Essentials of Technical Mathematics

CHERYL CLEAVES

Southwest Tennessee Community College

MARGIE HOBBS

The University of Mississippi

Upper Saddle River, New Jersey
Columbus, Ohio

Library of Congress Cataloging-in-Publication Data

Cleaves, Cheryl S.,
      Essentials of technical mathematics / Cheryl Cleaves, Margie Hobbs.
           p. cm.
      Includes index.
      ISBN 0-13-015672-8
        1. Mathematics. I. Hobbs, Margie J., II. Title.

      QA39.3 .C54 2002
      510--dc21   2001033199

Editor in Chief: Stephen Helba
Executive Editor: Frank Mortimer
Media Development Editor: Michelle Churma
Production Editor: Louise N. Sette
Production Supervision: Clarinda Publication Services
Design Coordinator: Robin G. Chukes
Cover Designer: Tom Mack
Cover art: © Photodisc
Production Manager: Brian Fox
Marketing Manager: Tim Peyton

This book was set in Times Roman and Futura by The Clarinda Company. It was printed and bound by Courier Kendallville, Inc. The cover was printed by The Lehigh Press, Inc.

Photo Credits: Susan Duke (Chapters 1, 4, 5, 6, 7, 10, 12, 16, 17, 18d) and Matthew Brown/Prentice Hall (Chapters 9, 11, 15, 17, 18a, 18b, 18c). Remaining photos were supplied by the authors.

Pearson Education Ltd., *London*
Pearson Education Australia Pty. Limited, *Sydney*
Pearson Education Singapore Pte. Ltd.
Pearson Education North Asia Ltd., *Hong Kong*
Pearson Education Canada, Ltd., *Toronto*
Pearson Educación de Mexico, S.A. de C.V.
Pearson Education—Japan, *Tokyo*
Pearson Education of Malaysia Pte. Ltd.
Pearson Education, *Upper Saddle River, New Jersey*

10 9 8 7 6 5 4 3 2 1
ISBN: 0-13-015672-8

Preface

To the Student

The mathematics you learn in this book will be helpful in your career. We have given much thought to the best way to teach mathematics and have done extensive research on how students learn. We suggest that you use the special features we have included in the text and supplementary materials to get the most from this book and your course. The following features are designed to help you learn the mathematics in this text.

Learning Outcomes.   The chapter opening pages list the learning outcomes for the chapter. Each section begins with its particular learning outcomes to show you what you will learn in that section. If you read and think about these outcomes before you begin the section, you will know what to look for as you work through the section.

Good Decisions through Teamwork.   Each chapter opens with a class project designed to promote teamwork. The projects incorporate a wide variety of team-building strategies. Each project engages your skills in a unique way—your computational skills, interpersonal skills, oral and written communication skills, organizational skills, research skills, critical-thinking and/or decision-making skills—skills that are highly valued by employers. You will prepare and present project reports for a variety of audiences, including instructors, peers, employers, and immediate supervisors.

Your instructor may use some or all of the projects, or he or she may organize teams within the class and have each team select a project from a different chapter. Even if a particular project is not used in your class, reading the projects will broaden your perception of the usefulness of mathematics.

Six-Step Approach to Problem Solving.   This approach gives you a system for solving a variety of math problems. You will learn how to organize the information given and how to develop a plan for solving the problem. You are asked to analyze and compare and also to estimate as you solve problems. Estimation helps you decide whether your answer is reasonable. You will learn to interpret the results of your calculations within the context of the problem, a skill you will use on your job.

Use of Color in the Text.   As you read the text and work through the examples notice the items shaded with color or with gray. These will help you follow the logic of working through the example. Color also highlights important items and boxed features such as the Tips!, Learning Strategies, and rules, procedures, and formulas.

Tip! Boxes.   These boxes give helpful hints and calculator strategies for doing mathematics, and they draw your attention to important generalizations or restrictions that you might otherwise overlook. Many of our students tell us that the tip boxes seem to anticipate and answer many of the questions they have when studying alone.

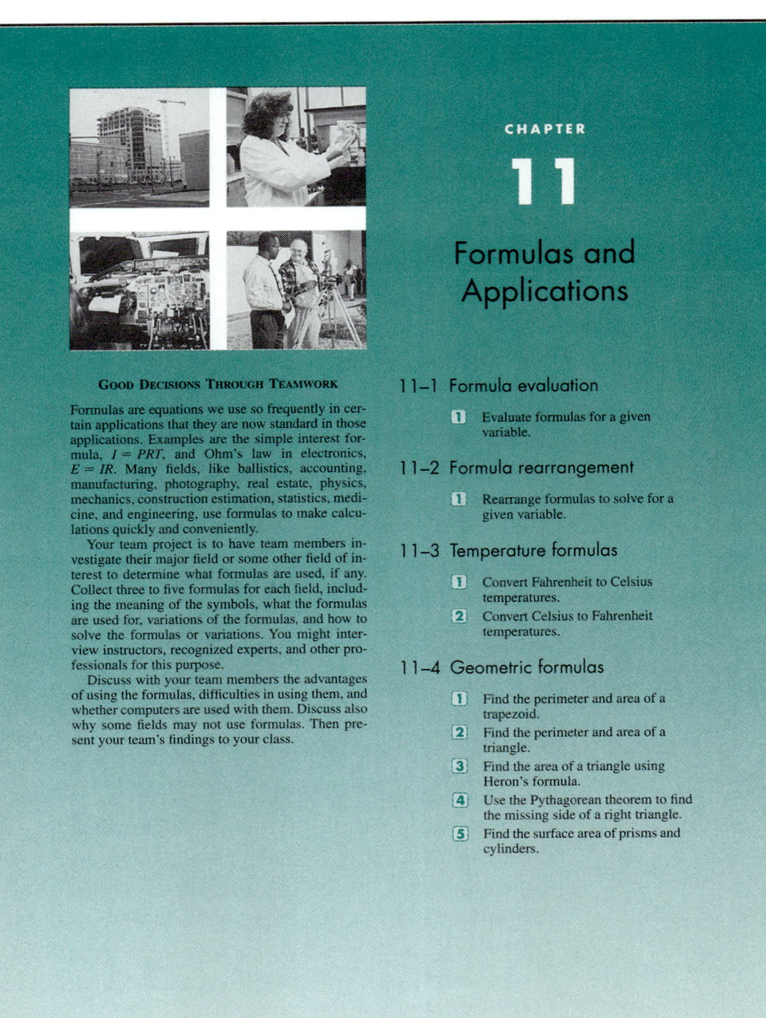

Learning Strategies. In each chapter you will find a number of learning strategies. These strategies can help you build a framework for successful learning. The strategies show ways to manage your learning of mathematics that you may not have thought of before. Use them to improve your "mathematical sense" and to give you a greater appreciation for the power of mathematics in your workplace and everyday life. You may find them useful in other areas of study also.

Using Your Calculator. Calculators are essential in all types of math, and especially in college math for technology. Some of the tips introduce useful and easy-to-follow calculator strategies. The tips show you how to analyze the procedure and set up a problem for a calculator solution; a sample series of keystrokes is often included. In addition, some tips give you strategies so you can determine how your calculator handles various operations.

Self-Study Exercises. These practice sets are keyed to the learning outcomes and appear at the end of each section. Use these exercises to check your understanding of the section. The answers to every problem are at the end of the text so you can get immediate feedback on your level of understanding of the material.

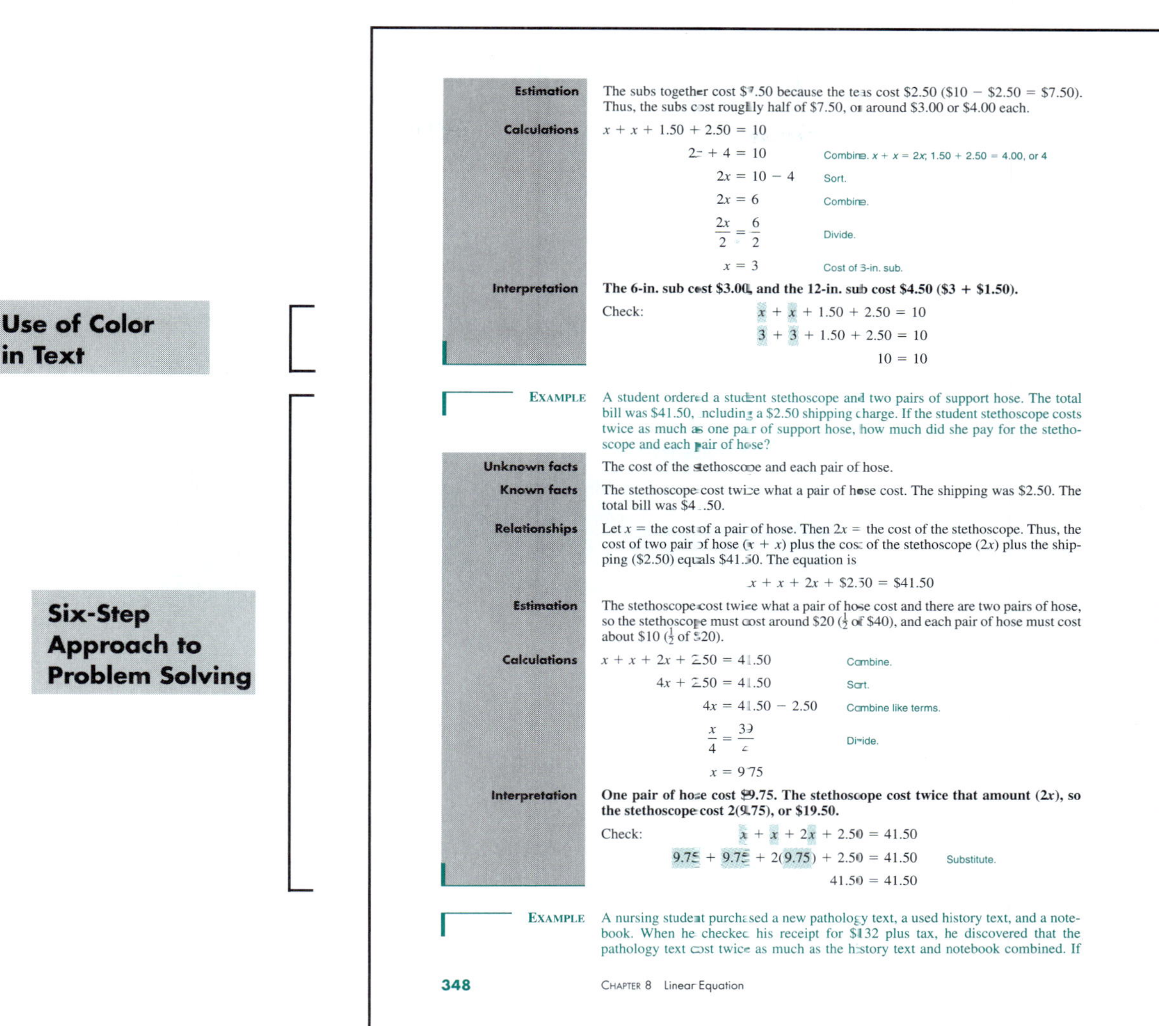

Estimation The subs together cost $7.50 because the teas cost $2.50 ($10 − $2.50 = $7.50). Thus, the subs cost roughly half of $7.50, or around $3.00 or $4.00 each.

Calculations

$$x + x + 1.50 + 2.50 = 10$$
$$2x + 4 = 10 \qquad \text{Combine. } x + x = 2x; 1.50 + 2.50 = 4.00, \text{ or } 4$$
$$2x = 10 - 4 \qquad \text{Sort.}$$
$$2x = 6 \qquad \text{Combine.}$$
$$\frac{2x}{2} = \frac{6}{2} \qquad \text{Divide.}$$
$$x = 3 \qquad \text{Cost of 3-in. sub.}$$

Interpretation **The 6-in. sub cost $3.00, and the 12-in. sub cost $4.50 ($3 + $1.50).**

Check:
$$x + x + 1.50 + 2.50 = 10$$
$$3 + 3 + 1.50 + 2.50 = 10$$
$$10 = 10$$

EXAMPLE A student ordered a student stethoscope and two pairs of support hose. The total bill was $41.50, including a $2.50 shipping charge. If the student stethoscope costs twice as much as one pair of support hose, how much did she pay for the stethoscope and each pair of hose?

Unknown facts The cost of the stethoscope and each pair of hose.

Known facts The stethoscope cost twice what a pair of hose cost. The shipping was $2.50. The total bill was $41.50.

Relationships Let x = the cost of a pair of hose. Then $2x$ = the cost of the stethoscope. Thus, the cost of two pair of hose ($x + x$) plus the cost of the stethoscope ($2x$) plus the shipping ($2.50) equals $41.50. The equation is

$$x + x + 2x + \$2.50 = \$41.50$$

Estimation The stethoscope cost twice what a pair of hose cost and there are two pairs of hose, so the stethoscope must cost around $20 ($\frac{1}{2}$ of $40), and each pair of hose must cost about $10 ($\frac{1}{2}$ of $20).

Calculations

$$x + x + 2x + 2.50 = 41.50 \qquad \text{Combine.}$$
$$4x + 2.50 = 41.50 \qquad \text{Sort.}$$
$$4x = 41.50 - 2.50 \qquad \text{Combine like terms.}$$
$$\frac{x}{4} = \frac{39}{4} \qquad \text{Divide.}$$
$$x = 9.75$$

Interpretation **One pair of hose cost $9.75. The stethoscope cost twice that amount ($2x$), so the stethoscope cost 2($9.75), or $19.50.**

Check:
$$x + x + 2x + 2.50 = 41.50$$
$$9.75 + 9.75 + 2(9.75) + 2.50 = 41.50 \qquad \text{Substitute.}$$
$$41.50 = 41.50$$

EXAMPLE A nursing student purchased a new pathology text, a used history text, and a notebook. When he checked his receipt for $132 plus tax, he discovered that the pathology text cost twice as much as the history text and notebook combined. If

348   Chapter 8   Linear Equation

Career Applications.   At the end of most chapters there is a career application. These applications resulted from interviews and research in the workplace, and they demonstrate how widespread math applications are in the workplace and the world around you. They provide opportunities to solve real-world problems, and demonstrate the ways that you regularly use the math concepts you are learning.

Assignment Exercises.   An extensive set of exercises appears at the end of each chapter so you can review all the learning outcomes presented in the chapter. You may be assigned these exercises, organized by section, as homework, or you may want to work them on your own for extra practice. Challenge problems are at the end of this set of exercises. The answers to the odd-numbered exercises are given at the end of the book, and worked-out solutions appear in a separate Student Solutions Manual. Your instructor has the solutions to the even-numbered exercises in the Instructor's Resource Manual.

Chapter Trial Test.   The trial test at the end of each chapter lets you check your understanding of the chapter concepts. You should be able to work each problem without referring to any examples in your text or your notes. Take this test before you take the class test to evaluate your understanding of the chapter material.

Tip

Learning Strategy

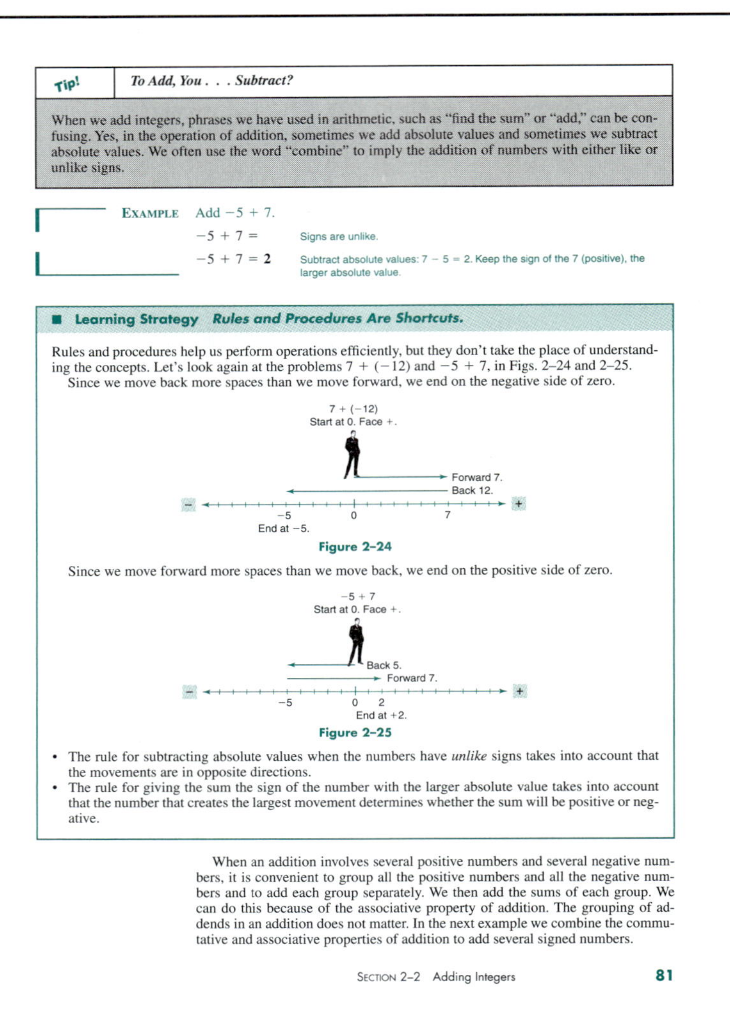

Tip! **To Add, You . . . Subtract?**

When we add integers, phrases we have used in arithmetic, such as "find the sum" or "add," can be confusing. Yes, in the operation of addition, sometimes we add absolute values and sometimes we subtract absolute values. We often use the word "combine" to imply the addition of numbers with either like or unlike signs.

EXAMPLE Add $-5 + 7$.

$-5 + 7 =$ Signs are unlike.

$-5 + 7 = 2$ Subtract absolute values: $7 - 5 = 2$. Keep the sign of the 7 (positive), the larger absolute value.

■ **Learning Strategy** *Rules and Procedures Are Shortcuts.*

Rules and procedures help us perform operations efficiently, but they don't take the place of understanding the concepts. Let's look again at the problems $7 + (-12)$ and $-5 + 7$, in Figs. 2–24 and 2–25.

Since we move back more spaces than we move forward, we end on the negative side of zero.

Figure 2–24

Since we move forward more spaces than we move back, we end on the positive side of zero.

Figure 2–25

• The rule for subtracting absolute values when the numbers have *unlike* signs takes into account that the movements are in opposite directions.
• The rule for giving the sum the sign of the number with the larger absolute value takes into account that the number that creates the largest movement determines whether the sum will be positive or negative.

When an addition involves several positive numbers and several negative numbers, it is convenient to group all the positive numbers and all the negative numbers and to add each group separately. We then add the sums of each group. We can do this because of the associative property of addition. The grouping of addends in an addition does not matter. In the next example we combine the commutative and associative properties of addition to add several signed numbers.

SECTION 2–2 Adding Integers   **81**

Answers to the odd-numbered problems appear at the end of the book, and their solutions appear in a separate Student Solutions Manual available in your bookstore. Your instructor has the solutions to the even-numbered problems in the Instructor's Resource Manual.

Glossary/Index. An extensive glossary/index makes this book a valuable reference. Use the index to cross-reference topics and to locate other topics that relate to the topic you are studying.

Table of Contents. The table of contents is your "roadmap" to this course. Study it carefully to determine how the topics are arranged. This will aid you in relating topics to each other.

Student Solutions Manual. This manual can be purchased at your bookstore. The manual contains worked-out solutions to the odd-numbered exercises in the Assignment Exercises and the Chapter Trial Test for each chapter of the text. Answers to these exercises appear in the back of your text, but using the manual to study the fully worked-out solutions can enhance your problem-solving skills and your understanding of the concepts covered.

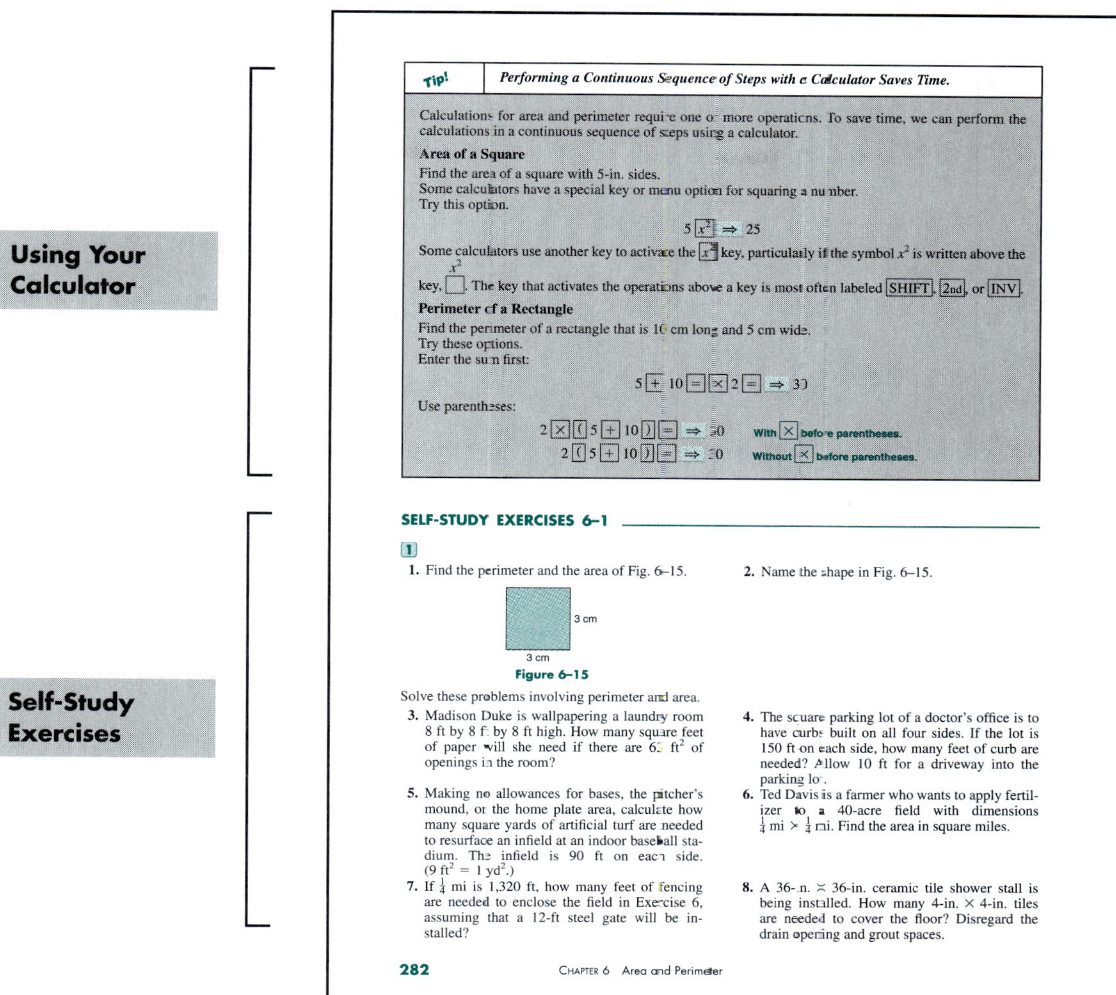

| Tip! | *Performing a Continuous Sequence of Steps with a Calculator Saves Time.* |

Calculations for area and perimeter require one or more operations. To save time, we can perform the calculations in a continuous sequence of steps using a calculator.

Area of a Square

Find the area of a square with 5-in. sides.
Some calculators have a special key or menu option for squaring a number.
Try this option.

$$5 \boxed{x^2} \Rightarrow 25$$

Some calculators use another key to activate the $\boxed{x^2}$ key, particularly if the symbol x^2 is written above the key, $\boxed{}$. The key that activates the operations above a key is most often labeled $\boxed{\text{SHIFT}}$, $\boxed{\text{2nd}}$, or $\boxed{\text{INV}}$.

Perimeter of a Rectangle

Find the perimeter of a rectangle that is 10 cm long and 5 cm wide.
Try these options.
Enter the sum first:

$$5 \boxed{+} 10 \boxed{=} \boxed{\times} 2 \boxed{=} \Rightarrow 30$$

Use parentheses:

$$2 \boxed{\times} \boxed{(} 5 \boxed{+} 10 \boxed{)} \boxed{=} \Rightarrow 30 \qquad \text{With } \boxed{\times} \text{ before parentheses.}$$
$$2 \boxed{(} 5 \boxed{+} 10 \boxed{)} \boxed{=} \Rightarrow 30 \qquad \text{Without } \boxed{\times} \text{ before parentheses.}$$

SELF-STUDY EXERCISES 6–1

1

1. Find the perimeter and the area of Fig. 6–15.

2. Name the shape in Fig. 6–15.

3 cm

3 cm

Figure 6–15

Solve these problems involving perimeter and area.

3. Madison Duke is wallpapering a laundry room 8 ft by 8 ft by 8 ft high. How many square feet of paper will she need if there are 65 ft² of openings in the room?

4. The square parking lot of a doctor's office is to have curbs built on all four sides. If the lot is 150 ft on each side, how many feet of curb are needed? Allow 10 ft for a driveway into the parking lot.

5. Making no allowances for bases, the pitcher's mound, or the home plate area, calculate how many square yards of artificial turf are needed to resurface an infield at an indoor baseball stadium. The infield is 90 ft on each side. $(9 \text{ ft}^2 = 1 \text{ yd}^2.)$

6. Ted Davis is a farmer who wants to apply fertilizer to a 40-acre field with dimensions $\frac{1}{4}$ mi × $\frac{1}{4}$ mi. Find the area in square miles.

7. If $\frac{1}{4}$ mi is 1,320 ft, how many feet of fencing are needed to enclose the field in Exercise 6, assuming that a 12-ft steel gate will be installed?

8. A 36-in. × 36-in. ceramic tile shower stall is being installed. How many 4-in. × 4-in. tiles are needed to cover the floor? Disregard the drain opening and grout spaces.

282   Chapter 6   Area and Perimeter

How to Study Technical Mathematics.   Your instructor can get free copies of this booklet, which describes various learning techniques you can use in class and to prepare for class that can make your learning of mathematics much more efficient and effective.

StudyWizard Software.   This software, which is packaged with the text, provides additional practice with the math concepts presented in the text. Each question contains a reference to the section and learning outcome number in the text where the concept first appears, making it easier to find the sections you want to review. Immediate feedback is provided to all questions, allowing you to strengthen your skills and test your knowledge of the concepts before a class test. The glossary included on the software allows you to review the terms and concepts presented in the text.

Companion Web Site.   This free web site, available at www.prenhall.com/cleaves, provides even more practice with the math concepts presented in the form of short quizzes for each section of the text. These quizzes are immediately graded, and you have the opportunity to send the results to your instructor via e-mail.

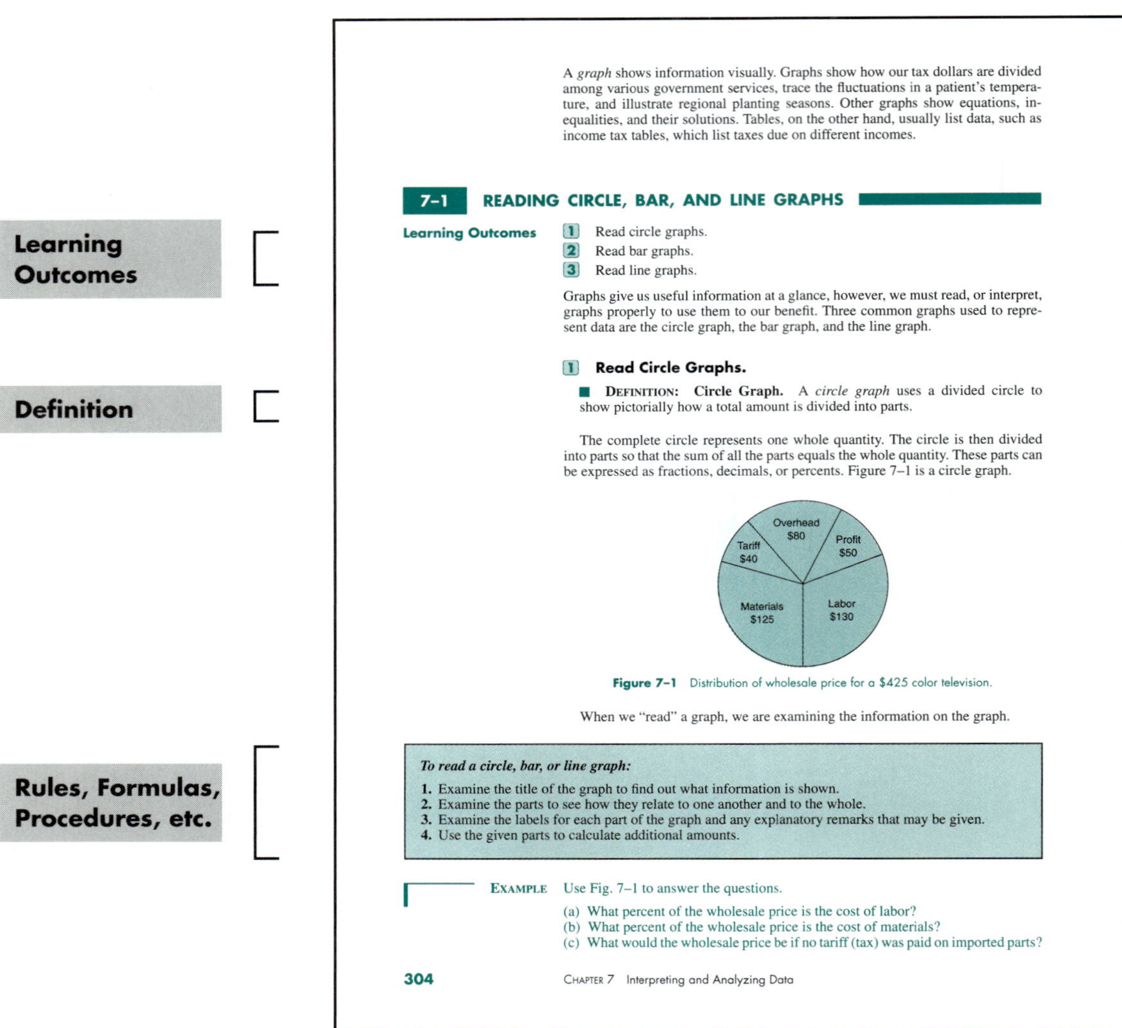

A *graph* shows information visually. Graphs show how our tax dollars are divided among various government services, trace the fluctuations in a patient's temperature, and illustrate regional planting seasons. Other graphs show equations, inequalities, and their solutions. Tables, on the other hand, usually list data, such as income tax tables, which list taxes due on different incomes.

7–1   READING CIRCLE, BAR, AND LINE GRAPHS

Learning Outcomes

1. Read circle graphs.
2. Read bar graphs.
3. Read line graphs.

Graphs give us useful information at a glance, however, we must read, or interpret, graphs properly to use them to our benefit. Three common graphs used to represent data are the circle graph, the bar graph, and the line graph.

1 Read Circle Graphs.

■ **DEFINITION: Circle Graph.** A *circle graph* uses a divided circle to show pictorially how a total amount is divided into parts.

The complete circle represents one whole quantity. The circle is then divided into parts so that the sum of all the parts equals the whole quantity. These parts can be expressed as fractions, decimals, or percents. Figure 7–1 is a circle graph.

Figure 7–1 Distribution of wholesale price for a $425 color television.

When we "read" a graph, we are examining the information on the graph.

To read a circle, bar, or line graph:

1. Examine the title of the graph to find out what information is shown.
2. Examine the parts to see how they relate to one another and to the whole.
3. Examine the labels for each part of the graph and any explanatory remarks that may be given.
4. Use the given parts to calculate additional amounts.

EXAMPLE Use Fig. 7–1 to answer the questions.

(a) What percent of the wholesale price is the cost of labor?
(b) What percent of the wholesale price is the cost of materials?
(c) What would the wholesale price be if no tariff (tax) was paid on imported parts?

304   CHAPTER 7 Interpreting and Analyzing Data

We wish you much success in your study of mathematics. Many of the features in this book were suggested by students such as yourself. If you have suggestions for improving the presentation, please give them to your instructor or e-mail the authors at ccleaves@bellsouth.net or mhobbs@watervalley.net.

Reading Your Math Textbook

In developing an effective study plan it is important to use all your available resources to their maximum advantage. The most accessible of these resources is your textbook. Incorporate an effective strategy for reading your textbook into your study plan.

Beginning a Chapter

1. Examine the chapter opening page or pages. Read the chapter title, section titles, and learning outcomes to determine what will be covered in the chapter.
2. Use the learning outcomes as a checklist to rate your initial knowledge of the topics presented in the chapter. This rating can be a numerical one. For example, 0 means you know nothing about this topic, 1 means you know a little but not much about this topic, 2 means you know quite a bit but there may be a few gaps, and 3 means you know this topic very well.

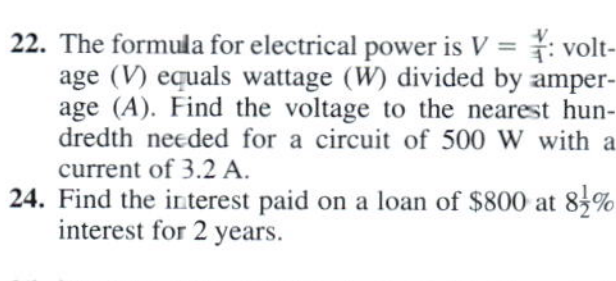

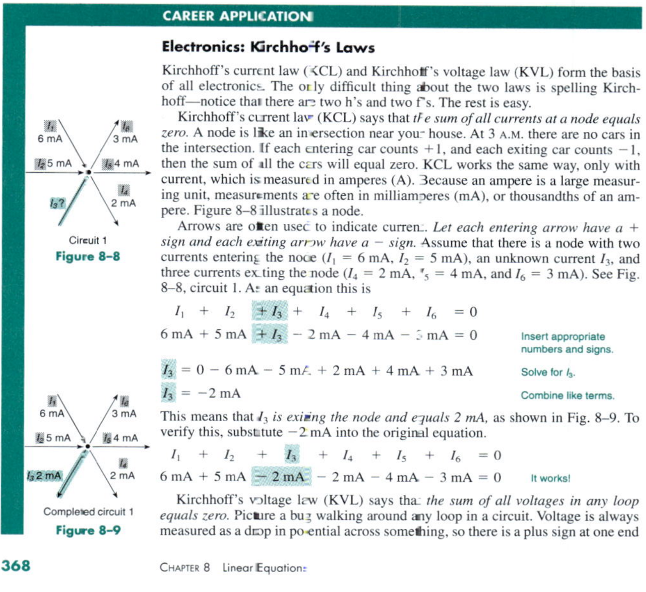

22. The formula for electrical power is $V = \frac{W}{A}$: voltage (V) equals wattage (W) divided by amperage (A). Find the voltage to the nearest hundredth needed for a circuit of 500 W with a current of 3.2 A.

23. Find the interest paid on a loan of $2,400 for 1 year at an interest rate of 11%.

24. Find the interest paid on a loan of $800 at $8\frac{1}{2}$% interest for 2 years.

25. Find the total amount of money (maturity value) that the borrower will pay back on a loan of $1,400 at $12\frac{1}{2}$% simple interest for 3 years.

26. Find the rate of interest on an investment of $2,500 made by Nurse Honda for a period of 2 years if she received $612.50 in interest.

27. Maddy Brown needed start-up money for her landscape service. She borrowed $12,000 for 30 months and paid $360 interest on the loan. What interest rate did she pay?

28. Raul Fletes needs money to buy lawn equipment. He borrows $500 for 7 months and pays $53.96 in interest. What is his rate of interest?

29. Linda Davis agrees to lend money to Alex Luciano at a special interest rate of 9%, on the condition that he borrow enough that he will pay her $500 in interest over a 2-year period. What is the minimum amount Alex can borrow?

30. Rob Thweatt needs money for medical school. He borrows $6,000 at 12% interest. If he pays $360 interest, what is the duration of the loan?

CAREER APPLICATION

Electronics: Kirchhoff's Laws

Kirchhoff's current law (KCL) and Kirchhoff's voltage law (KVL) form the basis of all electronics. The only difficult thing about the two laws is spelling Kirchhoff—notice that there are two h's and two f's. The rest is easy.

Kirchhoff's current law (KCL) says that *the sum of all currents at a node equals zero*. A node is like an intersection near your house. At 3 A.M. there are no cars in the intersection. If each entering car counts $+1$, and each exiting car counts -1, then the sum of all the cars will equal zero. KCL works the same way, only with current, which is measured in amperes (A). Because an ampere is a large measuring unit, measurements are often in milliamperes (mA), or thousandths of an ampere. Figure 8–8 illustrates a node.

Arrows are often used to indicate current. *Let each entering arrow have a $+$ sign and each exiting arrow have a $-$ sign.* Assume that there is a node with two currents entering the node ($I_1 = 6$ mA, $I_2 = 5$ mA), an unknown current I_3, and three currents exiting the node ($I_4 = 2$ mA, $I_5 = 4$ mA, and $I_6 = 3$ mA). See Fig. 8–8, circuit 1. As an equation this is

$$I_1 \ + \ I_2 \ + I_3 \ + \ I_4 \ + \ I_5 \ + \ I_6 \ = 0$$

$$6\text{ mA} + 5\text{ mA} + I_3 - 2\text{ mA} - 4\text{ mA} - 3\text{ mA} = 0 \qquad \text{Insert appropriate numbers and signs.}$$

$$I_3 = 0 - 6\text{ mA} - 5\text{ mA} + 2\text{ mA} + 4\text{ mA} + 3\text{ mA} \qquad \text{Solve for } I_3.$$

$$I_3 = -2\text{ mA} \qquad \text{Combine like terms.}$$

This means that I_3 is exiting the node and equals 2 mA, as shown in Fig. 8–9. To verify this, substitute -2 mA into the original equation.

$$I_1 \ + \ I_2 \ + \ I_3 \ + \ I_4 \ + \ I_5 \ + \ I_6 \ = 0$$

$$6\text{ mA} + 5\text{ mA} - 2\text{ mA} - 2\text{ mA} - 4\text{ mA} - 3\text{ mA} = 0 \qquad \text{It works!}$$

Kirchhoff's voltage law (KVL) says that *the sum of all voltages in any loop equals zero*. Picture a bug walking around any loop in a circuit. Voltage is always measured as a drop in potential across something, so there is a plus sign at one end

368   CHAPTER 8   Linear Equations

Another possible rating strategy can be a minus, check, plus system. Minus means you need to work on this topic, check means you know the topic moderately well, and plus means you know the topic very well.

Beginning a Section
1. Read the section title and the learning outcomes for the section.
2. Read the introductory paragraph.
3. Locate the Self-Study Exercises at the end of the section. Read the directions for each "clump" of exercises. This will give you an idea of the type of problems you will be working and what to look for as you read the section.
4. Begin reading the section. Make notes on concepts that you do not understand or examples for which you are not able to follow the explanation. This will be the basis for questions to ask in class.

Continuing through the Chapter
1. Work on one learning outcome at a time. After reading and studying one learning outcome, try some of the exercises for that outcome. Always check your answers with the text or StudyWizard and ask questions as appropriate. Assess your understanding of each outcome and practice or get help as you think necessary. Be realistic with your self-assessment!

Preface   ix

5. If a support has a depth three-fourths the size of the width, find its dimensions and volume.
6. This freeway will be in an earthquake fault zone, so state civil engineers decide to make the depth equal to the width of the support to withstand maximum vibration stress. Find the dimensions and volume of each support.
7. Why do the dimensions you found in Exercise 4 have a volume of exactly 200 ft^3, while the dimensions you found in Exercises 5 and 6 have volumes over and under the prescribed 200 ft^3?
8. A recent computer modeling simulation found that supports with the smallest cross-sectional perimeter per volume are best able to endure vibration stress, which means that supports in earthquake zones should be in the shape of a cylinder (a cylinder has a circular cross-sectional perimeter). Use $V = \pi r^2 h$ to find the radius of the support with a height of 16 ft and a volume of 200 ft^3.
9. Find the cross-sectional perimeters of the supports designed in Exercises 4, 5, 6, and 8. Does the circular cross section have the smallest perimeter?

Answers

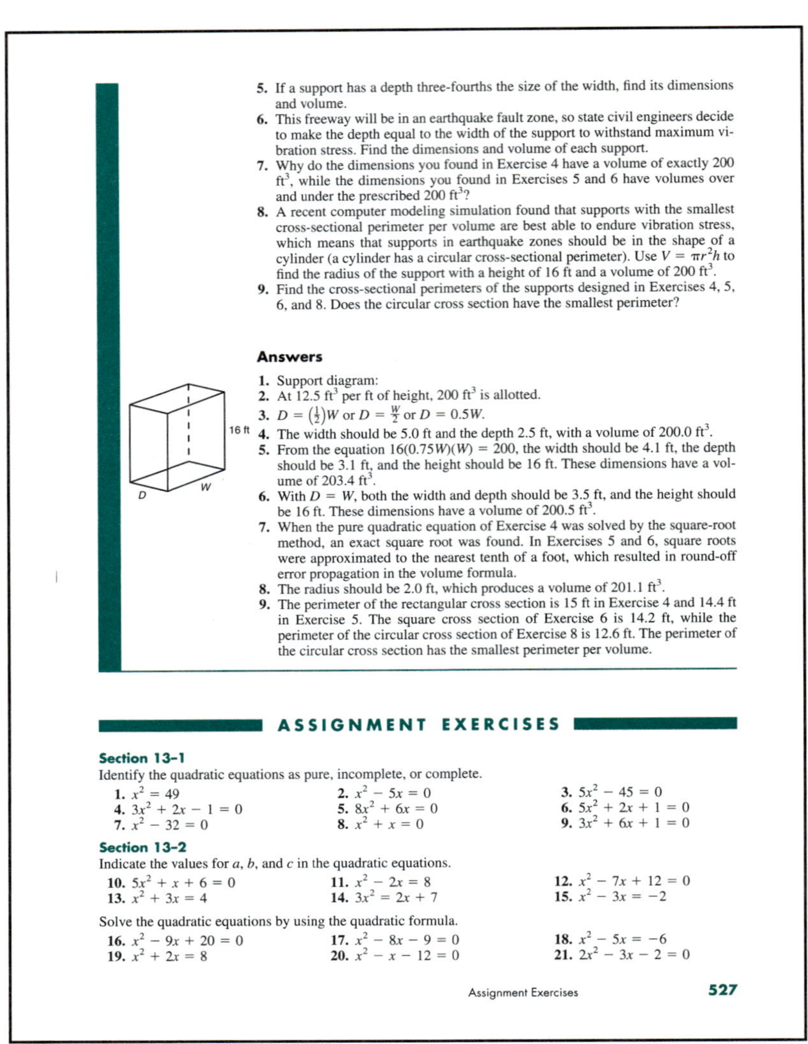

1. Support diagram:
2. At 12.5 ft^3 per ft of height, 200 ft^3 is allotted.
3. $D = \left(\frac{1}{2}\right)W$ or $D = \frac{W}{2}$ or $D = 0.5W$.
4. The width should be 5.0 ft and the depth 2.5 ft, with a volume of 200.0 ft^3.
5. From the equation $16(0.75W)(W) = 200$, the width should be 4.1 ft, the depth should be 3.1 ft, and the height should be 16 ft. These dimensions have a volume of 203.4 ft^3.
6. With $D = W$, both the width and depth should be 3.5 ft, and the height should be 16 ft. These dimensions have a volume of 200.5 ft^3.
7. When the pure quadratic equation of Exercise 4 was solved by the square-root method, an exact square root was found. In Exercises 5 and 6, square roots were approximated to the nearest tenth of a foot, which resulted in round-off error propagation in the volume formula.
8. The radius should be 2.0 ft, which produces a volume of 201.1 ft^3.
9. The perimeter of the rectangular cross section is 15 ft in Exercise 4 and 14.4 ft in Exercise 5. The square cross section of Exercise 6 is 14.2 ft, while the perimeter of the circular cross section of Exercise 8 is 12.6 ft. The perimeter of the circular cross section has the smallest perimeter per volume.

ASSIGNMENT EXERCISES

Section 13–1

Identify the quadratic equations as pure, incomplete, or complete.

1. $x^2 = 49$
2. $x^2 - 5x = 0$
3. $5x^2 - 45 = 0$
4. $3x^2 + 2x - 1 = 0$
5. $8x^2 + 6x = 0$
6. $5x^2 + 2x + 1 = 0$
7. $x^2 - 32 = 0$
8. $x^2 + x = 0$
9. $3x^2 + 6x + 1 = 0$

Section 13–2

Indicate the values for a, b, and c in the quadratic equations.

10. $5x^2 + x + 6 = 0$
11. $x^2 - 2x = 8$
12. $x^2 - 7x + 12 = 0$
13. $x^2 + 3x = 4$
14. $3x^2 = 2x + 7$
15. $x^2 - 3x = -2$

Solve the quadratic equations by using the quadratic formula.

16. $x^2 - 9x + 20 = 0$
17. $x^2 - 8x - 9 = 0$
18. $x^2 - 5x = -6$
19. $x^2 + 2x = 8$
20. $x^2 - x - 12 = 0$
21. $2x^2 - 3x - 2 = 0$

Assignment Exercises   **527**

---

2. Continue outcome by outcome, section by section, checking your understanding as you go.

Reviewing the Chapter

1. After finishing a chapter, thumb through the entire chapter, reading the Tip! boxes and Learning Strategies.
2. Look through the chapter again, this time reviewing new terminology. New terminology will appear in *italic* type or be set apart as a definition. Read the definitions and make a list of words that may need further review.
3. Read again the learning outcomes on the chapter opening pages and again rate your understanding of each outcome.
4. Work the Chapter Trial Test at the end of the chapter and check your answers. Review or get assistance as necessary.

Finishing the Chapter

1. Prepare for the test on the chapter. Ask your instructor which outcomes require mastery for testing purposes. Some outcomes may not require mastery, and others may even be optional.
2. Read the special features Good Decisions Through Teamwork, Mathematics in the Workplace, and Career Applications to gain some insight about where these concepts are used in real life.

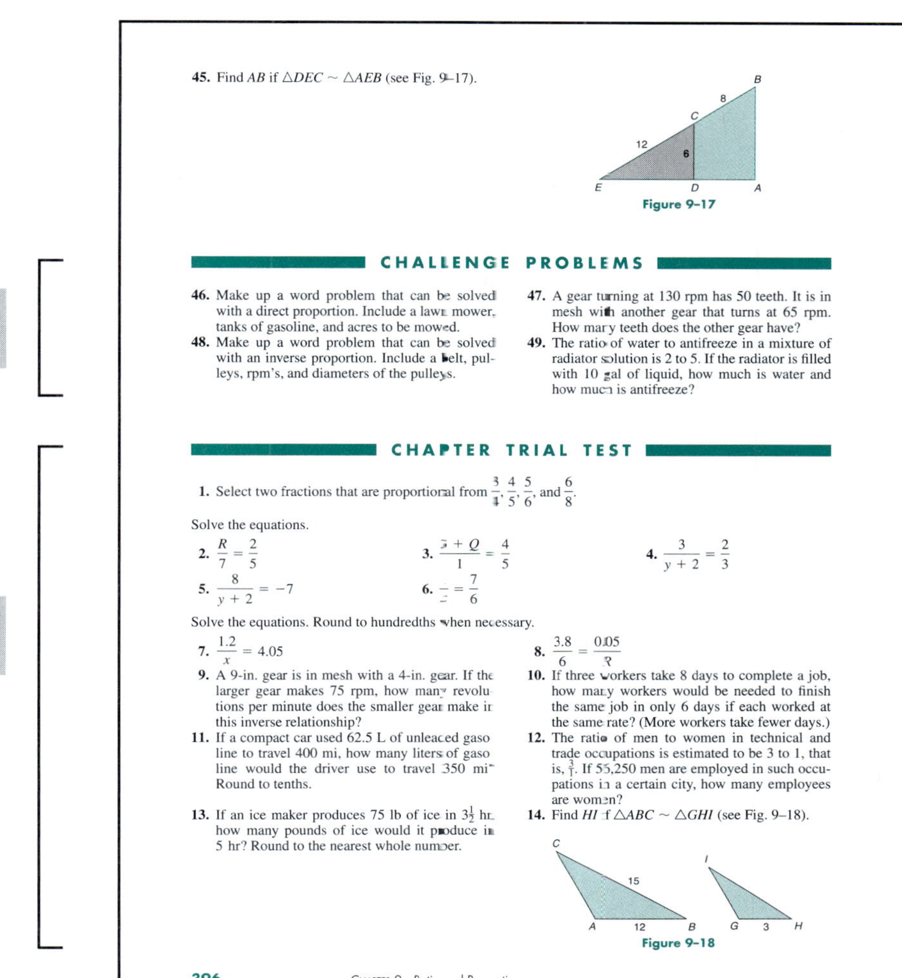

General Tips

1. Practice an outcome until you feel comfortable that you understand the concept. Abundant practice material is available to you that is specifically geared to your text (Self-Study Exercises, Assignment Exercises, StudyWizard, and Companion Web Site). Other practice is available through generic mathematics software and other texts. Only you know when you have practiced enough. Be realistic with the self-assessment of your understanding. Practice helps you retain the information for a longer period of time, but don't wear yourself out! Finding that appropriate balance is your goal.

2. Don't forget the Glossary/Index! As you move through the text you will forget definitions and concepts. Maybe you are not starting your study at the beginning of the text and need to review a few concepts that were in the chapters not covered. Examining the Glossary/Index should be your first step in accomplishing your review.

Good luck on your study of mathematics.

To the Instructor

In the development of the text we have tried to address a wide variety of teaching and learning styles and modes of instruction, including on-line course delivery. A holistic approach to student learning is our goal.

We suggest that you encourage your students to read the "To the Student" portion of the preface, having them pay particular attention to the suggestions provided in the section "Reading Your Math Textbook."

Commitment to Improving Mathematics Education.   The authors have been and continue to be active in implementing the standards of the American Mathematical Association of Two-Year Colleges (AMATYC), the National Council of Teachers of Mathematics (NCTM), and the Mathematical Association of America (MAA). We enthusiastically promote the standards and guidelines encouraged by these organizations and the SCANS document. The Instructor's Resource Manual gives specific references for implementing the Standards in your courses.

Calculator Usage.   Calculator tips that are appropriate for both scientific and graphing calculators are included. These tips are generic (that is, they do not pertain to specific models), and they help students determine how their calculator operates without referring to a user's manual.

We continue to emphasize the calculator as a tool that facilitates learning and understanding, but students' understanding of the mathematical concepts is even more important. To this end, we include assessment strategies throughout the text and supplementary materials that enable students to test their understanding of a concept independently of their calculator.

Study Strategies and Reference Features.   In our experiences as instructors, we are keenly aware of the need for students to develop good study habits and good independent learning skills. Students find a good reference text invaluable as they review mathematical concepts when the need arises. We take great pride in our students' praise of the usefulness of this text as a reference standard. For a detailed description of the features of the text and our suggestions for students, refer to the "To the Student" portion of the preface.

Additional Resources.   Several additional resources are available with the adoption of the text. These resources include the Instructor's Resource Manual (IRM). This manual includes notes and suggested activities for each chapter along with teaching tips as well as a variety of reproducible activities and worked-out solutions to even-numbered exercises. Also included are a Test Item File and a computerized test item file (PH Test Manager), a Student Solutions Manual, a "How to Study Technical Mathematics" booklet, StudyWizard software (packaged with the text), and a Companion Web site. Contact your Prentice Hall representative for more information.

Acknowledgments

A project such as this does not come together without help from lots of people. Our first avenue for input is through our students and fellow instructors at Southwest Tennessee Community College. Their comments and suggestions have been invaluable. The teaching of mathematics over time produces a wealth of knowledge about instructional strategies and specific content. We are grateful for the many valuable suggestions that we received in these areas. We wish to thank the following individuals who served as reviewers for *Essentials of Technical Mathematics*:

Terry B. Gaalswyk, Western Iowa Technical Community College
Edwin G. Landauer, Clackamas Community College (OR)
Karen Newson, Triangle Tech (PA)

Scott Randby, University of Akron (OH)
Behnaz Rouhani, Athens Area Technical Institute (GA)
David C. Shellabarger, Lane Community College (OR)
Joseph Sukta, Moraine Valley Community College (IL)
Jimmie A. Van Alphen, Ozarks Technical Community College (MO)

The photographs that open each chapter remind us that mathematics is a very human endeavor. Some of the photographs were taken by the authors, but the best ones were taken by Susan Duke and Matthew Brown. We appreciate the care they took in matching the photos to the projects.

We wish to express thanks to all the people who helped make this edition a reality. In particular, we thank Steve Helba, Editor in Chief, who has believed in our work and supported our ideas. Frank Mortimer, Executive Editor, contributed his extensive experience in technology and marketing to the success of this project. We also thank Louise Sette, Production Editor, and Michelle Churma, Media Development Editor, who patiently and carefully directed all ancillary materials for the text. Emily Autumn of Clarinda Publication Services was especially helpful.

Finally, our extended families continue to be our driving force, especially our spouses, Charles Cleaves and Allen Hobbs.

Cheryl Cleaves
Margie Hobbs

Contents

Choose a home business in your team's target career field and investigate the cost of starting the business. Decide whether the home business will market products, such as sports memorabilia, or services, such as desktop publishing. Once your team agrees on the product or service, investigate the required equipment, materials, and other items and their cost. For example, you might search business and trade magazines and interview a professional in the chosen business. Some items to consider purchasing are telephones, supplies, furniture, a computer system, fax capability, subscription to an on-line service, and, if necessary, the product inventory.

Now find the total cost to start your home business. Will you need a loan, or are other sources of funds available? If the budget is tight, can you eliminate or trim any nonessential expenses? Setting priorities early in the planning process is an important part of developing a realistic start-up budget.

Whole Numbers and Decimals

1–1 Whole numbers, decimals, and the place-value system

1. Identify place values in whole numbers.
2. Read and write whole numbers in words, standard notation, and expanded notation.
3. Identify place values in decimal numbers.
4. Read and write decimal numbers.
5. Write fractions with power-of-10 denominators as decimal numbers.
6. Compare decimal numbers.
7. Round a whole number or a decimal number to a place value.
8. Round a whole number or a decimal number to a number with one nonzero digit.

1–2 Adding whole numbers and decimal numbers

1. Add whole numbers.
2. Add decimal numbers.
3. Estimate and check addition.
4. Add using a calculator.

1-3 Subtracting whole numbers and decimals

1. Subtract whole numbers.
2. Subtract decimal numbers.
3. Estimate and check subtraction.
4. Subtract using a calculator.

1-4 Multiplying whole numbers and decimals

1. Multiply whole-number factors.
2. Multiply decimal factors.
3. Apply the distributive property.
4. Estimate and check multiplication.
5. Multiply using a calculator.

1-5 Dividing whole numbers and decimals

1. Use various symbols to indicate division.
2. Divide whole numbers.
3. Divide decimal numbers.
4. Estimate and check division.
5. Find the numerical average.
6. Divide using a calculator.

1-6 Exponents, roots, and powers of 10

1. Simplify expressions that contain exponents.
2. Square numbers and find the square roots of numbers.
3. Use powers of 10 to multiply and divide.
4. Use a calculator to find powers and roots.

1-7 Order of operations and problem solving

1. Apply the order of operations to a series of operations.
2. Use a calculator to perform a series of operations.
3. Solve applied problems using problem-solving strategies.

When we study a subject for the first time or review in some detail a subject we studied a while ago, we begin with the basics. Often, as we examine the basics of a subject, we discover—or rediscover—many pieces of useful information. In this sense, mathematics is no different from any other subject. We begin with a study of whole numbers and decimals, and the basic operations that we perform with them.

1-1 WHOLE NUMBERS, DECIMALS, AND THE PLACE-VALUE SYSTEM

Learning Outcomes

1. Identify place values in whole numbers.
2. Read and write whole numbers in words, standard notation, and expanded notation.
3. Identify place values in decimal numbers.
4. Read and write decimal numbers.
5. Write fractions with power-of-10 denominators as decimal numbers.
6. Compare decimal numbers.
7. Round a whole number or a decimal number to a place value.
8. Round a whole number or a decimal number to a number with one nonzero digit.

Our system of numbers, which is called the *decimal-number system,* uses 10 figures called *digits:* 0, 1, 2, 3, 4, 5, 6, 7, 8, 9. A *whole number* is made up of one or more digits. When a number contains two or more digits, each digit must be in the correct place for the number to have the value we intend it to have. If we mean "ninety-eight," we must place the 9 first and the 8 second to represent 98. If we change the places of these two digits by putting the 8 first and the 9 second, we get a new value (eighty-nine) and a new number (89).

1 Identify Place Values in Whole Numbers.

Each place a digit occupies in a number has a value called a *place value.* If we know the place value of each digit in a number, we can read the number and understand how much it means. Look at the chart of place values in Fig. 1–1. Notice that each place value *increases* as we move from *right to left* and that each increase is *10 times* the value of the place to the right. For example, the tens place is 10 times the ones place, the hundreds place is 10 times the tens place, and so on.

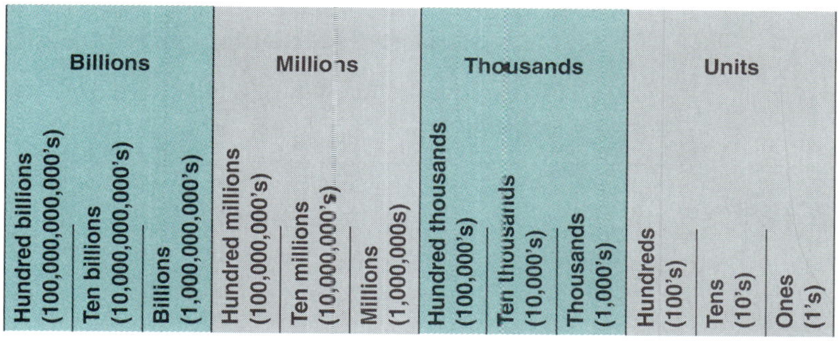

Figure 1–1 Whole-number place values.

The place values are arranged in *periods,* or groups of three, to make numbers easier to read. The first group of three is called *units,* the second group of three is called *thousands,* the third group is called *millions,* and the fourth group is called *billions.* Commas are used to mark off these groups. The commas make larger numbers easier to read because we can locate specific place values and interpret the number more easily. Each group of three digits will have a hundreds place, a tens place, and a ones place.

In four-digit numbers, the comma separating the units group from the thousands group is optional. Thus, 4,575 and 4575 are both correct.

To identify the place value of digits:

1. Mentally position the number on the place-value chart so that the last digit on the right aligns under the ones place.

2. Identify the place value of each digit according to its position on the chart.

EXAMPLE In the number 2,472,694,500, identify the place value of the digit 7.

We align the number on the place-value chart as shown in Fig. 1–2.

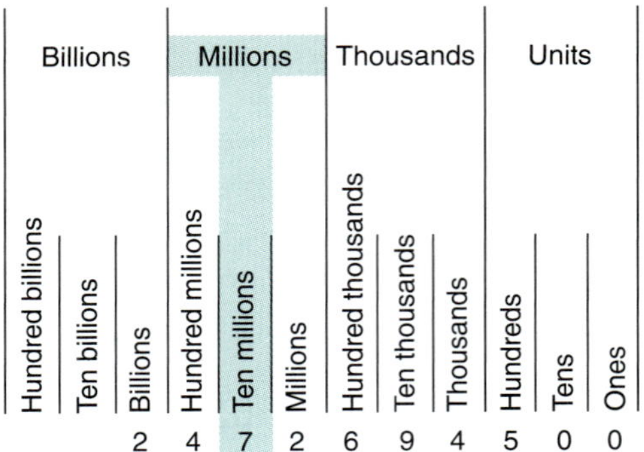

Figure 1–2 Whole-number place-value chart.

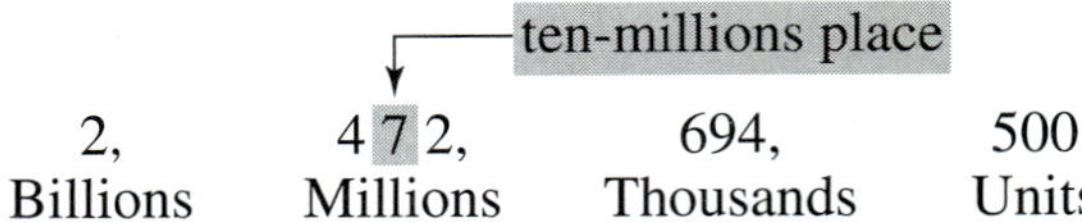

7 is in the ten-millions place.

2 Read and Write Whole Numbers in Words, Standard Notation, and Expanded Notation.

Now that we understand place values, we can read numbers using the following procedure.

> *To read numbers:*
>
> 1. Mentally position the number on the place-value chart so that the last digit on the right aligns under the ones place.
> 2. Examine the number from right to left, separating each group with commas.
> 3. Identify the leftmost group.
> 4. From the left, read the numbers in each group and the group name. (The group name *units* is not usually read. A group containing all zeros is not usually read.)

EXAMPLE Show how 7543026129 and 2000125 would be read by writing them in words.

Mentally align the digits on the place-value chart (Fig. 1–3). Starting at the right, separate each group of three digits with commas. Identify the leftmost group. The first number is seven *billion,* five hundred forty-three *million,* twenty-six *thousand,* one hundred twenty-nine.

The second number is two *million,* one hundred twenty-five. *Note:* Since the thousands group contains all zeros, it is *not* read.

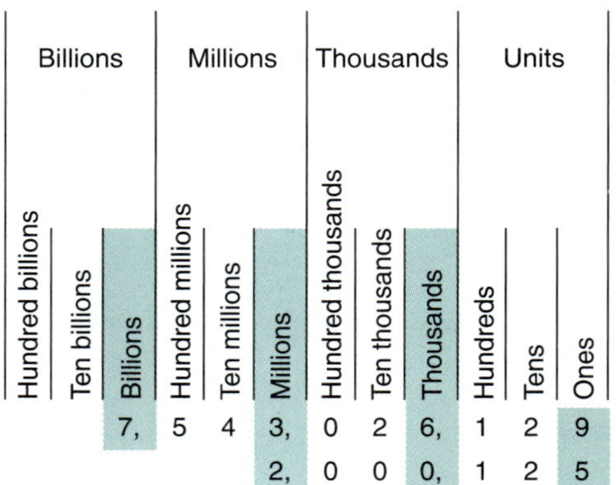

Figure 1–3 Group names.

<table>
<tr><td>**Tip!**</td><td>*Special Conventions with Reading and Writing Numbers:*</td></tr>
</table>

- A group name is inserted at each comma.
- The word *and* should not be used when reading whole numbers.
- The numbers from 21 to 99 (except 30, 40, 50, and so on) use a hyphen when they are written (forty-three, twenty-six, and so on).

Of course, there are times when we have to write spoken numbers on paper. Numbers written with digits in the appropriate place-value positions are numbers in *standard notation*.

To write numbers in standard notation:

1. Write the group names in order from left to right, starting with the first group in the number.
2. Fill in the digits in each group. Leave blanks for zeros if necessary.
3. Supply zeros as needed for each group. Each group except the leftmost group must have three digits.
4. Starting at the right, separate the groups with commas as needed.

EXAMPLE The exact cost of a group of airplanes is eight million, two hundred four thousand, twelve dollars. Write this amount as a number in standard notation.

	Millions	Thousands	Units
1.	Millions	Thousands	Units
2.	Millions	Thousands	Units
	8	204	12
3.	Millions	Thousands	Units
	8	204	012
4.	Millions	Thousands	Units
	8,	204,	012

The cost is $8,204,012.

To emphasize the value of a number, we may choose to write a number in *expanded notation*. Expanded notation shows the sum of each digit times its place value.

To write a number in expanded notation:

Write 516 in expanded notation.

1. Write each digit times its place value.

5×100
1×10
6×1

2. Add the expanded digits. $(5 \times 100) + (1 \times 10) + (6 \times 1)$

■ **Learning Strategy   Read and Illustrate Rules**

Sometimes the language of mathematics and the details in procedures and rules seem overwhelming. Here are some suggestions that may help.

- Illustrate the rule with an example.
- Make your example with reasonable numbers.

- Reword the rule into your own words.
- Discuss your rule and illustration with a classmate or study partner.
- Be sure you understand the concept of the rule or procedure rather than memorizing meaningless words and steps.

EXAMPLE A bank teller has only three types of currency (bills) in the cash drawer: $100, $10, and $1. Using the fewest number of bills, how many bills of each type are needed to cash a check for $243?

Mentally visualize 243 in expanded notation.

$$(2 \times 100) + (4 \times 10) + (3 \times 1)$$

Each place value in expanded notation represents a type of currency. Interpret the expanded notation as it relates to the type of money.

The cash includes two $100 bills, four $10 bills, and three $1 bills.

3 Identify Place Values in Decimal Numbers.

Numbers that are parts of a whole number are called *fractions*. In fraction notation, we write one number over another number. The bottom number, the *denominator,* represents the number of parts that a whole unit contains. The top number, the *numerator,* represents the number of parts being considered. We will use this notation to express division. Let's begin our examination of fractions with a special type of fraction called a *decimal fraction*. Other fractions will be examined in Chapter 3.

A decimal fraction is a fraction whose denominator is always 10 or some power of 10, such as 100 or 1000. For convenience, in this book we use the terms decimal fraction, decimal number, and decimal interchangeably. However, keep in mind that decimals are another way of writing fractions. In fraction notation, 3 out of 10 parts is written as $\frac{3}{10}$. In decimal notation, the denominator 10 is not written but is implied by position on the place-value chart. A decimal point (.) separates whole amounts on the left and fractional parts on the right. The fraction $\frac{3}{10}$ can be written in decimal notation as 0.3.

Examine the place-value chart shown in Fig. 1–4. We can use the place-value system to understand *decimal numbers.* In Fig. 1–4, to move from *left* to *right,* we divide by 10 to get the value of the next place. For example, to move from the hundreds place to the tens place, we have $100 \div 10 = 10$. To move from the tens place to the ones place, we have $10 \div 10 = 1$.

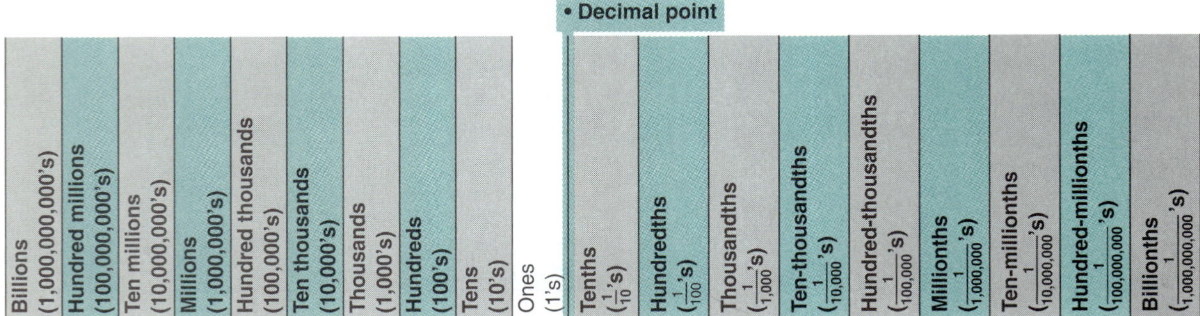

Figure 1–4 Place-value chart.

To extend the place-value chart on the right by moving from the ones place to the next place on the right, we have $1 \div 10 = \frac{1}{10}$. Thus, the place on the right of the ones place is called the *tenths place*. A period (.), called the *decimal point,* is placed between the ones place and tenths place to help us identify the place value of each digit.

The digits to the right of the ones place represent the numerator of the fraction. The place value of the rightmost digit indicates the denominator.

To identify the place value of digits in decimal fractions:

1. Mentally position the decimal number on the decimal place-value chart so that the decimal point of the number aligns with the decimal point on the chart.
2. Identify the place value of each digit according to its position on the chart.

EXAMPLE Identify the place value of each digit in 32.4675 using the rule.

Aligning the number 32.4675 with the place-value chart in Fig. 1–5, we see that 3 is in the tens place, 2 is in the ones place, 4 is in the tenths place, 6 is in the hundredths place, 7 is in the thousandths place, and 5 is in the ten-thousandths place.

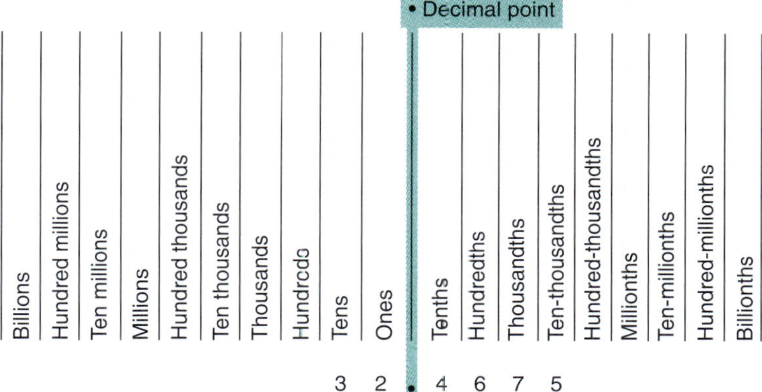

Figure 1–5 Decimal-number place-value chart.

The place-value chart can be extended on the right side for smaller fractions in the same manner that it is extended on the left for larger numbers.

4 Read and Write Decimal Numbers.

Examine the place-value chart in Fig. 1–5. Notice the *th* on the end of each place value to the right of the decimal point. When reading decimals, this *th* indicates a decimal number. The word *and* indicates the decimal point.

To read a decimal number, we first read the whole-number part, then we say *and* to indicate the decimal, then we read the decimal part. In reading the decimal part, we use the same procedure as for the whole-number part and end by reading the place value of the last digit in the decimal part. If the place value of the last digit in the decimal part is two words, these two words are hyphenated. For example, the ten-thousandths, hundred-thousandths, ten-millionths, and hundred-millionths places are all written with hyphens.

To read decimal numbers:

1. Mentally align the number on the decimal place-value chart so that the decimal point of the number is directly under the decimal point on the chart.
2. Read the whole-number part.
3. Use *and* for the decimal point only if there is a whole-number part.
4. Read the decimal part like the whole-number part.
5. End by reading the *place value* of the last digit in the decimal part.

EXAMPLE   Read 52.386 by writing it in words.

1. Align the number with the chart (see Fig. 1–6).
2. Read the whole-number part.

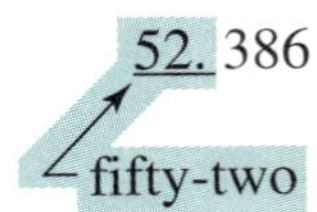

3. Use *and* for the decimal point because there is a whole-number part.

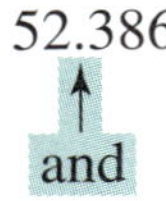

4. Read the decimal part like a whole-number part.

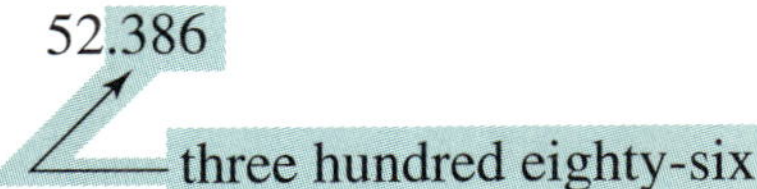

5. End by reading the *place value* of the last digit in the decimal part.

52.386 is "fifty-two and three hundred eighty-six thousandths."

Figure 1–6   Place-value chart.

■ **Learning Strategy**   ***Breaking a Process into Steps***

Sometimes processes that involve several steps can be overwhelming. Some strategies for handling multistepped processes are:

• Examine each step of the process to identify processes that are already familiar.

Look at Step 1 of reading a decimal. You experienced aligning numbers on a place-value chart in reading whole numbers. Step 2 refers to a skill that has already been learned. Step 3 gives specific instructions for connecting the steps. Step 4 applies the same skill as Step 2 to a different part of the number. Step 5 gives specific instructions for finishing the problem.

• Be sure that you can perform each step independently.

Since both Steps 2 and 4 involve reading whole numbers, review reading whole numbers if necessary.

• Examine how the steps are connected.

Steps 1 and 2 focus on the whole-number part of the number, and Steps 4 and 5 focus on the decimal part of the number. Step 3 connects the two parts.

<table>
<tr><td>Tip!</td><td>Informal Use of the Word Point:</td></tr>
<tr><td colspan="2">Informally, the decimal point is sometimes read as "point." Thus, 3.6 is read "three point six." The decimal 0.0162 can be read as "point zero one six two." This informal process is often used in communication to ensure that numbers are not miscommunicated.</td></tr>
</table>

EXAMPLE Read 0.0162 by writing it in words.

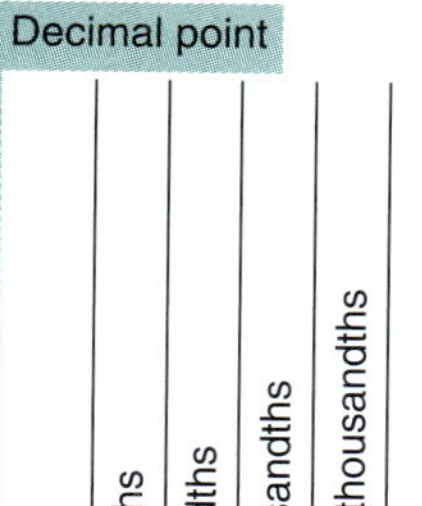

Figure 1–7 Place-value chart.

1. Align the number with the chart (see Fig. 1–7).
2. There is no whole-number part to read.

$$0.0162$$

3. Do *not* read the decimal point.

$$0.0162$$

4. Read the decimal part like a whole number.

$$0.0\ 162$$

one hundred sixty-two

5. End by reading the *place value* of the last digit in the decimal part.

$$0.016\ 2$$

ten-thousandths   Be sure to use the hyphen.

0.0162 is "one hundred sixty-two ten-thousandths."

<table>
<tr><td>Tip!</td><td>Unwritten Decimals</td></tr>
<tr><td colspan="2">When we write whole numbers using numerals we usually omit the decimal point; the decimal point is understood to be at the end of the whole number. Therefore, any whole number, such as 32, can be written without a decimal (32) or with a decimal (32.).</td></tr>
</table>

5 Write Fractions with Power-of-10 Denominators as Decimal Numbers.

On the job, we often encounter fractions with power-of-10 denominators like 10, 100, 1000, and so on. Examples of these are $\frac{1}{10}$ and $\frac{75}{100}$. These fractions are so quickly changed into decimals that many workers prefer to write them as decimals and perhaps use a calculator to arrive at the answer or solution to the problem. When we encounter fractions like $\frac{1}{10}$ or $\frac{75}{100}$, we may also want to use decimals instead. Here is a quick way to convert them.

Any fraction whose denominator is 10, 100, 1000, 10,000, and so on, can be written as a decimal number without making any calculations. The decimal number will have the same number of digits in its decimal part as the fraction has zeros in its power-of-ten denominator.

To write a fraction whose denominator is 10, 100, 1000, 10,000, and so on, as a decimal:

1. Use the denominator to find the number of decimal places.

Write $\frac{17}{1,000}$ as a decimal.

$0.\_\_\_$

$$10 \rightarrow 1 \text{ place}$$
$$100 \rightarrow 2 \text{ places}$$
$$1,000 \rightarrow 3 \text{ places}$$
$$10,000 \rightarrow 4 \text{ places}$$

2. Place the numerator so that the last digit is in the farthest place on the right.

$0.\_1\,7$

3. Fill in any blank spaces with zeros.

0.017

EXAMPLE Write $\frac{3}{10}$, $\frac{25}{100}$, $\frac{425}{100}$, and $\frac{3}{100}$ as decimal numbers.

$\frac{3}{10}$ is written **0.3.**

$\frac{25}{100}$ is written **0.25.**

$\frac{425}{100}$ is written **4.25.**

$\frac{3}{100}$ is written **0.03.**

Since $\frac{3}{10} = \frac{30}{100}$, then $0.3 = 0.30$. Similarly, $0.7 = 0.70 = 0.700$.

Tip!	***Do Zeros Change the Value of a Decimal Number?***

When we attach zeros on the *right* end of a decimal number, we do not change the value of the number.

$$0.5 = 0.50 = 0.500$$

6 Compare Decimal Numbers.

Our jobs may require us to work with close tolerances. Perhaps we are working with wire and need to compare the diameters of several sizes of wire to arrange them in order from largest to smallest or smallest to largest. To do this, we need to be able to compare decimals, because wires are often measured in decimals.

To make valid comparisons, we must compare like amounts. Tenths compare with tenths, thousandths compare with thousandths, and so on. One way to compare decimals is to make the decimal parts have the same number of places by attaching zeros at the right end. For instance, to compare 0.7 and 0.68, we change 0.7 to 0.70. Now, which is larger, 0.70 or 0.68? 0.70 is larger. We can also compare digits place by place.

> **To compare decimal numbers:**
>
> 1. Compare whole-number parts.
> 2. If the whole-number parts are equal, compare digits place by place, starting at the tenths place and moving to the right.
> 3. Stop when two digits in the same place are different.
> 4. The digit that is larger determines the larger decimal number.

EXAMPLE Compare the two numbers to see which is larger.

$$32.47 \qquad 32.48$$

1. Look at the whole-number parts: They are the same.
2. Look at the tenths place for each number. Both numbers have a 4 in the tenths place.
3. Look at the hundredths place. They are different.
4. **32.48 is the larger number** because 8 is larger than 7.

Two symbols for showing the relationship between two numbers are $<$ and $>$. These symbols are called *inequality symbols,* since they show that two numbers are *not* equal.

$5 < 7$	is read	*"5 is less than 7."*
$7 > 5$	is read	*"7 is greater than 5."*

EXAMPLE Write two inequalities for the pair of whole numbers 238 and 251.

Both numbers have the same number of digits.
Both numbers have a two in the hundreds place.
The digits in the tens place are different, with 5 being the larger digit. Therefore, 251 is the larger number.

$$238 < 251 \qquad \text{or} \qquad 251 > 238$$

EXAMPLE Write two inequalities for the numbers 0.4 and 0.07.

Since the whole-number parts are the same (0), we compare the digits in the tenths place. 0.4 is larger because 4 is larger than 0.

$$0.4 > 0.07 \qquad \text{or} \qquad 0.07 < 0.4$$

If we write both numbers in the preceding example so that they have the same number of digits in the decimal part, they may be easier to compare.

$$0.4 = 0.40$$

$$0.07 = 0.07$$

Since 40 is larger than 7, 0.4 is larger than 0.07. This is equivalent to the process used in comparing common fractions. A common denominator is found, and each fraction (or decimal fraction) is changed to an equivalent fraction with the common denominator.

Tip!	***Common Denominators in Decimals:***

Decimal fractions have a common denominator if they have the same number of digits to the right of the decimal point.

⑦ Round a Whole Number or a Decimal Number to a Place Value.

To check calculations made with a calculator or to make mental calculations, it is less cumbersome to use approximate numbers instead of exact numbers. To *round* a number is to express it as an *approximation*. Whenever a decimal (or any number) is rounded, it becomes less accurate than the original decimal or number. Thus, if we round $3.99 to $4.00, the $4.00 is less accurate than the original $3.99. However, the difference between $3.99 and $4.00 is slight, only 1 cent. On the other hand, if we round $4.25 to $4.00, the difference here is greater and the rounded decimal is much more inaccurate, with a difference of 25 cents. Some job applications require more accuracy in rounding than others.

In rounding, we are deciding that a number is closer to one of two numbers. For example, if 37 is rounded to the nearest ten, is it closer to 30 or 40? Look at a portion of the number line that includes the whole numbers from 30 to 40. The halfway point between 30 and 40 is 35. Locate 37 on the number line.

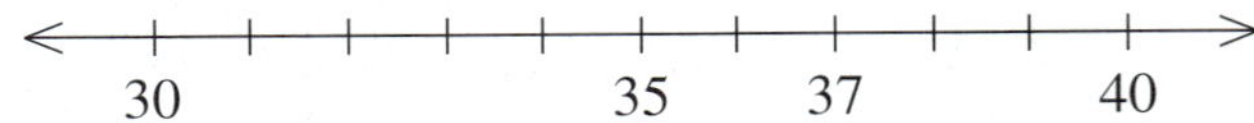

Is 37 closer to 30 or 40? 40. Thus, 40 is a better approximation to the nearest ten for 37. Another way to say this is that 37 rounded to the nearest ten is 40.

When rounding a number to a certain place value, we must make sure that we are as accurate as our employer wants us to be. Generally, the size of the number and its use dictates the decimal place to which it should be rounded.

■ **Learning Strategy**   *Procedures Do Not Substitute for Understanding Concepts*

Once you understand that rounding means to find the closest approximate number to a given number, you can use procedures or rules to facilitate rounding. Procedures are meaningless and are easily confused or forgotten if you do not understand the concept.

To round a whole or decimal number to a given place value:

1. Locate the digit that occupies the rounding place. Then examine the digit to the immediate right.
2. If the digit to the right of the rounding place is 0, 1, 2, 3, or 4, do not change the digit in the rounding place. If the digit to the right of the rounding place is 5, 6, 7, 8, or 9, add 1 to the digit in the rounding place.
3. Replace all digits to the *right* of the digit in the rounding place with zeros if they are to the left of the decimal point. Drop digits to the right of the digit in the rounding place *and* to the right of the decimal point.

EXAMPLE   Oregon has a land area of 96,187 square miles. What would be a reasonable approximate number for this land area?

96,187 rounds to the following approximate numbers:

96,190 to the nearest ten

96,200 to the nearest hundred

96,000 to the nearest thousand

100,000 to the nearest ten-thousand

Deciding to which place to round a value is a judgment depending on what use you will make of the rounded or approximate value. **Both 96,000 and 100,000 are reasonable approximations.**

If the directions do not give a specific rounding place, choose a place that is reasonable for the context of the problem. We illustrate this with two different situations:

If we are comparing the land area of Oregon (96,187 square miles) with the land area of Wyoming (96,988 square miles), Oregon has approximately 96,000 square miles, while Wyoming has approximately 97,000 square miles.

If we are comparing the land area of Oregon (96,187 square miles) with the land area of Texas (262,015 square miles), Oregon has approximately 100,000 square miles, while Texas has approximately 300,000 square miles.

Tip!	*Nine Plus One Still Equals Ten:*

When the digit in the rounding place is 9 and must be rounded up, it becomes 10. The 0 replaces the 9 and 1 is carried to the next place to the left.

EXAMPLE   Is 24.63 closer to 24.6 or 24.7?

We round to the tenths place. Applying the rounding rules, we circle the 6, which is in the tenths place, to indicate the rounding place.

$$24.\textcircled{6}3$$

Because the digit to the right of the 6 is 3, we do not change the 6.

$$24.\textcircled{6}3$$
$$6$$

Next, we write all digits to the left of the 6 as they are.

$$24.\textcircled{6}3$$
$$\mathbf{24.6}$$

Because the 3, which is to the right of 6, is to the right of the decimal, we drop it. **24.63 is closer to 24.6.**

EXAMPLE   Round 46.879 to the hundredths place.

46.8⑦9   7 is in the hundredths place.

46.8⑦9   The next digit to the right is 9, so add 1 to 7.
    8

46.8⑦9   Write all digits to the left of 7 as they are.
46.88

46.88   Drop the 9 because it is to the right of the rounding place and to the right of the decimal.

EXAMPLE   Round 32.6 to the tens place.

③2.6   3 is in the tens place.

③2.6   Leave 3 unchanged because 2, the next digit to the right, is less than 5.
3

30   Replace 2 with a zero because it is to the left of the decimal. Drop the 6 because it is to the right of the decimal.

EXAMPLE   Round 15.8 to the nearest whole number.

Rounding to the nearest whole number means rounding to the *ones* place.

1⑤.8   5 is in the ones place. Add 1 to 5 and drop the 8 because it is to the right of the decimal.

1 6

EXAMPLE   Round $293.48 to the nearest dollar.

When we round to the nearest dollar, we are rounding to the *ones* place.

$$\$29③.48$$

$293

EXAMPLE   Round $71.8986 to the nearest cent.

One cent is 1 hundredth of a dollar, so to round to the nearest cent is to round to the *hundredths* place.

$$\$71.8⑨86$$

$71.90

EXAMPLE   A motor has a horsepower rating of 30.497. Round this rating to the nearest whole number.

$$3⓪.497$$

30

8 Round a Whole Number or a Decimal Number to a Number with One Nonzero Digit.

We can also round so that the rounded number has only one digit that is not a zero. The only difference in this type of rounding and rounding to a specified place value is that the first nonzero digit determines the rounding place.

Look at the following numbers. Notice that the *first* digit on the left is a *nonzero digit* (not a zero); the other digits are all zeros.

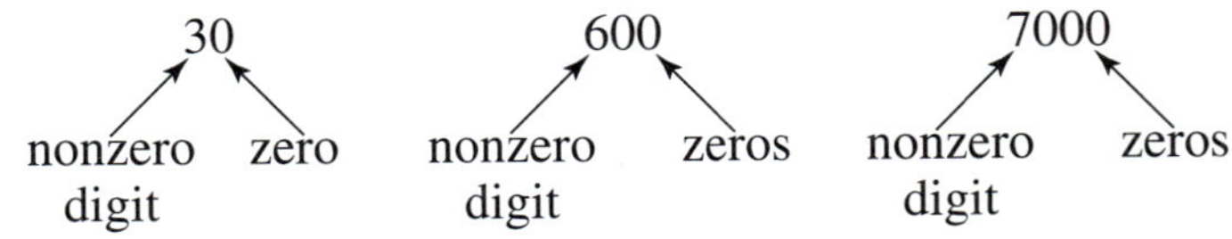

■ **Learning Strategy**   *Cumbersome Terminology*

Why do we use the terminology *one nonzero digit* or *first nonzero digit* instead of just saying "the first digit on the left"? All whole numbers begin with a number that is not a zero.

We want to introduce terms now that will be used later in similar procedures. In a decimal number like 0.00738, the *first nonzero digit*, 7, is not the first digit on the left.

To round to one nonzero digit:

1. Starting at the *left,* find the *first* digit that is not zero. This digit will be a 1, 2, 3, 4, 5, 6, 7, 8, or 9.
2. Round the number to the place value of the first nonzero digit.
3. Replace all digits to the right of the rounding place with zeros up to the decimal point. Drop any digits that are to the right of the digit in the rounding place *and* that follow the decimal point.

Rather than learn a new procedure in isolation, compare it with previously learned skills. How is the procedure for rounding decimals similar to the procedure for rounding whole numbers? In both cases, you

- Locate the rounding place.
- Examine the next digit to the right.
- Leave the digit in the rounding place the same if the next digit is 0, 1, 2, 3, or 4.
- Add one to the digit in the rounding place if the next digit is 5, 6, 7, 8, or 9.

How are the procedures different?

- In whole numbers, you replace digits to the right of the rounding place with zeros.
- In decimal numbers, you drop digits to the right of the rounding place that are also to the right of the decimal.

This strategy is often referred to as *making connections or linkages.*

EXAMPLE   Round 78.4 to one nonzero digit.

1. Find the first nonzero digit from the left.

78.4

The first nonzero digit is 7.

⑦8.4

2. Round to the place value of the first nonzero digit.

⑦8.4   Add 1 to 7 because the next digit to the right is 5 or more.
8

3. Replace each digit between the rounded digit and the decimal point with a zero. Drop all digits after the decimal point.

80   Replace the digit between 8 and the decimal point with a 0. Drop the digits after the decimal point.

EXAMPLE   Round 0.83 to one nonzero digit.

0.⑧3   Locate the first nonzero digit.
0.8

EXAMPLE   If the price of a typewriter is $497.95, estimate this cost to one nonzero digit.

④97.95
$500

Tip!	*Exact Amount Versus Approximate Amount:*

When an amount has been rounded, it is no longer an exact amount. The rounded amount is now an *approximate amount.*

1 In the number 2,304,976,186, identify the place value of these digits:

1. 3   **2.** 7   **3.** 1   **4.** 0   **5.** 2

In the number 8,972,069,143, identify the place value of these digits:

6. 0   **7.** 4   **8.** 7   **9.** 8   **10.** 6

2 Show how these numbers are read by writing them as words.

11. 6704   **12.** 89021   **13.** 662900714
14. 3000101   **15.** 15407294376   **16.** 150

Write these words as numbers in standard notation. Use commas when necessary.

17. Seven billion, four hundred

18. One million, six hundred twenty-seven thousand, one hundred six

19. Fifty-eight thousand, two hundred one

20. In a telephone conversation, a contractor submitted the following bid for a job: "one thousand six dollars." Write this bid as a number.

Write in expanded notation.

21. 718   **22.** 42   **23.** 1,983   **24.** 8,021   **25.** 52,010   **26.** 700

Write as ordinary numbers.

27. $(5 \times 100) + (3 \times 10) + (7 \times 1)$ **28.** $(9 \times 1,000) + (0 \times 100) + (8 \times 1)$
29. $(6 \times 10) + (5 \times 1)$
30. $(4 \times 10,000) + (0 \times 1,000) + (2 \times 100) + (1 \times 10) + (7 \times 1)$

3

31. What is the place value of 7 in 32.407?   **32.** What is the place value of 8 in 28.396?
33. What is the place value of 4 in 3.00254?   **34.** What is the place value of 3 in 457.2096532?

35. What is the place value of the 7 in 0.0387?

In the following problems, state what digit is in the place indicated.

36. Tens: 46.3079   **37.** Tenths: 2.0358
38. Thousandths: 520.0765   **39.** Ten-millionths: 3.002178356
40. Hundredths: 402.3786

4 Write the words for the decimal numbers.

41. 21.387   **42.** 420.059   **43.** 0.89
44. 0.0568   **45.** 30.02379   **46.** 21.205085

Write the digits for the word numbers.

47. Three and forty-two hundredths

48. Seventy-eight and one hundred ninety-five thousandths

49. Five hundred and five ten-thousandths

50. Seventy-five thousand thirty-four hundred-thousandths

5 Write the fractions as decimal numbers.

51. $\dfrac{5}{10}$ **52.** $\dfrac{23}{100}$ **53.** $\dfrac{7}{100}$ **54.** $\dfrac{683}{100}$ **55.** $\dfrac{79}{1000}$

56. $\dfrac{468}{1000}$ **57.** $\dfrac{587}{100}$ **58.** $\dfrac{108}{1000}$ **59.** $\dfrac{603}{100}$ **60.** $\dfrac{400}{100}$

6 Compare the number pairs and identify the larger number.

61. 3.72, 3.68   **62.** 7.08, 7.06   **63.** 0.23, 0.3
64. 0.56, 0.5   **65.** 2.75, 2.65   **66.** 0.157, 0.2

Compare the number pairs and identify the smaller number.

67. 8.9, 8.88   **68.** 0.25, 0.3   **69.** 0.913, 0.92

70. 0.761, 0.76   **71.** 5.983, 6.98   **72.** 1.972, 1.9735

Arrange the numbers in order from smallest to largest.

73. 0.23, 0.179, 0.314   **74.** 1.9, 1.87, 1.92   **75.** 72.1, 72.07, 73

76. Two micrometer readings are recorded as 0.837 in. and 0.81 in. Which is larger?

77. A micrometer reading for a part is 3.85 in. The specifications call for a dimension of 3.8 in. Which is larger, the micrometer reading or the specification?

78. A washer has an inside diameter of 0.33 in. Will it fit a bolt that has a diameter of 0.325 in.?

79. Aluminum sheeting can be purchased in thicknesses of 0.04 in. or 0.035 in. Which sheeting is thicker?

80. If No. 14 copper wire has a diameter of 0.064 in., and No. 10 wire has a diameter of 0.09 in., which has the larger diameter?

Write two inequalities for each pair of numbers.

81. 42 and 38   **82.** 168 and 160   **83.** 709 and 721

84. 5,297 and 984   **85.** 2,742 and 27,420   **86.** 150,000 and 134,812

87. Cleveland is 119 miles from Toledo, Ohio, and Detroit is 62 miles from Toledo. Write two inequalities that compare these distances. Write a statement about the relative distances of Detroit and Cleveland from Toledo. Compare your statement with those of your classmates.

88. The average July temperature for Laredo, Texas, is 96°F, and the average July temperature for Eugene, Oregon, is 82°F. Write two inequalities to compare the temperatures in these two cities. Write a statement in words to illustrate one of your inequalities.

89. Spokane, Washington, has an average January temperature of 26°F and an average July temperature of 70°F. Write two inequalities to compare Spokane's winter and summer temperatures.

90. San Diego's average January temperature is 57°F, and its average July temperature is 71°F. Write two inequalities to compare San Diego's winter and summer temperatures.

91. The Mexico City earthquake measured 8.1 on the Richter scale, and the 1906 San Francisco earthquake measured 8.3 on the Richter scale. Which earthquake had the greater Richter scale rating?

92. In 1980, the average annual population growth rate for the United States was 1.14, and in 1990, it was 1.02. From these two figures, would you say the growth rate for the United States appears to be rising or falling?

93. For the 1990–2000 decade, the average annual growth rate was 1.7 for Asia and 2.0 for the developing world. Is Asia growing faster or slower than the developing world?

94. If you roll a pair of dice, the probability of rolling a sum of 5 is 0.111 and the probability of rolling a sum of 6 is 0.139. Which sum has the greater probability of being rolled?

95. A 100-watt bulb that burns continuously for two minutes uses 0.003 kilowatt-hours (kWh) of electricity, and an 800-watt toaster uses 0.026 kilowatt-hours (kWh) for a piece of toast. Which uses the greater number of kilowatt-hours?

96. To change centimeters to inches, multiply by 0.394 and to change kilometers to miles, multiply by 0.621. Which factor is larger?

7 Round to the place value indicated.

97. Nearest hundred: 468   **98.** Nearest hundred: 6248

99. Nearest thousand: 8263   **100.** Nearest ten thousand: 429,207

101. Nearest thousand: 39,748   **102.** Nearest ten thousand: 39,748

103. Nearest million: 285,487,412   **104.** Nearest ten: 468

105. Nearest billion: 82,629,426,021   **106.** Nearest ten million: 297,384,726

The *Rand McNally Road Atlas and Travel Guide* gives the altitudes of the following cities:

Corpus Christi, Texas	35 feet above sea level
Denver, Colorado	5,280 feet above sea level
Jacksonville, Florida	19 feet above sea level
Indianapolis, Indiana	717 feet above sea level
Salt Lake City, Utah	4,260 feet above sea level

107. Use your knowledge of rounding to compare the approximate altitudes of Corpus Christi and Jacksonville. Indicate the rounding place you used.

108. To what place should you round to compare the approximate altitudes of Denver and Salt Lake City? Round to show the comparison of approximate altitudes.

109. To what place should you round to show the approximate comparison of altitudes of Indianapolis and Salt Lake City? Round to show the comparison of approximate altitudes.

110. Discuss your decisions for selecting a place value for rounding when comparing the approximate altitudes of the three pairs of cities in Exercises 107–109.

Round to the nearest whole number.

111. 42.7 **112.** 367.43 **113.** 7.983 **114.** 103.06 **115.** 2.9

Round to the nearest tenth.

116. 8.05 **117.** 12.936 **118.** 42.574 **119.** 83.23 **120.** 5.997

Round to the nearest hundredth.

121. 7.036 **122.** 42.065 **123.** 0.792 **124.** 3.198 **125.** 7.773

Round to the nearest thousandth.

126. 0.2173 **127.** 0.0196 **128.** 1.5085 **129.** 4.2378 **130.** 7.0039

Round to the nearest dollar.

131. $219.46 **132.** $82.93 **133.** $507.06 **134.** $2.83 **135.** $5.96

Round to the nearest cent.

136. $8.237 **137.** $0.291 **138.** $0.528 **139.** $5.796 **140.** $238.9238

141. A micrometer measure is listed as 0.7835 in. Round this measure to the nearest thousandth.

142. The diameter of an object is measured as 3.817 in. If specifications call for decimals to be expressed in hundredths, write this measure according to the specifications.

143. To the nearest tenth, what is the current of a 2.836-amp motor?

8 Round to numbers with one nonzero digit.

144. 483 **145.** 7.89 **146.** 62.5 **147.** 0.537 **148.** 0.0086
149. 0.095 **150.** 3.07 **151.** 52 **152.** 83.09 **153.** 52.8

154. If round steak costs $2.78 per pound, what is the cost per pound to the nearest dollar?

155. An estimate calls for converting 23.077 to an approximate number. What is the approximate number rounded to one nonzero digit?

156. Monthly rainfall (in inches) was identified for these months: January, 0.355; March, 1.785; May, 0.45; July, 1.409; September, 0.07; and December, 2.018. Identify the months with the most similar rainfall by rounding to one nonzero digit.

157. The average response times (in seconds) for drivers braking when they first see a road hazard were measured as follows: driver A, 0.0275; driver B, 0.0264; driver C, 0.0234; driver D, 0.0284; and driver E, 0.0379. Round each response time to one nonzero digit to identify the drivers whose response time was most similar.

Learning Outcomes

1 Add whole numbers.
2 Add decimal numbers.
3 Estimate and check addition.
4 Add using a calculator.

At one time or another, all employees add whole numbers on their jobs. Our purpose in this section is to understand addition and the basic properties that give us flexibility in the way we add. Also, we will review ways to gain speed and make fewer mistakes when adding whole numbers.

1 **Add Whole Numbers.**

Commutative and Associative Properties of Addition It is sometimes easier to discuss the procedures for working a problem if we name the parts of the problem. In an addition problem, the numbers being added are called *addends* and the answer is called the *sum* or *total*.

Two useful things to know about addition are that it is *commutative* and *associative*. By *commutative*, we mean that it does not matter in what *order* we add numbers. We can add 7 and 6 in any order and still get the same answer:

$$7 + 6 = 13 \qquad 6 + 7 = 13$$

■ **DEFINITION: Commutative Property.** The *commutative property* means that values being added (or multiplied) may be added (or multiplied) in any order.

By *associative*, we mean that we can *group* numbers together any way we want when we add and still get the same answer. To add $7 + 4 + 6$, we can group the $4 + 6$ to get 10; then we add the 10 to the 7:

$$7 + (4 + 6) = 7 + 10 = 17$$

Or we can group $7 + 4$ to get 11; then we add the 6 to the 11:

$$(7 + 4) + 6 = 11 + 6 = 17$$

Addition is a *binary operation;* that is, the rules of addition apply to adding *two* numbers at a time. The associative property of addition shows how addition is extended to more than two numbers.

■ **DEFINITION: Associative Property.** The *associative property* means that values being added (or multiplied) may be grouped in any manner.

The commutative and associative properties of addition can help us add whole numbers quickly and accurately. Look for these two aids in the examples that follow.

■ **Learning Strategy** *Symbolic Representation of Properties*

If we don't allow ourselves to be intimidated by symbols, we can develop a genuine appreciation for the symbolic representation of properties. Sometimes the simplest property can be very awkward or cumbersome to write in words.

To develop your mathematics communication skills, we suggest that as often as possible, write properties **both** symbolically and in words.

Using Symbols to Write Rules and Definitions

Many rules and definitions can be written symbolically. Symbolic representation allows a quick recall of the rule or definition.

Commutative Property of Addition: Two numbers may be added in any order and the sum remains the same.

$$a + b = b + a \quad \text{where } a \text{ and } b \text{ are numbers.}$$

Associative Property of Addition: Three numbers may be added using different grouping and the sum remains the same.

$$a + (b + c) = (a + b) + c \quad \text{where } a, b, \text{ and } c \text{ are numbers.}$$

The associative property of addition also allows other possible groupings and extends to more than three numbers.

$$7 + 4 + 6 = 13 + 4 = 17$$
$$3 + 5 + 7 + 9 = 8 + 16 = 24$$

EXAMPLE Add $7 + 3 + 6 + 8 + 2 + 4$ using the commutative and associative properties.

$$
\begin{array}{l}
7 \\
3 \\
6 \\
8 \\
2 \\
+4 \\
\hline
\mathbf{30}
\end{array}
$$

Group $7 + 3$, $8 + 2$, and $6 + 4$. Then add the three groups ($10 + 10 + 10 = 30$).

Another important property of addition is the *zero property of addition.*

Zero property of addition:

Adding zero to any number results in the same number.

$$n + 0 = n \quad \text{or} \quad 0 + n = n$$
$$5 + 0 = 5 \quad \text{or} \quad 0 + 5 = 5$$

When adding numbers of two or more digits, such as $238 + 456$, we must make sure that the numbers are placed in columns properly. This means that the same place values must be aligned under one another so that all the ones are in the far right column, all the tens in the next column, and so on.

To add numbers of two or more digits:

1. Arrange the numbers in columns so that the ones place values are in the same column.
2. Add the ones column, then the tens column, then the hundreds column, and so on, until all the columns have been added. *Carry* whenever the sum of a column is more than one digit.

EXAMPLE   Shipping fees are often charged by the total weight of the shipment. Find the total weight of this order: nails, 250 pounds (lb); tacks, 75 lb; brackets, 12 lb; and screws, 8 lb. Arrange in columns and add.

$$
\begin{array}{r}
{}^{1\,1} \\
250 \\
75 \\
12 \\
+\quad 8 \\
\hline
345
\end{array}
$$

The sum of the digits in the right column is 15. Record the 5 in the ones column and *carry* the 1 to the tens column.

The total weight is 345 lb.

2 Add Decimal Numbers.

We add whole numbers by adding in columns all digits in the ones place, then all digits in the tens place, and so on. To add decimal numbers, we will follow the same procedure. When decimal numbers are aligned in this manner, all decimal points fall in the same vertical line. Aligning decimal points has the same effect as using *like* denominators when adding (or subtracting) fractions. Like denominators are discussed in Chapter 3.

> *To add decimals:*
>
> Arrange the numbers so that the decimal points are in one vertical line. Then add each column. Align the decimal for the sum in the same vertical line.

EXAMPLE   Add 42.3 + 17 + 0.36.

$$
\begin{array}{r}
42.3 \\
17 \\
0.36 \\
\hline
\end{array}
$$

Note that the decimal in 17 is understood to be at the right end.

We should be very careful in aligning digits. Careless writing can cause unnecessary errors. To avoid difficulty in adding the columns, we may write each number so that all have the same number of decimal places by attaching zeros on the right.

$$
\begin{array}{r}
42.30 \\
17.00 \\
0.36 \\
\hline
59.66
\end{array}
$$

3 Estimate and Check Addition.

When you go to the grocery store and must stay within a budget, how do you know before you pay for your purchases if you have stayed within your budget? When you read the label on a paint can to determine how much paint you need for a remodeling project, how do you decide? You estimate. Estimating is an increasingly important skill for everyone to develop. Estimating depends in part on your number sense. You aren't born with number sense. It must be developed. We develop and strengthen our number sense by estimating and doing mental calculations.

   Estimating a sum before we perform the actual addition gives us an approximate answer. Estimating sums is important when we make mental calculations in the workplace or marketplace. It is also important when we use a calculator.

   When the difference between the exact sum and the estimated sum is large, the exact sum may be incorrect and we should add the numbers again.

It is also important to check calculations. In the workplace, it is often expected that calculations that are made on a calculator are to be done twice. This is one way to make fewer errors.

EXAMPLE Elston Home Renovators spent the following amounts on a job: $16,466.15, $23,963.10, and $5,855.20. Estimate the total amount by rounding to thousands. Then find the exact amount and check your answers.

Thousands Place	Estimate	Exact	Check
$16,466.15	$16,000	$16,466.15	$16,466.15
23,963.10	24,000	23,963.10	23,963.10
5,855.20	6,000	5,855.20	5,855.20
	$46,000	**$46,284.45**	**$46,284.45**

The estimate and exact answer are close. The exact answer is reasonable.

Tip!	***Accuracy of Estimates***

The accuracy of estimates varies depending on the place value to which the addends are rounded. The context of the problem will help in determining a reasonable rounding place.

4 Add Using a Calculator.

Knowing the properties of addition helps us estimate sums mentally or add small numbers mentally; however, we use calculators to add large numbers or to add several numbers. Many types of calculators are available, so we will help you determine how your calculator works. You will develop better calculator skills if you use the same calculator all the time. Labeling, key location, and sequences for entering numbers and operations vary widely from one calculator to another.

■ **Learning Strategy** *Test your calculator by entering a problem that you can do mentally.*

Add 3 + 5 on your calculator. Some options are

$$3 \boxed{+} 5 \boxed{=}$$

$$3 \boxed{+} 5 \boxed{+} \boxed{T}$$ $\boxed{T}$ represents Total.

$$3 \boxed{+} 5 \boxed{ENTER}$$

$$3 \boxed{+} 5 \boxed{EXE}$$ $\boxed{EXE}$ represents Execute or complete the instructions.

To add more than two numbers, does your calculator accumulate the total in the display as you enter numbers? Or, does your calculator wait and give the total after the $\boxed{\text{ENTER}}$, $\boxed{\text{EXE}}$, $\boxed{\text{T}}$, or $\boxed{=}$ key is pressed? We use the symbol $\Rightarrow$ to indicate that the digits showing in the calculator display will follow.

Tip! | ***Decimal Point and Zeros on the Calculator***

Only one new key has to be examined to perform any type of calculation with decimals on a calculator. The decimal key $\boxed{\cdot}$ is most often located near the number keys. This key is pressed when the decimal point appears in the number being entered.

Does the zero to the left of the decimal have to be entered before the decimal in a number like 0.5? Check it out: Add $3 + 0.2$ on the calculator.

$$\text{Options:} \quad 3\boxed{+}0\boxed{\cdot}2\boxed{=}$$
$$3\boxed{+}\boxed{\cdot}2\boxed{=}$$

Next, do zeros that follow decimals have to be entered?
Check it out: Add $3.00 + 1.50$ on the calculator.

$$\text{Options:} \quad 3\boxed{\cdot}00\boxed{+}1\boxed{\cdot}50\boxed{=}$$
$$3\boxed{+}1\boxed{\cdot}5\boxed{=}$$

EXAMPLE Estimate the sum, then use a calculator to find the sum.

$$2{,}345 + 3{,}894.745 + 758.05 \;=$$

Estimate:

$$2{,}000 + 4{,}000 + 1{,}000 = \mathbf{7{,}000}$$

Exact sum:

$$2{,}345\boxed{+}3{,}894\boxed{\cdot}745\boxed{+}758\boxed{\cdot}05\boxed{=} \;\Rightarrow\; \mathbf{6{,}997.795}$$

Tip! | ***Notation for Calculator Examples:***

The value following the symbol $\Rightarrow$ indicates the final answer in the calculator display.

■ **Learning Strategy** *Use Your Calculator to Verify Concepts*

The process of verifying a concept with a calculator does not constitute a formal proof, but it can be used to investigate speculations. Verify several examples to be sure you aren't making an improper generalization.

EXAMPLE Use your calculator to verify that $a + b = b + a$.

Try several examples. It helps to organize the results of your investigation in a table.

a	b	$a + b$	$b + a$
5	8	13	13
27	39	66	66
207	417	624	624

Are there any situations that might not work? Investigate them. What is your conclusion?

$$a + b = b + a \quad \text{for all whole numbers}$$

SELF-STUDY EXERCISES 1–2

1 Add.

1.	2.	3.	4.
4	6	3	6
1	9	5	4
5	7	2	5
3	4	4	6
+ 2	+ 1	4	7
		+ 1	+ 6

5. In an inventory, the following $\frac{1}{4}$-inch (in.) hex bolts of various lengths are counted: nine 1 in. long, four $1\frac{1}{4}$ in., seven $1\frac{1}{2}$ in., six 2 in., two $2\frac{1}{4}$ in., and nine $2\frac{1}{2}$ in. How many $\frac{1}{4}$-in. hex bolts are there in all?

6.
```
  1072
  6710
+ 1410
```

7.
```
  5273
  4001
+ 7682
```

8.
```
  59,718
+ 46,567
```

Write in columns and add.

9. $36 + 482 + 961 + 27 + 804$

10. $4,582 + 86,724 + 482 + 5,826$

11. A mechanic was paid the following for 5 days of work: $86, $124, $67, $85, and $94. How much was the mechanic paid for the 5 days?

12. A truck driver has the following weight slips on three loads of gravel: 8,114 lb, 8,027 lb. and 8,208 lb. What is the total weight of the gravel hauled by the driver?

13. During inventory, the following numbers of slotted-head screws were counted in four different boxes: 84, 63, 72, and 79. How many slotted-head screws were there all together?

14. Four bricklayers were working on the same job. In one day they laid the following numbers of bricks: 1,217, 1,103, 1,039, and 1,194. How many bricks did all four lay?

15. Canty Robbins, Director of Purchasing, placed an order for 15 gallons (gal) of white paint, 27 gal of black paint, 5 gal of crimson red paint, and 3 gal of canary yellow paint. How many gallons of paint were ordered?

2 Add.

16. $4.2 + 3.6 + 7.9$

17. $12.8 + 13.52 + 7.86$

18. $3.9 + 4.02 + 0.21$

19. $8.9 + 6.72 + 3.58 + 68.2$

20. $83.37 + 42 + 1.6 + 3$

21. $7 + 4.2 + 14.6 + 0.23$

22. $23.9 + 54.3$

23. $24.5 + 21.2$

24. $205.03 + 56.305$

25. $309.01 + 47.602$

26. $0.784 + 5 + 1.2$

27. $0.225 + 7 + 2.3$

28. $900.75 + 225.85$

29. $300.25 + 113.35$

30. $3.7 + 0.6 + 4.8 + 9$

31. $7.1 + 0.2 + 5.5 + 6$

32.
```
  0.70868
+ 0.10937
```

33.
```
  0.83967
+ 0.33675
```

34.
```
  51.006
+  4.507
```

35.
```
  82.005
+  3.406
```

36. $3.487 + 47.5 + 19$

37. A 0.103-in.-thick pipe has an inside diameter of 2.871 in. Find the outside diameter of the pipe.

38. The total current in amps (A) in a parallel circuit is found by adding the individual currents. If a circuit has individual currents of 3.98 A, 2.805 A, and 8.718 A, find the total current.

39. A part-time hourly worker earned $25.97 on Monday, $7.48 on Tuesday, $5.88 on Wednesday, $65.45 on Thursday, and $76.47 on Friday. Find the total week's wages.

40. A four-sided residential lot that measures 100.8 ft, 87.3 ft, 104.7 ft, and 98.6 ft is to be fenced. How many feet of fencing are required?

41. A technician purchased an ac voltage sensor for $11.95, a grounded outlet analyzer for $5.99, and a neon circuit tester for $1.99. How much did she pay for all three items?

42. Julio purchased a $\frac{3}{32}$-in. submini stereo plug adapter for $2.99 and a $\frac{3}{32}$-in. submini mono plug adapter for $1.76. If the sales tax was $0.29, what was his total bill?

43. A patient's normal body temperature registered at 98.2°F and his temperature rose 2.7°. What was his increased body temperature?

44. An investment portfolio is a listing of a person's investments. If you have an investment portfolio that totals $25,915.53 at the beginning of the year and it increases by $2,418.48 over the one-year period, what is the value of your portfolio at the end of the year?

45. eBay is an Internet auction house whose stock is traded publicly. When it was first traded, its stock sold at $8.4375 a share, then immediately increased by $215.8425 a share. What was the new price per share?

46. eBay had 1.2 million users in fall 1998, and by spring 1999, that number had increased by 2.6 million users. How many people were using eBay in early 1999?

47. If your college tuition is $1,564.50 and books cost $624.95, what is the total cost of your tuition and books?

48. Carlee's car odometer registered 36,263.7, then she drove 638.2 miles. What should her odometer have registered at the end of the trip?

49. Lucia purchased groceries costing $5.15, $1.84, $0.59, $7.21, and $1.47 for a party. What was the total cost of her purchases?

50. Caroline sets up a booth once a month at a flea market. For the most recent market, she accepted checks in the amounts of $85.50, $21.75, $13.83, $62.37, and $121.52. What is the total of these checks?

Estimate the sum by rounding to hundreds. Then find the exact sum and check.

51. 6,367.32
24,036.95
582.14

52. 532.73
1,823.14
2,384.93

53. 2,546.28
3,626.13
831.86
6,153.15

54. On a trip from Albuquerque, New Mexico, to Blowing Rock, North Carolina, Lynette and Tosha purchased 14.2, 12.9, 15.3, 9.2, 8.6, and 12.1 gallons of fuel. Estimate the amount of fuel they purchased by rounding to the nearest ten, then find the exact amount of fuel and compare the estimate with the exact amount.

3 Estimate the sum by rounding to hundreds. Then find the exact sum and check.

55. 4,256.65
3,892.10
576.46
8,293.00

56. 52,843
17,497
13,052
821

57. 24,003
5,874
319,467
52,855

58. Palmer Associates provided the following prices for items needed to build a sidewalk: concrete, $2,583.45; wire, $43.25; frame material, $18.90; labor, $798. Estimate the cost by rounding each amount to the nearest ten. Find the exact total cost.

59. Antonio Juarez expects to spend the following amounts for college next semester: food, $1,500; lodging, $1,285; books, $288; supplies, $130; transportation, $162. Estimate his expenditures for one semester by rounding each amount to the nearest hundred. Calculate the exact amount.

Estimate the sums by rounding addends to numbers with one nonzero digit. Then find the exact sum and check.

60.   940
+ 8,299

61.   478.125
+ 146.055

62.   1,901
+ 6,548

63.   149.25
+ 652.14

64.   16,259
+ 36,542

65.   32,501
+ 16,740

66. A hardware store filled the following order for nails: 25 lb, $2\frac{1}{2}$-in. common; 16 lb, 4-in. common; 12 lb, 2-in. siding; 24 lb, $2\frac{1}{2}$-in. floor brads; 48 lb, 2-in. roofing; and 34 lb, $2\frac{1}{2}$-in. finish. What was the total weight of the order?

68. A printer has three printing jobs that require the following numbers of sheets of paper: 185, 83, and 211. Will one ream of paper (500 sheets) be enough to finish the three jobs?

67. If four containers have a capacity of 12 gal, 27 gal, 55 gal, and 21 gal, can 100 gal of fuel be stored in these containers? (Find the total capacity of the containers first.)

69. How many feet of fencing are needed to enclose the area shown in Fig. 1–8?

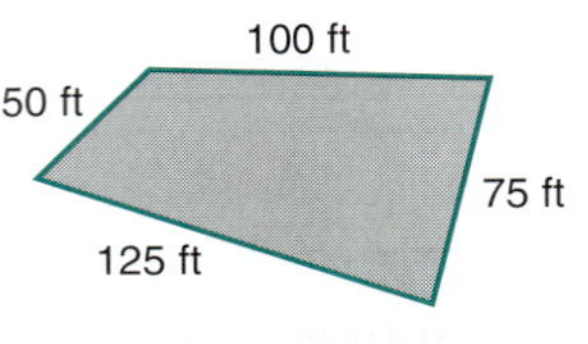

Figure 1–8

4 Estimate the sums by rounding addends to numbers with one nonzero digit. Then use a calculator to find the exact sum.

70. $47{,}287 + 33{,}409 + 81{,}496 + 28{,}594$

72. $31{,}592 + 8{,}584.6 + 13{,}215.05 + 968$

74. Mario's Restaurant had the following daily sales: Sunday, \$3,842.95; Monday, \$1,285.68; Tuesday, \$1,195.57; Wednesday, \$1,843.76; Thursday, \$1,526.47; Friday, \$2,984.89; and Saturday, \$4,359.72. Find the total sales for the week.

71. $387{,}483 + 879{,}583 + 592{,}801$

73. $1{,}328{,}591 + 35{,}803{,}502 + 10{,}387{,}921$

1–3 SUBTRACTING WHOLE NUMBERS AND DECIMALS

Learning Outcomes

1 Subtract whole numbers.
2 Subtract decimal numbers.
3 Estimate and check subtraction.
4 Subtract using a calculator.

Subtraction of whole numbers is another basic skill we use on the job. Subtraction is the *inverse operation* of addition. In addition, we add numbers to get their total (such as $5 + 4 = 9$), but to solve the subtraction problem $9 - 5 = ?$ we ask, "What number must be added to 5 to give us 9?" The answer is 4 because 4 added to 5 gives a total of 9. When we subtract two numbers, the answer is called the *difference* or *remainder*. The initial quantity is the *minuend*. The amount being subtracted from the initial quantity is the *subtrahend*.

1 Subtract Whole Numbers.

In addition it does not matter in which order numbers are added. But *in subtraction, order is important:* $8 - 3 = 5$, but $3 - 8$ does not equal 5; that is, $3 - 8 \neq 5$. Subtraction is *not* commutative, whereas addition is. In addition, grouping does not matter when a problem has three or more numbers. But *in subtraction, grouping is important.*

EXAMPLE Show that $9 - (5 - 1)$ does not equal $(9 - 5) - 1$.

$$9 - (5 - 1) = 9 - 4 = 5, \qquad \text{but} \qquad (9 - 5) - 1 = 4 - 1 = 3$$

If a subtraction problem does not contain parentheses to show a grouping, subtract the first two numbers on the left. Then subtract the next number from that difference.

EXAMPLE Subtract $8 - 3 - 1$.

$$8 - 3 - 1 = 5 - 1 = 4$$

Subtracting zero from a number results in the same number:

$$n - 0 = n, \qquad 7 - 0 = 7$$

Tip! **Subtraction and Zeros:**

Subtracting a number from zero is not the same as subtracting zero from a number; that is, $7 - 0 = 7$, but $0 - 7$ does not equal 7.

To subtract numbers of two or more digits:

1. Arrange the numbers in columns, with the minuend at the top and the subtrahend at the bottom.
2. Make sure the ones digits are in a vertical line on the right.
3. Subtract the ones column first, then the tens column, the hundreds column, and so on.
4. To subtract a larger digit from a smaller digit in a column, borrow 1 from the digit in the next column to the left. This is the same as borrowing *one* group of 10; thus, add 10 to the digit in the given column. Then, continue subtracting.

EXAMPLE Subtract $5,327 - 3,514$.

Arrange in columns.

$$\begin{array}{r} {}^{413} \\ 5,327 \\ -\ 3,514 \\ \hline \mathbf{1,813} \end{array}$$

Tip! **Words and Phrases That Imply Subtraction:**

These phrases indicate subtraction in applied problems:

how many are left	how many more
how much less	how much larger
how much smaller	

Also, some applied problems require more than one operation. Many problems that require more than one operation involve parts and a total. If we know the total and all the parts but one, we can add all the known parts and subtract the result from the total to find the missing part.

The Froehlichs left Memphis and drove 356 miles on the first day of their vacation. They drove 426 miles on the second day. If they are traveling to Albuquerque, which is 1,050 miles from Memphis, how many more miles do they have to drive?

The phrase, *how many more,* indicates subtraction.

$$1{,}050 \text{ miles} = \text{total miles}$$

$$356 + 426 + \text{miles left to drive} = 1{,}050 \text{ miles}$$

To find the miles left to drive, add $356 + 426$ and subtract the result from $1{,}050$.

$$356 + 426 = \boxed{782} \qquad 1{,}050 - \boxed{782} = \mathbf{268}$$

2 Subtract Decimal Numbers.

When we subtract decimals, we align the digits just like when we add decimals. Again, this is equivalent to finding common denominators for decimal fractions. Aligning the decimal points is sufficient.

To subtract decimals:

1. Arrange the numbers so that the decimal points align.
2. Subtract each column beginning at the right.
3. Interpret blank places as zeros.
4. Place the decimal in the difference in the same vertical line.

Subtract 8.29 from 13.76.

$$
\begin{array}{r}
13.76 \\
-\ \ 8.29 \\
\hline
\mathbf{5.47}
\end{array}
$$

Notice how the decimals are aligned.

Subtract 7.18 from 15.

In this problem, we must take care to align the decimals properly. Because 15 is a whole number, its decimal point is placed after the 5.

$$
\begin{array}{r}
15. \\
-\ \ 7.18
\end{array}
$$

To subtract, we put zeros in the tenths and hundredths places of 15 and then borrow.

$$
\begin{array}{r}
15.00 \\
-\ \ 7.18 \\
\hline
\mathbf{7.82}
\end{array}
$$

When a worker machines an object using a blueprint as a guide, a certain amount of variation from the blueprint specification is allowed for the machining process. This variation is called the *tolerance.* Thus, if a blueprint calls for a part to be 9.47 in. with a tolerance of ± 0.05, this means that the actual part can be 0.05 in. *more* or 0.05 in. *less* than the specification. To find the *largest* possible size of the object, we add $9.47 + 0.05 = 9.52$ in. To find the *smallest* possible size of the object, we subtract $9.47 - 0.05 = 9.42$ in. **The dimensions 9.52 in. and 9.42 in. are called the** *limit dimensions* **of the object.** That is, 9.52 in. is the largest acceptable measure, and 9.42 in. is the smallest acceptable measure.

CHAPTER 1 Whole Numbers and Decimals

 Find the limit dimensions of an object with a blueprint specification of 8.097 in. and a tolerance of ± 0.005. (This is often written 8.097 ± 0.005.)

8.097 in. = The blueprint specification for the dimension of an object.
± 0.005 = Tolerance of object's dimension.
Smallest dimension for object = Blueprint specification − Tolerance.
Largest dimension for object = Blueprint specification + Tolerance.

Estimation The tolerance of the part is very small—just five-thousandths, so the limit dimensions for the part should be only a few thousandths of an inch smaller or larger than the blueprint specification.

$8.097 - 0.005 = 8.092$
$8.097 + 0.005 = 8.102$

The smallest acceptable dimension for the object is 8.092 in. and the largest acceptable dimension is 8.102 in. That is, the limit dimensions are 8.092 in. and 8.102 in.

3 Estimate and Check Subtraction.

Estimating a subtraction problem is similar to estimating an addition problem. The numbers in the problem are rounded before the subtraction is performed. The accuracy of the estimate depends on the method used for rounding.

To estimate the difference:

1. Round each number to the indicated place value or to a number with one nonzero digit.
2. Subtract the rounded numbers.

To check subtraction, we can use the relationship between addition and subtraction. If $9 - 5 = 4$, then $4 + 5$ should equal 9.

To check a subtraction problem:

1. Add the subtrahend and difference.
2. Compare the result of Step 1 with the minuend. If the two numbers are equal, the subtraction is correct.

 Estimate by rounding to hundreds, then find the exact difference, and check.

$$427.45 - 125$$

	Estimate	Exact	Check
427.45	400	427.45	125.00
− 125.00	− 100	− 125.00	+ 302.45
	300	302.45	427.45

4 **Subtract Using a Calculator.**

Since order (commutative property) and grouping (associative property) *do* matter in subtraction, we must take care to enter subtraction problems in the proper order on our calculators. Note that commas are not entered into the calculator, and most calculator displays do not place commas in the answer.

EXAMPLE   Find the difference between $53,943.76 and $34,256.45.

One option:

53943 $\boxed{\cdot}$ 76 $\boxed{-}$ 34256 $\boxed{\cdot}$ 45 $\boxed{=}$ $\boxed{\text{ENTER}}$ or $\boxed{\text{EXE}}$ may replace $\boxed{=}$.

Calculator display: **19687.31**

■ **Learning Strategy   *What Happens If the Numbers Are Entered in the Wrong Order?***

Rework the previous example by entering the smaller number first.

$$34256.45 \boxed{-} 53943.76 \boxed{=}$$

The display shows -19687.31 as the difference. Is this correct? No. Why not? The larger number should be entered first.

Since the digit portion of the display is the same as the correct answer, is it acceptable to just ignore the negative sign? No. It will cause us to be careless in entering subtraction problems in the calculator. Poor habits are hard to break, and this practice will cause problems later on.

EXAMPLE   Two cuts are made from a 72-in. pipe (see Fig. 1–9). The two lengths cut from the pipe are 28 in. and 15 in. How much of the pipe is left after these cuts are made?

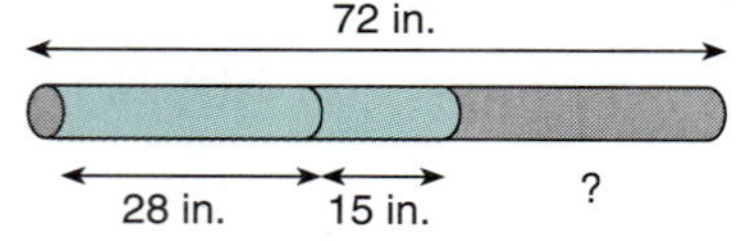

Figure 1–9   Lengths cut from pipe.

72 $\boxed{-}$ 28 $\boxed{-}$ 15 $\boxed{=}$

Calculator display: 29.

There is 29 in. of pipe left.

SELF-STUDY EXERCISES 1–3

1   Subtract.

1. $7 - 2$
2. $(8 - 3) - 4$
3. $8 - (5 - 4)$
4. $9 - 5 - 2$
5. $(7 - 1) - 3$
6. $8 - (2 - 1)$
7. $(8 - 2) - 1$
8. $9 - 3 - 4$
9. $8 - (5 - 1)$
10. $6 - 3 - 2$
11. $47 - 23$
12. $427 - 26$
13. $3672 - 2652$

14.

$$\begin{array}{r} 946 \\ -\ 831 \\ \hline \end{array}$$

15.
$$\begin{array}{r} 53{,}867 \\ -\quad 831 \\ \hline \end{array}$$

16. If a mason orders 75 bags of cement for a job and uses only 53, how many bags are left?

17. An inventory sheet shows that 468 outlet boxes were in stock on March 1. Sales during March were 127. How many outlet boxes were left at the end of the month?

18. A reel of cable contains 575 ft. The following amounts are used on three separate jobs: 112 ft, 101 ft, and 41 ft. How much cable is left on the reel?

2 Subtract.

19. $\begin{array}{r} 14.86 \\ -\ 7.93 \end{array}$
20. $\begin{array}{r} 20.07 \\ -\ 4.236 \end{array}$
21. $\begin{array}{r} 813.673 \\ -\ 9.98 \end{array}$
22. $\begin{array}{r} 84 \\ -\ 27.86 \end{array}$
23. $\begin{array}{r} 5.079 \\ -\ 0.985 \end{array}$

24. Subtract 8.8 from 12.7.

25. Subtract 24.38 from 316.2.

26. Subtract 13.5 from 21.

27. Subtract 67.2 from 378.

28. Find the difference between 42 and 37.6.

29. The Highway Loss Data Institute recently issued statistics on the frequency of insurance claims due to vehicle theft. One year, there were 15.2 claims per 1,000 insured vehicle-years, and eight years later, there were 2.5 claims per 1,000 insured vehicle-years. By how much did the number of claims drop over the 8-year period?

30. Nurse Jennings recorded a temperature of 103.6°F for her patient. If the patient's normal body temperature was 98.6°F, how many degrees of fever did the patient have?

31. One box of rivets weighs 52.6 lb and another box weighs 37.5 lb. How much more does the first box of rivets weigh?

32. According to a blueprint the length of an object is 12.09 in. If the tolerance is ± 0.01 in., what are the limit dimensions of the object?

33. Two lengths of copper tubing measure 63.6 cm and 3.77 cm. What is the difference in their lengths?

34. Find the limit dimensions of an object whose blueprint dimension is 4.195 in. ± 0.006 in.

35. Steel rods of 0.38 decimeter (dm) and 1.9 dm are cut from a steel rod 2.5 dm long. If cutting waste is ignored, how long is the piece that is left?

36. If a 3.95-A current is removed from a 15.5-A parallel circuit, find the remaining current in the circuit.

37. If a hardbound novel costs $27.50 and the softcover edition costs $18.75, how much can be saved by buying the softcover edition?

38. A student buys electronic supplies for an engineering technology class for $27.75 and pays with $30. How much is the change?

39. If the sheet music for a popular song costs $5.25, how much change will be returned if the purchaser pays with a $10 bill? (Disregard sales tax.)

3 Subtract, then check your answers.

40. $\begin{array}{r} 82 \\ -\ 37 \end{array}$
41. $\begin{array}{r} 961 \\ -\ 353 \end{array}$
42. $\begin{array}{r} 4,070 \\ -\ 2,497 \end{array}$
43. $\begin{array}{r} 30,021 \\ -\ 7,816 \end{array}$

Estimate the differences by rounding numbers to one nonzero digit. Then find the exact answer.

44. $\begin{array}{r} 589,760 \\ -\ 498,726 \end{array}$
45. $\begin{array}{r} 4,903.009 \\ -\ 2,814.125 \end{array}$
46. $\begin{array}{r} 3,070 \\ -\ 2,896 \end{array}$
47. $\begin{array}{r} \$\ 1,401.30 \\ -\ \ \ \ \ 802.25 \end{array}$

4 Estimate the differences by rounding numbers to one nonzero digit. Then use your calculator to find the exact answer.

48. $\begin{array}{r} 8,001 \\ -\ 3,604 \end{array}$
49. $\begin{array}{r} 3,804.621 \\ -\ 2,617.370 \end{array}$
50. $\begin{array}{r} 895,740 \\ -\ 387,465 \end{array}$

51. A fuel tank on one model automobile has a capacity of 75 liters (L). The tank on a different model has a capacity of 92 L. How much larger is the second tank?

52. A bricklayer laid 1,283 bricks on one day. A second bricklayer laid 1,097 bricks. How many more bricks did the first bricklayer lay?

53. A stockroom has 285.8 in. of bar stock and 173.5 in. of round stock. How many inches of bar stock remain after the object in Fig. 1–10 is made from this stock?

54. In Exercise 53, how many inches of round stock remain after the object in Fig. 1–10 is made?

55. Find the missing dimension in Fig. 1–11.

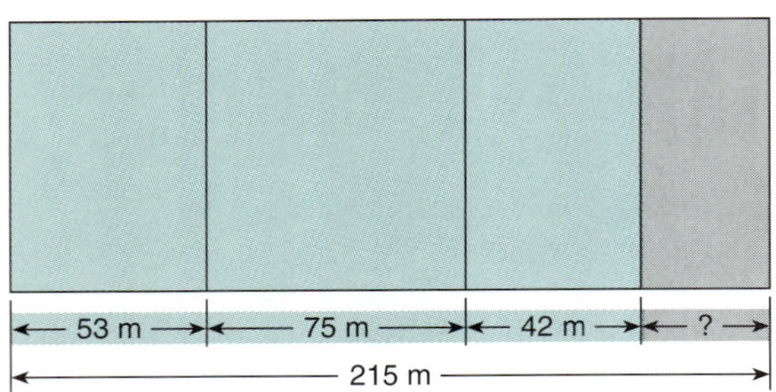

Figure 1–11

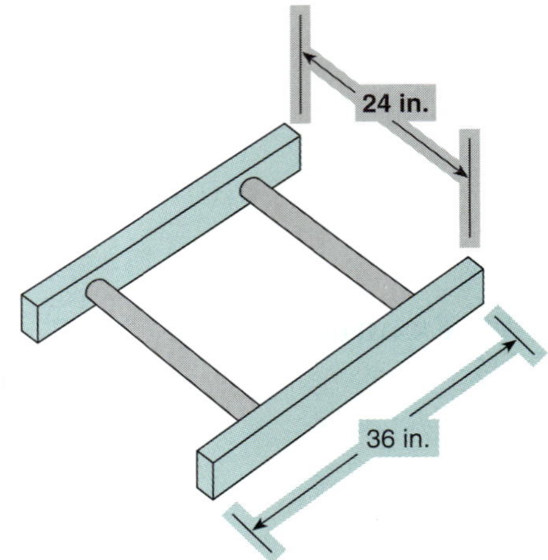

Figure 1–10

1–4   MULTIPLYING WHOLE NUMBERS AND DECIMALS

Learning Outcomes

1. Multiply whole-number factors.
2. Multiply decimal factors.
3. Apply the distributive property.
4. Estimate and check multiplication.
5. Multiply using a calculator.

1. Multiply Whole-Number Factors.

Multiplication is repeated addition. If we have three $10 bills, we have $10 + $10 + $10, or $30. Using multiplication, we see that this is the same as 3 times $10, or $30.

When we multiply two numbers, the first number is called the *multiplicand,* and the number we multiply by is called the *multiplier.* Either number is referred to as a *factor.* The answer or result of multiplication is called the *product.*

$$2 \quad \times \quad 3 \quad = \quad 6$$

    multiplicand   multiplier   product
     or factor   or factor

Tip!	***Various Notations for Multiplication:***

Besides the familiar $\times$ or "times" sign, the raised dot ($\cdot$), the asterisk ($*$), and parentheses ( ) are also used to show multiplication.

$$2 \cdot 3 = 6, \qquad 2 * 3 = 6, \qquad 2(3) = 6, \qquad (2)(3) = 6$$

Multiplication is commutative and associative, just like addition. The *commutative property of multiplication* permits two numbers to be multiplied in any order. In symbols, $a \times b = b \times a$.

$$4 \times 5 = 20, \qquad 5 \times 4 = 20$$

When more than two numbers are multiplied, the numbers must be grouped, and the *associative property of multiplication* permits the numbers to be grouped in any way. In symbols, $a \times (b \times c) = (a \times b) \times c$.

$$2 \times (3 \times 5) \qquad \text{or} \qquad (2 \times 3) \times 5$$
$$2 \times \quad 15 \qquad\qquad\qquad 6 \quad \times 5$$
$$30 \qquad\qquad\qquad\qquad 30$$

EXAMPLE Multiply $3 \times 2 \times 9$.

$3 \times 2 \quad \times 9$

$(3 \times 2) \quad \times 9$ Group any two factors.

$6 \quad\quad \times 9$ Multiply grouped factors.

$\mathbf{54}$ Multiply the factors: 6 and 9.

Another property is the *zero property of multiplication*.

The product of a number and zero is zero:

$$n \times 0 = 0, \qquad 0 \times n = 0, \qquad 4 \times 0 = 0, \qquad 0 \times 4 = 0$$

EXAMPLE Multiply $2 \times 5 \times 0 \times 7$.

$2 \times 5 \times 0 \times 7$

$(2 \times 5) \times (0 \times 7)$ Group factors.

$10 \quad \times \quad 0$ Multiply each group of factors.

$\mathbf{0}$ Multiply products.

■ **Learning Strategy** *How Important Is It to Learn the Basic Multiplication Facts?*

Some argue that the easy availability of the calculator makes learning basic multiplication facts less important. However, the basic facts are crucial to developing your number sense. Estimating, checking for the reasonableness of an answer, dividing, reducing fractions, and many other mathematical concepts are dependent on these basic multiplication facts.

What if you forget a multiplication fact, like the product of 7×8? Is there a fact involving 7 or 8 that you do remember? Suppose you remember that $7 \times 7 = 49$. Then, 7×8 is 7 more than 49, or $7 + 49 = 56$.

It is easier to learn patterns than unconnected facts!

Now, we will develop a procedure for multiplying factors of two or more digits.

To multiply factors of two or more digits:

1. Arrange the factors one under the other.
2. Multiply each digit in the multiplicand by each digit in the multiplier. The product of the multiplicand and each digit in the multiplier gives a *partial product.*
   a. Start with the ones digit in the multiplier and multiply the multiplicand from right to left.
   b. Align each partial product with its first digit directly under its multiplier digit.
3. Add the partial products.

EXAMPLE   Multiply 204×103.

$$
\begin{array}{r}
\overset{1}{204} \\
\times\ 103 \\
\hline
612 \\
0\ 00 \\
20\ 4 \\
\hline
21{,}012
\end{array}
$$

Multiply: $3 \times 204 = 612$. Align 612 under 3 in the multiplier.
Multiply: $0 \times 204 = 000$. Align 000 under 0 in the multiplier.
Multiply: $1 \times 204 = 204$. Align 204 under 1 in the multiplier.
Add the partial products as they are aligned.

Partial products 000 and 204 could be combined on a single line to be 2040, with the rightmost 0 aligned under the 0 in the multiplier and the 4 aligned under the 1 in the multiplier.

Some applied problems require more than one operation to answer the question. The following example requires both multiplication and addition.

EXAMPLE   John Paszel is taking inventory and finds 27 unopened boxes of candy. Each of the boxes contains 48 candy bars. In another part of the warehouse, John counts 33 unopened cases of the same candy. These cases contain 288 bars of candy each. How many candy bars should John show on his inventory report?

John must first calculate the number of candy bars in the 27 unopened boxes.

$$27 \times 48 = 1{,}296$$

Next, he must calculate the number of candy bars in the cases.

$$33 \times 288 = 9{,}504$$

Finally, the total number of candy bars is 1,296 + 9,504, or 10,800.

2 Multiply Decimal Factors.

Since decimals are fractions, multiplication of decimals will cause us to rethink or expand our basic number sense of multiplication. With decimals, the product of two numbers will not always result in a number that is larger than the factors being multiplied. For example, a part of a whole amount is smaller than the original amount. Also, a part of a part is a smaller part.

One way we can multiply decimal numbers is by writing them in their fractional form and then multiplying.

In Chapter 3, we will thoroughly examine situations and procedures for multiplying fractions. However, for now, let's examine the basic process:

- Numerator times numerator gives the numerator of the product.
- Denominator times denominator gives the denominator of the product.

$$\frac{a}{b} \times \frac{c}{d} = \frac{ac}{bd}$$

Note: ac means $a \times c$ and bd means $b \times d$.

To multiply 0.8×0.32, for example, we can write $\frac{8}{10} \times \frac{32}{100}$. Multiplying, we have $\frac{256}{1,000}$, which is written as 0.256 in decimal form.

Before continuing, let's use these three numbers to make some observations about decimal fractions and the place-value chart in Fig. 1–12. Recall that the number of zeros in the denominator is the same as the number of places to the right of the decimal point.

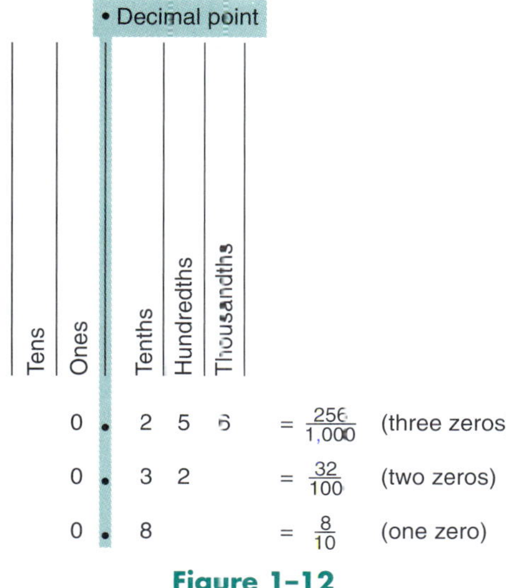

Figure 1–12

In 0.8×0.32, we see that 0.8 has one digit to the right of the decimal, while 0.32 has two digits to the right of the decimal. These two factors together have a total of three digits to the right of the decimal. If we compare this total with the number of digits to the right of the decimal in the product, 0.256, we see that they are the same. This fact allows us to multiply decimal numbers without converting them to fractions.

$$\begin{array}{rl} 0.32 & \text{2 places after decimal} \\ \underline{\times\quad 0.8} & \underline{\text{1 place after decimal}} \\ 0.256 & \text{3 places after decimal} \end{array}$$

$$0.256 = \frac{256}{1,000}$$

To multiply decimal numbers:

1. Align the numbers as if they were whole numbers and multiply.
2. Count the total number of digits to the right of the decimal in each factor.
3. Place the decimal in the product so that the number of decimal places is the sum of the number of decimal places in the factors.

EXAMPLE   Multiply 1.36×0.2.

$$\begin{array}{r} 1.36 \\ \underline{\times\quad 0.2} \\ \mathbf{0.272} \end{array}$$

Note that in multiplication the decimals do *not* have to be in a straight line. Place a zero in the ones place so that the decimal point will not be overlooked.

        0.309
   ×   0.17
        2163
        309
      0.05253   We did not have enough digits in the product, so we inserted a zero
                     on the *left* to give the appropriate number of decimal places.

EXAMPLE The outside diameter of a flower bed is 7.82 meters (m) (see Fig. 1–13). If the brick walk surrounding the bed is 1.56 m thick, find the inside diameter of the flower bed.

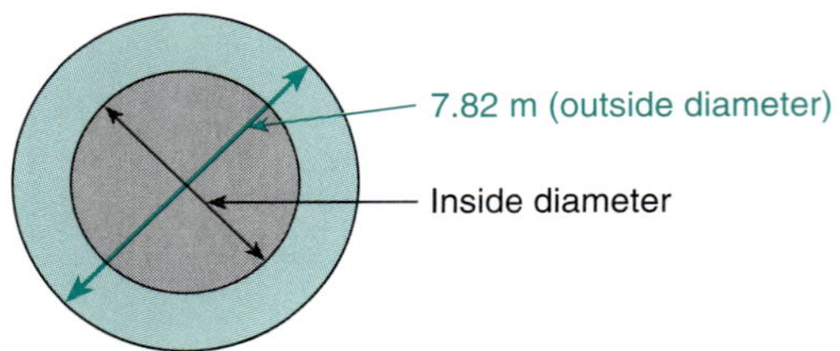

Figure 1–13

7.82 m = outside diameter of the flower bed
1.56 m = thickness of brick walk surrounding the flower bed
The diameter of a circle is the distance across the center of the circle.
Outside diameter − Two thicknesses (one on each end of the outside diameter) of the walk = Inside diameter

Estimation The outside diameter is nearly 8 m, and the total thickness to be subtracted is about 3 m, so the inside diameter should be about 5 m.
$7.82 − (2 × 1.56) = 7.82 − 3.12 = 4.70$
The inside diameter of the flower bed is 4.70 m.

3 Apply the Distributive Property.

Another property of multiplication is the distributive property. The *distributive property of multiplication* means that multiplying a sum or difference by a factor is equivalent to multiplying each term of the sum or difference by the factor.

Distributive Property of Multiplication:

1. Add or subtract the numbers within the grouping.
2. Multiply the result of Step 1 by the factor outside the grouping.

or

1. Multiply each number inside the grouping by the factor outside the grouping.
2. Add or subtract the products from Step 1.

Symbolically,

$$a \times (b + c) = a \times b + a \times c \qquad \text{or} \qquad a(b + c) = ab + ac$$
$$a \times (b - c) = a \times b - a \times c \qquad \text{or} \qquad a(b - c) = ab - ac$$

<table>
<tr><td>Tip!</td><td>Other Notations for Multiplication:</td></tr>
</table>

- Parentheses show multiplication when the distributive property is used: $a(b + c)$ means $a \times (b + c)$.
- The letters represent numbers.
- Letters written together with no operation sign between them imply multiplication: ab means $a \times b$; ac means $a \times c$.

EXAMPLE Multiply $3(2 + 4)$.

Multiplying first gives:	Adding first gives:
$3(2 + 4) =$	$3(2 + 4) =$
$3(2) + 3(4) =$	$3(6) = \mathbf{18}$
$6 + 12 = \mathbf{18}$	

EXAMPLE Multiply $2(6 - 5)$.

Multiplying first gives:	Subtracting first gives:
$2(6 - 5) =$	$2(6 - 5) =$
$2(6) - 2(5) =$	$2(1) = \mathbf{2}$
$12 - 10 = \mathbf{2}$	

The distributive property is found in many formulas. One example is the formula for the perimeter of a rectangle. A *rectangle* is a four-sided geometric shape whose opposite sides are equal in length, and each corner makes a square corner (see Fig. 1–14). The *perimeter* of a rectangle is the distance around the figure. We find perimeters when we fence a rectangular yard, install baseboard in a rectangular room, frame a picture, or outline a flower bed with landscaping timbers.

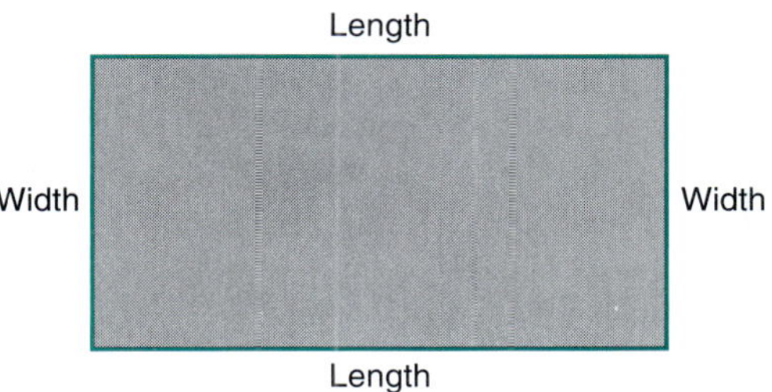

Figure 1–14 Perimeter of rectangle.

The formula for finding the perimeter of a rectangle is

$$P = 2(l + w) \qquad \text{or} \qquad P = 2l + 2w$$

EXAMPLE Find the number of feet of fencing needed to enclose a rectangular pasture that is 1,784.6 feet long and 847.3 feet wide.

$P = 2(l + w)$	or	$P = 2l + 2w$
$P = 2(1,784.6 + 847.3)$		$P = 2(1,784.6) + 2(847.3)$
$P = 2(2,631.9)$		$P = 3,569.2 + 1,694.6$
$P = 5,263.8$		$P = 5,263.8$

The amount of fencing needed is 5,263.8 feet.

4 Estimate and Check Multiplication.

To estimate the answer for a multiplication problem:

1. Round both factors to a chosen or specified place value or to one nonzero digit.
2. Multiply the rounded numbers.

To check a multiplication problem:

1. Multiply the numbers a second time and check the product.
2. Interchange the factors if convenient.

EXAMPLE Find the approximate and exact cost of 48 flower bulbs if each bulb costs $2.15. Estimate the cost by rounding each factor to a number with one nonzero digit. Then find the exact cost and check your work.

Total cost of flower bulbs
48 = Total number of flower bulbs
$2.15 = Cost of each bulb

Number of bulbs $\times$ Cost of each bulb = Total cost of flower bulbs

Estimation If 50 bulbs were purchased at $2 each, the total cost would be $100. So the exact cost should be close to $100.
48 $\times$ $2.15 = $103.20
The total cost of 48 flower bulbs is $103.20, which is approximately $100, as noted in the estimation.

We routinely estimate quantities, distances, and amounts in do-it-yourself projects before we calculate the exact amounts. One such example involves finding the area of a rectangular city house lot. The *area* of a geometric shape is the number of square units needed to cover the shape. We use area when we find the amount of carpet needed to cover a rectangular floor, the amount of paint needed for a wall, the amount of asphalt needed to pave a parking lot, the amount of fertilizer needed to treat a yard, or the amount of material needed to produce a rectangular sign. Area is measured in *square measures*. One square foot (1 ft^2) indicates a square that measures 1 foot on each side.

5 feet long by 3 feet wide = 15 square feet

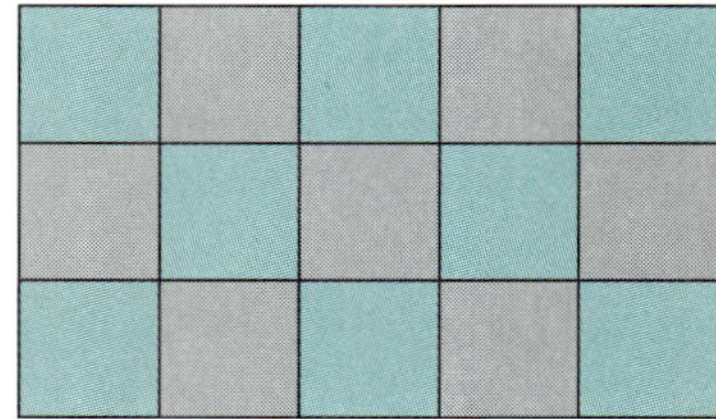

Figure 1–15 Area of a rectangle.

To find the area of a rectangle, we multiply the length by the width (see Fig. 1–15). We can state this in symbols using a *formula*.

$A = lw$ When two letters are written side-by-side with no operation sign, multiplication is implied.

EXAMPLE   Maintenance Consultants need to apply fertilizer to a customer's lawn. The lawn is 223.4 ft long and 132.8 ft wide. The fertilizer costs $0.004 per square foot to apply, what is the cost of applying the fertilizer?

Lawn is in the shape of a rectangle.
223.4 ft = Length of the lawn
132.8 ft = Width of the lawn
Fertilizer costs $0.004 per square foot to apply.
Area of a rectangle is the product of the length times the width ($A = lw$)
Total cost = Total number of square feet × Cost per square foot

Estimation   A lawn 200 ft by 100 ft contains 20,000 ft^2, and at a cost of $0.004 per square foot, the total cost is approximately $80. Note, the cost will be higher because we rounded both length and width *down*.

$A = lw$
$A = 223.4(132.8)$
$A = 29,667.52$ square feet

Cost = 29,667.52 × $0.004
Cost = $118.67

The total cost of applying fertilizer to the lawn is $118.67, which is higher than the estimated cost of $80, as we anticipated.

5 Multiply Using a Calculator.

To determine how your calculator handles multiplication, let's begin with some examples that you can do mentally. Notice whether your calculator accumulates the product as operation keys are pressed or whether the calculation is performed only after the $=$, ENTER, or EXE key is pressed.

$3 \times 4 = 12$

3 ⊠ 4 =   ENTER or EXE may be substituted for equal.

$2 \times 3 \times 4 = 24$

2 ⊠ 3 ⊠ 4 =

EXAMPLE   Find the product of the following numbers using a calculator.

$$3,283 \times 34.6 \times 34 \ =$$

3283 ⊠ 34 · 6 ⊠ 34 = ⟹ 3862121.2

The product is 3,862,121.2.

When either or both factors of a multiplication problem end in zeros, a shortcut process such as the one in the following example can be used.

EXAMPLE   Multiply 2,600 × 70.

1.   26|00   Separate the ending zeros from the other digits.
       × 7|0

2.   26|00
       × 7|0   Multiply the other digits as if the zeros were not there (26 × 7 = 182).
       ——
        182|

3.   26|00   Attach the zeros to the basic product. Note that the number of zeros affixed to
       × 7|0   the basic product is now the same as the sum of the number of zeros at the
       ——————   end of each factor.
      182|000

This process is sometimes necessary whenever a multiplication problem is too long to fit into a calculator. Many calculators have only an eight-digit display window. The problem $26,000,000 \times 3,000$ would not fit many basic calculators. This shortcut allows us to work the problem quickly, with or without a calculator.

EXAMPLE   Multiply $26,000,000 \times 3,000$.

$$
\begin{array}{r}
26,000,000 \\
\times \quad 3,000 \\
\hline
78,000,000,000
\end{array}
$$

Separate ending zeros and multiply 26×3.

Attach 9 zeros.

■ **Learning Strategy**   *The Mind Is Often Quicker Than the Fingers.*

Don't use your calculator as a crutch. It is a tool! When multiplying 2,500 times 30, you can multiply 25 times 3 mentally. $25 \times 3 = 75$. Then, attach three zeros to that product.

$$2,500 \times 30 = 75,000$$

This skill does not come automatically. You have to practice it! Like playing a musical instrument or mastering a sport, you don't develop skill by watching. You have to practice, practice, practice.

Performing multiplication on a calculator is similar to adding and subtracting. With some calculators, the operations are accumulated in the display as you press an operation key. With others, all operations are made after the equal or enter key is pressed. Try some examples with your calculator to assure yourself that you understand how your calculator performs multiplication.

Try this example on your calculator: $3 \times 4 \times 5 = 60$

SELF-STUDY EXERCISES 1–4

1   Multiply or answer the following.

1. Find the following products.
   (a) 5×3 **(b)** $7 * 8$
   (c) $(9)(7)$ **(d)** $4 \cdot 6$
3. What property of multiplication justifies the statement "$5(3) = 3(5)$"?

2. Jaime Oxnard has 9 wood boxes that he intends to sell for \$7 each. If Jaime sells 6 of the boxes, how much money will he make?
4. Explain the associative property of multiplication and give an example to illustrate the property.

Multiply.

5. $5 \cdot 3 \cdot 0 \cdot 6$

6. $3 * 7 * 9 * 2$

7. $7 \times 3 \times 4$

8. $(3)(2)(0)(8)$

9. $\begin{array}{r} 32 \\ \times \ 7 \\ \hline \end{array}$

10. $\begin{array}{r} 83 \\ \times \ 7 \\ \hline \end{array}$

11. $\begin{array}{r} 90 \\ \times \ 7 \\ \hline \end{array}$

12. 503×204

13. Margaret Johnston purchased 3 cases of candy bars. Each case contains 12 bags, and each bag contains 24 pieces of individually wrapped candy. How many individually wrapped candies did Margaret purchase?
15. Bill Wepner read 6 books a week over a period of 14 weeks. How many books did Bill read during this time period?

14. Chuckee Wright counted 8 unopened boxes of washers. Each box contained 512 washers. What total number of washers will be shown on the inventory sheet?
16. Each officer in the Public Safety office wrote, on average, 25 tickets a week. If there are 7 officers, how many tickets were written over a 4-week period?

17. Jossie Moore is planning to sell candy bars in her Smart Shop. She receives 12 boxes, and each box contains 24 candy bars. If Jossie sells the bars for \$1 each, how much money will she get for all the bars?
18. Tory Eaton teaches 5 classes, and each class has 46 students. Her sixth class has 19 students enrolled. How many students is Terry teaching?

   Multiply.

19. 53.4 $\times$ 0.29	**20.** 37.7 $\times$ 1.5	**21.** 9.27 $\times$ 0.35
22. 0.215 $\times$ 0.27	**23.** 0.271 $\times$ 0.32	**24.** 3.7 $\times$ 1.93

25. 2.78 $\times$ 0.03

26. 73.806 $\times$ 2.305

27. 1.9067 $\times$ 0.2013

28. 8.2037 $\times$ 0.602

29. 42 $\times$ 0.73

30. 81 $\times$ 7.37

31. 81.7 $\times$ 0.621

32. 72.6 $\times$ 0.532

33. 326 $\times$ 1.04

34. 261 $\times$ 2.07

35. 6.35 $\times$ 40

36. 9.21 $\times$ 20

37. 93.07 $\times$ 0.01

38. 76.02 $\times$ 0.01

39. 586.35 $\times$ 0.001

40. 732.65 $\times$ 0.001

41. 2.0037 $\times$ 4.6

42. 3.0062 $\times$ 3.8

43. 7.04 $\times$ 0.0025

44. 6.01 $\times$ 0.0045

45. 0.457 $\times$ 2.003

46. 0.387 $\times$ 1.009

47. 0.05 $\times$ 0.003

48. 0.07 $\times$ 0.008

49. A pipe has an outside diameter of 4.327 in. Find the inside diameter if the pipe wall is 0.185 in. thick.

50. A plastic pipe has an inside diameter of 4.75 in. Find the outside diameter if the pipe wall is 0.25 in. thick.

51. Electrical switches cost $5.70 wholesale. If the retail price is $7.99 each, how much would an electrician save over retail by buying 12 switches at the wholesale price?

52. How much No. 24 electrical wire is needed to make eight pieces each 18.9 in. long?

53. A retailer purchases 15 cases of potato chips at $8.67 per case. If the chips are sold at $12.95 per case, how much profit does the retailer make?

54. A contractor purchases 500 yd^3 of concrete at $21.00 per cubic yard, 24 yd^3 of sand at $5.00 per cubic yard, and 36 yd^3 of fill dirt at $3.00 per cubic yard. What is the total cost of the materials?

55. A micrometer is designed so that the screw thread makes the spindle advance 0.025 in. in one complete turn. How far will the spindle advance in 15 complete turns?

56. A 4 ft $\times$ 8 ft sheet of 1-in.-thick steel weighs 1,307 lb and costs $0.30 a pound. Find the cost of five of these sheets.

57. What is the total amperage (A) of a circuit if six 2.5-A devices are connected in parallel?

58. Sequoia orders 24 BNC T-adapters at $3.25 each. How much is the total order?

   Find the product by multiplying first, and check your answers by adding first.

59. 5(7 + 4)

60. 3(9 + 11)

61. 14(3 + 11)

62. 4(5 − 2)

63. 1.2(16 − 3)

64. 7(1.8 − 1)

Find the perimeter of the rectangles in Figs. 1–16 and 1–17.

65.

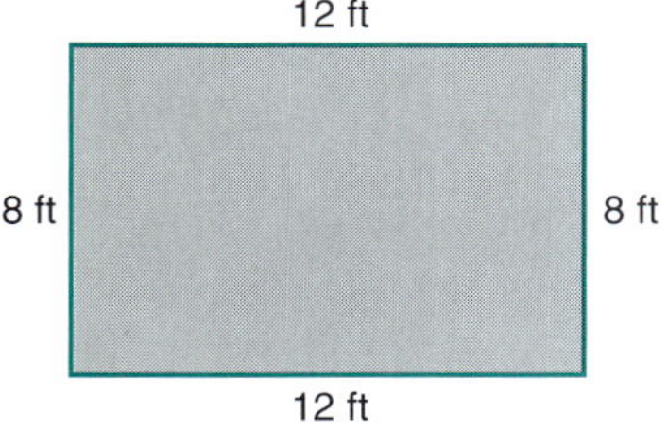

Figure 1–16

66.

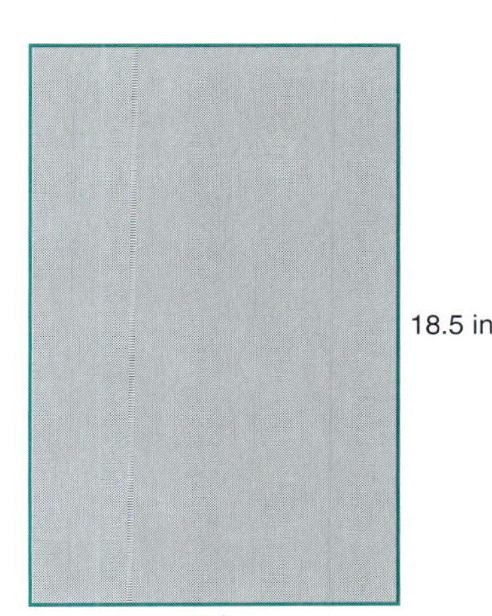

Figure 1–17

67. How much baseboard is required to surround a kitchen if the floor measures 15 ft by 13 ft and no allowance is made for door openings?

68. A dog pen is staked out in the form of a rectangle so that the width is 38 ft and the length is 47 ft. How much fencing must be purchased to build the pen?

4 Complete the multiplication problems as directed.

69. Find the approximate cost of 24 shirts if each shirt costs $32. What is the exact cost of the shirts?

Find the number of square feet in each rectangle illustrated in Figs. 1–18 and 1–19. Check your work.

70.

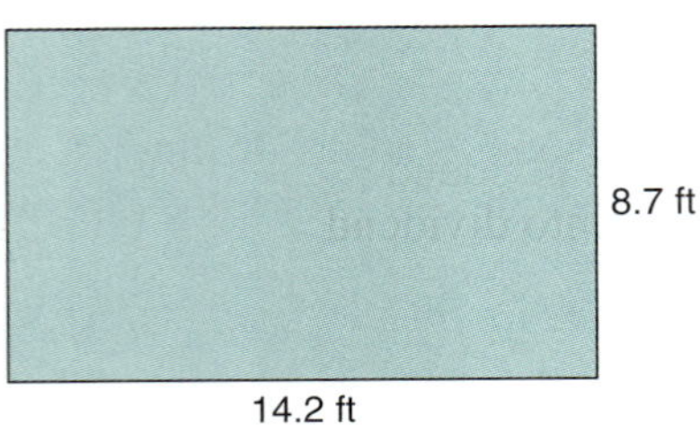

Figure 1–18

71.

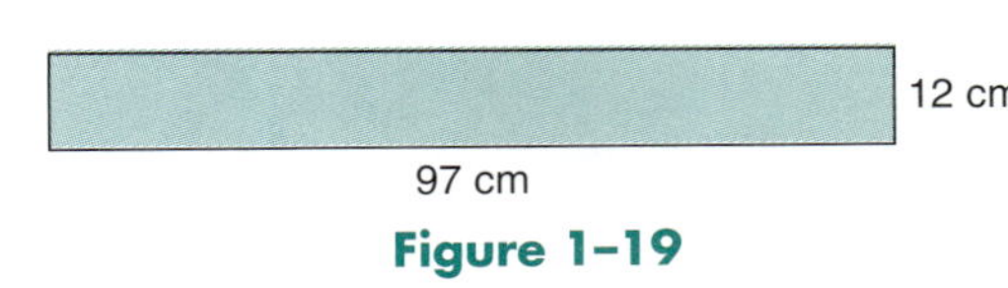

Figure 1–19

72. Estimate how many square feet of carpet should be purchased to cover a rectangular room that measures 42.3 ft by 32.5 ft. Find the exact area.

73. A rectangular table top is 12 ft long and 8 ft wide. How many square feet of glass are needed to cover it? Check your work.

5 Use a calculator to find the products.

74. $5896 \times 47.3 \times 12.83$

75. $52{,}834 * 15 * 29.5$

76. $832 \cdot 703 \cdot 960$

77. $583(294)(627)(0)(8)$

78. Paul Thomas earns $3,015 per month. What is his annual salary?

79. Ertha Brower earns $423.20 per week. Calculate her annual salary.

Perform the multiplications.

80. $3{,}500 \times 42{,}000{,}000$

81. $30{,}800 * 544{,}000$

82. $70{,}000(6{,}000)$

1–5 DIVIDING WHOLE NUMBERS AND DECIMALS

Learning Outcomes

1 Use various symbols to indicate division.
2 Divide whole numbers.
3 Divide decimal numbers.
4 Estimate and check division.
5 Find the numerical average.
6 Divide using a calculator.

1 Use Various Symbols to Indicate Division.

Division is the inverse operation of multiplication. Determining what number we should multiply 7 by to obtain a product of 28 would be equivalent to dividing 28 by 7. The result is 4. In division, the number we divide by is the *divisor.* The number being divided is the *dividend.* The answer is the *quotient.*

$$28 \div 7 = 4 \longleftarrow \text{quotient}$$

dividend —⏌   ⎣— divisor

Besides the "divided by" sign ($\div$), we also use the symbol $\overline{)}$. The problem $7\overline{)28}$ means $28 \div 7$.

$$7\overline{)28}^{\,4} \quad \longleftarrow \text{ quotient}$$

divisor ———↑ ↑——— dividend

We also use a bar to indicate division such as $\dfrac{28}{7}$, which also means $28 \div 7$.

$$\text{dividend} \longrightarrow \dfrac{28}{7} = 4 \longleftarrow \text{quotient} \longleftarrow \text{divisor}$$

These words indicate division:

1. "divide by" $\longrightarrow$ dividend divided by divisor
2. "goes into" $\longrightarrow$ divisor goes into dividend
3. "divide into" $\longrightarrow$ divisor divided into dividend

EXAMPLE Express "2 goes into 8" as a division using the symbol $\div$, the symbol $\overline{)}$, and the bar symbol —. Then solve the problem.

$$8 \div 2 = ? \quad \text{Because } 4 \times 2 = 8, \quad \mathbf{8 \div 2 = 4,} \quad 2\overline{)8}^{\,4}, \quad \dfrac{8}{2} = 4.$$

2 Divide Whole Numbers.

Some of the problems that arise in division involve *long division*. Study the following example. Long division, like long multiplication, involves using simple one-digit multiplication facts repeatedly to find the quotient.

EXAMPLE Divide $975 \div 12$.

$12\overline{)975}$ — Use long-division symbol.

$12\overline{)975}^{\,8}$ — 12 does not divide into 9, but it does go into 97 8 times. Write 8 over the 7, the last digit of 97.

$\begin{array}{r} 8 \\ 12\overline{)975} \\ \underline{96} \\ 15 \end{array}$ — Multiply $8 \times 12 = 96$. Write 96 under the 97 of the dividend.

Subtract $97 - 96 = 1$. This difference must be less than 12, the divisor. Bring down the next digit in the dividend, 5.

$\begin{array}{r} 8\,1 \\ 12\overline{)975} \\ \underline{96} \\ 15 \\ \underline{12} \\ 3 \end{array}$ — 12 divides into 15 one time. Write 1 over the 5 of the dividend.

Multiply $1 \times 12 = 12$. Write 12 under the 15.
Subtract $15 - 12 = 3$. This difference must be less than 12.

$\begin{array}{r} \mathbf{81} \ \ \mathbf{R3} \\ 12\overline{)975} \\ \underline{96} \\ 15 \\ \underline{12} \\ 3 \end{array}$ — 3 is the remainder.

<table>
<tr><td>**Tip!**</td><td>*Importance of Placing the First Digit Carefully:*</td></tr>
</table>

The correct placement of the first digit in the quotient is critical. If the first digit is out of place, all digits that follow will be out of place, giving the quotient too few or too many digits.

At times, after subtracting and bringing down the next digit, we find that the divisor will not divide into the resulting number at least one time. In this case, we place a zero in the quotient.

EXAMPLE Divide $5\overline{)2{,}535}$.

$$
\begin{array}{r}
5 \\
5\overline{)2{,}535} \\
\underline{2\,5} \\
03
\end{array}
$$

5 divides into 25 five times. Write 5 over the last digit of the 25. Subtract and bring down the 3.

$$
\begin{array}{r}
\mathbf{507} \\
5\overline{)2{,}535} \\
\underline{2\,5} \\
035 \\
\underline{35} \\
0
\end{array}
$$

5 divides into 3 zero times, so write 0 over the 3 of the dividend. Bring down the next digit, which is 5. Divide 5 into the 35: $35 \div 5 = 7$. Write the 7 over the 5 of the dividend. Continue as usual.

When dividing by a one-digit divisor, we can perform the division mentally. This process is called *short division*.

EXAMPLE Divide $7\overline{)180}$.

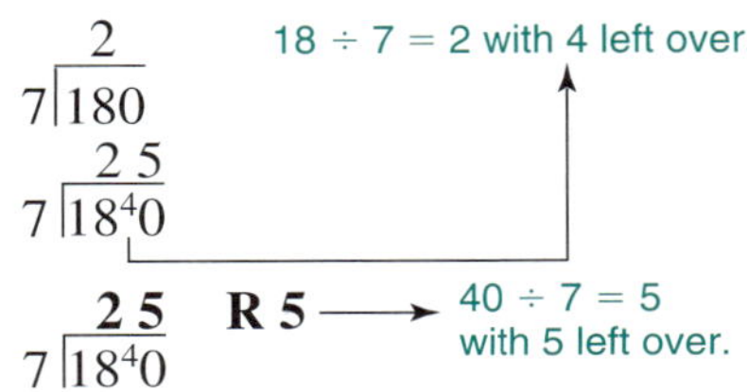

$18 \div 7 = 2$ with 4 left over.

$40 \div 7 = 5$ with 5 left over.

Special properties associated with division involve zeros, division of a number by itself, and division by 1.

> **The quotient of zero divided by a nonzero number is zero:**
>
> $$0 \div n = 0, \qquad n\overline{)0}, \qquad \frac{0}{n} = 0$$
>
> $$0 \div 5 = 0, \qquad 5\overline{)0}, \qquad \frac{0}{5} = 0$$

If a number divided by zero is to have an answer, then that answer times zero should equal the number. However, we learned in multiplication that zero times any number is zero. Therefore, a number divided by zero has *no* answer.

$$8 \div 0 = ? \qquad \text{What number} \times 0 = 8?$$ Because any number times $0 = 0$, $8 \div 0$ has no answer.

    CHAPTER 1 Whole Numbers and Decimals

The quotient of a division problem can be only *one* number. Consider the problem $0 \div 0$.

$0 \div 0 = ?$ What number $\times 0 = 0$? Because *any number* times $0 = 0$, $0 \div 0$ does not have just *one* number as its quotient; that is, the quotient is not *unique*. For this reason, we say that division of zero by zero has no answer, or that division of zero by zero is impossible.

The quotient of a number divided by zero is impossible (undefined):

$$\text{impossible}$$

$$n \div 0 \text{ is impossible,} \qquad 0\overline{)\,n\,}, \qquad \frac{n}{0} \text{ is impossible}$$

$$12 \div 0 \text{ is undefined, or impossible}$$

$$0 \div 0 \text{ is undefined, or impossible}$$

Dividing any nonzero number by itself yields 1:

$$n \div n = 1, \qquad \text{if } n \text{ is not equal to zero;} \qquad 12 \div 12 = 1$$

Dividing any number by 1 yields the same number:

$$n \div 1 = n, \qquad 5 \div 1 = 5$$

In division, as in subtraction, order is important. For example, $12 \div 4 = 3$; however, $4 \div 12$ is not 3. Therefore, division is *not* commutative. When dividing more than two numbers, as in subtraction, the way we group the numbers is important. For example, $(16 \div 4) \div 2 = 4 \div 2 = 2$. If we change the grouping to $16 \div (4 \div 2)$, we have $16 \div 2 = 8$. Notice the different answers: 2 and 8. Therefore, division is *not* associative. If there are no grouping symbols, we divide from left to right.

3 Divide Decimal Numbers.

Dividing decimals is similar to dividing whole numbers. However, as in multiplication, we must be careful about placing the decimal point in the quotient.

As with multiplying decimals, dividing by a decimal number will cause us to rethink the effects of division. With whole numbers, the quotient is the same as or smaller than the dividend. With decimals, this will not always be the case.

Suppose we have 3.6 lb of candy and we want to put 0.4 lb in a package. How many packages would we have? That is, 3.6 lb $\div$ 0.4 lb per package = how many packages?

This division can be written as the fraction $\frac{3.6}{0.4}$. If we multiply the numerator and denominator of the fraction by 10, we make each number a whole number. Because $\frac{10}{10} = 1$, we do not change the value of the fraction.

$$\frac{3.6}{0.4} \times \frac{10}{10} = \frac{3.6 \times 10}{0.4 \times 10} = \frac{36}{4} = 9$$

Now consider the same division problem in long-division form.

$$0.4\overline{)3.6}$$

The decimal point is shifted to the end of the divisor, 0.4, which in this case is one place. (Remember that multiplying by 10 results in moving the decimal one place to the right.) The dividend is also multiplied by 10, which shifts the decimal one place. The decimal in the quotient is written directly above the new decimal in the dividend. The decimal point should be placed in position *before* the division is started.

$$0.4\overline{)3.6}$$

The reason for moving the decimal point in the divisor is to make the divisor a *whole* number to simplify the division process. This changes the value of the divisor, so the decimal point in the dividend must be moved the same number of places to keep the value of the division problem unchanged.

To divide decimal numbers written in long-division form:

1. Move the decimal in the divisor so that it is on the right side of all digits. (By moving the decimal, you are multiplying by 10, 100, 1000, and so on.)
2. Move the decimal in the dividend to the right as many places as the decimal was moved in the divisor. (Attach zeros if necessary.)
3. Write the decimal point in the answer directly above the new position of the decimal in the dividend. (Do this *before* dividing.)
4. Divide as you would in whole numbers.

EXAMPLE Divide 4.8 ÷ 6.

$$6\overline{)4.8}\qquad \text{Insert a decimal point above the decimal point in the dividend.}$$

$$\begin{array}{r} 0.8 \\ 6\overline{)4.8} \end{array}\qquad \text{Divide.}$$

When the divisor is a whole number, the decimal is understood to be to the right of 6 and is not moved. When the divisor is a whole number, the decimal point in the dividend is not moved. The decimal is placed in the quotient directly above the decimal in the dividend.

EXAMPLE Divide 3.12 ÷ 1.2.

$$1.2\overline{)3.1\,2}\qquad \text{Move the decimal in the divisor and dividend.}$$

$$12\overline{)31.2}\qquad \text{Write the decimal in the quotient, then divide.}$$

$$\begin{array}{r} 2.6 \\ 12\overline{)31.2} \\ \underline{24} \\ 7\,2 \\ \underline{7\,2} \end{array}$$

Not all divisions have a remainder of zero. When divisions have a nonzero remainder, we have two choices. We can write the remainder as a fraction (place it over the divisor and add the fraction to the quotient), or we can round the quotient to a specified decimal place. Both choices give the same results; just remember to choose before you divide.

> **To express a remainder as a fraction:**
>
> 1. Divide to the specified place value.
> 2. Attach zeros to the dividend after the decimal if needed to carry out the division.
> 3. Write the remainder over the divisor as a fraction after the last place in the quotient. Reduce the fraction to lowest terms if appropriate (Section 3–5).

EXAMPLE   Divide to the hundredths place and write the remainder as a fraction.

$$3.2\overline{)15.27}$$

$$\begin{array}{r} 3.33\tfrac{23}{69} \\ 6.9\,\overline{)23.0\,00} \\ 20\ 7 \\ \hline 2\ 3\ 0 \\ 2\ 0\ 7 \\ \hline 2\ 30 \\ 2\ 07 \\ \hline 23 \end{array}$$

or $\mathbf{3.33\tfrac{1}{3}}$

We attach zeros so that you can carry the division to the decimal place specified, often hundredths.

> **To round a quotient to a place value:**
>
> 1. Divide to one place past the desired rounding place.
> 2. Attach zeros to the dividend after the decimal if necessary to carry out the division.
> 3. Round the quotient to the place specified.

EXAMPLE   Divide and round the quotient to the nearest tenth.

$$3.2\overline{)15.27}$$

$$\begin{array}{r} 4.77 \\ 3.2\,\overline{)15.2\,70} \\ 12\ 8 \\ \hline 2\ 4\ 7 \\ 2\ 2\ 4 \\ \hline 2\ 30 \\ 2\ 24 \\ \hline 6 \end{array}$$

Since you are rounding to the nearest tenth, divide to the hundredths place.

4.77 rounds to 4.8.

4   Estimate and Check Division.

Estimating a division is similar to estimating other operations. The numbers are rounded *before* the calculation is made.

> **To estimate division:**
>
> 1. Round the divisor and dividend to one nonzero digit.
> 2. Find the first digit of the quotient.
> 3. Attach a zero in the quotient for each remaining digit in the dividend.

Because division and multiplication are inverse operations, division is checked by multiplication.

As noted earlier, estimating and checking is an important process in extending our number sense to decimals.

EXAMPLE Estimate, find the exact answer, and check $913 \div 22$.

$$\text{Estimate:} \quad 20\overline{)900} \quad = 40$$

20 divides into 90 four whole times. Attach a zero after 4.

Exact:
$$22\overline{)913.0} = 41.5$$
$$\begin{array}{r} 88 \\ \hline 33 \\ 22 \\ \hline 110 \\ 110 \end{array}$$

Check:
$$\begin{array}{r} 41.5 \\ 2\,2 \\ \hline 83\,0 \\ 830 \\ \hline 913.0 \end{array}$$

The answer checks.

5 Find the Numerical Average.

Sometimes we need to use approximate numbers. Two commonly used approximations are averages and estimates. We use averages to make comparisons, such as when we compare the average mileage different cars get per gallon of gasoline. There are several types of averages. Here we examine the average known as the arithmetic average or mean. To get a quick but rough idea of the answer *before* we work a problem, we use estimates.

In most courses students take, their numerical grade is determined by a process called *numerical averaging*. If the grades are 92, 87, 76, 88, 95, and 96, we can find the average grade by adding the grades and dividing by the number of grades. Because we have six grades, we divide the sum by 6.

$$\frac{92 + 87 + 76 + 88 + 95 + 96}{6} = \frac{534}{6} = 89$$

EXAMPLE A car involved in an energy efficiency study had the following miles per gallon (mpg) listings for five tanks of gasoline: 21.7, 22.4, 26.9, 23.7, and 22.6 mpg. Find the average miles per gallon for the five tanks of gasoline.

Estimate: The low value is 21.7 and the high value is 26.9. The average will be between the two values.

$$\text{Exact:} \quad \frac{21.7 + 22.4 + 26.9 + 23.7 + 22.6}{5} = \frac{117.3}{5} = 23.46$$

$$= 23.5 \text{ mpg} \qquad \text{rounded}$$

The average is 23.5 mpg.

Tip!	***Rounding Answers When Averaging:***

If the numbers being averaged are expressed to the same place value, the answer is usually rounded to that place value.

6 Divide Using a Calculator.

When you divide on a calculator, remember that division is not commutative or associative. Thus, you must be careful when you enter division problems.

EXAMPLE Find the quotient of $9,264 \div 2.4$ using a calculator.

$$9264 \;\boxed{\div}\; 2 \;\boxed{\cdot}\; 4 \;\boxed{=} \;\Rightarrow\; \mathbf{3860.}$$

Tip!	***Division Notation on a Calculator:***

The display on some graphing calculators may use a slash (/) instead of the division symbol ($\div$).

Tip!	***Which Number Goes First in Division?***

One common error occurs when the numbers are entered on the calculator in the wrong order. To verify the correct order, use a division problem that you can work mentally. $2\overline{)10}$ or $10 \div 2$. The answer should be 5.

$$\text{Option 1:} \qquad 2 \div 10 \neq 5 \qquad \text{Incorrect.}$$
$$\text{Option 2:} \qquad 10 \div 2 = 5 \qquad \text{Correct.}$$

EXAMPLE Paul Vo shops regularly at Ajax Discount Superstore for items used in his business. In preparing his tax return for the year, Paul has receipts for these amounts: $67.48, $123.76, $47.53, $54.97, $94.50, $46.23, $112.09, and $47.58. What is the amount of his average purchase?

$$\$67.48 + \$123.76 + \$47.53 + \$54.97 + \$94.50 + \$46.23 + \$112.09 + \$47.58 = \$594.14$$
$$\$594.14 \div 8 = \$74.2675$$

His average purchase is $74.27.

Continuous Calculator Sequence

Averages can be found in a continuous sequence on a calculator. Remember to use the $=$ or ENTER key to find the sum *before* you divide by the number of measures. Let's use the calculator to work the previous example.

67 ⸱ 48 + 123 ⸱ 76 + 47 ⸱ 53 + 54 ⸱ 97 + 94 ⸱ 50 + 46 ⸱

23 + 112 ⸱ 09 + 47 ⸱ 58 = ÷ 8 = ⇒ **74.2675**   **$74.27, rounded**

SELF-STUDY EXERCISES 1–5

1   Write as divisions using the symbols ÷, ⌐, and − for each problem.

1. 4 into 8

2. 3 into 9

3. 6 divided into 24

4. 7 divided into 30

5. 6 divided by 2

6. 8 divided by 4

2   Perform the long-division problems.

7. 8⟌600

8. 9⟌207

9. 6⟌120

10. 7⟌21

11. 3⟌48

12. 8⟌96

13. 5⟌215

14. 37⟌1,739

15. 26⟌312

16. 4⟌93

17. 48⟌5,982

18. 133⟌7,528

19. In a building where 46 outlets are installed, 1472 ft of cable are used. What is the average number of feet of cable used per outlet?

20. Twelve water tanks are constructed in a welding shop at a total contract price of $14,940. What is the price per tank?

21. Five equally spaced holes are drilled in a piece of $\frac{1}{4}$-in. flat metal stock. The centers of the first and last holes are 2 in. from the end (see Fig. 1–20). What is the distance between the centers of any two adjacent holes? (*Caution:* How many equal center-to-center distances are there?)

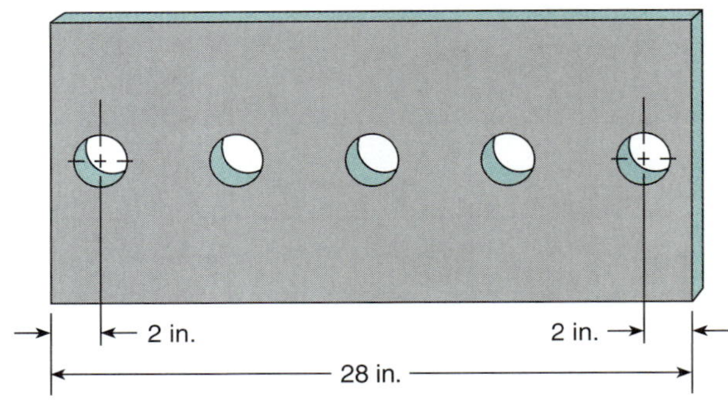

Figure 1–20

3   Divide.

22. 3⟌3.78

23. 26⟌80.34

24. 21⟌1.323

25. 1.2⟌342

26. 4.8⟌28.32

27. 7.6⟌342

28. 6.3⟌68.67

29. 0.23⟌0.0437

30. 0.09⟌0.0954

31. 0.41⟌0.1353

32. 0.32⟌8

33. 1449 ÷ 0.07

34. 8118 ÷ 0.09

35. 4066 ÷ 0.38

36. 8.28 ÷ 4.6

37. A 7.3-ft-long pipe weighs 43.8 lb. What is the weight of 1 ft of pipe?

38. A room requires 770.5 ft^2 of wallpaper, including waste. How many whole single rolls are needed for the job if a roll covers 33.5 ft^2 of surface?

39. The feed per revolution of a drill is 0.012 in. A hole 7.2 in. deep will require how many revolutions of the drill?

40. A piece of channel iron 5.6 ft long is cut into eight pieces. Assuming that there is no waste, what is the length of each piece?

41. If voltage is wattage (W) divided by amperage (A), find the voltage in a 300-W circuit drawing 3.2 A.

42. Exercises 37–41 involve measures. In some instances, the answer is a measure and in others, the answer is an amount telling how many. Explain when the answer will be a measure.

Divide to the hundredths place and write the remainder as a fraction.

43. $1.3\overline{)25.8}$ **44.** $2.4\overline{)4.37}$ **45.** $0.06\overline{)156.07}$ **46.** $41.7 \div 21$ **47.** $0.23 \div 2.9$

Divide and round the quotient to the place indicated.

48. Nearest tenth: $0.43\overline{)72.8}$

49. Nearest hundredth: $25\overline{)3.897}$

50. Nearest whole number: $4.1\overline{)34.86}$

51. Nearest cent: $5\overline{)\$4.823}$

52. Nearest dollar: $17\overline{)\$24.98}$

53. If 12 lathes cost $6895, find the cost of each lathe to the nearest dollar.

54. If 12 electrolytic capacitors cost $23.75, find the cost of one capacitor to the nearest cent.

55. To find the depth of an American Standard screw thread, the number of threads per inch is divided into 0.6495. Find the depth of thread of a screw that has 8 threads per inch. Round to the nearest ten-thousandth.

56. A stack of 40 sheets of fiberglass is 6.9 in. high. What is the thickness of one sheet to the nearest tenth of an inch?

4 Estimate by rounding to one nonzero digit. Then find the exact answer and check.

57. $7.2\overline{)904.32}$

58. $122\overline{)384,512}$

59. $221\overline{)824,604}$

60. $7\overline{)59.01}$

61. A landfill job needs 2294 yd^3 of dirt to be delivered. If one truck can carry 18 yd^3, how many loads will have to be made?

62. If a reel of wire contains 6362.5 ft of wire, how many 50-ft extension cords can be made?

5 Find the average, then round to the same place value used in the problems.

63. Temperature readings: 82.5°, 76.3°, 79.8°, 84.7°, 80.8°, 78.8°, 80.0°

64. Test scores: 86, 73, 95, 85

65. Weight of cotton bales: 515 lb, 468 lb, 435 lb, 396 lb

66. Monthly income: $873.46, $598.21, $293.85, $546.83, $695.83, $429.86, $955.34, $846.95, $1025.73, $1152.89, $957.64, $807.25

67. Average rainfall: 1.25 in., 0.54 in., 0.78 in., 2.35 in., 4.15 in., 1.09 in.

68. Amperes of current: 3.0 A, 2.5 A, 3.5 A, 4.0 A, 4.5 A

6 Work the division problems using your calculator. Round the answer to the nearest whole number.

69. $1.58\overline{)85.297}$ **70.** $4\overline{)1203}$ **71.** $5.6\overline{)224.178}$

72. $45\overline{)4,027,500}$ **73.** $32\overline{)1,286,400}$ **74.** $71\overline{)440,271}$

75. Three bricklayers laid 3210 bricks on a job in one day. What was the average number of bricks laid by each bricklayer?

76. A developer divides a tract of land into 14 equally valued parcels. If the tract is valued at $147,000, what is the value of each parcel?

77. A shipment of 150 machine parts costs $15,737.50. If this includes a $25 shipping charge, how much does each part cost if shipping charges are not included?

78. If a shop manager earns $23,400 annually, what is the monthly salary?

79. A machine operator earns an annual salary of $29,580. What is the operator's monthly salary?

80. A shipment of 288 headlights is billed out at $1329.60, which includes a $48 freight charge. What is the cost per headlight, excluding freight costs?

81. An invoice for the total cost of refrigerator replacement parts indicates $2116.80, which includes a $30 freight charge. The catalog price of the parts is $86.95 each. How many parts were ordered?

82. A job requires 95 ft of pipe, which will be cut into 9.3-ft lengths. How many 9.3-ft lengths of pipe will there be?

83. According to a real estate report in the Sunday newspaper, four homes in the Scenic Hills subdivision sold for $86,500, $68,750, $92,780, and $65,990. Find the average selling price of the homes in the subdivision.

84. Suprena Anderson paid the following amounts for her textbooks for the 2000 fall semester at Bolton Technical and Community College: English II, $67.50; general science, $92.75; psychology, $42.50; electronic technology I, $94.25; health, $18.50; and introduction to microcomputers, $42.75. Find the average cost of her textbooks for the semester.

85. The weather report listed the daily rainfall for the month of July as 0.2 in., 1.4 in., 0.5 in., 1.8 in., 0.6 in., 0.8 in., and 0.2 in. Find the average daily rainfall for the month to the nearest tenth of an inch.

1-6 EXPONENTS, ROOTS, AND POWERS OF 10

Learning Outcomes

1. Simplify expressions that contain exponents.
2. Square numbers and find the square roots of numbers.
3. Use powers of 10 to multiply and divide.
4. Use a calculator to find powers and roots.

1 Simplify Expressions That Contain Exponents.

The product of repeated factors can be written in shorter form using whole-number exponents. For example, $4 \times 4 \times 4 = 4^3$. The 4 is called the *base* and is the repeated factor. The 3 is the *exponent* and indicates how many times the factor is repeated. The expression 4^3 is read "four *cubed*" or "four to the third *power*" or "four raised to the third *power*." In the example $2 \times 2 \times 2 \times 2 \times 2 = 2^5$, the 2 is the base and the 5 is the exponent. The expression 2^5 is read "two to the fifth *power*." In expressions where 2 is an exponent, such as 4^2, the expression is usually read as "four *squared*"; however, it may also be read as "four to the second *power*."

To perform the multiplication of 2^5 (or $2 \times 2 \times 2 \times 2 \times 2) = 32$ is to *simplify* the expression. The expression 2^5 is written in *exponential notation*. The number 32 is written in *standard notation*.

$$\text{base} \rightarrow 2^5 \leftarrow \text{exponent}$$

Exponential notation is a shortcut way to write repeated multiplication.

> **Simplifying Expressions in Exponential Notation**
>
> 1. Use the base as a factor as many times as indicated by the exponent.
> 2. Perform the multiplication.

EXAMPLE Identify the base and exponent of the expressions, then simplify each expression.

(a) 5^3 (b) 1.5^2

(a) 5^3 5 is the base; 3 is the exponent.
$$5^3 = \boxed{5 \times 5 \times 5} = 125$$

(b) 1.5^2 1.5 is the base; 2 is the exponent.
$$1.5^2 = \boxed{1.5 \times 1.5} = 2.25$$

When 1 is the exponent, the 1 indicates that the number is used as a factor 1 time. For example, $5^1 = 5$. When the number does not show an exponent, the exponent is understood to be 1. For example, 8 means 8^1.

An exponent can be any number, including 1 and 0. Exponents that are not whole numbers are discussed later. Examine the following expressions in exponential notation.

$$4^3 = 4 \times 4 \times 4 \qquad \text{4 used as a factor 3 times}$$

$$4^2 = 4 \times 4 \qquad \text{4 used as a factor 2 times}$$

$$4^1 = 4 \qquad \text{4 used as a factor 1 time}$$

Any number with an exponent of 1 is the number itself:

$$a^1 = a, \qquad \text{for any base } a$$

EXAMPLE Write the exponential expressions as standard numbers.

(a) 3^1 (b) 8^1 (c) 5^1 (d) 2.3^1

(a) $3^1 = \mathbf{3}$ (b) $8^1 = \mathbf{8}$ (c) $5^1 = \mathbf{5}$ (d) $2.3^1 = \mathbf{2.3}$

EXAMPLE Express the standard numbers in exponential notation.

(a) 2   (b) 7   (c) 12   (d) 4.5

(a) $2 = \mathbf{2^1}$ (b) $7 = \mathbf{7^1}$ (c) $12 = \mathbf{12^1}$ (d) $4.5 = \mathbf{4.5^1}$

An exponent of 0 is a special case. When 0 is used as an exponent, we interpret the meaning of the exponent differently. Informally, we say the base is used as a factor 0 times. It does not mean that we have a factor of 0. In fact, we define a base with an exponent of 0 to be 1. If we translate our place-value chart into a base-10 number system, we see that this interpretation is consistent with our definitions of place values and exponents of 1 and 0.

$10^5 = 100,000$	$10^4 = 10,000$	$10^3 = 1,000$	$10^2 = 100$	$10^1 = 10$	$10^0 = 1$

Any number (except zero) with an exponent of 0 equals 1. The expression 0^0 is undefined.

$$a^0 = 1, \qquad \text{for any nonzero base } a$$

EXAMPLE Simplify the exponential expressions.

(a) 9^0 (b) 6^0 (c) 27^0 (d) 2.6^0

(a) $9^0 = \mathbf{1}$ (b) $6^0 = \mathbf{1}$ (c) $27^0 = \mathbf{1}$ (d) $2.6^0 = \mathbf{1}$

2 Square Numbers and Find the Square Roots of Numbers.

When a number has an exponent of two, we can call the standard form of the number a *square* number or a *perfect square*. That is, the result of multiplying a number times itself is a square number. This terminology evolved from a common formula for finding the area of a square. $A = s^2$, where A represents the area and s represents the length of one side. When we square 3, we write $3^2 = 3 \times 3 = 9$. We say 9 is the square of 3.

  Find the square.

(a) 2   (b) 7   (c) 3.2

(a) 2; $2^2 = 2 \times 2 = \mathbf{4}$
(b) 7; $7^2 = 7 \times 7 = \mathbf{49}$
(c) 3.2; $3.2^2 = 3.2 \times 3.2 = \mathbf{10.24}$

Just as addition and multiplication have the inverse operations of subtraction and division, respectively, squaring also has an inverse operation: taking the square root of a number. To take the square root of a number (a perfect square), we find the number (the *square root*) that was used as a factor twice to equal that perfect square. The square root of 9 is 3 because 3^2 or $3 \times 3 = 9$.

The *radical sign* $\sqrt{}$ indicates that the square root is to be taken of the number under the bar. This bar serves as a grouping symbol just like parentheses. The number under the bar is called the *radicand*. The entire expression is called a *radical expression*.

$$\text{radical sign} \longrightarrow \sqrt{\underset{\displaystyle\uparrow}{25}} = 5 \longleftarrow \text{square root}$$
$$\text{radicand} \longrightarrow$$

To find the square root of a perfect square by estimation:

1. Select a trial estimate of the square root.
2. Square the estimate.
3. If the square of the estimate is less than the original number, adjust the estimate to a larger number. If the square of the estimate is more than the original number, adjust the estimate to a smaller number.
4. Square the adjusted estimate from Step 3.
5. Continue the adjusting process until the square of the trial estimate is the original number.

■ **Learning Strategy** *Memorizing Versus Understanding.*

We often try to memorize facts rather than rely on our understanding of a concept. Memorized facts fade with time. For example, do you remember who was the 23rd president of the United States? (By the way, it was Benjamin Harrison.) If we understand the concept of a perfect square, we can generate a list whenever we need it. Of course, the more we use certain facts, the longer our memory retains them.

$11 \times 11 = 121$; $12 \times 12 = 144$; $13 \times 13 = 169$; $14 \times 14 = 196$; $15 \times 15 = 225$
$16 \times 16 = 256$; $17 \times 17 = 289$; $18 \times 18 = 324$; $19 \times 19 = 361$; $20 \times 20 = 400$

As we noted earlier, it is important to develop and strengthen our number sense.

  Find $\sqrt{256}$.

Select 15 as the estimated square root: 15^2 or $15 \times 15 = 225$. The number 225 is less than 256, so the square root of 256 must be larger than 15. We adjust the estimate to 17: 17^2 or $17 \times 17 = 289$. The number 289 is more than 256, so the square root of 256 must be smaller than 17. Now adjust the estimate to 16.

Because $16^2 = 256$, 16 is the square root of 256.

3 Use Powers of 10 to Multiply and Divide.

Powers of 10 are numbers whose only nonzero digit is 1. Thus, 10, 100, 1000, and so on, are powers of 10 because each value can be written in exponential form with a base of 10.

Compare the number of zeros in standard notation with the exponent in exponential notation.

One million	$1,000,000 = 10^6$	6 zeros
One hundred thousand	$100,000 = 10^5$	5 zeros
Ten thousand	$10,000 = 10^4$	4 zeros
One thousand	$1,000 = 10^3$	3 zeros
One hundred	$100 = 10^2$	2 zeros
Ten	$10 = 10^1$	1 zero
One	$1 = 10^0$	0 zeros

The exponents in powers of 10 indicate the number of zeros used in standard notation.

EXAMPLE Express as powers of 10.

(a) 10,000,000   (b) 100,000,000   (c) 100,000,000,000

(a) $10,000,000 = 10^7$
(b) $100,000,000 = 10^8$
(c) $100,000,000,000 = 10^{11}$

EXAMPLE Express in standard notation.

(a) 10^5 (b) 10^{13}

(a) $10^5 = 100,000$
(b) $10^{13} = 10,000,000,000,000$

In some applications, powers of 10 are used to simplify multiplication and division problems. Compare the examples to find the pattern for multiplying a number by a power of 10.

$5 \times 100 = 500$	$5 \times 10^2 = 500$	Decimal point moved 2 places to the right and 2 zeros attached.
$27 \times 1,000 = 27,000$	$27 \times 10^3 = 27,000$	Decimal point moved 3 places to the right and 3 zeros attached.

If 100 sheets of plywood cost $10.57 per sheet, we find the total cost by multiplying: $10.57 \times 100 = \$1,057$. When we solve problems involving multiplying by 10, 100, 1,000, and so on, we should be able to perform the calculations mentally by just moving the decimal point. Calculations such as the preceding one can be done more quickly mentally than with a calculator.

To multiply a number by a power of 10:

1. Move the decimal point in the number to the *right* as many places as the 10, 100, 1000, or so on, has zeros.
2. Attach zeros on the right if necessary.

EXAMPLE Multiply 237×100.

Using the shortcut rule, we attach two zeros to the *right* of the 7 in 237 and insert the appropriate comma.

$$237.00 \times 100 = 23,700$$

EXAMPLE Multiply $36.2 \times 1{,}000$.

There are three zeros in 1,000, so we move the decimal point three places to the *right*. Two zeros need to be attached.

$$36.200 \times 1{,}000 = \mathbf{36{,}200}$$

Dividing by numbers such as 10, 100, 1,000, and so on, can be done very quickly using a shortcut rule. Examine these divisions.

$32 \div 10 = 3.2$ Decimal point moved one place to the *left*.

$78.9 \div 100 = 0.789$ Decimal point moved two places to the *left*.

$52{,}900 \div 1{,}000 = 52.9$ Decimal point moved three places to the *left*.

If we compare the decimal point in the dividend with the decimal in the quotient, we notice that the decimal point in the quotient has been moved to the left as many places as the divisor (10, 100, 1,000, and so on) has zeros.

> **To divide a number by a power of 10:**
>
> **1.** Move the decimal to the *left* as many places as the divisor has zeros.
> **2.** Attach zeros to the left if necessary.
> **3.** You may drop zeros to the right of the decimal point if they follow the last nonzero digit of the quotient.

Compare this rule with the rule for multiplying decimal numbers by 10, 100, 1,000, and so on. For multiplication, the decimal shifts to the right; for division, the decimal shifts to the left.

EXAMPLE If 100 lb of floor cleaner costs \$63, what is the cost of 1 lb?

Use the shortcut rule to divide by 100. A whole number has a decimal point understood after its last digit.

$$63 \div 100 = 0.63 = \$0.63$$

The cost of 1 lb is \$0.63.

EXAMPLE If a stack of 100 concrete blocks weighs 1,080 lb, find the weight of one concrete block.

$$1{,}080 \div 100 = 10.80 \qquad \text{or} \qquad 10.8 \text{ lb}$$

One block weighs 10.8 lb.

4 Use a Calculator to Find Powers and Roots.

Some scientific and graphing calculators have specific power keys, such as keys for squares and cubes. They also have a "general power" key that can be used for all powers. Even though the "general power" key can be used to square and cube numbers, the "square" and "cube" keys require fewer keystrokes.

To find the square of a number, locate the $\boxed{x^2}$ key on your calculator. Note, some calculators use a $\boxed{\text{SHIFT}}$ or $\boxed{2^{\text{nd}}}$ or $\boxed{\text{ALT}}$ key to access calculator functions that are displayed above the keys. Try these examples, which you can do mentally, then compare with your calculator results.

$$3^2 = 9$$

Some calculator options are

$$3 \boxed{x^2}$$

$$3 \boxed{x^2} \boxed{\text{ENTER}}$$

$$3 \boxed{\text{SHIFT}} \boxed{\sqrt{}}^{\,x^2} \boxed{=}$$

Now, locate the "general power" key $\boxed{x^s}$ or $\boxed{y^x}$.

$$4^3 = 64$$

Options:

$$4 \boxed{y^x} 3 \boxed{=}$$

$$4 \boxed{x^y} 3 \boxed{\text{ENTER}}$$

$$4 \boxed{x^y} 3 \boxed{\text{EXE}}$$

Tip! | ***Labels on Calculator Keys Are Not Universal:***

To make calculator steps easier to follow, we identify a common label for a function in a box. The exact label will vary with the specific calculator model. Graphing calculators also provide many functions on menus instead of on specific keys. Even though we use a key notation in this text to show a calculator function, this function may actually appear on a menu. Check your calculator manual for exact location and labeling of functions.

EXAMPLE Use the calculator to find 35^2.

$$\text{Using the } x^2 \text{ function: } 25 \boxed{x^2} \boxed{=}$$

Note: Some calculators require the equal, enter, or execute key to be pressed. Others give the square immediately after pressing the $\boxed{x^2}$ function. Using the general power function, $35 \boxed{x^y} 2 \boxed{=}$, the exponent must be entered followed by the equal key.

1,225.

EXAMPLE Use the calculator to find 2.3^5.

$$2 \boxed{\cdot} 3 \boxed{x^y} 5 \boxed{=}$$

64.36343

Most calculators have a square root key $\boxed{\sqrt{}}$ or the square root symbol is above the key and can be accessed through the $\boxed{\text{SHIFT}}$, $\boxed{2^{\text{nd}}}$, or $\boxed{\text{ALT}}$ key. Try an example you can do mentally.

$$\sqrt{25} = 5$$

Some calculator options are

$$25 \boxed{\sqrt{}}$$

$$\boxed{\sqrt{}} 25 \boxed{=}$$

$$\boxed{2^{\text{nd}}} \boxed{x^2}^{\,\sqrt{}} 25 \boxed{\text{ENTER}}$$

$$\boxed{\sqrt{}} 25 \boxed{\text{EXE}}$$

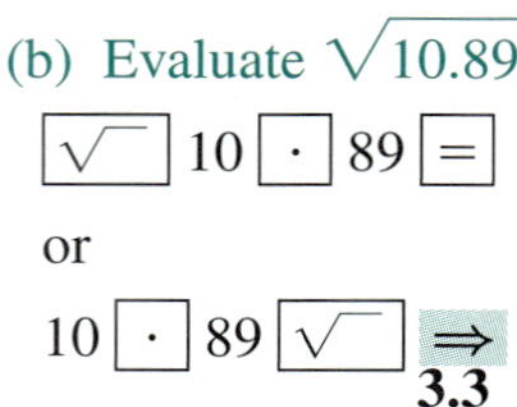

EXAMPLE (a) Evaluate $\sqrt{529}$.

$\boxed{\sqrt{}}\ \boxed{529}\ \boxed{=}$

or

$\boxed{529}\ \boxed{\sqrt{}} \Rightarrow 23.$

(b) Evaluate $\sqrt{10.89}$.

$\boxed{\sqrt{}}\ \boxed{10}\ \boxed{\cdot}\ \boxed{89}\ \boxed{=}$

or

$\boxed{10}\ \boxed{\cdot}\ \boxed{89}\ \boxed{\sqrt{}} \Rightarrow 3.3$

SELF-STUDY EXERCISES 1–6

1 Give the base and exponent of the expressions.

1. 4^3 **2.** 9^4 **3.** 2.7^9

Simplify the exponential expressions.

4. 1.5^3 **5.** 7^2 **6.** 10^3 **7.** 2^4
8. 3.4^2 **9.** 15^1 **10.** 8^1 **11.** 9^0

Express with an exponent of 1.

12. 8 **13.** 14.5 **14.** 12 **15.** 23
16. Explain what an exponent of 1 does to the base. **17.** Explain what an exponent of 0 does to the base.

2 Write in standard notation.

18. 8^2 **19.** 2^2 **20.** 3.5^2
21. 1.4^2 **22.** 13^2 **23.** 1^2
24. 100^2

Square the numbers.

25. 8 **26.** 9 **27.** 17
28. 18 **29.** 101 **30.** 22

Perform the operations.

31. $\sqrt{25}$ **32.** $\sqrt{49}$ **33.** $\sqrt{81}$
34. $\sqrt{196}$
35. What does it mean to square a number? **36.** What does it mean to take the square root of a number?

3 Multiply the whole numbers using powers of 10.

37. 10×10^2 **38.** $32 \times 1{,}000$ **39.** $3 * 10^4$
40. $2 * 10^4$ **41.** $102 * 100$ **42.** 22×10^3

Divide the whole numbers using powers of 10.

43. $250 \div 10$ **44.** $210 \div 10$ **45.** $\dfrac{300}{10^2}$

46. $\dfrac{900}{10^2}$ **47.** $2{,}500 \div 10$ **48.** $120 \div 10^2$

49. What does the exponent of a power of 10 mean when you are multiplying? **50.** What does the exponent of a power of 10 mean when you are dividing?

4 Use a calculator to evaluate.

51. 15^2 **52.** 7^3 **53.** 5^7
54. $\sqrt{324}$ **55.** $\sqrt{784}$ **56.** $\sqrt{1089}$

Learning Outcomes

1 Apply the order of operations to a series of operations.
2 Use a calculator to perform a series of operations.
3 Solve applied problems using problem-solving strategies.

1 Apply the Order of Operations to a Series of Operations.

Whenever several mathematical operations are performed, the proper *order of operations must be followed.*

The order of operations:

1. **Parentheses (grouping symbols):** Perform operations within parentheses (or other grouping symbols), beginning with the innermost set of parentheses, or apply the distributive property.
2. **Exponents and roots:** Evaluate exponential operations and find square roots in order from left to right.
3. **Multiply and divide** in order from left to right.
4. **Add and subtract** in order from left to right.

To summarize, use the following key words:

Parentheses (grouping), Exponents (and roots),

Multiplication and Division, Addition and Subtraction

Tip!	*A Memory Aid for the Order of Operations:*

To remember the order of operations, use the sentence, "Please Excuse My Dear Aunt Sally."

- Parentheses (grouping) • Exponents (roots) • Multiply/Divide • Add/Subtract

---

EXAMPLE   Evaluate $3(2 + 3)$.

Parentheses may show both a grouping and multiplication. When a number is multiplied by a sum or a difference, the distributive property can also be applied. Using the distributive property:

$$3(2 + 3) = 3(2) + 3(3) \qquad \text{Distribute.}$$
$$= 6 + 9 \qquad \text{Add.}$$
$$= 15$$

Using the order of operations:

$$3(2 + 3) = 3(5) \qquad \text{Do operations in parentheses first.}$$
$$= 15 \qquad \text{Multiply.}$$

EXAMPLE   Simplify $4^2 - 5(2) \div (4 + 6)$ by performing the operations in the correct order.

$$4^2 - 5(2) \div (4 + 6)$$
$$4^2 - 5(2) \div 10 \qquad \text{Do operations within parentheses first: } 4 + 6 = 10.$$
$$16 - 5(2) \div 10 \qquad \text{Evaluate exponentiation: } 4^2 = 16.$$

$$16 - \boxed{10} \div 10 \qquad \text{Then multiply: } 5(2) = 10.$$
$$16 - 1 \qquad \text{And divide: } 10 \div 10 = 1.$$
$$16 - 1 = \mathbf{15} \qquad \text{Subtract last.}$$

Tip!	***Parentheses Indicate Multiplication, the Distributive Property, or a Grouping:***

As in the previous example, parentheses can indicate multiplication or an operation that should be done first. If the parentheses contain an operation, they indicate a grouping. Otherwise, they indicate multiplication. The expression 5(2) indicates multiplication, while (4 + 6) indicates a grouping.

EXAMPLE Evaluate $5 \times \sqrt{16} - 5 + [15 - (3 \times 2)]$.

$$5 \times \sqrt{16} - 5 + [15 - \boxed{(3 \times 2)}] \qquad \text{Work innermost grouping: } 3 \times 2.$$
$$5 \times \sqrt{16} - 5 + [15 - \boxed{6}] \qquad \text{Work remaining grouping: } 15 - 6.$$
$$5 \times \boxed{\sqrt{16}} - 5 + 9 \qquad \text{Find square root: } \sqrt{16} = 4.$$
$$5 \times \boxed{4} - 5 + 9 \qquad \text{Multiply: } 5 \times 4.$$
$$\boxed{20 - 5} + 9 \qquad \text{Add and subtract from left to right.}$$
$$\boxed{15} + 9 = \mathbf{24}$$

EXAMPLE Evaluate $3.2^2 + \sqrt{21 - 5} \times 2$.

$$3.2^2 + \sqrt{\boxed{21 - 5}} \times 2 \qquad \text{Do operations within grouping first; the bar in the radical is a grouping symbol: } 21 - 5 = 16.$$
$$\boxed{3.2^2} + \sqrt{\boxed{16}} \times 2 \qquad \text{Evaluate exponent and square root from left to right: } 3.2^2 = 10.24; \; \sqrt{16} = 4.$$
$$\boxed{10.24} + \boxed{4 \times 2} \qquad \text{Then multiply: } 4 \times 2 = 8.$$
$$10.24 + \boxed{8} \qquad \text{Add last.}$$
$$10.24 + 8 = \mathbf{18.24}$$

2 Use a Calculator to Perform a Series of Operations.

Both scientific and graphing calculators are programmed to perform the order of operations in the proper order.

EXAMPLE Evaluate $2.4^2 - 5(2) \div (4 + 6)$.

$$2 \boxed{\cdot} 4 \boxed{x^2} \boxed{-} 5 \boxed{\times} 2 \boxed{\div} \boxed{(} 4 \boxed{+} 6 \boxed{)} \boxed{=} \Rightarrow \qquad \text{Use parentheses for groupings.}$$

4.76

Tip!	***Parentheses as Multiplication and Grouping***

Some calculators use the multiplication key to indicate the product of 5 and 2, or 5×2. With other calculators, 5(2) may be entered exactly as it appears *or* with the multiplication key. Parentheses keys, $\boxed{(}$ and $\boxed{)}$, are also used to show groupings.

EXAMPLE Evaluate $3^4 + \sqrt{77 - 13} \times 2[26 - (45 - 38)]$.

$$3\;\boxed{x^y}\;4\;\boxed{+}\;\boxed{\sqrt{}}\;\boxed{(}\;77\;\boxed{-}\;13\;\boxed{)}\;\boxed{\times}\;2\;\boxed{(}\;26\;\boxed{-}\;\boxed{(}\;45\;\boxed{-}\;38\;\boxed{)}\;\boxed{)}\;\boxed{=}\;\Rightarrow$$

385.

3 Solve Applied Problems Using Problem-Solving Strategies.

Problem solving is an important skill in the workplace and in everyday life. We begin our development of problem-solving strategies by introducing a systematic problem-solving plan. This plan gives you a framework for approaching problems, but we will investigate many strategies throughout this text.

It is important to keep a positive attitude when solving problems.

Work with a study partner or group.

Working with a partner or group develops confidence. As you work with others each of you verbalizes your understanding of the problem, and you can discuss and correct any misconceptions you may have.

You and your partners can examine several approaches to the problem and discuss their pros and cons. This discussion strengthens your investigative skills and gives you more options than you would probably come up with alone. You also may discover that there may be several *correct* approaches to solving the problem.

Be persistent.

Even the best problem solvers do not always use a correct approach on the first try. The mark of a good problem solver is persistence. Learn from failed attempts. Learn what doesn't work and why. This makes the correct approach easier to understand when you find it.

Don't make the "ruts" too deep.

After pursuing several dead ends, don't continue to the point of frustration. In other words, don't spin your wheels! The harder you work, the deeper the ruts get. Back away from the problem, and when you come back to it, you may think of a new approach you haven't tried before.

Another option is to seek help from others (additional classmates, tutors, your instructor). Just because a problem doesn't "click" right away, doesn't mean that you can't solve problems.

You will encounter many different plans for solving problems. They are all trying to help you organize the problem details so that you can find an appropriate procedure for solving the problem. Our plan, like others, is very structured. As you develop confidence and skill in solving problems you will probably use less structured and more intuitive approaches.

Six-Step Problem-Solving Plan

1. **Unknown Facts.** What facts are missing from the problem? What are you trying to find?
2. **Known Facts.** What relevant facts are known or given? What facts must you bring to the problem from your own background?
3. **Relationships.** How are the known facts and the unknown facts related? What formulas or definitions can you use to establish a model?
4. **Estimation.** What are some of the characteristics of a reasonable solution? For instance, should the answer be more than a certain amount or less than a certain amount?
5. **Calculations.** Perform the operations identified in the relationships.
6. **Interpretation.** What do the results of the calculations represent within the context of the problem? Is the answer reasonable? Have all unknown facts been found? Do the unknown facts check in the context of the problem?

EXAMPLE The 7th Inning buys baseball cards from eight different vendors. In November, the company purchased 8,832 boxes of cards. If an equal number of boxes was purchased from each vendor, how many boxes of cards were supplied by each vendor?

Unknown fact Number of boxes of cards supplied by each vendor

Known facts 8,832 = Total number of boxes purchased
       8 = Total number of vendors
An equal number of boxes were purchased from each vendor.

Relationships Total boxes purchased ÷ Number of vendors = Boxes purchased by each vendor

Estimation If 8,000 boxes were purchased in equal amounts from 8 vendors, then 1,000 were purchased from each vendor. Since more than 8,000 boxes were purchased, then more than 1,000 were purchased from each vendor.

Calculations
$$8 \overline{)8{,}832} = 1{,}104$$

Interpretation **Each vendor supplied 1,104 boxes of cards.**

In identifying the relationships of a problem or in developing your plan for solving a problem, it is often helpful if you look for key words or phrases that give you clues about the mathematical operations involved in the relationships. Key words give clues as to whether one quantity is added to, subtracted from, or multiplied or divided by another quantity. For example, if a problem tells you that Carol's salary in 2001 exceeds her 2000 salary by $2,500, you know that you should add $2,500 to her 2000 salary to find her 2001 salary. Table 1–1 summarizes important key words and what they generally imply when they are used in problems.

TABLE 1–1 Key Words and What They Generally Imply in Word Problems

Addition	Subtraction	Multiplication	Division	Equality
The sum of	Less than	Times	Divide(s)	Equals
Plus/total	Decreased by	Multiplied by	Divided by	Is/was/are
Increased by	Subtracted from	Of	Divided into	Is equal to
More/more than	Difference between	The product of	Half of (divided by 2)	The result is
Added to	Diminished by	Twice (2 times)	Third of (divided by 3)	What is left
Exceeds	Take away	Double (2 times)	Per	What remains
Expands	Reduced by	Triple (3 times)	How big is each part?	The same as
Greater than	Less/minus	Half of ($\frac{1}{2}$ times)	How many parts can be made from	Gives/giving
Gain/profit	Loss	Third of ($\frac{1}{3}$ times)		Makes
Longer	Lower			Leaves
Older	Shrinks			
Heavier	Smaller than			
Wider	Younger			
Taller	Slower			

In many applied problems, you must use more than one relationship to find the unknown facts. When this is the case, you usually need to read the problem several times to find all the relationships and plan your solution strategy.

EXAMPLE Carlee Anne McAnally needs to ship 78 crystal vases. With standard packing to prevent damage, 5 vases fit in each available box. How many boxes are required to pack the vases?

Unknown fact Number of boxes required to ship the vases

Known facts Total vases to be shipped = 78
Number of vases per box = 5

Relationships	Total boxes needed = Total number of vases ÷ Number per box
	Total boxes needed = $78 ÷ 5$
Estimation	$70 ÷ 5 = 14$ Round down the dividend.
	$80 ÷ 5 = 16$ Round up the dividend.
	Since 78 is between 70 and 80, the number of boxes needed is between 14 and 16.
Calculation	$78 ÷ 5 = 15 \text{ R } 3$
Interpretation	**16 boxes are needed;** 15 boxes will contain 5 vases each, and 1 box will contain 3 vases. The box with 3 vases will need extra packing.

■ Learning Strategy *Using Guess and Check to Solve Problems*

An effective strategy for solving problems involves guessing. Make a guess that you think may be reasonable and check to see if the answer is correct. If your guess is not correct, decide if it is too high or too low. Make another guess based on what you learned from your first guess. Continue until you find the correct answer.

Let's try guessing in the previous example. We found that we could pack 70 vases in 14 boxes and 80 vases in 16 boxes. Since we need to pack 78 vases, how many vases can we pack with 15 boxes? $15 \times 5 = 75$. Still not enough. Therefore, we will need 16 boxes, but the last box will not be full.

You can probably think of other ways to solve this problem. This is good. Any plan that leads to a correct solution is acceptable. Some plans will be more efficient than others, but you develop your problem-solving skills by pursuing a variety of strategies. As we progress through this text we will introduce other strategies that you may find helpful.

SELF-STUDY EXERCISES 1–7

1   Use the order of operations to evaluate each problem.

1. $5^2 + 4 - 3$

2. $4^2 + 6 - 4$

3. $4 \times 3 - 9 ÷ 3$

4. $5 \times 2.9 - 4 ÷ 2$

5. $25 ÷ 5 \times 4.8$

6. $64 ÷ 4 \times 2$

7. $6 \times \sqrt{36} - 2 \times 3$

8. $3 \times \sqrt{81} - 3 \times 4$

9. $4^2 \times 3^2 + (4 + 2) \times 2$

10. $2^2 \times 5^2 + (2 + 1) \times 3$

11. $54 - 3^3 - \dfrac{8}{2}$

12. $156 - 2^3 - \dfrac{9}{3}$

13. $32 - 2.05 \times 4^2 ÷ 2$

14. $5.2^2 - 3 \times 2^2 ÷ 6$

15. $4 + 15 ÷ 3 - \dfrac{0}{7}$

16. $3 + 8 ÷ 4 - \dfrac{0}{5}$

17. $3 - 2 + 3 \times 3 - \sqrt{9}$

18. $2 - 1 + 4 \times 4 - \sqrt{4}$

19. $2^4 \times (7 - 2) \times 2$

20. $3^4 \times (9 - 3) \times 3$

2   Use a calculator to perform the operations.

21. $6 \times 9^2 - \dfrac{12}{4}$

22. $7 \times 8^2 - \dfrac{10}{2}$

23. $4^3 + 14 - 8$

24. $2^3 + 12 - 7$

25. $2 \times \sqrt{16} + (8 - \sqrt{25})$

26. $4 \times \sqrt{49} + (9 - \sqrt{64})$

27. $3(2^2 + 1) - 30 ÷ 3$

28. $4(3^2 + 2) - 60 - 12$

29. $3 + 10 ÷ 5 + 2$

30. $4 + 12 ÷ 4 + 7$

31. $25 - 5^2 ÷ (7 - 2)$

32. $36 - 6^2 ÷ (8 - 2)$

33. $2(3.1^2 + 2) - \sqrt{7.29}$

34. $3^4 - 2 \times 4.6 ÷ \dfrac{0}{2}$

35. $24(3^2 + 1) - 48.25$

36. State the first operation in the order of operations.

37. State the last operation in the order of operations.

38. If you have 348 packages of Halloween candy to rebox for shipment to a discount store and you can pack 12 packages in each box, how many boxes will you need?

40. In a recent year, 21,960 people visited Bio Fach, Germany's biggest ecologically sound consumer goods trade fair. This figure was up from the 18,090 and 16,300 of the previous two years. What was the increase in visitors to Bio Fach over the two-year period?

39. If American Communications Network (ACN) has an annual payroll of $5,602,186 for its 214 employees, what is the average salary of an ACN employee?

41. The On-The-Square Card and Gift Shop buys cards from eight vendors. In November, the company purchased 8,832 boxes of cards. If the shop purchased an equal number of boxes from each vendor, how many boxes of cards did each vendor supply?

■ ASSIGNMENT EXERCISES ■

Section 1–1

Write as decimal numbers.

1. (a) $\dfrac{3}{10}$ **(b)** $\dfrac{15}{100}$ **(c)** $\dfrac{4}{100}$

2. (a) $\dfrac{75}{1,000}$ **(b)** $\dfrac{21}{10}$ **(c)** $\dfrac{652}{100,000}$

3. What is the place value of 6 in 21.836?

4. What is the place value of 5 in 13.0586?

5. In 13.7213, what digit is in the tenths place?

6. In 15.02167, what digit is in the ten-thousandths place?

7. Identify the place value of the digit 3:
   (a) 430 **(b)** 34,789 **(c)** 3,456,321

8. Identify the place value of the digit 2:
   (a) 2,785,901 **(b)** 45,923

9. Write 56,109,110 in words.

10. Write 61,201 in words.

11. Write one million, two hundred sixty-five thousand, four hundred one in standard notation.

12. Write thirty-two thousand, three hundred twenty-one in standard notation.

13. Write the words for 6.803.

14. Write the words for 0.0712.

15. The decimal equivalent of $\frac{5}{8}$ is six hundred twenty-five thousandths. Write this decimal using digits.

16. The thickness of a sheet of aluminum is forty-thousandths of an inch. Write this thickness as a decimal number.

17. Round to the indicated places.
   (a) 36 to the nearest tens
   (b) 74 to the nearest tens
   (c) 24.237 to the nearest whole number
   (d) $42.98 to the nearest dollar
   (e) 83.052 to tens
   (f) $8.9378 to the nearest cent
   (g) $0.9986 to the nearest cent
   (h) 0.097032 to hundred-thousandths

18. Round to the indicated places.
   (a) Nearest thousand: 65,763
   (b) Nearest thousand: 28,714
   (c) Nearest ten million: 497,283,016
   (d) Nearest hundred: 8,236
   (e) Nearest ten thousand: 248,217
   (f) Nearest hundredth: 7.0893
   (g) Nearest thousandth: 1.078834
   (h) Nearest tenth: 0.09783

19. Round to the indicated places.
   (a) Nearest ten: 324
   (b) Nearest thousand: 6,882
   (c) Nearest hundred: 468
   (d) Nearest ten thousand: 49,238
   (e) Nearest billion: 26,500,000,129
   (f) Nearest tenth: 41.378
   (g) Nearest hundredth: 6.8957
   (h) Nearest ten-thousandth: 23.46097

20. Round to numbers with one nonzero digit.
   (a) 98 **(b)** 94 **(c)** 25,786
   (d) 34,786 **(e)** 12.83 **(f)** 0.0736
   (g) 7.93 **(h)** 1.876

21. Which of these decimal numbers is larger: 4.783 or 4.79?

22. Which of these decimal numbers is smaller: 0.83 or 0.825?

23. Write these decimal numbers in order of size from smallest to largest: 0.021, 0.0216, 0.02.

24. Two measurements of an object are recorded. If the measures are 4.831 in. and 4.820 in., which is larger?

25. The decimal equivalent of $\frac{7}{8}$ is 0.875. The decimal equivalent of $\frac{6}{7}$ is approximately 0.857. Which fraction is larger?

26. Two parts are machined from the same stock. They measure 1.023 in. and 1.03 in. after machining. Which part has been machined more; that is, which part is now smaller?

Section 1–2
Add using a calculator as directed by your instructor.

27. **(a)** $6 + 9 + 3 + 5$
(b) $5 + 1 + 6 + 3 + 3$
(c)
```
   8
   5
   3
   6
   2
 + 4
```
(d)
```
   7
   4
   3
   2
   5
 + 4
```

28. **(a)** $3.47 + 42.32 + 3.82 + 4.09$
(b) $6.2 + 32.7 + 46.82 + 0.29 + 4.237$
(c) $86.3 + 9.2 + 70.02 + 3 + 2.7$
(d) $42 + 3.6 + 2.1 + 7.83$

29. An air conditioner uses 10.4 kW (kilowatts), a stove uses 15.3 kW, a washer uses 2.9 kW, and a dryer uses 6.3 kW. What is the total number of kilowatts used?

30. Add.

(a)	3,456.08 + 2,147.76	**(b)**	12,467 + 24,378
(c)	23,609 2,200 76 + 124	**(d)**	$43,045.36 5,047.47 87.10 + 213.08

31. Estimate by rounding to thousands, then find the exact answer. Check your answer.
(a) $16,742.83 + $12,349.26
(b) $17,402 + 18,646$

32. A do-it-yourself project requires $57.32 for concrete, $74.26 for fence posts, and $174.85 for fence boards. Estimate the cost by rounding to numbers with one nonzero digit, then find the exact cost. Check your answer.

Section 1–3
33. Subtract.
(a) $21.34 - 16.73$
(b) $15.934 - 12.807$
(c) $9 - 7$
(d) $5 - 0$
(e) $8 - 3 - 2 - 3$
(f) $284.73 - 79.831$
(g) $345 - 201$
(h) $13,342 - 1,202$

34. Estimate by rounding to hundreds, then find the exact answer. Check your answer.
(a) $12,346.87 - $4,468.63
(b) $3,495 - 3,090$
(c) $6,767 - 478$
(d) $293.86 - 148$

35. A blueprint calls for the length of a part to be 8.296 in. with a tolerance of ± 0.005 in. What are the limit dimensions of the part?

36. For a moving sale, a family sold a sofa for $75 and a table for $25. If a newspaper ad for the sale cost $12.75, how much did the family clear on the two items sold?

37. Planning a vacation, the Mendez family selected a scenic route covering 653 miles and a direct route covering 463 miles. Estimate the difference by converting to round numbers, then find the exact answer. Check your answer.

38. A new foreign car costs $12,677, and a comparable new American car costs $11,859. Estimate the difference by rounding to thousands, then find the exact answer. Check your answer.

Use Fig. 1–21 for Exercises 39–42.

39. Find the length of A if $D = 4.237$ in., $B = 1.861$ in., and $C = 1.946$ in.

40. What is the dimension of A if $E = 4.86$ in. and $B = 1.972$ in.?

41. Give the limit dimensions of D if $D = 8.935$ in. with a ± 0.005-in. tolerance.

42. What dimension should be listed for C if D measures 3.7 in. and E measures 1.6 in.?

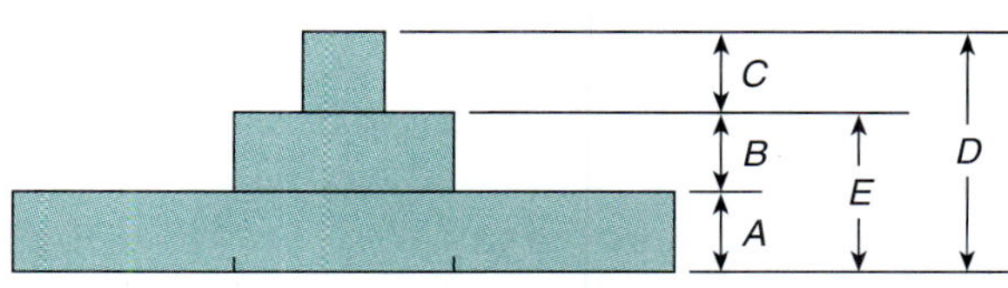

Figure 1–21

Multiply.

43. $2 \times 6 \times 7$ **44.** $6 \times 3 \times 2 \times 4$ **45.** 127×9

46. 76×5 **47.** 305×45 **48.** $236 \cdot 244$

49. $12{,}407(270)$ **50.** $527 * 342$ **51.** $56{,}002 \times 7{,}040$

52. A college bookstore sold 327 American history textbooks for \$39 each. How much did the bookstore receive for the books?

53. A mail-order supplier sells computer keyboards, for \$67 each. How much would a business pay for 21 keyboards?

54. An automotive tire dealer ran a special on heavy-duty, deluxe whitewall truck tires. If the dealer sold 105 tires for \$112 each, how much did the dealer take in on the sale?

55. If a wholesaler ordered 144 bare-bones computers for \$305 each, how much did the dealer pay for the order?

Perform the indicated operations.

56. $2.4(3 + 1)$ **57.** $5(6 - 2)$ **58.** $2(9.2 - 4)$

59. $6(3 + 7)$ **60.** $3(5 - 3)$ **61.** $8(7 + 3)$

62. What number is obtained if the sum of 5 and 2 is multiplied by 6? Use the distributive property.

63. A standard monitor for a computer sells for \$687, and a color printer sells for \$523. A multimedia monitor sells for two times the difference in cost between the monitor and the color printer. How much does the multimedia monitor cost?

64. A luxury car dealer pays a sound system installer \$33.25 per hour. How much is the sound system installer paid for 37 hours of work? Estimate by rounding to a number with one nonzero digit, then find the exact answer. Check the answer.

65. A worker is offered a job that pays \$365 per week. If the worker takes the job for 36 weeks, how much will the worker earn? Estimate by rounding to tens, then find the exact answer. Check your answer.

66. Find the area of a field 234.6 ft by 123.2 ft. Estimate the area by rounding to one nonzero digit, then find the exact answer. Check your answer. Express the area in square feet $(A = l \times w)$.

67. A parcel of land measures 1,940.7 ft by 620.4 ft. Estimate the area by rounding to hundreds, then find the exact area. Check your answer. Express the area in square feet $(A = l \times w)$.

68. A piecework employee averages 178.6 pieces per day. If the employee earns \$0.28 per item, how much is earned in 5 days?

69. If a steel tape expands 0.00014 in. for each inch when heated, how much will a tape 864 in. long expand?

Section 1–5

70. Express "3 divides into 4" using the symbols.

    (a) $\div$ **(b)** $\overline{}$ **(c)** —

71. Express "5 divided by 3" using the symbols.

    (a) $\div$ **(b)** $\overline{}$ **(c)** —

Divide.

72. $8 \div 8$ **73.** $1 \div 1$ **74.** $0 \div 3$

75. $7 \div 0$ **76.** $4 \div 1$ **77.** $8.43 \div 1.6$

78. $29.25 \div 3.6$ **79.** $325 \div 25$ **80.** $364.8 \div 6$

81. $30{,}126 \div 15$ **82.** $10{,}160 \div 20$

83. A group of 27 volunteers is seeking contributions to send a first-grade class to the circus. The group leader has 632 envelopes for the collection. If the envelopes are divided equally among the 27 volunteers, how many will each receive? How many will be left over?

84. A school marching band is in a formation of 7 rows, each with the same number of students. If the band has 56 members, how many are in each row?

85. Divide by short division: $46.7 \div 8$.

86. Divide by short division: $7{,}032 \div 9$.

Estimate, find the answer, and check.

87. $475 \div 86$ **88.** $35.16 \div 1.04$

89. If 27 volunteers collect \$287 to send a first-grade class to the circus, on average how much does each volunteer collect? Estimate the answer, then find and check the exact answer.

91. Find the average measure for 42.34 ft, 38.97 ft, 51.95 ft, and 61.88 ft. Round to the nearest hundredth.

90. A class of 16 students made 768 cookies for a school open house. How many cookies did each student make, assuming each made the same number? Estimate the answer, and then find and check the exact answer.

92. Five light fixtures cost \$74.98, \$23.72, \$51.27, \$125.36, and \$85.93. Find the average cost of the fixtures to the nearest cent.

Section 1–6

93. Give the base and exponent of each expression, then simplify.
(a) 7^3 (b) 2.3^4 (c) 8^4

95. Simplify.
(a) 0.9^1 (b) 35^1 (c) 1^1

97. Simplify.
(a) 1.7^0 (b) 1^0 (c) 8^0 (d) 149^0

99. Evaluate:
(a) 1^2 (b) 125^2 (c) 5.6^2 (d) 21^2

101. Express as powers of 10.
(a) 10 (b) 1,000 (c) 10,000
(d) 100,000

103. Divide by using powers of 10.
(a) $700 \div 100$ (b) $40.56 \div 1,000$
(c) $60.5 \div 100$ (d) $23,079 \div 10,000$
(e) $44,582 \div 1,000$

94. Give the base and exponent of each expression, then simplify.
(a) 5^6 (b) 1.2^2 (c) 10^6

96. Express with an exponent of 1.
(a) 904 (b) 76 (c) 0.3

98. Find the value of:
(a) 2^2 (b) 1.3^2 (c) 7^2 (d) 12^2

100. Find the square root.
(a) $\sqrt{2500}$ (b) $\sqrt{1.44}$ (c) $\sqrt{289}$
(d) $\sqrt{81}$

102. Multiply by using powers of 10.
(a) 3×100 (b) $75 \times 10,000$
(c) 2.2×1.000 (d) 5×100
(e) 40.6×10

Section 1–7

Evaluate.

104. $2 + (3 + 6) \div 3$

105. $4^2 \times (12 - 7) - 8 + 3$

106. $8^2 - (3 - 1.5) \times 5.2$

107. $4 + 5 - 2 \times 3$

108. $4 + \dfrac{8.6}{2} \times 2$

109. $3.1 \times 4 \times \sqrt{16} - 6^2$

110. $\sqrt{12.25} \times (4 - 2) + 8$

111. $5.2^3 - \sqrt{81} \times (2 + 1)$

112. $2^4 \div 2 - \sqrt{9}$

113. If you have 584 packages of hard candy to rebox for shipment to a discount store and you can pack 12 packages in each box, how many boxes will you need?

114. If European Internet Service (EIS) has an annual payroll of \$7,460,174,000,000 for its 194,582 employees, what is the average salary of an EIS employee?

115. In October, 42,670 people visited an aquarium in California. This figure was up from the 38,190 who visited in September and the 26,361 who visited in August. What was the increase in visitors from August to October?

116. The On-The-Square Card and Gift Shop buys magazines from 12 vendors. In August, the shop purchased 1,432 magazines. If an equal number of magazines were purchased from each vendor, how many magazines did each vendor supply?

117. Twin Towers Antique Mall has 88 booths that rent for \$1.40 per square foot. If 48 of the booths have 100 ft^2 and 40 booths have 110 ft^2, what is the total rental from the booths?

CHALLENGE PROBLEMS

118. Use a calculator to evaluate the powers.
(a) 0.03^2 (b) 0.07^2 (c) 0.005^2
(d) 0.009^2 (e) 0.02^3 (f) 0.004^3

119. You are designing a playground that has 2,160 yd^2 of space. If the playground is rectangular, determine how long and how wide it should be if one piece of equipment requires a space at least 15 yd long. Give whole-number solutions.

120. Of all the possible ways to design the playground in Exercise 119, which rectangle requires the least amount of fencing? Give whole-number solutions.

121. Use a calculator to find the square root of the decimals.
 (a) $\sqrt{0.64}$ (b) $\sqrt{0.09}$
 (c) $\sqrt{0.0009}$ (d) $\sqrt{0.0625}$
 (e) $\sqrt{0.000016}$ (f) $\sqrt{0.4}$

CHAPTER TRIAL TEST

1. Write 5,030,102 in words.

2. Write three hundred twenty-four thousand, five hundred twenty in standard notation.

3. Write the digits for seven and twenty-seven thousandths.

4. Write the digits for two hundred four millionths.

5. Round 2,743 to hundreds.

6. Round 34,988 to a number with one nonzero digit.

7. Which number is smaller, 5.09 or 5.1?

8. Round 4.018 to the nearest hundredth.

9. Round 48.3284 to the nearest tenth.

10. Round $4.834 to the nearest cent.

Perform the indicated operations.

11. $37 + 158 + 764 + 48$

12. $\$61,532 - \$47,245$

13. $\$13,207 \times 702$

14. $\$25,600 \div 12$

15. $3^2 + 5^3$

16. 46×10^3

17. $3 \times 6^2 - 4 \div 2$

18. $5^3 - (3 + 2) \times \sqrt{9}$

19. $86 \div 10^4$

20. If a stereo costs $495 and a stereo cabinet costs $269, use numbers with one nonzero digit to estimate the total cost. Find the exact cost.

21. If a man has a bill for $165 and his paycheck is $475, estimate how much of his paycheck is left after paying the bill by rounding to tens. Find the exact amount left.

22. A corporation buys 45 VCRs for $335 each. Use numbers with one nonzero digit to estimate the total cost. Find the exact cost.

23. A softball coach paid $126 for 9 pizzas for a party after a successful season. Estimate the cost of each pizza using numbers with one nonzero digit. Find the exact cost per pizza.

24. A mathematics professor promises to give 2 extra points for each time a student works a set of self-study exercises. If one student works 17 sets of self-study exercises, how many points should be given?

25. Explain the difference between the commutative property of addition and the associative property of addition. Give a numerical illustration to support your explanation.

26. Find the product: 4.08×0.05.

27. Find the product: $42.73 \times 1,000$.

28. Divide: $25\overline{)27.75}$.

29. Round answer to the nearest tenth: $7.2\overline{)83.41}$.

30. Divide: $52.38 \div 10,000$.

31. Find the average of these test scores: 82, 95, 76, 84, 72, and 91. Round to the nearest whole number.

32. Estimate the sum by rounding to the nearest whole number: $3.85 + 7.46$.

33. Estimate the difference by rounding each number to the nearest tenth: $0.87 - 0.328$.

34. Estimate the product by rounding each decimal to a number with one nonzero digit: 42.38×27.9.

35. Estimate the quotient by rounding the divisor and dividend to numbers with one nonzero digit: $37.2\overline{)2987.5}$

36. The blueprint specification for a machined part calls for its thickness to be 1.485 in. with a tolerance of ± 0.010 in. What are the limit dimensions of the part?

37. Heating oil costs $1.75 per gallon. What is the cost of 10,000 gal?

38. A construction job requires 16 pieces of reinforced steel, each 7.96 ft in length. What length of steel is needed?

2

Integers

GOOD DECISIONS THROUGH TEAMWORK

Each member of your five-member team will be vice president of operations for each office of your international company. Your business requires a weekly teleconference, and you each schedule working hours between 8:00 A.M. and 5:00 P.M., when possible. The company's home office is in Memphis, Tennessee, with offices in Toronto, Canada; Frankfurt, Germany; Tokyo, Japan; and San Juan, Puerto Rico.

You are all attending the annual meeting in Memphis, so this is a good time to schedule your weekly teleconference so that it fits within everyone's normal working hours. If this cannot be done, select a permanent weekly time or a rotation schedule that is convenient for most of you. To make these arrangements, assign the number 0 to the Memphis office and an appropriate integer to the remaining offices based on the difference in time between each location and Memphis. Time zone information can be found on the Internet or on a world globe, or in standard reference materials such as atlases or encyclopedias. Once you have assigned integers to each location, work together to establish a meeting time.

Prepare a report in which you describe your telecommunications plans, list some of the obstacles that needed to be resolved, and summarize what your team learned from this activity.

2-1 Natural numbers, whole numbers, and integers

1. Relate integers to natural numbers and whole numbers.
2. Compare integers.
3. Find the absolute value of integers.
4. Find the opposite of integers.
5. Locate points on a rectangular coordinate system.

2-2 Adding integers

1. Add integers with like signs.
2. Add integers with unlike signs.

2-3 Subtracting integers

1. Subtract integers.
2. Subtract with zero or opposites.
3. Combine addition and subtraction.

2-4 Multiplying integers

1. Multiply integers.
2. Multiply several integers.

3 Multiply with zero.
4 Evaluate powers of integers.

2–5 Dividing integers

1 Divide integers.
2 Divide with zero.

2–6 Order of operations

1 Use the order of operations for integers.
2 Use a calculator to evaluate operations with integers.

Historically, as the need for negative numbers became apparent, the number system was expanded to fill this need. Consider the following. When a temperature is colder than zero degrees, how can we express this temperature numerically? When the selling price of an item is less than the cost of an item, how can we express the profit made on the sale numerically? When a withdrawal on a bank account exceeds the amount of money in the account, how can we express the account balance numerically? These situations demonstrate a need to express values that are less than zero. The first type of number we explore that can have a value less than zero is the *integer.*

<table>
<tr><td>**2–1**</td><td>## NATURAL NUMBERS, WHOLE NUMBERS, AND INTEGERS</td></tr>
</table>

Learning Outcomes

1 Relate integers to natural numbers and whole numbers.
2 Compare integers.
3 Find the absolute value of integers.
4 Find the opposite of integers.
5 Locate points on a rectangular coordinate system.

1 Relate Integers to Natural Numbers and Whole Numbers.

The first set of numbers needed was the set of counting numbers, or *natural numbers.* Natural numbers begin with the numbers 1, 2, 3, 4, 5, 6, 7, and continue indefinitely. As the place-value system evolved, the number zero, 0, was added. The natural numbers and zero form a set of numbers called *whole numbers.* Figure 2–1 shows the whole numbers on a number line; note that zero is to the left of the number 1.

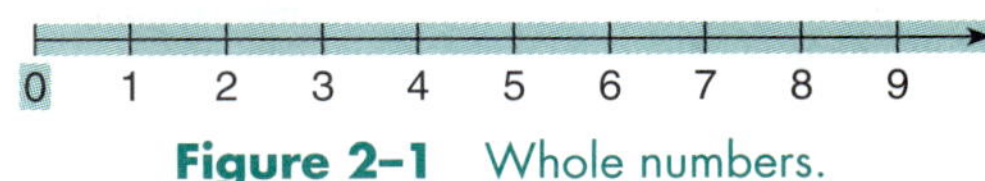

Figure 2–1   Whole numbers.

Many physical phenomena have values that are less than zero. For instance, when the temperature is 0°, that does not mean that the temperature cannot become any colder than zero. Another type of number is needed to express a temperature colder than zero degrees. When the opposite of each natural number is included, the set of whole numbers can be expanded to form the set of *integers.* Figure 2–2 shows how the set of whole numbers is extended to include all integers. A number line for integers continues indefinitely in both directions. Numbers get smaller as we proceed to the left and larger as we proceed to the right.

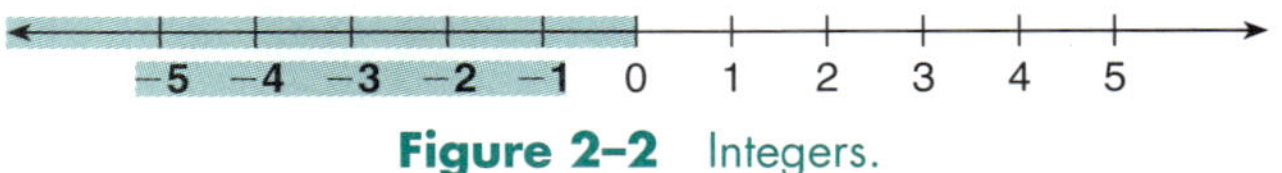

Figure 2–2   Integers.

The *opposite* of a number is the same number of units from zero, but in the opposite direction. Figure 2–3 shows a number line with whole numbers and their opposites.

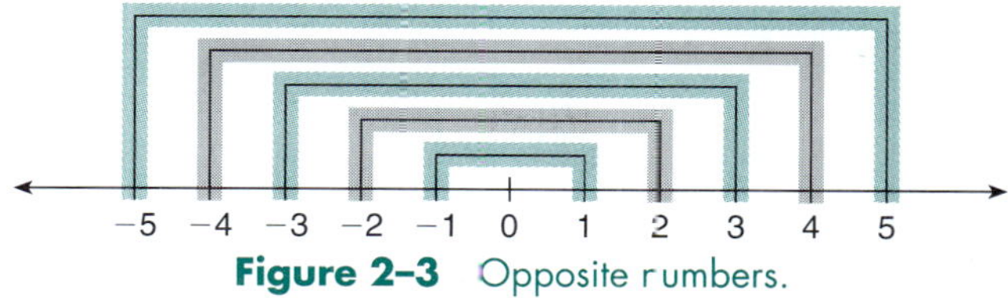

Figure 2–3 Opposite numbers.

Tip!	*Are Positive Signs Necessary?*

As you can see from the number line, it is not necessary to include the positive sign for positive values. You may sometimes want to include the sign to draw attention to it or to emphasize its direction. Negative signs, however, must always be included.

To see the relationship among the three sets of numbers—natural numbers, whole numbers, and integers—examine Fig. 2–4.

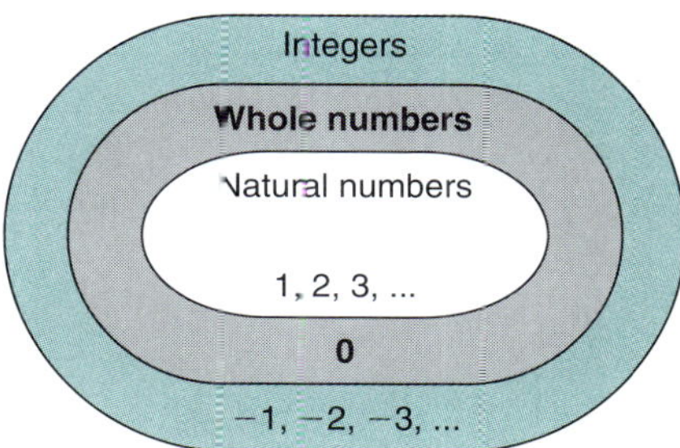

Figure 2–4 Relating natural numbers, whole numbers, and integers

EXAMPLE Answer the following questions.

(a) How is the set of natural numbers different from the set of whole numbers?
(b) How is the set of whole numbers different from the set of integers?
(c) Can a number be both an integer and a whole number? If so, give an illustration.
(d) Is any integer also a whole number?

(a) Natural numbers and whole numbers have all the same numbers except for the number zero. Natural numbers do not include zero; whole numbers do include zero.
(b) All whole numbers are included in the set of integers. In addition, integers include the opposite of every natural number.
(c) Yes, every whole number is also an integer. The number 3 is both a whole number and an integer.
(d) An integer is sometimes a whole number. The opposites of natural numbers (the negative integers) are integers but are not whole numbers. The integer -3 is not a whole number.

2 Compare Integers.

If a number is positioned to the left of another number on a number line, it represents a smaller value. Larger values are positioned to the right of a number on the number line. We show how two numbers compare in size by using two symbols: $<$ and $>$. The symbol $<$ is used when the first number is smaller than or "is less than" the second number. For example, "5 is less than 7" is written $5 < 7$. When the first number is larger than or "is greater than" the second number, the symbol $>$ is used. For example, "7 is greater than 5" is written $7 > 5$.

■ **Learning Strategy** ***Develop Strategies for Remembering Specific Facts.***

Memory Tips for Distinguishing Between $<$ and $>$:

Look at the number line in Fig. 2–5. You will see that numbers get smaller as we move to the left.

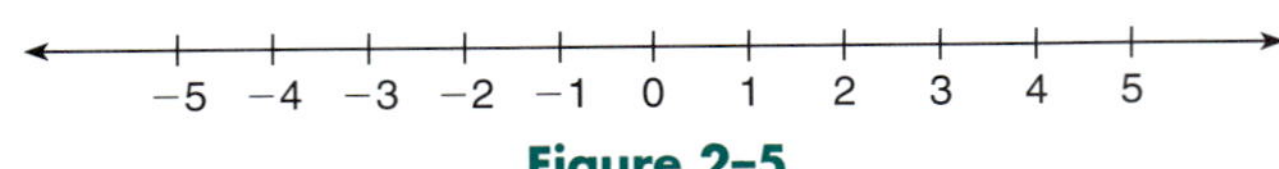

Figure 2–5

The arrow on the left end of the number line points in the same direction as the "less than" symbol. In other words, the arrow points toward the smaller number.

$$-4 < -1 \qquad -3 < 0 \qquad 0 < 4$$

On the other hand, numbers get larger as we move to the right. The arrow on the right end of the number line points in the same direction as the "greater than" symbol. Again, the arrow points toward the smaller number.

$$-1 > -4 \qquad 0 > -3 \qquad 4 > 0$$

EXAMPLE Use the symbols $<$ and $>$ to indicate whether the first number is less than or greater than the second number. (a) 7 ___ 9 (b) 0 ___ −1 (c) −4 ___ −2 (d) −2 ___ 0

(a) $7 < 9$ — 7 is smaller—to the left of 9 on the number line.
(b) $0 > -1$ — 0 is larger—to the right of −1 on the number line.
(c) $-4 < -2$ — −4 is smaller—to the left of −2 on the number line.
(d) $-2 < 0$ — −2 is smaller—to the left of 0 on the number line.

3 Find the Absolute Value of Integers.

The distance between each pair of consecutive integers is the same for all integers located on the number line. This concept of distance between numbers on the number line is related to the concept of *absolute value*. The absolute value of a number is often described as the number of units of distance the number is from zero. The symbol for absolute value is $| \ |$; $|3|$ is read "the absolute value of 3."

Distance is a physical property, so it cannot have a negative value. Thus, the absolute value of a number is always a nonnegative value. If a number is a positive number, then the absolute value of the number is the number itself. In symbols, if $a > 0$, then $|a| = a$. The absolute value of a negative number is the opposite of that number. In symbols, if $a < 0$, then $|a| = -a$.

The absolute value of zero is zero. In symbols, $|0| = 0$. Study the following examples of absolute values.

$$|-5| = 5 \qquad |27| = 27 \qquad |0| = 0$$

Tip! ***What Does Nonnegative Mean?***

Are positive numbers also nonnegative? Yes, every positive number is nonnegative.
Are there any nonnegative numbers that are not positive?
Yes, zero is nonnegative *and* nonpositive. Therefore, when we say nonnegative, we mean positive or zero.

Positive numbers are represented symbolically as $a > 0$, where a is any number greater than zero.

Negative numbers are represented by $a < 0$, which states that a is any number less than zero.

Nonnegative numbers (positives and zero) can be written symbolically as $a \geq 0$. The symbol $\geq$ means "is greater than or equal to."

Practice reading in words the symbolic statements that define the absolute value of a number.

■ **DEFINITION: Absolute Value**

$$|a| = a, \text{ for } a > 0$$

The absolute value of a positive number is equal to itself.

$$|a| = -a, \text{ for } a < 0$$

The absolute value of a negative number is equal to its opposite.

$$|0| = 0$$

The absolute value of zero is zero.

EXAMPLE   Give the absolute value of the following quantities: (a) $|9|$ (b) $|-4|$ (c) $|0|$

(a) $|9| = 9$ 9 is positive. Its absolute value is the number itself.
(b) $|-4| = 4$ −4 is negative. Its absolute value is the opposite of −4.
(c) $|0| = 0$ 0 is unsigned. Its absolute value is still zero.

Tip!	*Two Components of a Signed Number.*

Integers and other signed numbers tell us two things. For the integer -8, the negative sign tells us the number is to the left of zero and is referred to as the *directional sign* of the number. The 8 tells us how many units the number is from zero; it is the *absolute value* of the number.

4   Find the Opposite of Integers.

Every number except zero has an opposite. The opposite of a number is also called the *additive inverse* of the number. A number and its additive inverse (opposite) have the same absolute value but different directional signs.

EXAMPLE   Find the opposite of each number and show your answer on the number line.
(a) -8 (b) 6   (c) 0

(a) The opposite of -8 is 8, which is illustrated in Fig. 2–6.

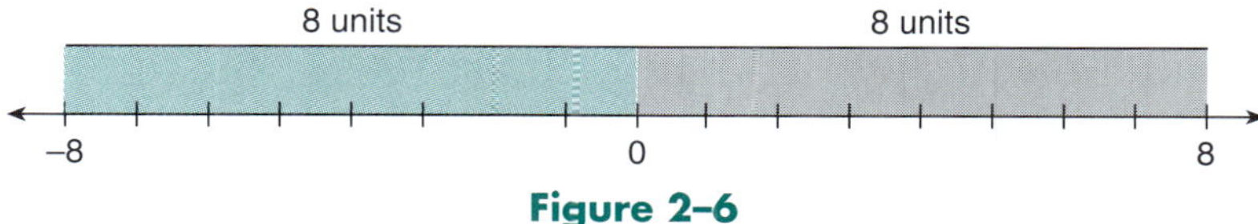

Figure 2–6

(b) The opposite of 6 is -6, which is illustrated in Fig. 2–7.

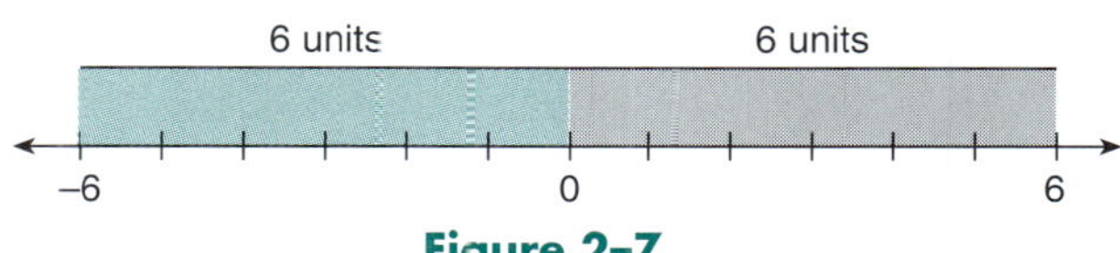

Figure 2–7

(c) Zero has no opposite, which is illustrated in Fig. 2–8.

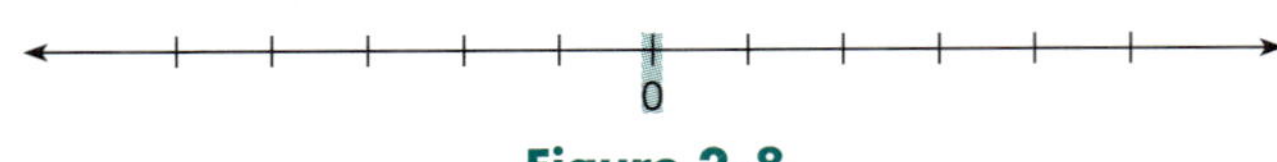

Figure 2–8

5 Locate Points on a Rectangular Coordinate System.

The number line shows a value pictorially. This kind of visual representation is called a *one-dimensional graph.* The one dimension represents the distance and direction that a value is from zero. The *rectangular coordinate system* gives a graphical representation of two-dimensional values. In the rectangular coordinate system, two number lines are positioned to form a right angle or square corner (see Fig. 2–9). The number line that runs from left to right is the *horizontal axis* and is represented by the letter x. The number line that runs from top to bottom is the *vertical axis* and is represented by the letter y.

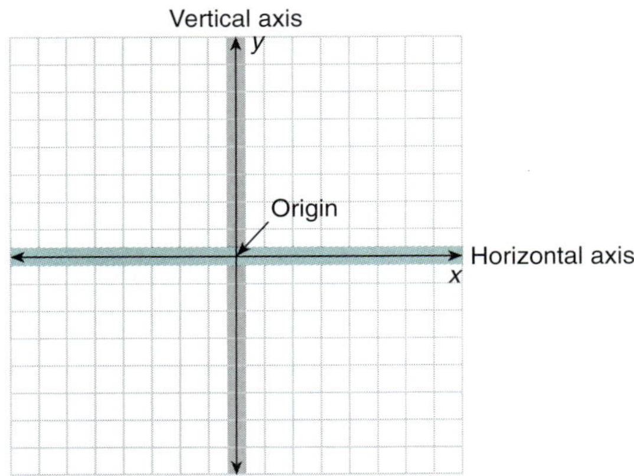

Figure 2–9

Zero on both number lines is located at the point where the two number lines cross. This point is called the *origin.* Other points are located by horizontal and vertical movements from the origin. Look at the points in the next example.

EXAMPLE Describe the location of the points in Fig. 2–10 by giving the amount of horizontal and vertical movement from the origin.

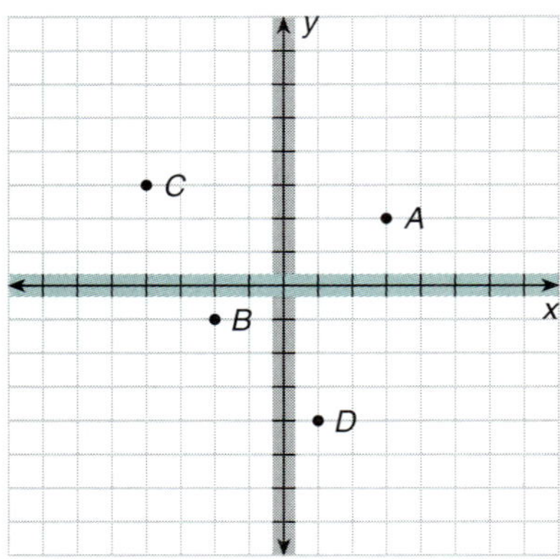

Figure 2–10

Point A: horizontal movement, $+3$; vertical movement, $+2$

Point B: horizontal movement, -2; vertical movement, -1

Point *C*: horizontal movement, -4; vertical movement, $+3$

Point *D*: horizontal movement, $+1$; vertical movement, -4

As is common in mathematics, the location of points on a rectangular coordinate system can be written in an abbreviated, symbolic form. Point notation uses two signed numbers to show the horizontal and vertical movement from the origin to a given point. The value that represents the horizontal movement is the *x-coordinate*, and the value that represents the vertical movement is the *y-coordinate*. These signed numbers are separated with a comma and enclosed in parentheses.

Tip! | ***Point Notation.***

Symbolically, we represent the location of a point as

(horizontal movement, vertical movement)   or (x, y)

where x is the horizontal movement from the origin, or the *x*-coordinate, and y is the vertical movement from the *x*-axis, or the *y*-coordinate.

EXAMPLE   Write the points in the preceding example using point notation.

Point A = (horizontal movement of $+3$, vertical movement of $+2$) or $(+3, +2)$

Point B = (horizontal movement of -2, vertical movement of -1) or $(-2, -1)$

Point C = (horizontal movement of -4, vertical movement of $+3$) or $(-4, +3)$

Point D = (horizontal movement of $+1$, vertical movement of -4) or $(+1, -4)$

To *plot* a point means to show its location on the rectangular coordinate system.

To plot a point on the rectangular coordinate system:

1. Start at the origin.
2. Count to the left or right the number of units of the first signed number.
3. From the ending point from Step 2, count up or down the number of units of the second signed number.

Tip! | ***Which Way Do I Go First?***

Proper use of point notation depends on knowing which direction applies to which number. Horizontal comes first, but think of other words that imply "horizontal movement" right/left, forward/back; for vertical movement, think of up/down.
   A memory strategy might be to think of the alphabetical order of some key words.

Alphabetically, *horizontal* comes before *vertical*. *Right* comes before *up*.

EXAMPLE   Plot these points: Point $A = (3, 1)$,   point $B = (-2, 5)$,   point $C = (-3, -2)$, point $D = (1, -3)$.

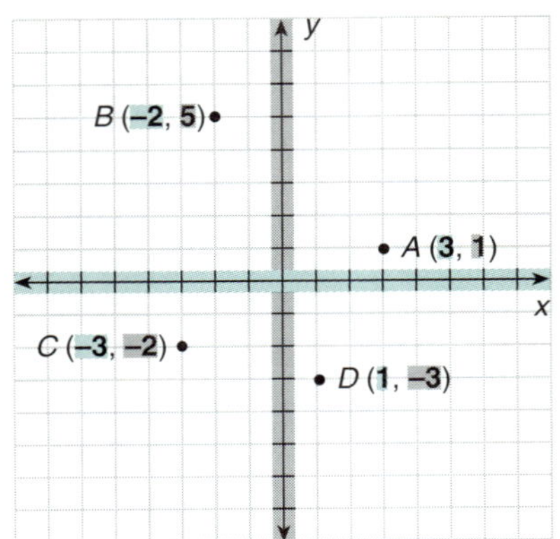

Figure 2–11

The solution is shown in Fig. 2–11.

A common mistake when plotting points is to move in the wrong direction. You can minimize mistakes by developing your spatial sense. To do this, consider the signs of the coordinates of points being plotted. Then, visualize in which part of the graph the point will fall.

Look again at the points $A = (3, 1)$; $B = (-2, 5)$; $C = (-3, -2)$; and $D = (1, -3)$.

Point A (3, 1) moves *right* and *up,* as shown in Fig. 2–12.

Point B (−2, 5) moves *left* and *up,* as shown in Fig. 2–13.

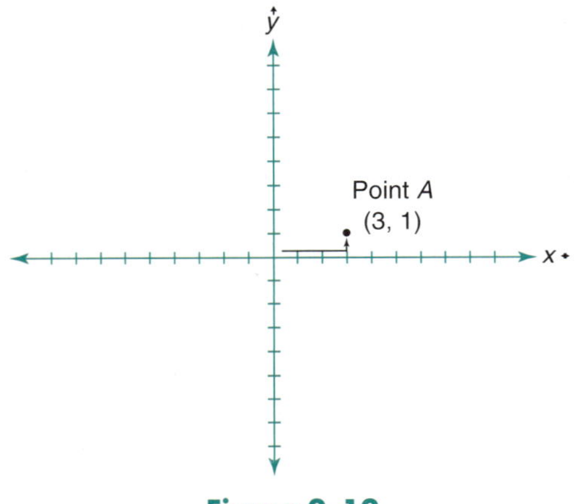

Figure 2–12

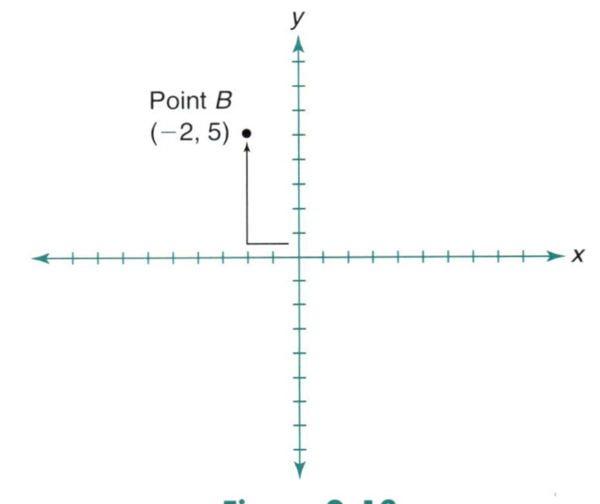

Figure 2–13

Point C $(-3, -2)$ moves *left* and *down,* as shown in Fig. 2–14.

Point D (1, −3) moves *right* and *down,* as shown in Fig. 2–15.

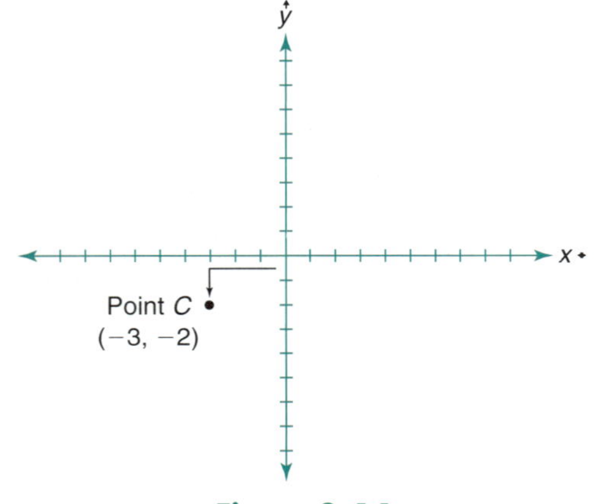

Figure 2–14

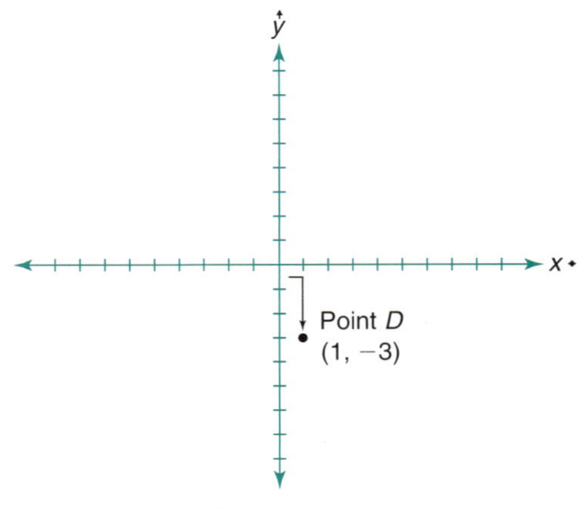

Figure 2–15

Suppose a point shows movement in only one direction. This point is written in point notation by using the number zero. Points on the horizontal number line have no vertical movement. Points on the vertical number line have no horizontal movement. Examine the points $(3, 0)$, $(-6, 0)$, $(0, 7)$, and $(0, -5)$ in Fig. 2–16.

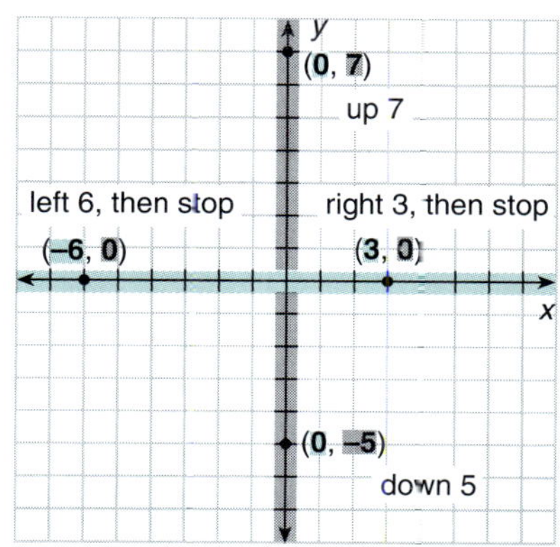

Figure 2–16

EXAMPLE   Plot the following points: $A = (2, 0)$, $B = (0, 2)$, $C = (-2, 0)$, $D = (0, -2)$.

Point $A = (2, 0)$ right 2, then stop
Point $B = (0, 2)$ up 2
Point $C = (-2, 0)$ left 2, then stop
Point $D = (0, -2)$ down 2

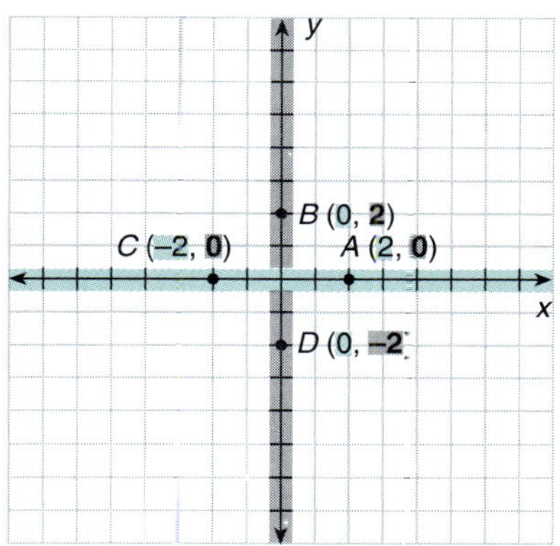

Figure 2–17

The solution is shown in Fig. 2–17.

SELF-STUDY EXERCISES 2–1

1   Exercises 1–8 refer to the number line in Fig. 2–18.

1. On which side of zero are the positive numbers?

2. On which side of zero are the negative numbers?

3. How long does the number line extend in each direction?

4. How many points exist on the number line for each distinct number?

5. What sign ($+$ or $-$) is assigned to zero?

6. Arrange these integers in order as they would appear on the number line: $-2, 3, 0, -1, -3, 2, 1$.

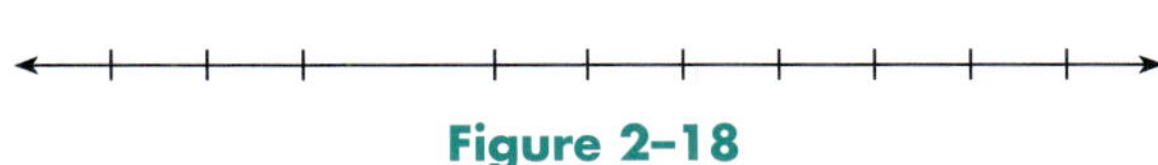

Figure 2–18

7. In which direction on the number line do the numbers increase in value?

9. Draw a thermometer and show 30° above zero, 20° above zero, 10° above zero, zero degrees, and 10° below zero.

8. In which direction on the number line do the numbers decrease in value?

10. There are 25 integer World Time Zones, which are assigned numbers from -12 through 0 to $+12$. Each time zone is measured relative to Greenwich, England (Greenwich Mean Time—GMT), which is designated as zero. New York City is -5:00 relative to GMT and Moscow, Russia is $+3$:00 GMT. Use a number line to show the times in New York City, Greenwich, and Moscow.

2 Use the "greater than" or "less than" symbol in the number pairs to make a true statement.

11. 5 __ 8

12. 0 __ -2

13. -4 __ 2

14. -2 __ 0

15. -5 __ -9

16. -5 __ 2

17. 0 __ -5

18. -3 __ 3

Rewrite the expressions using $<$ or $>$. When an expression contains one or more letters, the letters stand for numbers.

19. x is less than y

20. a is greater than b

21. $3 + 5$ is greater than 6

22. 9 is less than $18 - 6$

23. k is greater than t

24. r is less than s

3 Give the value.

25. $|23|$

26. $|0|$

27. $|-10|$

28. -17

29. $|-13|$

30. $|345|$

31. 67

32. $|-61|$

4 Illustrate the numbers and their opposites on a number line.

33. 7

34. -8

35. -4

36. 12

37. Which number has no opposite?

38. Describe the opposite of a positive integer. Describe the opposite of a negative integer.

5 Locate the points by giving the amount of horizontal and vertical movement from the origin in Fig. 2–19.

39. R

40. S

41. T

42. U

43. V

44. W

45. X

46. Y

47. Write the coordinates for the points on the graph in Fig. 2–20.

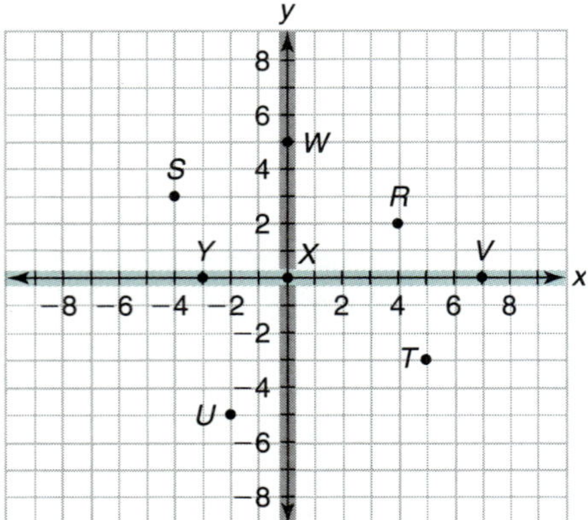

Figure 2–19

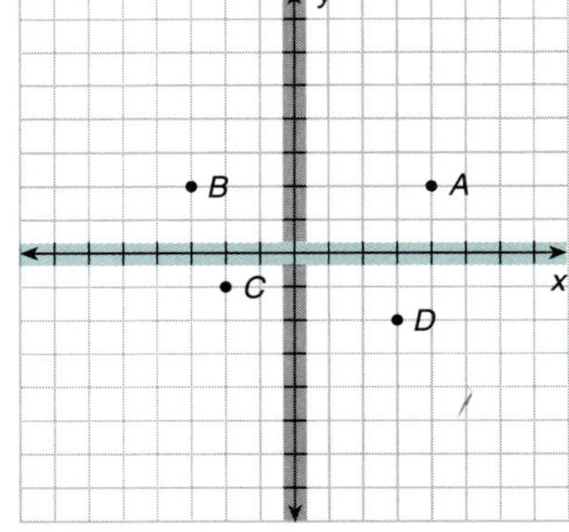

Figure 2–20

A = __ B = __
C = __ D = __

Draw a rectangular coordinate system and locate the points given in Exercises 48–53.

48. $A = (-7, 4)$

49. $B = (0, -10)$

50. $C = (-2, 0)$

51. $D = (-7, -7)$

52. $E = (3, 0)$

53. $F = (4, -3)$

54. What common property describes the coordinates of the points on the *x*-axis?

55. What common property describes the coordinates of the points on the *y*-axis?

56. Describe the signs of the coordinates of points that lie in the upper-left quarter of the graph.

57. Describe the signs of the coordinates of points that lie in the lower-right quarter of the graph.

2–2 ADDING INTEGERS

Learning Outcomes

1 Add integers with like signs.

2 Add integers with unlike signs.

Throughout history, numbers have been used for applications other than just counting and recording information. For example, if the low temperature for the day is 3 degrees below zero and it is expected to rise 24 degrees during the day, what will be the high temperature for the day?

1 Add Integers with Like Signs.

Since our earliest experiences with addition we have added positive numbers. In the addition problem $3 + 2 = 5$, all three numbers are positive. Numbers are understood to be positive when no sign is given. To emphasize that the numbers are positive, however, we can write the problem as $+3 + +2 = +5$. We can also write the problem using parentheses to separate the two plus signs, $3 + (+2) = +5$. The plus symbol serves two purposes in mathematical expressions. It is used as the directional sign for a positive number, and it is used to show addition. In the expression $+3 + (+2) = +5$, the plus sign in front of the parentheses shows the operation of addition. The plus sign inside the parentheses identifies the number as positive. Adding integers will build on our knowledge of addition of positive numbers. We will use the number line to illustrate addition of integers.

To add 3 and 2, we begin at zero and move three units to the right (positive direction) on the number line (see Fig. 2–21). This move takes us to $+3$. From the $+3$ position, we move two units to the right. This second move takes us to the $+5$ position. In other words, $3 + 2 = 5$.

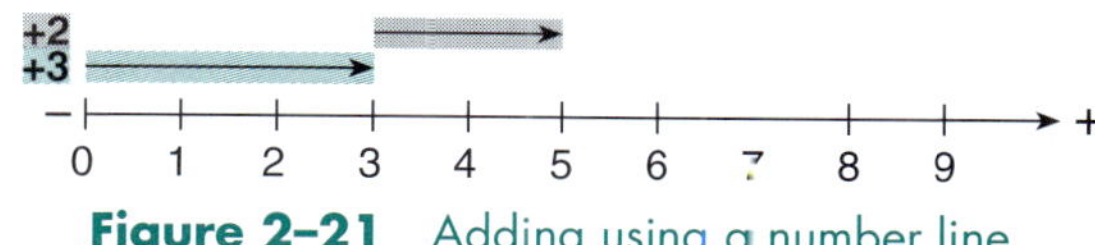

Figure 2–21 Adding using a number line.

To add two negative integers, say, $(-3) + (-2)$, we begin at zero and move three units to the left (negative direction) on the number line (see Fig. 2–22). This takes us to -3 on the number line. Then from -3 we move two units to the left. This second move takes us to -5. Thus, $-3 + (-2) = -5$.

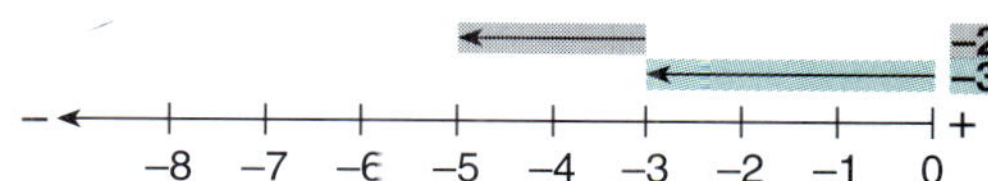

Figure 2–22 Add two negative numbers using a number line.

When adding the two positive numbers or the two negative numbers, notice that the sum could have been obtained using the following rule.

To add numbers with like signs:

1. Add the absolute values of the numbers.
2. Give the sum the common or like sign.

Add $3 + 6 + 8$.

$3 + 6 + 8$ Add absolute values. Keep common positive sign.

$9 + 8 = \mathbf{17}$

Before we look at additional examples, let's make some observations. First, addition is a *binary operation*. This means that only two numbers are used in the operation. If more than two numbers are involved, two are added and then the next number is added to the sum of the first two. Thus, rules developed for addition apply to two numbers at a time.

EXAMPLE   Add $-3 + (-4) + (-1)$.

$-3 + (-4) + (-1) =$ Add first two numbers.

$-7 + (-1) = \mathbf{-8}$ Add sum to remaining number.

■ **Learning Strategy**   *Using Precise Versus Casual Terminology.*

When adding numbers with *like* signs, we casually say "add and keep the *like* sign" when we actually mean "add *absolute values* and keep the *like* sign." Is it wrong to use casual language in mathematics? That's debatable. The important thing is to establish some ground rules with your classmates and instructor. You may decide casual language is acceptable in class discussions but precise language is preferred on assignments or tests, or your instructor may decide that you should use precise language at all times.

---

2   Add Integers with Unlike Signs.

Now let's examine the addition of numbers with unlike signs, which can also be illustrated on the number line. When adding numbers with unlike signs, we move in the direction indicated by the sign of each number, positive to the right and negative to the left. To add $-4 + 3$, we first move four units to the left of zero (see Fig. 2–23). This takes us to -4. Then from the -4 position, we move three units to the right. This takes us to -1. Thus, $-4 + 3 = -1$. This operation can also be expressed as a rule using absolute values.

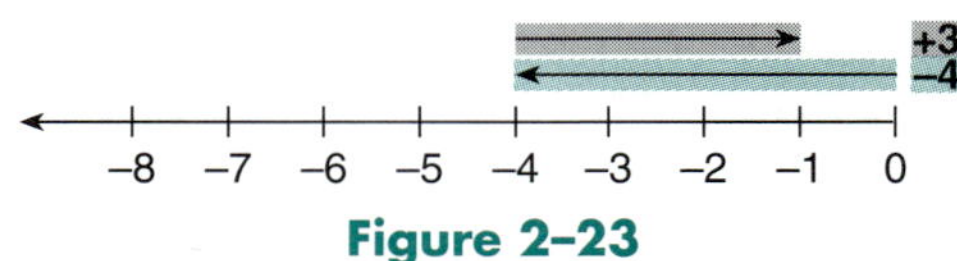

Figure 2–23

To add signed numbers with unlike signs:

1. *Subtract* the smaller absolute value from the larger absolute value.
2. Give the sum the sign of the number with the larger absolute value.

EXAMPLE   Add $7 + (-12)$.

$7 + (-12) =$ Signs are unlike.

$7 + (-12) = \mathbf{-5}$ Subtract absolute values: $12 - 7 = 5$. Keep the sign of the -12 (negative), the larger absolute value.

<table>
<tr><td>Tip!</td><td>To Add, You . . . Subtract?</td></tr>
</table>

When we add integers, phrases we have used in arithmetic, such as "find the sum" or "add," can be confusing. Yes, in the operation of addition, sometimes we add absolute values and sometimes we subtract absolute values. We often use the word "combine" to imply the addition of numbers with either like or unlike signs.

EXAMPLE   Add $-5 + 7$.

$-5 + 7 =$ Signs are unlike.

$-5 + 7 = 2$ Subtract absolute values: $7 - 5 = 2$. Keep the sign of the 7 (positive), the larger absolute value.

■ **Learning Strategy**   *Rules and Procedures Are Shortcuts.*

Rules and procedures help us perform operations efficiently, but they don't take the place of understanding the concepts. Let's look again at the problems $7 + (-12)$ and $-5 + 7$, in Figs. 2–24 and 2–25.
  Since we move back more spaces than we move forward, we end on the negative side of zero.

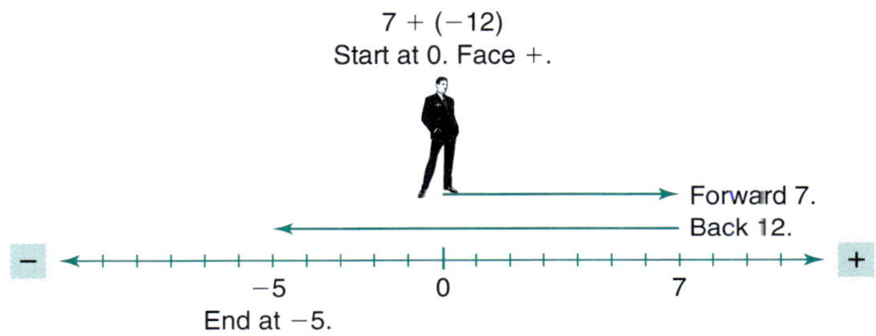

Figure 2–24

Since we move forward more spaces than we move back, we end on the positive side of zero.

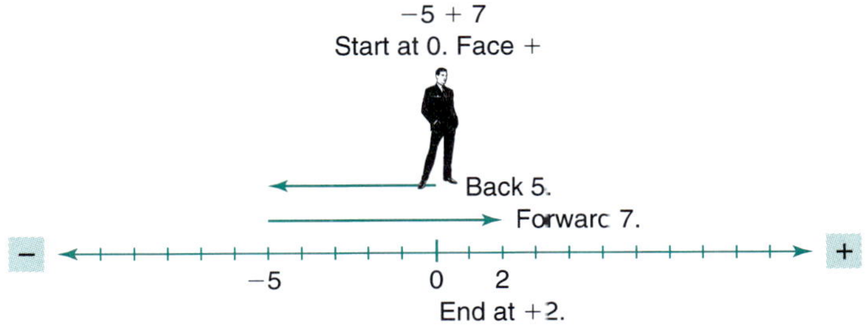

Figure 2–25

- The rule for subtracting absolute values when the numbers have *unlike* signs takes into account that the movements are in opposite directions.
- The rule for giving the sum the sign of the number with the larger absolute value takes into account that the number that creates the largest movement determines whether the sum will be positive or negative.

When an addition involves several positive numbers and several negative numbers, it is convenient to group all the positive numbers and all the negative numbers and to add each group separately. We then add the sums of each group. We can do this because of the associative property of addition. The grouping of addends in an addition does not matter. In the next example we combine the commutative and associative properties of addition to add several signed numbers.

EXAMPLE   Add $8 + (-9) + 13 + (-15)$.

$$8 + (-9) + 13 + (-15) =$$

Signs are unlike. Group numbers with like signs.

$$8 + 13 + [-9 + (-15)] =$$

The brackets, [ ], show the grouping and separate the operational plus sign from the directional minus sign of -9. Add groups of numbers with like signs.

$$21 + (-24) = -3$$

Add resulting sums using the rule for adding numbers with unlike signs.

The properties of addition that involve the number zero apply to integers as well as to whole numbers and decimals. When zero is added to a number, the number is not changed. We say that zero is the *additive identity*. The sum of a number and its opposite is zero. We call the opposite of a number the *additive inverse* of that number. We can also write these definitions using symbols.

■ **DEFINITION: Additive Identity**   Zero is the *additive identity* because $a + 0 = a$ for all values of a.

■ **DEFINITION: Additive Inverse (Opposite)**   The opposite of a number is the *additive inverse* of the number because $a + (-a) = 0$ for all values of a.

EXAMPLE   Add: (a) $-2 + 0$ (b) $5 + (-5)$ (c) $0 + (-3) + 4$
(d) $5 + (-2) + (-5)$

(a) $-2 + 0 = -2$ Additive identity.
(b) $5 + (-5) = 0$ Additive inverse.
(c) $0 + (-3) + 4 =$ Additive identity.
    $-3 + 4 = 1$ Add numbers with unlike signs.
(d) $5 + (-2) + (-5) =$ Commutative and associative properties.
    $5 + (-5) + (-2) =$ Additive inverse.
    $0 + (-2) = -2$ Additive identity.

SELF-STUDY EXERCISES 2–2

1   Add.

1. $7 + 10 + 12$
2. $-5 + (-8) + (-7)$
3. $12 + 87$
4. $-21 + (-38)$
5. $-32 + (-16)$
6. $(-58) + (-103)$

2   Add.

7. $-4 + (-2)$
8. $-3 + 7$
9. $4 + (-6)$
10. $-4 + 6$
11. $-18 + 8$
12. $32 + (-72)$
13. $21 + (-14)$
14. $17 + (-4) + 3 + (-1)$
15. $-3 + 2 + (-7)$
16. $47 + (-82) + 2$
17. $14 + (-6) + 1$
18. $-7 + (-3) + (-1)$
19. $4 + 2 + (-3) + 10$
20. $2 + (-1) + 8$
21. $-2 + 1 + (-8) + 12$
22. $15 + (-15)$
23. $-7 + 7$
24. $92 + (-92)$
25. $(-396) + (396)$
26. $503 + (-503)$
27. $0 + 9$
28. $-3 + 0$
29. $0 + (-7)$
30. $18 + 0$
31. $-8 + 0$
32. $0 + (-28)$

33. Write a sum that shows the additive inverse.

34. What is the additive identity?

35. Explain why the sum of two positive numbers is positive and the sum of two negative numbers is negative.

36. A stock priced at \$42 has the following changes in one week: $-3, +8, -6, -7, +2$. What is the value of the stock at the end of the week?

37. You open a bank account by depositing \$242. You then write checks for \$21, \$32, and \$123. What is your new account balance?

38. Paris, France, is assigned Greenwich Mean Time (GMT) of $+1:00$, and Jerusalem, Israel, is 1 time zone east (1 hour) of Paris. If east is represented by adding a positive number, what is the GMT for Jerusalem, Israel?

39. Memphis, Tennessee, is −6:00 GMT, and Portland, Oregon, is 2 time zones west of Memphis. Movement westward is represented by negative numbers. Find the GMT for Portland.

40. The temperature at the South Pole is recorded as −38°F at midnight. The temperature rises 15° by noon. Determine the temperature reading at noon.

2–3   SUBTRACTING INTEGERS

Learning Outcomes

1 Subtract integers.
2 Subtract with zero or opposites.
3 Combine addition and subtraction.

Before we discuss the subtraction of integers, let's examine two expressions involving the subtraction of integers, $6 - 9$ and $5 - (-3)$, and interpret the meaning of the notation. The first expression, $6 - 9$, means positive 9 is subtracted from positive 6. The sign between the 6 and the 9 means subtraction and is read as "6 subtract 9" or "6 minus 9," just as we did with whole numbers. The minus sign is the operational sign telling us to subtract positive 9 from positive 6. The expression can also be written as $6 - (+9)$.

The second expression, $5 - (-3)$, means that negative 3 is subtracted from positive 5. The first minus sign between 5 and 3 is the operational sign telling us to subtract. The second minus sign is the directional sign to show negative 3.

1   Subtract Integers.

Now let's focus on the subtraction of whole numbers. To show subtraction on the number line, we face the positive end of the number line and step forward or backward as determined by the sign of the first number. Then we turn and face the negative end of the number line and step forward or backward as determined by the sign of the second number (subtrahend), as shown in Figs. 2–26 and 2–27.

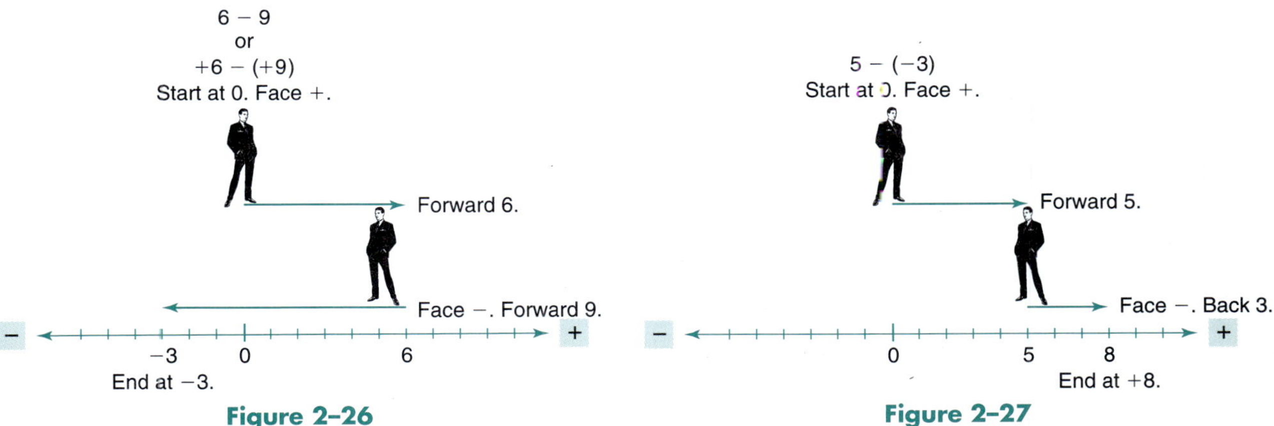

Figure 2–26   **Figure 2–27**

Now, let's examine the relationship between subtraction and addition so that we can subtract integers efficiently.

Suppose we say that subtracting a number is the *same as* adding the opposite of the number. How do we interpret this on the number line?

$$6 - 9$$

$$+6 - (+9) \qquad \text{Subtract } +9.$$

$$+6 + (-9) \qquad \text{Add } -9.$$

When we add on the number line, we always face the positive end of the number line, as shown in Figs. 2–28 and 2–29.

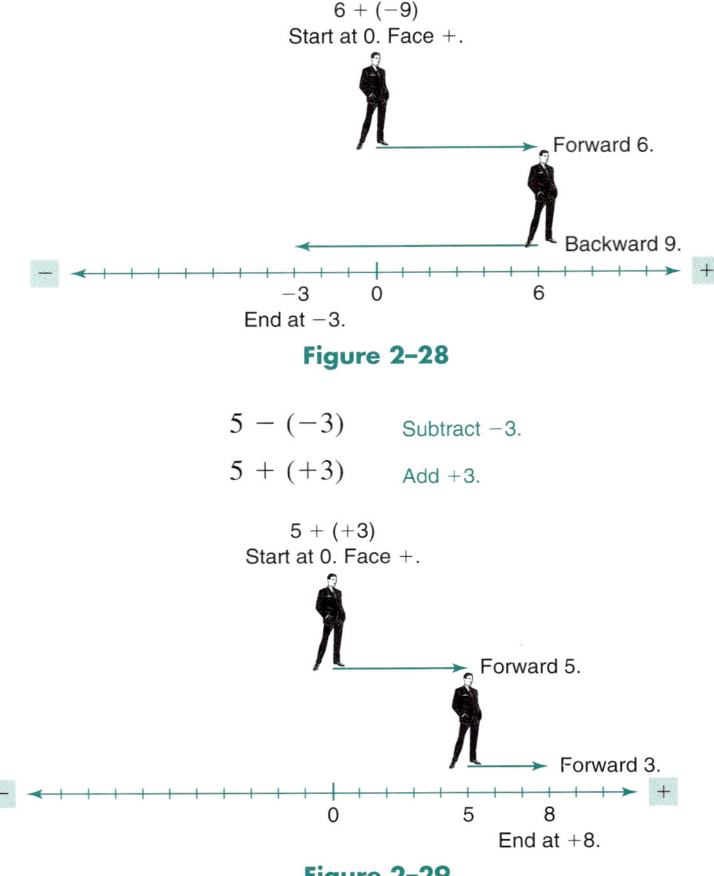

Figure 2–28

$$5 - (-3) \qquad \text{Subtract } -3.$$

$$5 + (+3) \qquad \text{Add } +3.$$

Figure 2–29

Figures 2–26 and 2–28 show how addition and subtraction are related. Similarly, Figures 2–27 and 2–29 relate addition and subtraction. Look at this relationship in words.

To subtract signed numbers:

1. Add the opposite of the subtrahend (second number) to the minuend (first number).
2. Apply the appropriate rule for adding signed numbers with like or unlike signs.

EXAMPLE Subtract: (a) $2 - 6$ (b) $-9 - 5$ (c) $12 - (-4)$ (d) $-7 - (-9)$

(a) $\qquad 2 - 6 =$ Subtract positive 6 from positive 2.

$\quad 2 - (+6) \ =$ Write sign of subtrahend.

$\quad 2 + (-6) \ =$ Add the opposite of 6, which is −6, to 2.

$\quad 2 + (-6) \ = \ \mathbf{-4}$ Apply the rule for adding numbers with unlike signs.

(b) $\qquad -9 - 5 =$ Subtract positive 5 from negative 9.

$\quad -9 - (+5)$ Write sign of subtrahend.

$\quad -9 + (-5) \ =$ Add the opposite of 5, which is −5, to −9.

$\quad -9 + (-5) \ = \ \mathbf{-14}$ Apply the rule for adding numbers with like signs.

$$
\begin{aligned}
\text{(c)} \quad 12 - (-4) &= \\
12 + (+4) &= \\
12 + 4 &= \mathbf{16}
\end{aligned}
$$

Subtract negative 4 from positive 12.
Add the opposite of -4, which is $+4$, to 12.
Apply the rule for adding numbers with like signs.

$$
\begin{aligned}
\text{(d)} \quad -7 - (-9) &= \\
-7 + (+9) &= \\
-7 + 9 &= 2
\end{aligned}
$$

Subtract negative 9 from negative 7.
Add the opposite of -9, which is $+9$, to -7.
Apply the rule for adding numbers with unlike signs.

Tip! | ***Writing Subtractions as Equivalent Additions.***

When writing a subtraction as an equivalent addition, you make *two* changes.

1. Change the operation from subtraction to addition.
2. Change the subtrahend to its opposite.

$$
\begin{array}{cccc}
2 - (+6) & -9 - (+5) & 12 - (-4) & -7 - (-9) \\
2 + (-6) & -9 + (-5) & 12 + (+4) & -7 + (+9) \\
\uparrow\ \uparrow & \uparrow\ \uparrow & \uparrow\ \uparrow & \uparrow\ \uparrow \\
\textbf{1. 2.} & \textbf{1. 2.} & \textbf{1. 2.} & \textbf{1. 2.}
\end{array}
$$

---

2 Subtract with Zero or Opposites.

Subtractions involving zero and opposites are still interpreted first as equivalent addition problems. Subtracting zero from a number is the same as adding zero: $5 - 0 = 5 + (0) = 5$. Subtracting a nonzero number from zero is the same as adding the opposite of the number to zero: $0 - 5 = 0 + (-5) = -5$. Subtracting an opposite from a number is the same as adding a number to itself: $5 - (-5) = 5 + (+5) = 10$.

EXAMPLE Evaluate: (a) $8 - 0$ (b) $0 - 15$ (c) $32 - (-32)$ (d) $-15 - 15$

$$
\begin{aligned}
\text{(a)} \quad 8 - 0 &= \\
8 + (0) &= \\
8 + (0) &= \mathbf{8}
\end{aligned}
$$

Subtract zero from positive 8.
Add zero to positive 8.
Zero added to any number is that number.

$$
\begin{aligned}
\text{(b)} \quad 0 - 15 &= \\
0 - (+15) &= \\
0 + (-15) &= \\
0 + (-15) &= \mathbf{-15}
\end{aligned}
$$

Subtract positive 15 from zero.
Write sign of subtrahend.
Add the opposite of 15, which is -15, to zero.
Any number added to zero is that number.

$$
\begin{aligned}
\text{(c)} \quad 32 - (-32) &= \\
32 + (+32) &= \\
32 + 32 &= \mathbf{64}
\end{aligned}
$$

Subtract negative 32 from positive 32.
Add the opposite of -32, which is $+32$, to 32.
Apply the rule for adding numbers with like signs.

$$
\begin{aligned}
\text{(d)} \quad -15 - (15) &= \\
-15 + (-15) &= \\
-15 + (-15) &= \mathbf{-30}
\end{aligned}
$$

Subtract positive 15 from negative 15.
Add the opposite of 15, which is -15, to -15.
Apply the rule for adding numbers with like signs.

---

3 Combine Addition and Subtraction.

When we express the sum and difference of integers with all the appropriate operational and directional signs, we have an expression with both an operational sign and a directional sign between every two numbers. We have four different possibilities when two signs are written together. The following Tip may reduce your confusion about the four possibilities.

Tip! | ***Omitting Signs.***

Writing mathematical expressions that include every operational and directional sign is cumbersome. In general practice, we omit as many signs as possible. When two signs are written between two numbers, we can write a simplified expression with only one sign.

Plus, Plus: $+3 + (+5)$ Add $+3$ and $+5$. $3 + 5$
Minus, Minus: $+3 - (-5)$ Change to addition.
$+3 + (+5)$ $3 + 5$
Plus, Minus: $+3 + (-5)$ Add $+3$ and -5. $3 - 5$
Minus, Plus: $+3 - (+5)$ Change to addition.
$+3 + (-5)$ $3 - 5$

To generalize, two like signs between integers, whether both plus or both minus, translate to adding a positive number. Use just one plus sign:

$$+ \; + \rightarrow + \qquad\qquad - \; - \rightarrow +$$

Two unlike signs between integers, either a plus/minus, or a minus/plus, translate to adding a negative number. Use just one minus sign:

$$+ \; - \rightarrow - \qquad\qquad - \; + \rightarrow -$$

The suggestions in the preceding Tip will simplify problems that combine both addition and subtraction of integers.

To add and subtract more than two numbers:

1. Rewrite the problem so that all integers are separated by only one sign.
2. Add a series of signed numbers.

EXAMPLE   Evaluate: (a) $3 - (-5) - 6$ (b) $-8 + 10 - (-7)$

(a) $3 - (-5) - 6 =$ Rewrite with only one sign between integers.
    $3 + 5 - 6 =$ Add numbers with like signs.
       $8 - 6 = \mathbf{2}$ Apply the rule for adding numbers with unlike signs.
(b) $-8 + 10 - (-7) =$ Rewrite with only one sign between integers.
    $-8 + 10 + 7 =$ Add numbers with like signs.
       $-8 + 17 = \mathbf{9}$ Apply the rule for adding numbers with unlike signs.

The key to solving applied problems with integers is to identify which numbers are positive and which are negative.

Positive amounts include profits, gains, money in the bank, temperatures above zero, receipts, income, winnings, and so on.

Negative amounts include losses, deficits, checks that cleared the bank, temperatures below zero, drops, declines, payments, and so on.

Once the positive and negative amounts are known, they are added or subtracted using the rules for adding or subtracting signed numbers.

EXAMPLE   A landscaping business makes a profit of $345 one week, has a loss of $34 the next week, and makes a profit of $235 the third week. What is the net profit?

The net profit is the sum of the profits and losses.

$\$345 - \$34 + \$235$ Interpret profits as positive and losses as negative.
profit   loss   profit

$345 + 235 - 34$ Rearrange and add positives.

$580 - 34 = 546$ Apply rule for adding numbers with unlike signs.

The net profit for the   Interpret answer.
three days was $546.

EXAMPLE In a recent year, 98°F was the highest temperature in Boston, and −2°F was the lowest temperature. What was the temperature range for the city that year? (The range is the difference between the highest and lowest values.)

$$98 - (-2)$$ Subtract −2 from 98.

$$98 + 2 = 100$$ Change to one sign only and apply the appropriate rule for adding integers.

The temperature range for Boston that year was 100°F. Interpret answer.

SELF-STUDY EXERCISES 2–3

1 Subtract.

1. $-3 - 9$
2. $8 - 2$
3. $9 - 15$
4. $-3 - (-7)$
5. $-11 - 14$
6. $-6 - (-3)$
7. $5 - (-3)$
8. $8 - 11$
9. $-8 - 1$
10. $11 - (-2)$
11. $(-8) - (-7)$
12. $(-15) - (-7)$

2 Subtract.

13. $15 - 0$
14. $0 - 8$
15. $-12 - 0$
16. $0 - (-8)$
17. $0 - (-7)$
18. $10 - 0$
19. $28 - (-28)$
20. $-46 - 46$
21. $7 - (-7)$
22. $-18 - 18$

3 Evaluate.

23. $-1 + 1 - 4$
24. $5 + 3 - 7$
25. $7 + 3 - (-4)$
26. $-8 - 2 - (-7)$
27. $-3 + 4 - 7 - 3$
28. $2 - 4 - 5 - 6 + 8$
29. $8 - 3 + 2 - 1 + 7$
30. $-5 - 3 + 8 - 2 + 4$
31. $6 - (-3) + 5 - 6 - 9$

32. The temperatures for Bowling Green, Kentucky, ranged from 102°F to −5°F. What was the temperature range for the city?

33. New Boston, Texas, registered −8°F as its lowest temperature one year and 99°F as its highest temperature for the same year. What was the temperature range for New Boston?

34. Computing Solutions records a profit of $28,296 one quarter (3 months), a loss of $1,896 for the second quarter, a profit of $52,597 for the third quarter, and a profit of $36,057 for the fourth quarter. What is their net profit for the year?

35. Explain the difference between subtracting zero from a number and subtracting a number from zero.

36. Time zones from Greenwich, England, west to the international date line are assigned negative numbers from −1 to −12. If Lima, Peru, is assigned a Greenwich Mean Time (GMT) of −5:00 and Los Angeles, California, which is west of Lima, is assigned −8:00 (GMT), express the time difference between Lima and Los Angeles as a positive or negative number.

2–4 MULTIPLYING INTEGERS

Learning Outcomes

1 Multiply integers.
2 Multiply several integers.
3 Multiply with zero.
4 Evaluate powers of integers.

Multiplying absolute values of integers is exactly the same as multiplying whole numbers. However, the rules for multiplying integers must also include the assignment of the proper sign to the product.

1 Multiply Integers.

Just like with whole numbers, multiplication of integers is commutative and associative; that is, factors can be multiplied in any order and grouped in any manner. These examples show how we assign signs to the products of two integers:

$$4(6) = 24, \quad (-3)(-7) = 21, \quad -10(2) = -20, \quad 8(-2) = -16$$

To multiply two integers:

1. Multiply the absolute values of the numbers as in whole numbers.
2. If the factors have like signs, the sign of the product is positive.
3. If the factors have unlike signs, the sign of the product is negative.

EXAMPLE Multiply: (a) $-12(-2)$ (b) $10 \cdot 3$ (c) $25(-3)$ (d) $-5 * 7$

(a) $-12(-2) = \mathbf{24}$ Like signs give a positive product.
(b) $10 \cdot 3 = \mathbf{30}$ Like signs give a positive product.
(c) $25(-3) = \mathbf{-75}$ Unlike signs give a negative product.
(d) $-5 * 7 = \mathbf{-35}$ Unlike signs give a negative product.

When the number 1 is multiplied by a number, the result is the number. Thus, 1 is the *multiplicative identity*.

■ **DEFINITION: Multiplicative Identity** 1 is the multiplicative identity because $a \cdot 1 = 1 \cdot a = a$ for all values of a.

2 Multiply Several Integers.

Multiplication, like addition, is a binary operation, so when we multiply three or more factors, we multiply two at a time. We apply the appropriate rule for signs each time we multiply. Multiplication is also commutative and associative, so we can change the order and grouping of the factors without affecting the final product.

EXAMPLE Multiply: (a) $4(-2)(6)$ (b) $-3(4)(-5)$ (c) $-2(-8)(-3)$
(d) $-2(-3)(-4)(-1)$

(a) $4(-2)(6)$ Multiply the first two factors and apply the rule for factors with unlike signs.

    $-8(6) = \mathbf{-48}$ Multiply and apply the rule for factors with unlike signs.

(b) $-3(4)(-5)$ Multiply the first two factors and apply the rule for factors with unlike signs.

    $-12(-5) = \mathbf{60}$ Multiply and apply the rule for factors with like signs.

(c) $-2(-8)(-3)$ Multiply the first two factors and apply the rule for factors with like signs.

    $16(-3) = \mathbf{-48}$ Multiply and apply the rule for factors with unlike signs.

(d) $-2(-3)(-4)(-1)$ Multiply the first two and the last two factors.

    $6(4) = \mathbf{24}$ Multiply and apply the rule for factors with like signs.

If we examine the multiplications in the preceding example more closely, we see that the number of negative factors affects the sign of the answer. Part a has one negative factor and the product is negative. Part b has two negative factors and the product is positive. Part c has three negative factors and the product is negative. Part d has four negative factors and the product is positive.

CHAPTER 2   Integers

> *To determine the sign of the product when multiplying three or more factors:*
>
> **1.** The sign of the product is *positive* if the number of negative factors is *even.*
> **2.** The sign of the product is *negative* if the number of negative factors is *odd.*

EXAMPLE Multiply: (a) $-2(6)(-1)(-3)$ (b) $(2)(-5)(1)(-3)$

(a) $-2(6)(-1)(-3) = -36$ Multiply absolute values. The odd number of negative factors makes the product negative.

(b) $(2)(-5)(1)(-3) = 30$ Multiply absolute values. The even number of negative factors makes the product positive.

3 Multiply with Zero.

Multiplications involving zero make use of a special property of zero that extends to integers and, in fact, to all real numbers. The product of zero and any number is zero: $x \cdot 0 = 0$. This is called the *zero property of multiplication.*

> **Tip!** | *Effect of a Zero Factor.*
>
> If we have two or more factors and one factor is zero, we can immediately write the product as zero without having to work through the steps.

EXAMPLE Multiply: (a) $3(-21)(2)(0)$ (b) $-9(-2)(8)(-1)$

(a) $3(-21)(2)(0) = 0$ Zero is a factor.
(b) $-9(-2)(8)(-1) = -144$ Zero is not a factor.

Applied problems often require multiplication of integers.

EXAMPLE In a three-week period, a technology stock declined approximately 2 points each week. How many points did the stock decline in the three weeks?

Because there are equal declines each week, we multiply the amount of weekly decline times the number of weeks: $-2(3) = -6$. Thus, **the stock declined (negative) a total of 6 points over the three-week period.**

4 Evaluate Powers of Integers.

Raising a number to a natural-number power is an extension of multiplication, so determining the sign of the result is similar to multiplying several integers. Observe the pattern for determining the sign of the result:

$$(+4)^2 = (+4)(+4) = +16 \qquad (-4)^2 = (-4)(-4) = +16$$
$$(+4)^3 = (+4)(+4)(+4) = +64 \qquad (-4)^3 = (-4)(-4)(-4) = -64$$

> *To raise integers to a natural-number power, use the following patterns:*
>
> **1.** A positive number raised to any natural-number power is positive.
> **2.** Zero raised to any natural-number power is zero.
> **3.** A negative number raised to an even natural-number power is positive.
> **4.** A negative number raised to an odd natural-number power is negative.

EXAMPLE Evaluate the powers: (a) 4^3 (b) 0^8 (c) $(-2)^4$ (d) $(-3)^5$

(a) $4^3 = 4(4)(4) = \mathbf{64}$
(b) $0^8 = \mathbf{0}$
(c) $(-2)^4 = (-2)(-2)(-2)(-2) = \mathbf{16}$
(d) $(-3)^5 = (-3)(-3)(-3)(-3)(-3) = \mathbf{-243}$

EXAMPLE Security systems sometimes use a four-digit code to activate the system. How many different codes can be made with four digits? Identify codes that may be impractical.

The number system has 10 digits: 0, 1, 2, 3, 4, 5, 6, 7, 8, and 9, so we can fill each one of the four slots of the code in 10 different ways. To find the total number of different codes, we multiply $10 \cdot 10 \cdot 10 \cdot 10$. This product can be written as 10^4, or 10,000. **There are 10,000 ways to make a four-digit security code. Some codes, such as 0000, may be impractical.**

Tip!	*Negative Base Versus an Opposite.*

$(-2)^4$ is not the same expression as -2^4.

$$(-2)^4 = (-2)(-2)(-2)(-2) = 16 \qquad -2^4 = -(2)(2)(2)(2) = -16$$

$(-2)^4$ is a negative base raised to a power. -2^4 is the opposite of 2^4.
Sometimes the values of two expressions are equal, but the interpretation is different. The expressions $(-3)^5$ and -3^5 give the same result because the exponent is an odd number.

$$(-3)^5 = (-3)(-3)(-3)(-3)(-3) = -243$$
$$-3^5 = -(3)(3)(3)(3)(3) = -243$$

As with whole numbers, raising a nonzero integer to the zero power gives 1: $x^0 = 1$.

EXAMPLE Evaluate the powers: (a) $(-2)^0$ (b) $(-7)^0$

(a) $(-2)^0 = \mathbf{1}$ (b) $(-7)^0 = \mathbf{1}$

SELF-STUDY EXERCISES 2–4

 Multiply.

1. $5 \cdot 8$
2. $-4(-3)$
3. $7 * 5$
4. $(-3)(-7)$
5. $-8(-3)$
6. $(-2)(-3)$
7. $5(-3)$
8. $(-2)(5)$
9. $-4 * 8$
10. $-3 \cdot 4$
11. $-7 * 8$
12. $6(-4)$

13. Holly Hobbs had four checks returned for insufficient funds, and her bank charged her a $28 service charge for each check. Use multiplication of signed numbers to show how these transactions affected her checking account balance.

14. Carolyn Luttrell made 7 withdrawals of $40 each from her checking account. Use signed numbers to show how these transactions affected her checking account balance.

15. Using the rules for adding signed numbers, you add or subtract the absolute values depending on whether the signs are alike or unlike. How are the multiplication rules for signed numbers similar or different?

2 Multiply.

16. $5(-2)(3)(2)$ **17.** $6(1)(-3)(-2)$ **18.** $4(0)(-12)(3)$ **19.** $15(-2)(-3)$
20. $5(2)(-3)(0)$ **21.** $-3(2)(-7)(-1)$ **22.** $9(-1)(3)(-2)$ **23.** $(-7)(-5)(-6)$
24. $(-3)(-9)(-12)(-7)$ **25.** $7(-3)(-10)(12)(-8)$

3 Multiply.

26. $-8(0)$ **27.** $5(0)$ **28.** $0(-12)$ **29.** $(-15)(0)$ **30.** $18(0)$
31. $0 \cdot 3$ **32.** $0(-15)$ **33.** $0(-17)$ **34.** $-28 \cdot 0$ **35.** $46 \cdot 0$
36. $5(1)(-3)(2)$ **37.** $-8(1)(3)(-7)$
38. Review the definitions for additive inverse and additive identity and write a similar definition for multiplicative inverse.

4 Evaluate.

39. $(-3)^2$ **40.** $(-2)^3$ **41.** $(-5)^2$ **42.** 0^{10} **43.** -2^3
44. $(-8)^0$ **45.** $(5)^4$ **46.** 3^4 **47.** 7^0 **48.** $(-42)^0$

49. How many seven-digit telephone numbers can be made with the 10 digits in the number system? Write the expression as a power and evaluate it.

50. Phone companies recently added a new three-digit, toll-free prefix—888. How many toll-free numbers are now in existence? (*Hint:* See Exercise 49 for the number of seven-digit phone numbers.)

51. Do you think all the phone numbers found in Exercise 49 can be used as legitimate phone numbers? Explain.

52. Many states have automobile license plates with three digits followed by three letters. How many different patterns can be formed with this sequence?

53. How many different license plates can be formed if each license plate has four digits only?

54. The prime meridian passes through Greenwich, England, and the time zone is designated as 0 GMT. Each time zone around the world encompasses 15° of longitude. Boston, Massachusetts, is assigned −5:00 GMT and San Francisco, California, is assigned −8:00 GMT. How many degrees of longitude (approximately) lie between Boston and San Francisco?

2–5   DIVIDING INTEGERS

Learning Outcomes

1 Divide integers.
2 Divide with zero.

Dividing the absolute values of the integers is exactly the same as dividing whole numbers. However, the rules for dividing integers must also include the assignment of the proper sign to the quotient.

1 Divide Integers.

The rules for determining the sign when dividing integers are similar to the rules for multiplying integers.

> *To divide two integers:*
>
> **1.** Divide the absolute values of the numbers as in whole numbers.
> **2.** If the values have like signs, the sign of the quotient is positive.
> **3.** If the values have unlike signs, the sign of the quotient is negative.

EXAMPLE Divide: (a) $\dfrac{-8}{-2}$ (b) $\dfrac{6}{3}$ (c) $\dfrac{10}{-2}$ (d) $\dfrac{-9}{1}$

(a) $\dfrac{-8}{-2} = \mathbf{4}$ Like signs give a positive quotient.

(b) $\dfrac{6}{3} = \mathbf{2}$ Like signs give a positive quotient.

(c) $\dfrac{10}{-2} = \mathbf{-5}$ Unlike signs give a negative quotient.

(d) $\dfrac{-9}{1} = \mathbf{-9}$ Unlike signs give a negative quotient.

2 Divide with Zero.

Division with zero works the same for integers as it does for whole numbers.

> **To evaluate division with zero:**
>
> Zero divided by any nonzero number is zero.
>
> $$\frac{0}{n} = 0; \quad n \neq 0 \qquad \frac{0}{-5} = 0$$
>
> Division by zero is undefined or impossible.
>
> $$\frac{n}{0} = \infty; \qquad \frac{-5}{0} = \text{undefined or impossible (or infinity)}$$

EXAMPLE Evaluate: (a) $\dfrac{-3}{0}$ (b) $\dfrac{0}{0}$ (c) $\dfrac{0}{-5}$

(a) $\dfrac{-3}{0} = \mathbf{undefined\ or\ impossible,\ or\ infinity}$

(b) $\dfrac{0}{0} = \mathbf{undefined\ or\ impossible,\ or\ infinity}$

(c) $\dfrac{0}{-5} = \mathbf{0}$

SELF-STUDY EXERCISES 2–5

1 Divide.

1. $\dfrac{-12}{-6}$ **2.** $-15 \div 5$ **3.** $\dfrac{8}{-2}$ **4.** $\dfrac{-24}{6}$ **5.** $-28 \div 7$

6. $\dfrac{-32}{-4}$ **7.** $\dfrac{-50}{-10}$ **8.** $\dfrac{36}{-6}$ **9.** $\dfrac{-48}{-6}$ **10.** $\dfrac{-25}{-5}$

11. You have decreased your house loan balance by $1800 over the past six months. Show this figure as a signed number, and find the average monthly decrease.

2 Divide.

12. $\dfrac{12}{0}$ **13.** $\dfrac{-15}{0}$ **14.** $\dfrac{0}{+3}$ **15.** $\dfrac{0}{-12}$ **16.** $\dfrac{0}{-8}$

17. $\dfrac{-20}{0}$ **18.** $\dfrac{-100}{0}$ **19.** $\dfrac{1}{0}$ **20.** $\dfrac{0}{8}$ **21.** $\dfrac{-350}{0}$

22. Which two operations have the same rules for handling signs when working with signed numbers?

23. A nor'easter storm blew into Green Bay, Wisconsin, and the temperature changed from 38°F to −22°F between 1 P.M. and 7 P.M. Represent the average hourly change in temperature with a signed number.

24. A business started the year with a net worth of −$4,852. During the first six months of the year, the business recovered and increased its net worth to a value of $15,983. Find the average monthly increase in net worth for the six-month period.

2–6 ORDER OF OPERATIONS

Learning Outcomes

1. Use the order of operations for integers.
2. Use calculators to evaluate operations with integers.

1 Use the Order of Operations for Integers.

Integers follow the same order of operations as whole numbers.

> *Perform operations in the following order as they appear from left to right:*
>
> **1.** Parentheses used as groupings and other grouping symbols.
> **2.** Exponents (powers and roots).
> **3.** Multiplications and divisions.
> **4.** Additions and subtractions.

EXAMPLE Evaluate $(4 + 3) - (3 + 1)$.

$(4 + 3) - (3 + 1)$ Perform operations inside parentheses.

$7 \quad - \quad 4$ Add or subtract last.

$7 - 4 = 3$

EXAMPLE Evaluate $8(4 + 6)$.

Work within the grouping symbols first:

$8(4 + 6)$ Add $4 + 6 = 10$.

$8(10)$ Multiply 8 by 10.

80

Or, use the distributive principle:

$8(4 + 6)$ Multiply each term by the factor 8.

$8(4) + 8(6)$

$32 + 48 = 80$ Add $32 + 48$.

Symbols for division are also grouping symbols.

EXAMPLE Evaluate $\dfrac{3-5}{2} - \dfrac{9}{2+1} + 4(5)$.

$\dfrac{3-5}{2} - \dfrac{9}{2+1} + 4(5)$ Perform operations grouped by the fraction bar.

$\dfrac{-2}{2} - \dfrac{9}{3} + 4(5)$ Multiply and divide.

$-1 - 3 + 20$ Add and subtract last.

$-4 + 20 = \mathbf{16}$

EXAMPLE Evaluate $-12 \div 3 - (2)(-5)$.

$-12 \div 3 - (2)(-5)$

$-12 \div 3 - (2)(-5)$ Multiply and divide first.

$-4 - (-10)$ Change to a single sign between the numbers.

$-4 + 10$ Add integers with unlike signs.

$-4 + 10 = \mathbf{6}$

The problem in the preceding example can also be written without parentheses around the 2. The problem is then expressed as

$$-12 \div 3 - 2(-5)$$

Again, we multiply and divide first, but notice how we perform the multiplication.

$-12 \div 3 - 2(-5)$ Consider the minus sign as the sign of the 2. Add the results of the division and the multiplication.

$-4 \qquad +10$

$-4 + 10 = \mathbf{6}$ Add integers with unlike signs.

EXAMPLE Evaluate $10 - 3(-2)$.

$10 - 3(-2)$ Think of the multiplication as -3 times -2.

$10 \quad +6$

$10 + 6 = \mathbf{16}$ Add integers with like signs.

Only after parentheses and all multiplications and/or divisions are taken care of do we perform the final additions and/or subtractions from left to right.

EXAMPLE Evaluate $4 + 5(2 - 8)$.

$4 + 5(2 - 8)$ First, perform operations in parentheses.

$4 + 5(-6)$ Multiply.

$4 - 30$ Finally, add integers with unlike signs.

$4 - 30 = \mathbf{-26}$

<table>
<tr><td>Tip!</td><td>The Order of Operations Is Important.</td></tr>
</table>

Note what happens if we proceed *out of order* in the preceding example!

Incorrectly Worked

$4 + 5(2 - 8)$ **Incorrectly add first instead of last.**

$9(2 - 8)$ **Perform operation in parentheses second instead of first.**

$9(-6)$ **Multiply last instead of second.**

$9(-6) = -54$ **Incorrect solution.**

We get a wrong answer. *The order of operations must be followed to arrive at a correct solution.*

EXAMPLE   Evaluate $5 + (-2)^3 - 3(4 + 1)$.

$5 + (-2)^3 - 3\,(4 + 1)$ Perform operation in parentheses.

$5 + (-2)^3 - 3\,(5)$ Raise to a power.

$5 + (-8) - 3(5)$ Multiply.

$5 + (-8) - 15$ Change to one sign only between numbers.

$5 - 8 - 15$ First addition of integers.

$-3 - 15 =$ Remaining addition of integers.

-18

---

2   Use a Calculator to Evaluate Operations with Integers.

Since negative numbers are such an integral part of our everyday lives, practically all types of calculators, even the basic calculator, can deal with negative values. However, the notation for negatives and the process for entering negatives varies widely from calculator to calculator. We will look at the most common options.

Some calculators use the subtraction key for both subtraction and entering negative numbers. Others have a special key for entering a negative sign that is labeled as a negative sign enclosed in parentheses $\boxed{(-)}$ A common option with basic calculators is the *sign-change key* $\boxed{+/-}$. This key is a *toggle key* and changes the sign of whatever is in the display to the opposite sign.

We continue our basic calculator strategy of examining options with examples we can work mentally. First, see how negatives are displayed on your calculator by entering the problem $3 - 4$. Next, examine your options for changing the sign in the display to its opposite. The *sign-change* may be an option, but if it is not available, try this: Enter a negative sign, then enter the previous answer $\boxed{\text{ANS}}$ followed by the $\boxed{=}$, $\boxed{\text{ENTER}}$, or $\boxed{\text{EXE}}$ key.

Now, try a variety of operations until you are comfortable and confident with your calculator and negative numbers.

EXAMPLE   Use a calculator to evaluate.

(a) $2 - 7$ (b) $-7 + 2$

(c) $(-7)(2)$ (d) $\dfrac{-4}{2} + 14$

The most common options:

(a) $2 - 7$ $2\ \boxed{(-)}\ 7\ \boxed{=}\ \Rightarrow\ -5$ $\boxed{=}$ may be labeled $\boxed{\text{ENTER}}$ or $\boxed{\text{EXE}}$.

(b) $-7 + 2$ $\boxed{(-)}\ 7\ \boxed{+}\ 2\ \boxed{=}\ \Rightarrow\ -5$

(c) $(-7)(2)$ $\boxed{(-)}\ 7\ \boxed{\times}\ 2\ \boxed{=}\ \Rightarrow\ -14$

or

$\boxed{(-)}\ 7\ \boxed{(}\ 2\ \boxed{)}\ \boxed{=}\ \Rightarrow\ -14$ Some calculators interpret parentheses as multiplication.

(d) $\dfrac{-4}{2} + 14$ $\boxed{(-)}\ 4\ \boxed{\div}\ 2\ \boxed{=}\ \boxed{+}\ 14\ \boxed{=}\ \Rightarrow\ 12$

or

$\boxed{(-)}\ 4\ \boxed{/}\ 2\ \boxed{=}\ \boxed{+}\ 14\ \boxed{=}\ \Rightarrow\ 12$ A slash may be used for division.

Other options:

(a) $2 - 7$ $2\ \boxed{-}\ 7\ \boxed{=}\ \Rightarrow\ -5$

(b) $-7 + 2$ $7\ \boxed{+/-}\ \boxed{+}\ 2\ \boxed{=}\ \Rightarrow\ -5$

(c) $(-7)(2)$ $7\ \boxed{+/-}\ \boxed{\times}\ 2\ \boxed{=}\ \Rightarrow\ -14$

(d) $\dfrac{-4}{2} + 14$ $4\ \boxed{+/-}\ \boxed{\div}\ 2\ \boxed{+}\ 14\ \boxed{=}\ \Rightarrow\ 12$

You may need to impose the appropriate order of operations by using the $\boxed{=}$ key when you perform a series of calculations. This is necessary with a basic calculator when no parentheses keys are available.

When a bar is used as both a grouping and division symbol on a calculator, we must instruct the calculator to work the grouping first by enclosing the grouping in parentheses.

EXAMPLE   Evaluate:

(a) $\dfrac{5 + 1}{3} + 2$ (b) $2(4 - 1) + 3$

Most common calculator options:

(a) $\dfrac{5 + 1}{3} + 2$ $\boxed{(}\ 5\ \boxed{+}\ 1\ \boxed{)}\ \boxed{\div}\ 3\ \boxed{+}\ 2\ \boxed{=}\ \Rightarrow\ 4$

(b) $2(4 - 1) + 3$ $2\ \boxed{(}\ 4\ \boxed{-}\ 1\ \boxed{)}\ \boxed{+}\ 3\ \boxed{=}\ \Rightarrow\ 9$

Other calculator options:

(a) $\dfrac{5 + 1}{3} + 2$ $5\ \boxed{+}\ 1\ \boxed{=}\ \boxed{\div}\ 3\ \boxed{+}\ 2\ \boxed{=}\ \Rightarrow\ 4$

(b) $2(4 - 1) + 3$ $2\ \boxed{\times}\ \boxed{(}\ 4\ \boxed{-}\ 1\ \boxed{)}\ \boxed{+}\ 3\ \boxed{=}\ \Rightarrow\ 9$

SELF-STUDY EXERCISES 2–6

1   Evaluate, paying careful attention to the order of operations.

1. $-12 + 8$ **2.** $6(-5)$ **3.** $3(5 - 8)$

4. $8 - (-9)$ **5.** $(7 + 8) - (2 - 7)$ **6.** $3(7) + 9 \div 3$

7. $-6(-2) - 4$

8. $-6(-4)$

9. $7(1 - 4) \div 3 + 2$

10. $\dfrac{2 - 8}{3} + 5(-2) \div 2$

11. $\dfrac{5 - 9}{4} + 4(-3) \div 6$

12. $4 \times 8 - 7 \times 3^2 + 18 \div 6$

2 Use a scientific or graphing calculator to evaluate.

13. $5(3 - 4) - 7(2 - 5) \div (-3)$

14. $142(3 - 21) + 48(27)$

15. $24 - 3(2 + 7) \div 3 + 12$

16. $\dfrac{12 - 18}{3} + 15 - 81 \div 3$

17. $14 - \dfrac{3 + 7}{5} \div 2 + 5(3)$

18. $2(3 - 12) \div 4 \times 6 - 8$

19. $2(4 - 3) - 7 + 4(2 - 8)$

20. $7 - 21 + 138 - 256$

21. Use your calculator to verify that division by zero is impossible. Describe the calculator display that results from this division.

22. Use your calculator to evaluate several addition problems, then change the order of the addends and evaluate. Write a statement about the order of addition.

23. Evaluate several subtraction problems, then change the order of each problem and evaluate. Compare the results for each problem. Write a statement about the order of subtraction.

24. After testing several multiplication examples, write a statement about the order of multiplication.

25. After testing several division examples, write a statement about the order of division.

CAREER APPLICATION

Consumer Interest: Stock Market Prices

Many savvy savers have recently discovered that over the long term their principal earns much more invested in the stock market than invested in savings accounts, certificates of deposit (CDs), mutual funds, or bonds. This trend toward investing in the market has been fueled by the practice of many discount stock brokerage firms to charge a fraction of what other firms charge per transaction. The basic idea is simple: You buy shares of publicly traded companies at a specified price, hold them, then sell them at a specified price. Ideally you buy at a low price and sell at a high price. Investors who can hold the stock a long time (say, 40 years until retirement, or 18 years until a child goes to college), usually make the most profit. Not only do stock prices usually increase over time, but the stock may "split." This means that one share becomes two shares at half the current price, but the price often rebounds quickly to its presplit price.

The stock prices you see scrolling along the bottom of financial television channels, news programs, or computer screens report only the change in the price of the stock from the previous day's closing price. This change can be positive, negative, or zero, depending on many factors within the individual company, on politics, or on the financial network.

Let's assume that your child is born in the year 2000, and you sell your restored 1965 Chevrolet Corvair for $10,000 profit to invest for your child's college education. With advice from your certified financial planner and discount stock broker, you decide to invest in Coca-Cola common stock, which sells for $80 per share. You hold these shares for 18 years until your daughter is about to attend college. Your broker advises you to sell within the next month (October 2018) because fewer soft drinks are consumed in the winter, and the stock price could fall. The closing price on Friday, September 28, 2018, is $341. You record the daily Coca-Cola price *change* in dollars for the next two weeks (October 1–12, 2018; Monday through Friday):

$$-2, +4, -1, 0, +1, -2, -1, 0, +3, +1$$

Exercises

Use the preceding information to solve the exercises. Round all answers to the nearest tenth.

1. Assuming no splits, how many shares of Coca-Cola common stock do you own in 2018?
2. If you sell your stock at the start of the day on October 1, how much money will you receive, and how much profit will you have made? Find the simple interest earned for 18 years by dividing the profit by the principal invested, and multiplying by 100. Dividing this interest by 18 years gives you the simple annual interest. Find it.
3. What is the closing price on October 1, 2018? What will be the difference in the amount of money received if you sell at the end of the day on October 1 instead of at the beginning?
4. Check the price changes given and guess which day would have been the best to sell on and the worst to sell on. Now, list the closing prices on those days and see if your guesses are correct. For the best and worst selling days, what is the difference in price per share, and difference in the amount received from the sale of 125 shares?
5. Assume that when you call in your request to sell on October 12, your broker advises you to hold because she has heard a rumor that the stock may soon split. It does split on October 15 at the closing price of $346. Now how many shares do you hold at what price?
6. If you sell at the beginning of the day on October 16, how much do you receive? If your broker advises you to wait until the end of the month, and the closing price on October 31 is $185 per share, how much more will you receive?
7. You learn that your smart daughter has received a full academic National Merit scholarship. You decide to hold your stock another year, then sell it and build the vacation/retirement home of your dreams on lakefront property that you own. If the closing price on October 31, 2019, is $238, will you have enough to pay for an $80,000 A-frame home?
8. If you sell your stock on October 31, 2020, and the closing price is $329, will you have enough for your home?
9. Assuming you had to pay $722.50 in broker fees, find the profit earned on your 20-year investment of $10,000 and how much annual simple interest you earned.
10. Compare your interest rate with current savings account, certificates of deposit, money market mutual fund, and bond interest rates. What is the biggest disadvantage to investing in the stock market?

Answers

1. $10,000 divided by $80 per share $=$ 125 shares.
2. You will receive $42,625, or a profit of $32,625, which amounts to 326% simple interest for 18 years (or 18% per year).
3. The closing price on October 1 is $339, which means you will receive $250 less.
4. Closing prices are October 1: $339, October 2: $343, October 3: $342, October 4: $342, October 5: $343, October 8: $341, October 9: $340, October 10: $340, October 11: $343, October 12: $344. The best day to sell on would have been at the end of October 12, and the worst day to sell on would have been at the end of October 1, for a difference of $5 per share, or $625 for 125 shares.
5. You now have 250 shares at $173 per share.
6. At $173, 250 shares gives $43,250, while at $185, 250 shares gives $46,250, or $3,000 more.
7. At $238, 250 shares gives $59,500, so you will not have enough.

8. At \$329, 250 shares gives \$82,250, so you will have enough.
9. Your profit of \$71,527.50 (\$72,250 − \$722.50) earned 40% annual simple interest.
10. This interest rate is much higher than if you had invested in savings accounts, certificates of deposit, money market mutual funds, or bonds. The disadvantage is that you are taking a risk that is uninsured by the Federal Deposit Insurance Corporation (FDIC). If the price of the stock falls, you can lose all or part of your investment.

ASSIGNMENT EXERCISES

Section 2-1

The number line in Fig. 2–30 shows 14 positions lettered A–N. Give the letter that corresponds to each number.

1. -1
2. $+1$
3. -2
4. $+2$
5. -3
6. $+3$
7. -7
8. 4
9. 5
10. -5
11. 6
12. -6

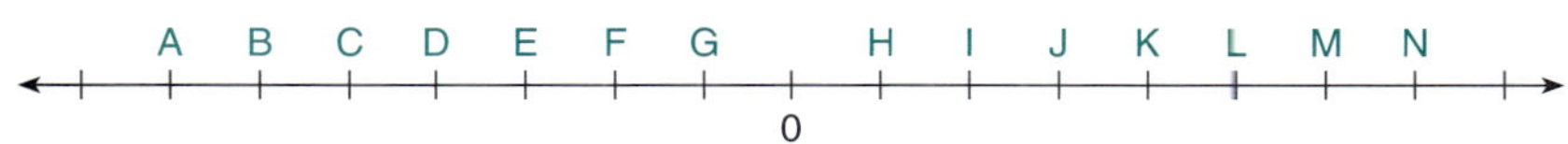

Figure 2-30

13. Write three examples of natural numbers.
14. Write three examples of integers that are not natural numbers.
15. Is there any number that is a whole number and an integer but not a natural number? If so, what is it? If not, explain why.
16. What is the smallest whole number?
17. What is the smallest natural number?
18. What is the smallest integer?

Use the symbols $>$ and $<$ to show the relationships of the numbers.

19. Is 72°F more than or less than -80°F?
20. Is 0 more than or less than -3?
21. Is 7 more than or less than -5?
22. Is 5 more than or less than 8?
23. Is -9 more than or less than 5?
24. Is -12 more than or less than -8?

Give the value of each number:

25. $|5|$
26. $|-8|$
27. $|-3|$
28. $|0|$
29. $|+7|$
30. $|+8|$
31. $|-11|$
32. $|-52|$

Give the opposite of each number:

33. -12
34. 8
35. 15
36. 0
37. -2
38. -13
39. -42
40. 156
41. 87
42. -19

Draw a rectangular coordinate system and locate the points.

43. $A = (5, -2)$
44. $B = (-8, -3)$
45. $C = (0, -4)$
46. $D = (3, 7)$
47. $E = (-3, 2)$
48. $F = (-3, 0)$
49. What are the coordinates of the origin?
50. Which of the four quarters of the coordinate system is used to plot coordinates that are both negative?

51. Write the coordinates for the points on the graph in Fig. 2–31.

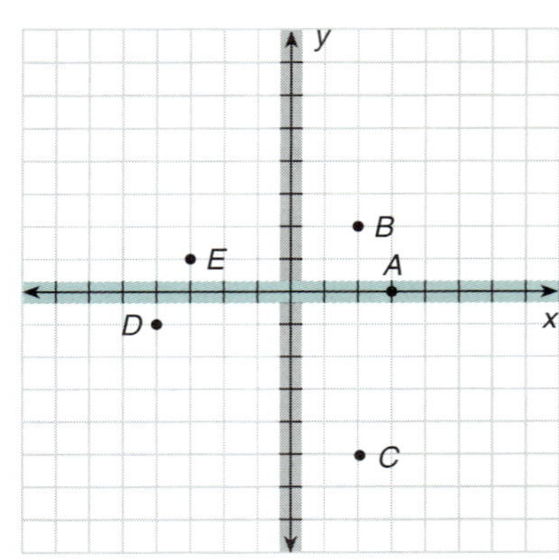

Figure 2–31

Section 2–2
Add.

52. $-3 + (-8)$ **53.** $5 + 12$ **54.** $-7 + 12$ **55.** $(-15) + 8$
56. $7 + (-11)$ **57.** $-5 + (-8)$ **58.** $-6 + 6$ **59.** $8 + (-8)$
60. $7 + 0$ **61.** $0 + (-8)$ **62.** $(-7) + 7$ **63.** $-25 + 0 + 12 + 7$

64. A publicly traded company has a profit of $256,872 for one year and a loss of $38,956 for the following year. What is the net profit over the two-year period?

65. A football team gained and lost the following yardage during a series of plays beginning with first down: $+4$, -5, $+9$. If you are the coach, what is your next play? Why?

66. Because of stormy weather, a pilot flying at 35,000 feet descends 8,000 feet. What is his new altitude?

67. Agnes opens a checking account by depositing $500. She then writes checks for $42, $18, and $21. What is her balance after depositing another $150?

68. You owe your friend $18 and borrow an additional $42. Use signed numbers to express your debt.

69. Explain the difference between additive identity and additive inverse.

Section 2–3
Evaluate.

70. $8 - 5$ **71.** $-9 - 4$ **72.** $-7 - (-2)$
73. $11 - (-3)$ **74.** $12 + 3 + (-8) - 5$ **75.** $-6 + 3 - 5 - 7$
76. $8 - 0$ **77.** $-5 - 0$ **78.** $0 - 3$
79. $0 - (-2)$ **80.** $-5 - 5$ **81.** $18 - (-18)$

82. Temperatures in northern Canada ranged as high as $37°F$ one summer. That same year the lowest temperature was $-28°F$. What was the range of temperatures for the year?

83. What is the difference (or range) in temperatures of $43°$ above zero and $27°$ below zero?

84. What is the difference in temperatures $47°$ below zero and $28°$ below zero?

85. Two successive recordings for a surgery patient's temperature were $103.2°F$ and $97.8°F$. Express the temperature change with a signed number.

86. A 10-ft-long fence post is placed in a hole that is 3 ft deep. How much of the post is above the ground?

Section 2–4
Evaluate.

87. $5(8)$ **88.** -3×-7 **89.** 7×-2 **90.** $(-3)(+2)$
91. $6(-2)$ **92.** $-7(3)$ **93.** $-8(0)$ **94.** $0(5)$
95. $2(3)(-7)(0)$ **96.** $5(-2)(-1)(-3)$ **97.** $4(3)(-2)(7)$ **98.** $(-3)^2$
99. $(7)^3$ **100.** $(-4)^3$ **101.** -4^2 **102.** -2^3

103. On one winter day, the temperature dropped 2° each hour for 5 hours. What was the total drop in temperature?

104. If the temperature in Exercise 103 was 8° originally, what was the temperature at the end of the 5-hour period?

105. You research stock prices and find an article that says, "XYZ stock has dropped 4 points each week for the past 5 weeks." Was the stock price higher or lower 5 weeks ago than it is today? How much higher or lower? Use negative numbers to express drops in prices. Use a signed number to express the amount the stock price decreased or increased.

Section 2–5

Divide.

106. $-8 \div (-4)$

107. $12 \div 3$

108. $\dfrac{18}{9}$

109. $\dfrac{-20}{-5}$

110. $\dfrac{16}{-4}$

111. $\dfrac{-24}{6}$

112. $\dfrac{0}{-8}$

113. $\dfrac{-7}{0}$

114. $\dfrac{-51}{3}$

115. Greenwich Mean Time (GMT) is known as Zulu time or time zone Z, and its GMT number is 0. The Central Standard Time Zone is called Foxtrot or F and its GMT number is -6. If there are approximately 90° of longitude between these two time zones, and a negative GMT indicates locations west of Greenwich, England, approximately how many degrees of longitude are in each time zone between Zulu and Foxtrot time zones?

116. A company records the following gains and losses in net profit for a six-month period. $22,973; $-$12,357; $-$2,791; $32,872; $18,930; and $2,093. Find the net gain or loss for the six-month period.

Section 2–6

Use the order of operations to evaluate the following. Verify the results with a calculator.

117. $7(3 + 5)$

118. $-2(3 - 1)$

119. $\dfrac{15 - 7}{8}$

120. $-20 \div 4 - 3(-2)$

121. $12 - 8(-3)$

122. $7 + 3(4 - 6)$

123. $4 + (-3)^4 - 2(5 + 1)$

124. $(-3)^3 + 1 - 8$

125. $-3 + 2^3 - 7$

126. $4(-6 - 2) - \dfrac{8 + 2}{7 - 5}$

127. $296 - 382(-4)^5$

128. $-71 + 3(-19)$

CHALLENGE PROBLEMS

129. The temperature at 8:00 A.M. is recorded as -3°C. Calculate the temperature at each hour as recorded by the increases (inc), decreases (dec), and no change (n.c.).

9:00 A.M.: (inc) 2°	2:00 P.M.: (inc) 3°
10:00 A.M.: (inc) 1°	3:00 P.M.: (dec) 4°
11:00 A.M.: (inc) 0°	4:00 P.M.: (dec) 7°
12:00 P.M.: (inc) 1°	5:00 P.M.: (dec) 8°
1:00 P.M.: (n.c.)	6:00 P.M.: (dec) 12°

130. How many automobile license plates can be formed if the pattern is 3 digits and 3 letters? 6 letters?

Use the symbols $>$ and $<$ to write the following as *true* statements.

1. -8 is more than 0

2. 2 is less than 3

3. -5 is more than -1

Answer the questions.

4. What is the value of $|-12|$?

5. What is the opposite of 8?

Perform the operations.

6. $-8 - 2$

7. $-3 + 7$

8. $\dfrac{8}{-2}$

9. $2(6)(-4)$

10. $\dfrac{-6}{-2}$

11. $8 + 4 + (-2) + (-7)$

12. $7 - (-3)$

13. $2(-1)(-4)$

14. $-8 + (-3)$

15. $(-8)(3)(0)(-1)$

16. $\dfrac{0}{3}$

17. $0 - 7$

18. $-3 + 5 + 0 + 2 + (-5)$

19. $\dfrac{-7}{0}$

20. $4(-13)$

21. $5 + (-7) - (-3) + 2 - 6$

22. $\dfrac{6}{-1}$

23. $\dfrac{4}{-2}$

24. $8(3) - 2 + 6(7)$

25. $2(3 - 9) \div 2^2 + 7$

26. $\dfrac{10 - 4}{3} - 3$

Use a calculator to perform the operations.

27. $5(6) - 2 + 10 \div 2$

28. $5(2 - 3) + 16 \div 4$

29. $-2 - 3(-4) + (4)(3)$

30. $\dfrac{-18}{3}$

31. $\dfrac{4 - 2}{2} - 6^2$

32. Write two signed numbers that are opposites. Describe the relationship between the two numbers.

33. How are the multiplicative identity and the additive identity alike? Give examples.

34. Write the sum of two signed numbers that have unlike signs and whose addition results in a negative number.

35. Temperatures around the world may range from 135°F in Seville, Spain, to -40°F in Fairbanks, Alaska. What is the range (difference) of temperatures?

36. In Grand Rapids, Minnesota, the temperature ranged from 42°F at 12 noon to -14°F at 7 P.M. Represent the change in temperature with a signed number.

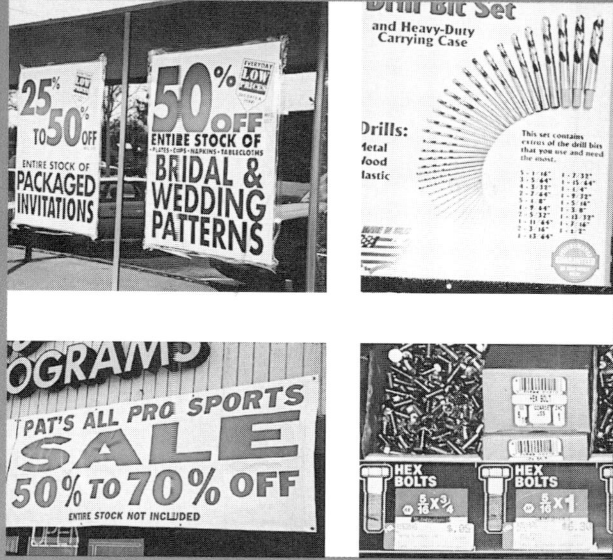

Fractions and Percents

GOOD DECISIONS THROUGH TEAMWORK

Over the course of one week, note each time you see a fraction or percent used outside the classroom. Describe the situation and the use made of the fraction or percent. For instance, a computer store advertises a half-off sale on selected peripherals, your bank charges $9\frac{1}{2}\%$ interest on 60-month car loans, or a painter adds $4\frac{1}{2}$ ounces of TSP to a solution to clean a wall before painting it.

With your team, make a master list of situations, eliminating duplications. For each situation on your master list, discuss why a fraction or a percent is used rather than a whole number or a decimal. Also discuss why a fraction is used rather than a percent. On the basis of your discussion, identify major categories for the use of fractions and percents. How many of these categories apply to technical settings, such as in science or industry? Give a technology-related example for each category your team judges to be related to a technical setting. Choose a team member to share the results of your discussion with the class.

3–1 Fraction terminology

1 Identify fraction terminology.

3–2 Multiples, divisibility, and factor pairs

1 Find multiples of a natural number.
2 Determine the divisibility of a number.
3 Find all factor pairs of a natural number.

3–3 Prime and composite numbers

1 Factor prime and composite numbers.
2 Determine the prime factorization of composite numbers.

3–4 Least common multiple and greatest common factor

1 Find the least common multiple of two or more numbers.
2 Find the greatest common factor of two or more numbers.

3–5 Equivalent fractions and decimals

1 Write equivalent fractions with higher denominators.
2 Write equivalent fractions with lowest denominators.

3. Change decimals to fractions.
4. Change fractions to decimals.

3–6 Improper fractions and mixed numbers

1. Convert improper fractions to whole or mixed numbers.
2. Convert mixed numbers and whole numbers to improper fractions.

3–7 Finding common denominators and comparing fractions and decimals

1. Find common denominators.
2. Compare fractions and decimals.

3–8 Adding fractions and mixed numbers

1. Add fractions.
2. Add mixed numbers.

3–9 Subtracting fractions and mixed numbers

1. Subtract fractions.
2. Subtract mixed numbers.

3–10 Multiplying fractions and mixed numbers

1. Multiply fractions.
2. Multiply mixed numbers.

3–11 Dividing fractions and mixed numbers

1. Find reciprocals.
2. Divide fractions.
3. Divide mixed numbers.
4. Simplify complex fractions.

3–12 Signed fractions and decimals

1. Change a signed fraction to an equivalent signed fraction.
2. Perform basic operations with signed fractions.
3. Perform basic operations with signed decimals.

3–13 Calculators with fraction key

1. Use the fraction key on a calculator to perform operations with fractions.

3–14 Finding number and percent equivalents

1. Change any number to its percent equivalent.
2. Change any percent to its numerical equivalent.

3–1   FRACTION TERMINOLOGY

Learning Outcome

1. Identify fraction terminology.

The language of mathematics is important in the study of mathematics. Mathematical symbols help us describe situations in shortcut fashion, and terminology helps us understand mathematical concepts. Thus far, we have focused on the terminology of whole numbers and decimals. However, we must often work with numbers called fractions (such as $\frac{1}{2}$, $\frac{1}{4}$, $1\frac{1}{4}$, and $1\frac{1}{2}$) and percents. In this chapter we study how fractions and percents relate to whole numbers and decimals, and how we add, subtract, multiply, and divide fractions. Learning the terminology of fractions is our first step.

In Chapter 1, we examined a special type of fraction called a decimal fraction. Now we look more closely at other types of fractions. A *fraction* is a value that can be expressed as the quotient of two integers.

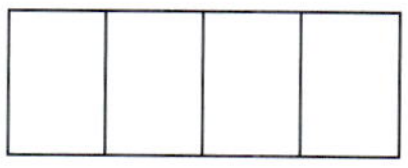

If one unit (or amount) is divided into four parts, we can write the fraction $\frac{4}{4}$ to represent this single unit. Figure 3–1 illustrates a unit divided into four equal parts.

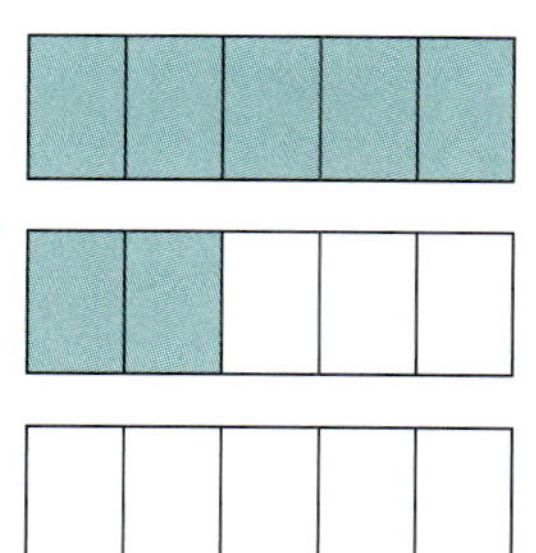

Figure 3–1

The fraction $\frac{4}{4}$ is an example of a *common fraction*. A common fraction consists of two whole numbers. The bottom number, the *denominator*, indicates *how many equal parts* one whole unit has been divided into. The top number, the *numerator*, tells *how many of these parts* are being considered. Thus, in Fig. 3–2, one of the four equal parts is shaded and represents the fraction $\frac{1}{4}$. Similarly, the shaded part in Fig. 3–3 is written as the fraction $\frac{3}{7}$.

Figure 3–2 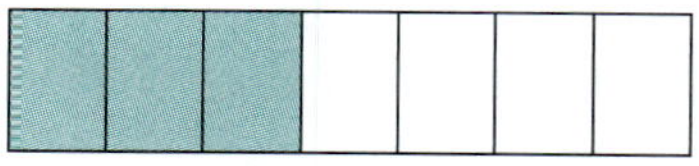   **Figure 3–3**

The unit is always our *standard* amount when writing fractions. Thus, $\frac{4}{4}$ and $\frac{3}{3}$ represent one unit each, or simply the number 1. Some fractions represent less than one unit (less than 1), for example, $\frac{1}{4}$ or $\frac{3}{4}$. These fractions are called *proper fractions*. Other fractions represent one or more units, for example, $\frac{7}{5}$. These fractions are called *improper fractions*.

Suppose that three units are divided into five equal parts each. If we have seven of these parts, the fraction $\frac{7}{5}$ represents this amount. This fraction is greater than one unit (see Fig. 3–4) and is thus an improper fraction.

Similarly, if we have 10 of these parts, the fraction $\frac{10}{5}$ represents this amount. If five parts equal one unit, then 10 parts equal two units (see Fig. 3–5). Fifteen parts, $\frac{15}{5}$, equal three units (see Fig. 3–6).

Figure 3–4

Fractions and division are related. We can see this in the fractions $\frac{10}{5}$ (two units) and $\frac{15}{5}$ (three units) because $10 \div 5 = 2$ and $15 \div 5 = 3$. This relationship to division is important to our understanding of fractions. Now, let's formally define the terms we have been discussing.

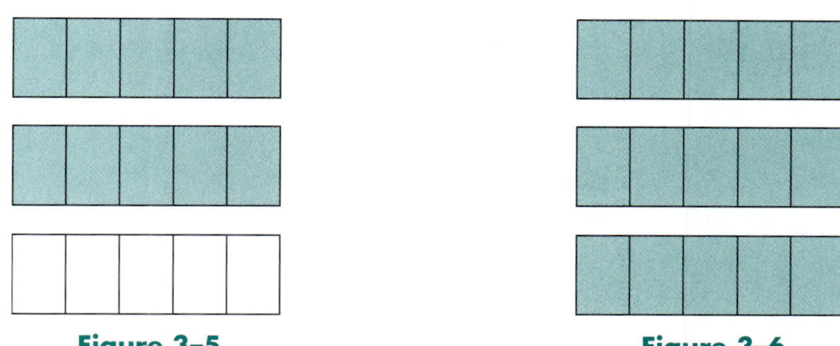

Figure 3–5   **Figure 3–6**

■ **DEFINITION: Common Fraction.** A *common fraction* is the division of one integer by another integer.

■ **DEFINITION: Denominator.** The *denominator* of a fraction is the number of parts one unit has been divided into. It is the *bottom* number of a fraction or the *divisor* of the indicated division.

■ **DEFINITION: Numerator.** The *numerator* of a fraction is the number of parts being considered. It is the *top* number of a fraction, or the *dividend* of the indicated division.

The numerator and denominator of a fraction are usually separated by a horizontal line, although sometimes a slash is used. This is the *fraction line,* and it also serves as a division symbol.

Another term for a fraction is *rational number.* A rational number can be written in various forms. So far we have looked at several forms for writing rational numbers: natural numbers, whole numbers, fractions, and decimals. However, we can use the fraction form of numbers to define a rational number.

■ **DEFINITION: Rational Number.** Any number that can be written in the form of a fraction, with an integer for a numerator and a nonzero integer for a denominator, is a rational number.

To interpret natural numbers, whole numbers, and integers as rational numbers, consider one whole unit as one part. Thus, 5 can be written in fraction form as $\frac{5}{1}$. Look at the fraction $\frac{2}{3}$. This fraction can be visualized as two of three parts (Fig. 3–7).

Let's discuss the fraction $\frac{2}{3}$ using the relationships and terminology we have used so far with fractions.

Figure 3–7

EXAMPLE Answer each question as it relates to $\frac{2}{3}$.

(a) One unit is divided into how many parts?
   The denominator shows one whole is divided into three parts.
(b) How many of these parts are used?
   The numerator shows that two parts are used.
(c) The numerator of the fraction is ________.
   The numerator is the top number, so the numerator of $\frac{2}{3}$ is 2.
(d) The denominator of the fraction is ________.
   The denominator is the bottom number, so the denominator of $\frac{2}{3}$ is 3.
(e) The fraction can be read as ________ divided by ________.
   The numerator is divided by the denominator, so **2 is divided by 3.**
(f) ________ is the divisor of the division.
   The divisor is the number we are dividing by, so **3 is the divisor.**
(g) ________ is the dividend of the division.
   The dividend is the number being divided, so **2 is the dividend.**
(h) Is this common fraction proper or improper?
   A proper fraction is less than one whole amount. In $\frac{2}{3}$, the whole is divided into three parts and 2 is less than 3. Therefore, $\frac{2}{3}$ **is a proper fraction.**
(i) This fraction represents ________ out of ________ parts.
   $\frac{2}{3}$ is *two* **out of** *three* **parts.**
(j) Does this fraction have a value less than, more than, or equal to 1?
   Since two parts is less than the three parts needed for one whole, then $\frac{2}{3}$ **is less than 1.**

Now, let's focus on the division property of fractions. We will come back to this concept again and again as we progress through this book. Look at the ways we have seen so far for writing division. Consider 6 divided by 2.

$$2\overline{)6} \qquad 6 \div 2 \qquad \frac{6}{2} \qquad 6/2$$

Using fraction terminology to describe division, remember that in all these cases the numerator is divided by the denominator.

$$\text{denominator}\,\overline{)\text{numerator}} \qquad \text{numerator} \div \text{denominator} \qquad \frac{\text{numerator}}{\text{denominator}}$$

■ Learning Strategy *Learning Definitions.*

Learning new definitions can sometimes be overwhelming. A key strategy for learning definitions is to understand the basic concepts, to connect the new words to other words or concepts you already know, and to use memory devices.

Learning terminology is important in developing your math survival skills. When you forget the details of a procedure (and you probably will if you don't use the procedure often), the term will help you use the index or table of contents to find the topic in the text that will refresh your skills.

The words *numerator* and *denominator* are used in so many rules and definitions that we need to concentrate on distinguishing between these two words. Relate denominator to denomination. When examining money, the denomination describes how much the money is worth. For example, we have currency in denominations of \$1, \$5, \$10, \$20, and so on.

In fractions, the denominator describes how a whole amount is divided into parts. Now, let's relate the word numerator to number. The numerator tells the number of parts that are being considered.

Another way to remember is to create a visual image in words and as in Fig 3–8.

Figure 3–8

$$\frac{\text{numerator}}{\text{denominator}} \qquad \frac{N}{D} \qquad \frac{\text{top}}{\text{bottom}}$$

$$\text{3 out of 4} \qquad \frac{3}{4} \qquad \frac{\text{part}}{\text{total}}$$

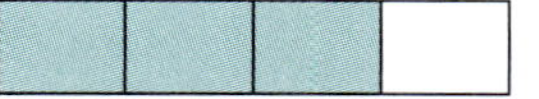

Now, let's continue our discussion of common fractions.

There are two types of common fractions: those whose value is less than 1 and those whose value is equal to or greater than 1.

■ **Definition: Proper Fraction.** A *proper fraction* is a common fraction whose value is less than one unit; that is, the numerator is less than the denominator. *Examples:*

$$\frac{2}{5}, \quad \frac{3}{7}, \quad \frac{15}{16}, \quad \frac{1}{3}, \quad \frac{7}{10}$$

■ **Definition: Improper Fraction.** An *improper fraction* is a common fraction whose value is equal to or greater than one unit; that is, the numerator is equal to or greater than the denominator. *Examples:*

$$\frac{4}{4}, \quad \frac{7}{3}, \quad \frac{8}{4}, \quad \frac{100}{10}, \quad \frac{17}{5}$$

When the denominator divides evenly into the numerator, the improper fraction can be written as a whole number, such as $\frac{8}{4} = 2$; but when the value is more than one unit and the denominator cannot divide evenly into the numerator, the improper fraction can be written as a combination of a whole number and a fractional part, such as $\frac{9}{4} = 2\frac{1}{4}$. Such a combination is called a *mixed number.*

■ **Definition: Mixed Number.** A *mixed number* consists of both a whole number and a fraction. The whole number and fraction are added together. *Example:*

$$3\frac{2}{5} \text{ means 3 whole units and } \frac{2}{5} \text{ of another unit} \qquad \text{or} \qquad 3\frac{2}{5} = 3 + \frac{2}{5}$$

Sometimes computations in technical applications are more complex than the fractions and mixed numbers we have looked at so far. We have seen that a fraction like $\frac{2}{7}$ means that a unit has been divided into seven equal parts and we are considering only two of those seven parts. But suppose our job requires us to consider $2\frac{1}{3}$ of those seven parts instead of just two. In this instance, our fraction would be

$$2\frac{\frac{1}{3}}{7}$$

which is a *complex fraction.*

■ **DEFINITION: Complex Fraction.** A *complex fraction* has a fraction or mixed number in its numerator or in its denominator, or in both.

Examples of complex fractions:

$$\frac{\frac{1}{2}}{2}, \quad \frac{7}{\frac{3}{4}}, \quad \frac{3\frac{1}{3}}{7}, \quad \frac{5}{4\frac{1}{2}}, \quad \frac{\frac{1}{8}}{\frac{5}{16}}, \quad \frac{6\frac{3}{8}}{4\frac{1}{2}}$$

It is very important to understand that fractions indicate division. The fraction is read from *top to bottom,* with the fraction line read as "divided by."

$$\frac{\frac{1}{2}}{2} \quad \text{is read as } \frac{1}{2} \text{ divided by 2.}$$

$$\frac{7}{\frac{3}{4}} \quad \text{is read as 7 divided by } \frac{3}{4}.$$

As you learned in Chapter 1, many job-related applications make use of another kind of fraction, whose denominator is 10 or a power of 10, such as 100, 1,000, and so on. Instead of writing $\frac{3}{10}$, we may write 0.3 (with the zero indicating no whole number, the period or *decimal point* indicating the beginning of the fraction, and the 3 indicating how many $\frac{1}{10}$'s, in this case 3). The mixed number $2\frac{17}{100}$ can be written as 2.17 (with the 2 indicating the whole number, the decimal point indicating the beginning of the fraction, and the 17 indicating how many $\frac{1}{100}$'s, in this case 17).

Tip!	***Relate the Number of Decimal Places and the Zeros in the Denominator to the Place Value.***

The number of places after the decimal point indicates the number of zeros in the denominator of the power of 10.

$$0.015 = \frac{15}{1,000} \qquad 2.43 = 2\frac{43}{100}$$

Note that the number after the decimal point indicates the numerator of the fraction.

■ **DEFINITION: Decimal Fraction.** A *decimal fraction* is a fractional notation that uses the decimal point and the place values to its right to represent a fraction whose denominator is 10 or some power of 10, such as 100, 1,000, and so on. A decimal fraction is also referred to as a decimal, a decimal number, or a number using decimal notation.

■ **DEFINITION: Mixed-Decimal Fraction.** A *mixed-decimal fraction* is a notation used to represent a decimal fraction that contains a whole number part as well as a fractional part.

EXAMPLE Rewrite the decimal fractions and mixed-decimal fractions as proper fractions or mixed numbers:

$$0.4, \quad 0.18, \quad 0.014, \quad 3.7, \quad 4.103$$

The number of places to the right of the decimal indicates the number of zeros in the denominator of the fraction.

$$0.4 = \frac{4}{10} \qquad\qquad 3.7 = 3\frac{7}{10}$$

$$0.18 = \frac{18}{100} \qquad\qquad 4.103 = 4\frac{103}{1,000}$$

$$0.014 = \frac{14}{1,000}$$

We summarize our study of rational numbers or fractions in Fig. 3–9.

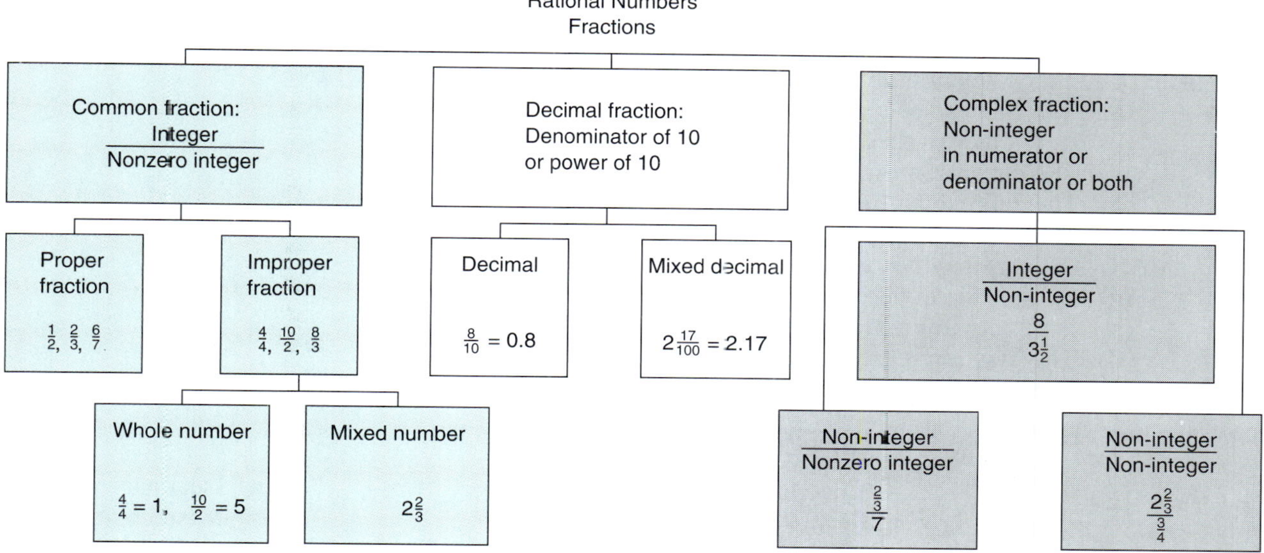

Figure 3–9

Fractions extend the numbers we have studied so far to another level. We often broadly interpret fractions as *rational* numbers. Rational numbers include whole numbers, natural numbers, integers, fractions, and decimal numbers because all these types of numbers can be expressed in fraction form. In case you were wondering, yes, fractions can be negative numbers. We will look at signed fractions and decimals later in this chapter.

SELF-STUDY EXERCISES 3–1

1 Write a fraction to represent the shaded portion in Exercises 1–7 (Figs. 3–10 to 3–16).

1.   **2.**   **3.**

Figure 3–10

Figure 3–11

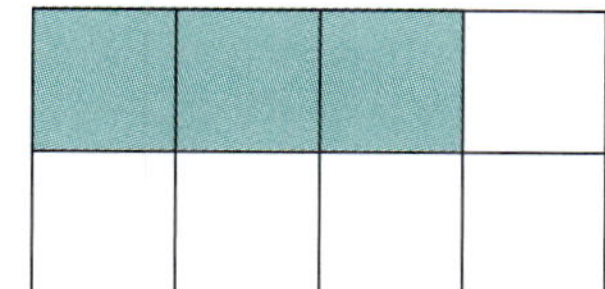

Figure 3–12

4.

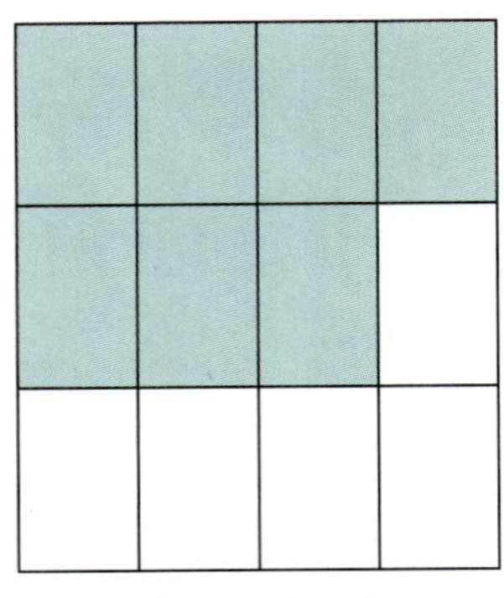

Figure 3–13

5.

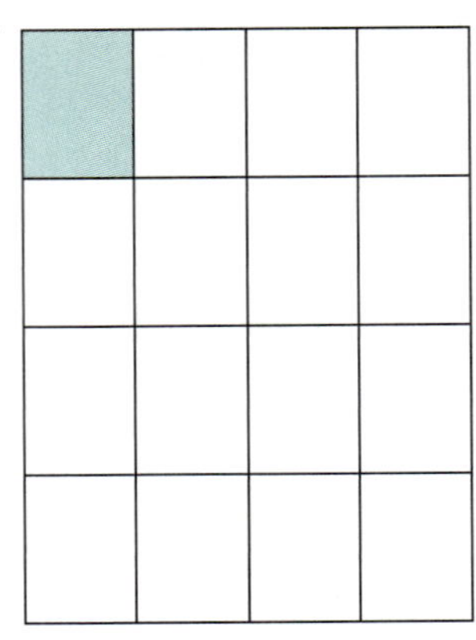

Figure 3–14

6.

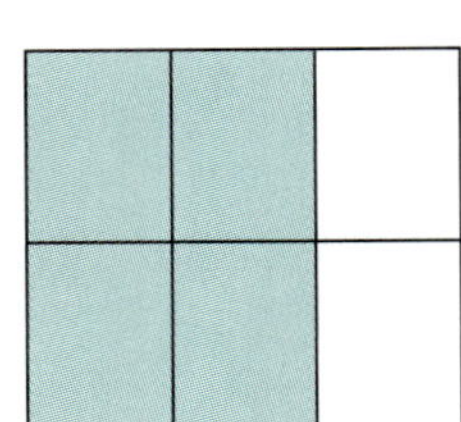

Figure 3–15

7.

Figure 3–16

Examine each common fraction in Exercises 8–11 and answer the following questions.

8. $\dfrac{5}{6}$

9. $\dfrac{8}{8}$

10. $\dfrac{11}{5}$

11. $\dfrac{12}{3}$

(a) One unit has been divided into how many parts?
(b) How many of these parts are used?
(c) The numerator of the fraction is ______.
(d) The denominator of the fraction is ______.
(e) The fraction can be read as ______ divided by ______.
(f) ______ is the divisor of the division.
(g) ______ is the dividend of the division.
(h) Is this common fraction proper or improper?
(i) This fraction represents ______ out of ______ parts.
(j) Does this fraction have a value less than, more than, or equal to 1?

Match the best description with the number in each exercise.

______ **12.** $2\dfrac{3}{7}$

______ **13.** 0.689

______ **14.** $\dfrac{8}{3}$

______ **15.** $14\dfrac{3}{16}$

______ **16.** 6.27

______ **17.** 3.7

______ **18.** 0.3

______ **19.** $\dfrac{3}{7}$

______ **20.** $\dfrac{\frac{3}{1}}{\frac{1}{2}}$

______ **21.** $\dfrac{1\frac{1}{2}}{3}$

______ **22.** $\dfrac{\frac{3}{4}}{\frac{2}{7}}$

(a) Decimal fraction
(b) Mixed number
(c) Complex fraction
(d) Mixed-decimal fraction
(e) Proper fraction
(f) Improper fraction

Rewrite the fractions as proper fractions or mixed numbers.

23. 6.27

24. 0.3

25. 3.7

26. 1.53

3-2 MULTIPLES, DIVISIBILITY, AND FACTOR PAIRS

Learning Outcomes

1 Find multiples of a natural number.
2 Determine the divisibility of a number.
3 Find all factor pairs of a natural number.

As we mentioned earlier, fractions indicate division, and multiplication and division are inverse operations. Before we examine computations with fractions, let's look at some relationships involving multiplication and division.

1 **Find Multiples of a Natural Number.**

If we count by threes, such as 3, 6, 9, 12, 15, 18. we obtain natural numbers that are *multiples* of 3. Each of the preceding numbers is a multiple of 3 because each is the product of 3 and a natural number; that is,

$$3 = 3 \times 1, \qquad 6 = 3 \times 2, \qquad 9 = 3 \times 3$$
$$12 = 3 \times 4, \qquad 15 = 3 \times 5, \qquad 18 = 3 \times 6$$

■ **DEFINITION: Multiple.** A *multiple* of a natural number is the product of that number and a natural number.

EXAMPLE Show that 2, 4, 6, 8, and 10 are multiples of 2.

$$2 = 2 \times 1, \qquad 4 = 2 \times 2, \qquad 6 = 2 \times 3$$
$$8 = 2 \times 4, \qquad 10 = 2 \times 5$$

■ **DEFINITION: Even Number.** Natural numbers that are multiples of 2 are *even numbers.*

■ **DEFINITION: Odd Number.** Natural numbers that are not multiples of 2 are *odd numbers.*

EXAMPLE When we organize documents to be printed on both sides of a paper, we normally put odd-numbered pages on the front of the paper and even-numbered pages on the back. Also, it is common practice to start each new unit or chapter on the front of a sheet of paper, even if this creates a preceding blank page. Assign page numbers to the document based on these guidelines.

> Chapter 1, 15 pages;   Chapter 2, 17 pages;
> Chapter 3, 24 pages;   Chapter 4, 15 pages

(a) What page number will start Chapter 2?
(b) How many pages will the entire document include?
(c) How many blank pages will be included in the document?

(a) Chapter 1 has an odd number of pages, so page 16 will be blank. Chapter 2 will begin on page 17.
(b) Chapter 1, 16 pages   including 1 blank page
    Chapter 2, 18 pages   including 1 blank page
    Chapter 3, 24 pages   including no blank pages
    Chapter 4, 16 pages   including 1 blank page
(c) There will be three blank pages in the total document, 1 each in Chapters 1, 2, and 4.

2 **Determine the Divisibility of a Number.**

As we continue to develop our number sense an important skill to acquire is to be able to determine when a number will divide into another number with no remainder. We say that a number is *divisible* by another number if the quotient has no remainder or if the dividend is a multiple of the divisor.

In developing this skill, we want to be able to determine divisibility by *inspection*. This means that we can examine the number being divided (dividend) and decide if it is divisible by a divisor without actually having to perform the division.

Is 35 divisible by 7?

35 is divisible by 7 if $35 \div 7$ has no remainder or if 35 is a multiple of 7.

$$35 \div 7 = 5 \qquad \text{or} \qquad 35 = 5 \times 7$$

Yes, 35 is divisible by 7.

A holiday party is planned for 47 first-grade students. The teachers have a gross (144) of assorted party favors. Will each child receive an equal number of favors with none left over?

This problem is really asking whether 144 is divisible by 47.

$$144 \div 47 = 3R3 \qquad \text{or} \qquad 144 \div 47 = 3.063829787 \qquad \text{By calculator.}$$

$144 \div 47$ has a remainder, and the calculator answer is not a natural number; therefore, 144 is not divisible by 47. **No, there will be 3 party favors left over.**

A number of rules or tests can help us decide by inspection if certain numbers are divisible by other numbers.

Tests for divisibility:

A number is divisible by

1. 2 if the last digit is an even number (0, 2, 4, 6, or 8).
2. 3 if the sum of its digits is divisible by 3.
3. 4 if the last two digits form a number that is divisible by 4.
4. 5 if the last digit is 0 or 5.
5. 6 if the number is divisible by *both 2 and* 3.
6. 7 if the division has no remainder.
7. 8 if the last three digits form a number divisible by 8.
8. 9 if the sum of its digits is divisible by 9.
9. 10 if the last digit is 0.

Use the tests for divisibility to identify which number in each pair is divisible by the given divisor.

Numbers	Divisor	Answer
874 or 873	2	**874;** the last digit is an even digit (4).
275 or 270	2	**270;** the last digit is an even digit (0).
427 or 423	3	**423;** the sum of the digits is divisible by 3: $4 + 2 + 3 = 9$.
5,912 or 5,913	4	**5,912;** the last two digits form a number divisible by 4: $12 \div 4 = 3$.
80 or 82	5	**80;** the last digit is 0.
56 or 65	5	**65;** the last digit is 5.
804 or 802	6	**804;** the last digit is even and the sum of the digits is divisible by 3.
58 or 56	7	**56;** it divides by 7 with no remainder.
3,160 or 3,162	8	**3,160;** the last three digits form a number divisible by 8: $160 \div 8 = 20$.
477 or 475	9	**477;** the sum of the digits is divisible by 9: $4 + 7 + 7 = 18$.
182 or 180	10	**180;** the last digit is 0.

CHAPTER 3   Fractions and Percents

3   **Find All Factor Pairs of a Natural Number.**

The *natural numbers* are the counting numbers—1, 2, 3, 4, 5, 6, 7, 8, 9, 10, 11, 12, and so forth. The natural numbers can also be described as the whole numbers excluding zero.

Any natural number can be expressed as the product of two natural numbers. These two natural numbers are called a *factor pair* of the number.

> ■   **DEFINITION: Factor Pair.**   A *factor pair* of a natural number consists of two natural numbers whose product equals the given natural number.

Every natural number greater than 1 has at least one factor pair, 1 and the number itself.

$$1 \text{ and } 3 \text{ form a factor pair for } 3: 1 \times 3 = 3.$$

$$1 \text{ and } 5 \text{ form a factor pair for } 5: 1 \times 5 = 5.$$

Many natural numbers have more than one factor pair. To find all the factor pairs of a number, we initially examine each natural number that is less than the given number. For example, to list all number pairs of the number 12, we start with the pair 1×12. Every number has a factor pair of the number and 1. Next, we examine each number from 2 to 11.

2	12 is divisible by 2	$12 \div 2 = 6$	2×6 is a factor pair of 12.
3	12 is divisible by 3	$12 \div 3 = 4$	3×4 is a factor pair of 12.
4	12 is divisible by 4	$12 \div 4 = 3$	4×3 is a factor pair of 12.
5	12 is not divisible by 5		
6	12 is divisible by 6	$12 \div 6 = 2$	6×2 is a factor pair of 12.
7	12 is not divisible by 7		
8	12 is not divisible by 8		
9	12 is not divisible by 9		
10	12 is not divisible by 10		
11	12 is not divisible by 11		

Are the factor pairs 2×6 and 6×2 different pairs? No. Both pairs have the same numbers and since multiplication is commutative, we count that as one pair. Similarly, 3×4 and 4×3 are the same factor pair. Is it really necessary to examine every number less than 12? No. Once we get the first repeat, 4×3, we can assume we have found all the factor pairs of the number.

$$\text{The factor pairs of 12 are } 1 \times 12, 2 \times 6, \text{ and } 3 \times 4.$$

Here is a strategy for finding all factor pairs of a natural number.

To find all factor pairs of a natural number:

1. Write the factor pair of 1 and the number.
2. Check to see if the number is divisible by 2. If so, write the factor pair of 2 and the quotient of the beginning natural number and 2.
3. Check each natural number in turn for divisibility until you reach a number that has already been found as a quotient in a previous factor pair.

Once we have listed all factor pairs of a number, we can then list all factors of a number. From the factor pairs, we list every different factor that appears in any factor pair. The factors of 12 are 1, 2, 3, 4, 6, and 12.

EXAMPLE List all the factor pairs of 18, then write each distinct factor in order from smallest to largest.

1×18
2×9 18 is divisible by 2: $18 \div 2 = 9$.
3×6 18 is divisible by 3: $18 \div 3 = 6$.
         18 is not divisible by 4 or 5.
         18 is divisible by 6; 6 was found in the factor pair 3×6, so we stop.

Factors: 1, 2, 3, 6, 9, 18

EXAMPLE List all the factor pairs of 48, then write each distinct factor in order from smallest to largest.

1×48
2×24 48 is divisible by 2: $48 \div 2 = 24$.
3×16 48 is divisible by 3: $48 \div 3 = 16$.
4×12 48 is divisible by 4: $48 \div 4 = 12$.
         48 is not divisible by 5.
6×8 48 is divisible by 6: $48 \div 6 = 8$.
         48 is not divisible by 7.
         48 is divisible by 8; 8 was found in the factor pair 6×8, so we stop.

Factors: 1, 2, 3, 4, 6, 8, 12, 16, 24, 48

EXAMPLE Telephone message pads are purchased in bulk as cases of 144 pads. To sell the pads in smaller quantities, an office supply store needs to repackage them, with no unpackaged pads remaining. What packaging options does the store have?

Find all factors of 144 by first finding all factor pairs.

1×144
2×72
3×48
4×36
5 144 is not divisible by 5.
6×24
7 144 is not divisible by 7.
8×18
9 144 is not divisible by 9.
10 144 is not divisible by 10.
11 144 is not divisible by 11.
12×12

The factors of 144 are 1, 2, 3, 4, 6, 8, 12, 18, 24, 36, 48, 72, 144.

The factors of 144 represent the choices for packaging the telephone message pads; however, some options may be considered impractical.

SELF-STUDY EXERCISES 3–2

1 Show that each number is a multiple of the first number by listing its factorization with the first number.

1. 5, 10, 15, 20, 25, 30   **2.** 6, 12, 18, 24, 30, 36   **3.** 8, 16, 24, 32, 40, 48
4. 9, 18, 27, 36, 45, 54   **5.** 10, 20, 30, 40, 50, 60   **6.** 30, 60, 90, 120, 150, 180

Find 5 multiples of the numbers by listing their factorizations, starting with the given number times 1.

7. 5   **8.** 12   **9.** 7
10. 3   **11.** 50   **12.** 4

2   Are these numbers divisible by the given number? Explain.

13. 2,434 by 6   **14.** 230 by 5   **15.** 2,434 by 4
16. 1,221 by 3   **17.** 756 by 7   **18.** 920 by 8
19. 621 by 3   **20.** 426 by 6   **21.** 1,232 by 2

3   Find all the factor pairs for each number.

22. 24   **23.** 36   **24.** 45   **25.** 32   **26.** 16
27. 27   **28.** 20   **29.** 30   **30.** 12   **31.** 8
32. 4   **33.** 15   **34.** 81   **35.** 64   **36.** 38
37. 46   **38.** 51   **39.** 18   **40.** 72

3-3   PRIME AND COMPOSITE NUMBERS

Learning Outcomes

1   Factor prime and composite numbers.
2   Determine the prime factorization of composite numbers.

1   Factor Prime and Composite Numbers.

When all the factor pairs of natural numbers are listed, some numbers will have only one pair of factors, the number itself and 1. These numbers form a special group of numbers called prime numbers.

> ■ **DEFINITION: Prime Number.** A *prime number* is a whole number greater than 1 that has only one factor pair, the number itself and 1.

Note that 1 is not a prime number because it has only one factor.

EXAMPLE   Identify the prime numbers by examining the factor pairs.

(a) 8   (b) 1   (c) 3   (d) 9   (e) 7

(a) **8 is *not* a prime number** because its factor pairs are 1×8 and 2×4.
(b) **1 is *not* a prime number** because a prime must be greater than 1 or because it does not have a factor pair.
(c) **3 is a prime number** because it has one factor pair, 1×3.
(d) **9 is *not* a prime number** because its factor pairs are 1×9 and 3×3.
(e) **7 is a prime number** because it has one factor pair, 1×7.

> ■ **DEFINITION: Composite Number.** A *composite number* is a whole number greater than 1 that is not a prime number.

In the preceding example, 8 and 9 are composite numbers. A composite number has at least one factor pair other than itself and 1.

EXAMPLE   Identify the composite numbers by examining the factor pairs of the numbers.

(a) 4   (b) 10   (c) 13   (d) 12   (e) 5

(a) **4 is a composite number** because its factor pairs are 1×4 and 2×2.
(b) **10 is a composite number** because its factor pairs are 1×10 and 2×5.
(c) **13 is *not* a composite number** because its only factor pair is 1×13. It is a prime number.
(d) **12 is a composite number** because its factor pairs are 1×12, 2×6, and 3×4.
(e) **5 is *not* a composite number** because its only factor pair is 1×5. It is a prime number.

<table>
<tr><td>Tip!</td><td>Prime Numbers Less Than 50.</td></tr>
</table>

We can find all the prime numbers that are 50 or less using an ancient technique developed by the mathematician Eratosthenes.

Step 1. List the numbers from 1 through 50.

Step 2. Eliminate numbers that are not prime using the systematic process:

(a) 1 is not prime. Eliminate 1.   (b) 2 is prime. Eliminate all multiples of 2.
(c) 3 is prime. Eliminate all multiples of 3.   (d) 4 has already been eliminated.
(e) 5 is prime. Eliminate all multiples of 5.   (f) 6 has already been eliminated.
(g) 7 is prime. Eliminate all multiples of 7.

Step 3. Circle remaining numbers as prime numbers.

1	②	③	4	⑤	6	⑦	8	9	10
⑪	12	⑬	14	15	16	⑰	18	⑲	20
21	22	㉓	24	25	26	27	28	㉙	30
㉛	32	33	34	35	36	�37	38	39	40
㊀41	42	㊀43	44	45	46	㊀47	48	49	50

All numbers not already eliminated are prime. Why? The numbers 8, 9, and 10 have already been eliminated as multiples of 2, 3, and 5, respectively. Multiples of 11 that are less than 50 have already been eliminated: $11 \times 2 = 22$, $11 \times 3 = 33$, $11 \times 4 = 44$. $11 \times 5 = 55$ is greater than 50. Similarly, all other composite numbers have already been eliminated.

2 Determine the Prime Factorization of Composite Numbers.

A composite number can be expressed as a product of prime numbers. *Prime factorization* refers to writing a composite number as the product of *only* prime numbers. In this case, the factors are *prime factors*.

To find the prime factors of a composite number:

1. Test each prime number to see if the composite number is divisible by the prime.
2. Make a factor pair using the first prime number that passes the test in Step 1.
3. Carry forward the prime factors and test the remaining factors by repeating Steps 1 and 2.

EXAMPLE Find the prime factorization of 30.

$$30 = 2 \times 15$$

first prime ⎯⎯⎯⎯

30 is divisible by 2. Factor 30 into a factor pair using its smallest prime factor, 2.

$$30 = 2 \times 3 \times 5$$

last primes ⎯⎯⎯⎯

Carry the prime factor 2 forward. Factor the composite 15 using its smallest prime factor, 3. Because 5 is also prime, the factoring is complete.

The prime factorization of 30 is $2 \times 3 \times 5$.

EXAMPLE Find the prime factorization of 16.

$$16 = 2 \times 8$$

first prime ⟶

Factor 16 into two factors using its smallest prime factor.

$$16 = 2 \times 2 \times 4$$

second prime ⟶

Factor 8 into two factors using its smallest prime factor.

$$16 = 2 \times 2 \times 2 \times 2$$

last two primes ⟶

Factor 4 into two factors using its smallest prime factor.

The prime factorization of 16 is 2 × 2 × 2 × 2. We can write this expression in exponential notation as 2^4.

SELF-STUDY EXERCISES 3–3

1 List all factor pairs, then write the factors in order from smallest to largest.

1. 14 **2.** 22 **3.** 11 **4.** 17 **5.** 18 **6.** 24

2 Find the prime factorization.

7. 50 **8.** 52 **9.** 225 **10.** 125 **11.** 100 **12.** 200
13. 65 **14.** 75 **15.** 121 **16.** 144

Write the prime factorization using exponential notation.

17. 568 **18.** 112 **19.** 124 **20.** 164 **21.** 72 **22.** 900

3–4 LEAST COMMON MULTIPLE AND GREATEST COMMON FACTOR

Learning Outcomes

1 Find the least common multiple of two or more numbers.
2 Find the greatest common factor of two or more numbers.

1 Find the Least Common Multiple of Two or More Numbers.

Multiples, especially the *least common multiple*, are useful when we add and subtract fractions with different denominators. The least common multiple is the lowest common denominator.

If we count by threes, we obtain natural numbers that are multiples of 3 ($3 \times 1 = 3, 3 \times 2 = 6$, and so on). If we count by fives, we obtain natural numbers that are multiples of 5 ($5 \times 1 = 5, 5 \times 2 = 10$, and so on).

Multiples of 3:   3, 6, 9, 12, 15, 18, 21, 24, 27, 30, 33, 36, 39, . . .

Multiples of 5:   5, 10, 15, 20, 25, 30, 35, 40, . . .

The common factors in these lists that are less than 40 are 15 and 30. The smallest multiple common to both sets of numbers is 15, so 15 is the *least common multiple* of 3 and 5. It is the smallest number divisible by both 3 and 5.

■ **DEFINITION: Least Common Multiple.** The *least common multiple (LCM)* of two or more natural numbers is the smallest number that is a multiple of each number. It is divisible by each number.

Prime factorization can also be used to find the least common multiple of two or more numbers.

The least common multiple of numbers can be found by using the prime factorization of the numbers:

1. List the prime factorization of each number using exponential notation.
2. List the prime factorization of the least common multiple by including the prime factors appearing in *each* number. If a prime factor appears in more than one number, use the factor with the *largest* exponent.
3. Write the resulting expression in standard notation.

EXAMPLE   Find the least common multiple of 12 and 40 by prime factorization.

$$12 = 2 \times 2 \times 3 \qquad = 2^2 \times 3$$ Prime factorization of 12.

$$40 = 2 \times 2 \times 2 \times 5 = 2^3 \times 5$$ Prime factorization of 40.

$$\text{LCM} = 2^3 \times 3 \times 5$$ Prime factorization of LCM.

$$\text{LCM} = 120$$ LCM in standard notation.

2 Find the Greatest Common Factor of Two or More Numbers.

A *common factor* is a factor common to two or more numbers or products. Common factors, especially the *greatest common factor,* are useful in simplifying or reducing fractions.

■ **DEFINITION: Greatest Common Factor.** The *greatest common factor (GCF)* of two or more numbers is the largest factor common to each number. Each number is divisible by the GCF.

Let's take the numbers 30 and 42. The prime factors are

$$30 = 2 \times 3 \times 5, \qquad 42 = 2 \times 3 \times 7$$

The *common* prime factors of both 30 and 42 are 2 and 3, which represent the composite factor 6. The *greatest* common factor is the product of the common prime factors, $2 \times 3 = 6$. So 6 is the GCF of 30 and 42. Each number is divisible by 6; that is, 6 is the largest number that divides evenly into both 30 and 42.

To find the greatest common factor (GCF) of two or more natural numbers:

1. List the prime factorization of each number using exponential notation when appropriate.
2. List the prime factorization of the greatest common factor by including each prime factor appearing in *every* number. If a prime factor appears in more than one number, use the factor with the *smallest* exponent. If there are no common prime factors, the GCF is 1.
3. Write the resulting expression in standard notation.

EXAMPLE   Find the greatest common factor of 15, 30, and 45.

$$15 = 3 \times 5 \qquad = 3 \times 5$$ Prime factorization of 15.

$$30 = 2 \times 3 \times 5 = 2 \times 3 \times 5$$ Prime factorization of 30.

$$45 = 3 \times 3 \times 5 = 3^2 \times 5$$ Prime factorization of 45.

$$\text{GCF} = 3 \times 5$$ Common prime factors.

$$\text{GCF} = 15$$ GCF in standard notation.

EXAMPLE   Find the greatest common factor of 10, 12, and 13.

$$10 = 2 \times 5 \qquad\quad = 2 \times 5$$ Prime factorizat on of 10.
$$12 = 2 \times 2 \times 3 = 2^2 \times 3$$ Prime factorizat on of 12.
$$13 = 13 \qquad\qquad = 13$$ Prime factorization of 13.

GCF = 1   No common prime factors.

Tip!	*LCM Versus GCF.*

The least common multiple (LCM) and greatest common factor (GCF) are easily confused. Because multiples of a number are the products of the number and any natural number, multiples will be as large as or larger than the original number. The LCM is the *smallest* of the "large or larger" multiples of the original number.

Because factors of a number are the same as or smaller than the given number, the GCF is the *largest* of the "small or smaller" factors of the original number.

SELF-STUDY EXERCISES 3–4

1   Find the least common multiple.

1. 2 and 3
2. 5 and 6
3. 7 and 8
4. 3 and 4
5. 18 and 60
6. 10 and 12
7. 12 and 24
8. 9 and 18
9. 4, 8, and 12
10. 20, 25, and 35
11. 3, 9, and 27
12. 2, 8, and 16
13. 6, 15, and 18
14. 20, 24, and 30
15. 12, 18, and 20
16. 6, 10, and 12
17. 8, 12, and 32
18. 8, 12, and 18
19. 10, 15, and 20
20. 30, 50, and 60
21. 6, 11, and 33
22. 8, 13, and 39

2   Find the greatest common factor.

23. 18 and 24
24. 15 and 25
25. 10, 11, and 14
26. 12, 13, and 16
27. 6, 8, and 14
28. 4, 10, and 18
29. 30 and 45
30. 40 and 55
31. 36, 60, and 216
32. 18, 30, and 108

MATHEMATICS IN THE WORKPLACE

Astronomy: Planet Alignment

Every body in the universe exerts a gravitational pull on other objects. The more mass a body has, the greater the tug. For example, the Earth has more mass than the Moon, so its gravitational pull on the Moon is stronger than the Moon's gravitational pull on the Earth. As a result, the Moon stays in orbit around the Earth. Planets have huge gravitational forces, and when planets are in alignment, their combined forces of gravity are tremendous. The strength of their combined gravities on another object (say, an asteroid) depends on four factors:

1. mass of the planets
2. planets' distance apart
3. closest planet's distance from the asteroid
4. planet's and asteroid's orbital angle from the ecliptic plane

To be in alignment means that the planets and the Sun lie on a straight line in a bird's-eye view of the ecliptic plane. Planets are usually closest to each other when in alignment. However, some planets' orbits tilt a few degrees out of the ecliptic plane, so their minimal distance does not occur when those planets are in alignment.

The least common multiple (LCM) of the planets' orbital periods around the Sun is used to calculate how often this alignment occurs. Venus and Earth are in alignment about every 45.0 years. This alignment is computed by first finding the prime factorization of their orbital periods. For Earth, 365 days = $5 \cdot 73$. For Venus, 225 days = $3^2 \cdot 5^2$. The LCM of 365 and 225 is $3^2 \cdot 5^2 \cdot 73 = 16{,}425$ days = 45.0 years. Thus, about every 45 years Earth, Venus, and the Sun are collinear in the ecliptic plane.

Planet alignment is important when NASA and other space agencies decide on a spacecraft's launch-date window and its trajectory. The combined gravitational forces of aligned planets could greatly affect the spacecraft's flight path and could even cause a spacecraft to hit a planet.

Space agencies also use planet alignment LCM to conserve spacecraft fuel, flight distance, and time by launching when the earth is closest to the target planet. For example, NASA launched its Mars *Global Surveyor* and Mars *Pathfinder* spacecraft on November 7 and December 4, 1996, respectively, to take advantage of the Earth–Mars alignment that occurs about every 2 years. Planet alignment also played a part in choosing the October 6, 1997 launch date and path of the *Cassini* spacecraft bound for Saturn.

When larger planets align, the combined gravitational effect is *enormous*. Some astronomers believe that Pluto and its moon Charon may have been tugged into orbit around the Sun by the alignment of any two or three of the large outer planets: Neptune, Uranus, Saturn, and Jupiter. Because planets far away from the sun have long orbital periods, alignment of two planets is rare, and alignment of three is even more rare. *Voyager II* took a rare photo as it left our solar system: seven of our nine planets in the same camera frame. Although they were not in perfect alignment, even a close alignment of seven planets happens once in a lifetime.

<table><tr><td>3–5</td><td>EQUIVALENT FRACTIONS AND DECIMALS</td></tr></table>

Learning Outcomes

1 Write equivalent fractions with higher denominators.

2 Write equivalent fractions with lowest denominators.

3 Change decimals to fractions.

4 Change fractions to decimals.

Let's look at numbers in fractional form once more. There are many different ways to express the same value in fractional form. For example, the whole number 1 can be written as $\frac{1}{1}, \frac{4}{4}, \frac{7}{7}, \frac{15}{15}$, and so on.

Fractions are equivalent if they represent the same value. Let's compare the illustrations in Fig. 3–17. In Fig. 3–17, line a is one whole unit divided into only 1 part ($\frac{1}{1}$). Line b is one unit divided into 2 parts ($\frac{2}{2}$). Lines c and d are divided into 4 parts ($\frac{4}{4}$), line e is divided into 8 parts ($\frac{8}{8}$), and line f is divided into 16 parts ($\frac{16}{16}$).

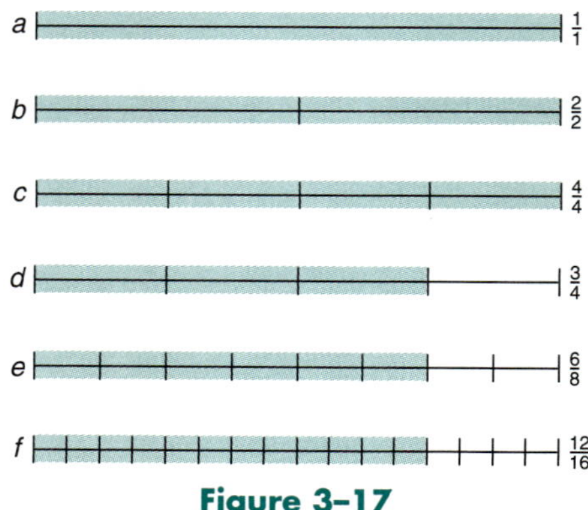

Figure 3–17

Now, look at lines d, e, and f. Line d is divided into 4 parts and 3 of them are shaded. Line e is divided into 8 parts and 6 of them are shaded. Line f is divided into 16 parts and 12 of them are shaded.

Look at the shaded portions of lines d, e, and f. They are the same length, even though they are divided into a different number of parts. Thus, $\frac{3}{4}$, $\frac{6}{8}$, and $\frac{12}{16}$ are equivalent fractions since they represent the same shaded part of one whole unit. That is,

$$\frac{3}{4} = \frac{6}{8} = \frac{12}{16}$$

1 Write Equivalent Fractions with Higher Denominators.

Equivalent fractions are in the same "family of fractions." The first member of the "family" is the fraction in lowest terms; that is, no whole number divides evenly into *both* the numerator and denominator except the number 1. Other "family members" are found by multiplying both the numerator and denominator by the same number.

EXAMPLE   Find five fractions that are equivalent to $\frac{1}{2}$.

$$\frac{1 \times 2}{2 \times 2} \quad \text{or} \quad \frac{1}{2} \times \frac{2}{2} = \frac{2}{4} \qquad\qquad \frac{1 \times 5}{2 \times 5} \quad \text{or} \quad \frac{1}{2} \times \frac{5}{5} = \frac{5}{10}$$

$$\frac{1 \times 3}{2 \times 3} \quad \text{or} \quad \frac{1}{2} \times \frac{3}{3} = \frac{3}{6} \qquad\qquad \frac{1 \times 6}{2 \times 6} \quad \text{or} \quad \frac{1}{2} \times \frac{6}{6} = \frac{6}{12}$$

$$\frac{1 \times 4}{2 \times 4} \quad \text{or} \quad \frac{1}{2} \times \frac{4}{4} = \frac{4}{8}$$

Tip!	*Multiplication, Division, and 1.*

In the preceding example $\frac{1}{2}$ is multiplied by a fraction whose value is 1, and 1 times any number does not change the value of that number. Written symbolically,

$$\frac{n}{n} = 1 \qquad \text{and} \qquad 1 \times n = n, \, n \neq 0$$

Fractions in the same family can be generated by multiplying the fraction by 1 in the form of $\frac{2}{2}$, $\frac{3}{3}$, $\frac{4}{4}$, and so on.

We frequently need to find a fraction with a specific denominator that is equivalent to our given fraction. In this case, we first decide what number multiplied by the original denominator will give the new denominator, then we multiply the numerator and denominator by that number. This is allowed because of the **fundamental principle of fractions.**

■ **DEFINITION: Fundamental Principle of Fractions.** If the numerator and denominator of a fraction are multiplied by the same nonzero number, the value of the fraction remains unchanged.

EXAMPLE   Find a fraction equivalent to $\frac{2}{3}$ that has a denominator of 12.

3 times what number is 12? Or 12 divided by 3 is what number? The answer is 4. Then

$$\frac{2 \times 4}{3 \times 4} \qquad \text{or} \qquad \frac{2}{3} \times \frac{4}{4} = \frac{8}{12}$$

To change a fraction to an equivalent fraction with a larger denominator:

1. Divide the larger denominator by the original denominator.
2. Multiply the original numerator and denominator by the quotient found in Step 1.

EXAMPLE Change $\frac{5}{8}$ to an equivalent fraction whose denominator is 32.

$$\frac{5}{8} = \frac{?}{32}, \qquad 32 \div 8 = 4 \text{ and } 4 \times 5 = 20$$

$$\frac{5}{8} = \frac{20}{32} \qquad \text{or} \qquad \frac{5}{8} \times \frac{4}{4} = \frac{20}{32}$$

2 Write Equivalent Fractions with Lowest Denominators.

Because each fraction has an unlimited number of equivalent fractions, we usually work with decimal equivalents or fractions in *lowest terms.*

By *lowest terms,* we mean that no *whole* number other than 1 divides evenly into both the numerator and denominator. Another way of saying this is that the numerator and denominator have no common factors other than 1. When we find an equivalent fraction with smaller numbers and no common factors in the numerator and denominator, we have *reduced to lowest terms.*

To change a fraction to an equivalent fraction with a smaller denominator or to reduce a fraction to lowest terms:

1. Find a common factor greater than 1 for the numerator and denominator.
2. Divide both the numerator and denominator by this common factor.
3. Continue until the fraction is in lowest terms or has the desired smaller denominator.

Note: To reduce to lowest terms in the fewest steps, find the greatest common factor (GCF) in Step 1.

EXAMPLE Reduce $\frac{8}{10}$ to lowest terms.

Prime factors of 8: $2 \times 2 \times 2$ or 2^3
Prime factors of 10: 2×5
The greatest common factor (GCF) is 2.

$$\frac{8 \div 2}{10 \div 2} \qquad \text{or} \qquad \frac{8}{10} \div \frac{2}{2} = \frac{4}{5}$$

Tip! | *Reducing and the Properties of 1.*

To reduce the fraction $\frac{8}{10}$, we divide by the whole number 1 in the form of $\frac{2}{2}$. $\frac{8}{10} \div \frac{2}{2} = \frac{4}{5}$. A nonzero number divided by itself is 1, and to divide a number by 1 does not change the value of the number. Symbolically,

$$\frac{n}{n} = 1 \qquad \text{and} \qquad n \div 1 = n;\, n \neq 0$$

EXAMPLE   Reduce $\frac{18}{24}$ to lowest terms.

Prime factors of 18: $2 \times 3 \times 3$ or 2×3^2
Prime factors of 24: $2 \times 2 \times 2 \times 3$ or $2^3 \times 3$
The GCF is 2×3 or 6.

$$\frac{18 \div 6}{24 \div 6} \qquad \text{or} \qquad \frac{18}{24} \div \frac{6}{6} = \frac{3}{4}$$

Tip!	***Do You Have to Use the GCF to Reduce to Lowest Terms?***

A fraction can be reduced to lowest terms in the fewest steps by using the *greatest common factor;* however, it can still be reduced correctly using any common factor; this just takes a few more steps.

$$\frac{18 \div 2}{24 \div 2} = \frac{9}{12}$$

$$\frac{9 \div 3}{12 \div 3} = \frac{3}{4}$$

3   Change Decimals to Fractions.

When a decimal is written as a fraction, the number of digits that follow the decimal point determines the denominator of the fraction. In the decimal 0.23, two digits follow the decimal, so the denominator will be 100 (100 has two zeros). Decimal numbers with one digit after the decimal have one zero in the denominator. Decimal numbers with three digits after the decimal have three zeros in the denominator, and so on.

To convert a decimal number to a fraction or mixed number in lowest terms:

1. Write the digits without the decimal point as the numerator.
2. Write the denominator as a power of 10 with as many zeros as there are places after the decimal point.
3. Reduce and, if the fraction is improper, convert to a mixed number.

EXAMPLE   Write as a fraction 0.4, and 0.075.

$$0.4 = \frac{4}{10} = \frac{2}{5} \qquad \text{Tenths indicates a denominator of 10.}$$

$$0.075 = \frac{75}{1,000} = \frac{3}{40} \qquad \text{Thousandths indicates a denominator of 1,000.}$$

4   Change Fractions to Decimals.

With increased calculator use, we find it convenient to work with decimals. Therefore, we often change fractions to decimals when we make certain computations.

The bar separating the numerator and denominator of a fraction indicates division: $\frac{2}{5}$ also means $2 \div 5$ or $5\overline{)2}$.

If we write a decimal after the 2 and attach a zero, we can divide.

$$\begin{array}{r} 0.4 \\ 5\overline{)2.0} \end{array}$$

Therefore, $\frac{2}{5} = 0.4$.

To convert a fraction to a decimal number:

1. Place a decimal point after the numerator.
2. Divide the numerator by the denominator using long division.
3. Attach zeros to the numerator as needed for division.

EXAMPLE   Change $\frac{7}{8}$ to a decimal.

$$7 \div 8 \qquad \text{or} \qquad \begin{array}{r} 0.875 \\ 8\overline{)7.000} \\ \underline{6\,4} \\ 60 \\ \underline{56} \\ 40 \\ \underline{40} \end{array}$$

Attach zeros until the division terminates; that is, it has no remainder.

To convert a mixed number to a decimal number:

1. The whole-number part remains the same.
2. Convert only the fraction part using the same procedure as before.

EXAMPLE   Change $3\frac{2}{5}$ to a mixed decimal.

$$\frac{2}{5} = \begin{array}{r} 0.4 \\ 5\overline{)2.0} \end{array} \qquad \text{or} \qquad 0.4$$

Then, $3\frac{2}{5} = 3.4$.

When we change some fractions to decimals, the quotient does not terminate.

EXAMPLE   Change $\frac{1}{3}$ and $\frac{4}{11}$ to decimals.

$$1 \div 3 \qquad \text{or} \qquad \begin{array}{r} 0.333 \\ 3\overline{)1.000} \\ \underline{9} \\ 10 \\ \underline{9} \\ 10 \\ \underline{9} \\ 1 \end{array} \qquad \text{or} \qquad \mathbf{0.33\frac{1}{3}}$$

$$4 \div 11 \qquad \text{or} \qquad \begin{array}{r} 0.363636 \\ 11\overline{)4.000000} \\ \underline{3\,3} \\ 70 \\ \underline{66} \\ 40 \\ \underline{33} \\ 70 \\ \underline{66} \\ 40 \\ \underline{33} \\ 70 \\ \underline{66} \\ 4 \end{array} \qquad \text{or} \qquad \mathbf{0.36\frac{4}{11}}$$

When fractions are changed into decimals that do not terminate, the decimals are called *repeating decimals*. Repeating decimals can be written with a line over the digit(s) that repeat or three dots at the end to indicate that they repeat. The decimal equivalent can be rounded to any desirable place.

$$\frac{1}{3} = 0.\overline{3} \quad \text{or} \quad 0.333333 \ldots \qquad \frac{4}{11} = 0.\overline{36} \quad \text{or} \quad 0.3636 \ldots$$

Tip!	***Changing Fractions to Decimals on the Calculator.***

To convert fractions to decimal numbers on your calculator divide the numerator by the denominator (divisor). Always enter the dividend (numerator) *first*. Let's convert the fractions in the preceding example to decimals using our calculator.

$$\text{Options:} \quad 1 \div 3 = \Rightarrow 0.333333333$$
$$4 \div 11 = \Rightarrow 0.36363636$$

SELF-STUDY EXERCISES 3–5

1

1. Find five fractions that are equivalent to $\frac{4}{5}$.

2. Find five fractions that are equivalent to $\frac{7}{10}$.

3. Find a fraction that is equivalent to $\frac{3}{4}$ and has a denominator of 24.

Find the equivalent fractions with the indicated denominators.

4. $\dfrac{3}{8} = \dfrac{?}{16}$

5. $\dfrac{4}{7} = \dfrac{?}{21}$

6. $\dfrac{9}{11} = \dfrac{?}{44}$

7. $\dfrac{1}{3} = \dfrac{?}{15}$

8. $\dfrac{5}{6} = \dfrac{?}{24}$

9. $\dfrac{7}{8} = \dfrac{?}{24}$

10. $\dfrac{2}{5} = \dfrac{?}{30}$

2 Reduce the fractions to lowest terms.

11. $\dfrac{4}{8}$

12. $\dfrac{6}{10}$

13. $\dfrac{12}{16}$

14. $\dfrac{10}{32}$

15. $\dfrac{16}{32}$

16. $\dfrac{28}{32}$

17. $\dfrac{20}{64}$

18. $\dfrac{2}{8}$

19. $\dfrac{8}{32}$

20. $\dfrac{12}{50}$

21. $\dfrac{10}{16}$

22. $\dfrac{4}{16}$

23. $\dfrac{24}{32}$

24. $\dfrac{12}{64}$

25. $\dfrac{14}{64}$

3 Change the decimals to their fraction or mixed-number equivalents, and reduce answers to lowest terms.

26. 0.5

27. 0.1

28. 0.2

29. 0.7

30. 0.25

31. 0.025

32. 3.9

33. 4.8

34. 0.378

35. 0.875

36. 0.375

37. 0.625

38. A measure of 0.75 in. represents what fractional part of an inch?

39. What fraction represents 0.1875 ft?

40. The length of a screw is 2.375 in. Represent this length as a mixed number.

41. An instrument weighs $0.83\frac{1}{3}$ lb. Write this as a fraction of a pound.

42. Some sheet metal is 0.3125 in. thick. What is the thickness expressed as a fraction?

43. A predrilled PC board is 3.125 in. long. Write this length as a mixed number.

4 Change these fractions and mixed numbers to decimals.

44. $\dfrac{2}{5}$ **45.** $\dfrac{3}{10}$ **46.** $\dfrac{7}{8}$ **47.** $\dfrac{5}{8}$ **48.** $\dfrac{9}{20}$

49. $\dfrac{49}{50}$ **50.** $\dfrac{21}{100}$ **51.** $3\dfrac{7}{8}$ **52.** $1\dfrac{7}{16}$ **53.** $4\dfrac{9}{16}$

Write these fractions as repeating decimals.

54. $\dfrac{2}{3}$ **55.** $\dfrac{3}{11}$ **56.** $\dfrac{7}{9}$ **57.** $\dfrac{5}{13}$

Write these fractions as decimals to the hundredths place. Express the remainder as a fraction.

58. $\dfrac{5}{6}$ **59.** $\dfrac{7}{12}$

60. An aerial map shows a building measuring $2\dfrac{3}{64}$ in. on one side. What is the measure of the side of the building in decimal numbers?

61. The property tax rate in one state is $4\dfrac{1}{2}\%$ of the assessed value. Express the tax rate as a mixed-decimal percent.

62. A plan allows a gap of $\dfrac{1}{8}$ in. between vinyl flooring and the wall for expansion. What is the gap measure in decimal notation?

3–6   IMPROPER FRACTIONS AND MIXED NUMBERS

Learning Outcomes

1 Convert improper fractions to whole or mixed numbers.

2 Convert mixed numbers and whole numbers to improper fractions.

1 Convert Improper Fractions to Whole or Mixed Numbers.

In Section 3–1, we noted that an improper fraction is a fraction whose value is equal to or greater than one unit, such as $\dfrac{6}{3}$, $\dfrac{15}{7}$, or $\dfrac{5}{5}$. Problems involving improper fractions are sometimes easier to solve if we change the improper fractions into whole numbers or mixed numbers.

> **To convert an improper fraction to a whole or mixed number:**
>
> Perform the division indicated. Express any remainder of the division as a fraction or decimal equivalent.

EXAMPLE   Convert $\dfrac{15}{3}$ to a whole or mixed number.

$$\dfrac{15}{3} \quad \text{means} \quad 15 \div 3 \quad \text{or} \quad 3\overline{)\begin{array}{c}5\\15\\\underline{15}\\0\end{array}}; \quad \dfrac{15}{3} = 5$$

If there is no remainder from the division, the improper fraction converts to a whole number.

Tip!	***Converting to a Whole or Mixed Number Is Different from Reducing.***

Do not confuse converting an improper fraction to a whole or mixed number with reducing to lowest terms. An improper fraction is in lowest terms if its numerator and denominator have no common factor. Therefore, the improper fraction $\dfrac{10}{7}$ is in lowest terms. The improper fraction $\dfrac{10}{4}$ is not in lowest terms. It will reduce to $\dfrac{5}{2}$, which is in lowest terms.

When converting an improper fraction to a whole or mixed number, we must make sure that the fractional part is in lowest terms. We can reduce to lowest terms either before dividing or after dividing.

EXAMPLE   Convert $\frac{28}{8}$ to a mixed number.

$$\frac{28}{8} = \frac{7}{2} \qquad 2\overline{)7} = 3\frac{1}{2}$$

or

$$\frac{28}{8} \qquad 8\overline{)28} = 3\frac{4}{8} = 3\frac{1}{2}$$

2 Convert Mixed Numbers and Whole Numbers to Improper Fractions.

To solve certain problems, it's sometimes useful to convert mixed numbers to improper fractions. In improper fractions, the numerator indicates the total number of parts considered. The denominator indicates the number of parts one unit has been divided into.

> *To convert a mixed number to an improper fraction:*
>
> 1. Multiply the denominator of the fractional part by the whole number.
> 2. Add the numerator of the fractional part to the product; this sum becomes the numerator of the improper fraction.
> 3. The denominator of the improper fraction is the same as the denominator of the fractional part of the mixed number.

EXAMPLE   Change $6\frac{2}{3}$ to an improper fraction.

$$6\frac{2}{3} = \frac{(3 \times 6) + 2}{3} = \frac{20}{3}$$

Whole numbers can also be written as improper fractions. First, write the whole number as a fraction with a denominator of 1. Then, change the whole-number fraction divided by 1 to an improper fraction having any denominator by using the procedures described in Section 3–5.

EXAMPLE   Change 8 to fifths.

$$8 = \frac{8}{1} = \frac{8 \times 5}{1 \times 5} = \frac{40}{5}$$

SELF-STUDY EXERCISES 3–6

1   Convert the improper fractions to whole or mixed numbers.

1. $\dfrac{12}{5}$ 2. $\dfrac{10}{7}$ 3. $\dfrac{12}{12}$ 4. $\dfrac{32}{7}$ 5. $\dfrac{24}{6}$

6. $\dfrac{15}{7}$ 7. $\dfrac{23}{9}$ 8. $\dfrac{47}{5}$ 9. $\dfrac{86}{9}$ 10. $\dfrac{38}{21}$

11. $\dfrac{57}{15}$ 12. $\dfrac{64}{4}$ 13. $\dfrac{72}{10}$ 14. $\dfrac{19}{2}$ 15. $\dfrac{36}{4}$

2   Change the mixed numbers to improper fractions.

16. $2\dfrac{1}{3}$ 17. $3\dfrac{1}{8}$ 18. $1\dfrac{7}{8}$ 19. $6\dfrac{5}{12}$ 20. $9\dfrac{5}{8}$

21. $3\dfrac{7}{8}$ 22. $7\dfrac{5}{12}$ 23. $6\dfrac{7}{16}$ 24. $8\dfrac{1}{32}$ 25. $1\dfrac{5}{64}$

26. $7\dfrac{3}{10}$ 27. $8\dfrac{2}{3}$ 28. $33\dfrac{1}{3}$ 29. $66\dfrac{2}{3}$ 30. $12\dfrac{1}{2}$

Change the whole numbers to an equivalent fraction with the given denominator.

31. $5 = \dfrac{?}{3}$ 32. $9 = \dfrac{?}{2}$ 33. $7 = \dfrac{?}{8}$ 34. $8 = \dfrac{?}{4}$ 35. $3 = \dfrac{?}{16}$

3–7   FINDING COMMON DENOMINATORS AND COMPARING FRACTIONS AND DECIMALS

Learning Outcomes

1   Find common denominators.

2   Compare fractions and decimals.

1   Find Common Denominators.

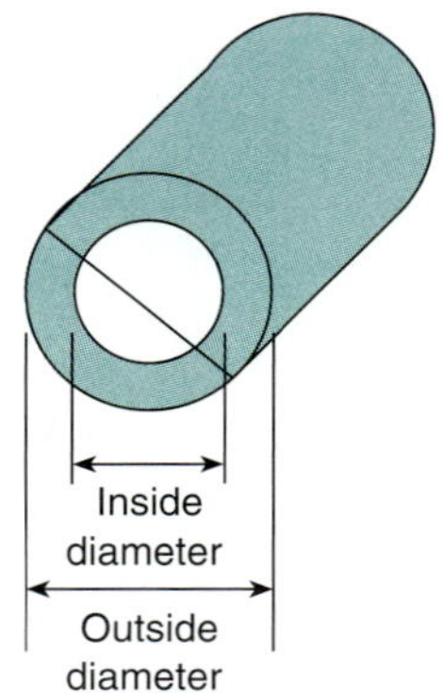

Figure 3–18

Is it possible to have a pipe with an outside diameter of $\frac{5}{8}$ in. and an inside diameter of $\frac{21}{32}$ in. (see Fig. 3–18)? To answer this question, we need to compare two fractions. To do this, we first select a *common denominator;* that is, we find a denominator that each denominator will divide into evenly. A common denominator can always be found by multiplying the two denominators together. However, many times a smaller number can be used as a common denominator. We should use the least common multiple (LCM) as our *least common denominator* (LCD). The LCM or LCD is equal to or greater than the larger factor. Because 8 divides evenly into 32, we can use 32 for our common denominator. Next, we change $\frac{5}{8}$ to an equivalent fraction with a denominator of 32.

$$\frac{5}{8} = \frac{5 \times 4}{8 \times 4} = \frac{20}{32}$$

To answer our original question, is $\frac{20}{32}$, which is equivalent to $\frac{5}{8}$, larger than $\frac{21}{32}$? No, so $\frac{5}{8}$ in. cannot be the outside diameter of a pipe with an inside diameter of $\frac{21}{32}$ in.

    With many fractions, the least common denominator can be found by inspection. By *inspection* we mean to inspect each denominator and intuitively (or mentally) determine the LCD. For fractions with larger denominators, you may need to use the procedure for finding the LCM discussed in Section 3–4.

To find the least common denominator (LCD) of two or more fractions:

Find the least common multiple (LCM) of the denominators of the fractions.

EXAMPLE Find the least common denominator for the fractions: $\frac{5}{12}, \frac{4}{15}, \frac{3}{8}$. Write the prime factorization of each denominator.

$$
\begin{array}{ccc}
\underline{12} & \underline{15} & \underline{8} \\
2 \times 6 & 3 \times 5 & 2 \times 4 \\
2 \times 2 \times 3 & & 2 \times 2 \times 2 \\
2^2 \times 3 & & 2^3
\end{array}
$$

$$\text{LCM or LCD} = 2^3 \times 3 \times 5$$
$$= 8 \times 3 \times 5$$
$$= 120$$

Tip! | ***Alternative Procedure for Finding the LCM or LCD.***

We can also find the LCM or LCD of several fractions by dividing duplicated factors and then, multiplying.

1. Arrange the denominators horizontally.
2. Divide by any prime factor that divides evenly into at least two denominators.
3. The LCM or LCD is the product of the primes and remaining factors.

Look at the denominators from the preceding example.

2	8	12	15	**Divide by a factor of 2.**
2	4	6	15	**Divide by another factor of 2.**
3	2	3	15	**Divide by a factor of 3.**
	2	1	5	

Primes: 2, 2, 3
Remaining factors: 2, 1, 5
LCM $= 2 \cdot 2 \cdot 3 \cdot 2 \cdot 1 \cdot 5$
$\quad\;\, = 120$

---

2 **Compare Fractions and Decimals.**

To compare quantities, we must compare like amounts. Thus, if we are to compare fractions, the fractions must have the same denominator. To compare $\frac{3}{7}$ and $\frac{5}{7}$, which have the same denominators, we compare the numerators, and $\frac{3}{7}$ is smaller than $\frac{5}{7}$.

To compare fractions with different denominators, we first change the fractions to equivalent fractions, which have a common denominator.

To compare fractions:

1. Find the least common denominator (LCD) that is also the least common multiple (LCM).
2. Change each fraction to an equivalent fraction with the least common denominator (LCD) as its denominator.
3. Compare the numerators.

Tip! | ***Comparing Fractions.***

We can compare fractions by changing them to equivalent fractions with *any* common denominator, but using the least common denominator gives us smaller numbers with which to work.

EXAMPLE Two drill bits have diameters of $\frac{3}{8}$ in. and $\frac{5}{16}$ in., respectively. Which drill bit makes the larger hole?

The least common denominator is 16.

$$\frac{3}{8} = \frac{6}{16} \qquad \frac{5}{16} = \frac{5}{16}$$

Now that each fraction has been changed to an equivalent fraction with the same denominator, we compare the numerators: $\frac{6}{16}$ is larger than $\frac{5}{16}$ because 6 is larger than 5, so $\frac{3}{8}$, which is equivalent to $\frac{6}{16}$, is larger than $\frac{5}{16}$. **The drill bit with a $\frac{3}{8}$-in. diameter will drill the larger hole.**

Sometimes we are asked to compare a fraction and a decimal. To do this, we change one number so that both numbers are either in fraction format or decimal format. Unless otherwise specified, you may use either format.

EXAMPLE An engineering rule is used to measure the length of a computer motherboard as 15.8 in. The computer case that will hold the board is $15\frac{7}{8}$ in. Will the board fit inside the case?

Change $15\frac{7}{8}$ to a decimal.

$$7 \div 8 = 0.875$$

$$15\frac{7}{8} = 15.875$$

The case measurement, 15.875, is greater than 15.80, so the board will fit into the case.

Our jobs sometimes require us to work with close tolerances. For instance, we might work with wire and need to compare the diameters of several sizes of wire to arrange them in order from largest to smallest or smallest to largest. To do this, we need to be able to compare decimals, because wire size is often measured in decimals.

To compare decimal numbers:

1. Compare whole-number parts.
2. If whole-number parts are equal, compare digits place by place, starting at the tenths place and moving to the right.
3. Stop when two digits in the same place are different.
4. The larger digit determines the larger decimal number.

EXAMPLE Compare the two numbers to see which is larger.

$$32.47 \qquad 32.48$$

We first look at the whole-number parts of the two numbers. They are the same so we look at the tenths place for each number. Both numbers have a 4 in the tenths place. **In the hundredths place, we see that 8 is larger than 7, so 32.48 is the larger number.**

EXAMPLE Compare 0.4 and 0.07 to see which is larger.

Since the whole-number parts are the same (0), we compare the digits in the tenths place: **4 is larger than 0, so 0.4 is larger.**

If we write both numbers in the preceding example so that they have the same number of digits in the decimal part, they are easier to compare.

$$0.4 = 0.40$$

$$0.07 = 0.07$$

Because 40 is larger than 7, 0.4 is larger than 0.07. This is equivalent to the process we used to compare common fractions. A common denominator is found, and each fraction (or decimal fraction) is changed to an equivalent fraction having the common denominator.

Tip!	***Decimal Fractions with Common Denominators.***

Decimal fractions have a common denominator if they have the same number of digits after the decimal point.

EXAMPLE A micrometer is used to measure two objects. The measurements are 0.386 in. and 0.388 in. Which object is larger?

To compare these measurements, we note that the digits in the tenths and hundredths places are the same. **Comparing the digits in the thousandths places, we see that the object measuring 0.388 in. is larger.**

SELF-STUDY EXERCISES 3–7

1 Find the least common denominator.

1. $\dfrac{5}{8}, \dfrac{4}{9}$
    2. $\dfrac{3}{10}, \dfrac{4}{15}$
    3. $\dfrac{9}{10}, \dfrac{4}{25}$
    4. $\dfrac{7}{12}, \dfrac{9}{16}, \dfrac{5}{8}$
    5. $\dfrac{2}{3}, \dfrac{5}{12}, \dfrac{7}{8}$

2 Show which fraction is larger.

6. $\dfrac{2}{3}, \dfrac{3}{5}$
    7. $\dfrac{5}{12}, \dfrac{7}{16}$
    8. $\dfrac{8}{9}, \dfrac{7}{8}$
    9. $\dfrac{5}{8}, \dfrac{11}{16}$
    10. $\dfrac{15}{32}, \dfrac{29}{64}$

11. $\dfrac{7}{12}, \dfrac{9}{16}$
    12. $\dfrac{3}{8}, \dfrac{4}{5}$
    13. $\dfrac{7}{11}, \dfrac{9}{10}$
    14. $\dfrac{4}{15}, \dfrac{3}{16}$
    15. $\dfrac{1}{2}, \dfrac{7}{16}$

16. Is the thickness of a $\frac{3}{16}$-in. sheet of metal greater than the length of a $\frac{15}{64}$- in. sheet metal screw?

17. Will a pipe with a $\frac{5}{16}$-in. outside diameter fit inside a pipe with a $\frac{3}{8}$-in. inside diameter?

18. A hollow-wall fastener has a grip range up to $\frac{7}{16}$ in. Is it long enough to fasten a thin sheet metal strip to a plywood wall if the combined thickness is $\frac{3}{8}$ in.?

19. A range top is $29\frac{1}{8}$ in. long by $19\frac{1}{2}$ in. wide. Is it smaller than an existing opening $29\frac{3}{16}$ in. long by $19\frac{9}{16}$ in. wide?

20. A wrench is marked $\frac{5}{8}$ at one end and $\frac{19}{32}$ at the other. Which end is larger?

21. Charles Bryant has a wrench marked $\frac{25}{32}$, but it is too large. Would a $\frac{7}{8}$-in. wrench be smaller?

22. For a do-it-yourself project, Brenda Jinkins needs to cut a piece of sheet metal slightly longer than the $10\frac{21}{32}$ in. required in the plans and trim it down to size. Brenda cuts the piece $10\frac{3}{4}$ in. Is it too short?

23. A plastic anchor for a No. 6 $\times$ 1-in. screw requires that at least a $\frac{3}{16}$-in.-diameter hole be drilled. Will a $\frac{1}{4}$-in. drill bit be large enough?

24. The plastic anchor in Exercise 23 requires a minimum hole depth of $\frac{7}{8}$ in. Is a $\frac{3}{4}$-in. hole deep enough?

25. Is a $\frac{3}{8}$-in. wrench too large or too small for a $\frac{1}{2}$-in. bolt?

Show which decimal is larger.

26. 0.023 or 0.03

27. 0.38 or 0.392

28. 5.38 or 5.375

29. 4.207 or 4.71

30. A nurse records temperatures of 96.5°, 97.3°, 98.6°. Which temperature is highest?

31. Two quality-control technicians measure and record two hip replacement medical implants as 0.0256 cm and 0.0249 cm. Which implant is larger?

32. Two cities base the property tax rate on mills. One city has a rate of 0.973 mills, and the other has a rate of 0.975 mills. Which city has the lower tax rate?

33. Describe an instance when you needed to compare two or more decimals or fractions. How did you do this?

3–8   ADDING FRACTIONS AND MIXED NUMBERS

Learning Outcomes

 Add fractions.

 Add mixed numbers.

Thus far, we have studied what fractions and mixed numbers represent and how to convert them to their equivalents. Now let's put that knowledge to use and solve problems encountered on the job. Let's look first at problems using addition.

 Add Fractions.

The most important rule to remember when adding fractions is that all fractions to be added must have the same denominator. Trying to add unlike fractions is like trying to add unlike objects or measures. Before adding unlike fractions, we change the fractions to equivalent fractions with a common denominator and then add the fractions. We use the following rule.

To add fractions:

1. If the denominators are not the same, find the least common denominator.
2. Change each fraction not already expressed in terms of the common denominator to an equivalent fraction with the common denominator.
3. Add the numerators only.
4. The common denominator is the denominator of the sum.
5. Reduce the answer to lowest terms and change improper fractions to whole or mixed numbers.

EXAMPLE   Find the sum of $\dfrac{3}{8} + \dfrac{1}{8}$.

Because the denominators are the same, start with Step 3 of the addition procedure.

$$\frac{3}{8} + \frac{1}{8} = \frac{4}{8} = \frac{1}{2}$$

■ **Learning Strategy**   *Shortening Written Steps.*

We are beginning to work some basic steps mentally. For example, $\frac{4}{8} = \frac{4 \div 4}{8 \div 4} = \frac{1}{2}$ was written as $\frac{4}{8} = \frac{1}{2}$. Understanding basic principles is important; however, we don't want to burden ourselves with unnecessary written procedures when we can do steps mentally.

EXAMPLE   Add $\dfrac{5}{32} + \dfrac{3}{16} + \dfrac{7}{8}$.

The least common denominator may be found by inspection. Both 8 and 16 divide evenly into 32, so we use 32 as the common denominator. Next, we change each fraction to an equivalent fraction whose denominator is 32.

$$\frac{5}{32} = \frac{5}{32}, \qquad \frac{3}{16} = \frac{6}{32}, \qquad \frac{7}{8} = \frac{28}{32}$$

Then we add the numerators.

$$\frac{5}{32} + \frac{6}{32} + \frac{28}{32} = \frac{39}{32}$$

Finally, we change to a mixed number.

$$\frac{39}{32} = 1\frac{7}{32}$$

Tip!	***Using Several Skills in One Problem.***

It is important to keep your skills fresh and readily accessible. When you need to review, your independent learning skills will be helpful. In a problem like the preceding example, nearly all fraction skills are used.

EXAMPLE   A plumber uses a $\frac{9}{16}$-in.-diameter copper tube wrapped with $\frac{5}{8}$-in. insulation. What size hole must he bore in the stud to install the insulated pipe?

From Fig. 3–19, we see that the $\frac{5}{8}$-in. insulation increases the diameter on each side of the pipe. To get the total diameter of the pipe and insulation, we add $\frac{9}{16} + \frac{5}{8} + \frac{5}{8}$. The thickness of the insulation is added twice because it counts in the total diameter of the pipe and insulation two times.

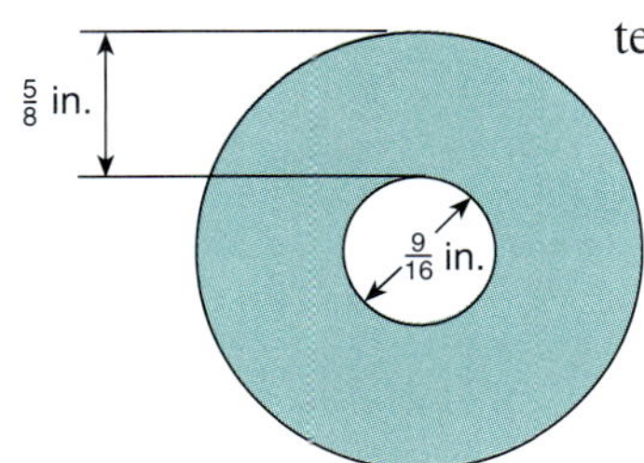

Figure 3–19

$$\frac{9}{16} = \frac{9}{16}$$

$$\frac{5}{8} = \frac{10}{16}$$

$$+ \ \frac{5}{8} = \frac{10}{16}$$

$$\frac{29}{16} = 1\frac{13}{16}$$

The LCD is 16.

The total diameter is $1\frac{13}{16}$ in., and the diameter of the hole must be at least this large.

2 Add Mixed Numbers.

There are two popular methods for adding mixed numbers. In one method, we add the whole numbers and fractions separately and then combine the sums. In the second method, we change each mixed number to an improper fraction and then follow the rules for adding fractions. The first method is more appropriate for most situations.

EXAMPLE Add $5\frac{2}{3} + 7\frac{3}{8} + 4\frac{1}{2}$.

It is usually more convenient to arrange the problem vertically when adding. Using the first method, we have

$$5\frac{2}{3} = 5\frac{16}{24} \qquad \text{The LCD is 24.}$$

$$7\frac{3}{8} = 7\frac{9}{24}$$

$$+\ 4\frac{1}{2} = 4\frac{12}{24}$$

$$16\frac{37}{24} \qquad \frac{37}{24} = 1\frac{13}{24}$$

$$16 + 1\frac{13}{24} = 17\frac{13}{24}$$

Using the second method, we have

$$5\frac{2}{3} = \frac{17}{3} = \frac{136}{24}$$

$$7\frac{3}{8} = \frac{59}{8} = \frac{177}{24}$$

$$+\ 4\frac{1}{2} = \frac{9}{2} = \frac{108}{24}$$

$$\frac{421}{24} = 17\frac{13}{24}$$

EXAMPLE A carpenter cuts the following pieces from a metal rod: $3\frac{3}{8}$ in., $9\frac{5}{16}$ in., and $7\frac{3}{4}$ in. What is the total length cut from the rod?

To find the total length cut from the rod, we add the mixed numbers.

$$3\frac{3}{8} = 3\frac{6}{16} \qquad \text{The LCD is 16.}$$

$$9\frac{5}{16} = 9\frac{5}{16}$$

$$+\ 7\frac{3}{4} = 7\frac{12}{16}$$

$$19\frac{23}{16} = 20\frac{7}{16} \qquad \frac{23}{16} = 1\frac{7}{16}, \ 19 + 1\frac{7}{16} = 20\frac{7}{16}$$

$20\frac{7}{16}$ in. was cut from the rod.

| **Tip!** | **Writing Whole Numbers in Mixed-Number Form.** |

When you add mixed numbers and whole numbers, think of the whole number as a mixed number with zero as the numerator in the fraction.

$$5 + 3\frac{1}{3} = \quad 5\frac{0}{3}$$
$$+ \ 3\frac{1}{3}$$

Recall the addition property of zero. Adding zero to any number does not change the value of the number. Thus, $5 + 3\frac{1}{3} = 8\frac{1}{3}$.

EXAMPLE Find the largest permissible measurement of a part if the blueprint calls for the part to be 2 in. long and the tolerance is $\pm\frac{1}{8}$ in.

Tolerance is the amount a part can vary from the blueprint specification. To find the largest permissible measure, we add.

$$2 + \frac{1}{8} = 2\frac{1}{8}$$

The largest permissible measure is $2\frac{1}{8}$ in.

| **Tip!** | **What Are Tolerance and Limit Dimensions?** |

Tolerance is an often-used concept in technical applications. The "plus or minus" symbol ($\pm$) indicates that the actual measure of an object can vary, but it must be no more than the specified amount (plus) nor no less than the specified amount (minus). If you are asked to find the largest possible acceptable measure, *add* the specified amount. If you are asked to find the smallest possible acceptable measure, *subtract* the specified amount. If you are asked to find the *limits* or *limit dimensions,* find *both* the largest and smallest acceptable measures.

SELF-STUDY EXERCISES 3–8

1 Add; reduce answers to lowest terms and convert any improper fractions to whole or mixed numbers.

1. $\dfrac{5}{16} + \dfrac{1}{16}$

2. $\dfrac{1}{2} + \dfrac{1}{8} + \dfrac{3}{4}$

3. $\dfrac{1}{8} + \dfrac{1}{2}$

4. $\dfrac{3}{8} + \dfrac{5}{32} + \dfrac{1}{4}$

5. $\dfrac{5}{16} + \dfrac{1}{4}$

6. $\dfrac{15}{16} + \dfrac{1}{2}$

7. $\dfrac{3}{32} + \dfrac{5}{64}$

8. $\dfrac{7}{8} + \dfrac{3}{5}$

9. $\dfrac{3}{4} + \dfrac{8}{9}$

10. $\dfrac{7}{8} + \dfrac{5}{24}$

11. What is the thickness of a countertop made of $\frac{7}{8}$-in. plywood and $\frac{1}{16}$-in. Formica?

12. Three pieces of steel are joined together. What is the total thickness if the pieces are $\frac{1}{2}$ in., $\frac{7}{16}$ in., and $\frac{29}{32}$ in.?

13. Three books are placed side by side. They are $\frac{5}{16}$ in., $\frac{7}{8}$ in., and $\frac{3}{4}$ in. wide. What is the total width of the books if they are polywrapped in one package?

14. Find the outside diameter of a pipe (see Fig. 3–20) whose wall is $\frac{1}{2}$ in. thick if its inside diameter is $\frac{7}{8}$ in.
15. What length bolt is needed to fasten two pieces of metal each $\frac{7}{16}$ in. thick if a $\frac{1}{8}$-in. lock washer is used and a $\frac{1}{4}$-in. nut is used.

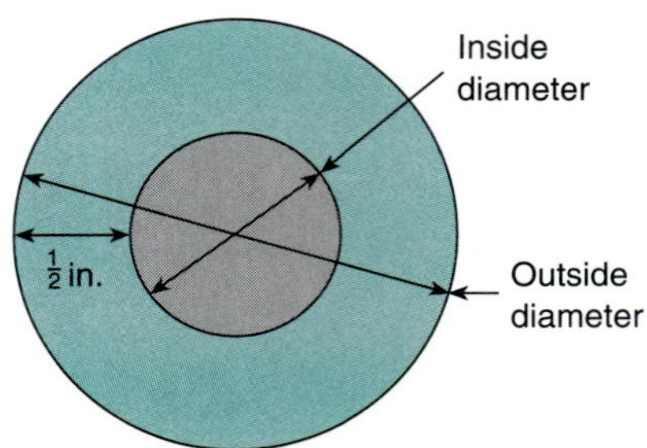

Figure 3–20

2 Add; reduce answers to lowest terms and convert any improper fractions to whole or mixed numbers.

16. $2\frac{3}{5} + 4\frac{1}{5}$

17. $1\frac{5}{8} + 2\frac{1}{2}$

18. $3\frac{3}{4} + 7\frac{3}{16} + 5\frac{7}{8}$

19. $\frac{1}{6} + \frac{7}{9} + \frac{2}{3}$

20. $2\frac{1}{4} + 2\frac{9}{16}$

21. $1\frac{5}{16} + 4\frac{7}{32}$

22. $3\frac{1}{4} + 1\frac{7}{16}$

23. $4\frac{1}{2} + 9$

24. $\frac{2}{3} + 3\frac{4}{5}$

25. $5\frac{1}{8} + 3 + 4\frac{9}{16}$

26. The studs of an outside wall are $5\frac{3}{4}$ in. thick. The inside wallboard is $\frac{7}{8}$ in. thick, and the outside covering is $2\frac{3}{16}$ in. thick. What is the total thickness of the wall?

27. A blueprint calls for a piece of bar stock $3\frac{7}{8}$ in. long. If a tolerance of $\pm\frac{1}{16}$ in. is allowed, what is the longest permissible measurement for the bar stock?

28. If $4\frac{3}{8}$ gal of water are used to dilute $7\frac{1}{4}$ gal of acid, how many gallons are in the mixture?

29. How much bar stock is needed to make bars of the following lengths: $10\frac{1}{4}$ in., $8\frac{7}{16}$ in., $5\frac{15}{32}$ in.? Disregard waste.

30. Three pieces of I-beam each measuring $7\frac{5}{8}$ in. are needed to complete a job. How much I-beam is needed?

3–9   SUBTRACTING FRACTIONS AND MIXED NUMBERS

Learning Outcomes

1 Subtract fractions.
2 Subtract mixed numbers.

1 Subtract Fractions.

The steps for subtracting fractions and mixed numbers are very similar to the steps for adding fractions and mixed numbers.

To subtract fractions:

1. If the denominators are not the same, find the least common denominator.
2. Change each fraction not expressed in terms of the common denominator to an equivalent fraction having the common denominator.
3. Subtract the numerators.
4. The common denominator will be the denominator of the difference.
5. Reduce the answer to lowest terms.

Fractions must have the same denominator before they can be subtracted. Chapter 1 emphasized a very important contrast between addition and subtraction. In addition the order of the terms (addends) does not matter; that is, $3 + 4$ is the same as $4 + 3$. In subtraction, however, the order of the terms is very important because subtraction is not commutative. We must be careful when arranging a subtraction problem.

EXAMPLE Subtract $\dfrac{3}{8} - \dfrac{7}{32}$.

$$\dfrac{3}{8} = \dfrac{12}{32} \qquad \text{Change } \tfrac{3}{8} \text{ to an equivalent fraction with a denominator of 32.}$$

$$-\dfrac{7}{32} = \dfrac{7}{32} \qquad \text{Subtract numerators and keep the common denominator.}$$

$$\dfrac{5}{32}$$

2 Subtract Mixed Numbers.

As in addition, there is more than one method for subtracting mixed numbers. Because the method that requires changing each mixed number to an improper fraction is generally not the most practical method, the following examples use the method that considers the whole numbers and fractional parts separately.

EXAMPLE Subtract $15\dfrac{7}{8} - 4\dfrac{1}{2}$.

$$15\dfrac{7}{8} = 15\dfrac{7}{8}$$

$$-\ 4\dfrac{1}{2} = 4\dfrac{4}{8}$$

$$11\dfrac{3}{8}$$

When the fraction in the subtrahend (number being subtracted) is larger than the fraction in the minuend (first number), we borrow from the whole-number part of the minuend.

> **To borrow when subtracting mixed numbers:**
>
> 1. If the fractional parts of the mixed numbers don't have the same denominator, change them to equivalent fractions with a common denominator.
> 2. When the fraction in the subtrahend is larger than the fraction in the minuend, borrow one whole number from the whole-number part of the minuend. This makes the whole number 1 less.
> 3. Change the whole number borrowed to an improper fraction with the common denominator. For example, $1 = \dfrac{3}{3}$, $1 = \dfrac{8}{8}$, $1 = \dfrac{n}{n}$, where n is the common denominator.
> 4. Add the borrowed fraction ($\tfrac{n}{n}$) to the fraction already in the minuend.
> 5. Subtract the fractional parts and the whole-number parts.
> 6. Reduce the answer to lowest terms.

EXAMPLE Subtract $15\dfrac{3}{4}$ from $18\dfrac{1}{2}$.

$$18\dfrac{1}{2} = 18\dfrac{2}{4} = 17\dfrac{4}{4} + \dfrac{2}{4} = 17\dfrac{6}{4} \qquad 18 - 1 = 17,\ 1 = \dfrac{4}{4},\ \dfrac{4}{4} + \dfrac{2}{4} = \dfrac{6}{4}$$

$$-\ 15\dfrac{3}{4} = 15\dfrac{3}{4} = 15\dfrac{3}{4} \qquad = 15\dfrac{3}{4}$$

$$2\dfrac{3}{4}$$

EXAMPLE A metal block weighing $127\frac{1}{2}$ lb is removed from a flatbed truck with a payload of $433\frac{3}{8}$ lb. How many pounds remain on the truck?

$$433\frac{3}{8} = 433\frac{3}{8} = 432\frac{11}{8} \qquad 433 - 1 = 432,\ 1 = \frac{8}{8},\ \frac{8}{8} + \frac{3}{8} = \frac{11}{8}$$

$$- 127\frac{1}{2} = 127\frac{4}{8} = 127\frac{4}{8}$$

$$305\frac{7}{8}$$

$305\frac{7}{8}$ lb remains on the truck.

Tip! | ***Think of Whole Numbers in Mixed-Number Form Before Subtracting.***

As in addition, when subtracting whole numbers and mixed numbers, consider the whole number to have zero fractional parts. Then follow the same procedures as before. Borrow when necessary.

EXAMPLE Subtract 27 from $45\frac{1}{3}$.

$$45\frac{1}{3} = 45\frac{1}{3}$$

$$- 27\ \ = 27\frac{0}{3}$$

$$18\frac{1}{3}$$

EXAMPLE How many feet of wire are left on a 100-ft roll if $27\frac{1}{4}$ ft are used from the roll?

$$100 = 99\frac{4}{4}$$

$$- 27\frac{1}{4} = 27\frac{1}{4}$$

$$72\frac{3}{4}$$

$72\frac{3}{4}$ ft of wire is left on the roll.

EXAMPLE Three lengths measuring $5\frac{1}{4}$ in., $7\frac{3}{8}$ in., and $6\frac{1}{2}$ in. are cut from a 64-in. bar of angle iron. If $\frac{3}{16}$ in. is wasted on each cut, how many inches of angle iron remain?

Visualize the problem by making a sketch (see Fig. 3–21). Then find the total amount of angle iron used. This includes the three lengths and the waste for three cuts.

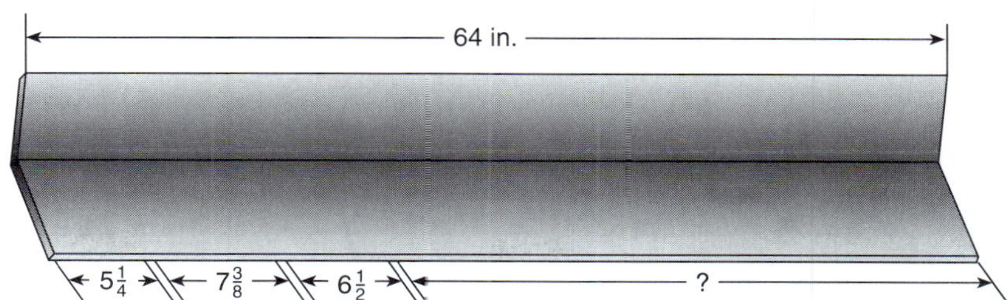

Figure 3–21

$$5\frac{1}{4} + 7\frac{3}{8} + 6\frac{1}{2} + \frac{3}{16} + \frac{3}{16} + \frac{3}{16} =$$

$$5\frac{4}{16} + 7\frac{6}{16} + 6\frac{8}{16} + \frac{3}{16} + \frac{3}{16} + \frac{3}{16} = 18\frac{27}{16}$$

$$18\frac{27}{16} = 18 + \frac{27}{16} = 18 + 1\frac{11}{16} = 19\frac{11}{16}$$ The total amount removed is $19\frac{11}{16}$ in.

$$\text{Amount of angle iron remaining} = \frac{\text{Beginning}}{\text{length}} - \frac{\text{Total iron removed}}{\text{and wasted}}$$

$$= 64 - 19\frac{11}{16}$$

$$= 63\frac{16}{16} - 19\frac{11}{16} = 44\frac{5}{16}$$

Thus, $44\frac{5}{16}$ in. of angle iron remains.

SELF-STUDY EXERCISES 3–9

1 Subtract; reduce when necessary.

1. $\dfrac{7}{8} - \dfrac{5}{8}$

2. $\dfrac{9}{16} - \dfrac{3}{8}$

3. $\dfrac{7}{16} - \dfrac{3}{8}$

4. $\dfrac{5}{8} - \dfrac{1}{2}$

5. $\dfrac{5}{32} - \dfrac{1}{64}$

6. $\dfrac{7}{8} - \dfrac{3}{4}$

2

7. $9\dfrac{11}{16} - 5$

8. $23\dfrac{3}{16} - 5\dfrac{7}{16}$

9. $9\dfrac{1}{4} - 4\dfrac{5}{16}$

10. $9\dfrac{1}{32} - 3\dfrac{3}{8}$

11. A length of bar stock $16\frac{3}{8}$ in. long is cut so that a piece only $7\frac{9}{16}$ in. long remains. What is the length of the cutoff piece? Disregard waste.

12. A concrete foundation includes $7\frac{7}{8}$ in. of base fill. If the foundation is to be 18 in. thick, how thick must the concrete be?

13. A casting is machined so that $22\frac{1}{5}$ lb of metal remains. If the casting weighed $25\frac{3}{10}$ lb, how many pounds were removed by machine?

14. A bolt $2\frac{5}{8}$ in. long fastens two pieces of wood 1 in. and $1\frac{7}{32}$ in. thick. If a $\frac{3}{32}$-in.-thick lockwasher and a $\frac{1}{8}$-in.-thick washer are used, what thickness is the nut if it is flush with the bolt? The measure of a bolt length does not include the bolt head.

15. Find the missing length in Fig. 3–22.

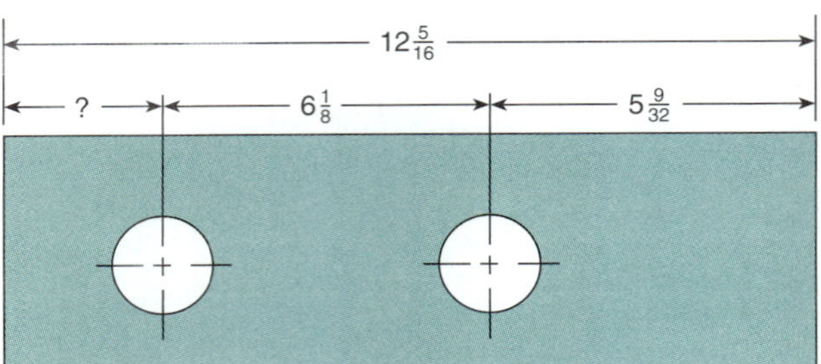

Figure 3–22

3-10 MULTIPLYING FRACTIONS AND MIXED NUMBERS

Learning Outcomes

1 Multiply fractions.
2 Multiply mixed numbers.

We saw in Chapter 1 that multiplying whole numbers is a shortcut for addition and therefore a time-saver on the job. For example, we saw that multiplying 2×4 to get 8 is quicker and more efficient than adding $2 + 2 + 2 + 2$ to get the same number 8. When multiplying a whole number by a whole number, we are increasing a given number of units by a certain number of times. Thus, in $2 \times 7 = 14$, we are increasing 2 by 7 times. However, when multiplying a fraction by a fraction, we are doing something different.

When multiplying a fraction by a fraction, we are finding *a part of a part*. For instance, $\frac{1}{2} \times \frac{1}{2}$ is $\frac{1}{2}$ of $\frac{1}{2}$. The word "of" is the clue that we must multiply to find the part we are looking for. The examples in Fig. 3–23 illustrate what we mean by finding a part of a part.

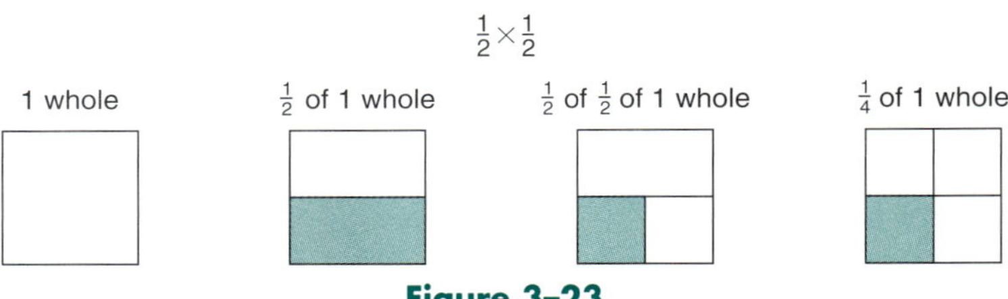

Figure 3–23

When we add or subtract fractions and mixed numbers, we need a common denominator. Otherwise, we cannot add or subtract the fractional parts. However, this is not the case when we multiply. In multiplying fractions, we do *not* change fractions to equivalent fractions with a common denominator. Look at two more examples of taking a part of a part (Figs. 3–24 and 3–25).

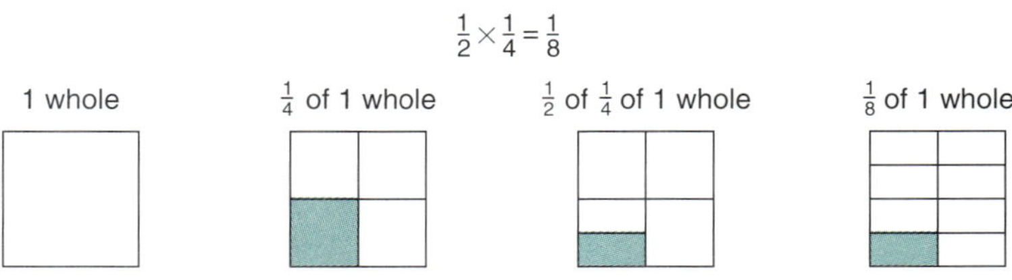

Figure 3–24

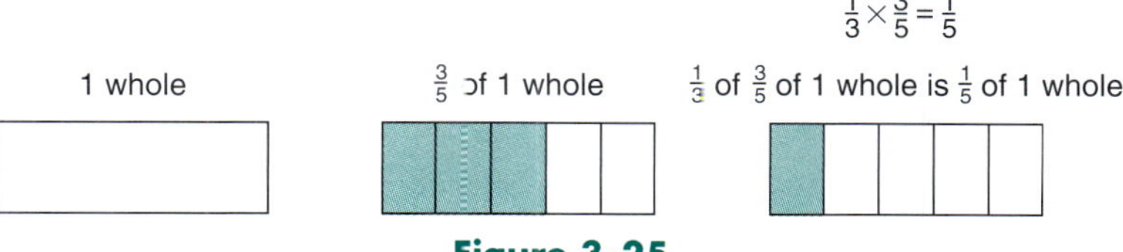

Figure 3–25

1 Multiply Fractions.

Figures 3–23, 3–24, and 3–25 show visually how we can multiply a fraction by a fraction, that is, take a part of a part. On the job, however, we may not want to stop and draw pictures or figures. Instead, we use rules to compute quickly.

> ***To multiply fractions:***
>
> **1.** Multiply the numerators of the fractions to get the numerator of the product.
> **2.** Multiply the denominators to get the denominator of the product.
> **3.** Reduce the product to lowest terms.

EXAMPLE   Find $\frac{1}{2}$ of $\frac{1}{4}$.

$$\frac{1}{2} \times \frac{1}{4} = \frac{1}{8}$$

EXAMPLE   Find $\frac{1}{3}$ of $\frac{3}{5}$.

$$\frac{1}{3} \times \frac{3}{5} = \frac{3}{15} = \frac{1}{5}$$

Tip! | ***Reduce or Cancel Before Multiplying.***

In the preceding example, reducing is possible. When multiplying fractions, we reduce common factors before multiplying them.

$$\frac{1}{3} \times \frac{3}{5} = \frac{1 \times \overset{1}{\cancel{3}}}{\underset{1}{\cancel{3}} \times 5} = \frac{1}{5}$$

In this example a numerator and a denominator both have a common factor of 3, so the common factor can be reduced before multiplying. Reducing applies the principles $\frac{n}{n} = 1$ and $1 \times n = n$. This process is also referred to as *canceling*.

EXAMPLE   Find $\frac{3}{9}$ of $\frac{2}{7}$.

$$\frac{\overset{1}{\cancel{3}}}{\underset{3}{\cancel{9}}} \times \frac{2}{7} = \frac{2}{21}$$

3 is a common factor of both a numerator and a denominator. Reduce before multiplying.

EXAMPLE Multiply $\dfrac{2}{3} \times \dfrac{5}{9} \times \dfrac{1}{6}$.

$$\dfrac{\overset{1}{2}}{3} \times \dfrac{5}{9} \times \dfrac{1}{\underset{3}{6}} = \dfrac{5}{81}$$

2 is a common factor of both a numerator and a denominator. Reduce before multiplying.

Tip!	***Cancel from Any Numerator to Any Denominator.***

Common factors that are reduced can be diagonal to each other, one above the other, or separated by another fraction, but one factor *must* be in the numerator and the other in the denominator.

EXAMPLE Multiply $\dfrac{2}{5} \times \dfrac{10}{21} \times \dfrac{6}{12}$.

$$\dfrac{\overset{1}{2}}{\underset{1}{5}} \times \dfrac{\overset{2}{10}}{21} \times \dfrac{\overset{1}{6}}{\underset{\underset{1}{2}}{12}} = \dfrac{2}{21}$$

5 and 10 are diagonal to each other.
6 is above 12.
2 and 2 are separated by another fraction.

Other patterns of reducing could also have been used.

2 Multiply Mixed Numbers.

We use the same rules to multiply mixed numbers or combinations of whole numbers, fractions, and mixed numbers as we use to multiply fractions. When multiplying mixed numbers or combinations of whole numbers, fractions, and mixed numbers, we first change each whole number or mixed number to an improper fraction. Then we proceed as in multiplying fractions.

To multiply mixed numbers, fractions, and whole numbers:

1. Change each mixed number or whole number to an improper fraction.
2. Reduce as much as possible.
3. Multiply numerators.
4. Multiply denominators.
5. Convert the answer to a whole or mixed number if possible.

EXAMPLE Multiply $2\dfrac{1}{2} \times 5\dfrac{1}{3}$.

$$2\dfrac{1}{2} \times 5\dfrac{1}{3} = \dfrac{5}{2} \times \dfrac{16}{3} = \dfrac{5}{\underset{1}{2}} \times \dfrac{\overset{8}{16}}{3} = \dfrac{40}{3} = 13\dfrac{1}{3}$$

EXAMPLE An engineering technician needs 10 pieces of wire each $3\dfrac{5}{8}$ in. long. What total length will she need for the wire pieces?

$$3\dfrac{5}{8} \times 10 = \dfrac{29}{\underset{4}{8}} \times \dfrac{\overset{5}{10}}{1} = \dfrac{145}{4} = 36\dfrac{1}{4}$$

$36\dfrac{1}{4}$ in. of wire will be needed.

   CHAPTER 3   Fractions and Percents

As we have noted before, formulas give us a plan for performing operations. Common formulas are found in many texts and reference materials. Others we develop to solve specific problems. The great benefit in working with formulas is that we use our critical thinking skills to develop a formula or select an appropriate formula. Then, in the calculations we focus on the necessary manipulation skills.

EXAMPLE Bricks that are $2\frac{1}{4}$ in. thick form a brick wall with $\frac{3}{8}$-in. mortar joints (Fig. 3–26). What is the height of the wall above the foundation after nine courses?

Use the Six-Step Problem-Solving Plan.

Unknown facts Height of the wall after nine courses of brick have been laid

Known facts $2\frac{1}{4}$ in.   thickness of each brick

$\frac{3}{8}$ in.   thickness of each mortar joint

9   number of courses (or rows) of brick and mortar joints

Relationships Height of wall = Thickness of each mortar joint × Number of mortar joints + Thickness of each brick × Number of rows of brick

Estimation Each brick is a little more than 2 in. thick, so the wall should be at least 2 × 9 or 18 in. high. Since the mortar joint is not quite $\frac{1}{2}$ in. and the fractional portion of the brick's thickness is less than $\frac{1}{2}$ in., the combined thickness of the brick and mortar joint must be less than 3 in. So the total height of the wall should be less than 3 × 9 or 27 in. Thus, we estimate the wall height to be between 18 and 27 inches.

Calculation

$$\left(9 \times 2\frac{1}{4}\right) + \left(9 \times \frac{3}{8}\right)$$

$$\left(\frac{9}{1} \times \frac{9}{4}\right) + \left(\frac{9}{1} \times \frac{3}{8}\right)$$

$$\frac{81}{4} + \frac{27}{8}$$

$$20\frac{1}{4} + 3\frac{3}{8}$$

$$20\frac{2}{8} + 3\frac{3}{8} = 23\frac{5}{8}$$

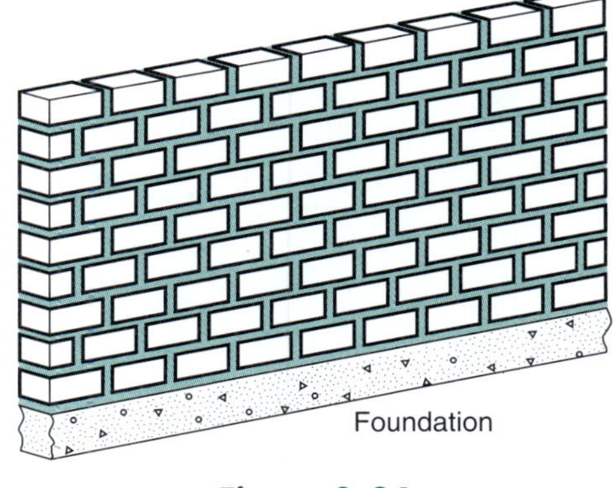

Figure 3–26

Interpretation **The wall will be $23\frac{5}{8}$ in. high.**

SELF-STUDY EXERCISES 3–10

1 Multiply and reduce answers to lowest terms.

1. $\dfrac{3}{4} \times \dfrac{1}{8}$ **2.** $\dfrac{1}{2} \times \dfrac{7}{16}$ **3.** $\dfrac{5}{8} \times \dfrac{7}{10}$ **4.** $\dfrac{2}{3} \times \dfrac{7}{8}$

5. $\dfrac{1}{2} \times \dfrac{3}{4} \times \dfrac{8}{9}$ **6.** $\dfrac{3}{8} \times \dfrac{5}{6} \times \dfrac{1}{2}$ **7.** $\dfrac{7}{8} \times \dfrac{2}{5} \times \dfrac{4}{21}$ **8.** $\dfrac{11}{12} \times \dfrac{9}{10} \times \dfrac{8}{15}$

9. $\dfrac{3}{10} \times \dfrac{4}{15}$ **10.** $\dfrac{15}{16} \times \dfrac{7}{10}$

2 Multiply and reduce answers to lowest terms. Convert improper fractions to whole or mixed numbers.

11. $7 \times 3\frac{1}{8}$

12. $\frac{3}{5} \times 125$

13. $2\frac{3}{4} \times 1\frac{1}{2}$

14. $9\frac{1}{2} \times 3\frac{4}{5}$

15. $\frac{1}{5} \times 7\frac{5}{8}$

16. $\frac{2}{3} \times 3\frac{1}{4}$

17. A fuel tank that holds 75 liters (L) of fuel is $\frac{1}{4}$ full. How many liters of fuel are in the tank?

18. If steps are 12 risers high and each riser is $7\frac{1}{2}$ in. high, what is the total rise of the steps?

19. A piece of sheet metal is $\frac{1}{16}$ in. thick. How thick is a stack of 250 pieces?

20. An alloy, which is a substance composed of two or more metals, is $\frac{11}{16}$ copper, $\frac{7}{32}$ tin, and $\frac{3}{32}$ zinc. How many kilograms (kg) of each metal are needed to make 384 kg of alloy?

3–11   DIVIDING FRACTIONS AND MIXED NUMBERS

Learning Outcomes

1   Find reciprocals.

2   Divide fractions.

3   Divide mixed numbers.

4   Simplify complex fractions.

Once we can multiply fractions, we can easily handle the division of fractions and mixed numbers. The reason for this is that multiplication and division are inverse operations. However, the real focus is to understand when to multiply and when to divide.

Let's make a comparison. If six units are divided into two equal parts, how many units will be in each part? $6 \div 2 = ?$ The answer is 3. But how many units are $\frac{1}{2}$ of 6 units? $\frac{1}{2} \times 6 = ?$ Again, the answer is 3. In these two examples $6 \div 2$ and $\frac{1}{2} \times 6$ represent three units. To divide by 2, then, is the same as taking half of a quantity.

1   Find Reciprocals.

If we compare $12 \div 3$ and $\frac{1}{3} \times 12$, we find that both answers are 4. That is, $12 \div 3 = \frac{1}{3} \times 12$ or $12 \times \frac{1}{3}$. So, not only is there a relationship between multiplication and division, there is also a relationship between 3 and $\frac{1}{3}$. Pairs of numbers like $\frac{1}{2}$ and 2 or $\frac{1}{3}$ and 3 are called *reciprocals*.

■   **DEFINITION: Reciprocals.** Two numbers are *reciprocals* if their product is 1. Thus, $\frac{1}{2}$ and 2 are reciprocals because $\frac{1}{2} \times 2 = 1$, and $\frac{2}{3}$ and $\frac{3}{2}$ are reciprocals because $\frac{2}{3} \times \frac{3}{2} = 1$.

To find the reciprocal of a number:

1. Write the number in fractional form.
2. Interchange the numerator and denominator so that the numerator is the denominator and the denominator is the numerator.

Tip!	*Reciprocals and Inverting.*

Interchanging the numerator and denominator of a fraction is commonly called *inverting* the fraction.

EXAMPLE   Find the reciprocal of $\frac{2}{3}, \frac{4}{7}, \frac{1}{5}, 3, 1, 2\frac{1}{2}, 0.8, 3.5$, and 0.

The reciprocal of $\frac{2}{3}$ is $\frac{3}{2}$.

The reciprocal of $\frac{4}{7}$ is $\frac{7}{4}$.

The reciprocal of $\frac{1}{5}$ is $\frac{5}{1}$ or **5**.

The reciprocal of 3 is $\frac{1}{3}$. $(3 = \frac{3}{1})$

The reciprocal of 1 is **1**. $(1 = \frac{1}{1})$

The reciprocal of $2\frac{1}{2}$ is $\frac{2}{5}$. $(2\frac{1}{2} = \frac{5}{2})$

Write 0.8 as a fraction: $\frac{8}{10} = \frac{4}{5}$. The reciprocal is $\frac{5}{4}$ or **1.25**.

Write 3.5 as an improper fraction: $3\frac{5}{10} = 3\frac{1}{2} = \frac{7}{2}$. The reciprocal is $\frac{2}{7}$ or **0.285714285**.

0 has no reciprocal because division by 0 is impossible. $0 = \frac{0}{1}, \frac{1}{0}$ is undefined.

2 Divide Fractions.

Let's review the terminology of division.

$$15 \quad \div \quad 3 \quad = \quad 5$$
$$\text{dividend} \quad \text{divisor} \quad \text{quotient}$$

To identify the divisor, remember that the symbol $\div$ is always read "divided by."

To divide fractions:

1. Change the division to an equivalent multiplication by replacing the divisor with its reciprocal and replacing the division sign ($\div$) with a multiplication sign ($\times$).
2. Perform the resulting multiplication.

EXAMPLE   Find $\dfrac{5}{8} \div \dfrac{2}{3}$.

$$\frac{5}{8} \div \frac{2}{3} = \frac{5}{8} \times \frac{3}{2} = \frac{15}{16}$$

■ **Learning Strategy** *Put Rules into Your Own Words.*

We commonly state the division of fractions rule as

"Invert the divisor and multiply."

We could also say

"Invert the second number and multiply."

or

"Invert the number after the division sign and multiply."

A rule in your own words is often easier for you to remember. Just be sure the words guide you to an appropriate process.

EXAMPLE   Divide $\dfrac{7}{8}$ by $\dfrac{14}{15}$.

$$\frac{7}{8} \div \frac{14}{15} = \frac{\overset{1}{7}}{8} \times \frac{15}{\underset{2}{14}} = \frac{15}{16}$$

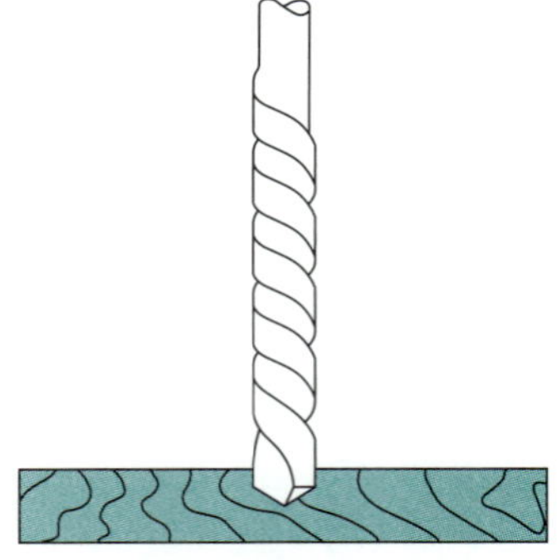

EXAMPLE An auger bit advances $\frac{1}{16}$ in. for each turn (see Fig. 3–27). How many turns are needed to drill a hole $\frac{5}{8}$ in. deep? ($\frac{5}{8}$ in. can be divided into how many $\frac{1}{16}$-in. parts?)

$$\frac{5}{8} \div \frac{1}{16} = \frac{5}{\overset{}{8}} \times \frac{\overset{2}{\cancel{16}}}{1} = 10$$

Figure 3–27

Ten turns are needed.

3 Divide Mixed Numbers.

To divide mixed numbers and whole numbers, we first write the mixed numbers or whole numbers as improper fractions and then follow the rules for dividing fractions.

> **To divide mixed numbers, fractions, and whole numbers:**
>
> 1. Change each mixed number or whole number to an improper fraction.
> 2. Convert to an equivalent multiplication problem using the reciprocal of the divisor.
> 3. Multiply according to the rule for multiplying fractions.

EXAMPLE Find $2\frac{1}{2} \div 3\frac{1}{3}$.

$$2\frac{1}{2} \div 3\frac{1}{3} = \frac{5}{2} \div \frac{10}{3} = \frac{\overset{1}{\cancel{5}}}{2} \times \frac{3}{\cancel{10}} = \frac{3}{4}$$

EXAMPLE Find $5\frac{3}{8} \div 3$.

$$5\frac{3}{8} \div 3 = \frac{43}{8} \div \frac{3}{1} = \frac{43}{8} \times \frac{1}{3} = \frac{43}{24} = 1\frac{19}{24}$$

EXAMPLE A developer subdivides $5\frac{1}{4}$ acres into lots; each lot is $\frac{7}{10}$ of an acre. How many lots are made?

$$5\frac{1}{4} \div \frac{7}{10} = \frac{21}{4} \div \frac{7}{10} = \frac{\overset{3}{\cancel{21}}}{\underset{2}{\cancel{4}}} \times \frac{\overset{5}{\cancel{10}}}{\underset{1}{\cancel{7}}} = \frac{15}{2} = 7\frac{1}{2}$$

Seven lots are made so that each is $\frac{7}{10}$ of an acre. The $\frac{1}{2}$ lot is left over.

EXAMPLE

How many pieces of cable, each $1\frac{3}{4}$ ft long, can be cut from a reel containing 100 ft of cable? See Fig. 3–28.

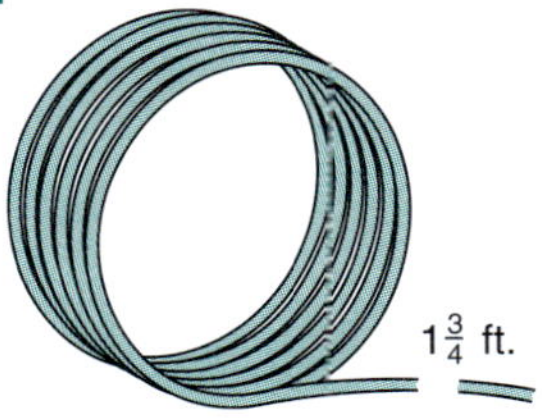

$$100 \div 1\frac{3}{4} = \frac{100}{1} \div \frac{7}{4} = \frac{100}{1} \times \frac{4}{7} = \frac{400}{7} = 57\frac{1}{7}$$

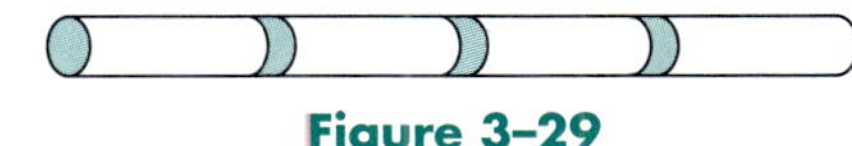

Figure 3–28

Fifty-seven pieces of cable can be cut to the desired length. The extra $\frac{1}{7}$ of the desired length is considered waste.

EXAMPLE

A pipe that is $21\frac{1}{2}$-in.-long is cut into four equal pieces (see Fig. 3–29). If $\frac{3}{16}$ in. is wasted on each cut, how long is each piece?

Figure 3–29

Use the Six-Step Problem-Solving Plan.

Unknown facts   Length of cuts to be made

Known facts

4   number of pieces needed

3   number of cuts to be made

$\frac{3}{16}$ in.   waste for each cut

$21\frac{1}{2}$ in.   length of pipe

Relationships   Length of each piece = (Total length − 3 × Amount wasted for each cut) ÷ 4

Estimation   If the pipe is 20 in. long and cut into 4 pieces, each piece will be 5 in.

Calculation   Three cuts are to be made and each cut wastes $\frac{3}{16}$ in., so first find the amount wasted.

$$\frac{3}{16} \times 3 = \frac{9}{16} \qquad \text{Total waste.}$$

Next, find how much pipe will be left to divide equally into four pieces.

$$21\frac{1}{2} - \frac{9}{16}$$

$$21\frac{1}{2} = 21\frac{8}{16} = 20\frac{24}{16} \qquad \text{Align vertically.}$$

$$- \quad \frac{9}{16} = \frac{9}{16} = \frac{9}{16}$$

$$\overline{\qquad\qquad\qquad\qquad 20\frac{15}{16} \text{ in.}}$$

Now find the length of each piece.

$$20\frac{15}{16} \div 4 = \frac{335}{16} \div \frac{4}{1} = \frac{335}{16} \times \frac{1}{4} = \frac{335}{64} = 5\frac{15}{64}$$

Interpretation   **Each piece is $5\frac{15}{64}$ in. long.**

4 Simplify Complex Fractions.

Recall from Section 3–1 that a complex fraction is one in which either the numerator or the denominator or both contain a fraction. Examples of complex fractions are

$$\frac{\frac{3}{4}}{2}, \qquad \frac{6}{\frac{1}{2}}, \qquad \frac{\frac{2}{3}}{\frac{1}{5}}, \qquad \frac{2\frac{1}{2}}{7}, \qquad \frac{1\frac{2}{3}}{3\frac{5}{8}}, \qquad \frac{4}{1\frac{1}{2}}$$

Recall also that fractions indicate division as we read from top to bottom. The large fraction line is read as "divided by." For example,

$$\frac{2\frac{1}{2}}{7} \quad \text{is read} \quad \text{"}2\frac{1}{2}\text{ divided by 7"}$$

and

$$\frac{4}{1\frac{1}{2}} \quad \text{is read} \quad \text{"}4\text{ divided by }1\frac{1}{2}\text{"}$$

If we think of fractions as divisions, then complex fractions are merely another way of writing problems.

To simplify a complex fraction:

Perform the indicated division.

EXAMPLE Simplify $\dfrac{6\frac{3}{8}}{4\frac{1}{2}}$.

$$\frac{6\frac{3}{8}}{4\frac{1}{2}} = 6\frac{3}{8} \div 4\frac{1}{2} = \frac{51}{8} \div \frac{9}{2} = \frac{\overset{17}{\cancel{51}}}{\underset{4}{\cancel{8}}} \times \frac{\overset{1}{\cancel{2}}}{\underset{3}{\cancel{9}}} = \frac{17}{12} = 1\frac{5}{12}$$

EXAMPLE Simplify $\dfrac{\frac{4}{5}}{6}$.

$$\frac{\frac{4}{5}}{6} = \frac{4}{5} \div 6 = \frac{4}{5} \div \frac{6}{1} = \frac{\overset{2}{\cancel{4}}}{5} \times \frac{1}{\underset{3}{\cancel{6}}} = \frac{2}{15}$$

We sometimes use complex fractions to change mixed decimals to fractions. For example, $0.33\frac{1}{3}$ contains the fraction $\frac{1}{3}$. To convert a decimal number to a fraction, we write it as a complex fraction.

EXAMPLE Write $0.33\frac{1}{3}$ as a fraction.

$$0.33\frac{1}{3} = \frac{33\frac{1}{3}}{100} = 33\frac{1}{3} \div 100$$

Count places for digits only; that is, do not count the fraction $\frac{1}{3}$ as a place.

Following our rule for dividing mixed numbers, write $33\frac{1}{3}$ as an improper fraction. Then invert the divisor 100 and multiply.

$$33\frac{1}{3} \div 100 = \frac{100}{3} \div \frac{100}{1} = \frac{\overset{1}{\cancel{100}}}{3} \times \frac{1}{\underset{1}{\cancel{100}}} = \frac{1}{3}$$

Tip!

What Place Does $\frac{1}{3}$ Hold If the Decimal Is $0.33\frac{1}{3}$?

When changing decimals to fractions, we divide by the place value of the last digit. Is $\frac{1}{3}$ in $0.33\frac{1}{3}$ in the hundredths or thousandths place? Hundredths. The fraction attaches to the last digit.

$$0.12\frac{1}{2} \text{ is read "twelve and one-half hundredths"}$$

$$0.008\frac{1}{3} \text{ is read "eight and one-third thousandths"}$$

SELF-STUDY EXERCISES 3–11

1 Give the reciprocal.

1. $\frac{3}{7}$ **2.** 8 **3.** $2\frac{1}{5}$ **4.** $\frac{1}{5}$ **5.** 7

2 Divide and reduce answers to lowest terms. Convert improper fractions to whole or mixed numbers.

6. $\frac{4}{5} \div \frac{8}{9}$ **7.** $\frac{11}{32} \div \frac{3}{8}$ **8.** $\frac{3}{4} \div \frac{3}{8}$ **9.** $\frac{7}{10} \div \frac{2}{3}$

10. $\frac{5}{6} \div \frac{2}{5}$ **11.** $\frac{7}{8} \div \frac{3}{16}$ **12.** $\frac{1}{5} \div \frac{1}{2}$

13. One-half inch is divided into $\frac{1}{16}$-in. segments. How many $\frac{1}{16}$-in. segments are in $\frac{1}{2}$ in.?

3 Divide and reduce answers to lowest terms. Convert improper fractions to whole or mixed numbers.

14. $10 \div \frac{3}{4}$ **15.** $8 \div \frac{1}{4}$ **16.** $2\frac{1}{2} \div 4$ **17.** $7\frac{3}{8} \div \frac{1}{2}$ **18.** $30\frac{1}{3} \div 4\frac{1}{3}$

19. Lumber is given in rough dimensions. Rough lumber that is 2 in. thick dresses out to $1\frac{5}{8}$ in. How many dressed 2×4's are in a stack that is $29\frac{1}{4}$ in. high?

20. How many $4\frac{5}{8}$-ft lengths can be cut from a 50-ft length of conduit?

21. A truck will hold 21 yd^3 (cubic yards) of gravel. If an earth mover has a shovel capacity of $1\frac{3}{4}$ yd^3, how many shovelfuls are needed to fill the truck?

22. Three shelves of equal length are cut from a 72-in. board. If $\frac{1}{8}$ in. is wasted on each cut, what is the maximum length of each shelf? (Two cuts are made to divide the entire board into three equal lengths.)

23. If $\frac{1}{8}$ in. represents 1 ft on a drawing, find the dimensions of a room that measures $2\frac{1}{2}$ in. by $1\frac{7}{8}$ in. on the drawing. (How many $\frac{1}{8}$'s are there in $2\frac{1}{2}$; how many $\frac{1}{8}$'s are there in $1\frac{7}{8}$?)

24. A segment of I-beam is $10\frac{1}{2}$ ft long. Into how many whole $2\frac{1}{4}$-ft pieces can it be divided? Disregard waste.

25. How many $17\frac{5}{8}$-in. strips of quarter-round molding can be cut from a piece $132\frac{3}{4}$ in. long? Disregard waste.

26. How many $9\frac{1}{4}$-in. drinking straws can be cut from a $216\frac{1}{2}$-in. length of stock? How much stock is left over?

27. It takes $12\frac{1}{2}$ minutes to sand a piece of oak stock. If Isom Tibbs works 100 minutes, how many pieces does he sand?

4   Divide and reduce answers to lowest terms. Convert improper fractions to whole or mixed numbers.

28. $\dfrac{\frac{1}{2}}{\frac{2}{5}}$ **29.** $\dfrac{\frac{2}{2}}{\frac{2}{3}}$ **30.** $\dfrac{\frac{7}{1}}{1\frac{1}{4}}$ **31.** $\dfrac{2\frac{2}{5}}{5}$ **32.** $\dfrac{3\frac{1}{4}}{9\frac{3}{4}}$

33. $\dfrac{12\frac{1}{2}}{2\frac{1}{2}}$ **34.** $\dfrac{\frac{7}{1}}{1\frac{1}{4}}$ **35.** $\dfrac{1\frac{1}{3}}{\frac{5}{6}}$ **36.** $\dfrac{10}{3\frac{1}{3}}$ **37.** $\dfrac{33\frac{1}{3}}{100}$

Write as a fraction.

38. $0.16\frac{2}{3}$ **39.** $0.83\frac{1}{3}$ **40.** $0.66\frac{2}{3}$

3–12   SIGNED FRACTIONS AND DECIMALS

Learning Outcomes

  1   Change a signed fraction to an equivalent signed fraction.
  2   Perform basic operations with signed fractions.
  3   Perform basic operations with signed decimals.

1   **Change a Signed Fraction to an Equivalent Signed Fraction.**

A fraction has three basic signs, the sign of the fraction, the sign of the numerator, and the sign of the denominator. The fraction $\frac{2}{3}$ expressed as a *signed fraction* is $+\frac{+2}{+3}$. When a signed fraction has negative signs, we often change the signed fraction to an equivalent signed fraction.

The rules for operating with integers can be extended to apply to signed fractions. Applying the rules of subtraction and division allows us to manipulate the signs of a fraction.

To find an equivalent signed fraction:

Change any two of the three signs to the opposite sign.

EXAMPLE   Change $-\dfrac{-2}{-3}$ to three equivalent signed fractions.

$$-\frac{-2}{-3} = +\frac{+2}{-3}$$ Change the sign of the fraction and the sign of the numerator.

$$-\frac{-2}{-3} = +\frac{-2}{+3}$$ Change the signs of the fraction and the denominator.

$$-\frac{-2}{-3} = -\frac{+2}{+3}$$ Change the signs of the numerator and the denominator.

Changing the signs of a fraction is a manipulation tool that simplifies our work when we perform basic operations with signed fractions.

Tip! | ***Why Would We Want to Change the Signs of a Fraction?***

Sometimes we get so wrapped up in the flexibility of mathematics that we lose sight of why a manipulation is useful. When changing any two signs of a fraction, we can accomplish these desirable outcomes.

- Avoid dealing with negatives.

$$-\frac{-3}{+4} = +\frac{+3}{+4}$$

$$+\frac{-6}{-7} = +\frac{+6}{+7}$$

$$-\frac{+5}{-9} = +\frac{+5}{+9}$$

- Change subtraction to addition.

$$-\frac{+5}{+6} = +\frac{-5}{+6}$$

- Avoid negative denominators.

$$-\frac{+3}{-4} = +\frac{+3}{+4}$$

$$+\frac{+2}{-5} = +\frac{-2}{+5}$$

- Deal with fewer negatives.

$$-\frac{-5}{-8} = +\frac{-5}{+8}$$

---

2 **Perform Basic Operations with Signed Fractions.**

We use the same rules for performing the basic operations with integers as we use for signed fractions. In applying these rules, we change the signs of a fraction whenever it gives us a simpler problem to work.

EXAMPLE   Add $\dfrac{-3}{4} + \dfrac{5}{-8}$.

$$\frac{-3}{4} + \frac{-5}{8}$$ Change the signs of the numerator and denominator in the second fraction so that both denominators are positive.

$$\frac{-6}{8} + \frac{-5}{8}$$ Change to equivalent fractions with a common denominator.

$$\frac{-11}{8}$$ Add numerators, applying the rule for adding numbers with like signs.

$$-1\frac{3}{8}$$ Change to mixed number. The sign of the mixed number is determined by the rule for dividing numbers with unlike signs.

$\boxed{\qquad}$ **EXAMPLE**   Subtract $\dfrac{-3}{7} - \dfrac{-5}{7}$.

$$\dfrac{-3}{7} + \dfrac{5}{7}$$ Change the signs of the second fraction by changing the signs of the fraction and the numerator.

$$\dfrac{2}{7}$$ Apply the rule for adding numbers with unlike signs.

Tip!	*Manipulating Signs.*

Manipulating the signs of a fraction is used to simplify the steps of a problem. Generally, we use this tool to create an equivalent problem with fewer negative signs.

EXAMPLE   Multiply $\left(\dfrac{-4}{5}\right)\left(\dfrac{3}{-7}\right)$.

$$\dfrac{-4}{5} \cdot \dfrac{3}{-7}$$ Multiply numerators and denominators.

$$\dfrac{-12}{-35}$$ Apply rule for dividing numbers with like signs.

$$\dfrac{12}{35}$$

EXAMPLE   Simplify $\left(\dfrac{-2}{3}\right)^3$.

$$\left(\dfrac{-2}{3}\right)^3$$ Cube numerator and cube denominator.

$$\dfrac{(-2)^3}{3^3}$$ Apply rule for raising a negative number to an odd power.

$$\dfrac{-8}{27} \quad \text{or} \quad -\dfrac{8}{27}$$

3 Perform Basic Operations with Signed Decimals.

The same rules for performing the basic operations with integers can be used with signed decimals.

EXAMPLE   Add -5.32 and -3.24.

$$\begin{array}{r} -5.32 \\ -3.24 \\ \hline -8.56 \end{array}$$ Align decimals and use the rule for adding numbers with like signs.

EXAMPLE   Subtract -3.7 from 8.5.

$$8.5 - (-3.7) = 8.5 + 3.7 = \mathbf{12.2}$$

EXAMPLE Multiply 3.91 and -7.1.

$$
\begin{array}{r}
3.91 \\
-\ 7.1 \\
\hline
391 \\
27\ 37 \\
\hline
-27.761
\end{array}
$$

Apply rule for multiplying numbers with unlike signs.

Or use your calculator.

Calculator with sign change key:

$$3\ \boxed{\cdot}\ 9\ 1\ \boxed{\times}\ 7\ \boxed{\cdot}\ 1\ \boxed{+/-}\ \boxed{=}\ \Rightarrow\ -27.761$$

Calculator with negative key:

$$3\ \boxed{\cdot}\ 9\ 1\ \boxed{\times}\ \boxed{(-)}\ 7\ \boxed{\cdot}\ 1\ \boxed{=}\ \Rightarrow\ -27.761$$

EXAMPLE Divide: $(-1.2) \div (-0.4)$.

Use the rule for dividing signed numbers to determine the sign of the quotient. Division of two negative numbers results in a positive quotient. Perform the division of decimals.

$$0.4\,\overline{)1.2}^{\ 3.}$$

Shift decimal points.

Tip!	*The Rules of Signed Numbers Apply to All Types of Numbers.*

The rules of signed numbers apply to all types of numbers. First, determine the appropriate sign of a problem, then perform the necessary calculations. You will make *fewer* mistakes when you apply the sign rules mentally and use the calculator only to perform calculations with the absolute values of the numbers.

SELF-STUDY EXERCISES 3–12

1 Change the fractions to three equivalent signed fractions.

1. $+\dfrac{+5}{+8}$ **2.** $-\dfrac{3}{4}$ **3.** $\dfrac{-2}{-5}$ **4.** $-\dfrac{-7}{-8}$ **5.** $\dfrac{7}{8}$

2 Perform the operations.

6. $\dfrac{-7}{8} + \dfrac{5}{8}$ **7.** $\dfrac{-4}{5} + \dfrac{-3}{10}$ **8.** $\dfrac{1}{2} - \dfrac{-3}{5}$ **9.** $\dfrac{-3}{5} \times \dfrac{10}{-11}$ **10.** $-\dfrac{5}{8} \div \dfrac{4}{5}$

3 Perform the operations with signed decimals.

11. $5.823 - 32.12$ **12.** $-8.32 + 7.21$ **13.** $-84.23 - 7.21$ **14.** $34.6 \times (-3.2)$
15. -7.2×8.2 **16.** $-83.1 \times (-4.1)$ **17.** $83.2 \div (-3)$ **18.** $-0.826 \div -2$
19. $-3.2 + 7.8(-3.2 + 0.2)$ **20.** $4.23 - 4.2/(-1.2)$

Learning Outcome

1 Use the fraction key on a calculator to perform operations with fractions.

1 Use the Fraction Key on a Calculator to Perform Operations with Fractions.

Some calculators have a special key for computing with fractions. On these calculators, numbers are entered as fractions, and the results are displayed either in fraction or decimal form (or mixed-number form or improper-fraction form). The *fraction key* is generally labeled $\boxed{a\,b/c}$. The numerator and denominator are separated with a special symbol. The fraction $\frac{2}{3}$ appears as 2 ⌐ 3 or 2/3. The mixed number $3\frac{1}{5}$ appears as 3 ⌐ 1 ⌐ 5.

To reduce a fraction using a calculator:

To reduce the fraction, enter the numerator, press the fraction key, and then enter the denominator. To display the fraction in lowest terms, press the $\boxed{=}$ key or the appropriate entry key.

EXAMPLE Reduce $\frac{30}{36}$.

$$30 \;\boxed{a\,b/c}\; 36 \;\boxed{=}\; \Rightarrow\; 5 \;⌐\; 6$$

Although we have already discussed finding the decimal equivalent of a fraction by dividing, now let's find decimal equivalents with the fraction key.

To find the decimal equivalent of a fraction with a calculator:

When a fraction or mixed number is displayed on the calculator and after the $\boxed{=}$ key has been pressed, press the fraction key, $\boxed{a\,b/c}$.

EXAMPLE Use the fraction key to show the decimal equivalent of $\frac{5}{6}$.

$$5 \;\boxed{a\,b/c}\; 6 \;\boxed{=}\; \boxed{a\,b/c}\; \Rightarrow\; 0.833333333$$

Improper fractions are found by accessing the function $\boxed{d/c}$ above the fraction key, which appears on the calculator as

$$\begin{array}{c} d/c \\ \boxed{a\,b/c} \end{array}$$

Tip! | ***Calculator Shift Keys.***

A special key is used to access calculator functions that are written above the keys. This key is labeled differently on various calculators. Frequently used labels are $\boxed{\text{SHIFT}}$ $\boxed{\text{INV}}$ and $\boxed{\text{2nd}}$. Many times, these keys and the labels above the keys are color coded to make the functions and the access key easier to locate.

EXAMPLE Change $2\frac{7}{8}$ to an improper fraction using the improper-fraction key.

2 $\boxed{a\ b/c}$ 7 $\boxed{a\ b/c}$ 8 $\boxed{\text{SHIFT}}$ $\boxed{d/c}$ $\Rightarrow$ Shift key may be labeled differently.

$$\frac{23}{8}$$

Tip!	**Calculator Toggle Keys.**

Some keys on a calculator act as a "toggle" in certain situations. With $\frac{23}{8}$ in the display, press the fraction key several times. The display alternates among the improper-fraction form, the decimal equivalent, and the mixed number. The $\boxed{a\ b/c}$ key is a toggle key in this instance.

The fraction key can also be used to add, subtract, multiply, or divide fractions.

To perform calculations of fractions on the calculator:

1. Enter the fraction or mixed number using the fraction key.
2. Use the operation keys as usual.

EXAMPLE Simplify $\frac{3}{4} - \frac{2}{3} + 3\frac{1}{2}$.

3 $\boxed{a\ b/c}$ 4 $\boxed{-}$ 2 $\boxed{a\ b/c}$ 3 $\boxed{+}$ 3 $\boxed{a\ b/c}$ 1 $\boxed{a\ b/c}$ 2 $\boxed{=}$ $\Rightarrow$

$$3\frac{7}{12}$$

SELF-STUDY EXERCISES 3–13

1 Use the fraction key on your calculator to perform the operations.

1. $\dfrac{5}{8} + \dfrac{7}{12}$

2. $3\dfrac{1}{7} - 5\dfrac{3}{5}$

3. $2\dfrac{7}{8} \times 5\dfrac{1}{12}$

4. $\dfrac{3}{5} \div \dfrac{12}{35}$

5. $3\dfrac{1}{5} \div 4$

6. $\dfrac{-1}{4} + \dfrac{3}{4}$

7. $\dfrac{-5}{8} + \dfrac{7}{12}$

8. $15\dfrac{3}{5} \times 18\dfrac{1}{12}$

9. $3\dfrac{4}{5} + 7\dfrac{3}{8}$

10. $-\dfrac{7}{8} \div -\dfrac{2}{3}$

3-14 FINDING NUMBER AND PERCENT EQUIVALENTS

Learning Outcomes

1 Change any number to its percent equivalent.

2 Change any percent to its numerical equivalent.

In our study of fractions and decimals, we have learned some very useful ways to express parts of quantities and to compare quantities. In this section we standardize our comparison of parts of quantities. We do this by expressing quantities in relation to a standard unit of 100. This relationship, called a *percent*, helps us solve many types of technical problems.

The word *percent* means "per hundred" or "for every hundred." Thus, 35 percent means 35 per hundred, or 35 out of every hundred, or $\frac{35}{100}$, and 100 percent means 100 out of 100 parts, or $\frac{100}{100}$, or 1 whole quantity. The symbol % is used to represent "percent."

■ **DEFINITION: Percent.** A *percent* is a fractional part of 100 expressed with a percent sign (%).

1 Change Any Number to Its Percent Equivalent.

We often need to use percents on the job, but the numbers we are given are not expressed as percents. For example, suppose an electronic parts supply house always keeps its inventory of general-purpose soldering irons at 1,200. After a large order is filled, the inventory drops to 800 soldering irons, and the parts manager orders one-half of that amount to bring the inventory back up to 1,200. To express this $\frac{1}{2}$ as a percent, we multiply by 1 in the form of 100%.

$$\frac{1}{2} \times 100\% = \frac{1}{\overset{}{\underset{1}{2}}} \times \frac{\overset{50}{\cancel{100}}\%}{1} = 50\%$$

> *To change any number to its percent equivalent:*
>
> Multiply by 1 in the form of 100%.

We can use this rule to change any type of number—such as a fraction, decimal, whole number, or mixed number—to a percent equivalent.

Tip!	***Multiplying by 1 Can Take Many Forms.***

In working with whole numbers, we studied the multiplicative identity, which states that $n \times 1 = n$. In working with fractions, we use the multiplicative identity again to change the fraction to an equivalent fraction. This time we use the number 1 in the form of $\frac{n}{n}$.

$$\frac{1}{2} \times \frac{3}{3} = \frac{3}{6}$$

Now, we use the property with the number 1 in the form of 100%.

$$100\% = 1$$

EXAMPLE Change the fractions to percent equivalents: $\frac{1}{4}, \frac{1}{3}, \frac{3}{8}, \frac{4}{7}, \frac{1}{200}$, and $\frac{3}{1,000}$.

$$\frac{1}{4} \times 100\% = \frac{1}{\underset{1}{\cancel{4}}} \times \frac{\overset{25}{\cancel{100}}\%}{1} = \mathbf{25\%}$$

$$\frac{1}{3} \times 100\% = \frac{1}{3} \times \frac{100\%}{1} = \frac{100\%}{3} = \mathbf{33\tfrac{1}{3}\%}$$

$$\frac{3}{8} \times 100\% = \frac{3}{\underset{2}{\cancel{8}}} \times \frac{\overset{25}{\cancel{100}}\%}{1} = \frac{75\%}{2} = \mathbf{37\tfrac{1}{2}\%}$$

$$\frac{4}{7} \times 100\% = \frac{4}{7} \times \frac{100\%}{1} = \frac{400\%}{7} = 57\frac{1}{7}\%$$

$$\frac{1}{200} \times 100\% = \frac{1}{\underset{2}{200}} \times \frac{\overset{1}{100\%}}{1} = \frac{1}{2}\%$$

$\frac{1}{2}\%$ means $\frac{1}{2}$ of every hundredth, or $\frac{1}{2}$ of 1%.

$$\frac{3}{1,000} \times 100\% = \frac{3}{\underset{10}{1,000}} \times \frac{\overset{1}{100\%}}{1} = \frac{3}{10}\%$$

$\frac{3}{10}\%$ means $\frac{3}{10}$ of every hundredth, or $\frac{3}{10}$ of 1%.

EXAMPLE Change the decimals to percent equivalents: 0.3, 0.25, 0.07, 0.006, and 0.0025.

Let's use the shortcut procedure to multiply by 100: Move the decimal point two places to the right.

$$0.3 \times 100\% = 030.\% = \mathbf{30\%}$$
$$0.25 \times 100\% = 025.\% = \mathbf{25\%}$$
$$0.07 \times 100\% = 007.\% = \mathbf{7\%}$$
$$0.006 \times 100\% = 000.6\% = \mathbf{0.6\%}$$

0.6% means 0.6 of every hundredth, or 0.6 of 1%.

$$0.0025 \times 100\% = 000.25\% = \mathbf{0.25\%}$$

0.25% means 0.25 of every hundredth, or 0.25 of 1%.

EXAMPLE Change the whole numbers to their percent equivalents: 1, 3, 7.

$$1 \times 100\% = \mathbf{100\%}$$ 100 out of 100 or all of something.

$$3 \times 100\% = \mathbf{300\%}$$

When we talk of more than 100%, we are talking about more than one whole quantity. Thus, 300% is three whole quantities, or three times a quantity.

$$7 \times 100\% = \mathbf{700\%}$$ 7 whole quantities, or 7 times a quantity.

EXAMPLE Change the mixed numbers and decimals to their percent equivalents: $1\frac{1}{4}$, $3\frac{2}{3}$, 5.3, and 5.12.

$$1\frac{1}{4} \times 100\% = \frac{5}{\underset{1}{4}} \times \frac{\overset{25}{100\%}}{1} = \mathbf{125\%}$$

$$3\frac{2}{3} \times 100\% = \frac{11}{3} \times \frac{100\%}{1} = \frac{1,100\%}{3} = \mathbf{366\frac{2}{3}\%}$$

$$5.3 \times 100\% = 530.\% = \mathbf{530\%}$$

$$5.12 \times 100\% = 512.\% = \mathbf{512\%}$$

2 Change Any Percent to Its Numerical Equivalent.

Percents are a convenient way to express the relationship of any quantity to 100. They are excellent time-savers when we make comparisons or state problems on the job. However, we cannot use percents to solve a problem. Instead, we first convert the percents to fraction, decimal, whole, or mixed-number equivalents.

Tip! | ***Dividing by 1 Can Take Many Forms.***

With whole numbers $\frac{n}{n} = 1$. To reduce fractions, we use this property again as

$$\frac{6}{10} \div \frac{2}{2} = \frac{6 \div 2}{10 \div 2} = \frac{3}{5}$$

Now, with percents we apply this property using 1 in the form of 100%.

The numerical equivalent of a percent can be expressed in fraction or decimal form as shown in the following example.

EXAMPLE Change the percents to their fraction and decimal equivalents: 75%, 38%, and 5%.

	Fraction equivalent	**Decimal equivalent**
75%	75% ÷ 100%	75% ÷ 100%

$$\frac{\overset{3}{\cancel{75\%}}}{1} \times \frac{1}{\underset{4}{\cancel{100\%}}} = \frac{3}{4} \qquad .75 = \mathbf{0.75}$$

Use the shortcut procedure for dividing by 100%: move the decimal point two places to the left.

38%	38% ÷ 100%	38% ÷ 100%

$$\frac{\overset{19}{\cancel{38\%}}}{1} \times \frac{1}{\underset{50}{\cancel{100\%}}} = \frac{19}{50} \qquad .38 = \mathbf{0.38}$$

5%	5% ÷ 100%	5% ÷ 100%

$$\frac{\overset{1}{\cancel{5\%}}}{1} \times \frac{1}{\underset{20}{\cancel{100\%}}} = \frac{1}{20} \qquad .05 = \mathbf{0.05}$$

Tip! | ***Division Expressed as Multiplication.***

As with fractions, we see that it is again convenient to change division to an equivalent multiplication. Is dividing by 100% the same as multiplying by $\frac{1}{100\%}$? Yes.

$$\text{A percent} \div 100\% = \text{A percent} \div \frac{100\%}{1} = \text{A percent} \times \frac{1}{100\%}$$

Some percents will change more conveniently to one numerical equivalent than to another. In solving problems, we normally change the percent to the most convenient equivalent for the problem. In the following examples, both fraction and decimal equivalents are given, and you can judge for yourself when fraction equivalents are more convenient than decimal equivalents, and vice versa.

 Change the percents to their fraction and decimal equivalents: $33\frac{1}{3}\%$, $37\frac{1}{2}\%$.

Fraction equivalent   **Decimal equivalent**

$$33\frac{1}{3}\% \div 100\% \qquad 33\frac{1}{3}\% \div 100\%$$

$$\frac{\overset{1}{\cancel{100\%}}}{3} \times \frac{1}{\underset{1}{\cancel{100\%}}} = \frac{1}{3} \qquad .33\frac{1}{3} = \mathbf{0.33\frac{1}{3}}$$

Remember, a decimal point separates whole quantities from the fraction parts. Therefore, there is an *unwritten* decimal between 33 and $\frac{1}{3}$. Because $\frac{1}{3}$ does not change to a terminating decimal equivalent, using the decimal equivalent of $33\frac{1}{3}\%$ will create more extensive calculations.

Fractional equivalent   **Decimal equivalent**

$$37\frac{1}{2}\% \div 100\% \qquad 37\frac{1}{2}\% \div 100\% = \mathbf{0.37\frac{1}{2}}$$

$$\frac{\overset{3}{\cancel{75}}}{2}\% \times \frac{1}{\underset{4}{\cancel{100\%}}} = \frac{3}{8}$$

Decimal equivalents are desirable when we use calculators. However, the decimal $0.37\frac{1}{2}$ would still not be adaptable to most calculators because of the fraction that remains. For that reason, when changing a mixed-number percent to decimal form, we first change the fractional part of the mixed number to its decimal equivalent.

$$\frac{1}{2} = 0.5 \qquad 2\overline{)1.0}^{\,0.5}$$

Thus, $37\frac{1}{2}\% = 37.5\%$. Then, divide by 100%.

$$37.5\% \div 100\% = \mathbf{0.375}$$

Now the decimal equivalent may be used on a calculator.

 Change 5.25% to its fractional and decimal equivalents.

Fractional equivalent   **Decimal equivalent**

$$5.25\% = 5\frac{25}{100}\% = 5\frac{1}{4}\% \qquad\qquad 5.25\% = 5.25\% \div 100\% = \mathbf{0.0525}$$

$$5\frac{1}{4}\% = 5\frac{1}{4}\% \div 100\%$$

$$= \frac{21}{4}\% \times \frac{1}{100\%}$$

$$= \mathbf{\frac{21}{400}}$$

SECTION 3–14   Finding Number and Percent Equivalents

What Happens to the % (Percent) Sign?

Remember from multiplying fractions that we can reduce or cancel common factors from a numerator to a denominator. Percent signs and other types of labels can also cancel.

$$\frac{\cancel{\%}}{1} \times \frac{1}{\cancel{\%}} = 1$$

EXAMPLE Change the percents to their fractional and decimal equivalents: $\frac{1}{2}\%$ and 0.25%.

Fractional equivalent

$$\frac{1}{2}\% = \frac{1}{2}\% \div 100\% = \frac{1}{2}\% \times \frac{1}{100\%} = \frac{1}{200}$$

$$0.25\% = \frac{25}{100}\% = \frac{1}{4}\%$$

$$\frac{1}{4}\% = \frac{1}{4}\% \div 100\%$$

$$= \frac{1}{4}\% \times \frac{1}{100\%}$$

$$= \frac{1}{400}$$

Decimal equivalent

$$\frac{1}{2}\% = 0.5\% = 0.5\% \div 100\% = \mathbf{0.005}$$

$$0.25\% = 0.25\% \div 100\% = \mathbf{0.0025}$$

When a quantity is 100% or more, the fractional and decimal equivalents will be equal to or more than the whole number 1. The numerical equivalents will be whole numbers, mixed numbers, or mixed decimals. Again, when solving problems, use the more convenient equivalent.

EXAMPLE Change the percents to their whole-number equivalents or to both their mixed-number and mixed-decimal equivalents: 700%, 375%, $233\frac{1}{3}\%$, and $462\frac{1}{2}\%$.

$$700\% \div 100\% = \frac{\cancel{700\%}^{7}}{1} \times \frac{1}{\cancel{100\%}_{1}} = \mathbf{7}$$

or

$$700\% \div 100\% = \mathbf{7}$$

Mixed-number equivalent

$$375\% \div 100\%$$

$$\frac{\cancel{375\%}^{15}}{1} \times \frac{1}{\cancel{100\%}_{4}} = \frac{15}{4} = \mathbf{3\frac{3}{4}}$$

Mixed-decimal equivalent

$$375\% \div 100\% = \mathbf{3.75}$$

Mixed-number
equivalent

$$233\frac{1}{3}\% \div 100\% = \frac{\overset{7}{\cancel{700}\%}}{3} \times \frac{1}{\cancel{100}\%} = \frac{7}{3} = 2\frac{1}{3}$$

To allow more mental calculation, consider $233\frac{1}{3}\%$ to be $200\% + 33\frac{1}{3}\%$.

$$200\% = 2 \qquad \text{and} \qquad 33\frac{1}{3}\% = \frac{1}{3}$$

Thus, $200\% + 33\frac{1}{3}\% = 2 + \frac{1}{3} = \mathbf{2\frac{1}{3}}$.

Mixed-decimal
equivalent

$$233\frac{1}{3}\% \div 100\% = \mathbf{2.33\frac{1}{3}}$$

Mixed-number
equivalent

$$462\frac{1}{2}\% \div 100\% = \frac{\overset{37}{\cancel{925}}}{2} \times \frac{1}{\cancel{100}} = \frac{37}{8} = 4\frac{5}{8}$$

Again, it is easier to consider $462\frac{1}{2}\%$ to be $400\% + 62\frac{1}{2}\%$.

$$400\% = 4 \qquad \text{and} \qquad 62\frac{1}{2}\% \div 100\% = \frac{\overset{5}{\cancel{125}}}{2} \times \frac{1}{\cancel{100}} = \frac{5}{8}$$

Then, $462\frac{1}{2}\% = 4 + \frac{5}{8} = \mathbf{4\frac{5}{8}}$.

Mixed-decimal
equivalent

$$462\frac{1}{2}\% = 462.5\%$$

$$462.5\% \div 100\% = \mathbf{4.625}$$

Because many percents, fractions, and decimals are frequently used, it would be helpful to know these equivalents from memory. Knowing these will save time on the job. Table 3–1 lists the most commonly used percents and equivalents. The equivalents were determined by the procedures used in this section.

Percent	Fraction	Decimal	Percent	Fraction	Decimal
10%	$\frac{1}{10}$	0.1	60%	$\frac{3}{5}$	0.6
20%	$\frac{1}{5}$	0.2	$66\frac{2}{3}\%$	$\frac{2}{3}$	$0.66\frac{2}{3}$ or 0.667^{a}
25%	$\frac{1}{4}$	0.25	70%	$\frac{7}{10}$	0.7
30%	$\frac{3}{10}$	0.3	75%	$\frac{3}{4}$	0.75
$33\frac{1}{3}\%$	$\frac{1}{3}$	$0.33\frac{1}{3}$ or 0.333^{a}	80%	$\frac{4}{5}$	0.8
40%	$\frac{2}{5}$	0.4	90%	$\frac{9}{10}$	0.9
50%	$\frac{1}{2}$	0.5	100%	$\frac{1}{1}$	1.0

[a]These decimals can be expressed with fractions or rounded decimals. Their fraction equivalents are exact amounts, and rounded decimals are approximate amounts.

To help us remember these equivalents, let's group them differently and suggest some mental calculations as memory aids.

$$50\% = \frac{1}{2} = 0.5$$

This is one of the easiest to learn because of its direct comparison to money. One dollar is 100 cents. One-half dollar is 50 cents. One-half dollar is $0.50 in dollar-and-cent notation.

$$25\% = \frac{1}{4} = 0.25$$

$$75\% = \frac{3}{4} = 0.75$$

Again, relating these to money, one-fourth of a dollar is 25 cents or $0.25, and three-fourths of a dollar is three times as much or 75 cents or $0.75.

Both 10% and multiples of 10% are usually easy to remember.

EXAMPLE Write the percents as decimals. 10%, 20%, 30%, 40%, 50%, 60%, 70%, 80%, 90%, 100%, $33\frac{1}{3}\%$, and $66\frac{2}{3}\%$.

$$10\% = \frac{1}{10} = \mathbf{0.1} \qquad 60\% = \frac{6}{10} = \frac{3}{5} = \mathbf{0.6}$$

$$20\% = \frac{2}{10} = \frac{1}{5} = \mathbf{0.2} \qquad 70\% = \frac{7}{10} = \mathbf{0.7}$$

$$30\% = \frac{3}{10} = \mathbf{0.3} \qquad 80\% = \frac{8}{10} = \frac{4}{5} = \mathbf{0.8}$$

$$40\% = \frac{4}{10} = \frac{2}{5} = \mathbf{0.4} \qquad 90\% = \frac{9}{10} = \mathbf{0.9}$$

$$50\% = \frac{5}{10} = \frac{1}{2} = \mathbf{0.5} \qquad 100\% = \frac{1}{1} = \mathbf{1}$$

When 3 is divided into 100, the result is $33\frac{1}{3}$.

$$3\overline{)100} \quad = \quad 33\frac{1}{3}$$

$$33\frac{1}{3}\% = \frac{1}{3} = 0.33\frac{1}{3}$$

Two times $33\frac{1}{3}$ is $66\frac{2}{3}$.

$$66\frac{2}{3}\% = \frac{2}{3} = 0.66\frac{2}{3}$$

Tip! | ***Estimating Percents.***

Find a relationship between the given fraction and some common fraction and percent equivalent that you know from memory. Then, use that common equivalent to estimate the percent equivalent of the given fraction.

- Approximately what percent is $\frac{1}{6}$ of a quantity? $\frac{1}{6} = \frac{1}{2}$ of $\frac{1}{3}$ and $\frac{1}{3}$ equals $33\frac{1}{3}\%$. What is $\frac{1}{2}$ of $33\frac{1}{3}\%$?

  So $\frac{1}{6}$ of a quantity is approximately 16% or 17%.

- Approximately what percent is $\frac{3}{8}$ of a quantity? $\frac{3}{8}$ is halfway between $\frac{2}{8}$ or $\frac{1}{4}$ and $\frac{4}{8}$ or $\frac{1}{2}$. Then, $\frac{3}{8}$ is halfway between 25% and 50% or $37\frac{1}{2}\%$. So, $37\frac{1}{2}\%$ is the exact equivalent rather than an estimated equivalent. An estimate that $\frac{3}{8}$ is between 35% and 40% is a good estimate.

EXAMPLE Estimate the percent equivalent of $\frac{3}{20}$ and $\frac{4}{7}$.

Approximately what percent is $\frac{3}{20}$?

$$\frac{1}{20} = \frac{1}{2} \text{ of } \frac{1}{10} = \frac{1}{2} \text{ of } 10\%$$

So $\dfrac{1}{20} = \dfrac{1}{2}$ of $10\% = 5\%$

Then $\dfrac{3}{20} = 3 \times \dfrac{1}{20} = 3 \times 5\% = \mathbf{15\%}$ This is an exact equivalent.

Approximately what percent is $\frac{4}{7}$?

$$\frac{3}{7} < \frac{1}{2} \quad \text{and} \quad \frac{4}{7} > \frac{1}{2}$$

So $\dfrac{4}{7} > 50\%$

55% or **60%** would be a good estimate.

<table>
<tr><td>Tip!</td><td>Do Part of the Calculation on the Calculator and the Other Part Mentally.</td></tr>
</table>

When you need to find the percent equivalent of a fraction, why not change the fraction to a decimal equivalent by dividing? Then, change the decimal to a percent mentally by multiplying by 100%.

$$\frac{4}{7} = 4 \div 7 = 0.571428571$$

Then, $\frac{4}{7} \approx 57.14\%$ **Mentally move the decimal two places to the right.**

SELF-STUDY EXERCISES 3–14

1

1. If $\frac{2}{5}$ of the electricians in a city are self-employed, what percent are self-employed?

2. If $\frac{7}{10}$ of the bricklayers in a city are male, what percent are male?

Change the numbers to their percent equivalents.

3. $\frac{5}{8}$ **4.** $\frac{7}{9}$ **5.** $\frac{7}{1000}$ **6.** $\frac{1}{350}$ **7.** 0.2

8. 0.14   **9.** 0.007   **10.** 0.0125   **11.** 5   **12.** 8

13. $1\frac{1}{3}$ **14.** $3\frac{1}{2}$ **15.** $4\frac{3}{10}$ **16.** $2\frac{1}{5}$ **17.** 3.05

18. 7.2   **19.** 15.1   **20.** 36.25

2 Change to both fractional and decimal equivalents.

21. 36%   **22.** 45%   **23.** 20%

24. 75%   **25.** $6\frac{1}{4}\%$ **26.** 62.5%

27. $66\frac{2}{3}\%$ **28.** 0.6%   **29.** $\frac{1}{5}\%$

30. 0.05%   **31.** $8\frac{1}{3}\%$ **32.** 18.75%

Change to equivalent whole numbers.

33. 800%   **34.** 400%

Change to both mixed-number and mixed-decimal equivalents.

35. 250%   **36.** 425%   **37.** 176%

38. 380%   **39.** $137\frac{1}{2}\%$ **40.** 387.5%

Change to mixed-number equivalents.

41. $166\frac{2}{3}\%$ **42.** $316\frac{2}{3}\%$

Change to mixed-decimal equivalents.

43. 115.3%   **44.** $212\frac{1}{2}\%$ **45.** $106\frac{1}{4}\%$

Fill in the missing percents, fractions, or decimals *from memory*. Work this exercise only after memorizing the equivalents in Table 3–1.

Percent	Fraction	Decimal	Percent	Fraction	Decimal	Percent	Fraction	Decimal
46. 10%			**47.**	$\frac{1}{4}$		**48.**		0.2
49.	$\frac{1}{3}$		**50.** 50%			**51.**	$\frac{4}{5}$	
52.		0.75	**53.** $66\frac{2}{3}\%$			**54.**		1
55.	$\frac{3}{10}$		**56.** 40%			**57.**		0.7
58.	$\frac{9}{10}$		**59.** 60%			**60.**	$\frac{1}{1}$	
61.		0.25	**62.**	$\frac{2}{3}$		**63.** 70%		
64.		0.5	**65.**	$\frac{3}{5}$		**66.**		0.1
67. 20%			**68.**		$0.33\frac{1}{3}$	**69.**		$0.66\frac{2}{3}$
70.	$\frac{3}{4}$		**71.** 30%			**72.**		0.9
73.	$\frac{1}{5}$		**74.**		0.6	**75.** 100%		

Estimate the percent equivalents for the fractions.

76. $\frac{3}{5}$ **77.** $\frac{3}{7}$ **78.** $\frac{7}{10}$ **79.** $\frac{5}{8}$ **80.** $\frac{9}{10}$

Aerospace Technology: Asteroid and Comet Impacts on Earth

In January 1997, NASA tracked 115+ asteroids in resonance with Earth, 7,000+ numbered asteroids (well-defined orbits), and 15,000+ unnumbered asteroids. Most asteroids revolve around the Sun in a more elliptical (oval) orbit than Earth's nearly circular orbit, which means that some asteroids from the asteroid belt cross Earth's path. These Earth-crossing asteroids are called Apollo asteroids. An asteroid is said to be "in resonance" with the Earth if it is periodically "close" to Earth. A collision could eventually result because Earth's gravitational pull would tug the asteroid even closer.

For obvious reasons, predicting when an asteroid or comet might hit the Earth is of great interest to NASA, the U.S. National Security Council, and the general public. Recent technological developments in computer simulation software, electronic CCD cameras (mounted onto telescopes), and electronic transmission of data have made it possible to detect asteroids more easily, define their orbits more quickly, and predict with reasonable confidence any close approaches to Earth. A close approach is generally regarded as 0.15 astronomical unit (AU). One AU is the average distance of Earth from the Sun, approximately 93,000,000 miles. Thus, a close approach is approximately 0.15 × 93,000,000 or 13,950,000 miles.

You can mathematically predict how often an Apollo asteroid will come close to Earth (and possibly collide with Earth) by finding the least common multiple

(LCM) of their orbiting periods around the Sun. For example, an asteroid named Toro in a highly elliptical orbit around the Sun comes close periodically to Venus's almost circular orbit. Toro orbits the Sun in about 584 days (approximately 1.6 years or ~1.6 years), and Venus orbits in about 224 days (~0.6152 years). The LCM of these orbiting periods is 2,920 days (~8 years). In 2,920 days, Toro orbits the Sun 5 times while Venus orbits the Sun 13 times. We say Toro is in "5 to 13 resonance" with Venus, and collision is possible. Whether the collision will occur depends on such factors as their minimal distance apart in their current orbits, Toro's orbital inclination to the ecliptic plane, and other planets in resonance with Toro that might intersect it first.

Exercises

For these exercises, assume that 1 year = 365 days.

1. Toro is also in resonance with Earth. Find the LCM of Toro's and Earth's orbiting periods (584 and 365 days, respectively).
2. How many orbits will Toro and Earth complete each during this time? State the resonance of Toro with Earth.
3. Earth's collision with Toro is not expected for at least 1 million years because Toro's orbit is tipped slightly out of the ecliptic plane. Diagram the orbits of Earth and Toro around the Sun looking down on the ecliptic plane, then show the same information but from a side view of the ecliptic plane on a second diagram.
4. NASA's definition of a *close approach* is when an asteroid or comet comes within 13,950,000 miles (0.15 AU) of Earth. Comet Toutaki is predicted to be ~3,255,000 miles (0.035 AU) from Earth on December 30, 2000. It will have its next close approach to Earth in 2004, which means Toutaki is in a 1 to 4 resonance with Earth. What is the LCM of Toutaki's and Earth's orbiting periods in years? In days?
5. Do you have enough information to find Toutaki's orbiting period around the Sun? If so, what is its period in years? In days? Is this the only possible answer?
6. NASA's Near Earth Asteroid Rendezvous (NEAR) Mission reached Eros, a 22-mi-long asteroid, in February 2000 and will orbit it for at least 6 months. Scientists hope to learn its dimensions, mass, density, spin, chemical composition, and geology. Eros orbits the Sun 4 times in the time Earth orbits the Sun 7 times, so Eros is in a 4:7 resonance with Earth. Find Eros's orbiting period in days and in years.
7. If NASA wants to reach Eros again, when is the next time Eros and Earth will be positioned about the same as in 2000?
8. Scientists have found gaps in the asteroid belt corresponding to missing asteroids whose mathematically calculated orbiting periods would have predicted them to be in resonance with Jupiter. Give one possible explanation of what happened to the missing asteroids.
9. If a missing asteroid's orbiting period is predicted to be ~3,650 days, and Jupiter's orbiting period is ~12 years, find the LCM that predicts how often the asteroid and Jupiter will be close to each other.
10. Use Exercise 9 to find the resonance of the asteroid with Jupiter.

Answers

1. LCM = 2,920 days, or 8 years.
2. Toro will complete 5 orbits, and Earth will complete 8 orbits. Toro is in a 5:8 resonance with Earth.

3. Top view: Side view:

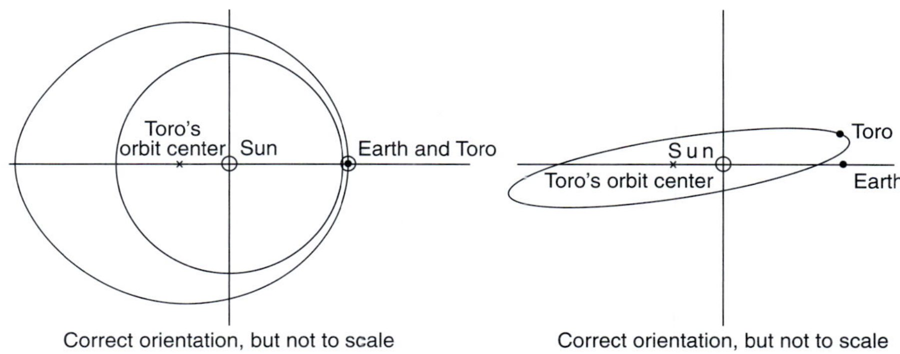

4. LCM = 4 years = ~1460 days.
5. Yes. Comet Toutaki's orbiting period is 4 years, or 1,460 days. This is the only possible answer.
6. Eros's orbiting period is ~639 days, or ~1.75 years.
7. The next time will be in 2008.
8. The missing asteroids were predicted to be in resonance with Jupiter, so they were frequently close to Jupiter. Jupiter's gravitational pull would have eventually sucked in the missing asteroids, leaving "negative rings," or gaps, in the asteroid belt.
9. 3650 days = 10 years. The LCM of 10 years and 12 years is 60 years. So the asteroid and Jupiter would have been close every 60 years.
10. In 60 years, the asteroid orbited the Sun 6 times and Jupiter orbited the Sun 5 times, so the asteroid was in a 5:6 resonance with Jupiter.

ASSIGNMENT EXERCISES

Section 3–1
Write a fraction that represents the shaded portion of Figs. 3–30 to 3–34.

1.

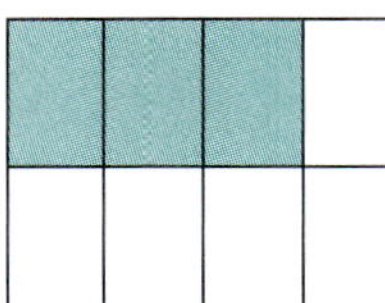

Figure 3–30

2.

Figure 3–31

3.

Figure 3–32

4.

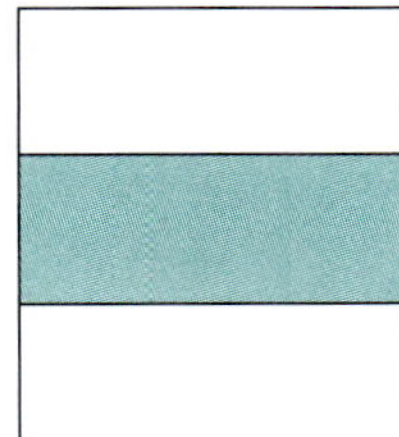

Figure 3–33

5.

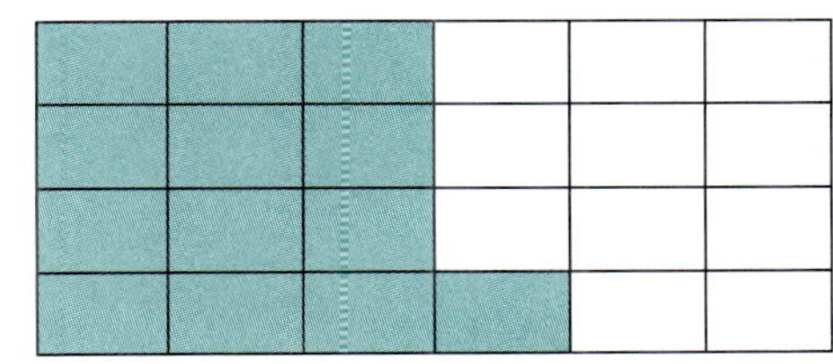

Figure 3–34

Use the numbers in Exercises 6–8 to answer the questions.

6. $\dfrac{3}{7}$

7. $\dfrac{9}{4}$

8. $\dfrac{2}{2}$

(a) What is the numerator of the fraction?
(b) What is the denominator of the fraction?
(c) One unit has been divided into how many parts?
(d) How many of these parts are used?
(e) The fraction can be read as _____ divided by _____.
(f) What is the divisor of the division?
(g) What is the dividend of the division?
(h) Is this fraction proper or improper?
(i) Does this fraction have a value less than, more than, or equal to 1?
(j) This fraction represents _____ out of _____ parts.

Match the best description with each fraction.

_____ **9.** $\dfrac{5}{2}$

_____ **10.** 0.172

_____ **11.** $6\dfrac{1}{7}$

_____ **12.** 4.59

_____ **13.** $\dfrac{7}{11}$

_____ **14.** $\dfrac{\frac{2}{3}}{\frac{1}{4}}$

_____ **15.** $33\dfrac{1}{3}$

(a) Mixed-decimal fraction
(b) Decimal fraction
(c) Complex fraction
(d) Mixed number
(e) Proper fraction
(f) Improper fraction

Rewrite the decimals as proper fractions or mixed numbers.

16. 4.273 **17.** 0.87 **18.** 1.002 **19.** 2.03 **20.** 1.1

Section 3–2
Find five multiples of each number.

21. 2 **22.** 9 **23.** 21 **24.** 14
25. 7 **26.** 11 **27.** 8 **28.** 15

Is the number divisible by the given number? Explain.

29. 153 by 3 **30.** 8,234 by 4 **31.** 8,726 by 6 **32.** 5,986 by 5

List all factor pairs for each number, then write the factors in order from smallest to largest.

33. 48 **34.** 50 **35.** 51 **36.** 63

Section 3–3
Find the prime factorization of each number. Write in factored form and then in exponential notation.

37. 44 **38.** 128 **39.** 216 **40.** 98

Section 3–4
Find the least common multiple.

41. 18 and 40 **42.** 12 and 18 **43.** 12, 18, and 30 **44.** 6, 10, and 12

Find the greatest common factor.

45. 10 and 12 **46.** 12 and 18 **47.** 12, 18, and 30 **48.** 4, 9, and 16

Section 3–5
Find the equivalent fractions using the indicated denominators.

49. $\dfrac{5}{8} = \dfrac{?}{24}$ **50.** $\dfrac{3}{7} = \dfrac{?}{35}$ **51.** $\dfrac{5}{12} = \dfrac{?}{60}$ **52.** $\dfrac{4}{5} = \dfrac{?}{40}$ **53.** $\dfrac{2}{3} = \dfrac{?}{15}$

54. $\dfrac{4}{9} = \dfrac{?}{18}$ **55.** $\dfrac{3}{4} = \dfrac{?}{32}$ **56.** $\dfrac{1}{6} = \dfrac{?}{30}$ **57.** $\dfrac{1}{5} = \dfrac{?}{55}$ **58.** $\dfrac{7}{8} = \dfrac{?}{64}$

Reduce to lowest terms.

59. $\dfrac{6}{12}$ **60.** $\dfrac{8}{10}$ **61.** $\dfrac{4}{32}$ **62.** $\dfrac{26}{64}$

63. $\dfrac{2}{8}$ **64.** $\dfrac{8}{32}$ **65.** $\dfrac{34}{64}$ **66.** $\dfrac{16}{64}$

67. $\dfrac{12}{32}$ **68.** $\dfrac{45}{90}$ **69.** $\dfrac{6}{8}$ **70.** $\dfrac{75}{100}$

Change the decimals to fractions in lowest terms.

71. 0.7 **72.** 0.83 **73.** 0.95 **74.** 0.25

75. 0.872 **76.** 0.081 **77.** 0.02 **78.** 0.005

Change the fractions to decimals. If necessary, round to the nearest thousandth.

79. $\dfrac{1}{5}$ **80.** $\dfrac{1}{10}$ **81.** $\dfrac{5}{8}$ **82.** $\dfrac{3}{7}$ **83.** $\dfrac{9}{11}$

Section 3–6

Write the improper fractions as whole or mixed numbers.

84. $\dfrac{35}{7}$ **85.** $\dfrac{18}{5}$ **86.** $\dfrac{27}{6}$ **87.** $\dfrac{39}{8}$ **88.** $\dfrac{21}{15}$

89. $\dfrac{43}{8}$ **90.** $\dfrac{22}{7}$ **91.** $\dfrac{175}{2}$ **92.** $\dfrac{135}{3}$

Write the whole or mixed numbers as improper fractions.

93. 8 **94.** $10\dfrac{1}{2}$ **95.** $7\dfrac{1}{8}$ **96.** $5\dfrac{7}{12}$ **97.** $9\dfrac{3}{16}$

98. $7\dfrac{8}{17}$ **99.** $4\dfrac{3}{5}$ **100.** $9\dfrac{1}{9}$ **101.** 12 **102.** $16\dfrac{2}{3}$

Change the whole number to an equivalent fraction using the indicated denominator.

103. $2 = \dfrac{?}{10}$ **104.** $6 = \dfrac{?}{4}$ **105.** $11 = \dfrac{?}{3}$ **106.** $7 = \dfrac{?}{5}$

Section 3–7

Find the least common denominator.

107. $\dfrac{1}{4}, \dfrac{1}{3}, \dfrac{1}{5}$ **108.** $\dfrac{7}{8}, \dfrac{2}{3}$ **109.** $\dfrac{3}{4}, \dfrac{1}{16}$

110. $\dfrac{1}{12}, \dfrac{3}{4}$ **111.** $\dfrac{5}{12}, \dfrac{3}{10}, \dfrac{13}{15}$ **112.** $\dfrac{1}{12}, \dfrac{3}{8}, \dfrac{15}{16}$

113. Is a $\frac{5}{8}$-in. wrench larger or smaller than a $\frac{9}{16}$-in. wrench?

114. An alloy contains $\frac{2}{3}$ metal A, and the same quantity of another alloy contains $\frac{3}{5}$ metal A. Which alloy contains more of metal A?

115. Is a $\frac{3}{8}$-in.-thick piece of plasterboard thicker than a $\frac{1}{2}$-in.-thick piece?

116. A $\frac{9}{16}$-in. tube must pass through a wall. Is a $\frac{3}{4}$-in.-diameter hole large enough?

117. Is a $\frac{19}{32}$-in. wrench larger or smaller than a $\frac{7}{8}$-in. bolt head?

Which fraction is smaller?

118. $\dfrac{3}{7}, \dfrac{1}{9}$ **119.** $\dfrac{3}{8}, \dfrac{4}{8}$ **120.** $\dfrac{5}{9}, \dfrac{4}{9}$ **121.** $\dfrac{1}{4}, \dfrac{3}{16}$

122. $\dfrac{5}{8}, \dfrac{11}{16}$ **123.** $\dfrac{7}{8}, \dfrac{27}{32}$ **124.** $\dfrac{7}{64}, \dfrac{1}{4}$ **125.** $\dfrac{1}{2}, \dfrac{9}{19}$

Section 3–8

Add; reduce sums to lowest terms and convert improper fractions to mixed numbers or whole numbers.

126. $\dfrac{1}{8} + \dfrac{5}{16}$ **127.** $\dfrac{3}{16} + \dfrac{9}{64}$ **128.** $\dfrac{3}{14} + \dfrac{5}{7}$

129. $\dfrac{3}{5} + \dfrac{5}{6}$

130. $3\dfrac{7}{8} + 7 + 5\dfrac{1}{2}$

131. $2\dfrac{7}{16} + 6\dfrac{5}{32}$

132. $9\dfrac{7}{8} + 5\dfrac{3}{4}$

133. $3\dfrac{7}{8} + 5\dfrac{3}{16} + 1\dfrac{7}{32}$

134. $2\dfrac{1}{4} + 3\dfrac{7}{8}$

135. A pipe is cut into two pieces measuring $7\dfrac{5}{8}$ in. and $10\dfrac{7}{16}$ in. How long was the pipe before it was cut? Disregard waste.

136. Find the total thickness of a wall if the outside covering is $3\dfrac{7}{8}$ in. thick, the studs are $3\dfrac{7}{8}$ in., and the inside covering is $\dfrac{5}{16}$-in. paneling.

137. If $7\dfrac{5}{16}$ in. of a piece of square bar stock is turned so that it is cylindrical and $5\dfrac{9}{32}$ in. remains square, what is the total length of the bar stock?

138. Two castings weigh $27\dfrac{1}{2}$ lb and $20\dfrac{3}{4}$ lb. What is the total weight of the two castings?

139. Three metal rods $3\dfrac{1}{8}$ in., $5\dfrac{3}{32}$ in., and $7\dfrac{9}{16}$ in. are welded together end to end. How long is the welded rod?

140. In Fig. 3–35, what is the length of side A?

141. In Fig. 3–35, what is the length of side B?

142. A hollow-wall fastener has a grip range up to $\dfrac{3}{4}$ in. Is it long enough to fasten three sheets of metal $\dfrac{5}{16}$ in. thick, $\dfrac{3}{8}$ in. thick, and $\dfrac{1}{16}$ in. thick?

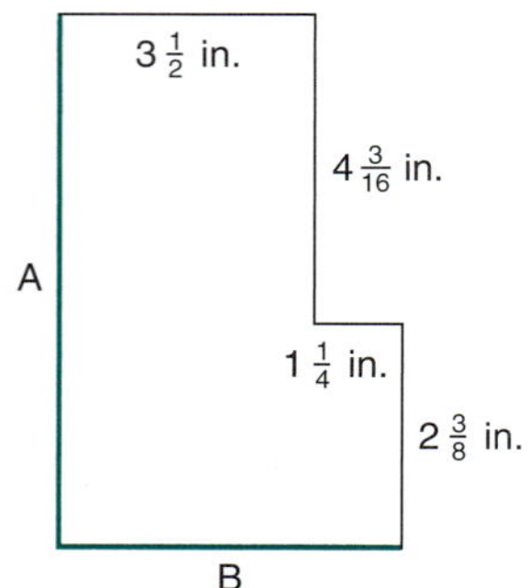

Figure 3–35

143. Figure 3–36 shows $\dfrac{1}{2}$-in. copper tubing wrapped in insulation. What is the distance across the tubing and insulation?

144. In Exercise 143, what would be the overall distance across the tubing and insulation if $\dfrac{3}{8}$-in.-ID (inside diameter) tubing were used?

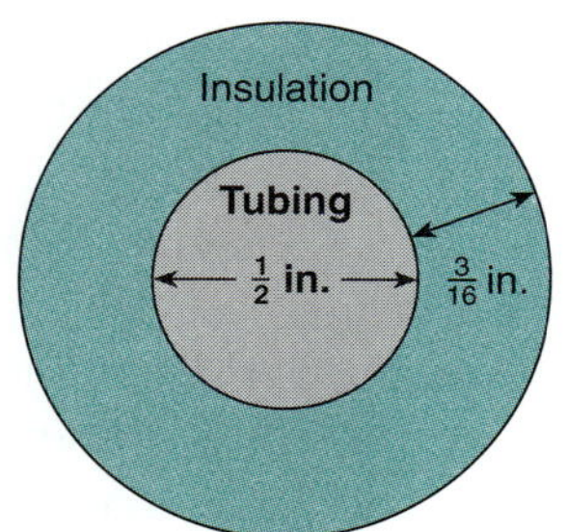

Figure 3–36

Section 3–9

Subtract; reduce to lowest terms when necessary.

145. $\dfrac{5}{9} - \dfrac{2}{9}$

146. $\dfrac{11}{32} - \dfrac{5}{64}$

147. $3\dfrac{5}{8} - 2$

148. $7 - 4\dfrac{3}{8}$

149. $8\dfrac{7}{8} - 2\dfrac{29}{32}$

150. $7 - 2\dfrac{9}{16}$

151. $12\dfrac{11}{16} - 5$

152. $48\dfrac{5}{12} - 12\dfrac{11}{15}$

153. A bolt 2 in. long fastens a piece of $\dfrac{7}{8}$-in.-thick wood to a piece of metal. If a $\dfrac{3}{16}$-in.-thick lock washer, a $\dfrac{1}{16}$-in. washer, and a $\dfrac{7}{16}$-in.-thick nut are used, what is the thickness of the metal if the nut is flush with the bolt after tightening?

154. Pins of $2\dfrac{3}{8}$ in. and $3\dfrac{7}{16}$ in. are cut from a drill rod 12 in. long. If $\dfrac{1}{16}$ in. of waste is allowed for each cut, how many inches of drill rod are left?

155. A piece of tapered stock has a diameter of $2\frac{5}{16}$ in. at one end and a diameter of $\frac{55}{64}$ in. at the other end. What is the difference in the diameters?

156. Four lengths measuring $6\frac{1}{4}$ in., $9\frac{3}{16}$ in., $7\frac{1}{8}$ in., and $5\frac{9}{32}$ in. are cut from 48 in. of copper tubing. How much copper tubing remains? Disregard waste.

Section 3–10

Multiply and reduce answers to lowest terms. Convert improper fractions to whole or mixed numbers.

157. $\frac{1}{3} \times \frac{7}{8}$

158. $\frac{2}{5} \times \frac{7}{10}$

159. $\frac{7}{9} \times \frac{3}{8}$

160. $\frac{2}{3} \times \frac{5}{8} \times \frac{3}{16}$

161. $\frac{15}{16} \times \frac{4}{5} \times \frac{2}{3}$

162. $5 \times \frac{3}{4}$

163. $\frac{7}{16} \times 18$

164. $\frac{3}{16} \times 184$

165. $1\frac{1}{2} \times \frac{4}{5}$

166. In a concrete mixture, $\frac{4}{7}$ of the total volume is sand. How much sand is needed for 135 cubic yards (yd^3) of concrete?

167. Concrete blocks are 8 in. high. If a $\frac{3}{8}$-in. mortar joint is used, how high will a wall of 12 courses of concrete blocks be? (*Hint:* There are 12 rows of mortar joints.)

168. An adjusting screw will move $\frac{3}{64}$ in. for each full turn. How far will it move in four turns?

169. A chef is making a dessert that is $\frac{3}{4}$ the original recipe. How much flour should be used if the original recipe calls for $3\frac{2}{3}$ cups of flour?

170. If an alloy is $\frac{3}{5}$ copper and $\frac{2}{5}$ zinc, how many pounds of each metal are in a casting weighing $112\frac{1}{2}$ lb?

Section 3–11

Divide and reduce answers to lowest terms. Convert improper fractions to whole or mixed numbers.

171. $\frac{7}{8} \div \frac{3}{4}$

172. $\frac{4}{9} \div \frac{5}{16}$

173. $\frac{7}{8} \div \frac{3}{32}$

174. $8 \div \frac{2}{3}$

175. $18 \div \frac{3}{4}$

176. $35 \div \frac{5}{16}$

177. $5\frac{1}{10} \div 2\frac{11}{20}$

178. $27\frac{2}{3} \div \frac{2}{3}$

179. $7\frac{1}{5} \div 12$

180. On a house plan, $\frac{1}{4}$ in. represents 1 ft. Find the dimensions of a porch that measures $4\frac{1}{8}$ in. by $6\frac{1}{2}$ in. on the plan. (How many $\frac{1}{4}$'s are there in $4\frac{1}{8}$; how many $\frac{1}{4}$'s are there in $6\frac{1}{2}$?)

181. A pipe that is 12 in. long is cut into four equal parts. If $\frac{3}{16}$ in. is wasted per cut, what is the maximum length of each pipe? (It takes three cuts to divide the entire length into four equal parts.)

182. A stack of $\frac{5}{8}$-in. plywood is $21\frac{7}{8}$ in. high. How many sheets of plywood are in the stack?

183. A rod $1\frac{1}{8}$ yd long is cut into 6 equal pieces. What is the length of each piece? Disregard waste.

184. If $7\frac{1}{2}$ gallons of liquid are distributed equally among five containers, what is the average number of gallons per container?

Divide and reduce answers to lowest terms. Convert improper fractions to whole or mixed numbers.

185. $\dfrac{\frac{1}{3}}{6}$

186. $\dfrac{4}{4\frac{4}{5}}$

187. $\dfrac{8}{1\frac{1}{2}}$

188. $\dfrac{3\frac{1}{4}}{5}$

189. $\dfrac{2\frac{1}{5}}{8\frac{4}{5}}$ **190.** $\dfrac{16\frac{2}{3}}{3\frac{1}{3}}$ **191.** $\dfrac{12\frac{1}{2}}{100}$ **192.** $\dfrac{37\frac{1}{2}}{100}$

Section 3–12

Change each fraction to three equivalent signed fractions.

193. $-\dfrac{3}{8}$ **194.** $\dfrac{-5}{9}$ **195.** $\dfrac{-7}{-8}$

Perform the indicated operations.

196. $\dfrac{-7}{8} + \dfrac{-3}{8}$ **197.** $\dfrac{5}{9} - \dfrac{-3}{7}$ **198.** $\dfrac{-5}{8} * \dfrac{-2}{3}$ **199.** $\dfrac{-4}{5} \div -\dfrac{7}{15}$

200. $3.23 + (-4.61)$ **201.** 0.27×0.13 **202.** $-4.36 + (-7.23)$ **203.** $-12.4 \div 0.2$

Section 3–13

Use your calculator to perform the operations.

204. $\dfrac{-11}{12} - \dfrac{-7}{8}$ **205.** $-\dfrac{7}{8} + \left(-\dfrac{5}{12}\right)$ **206.** $1\frac{3}{5} \div \left(-7\frac{5}{8}\right)$ **207.** $-2\frac{5}{8} \times 4\frac{1}{2}$

Section 3–14

Change to percent equivalents.

208. $\dfrac{87}{100}$ **209.** 0.7 **210.** $4\frac{1}{3}$ **211.** 125

212. $\dfrac{5}{6}$ **213.** 17.3 **214.** 18

Change to both decimal and fraction equivalents.

215. 72% **216.** 40% **217.** $12\frac{1}{2}\%$ **218.** $16\frac{2}{3}\%$ **219.** $\frac{2}{3}\%$

220. $\frac{3}{5}\%$ **221.** 275% **222.** 124% **223.** $112\frac{1}{2}\%$ **224.** $183\frac{1}{3}\%$

Change to decimal equivalents.

225. 227.2% **226.** 73.8% **227.** 9.275% **228.** 275% **229.** 340%

CHALLENGE PROBLEM

230. Len Smith has 180 ft of fencing and needs to build two square or rectangular holding yards for his two sheltie dogs. How should he design the two holding yards to get the largest area for each dog?

CHAPTER TEST

Represent as fractions.

1. 3 out of 4 people in a survey **2.** $7 \div 9$

Convert to mixed or whole numbers.

3. $\dfrac{9}{3}$ **4.** $\dfrac{14}{9}$

Convert to improper fractions.

5. $4\dfrac{6}{7}$

6. $3\dfrac{1}{10}$

Write the prime factors.

7. 96

8. 132

Perform the operations. When possible, simplify first. Make sure your answers are in lowest terms.

9. $\dfrac{5}{6} \times \dfrac{3}{10}$

10. $\dfrac{3}{7} \times \dfrac{2}{9}$

11. $2\dfrac{2}{9} \times 1\dfrac{3}{4}$

12. $7 \times \dfrac{1}{3}$

13. $7\dfrac{1}{2} \div \dfrac{5}{9}$

14. $\dfrac{4\dfrac{2}{3}}{2\dfrac{1}{2}}$

15. $\dfrac{7}{12} + \dfrac{5}{6}$

16. $\dfrac{5}{12} \div \dfrac{5}{6}$

17. $2\dfrac{3}{7} + 5 + \dfrac{1}{2}$

18. $\dfrac{3}{32} + 4 + 1\dfrac{3}{4}$

19. $\dfrac{7}{9} - \dfrac{2}{3}$

20. $6\dfrac{1}{4} - 2\dfrac{3}{4}$

21. $\dfrac{5\dfrac{2}{3}}{1\dfrac{1}{9}}$

Determine which is larger. Show your work.

22. $\dfrac{7}{8}, \dfrac{11}{12}$

23. $\dfrac{7}{32}, \dfrac{5}{16}$

Arrange the fractions in order, beginning with the smallest. Show your work.

24. $\dfrac{5}{7}, \dfrac{10}{21}, \dfrac{3}{4}$

Write the percent equivalent.

25. $\dfrac{3}{5}$

26. $\dfrac{5}{8}$

Solve the problems.

27. Two of the seven security employees at the local community college received safety awards from the governor. Represent the part of the total number of employees who received an award as a fraction.

28. A candy-store owner mixes $1\dfrac{1}{2}$ lb of caramels, $\dfrac{3}{4}$ lb of chocolates, and $\dfrac{1}{2}$ lb of candy corn. What is the total weight of the mixed candy?

29. A homemaker has $5\dfrac{1}{2}$ cups of sugar on hand to make a batch of cookies requiring $1\dfrac{2}{3}$ cup of sugar. How much sugar is left?

30. If $6\dfrac{1}{4}$ ft of wire is needed to make one electrical extension cord, how many extension cords can be made from $68\dfrac{3}{4}$ ft of wire?

31. A costume maker figures one costume requires $2\dfrac{2}{3}$ yd of red satin. How many yards of red satin are needed to make three costumes?

32. Will a $\dfrac{5}{8}$-in.-wide drill bit make a hole wide enough to allow a $\dfrac{1}{2}$-in. (outside diameter) copper tube to pass through?

Problem Solving with Percents

Good Decisions Through Teamwork

You work in a department that has 6 members, and you are to distribute annual raises that must average 4% per department. No employee can get exactly 4%, and the raise must be at least 2% and no more than 6%. You may establish your own criteria for distributing the raises, but you have also been given the years of experience and the annual performance review ratings for each member. A performance rating of 1 is the lowest rating possible and a rating of 5 is the highest. Each member should assume the role of one of the following employees for purposes of this project: Employee 1 has 4 years experience, a performance rating of 4, and a salary of $28,500; employee 2 has 3 years experience, a performance rating of 3, and a salary of $22,800; employee 3 has 10 years experience, a performance rating of 4, and a salary of $32,700; employee 4 has 7 years experience, a performance rating of 3, and a salary of $31,400; employee 5 has 15 years experience, a performance rating of 3, and a salary of $34,600; employee 6 has 12 years experience, a performance rating of 5, and a salary of $32,400.

As a team, decide the amount of increase each of you will recommend to your supervisor. Prepare a report for your supervisor that justifies your recommendations; a table showing the original salary, the amount of increase, the new salary, and the percent of increase for each employee; and calculations to verify that the total increase is exactly 4% of the total original salaries except for rounding discrepancies. Will the percents of change also average 4%? Why or why not?

4–1 The percentage proportion

1. Solve a proportion for any missing element.
2. Identify the rate, base, and percentage in percent problems.
3. Solve the percentage proportion for any missing element.

4–2 Increases and decreases

1. Find the amount of increase or decrease in percent problems.
2. Find the new amount directly in percent problems.
3. Find the rate or the base in increase or decrease problems.

Problems involving percents are one of the most common uses of mathematics on the job and in everyday life. Everyone needs to understand percents to be a well-informed employee and citizen.

4–1   THE PERCENTAGE PROPORTION

Learning Outcomes

1. Solve a proportion for any missing element.
2. Identify the rate, base, and percentage in percent problems.
3. Solve the percentage proportion for any missing element.

One of the most practical methods for solving percent problems involves solving the percentage proportion. We use this method at this point in the text because it applies to numerous situations and requires very few formula manipulation skills. As we increase our experience with manipulating formulas we will examine other methods.

1   Solve a Proportion for Any Missing Element.

The equivalent fractions $\frac{5}{10}$ and $\frac{1}{2}$ can be written as the proportion $\frac{5}{10} = \frac{1}{2}$. Each fraction is also called a *ratio*, and two ratios that are equal or equivalent form a *proportion*.

■ **DEFINITION: Ratio.**   A *ratio* is a fraction that compares a quantity or measure in the numerator to a quantity or measure in the denominator.

■ **DEFINITION: Proportion.**   A *proportion* is a mathematical statement that shows two fractions or ratios are equal.

To solve proportions with a missing element, we use the property that the *cross products* in a proportion are equal.

■ **DEFINITION: Cross Products.**   In a proportion, the *cross products* are the product of the numerator of the first fraction times the denominator of the second, and the product of the denominator of the first fraction times the numerator of the second. In the proportion $\frac{a}{b} = \frac{c}{d}$, the cross products are $a \times d$ and $b \times c$.

Property of proportions:

The cross products in a proportion are equal. Symbolically, if $\frac{a}{b} = \frac{c}{d}$, then $a \times d = b \times c$, provided that b and d are not equal to zero. Also, if $a \times d = b \times c$, then $\frac{a}{b} = \frac{c}{d}$.

Let's use this property to show that $\frac{5}{10} = \frac{1}{2}$ is a true statement. The first cross product is $5 \times 2 = 10$. The second cross product is $10 \times 1 = 10$. The cross products are equal; thus, the fractions are equal.

This property can also help us find a missing element when three of the four elements are known.

EXAMPLE   Find the value for a in $\dfrac{a}{12} = \dfrac{3}{9}$.

$$\frac{a}{12} = \frac{3}{9}$$

$$a \times 9 = 12 \times 3 \qquad \text{Find the cross products.}$$

$$a \times 9 = 36$$

$$a = \frac{36}{9}$$

$$a = 4$$

Thus, $\dfrac{4}{12} = \dfrac{3}{9}$.

2   Identify the Rate, Base, and Percentage in Percent Problems.

We can use our previous knowledge of fractions and decimals to solve percent problems.

You probably already have some experience doing this when shopping for items on sale. Suppose an advertisement reads "50% off sale." Does this mean that you pay only 50% of the original price? Yes. So instead of using a calculator to find 50% of an item originally priced at $40, you mentally calculate half of $40, and realize that you will pay $20 for the item on sale.

We will come back to estimating prices later, but for now, let's look at some structured procedures for solving percent problems.

All problems involving percents have three basic elements: the rate, the base, and the percentage. Knowing what each element is and how all three are related helps us solve problems with percents.

The *rate R* is the percent; the *base B* is the original or total amount; the *percentage P* is part of the base. In the statement 50% of 80 is 40, the rate is 50%, the base is 80, and the percentage is 40.

EXAMPLE   Identify the given and missing elements for

(a) 20% of 75 is what number?
(b) What percent of 50 is 30?
(c) Eight is 10% of what number?

Use the identifying key words for rate (*percent* or %), base (*total, original,* associated with the word *of*), and percentage (*part,* associated with the word *is*).

$$\begin{array}{ccc} R & B & P \end{array}$$
(a) 20% of 75 is what number?
    percent   total   part

$$\begin{array}{ccc} R & B & P \end{array}$$
(b) What percent of 50 is 30?
     percent   total part

$$\begin{array}{ccc} P & R & B \end{array}$$
(c) Eight is 10% of what number?
    part   percent   total

3   Solve the Percentage Proportion for Any Missing Element.

Traditionally, percent problems are solved using three separate formulas, one for the missing percentage, one for the missing rate, and one for the missing base.

$$\text{Percentage} = \text{Rate} \times \text{Base} \quad \text{or} \quad P = RB$$

$$\text{Rate} = \frac{\text{Percentage}}{\text{Base}} \quad \text{or} \quad R = \frac{P}{B}$$

$$\text{Base} = \frac{\text{Percentage}}{\text{Rate}} \quad \text{or} \quad B = \frac{P}{R}$$

When we use a calculator, there may be some advantage in using separate formulas, but for now, let's use one formula or equation that will work for all three situations.

We can consolidate all three elements into one formula called the *percentage proportion*. The relationships among the rate, percentage, base, and the standard unit of 100 are represented by two fractions that are equal to each other.

Percentage proportion formula:

$$\frac{R}{100} = \frac{P}{B}$$

where R = rate or percent

P = percentage or part

B = base or total

In a percent problem, if we know any two of the elements, we can find the third element using the property of proportions and cross multiplication. This is called *solving* the proportion.

To solve the percentage proportion:

1. Cross multiply to find the cross products.
2. Divide the product of the two known factors by the factor with the letter.

EXAMPLE What is 20% of 75?

The rate is 20%, the base is 75, and the percentage is missing.

$$\frac{R}{100} = \frac{P}{B}$$
Set up the proportion.

$$\frac{20}{100} = \frac{P}{75}$$
Substitute the known elements.

$$20 \times 75 = 100 \times P$$
Cross multiply.

$$1,500 = 100 \times P$$

$$\frac{1,500}{100} = P$$
Divide to find P.

$$15 = P$$

Therefore, the percentage is 15.

Tip!	**Multiply, Then Divide.**

Solving percentage proportions always involves one multiplication and one division. One cross product will be two numbers—this is the multiplication. The other cross product will be one number and one letter—this is the division. We divide the cross product with two numbers by the one number in the other cross product.

(continued)

$$\frac{20}{100} = \frac{P}{75}$$

$$20 \times 75 = 100 \times P \qquad \textbf{Cross products.}$$

$$1{,}500 = 100P$$

$$1{,}500 \div 100 = P \qquad \textbf{Now we divide.}$$

$$15 = P$$

Tip! | ***Proportions as a Continuous Series of Steps.***

To solve proportions with a calculator, we follow a continuous series of steps.

$$\frac{20}{100} = \frac{P}{75}$$

$$20 \;\boxed{\times}\; 75 \;\boxed{\div}\; 100 \;\boxed{=}\; \Rightarrow \; 15$$

The next example shows how to use the fractional equivalent of a mixed-number percent.

EXAMPLE $33\frac{1}{3}\%$ of 282 is what number?

The rate is $33\frac{1}{3}\%$. The key word *of* tells us that 282 is the base. The percentage is missing.

$$\frac{33\frac{1}{3}}{100} = \frac{P}{282}$$

If you know the fractional equivalent for $33\frac{1}{3}\%$ you can simplify the calculations by substituting it into the proportion. ($33\frac{1}{3}\% = \frac{33\frac{1}{3}}{100} = \frac{1}{3}$; see the first example on page 159.)

$$\frac{1}{3} = \frac{P}{282}$$

$$1 \times 282 = 3 \times P \qquad \text{Cross multiply.}$$

$$\frac{282}{3} = P \qquad \text{Divide.}$$

$$94 = P$$

$33\frac{1}{3}\%$ **of 282 is 94.**

Tip! | ***How Many Digits Do You Use?***

Even with a calculator, you may prefer to use the fractional equivalent because the decimal equivalent of $33\frac{1}{3}\%$ is a repeating decimal, and requires rounding. Examine the effect of rounding to various places.

$$0.3 \times 282 = 84.6$$

$$0.33 \times 282 = 93.06$$

(continued)

   Chapter 4   Problem Solving with Percents

$$0.333 \times 282 = 93.906$$

$$0.3333 \times 282 = 93.9906$$

$$0.333333333 \times 282 = 93.999999906$$

Because the desired degree of accuracy varies depending on the problem, it is often advisable to find the exact answer on the calculator using the fractional equivalent. A close approximate answer can be found by using the full calculator value for nonterminating decimal equivalents.

To solve for percents that are less than 1% or more than 100%, we follow the same procedures and cautions as with other percents. However, we can use some time-savers to make our work easier and faster.

It is always advisable to anticipate the approximate size of an answer before you make any calculations. This approximation helps you discover many careless mistakes, especially when you are using your calculator. For example, 1% of any number can be found by making mental calculations.

$$1\% \text{ of } 175 \text{ is} \qquad \frac{1}{100} = \frac{P}{175}$$

$$175 = 100 \times P$$

$$\frac{175}{100} = P$$

$$1.75 = P$$

Tip!	*Mentally Find 1% of a Number.*

To find 1% of a number, divide the number by 100 or move the decimal two places to the left.

Because 1% of a number can be found mentally, when we work with a percent that is less than 1%, we first find 1% of the number. The answer we seek must be less than 1% of the number. This estimating procedure is very useful in checking the decimal placement. For instance, let's find $\frac{1}{4}\%$ of 875:

$\frac{1}{4}\%$ means $\frac{1}{4}$ of 1%

1% of 875   is   8.75   Move decimal two places to the left.

$\frac{1}{4}\%$ of 875 is $\frac{1}{4}$ of 8.75, or about 2

Let's use this estimate to check our work in the following example.

EXAMPLE $\frac{1}{4}\%$ of 875 is what number?

$\frac{1}{4}\%$ and 0.25% are equivalent. Either can be used to solve this problem. We are given the rate and the base, so we need to find the percentage.

Option 1   **Option 2**

$$\frac{\frac{1}{4}}{100} = \frac{P}{875} \qquad\qquad \frac{0.25}{100} = \frac{P}{875}$$

$$\frac{1}{4} \times 875 = 100 \times P \qquad 0.25 \times 875 = 100 \times P \qquad \text{Cross multiply.}$$

$$\frac{875}{4} = 100 \times P \qquad\qquad 218.75 = 100 \times P$$

$$\frac{\frac{875}{4}}{100} = P \qquad\qquad \frac{218.75}{100} = P \qquad\qquad \text{Divide by 100.}$$

$$\frac{875}{4} \div 100 = P \qquad\qquad 2.1875 = P$$

$$\frac{\overset{35}{\cancel{875}}}{4} \times \frac{1}{\underset{4}{\cancel{100}}} = P$$

$$\frac{35}{16} = P$$

$$2\frac{3}{16} = P$$

$\frac{1}{4}\%$ **of 875 is** $2\frac{3}{16}$**, or 2.1875.** Use your calculator to verify that $\frac{3}{16} = 0.1875$ ($3 \div 16 = 0.1875$).

Recall that 100% of a number is 1 times the number, or the number itself. When working with percents that are larger than 100%, we can also roughly estimate the answer.

For instance, 325% of 86 is more than 3 times 86. 100% of $86 = 86$. Thus, 325% of 86 must be more than 3 times 86, or more than 258. Thus, our estimate for the preceding example is more than 258.

EXAMPLE   325% of 86 is what number?

Again, we are finding the percentage. The rate is 325% and the base is 86.

Option 1

$$\frac{325}{100} = \frac{P}{86}$$

$$325 \times 86 = 100 \times P$$

$$\frac{27{,}950}{100} = P$$

$$279.5 = P$$

Option 2

$$\frac{13}{4} = \frac{P}{86} \quad \left(325\% = 3\frac{1}{4} = \frac{13}{4}\right)$$

$$13 \times 86 = 4 \times P \qquad \text{Cross multiply.}$$

$$\frac{1{,}118}{4} = P \qquad \text{Divide.}$$

$$279.5 = P$$

325% of 86 is 279.5 or $279\frac{1}{2}$**.**

Tip!	***Using the*** $\boxed{\%}$ ***Key on the Calculator.***

Some calculators have a $\boxed{\%}$ key, but it is very common to work percent problems without it. To solve percent problems, enter the percent in decimal notation when finding the percentage or base, or change the calculator display to percent notation when finding the rate. Even when a calculator has a $\boxed{\%}$ key, its use may vary. The most common practice is to mentally convert percent notation to decimal notation, or vice versa. **The** $\boxed{\%}$ **key most often functions as the** $\boxed{=}$ **key.** In addition to functioning as the $\boxed{=}$ key, the $\boxed{\%}$ key also uses the decimal equivalent of the rate when the rate is used in the problem.

28% of 37 is what number?

(continued)

Using the $\boxed{\%}$ *key:* The $\boxed{\%}$ key automatically converts the rate to a decimal.

$$28 \boxed{\times} 37 \boxed{\%} \Rightarrow 10.36$$

Using the $\boxed{=}$ *key:* Mentally convert 28% to 0.28.

$$\boxed{\cdot} 28 \boxed{\times} 37 \boxed{=} \Rightarrow 10.36$$

15 is what percent of 52?

Using the $\boxed{\%}$ *key:* The $\boxed{\%}$ key automatically converts the decimal answer to a percent.

$$15 \boxed{\div} 52 \boxed{\%} \Rightarrow 28.84615385 \qquad \text{\textbf{A percent.}}$$

15 is approximately 28.8% of 52.

Using the $\boxed{=}$ *key:* The $\boxed{=}$ key gives a decimal answer. You must change the decimal to a percent by mentally multiplying by 100%.

$$15 \boxed{\div} 52 \boxed{=} \Rightarrow 0.2884615385 \qquad \text{\textbf{A decimal equivalent.}}$$

0.2884615385 is approximately 28.8%.

The rate is the easiest of the three parts to identify in a percentage problem. When neither of the given numbers in a problem has a percent symbol or is followed by the word *percent,* you know the rate is missing.

EXAMPLE   What percent of 48 is 24?

The missing element in this problem is the rate. Is 48 the percentage (part) or the base (total)? 48 follows the key word *of* and is the base. 24 follows the key word *is* and is the percentage.
Set up the proportion.

$$\frac{R}{100} = \frac{24}{48}$$

$$R \times 48 = 100 \times 24 \qquad \text{Cross multiply.}$$

$$R \times 48 = 2{,}400$$

$$R = \frac{2{,}400}{48} \qquad \text{Divide.}$$

$$R = 50$$

R is the rate, so 50% of 48 is 24.

EXAMPLE   What percent is 2 out of 600?

The rate is missing. The 2 represents the part or percentage and 600 is the base.

$$\frac{R}{100} = \frac{2}{600}$$

$$R \times 600 = 2 \times 100 \qquad \text{Cross multiply.}$$

$$R \times 600 = 200$$

$$R = \frac{200}{600} \qquad \text{Divide or reduce.}$$

$$R = \frac{1}{3}$$

2 out of 600 is $\frac{1}{3}$%.

 261 is what percent of 87?

The rate is missing in this problem. Does 261 represent the percentage or the base? The key word here is *is*, so 261 is the percentage or part. The word *of* indicates that 87 is the base (one total amount). The percentage is larger than the base, so the rate will be more than 100%.

$$\frac{R}{100} = \frac{261}{87}$$

$26{,}100 = 87 \times R$ Cross multiply.

$\dfrac{26{,}100}{87} = R$ Divide.

$300 = R$

261 is 300% of 87.

The base is the original or whole amount. The base is larger than the percentage when the rate is less than 100%.

EXAMPLE 20% of what number is 45?

This time we know the rate, 20%, and the percentage, 45. We are looking for the base, as indicated by the key word *of.*

Option 1	**Option 2**	**Option 3**
$\dfrac{20}{100} = \dfrac{45}{B}$	$\dfrac{1}{5} = \dfrac{45}{B}$	$\dfrac{0.2}{1} = \dfrac{45}{B}$
$20 \times B = 100 \times 45$	$B = 5 \times 45$	$45 = 0.2 \times B$
$20 \times B = 4{,}500$	$B = 225$	$\dfrac{45}{0.2} = B$
$B = \dfrac{4{,}500}{20}$		$225 = B$
$B = 225$		

20% of 225 is 45.

EXAMPLE $\frac{3}{4}$% of what number is 11.25?

The rate is $\frac{3}{4}$% and the percentage is 11.25, as signaled by the key word *is.* The base is missing.

Option 1	**Option 2**	
$\dfrac{\frac{3}{4}}{100} = \dfrac{11.25}{B}$	$\dfrac{0.75}{100} = \dfrac{11.25}{B}$	
$\dfrac{3}{4} \times B = 1{,}125$	$0.75 \times B = 1{,}125$	Cross multiply.
$B = \dfrac{1{,}125}{\frac{3}{4}}$	$B = \dfrac{1{,}125}{0.75}$	Divide.
	$B = 1{,}500$	

$$B = \frac{\overset{375}{\cancel{1{,}125}}}{1} \times \frac{4}{\underset{1}{\cancel{3}}}$$

$$B = 1{,}500$$

$\frac{3}{4}\%$ of 1,500 is 11.25.

To mentally check that the answer is reasonable, find 1% of 1,500: 1% of 1,500 = 15, so $\frac{3}{4}\%$ of 1,500 should be less than 15. The answer 11.25 is less than 15, so it is reasonable.

In the next example the rate is more than 100% so we expect the base to be less than the percentage.

EXAMPLE   **398.18 is 215% of what number?**

Here, we are looking for the base, as indicated by the key word *of*. We are given the percentage and the rate.

$$\frac{215}{100} = \frac{398.18}{B}$$

$$215 \times B = 39{,}818 \qquad\qquad \textsf{\textbf{\textcolor{teal}{Cross multiply.}}}$$

$$B = \frac{39{,}818}{215} \qquad\qquad \textsf{\textbf{\textcolor{teal}{Divide.}}}$$

$$B = 185.2 \quad \text{or} \quad 185\frac{1}{5}$$

398.18 is 215% of 185.2 or $185\frac{1}{5}$.

Because the rate is more than 100%, we expected the base to be smaller than the percentage. Thus, the answer is reasonable.

4 Use the Percentage Proportion to Solve Applied Problems.

When solving applied problems, our most difficult task is identifying the two given parts and determining which part is missing. Then the proportion can be set up and solved. Let's examine several applied problems involving percents.

EXAMPLE   **If a type of solder contains 55% tin, how many pounds of tin are needed to make 10 lb of solder?**

First, let's be sure we understand the word *solder.* Solder is a mixture of metals. In this problem, the *total amount* or the *base* is the 10 lb of solder, and it is made of tin and other metals.

Known facts
55% of 10 lb of solder is tin.
Rate: Percent of tin = 55%
Base: Amount of solder = 10 lb

Unknown fact
Percentage or number of pounds of tin = P

Relationships
Percentage proportion

$$\frac{R}{100} = \frac{P}{B}$$

$$\frac{55}{100} = \frac{P}{10 \text{ lb}}$$

Estimation	To estimate, 55% is more than $\frac{1}{2}$. $\frac{1}{2}$ of $10 = 5$. So the amount should be more than 5 lb.
Calculations	

$$\frac{55}{100} = \frac{P}{10}$$

$$55 \times 10 = 100 \times P \qquad \text{Cross multiply.}$$

$$550 = 100 \times P$$

$$\frac{550}{100} = P \qquad \text{Divide by 100.}$$

$$5\frac{1}{2} = P$$

Interpretation

Thus, $5\frac{1}{2}$ lb of tin is needed to make 10 lb of solder.

To check the reasonableness of the answer, $5\frac{1}{2}$ lb is a little more than 5.

EXAMPLE If a 150-horsepower (hp) engine delivers only 105 hp to the driving wheels of a car, what is the efficiency of the engine?

Efficiency means the *percent* the output (105 hp) is of the total amount (150 hp) the engine is capable of delivering. Thus, the base amount is 150 hp, the part or percentage delivered is 105 hp, and the percent of 150 represented by 105 is the rate or efficiency.

Known facts

Base: Total amount of horsepower = 150 hp
Percentage: Amount of horsepower engine delivers = 105 hp

Unknown facts

Efficiency or percent of the total horsepower

Relationships

Percentage proportion

$$\frac{R}{100} = \frac{P}{B}$$

$$\frac{R}{100} = \frac{105}{150}$$

Estimation

Because the engine is not operating at full capacity (150 hp), we expect the efficiency to be less than 100%. Since 105 is more than $\frac{1}{2}$ of 150, the percent or rate will be more than 50%. A more precise estimate is

$$\frac{100}{150} = \frac{2}{3} = 66\frac{2}{3}\%$$

$$\text{Since } 105 > 100 \qquad \text{> is read "is greater than."}$$

$$\text{Rate} > 66\frac{2}{3}\%$$

Calculations

$$\frac{R}{100} = \frac{105}{150}$$

$$R \times 150 = 100 \times 105$$

$$R \times 150 = 10{,}500$$

$$R = \frac{10{,}500}{150}$$

$$R = 70$$

Interpretation

The engine is 70% efficient.

CHAPTER 4 Problem Solving with Percents

EXAMPLE The effective value of current or voltage in an ac circuit is 71.3% of the maximum voltage. If a voltmeter shows a voltage of 110 volts (V) in a circuit, what is the maximum voltage?

71.3% of the maximum voltage is 110 V. The maximum voltage is the *base,* and the amount of voltage shown in the voltmeter is 110 V, which is the *percentage.*

Estimation The rate is less than 100%. Therefore, the base is larger than the percentage. $B > 110$.

$$\frac{71.3}{100} = \frac{110}{B}$$

$$71.3 \times B = 100 \times 110$$

$$71.3 \times B = 11{,}000$$

$$B = \frac{11{,}000}{71.3}$$

$$B = 154.2776999$$

or

$$B = 154 \text{ V} \qquad \text{To the nearest volt.}$$

Interpretation **The maximum voltage is 154 V.**

Using percentage proportions gives us several advantages. The obvious advantage is that we need only one formula to solve problems for percentage, rate, and base. Another advantage is that we can standardize our approach to percentage problems—one basic solution procedure works in all cases.

This general approach to percentage problems has other advantages, also. One is that the standard 100 takes care of having to convert decimal numbers or fractions to percents or to convert percents to decimal or fraction equivalents, as is done in traditional approaches. This approach also simplifies the steps. For example, all problems involve one multiplication (cross multiplication) and one division step (division by whichever number the letter is multiplied by). This approach also includes either multiplication by 100 or division by 100, which can be done mentally.

SELF-STUDY EXERCISES 4–1

1 Find the value of the letter in each proportion.

1. $\dfrac{x}{10} = \dfrac{1}{2}$

2. $\dfrac{b}{16} = \dfrac{2}{8}$

3. $\dfrac{3}{a} = \dfrac{4}{12}$

4. $\dfrac{4}{c} = \dfrac{2}{5}$

5. $\dfrac{2}{3} = \dfrac{x}{6}$

6. $\dfrac{2}{5} = \dfrac{a}{20}$

7. $\dfrac{1}{2} = \dfrac{3}{x}$

8. $\dfrac{2}{4} = \dfrac{9}{b}$

9. $\dfrac{d}{3} = \dfrac{1}{2}$

10. $\dfrac{2}{x} = \dfrac{3}{5}$

11. $\dfrac{4}{9} = \dfrac{c}{2}$

12. $\dfrac{8}{3} = \dfrac{2}{b}$

2 Identify the given and missing elements as R (rate), B (base), and P (percentage).

13. What percent of 10 is 2?
14. Two is 20% of what number?
15. 20% of 10 is what number?
16. Three books is what percent of four books?
17. 15% of how many dollars is $9?
18. What percent of 25 students is 5 female students?
19. 6 of 15 motorists is what percent?
20. How many nurses is 20% of the 15 nurses on duty?
21. 35% of how many pieces of sod is 70 pieces?
22. What percent of a total bill of $45 is $3.15?

3 Solve the percentage proportion.

23. 20% of 375 is what number?
24. 75% of 84 is what number?
25. $66\frac{2}{3}$% of 309 is what number?
26. 34.5% of 336 is what number?
27. $\frac{3}{4}$% of 90 is what number?
28. 0.2% of 470 is what number?
29. What number is 134% of 115?
30. Find 275% of 84.
31. 400% of 231 is what number?
32. $37\frac{1}{2}$% of 920 is what number?
33. What percent of 348 is 87?
34. What percent of 350 is 105?
35. 72 is what percent of 216?
36. 28 is what percent of 85 (to the nearest tenth percent)?

37. 37.8 is what percent of 240?
38. What percent of 175 is 28?
39. 32 is what percent of 4,000?
40. What percent is 2 out of 300?
41. What percent of 125 is 625?
42. 173.55 is what percent of 156?
43. 50% of what number is 36?
44. 60% of what number is 30?
45. $12\frac{1}{2}$% of what number is 43?
46. 15.87 is 34.5% of what number?
47. $\frac{2}{3}$% of what number is $2\frac{2}{5}$?
48. 0.3% of what number is 0.825?
49. 150% of what number is $112\frac{1}{2}$?
50. 43% of what number is 107.5?
51. 92 is 500% of what number?
52. $133\frac{1}{3}$% of what number is 348?
53. Find the percentage if the base is 75 and the rate is 5%.
54. Find the percentage if the base is 25 and the rate is 2.5%.
55. What is the rate if the base is 10.5 and the percentage is 7?
56. Find the rate when the base is 80 and the percentage is 30.
57. If the percentage is 4.75 and the rate is $33\frac{1}{3}$%, find the base.
58. Find the base when the rate is 15% and the percentage is 52.5.
59. If the rate is $12\frac{1}{2}$% and the base is 75, find the percentage.
60. If the percentage is 11 and the rate is 5%, find the base.
61. Find the rate when the percentage is 15 and the base is 75.
62. What is the base if the percentage is 35 and the rate is 17.5%?

4 Solve.

63. Cast iron contains 4.25% carbon. How much carbon is contained in a 25-lb bar of cast iron?
64. 3,645 rolls of landscape fabric are manufactured during one day. After being inspected, 121 of these rolls are rejected as imperfect. What percent of the rolls is rejected? (Round to the nearest whole percent.)

65. An engine operating at 82% efficiency transmits 164 hp. What is the engine's maximum capacity in horsepower?
66. If wrought iron contains 0.07% carbon, how much carbon is in a 30-lb bar of wrought iron?
67. The voltage of a generator is 120 V. If 6 V is lost in a supply line, what is the rate of voltage loss?
68. A contractor figures it costs $\frac{1}{2}$% of the total cost of a job to make a bid. What would be the cost of making a bid on a $115,000 job?

69. A certain ore yields an average of 67% iron. How much ore is needed to obtain 804 lb of iron?
70. A contractor makes a profit of $12,350 on a $115,750 job. What is the percent of profit? (Round to the nearest whole percent.)

71. 385 defective alcohol swabs were produced during a day. If 4% of the alcohol swabs produced were defective, how many alcohol swabs were produced in all?
72. In a welding shop, 104,000 welds are made. If 97% of them are acceptable, how many are acceptable?

Learning Outcomes

1. Find the amount of increase or decrease in percent problems.
2. Find the new amount directly in percent problems.
3. Find the rate or the base in increase or decrease problems.

Percents are often used in problems dealing with increases or decreases. For instance, if a TV repair shop is advised that its recent order for 250 power cords will be reduced by 14% because of a shortage of copper wire, the shop needs to find the number of power cords this amounts to. The shop may have to order additional cords from another supplier.

1 Find the Amount of Increase or Decrease in Percent Problems.

When we work with increases or decreases, the *original amount* (250 power cords) is the *base*. The *percentage* (power cords *not* received) is the *amount of change (increase* or *decrease*). The *new amount* is the original amount plus or minus the amount of change.

EXAMPLE   Pipefitters are to receive a 9% increase in wages per hour. If they were making $9.25 an hour, what is the *amount of increase per hour* (to the nearest cent)? Also, what is the *new wage per hour*?

The original wage per hour is the base, and we want to find the amount of increase (percentage).

Estimation   10% of $9.25 is $.92. The increase will be less than $.92. The new wage per hour will be approximately $10.

$$\frac{9}{100} = \frac{P}{9.25}$$

P represents amount of increase, and 9% is the rate of increase.

$$9 \times 9.25 = 100 \times P$$

$$83.25 = 100 \times P$$

$$\frac{83.25}{100} = P$$

$$\$0.8325 = P$$

$0.83 to the nearest cent is the amount of increase.

Interpretation   **The pipefitters will receive an $0.83 per hour increase in wages.**

$$\$9.25 + \$0.83 = \$10.08$$

New amount = Original amount + Amount of increase.

Their new hourly wage will be $10.08.

EXAMPLE   Molten iron shrinks 1.2% while cooling. What is the cooled length of a piece of iron if it is cast in a 24-cm pattern?

First, we find the amount of shrinkage (amount of decrease, percentage). The original amount, 24 cm, is the base.

Estimation   The cooled piece will be less than 24 cm.

$$\frac{1.2}{100} = \frac{P}{24}$$

P is the amount of decrease, and 1.2% is the rate of decrease.

$$1.2 \times 24 = 100 \times P$$

$$28.8 = 100 \times P$$

$$\frac{28.8}{100} = P$$

$$0.288 = P$$

The amount of shrinkage is 0.288 cm, so the length of the cooled piece (new amount) is

24 − 0.288 = 23.712 cm   New amount = Original amount − Amount of decrease

EXAMPLE   Uncut earth is hard, packed soil. As it is dug the volume increases or swells. A contractor figures that there will be a 20% earth swell when a mixture of uncut loam and clay soil is excavated. If 150 cubic yards (yd^3) of uncut earth is to be removed, taking into account the earth swell, how many cubic yards will have to be hauled away?

To find the amount of earth swell (amount of increase, percentage), find 20% of 150 yd^3, the original amount or base.

Estimation   10% of 150 = 15, so 20% of 150 = 30. The amount of earth to be hauled away is 150 + 30 = 180 yd^3

Option 1

$$\frac{20}{100} = \frac{P}{150}$$

$$20 \times 150 = 100 \times P$$

$$3{,}000 = 100 \times P$$

$$\frac{3{,}000}{100} = P$$

$$30 = P$$

Option 2

$$\frac{1}{5} = \frac{P}{150}$$

$$150 = 5 \times P$$

$$\frac{150}{5} = P$$

$$30 = P$$

P is the amount of increase, and 20%, or $\frac{1}{5}$, or 0.2, is the rate of increase.

Option 3

$$\frac{0.2}{1} = \frac{P}{150}$$

$$0.2 \times 150 = P$$

$$30 = P$$

Thus, the earth will swell 30 yd^3 when cut. To find the amount of earth to be hauled away, add 150 yd^3 and 30 yd^3.

150 + 30 = 180 yd^3 to be hauled away   New amount = Original amount + Amount of increase

2 Find the New Amount Directly in Percent Problems.

When knowing the amount of increase is not necessary, the new amount can be figured directly. Remember, all of a quantity is 100%. If a quantity is to be increased by 20%, then the new amount will be 100% + 20%, or 120% of the original amount.

If 150 yd^3 of uncut earth (base) increases by 20% when cut, the new amount (percentage) will be 120% of 150 yd^3.

Option 1

$$\frac{120}{100} = \frac{P}{150}$$

$$\frac{6}{5} = \frac{P}{150}$$

$$6 \times 150 = 5 \times P$$

$$900 = 5 \times P$$

$$\frac{900}{5} = P$$

$$180 = P$$

Option 2

$$\frac{1.2}{1} = \frac{P}{150}$$

$$1.2 \times 150 = P$$

$$180 = P$$

P is the new amount because 120% is the new rate; that is, 120% of 150 is the new amount.

EXAMPLE A drying process causes a 2% weight loss in a casting. If the wet casting weighs 130 kg, how much will the dried casting weigh?

The amount of weight loss (amount of decrease) is 2% of 130 kg.

Estimation The dried casting will weigh slightly less than the wet casting.

Option 1

$$\frac{2}{100} = \frac{P}{130}$$

P is the amount of decrease because 2% is the rate or percent of decrease.

$$2 \times 130 = 100 \times P$$

$$260 = 100 \times P$$

$$\frac{260}{100} = P$$

$$2.6 = P$$

Interpretation Thus, the casting will have a weight loss of 2.6 kg. **The dried casting will weigh**

$$130 - 2.6 = \textbf{127.4 kg}$$ New amount = Original amount − Amount of decrease

Knowing the amount of weight loss is not necessary, so we could have calculated the dried casting weight directly. The original weight equals 100%. There will be 2% weight loss, so the dried weight will be 100% − 2%, or 98% of the original weight.

Option 2

$$\frac{98}{100} = \frac{P}{130}$$

P is the new amount because the new amount is 98% of the original weight; that is, the new amount is 98% of 130.

$$98 \times 130 = 100 \times P$$

$$\frac{12,740}{100} = P$$

$$127.4 = P$$

The dried casting weighs 127.4 kg.

EXAMPLE A 3% error is acceptable for a machine part to be usable. If the part is intended to be 57 cm long, what is the range of measures that is acceptable for this part?

The machine part can be ±3% from the ideal length of 57 cm. The range of acceptable measures is found by calculating the smallest acceptable measure and the largest acceptable measure. The symbol ± is read "plus or minus." It means that the measure can be more (+) or less (−) than the designated amount. In this case, the part can be 3% longer than or shorter than 57 cm and still be usable. The smallest acceptable value is 97% of the ideal length (100% − 3%).

$$\frac{97}{100} = \frac{P}{57}$$

P is the smallest acceptable amount because 97% is the *smallest acceptable percent.*

$$97 \times 57 = 100 \times P$$

$$\frac{5{,}529}{100} = P$$

$$55.29 = P$$

Interpretation

The smallest acceptable measure is 55.29 cm.
The largest acceptable value is 103% of the ideal length (100% + 3%).

$$\frac{103}{100} = \frac{P}{57}$$

P is the largest acceptable amount because 103% is the *largest acceptable percent.*

$$5{,}871 = 100 \times P$$

$$\frac{5{,}871}{100} = P$$

$$58.71 = P$$

Interpretation

The largest acceptable value is 58.71 cm. The range of acceptable measures is from 55.29 cm to 58.71 cm.
We can also work this problem by finding the amount of acceptable error. Then we subtract this amount from the intended length of the machine part to get the smallest acceptable measure and add to get the largest acceptable measure.
3% of 57 is the amount of acceptable error.

$$\frac{3}{100} = \frac{P}{57}$$

P is the acceptable error because 3% is the *acceptable error percent.*

$$171 = 100 \times P$$

$$\frac{171}{100} = P$$

$$1.71 = P$$

Thus, 57 ± 1.71 represents the range of acceptable measures.

$$57 - 1.71 = 55.29$$

Smallest acceptable amount = Original amount − Amount of acceptable error

The smallest acceptable value is 55.29 cm.

$$57 + 1.71 = 58.71$$

Largest acceptable amount = Original amount + Amount of acceptable error

The largest acceptable value is 58.71 cm.

3 Find the Rate or the Base in Increase or Decrease Problems.

Many kinds of increase or decrease problems involve finding either the rate or the base.

The rate is often called the *percent of change* or the *percent of increase or decrease.* The base is called the *original amount.*

EXAMPLE   A worn brake lining is measured to be $\frac{3}{32}$ in. thick. If the original thickness was $\frac{1}{4}$ in., what is the percent of wear?

First, the amount of wear is $\frac{1}{4} - \frac{3}{32}$.

$$\frac{8}{32} - \frac{3}{32} = \frac{5}{32}$$ Find the common denominator. Subtract.

The amount of wear (decrease) is the percentage. The base is the original amount.

$$\frac{R}{100} = \frac{\frac{5}{32}}{\frac{1}{4}}$$

R is the percent of wear.
$\frac{5}{32}$ is the amount of wear.

$$\frac{1}{4} \times R = \frac{\overset{25}{\cancel{100}}}{1} \times \frac{5}{\underset{8}{\cancel{32}}}$$

$$\frac{1}{4} \times R = \frac{125}{8}$$

$$R = \frac{125}{8} \div \frac{1}{4}$$

$$R = \frac{125}{\underset{2}{\cancel{8}}} \times \frac{\overset{1}{\cancel{4}}}{1}$$

$$R = 62\frac{1}{2}$$

Interpretation   **The percent of wear is $62\frac{1}{2}\%$.**

EXAMPLE   During the month of May, an electrician made a profit of $1,525. In June, she made a profit of $1,708. What is the percent of increase?

The amount of increase is $1,708 − $1,525 = $183, so $183 is what percent of the *original* amount?

$$\frac{R}{100} = \frac{183}{1,525}$$

R is the percent of increase.
183 is the amount of increase.

$$R \times 1,525 = 18,300$$

$$R = \frac{18,300}{1,525}$$

$$R = 12$$

Interpretation   **The percent of increase is 12%.**

From the applications in this section, we can see how important it is to understand what 100% of something really means. Many applications of percents use the fact that 100% of a quantity is the entire quantity. This knowledge lets us compute many quantities directly, without having to first figure the increase or decrease separately.

1 Solve.

1. Find the amount of increase if 432 is increased by 25%.
2. If 78 is increased by 40%, what is the new amount?
3. Find the amount of decrease if 68 is decreased by 15%.
4. If 135 is decreased by 75%, what is the new amount?

2 Solve.

5. Steel rods shrink 10% when cooled from furnace temperature to room temperature. If a tie rod is 30 in. long at furnace temperature, how long is the cooled tie rod?
6. A contractor needs 1,650 board feet of 1-in. × 8-in. common boards to subfloor a house. If she needs 10% extra flooring to allow for waste when the boards are laid square, how much flooring should she order?
7. If 17% extra flooring is needed to allow for waste when the boards are laid diagonally, how much flooring should be ordered to cover 2,045 board feet of floor? Answer to the nearest whole board foot.
8. A construction company requires 25,400 bricks for a job. If they allow 2% more bricks for breakage, how many bricks must they order?
9. When making an estimate on a job, a contractor wants to make a 10% profit. If all the estimated costs are $15,275, what is the total bid of cost and profit for the job?
10. Rock must be removed from a highway right-of-way. If 976 cubic yards (yd^3) of unblasted rock is to be removed, how many cubic yards is this after blasting? Blasting causes a 40% swell in volume.

3 Solve.

11. The cost of a pound of nails increased from $2.36 to $2.53. What is the percent of increase to the nearest whole-number percent?
12. A landscape contractor estimated that materials, shrubs, saplings, and labor for a job would cost $5,385. An estimate one year later for the same job was $7,808, due to inflation. Find the percent of increase due to inflation to the nearest whole number.
13. A chicken farmer bought 2,575 baby chicks. Of this number, 2,060 lived to maturity. What percent loss was experienced by the chicken farmer?
14. An electrician recorded costs of $1,297 for a job. If he received $1,232 for the job, what was the percent of money lost on the job? Round to the nearest whole number.
15. An engine that has a 4% loss of power has an output of 336 hp. What is the input (base) horsepower of the engine?
16. A contractor figures that 10 yd^3 of sand is needed for a job. If a 5% allowance for waste must be included, how much sand must be ordered?
17. A floor in a doctor's office that would normally require 2,580 board feet is to be laid diagonally. If a 17% waste allowance is necessary for flooring laid diagonally, how much flooring must be ordered? (Round to the nearest whole number.)
18. A shop manager records a 14% loss on rivets for waste. If the shop needs 25 lb of rivets, how many pounds must be ordered to compensate for loss due to waste?
19. Steel bars shrink 10% when cooled from furnace temperature to room temperature. If a cooled steel bar is 36 in. long, how long was it when it was formed?
20. The cost of No. 1 pine studs increased from $3.85 each to $4.62 each. Find the percent of increase.
21. A 141-hp output is required for an engine. If there is a 6% loss of power, what amount of input horsepower (or base) is needed?

Electronics: Percentage Error

Technicians are often asked to find the percentage error in components and in parts of circuits.

When you calculate a value (called the *theoretical value* or the *predicted value*) and then measure that value (the *measured* or *actual value*), the results are seldom exactly the same. It is not enough to say that measured values were close or not close to the predicted values. You need to give a percentage answer. This is the percentage error or percentage difference.

$$\% \text{ error} = \frac{\text{Measured value} - \text{Theoretical value}}{\text{Theoretical value}} \times 100\%$$

Suppose you have a circuit in which you have calculated and then measured three voltage drops, or potential differences in voltage. These voltage drops are called V_1 and V_2 and V_3, and they are read as "V sub 1" and "V sub 2" and "V sub 3." They mean "voltage drop number one," "voltage drop number two," and "voltage drop number three." The subscripts are written below the line and are there only for identification purposes. Subscripts make complicated sets of numbers easier to read. Using sequential numbers, we can arrange the voltage drops in order.

As long as you understand what is meant, you can also write $V1$ and $V2$ and $V3$. This convention means the number following the letter is the identifier. This practice is particularly important when you program computers.

Suppose your calculations show that you should get

$$V_1 = 20 \text{ V}, \qquad V_2 = 30 \text{ V}, \qquad V_3 = 40 \text{ V}$$

When you make the measurements with a voltmeter, the results are

$$M_1 = 21 \text{ V}, \qquad M_2 = 29 \text{ V}, \qquad M_3 = 43 \text{ V}$$

First consider V_1. Since $M_1 = 21$ V and $V_1 = 20$ V,

$$\% \text{ error} = \frac{21 \text{ V} - 20 \text{ V}}{20 \text{ V}} \times 100\% = \frac{1}{20} \times 100\% = \frac{100\%}{20} = +5\%$$

Next consider V_2. Since $M_2 = 29$ V and $V_2 = 30$ V,

$$\% \text{ error} = \frac{29 \text{ V} - 30 \text{ V}}{30 \text{ V}} \times 100\% = \frac{-1}{30} \times 100\% = \frac{-100\%}{30} = -3.3\%$$

Next consider V_3. Since $M_3 = 43$ V and $V_3 = 40$ V,

$$\% \text{ error} = \frac{43 \text{ V} - 40 \text{ V}}{40 \text{ V}} \times 100\% = \frac{300\%}{40} = 7.5\%$$

Your report should say that all measurements are within 10% of what was predicted.

Now, calculate the % error for three more voltage drops, each with an actual difference of 3 V. What is the % error if $M_4 = 403$ V and $V_4 = 400$ V?

$$\% \text{ error} = \frac{403 \text{ V} - 400 \text{ V}}{400 \text{ V}} \times 100\% = \frac{300\%}{400} = 0.75\%, \text{ or less than } 1\%$$

Close again.

What is the % error if $M_5 = 3{,}997$ V and $V_5 = 4{,}000$ V? The actual measurement difference is only 3 V. But

$$\% \text{ error} = \frac{3{,}997 \text{ V} - 4{,}000 \text{ V}}{4{,}000 \text{ V}} \times 100\% = \frac{-300\%}{4{,}000}$$

$$= -0.075\%, \text{ or less than } 1\%$$

which is extremely close.

What is the % error if $M_6 = 1$ V and $V_6 = 4$ V? The actual measurement difference is only 3 V.

$$\% \text{ error} = \frac{1 \text{ V} - 4 \text{ V}}{4 \text{ V}} \times 100\% = \frac{-300\%}{4} = -75\%, \text{ which is extremely high.}$$

Notice that when the measured value is more than the theoretical value, as with V_1, V_3, and V_4, the % error is positive. When the measured value is less than the theoretical value, as with V_2, V_5, and V_6, the % error is negative.

Exercises

Fill in the % errors in the data table and then decide if each error is high, low, or in the middle. Low is between -5% and $+5\%$, middle is from -10% to -5% or 5% to 10%, and high is less than -10% or greater than $+10\%$.

Sample Data Table

	Theoretical Value (V)	Measured Value (M)	% Error	High, Low, or Middle?
1.	50	55		
2.	50	46		
3.	500	505		
4.	500	480		
5.	250	200		
6.	250	260		
7.	25	26		
8.	25	23		
9.	2	1.5		
10.	2	2.2		

Answers for Exercises

	Theoretical Value (V)	Measured Value (M)	% Error	High, Low, or Middle?
1.	50	55	10	Middle
2.	50	46	−8	Middle
3.	500	505	1	Low
4.	500	480	−4	Low
5.	250	200	−20	High
6.	250	260	4	Low
7.	25	26	4	Low
8.	25	23	−8	Middle
9.	2	1.5	−25	High
10.	2	2.2	10	Middle

Section 4-1

Find the value of the letter in each proportion.

1. $\dfrac{1}{2} = \dfrac{a}{9}$
2. $\dfrac{x}{7} = \dfrac{3}{4}$
3. $\dfrac{7}{16} = \dfrac{21}{y}$
4. $\dfrac{3}{a} = \dfrac{2}{5}$

5. $\dfrac{2}{9} = \dfrac{x}{6}$
6. $\dfrac{3}{16} = \dfrac{a}{4}$
7. $\dfrac{2}{3} = \dfrac{4}{c}$
8. $\dfrac{1}{4} = \dfrac{2}{a}$

9. $\dfrac{d}{5} = \dfrac{1}{2}$
10. $\dfrac{3}{x} = \dfrac{5}{4}$
11. $\dfrac{2}{9} = \dfrac{c}{3}$
12. $\dfrac{7}{3} = \dfrac{10}{x}$

Identify the given and missing elements as R (rate), B (base), and P (percentage).

13. What percent of 25 is 5?
14. Six is 40% of what number?
15. 5% of 180 is what number?
16. $15 is what percent of $120?
17. 45% of how many dollars is $36?
18. What percent of 10 syringes is 2 syringes?
19. Six is what percent of 25 sacks of grass seed?
20. How many landscape contractors is 15% of 40 landscape contractors?
21. 18% of 150 pieces of sod is how many?
22. What percent of a total of 28 students is 8 students?

Solve.

23. 5% of 480 is what number?
24. $62\frac{1}{2}$% of 120 is what number?
25. $\frac{1}{4}$% of 175 is what number?
26. $233\frac{1}{3}$% of 576 is what number?
27. 39 is what percent of 65?
28. What percent of 118 is 42.48?
29. What percent of 65 is 162.5?
30. 80% of what number is 116?
31. 24% of what number is 19.92?
32. 7.56 is $6\frac{3}{4}$% of what number?
33. 260% of what number is 395.2?
34. 3 is 0.375% of what number?
35. 38.25 is what percent of 250?
36. 83% of 163 is what number?
37. What percent of 26 is 130?
38. 4.75% of 348.2 is what number?
39. $10\frac{1}{3}$% of what number is 8.68?
40. Specifications for bronze call for 80% copper. How much copper is needed to make 300 lb of bronze?

41. Zinc makes up 84 lb of an alloy that weighs 224 lb. What percent of the alloy is zinc?
42. When a subfloor using common 1-in. × 8-in. boards is laid diagonally, 17% is allowed for waste. How many board feet will be wasted out of 1,250 board feet?

43. Of the 2,374 pieces produced by a particular machine 27 were defective. What percent (to the nearest hundredth of a percent) were defective?
44. The voltage loss in a line is 2.5 V. If this is 2% of the generator voltage, what is the generator voltage?

45. If a family spends 28% of its income on food, how much of a $950 paycheck goes for food?
46. If a freshman class of 1,125 college students is made up of 8% international students, how many international students are in the class?

47. A city prosecuted 1,475 individuals with traffic citations. If 36,875 individuals received traffic citations, what percent were prosecuted?
48. A survey studied 600 people for their views on nuclear power plants near their towns. Of these, 75 people said that they approved of nuclear power plants near their towns. What percent approved?

49. In one college, 67 students made the dean's list. If this was 33.5% of the student body, what was the total number of students in the college?
50. It is estimated that only 19% of the licensed big game hunters on a state wildlife management area are successful. If 95 big game hunters were successful, what was the total number of big game hunters in a particular hunting season?

51. A rough casting weighs 32.7 kg. After finishing on a lathe, it weighs 29.3 kg. Find the percent of weight loss to the nearest whole-number percent.

52. A steel beam expands 0.01% of its length when exposed to the sun. If a beam measures 49.995 ft after being exposed to the sun, what is its cooled length?

53. A contractor ordered 800 board feet of lumber for a job that required 750 board feet. What percent, to the nearest whole number, of the required lumber was ordered for waste?

54. A brickmason received a 12% increase in wages, amounting to $35.40. Find the amount of wages received before the increase. Find the amount of wages received after the increase.

55. A lathe costing $600 was sold for $516. What was the percent of decrease in the price of the lathe?

56. According to specifications, a machined part may vary from its specified measure by $\pm 0.4\%$ and still be usable. If the specified measure of the part is 75 in. long, what is the range of measures acceptable for the part?

57. A wet casting weighing 145 kg has a 2% weight loss in the drying process. How much does the dried casting weigh?

58. A mixture of uncut loam and clay soil will have a 20% earth swell when it is excavated. If 300 yd^3 of uncut earth is to be removed, how many cubic yards will have to be hauled away if the earth swell is taken into account?

59. A blueprint specification for a part lists its overall length as 62.5 cm. If a tolerance of $\pm 0.8\%$ is allowed, find the limit dimensions for the part's length.

60. A book that did sell for $18.50 now sells for 20% more. How much does the book now sell for?

61. A computer disk that once sold for $2.25 now sells for 25% less. How much does the computer disk sell for?

62. A paperback dictionary originally sold for $4. It now sells for $1 more. What is the percent of the increase.

63. A laptop computer originally priced at $2,400 now sells for $300 more. What is the percent of the increase?

64. When earth is dug, it usually increases in volume or expands by 20%. How much earth will a contractor have to haul away if 150 ft^3 is dug?

65. A shipping carton is rated to hold 50 lb. A stronger carton that holds 15% more weight will be used for added safety. How much weight will the new carton hold?

66. A 20-inch bar of iron measures 20.025 in. when it is heated. What is the percent of increase?

67. An 18-in. pearl necklace is exchanged for a 24-in. one. What is the percent of increase?

68. A casting weighed 130 ounces (oz) when first made. After it dried, it weighed 127.4 oz. What was the percent of weight loss caused by drying?

69. A dieter went from 168 lb to 160 lb in one week. To the nearest tenth of a percent, what was the percent of weight loss?

70. Workers took a 10% pay cut to help their company stay open during economic hard times. What is the reduced annual salary of a worker who originally earned $35,000?

71. A motorist trades an older car with 350 hp for a new car with 17.4% less horsepower. What is the horsepower of the new car to the nearest whole number?

CHALLENGE PROBLEMS

72. A homeowner with an annual family income of $35,500 spends in a year $6,900 for a home mortgage, $950 for property taxes, $380 for homeowner's insurance, $2,400 for utilities, and $200 for maintenance and repair. To the nearest tenth, what percent of the homeowner's annual income is spent for housing?

73. A motorist with an annual income of $18,250 spends each year $3,420 on automobile financing, $652 on gasoline, $625 on insurance, and $150 on maintenance and repair. To the nearest tenth, what percent of the motorist's annual income is used for automotive transportation?

74. A 78-lb alloy of tin and silver contains 69.3 lb of tin. Find the percent of silver in the alloy to the nearest tenth of a percent.

75. There are 25 women in a class of 35 students. Find the percent of men in the class to the nearest tenth of a percent.

76. A librarian checked out 25 books on interlibrary loan. The books included 5 mysteries, 2 science fiction novels, 8 biographies, and 10 classics. Mysteries and biographies accounted for what percent of the books checked out?

CHAPTER TRIAL TEST

Find the value of the letter in each proportion.

1. $\dfrac{a}{10} = \dfrac{4}{5}$

2. $\dfrac{2}{8} = \dfrac{3}{x}$

3. $\dfrac{2}{c} = \dfrac{1}{6}$

Identify the given and missing elements as R (rate), B (base), and P (percent).

4. $10 is what percent of $35?

5. 40% of 10 x-ray technicians is how many?

6. What percent of 24 syringes is 8?

7. 9 is what percent of 27 dogwood trees?

8. How many books is 30% of 40 books?

9. 12% of 50 grass plugs is how many?

Solve.

10. 10% of 150 is what number?

11. What number is $6\frac{1}{4}\%$ of 144?

12. What percent of 275 is 33?

13. 45.75 is 15% of what number?

14. 55 is what percent of 11?

15. 250% of what number is 287.5?

16. What percent of 360 is 1.2?

17. 245% of what number is 164.4? Round to the nearest hundredth.

18. A casting measuring 48 cm when poured shrinks to 47.4 cm when cooled. What is the percent of decrease?

19. An electronic parts salesperson earned $175 in commission. If the commission is 7% of sales, how much did the salesperson sell?

20. Electronic parts increased 15% in cost during a certain period, amounting to an increase of $65.15 on one order. How much would the order have cost before the increase? Round to the nearest cent.

21. In 2000, an area vocational school had an enrollment of 325 men and 123 women. In 2001, there were 149 women. What was the percent of increase of women students? Round to the nearest hundredth.

22. The payroll for an electrical shop for one week is $1,500. If federal income and FICA taxes average 28%, how much is withheld from the $1,500?

23. During one period, a bakery rejected 372 items as unfit for sale. In the following period, the bakery rejected only 323 items, a decrease in unfit bakery items. What was the percent of decrease? Round to the nearest hundredth.

24. Materials to landscape a new home cost $643.75. What is the amount of tax if the rate is 6%? Round to the nearest cent.

25. A casting weighed 36.6 kg. After milling, it weighs 34.7 kg. Find the percent of weight loss to the nearest whole percent.

26. After soil was excavated for a project, it swelled 15%. If 275 yd^3 was excavated, how many cubic yards of soil was there after excavation?

27. The total bill for machinist supplies was $873.92 before a discount of 12%. How much was the discount? Round to the nearest cent.

28. A business paid a $5.58 finance charge on a monthly balance of $318.76. What was the monthly rate of interest? Round to the nearest hundredth.

5

Direct Measurement

5–4 Introduction to the metric system

1 Identify uses of metric measures of length, weight, and capacity.

2 Convert from one metric unit of measure to another.

3 Make calculations with metric measures.

5–5 Metric–U.S. customary comparisons

1 Convert between U.S. customary measures and metric measures.

5–6 Time

1 Convert from one unit of time to another.

2 Make calculations with measures of time.

3 Equate time of day among time zones.

5–7 Reading measuring instruments

1 Determine the percent of error of a measuring instrument.

2 Find the precision and greatest possible error.

3 Read the U.S. customary rule.

4 Read the metric rule.

5 Read the vernier caliper.

6 Read the micrometer.

7 Find the relative error and the percent error of a measurement.

8 Read circular, uniform, and nonuniform scales.

We routinely use measurements in business, industry, health care, and in our everyday lives. International trade created a need for a worldwide standard of measurements. The International System of Units (SI), more commonly called the *metric system*, has been adopted by most nations of the world, including the United States, and a conversion to this system is gradually taking place.

However, the United States has used a nonmetric system of measurement for many years, and this system is still the customary system of measurement for many businesses and industries. This system is called the *English* or the *U.S. customary system of measurement*, and we begin our study of measurements with it.

5–1 THE U.S. CUSTOMARY SYSTEM OF MEASUREMENT

Learning Outcomes

1 Identify uses of U.S. customary system measures of length, weight, and capacity.

2 Write equivalent measures as unity ratios.

3 Convert one U.S. customary unit of measure to another using unity ratios.

4 Convert from one U.S. customary unit of measure to another using conversion factors.

5 Write mixed U.S. customary measures in standard notation.

The *U.S. customary system of measurement* continues to be used in the United States. Many of the units in this system are now obsolete. In our discussion of the U.S. customary system of measurement, we will include only selected measures.

Length

Four basic units in the U.S. customary system are commonly used to measure length. They are the inch, the foot, the yard, and the mile. Table 5–1 gives the relationships among these measurements of length.

TABLE 5–1 U.S. Customary Units of Length or Distance

12 inches (in.)^a = 1 foot (ft)^b	36 inches (in.) = 1 yard (yd)
3 feet (ft) = 1 yard (yd)	5,280 feet (ft) = 1 mile (mi)

^{a}The symbol ″ means inches (8″ = 8 in.) or seconds (60″ = 60 seconds).

^{b}The symbol ′ means feet (3′ = 3 ft) or minutes (60′ = 60 minutes).

Figure 5–1

Inch: An inch is slightly less than the width of a quarter coin (Fig. 5–1). An inch once was considered to be the width of a man's thumb. The width of a U.S. quarter is a good estimate for 1 inch. Lengths measured in inches include the sizes of men's trousers, belts, shirts, and jackets. The diagonal distance in inches from corner to opposite corner of a television screen determines its size, such as 19-in. or 36-in. television screens. Photographs and picture frames, such as 5 × 7 or 8 × 10 are measured in inches. Lumber, like 2 × 4's or 2 × 12's, is measured in inches.

Foot: A foot is 12 inches, or about the width of a placemat for a dining table (Fig. 5–2). This measure was originally based on the length of a human foot. We use feet to measure lengths varying in size by increments of 12 inches. Lumber, for instance, is sold by the foot. A 2 × 4 is usually sold in lengths of 8 ft, 12 ft, 16 ft, or 20 ft. Human heights are also often measured in feet (and inches), such as 5 ft, 6 in.; 5 ft, 11 in.; or 6 ft, 2 in. Elevation, such as the height of mountains and the altitude of airplanes, is designated in feet.

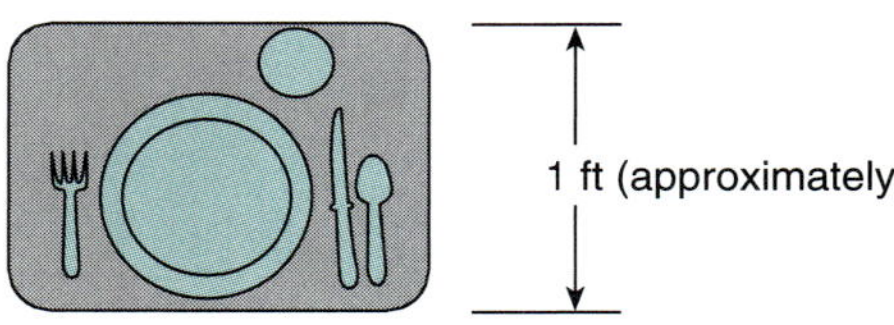

Figure 5–2

Mile: A mile is 5,280 feet, or the length of approximately eight city blocks (Fig. 5–3). The mile was originally used by the Romans, who considered a mile to be 1,000 paces of 5 feet each. Distances between cities and the lengths traveled on a road, street, or highway are measured in miles. The mile is also used to designate the speed of a vehicle and speed limits, such as 55 miles per hour (mph or mi/hr) or 30 mph.

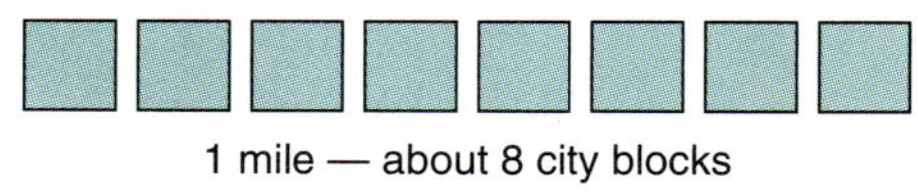

Figure 5–3

Another measure of length is the *yard.* The yard is 3 feet or 36 inches. Yards are sometimes used instead of feet. For example yards are used to measure fabric, carpet, and football fields.

Weight or Mass

Many goods are exchanged or sold according to the amount of their weight or mass. The terms weight and mass are commonly used interchangeably; however, in technical, engineering, and scientific applications a distinction is sometimes made. The *mass* of an object is the quantity of material that makes up the object. The *weight* of an object is a measure of the Earth's gravitational pull on the object. Three commonly used measuring units for weight or mass in the U.S. customary system are the ounce, pound, and ton. Table 5–2 gives the relationships among these measures of weight or mass.

TABLE 5–2 U.S. Customary Units of Weight or Mass

16 ounces (oz) = 1 pound (lb)
2,000 pounds (lb) = 1 ton (T)

Ounce: The ounce is $\frac{1}{16}$ of a pound and is used to measure objects or products that vary in weight by increments of one ounce. The weight of three individual packages of artificial sweetener is approximately one ounce (Fig. 5–4). (The troy ounce, which will not be used in this text, is used to measure precious metals and is $\frac{1}{12}$ of a pound. *Ounce* is derived from a Latin word for $\frac{1}{12}$.) Ounces are used to measure lightweight items like first-class letters, tubes of toothpaste or medicinal ointments, canned goods, and dry-packaged foods like pasta, candy, and gravy mixes.

Pound: A pound is equal to 16 ounces. It is used for items that vary by increments of one pound. The term is derived from an old English word for weight. A pound is the weight of a 4.25-in.-tall can of cut green beans or a paper container of coffee (Fig. 5–5). Many small, medium, and large items of merchandise are measured in pounds, like grocery and market items, people's weights, and air pressure in automotive tires. The pound is also used to calculate shipping charges for merchandise and parcel-post packages.

Ton: A ton is 2,000 pounds. The word originally meant a weight or measure. The typical American-made SUV weighs 3 tons and may help you visualize this measure (Fig. 5–6). The ton is used for extremely heavy items, such as very large volumes of grain, the weights of huge animals like elephants (4–7 tons), and the weights by which coal and iron are sold.

Capacity or Volume

The U.S. customary system of measurement has several units for capacity or volume. The system includes units for both liquid and dry capacity measures; however, the dry capacity measures are seldom used. It is more common to express dry measures in terms of weight than in terms of capacity.

The U.S. customary units of measure for capacity or volume are the ounce, cup, pint, quart, and gallon. Table 5–3 gives the relationships among the liquid measures for capacity or volume. In the U.S. customary system the term *ounce* represents both weight and liquid capacity. The measures are different and have no common relationship. The context of the problem will suggest whether the unit for weight or capacity is meant.

TABLE 5–3 U.S. Customary Units of Liquid Capacity or Volume

8 ounces (oz) = 1 cup (c)	4 cups (c) = 1 quart (qt)
2 cups (c) = 1 pint (pt)	4 quarts (qt) = 1 gallon (gal)
2 pints (pt) = 1 quart (qt)	

Ounce: A liquid ounce is a volume, not a weight. It is about the volume of two small bottles of fingernail polish or perfume (Fig. 5–7). Liquid ounces are used for

1 ounce — approximately three packets of artificial sweetener

Figure 5–4

1 lb — about one can of cut green beans

Figure 5–5

3 tons – approximately

Figure 5–6

1 ounce — approximately two small bottles of fingernail polish

Figure 5–7

1 cup — approximately one large coffee cup

Figure 5–8

1 pint – one large single-serving container of milk

Figure 5–9

bottled medicine, canned or bottled carbonated beverages, baby bottles and formula, and similar-sized quantities.

Cup: A cup is equal to 8 ounces. The term is derived from an Old English word for tub, a kind of container. This measure is about the volume of a coffee cup (Fig. 5–8). Cups are most often used for 8-oz quantities in cooking. Measuring cups used for cooking usually divide their contents into cups and portions of cups.

Pint: A pint is two cups. Its name comes from an Old English word meaning the spot that marks a certain level in a measuring device. It is the quantity of a *medium*-size paper container of milk (Fig. 5–9). Liquids like milk and automobile motor additives are packaged in pint containers.

Quart: A quart is two pints, or four cups, or 32 ounces. The term is derived from an Old English word for fourth. It is a fourth of a gallon. Milk and various citrus juices are often sold in quart containers (Fig. 5–10). Motor oil and bottled water are also packaged in quart containers. Quarts are used to measure liquid quantities in cooking and also the capacities of cookware like mixing bowls, pots, and casserole dishes.

Gallon: A gallon is four quarts. It is based on the "wine gallon" of British origin. Paint, varnish, and stain are often sold in gallon cans (Fig. 5–11). Motor fuel and heating oil are usually sold by the gallon, as are large quantities of liquid propane and various chemicals. Often, statistics on liquid consumption, such as water or alcoholic beverages, are reported in gallons consumed per individual during a specified time period.

1 quart — one container of milk

Figure 5–10

1 gallon — one large can of paint

Figure 5–11

EXAMPLE   Select a letter that indicates the most reasonable unit:

1. Package of rice	a. Pounds or ounces
2. Height of Doctor Washington	b. Feet, or feet and inches
3. A hippopotamus	c. Tons or pounds
4. Sugar for a cake	d. Cups
5. Expensive French perfume	e. Ounces
6. Distance between Memphis and New Orleans	f. Miles
7. Bottle of suntan lotion	g. Ounces
8. Container of eggnog	h. Gallon, quart, or pint
9. Sherbet in grocery store freezer	i. Gallon or pint
10. Tank of pesticide	j. Gallons
11. Man's dress shirt size	k. Inches
12. Material for draperies	l. Yards

2   **Write Equivalent Measures as Unity Ratios.**

Two useful properties of numbers are $\frac{n}{n} = 1$ and $1 \times n = n$. That is, a number divided by itself equals 1, and 1 times any number equals the number. Using the relationship between two units of measure, we can form a ratio that has a value of 1 in two different ways. We call this type of ratio a *unity ratio*.

■ **DEFINITION:** **Unity Ratio.** A *unity ratio* is a ratio of measures that has a value of 1.

A ratio is a fraction. A unity ratio, then, is a fraction with one unit of measure in the numerator and a different, but equivalent, unit of measure in the denominator. Some examples of unity ratios are

$$\frac{12 \text{ in.}}{1 \text{ ft}}, \qquad \frac{1 \text{ ft}}{12 \text{ in.}}, \qquad \frac{3 \text{ ft}}{1 \text{ yd}}, \qquad \frac{1 \text{ mi}}{5{,}280 \text{ ft}}$$

In each unity ratio, the value of the numerator equals the value of the denominator. When we have a ratio with the numerator and denominator equal, the value of the ratio is 1. We call this ratio a *unity ratio* because its value is 1. The word *unity* means 1. When we convert from one unit of measure to another, we use a unity ratio that contains the original unit and the new unit.

EXAMPLE Write two unity ratios that relate the pair of measures.

(a) ounces and pounds　　　　　　　　(b) cups and pints

(a) The relationship between ounces and pounds is 1 lb contains 16 oz. The unity ratios involving ounces and pounds are

$$\frac{1 \text{ lb}}{16 \text{ oz}} \quad \text{and} \quad \frac{16 \text{ oz}}{1 \text{ lb}}$$

(b) The relationship between cups and pints is 1 pint contains 2 cups. The unity ratios involving cups and pints are

$$\frac{1 \text{ pint}}{2 \text{ cups}} \quad \text{and} \quad \frac{2 \text{ cups}}{1 \text{ pint}}$$

3 Convert One U.S. Customary Unit of Measure to Another Using Unity Ratios.

Unity ratios help us convert from one unit of measure to another. This is an alternative process to the proportion process and will be useful in formulas and dimension analysis.

To change from one U.S. customary unit of measure to another using unity ratios:

1. Set up the original amount as a fraction with the original unit of measure in the numerator.
2. Multiply this by a unity ratio with the original unit in the denominator and the new unit in the numerator.
3. Reduce like units of measure and all numbers wherever possible.

EXAMPLE Find the number of inches in 5 ft.

To make this conversion, we multiply 5 ft by a unity ratio that contains both inches and feet.

Because 5 ft is a whole number, we write it with 1 as the denominator. We place the original unit with the 5 in the *numerator* of the first fraction.

$$\frac{5 \text{ ft}}{1} \left(\frac{}{} \right)$$

We are changing *from* (feet), so we place ft in the *denominator* of the unity ratio, which is shown in parentheses. This allows us to reduce the units later.

$$\frac{5 \text{ ft}}{1} \left(\frac{}{\text{ft}} \right)$$

We are changing *to* (inches), so we place in. in the *numerator* of the unity ratio.

$$\frac{5 \text{ ft}}{1}\left(\frac{\text{in.}}{\text{ft}}\right)$$

Now we place in the unity ratio the numerical values that make these two units of measure equivalent (1 ft = 12 in.). Then we complete the calculation, reducing wherever possible.

$$\frac{5 \text{ ft}}{1}\left(\frac{12 \text{ in.}}{1 \text{ ft}}\right) = 60 \text{ in.}$$

Thus, 5 ft = 60 in.

EXAMPLE   How many pints are in 4.5 quarts?

$$\frac{4.5 \text{ qt}}{1}\left(\frac{2 \text{ pt}}{1 \text{ qt}}\right) = 4.5 \,(2 \text{ pt}) = 9 \text{ pt} \qquad \text{From quart (denominator) to pint (numerator).}$$

Thus, 4.5 qt = 9 pt.

When we are working with units of measure, it is very important to include the measuring unit in our analysis. Measurements are also referred to as *dimensions,* and the systematic examination of the appropriate measuring units of a solution is referred to as *dimension analysis.* This analysis is very important in the study of chemistry, physics, engineering, nursing, and many other fields of study.

Sometimes it is necessary to convert a U.S. customary unit to a U.S. unit that is *not* the next larger or smaller unit of measure. Suppose a dressmaker needs to know how many yards are in a certain number of inches of fabric. We solve this problem by converting inches to feet with one unity ratio and then converting feet to yards with another unity ratio.

Tip!	***Changing to Any Larger or Smaller Unit.***

To change from a U.S. customary unit to one other than the next larger or smaller unit, proceed as before, but multiply the original amount by as many unity ratios as needed to attain the new U.S. customary unit.

For instance, to change from yards to inches:

$$\text{yards} \rightarrow \text{feet} \rightarrow \text{inches}$$

To change from gallons to ounces:

$$\text{gallons} \rightarrow \text{quarts} \rightarrow \text{pints} \rightarrow \text{cups} \rightarrow \text{ounces}$$

EXAMPLE   How many inches are in $2\frac{1}{3}$ yd?

Write $2\frac{1}{3}$ as an improper fraction.

$$2\frac{1}{3} \text{ yd} = \frac{7}{3} \text{ yd}$$

Multiply the improper fraction by two unity ratios. To change from yards to inches, first change from yards to feet, then from feet to inches. The first unity ratio converts yards to feet, and the second one converts feet to inches. Place the original unit in the *numerator* of the improper fraction, $\frac{7}{3}$.

$$\frac{7 \text{ yd}}{3}\left(\frac{\text{ft}}{\text{yd}}\right)\left(\frac{\text{in.}}{\text{ft}}\right)$$

Insert the correct numerical values for each unity ratio (3 ft = 1 yd; 12 in. = 1 ft), then complete the calculation, reducing wherever possible.

$$\frac{7 \text{ yd}}{\cancelto{}{3}} \left(\frac{\cancelto{1}{3} \text{ ft}}{1 \text{ yd}}\right)\left(\frac{12 \text{ in.}}{1 \text{ ft}}\right) = 84 \text{ in.}$$

Thus, 84 in. $= 2\frac{1}{3}$ yd.

Alternative method

If we use the relationship for inches and yards, 36 in. = 1 yd, we need only one unity ratio for the calculation. That is, we convert $2\frac{1}{3}$ yd ($\frac{7}{3}$ yd) to inches as follows:

$$\frac{7 \text{ yd}}{3}\left(\frac{36 \text{ in.}}{1 \text{ yd}}\right)$$

Set up the unity ratio using 36 in. = 1 yd.

$$\frac{7 \text{ yd}}{\cancelto{1}{3}}\left(\frac{\cancelto{12}{36} \text{ in.}}{1 \text{ yd}}\right) = 84 \text{ in.}$$

Reduce units and numbers; then multiply.

Thus, $2\frac{1}{3}$ yd $= 84$ in.

■ Learning Strategy *Focus on One Thing at a Time.*

Sometimes the steps in a multistepped problem can be overwhelming. It is often helpful to focus on one aspect of the problem at a time. In the example changing yards to inches, focus first on just the units of measure or dimensions.

$$\frac{\text{yd}}{1}\left(\frac{\text{ft}}{\text{yd}}\right)\left(\frac{\text{in.}}{\text{ft}}\right)$$

Then, reduce as appropriate. Yards reduce to 1. Feet reduce to 1. The only measuring unit left is inches. Therefore, the result will be in inches.

Next, focus on the numbers.

$$\frac{7 \text{ yd}}{3}\left(\frac{3 \text{ ft}}{1 \text{ yd}}\right)\left(\frac{12 \text{ in.}}{1 \text{ ft}}\right) \rightarrow \frac{7}{3}\left(\frac{3}{1}\right)\left(\frac{12}{1}\right) = 84$$

Now, putting the number and unit together, you have 84 in.

EXAMPLE Find the number of ounces in 2 gallons.

To change from gallons to ounces, we use the following conversions:

$$\text{gallons} \rightarrow \text{quarts} \rightarrow \text{pints} \rightarrow \text{cups} \rightarrow \text{ounces}$$

Thus, we need four unity ratios to convert gallons to quarts, quarts to pints, pints to cups, and cups to ounces. First, examine the units of measure and set up the units in ratios. Insert numbers that relate the units to complete unity ratios. Reduce the units. The values are listed in Table 5–3.

$$\frac{2 \text{ gal}}{1}\left(\frac{\text{qt}}{\text{gal}}\right)\left(\frac{\text{pt}}{\text{qt}}\right)\left(\frac{\text{c}}{\text{pt}}\right)\left(\frac{\text{oz}}{\text{c}}\right) =$$

$$\frac{2 \text{ gal}}{1}\left(\frac{4 \text{ qt}}{1 \text{ gal}}\right)\left(\frac{2 \text{ pt}}{1 \text{ qt}}\right)\left(\frac{2 \text{ c}}{1 \text{ pt}}\right)\left(\frac{8 \text{ oz}}{1 \text{ c}}\right) =$$

Reduce measures.

$$2(4)(2)(2)(8) \text{ oz} = 256 \text{ oz}$$

Multiply.

Thus, 2 gal $= 256$ oz.

We can work this same problem with fewer unity ratios if we make some preliminary calculations using the relationships in Table 5–3. For example, $4\text{ c} = 1\text{ qt}$ and $8\text{ oz} = 1\text{ c}$, so we multiply 4 (cups) by 8 (ounces per cup) to find the number of ounces in a quart; that is, $4 \times 8 = 32\text{ oz}$. Thus, $32\text{ oz} = 1\text{ qt}$. We can now find the number of ounces in 2 gallons with fewer unity ratios.

$$\frac{2\text{ gal}}{1}\left(\frac{4\text{ qt}}{1\text{ gal}}\right)\left(\frac{32\text{ oz}}{1\text{ qt}}\right) \qquad \text{Set up unity ratios.}$$

$$\frac{2\text{ gal}}{1}\left(\frac{4\text{ qt}}{1\text{ gal}}\right)\left(\frac{32\text{ oz}}{1\text{ qt}}\right) = 256\text{ oz} \qquad \text{Reduce and multiply.}$$

Thus, 2 gal = 256 oz.

Tip!	*Estimation and Dimension Analysis.*

When estimating unit conversions, first see if the new unit is larger or smaller than the original unit.

Larger to smaller

Each larger unit can be divided into smaller units. Thus, larger-to-smaller conversions mean *more* smaller units. *More* implies multiplication.

To convert a U.S. customary unit to a desired *smaller* unit: *Multiply* the number of larger units by the number of smaller units that equals 1 larger unit.

$2\text{ yd} = \underline{\qquad}\text{ ft}$ **The smaller unit is feet: 3 ft = 1 yd.**

$2 \times 3\text{ ft} = 6\text{ ft}$ **Multiply number of yards by 3 ft.**

Dimension analysis: $\dfrac{2\text{ yd}}{1} \times \dfrac{3\text{ ft}}{1\text{ yd}} = 6\text{ ft}$

Thus, $2\text{ yd} = 6\text{ ft}$.

Let's look at the key word clues.

$$\text{Larger to smaller unit} \rightarrow \text{obtain more units} \rightarrow \text{multiply}$$

Smaller to larger

Several small units combine to make one large unit. Thus, smaller-to-larger conversions mean *fewer* large units. *Fewer* implies division.

To convert a U.S. customary unit to a *larger* unit: *Divide* the original unit by the number of smaller units that equals 1 desired larger unit.

$12\text{ ft} = \underline{\qquad}\text{ yd}$ **The larger unit is yards: 1 yd = 3 ft.**

$12 \div 3 = 4$ **Divide by 3 ft to get yards.**

Thus, $12\text{ ft} = 4\text{ yd}$.

Dimension analysis: $\dfrac{12\text{ ft}}{1} \times \dfrac{1\text{ yd}}{3\text{ ft}} = 4\text{ yd}$

Using key word clues:

$$\text{Smaller to larger unit} \rightarrow \text{fewer units} \rightarrow \text{divide}$$

Here's a quick way to estimate: Decide if the end result will be more or fewer units than you started with. This type of estimation can catch errors in setting up the problem, but it will not likely catch calculation errors.

---

4   **Convert from One U.S. Customary Unit of Measure to Another Using Conversion Factors.**

Unity ratios help us develop conversion factors. With conversion factors, we *always* multiply. To develop conversion factors, look again at reciprocals and the relationship between multiplication and division.

To change from feet to inches, we use the unity ratio $\dfrac{12 \text{ in.}}{1 \text{ ft}}$. The numerical portion of the ratio reduces to the whole number 12. Therefore, the conversion factor for changing from feet to inches is 12.

To change from inches to feet, we use the unity ratio $\dfrac{1 \text{ ft}}{12 \text{ in.}}$. The numerical portion of the ratio is $\dfrac{1}{12}$. From the relationship between multiplication and division, we can interpret this as multiplying by $\dfrac{1}{12}$ or dividing by 12.

Multiplying by $\dfrac{1}{12}$ is the same as multiplying by the decimal equivalent of $\dfrac{1}{12}$, $1 \div 12 = 0.08\overline{3}$, so the conversion factor for converting inches to feet is $0.08\overline{3}$.

EXAMPLE Develop conversion factors for the following units.

(a) cups and pints

$$\dfrac{\text{cups}}{1}\left(\dfrac{\text{pints}}{\text{cups}}\right)$$

$$\dfrac{\text{cups}}{1}\left(\dfrac{1 \text{ pint}}{2 \text{ cups}}\right)$$

$$\dfrac{1}{2} = 0.5$$

cups $\times$ 0.5 = pints

$$\dfrac{\text{pounds}}{1}\left(\dfrac{\text{ounces}}{\text{pounds}}\right)$$

$$\dfrac{\text{pounds}}{1}\left(\dfrac{16 \text{ ounces}}{1 \text{ pound}}\right)$$

$$\dfrac{16}{1} = 16$$

pounds $\times$ 16 = ounces

(b) pounds and ounces

$$\dfrac{\text{pints}}{1}\left(\dfrac{\text{cups}}{\text{pints}}\right)$$

$$\dfrac{\text{pints}}{1}\left(\dfrac{2 \text{ cups}}{1 \text{ pint}}\right)$$

$$\dfrac{2}{1} = 2$$

pints $\times$ 2 = cups

$$\dfrac{\text{ounces}}{1}\left(\dfrac{\text{pounds}}{\text{ounces}}\right)$$

$$\dfrac{\text{ounces}}{1}\left(\dfrac{1 \text{ pound}}{16 \text{ ounces}}\right)$$

$$1 \div 16 = 0.0625$$

ounces $\times$ 0.0625 = pounds

EXAMPLE Use a conversion factor to convert 54 ounces to pounds.

ounces $\times$ 0.0625 = pounds   Conversion factor for ounces to pounds

$56 \times 0.0625 = 3.5$ pounds

56 ounces is 3.5 pounds.

5 Write Mixed U.S. Customary Measures in Standard Notation.

We express some measures with two or more different units. For example, the weight of an object is 5 lb 3 oz.

■ **DEFINITION: Mixed Measures.** Measures that use two or more units are called *mixed measures*.

■ **DEFINITION: Standard Notation.** A mixed measure is in standard notation if the number associated with each unit of measure is smaller than the number required to convert to the next larger unit. The number in the largest unit of measure given may or may not be converted as desired.

Therefore, the 15 in. in 3 ft 15 in. converts to 1 ft 3 in., and we add 1 ft to the 3 ft as follows:

$$3 \text{ ft } \boxed{15 \text{ in.}} = 3 \text{ ft } + \boxed{12 \text{ in.} + 3 \text{ in.}} = \boxed{3 \text{ ft } + 1 \text{ ft}} + 3 \text{ in.} = \boxed{4 \text{ ft}} \, 3 \text{ in.}$$

Thus, standard notation for 3 ft 15 in. is 4 ft 3 in. We could also write this as 1 yd 1 ft 3 in.

EXAMPLE   Express (a) 8 lb 20 oz and (b) 1 gal 5 qt in standard notation.

(a) 8 lb 20 oz   20 oz = 1 lb 4 oz

8 lb 20 oz = 8 lb + 1 lb 4 oz = **9 lb 4 oz**   Standard notation.

(b) 1 gal 5 qt   5 qt = 1 gal 1 qt

1 gal 5 qt = 1 gal + 1 gal 1 qt = **2 gal 1 qt**   Standard notation.

If the mixed measure contains three or more different units, then conversion to standard notation may require two steps.

EXAMPLE   Express 2 yd 4 ft 16 in. in standard notation.

$$2 \text{ yd } 4 \text{ ft } \boxed{16 \text{ in.}} = 2 \text{ yd } 4 \text{ ft } \boxed{12 \text{ in.} + 4 \text{ in.}} = 2 \text{ yd } 4 \text{ ft } + \boxed{1 \text{ ft} + 4 \text{ in.}}$$

$$= 2 \text{ yd } \boxed{5 \text{ ft}} \, 4 \text{ in.} = 2 \text{ yd } \boxed{3 \text{ ft} + 2 \text{ ft}} \, 4 \text{ in.}$$

$$= 2 \text{ yd } + \boxed{1 \text{ yd } 2 \text{ ft}} \, 4 \text{ in.} = \textbf{3 yd 2 ft 4 in.} \qquad \text{Standard notation.}$$

Tip!	***Standard Conventions.***

In one illustration, we said that 3 ft 15 in. is 4 ft 3 in. in standard notation. Why not change 4 ft 3 in. to 1 yd 1 ft 3 in.? There may be situations when 1 yd 1 ft 3 in. is the desirable form; however, in general, we keep the same units of measure in standard notation. 3 qt 3 pt = 4 qt 1 pt (instead of 1 gal 1 pt).

SELF-STUDY EXERCISES 5–1

1   Identify the appropriate U.S. customary measure.

1. Shipping weight of a sofa
2. Liquid medicine in a bottle
3. Package of dried beans
4. Large home aquarium
5. Container of motor oil
6. Shipment of coal
7. Sack of potatoes
8. Size of a casserole dish
9. Height of a hospital patient
10. Milk for a cake recipe
11. Man's belt size
12. Cloth for a Mardi Gras costume
13. Hourly speed of a train
14. Weight of a first-class letter
15. Shaving lotion
16. Distance from home to a doctor's office across town
17. Package of macaroni
18. Pork roast
19. Parcel-post package
20. Tube of ointment

2   Write two unity ratios that relate the *given* pair of measures.

21. pints and quarts
22. feet and miles
23. inches and feet
24. feet and yards
25. days and weeks
26. quarts and gallons

3   Use unity ratios to convert the units.

27. 4 ft = _______ in.
28. 7 yd = _______ ft
29. $2\frac{1}{2}$ mi = _______ yd
30. Find the number of yards in 28 ft.
31. If a car with 0 miles on the odometer is driven 10,560 ft, how many miles will register on the odometer?

32. How many ounces are in 3 lb?

34. How many ounces are in 45.8 lb?

33. Find the number of pounds in $36\frac{4}{5}$ oz.

35. A can of vegetables weighs 1.2 lb. How many ounces is this?

36. The net weight of a can of corn is 17 oz. If a case contains 24 cans, what is the net weight of a case in ounces? In pounds?

37. How many quarts are in 5 gal?

38. How many pints are in $6\frac{1}{2}$ qt?

39. Find the number of gallons in 6 qt.

40. How many pints are in $7\frac{1}{2}$ gal?

41. How many gallons are in 72 pt?

42. How many ounces are in 2 gal?

43. How many yards are in 54 in.?

44. How many inches are in 3 yd?

45. How many feet are in $1\frac{1}{3}$ mi?

46. How many cups are in $1\frac{1}{2}$ gal?

4 Develop conversion factors for the pairs of units.

47. feet and yards

48. quarts and gallons

49. quarts and pints

50. ounces and cups

Use conversion factors to convert the units of measure.

51. 460 oz is equivalent to how many pints?

52. How many pints are in 46 qt?

53. A 50-ft garden hose is how many yards long?

54. How many gallons are in 200 pt?

55. How many pounds are in 580 oz?

56. If a cabinet is $12\frac{3}{4}$ ft long, how many inches is this?

57. A bolt of fabric contains about 20 yd. How many inches is this?

58. A stack of plywood is 56 in. high. How many feet is this?

59. How many feet are in 1.5 miles?

60. How many quarts are in 7 gal?

5 Express the measures in standard notation.

61. 2 ft 20 in.

62. 1 mi 6,375 ft

63. 2 lb $19\frac{1}{2}$ oz

64. 1 gal 5 qt

65. 1 gal 3 qt 2 pt

66. 1 T 2,500 lb

67. 2 yd 1 ft 23 in.

68. 1 qt 5 c 10 oz

69. 2 ft 10 in.

70. 5 lb 25 oz

71. 3 gal 5 qt 48 oz

72. 6 qt 20 oz

73. 1 mi 265 yd 4,500 ft 25 in.

74. 2 ft 40 in.

75. 2 T 2,600 lb 15 oz

76. 3 lb 21 oz

77. 2 lb 5 oz

78. 1 yd 4 ft 16 in.

79. 2 yd 2 ft 11 in.

80. 1 mi 875 ft 6 in.

5-2   ADDING AND SUBTRACTING U.S. CUSTOMARY MEASURES

Learning Outcomes

1 Add U.S. customary measures.

2 Subtract U.S. customary measures.

Once we can convert from one U.S. customary unit to another, we can perform addition, subtraction, multiplication, and division.

1   Add U.S. Customary Measures.

We can add U.S. customary measures *only* when their units are the same; that is, measures with the same units are *like* measures. *Unlike* measures are measures with different units. Before we can add 3 ft and 2 in., we need to change one of the measures so they are both expressed as inches or as feet.

To add unlike U.S. customary measures:

1. Convert to a common U.S. customary unit.

2. Add.

   Add 3 ft + 2 in.

Because 2 in. is a fraction of a foot, we can avoid working with fractions by converting the larger unit (feet) to the smaller unit (inches).

$$3 \text{ ft} = \frac{3 \text{ ft}}{1}\left(\frac{12 \text{ in.}}{1 \text{ ft}}\right) = 36 \text{ in.}$$

Because 3 ft = 36 in., then 3 ft + 2 in. =

36 in. + 2 in. = 38 in.

■ **Learning Strategy**   *Relate to Previous Processes.*

We are using a process that requires us to add *like quantities*. When have we dealt with *likes* before? *Like* signs in adding integers, *like* denominators when adding fractions.

What did we do when we had *unlike quantities?*

With integers having *unlike* signs, we use a different rule. With fractions having *unlike* denominators, we change to fractions with a common denominator. Which process will we pattern after here?

In many instances, unlike units can be changed to a common unit. To add 3 ft + 2 in., we change 3 ft to inches. Are there situations when we cannot change to a common unit? Yes, but sometimes in our investigation of possible solutions, we may temporarily consider adding unlike measures.

Can we add 3 ft + 2 lb? No. Why not? Feet and pounds measure different types of units: length and weight.

To add mixed U.S. customary measures:

1. Align the measures vertically so the common units are written in the same vertical column.
2. Add.
3. Express the sum in standard notation.

EXAMPLE   Add and write the answer in standard form: 6 lb 7 oz and 3 lb 13 oz.

```
   6 lb  7 oz
+  3 lb 13 oz
   9 lb 20 oz      20 oz = 1 lb 4 oz
```

Thus, 9 lb 20 oz = 10 lb 4 oz.

2 Subtract U.S. Customary Measures.

To add measures, we must express different units as a common unit. To subtract measures, we must also find a common unit.

To subtract unlike U.S. customary units:

1. Convert to a common U.S. customary unit.
2. Subtract.

EXAMPLE   Subtract 15 in. from 2 ft.

Changing 15 in. to feet gives us a mixed number, so it is easier to convert 2 ft to inches.

$$2 \text{ ft} = 24 \text{ in.}$$

$$24 \text{ in.} - 15 \text{ in.} = \textbf{9 in.}$$

To subtract mixed U.S. customary measures:

1. Align the measures vertically so that the common units are written in the same vertical column.
2. Subtract.

EXAMPLE Subtract 5 ft 3 in. from 7 ft 4 in.

Align like measures in a vertical line, then subtract.

$$\begin{array}{r} 7\text{ ft }4\text{ in.} \\ -\ 5\text{ ft }3\text{ in.} \\ \hline \mathbf{2\text{ ft }1\text{ in.}} \end{array}$$

Sometimes when we subtract mixed measures, we subtract a larger unit from a smaller one. In this case, we use our knowledge of borrowing.

To subtract a larger U.S. customary unit from a smaller unit in mixed measures:

1. Align the common measures in vertical columns.
2. Borrow one unit from the next larger unit of measure in the minuend, convert to the equivalent smaller unit, and add it to the smaller unit.
3. Subtract the measures.

EXAMPLE Subtract 3 lb 12 oz from 7 lb 8 oz.

$$\begin{array}{r} 7\text{ lb }\ 8\text{ oz} \\ -3\text{ lb }12\text{ oz} \end{array}$$

We always begin subtraction with the smallest unit, which should be on the *right*. In this example, notice that 12 oz cannot be subtracted from 8 oz, so we rewrite 7 lb as 6 lb 16 oz. Then, we add 16 oz to 8 oz to get 24 oz.

$$\begin{array}{rcl} 7\text{ lb }\ 8\text{ oz} = 6\text{ lb }16\text{ oz} + 8\text{ oz} = & & 6\text{ lb }24\text{ oz} \\ -\ 3\text{ lb }12\text{ oz} = & & \underline{-\ 3\text{ lb }12\text{ oz}} \\ & & \mathbf{3\text{ lb }12\text{ oz}} \end{array}$$

SELF-STUDY EXERCISES 5–2

1 Add and write the answer in standard notation.

1. 5 oz + 2 lb

2. 4 ft + 7 in.

3.
$$\begin{array}{r} 8\text{ lb }2\text{ oz} \\ +\ 7\text{ lb }9\text{ oz} \end{array}$$

4.
$$\begin{array}{r} 5\text{ ft }45\text{ in.} \\ +\ 7\text{ ft }30\text{ in.} \end{array}$$

5.
$$\begin{array}{r} 5\text{ qt }1\ \text{pt} \\ +\ 2\text{ qt }1\tfrac{1}{2}\text{ pt} \end{array}$$

6.
$$\begin{array}{r} 8\text{ gal }3\text{ qt} \\ +\ 5\text{ gal }2\text{ qt} \end{array}$$

7.
$$\begin{array}{r} 7\text{ yd }2\text{ ft} \\ +\ 1\text{ yd }2\text{ ft} \end{array}$$

8.
$$\begin{array}{r} 3\text{ c }6\text{ oz} \\ +\ 1\text{ c }3\text{ oz} \end{array}$$

9.
$$\begin{array}{r} 1\text{ ft }13\text{ in.} \\ +\ 2\text{ ft }15\text{ in.} \end{array}$$

10.
$$\begin{array}{r} 4\text{ yd }2\text{ ft }\ 7\text{ in.} \\ +\ 2\text{ yd }1\text{ ft }10\text{ in.} \end{array}$$

Solve the problems. When necessary, express the answers in standard notation.

11. A plumber has a 2-ft length of copper tubing and a 7-in. length of copper tubing. What is the total?

12. How long are two extension cords together if one is 6 ft and the other is 60 in.?

13. A mixture for hamburgers contains 2 lb 8 oz of ground round steak and 3 lb 7 oz of regular ground beef. How much does the hamburger mixture weigh?

14. According to hospital maternity records, one infant twin weighed 6 lb 1 oz and the other weighed 5 lb 15 oz at birth. What was their total weight?

2   Subtract and write the answer in standard form.

15. 2 ft − 18 in.

16. 2 qt − 3 pt

17. 3 yd − 7 ft

18. 6 lb − 18 oz

19.
$$\begin{array}{r} 3 \text{ lb } 12 \text{ oz} \\ -\ 2 \text{ lb }\ 6 \text{ oz} \\ \hline \end{array}$$

20.
$$\begin{array}{r} 12 \text{ lb }\ 7 \text{ oz} \\ -\ \ 5 \text{ lb } 12 \text{ oz} \\ \hline \end{array}$$

21.
$$\begin{array}{r} 3 \text{ ft } 8 \text{ in.} \\ -\ 2 \text{ ft } 9 \text{ in.} \\ \hline \end{array}$$

22.
$$\begin{array}{r} 19 \text{ lb } 12 \text{ oz} \\ -\ \ 8 \text{ lb } 15 \text{ oz} \\ \hline \end{array}$$

23.
$$\begin{array}{r} 2 \text{ ft } 30 \text{ in.} \\ -\ 1 \text{ ft } 40 \text{ in.} \\ \hline \end{array}$$

24.
$$\begin{array}{r} 5 \text{ gal } 3 \text{ qt }\ 1 \text{ pt} \\ -\ 1 \text{ gal } 3 \text{ qt } 1\frac{1}{2} \text{ pt} \\ \hline \end{array}$$

25. A seamstress has a length of cloth 5 ft long. What is its length after 10 in. is cut off?

26. A cabinet maker cut 9 in. from a 3-ft shelf in a medical lab. How long was the shelf after it was cut?

27. If a computer with a Pentium® processor sorts a list of names in 1 min 30 sec and a computer with a math coprocessor does the same job in 45 sec, how much time is saved by using the computer with the math coprocessor?

28. A car with a 2.0-L engine accelerates a certain distance in 1 min 27 sec. A car with a 5.0-L engine accelerates the same distance in 48 sec. How much faster is the car with the 5.0-L engine?

29. A package containing a laser printer weighs 74 lb 3 oz. The container and packing material weighs 4 lb 12 oz. How much does the laser printer weigh?

30. To weigh a pet poodle, Stacey held the dog while standing on a scale. If the scale showed 132 lb 6 oz and Stacey weighs 115 lb 8 oz, how much does the poodle weigh?

<table><tr><td>**5-3**</td><td>**MULTIPLYING AND DIVIDING U.S. CUSTOMARY MEASURES**</td></tr></table>

Learning Outcomes

1 Multiply a U.S. customary measure by a number.
2 Multiply a U.S. customary measure by a measure.
3 Divide a U.S. customary measure by a number.
4 Divide a U.S. customary measure by a measure.
5 Change from one U.S. customary rate measure to another.

1   **Multiply a U.S. Customary Measure by a Number.**

In addition to adding and subtracting U.S. customary measures, we often need to multiply or divide those measures by some number. Suppose a carpenter buys cedar siding in lengths of 18 ft 6 in. To find the total length of six pieces of siding, we multiply 18 ft 6 in. by 6. We do this by multiplying 6 by the numbers associated with each unit of measure.

> *To multiply a U.S. customary measure by a number:*
>
> **1.** Multiply the numbers associated with each unit of measure by the given number.
> **2.** Write the resulting measure in standard notation.

EXAMPLE   A container holds 21 gal 3 qt of weed killer. How much do eight containers hold?

$$\begin{array}{r} 21 \text{ gal } 3 \text{ qt} \\ \times\ \ \ \ \ \ \ 8 \\ \hline 168 \text{ gal } 24 \text{ qt} \end{array} \qquad 24 \text{ qt} = 6 \text{ gal}$$

$$168 \text{ gal } 24 \text{ qt} = 168 \text{ gal} + 6 \text{ gal} = 174 \text{ gal in standard notation}$$

The eight containers hold 174 gal.

  Find the total length of six pieces of siding that are each 18 ft 6 in. long.

$$18 \text{ ft } 6 \text{ in.}$$
$$\times \quad 6$$
$$\overline{108 \text{ ft } 36 \text{ in.}} \qquad 36 \text{ in} = 3 \text{ ft}$$
$$108 \text{ ft} + 3 \text{ ft} = 111 \text{ ft}$$

The total length of siding is 111 ft.

2 Multiply a U.S. Customary Measure by a Measure.

Length measures multiplied by like length measures produce square units. Square measures indicate areas (Fig. 5–12).

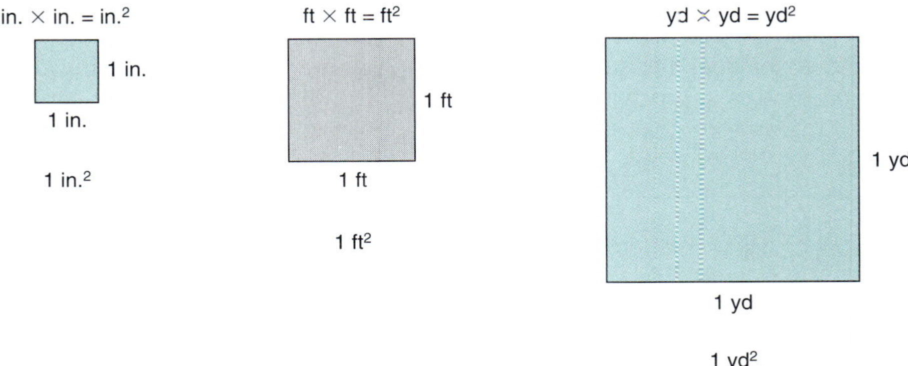

Figure 5–12

Area gives us a different perspective on units of measure. Units of area measure surfaces instead of lengths.

$$\text{length} \times \text{length} = \text{length}^2 \text{ or area}$$

To multiply a length measure by a like length measure:

1. Multiply the numbers associated with each like unit of measure.
2. Write the product as a square unit of measure.

  A desktop is 2 ft × 3 ft (Fig. 5–13). What is the number of square feet in the surface?

$$2 \text{ ft} \times 3 \text{ ft} = \mathbf{6 \text{ ft}^2}$$

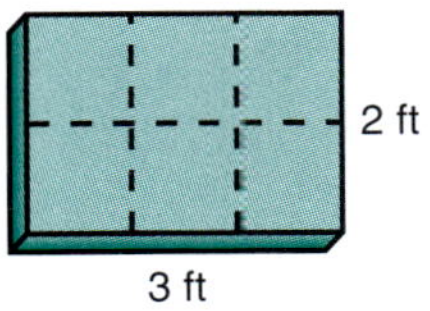

Figure 5–13

3 Divide a U.S. Customary Measure by a Number.

We are frequently required to divide measures by a number. For example, if a recipe calls for 2 gal 2 qt of liquid, and we half the recipe (divide by 2), how much liquid should be used? To solve this problem, we divide the measure by 2.

> ***To divide a U.S. customary measure by a number that divides evenly into each measure:***
>
> **1.** Divide the numbers associated with each unit of measure by the given number.
> **2.** Write the resulting measure in standard notation.

EXAMPLE   How much milk is needed for a half-recipe if the original recipe calls for 2 gal 2 qt?

$$2\overline{)2\text{ gal }2\text{ qt}}$$ Divide each measure by 2.

$$\begin{array}{c} 1\text{ gal }1\text{ qt} \\ 2\overline{)2\text{ gal }2\text{ qt}} \end{array}$$

The half-recipe requires 1 gal 1 qt of milk.

If a given number does not divide evenly into each measure, there will be a remainder. For instance, if we divide 5 gal by 2, the 2 divides into the 5 gal 2 times with 1 gal left over as a remainder.

> ***To divide U.S. customary measures by a number that does not divide evenly into each measure:***
>
> **1.** Set up the problem and proceed as in long division.
> **2.** When a remainder occurs after subtraction, convert the remainder to the same unit used in the next smaller measure and add it to the quantity in the next smaller measure.
> **3.** Divide the given number into this next smaller unit.
> **4.** If a remainder occurs when the smallest unit is divided, express the remainder as a fractional part of the smallest unit.

EXAMPLE   Divide 5 gal 3 qt 1 pt by 3.

$$\begin{array}{llll}
& 1\text{ gal} & 3\text{ qt} & 1\tfrac{2}{3}\text{ pt} \\
3\overline{)}\!\!\! & 5\text{ gal} & 3\text{ qt} & 1\text{ pt} \\
& 3\text{ gal} & & \\
\cline{2-2}
& 2\text{ gal} = & 8\text{ qt} & \\
& & 11\text{ qt} & \\
& & 9\text{ qt} & \\
\cline{3-3}
& & 2\text{ qt} = & 4\text{ pt} \\
& & & 5\text{ pt} \\
& & & 3\text{ pt} \\
\cline{4-4}
& & & 2\text{ pt}
\end{array}$$

5 gal ÷ 3 = 1 gal, remainder 2 gal
2 gal = 8 qt
3 qt + 8 qt = 11 qt
11 qt ÷ 3 = 3 qt, remainder 2 qt
2 qt = 4 pt
1 pt + 4 pt = 5 pt
5 pt ÷ 3 = 1 pt, remainder 2 pt

Write the final remainder, 2, as a fraction $\frac{2}{3}$, and add to 1 pint to get $1\frac{2}{3}$ pints.

Thus, 1 gal 3 qt $1\frac{2}{3}$ pt is the solution.

4   Divide a U.S. Customary Measure by a Measure.

We often need to divide a measure by a measure. For instance, if tubing is manufactured in lengths of 8 ft 4 in. and a part is 10 in. long, how many parts can be cut from the length of tubing if we do not account for waste? To solve such problems, we express both measures in the same unit, just as in adding and subtracting measures. We generally convert to the *smallest* unit used in the problem.

To divide a U.S. customary measure by a U.S. customary measure:

1. Convert both measures to the same unit if they are different.
2. Write the division as a fraction, including the common unit in the numerator and the denominator.
3. Reduce the units and divide the numbers.

EXAMPLE If tubing is manufactured in lengths of 8 ft 4 in. and a part is 10 in. long, how many parts can be cut from the length of tubing if we do not account for waste? Divide 8 ft 4 in. by 10 in.

Converting 8 ft 4 in. to inches, we have

$$8 \text{ ft} = \frac{8 \text{ ft}}{1}\left(\frac{12 \text{ in.}}{1 \text{ ft}}\right) = 96 \text{ in.}$$

$$8 \text{ ft } 4 \text{ in.} = 96 \text{ in.} + 4 \text{ in.} = 100 \text{ in.}$$

Dividing, we have 100 in. $\div$ 10 in.

If we write this division in fraction form, we can see more easily that the common units reduce. In other words, the answer will be a number (not a measure) telling *how many times* one measure divides into another.

$$\frac{100 \text{ in.}}{10 \text{ in.}} = 10$$

Therefore, 10 parts of equal length can be cut from the tubing.

5 Change from One U.S. Customary Rate Measure to Another.

A *rate measure* is a ratio of two different kinds of measures. It is often referred to simply as a *rate*. Some examples of rates are 55 miles per hour, 20 cents per mile, and 3 gallons per minute. In each of these rates, the word *per* means *divided by*.

The rate 55 miles per hour means 55 miles $\div$ 1 hour or $\frac{55 \text{ mi}}{1 \text{ hr}}$. The rate 20 cents per mile means 20 cents $\div$ 1 mile or $\frac{20 \text{ cents}}{1 \text{ mi}}$. The rate 3 gallons per minute means 3 gallons $\div$ 1 minute or $\frac{3 \text{ gal}}{1 \text{ min}}$.

Notice that rate measures require two units of measure, one in the numerator and one in the denominator. Sometimes we need to express the numerator and/or denominator of a rate in units other than the ones given. We do this by using unity ratios.

Many rate measures involve measures of time. The units we use to measure time are universally accepted and used around the world. The basic units of time are the year, month, week, day, hour, minute, and second. Table 5–4 gives the relationships among the units for time. These units are often used when working with rates.

TABLE 5–4 Units of Time

1 year (yr) = 12 months (mo)	1 day (da) = 24 hours (hr)
1 year (yr) = 365 days (da)	1 hour (hr) = 60 minutes (min)[a]
1 week (wk) = 7 days (da)	1 minute (min) = 60 seconds (sec)[b]

[a] The symbol ′ means feet (3′ = 3 ft) or minutes (60′ = 60 minutes).
[b] The symbol ″ means inches (8″ = 8 in.) or seconds (60″ = 60 seconds).

To convert one U.S. customary rate measure to another:

1. Compare the units of both numerators and both denominators to determine which units will change.
2. Multiply each unit that changes by a unity ratio containing the new unit so that the unit to be changed will reduce.

EXAMPLE   Change $8\dfrac{\text{pt}}{\text{min}}$ to $\dfrac{\text{qt}}{\text{min}}$.

Examine the rates:

> Numerators—pints to quarts
> Denominators—no change

Develop an appropriate unity ratio:

$$\frac{\cancel{\text{pt}}}{\text{min}}\left(\frac{\text{qt}}{\cancel{\text{pt}}}\right)=\frac{\text{qt}}{\text{min}}$$

Estimation   Pints to quarts is *smaller* to *larger,* so there will be fewer quarts.

$$\frac{\overset{4}{\cancel{8}}\ \cancel{\text{pt}}}{\text{min}}\left(\frac{1\ \text{qt}}{\underset{1}{\cancel{2}}\ \cancel{\text{pt}}}\right)=\frac{4\ \text{qt}}{\text{min}}\qquad\text{Multiply.}$$

Interpretation   Thus, $8\dfrac{\text{pt}}{\text{min}}$ equals $4\dfrac{\text{qt}}{\text{min}}$.

A separate unity ratio is used for each unit in the rate that changes. For example, when both the numerator and denominator of a rate measure change, we multiply by at least two unity ratios to make the conversion: First, we multiply by a unity ratio that changes the unit in the numerator of the original rate measure. Then, we multiply by a unity ratio that changes the unit in the denominator of the original rate measure.

EXAMPLE   Change $\dfrac{3\ \text{pt}}{\text{sec}}$ to $\dfrac{\text{qt}}{\text{min}}$.

Examine the rates:

> Numerators—pints to quarts
> Denominators—seconds to minutes

Develop appropriate unity ratios:

We are converting the numerator from pints to quarts, so we use a unity ratio with quarts in the numerator and pints in the denominator. This allows us to reduce pints.

$$\frac{3\ \text{pt}}{\text{sec}}\left(\frac{1\ \text{qt}}{2\ \text{pt}}\right)$$

To complete the conversion, we multiply by the unity ratio that changes *seconds* to *minutes.* Seconds must be reduced, so our unity ratio has seconds in the *numerator.*

$$\frac{3\ \text{pt}}{\text{sec}}\left(\frac{1\ \text{qt}}{2\ \text{pt}}\right)\left(\frac{60\ \text{sec}}{1\ \text{min}}\right)$$

$$\frac{3\ \cancel{\text{pt}}}{\cancel{\text{sec}}}\left(\frac{1\ \text{qt}}{\underset{1}{\cancel{2}}\ \cancel{\text{pt}}}\right)\left(\frac{\overset{30}{\cancel{60}}\ \cancel{\text{sec}}}{1\ \text{min}}\right)=90\frac{\text{qt}}{\text{min}}\qquad\text{Reduce and multiply.}$$

Thus, $3\frac{\text{pt}}{\text{sec}}=90\frac{\text{qt}}{\text{min}}$.

SELF-STUDY EXERCISES 5–3

1   Multiply and write the answers for mixed measures in standard notation.

1. 12 mi	**2.** 18 gal	**3.** 21 lb	**4.** 3 qt 1 pt
$\underline{\quad 5\quad}$	$\underline{\quad 6\quad}$	$\underline{\quad 12\quad}$	$\underline{\quad 4\quad}$

5. 7 lb 3 oz
$\underline{\quad 8 \quad}$

6. 5 gal 2 qt
$\underline{\quad 7 \quad}$

7. 8 gal 3 qt
$\underline{\quad 5 \quad}$

8. 7 ft 3 in.
$\underline{\quad 8 \quad}$

9. Tuna is packed in 1 lb 8 oz cans. If a case contains 24 cans, how much does a case weigh?

10. A car used 1 qt 1 pt of oil each month for 5 days. Find the total oil used.

2 Multiply.

11. 5 in. × 7 in.

12. 12 ft × 9 ft

13. 15 yd × 12 yd

14. 4 mi × 27 mi

15. A room is to be covered with square linoleum tiles that are 1 ft by 1 ft. If the room is 18 ft by 21 ft, how many tiles (square feet) are needed?

16. A horticulturist stores a stock solution of fertilizer in two tanks, each with a capacity of 23 gal 9 oz. How much liquid fertilizer is needed to fill both tanks?

17. Latonya has three containers, each containing 1 qt 3 pt of photographic solution. How much total photographic solution is in all three containers?

18. A package of nails weighs 1 lb 4 oz. How much would five packages weigh?

19. If a history textbook weighs 2 lb 8 oz, how much would three of the same books weigh?

3 Divide.

20. 12 gal ÷ 2

21. 3 days 6 hr ÷ 2

22. 20 yd 2 ft 6 in. ÷ 2

23. 4 yd 1 ft 9 in. ÷ 3

24. 4 gal 3 qt 1 pt ÷ 6

25. 60 gal ÷ 9 (express in gallons only)

26. 5 hr ÷ 3 (express in hours and minutes)

27. If eight trucks require 42 qt of oil for each to get a complete oil change, how many quarts are required for each truck?

28. Sixty feet of wire is required to complete eight jobs. If each job requires an equal amount of wire, find the amount of wire required for one job. (Express in feet and inches.)

29. A vat holding 10 gal 2 qt of defoliant is emptied equally into three tanks. How many gallons and quarts are in each tank?

30. How many pieces of $\frac{1}{2}$-in. OD (outside diameter) plastic pipe 8 in. long can be cut from a piece 72 in. long?

31. A roll of No. 14 electrical cable 150 ft long is divided into 30 equal sections. How long is each section?

32. A greenhouse attendant has a container with 6 gal 2 qt 10 oz of potassium nitrate solution that will be stored in two smaller containers. How much solution will be stored in each smaller container?

33. For a family picnic, Mr. Sonnier prepared 96 lb 12 oz of boiled crawfish. He brought the crawfish to the picnic site in four containers containing equal amounts. How much did the crawfish in each container weigh?

4 Divide.

34. 36 ft ÷ 12 ft

35. 51 in. ÷ 3 in.

36. 2 ft 8 in. ÷ 4 in.

37. 6 lb 12 oz ÷ 6 oz

38. 2 ft 6 in. ÷ 10 in.

39. How many 6-in. pieces can be cut from 48 in. of pipe?

40. How many 2-lb boxes can be filled from 18 lb of nails?

41. How many 15-oz cans are in a case if the case weighs 22 lb 8 oz?

42. How many 8-dollar tickets can be purchased for 72 dollars?

43. How many pieces of wood 5 in. long can be cut from a piece 45 in. long?

5 Work the problems.

44. $\dfrac{60 \text{ qt}}{\text{sec}} = \dfrac{\text{gal}}{\text{sec}}$

45. $\dfrac{45 \text{ lb}}{\text{hr}} = \dfrac{\text{lb}}{\text{min}}$

46. $\dfrac{3 \text{ mi}}{\text{hr}} = \dfrac{\text{ft}}{\text{hr}}$

47. $\dfrac{144 \text{ lb}}{\text{min}} = \dfrac{\text{oz}}{\text{min}}$

48. $\dfrac{30 \text{ gal}}{\text{min}} = \dfrac{\text{qt}}{\text{sec}}$

49. $\dfrac{30 \text{ lb}}{\text{day}} = \dfrac{\text{oz}}{\text{hr}}$

50. $\dfrac{8 \text{ in}}{\text{sec}} = \dfrac{\text{ft}}{\text{min}}$

51. $\dfrac{5 \text{ mi}}{\text{min}} = \dfrac{\text{ft}}{\text{sec}}$

52. A car traveling at the rate of 30 mph (miles per hour) is traveling how many feet per second?

53. A pump that can pump $45 \frac{\text{gal}}{\text{hr}}$ can pump how many quarts per minute?

54. A pump can dispose of sludge at the rate of $3,200 \frac{\text{lb}}{\text{hr}}$. How many pounds can be disposed of per minute?

55. If water flows through a pipe at the rate of $50 \frac{\text{gal}}{\text{min}}$, how many gallons will flow per second?

5–4   INTRODUCTION TO THE METRIC SYSTEM

Learning Outcomes

1   Identify uses of metric measures of length, weight, and capacity.

2   Convert from one metric unit of measure to another.

3   Make calculations with metric measures.

In the *metric system,* or the International System of Units (SI), there is a standard unit for each type of measurement. The *meter* is used for length or distance, the *gram* is used for weight or mass, and the *liter* is used for capacity or volume. A series of prefixes are affixed to the standard unit to indicate measures greater than the standard unit or less than the standard unit.

Many consider the metric system easier to use than the U.S. customary system. One reason is that all units greater than or less than the standard unit are expressed in powers of 10. Changing from larger to smaller or from smaller to larger units requires only multiplying or dividing by 10 or a power of 10. The metric system is also easier to use because the prefixes represent the power of 10 by which the standard unit is divided or multiplied.

■ **DEFINITION: Metric System.** The *metric system* is an international system of measurement that uses standard units and prefixes to indicate powers of 10.

1   **Identify Uses of Metric Measures of Length, Weight, and Capacity.**

Before we start our study of metric measurements, let's look at some prefixes used in the metric system. Keep in mind that the prefix has the same meaning no matter which unit (meter, gram, or liter) the prefix is attached to. The prefixes used in this text for units *smaller* than the standard unit are:

$$\textbf{deci-} \ \frac{1}{10} \text{ of} \qquad \textbf{centi-} \ \frac{1}{100} \text{ of} \qquad \textbf{milli-} \ \frac{1}{1,000} \text{ of}$$

The prefixes used in this text for units *larger* than the standard unit are

$$\textbf{deka-} \ 10 \text{ times} \qquad \textbf{hecto-} \ 100 \text{ times} \qquad \textbf{kilo-} \ 1,000 \text{ times}$$

Other prefixes for larger and smaller measures exist but are not discussed until later (Career Application, pp. 435–436).

We can relate the prefixes to our decimal system. Let's compare our decimal place-value chart with these prefixes (Fig. 5–14). The standard unit (whether meter, gram, or liter) corresponds to the *ones* place. All the places to the left are powers of the standard unit. That is, the value of *deka-* (some textbooks use *deca-*) is 10 times the standard unit; the value of *hecto-* is 100 times this unit; the value of *kilo-* is 1,000 times this unit; and so on. All the places to the right of the standard unit are subdivisions of the standard unit. That is, the value of *deci-* is $\frac{1}{10}$ of the standard unit; the value of *centi-* is $\frac{1}{100}$ of the standard unit; the value of *milli-* is $\frac{1}{1,000}$ of the standard unit; and so on.

Thousands (1000)	Hundreds (100)	Tens (10)	Units or ones (1) STANDARD UNIT	Tenths ($\frac{1}{10}$)	Hundredths ($\frac{1}{100}$)	Thousandths ($\frac{1}{1000}$)
Kilo-	Hecto-	Deka-		Deci-	Centi-	Milli-

Figure 5-14

EXAMPLE Give the value of the metric units using the standard unit (gram, liter, or meter).

(a) Kilogram (kg) = 1,000 times 1 gram or 1000 g
(b) Deciliter (dL) = $\frac{1}{10}$ of 1 liter or 0.1 L
(c) Hectometer (hm) = 100 times 1 meter or 100 m
(d) Dekaliter (dkL) = 10 times 1 liter or 10 L
(e) Milliliter (mL) = $\frac{1}{1,000}$ of 1 liter or 0.001 L
(f) Centigram (cg) = $\frac{1}{100}$ of 1 gram or 0.01 g

Remember that all metric units of length, weight, and volume are expressed either as a standard unit or as a standard unit with a prefix. Let's examine some common metric units to develop an intuitive sense of their size.

Length

Meter: The *meter* is the standard unit for measuring length. Both *meter* and *metre* are acceptable spellings for this unit of measure; however, we use the spelling *meter* throughout this book. A meter is about 39.37 in., or 3.37 in. (approximately $3\frac{1}{3}$ in.) longer than a yard (Fig. 5–15). We use the meter to measure lengths and distances like room dimensions, land dimensions, lengths of poles, heights of mountains, and heights of buildings. The abbreviation for meter is m.

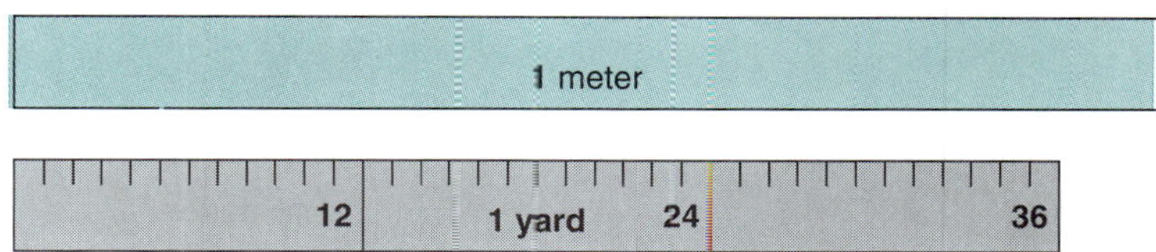

1 meter – slightly longer than a yardstick
(36 inches = 1 yard) (39.37 inches = 1 meter)

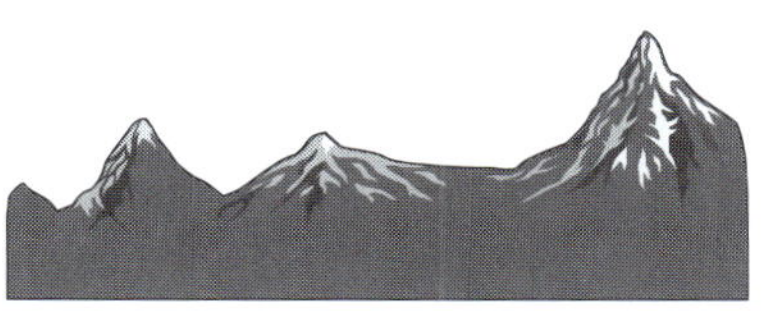
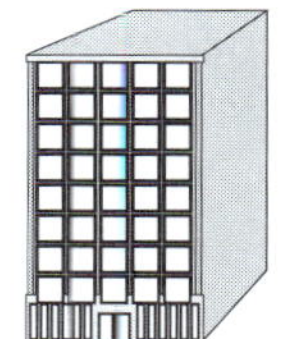

Figure 5-15

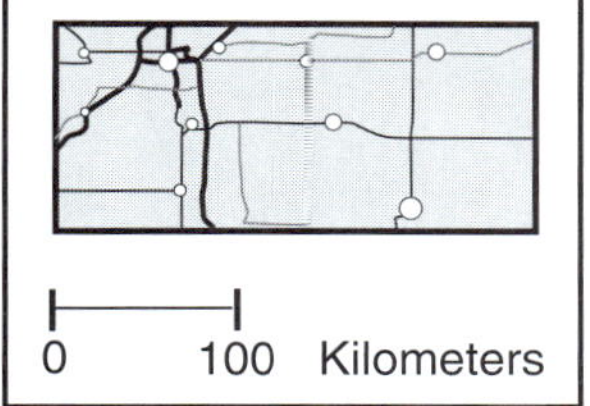

Figure 5-16

Kilometer: A *kilometer* is 1,000 meters and is used for longer distances (Fig. 5–16). The abbreviation for kilometer is km. The prefix *kilo* means 1,000. We measure the distance from one city to another, one country to another, or one landmark to another in kilometers. Driving at a speed of 55 mi (or 90 km) per hour, we

1 kilometer — about five city blocks
1 mile — about eight city blocks

Figure 5–17

would travel 1 km in about 40 sec (Fig. 5–17). An average walking speed is 1 km in about 10 min.

Centimeter: To measure objects less than 1 m long, we commonly use the *centimeter* (cm). The prefix *centi* means "$\frac{1}{100}$ of," and a centimeter is one hundredth of a meter. A centimeter is about the width of a thumbtack head, somewhat less than $\frac{1}{2}$ in. (Fig. 5–18). We use centimeters to measure medium-sized objects such as tires, clothing, textbooks, and television pictures.

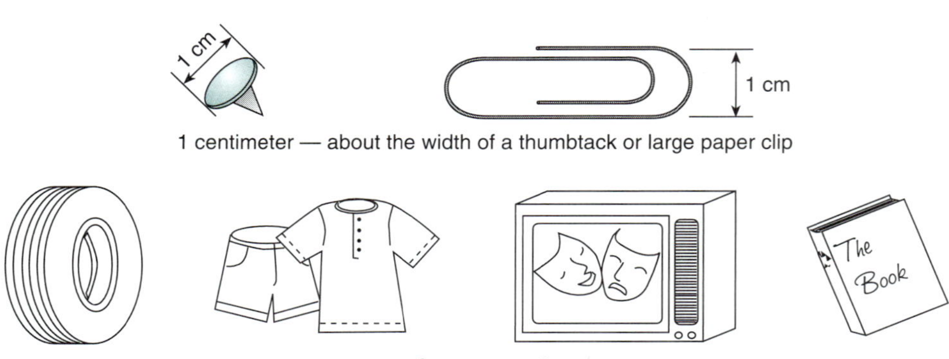

Figure 5–18

Millimeter: Many objects are too small to be measured in centimeters, so we use a *millimeter* (mm), which is "$\frac{1}{1,000}$ of" a meter. It is about the thickness of a plastic credit card or a dime (Fig. 5–19). Certain film sizes, bolt and nut sizes, the length of insects, and similar items are measured in millimeters.

Other units and their abbreviations are decimeter, dm; dekameter, dkm; and hectometer, hm.

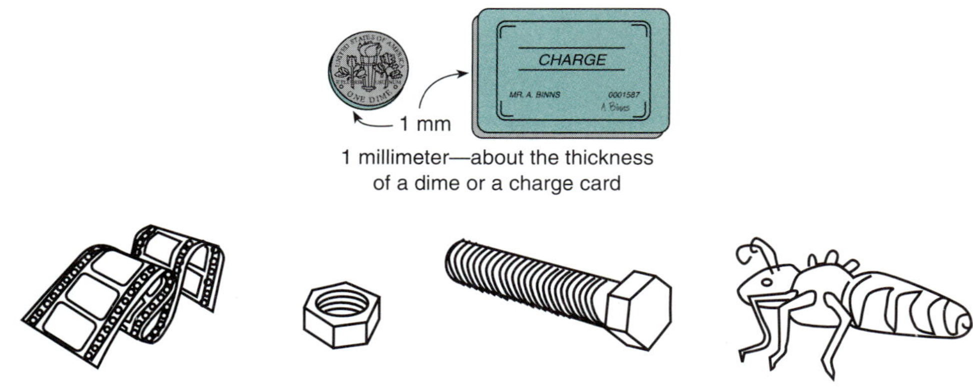

Figure 5–19

Weight or Mass

Gram: The standard unit for measuring mass in the metric system is the gram (Fig. 5–20). A gram is the mass of 1 cubic centimeter (cm^3) of water at its maximum density. A cubic centimeter is a cube whose edges are each 1 centimeter long. It is a little smaller than a sugar cube. The abbreviation g is used for gram. We use grams to measure small or light objects such as paper clips, cubes of sugar,

CHAPTER 5   Direct Measurement

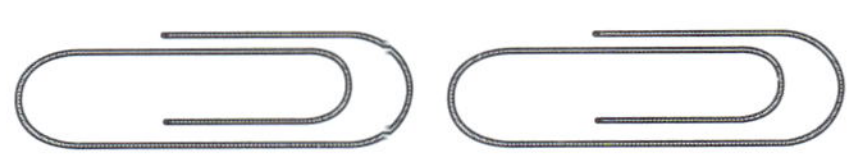
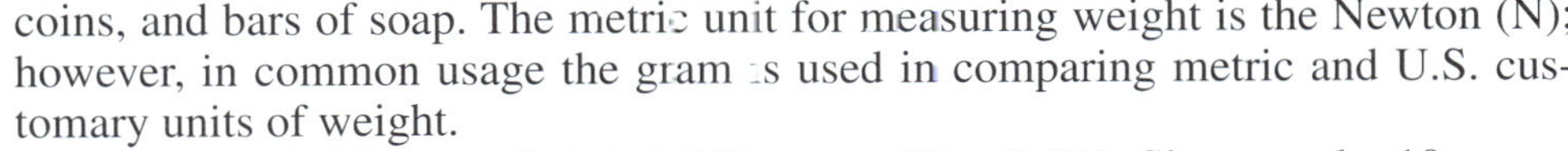

1 gram — about the weight of two paper clips

Figure 5–20

coins, and bars of soap. The metric unit for measuring weight is the Newton (N); however, in common usage the gram is used in comparing metric and U.S. customary units of weight.

Kilogram: A *kilogram* (kg) is 1,000 grams (Fig. 5–21). Since a cube 10 cm on each edge can be divided into 1,000 cm^3, the weight of water required to fill this cube is 1,000 grams or 1 kilogram. A kilogram is approximately 2.2 lb. The kilogram is probably the most often used unit for weight. We use it to weigh people, meat, sacks of flour, automobiles, and so on.

1 kilogram — about the weight of an average book with a one-inch spine

Figure 5–21

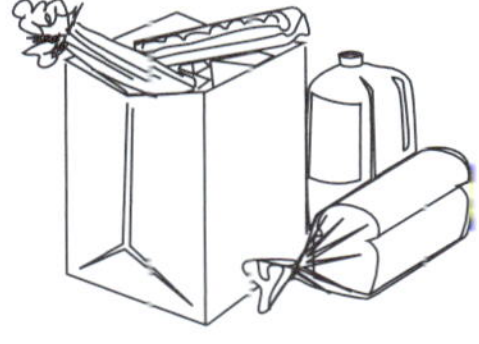

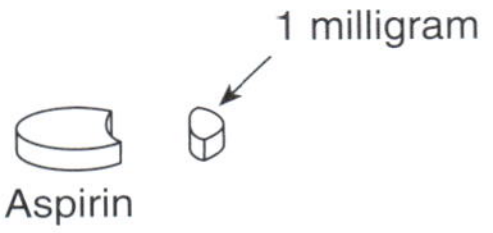

1 milligram

Aspirin

Figure 5–22

Milligram: The gram is used to measure small objects, and the *milligram* (1/1,000 of a gram) is used to measure *very* small objects. Milligrams are much too small for ordinary uses; however, pharmacists use milligrams (mg) to measure small amounts of drugs, vitamins, and medications (Fig. 5–22).

Other units and their abbreviations are decigram, dg; centigram, cg; dekagram, dkg; and hectogram, hg.

Capacity or Volume

Liter: A *liter* (L) is the volume of a cube 10 cm on each edge. It is the standard metric unit of capacity (Fig. 5–23). Like the meter, it may be spelled *liter* or *litre*, but we use the spelling *liter*. A cube 10-cm on each edge filled with water weighs approximately 1 kg, so 1 L of water weighs about 1 kg. One liter is just a little larger than a liquid quart. Soft drinks are often sold in 2-liter bottles, gasoline is sold by the liter at some service stations, and numerous other products are sold in liter containers.

Milliliter: A liter is 1,000 cm^3, so $\frac{1}{1,000}$ of a liter, or a *milliliter,* has the same volume as a cubic centimeter (Fig. 5–24). Most liquid medicine is labeled and sold in milliliters (mL) or cubic centimeters (cc or cm^3). Medicines, perfumes, and other very small quantities are measured in milliliters.

1 liter — about the volume of a quart of milk or a soft drink in a plastic bottle

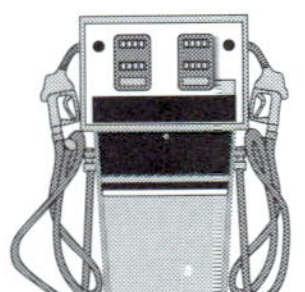

Figure 5–23

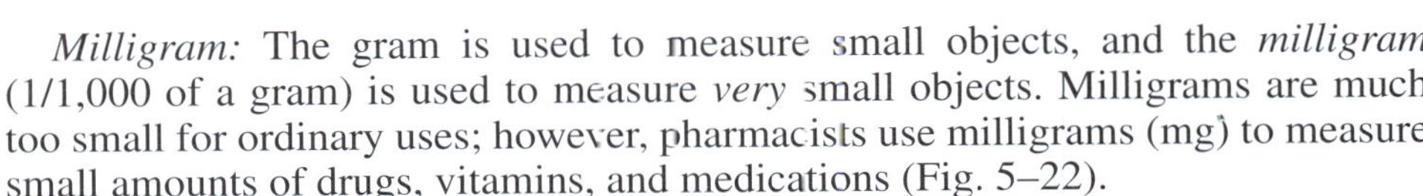

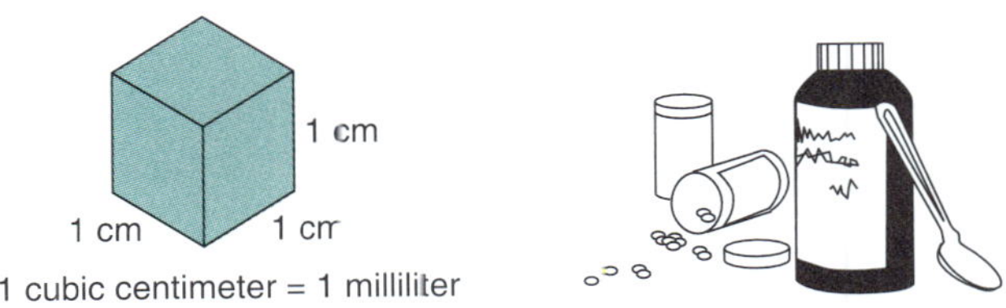

1 cubic centimeter = 1 milliliter

Figure 5–24

Other units and their abbreviations are deciliter, dL; centiliter, cL; dekaliter, dkL; hectoliter, hL; and kiloliter, kL.

Other multiples and standard units in the metric system exist, but those discussed here are the ones frequently used.

Choose the most reasonable metric measure.

1. Distance from Jackson, Mississippi, to New Orleans, Louisiana
   (a) 322 m   (b) 322 km   (c) 322 cm   (d) 322 mm
2. Weight of an adult woman
   (a) 56 g   (b) 56 mg   (c) 56 kg   (d) 56 dkg
3. Width of a roll of color film
   (a) 35 cm   (b) 35 m   (c) 35 mm   (d) 35 km
4. Bottle of eye drops
   (a) 30 dL   (b) 30 dkL   (c) 30 L   (d) 30 mL
5. Weight of a pill
   (a) 352 mg   (b) 352 dg   (c) 352 g   (d) 352 kg

1. (b) 322 km (about 200 mi)
2. (c) 56 kg (about 120 lb)
3. (c) 35 mm (roll of film for automatic cameras)
4. (d) 30 mL (about 1 oz)
5. (a) 352 mg (one regular-strength aspirin)

2 Convert from One Metric Unit of Measure to Another.

To understand how we change one metric unit to another, let's arrange the units into a place-value chart like the one we used for decimals (Fig. 5–25). The units are arranged from left to right and from largest to smallest.

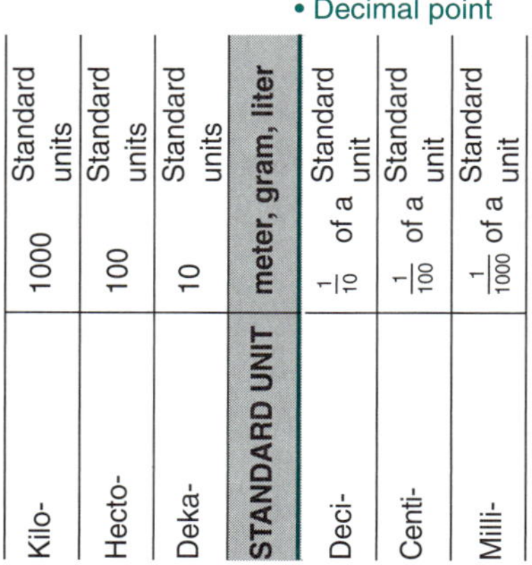

Figure 5–25   Metric prefixes

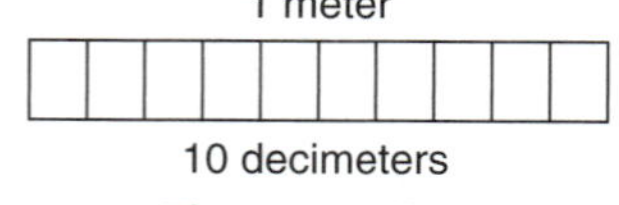

Figure 5–26

As we move from any place in the chart one place to the *right,* the metric unit changes to the next smaller unit. The larger unit is broken down into 10 smaller units, so we are *multiplying* the larger unit by 10 when we move one place to the right (Fig. 5–26).

Suppose we want to change 2 m to decimeters. A decimeter is $\frac{1}{10}$ of a meter, so there must be 10 dm in 1 meter. If 10 dm make 1 m, how many decimeters make 2 m? Twice as many:

$$2 \times 10 = 20$$

Therefore, 2 m = 20 dm.

Instead of multiplying by 10, let's use the metric-value chart as directed in the following rule, because when we change to a smaller unit, we always move to the right on the chart. Suppose we need to change 4 L to centiliters (Figure 5–27). The first step to the right changes 4 L to 40 dL. The next step to the right changes the 40 dL to 400 cL.

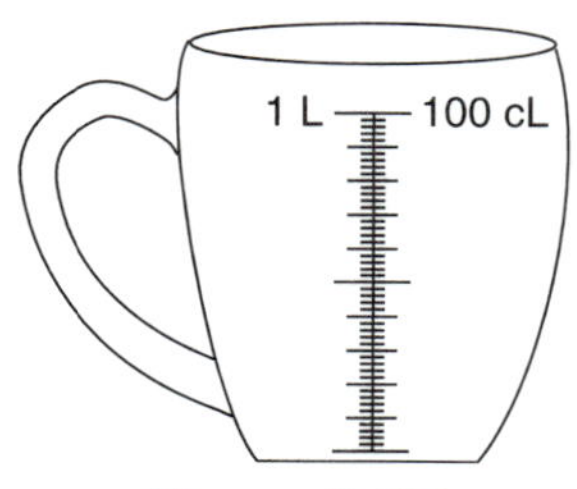

Figure 5–27

CHAPTER 5   Direct Measurement

To change from one metric unit to any smaller metric unit:

1. Mentally position the measure on the metric-value chart so that the decimal immediately follows the original measuring unit.

2. Move the decimal to the right so that it *follows* the new measuring unit. (Attach zeros if necessary.)

EXAMPLE 43 dkm = _____ cm?

Place 43 dkm on the metric-value chart so that the last digit is in the dekameters place; that is, the understood decimal that follows the 3 will be *after* the dekameters place (see Fig. 5–28). To change to centimeters, move the decimal point so that it follows the centimeters place. Fill in the empty places with zeros (see Fig. 5–29). Recall the shortcut to multiplication by powers of 10: Move the decimal point one place to the right for each time 10 is used as a factor.

Thus, 43 dkm = 43,000 cm.

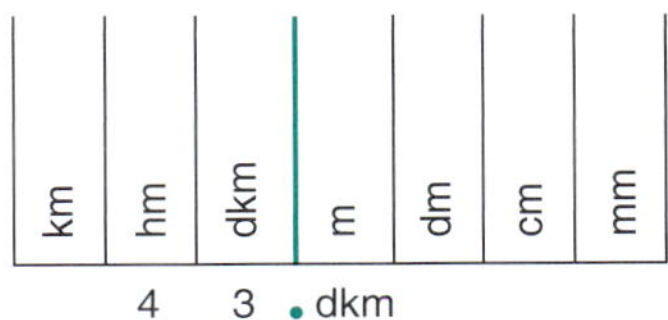

Figure 5–28

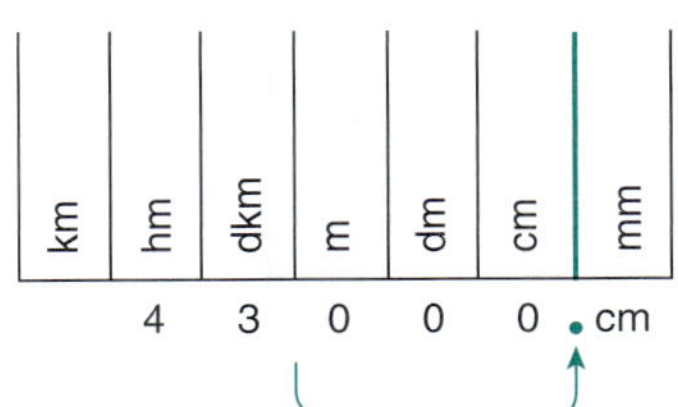

Figure 5–29

EXAMPLE 2.5 dg = _____ mg? (Note the decimal location.)

Place the number 2.5 on the chart so that the decimal point follows the decigrams place (see Fig. 5–30). (Note the decimal after the decigrams place.) To change to milligrams, shift the decimal two places to the right so that it follows the milligrams place (see Fig. 5–31). (Note the decimal after the milligrams place.)

Thus, 2.5 dg = 250 mg.

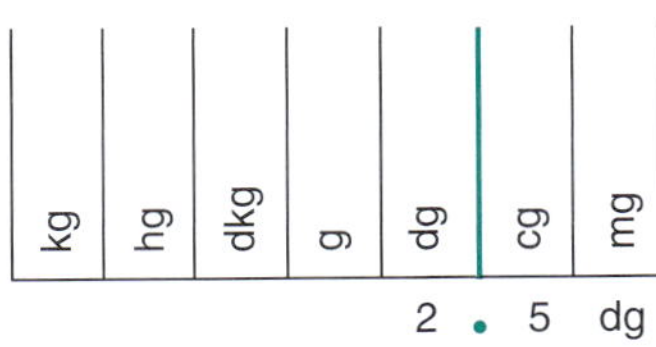

Figure 5–30

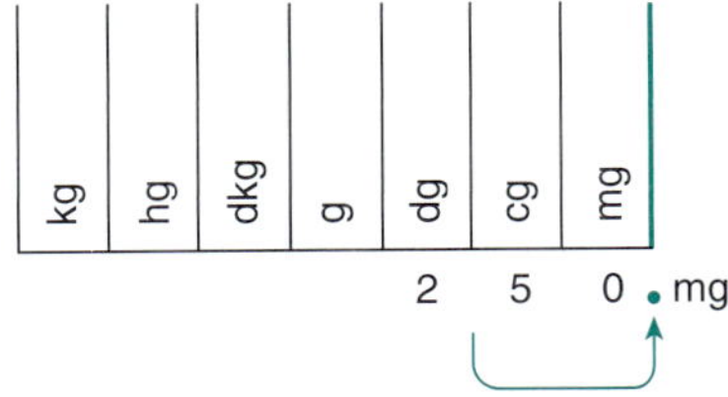

Figure 5–31

As we move one place to the *left* on the metric-value chart, the metric unit changes to the next larger unit. The smaller units are combined into one larger unit 10 times larger than each smaller unit, so we are *dividing* the smaller unit by 10 when we move one place to the left.

Suppose we want to change 20 mm to centimeters. One millimeter is $\frac{1}{10}$ of a centimeter, so there are 10 mm in 1 cm. If 1 cm equals 10 mm, how many centimeters are there in 20 mm? There are as many centimeters as there are 10s in 20 ($20 \div 10 = 2$). Therefore, 20 mm = 2 cm.

Because we are dividing by 10, we can use the metric-value chart as directed in the following rule. When we change to a larger unit, we always move to the left on the chart.

EXAMPLE   3,495 L = _____ kL?

Place the number on the chart so the digit 5 is in the liters place; that is, the decimal point follows the liters place (see Fig. 5–32). (Note the understood decimal point after the liters place.) To change to kiloliters, move the decimal *three* places to the left so the decimal follows the kiloliters place (see Fig. 5–33). Recall the shortcut to division by powers of 10: Move the decimal point one place to the left for each time 10 is used as a divisor.
Thus, 3,495 L = 3.495 kL.

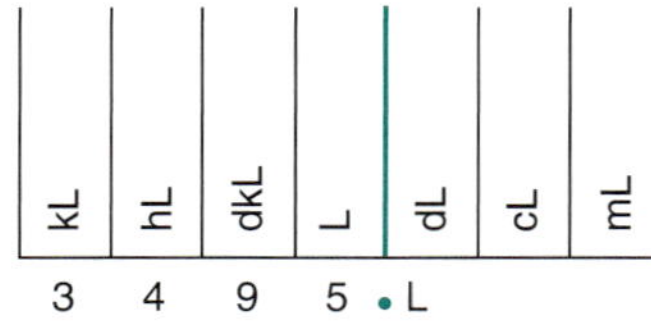

Figure 5–32

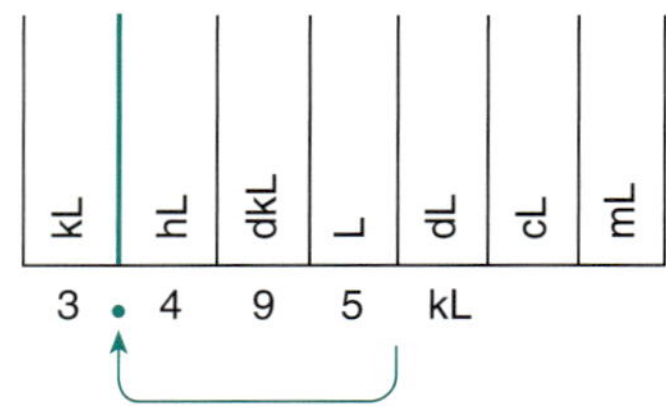

Figure 5–33

EXAMPLE   2.78 cm = _____ dkm?

Place the number on the chart so that the decimal follows the centimeters place (see Fig. 5–34). (Note the decimal position after the centimeters place.) Move the decimal three places to the left so that it follows the dekameters place (see Fig. 5–35). (Note the decimal position after the dekameters place.)
Therefore, 2.78 cm = 0.00278 dkm.

Figure 5–34

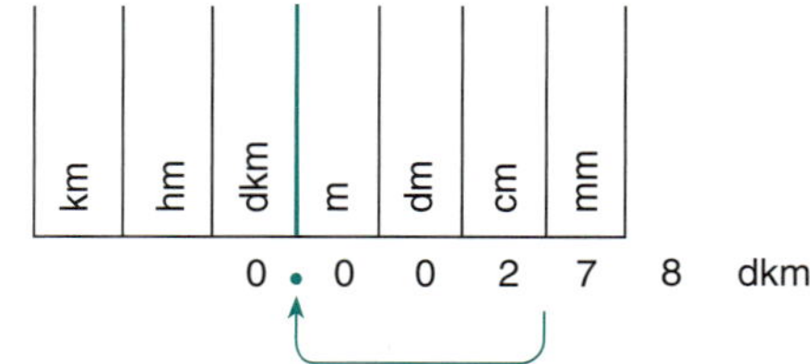

Figure 5–35

To summarize, we may change from one metric unit to another metric unit of the same type by positioning the measure on the metric-value chart so that the decimal immediately follows the original measuring unit. Then, we move the decimal to the right or left as necessary so that it *follows* the new measuring unit. We attach zeros if necessary.

We can also change from one metric unit to another by mentally locating the original *unit* on the metric-value chart and moving to the desired new *unit*. As we move we carefully note the number of units we move past and the direction in which we move. Then, we locate the decimal point in the original metric measure and shift the decimal in the same direction and the same number of units.

<table>
<tr><td>Tip!</td><td>How Far and Which Way?</td></tr>
</table>

To determine the movement of the decimal point when changing from one metric unit to another, answer these questions:

1. *How far* is it from the original unit to the new unit (how many places)?
2. *Which way* is the movement on the chart (left or right)?

Change 28.392 cm to m.

How far is it from cm to m (Fig. 5–36)?   Two places
Which way?   Left

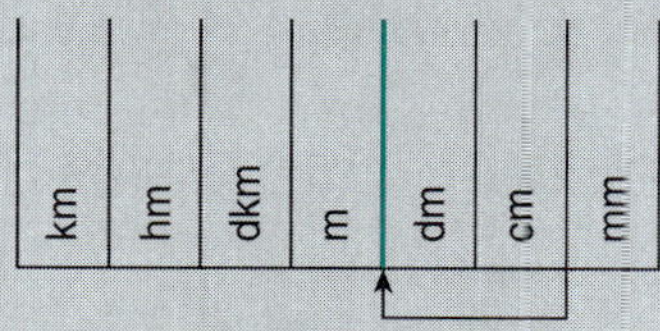

Figure 5–36

Move the decimal in the original measure *two places to the left*.

$$28.392 \text{ cm} = 0.28392 \text{ m}$$

3 Make Calculations with Metric Measures.

Thus far, we have compared metric units and changed to a larger or smaller unit. Now let's use metric units in problems involving addition, subtraction, multiplication, and division. Our jobs often require these basic mathematical skills.

When adding or subtracting measures, we must remember to add or subtract only *like* or *common* measures. By this we mean that the same measuring unit is used. For instance, 5 cm and 3 cm are *like* measures. In contrast, 17 kg and 4 hg are *unlike* measures.

---

*To add or subtract **like** or **common** metric measures:*

1. Add or subtract the numerical values.

$$7 \text{ cm} + 4 \text{ cm} = 11 \text{ cm}$$

2. Give the answer in the common unit of measure.

$$15 \text{ dkg} - 3 \text{ dkg} = 12 \text{ dkg}$$

---

*To add or subtract **unlike** metric measures:*

1. Change the measures to a common unit of measure.
2. Add or subtract the numerical values.
3. Give the answer in the common unit of measure.

EXAMPLE   Add 9 mL + 2 cL.

Change cL to mL; that is, 2 cL = 20 mL. Thus, 9 mL + 20 mL = **29 mL**.
Look at this example again:

$$9 \text{ mL} + 2 \text{ cL}$$

We could have changed milliliters to centiliters:

$$9 \text{ mL} = 0.9 \text{ cL}$$

Then,

$$0.9 \text{ cL} + 2 \text{ cL} =$$

    0.9 cL   Note alignment of decimals.

    <u>2   cL</u>   An understood decimal follows the addend 2.

    2.9 cL

Does 2.9 cL = 29 mL? Yes. Therefore, our answers are equivalent.

EXAMPLE   Subtract: 14 km − 34 hm.

Change hm to km; that is, 34 hm = 3.4 km.

    14.0 km

<u>−   3.4 km</u>   Caution: Watch alignment of decimals.

    10.6 km

The difference is 10.6 km, or 106 hm.

Tip!	*Incompatible Measures.*

Look at this problem: 2 cm + 5 g. Can 5 g be added to 2 cm? Can a common unit be found for centimeters and grams? No, centimeters measure length, and grams measure weight (Fig. 5–37), so there is no common unit. Thus, we cannot add 5 g + 2 cm.

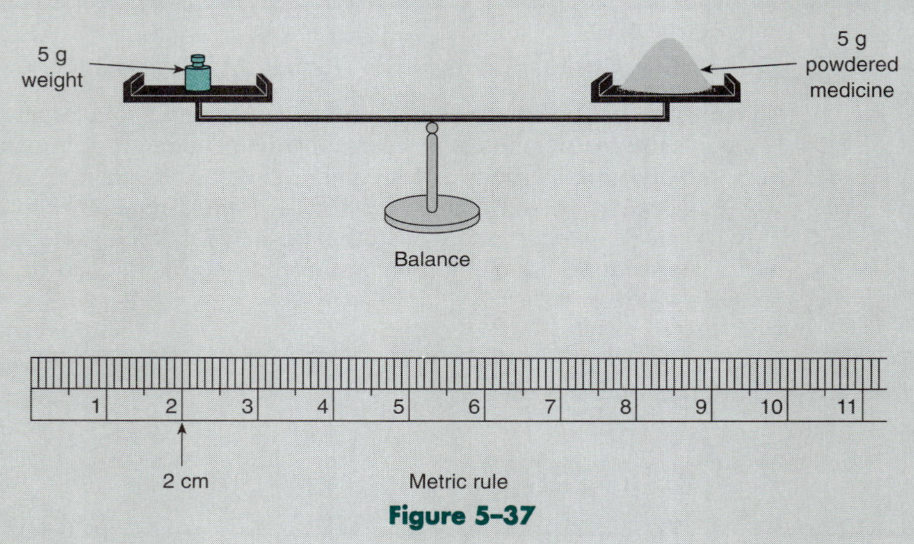

Figure 5–37

Multiplication is a shortcut for repeated addition. If we want to find the total length of three pieces of landscape timber that are each 2.7 m long, we can multiply 2.7 m by 3; that is, 2.7 m × 3 = 8.1 m.

To multiply a metric measure by a number:

1. Multiply the numerical values.
2. Give the answer in the same unit as the original measure.

Suppose we have a tube that is 30 cm long and we need it cut into five equal parts (Fig. 5–38). How long is each part? To solve this problem, we divide 30 cm by 5.

$$\frac{30 \text{ cm}}{5} = 6 \text{ cm}$$

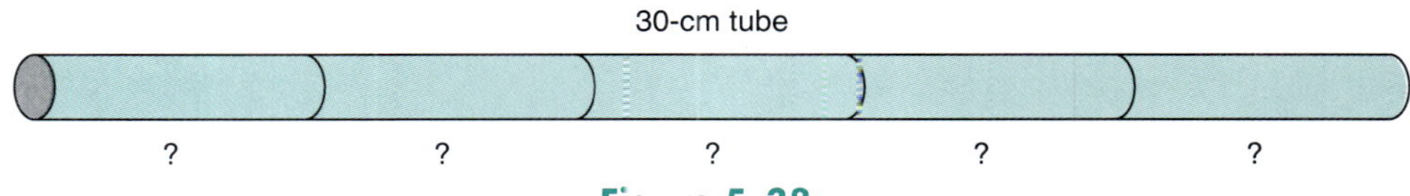

Figure 5–38

To divide a metric measure by a number:

1. Divide the numerical values.
2. Give the answer in the same unit as the original measure.

We can divide all measures by a number because we are dividing a quantity into parts.

In some situations, we need to divide a measure by a measure. Suppose we have to administer a 250-mg dose of ascorbic acid in 100-mg tablets. How many tablets are needed per dose?

$$250 \text{ mg} \div 100 \text{ mg} \qquad \text{or} \qquad \frac{250 \text{ mg}}{100 \text{ mg}} = 2.5 \qquad \text{Units reduce.}$$

Notice that the answer, 2.5, does not have a units label. That is because we are looking for *how many* tablets are needed.

To divide a metric measure by a like measure:

1. Divide the numerical values.
2. Give the answer as a number that tells how many parts there are.
3. If unlike measures are used, first change them to like measures.

EXAMPLE   Solve.

(a) Four micrometers weighing 752 g each will fit into a shipping carton. If the shipping carton and filler weigh 217 g, what is the total weight of the shipment?
(b) A 5-m-long board is to be cut into four equal parts. How long is each part?
(c) How many 25-g packages of seed can be made from 2 kg of seed?

(a) Total weight = Carton and filler + Four micrometers

$$= 217 \text{ g} + (4 \times 752 \text{ g})$$

$$= 217 \text{ g} + 3,008 \text{ g}$$

$$= 3,225 \text{ g, or } 3.225 \text{ kg}$$

The total weight is 3,225 g, or 3.225 kg.

(b) $\dfrac{5 \text{ m}}{4} = 1.25 \text{ m}$ Divide length by number of parts.

Each part is 1.25 m long.

(c) $\dfrac{2 \text{ kg}}{25 \text{ g}} = \dfrac{2,000 \text{ g}}{25 \text{ g}}$ Convert kg to g, then divide. Reduce units.

$$= 80$$

80 packages of seed can be made.

1

1. Give the value of the metric units.
   (a) kilometer (km)
   (b) dekaliter (dkL)
   (c) decigram (dg)
   (d) millimeter (mm)
   (e) hectogram (hg)
   (f) centiliter (cL)

Choose the most reasonable metric measure.

2. Height of a 6-year-old pediatric patient
   (a) 1.5 km
   (b) 1.5 m
   (c) 1.5 cm
   (d) 1.5 mm
3. Diameter of a dime
   (a) 1.5 m
   (b) 1.5 cm
   (c) 1.5 mm
   (d) 1.5 km
4. Distance from Los Angeles to San Francisco
   (a) 800 m
   (b) 800 mm
   (c) 800 km
   (d) 800 cm
5. Length of a pencil
   (a) 20 km
   (b) 20 cm
   (c) 20 m
   (d) 20 mm
6. Overnight accumulation of snowfall
   (a) 8 cm
   (b) 8 m
   (c) 8 km
7. Width of home videotape
   (a) 8 m
   (b) 8 km
   (c) 8 cm
   (d) 8 mm
8. Weight of the average male adult
   (a) 70 mg
   (b) 70 kg
   (c) 70 g
9. Weight of a can of tuna
   (a) 184 mg
   (b) 184 kg
   (c) 184 g
10. Weight of a teaspoonful of sugar
    (a) 2 g
    (b) 2 kg
    (c) 2 mg
11. Weight of a vitamin C tablet
    (a) 250 g
    (b) 250 mg
    (c) 250 kg
12. Weight of a dinner plate
    (a) 350 kg
    (b) 350 mg
    (c) 350 g
13. Weight of a bag of dog food
    (a) 25 g
    (b) 25 kg
    (c) 25 mg
14. Volume of a tank of pesticide
    (a) 60 L
    (b) 60 mL
15. Dose of cough syrup
    (a) 100 L
    (b) 100 mL
16. Glass of juice
    (a) 0.12 mL
    (b) 0.12 L

2 Change to the measure indicated. When using the metric-value chart, place the decimal immediately *after* the measuring unit.

17. 4 m = _____ dm
18. 7 g = _____ dg
19. 58 km = _____ hm
20. 8 hL = _____ dkL
21. 0.25 km = _____ hm
22. 21 dkL = _____ L
23. 8.5 cm = _____ mm
24. 14.2 dg = _____ cg
25. How many milliliters are there in 15.3 cL?
26. How many meters are there in 46 dkm?
27. How many dekagrams are there in 7.5 hg?
28. How many millimeters are there in 16 cm?

Change each measure to the indicated unit of measure. When using the metric-value chart, place the decimal immediately *after* the measuring unit.

29. 4L = _____ cL
30. 8 m = _____ mm
31. 58 km = _____ m
32. 8 hg = _____ g
33. 0.25 km = _____ m
34. 21 dkg = _____ dg
35. 10.25 hm = _____ cm
36. 8.33 L = _____ mL
37. 2 km = _____ mm
38. 0.7 g = _____ cg
39. Change 2.36 hL to liters.
40. Change 0.467 dkm to centimeters.
41. Change 3.8 kg to decigrams.
42. Change 13 dkm to centimeters.

Change each measure to the unit indicated. When using the metric-value chart, place the decimal immediately *after* the unit.

43. 28 m = _____ dkm
44. 238 hL = _____ kL
45. 101 mg = _____ cg
46. 60 hm = _____ km
47. 29 dkL = _____ hL
48. 192.5 g = _____ dkg
49. 17 cm = _____ dm
50. How many decimeters are in 4,389 cm?
51. How many grams are in 47 dg?
52. How many deciliters are in 2.25 cL?

Change each measure to the measure indicated. When using the metric-value chart, place the decimal immediately *after* the unit.

53. 2,743 mm = _____ m

54. 385 g = _____ kg

55. 15 dkm = _____ km

56. 8 cL = _____ L

57. 296,484 m = _____ hm

58. 29.83 dg = _____ dkg

59. 0.3 cm = _____ dkm

60. 40 dL = _____ kL

61. 2,857 mg = _____ kg

62. 15,285 m = _____ km

63. Change 297 cm to hectometers.

64. Change 0.03 mL to liters.

3 Add or subtract these measures.

65. 3 m + 8 m

66. 7 hL + 5 hL

67. 15 cg − 9 cg

68. 2 dm + 4 cm

69. 5 cL + 9 mL

70. 4 m + 2 L

71. 14 kL − 39 hL

72. 1 g − 45 cg

73. 3 cg − 5 mL

74. 7 km + 2 m

75. A patient absorbs 175 mL of fluid through an IV. If the IV bag has 825 mL left, how much fluid was in the bag to begin with?

76. 653 dkL of orange juice concentrate is removed from a vat containing 8 kL of the concentrate. How much concentrate remains in the vat?

Multiply.

77. 43 m × 12

78. 3.4 m × 12

79. 50.32 dm × 3

80. A plot of ground is divided into seven plots, each with road frontage of 138.5 m. What is the total road frontage of the plot of ground?

Divide.

81. 39 m ÷ 3

82. $\dfrac{54\ cL}{6}$

83. A block of silver weighing 978 g is cut into six equal pieces. How much does each piece weigh?

84. Two pieces of steel each 12 m long are cut into a total of 60 equal pieces. How long is each piece?

85. 2.5 cg ÷ 0.5 cg

86. 3 m ÷ 10 cm

87. How many 250-mL prescriptions can be made from a container of 4 L of decongestant?

88. How many 500-g containers are needed to hold 40 kg of grass seed?

Solve.

89. Add 4.6 cL + 5.28 dL of photographic developer.

90. Add 3 m + 2 dkm of fabric for draperies.

91. Subtract 19.8 km − 32.3 hm of paved highway.

92. Subtract 13 kL − 39 hL of stored liquid.

93. Multiply 0.25 cL of cologne by 5.

94. Multiply a 35-mm film size by 2.

95. A length of satin fabric 30 dm long is cut into 4 equal pieces. How long is each piece?

96. An IV bag holding 250 ml of an antibiotic is calibrated (marked off) into 5 equal sections. How many milliliters are represented by each section?

97. How many containers of jelly can be made from 8,500 L of jelly if each container holds 4 dL of jelly?

98. How many 2-kg vials of hydrochloric acid (HC1) can be obtained from 38 kg of HC1?

5-5 METRIC–U.S. CUSTOMARY COMPARISONS

Learning Outcome

1 Convert between U.S. customary measures and metric measures.

Learning to convert from the U.S. customary system to the metric system, and vice versa, is important because both systems are used in the United States. Converting from U.S. customary to metric or metric to U.S. customary is more challenging than working entirely within one system. Certain industries use one system exclusively, making conversions from one system to another uncommon. However, an industry that uses the U.S. customary system in the United States often needs to convert to the metric system to market its products in other countries.

1 Convert Between U.S. Customary Measures and Metric Measures.

To convert measures from the U.S. customary system to the metric system, and vice versa, we need only one conversion factor for each type of measure (length, weight, and capacity). The following factors have been rounded to the nearest hundredth of a unit.

For length:	1 m = 1.09 yd
For weight:	1 kg = 2.20 lb
For capacity:	1 L = 0.91 dry qt (dry measure)
	1 L = 1.06 liquid qt (liquid measure)

Dry quarts are less common than liquid quarts; if the type of quart (dry or liquid) is not specified, use liquid quarts.

Additional conversion factors, such as centimeters to inches and kilometers to miles, can be derived from these factors using unity ratios. Naturally, the more conversion factors we have before us, the better, but one per measurement type is all that is necessary.

Other conversion factors are listed in Table 5–5. The numbers in parentheses are rounded to ten-thousandths and can be used if greater accuracy is desired. Calculators sometimes have conversion factors with 10 or more decimal places programmed into the calculator for even greater accuracy.

TABLE 5–5 Metric–U.S. Customary Conversion Factors

Measures of Length	
Metric to U.S. customary units	*U.S. customary to metric units*
1 meter = 39.37 inches	1 inch = 25.4 millimeters
1 meter = 3.28 feet (3.2808)	1 inch = 2.54 centimeters
1 meter = 1.09 yards (1.0936)	1 inch = 0.0254 meter
1 centimeter = 0.39 or 0.4 inch (0.3937)	1 foot = 0.3 meter (0.3048)
1 millimeter = 0.04 inch (0.03937)	1 yard = 0.91 meter (0.9144)
1 kilometer = 0.62 mile (0.6214)	1 mile = 1.61 kilometers (1.6093)

Capacity: Liquid Measures	
Metric to U.S. customary units	*U.S. customary to metric units*
1 liter = 1.06 liquid quarts (1.0567)	1 liquid quart = 0.95 liter (0.9463)

Capacity: Dry Measures	
Metric to U.S. customary units	*U.S. customary to metric units*
1 liter = 0.91 dry quart (0.9081)	1 dry quart = 1.1 liters (1.1012)

Measures of Weight	
Metric to U.S. customary units	*U.S. customary to metric units*
1 gram = 0.04 ounce (0.0353)	1 ounce = 28.35 grams (28.3495)
1 kilogram = 2.2 pounds (2.2046)	1 pound = 0.45 kilogram (0.4536)

**To convert metric measures to U.S. customary measures or U.S. customary measures to metric measures:**

1. Select the appropriate conversion factor. (For convenience, select the conversion factor that places the new unit in the _numerator_ and a 1 in the denominator.)
2. Set up a unity ratio so that the original unit reduces and the new unit remains.

EXAMPLE   A hospital patient weighs 90 lb. How many kilograms does the patient weigh?

$$\frac{90 \text{ lb}}{1}\left(\frac{0.45 \text{ kg}}{1 \text{ lb}}\right)$$

Using 1 kg = 2.2 lb involves division. Answer will vary slightly.

$$\frac{90 \text{ lb}}{1}\left(\frac{0.45 \text{ kg}}{1 \text{ lb}}\right) = 90(0.45 \text{ kg}) = 40.5 \text{ kg}$$

The patient weighs about 40.5 kg.

EXAMPLE   A vinyl landscape pond that has a capacity of 76 liters is about to be filled with water. How many gallons is its capacity?

$$\frac{76 \text{ L}}{1}\left(\frac{1.06 \text{ qt}}{1 \text{ L}}\right)\left(\frac{1 \text{ gal}}{4 \text{ qt}}\right)$$

Table 5–5 does not give a conversion factor for gallons, so use two unity ratios.

$$\frac{\overset{19}{76} \text{ L}}{1}\left(\frac{1.06 \text{ qt}}{1 \text{ L}}\right)\left(\frac{1 \text{ gal}}{\underset{1}{4} \text{ qt}}\right) = 20.14 \text{ gal or } 20 \text{ gal}$$

The pond is a 20-gal pond.

SELF-STUDY EXERCISES 5–5

1   Change to the units indicated.

1. 9 m to inches

2. 120 m to yards

3. 42 km to miles

4. 6 L to liquid quarts

5. 10 dry qts to liters

6. 27 kg to pounds

7. 50 lb to kilograms

8. 7 in. to centimeters

9. 18 ft to meters

10. 39 mL to ounces (1 liquid qt = 32 oz.)

11. $5\frac{3}{4}$ gal to liters (1 gal = 4 qt)

12. A spool of wire contains 100 ft of wire. How many meters of wire are on the spool?

13. A 60-lb sheet of metal weighs how many kilograms?

14. Two cities 150 mi apart are how many kilometers apart?

15. A 30-m-wide road is how many yards wide?

16. A tourist in Europe traveled 200 km, 60 km, and 120 km by car. How many total miles was this?

17. A patient in therapy jogged 5 km, 4 km, and 3 km. How many miles did the patient jog?

18. A container holds 12 dry quarts. How many liters will the container hold?

19. A spool of electrical wire contains 100 m of wire. How many feet of wire are on the spool?

Health Sciences: Body Mass Index

Overweight Americans now outnumber those who are not overweight, according to the National Center for Health Statistics. Body mass index (BMI) is the standard unit for measuring a person's degree of obesity or emaciation. BMI is body weight in kilograms (kg) divided by height in meters squared, or $BMI = w/h^2$.

According to federal guidelines, a BMI greater than 25 means that you are overweight. Surveys show that 59% of men and 49% of women have BMIs greater than 25. People in their fifties tend to be the fattest—73% of men and 64% of women in this group have BMIs of 25+. Extreme obesity is defined as a BMI greater than 40. To calculate your BMI:

1. Multiply your weight by 0.45 to convert to kilograms.
2. Convert your height to inches.
3. Multiply the inches by 0.025 to get meters.
4. Square this number.
5. Divide your weight in kilograms by this number, and round to the nearest whole number. The result is your BMI.

People with high or low BMIs can expect to pay higher life and health insurance premiums. They may even be denied life and health insurance. BMI is used in many medical applications. For example, overweight people who need certain surgeries (especially orthopedic) may be required to lose weight before they undergo surgery due to the health risk.

BMI is also used to find a person's body surface area (BSA) in square meters. People with heart problems are tested for heart efficiency. A person's cardiac index (CI) measures heart efficiency through volume of blood rate flow and their body surface area according to this formula:

$$CI = \frac{\text{cardiac output}}{\text{BSA}}$$

$$= \frac{\text{liters of blood per minute}}{\text{BSA}} \; (\text{L/min/m}^2)$$

For a patient with a low cardiac index, doctors usually operate to correct the low blood flow problem, and patients are typically instructed to lose weight. A low CI can result from low cardiac output, high body surface area, or both. A very low CI can be life threatening and may require treatment.

For these weight-related problems and many others, health officials strongly recommend maintaining a reasonable weight throughout your life.

5–6 TIME

Learning Outcomes

1. Convert from one unit of time to another.
2. Make calculations with measures of time.
3. Equate time of day among time zones.

As we examined rate measures we introduced the units of time. We convert from one unit of time to another just as we convert other compatible units.

1 Convert from One Unit of Time to Another.

The relationships among various units of time are given in Table 5–6.

TABLE 5–6 Units of Time

1 year (yr) = 12 months (mo)	1 day (da) = 24 hours (hr)
1 year (yr) = 365 days (da)	1 hour (hr) = 60 minutes (min)[a]
1 week (wk) = 7 days (da)	1 minute (min) = 60 seconds (sec)[b]

[a]The symbol ′ means feet (3′ = 3 ft) or minutes (60′ = 60 minutes).

[b]The symbol ″ means inches (8″ = 8 in.) or seconds (60″ = 60 seconds).

We can use either unity ratios or conversion factors to convert from one unit of time to another.

Convert 3 hours to minutes.

Hours to minutes → larger to smaller → more minutes

Use a unity ratio to make the conversion:

$$3 \text{ hours} \left(\frac{\text{min}}{\text{hr}} \right) \qquad \text{Use an appropriate unity ratio.}$$

$$3 \text{ hr} \left(\frac{60 \text{ min}}{1 \text{ hr}} \right) = 3 \times 60 \text{ min} = 180 \text{ minutes}$$

There are 180 minutes in 3 hours.

Or, use a conversion factor to make the conversion:

1 hour = 60 minutes

3 hr × 60 min per hour = 180 min

2 Make Calculations with Measures of Time.

Suppose we have a 1-hr meeting and five items to be covered on our agenda. If we are to devote the same amount of time to each item, how much time should we give to each item?

$$1 \text{ hr} \div 5 = \frac{1}{5} \text{ hr}$$

How much is $\frac{1}{5}$ hr? Is this the usual way to express an amount of time? No. When less than 1 hr is involved, we often change to a smaller unit of time.

$$\frac{1}{\overset{}{5}} \text{ hr} \left(\frac{\overset{12}{60} \text{ min}}{1 \text{ hr}} \right) = 12 \text{ min}$$

A technician can assemble one precut picture frame in 7 minutes. How long will it take to assemble 25 frames?

If the frame could be assembled in 6 minutes, then the technician could assemble 10 per hour (6 min × 10 = 60 min = 1 hr). Thus, it would take 2 hours to assemble 20 frames and $2\frac{1}{2}$ hours to assemble 25 frames. Since it actually takes 7 minutes per frame, it will take more than $2\frac{1}{2}$ hours.

$$\frac{7 \text{ min}}{\text{frame}} \times 25 \text{ frames} = 175 \text{ min}$$

$$175 \text{ min} \times \frac{1 \text{ hr}}{60 \text{ min}} = 2.91\overline{6} \text{ hr}$$

Or, if we want to know how many hours and minutes, 2 hours = 120 minutes

$$175 \text{ min} - 120 \text{ min} = 55 \text{ min}$$

It will take 2 hr 55 min to assemble 25 frames.

To calculate from the decimal part of an hour, use the decimal portion of $2.91\overline{6}$ hr:

$$0.91\overline{6} \text{ hr} \times \frac{60 \text{ min}}{1 \text{ hr}} = 54.99\overline{9} \text{ or } 55 \text{ min}$$

3 Equate Time of Day Among Time Zones.

As our business operations become more global we have to account for the different time zones around the world.

Figure 5–39

Since 1 day is 24 hours, Earth is divided into *time zones* based on a 24-hr day (Fig. 5–39).

To understand how time zones work, let's look at how locations are defined on a globe of Earth. The grid system has similarities to the rectangular coordinate system, but since Earth is approximately a sphere (a ball-like shape), there are also differences.

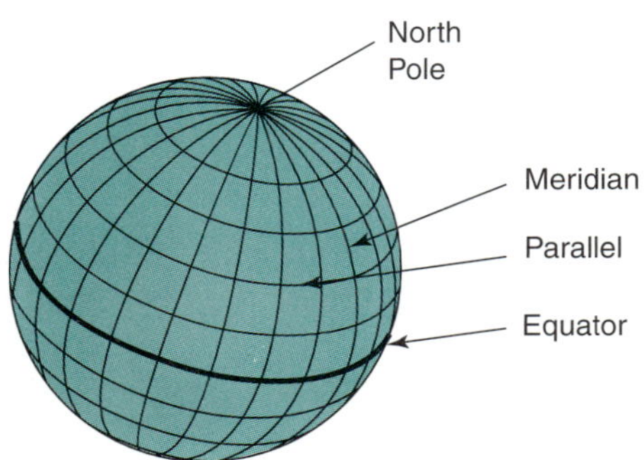

Figure 5–40

The imaginary horizontal line we call the *equator* divides Earth into two halves called *hemispheres.* Additional imaginary horizontal circles or rings are called *parallels of latitude.* Since the equator is our north/south reference, we define it as 0° on the scale. The parallels number from 0° to 90° both north and south. The North Pole is at 90° N, and the South Pole is at 90° S. We can relate this scale to the rectangular coordinate system and the vertical axis. Northern latitudes correspond to positive *y*-values on the rectangular coordinate system. Southern latitudes correspond to negative *y*-values.

Now, let's complete our spherical grid system with east/west measures. Imaginary circles drawn around the globe that pass through both the North and South Poles are called *meridians of longitude.* Figure 5–40 shows the equator, parallels, and meridians. Unlike the parallels, meridians are farthest apart at the equator and closest together at the North and South Poles. They roughly define time zones.

One complete rotation is 360°, so one time zone covers approximately 15° of Earth's rotation (360° ÷ 24 = 15°). We say *approximately* because time zone boundaries are not defined strictly by meridians. Actual time zone boundaries are defined by country, state, or province boundaries that are near a particular meridian. A few countries or small islands that are close to halfway between time zone boundaries define their time as a $\frac{1}{2}$-hour difference from a neighboring time zone. To be sure how a country defines its time, consult an atlas or other reference. Many software programs such as Microsoft Outlook™ give this information. Just as the equator is the longest parallel and is the zero line that separates north and south measures, one meridian must serve as zero for the east and west measures. All meridians are the same length, so one was arbitrarily selected; it is the one that passes through Greenwich, England, and is called the *prime meridian.*

    CHAPTER 5   Direct Measurement

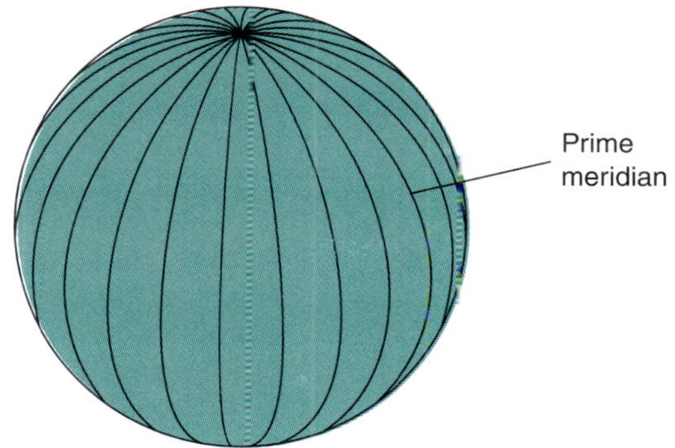

Figure 5–41

As the prime meridian (Fig. 5–41) passes through the North and South Poles and becomes the 180° meridian on the side of Earth opposite Greenwich, England, it is called the *international date line.* In references that show the appropriate time zone for a location (city, state, or country), the time is given as plus or minus a specified number of hours from *Greenwich Mean Time (GMT).* For example, Memphis, Tennessee, is defined as −6 hours GMT, written −6 GMT (Table 5–7). Mount Angel, Oregon, is defined as −8 GMT. What is the difference between Mount Angel time and Memphis time? $-8 - (-6) = -8 + 6 = -2$. Thus, Mount Angel is 2 hours earlier than Memphis time.

TABLE 5–7 Time Zone Chart

GMT	Military	Phonetic	Civilian Time Zones	Cities/Country
+0:00	z	Zulu	GMT—Greenwich Mean UT or UTC—Universal (Coordinated) WET—Western European	London, England Dublin, Ireland Edinburgh, Scotland Lisbon, Portugal Reykjavik, Iceland Casablanca, Morocco
−1:00	a	Alpha	WAT—West Africa	Azores, Cape Verde Islands
−2:00	b	Bravo	AT—Azores	
−3:00	c	Charlie		Brasilia, Brazil Buenos Aires, Argentina Georgetown, Guyana
−4:00	d	Delta	AST—Atlantic Standard	Caracas, Venezuela La Paz, Bolivia
−5:00	e	Echo	EST—Eastern Standard	Bogota, Colombia Lima, Peru New York, NY, USA
−6:00	f	Foxtrot	CST—Central Standard	Mexico City, Mexico Chicago, IL, USA Saskatchewan, Canada
−7:00	g	Golf	MST—Mountain Standard	Denver, CO, USA
−8:00	h	Hotel	PST—Pacific Standard	Los Angeles, CA, USA
−9:00	j	Juliet	YST—Yukon Standard	Whitehorse, Yukon, Canada Juneau, Alaska, USA
−10:00	k	Kilo	AHST—Alaska–Hawaii Standard CAT—Central Alaska HST—Hawaii Standard EAST—East Australian Standard	Honolulu, Hawaii, USA Anchorage, Alaska, USA Fairbanks, Alaska, USA
−11:00	l	Lima	NT—Nome	Nome, Alaska, USA American Samoa Aleutian Islands

(continued)

GMT	Military	Phonetic	Civilian Time Zones	Cities/Country
−12:00 +12:00	m y	Mike Yankee	IDLW—**International Date Line West** IDLE—**International Date Line East** NZST—New Zealand Standard NZT—New Zealand	Attu, Alaska, USA Wellington, New Zealand Fiji Marshall Islands
+11:00	x	X-ray	Russia Zone 10	Tasmania; Sydney, Australia; Micronesia
+10:00	w	Whiskey	GST—Guam Standard, Russia Zone 9	Tamuning, Guam Vladivostok; Papua, New Guinea
+9:00	v	Victor	JST—Japan Standard, Russia Zone 8	Korea, Singapore, Saipan, Malaysia
+8:00 +7:00	u t	Uniform Tango	CCT—China Coast, Russia Zone 7 WAST—West Australian Standard, Russia Zone 6	Mongolia; Perth, Australia Christmas Island, Australia; Java, Indonesia, Vietnam, Laos
+6:00	s	Sierra	ZP6—Chesapeake Bay, Russia Zone 5	Omsk, Kazakhstan (East), Kyrgyzstan
+5:00	r	Romeo	ZP5—Chesapeake Bay, Russia Zone 4	Perm, Pakistan; Kazakhstan (West)
+4:00	q	Quebec	ZP4—Russia Zone 3	Mauritius/Abu Dhabi, UAE Muscat, Aman Tblisi, Republic of Georgia Volgograd, Russia Kabul, Afghanistan
+3:00	p	Papa	BT—Baghdad, Russia Zone 2	Kuwait Nairobi, Kenya Riyadh, Saudi Arabia Moscow, Russia Tehran, Iran
+2:00	o	Oscar	EET—Eastern European, Russia Zone 1	Athens, Greece Helsinki, Finland Istanbul, Turkey Jerusalem, Israel Botswana, Zambia, Mozambique Harare, Zimbabwe
+1:00	n	November	CET—Central European FWT—French Winter MET—Middle European MEWT—Middle European Winter SWT—Swedish Winter	Paris, France Berlin, Germany Amsterdam, The Netherlands Brussels, Belgium Vienna, Austria Madrid, Spain Rome, Italy Bern, Switzerland Stockholm, Sweden Oslo, Norway

EXAMPLE Time in Madrid, Spain, is defined as +1 GMT. Time in Phoenix, Arizona, is defined as −7 GMT. When it is 8:00 A.M. in Madrid, what time is it in Phoenix?

−7 − (+1) = −8. So Phoenix is 8 hours earlier than Madrid. Therefore, when it is 8:00 A.M. in Madrid, we should subtract 8 hours to get the time in Phoenix. Eight hours earlier than 8:00 A.M. is 12 A.M. (midnight). So when it is 8:00 A.M. in Madrid, it is 12:00 midnight in Phoenix.

EXAMPLE Students in Boston, Massachusetts; Ogden, Utah; and Honolulu, Hawaii, are setting up an interactive chat on the Internet. Is there a time that this chat can be arranged between 8:00 A.M. and 3 00 P.M. at each site? At what time would each student need to schedule the session according to the local time?

Boston is -5 GMT; Ogden, Utah, is -7 GMT; and Honolulu, Hawaii, is -10 GMT.

$$-10 - (-5) = -10 + 5 = -5$$

Honolulu is 5 hours earlier than Boston. When it is 3:00 P.M. in Boston, it is 10 A.M. in Honolulu. When it is 1:00 P.M. in Boston, it is 8 A.M. in Honolulu. Thus, the chat could be arranged between 1:00 P.M. and 3:00 P.M. by Boston's time, which would be 8:00 A.M. to 10:00 A.M. by Honolulu's time. Since Ogden, Utah, is between the time zones, the session would be between 8:00 A.M. and 3:00 P.M. for them.

$$-7 - (-5) = -7 + 5 = -2$$

Odgen is 2 hours earlier than Boston, so the students in Ogden should **schedule the chat between 11:00 A.M. and 1:00 P.M.**

Tip!	*Subtracting a Negative.*

Dealing with time zones is a practical application for signed number skills. When we calculated the difference between Mount Angel time and Memphis time, we got $-8 - (-6) = -8 + 6 = -2$.

Our answer of -2 means that it is 2 hours *earlier* in Mount Angel than it is in Memphis. If we instead find the difference between Memphis time and Mount Angel time ($-6 - (-8) = -6 + 8 = +2$), our answer of $+2$ means that it is 2 hours *later* in Memphis than it is in Mount Angel.

SELF-STUDY EXERCISES 5–6

1   Use ratios or conversion factors to convert the measures of time.

1. How many days are in 36 hrs?

2. Find the number of minutes in 580 sec.

3. How many minutes are in 2.5 hr?

4. A physical test took 5.3 min to perform. How many seconds is this?

5. If you studied for 3.5 hr, how many minutes did you study?

6. A process takes 182 sec to perform. How many minutes is this?

7. Convert 7.2 hr to minutes.

8. Convert 88 sec to minutes.

2   Make the conversions.

9. A conveyor belt can move 72 lb of rice per minute. How many pounds can be moved on the conveyor belt in one hour?

10. Gasoline moves through a pipeline at 40 gallons per minute. How many gallons per hour can move through the pipeline?

11. A dog kennel uses 30 lb of dog food per day. How many pounds are used in a year?

12. A toll station can accommodate on average 28 vehicles per minute. How many vehicles can be accommodated in an hour?

13. A car traveling at the rate of 60 mph (miles per hour) is traveling how many miles per second?

14. A pump that can pump $125 \frac{\text{gal}}{\text{hr}}$ can pump how many gallons per minute?

15. A pump can dispose of sludge at the rate of $3,600 \frac{\text{lb}}{\text{hr}}$. How many pounds can be disposed of per minute?

16. If water flows through a pipe at the rate of $96 \frac{\text{gal}}{\text{min}}$, how many gallons will flow per second?

17. Time in Birmingham, Alabama, is defined as -6 GMT, and time in Athens, Greece, is defined as $+2$ GMT. If it is 2:00 A.M. in Athens, what time is it in Birmingham?

18. Business executives in New York City (-5 GMT), St. Louis, Missouri (-6 GMT), and London, England (0 GMT) want to schedule a conference call between 8 A.M. and 4 P.M. What range of hours can be used to schedule the meeting according to the local time in each city?

19. If you are visiting Paris, France ($+1$ GMT) and plan to call home to Stanford, Kentucky (-5 GMT), what time should you make the call using Paris local time?

20. What is the time difference between London, England, and Paris, France?

21. Suppose the Super Bowl football game is being played at 4:00 P.M. in New Orleans, Louisiana (-6 GMT), and you plan to watch it live from your hotel room in Vienna, Austria ($+1$ GMT). What time will it be on TV Vienna local time?

22. Your airplane leaves Detroit, Michigan (-6 GMT) at 5:45 P.M. and arrives in Helsinki, Finland ($+2$ GMT) at 11:45 A.M. How many hours was your flight?

5–7 READING MEASURING INSTRUMENTS

Learning Outcomes

1 Determine the percent of error of a measuring instrument.
2 Find the precision and greatest possible error.
3 Read the U.S. customary rule.
4 Read the metric rule.
5 Read the slide caliper.
6 Read the micrometer.
7 Find the relative error and the percent error of a measurement.
8 Read circular, uniform, and nonuniform scales.

Many measuring instruments are available for making measurements of all types. On some measuring instruments we read the measurement directly from a scale, whereas on other measuring instruments that have electronic sensors we read the measurement on a digital display. Whether the measurement is made manually or electronically, all measurements are *approximate* values.

1 Determine the Percent of Error of a Measuring Instrument.

Approximate numbers that represent measured values may have varying degrees of accuracy. The *accuracy* of a measurement refers to how close the measured value is to the true or accepted value. It may be difficult to determine the accuracy of a measure, since ordinarily the true value is not known. The accuracy of a measuring instrument can be determined by comparing a known or standard amount with a measured amount. For example, the accuracy of a device that measures weight can be determined by placing a known or standard weight on the device. If a 100-g weight is placed on a scale and it measures 96.9 g, the scale can be adjusted or calibrated to read 100 g. To *calibrate* a measuring device is to check, adjust, or systematically standardize the graduations of the device. After a device has been calibrated, objects of unknown weight can be measured more accurately. With extended use, many measuring devices will lose accuracy. These types of devices are calibrated often so that consistently accurate measurements can be obtained.

EXAMPLE A 1-oz standard weight produces a reading of 1.05 oz on a measuring device. What is the percent of error of the device?

$$\text{Amount of error} = \text{Measured value} - \text{Standard value}$$

$$= 1.05 \text{ oz} - 1 \text{ oz}$$

$$= 0.05 \text{ oz}$$

$$\frac{\text{percent of error}}{100} = \frac{\text{amount of error}}{\text{standard value}} \qquad \frac{R}{100} = \frac{P}{B}$$

$$\frac{R}{100} = \frac{0.05}{1}$$

$$R = 100(0.05)$$

$$R = 5\%$$

2 Find the Precision and Greatest Possible Error.

Precision is the degree to which several measurements provide values very close to one another. It is common to think of accuracy and precision as almost the same thing, but in technology and science, these terms are used in different ways. To illustrate the difference consider the results of three archers (Fig. 5–42).

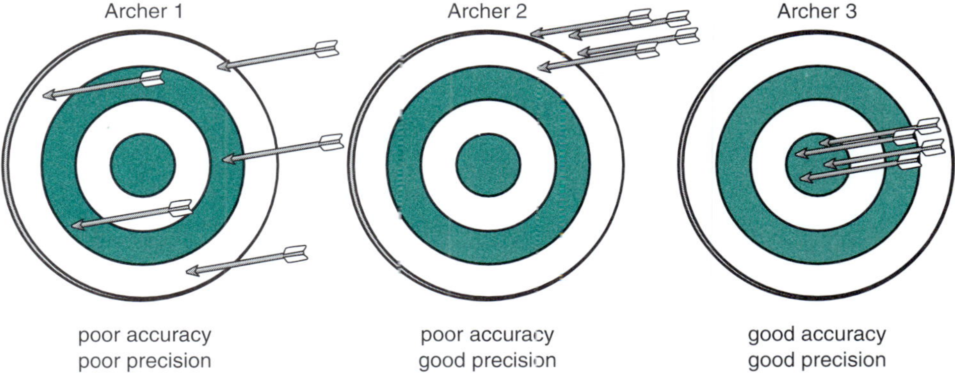

Figure 5–42 Accuracy versus precision

Archer 1's shots did not get close to the bull's-eye and did not hit in the same area of the target. Archer 2's shots did not get close to the bull's-eye but they were clustered. Archer 3's shots were both close to or in the bull's-eye and close together. It is desirable for measurements to be both accurate and precise.

One indicator of the degree of precision of a measurement is the number of *significant digits* in a measure. The numbers 2,500; 250; 25; 2.5; 0.25 and 0.025 all have two significant digits. When a number contains no zeros, all the digits are significant. However, zeros are special digits. Sometimes zeros are significant digits, and sometimes they are just placeholders.

> ### To determine the significant digits of a number:
>
> **For whole numbers:**
>
> 1. Start with the leftmost nonzero digit.
> 2. Count each digit through the rightmost nonzero digit.
>
> **For decimals and mixed decimals:**
>
> 1. Start with the leftmost nonzero digit.
> 2. Count each digit through the last digit.
>
> Determine the number of significant digits in the following:
>
> 200 has 1 significant digit.
> 28 has 2 significant digits.
> 1,320 has 3 significant digits.
> 2,005 has 4 significant digits.
>
> 0.07 has 1 significant digit.
> 2.0 has 2 significant digits.
> 1.20 has 3 significant digits.
> 0.01250 has 4 significant digits.

In other words, ending zeros on whole numbers are not significant, as 200 has only one significant digit. Beginning zeros in decimal numbers are not significant, as 0.07 has only one significant digit. Zeros between nonzero digits are significant, as 2,005 has four significant digits. Ending zeros to the right of the decimal are significant, as 1.20 has three significant digits.

Occasionally, an exception to this convention may be necessary. For example, a measure or approximate number to the nearest whole number may actually be 20 (two significant digits). This type of exception is sometimes clarified in words as "20 to the nearest whole number" and indentified with an overbar as $2\overline{0}$.

EXAMPLE Indicate the number of significant digits in the numbers.

(a) 203,000   (b) 0.0012   (c) 2.03   (d) 4.300   (e) $54\overline{0}$

(a) 203,000 has 3 significant digits. $\overline{203{,}000}$
(b) 0.0012 has 2 significant digits. $0.00\underline{12}$
(c) 2.03 has 3 significant digits. $\overline{2.03}$
(d) 4.300 has 4 significant digits. $\overline{4.300}$
(e) $54\overline{0}$ has 3 significant digits. $\underline{540}$

The precision of a measure or of an approximate number in decimal notation is the place value of the rightmost significant digit. The precision of 230 is tens, of 3.17 is hundredths, and of 4.0 is tenths.

The precision of a measure or of an approximate number in fraction notation is determined by the denominator of the fraction. The precision of $2\frac{5}{8}$ is $\frac{1}{8}$ and of $\frac{7}{16}$ is $\frac{1}{16}$.

Tip! | ***Do exact numbers have a degree of precision?***

The concept of precision is associated with approximate numbers. Exact numbers have infinite precision. For example, the average measurement of 2.3 cm, 4.25 cm, and 5.125 cm can be no more precise than the least precise of the measurements (tenths).

For the calculations $\frac{2.3 \ + \ 4.25 \ + \ 5.125}{3}$, we would round the results to the nearest tenth. The number 3 is an exact number (exactly three measures) and has an infinite precision so tenths is still the least precise.

$$\frac{2.3 + 4.25 + 5.125}{3} = \frac{11.675}{3} = 3.89166666 = 3.9 \text{ (rounded to tenths)}$$

The *greatest possible error* of a measurement is half the precision.

> **Find the greatest possible error of a measurement.**
>
> $$1\frac{3}{4}$$
>
> 1. Determine the precision of the measurement. $\qquad$ precision $= \dfrac{1}{4}$
>
> 2. Find half of the precision. $\qquad \dfrac{1}{2} \times \dfrac{1}{4} = \dfrac{1}{8}$

EXAMPLE Find the greatest possible error of the measurements.

(a) $2\dfrac{5}{8}$ in. (b) 3.5 cm

(a) $2\dfrac{5}{8}$ in. $\quad$ Precision $= \dfrac{1}{8}$.

Greatest possible error $= \dfrac{1}{2} \times \dfrac{1}{8} = \dfrac{1}{16}$.

(b) 3.5 cm $\qquad$ Precision $= 0.1$.
Greatest possible error $= 0.5(0.1) = 0.05$.

3 Read the U.S. Customary Rule.

Most of us are familiar with the *U.S. customary rule;* in elementary and high school we called it a *ruler*. It is divided into inches and parts of an inch, and although we use it for drawing straight lines, it is also used to measure length.

The U.S. customary rule uses an inch as the standard unit. Each inch is subdivided into fractional parts, usually 8, 16, 32, or 64. Let's examine the rule illustrated in Fig. 5–43.

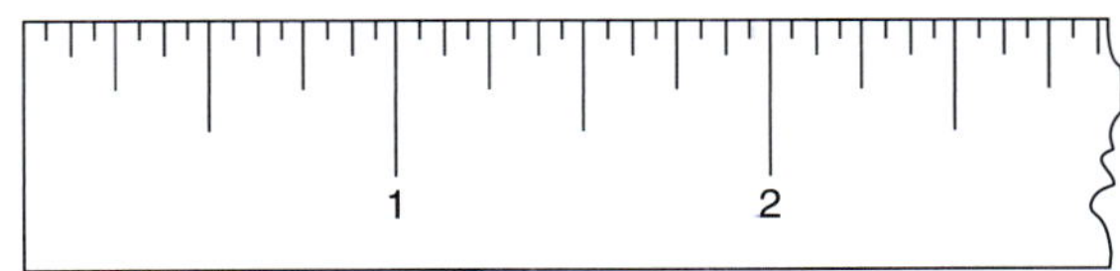

Figure 5–43 U.S. customary rule

Each inch is divided into 16 equal parts, so each part is $\frac{1}{16}$ in.; that is, the first mark from the left edge represents $\frac{1}{16}$ in. The left end of the rule represents zero (0). (Some rules leave a small space between zero and the end of the rule.)

The second mark from the left edge represents $\frac{2}{16}$ or $\frac{1}{8}$ in. This mark is slightly longer than the first mark. Look at Fig. 5–44, which labels each division mark, for one inch.

The fourth mark from the left is labeled $\frac{1}{4}$; that is $\frac{4}{16} = \frac{1}{4}$. In each case, fractions are always reduced to lowest terms. Notice that the $\frac{1}{4}$ mark is slightly longer than the $\frac{1}{8}$ mark.

The division marks are different lengths to make the rule easier to read. The shortest marks represent fractions that, in lowest terms, are sixteenths ($\frac{1}{16}$, $\frac{3}{16}$, $\frac{5}{16}$, $\frac{7}{16}$, $\frac{9}{16}$, $\frac{11}{16}$, $\frac{13}{16}$, $\frac{15}{16}$). Thus, the fractions that reduce to eighths are slightly longer than the sixteenths marks ($\frac{1}{8}$, $\frac{3}{8}$, $\frac{5}{8}$, $\frac{7}{8}$). Next, the marks representing fractions that reduce to fourths are slightly longer than the eighths ($\frac{1}{4}$, $\frac{3}{4}$). The marks for one-half ($\frac{1}{2}$) are longer than the fourths, and the inch marks are the longest. Look at Fig. 5–44 again and pay particular attention to the lengths of the division marks.

In many of the examples and exercises that follow, we will measure *line segments* whose beginning and end are identified by capital letters, such as line segment *AB* in the next example.

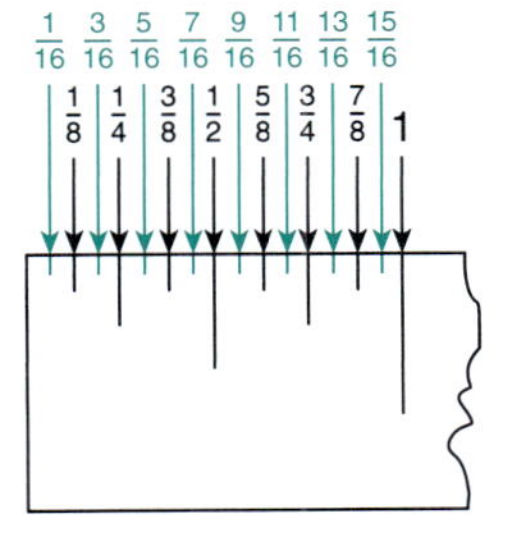

Figure 5–44 One inch

EXAMPLE   Measure line segment *AB* (Fig. 5–45).

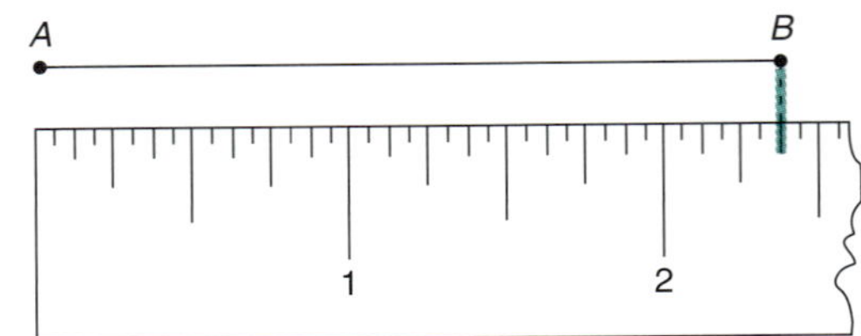

Figure 5–45   Line segment *AB*

Align point *A* with zero. Line segment *AB* goes past the 2-in. mark but not up to the 3-in. mark. Therefore, the measure of *AB* will be a mixed number between 2 and 3. Point *B* is $\frac{3}{8}$ in. past 2. **Thus, *AB* is $2\frac{3}{8}$ in.**

Tip!	***Judge to the Closest Mark.***

A line segment may not always align exactly with a division mark. If this is the case, use eye judgment to decide which mark is closer to the end of the line segment.

EXAMPLE   Measure line segment *CD* (Fig. 5–46).

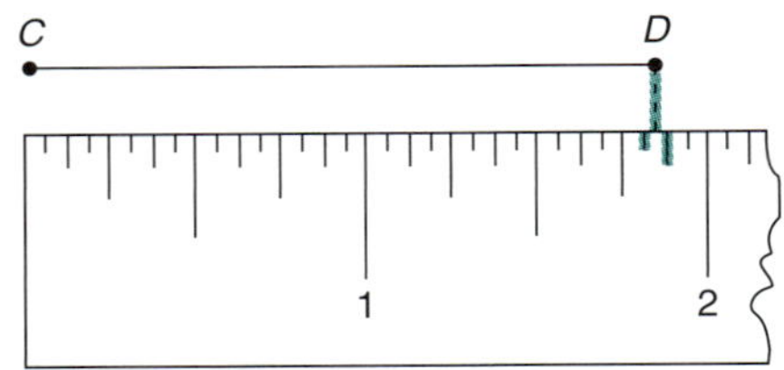

Figure 5–46

Point *D* aligns between $1\frac{13}{16}$ and $1\frac{7}{8}$. Recall that measurements are always approximations; using our best eye judgment, we determine that point *D* seems closer to $1\frac{13}{16}$ than to $1\frac{7}{8}$.
We will say *CD* is $1\frac{13}{16}''$ to the nearest sixteenth of an inch.

In the preceding example, $1\frac{7}{8}$ would also be a reasonable measure for line segment *CD*.

In practice, measurements are considered acceptable if they are within a desired *tolerance*. In the preceding example, the smallest division is $\frac{1}{16}$, so the desired tolerance would normally be plus or minus one-half of one-sixteenth, or $\pm\frac{1}{32}$ ($\frac{1}{2}$ of $\frac{1}{16} = \frac{1}{32}$). That is, the acceptable measure can be $\frac{1}{32}$ more than or $\frac{1}{32}$ less than the ideal measure.

If the ideal measure is $1\frac{13}{16}$ in. and the tolerance is $\frac{1}{32}$ in., the *range* of acceptable values is from $1\frac{13}{16} - \frac{1}{32}$ to $1\frac{13}{16} + \frac{1}{32}$. The acceptable range is from $1\frac{25}{32}$ to $1\frac{27}{32}$.

If the desired tolerance is $\pm\frac{1}{16}$ inch, then $1\frac{13}{16}$ and $1\frac{7}{8}$ would both be within tolerance. ($1\frac{12}{16}$ to $1\frac{14}{16}$)

4   Read the Metric Rule.

Many standard rulers have both a U.S. customary scale and a metric scale. The metric rule is usually calibrated in centimeters or millimeters. The *metric rule* illustrated in Fig. 5–47 shows centimeters (cm) as the major divisions, represented by the longest lines. Each centimeter is divided into 10 millimeters (mm), which are the shortest lines. A line slightly longer than the millimeter line divides each centimeter into two equal parts of 5 mm each.

CHAPTER 5   Direct Measurement

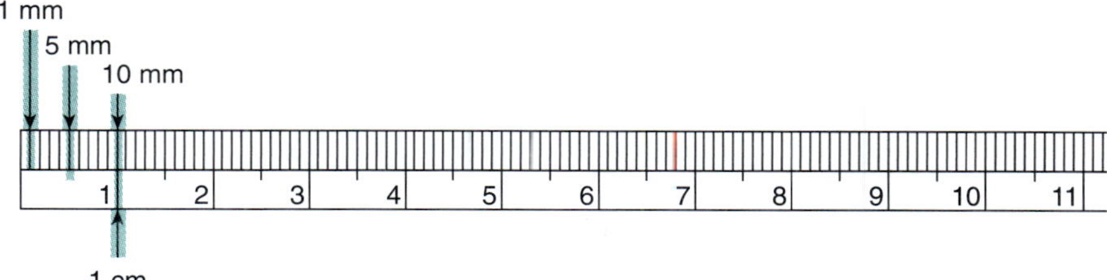

Figure 5–47 The metric rule

The metric rule is read like the U.S. customary rule with the exception that only two measures are indicated, centimeters and millimeters. Other measures are calculated in relation to millimeters or centimeters.

Since 1 millimeter is $\frac{1}{10}$ of a centimeter, we can write metric measures in decimals. For example, a measure of 3 centimeters and 4 millimeters is written as either 3.4 centimeters or 34 millimeters.

EXAMPLE   Find the length of line segment AB (Fig. 5–48) to the nearest millimeter.

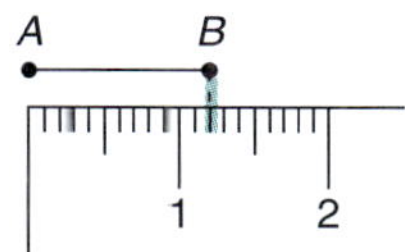

Figure 5–48

The line segment AB extends two marks past 1 cm, **so line segment AB is 12 mm or 1.2 cm to the nearest millimeter.**

EXAMPLE   Find the length of the line segment CD (Fig. 5–49).

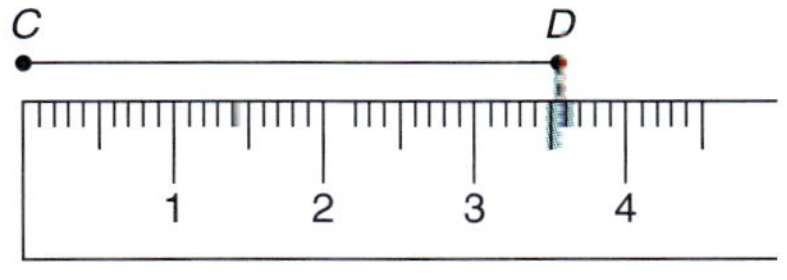

Figure 5–49

The end of line segment CD falls approximately halfway between the 35-mm mark and the 36-mm mark, by eye judgment, measuring about 35.5 mm. Both 35 and 36 mm would be acceptable measures of the line segment CD. **Thus, 35 mm, 35.5 mm, and 36 mm are acceptable approximations for line segment CD.**

5 Read the Slide Caliper.

The U.S. customary and metric rules are used to make measurements of items that do not require a high degree of precision. The *slide caliper* is just one of several measuring instruments used to make precise measurements.

One of the oldest precision measuring tools that measures by "feel" is a caliper. To find the measure of an object using a slide caliper, we may make the reading from a digital display, a dial, or a scale. Figure 5–50 shows an electronic digital caliper. Measurements are calculated automatically by selecting the desired measuring system (U.S. customary or metric) and reading the digital display.

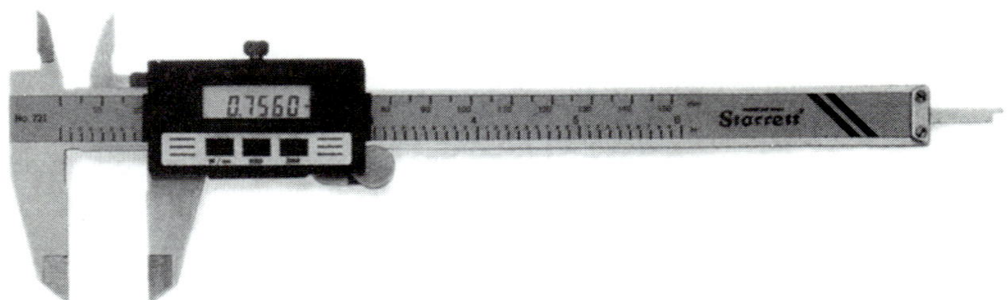

Figure 5–50 Digital Caliper

A dial caliper is specifically purchased for making measurements in the U.S. customary or metric system (Figure 5–51). The procedures for reading dials are given in Learning Outcome 8.

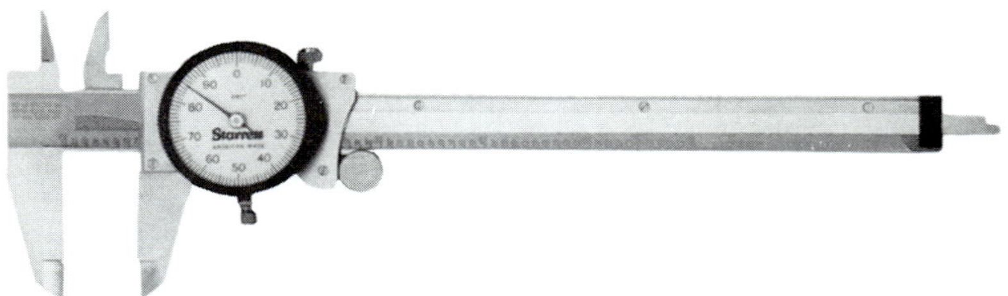

Figure 5–51 Dial Caliper

A slide caliper that requires the user to read the measure from a scale is available with scales of different graduations. Scales graduated in 32nds of an inch, 64ths of an inch or metric units are the most common (Figure 5–52).

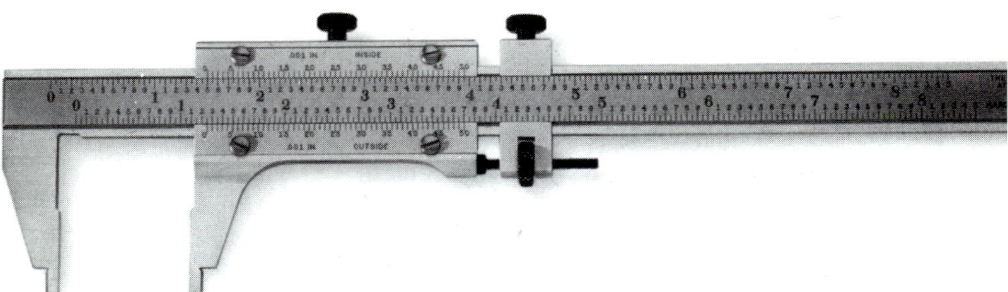

Figure 5–52 Slide Caliper

For a higher degree of precision the vernier caliper is used. A vernier caliper has an additional graduated scale that allows reading to the nearest thousandth of an inch in the U.S. customary system or to the nearest hundredth of a millimeter in the metric system. To get a good reading with any precision tool involves skill and experience. Figure 5–53 shows a vernier caliper with both U.S. customary and metric scales.

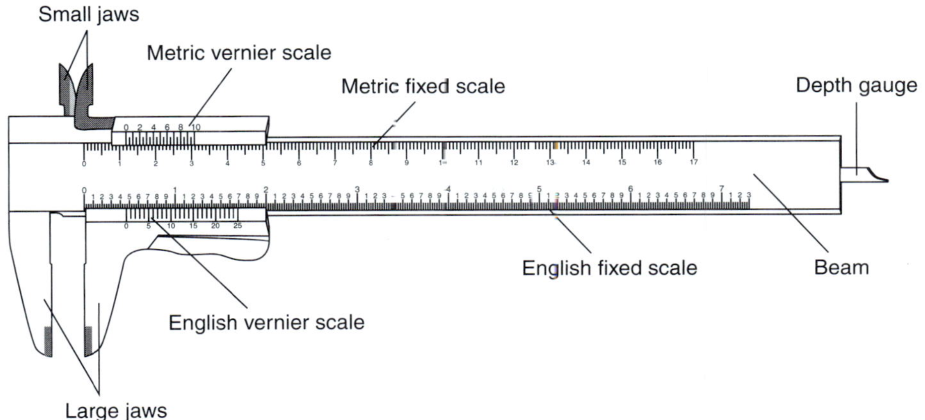

Figure 5–53   Vernier Caliper

The vernier caliper can be used to take inside, outside, or depth measurements. Outside measurements are made by closing the large jaws of the tool around the outside of the object. To measure the inside of an object, place the small jaws inside the object. The depth of an object is measured by inserting the depth gauge into the object.

One type of vernier caliper has two metric scales and two U.S. customary scales. Figure 5–54 shows the two metric scales on the upper part of the beam. The fixed scale is graduated in centimeters and subdivided into millimeters. Measurements taken with the metric scales are recorded in millimeters (mm). Each numbered graduation on the fixed scale represents 10 mm, and the unnumbered graduations represent 1 mm. The other metric scale, called the *vernier scale,* has 11 long marks (0, 1, 2, . . ., 9, 10) graduated in tenths of millimeters (0.10 mm) that are subdivided into halves of tenths $\left(\frac{1}{2} \times \frac{1}{10} = \frac{1}{20}\right)$, or five hundredths (0.05). The precision of the metric vernier scale is 0.05 mm.

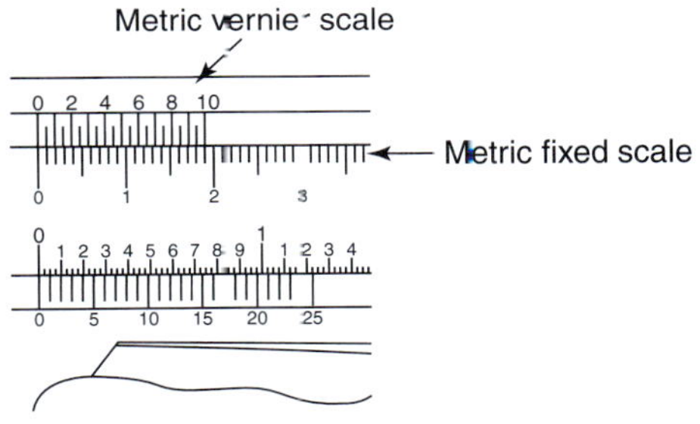

Figure 5–54

To read a vernier caliper metric scale:

1. On the fixed metric scale (top part of the fixed beam) determine the numbered graduation that is the tens place of the measurement to the left of the zero graduation on the vernier metric scale. This number is the tens place of the measurement.
2. Determine the number of unmarked graduations between the numbered graduation on the fixed scale and the zero (0) mark on the vernier scale. This is the ones place of the measurement.
3. If the zero graduation on the vernier metric scale is aligned directly above a line on the fixed scale, this indicates there are 0.00 (zero hundredths), so read the measurement from the fixed scale. Write the measurement as the whole number read from the fixed scale followed by .00 (zero hundredths).

(continued)

<table>
<tr><td>

3a. If the zero graduation on the vernier scale is not directly aligned with a graduation on the fixed scale, record the number of whole millimeters in the measurement, then find the graduation on the vernier scale that most nearly aligns with *any* graduation on the fixed scale. If a long mark represents the best alignment, record the number of tenths of millimeters and a zero in the hundredths place.

3b. If a short mark on the vernier scale represents the best alignment, record the number of tenths of millimeters represented by the long mark to the left and a five in the hundredths place.

</td></tr>
</table>

Tip!	***One Digit per Column.***

The vernier caliper is easier to read if you read one digit per column for each of the four places—tens, ones, tenths, hundredths.

- Tens place (or tens and hundreds): read from the numbered graduation on the fixed beam.
- Ones place: count the number of graduations on the fixed beam between the numbered graduation and zero on the vernier scale.
- Tenths place: read from the numbered graduation on the vernier scale that most closely aligns with *any* graduation on the fixed scale.
- Hundredths place: read as zero if the two beams align on a long graduation of the vernier scale, and read as 5 if the two beams align on a short graduation of the vernier scale.

EXAMPLE Read the measurement in millimeters (to the nearest hundredth) on the vernier caliper in Fig. 5–55.

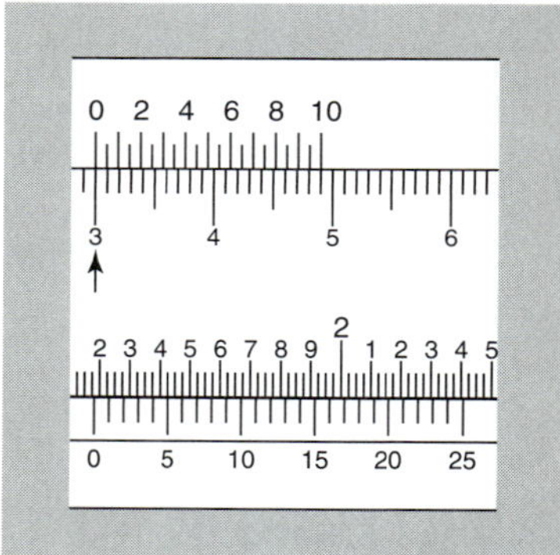

Figure 5–55

The zero graduation on the vernier scale is the graduation that most nearly aligns with a graduation on the fixed beam, and it aligns with the long mark numbered 3. **The measurement is 30.00 mm.**

EXAMPLE Read the measurement in millimeters (to the nearest hundredth) on the vernier caliper in Fig. 5–56.

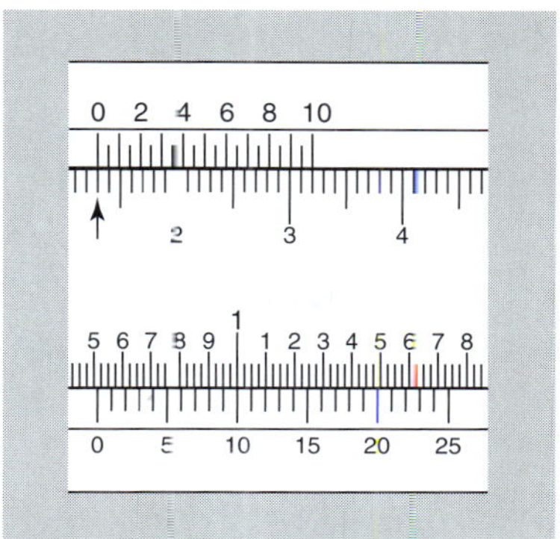

Figure 5-56

The zero graduation on the vernier scale is the graduation that most nearly aligns with a graduation on the fixed beam, and it aligns with the short mark 3 units to the right of the graduation marked 1. **The measurement is 13.00 mm.**

EXAMPLE   Read the measurement in millimeters (to the nearest hundredth) on the vernier caliper in Fig. 5–57.

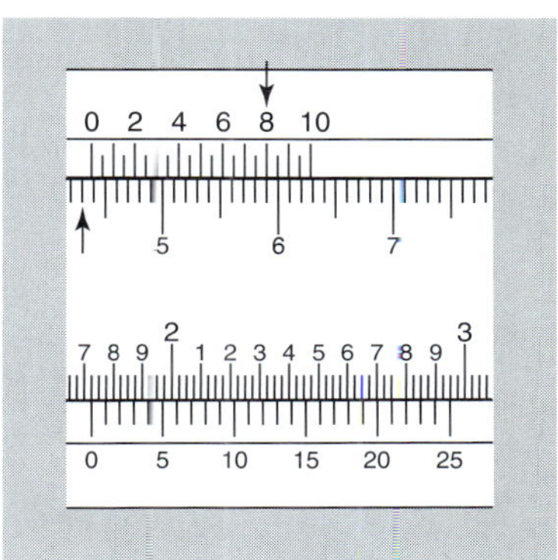

Figure 5-57

4_.__ mm	The zero graduation on the vernier scale is between the numbers 4 and 5 on the fixed beam.
43.__ mm	The first graduation to the left of the zero is 3 marks to the right of 4.
43.8_ mm	The graduation on the vernier scale that best aligns with a mark on the fixed beam is 8.
43.80 mm	Since 8 is a long mark, the hundredths place is zero.

The measurement is 43.80 mm.

EXAMPLE   Read the measurement in millimeters (to the nearest 0.05 mm) on the vernier caliper in Fig. 5–58.

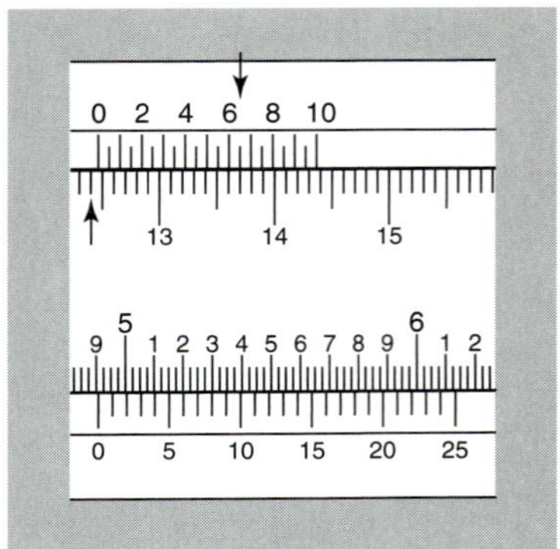

Figure 5–58

124.__ mm   The zero on the vernier scale is 4 graduations to the right of 12.
124.65 mm   The short mark to the right of the 6 on the vernier scale aligns best with a mark on the fixed scale.

The measurement is 124.65 mm.

The U.S. customary scales on the vernier caliper are located at the bottom of the scale (Fig. 5–59). The fixed scale is on the lower part of the fixed beam and is divided into whole inches and tenths (0.100 in., 0.200 in., . . .) of an inch. These graduations are numbered. Each tenth is subdivided into 4 parts, so each part represents $\frac{1}{4}$ of a tenth-inch $\left(\frac{1}{4} \times \frac{1}{10} = \frac{1}{40} \text{ or } 0.025 \text{ in.}\right)$. The vernier scale (movable scale at the bottom) is divided into 25 parts, and each part represents one thousandth of an inch. The precision of the U.S. customary scale is 0.001 in.

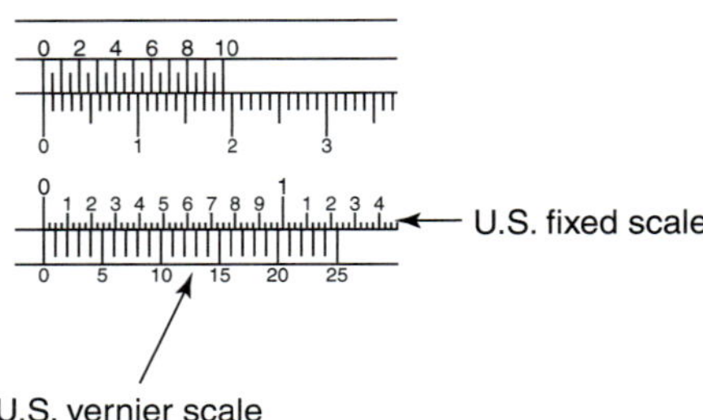

Figure 5–59

> *To read a vernier caliper U.S. customary scale:*
>
> **1.** Determine the number of whole inches by reading the number on the longest mark to the immediate left of the zero graduation on the vernier scale.
> **2.** Read the number of tenths of inches by reading the number on the short mark that is immediately to the left of the zero mark on the vernier scale. Read each unnumbered graduation as 0.025 in.
> **3.** Read each mark on the vernier scale as one-thousandth of an inch (0.001 in.) and add its value to the thousandths read from the fixed beam.
> **4a.** If the zero graduation on the vernier scale directly aligns with a graduation on the fixed scale, read the measurement from the fixed scale and attach zeros in the hundredths and thousandths places to show the precision level.
> **4b.** If the zero graduation on the vernier scale does not align with a graduation on the fixed scale, find the graduation on the vernier scale that most nearly aligns with *any* graduation on the fixed scale. Add that value in thousandths of an inch to the measurement.

EXAMPLE   Read the measurement in inches shown on the vernier caliper in Fig. 5–60.

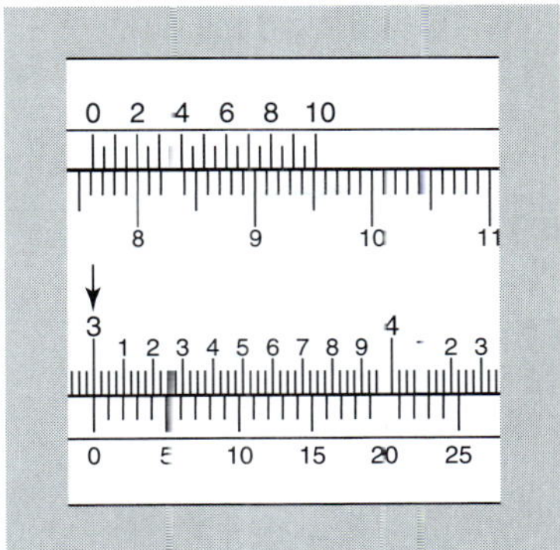

Figure 5–60

Zero aligns best with the mark numbered 3, **so the measurement is 3.000 in.**

EXAMPLE   Read the measurement in inches shown on the vernier caliper in Fig. 5–61.

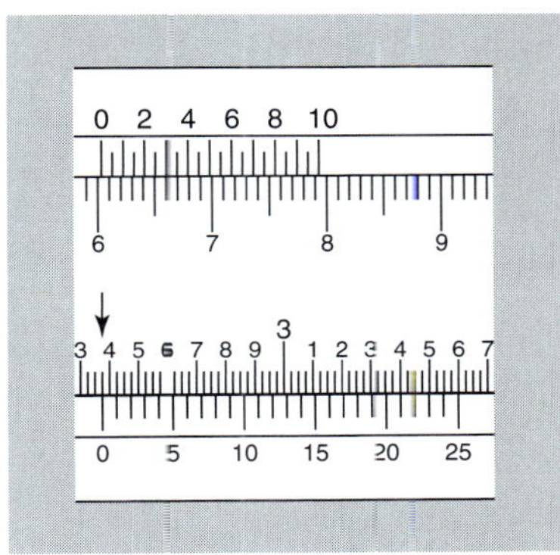

Figure 5–61

2.	in.	The first longest mark to the left of zero is numbered 2.
0.3	in.	The shorter numbered mark is 3.
0.075	in.	The number of 0.025-in. graduations is 3, so add 3 × 0.025 in. = 0.075 in.
0.000	in.	Zero on the vernier scale aligns directly with a mark on the fixed beam, so read 0.000 in. from the vernier scale.
2.375	in.	(Note the vertical alignment of decimal points.)

The measurement is 2.3075 in.

EXAMPLE   Read the measurement in inches shown on the vernier caliper in Fig. 5–62.

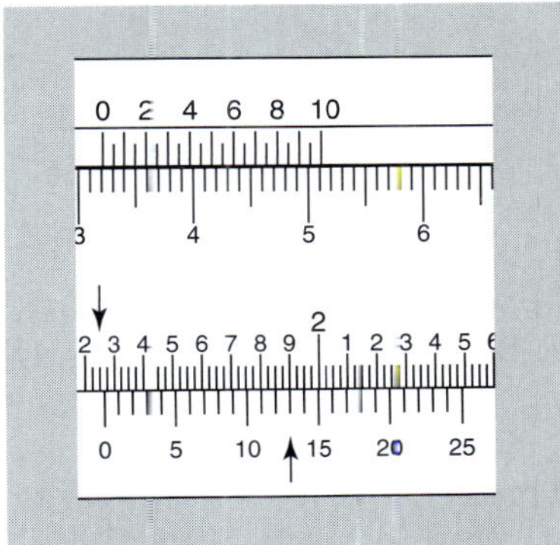

Figure 5–62

1.2 in.	Read 1.2 in. from the numbered graduations on the fixed beam.
0.050 in.	There are 2 unnumbered graduations to the left of zero, so add 2 × 0.025 in. = 0.050 in.
0.013 in.	The 13 mark on the vernier scale best aligns with a mark on the fixed scale, so add 0.013 in.
1.263 in.	*(Note the vertical alignment of decimal points.)*

The measurement is 1.263 in.

Proper Use of a Vernier Caliper.

- Hold the caliper so that it is perpendicular (at right angles) to the object being measured.
- When measuring the diameter of a round object, be sure to position the caliper so the greatest distance across the piece is measured.
- When measuring round parts or to ensure uniform thickness, take two or more measurements, then calculate and use their average.

6 Read the Micrometer.

The *micrometer* is a type of caliper used in technical fields that require more precise measurements than can be obtained with a vernier caliper. Some industries use computerized micrometers that provide measurements as digital readouts.

A micrometer is designed to make measurements in *either* metric or U.S. customary units of measure. The metric micrometer is graduated to provide readings to the nearest hundredth of a millimeter (0.01 mm). The U.S. customary "mike" provides readings to the nearest thousandth of an inch (0.001 in.).

Figure 5–63 illustrates the parts of an outside micrometer. Familiarity with the parts of the micrometer aids our understanding of the procedure for measuring with a micrometer. Examine the parts of the micrometers in Fig. 5–63 carefully.

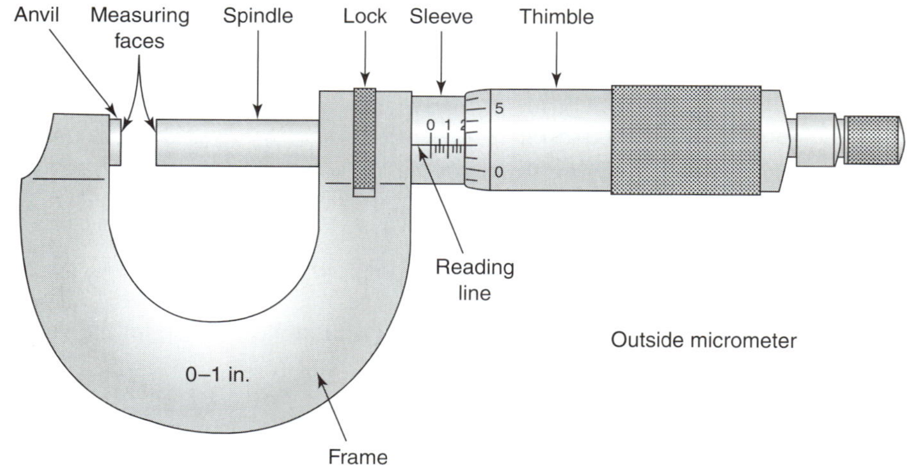

Figure 5–63

The frame of the micrometer is the curved rigid portion. Micrometers come in various sizes, and the size of the micrometer is generally identified on the frame. The anvil is located on the upper end of the frame. The spindle is the movable part that is adjusted to the size of the object being measured. The object being measured fits snugly between the anvil and the spindle.

The sleeve is the graduated portion of the micrometer that is attached to the frame. It has 40 vertical graduation marks, with every fourth mark labeled with a number from 1 to 10. Each labeled graduation represents one-tenth of an inch

   CHAPTER 5   Direct Measurement

(0.1 in.). Each vertical mark represents 0.025 in. Thus, the three marks between each labeled vertical mark represent 0.025, 0.050, and 0.075 in., respectively.

The thimble of the micrometer is graduated into 25 parts, each representing one-thousandth of an inch (0.001 in.). These marks are labeled every five marks for ease in reading.

The micrometer is placed so the object to be measured is between the anvil and the spindle. The thimble is turned slowly until it fits snugly on the object. *The threads of the spindle located inside the thimble are very delicate and can be damaged if the thimble is forced.* Some micrometers use a ratchet design that prevents the thimble from being turned with damaging force. Some micrometers have a lock that keeps the micrometer from slipping while the measurement is read.

To read an outside micrometer:

1. Place the object to be measured snugly between the anvil and the spindle.
2. Determine the inch portion of the measure by reading the smallest number given in the range of the micrometer. This range is found on the frame.
3. Read the tenth portion of the measurement by reading the last number that is showing on the sleeve of the micrometer.
4. Determine *part* of the thousandths measure by reading the number of marks that are showing to the right of the tenths number. Each mark represents 0.025.
5. Read *additional* thousandths from the thimble of the micrometer. The reading is made where the horizontal line on the sleeve meets the thimble. Give the reading to the closest mark.
6. Add the readings in 2 through 5. (Steps 2 and 3 represent whole numbers and tenths, respectively. Steps 4 and 5 represent thousandths and will fall in the same columns when the numbers are added.)

EXAMPLE   Determine the reading on the micrometer in Fig. 5–64.

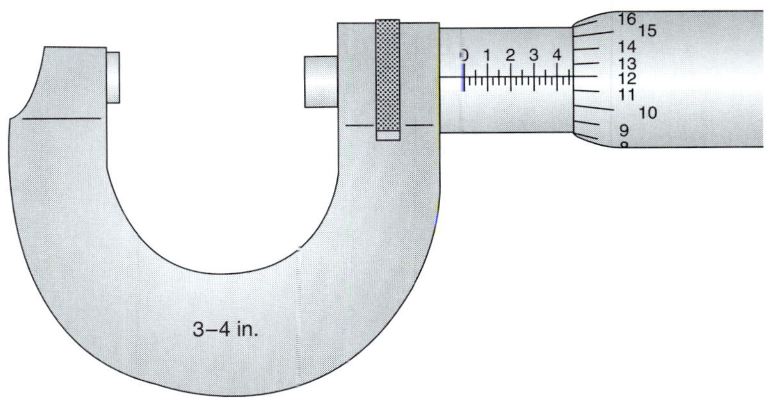

Figure 5–64

3.	The inch measurement is given to be 3 in.
0.4	The tenths measurement is 0.4 in.
0.050	The sleeve reading is two marks (0.050) past the 4.
<u>0.012</u>	The thimble reading is 12 (0.012).
3.462 in.	*(Note the vertical alignment of decimal points.)*

The measurement is 3.462 in.

A vernier micrometer has a vernier scale that allows readings to the nearest ten-thousandth of an inch (0.0001 in.). The graduations of the vernier scale are located on the top of the sleeve or barrel of the micrometer, and they run lengthwise on the barrel. There are 10 spaces indicated on the vernier scale, and each space represents one-tenth of a space on the thimble, or 0.0001 in. (One-tenth of one thousandth is one ten-thousandth.) Locate the vernier scale on Fig. 5–65.

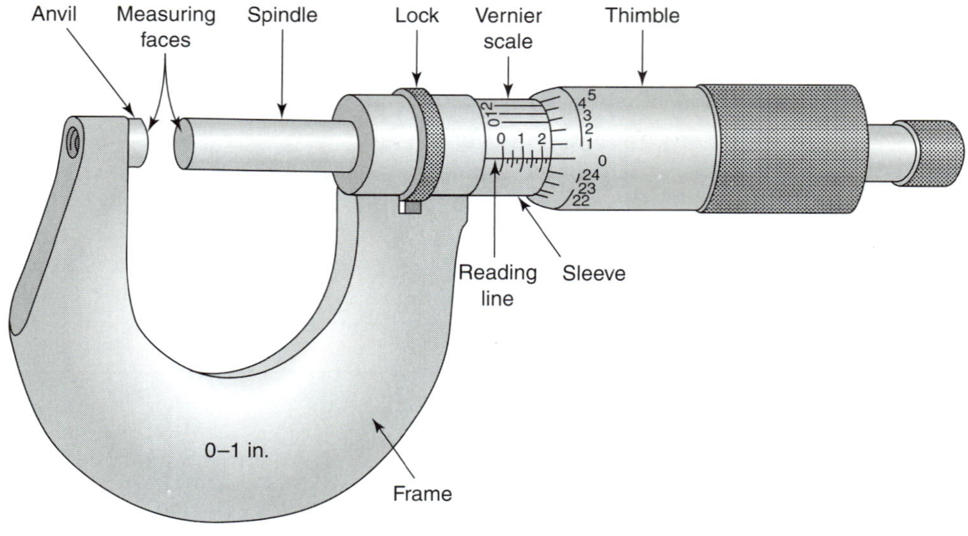

Outside vernier micrometer

Figure 5–65

To read an outside vernier micrometer:

1. Follow the first four steps for reading an outside micrometer.
2. When determining the thimble reading, use the smaller number that the measure falls between.
3. Locate the number on the vernier scale that aligns with *any* mark on the thimble. This indicates the number of ten-thousandths.
4. Add the readings in each step.

EXAMPLE   Determine the reading on the 2- to 3-in. vernier micrometer in Fig. 5–66.

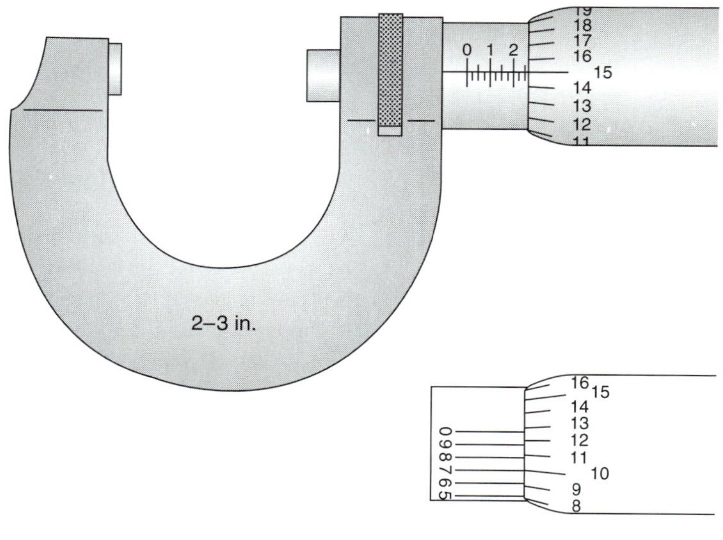

Vernier scale

Figure 5–66

2.	The inch measurement is given to be 2 in.
0.2	The tenth reading is 0.2 in.
0.050	The sleeve reading is two marks past the 2; thus, it is 0.050 in.
0.015	The thimble reading is 15; thus, it is 0.015.
0.0007	On the vernier scale the number 7 aligns with the 10 on the thimble; thus, it is 0.0007.
2.2657 in.	(Note the vertical alignment of decimal points.)

The measurement is 2.2657 in.

Several measuring instruments have scales similar to those on the outside micrometer and the outside vernier micrometer. Two are shown in Figure 5–67. The basic principles for reading these instruments are similar to those for reading the micrometer. Figure 5–67 illustrates a depth micrometer and an inside micrometer.

Depth micrometer

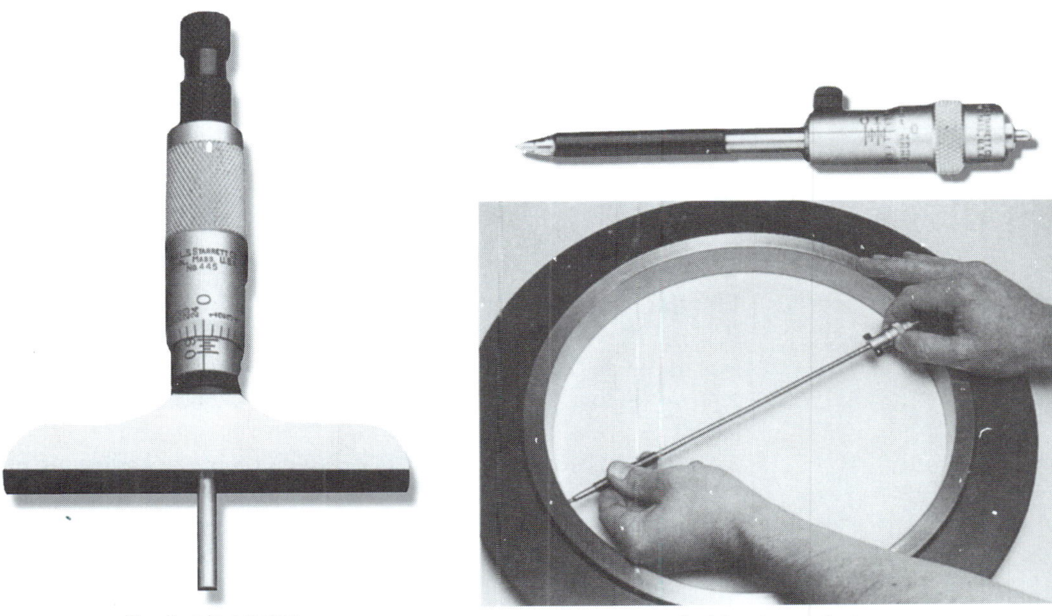

Depth micrometer

Inside micrometers

Figure 5–67 Other types of micrometers

7 Find the Relative Error and the Percent Error of a Measurement.

Because factories and other industrial concerns are extremely interested in issues related to quality control, most regularly measure parts to ensure that they fall within preestablished tolerance limits for quality. Every measurement taken has some amount of error. There are three ways of reporting the error: as an absolute error, a relative error, and a percent of error. Reporting the error as an absolute error does not give much information, so relative error and percent error are frequently calculated.

■ **DEFINITION: Absolute Error.**   The absolute value of the difference between the true value and the observed measurement is the *absolute error.*

■ **DEFINITION: Relative Error.**   The quotient of the absolute error and the observed measurement is the *relative error.*

■ **DEFINITION: Percent Error.**   The relative error converted to a percent is the *percent error.*

EXAMPLE The blueprint for a part calls for it to be 32.155 mm. A measurement of an actual part is recorded as 32.112 mm. Find the absolute error, relative error, and percent error.

$$\text{Absolute error} = |\text{Observed value} - \text{True value}|$$

$$\text{Absolute error} = |32.155 - 32.112|$$

Absolute error = 0.043

$$\text{Relative error} = \frac{\text{Absolute error}}{\text{True value}}$$

$$\text{Relative error} = \frac{0.043}{32.155}$$

Relative error = 0.001337

$$\text{Percent error} = \text{Relative error} \times 100\%$$

$$\text{Percent error} = 0.001337 \times 100\%$$

Percent error = 0.134%

8 Read Circular, Uniform, and Nonuniform Scales.

Circular scales are prevalent in our environment and workplace. Whether reading a water or gas meter dial or a dial used to test the depth of cut on a flat metal bar, the first thing we need to determine is the scale being used and the basic unit for that scale. For example, a water meter may be composed of six circular dials such as the one in Fig. 5–68. Notice that each dial has ten graduations on its scale and represents a power of ten. To read the dial, start with the 100,000 scale and continue clockwise to the 1-cubic-foot scale. If a dial indicator is between digits, read the smaller digit. The reading for the water meter in Fig. 5–68 is 381,835 cubic feet (ft^3).

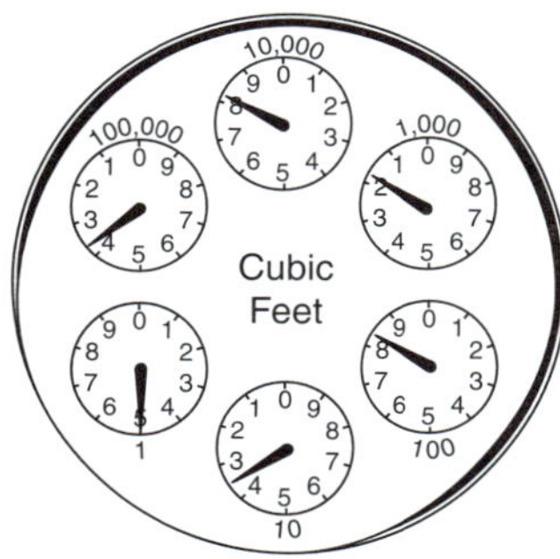

Figure 5–68 Water meter

EXAMPLE Read the water meter shown in Fig. 5–69.

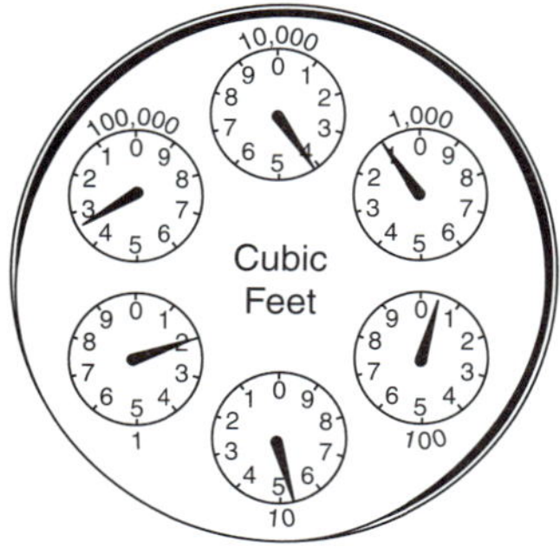

Figure 5–69 Water meter

Begin reading with the 100,000 scale and read clockwise through the scale marked 1.

The reading is 341,052 ft³.

An electric meter is another example of a circular scale. The electric meter in Fig. 5–70 has four dials; the first reads thousands, the next is hundreds, the third is tens, and the fourth is units (or ones). The measurement is given in kilowatt hours (kWh).

Figure 5–70 Electric meter

EXAMPLE Read the electric meter shown in Fig. 5–71.

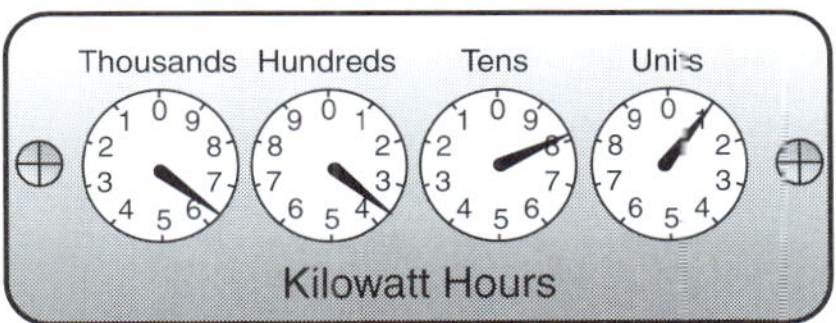

Read and record the thousands dial. 6
Read and record the hundreds dial. 3
Read and record the tens dial. 8
Read and record the ones dial. 1

Figure 5–71 Electric meter

The meter reads 6,381 kWh.

Some dial gauges have a large circular scale and a smaller circular scale on the inside face. Such a metric gauge is shown in Fig. 5–72. Each graduation of the large dial represents 0.01 mm. Thus, when the needle registers at or below 35 on the right of zero, the object being measured is 35 × 0.01 mm. The small needle records the number of complete revolutions made by the large needle when making the measurement. A complete revolution corresponds to 1.00 mm. The small needle is usually read first, followed by the large needle.

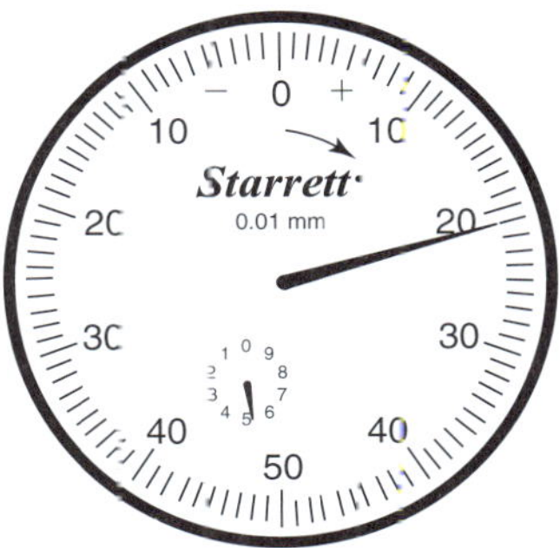

Figure 5–72

EXAMPLE Read the metric dial in Figure 5–72.

The small needle indicates +5:	5 × 1.00 mm = +5.00 mm
The large needle indicates +21:	21 × 0.01 mm = +0.21 mm
Total reading:	**+5.21 mm**

A U.S. customary dial is read much like the metric dial. Examine the U.S. customary dial in Fig. 5–73. Each complete revolution of the large needle corresponds to 0.100 in. and the small needle on the left indicates the number of complete revolutions made by the large needle. Each graduation on the large scale represents 0.001 in.

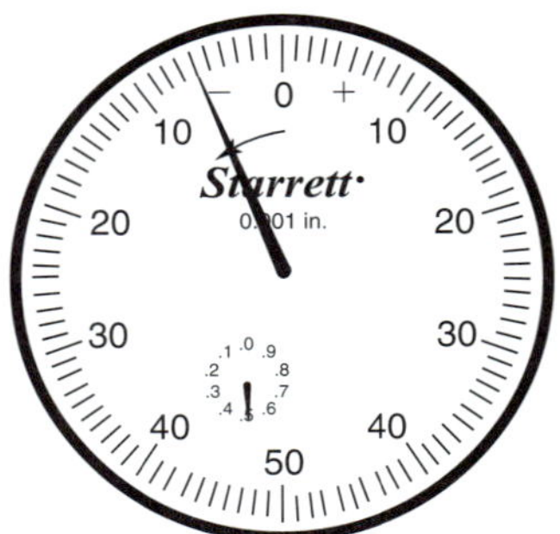

Figure 5–73

EXAMPLE   Read the U.S. customary dial in Fig. 5–73.

The small needle reads 5:	5×0.100 in. $= 0.500$ in.
The large needle reads −6:	-6×0.001 in. $= -0.006$ in.
Total reading:	0.494 in.

The measurement is 0.494.

The volt–ohm meter (VOM) is an example of a uniform scale. This instrument is used in electrical circuits to measure voltage in volts (V), and resistance in ohms (V). The voltage scales are uniform, but the resistance scale is nonuniform. On a *uniform scale* all the graduations are equally spaced, and each subdivision represents the same number of units.

EXAMPLE   Read the voltage scale shown in Fig. 5–74.

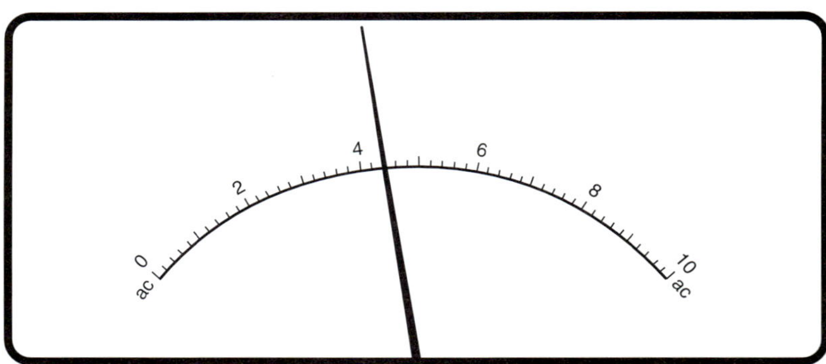

Figure 5–74   Volt–ohm meter (VOM) scales

Each long unnumbered mark represents an odd whole number of volts. Each short unnumbered mark represents 0.2 V. The needle is on the second graduation to the right of 4 ($2 \times 0.2 = 0.4$).

The reading is 4.4 V.

An ohmmeter is an example of a *nonuniform scale,* that is, the spacing between numbered graduations is not the same. In Fig. 5–75, each unit between 0 and 5 has 5 subdivisions. Therefore, each subdivision represents $\frac{1}{5} \times 1$ Ω, or 0.2 Ω. Between 5 and 10, each large division is subdivided into 2 parts, so each subdivision represents $\frac{1}{2} \times 1$ Ω, or 0.5 Ω. Between 10 and 20, each division represents 1 Ω. The values of the subdivisions for the other divisions of the scale are determined in a similar manner. The ranges and value of each subdivision are shown in the table.

CHAPTER 5   Direct Measurement

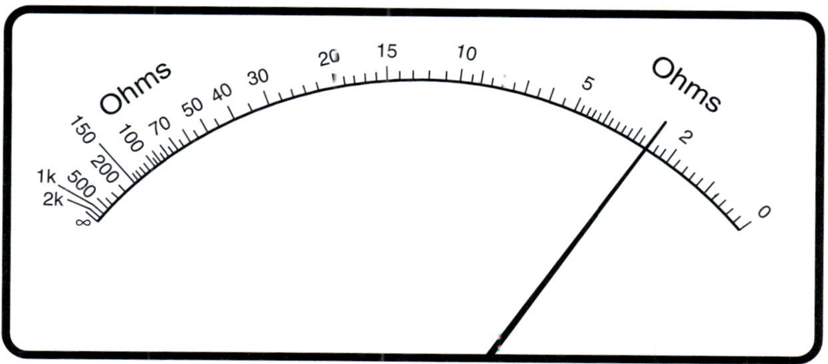

Figure 5-75

Range	Value of each subdivision
0–5 Ω	0.2 Ω
5–10 Ω	0.5 Ω
10–20 Ω	1 Ω
20–30 Ω	2 Ω
30–100 Ω	5 Ω
100–150 Ω	10 Ω
200–500 Ω	100 Ω

EXAMPLE Read the scale shown in Fig. 5–75.

The needle is on the third subdivision to the left of 2. Each subdivision in that range represents 0.2 Ω.
The scale reads 2.6 Ω.

Thickness gauges come in sets of various thicknesses of metal, with the measurements indicated on the gauge. These sets of gauges are available in different combinations of sizes and numbers and to varying degrees of precision. The gauges can be produced to allow measurements to the nearest two-millionth of an inch (0.000002 in.). Figure 5–76 illustrates two different sets of thickness gauges. These gauges are used in different combinations to determine the appropriate measurement. The sum of the measures of the gauges used represents the measurement of the object.

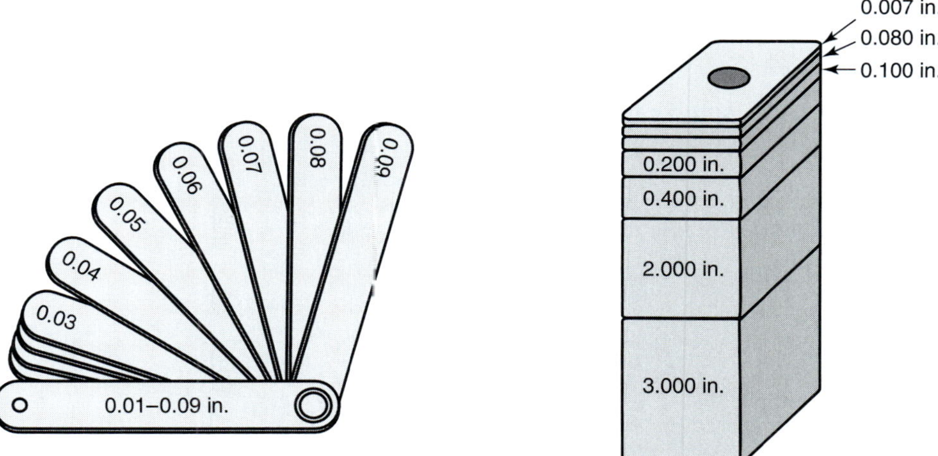

Figure 5-76

Suppose that a set of thickness gauges comes in series of ten-thousandths (0.0001 to 0.0009), thousandths (0.001 to 0.009), hundredths (0.01 to 0.09), and tenths (0.1 to 0.9). A measurement of 0.2493 in. can be made by selecting the 0.2, 0.04, 0.009, and 0.0003 gauges.

To measure an object that is less than 1 in., select the largest tenth-inch gauge that is less than the object. Then combine the tenth-inch gauge with the largest gauge from the hundredth-inch series that is less than the object. Continue with the largest appropriate thousandth-inch and ten-thousandth-inch gauges. To determine the measure of the object, add the measures of each of the gauges used.

EXAMPLE The following gauges were used to measure an object to the nearest hundred-thousandth of an inch: 0.2, 0.04, 0.006, 0.0001, and 0.00008. What is the measure of the object?

To find the measure, add the measures of the gauges used.

$$
\begin{array}{r}
0.2 \\
0.04 \\
0.006 \\
0.0001 \\
\underline{0.00008} \\
\mathbf{0.24618 \text{ in.}}
\end{array}
$$

SELF-STUDY EXERCISES 5–7

1 Determine the percent error of the measuring devices.

1. A 1-m standard measuring tool graduated in millimeters produces a reading of 970 mm. What is the percent error?

2. A standard 16-oz weight is placed on a scale, and the scale reads 15 oz. What is the percent error?

3. A gas pump registers 5.06 gal when the measuring standard of 5.00 gal is used. What is the percent error?

2 Indicate the number of significant digits in the numbers.

4. 583,000
5. 702,500
6. 0.0057
7. 82.07
8. 7.2000
9. 86$\bar{0}$

Find the greatest possible error of the measurements.

10. $7\dfrac{3}{16}$ in.
11. 7.2 mm
12. $5\dfrac{1}{4}$ in.
13. 19 oz
14. 8 L

3 Measure line segments 15–24 in Fig. 5–77 to the nearest sixteenth of an inch (tolerance $= \pm\frac{1}{32}$ in.).

   CHAPTER 5   Direct Measurement

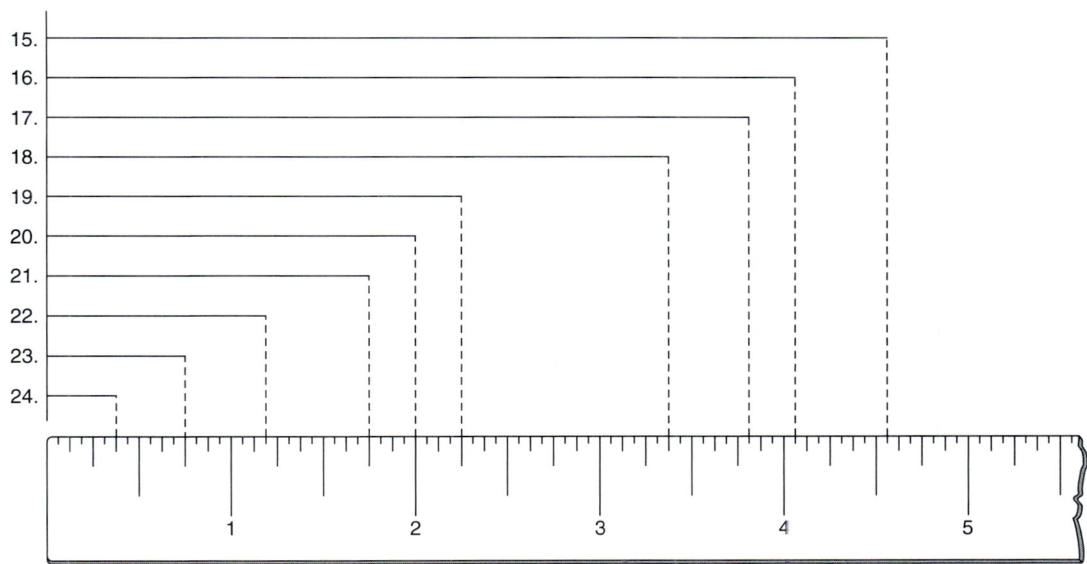

Figure 5-77

Measure line segments 25–28 in Fig. 5–78 to the nearest hundredth of an inch (tolerance = ±0.005 in.).

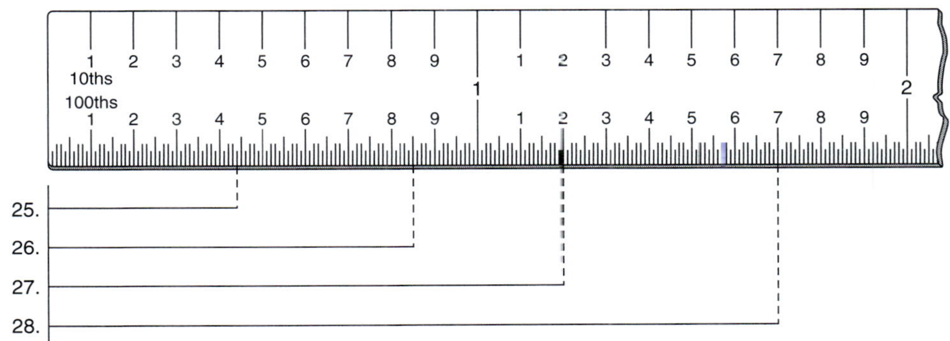

Figure 5-78

4 Measure line segments 29–38 in Fig. 5–79 to the nearest millimeter. Express answers in millimeters or centimeters.

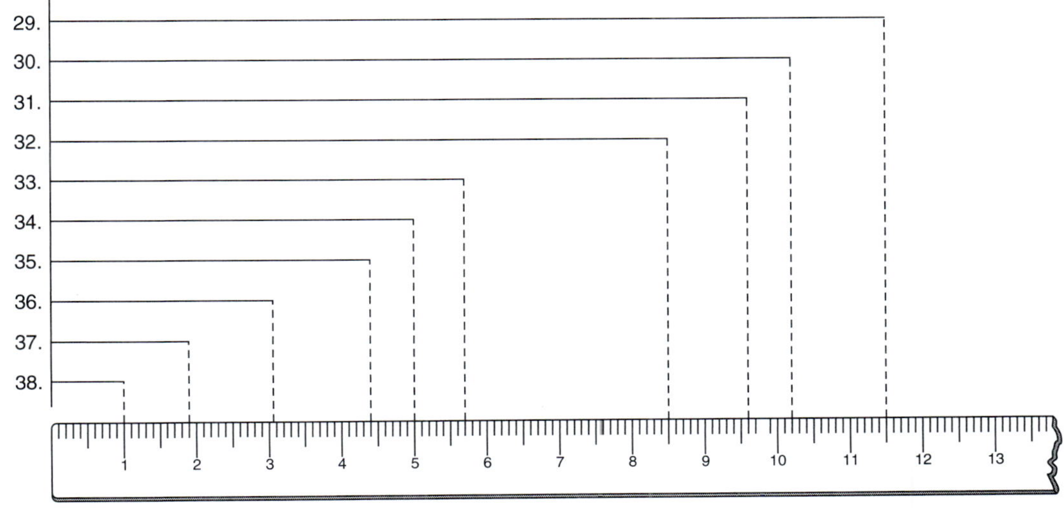

Figure 5-79

 Read the measurement shown on the vernier calipers in Figs. 5–80 to 5–83 in millimeters and in inches.

39.

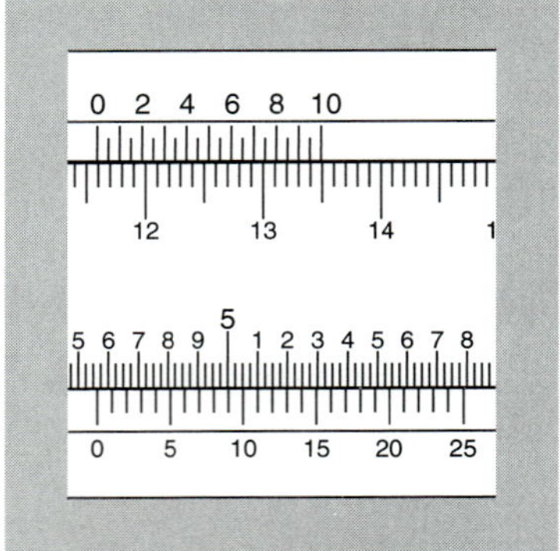

Figure 5–80

40.

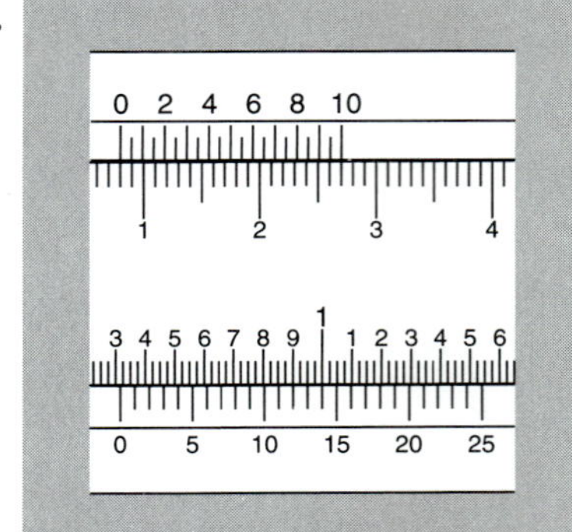

Figure 5–81

41.

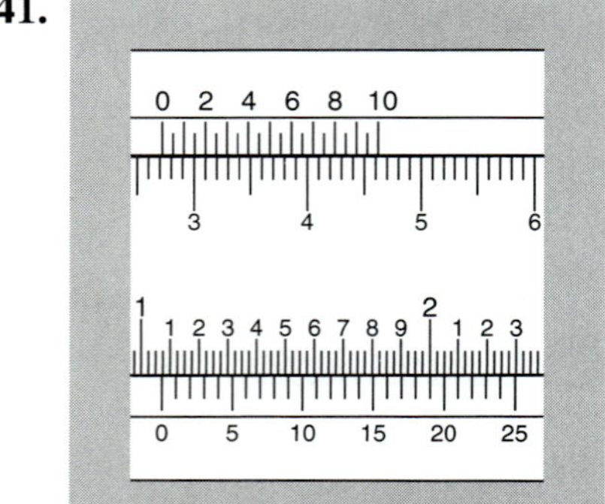

Figure 5–82

42.

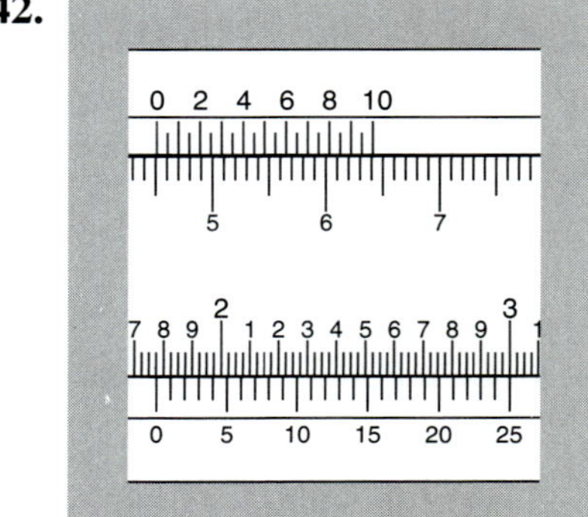

Figure 5–83

 Determine the measurements indicated on the micrometer readings in Figs. 5–84 to 5–88.

43.

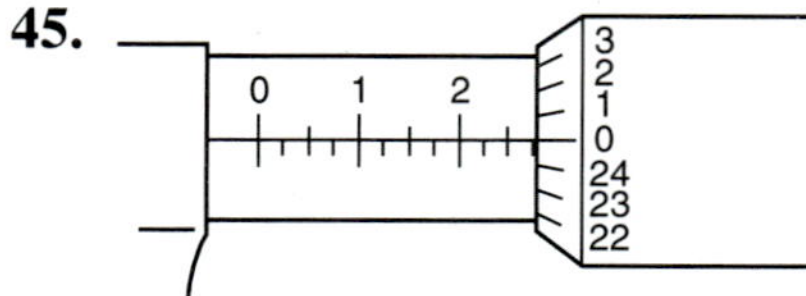

Figure 5–84
Frame size 0–1 in.

44.

Figure 5–85
Frame size 4–5 in.

45.

Figure 5–86
Frame size 2–3 in.

46.

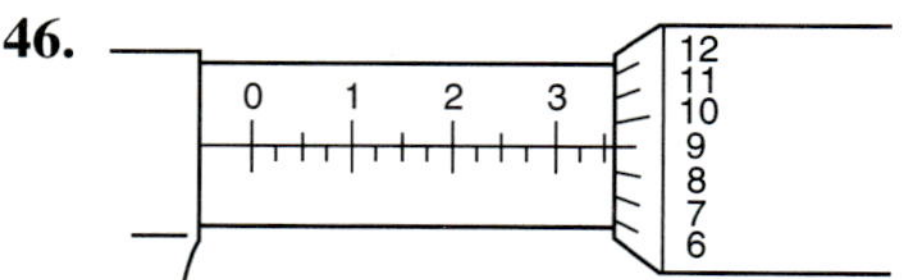

Figure 5–87
Frame size 4–5 in.

47.

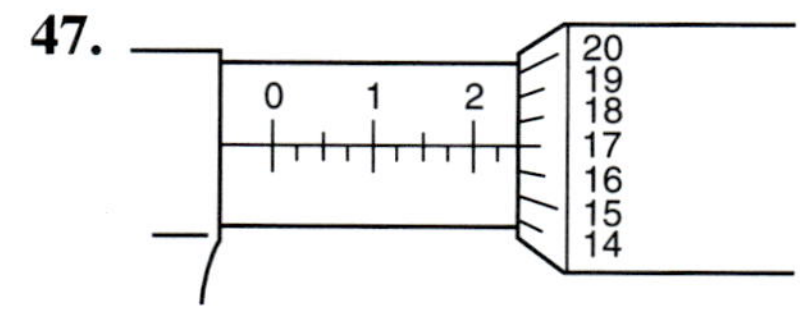

Figure 5–88
Frame size 1–2 in.

Determine the measurements indicated on the vernier micrometers in Figs. 5–89 to 5–93.

48.

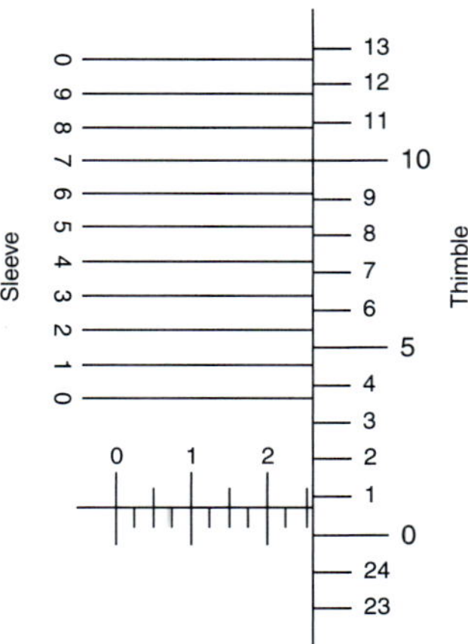

Figure 5–89
Frame size 0–1 in.

49.

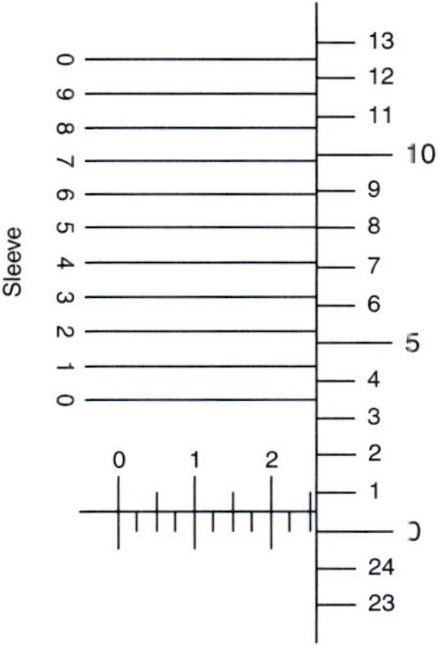

Figure 5–90
Frame size 2–3 in.

50.

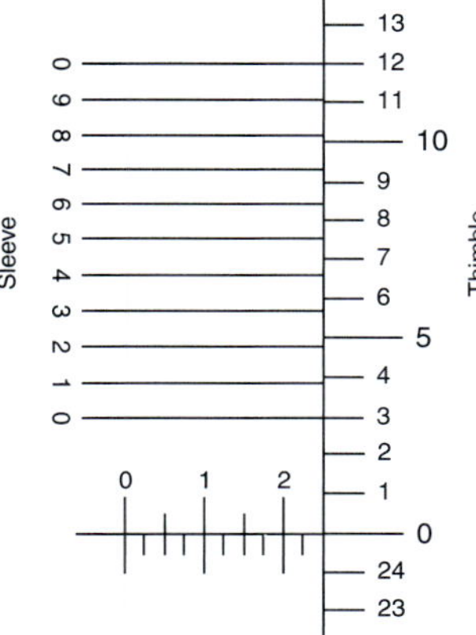

Figure 5–91
Frame size 1–2 in.

51.

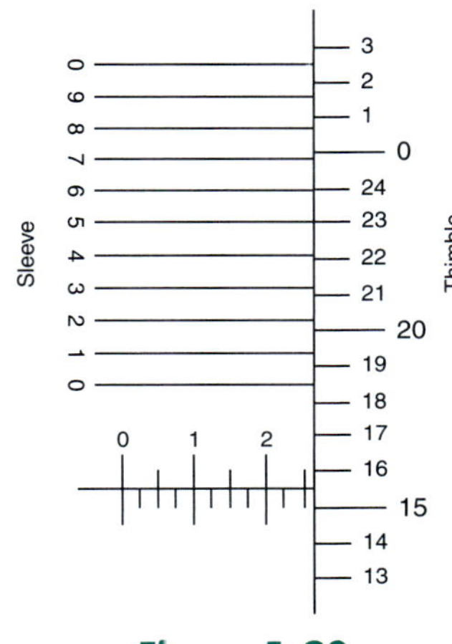

Figure 5–92
Frame size 4–5 in.

52.

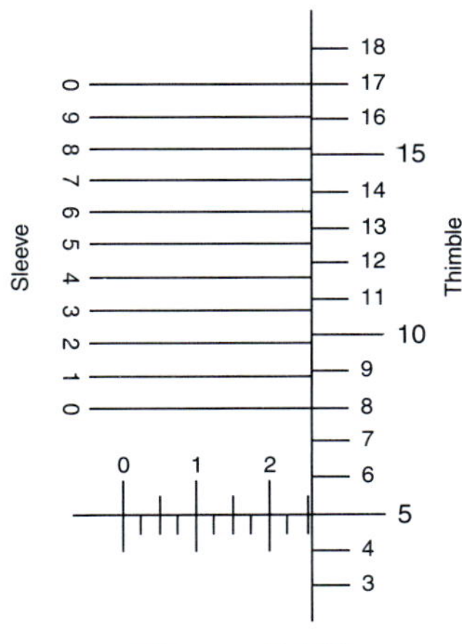

Figure 5–93
Frame size 2–3 in.

 **7**

53. A measurement is observed to be 52.02 cm, against a true measure of 52 cm. Find the absolute error, relative error, and percent error.

54. A blueprint for a machine part specifies a measure of 52.4 mm. An actual part has a measurement of 52.2 mm. Find the absolute error, relative error, and percent error.

8 Write the readings for the scales in Figs. 5–94 to 5–109.

Read each water meter:

55.

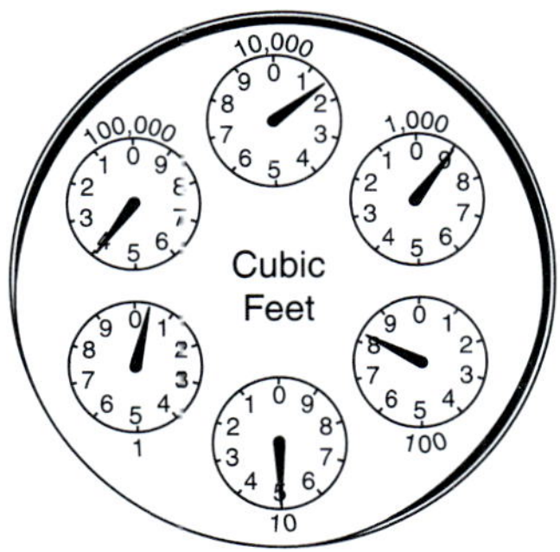

Figure 5–94 December 19

56.

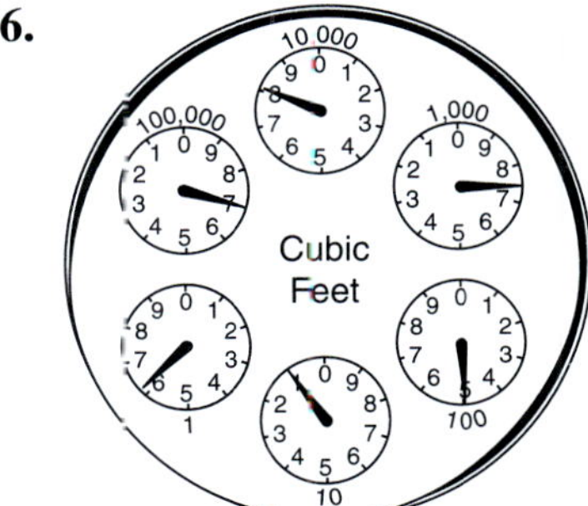

Figure 5–95 August 12

Figure 5–96   June 13

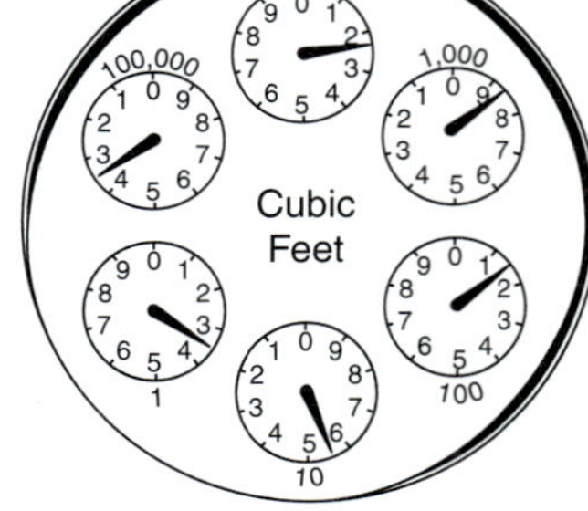

Figure 5–97   October 18

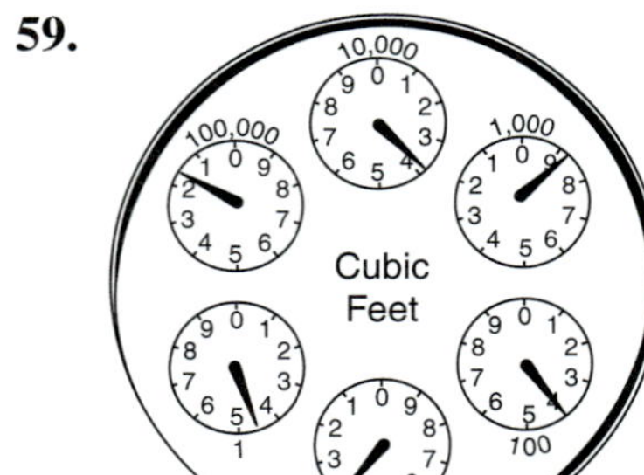

Figure 5–98   November 5

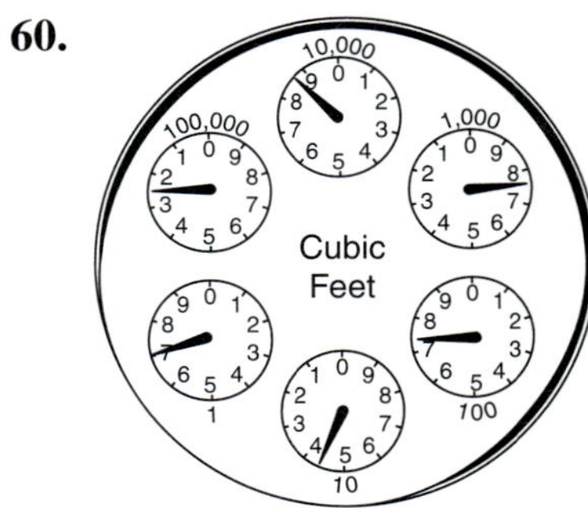

Figure 5–99   January 3

Read each electric meter:

61.

Figure 5–100

62.

Figure 5–101

Read each metric dial.

63.

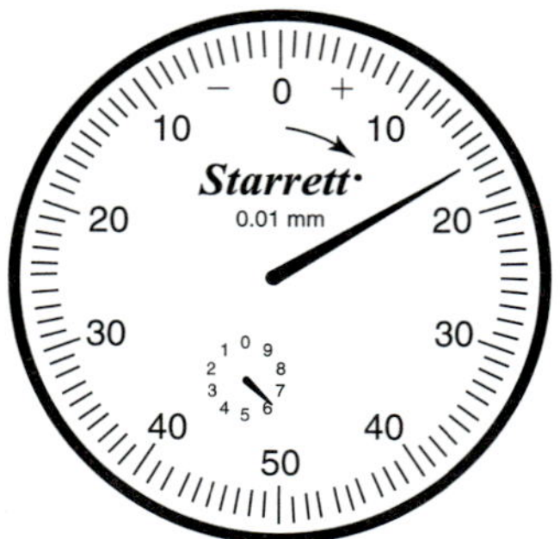

Figure 5–102

64.

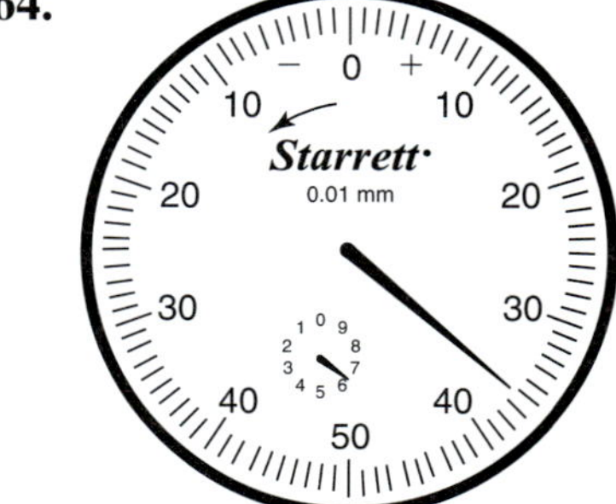

Figure 5–103

65.

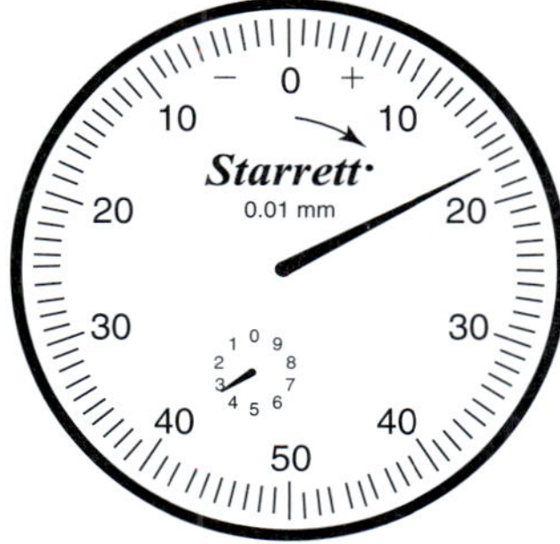

Figure 5–104

66.

Figure 5–105

Read each voltmeter.

67.

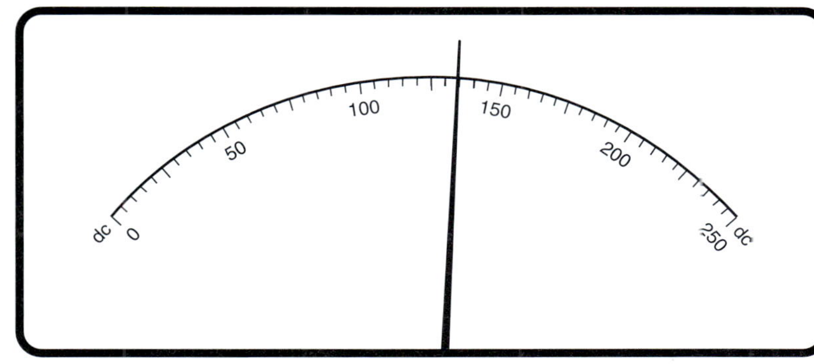

Figure 5–106

68.

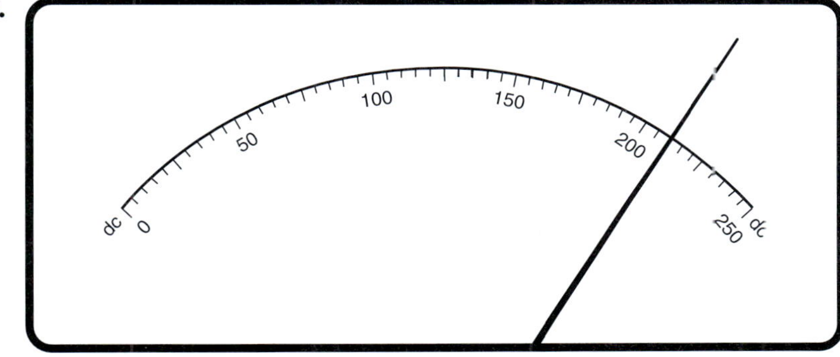

Figure 5–107

Read each ohmmeter.

69.

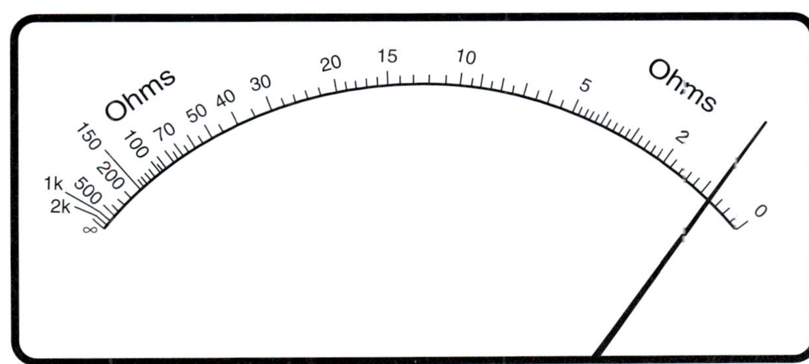

Figure 5–108

70. 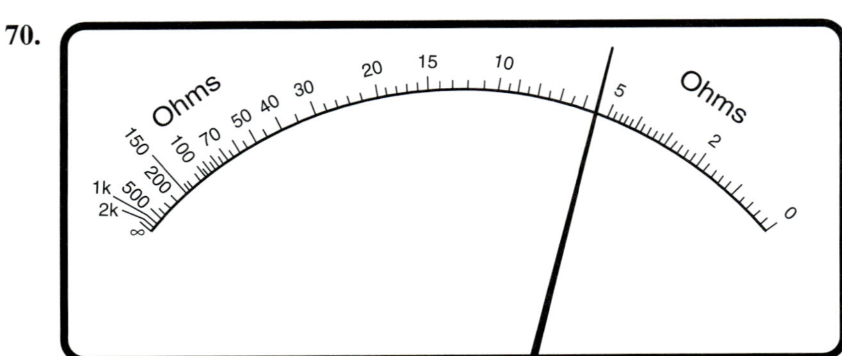

Figure 5–109

71. Find the measure of an object that is flush (even in height) with the following thickness gauges:

0.4 in., 0.003 in., 0.000002 in.

72. What thickness gauges would be needed to make a measure of 1.034506 in.?

CAREER APPLICATION

Nursing: Drug Dosage and IV Concentration

When nurses administer patient drugs, they spend most of their time calculating correct dosage and concentration, both of which depend on the patient's body weight. Even minor errors in dosage or concentration can have drastic (or even life-threatening) consequences, so it is imperative that the dosages be calculated correctly. Drug manufacturers recommend dosages based on milligrams of medicine per kilogram of body weight (mg/kg).

For example, the recommended dosage of *Lasix,* a common drug used in congestive heart failure, is 1 mg of *Lasix* per kg of body weight (1 mg/kg). A patient's weight is typically measured in pounds, so nurses must convert pounds to kilograms by dividing the weight in pounds by 2.2 (1 kg = 2.2 lb). Conversely, to change a patient's weight from kilograms to pounds, nurses multiply the weight in kilograms by 2.2.

Although nurses administer metric doses in hospitals, they usually convert to the U.S. customary system for patient at-home-care instructions. For example, *Children's Tylenol,* a common drug used to treat fever and pain, contains 80 mg of medicine per 5 cubic centimeters (cc) of volume. Knowing that 5 cc equals 1 teaspoon (tsp), nurses calculate how many tsp should be given at home to a patient whose body weight requires a stronger or weaker dose of *Tylenol.*

Some medicines are given through an intravenous (IV) drip. The rate of the IV drip determines the concentration of medicine given. For example, an infusion of *Dobutamine,* a cardiac drug used to improve a patient's heart muscle contractility, at 1.0 cc/hr equals a concentration of 10 micrograms of medicine per kilogram of body weight per minute (10 mcg/kg/min). Nurses then calculate how fast the drip needs to be if the doctor increases the dosage to 14 mcg/kg/min. Doctors often change the IV concentration order, which means the amount of medicine dissolved in the IV liquid is increased (for a higher concentration) or decreased (for a lower concentration).

Exercises

Use the preceding information for the exercises. Round all answers to the nearest tenth.

1. A premature baby boy weighing 4.5 lb needs a dose of *Lasix.* Convert his weight to kg and calculate how many mg of the drug he should receive.

   CHAPTER 5   Direct Measurement

2. Find the dosage of *Lasix* needed for his 3.4-lb twin sister.
3. Calculate the dosage of *Lasix* needed for a 67-year-old man weighing 192 lb.
4. The recommended dosage of *Children's Tylenol* is 10 mg/kg. How many mg of *Tylenol* should be given to an 18-month-old child who weighs 24 lb and has an ear infection?
5. Use your answer from Exercise 4 to tell the parent this child's dosage in teaspoons.
6. If a *Dobutamine* IV drip is flowing at 1.0 cc/hr, which equals 10 mcg/kg/min, how fast would it need to flow if the doctor ordered it increased to 20 mcg/kg/min?
7. For a *Dobutamine* IV drip flowing at 1.0 cc/hr (10 mcg/kg/min), how fast would it need to flow if the doctor ordered it decreased to 8 mcg/kg/min?
8. If a doctor ordered the concentration of a *Dobutamine* IV drip doubled, how would a nurse change the amount of medicine added to a bag of IV liquid?

Answers

1. For a weight of about 2 kg, 2 mg of *Lasix* should be given.
2. For a weight of about 1.5 kg, 1.5 mg of *Lasix* should be given.
3. For about 87.3 kg of weight, 87.3 mg of *Lasix* should be given.
4. For about 10.9 kg of weight, 109.0 mg of *Tylenol* should be given.
5. *Children's Tylenol* comes 80 mg/1 tsp, so about 1.4 or $1\frac{1}{2}$ tsp should be given.
6. The concentration has been doubled, so the IV should drip at twice the rate, or 2.0 cc/hr.
7. The IV should drip at 0.8 cc/hr.
8. The amount of medicine added to the bag of IV liquid should be doubled.

ASSIGNMENT EXERCISES

Section 5–1

Identify the appropriate U.S. customary measure for each item.

1. Package of spaghetti
2. Tank of gasoline
3. Container of motor oil
4. Distance from work to the hospital
5. Package of taco shells
6. Porterhouse steak
7. Shipment of iron
8. Bag of ammonium nitrate
9. Size of an aluminum pot
10. Height of a tree
11. Sugar for a pie recipe
12. Man's shirt size
13. Cloth for a pair of kitchen curtains
14. Hourly speed of an aircraft

Write two unity ratios that relate the given pair of measures.

15. cups and quarts
16. hours and days
17. pounds and tons
18. yards and miles

Using unity ratios, convert the given measures to the new units.

19. 12 ft = _____ yd
20. 11 yd = _____ ft
21. $1\frac{1}{5}$ mi = _____ ft

22. How many feet of wire are needed to put a fence along a property line $2\frac{1}{4}$ mi long?
23. How many ounces are in 5 lb?
24. An object weighing $57\frac{3}{5}$ lb weighs how many ounces?
25. Find the number of pounds in 680 oz.
26. A can of fruit weighs 22.4 oz. How many pounds is this?
27. The net weight of a can of peas is 19 oz. If a case contains 16 cans, what is the net weight of a case in ounces? In pounds?
28. How many quarts are in 8 pt?
29. How many pints are in $7\frac{1}{2}$ qt?
30. Find the number of gallons in 15 qt.
31. Find the number of pints in 3 gal.

32. How many gallons are in 36 pt?

33. How many feet of wire are needed to fence a property line $1\frac{1}{4}$ mi long?

34. A cook has 1 qt of vegetable oil. His recipe requires 2 c of oil. How many recipes can be made from the quart of oil?

Express the measures in standard notation.

35. 6 ft 17 in.

36. 1 mi 5,375 ft

37. 12 lb $17\frac{1}{2}$ oz

38. 2 gal 7 qt

39. 1 gal 2 qt 5 pt

40. 2 T 3,100 lb

41. 3 yd 2 ft 16 in.

42. 1 qt 3 c 12 oz

43. $3\frac{1}{4}$ ft 10 in.

44. 2 lb 21 oz

45. 1 gal 3 qt 48 oz

Section 5–2

Add or subtract. Write answers in standard form.

46. 12 oz + 2 lb

47. 8 ft − 49 in.

48. 4 gal + 3 qt

49.
$$\begin{aligned}&\ \ 2\text{ ft }9\text{ in.}\\ +\ &\ 8\text{ ft }2\text{ in.}\end{aligned}$$

50.
$$\begin{aligned}&\ 7\text{ lb }8\text{ oz}\\ +\ &\ 5\text{ lb }9\text{ oz}\end{aligned}$$

51.
$$\begin{aligned}&\ 5\text{ gal }3\text{ qt}\\ +\ &\ 2\text{ gal }3\text{ qt}\end{aligned}$$

52.
$$\begin{aligned}&\ 7\text{ ft }9\text{ in.}\\ -\ &\ 4\text{ ft }6\text{ in.}\end{aligned}$$

53.
$$\begin{aligned}&\ 4\text{ lb }\ 9\text{ oz}\\ -\ &\ 3\text{ lb }11\text{ oz}\end{aligned}$$

54.
$$\begin{aligned}&\ 4\text{ yd }1\text{ ft }\ 8\text{ in.}\\ -\ &\ 2\text{ yd }2\text{ ft }11\text{ in.}\end{aligned}$$

55. A rug 12 ft 6 in. long must fit in a room whose length is 10 ft 9 in. How much should be trimmed from the rug to make it fit the room?

56. Two packages to be sent air express weigh 5 lb 4 oz each. What is the shipping weight of the two packages?

57. A water hose purchased for an RV was 2 ft long. What was its length after 7 in. was cut off?

58. A vinyl flooring installer cut 19 in. from a piece of vinyl 13 ft long. How long was the vinyl piece after it was cut?

Section 5–3

Multiply and write answers for mixed measures in standard form.

59.
$$\begin{aligned}42\text{ ft}\\ \underline{12\text{ ft}}\end{aligned}$$

60.
$$\begin{aligned}8\text{ lb }3\text{ oz}\\ \underline{\qquad 9}\end{aligned}$$

61.
$$\begin{aligned}9\text{ in.}\\ \underline{7\text{ in.}}\end{aligned}$$

62.
$$\begin{aligned}10\text{ gal }3\text{ qt}\\ \underline{\qquad 7}\end{aligned}$$

Divide.

63. 20 yd 2 ft 6 in. ÷ 2

64. 5 gal 3 qt 2 pt ÷ 6

65. 65 ft ÷ 12 (Write answer in feet.)

66. 21 ft ÷ 4 (Write answer in feet and inches.)

67. If 18 lb of candy is divided equally into four boxes, express the weight of the contents of each box in pounds and ounces.

68. If 32 equal lengths of pipe are needed for a job, and each length is to be 2 ft 8 in., how many feet of pipe are needed for the job?

Work the problems.

69. 14 ft ÷ 4 ft

70. 2 mi 120 ft ÷ 15 ft

71. 400 lb ÷ 90 lb

72. 6 yd 2 ft ÷ 5 ft

73. $5\dfrac{\text{mi}}{\text{min}} = \underline{\qquad}\dfrac{\text{mi}}{\text{hr}}$

74. $2{,}520\dfrac{\text{gal}}{\text{hr}} = \underline{\qquad}\dfrac{\text{qt}}{\text{hr}}$

75. $88\dfrac{\text{ft}}{\text{sec}} = \underline{\qquad}\dfrac{\text{mi}}{\text{hr}}$

76. $18\dfrac{\text{mi}}{\text{gal}} = \underline{\qquad}\dfrac{\text{ft}}{\text{gal}}$

77. A pump that moves water at the rate of $75\frac{\text{gal}}{\text{hr}}$ can move how many gallons per minute?

78. A plane that travels at the rate of 240 mph is traveling how many feet per second?

79. How many quarts of milk are needed for a recipe that calls for 3 pt of milk?

80. How many $\frac{1}{2}$-oz servings of jelly can be made from a $1\frac{1}{2}$-lb container of jelly?

Section 5–4

Give the prefix that relates each number to the standard unit in the metric system.

81. 1,000 times

82. $\dfrac{1}{10}$ of

83. $\dfrac{1}{1{,}000}$ of

84. 10 times

85. $\dfrac{1}{100}$ of

86. 100 times

Give the value of the prefixes based on a standard measuring unit.

87. dekameter (dkm) **88.** hectogram (hg) **89.** milligram (mg)
90. centigram (cg) **91.** kiloliter (kL) **92.** deciliter (dL)

Choose the most reasonable answer.

93. Height of the Washington Monument
   (a) 200 m **(b)** 200 cm
   (c) 200 mm **(d)** 200 km

94. Height of Mount Rushmore
   (a) 1.6 km **(b)** 1.6 m
   (c) 1.6 cm **(d)** 1.6 mm

95. Weight of an egg
   (a) 50 g **(b)** 50 kg
   (c) 50 mg

96. Weight of a saccharin tablet
   (a) 50 kg **(b)** 50 mg
   (c) 50 g

97. Weight of a man's shoe
   (a) 0.25 g **(b)** 0.25 mg
   (c) 0.25 kg

98. Carton of milk
   (a) 4 L **(b)** 4 mL

99. Bottle of medicine
   (a) 50 L **(b)** 50 mL

Change to the unit indicated.

100. 0.4 dkm = _____ hm
101. 67.1 m = _____ dkm
102. 4 m = _____ dm
103. 2.3 m = _____ mm
104. 5 cm = _____ mm
105. 0.123 hm = _____ mm
106. How many millimeters are in 0.432 km?
107. 23 dkm = _____ mm
108. 42.7 cm = _____ dkm
109. 41,327 dkm = _____ km
110. A board is 1.82 m long. How many centimeters long is the board?
111. 394.5 g = _____ hg
112. 2.7 hg = _____ dg
113. 3,000,974 cg = _____ kg

Perform the operations indicated.

114. 25 mm − 14 mm
115. 12 g + 5 m
116. 17 mg − 8 mL
117. 8 g − 52 cg
118. 43 dkg × 7
119. 6.83 cg × 9
120. $\dfrac{18 \text{ cm}}{9}$
121. 7.5 kg ÷ 0.5 kg
122. $\dfrac{8 \text{ hL}}{20 \text{ L}}$
123. 34 hL ÷ 4
124. 2.4 m ÷ 5 cm

125. Fabric must be purchased to make seven garments, each requiring 2.7 m of fabric. How much fabric must be purchased?

126. Candy weighing 526 g is mixed with candy weighing 342 g. What is the weight of the mixture?

127. A recipe calls for 5 mL of vanilla flavoring and 24 cL of milk. How much liquid is this?

128. Twenty boxes, each weighing 42 kg, are to be moved. How much weight must be moved?

129. A metal rod 42 m long is cut into seven equal pieces. How long is each piece?

130. Thirty-two kilograms of a chemical is distributed equally among 16 chemistry students. How many kilograms of chemical does each student receive?

131. A serving of punch is 25 cL. How many servings can be obtained from 25 L of punch?

132. How many containers of jelly can be made from 8,548 L of jelly if each container holds 4 dL of jelly?

133. A bolt of fabric contains 6.8 dkm. If a shirt requires 1.7 m, how many shirts can be made from the bolt?

Section 5–5

Make the conversions.

134. 7 m = _____ inches
135. 215 m = _____ yards
136. 69 km = _____ miles
137. 15 L = _____ liquid quarts
138. 12 dry qt = _____ liters
139. 32 kg = _____ pounds
140. 10 lb = _____ kilograms
141. 9 in. = _____ centimeters
142. 21 ft = _____ meters
143. 14.8 dkL = _____ quarts
144. $3\frac{1}{2}$ gal = _____ liters

145. How many meters long is 200 ft of pipe?

146. Concrete weighing 90 lb weighs how many kilograms?

147. Two cities 175 mi apart are how many kilometers apart?

148. A room 10 m wide is how many feet wide?

Section 5–6

Use ratios or conversion factors to convert the measures of time.

149. How many days are in 72 hr?

150. How many minutes are in 2.4 hr?

151. Convert 158 min to hr.

152. A doctor ordered a surgery patient not to drive for 3 weeks. How many days is this?

153. A heart surgery patient was held in intensive care for 96 hr. How many days is this?

154. A bone marrow patient remained hospitalized for 72 days. How many weeks is this?

155. If you have worked for your employer for 39 months, how many years have you worked?

156. There are 96 days remaining in a year. How many weeks remain?

Section 5–7

157. A gas pump registers 4.95 gal when the measuring standard of 5.00 gal is used. What is the percent error?

158. A pharmacy balance reads 2.03 g when a standard 2.00-g weight is used in calibration. What is the percent error?

159. How many significant digits are there in 0.5010?

160. How many significant digits are there in 203.07?

161. What is the greatest possible error of the measurement 4.7 cg?

162. What is the greatest possible error of the measurement $2\frac{5}{32}$ in.?

Measure line segments 163–172 in Fig. 5–110 (tolerance $= \pm\frac{1}{32}$ in.).

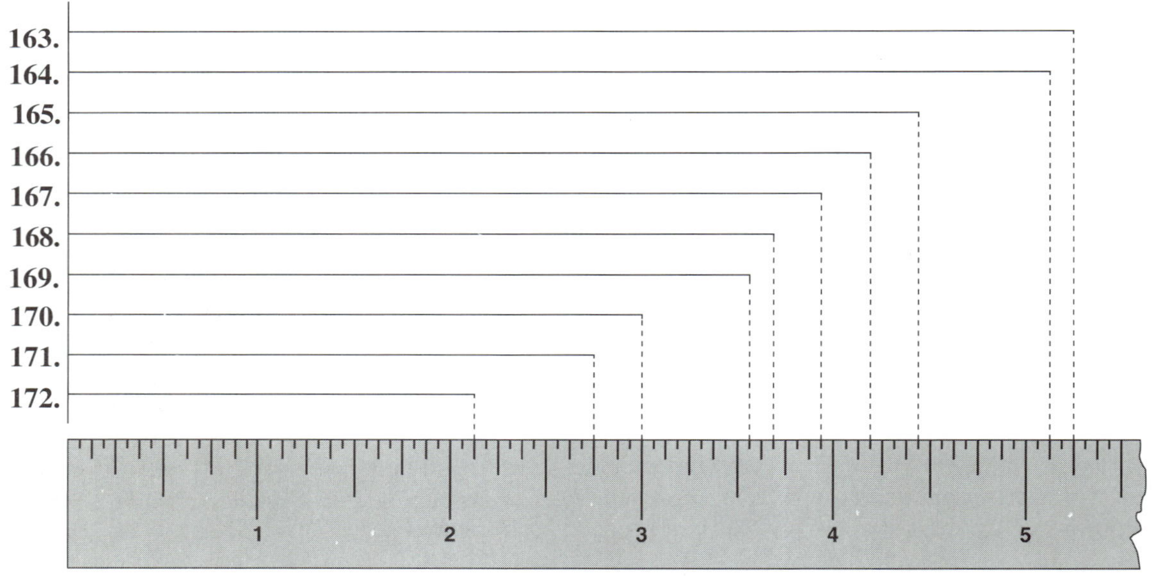

Figure 5–110

Measure line segments 173–182 in Fig. 5–111 to the nearest millimeter.

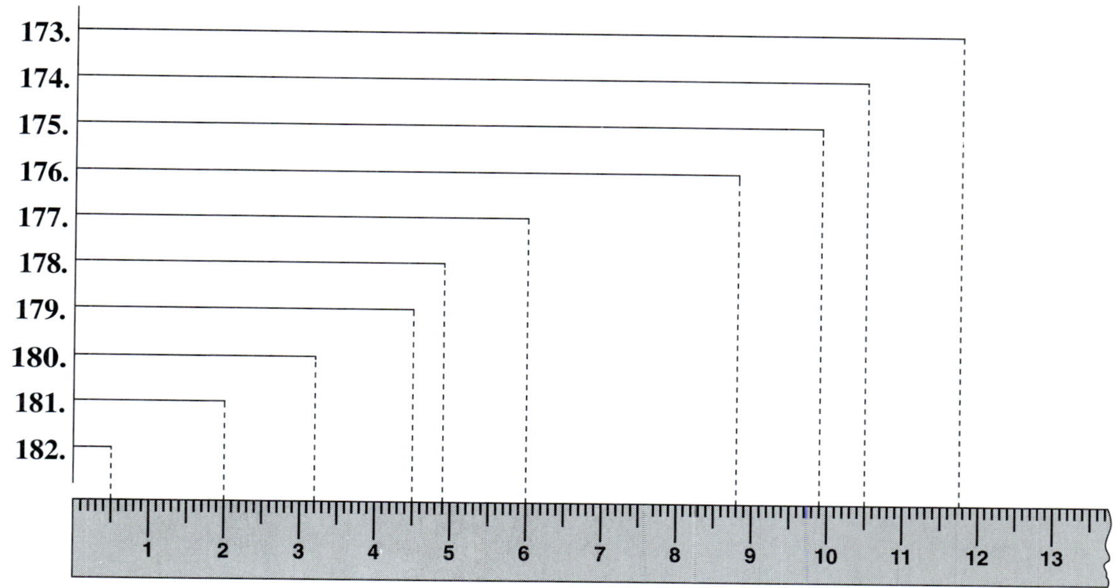

Figure 5–111

Read the measurements shown on metric and U.S. customary scales on the vernier calipers in Figs. 5–112 and 5–113.

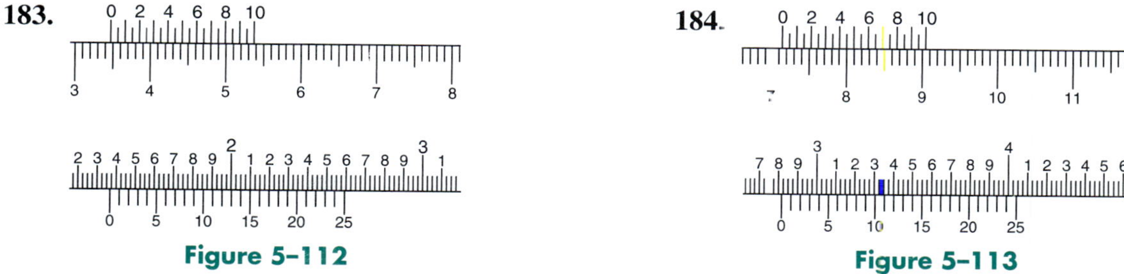

183.

184.

Figure 5–112

Figure 5–113

Determine the measurement indicated on the micrometer readings in Figs. 5–114 to 5–116.

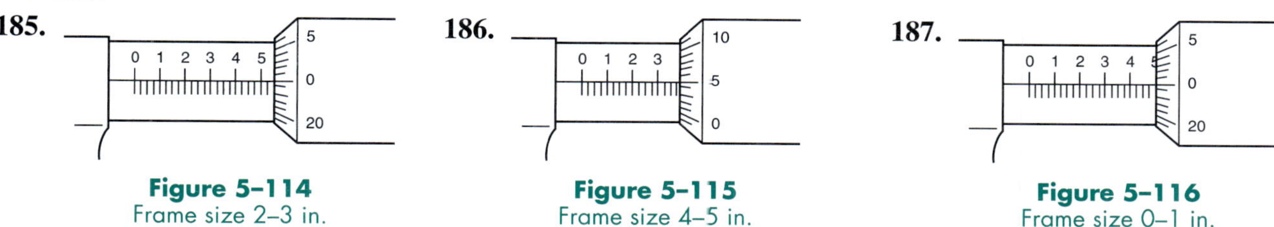

185.

186.

187.

Figure 5–114
Frame size 2–3 in.

Figure 5–115
Frame size 4–5 in.

Figure 5–116
Frame size 0–1 in.

Determine the measurements indicated on the vernier micrometer readings in Figs. 5–117 to 5–119.

188.

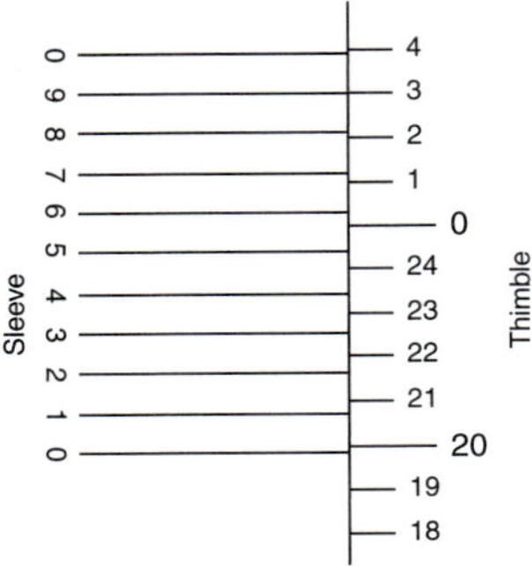

189.

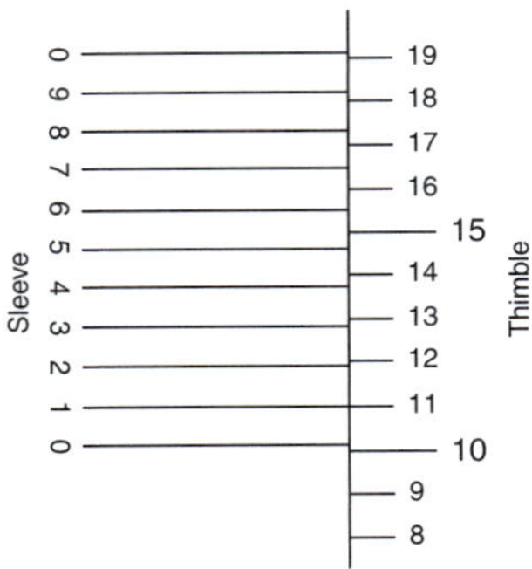

190.

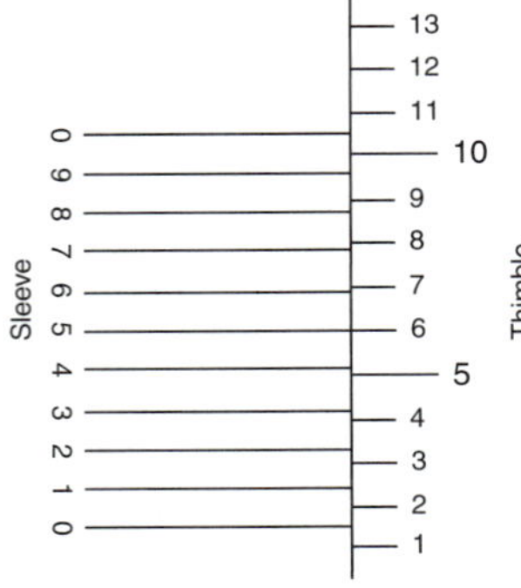

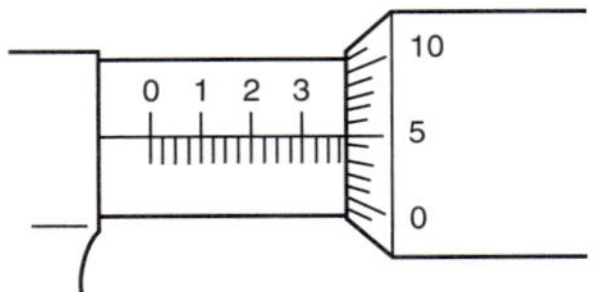

Figure 5–117

Figure 5–118

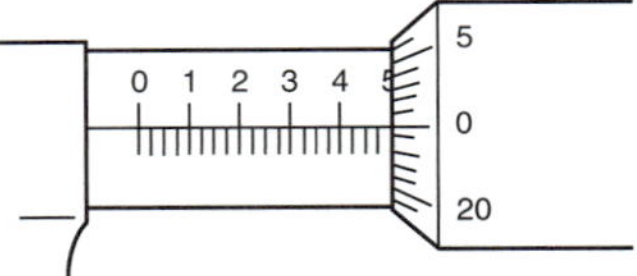

Figure 5–119

191. A blueprint reading indicates the thickness of a casting to be 12.40 cm. The actual part is measured and found to be 11.92 cm. Find the absolute error, relative error, and percent error.

192. The true measure of a machine tool is 18.0 in., but its measurement is recorded as 18.09 in. Find the absolute error, relative error, and percent error.

193. Read the water meter in Fig. 5–120.

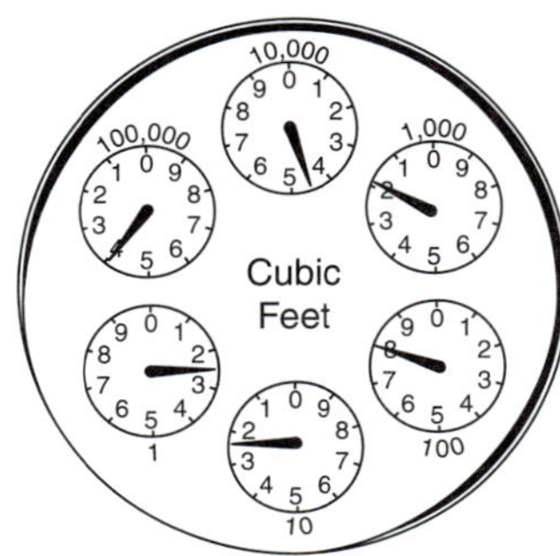

Figure 5–120

Read the electric meters in Figs. 5–121 and 5–122.

194.

Figure 5–121

195.

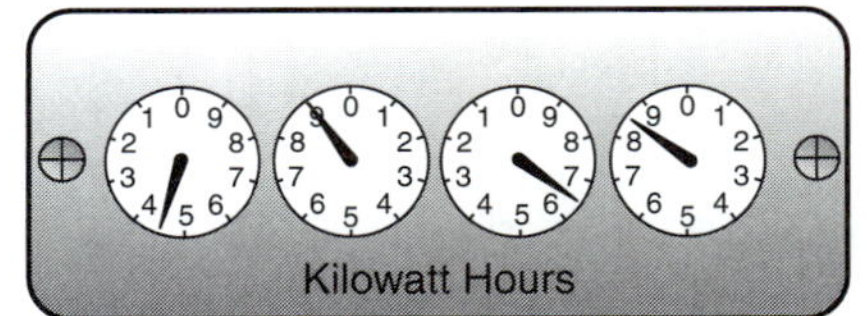

Figure 5–122

Read the metric dials in Figs. 5–123 and 5–124 (the arrow near the zero indicates the initial deflection of the needle).

196.

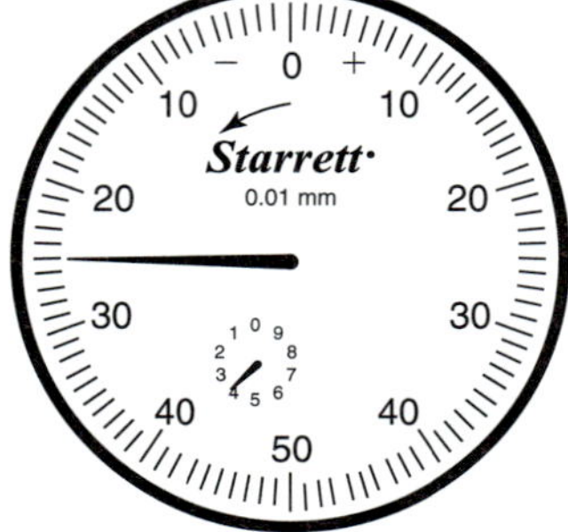

Figure 5–123

197.

Figure 5–124

CHAPTER 5 Direct Measurement

Read the voltmeters in Figs. 5–125 and 5–126.

198.

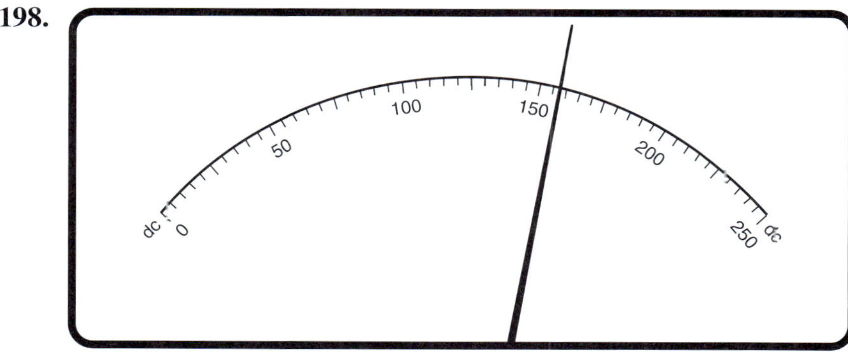

Figure 5–125

199.

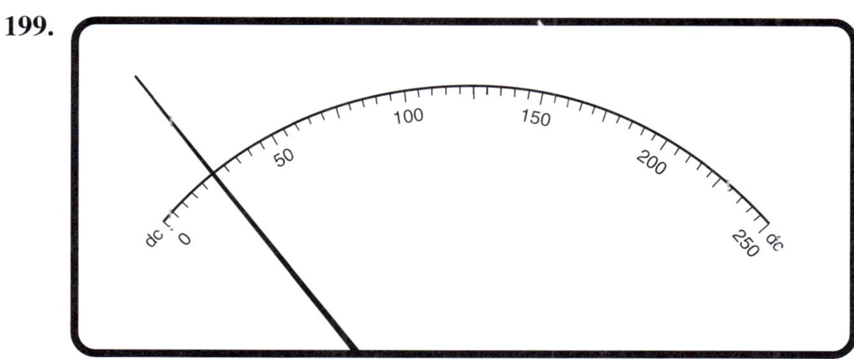

Figure 5–126

Read the ohmmeters in Figs. 5–127 and 5–128 (in ohms Ω).

200.

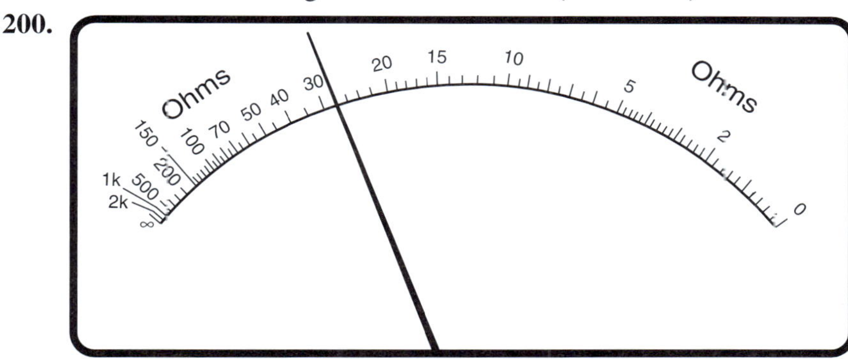

Figure 5–127

201.

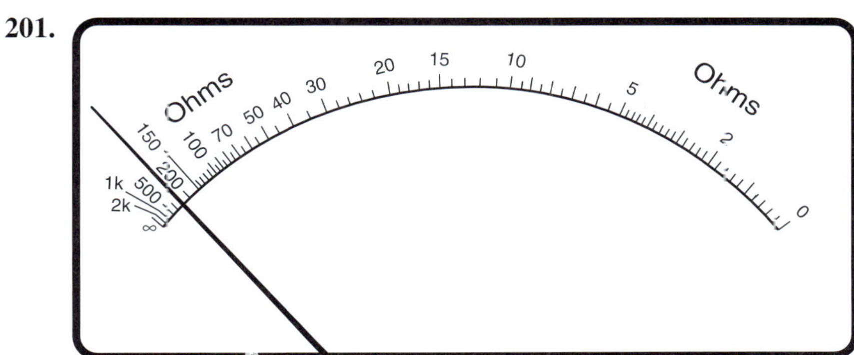

Figure 5–128

Change to the measure indicated (see text for conversion factors).

1. 3 ft = _____ in.

2. 36 oz = _____ lb

3. 32 qt = _____ gal

4. Add 2 yd 6 ft 10 in. + 3 ft 7 in. Write your answer in standard notation.

5. $60\dfrac{\text{gal}}{\text{min}} = \underline{\qquad}\dfrac{\text{qt}}{\text{sec}}$

6. 21 in. ÷ 3 in.

7. A 495-ft section of highway is to be resurfaced. How many yards is this?

8. How many quarts are contained in a 55-gal drum?

9. If an automobile travels at 55 mph, how many feet does it travel per second?

10. A spark plug wire kit contains $18\frac{1}{2}$ ft of wire in a coil. The directions call for cutting off individual lengths of 28 in. each. How many 28-in. wires can be cut from the coil?

11. Find the measure of the line segment AB in Fig. 5–129 (tolerance = $\pm\frac{1}{32}$ in.).

12. Select the appropriate U.S. customary measure for the contents of a swimming pool.

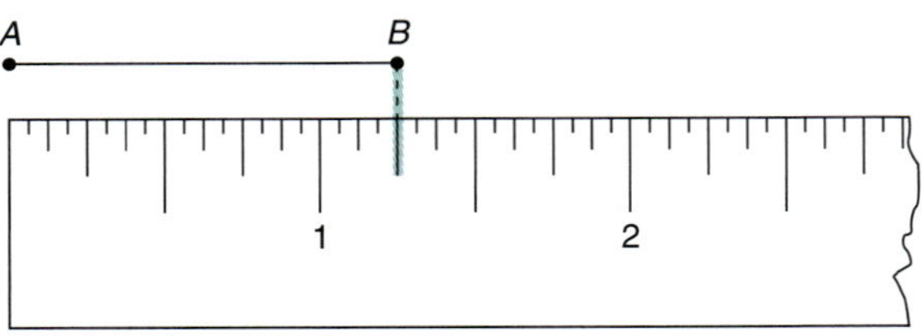

Figure 5–129

13. Write 1 gal 3 qt 6 c 20 oz in standard notation.

14. Which is the appropriate metric measure for eye drops: 28 mL, 28 dL, 28 L, or 28 kL?

15. If Lisbon, Portugal, is 0 GMT and Bern, Switzerland, is +1 GMT, how many hour's time difference is there between the two cities?

16. Grayson Lee traveled to Wellington, New Zealand, to do research on geological formations. He arrived at 3:00 P.M. local time (NZST) and needed to contact his office located in Los Alamos, New Mexico (MST). What time was it in Los Alamos?

Give the metric prefix.

17. $\dfrac{1}{10}$ of standard unit

18. 10 × standard unit

Change to the metric unit indicated.

19. 298 m = _____ km

20. 8 dm = _____ mm

21. 5.2 dL of liquid is poured from a container holding 10 L. How many liters of liquid remain in the container?

22. A bar of soap weighs 175 g. How much do 15 bars of soap weigh (in kilograms)?

23. 75 mi = _____ km

24. 25 kg = _____ lb

25. 4 L = _____ pt

26. How many liters of weed killer are contained in a 55-gal drum?

27. A .243 caliber bullet travels $2{,}450\frac{\text{ft}}{\text{sec}}$ for the first 200 yd. Using conversion factors, change $2{,}450\frac{\text{ft}}{\text{sec}}$ to $\frac{\text{m}}{\text{sec}}$.

28. Find the relative error and percent error to the nearest hundredth percent for a measurement of 15.60 cm for a blueprint specification of 15.65 cm.

29. Determine the measurement indicated on the micrometer in Fig. 5–130.

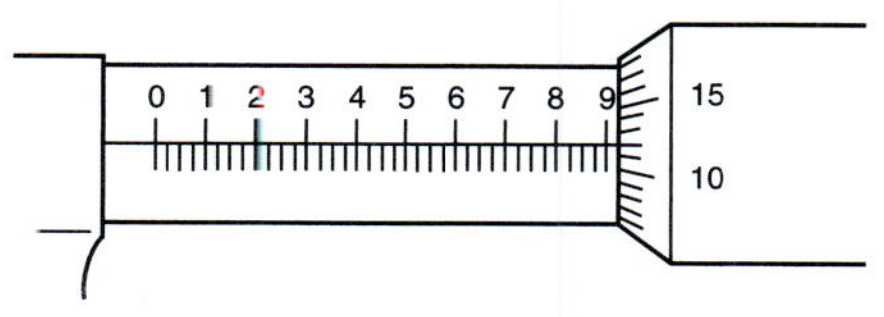

Figure 5–130
Frame size 2–3 in.

30. Read the electric meter in Fig. 5–131.

Figure 5–131

31. Read the U.S. customary dial in Fig. 5–132.

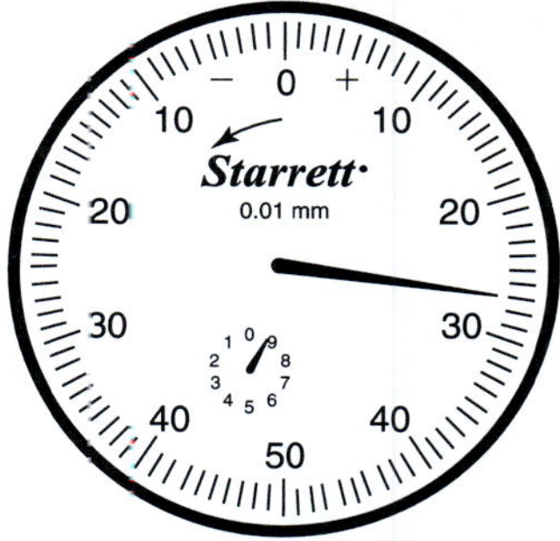

Figure 5–132

32. Read the voltmeter in Fig. 5–133.

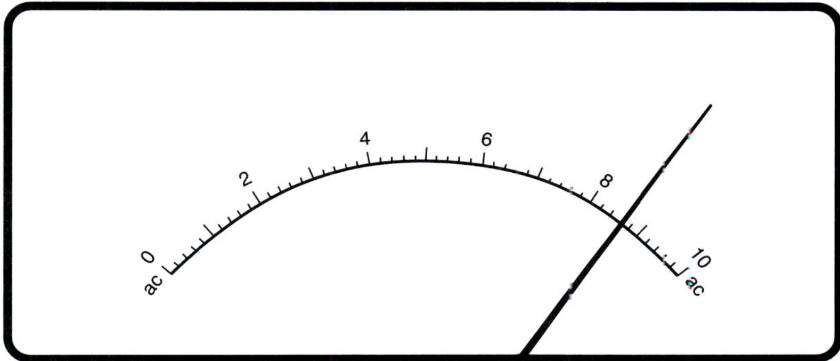

Figure 5–133

33. Discuss the similarities and differences between a uniform scale and a nonuniform scale. Give an example of each.

Area and Perimeter

Good Decisions Through Teamwork

You and your team are designing options for placing a single-family home on a lot your customer has purchased. The subdivision covenants where the lot is located require that permanent structures be located no closer than 25 feet from the front curb and 10 feet from the side and back property lines. The minimum living space must be 2,800 square feet, with at least 1,800 square feet on the ground level. The lot measures $103\frac{1}{2}$ ft wide $\times$ $121\frac{3}{4}$ ft deep.

Make three different scale drawings of the lot and a house, as viewed from above, that meet the subdivision covenants. Prepare a written document for your customer that presents the strengths and weaknesses of each option.

Pair with another team and have each team alternately play the role of the customer. In an open discussion, the customer team will decide which, if any, of the options is suitable for its needs. If no option presented meets the needs of the customer, construct a new scale drawing that will meet the needs of the customer, or show the customer why his or her needs cannot be accommodated within the requirements of the subdivision covenants.

6–1 Squares, rectangles, and parallelograms

1. Find the perimeter and area of a square.
2. Find the perimeter and area of a rectangle.
3. Find the perimeter and area of a parallelogram.

6–2 Area and circumference of a circle

1. Find the circumference of a circle.
2. Find the area of a circle.
3. Find the perimeter and area of composite figures.

In everyday life and in almost every occupation, we sometimes need to figure how much area a straight-sided, flat figure called a polygon occupies or what is the combined length of its sides. Most of us work with such geometric figures and their measurements at one time or another. Familiarity with polygons and their formulas saves both time and money on the job and at home.

6–1   SQUARES, RECTANGLES, AND PARALLELOGRAMS

Learning Outcomes

1. Find the perimeter and area of a square.
2. Find the perimeter and area of a rectangle.
3. Find the perimeter and area of a parallelogram.

A *polygon* is a plane or flat closed figure described by straight line segments and angles. Among the most familiar and easy-to-work-with polygons are the *parallelogram*, the *rectangle*, and the *square.*

■ **DEFINITION: Parallelogram.** A *parallelogram* is a four-sided polygon whose opposite sides are parallel (Fig. 6–1).

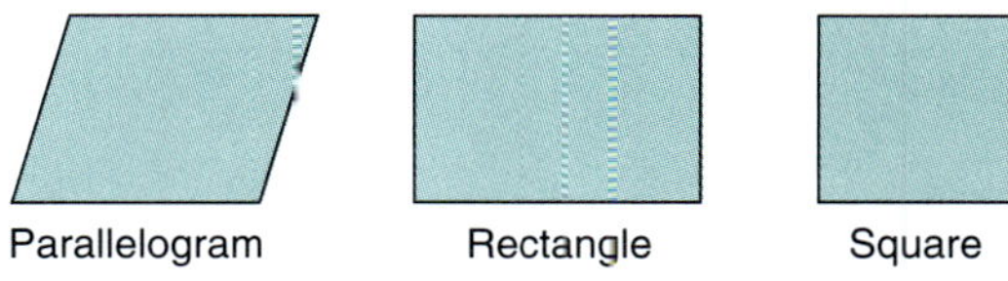

Figure 6–1

■ **DEFINITION: Rectangle.** A *rectangle* is a parallelogram whose angles are all right angles (Fig. 6–1).

■ **DEFINITION: Square.** A *square* is a parallelogram whose sides are all of equal length and whose angles are all right angles (Fig. 6–1).

1   Find the Perimeter and Area of a Square.

Two of the most useful concepts associated with polygons are *area* and *perimeter.* It is important to distinguish between these concepts.

■ **DEFINITION: Perimeter.** The *perimeter* is the total length of the sides of a plane figure (Fig. 6–2).

■ **DEFINITION: Area.** The *area* is the amount of surface of a plane figure (Fig. 6–3).

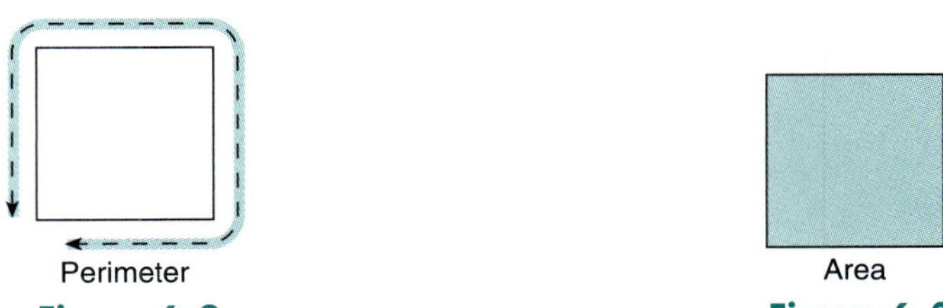

Figure 6–2   **Figure 6–3**

When do you need to find the perimeter of a polygon and when do you need to find the area? If a carpenter is installing baseboard molding in a newly constructed den or family room, the carpenter needs to know how many feet of baseboard

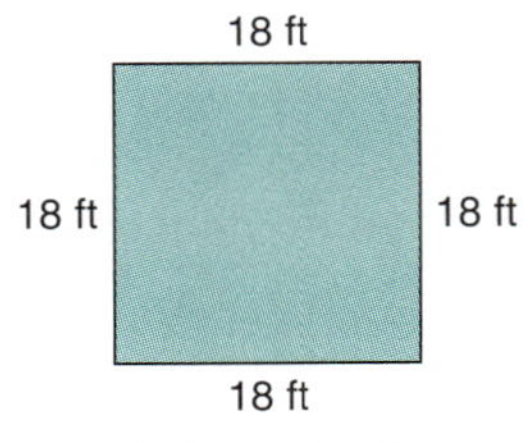

Figure 6-4

molding are needed. The carpenter also needs to know the distance around the room, that is, the perimeter of the room. According to the plans, the room is square and measures 18 ft on each side (Fig. 6–4).

The perimeter, or distance around the room, is figured by adding the lengths of each wall: 18 ft + 18 ft + 18 ft + 18 ft = 72 ft. All sides are equal and there are four sides, so we multiply 4×18.

> ### Formula for perimeter of a square:
>
> $$P_{square} = 4s$$
>
> where s is the length of one side.

An expression like $4s$ means 4 times the value represented by s.

Tip!	*Using Subscripts.*

Subscripted words, letters, or numbers are a handy way to provide additional information. Because we will examine the formulas for the perimeter and area of several different shapes, we sometimes use subscripts to distinguish among them. For instance, the perimeter of a square, rectangle, or parallelogram can be indicated as P_{square}, $P_{rectangle}$, or $P_{parallelogram}$, respectively.

EXAMPLE How much aluminum edge molding is needed to surround a stainless steel kitchen sink that measures 40 cm on each side (Fig. 6–5)?

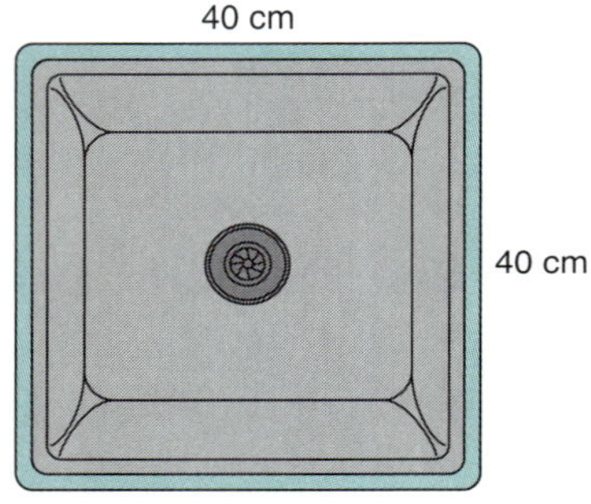

Figure 6-5

$P_{square} = 4s$ The figure is a square.

$P_{square} = 4(40 \text{ cm})$ Substitute measurement of one side. Multiply.

$P_{square} = 160 \text{ cm}$

The sink requires 160 cm of molding.

In contrast, the area of a square is the entire amount of surface that the square occupies.

Let's say we have a square that measures 6 in. on a side (Fig. 6–6). Each small square in the larger figure is a *square inch*, that is, a square measuring 1 in. on each side. If we count the number of square inches in the larger figure, we get 36. This is the area of the square in question. The area is expressed as 36 sq in. or 36 in.^{2} Instead of counting each square inch in the larger square, we could have arrived at 36 sq in. or 36 in.^{2} by multiplying one side by another (6 in. $\times$ 6 in. = 36 in.^{2}). From this equivalency we get the formula for the area of a square.

Figure 6-6

CHAPTER 6   Area and Perimeter

<table>
<tr><td>Tip!</td><td>Writing Square Units.</td></tr>
</table>

The exponent 2 following a unit of measure indicates *square measure* or area. Thus, $5 \text{ in.}^2 = 5$ square inches, $23 \text{ ft}^2 = 23$ square feet, $120 \text{ cm}^2 = 120$ square centimeters, and so on. This is a shortcut way to express a measure that has *already* been "squared."

Formula for area of a square:

$$A_{\text{square}} = s^2$$

where s is the length of one side.

The expression s^2 means $s \times s$, just as 5^2 means 5×5.

EXAMPLE Lynn Fly, a vinyl floor installer, discovers that one of the squares in the flooring pattern is damaged. He decides to cut out the damaged square and replace it. The damaged square measures 20 cm on each side. What is the area of the square to be replaced?

$A_{\text{square}} = s^2$

$A_{\text{square}} = (20 \text{ cm})^2$ Substitute measurement of one side and square it.

$A_{\text{square}} = 400 \text{ cm}^2$

The vinyl square measures 400 cm^2.

<table>
<tr><td>Tip!</td><td>Dimension Analysis:</td></tr>
</table>

When formulas are evaluated, the measuring units are sometimes omitted from the written steps. However, it is very important to use the correct unit in your calculations so that the unit in the solution is correct. Look at the preceding two examples.

Example (molding for sink)

$P_{\text{square}} = 4s$

$P_{\text{square}} = 4(40)$ **The measuring unit is centimeters; the measure is multiplied by a number.**

$P_{\text{square}} = 160 \text{ cm}$ **The measuring unit in the solution is centimeters.**

Perimeter is always a linear measure, which means the measuring unit should be to the first power.

Example (vinyl flooring)

$A_{\text{square}} = s^2$

$A_{\text{square}} = (20)^2$ **The measuring unit is centimeters; a measure is squared or a measure is multiplied by a measure.**

$A_{\text{square}} = 400 \text{ cm}^2$ **The measuring unit in the solution is square centimeters.**

Area is always a square measure, which means that measures of area are to the second power or have an exponent of 2.

2   Find the Perimeter and Area of a Rectangle.

Finding the perimeter of a rectangle is similar to finding the perimeter of a square. We add the measures of the four sides; the longer sides are the *length* (*l*) and the shorter sides are the *width* (*w*), as shown in Fig. 6–7.

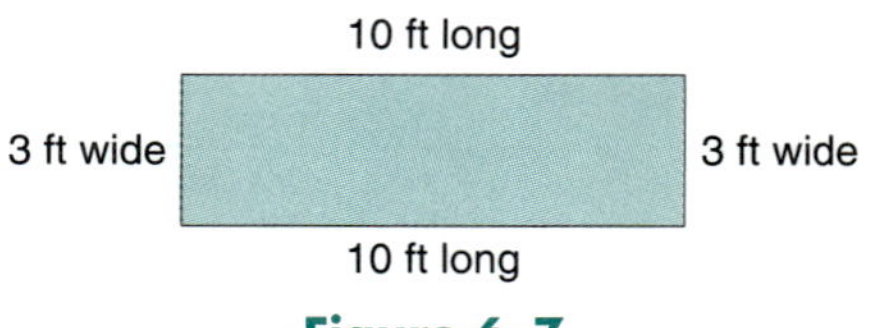

Figure 6–7

$$P = 10 + 3 + 10 + 3 = 26 \text{ ft}$$

Notice that the perimeter is composed of two lengths and two widths:

$$P = 2l + 2w$$

The distributive property shows that $2(l + w)$ and $2l + 2w$ are equivalent.

> **Formula for perimeter of a rectangle:**
>
> $$P_{\text{rectangle}} = 2(l + w) \quad \text{or} \quad P_{\text{rectangle}} = 2l + 2w$$
>
> where l is the length and w is the width.

The formula shows that *both* the length *and* the width are multiplied by 2.

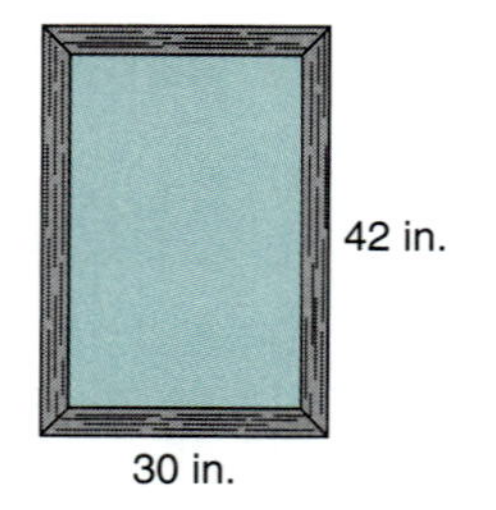

Figure 6–8

EXAMPLE A shop that makes custom picture frames has an order for a frame whose outside measurements are 42 in. by 30 in. (Fig. 6–8). How many inches of picture frame molding are needed for the job?

$$P_{\text{rectangle}} = 2(l + w)$$

$$P_{\text{rectangle}} = 2(42 + 30) \qquad \text{Substitute and perform operations.}$$

$$P_{\text{rectangle}} = 2(72)$$

$$P_{\text{rectangle}} = 144$$

When we analyze the dimensions, we add inches to inches, so the result is written in inches. We then multiply the inches by a number, and the final result is inches. **Thus, 144 in. are needed for the project.**

In some applications, we need to decrease the total perimeter to account for doorways and other openings in the perimeter.

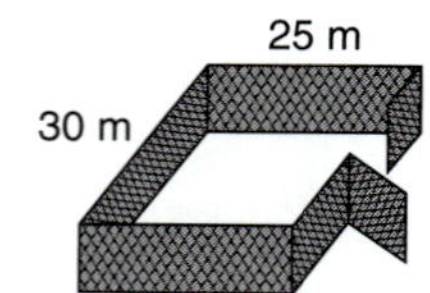

Figure 6–9

EXAMPLE A chain-link fence is to be installed around a yard measuring 25 m by 30 m (Fig. 6–9). A gate 3 m wide will be installed over the driveway. The gate comes pre-assembled from the manufacturer. How much fencing does the installer need to put up the fence?

$$P_{\text{rectangle}} = 2(l + w) - 3 \qquad \text{Subtract the length of the gate.}$$

$$P_{\text{rectangle}} = 2(30 + 25) - 3 \qquad \text{Substitute and perform operations.}$$

$$P_{\text{rectangle}} = 2(55) - 3$$

$$P_{\text{rectangle}} = 110 - 3$$

$$P_{\text{rectangle}} = 107 \text{ m} \qquad \text{Perimeter is a linear measure.}$$

The job requires 107 m of fencing.

The area of a rectangle is found by dividing it into smaller squares and counting the number of squares (Fig. 6–10). There are 15 square meters, or 15 m^2, in this rectangle. Multiplying length by width also gives us 15 m^2.

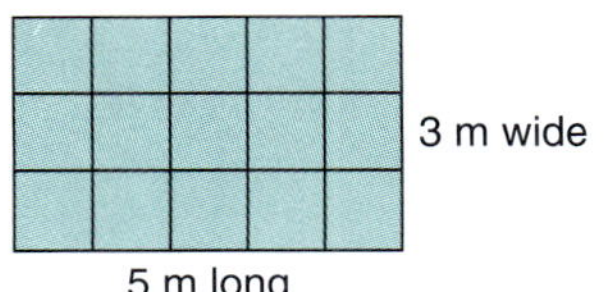

Figure 6–10

The expression lw represents the product of the length and width.

EXAMPLE A carpet installer is carpeting a room measuring 16 ft by 20 ft. Projecting out from one wall is a fireplace whose hearth measures 3 ft by 6 ft (Fig. 6–11). How many square yards of carpet does the installer require for the job? How much is wasted?

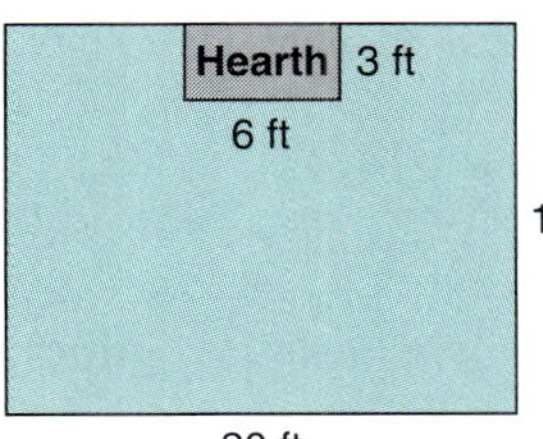

Figure 6–11

This problem has several steps. First, we compute the area of the room without considering the area of the hearth. This is the amount of carpet needed. Second, we compute the area of the fireplace hearth. This is the amount wasted. Finally, we analyze the dimensions to see if the result is expressed in the desired unit. If not, we will need to convert to the desired unit.

Let A_{room} = the area of the room and A_{hearth} = the area of the hearth.

$$A_{room} = lw$$

$$A_{room} = 20 \times 16 \qquad \text{Substitute and multiply.}$$

$$A_{room} = 320 \text{ ft}^2 \qquad \text{Area is a square measure.}$$

$$A_{hearth} = lw$$

$$A_{hearth} = 3 \times 6 \qquad \text{Substitute and multiply.}$$

$$A_{hearth} = 18 \text{ ft}^2 \qquad \text{Area is a square measure.}$$

Because carpet is sold in square yards, we convert the square footage to square yards. Since $9 \text{ ft}^2 = 1 \text{ yd}^2$, we can use a *unity ratio*.

Recall from Chapter 5 that a unity ratio is a fraction with one unit of measure in the numerator and a different, but equivalent, unit of measure in the denominator. A unity ratio contains the original unit and the new unit. Let's complete the problem using the unity ratio $\dfrac{1 \text{ yd}^2}{9 \text{ ft}^2}$.

$$\frac{320 \text{ ft}^2}{1} \times \frac{1 \text{ yd}^2}{9 \text{ ft}^2} = \frac{320 \text{ yd}^2}{9} = 35\frac{5}{9} \text{ yd}^2 \text{ carpet needed,}$$

$$\frac{\overset{2}{18 \text{ ft}^2}}{1} \times \frac{1 \text{ yd}^2}{9 \text{ ft}^2} = 2 \text{ yd}^2 \text{ carpet wasted}$$

This job requires $35\frac{5}{9}$ yd^2 of carpet. Of this, 2 yd^2 is waste that can be used for carpeting smaller areas, such as closets.

The preceding example brings up some interesting questions. Can a fraction of a yard of carpet be purchased? If carpet can be purchased only in 12-ft or 15-ft widths, will there be additional waste? Is it more practical to purchase extra carpet that will be wasted or to spend extra labor costs to install the carpet with several seams? You may want to investigate common practices in the industry.

In other applications, we will find that additional calculations are necessary for finding one or more areas. For instance, when we estimate amounts of paint, wall covering, tongue-and-groove flooring, roofing, shingles, siding, floor tiles, bricks, and similar building materials, we must consider the decrease or increase in area coverage of these materials on installation. Generally, coverage information is standardized by the industry and made known to the user. For example, paint manufacturers indicate on the product label different product coverages on bare wood and on previously painted surfaces. Roofing shingle manufacturers indicate how many bundles of shingles are needed per 100 ft^2 of roof depending on the amount of shingle overlap desired.

EXAMPLE If a $\frac{1}{2}$-in. mortar joint is used in a wall of 2-in. × 4-in. × 8-in. bricks, six bricks cover 1 square foot. If a $\frac{1}{4}$-in. mortar joint is used, seven bricks cover 1 square foot. How many bricks are needed to build a 50-ft × 14-ft wall with $\frac{1}{2}$-in. joints? With $\frac{1}{4}$-in. joints?

$$A_{\text{wall}} = 50 \times 14$$

$$A_{\text{wall}} = 700 \text{ ft}^2$$

$$\frac{700 \text{ ft}^2}{1} \times \frac{6 \text{ bricks}}{1 \text{ ft}^2} = 4{,}200 \text{ bricks}$$ Find number of bricks at 6 per ft^2, using a unity ratio.

$$\frac{700 \text{ ft}^2}{1} \times \frac{7 \text{ bricks}}{1 \text{ ft}^2} = 4{,}900 \text{ bricks}$$ Find number of bricks at 7 per ft^2, using a unity ratio.

With $\frac{1}{2}$-in. mortar joints, 4,200 bricks are needed. With $\frac{1}{4}$-in. mortar joints, 4,900 bricks are needed.

EXAMPLE Asphalt roofing shingles are sold in bundles. The number of bundles needed to cover a square (100 ft^2) depends on the overlap when the shingles are installed. If the overlap allows 4 in. of each shingle to be exposed, then 4 bundles are needed per square. With a 5-in. exposure, 3.2 bundles are needed per square. Figure the number of bundles for a 4-in. exposure and for a 5-in. exposure for a roof measuring 30 ft × 20 ft.

$$A_{\text{roof}} = 30 \times 20$$

$$A_{\text{roof}} = 600 \text{ ft}^2$$

$$\frac{\overset{6}{\cancel{600}} \text{ ft}^2}{1} \times \frac{1 \text{ square}}{\underset{1}{\cancel{100}} \text{ ft}^2} = 6 \text{ squares}$$ Find the number of squares (100-ft^2) using a unity ratio.

$$6 \times 4 = 24 \text{ bundles}$$ Find the number of bundles for a 4-in. exposure.

$$6 \times 3.2 = 19.2 \text{ } or \text{ } 20 \text{ bundles}$$ Find the number of bundles for a 5-in. exposure.

Therefore, 24 bundles of shingles are required for a 4-in. exposure, and 20 bundles (from 19.2 bundles) are required for a 5-in. exposure.

3 **Find the Perimeter and Area of a Parallelogram.**

The perimeter of a parallelogram, like the perimeter of a rectangle, is the sum of its four sides. The sides of the parallelogram are called *base* and *adjacent side* (instead of length and width). Notice the locations of the base and adjacent side in Fig. 6–12.

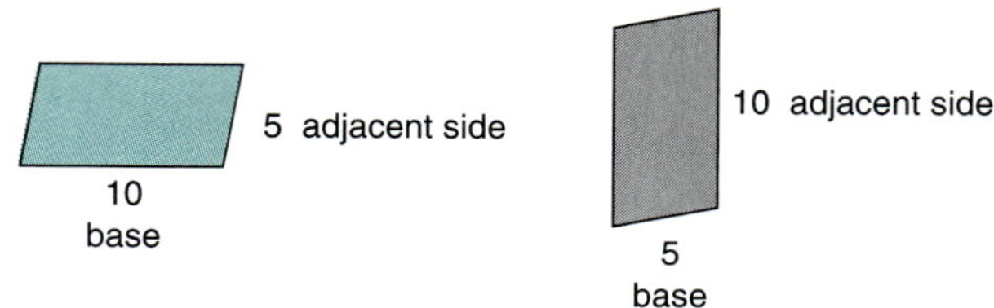

Figure 6–12

CHAPTER 6   Area and Perimeter

■ **DEFINITION: Base.** The *base* of any polygon is the horizontal side or a side that would be horizontal if the polygon's orientation is modified.

■ **DEFINITION: Adjacent Side.** The *adjacent side* of any polygon is the side that has an end point in common with the base.

The formula for the perimeter of a parallelogram is twice the sum of the base and adjacent side.

> **Formula for perimeter of a parallelogram:**
>
> $$P_{\text{parallelogram}} = 2(b + s) \qquad \text{or} \qquad P_{\text{parallelogram}} = 2b + 2s$$
>
> where b is the base and s is the adjacent side.

The preceding formula is similar to the formula for the perimeter of a rectangle. However, the formula for finding the *area* of a parallelogram is somewhat different from the one for the area of a rectangle. Study the parallelogram in Fig. 6–13.

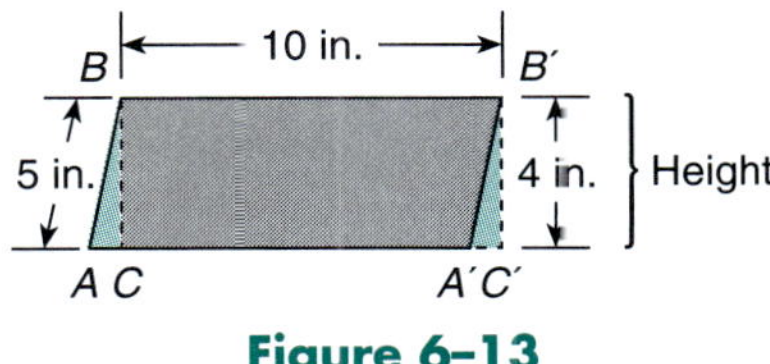

Figure 6–13

The *height* of a parallelogram or of any other polygon is an important dimension. Height is also called *altitude*.

■ **DEFINITION: Height.** The *height* of a polygon is the perpendicular distance from the base to the highest point of the polygon above the base.

The height of the parallelogram in Fig. 6–13 is 4 in. Notice that the height and the adjacent side are different lengths, and the color portions of Fig. 6–13 are triangles.

If triangle ABC ($\triangle ABC$) in Fig. 6–13 was transposed to the right side of the parallelogram ($\triangle A'B'C'$), we would have a rectangle. The area, therefore, is the product of the *base* times the *height;* that is, $10 \times 4 = 40$ in.^{2}.

> **Formula for area of a parallelogram:**
>
> $$A_{\text{parallelogram}} = bh$$
>
> where b is the base and h is the height.

EXAMPLE Find the perimeter and the area of a parallelogram with a base of 16 in., an adjacent side of 8 in., and a height of 7 in. (Fig. 6–14).

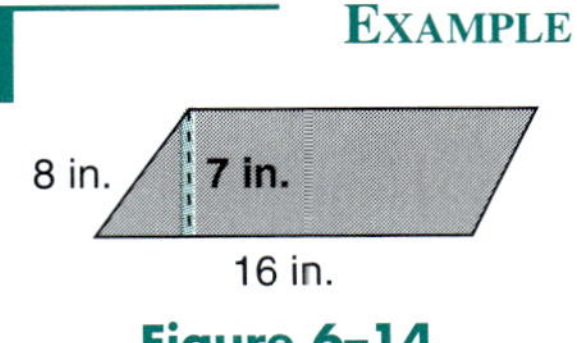

Figure 6–14

Visualize the parallelogram.

$P_{\text{parallelogram}} = 2(b + s)$ $\qquad$ $A_{\text{parallelogram}} = bh$

$P_{\text{parallelogram}} = 2(16 + 8)$ $\qquad$ $A_{\text{parallelogram}} = 16 \times 7$ $\qquad$ Substitute.

$P_{\text{parallelogram}} = 2(24)$ $\qquad$ $A_{\text{parallelogram}} = 112$ in.^{2}

$P_{\text{parallelogram}} = 48$ in.

<table>
<tr><td>Tip!</td><td>Performing a Continuous Sequence of Steps with a Calculator Saves Time.</td></tr>
</table>

Calculations for area and perimeter require one or more operations. To save time, we can perform the calculations in a continuous sequence of steps using a calculator.

Area of a Square

Find the area of a square with 5-in. sides.
Some calculators have a special key or menu option for squaring a number.
Try this option.

$$5 \boxed{x^2} \Rightarrow 25$$

Some calculators use another key to activate the $\boxed{x^2}$ key, particularly if the symbol x^2 is written above the

key, $\boxed{}$. The key that activates the operations above a key is most often labeled $\boxed{\text{SHIFT}}$, $\boxed{\text{2nd}}$, or $\boxed{\text{INV}}$.

Perimeter of a Rectangle

Find the perimeter of a rectangle that is 10 cm long and 5 cm wide.
Try these options.
Enter the sum first:

$$5 \boxed{+} 10 \boxed{=} \boxed{\times} 2 \boxed{=} \Rightarrow 30$$

Use parentheses:

$$2 \boxed{\times} \boxed{(} 5 \boxed{+} 10 \boxed{)} \boxed{=} \Rightarrow 30 \qquad \textbf{With } \boxed{\times} \textbf{ before parentheses.}$$

$$2 \boxed{(} 5 \boxed{+} 10 \boxed{)} \boxed{=} \Rightarrow 30 \qquad \textbf{Without } \boxed{\times} \textbf{ before parentheses.}$$

SELF-STUDY EXERCISES 6–1

1. Find the perimeter and the area of Fig. 6–15.

2. Name the shape in Fig. 6–15.

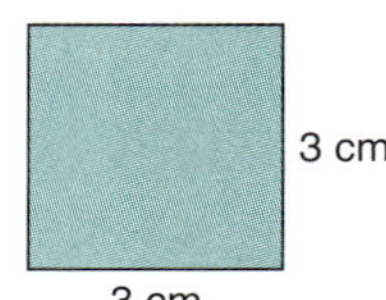

Figure 6–15

Solve these problems involving perimeter and area.

3. Madison Duke is wallpapering a laundry room 8 ft by 8 ft by 8 ft high. How many square feet of paper will she need if there are 63 ft^2 of openings in the room?

4. The square parking lot of a doctor's office is to have curbs built on all four sides. If the lot is 150 ft on each side, how many feet of curb are needed? Allow 10 ft for a driveway into the parking lot.

5. Making no allowances for bases, the pitcher's mound, or the home plate area, calculate how many square yards of artificial turf are needed to resurface an infield at an indoor baseball stadium. The infield is 90 ft on each side. ($9 \text{ ft}^2 = 1 \text{ yd}^2$.)

6. Ted Davis is a farmer who wants to apply fertilizer to a 40-acre field with dimensions $\frac{1}{4}$ mi $\times$ $\frac{1}{4}$ mi. Find the area in square miles.

7. If $\frac{1}{4}$ mi is 1,320 ft, how many feet of fencing are needed to enclose the field in Exercise 6, assuming that a 12-ft steel gate will be installed?

8. A 36-in. $\times$ 36-in. ceramic tile shower stall is being installed. How many 4-in. $\times$ 4-in. tiles are needed to cover the floor? Disregard the drain opening and grout spaces.

9. A border of 4-in. × 4-in. wall tiles surrounds the floor of the shower in Exercise 8. How many tiles are needed for this border? Disregard spaces for grout.

10. A 20-in. × 20-in. central heating and air conditioning return air vent is being installed in a wall. Find the perimeter and area of the wall opening.

11. Find the perimeter and the area of the shape in Fig. 6–16.

12. Name the shape in Fig. 6–16.

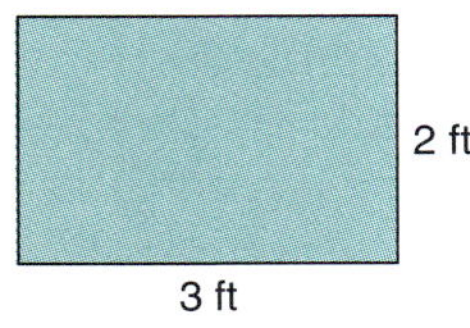

Figure 6–16

Solve the problems involving perimeter and area.

13. A rectangular parking lot is 340 ft by 125 ft. Find the number of square feet in the parking lot.

14. A room is 15 ft by 12 ft. How many square feet of flooring are needed for the room?

15. A small kitchen 9 ft by 10 ft by 8 ft high is being wallpapered. How many square feet of paper are needed if there are 63 ft^2 of openings in the kitchen?

16. How many feet of quarter-round molding are needed to finish around the baseboard after sheet vinyl flooring is installed if the room is 16 ft by 18 ft and there are three 3-ft-wide doorways?

17. The dimensions of a sun porch are 9 ft 6 in. by 15 ft. How many board feet of 1-in.-thick tongue-and-groove flooring are needed to build the porch? (Board feet for 1-in.-thick lumber is the same number as ft^2.) Disregard waste.

18. The swimming pool in Fig. 6–17 measures 32 ft by 18 ft. How much fencing is needed, including material for a gate, if the fence is to be built 7 ft from each side of the pool?

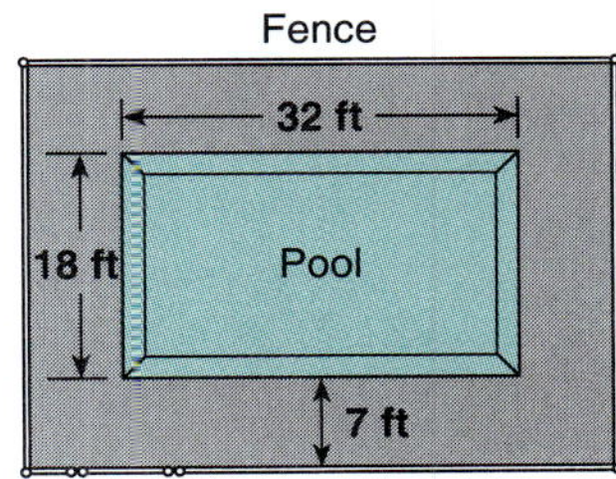

Figure 6–17

19. A Formica tabletop measures 40 in. by 62 in. How many feet of edge trim are needed (12 in. = 1 ft)?

20. A countertop requires rolled edging to be installed on all four sides. How much rolled edging material is needed if the countertop measures 25 in. × 40 in.?

21. Find the perimeter and area of the shape in Fig. 6–18.

22. Name the shape in Fig. 6–18.

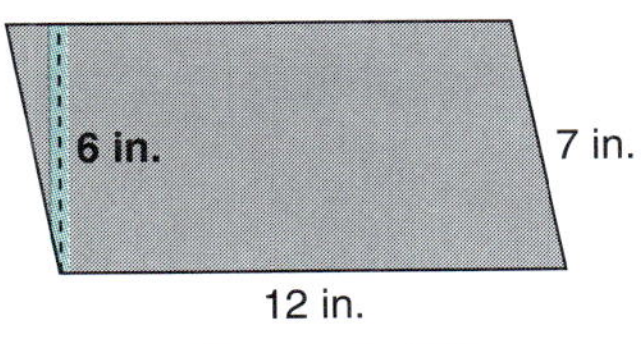

Figure 6–18

Solve the problems involving perimeter and area.

23. An illuminated sign in the main entrance of a hospital is a parallelogram with a base of 48 in. and an adjacent side of 30 in. How many feet of aluminum molding are needed to frame the sign?

24. Find the area of a parking lot that is a parallelogram 200 ft long if the perpendicular distance between the sides is 45 ft (Fig. 6–19).

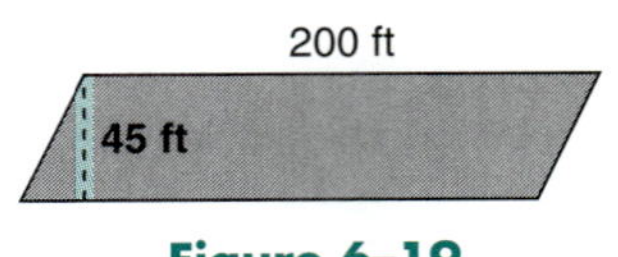

Figure 6–19

25. A customized van has a window cut in each side in the shape of a parallelogram with a base of 20 in. and an adjacent side of 11 in. How many inches of trim are needed to surround the two windows?

26. A design in a marble floor features parallelograms with a base of 48 in. and a height of 24 in. If the figure is made of 8-in. × 8-in. marble tiles, how many tiles are needed for each figure, assuming no waste?

27. A machine cuts sheet metal into parallelograms measuring 6 cm along the base and 3 cm in height. How many can be cut from a piece of sheet metal 100 cm × 200 cm, assuming no waste?

28. A contemporary building has a window in the shape of a parallelogram with a base of 50 in. and an adjacent side of 30 in. How many inches of trim are needed to surround the window?

29. A table for a reading lab has a top in the shape of a parallelogram with a base of 36 in. and an adjacent side of 18 in. How many inches of edge trim are needed to surround the tabletop?

30. Signs in the shape of parallelograms with a base of 24 in. and a height of 10 in. will be cut from sheet metal. How many whole signs can be cut from a piece of sheet metal 48 in. × 48 in.? Illustrate your answer with a drawing.

MATH IN THE WORKPLACE

Engineering: Computer Analysis of a Tornado

Engineering Analysis, Inc., specializes in severe weather threat assessment for business and government. According to a new software program written by engineers at Engineering Analysis, Inc., Mississippi ranks first as the nation's most tornado-prone state. Mississippi beats out states traditionally perceived as tornado-prone, such as Kansas, Nebraska, Oklahoma, and Indiana, which round out the top five. Arkansas is sixth, followed by Iowa, Alabama, Illinois, and Wisconsin. Texas ranks twentieth. The least likely tornado targets are Idaho and Hawaii.

The software ranks states according to the amount of *land area* damaged by twisters in each state. Instead of just counting the number of tornadoes per state each year, this software divides the annual total land area damaged in each state by the state's total land area. The resulting number is the average amount of land damaged by tornadoes in that state over the previous year.

Engineering Analysis, Inc., obtained National Weather Service data that show the starting and ending points and width of every U.S. tornado since 1950. Using *geometry* and *trigonometry,* the program calculates the distance each tornado traveled on land, and multiplies the distance by its width ($A = l \times w$) to determine how much area the twister passed over. Developed by a National Weather Service forecaster, this index is a reliable indicator of tornado threat because it considers both the number of tornadoes and the severity of each storm.

6–2   AREA AND CIRCUMFERENCE OF A CIRCLE

Learning Outcomes

1 Find the circumference of a circle.

2 Find the area of a circle.

3 Find the perimeter and area of composite figures.

Tanks, rings, MRI equipment, pipes, pillars, arches, gears, pulleys, wheels, and similar items are all based on the *circle*. The circle or some part of the circle is often found in composite figures in combination with one or more polygons. We now look at the circumference and area of circles.

Most of us are familiar with circles and some of their properties. However, let's clarify what circles are and what some of their properties are. Look at Fig. 6–20.

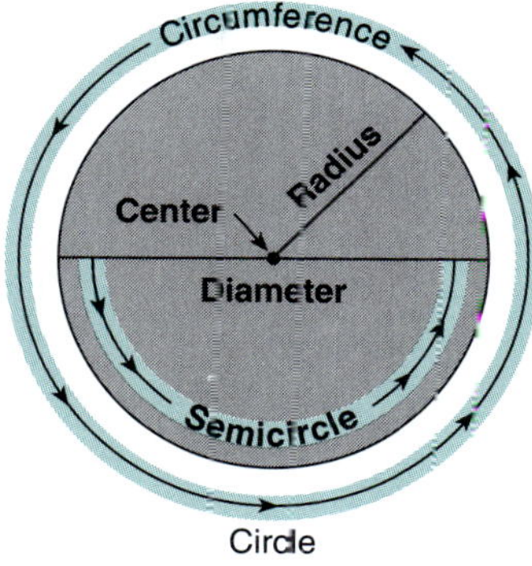

Figure 6–20

■ **DEFINITION: Circle.** A *circle* is a closed curved line whose points lie in a plane and are the same distance from the *center* of the figure.

■ **DEFINITION: Center.** The *center* of a circle is the point that is the same distance from every point on the circumference of the circle.

■ **DEFINITION: Radius.** A *radius* (plural: *radii,* pronounced "ray · dē · ī") is a straight line segment from the center of a circle to a point on the circle. It is half the diameter.

■ **DEFINITION: Diameter.** The *diameter* of a circle is the straight line segment from a point on the circle through the center to another point on the circle.

■ **DEFINITION: Circumference.** The *circumference* of a circle is the perimeter or length of the closed curved line that forms the circle.

■ **DEFINITION: Semicircle.** A *semicircle* is half a circle and is created by drawing a diameter.

1 Find the Circumference of a Circle.

The circle is a geometric form with a special relationship between its circumference and its diameter. If we divide the circumference of any circle by its diameter, the quotient is always the same number.

$$\pi = \frac{\text{circumference}}{\text{diameter}} = \frac{C}{d}$$

This number is a nonrepeating, nonterminating decimal approximately equal to 3.1415927 to seven decimal places. The Greek letter π (pronounced "pie") represents this value. Convenient approximations often used in calculations involving π are $3\frac{1}{7}$ and 3.14.

<table>
<tr><td>Tip!</td><td>What Is π?</td></tr>
</table>

Here is a way to visualize π.

Find a circular object (Fig. 6–21). Use a tape measure to measure the circumference of the object. If a tape measure is not available, wrap a string around the object, then stretch the string along a rule to measure the object.

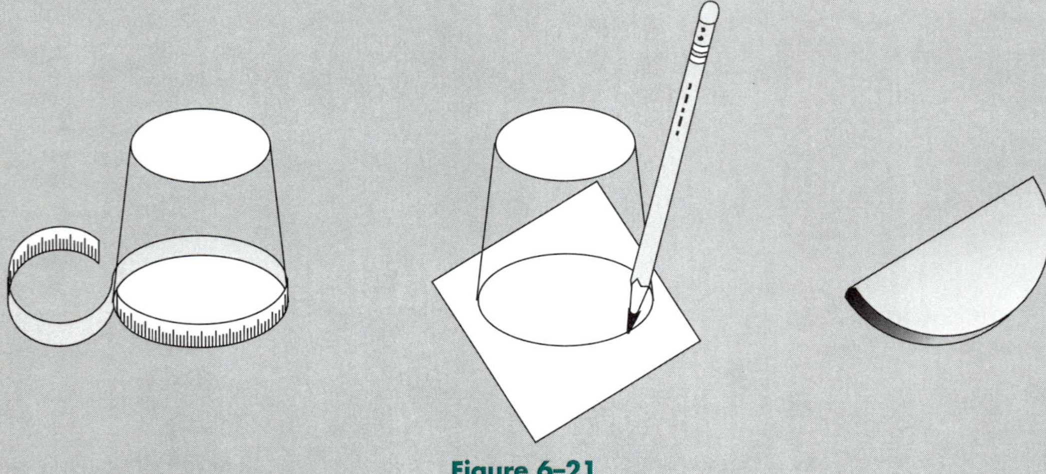

Figure 6–21

Next, trace the object on a sheet of paper and cut out the circle. Fold the circle in half and measure the straight edge (diameter).

Now, divide the circumference by the diameter $\frac{C}{d}$. Since measurements are approximate, the quotient will probably not be exactly 3.14, but it should be slightly over 3.

Repeat the activity again with different-sized circular objects. The larger the circle, the closer your approximation should be to π.

The quotient of the circumference and the diameter is π, and division and multiplication are inverse operations, so the circumference equals π times the diameter. We can also show the relationship using a proportion:

$$\pi = \frac{C}{d}$$

$$\frac{\pi}{1} = \frac{C}{d} \qquad \text{Cross multiply.}$$

$$C = \pi d$$

Because the diameter is twice the radius, the circumference is also equal to π times twice the radius.

Formula for circumference of a circle:

$$C = \pi d \qquad \text{or} \qquad C = 2\pi r$$

where d is the diameter and r is the radius.

Calculator Values of π:

Calculations involving π are always approximations. Many calculators include a π function or menu option whereby π to seven or more decimal places is computed by pressing a single key. Other calculators require the use of more than one key to activate the $\boxed{\pi}$ key. However, for computations by hand or a calculator without the $\boxed{\pi}$ function, 3.14 is sometimes adequate. **We use the calculator value 3.141592654 for π in all examples and exercises.**

EXAMPLE What is the circumference of a circle that has a diameter of 1.3 m (Fig. 6–22)? Round to tenths.

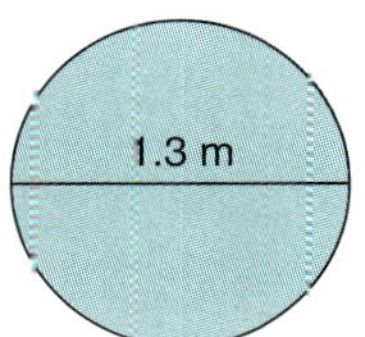

Figure 6–22

$$C = \pi d = \pi(1.3) = 4.08407045$$ Use the $\boxed{\pi}$ key on your calculator for π.

The circumference is 4.1 m (rounded).   Circumference is a linear measure.

Fractions versus Decimals.

When you use your calculator, you will find it easier if you convert mixed U.S. customary linear measurements to their decimal equivalents. For instance, if a diameter is 7 ft 6 in., convert it to 7.5 ft (from $7\frac{6}{12}$ ft, in which $\frac{6}{12} = 0.5$).

EXAMPLE Find to the nearest hundredth the circumference of a circle with a radius of 1 ft 9 in. (Fig. 6–23).

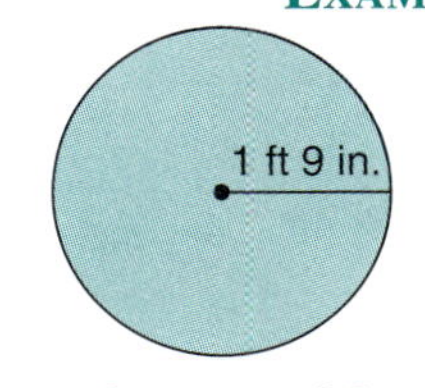

Figure 6–23

$C = 2\pi r$

$C = 2\pi(1.75)$ Substitute for π and r: $1\frac{9}{12}$ ft = 1.75 ft

$C = 10.99557429$

$C = 11.00$ ft   (rounded)   Circumference is a linear measure.

2 Find the Area of a Circle.

The area of a circle, like the circumference, is obtained from relationships within the circle itself. If we divide a circle into two semicircles and then subdivide each semicircle into pie-shaped pieces, we get something like Fig. 6–24a. If we then spread out the upper and lower pie-shaped pieces, we get Fig. 6–24b. Now, if we push the upper and lower pieces together, the result approximates the rectangle in Fig. 6–24c,

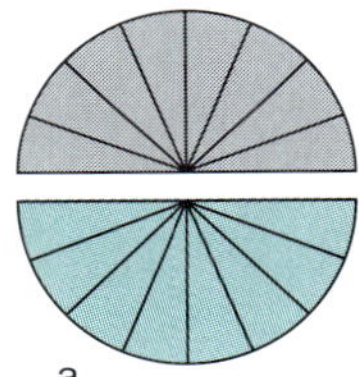

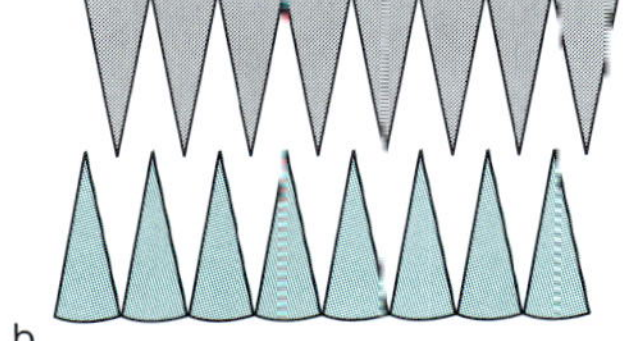

Figure 6–24

whose length is half the circumference and whose width is the radius. Thus, the area of the circle is approximately the area of a rectangle, that is, length times width. Since the length of the rectangle is half the circumference, and the width is the radius, the area of a circle equals half the circumference times the radius.

$$A = \frac{1}{2}C \times r$$
The formula for circumference is $C = 2\pi r$, so we substitute $2\pi r$ for C.

$$A = \frac{1}{2}(2\pi r)(r)$$

$$A = \frac{1}{\cancel{2}}(\cancel{2}\pi r)(r)$$
Multiply. Reduce where possible.

$$A = \pi r^2$$

> **Formula for area of a circle:**
>
> $$A_{\text{circle}} = \pi r^2$$
>
> where r is the radius.

Recall that r^2 means $r \times r$.

EXAMPLE Find the area of a circle whose radius is 8.5 m (Fig. 6–25). Round to tenths.

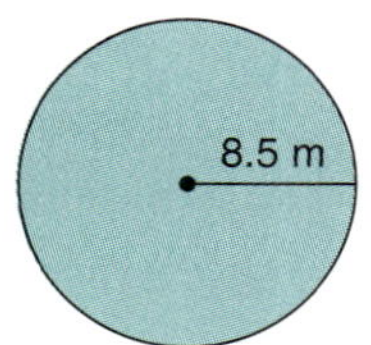

Figure 6–25

$$A = \pi r^2 = \pi(8.5)^2$$

$$A = \pi(72.25) = 226.9800692 \text{ m}^2$$
Area is a square measure.

The area of the circle is 227.0 m².

Recall that in the order of operations powers precede multiplication.

Tip!	*Use a Continuous Calculator Sequence for Accuracy.*

When mixed U.S. customary measures do not convert to convenient, terminating decimal equivalents, a continuous calculator sequence may be preferred for accuracy. For example,

$$3 \text{ ft } 4 \text{ in.} = 3 + \frac{4}{12} \text{ ft}$$

We can reduce and calculate with the mixed number or the improper fraction equivalent,

$$3 + \frac{4}{12} = 3 + \frac{1}{3} = 3\frac{1}{3} = \frac{10}{3}$$

or we can make a continuous sequence using $3 + \frac{4}{12}$.

$$3 \boxed{+} 4 \boxed{\div} 12 \boxed{=} \Rightarrow 3.333333333$$

Remember, scientific and graphing calculators always apply the order of operations.

EXAMPLE Find the area to the nearest hundredth of the top of a circular tank with a diameter of 12 ft 8 in.

If we are using a calculator, we can calculate the area using continuous steps as shown in the tip following the example.

$$A = \pi r^2$$

$$A = (\pi)\left(\frac{12 + \frac{8}{12}}{2}\right)^2$$

Follow the order of operations. 8 in. = $\frac{8}{12}$ ft.
The diameter, $12 + \frac{8}{12}$, divided by 2 is the radius.

$$A = 126.012772 \text{ ft}^2$$

Calculator result

$$A = 126.01 \text{ ft}^2 \quad \text{(rounded)}$$

Area is a square measure.

The area of the top of the tank is 126.01 ft² or 125.95 ft² (basic calculator using $\pi = 3.14$).

Tip!	***Develop Your Calculator Proficiency.***

It is desirable to perform continuous steps in your calculator. Continuous operations give more accurate results; however, be sure you know how your calculator performs continuous operations.

Let's try a problem we have already worked so we know what the correct answer should be. We will rework the preceding example.

Option 1:

$$12 \boxed{+} 8 \boxed{\div} 12 \boxed{=} \boxed{\div} 2 \boxed{=} \boxed{x^2} \boxed{\times} \boxed{\pi} \boxed{=} \Rightarrow 126.012772$$

Option 2:

$$12 \boxed{+} 8 \boxed{\div} 12 \boxed{=} \boxed{\div} 2 \boxed{=} \boxed{x^2} \boxed{\times} \boxed{\pi} \boxed{=} \Rightarrow 126.0127721$$

An alternative sequence of keystrokes is required if the parentheses keys are used. Experiment with your calculator to find other sequences that work.

Option 3:

$$\boxed{\pi} \boxed{\times} \boxed{(} \boxed{(} 12 + 8 \div 12 \boxed{)} \boxed{\div} 2 \boxed{)} \boxed{x^2} \boxed{=} \Rightarrow 126.0127721$$

Work several problems both continuously and by recording and reentering intermediate calculations to learn how your calculator works and to develop your skills and confidence.

---

3 Find the Perimeter and Area of Composite Figures.

As we look about us we see geometric shapes that are not squares, rectangles, parallelograms, or circles, yet our work may require us to calculate the perimeter and area of these shapes. In such cases, we can use what we already know to find both perimeter and area. These shapes are called *composite* figures.

■ **DEFINITION: Composite Figure.** A *composite figure* is a geometric figure made up of two or more geometric figures.

For instance, an L-shaped slab foundation for a building is really a composite of two rectangles. As shown in Fig. 6–26, a composite figure can be considered in more than one way. Layout I is partitioned vertically, and layout II is partitioned horizontally.

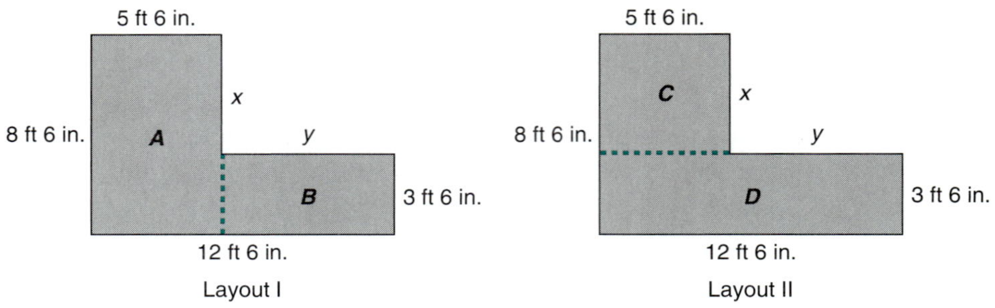

Figure 6–26

Note that such a layout may not always have all its dimensions indicated; however, we can calculate the missing dimensions from our knowledge of polygons and the given dimensions.

EXAMPLE   Find the missing dimensions x and y on the slab foundation in Fig. 6–26.

In layout I, the side of B opposite its $3'6''$ side is also $3'6''$ because opposite sides of a rectangle are equal. The side of A opposite its $8'6''$ side is, for the same reason, $8'6''$. Dimension x must therefore be the difference between $8'6''$ and $3'6''$.

$$x = 8'6'' - 3'6''$$

$$x = 5'$$

If we think of layout II as two horizontal rectangles, we can find dimension y. The side opposite the $5'6''$ side of C must be $5'6''$. The side of D opposite the $12'6''$ side must also be $12'6''$. Dimension y must therefore be the difference between $12'6''$ and $5'6''$.

$$y = 12'6'' - 5'6''$$

$$y = 7'$$

The missing dimensions are $x = 5'$ and $y = 7'$.

The perimeter is the sum of the lengths of the sides of a figure. The number and the length of the sides vary from one composite figure to the next, so no formula covers the entire variety of composite shapes that exist; however, we can use a general formula.

> ***Formula for perimeter of a composite figure***
>
> $$P = a + b + c + \cdots$$

EXAMPLE   Find the number of feet of 4-in. stock needed for the base plates of a room that has the layout shown in Fig. 6–27. Make no allowances for openings when estimating the linear footage of the base plates.

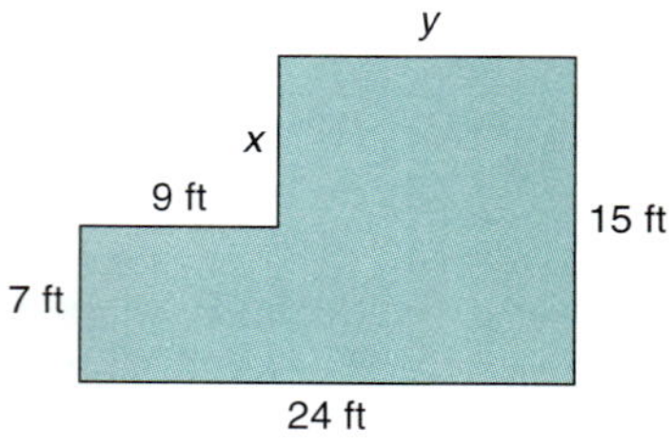

Figure 6–27

   CHAPTER 6   Area and Perimeter

1. Find the missing dimensions.

$$x = 15 \text{ ft} - 7 \text{ ft} \qquad y = 24 \text{ ft} - 9 \text{ ft}$$
$$x = 8 \text{ ft} \qquad\qquad y = 15 \text{ ft}$$

2. Apply the general formula for the perimeter of a polygon.

$$P = a + b + c + \cdots$$
$$P = 7 \text{ ft} + 24 \text{ ft} + 15 \text{ ft} + 15 \text{ ft} + 8 \text{ ft} + 9 \text{ ft} = 78 \text{ ft}$$

3. Count the number of sides on the layout to make sure that each is substituted into the formula.

The room needs 78 ft of 4-in. stock for the base plates.

EXAMPLE   Find the number of square yards of carpeting required for the room in the preceding example ($9 \text{ ft}^2 = 1 \text{ yd}^2$).

1. We divide the composite shape into two polygons with areas we can compute (Fig. 6–28). In this case, A is a rectangle and B is a square.

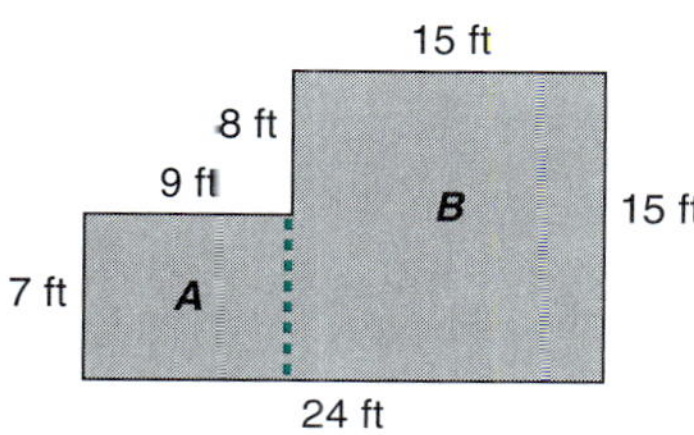

Figure 6–28

2. We find the areas of the smaller polygons and add them.

Rectangle A	**Square B**
$A_1 = lw$	$A_2 = s^2$
$A_1 = 9 \times 7$	$A_2 = 15^2$
$A_1 = 63 \text{ ft}^2$	$A_2 = 225 \text{ ft}^2$

$$A_1 + A_2 = \text{total area}$$
$$63 + 225 = 288 \text{ ft}^2$$

3. We convert square feet to square yards using a unity ratio.

$$\frac{\overset{32}{\cancel{288} \text{ ft}^2}}{1} \times \frac{1 \text{ yd}^2}{\underset{1}{\cancel{9} \text{ ft}^2}} = 32 \text{ yd}^2$$

The room requires 32 yd^2 of carpeting.

EXAMPLE   A 15-in.-diameter wheel has a 3-in. hole in the center. Find the area of a side of the wheel to the nearest tenth (Fig. 6–29).

We are asked to find the area of the color portion of the wheel in Fig. 6–29. To do so, we find A_{outside}, the area of the larger circle (diameter 15 in.), and *subtract* from it the area of A_{inside}, the smaller circle (diameter 3 in.). The color portion is called a *ring*.

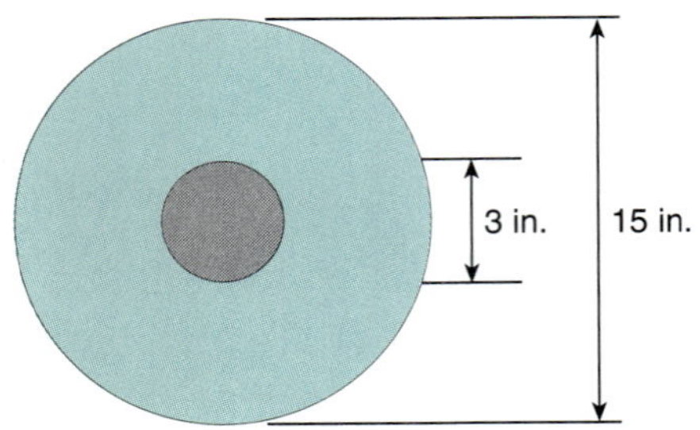

Figure 6–29

$$A_{\text{outside}} = \pi r^2 \quad \left(r = \frac{15}{2} = 7.5\right) \qquad A_{\text{inside}} = \pi r^2 \quad \left(r = \frac{3}{2} = 1.5\right)$$

$$A_{\text{outside}} = \pi(7.5)^2 \qquad\qquad A_{\text{inside}} = \pi(1.5)^2$$

$$A_{\text{outside}} = \pi(56.25) \qquad\qquad A_{\text{inside}} = \pi(2.25)$$

$$A_{\text{outside}} = 176.7145868 \text{ in.}^2 \qquad A_{\text{inside}} = 7.068583471 \text{ in.}^2$$

Area of wheel (ring) $= A_{\text{outside}} - A_{\text{inside}}$

$$A_{\text{wheel}} = 176.7145868 - 7.068583471 = 169.6460033 \qquad \text{or} \qquad 169.6 \text{ in.}^2$$

The area of the wheel (ring) is 169.6 in.^{2}

EXAMPLE A bandsaw has two 25-cm wheels spaced 90 cm between centers (Fig. 6–30). Find the length of the saw blade.

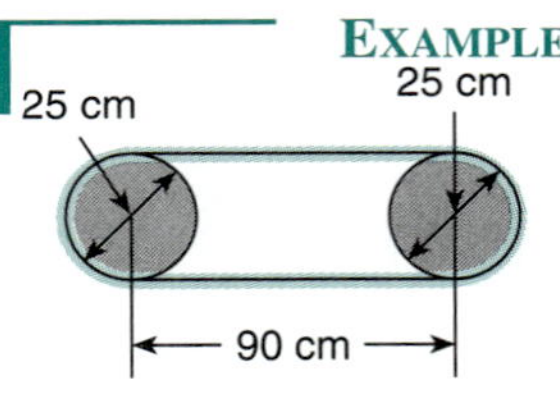

Figure 6–30

This layout is a composite figure that consists of a semicircle at each end and a rectangle in the middle. It is called a *semicircular-sided* figure. The two semicircles equal one whole circle, so we need to find the circumference of one circle (wheel) and add it to the lengths of the two sides of the rectangle.

$C = \pi d$ Total length of blade $= C + 2l$

$C = \pi(25)$ Total length of blade $= 78.53981634 + 2(90)$

$C = 78.53981634$ cm   Total length of blade $= 258.5$ cm   (rounded)

The bandsaw blade is 258.5 cm in length.

Specific applications for a particular industry or career often use the formulas for area, perimeter, or circumference.

EXAMPLE Find the cutting speed of a lathe if a piece of work 7 in. in diameter turns on a lathe at 75 revolutions per minute (rpm) (Fig. 6–31).

The *cutting speed* is the speed of a tool that passes over the work, such as the speed of a lathe or a sander as it sands (passes over) a piece of wood. If the cutting speed is too fast or too slow, safety and quality are impaired. The formula for cutting speed is CS $= C$ (in feet) $\times$ rpm.

$$\text{CS} = \text{cutting speed}$$

$$C = \text{circumference or one revolution (in feet)}$$

$$\text{rpm} = \text{revolutions per minute}$$

Cutting speed is measured in *feet per minute* (ft/min).

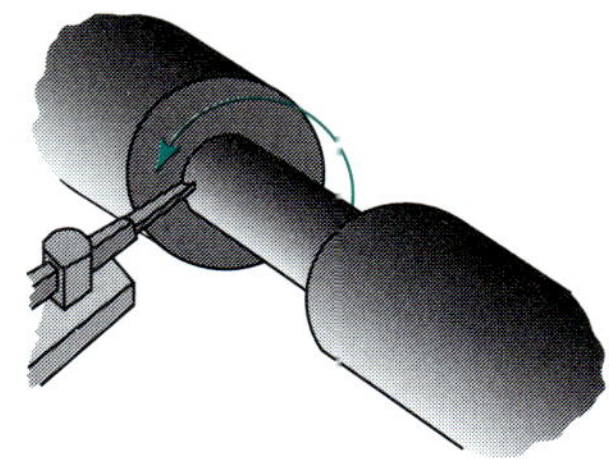

Figure 6–31

$$C = \pi d$$
$$C = \pi(7)$$
$$C = 21.99114858 \text{ in.}$$
$$C = 1.832595715 \text{ ft}$$

Convert 21.99114858 in. into feet using a unity ratio.

$$\frac{21.99114858 \text{ in.}}{1} \times \frac{1 \text{ ft}}{12 \text{ in.}} = 1.832595715 \text{ ft}$$

$$CS = C \times \text{rpm} = 1.832595715 \times 75 = 137 \text{ ft/min.} \qquad \text{Rounded.}$$

The cutting speed of the lathe is approximately 137 ft/min.

SELF-STUDY EXERCISES 6–2

1 Find the circumference of circles with the following dimensions. Round to tenths.

1. Diameter = 8 cm
2. Radius = 3 in.
3. Radius = 1.5 ft
4. Diameter = 5.5 m

2 Find the area of circles with the following dimensions. Round to tenths.

5. Diameter = 8 cm
6. Radius = 3 in.
7. Radius = 1.5 ft
8. Diameter = 5.5 m

3 Find the missing dimensions x and y in Figs. 6–32 and 6–33.

9.

10.

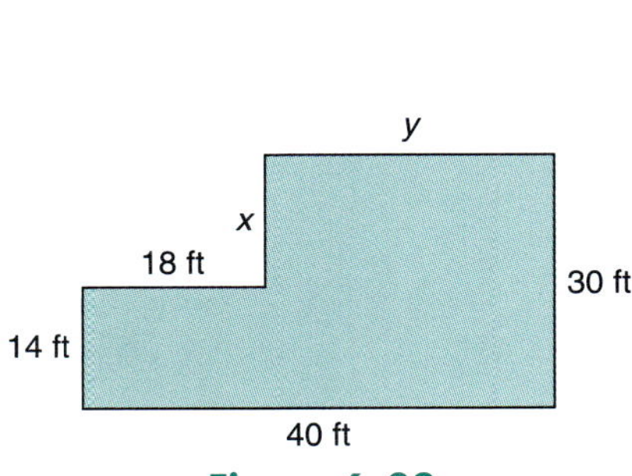

Figure 6–32

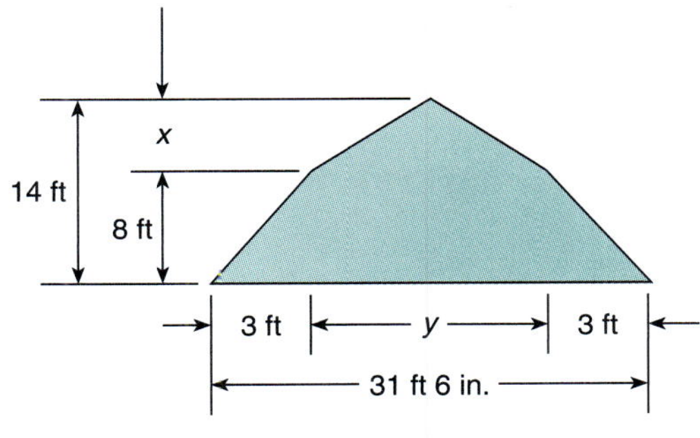

Figure 6–33

Find the color area of Figs. 6–34 through 6–37 to the nearest tenth.

11.

12.

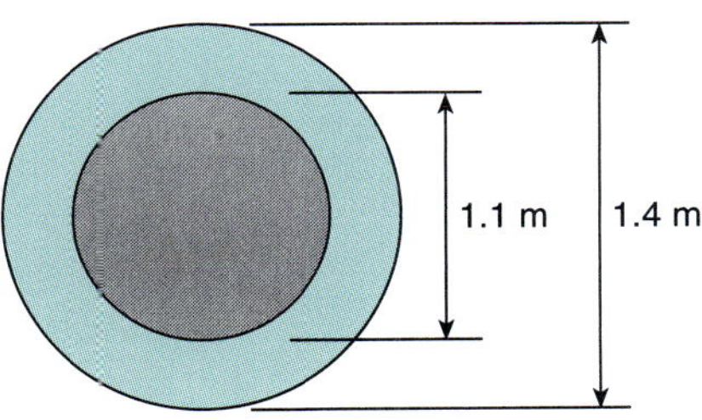

Figure 6–34

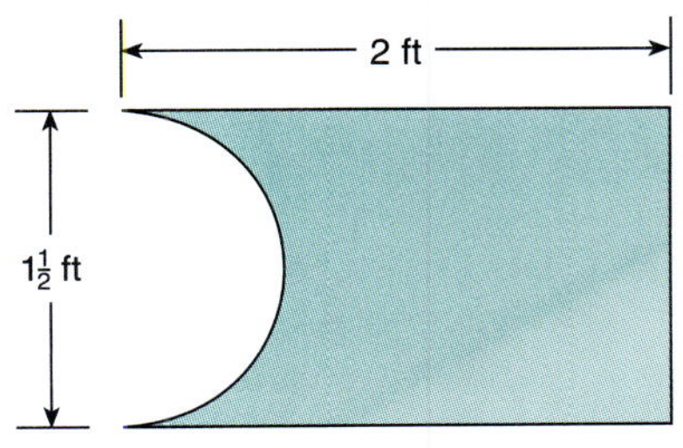

Figure 6–35

13.

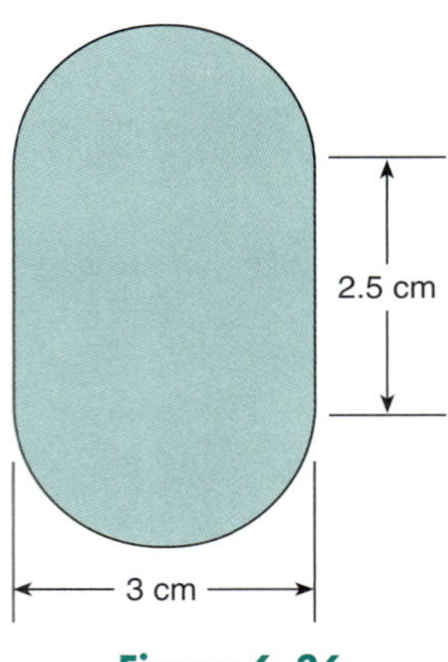

2.5 cm

3 cm

Figure 6–36

14.

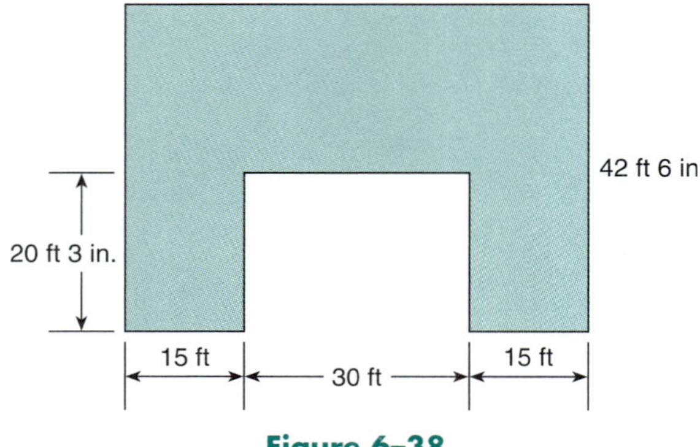

$\frac{1}{2}$ in.

$1\frac{1}{4}$ in.

Figure 6–37

15. Find the perimeter of the layout in Fig. 6–38.

16. Find the area of the layout in Fig. 6–38.

42 ft 6 in.

20 ft 3 in.

15 ft

30 ft

15 ft

Figure 6–38

Solve the problems involving perimeter and area of composite figures.

17. A house features a den with the layout shown in Fig. 6–39. Find the number of feet of 4-in. stock needed for the base plates around the perimeter. Make no allowances for doorways or other openings.

18. If the den in Exercise 17 is to be covered with roll vinyl that costs $16.50 per square yard for materials and labor, how much will it cost to install the vinyl? (*Note:* Round to the next whole square yard before figuring cost.)

19. The metal piece in Fig. 6–40 has the dimensions indicated. If the 18-gauge steel weighs 2 lb per square foot, how much does the piece weigh to the nearest pound? ($144 \text{ in.}^2 = 1 \text{ ft}^2$)

20. The metal used in Exercise 19 costs $2.78 per pound. Any fraction of a pound is rounded to the next pound. How much will the material cost for making the metal piece?

8 ft 6 in.

4 ft 3 in.

12 ft 8 in.

16 ft 8 in.

Figure 6–39

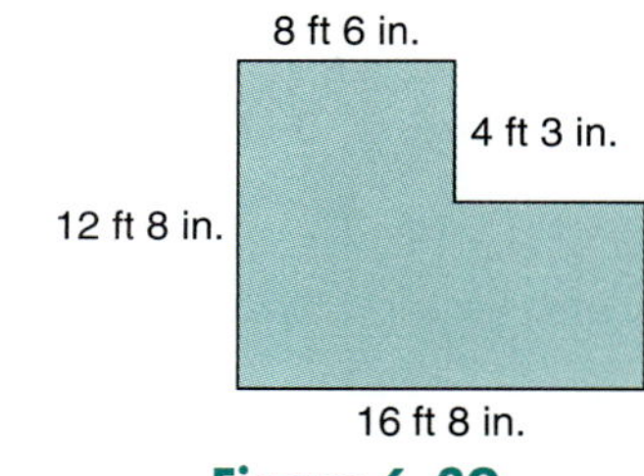

$72\frac{1}{2}$ in.

15 in.

37 in.

30 in.

Figure 6–40

CHAPTER 6   Area and Perimeter

21. A swimming pool is in the form of a semicircular-sided figure. Its width is 20 ft and the parallel portions of the sides are each 20 ft (Fig. 6–41). What is the area of a 5-ft-wide walk surrounding the pool?

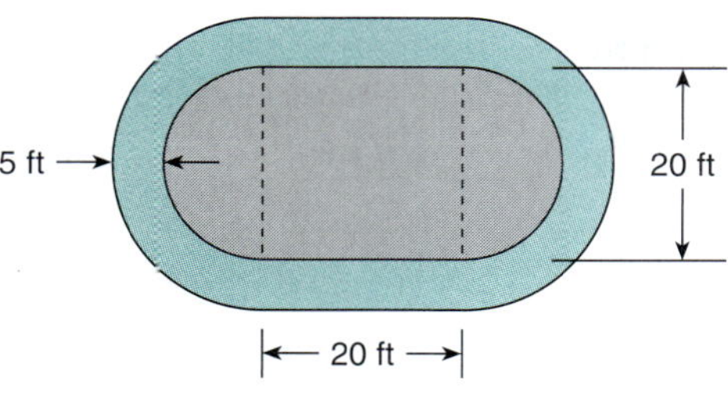

Figure 6–41

22. A belt connecting two 9-in.-diameter drums on a conveyor system needs replacing. How many inches must the new belt be if the centers of the drums are 10 ft apart (Fig. 6–42)?

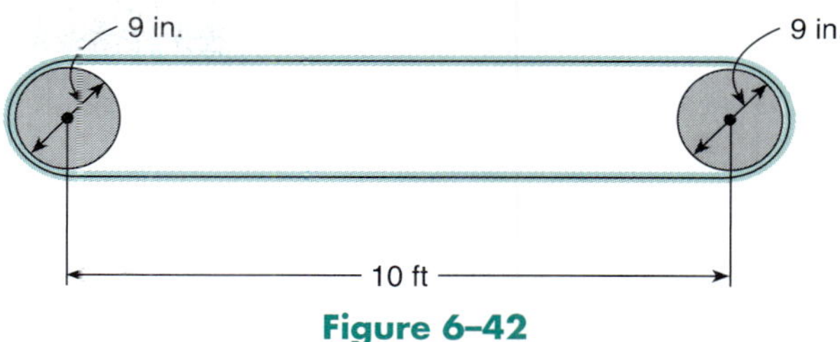

Figure 6–42

23. Find the area of the ring formed by the crosscut section of Fig. 6–43.

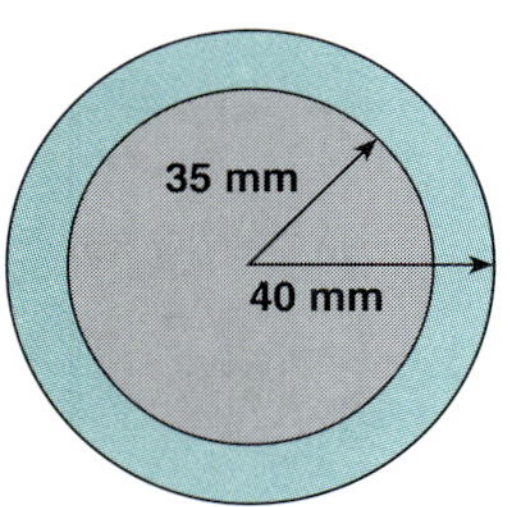

Figure 6–43

24. The wall of a galvanized water pipe is 3.5 mm thick. If the outside circumference is 68 mm, what is the area (to the nearest tenth) of the colored interior crosscut section (Fig. 6–44)?

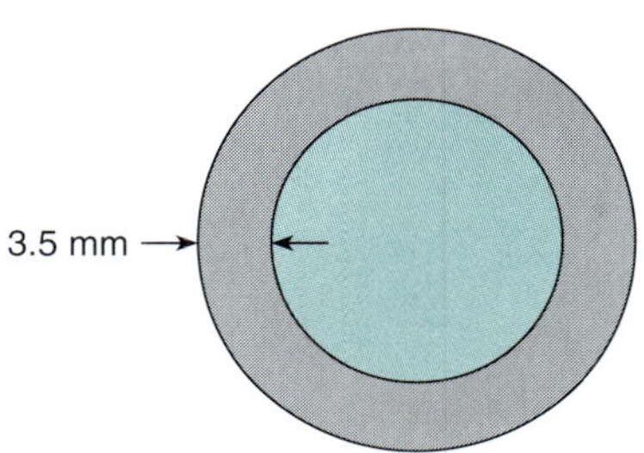

Figure 6–44

25. A 2-in.-inside-diameter pipe and a 4-in.-inside-diameter pipe empty into a third pipe whose inside diameter is 5 in. (Fig. 6–45). Is the third pipe large enough for the combined flow? (Justify your answer.)

Figure 6–45

26. A large pipe whose interior cross-sectional area is 20 in.^{2} empties into two smaller pipes that have an interior diameter of 4 in. each (Fig. 6–46). Are the smaller pipes together large enough to carry off the flow from the larger pipe? (Justify your answer.)

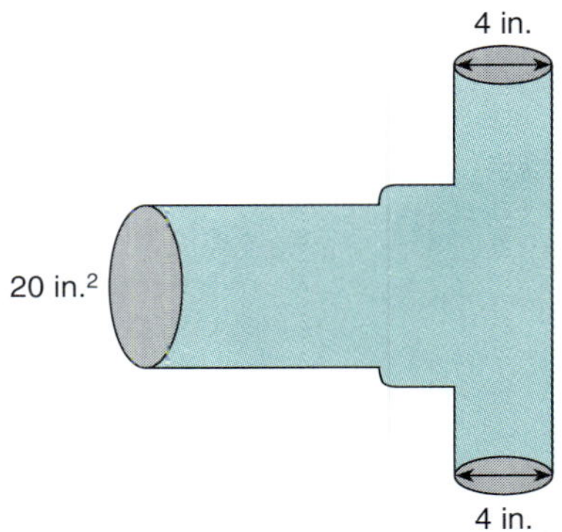

Figure 6–46

27. Cutting speed, when applied to a grinding wheel, is called *surface speed*. What is the surface speed in feet per minute of a 9-in.-diameter grinding wheel revolving at 1,200 rpm? (Surface speed = circumference in feet × rpm.)

28. A 12-in.-diameter polishing wheel revolves at 500 rpm. What is the surface speed? (See Exercise 27 for formula.)

29. A lamp effectively lights up an area 10′ in diameter (Fig. 6–47). How many square feet does the lamp light effectively?

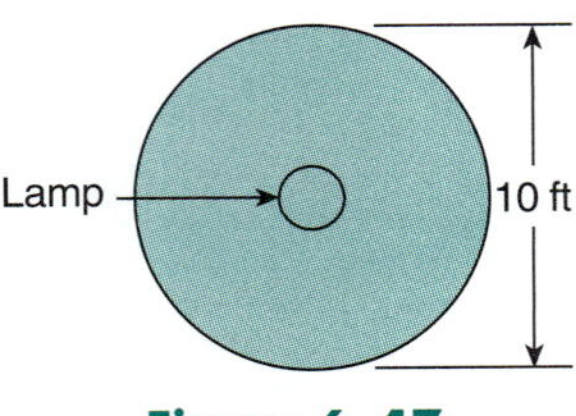

Figure 6–47

30. A steel sleeve with a 3-in. outside diameter measures $2\frac{1}{2}$ in. across the inside. The sleeve will be babbitted to fit a 1.604-in.-diameter motor shaft. What must the thickness of the babbitt be (in hundredths)? The babbitt is the colored ring in Fig. 6–48. A babbitt is a lining of babbitt metal, a soft antifriction alloy, that allows two metal parts to fit snugly together.

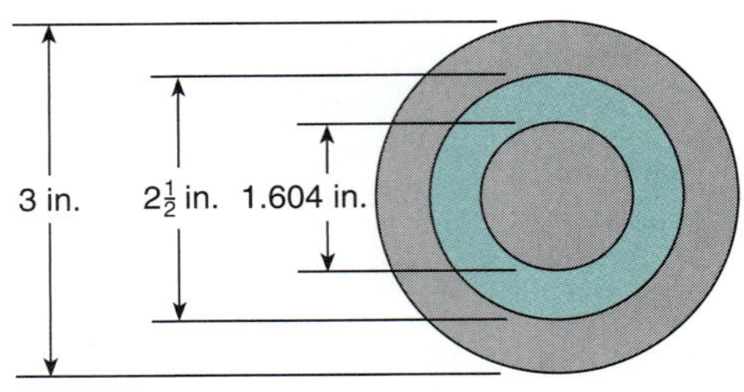

Figure 6–48

31. Two 12-cm-diameter drums are connected by a belt to form a conveyor system. The centers of the drums are 2 m apart (Fig. 6–49). How long must a replacement belt be (to the nearest tenth of a meter)?

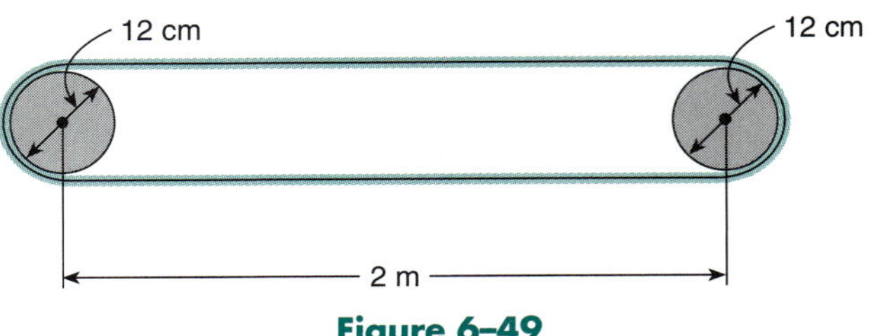

Figure 6–49

Interior Design: Paint, Wallpaper, and Border

Suppose you work for an interior design company. After the decorator helps the customer choose paint, wallpaper, and border, your job is to go to the site, measure the room(s), and order the correct amounts. Your company sometimes orders an extra gallon of paint, one extra single roll of wallpaper, and 20 extra feet of wallpaper border.

Your first job is a child's rectangular room measuring 12 ft by 16 ft with 8-ft ceilings. Three walls will be wallpapered and the fourth short wall will be painted. A wallpaper border 6 in. wide will circle the room. The painted wall and one long wall both have a centered window 6 ft wide by 5 ft tall. The other long wall has a 30-in. × 80-in. entry door and a 60-in. × 80-in. closet door. The doors will not be papered or painted.

Exercises

Use the given measurements to find the answers. Show computations. Round all answers to the next whole unit.

1. Using a ruler, draw a two-dimensional diagram of this room looking down from the ceiling. Label the length of each wall, and place the doors and windows correctly.
2. Using a ruler, draw a two-dimensional diagram of the two short walls looking at the walls while standing inside the room. Label the length of each side, and place the window correctly.

3. Using a ruler, draw a two-dimensional diagram of the two long walls looking at the walls while standing inside the room. Label the length of each side, and place the doors and the window correctly.
4. Using the diagram from Exercise 2, calculate the amount of paint needed to give the short wall without the window two coats. Each gallon of paint covers 300 ft².
5. Using the diagram from Exercise 1, calculate the length of wallpaper border needed to circle the entire room. Keep in mind that the doors and windows do not reach the ceiling. If the wallpaper border comes only in 20-ft rolls, how many rolls will you need? If the customer wants border seams only at the corners of the room, how many rolls will it take?
6. Use the diagrams from Exercises 2 and 3 to calculate the area to be covered with wallpaper on the three walls. The wallpaper is put up first, then the border covers the wallpaper at the ceiling. Don't forget to subtract the area of the doors and windows. (*Note:* 1 ft² = 144 in.².)
7. The customer has chosen a uniform wallpaper pattern that does not require matching. How many rolls of wallpaper will it take if each single roll of wallpaper covers 28 ft²? If this pattern is sold only in double rolls, how many double rolls will you need?
8. Most wallpaper comes in 27-in.-wide rolls. Paperers cut panels equal to the height of the room and cut out doors and windows from each panel. To find the number of single rolls needed, professional wallpaper hangers typically divide the total area to be covered by 22 ft². This technique ensures the minimum number of seams after allowing for door and window waste and pattern matching. How many single and double rolls will be needed using this method of calculation?
9. Place the order for the paint, rolls of wallpaper border, and double rolls of wallpaper.

Answers

1. Ceiling-view diagram:

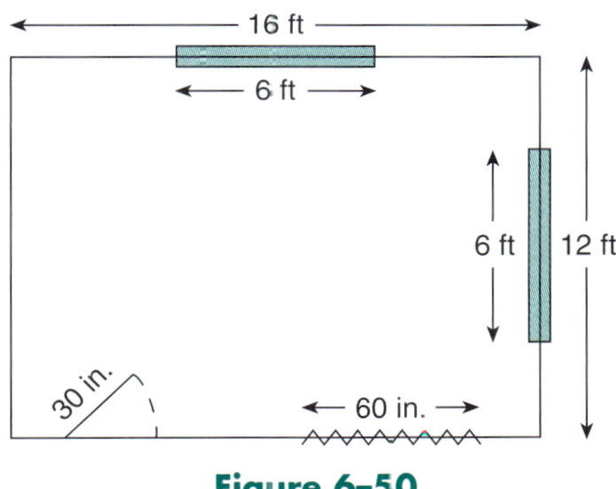

Figure 6–50

2. Short-wall diagrams:

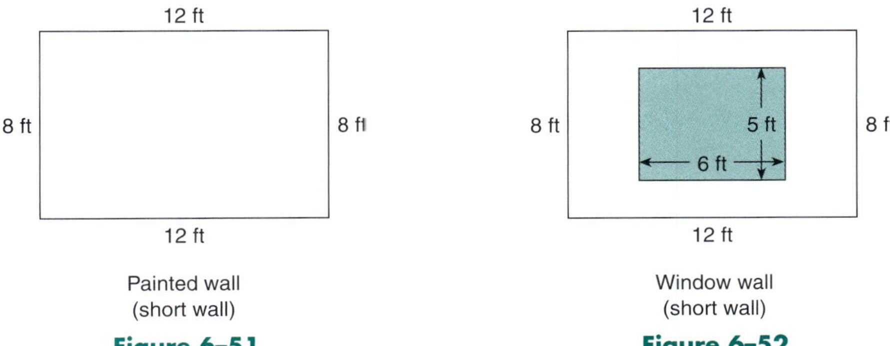

Figure 6–51

Figure 6–52

3. Long-wall diagrams:

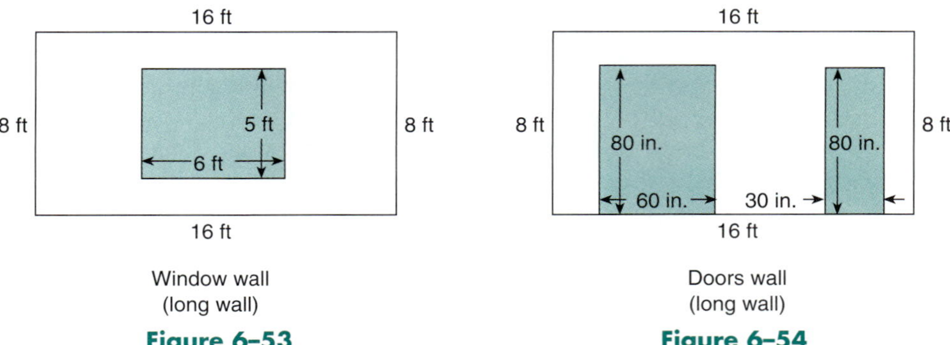

Window wall
(long wall)

Figure 6–53

Doors wall
(long wall)

Figure 6–54

4. The painted wall's area is 96 ft^2. Double that for two coats equals 192 ft^2, so 1 gal is plenty.

5. a. 56 ft of wallpaper border are needed.
    b. 3 rolls are needed for seams anywhere.
    c. 4 rolls are needed for corner seams only.

6. Short wall with window: 66 ft^2.
Long wall with window: 98 ft^2.
Long wall with doors: 78 ft^2.
Total wallpaper area: 242 ft^2.

7. 9 single rolls; 5 double rolls

8. 11 single rolls; 6 double rolls

9. Order 2 gal of paint, 5 rolls of wallpaper border, and 6 double rolls of wallpaper.

Thanks to Bettye Yates, owner of Interior Creations, Collierville, TN.

ASSIGNMENT EXERCISES

Section 6–1

Find the perimeter and the area of Figs. 6–55 through 6–57.

1.

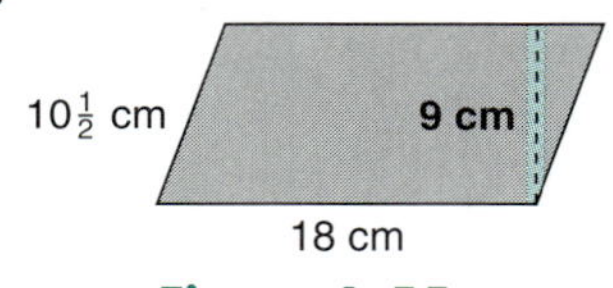

Figure 6–55

2.

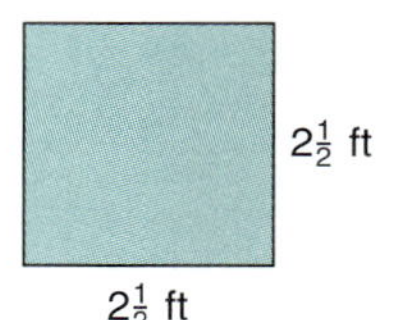

Figure 6–56

3.

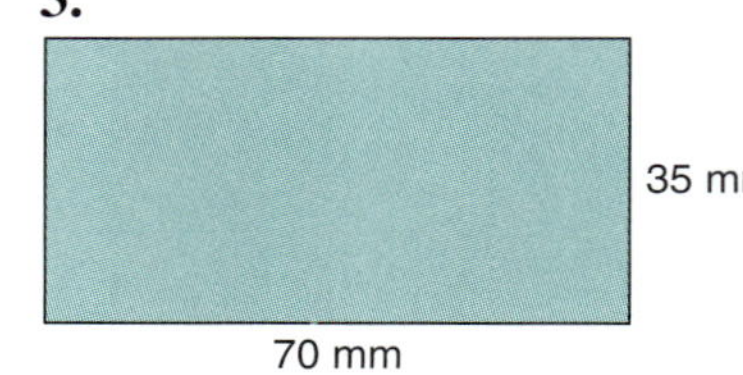

Figure 6–57

4. If a parking lot for a new hospital measures 275 ft by 150 ft, how many square feet need to be blacktopped?

5. In Exercise 4, find the amount of curbing needed to surround the parking lot if 14 ft is allowed for a driveway.

6. A hall wall with no windows or doors measures 25 ft long by 8 ft high. Find the number of square feet to be covered if paneling is installed on the two walls.

7. A den 18 ft by $16\frac{1}{2}$ ft is to be carpeted. How many square yards of carpeting are needed?

8. Vincent Ores, a contractor, is to brick the storefront of a landscape service that has a doorway measuring 7 ft by 6 ft. How many bricks are needed if the storefront is 20 ft by 12 ft and the bricks cover at the rate of 6 per square foot using $\frac{1}{2}$-in. mortar joints?

9. A roof measuring 16 ft by 20 ft is to be covered with asphalt roofing cement. How much will the project cost if the asphalt roofing cement spreads at the rate of 150 ft^2 per gallon and costs $4.75 per gallon? The cement is purchased by the gallon only.

10. Estimate the number of feet of roll fencing needed to fence a square storage area measuring $15\frac{1}{2}$ ft on a side. A preassembled gate 4 ft wide will be installed.

12. A square area of land is 10,000 m². If a baseball field must be at least 99.1 m along each foul line to the park fence, is the square adequate for regulation baseball?

11. A dining room wall 12 ft long by 8 ft high will be wallpapered. How many single rolls (24 in. × 20 ft) of wallpaper will be needed? Assume there is no waste.

13. Debbie Murphy is building a contemporary home with four front windows, each in the form of a parallelogram. If each window has a 5-ft base and is 2 ft high, how many square feet of the 25 ft × 11 ft wall will require stain?

14. If the stain used on the wall in Exercise 13 is applied at a cost of $2.75 per square yard, find the cost to the nearest dollar of staining the front wall.

Section 6–2

Find the perimeter (or circumference) and the area of Figs. 6–58 through 6–61. Round to hundredths.

15.

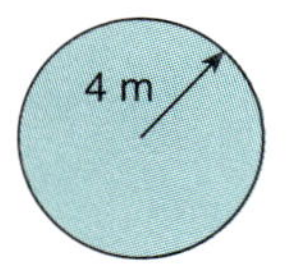

Figure 6–58

16.

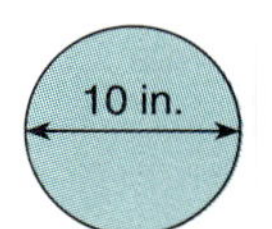

Figure 6–59

17.

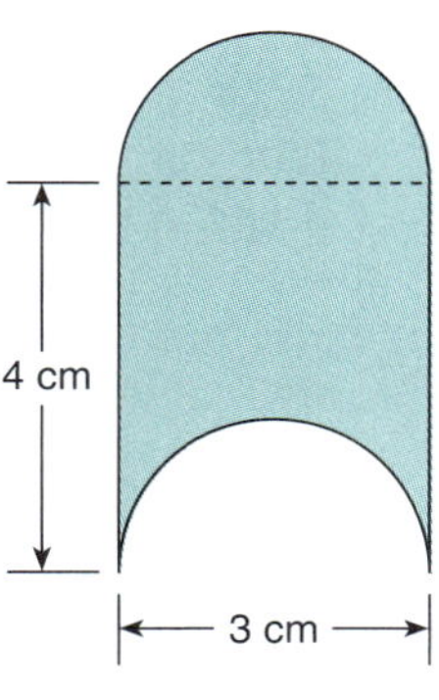

Figure 6–60

18.

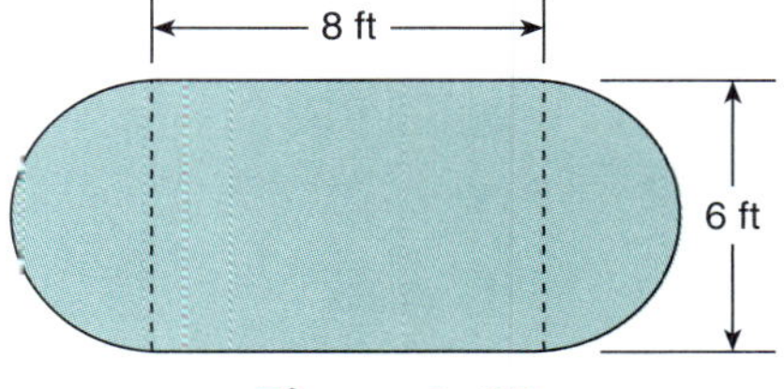

Figure 6–61

19. A circle has a diameter of 2.5 m. If its diameter is increased 1.2 m, what will be the difference in the areas (to the nearest tenth) of the two circles?

21. Three water hoses, with interior cross-sectional areas of 1 in.², $1\frac{1}{2}$ in.², and 2 in.², empty into a larger hose. What is the inside diameter of the larger hose if it carries the combined flow of the three smaller hoses? Round to tenths.

23. A $\frac{1}{4}$-in. electric drill with variable speed control turns as slowly as 25 rpm. If an abrasive disk with a 5-in. diameter is attached to the drill drive shaft, what is the disk's slowest cutting speed in ft/min? Round to the nearest whole number. (Cutting speed = circumference measured in feet × rpm.)

20. If a circular flower bed has a radius of 24 in., how many $5\frac{3}{4}$-in.-long flat bricks are needed to enclose it at the rate of two bricks per 12 in.?

22. Two 35-mm-diameter drums connect a conveyor belt. If the centers of the drums are 70 cm apart, how many centimeters long is the conveyor belt? Round to tenths.

24. What is the cross-sectional area of the opening in a round flue tile whose inside diameter is 8 in.? Round any part of an inch to the next tenth of an inch.

25. Triple-wall galvanized chimney pipe for prefabricated sheet-metal fireplaces has a circumference of 47 in. What is the inside diameter of the fire-stop spacer through which the chimney pipe passes as it goes through the ceiling? Round to the nearest tenth.

27. Find the area of the colored portion of the sidewalk in Fig. 6–62. The colored portion is one-fourth of a circle. Express the answer in square yards rounded to tenths.
(144 in.^{2} = 1 ft^2, 9 ft^2 = 1 yd^2.)

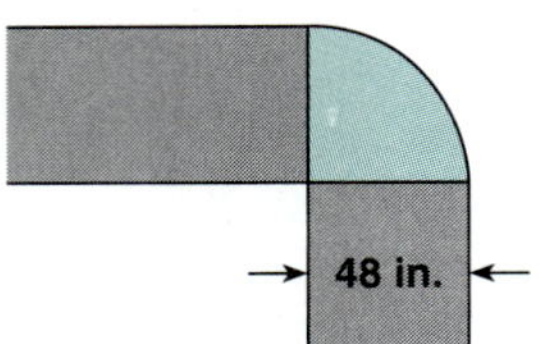

Figure 6–62

29. If a cogwheel makes one complete revolution and its radius is 9.4 in., how long is the path traveled by any one point on the cogwheel? Round to tenths.

26. Find the area of the cross section of a wire whose diameter is $\frac{1}{16}$ in. (Express your answer to the nearest thousandth.)

28. Find the area of the composite layout shown in Fig. 6–63. Round the answer to the nearest tenth.

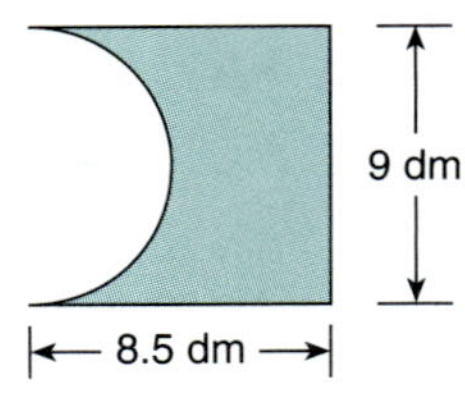

Figure 6–63

CHALLENGE PROBLEMS

30. Lou Ferrante plans to build a dog pen. He has 360 ft of fencing and would like to enclose as much area as possible for his dog. What length and width should he make the dog pen?

32. Given that the formula for finding the area of a triangle is $A = \frac{1}{2}bh$, find the area of the triangle in Fig. 6–64.

31. Tosha May, an interior designer, needs to order a tablecloth to cover a circular table that is 24 in. from the floor and 18 in. across the top. The cloth must drape the table so that it just touches the floor. Determine the shape of the tablecloth. Find the number of square feet of fabric needed for the tablecloth.

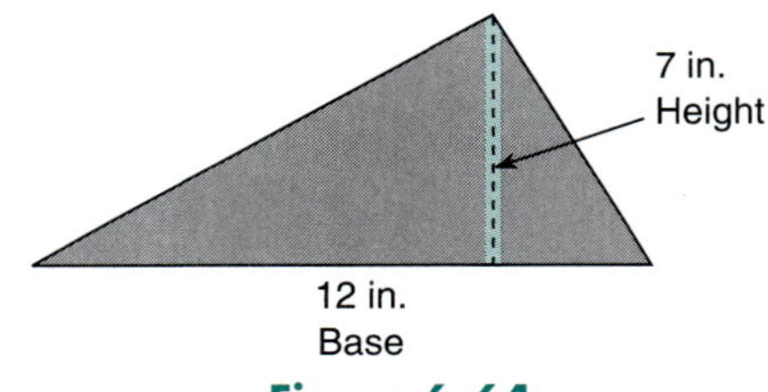

Figure 6–64

CHAPTER TRIAL TEST

Find the perimeters and areas of Figs. 6–65 through 6–68.

1.

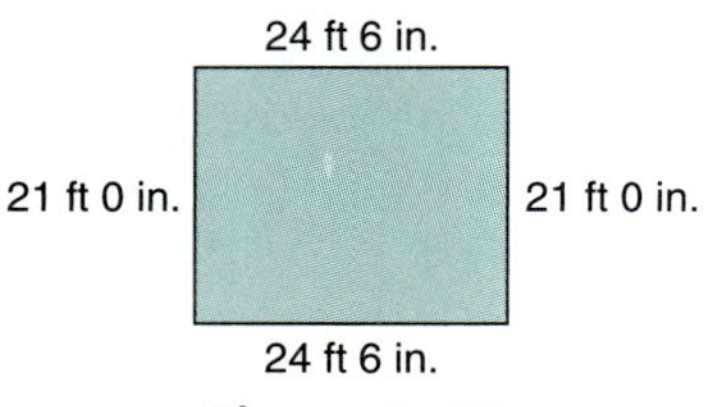

Figure 6–65

2.

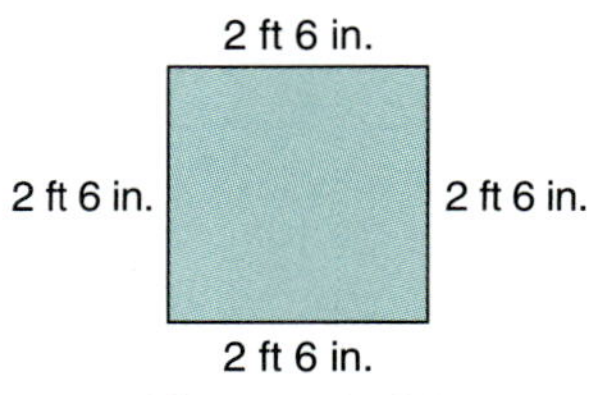

Figure 6–66

3.

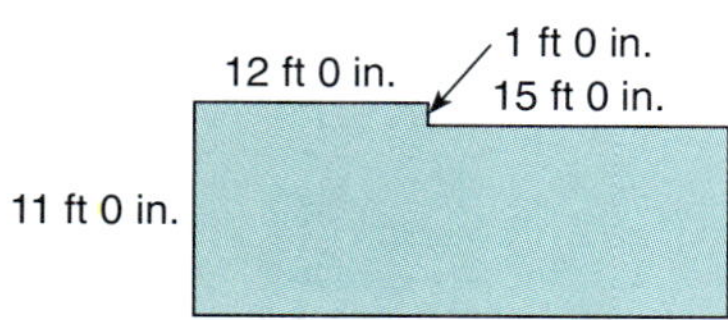

3.8 in.

2.3 in.   **1.6 in.**   2.3 in.

3.8 in.

Figure 6–67

4.

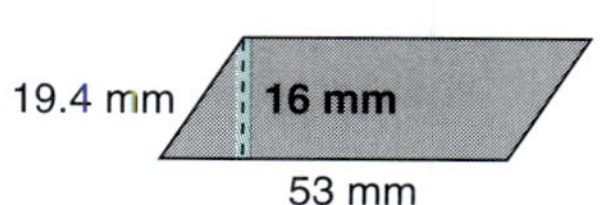

19.4 mm   **16 mm**

53 mm

Figure 6–68

5. A company charges $3.00 labor per square yard to install carpet and padding. If a dining room is 12 ft × 13 ft, what is the labor cost to install the padding and carpet ($9 \text{ ft}^2 = 1 \text{ yd}^2$)?

6. How many squares (100 ft^2) of siding are needed for four sides of a garage 18 ft × 15 ft × 9 ft high if there is 125 ft^2 of openings? Allow one-fourth extra siding for overlap and waste.

7. How many square feet of floor space are there in the plan shown in Fig. 6–69?

12 ft 0 in.   1 ft 0 in.

15 ft 0 in.

11 ft 0 in.

Figure 6–69

8. A machine stamps 1-in. squares from sheet metal. How many can it cut from a 3-ft × 4-ft sheet of metal?

Use Fig. 6–70 to identify the numbered parts of the circle.

9. _____

10. _____

11. _____

12. _____

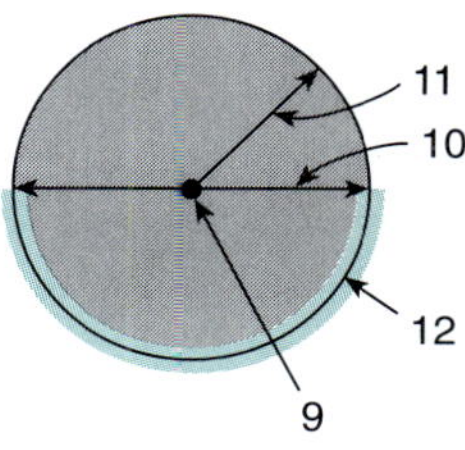

Figure 6–70

Solve these problems and when necessary, round to hundredths.

13. Find the circumference of a circle whose radius is 6 in.

14. Find the area of the circle in Exercise 13.

15. What is the perimeter of a roller-skating rink with the dimensions shown in Fig. 6–71?

16. Find the area to the nearest square foot of the skating rink in Exercise 15.

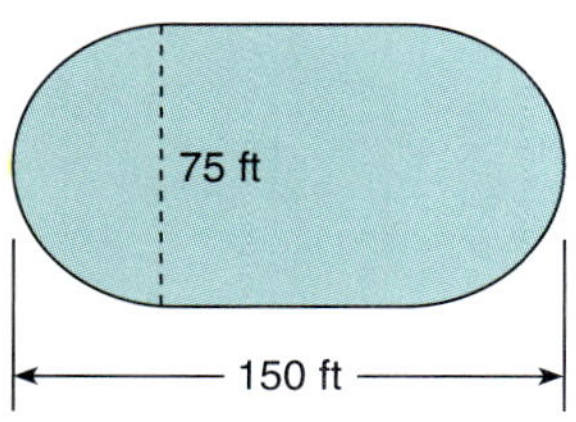

Figure 6–71

17. A water pipe with an inside diameter of $\frac{5}{8}$ in. and an outside diameter of $\frac{7}{8}$ in. is cut off at one end. What is the cross-sectional area of the ring formed by the wall of the pipe?

18. Find the area of the colored portion of the tiled walk that surrounds a rectangular swimming pool (Fig. 6–72).

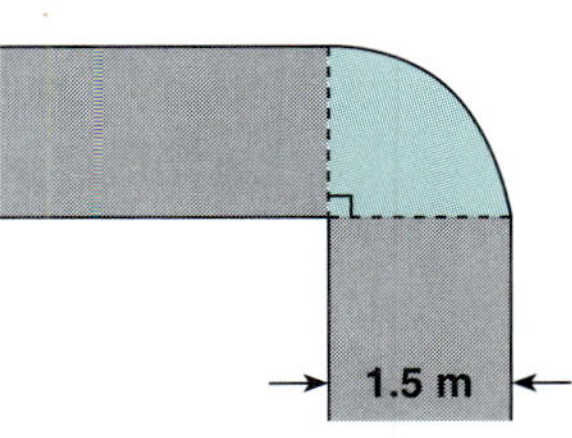

Figure 6–72

19. Find the area of a washer with an inside diameter of $\frac{1}{4}$ in. and an outside diameter of $\frac{1}{2}$ in.

20. If it costs \$1.25 to sod each square foot of a circular lawn (Fig. 6–73), how much will it cost, to the nearest cent, if the lawn is $6\frac{1}{2}$ ft wide and surrounds a round goldfish pond with a diameter of 12 ft?

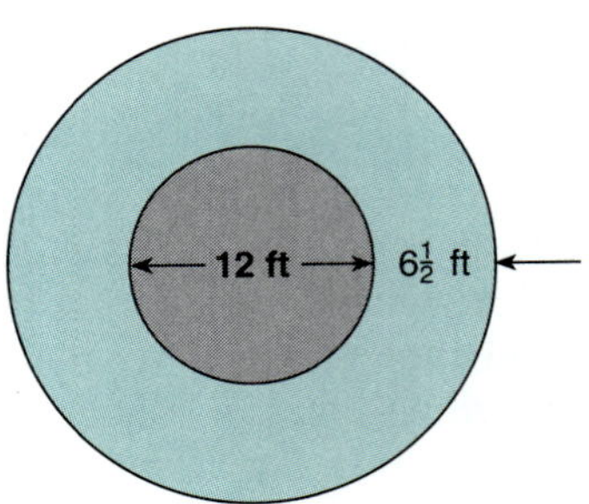

Figure 6–73

Interpreting and Analyzing Data

GOOD DECISIONS THROUGH TEAMWORK

Your team has been hired to conduct market research for a major consumer magazine. Your assignment is to choose an area of interest, conduct a survey, and prepare a report of your findings. Pay careful attention to the content of your survey: How long should it be? Should the questions be formatted to give yes or no answers, or do you want the respondents to rate the product or opinion topic? Two examples of suitable multiple-choice survey questions are: Which long-distance telephone company do you prefer? What single piece of exercise equipment would you most want to purchase? Then identify 5–10 possible responses, including "None of the above."

Next, determine your team's survey methods. How many responses do you feel will be adequate to make the results reliable? When and where will you survey people? On campus? At a mall? Through the mail? Discuss how the time of day and location can affect survey results.

Conduct the survey and record the responses, then tabulate the total number of respondents and the number choosing each possible response. Use a circle, a bar, and a line graph to plot your findings.

Write a report documenting your methods, results, and conclusions. Include the tabulation of responses and the summary graphs. Keep in mind that a high-quality report could mean another high-paying market research project for your team.

7–1 Reading circle, bar, and line graphs

1. Read circle graphs.
2. Read bar graphs.
3. Read line graphs.

7–2 Frequency distributions, histograms, and frequency polygons

1. Interpret and make frequency distributions.
2. Interpret and make histograms.
3. Interpret and make frequency polygons.

7–3 Finding statistical measures

1. Find the arithmetic mean or arithmetic average.
2. Find the median, the mode, and the range.
3. Find the standard deviation of a set of data.

A *graph* shows information visually. Graphs show how our tax dollars are divided among various government services, trace the fluctuations in a patient's temperature, and illustrate regional planting seasons. Other graphs show equations, inequalities, and their solutions. Tables, on the other hand, usually list data, such as income tax tables, which list taxes due on different incomes.

7-1 READING CIRCLE, BAR, AND LINE GRAPHS

Learning Outcomes

1 Read circle graphs.
2 Read bar graphs.
3 Read line graphs.

Graphs give us useful information at a glance, however, we must read, or interpret, graphs properly to use them to our benefit. Three common graphs used to represent data are the circle graph, the bar graph, and the line graph.

1 Read Circle Graphs.

■ **DEFINITION: Circle Graph.** A *circle graph* uses a divided circle to show pictorially how a total amount is divided into parts.

The complete circle represents one whole quantity. The circle is then divided into parts so that the sum of all the parts equals the whole quantity. These parts can be expressed as fractions, decimals, or percents. Figure 7–1 is a circle graph.

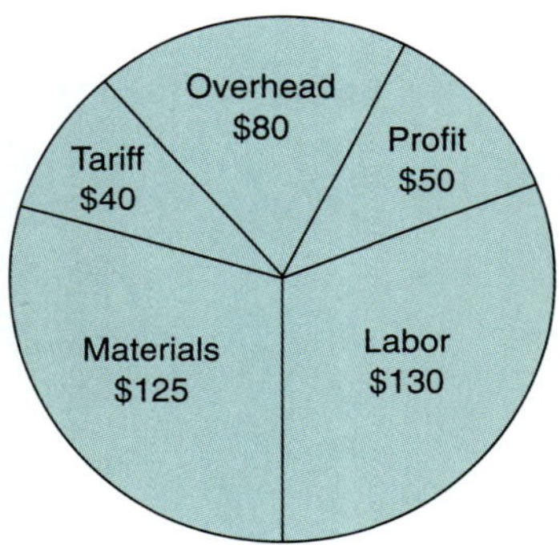

Figure 7–1 Distribution of wholesale price for a $425 color television.

When we "read" a graph, we are examining the information on the graph.

To read a circle, bar, or line graph:

1. Examine the title of the graph to find out what information is shown.
2. Examine the parts to see how they relate to one another and to the whole.
3. Examine the labels for each part of the graph and any explanatory remarks that may be given.
4. Use the given parts to calculate additional amounts.

EXAMPLE Use Fig. 7–1 to answer the questions.

(a) What percent of the wholesale price is the cost of labor?
(b) What percent of the wholesale price is the cost of materials?
(c) What would the wholesale price be if no tariff (tax) was paid on imported parts?

(a) $$\frac{R}{100} = \frac{130}{425}$$ R is the percent of wholesale price that is labor cost.

$$R \times 425 = 13{,}000$$

$$R = \frac{13{,}000}{425}$$

$$R = 30.58823529$$

$$\mathbf{R = 30.6\% \ (labor)}$$

(b) $$\frac{R}{100} = \frac{125}{425}$$ R is the percent of wholesale price that is materials cost.

$$R \times 425 = 12{,}500$$

$$R = \frac{12{,}500}{425}$$

$$R = 29.41176471$$

$$\mathbf{R = 29.4\% \ (materials)}$$

(c) **Price $-$ tariff $= 425 - 40 = \$385$ (cost without tariff)**

2 Read Bar Graphs.

Different types of graphs allow us to access different types of information. The bar graph is no exception.

> ■ **DEFINITION: Bar Graph.** A *bar graph* uses two or more bars to compare two or more amounts.

The bar lengths represent the amounts being compared. Bars can be drawn either horizontally or vertically.

The *axis* or *reference line* that runs along the length of the bars is a scale of the amounts being compared; in Fig. 7–2, this line is horizontal. The other reference line (vertical in this case) labels the bars. Figure 7–2 is a horizontal bar graph.

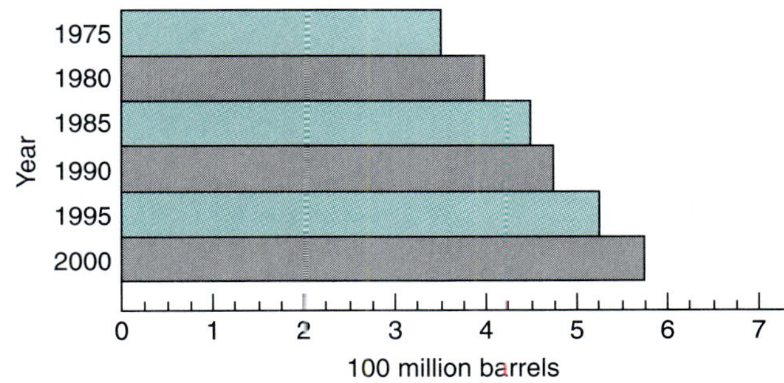

Figure 7–2 Company oil production.

EXAMPLE Use Fig. 7–2 to answer the questions.

(a) How many more 100 million barrels of oil are indicated for the company in 2000 than in 1980?

(b) Judging from the graph, should company oil production in 2005 be more or less than in 2000?

(c) How many 100 million barrels of oil did the company produce in 1980?

(a) 2000 production − 1980 production = 5.75 − 4.00 = 1.75 hundred million barrels.
(b) More, because the trend has been toward greater production.
(c) Four hundred million barrels in 1980.

3 Read Line Graphs.

A third kind of graph we encounter in industrial reports, handbooks, and the like, is the line graph.

■ **DEFINITION: Line Graph.**   A *line graph* uses one or more lines to show changes in data.

The horizontal axis or reference line on a line graph usually represents periods of time or specific times. The vertical axis or reference line is usually scaled to represent numerical amounts. Line graphs show trends in data and high values and low values at a glance. Figure 7–3 is a line graph.

EXAMPLE   Use Fig. 7–3 to answer the questions regarding a patient's temperature.

(a) On what date and time of day did the patient's temperature first drop to within 0.2° of normal (98.6°F)?
(b) On which post-op (postoperative) days was the patient's temperature within 0.2° of normal?
(c) What was the highest temperature recorded for the patient?

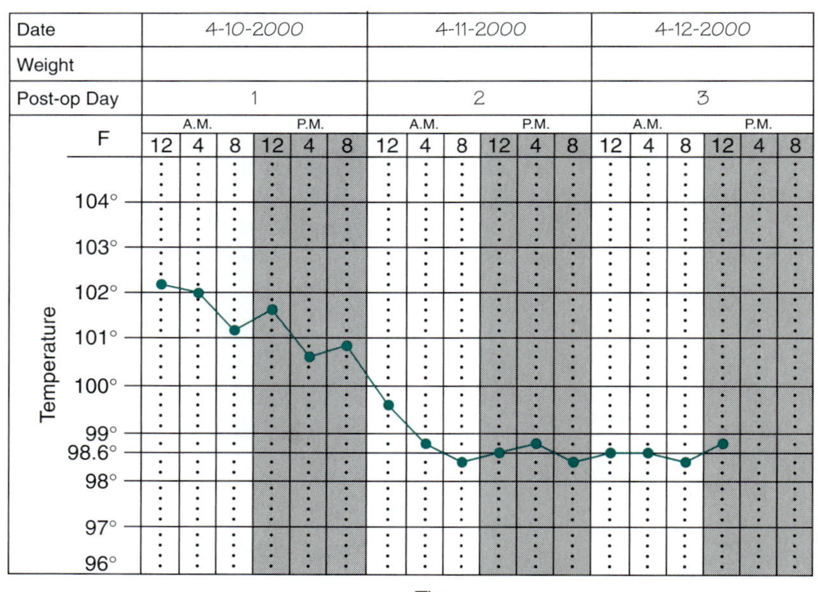

Figure 7–3   Graphic temperature chart.

(a) **4-11-00 at 4 A.M.** (Each "dot" is 0.2°, so the temperature was 98.8°F.)
(b) **Post-op days 2 and 3** (beginning at 4 A.M. on day 2)
(c) **102.2°** (recorded at 12 A.M. on 4-10-00)

1 Use Fig. 7–4 to answer Exercises 1–3.

1. What percent of the gross salary goes into savings? Round to tenths.
2. What percent of the take-home pay is federal income tax? Round to tenths.
3. What percent of the gross pay is the take-home pay? Round to tenths.

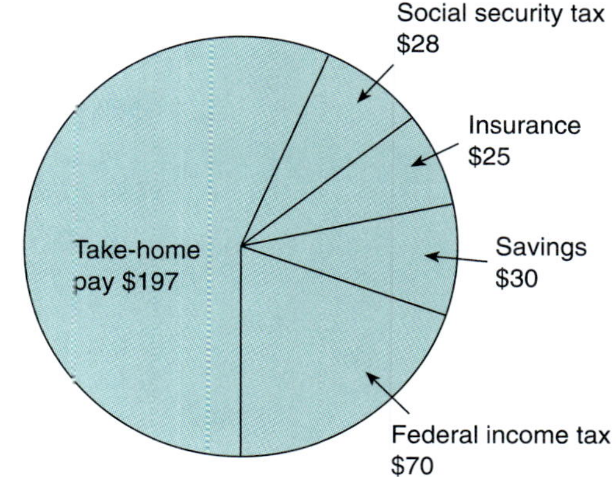

Figure 7–4   Distribution of weekly salary of $350.

2 Use Fig. 7–5 to answer Exercises 4–6.

4. What expenditure is expected to be the same next year as this year?
5. What two expenditures are expected to increase next year?
6. What two expenditures are expected to decrease next year?

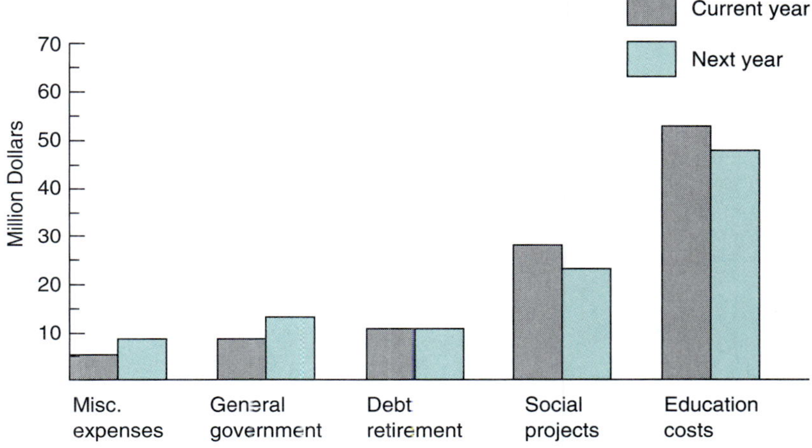

Figure 7–5   Distribution of local tax dollars.

3 Use Fig. 7–6 to answer Exercises 7–10.

7. How many amperes of current are produced by 50 V when the resistance is 10 ohms?
8. How many volts are needed to produce 2 A of current when the resistance is 25 ohms?
9. Approximately how many volts are required to produce a current of 3.5 A when the resistance is 10 ohms?
10. Find the resistance when 100 V is needed to produce a current of 4 A.

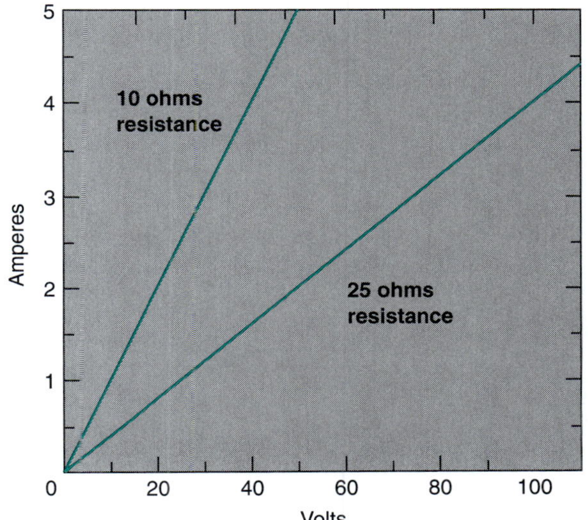

Figure 7–6   Amperage produced by voltage across two resistances.

FREQUENCY DISTRIBUTIONS, HISTOGRAMS, AND FREQUENCY POLYGONS

Learning Outcomes

1. Interpret and make frequency distributions.
2. Interpret and make histograms.
3. Interpret and make frequency polygons.

1 Interpret and Make Frequency Distributions.

Suppose for a class of 25 students the instructor records the following grades:

$$76 \quad 91 \quad 71 \quad 83 \quad 97 \quad 87 \quad 77 \quad 88 \quad 93 \quad 77 \quad 93 \quad 81 \quad 63$$

$$79 \quad 74 \quad 77 \quad 76 \quad 97 \quad 87 \quad 89 \quad 68 \quad 90 \quad 84 \quad 88 \quad 91$$

It is difficult to make sense of all these numbers as they appear here, but the instructor can arrange the scores into several smaller groups, called *class intervals*. The word *class* means a special category.

These scores can be grouped into class intervals of 5, such as 60–64, 65–69, 70–74, 75–79, 80–84, 85–89, 90–94, and 95–99. Each class interval has an odd number of scores. The *middle score* of each interval is a *class midpoint*.

The instructor can now *tally* the number of scores that fall into each class interval to get a *class frequency,* the number of scores in each class interval.

A compilation of class intervals, midpoints, tallies, and class frequencies is called a *frequency distribution.*

EXAMPLE Examine the frequency distribution in Table 7–1 and answer the following questions.

(a) How many students scored 70 or above?

$$2 + 6 + 3 + 5 + 5 + 2 = 23$$

23 students scored 70 or above.

(b) How many students made A's (90 or higher)?

$$5 + 2 = 7$$

7 students made A's (90 or higher).

(c) What percent of the total grades were A's (90's)?

$$\frac{7 \text{ A's}}{25 \text{ total}} = \frac{7}{25} = 0.28 = \textbf{28\% A's}$$

(d) Were the students prepared for the test, or was the test too difficult?

The relatively high number of 90's (7) compared with the relatively low number of 60's (2) suggests that **in general, most students were prepared for the test.**

TABLE 7–1 Frequency Distribution of 25 Scores

Class Interval	Midpoint	Tally	Class Frequency
60–64	62	/	1
65–69	67	/	1
70–74	72	//	2
75–79	77	++++ /	6
80–84	82	///	3
85–89	87	++++	5
90–94	92	++++	5
95–99	97	//	2

(e) What is the ratio of F's (60's) to A's (90's)?

$$\frac{2 \text{ F's}}{7 \text{ A's}} = \frac{2}{7}$$

The ratio is $\frac{2}{7}$.

EXAMPLE Students in a history class reported their credit-hour loads as shown. Make a frequency distribution of their credit hours. Credit hours carried: 3, 12, 15, 3, 6, 6, 12, 9, 12, 9, 6, 3, 12, 18, 6, 9.

To establish a class interval with an easy-to-find midpoint, use an odd number of points in the interval. Here, an interval of 5 is used; that is, 0–4 contains five possibilities: 0, 1, 2, 3, and 4. The middle number is the midpoint, 2. Make a tally mark for each time the credit hours of a student falls in the interval. Then count the tally marks to get the class frequency (Table 7–2).

TABLE 7–2

Class Interval	Midpoint	Tally	Class Frequency
0–4	2	///	3
5–9	7	+++ //	7
10–14	12	////	4
15–19	17	//	2

2 Interpret and Make Histograms.

A *histogram* is a bar graph that uses two scales, one for class intervals and one for class frequencies. The frequency distribution in Table 7–2 can be made into a histogram. The frequencies in this histogram form the vertical scale. The class intervals form the horizontal scale.

EXAMPLE Use the histogram in Figure 7–7 to answer the questions.

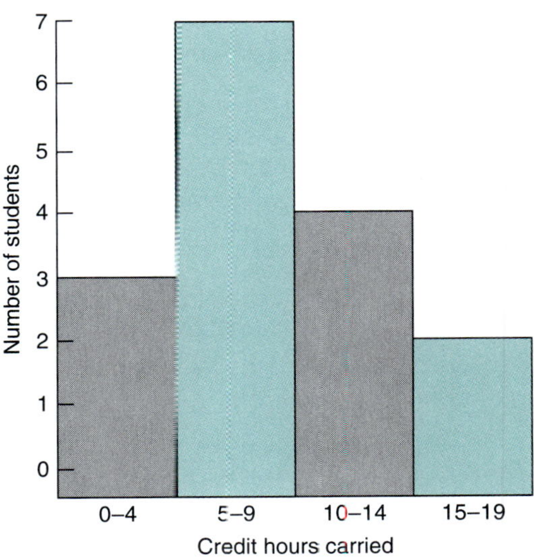

Figure 7–7 Credit hours carried by 16 students in history class.

(a) How many students carried 5–9 hours?

7 carried 5–9 hours.

(b) How many students carried less than 15 hours?

$3 + 7 + 4 =$ **14 carried less than 15 hours.**

(c) What percent of the total carried 10–14 hours?

$$\frac{4}{16} = \frac{1}{4} = 0.25 = \mathbf{25\%}$$

(d) What is the ratio of students carrying 0–4 hours to those carrying 5–9 hours?

$$\frac{3 \text{ with } 0\text{–}4}{7 \text{ with } 5\text{–}9} = \frac{3}{7}$$

The ratio is $\frac{3}{7}$.

EXAMPLE A hospital department bases employee vacation leave on years of service. Employees fall into four categories: 0–2 years, 8 employees; 3–5 years, 6 employees; 6–8 years, 4 employees; 9–11 years, 2 employees. Make a histogram showing this information.

The years of service are already arranged in class intervals and so the intervals may be used as given. The class frequency or number of employees in each interval is also provided and so the frequencies may also be used as given. Therefore, it is *not* necessary to make a frequency distribution. The class intervals form the horizontal scale, and the frequencies form the vertical scale (see Fig. 7–8).

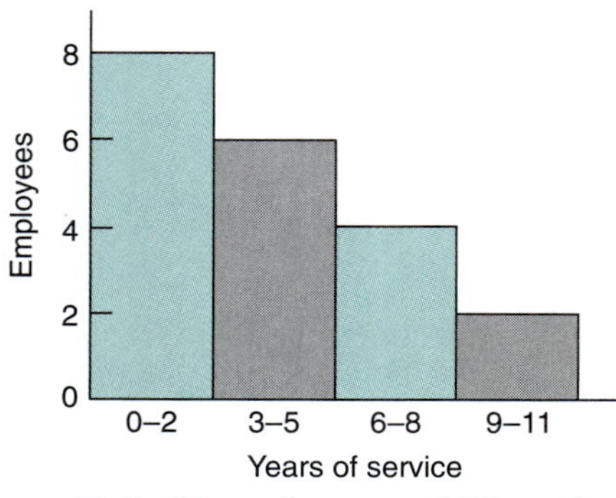

Figure 7–8 Years of service of 20 employees.

3 Interpret and Make Frequency Polygons.

To make a *frequency polygon,* we start with a histogram, identify each class midpoint, mark it with a dot, and connect the dots to make a line graph. Let's use Fig. 7–8 to make a frequency polygon. The midpoints may be considered an average for each class interval.

EXAMPLE Make a frequency polygon for the histogram in Fig 7–8.

First, identify the class midpoints with a dot, as in Fig. 7–9. Connect the dots with a line, as in Fig. 7–10. Write the line graph without the bars and connect to the horizontal axis at the smallest value of the first interval and the largest value of the last interval, as in Fig. 7–11.

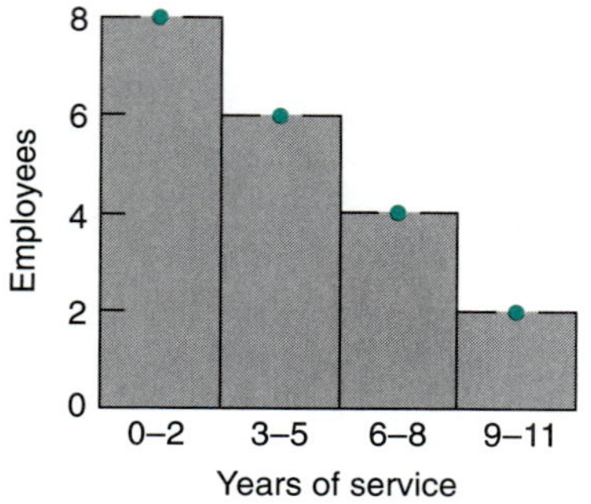

Figure 7–9 Years of service of 20 employees.

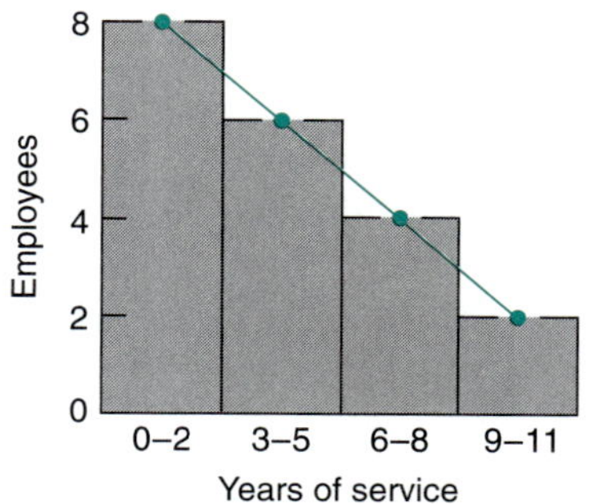

Figure 7–10 Years of service of 20 employees.

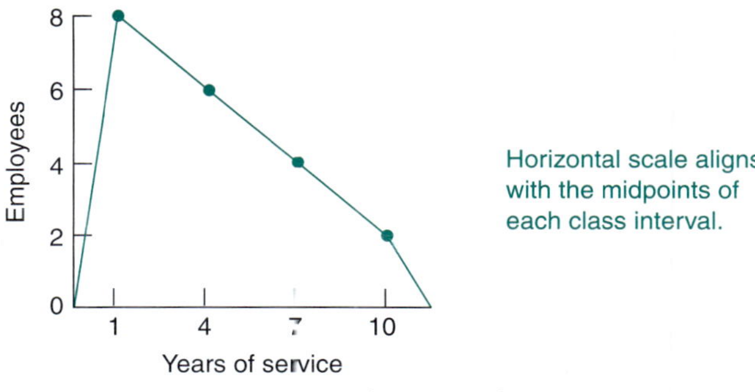

Figure 7–11 Years of service of 20 employees.

EXAMPLE Use the frequency polygon in Fig. 7–11 to answer the questions.

(a) How many employees have 5 or fewer years of service?

$$8 + 6 = \textbf{14 with 5 or fewer years of service}$$

(b) What is the ratio of employees with 2 or fewer years of service to employees with 9 or more years of service?

$$\frac{8 \text{ with 0 to 2 years}}{2 \text{ with 9 or more}} = \frac{8}{2} = \frac{4}{1}$$

The ratio is $\frac{4}{1}$.

(c) What percent of the number of employees with 2 or fewer years of service is the number of employees with 9 or more years of service?

$$\frac{2}{8} = \frac{1}{4} = 0.25 = \textbf{25\%}$$

(d) What is the average number of years of service for employees in the 6 to 8 years of service bracket?

7 is the midpoint or average for the interval.

SELF-STUDY EXERCISES 7–2

1 Use Table 7–3 to answer the questions. The frequency distribution shows the ages of 25 college students in a landscaping class.

1. How many students are 22 or younger?
2. How many students are older than 34?
3. What is the ratio of the number of students 38–40 to the number of students 17–19?
4. What is the ratio of the smallest class frequency to the largest class frequency?
5. What percent of the total class are students age 17–19?
6. What percent of the total class are students age 20–22?
7. What two age groups make up the smallest number of students in the class?
8. What two age groups make up the largest number of students in the class?

TABLE 7–3 Frequency Distribution of 25 Ages

Class Interval	Midpoint	Tally	Class Frequency
38–40	39	/	1
35–37	36	/	1
32–34	33	//	2
29–31	30	///	3
26–28	27	//	2
23–25	24	//// /	6
20–22	21	//// //	7
17–19	18	///	3

9. How many students are over age 28?
10. How many students are under age 26?

Use the given hourly pay rates (rounded to the nearest whole dollar) for 33 support employees in a private college to complete a frequency distribution using the format shown in Table 7–4.

TABLE 7–4 Pay Rates of 33 Support Employees

Class Interval	Midpoint	Tally	Class Frequency
11. $14–16	_______	_______	_______
12. $11–13	_______	_______	_______
13. $8–10	_______	_______	_______
14. $5–7	_______	_______	_______

$6	$6	$10	$7	$6	$6	$6
$6	$7	$7	$8	$8	$6	$6
$11	$10	$7	$11	$8	$16	$6
$6	$9	$6	$7	$9	$6	$6
$12	$13	$7	$15	$5		

Use the given 40 test scores of two physics classes to complete a frequency distribution using the format in Table 7–5.

TABLE 7–5 Test Scores of 40 Physics Students

Class Interval	Midpoint	Tally	Class Frequency
15. 91–95	_______	_______	_______
16. 86–90	_______	_______	_______
17. 81–85	_______	_______	_______
18. 76–80	_______	_______	_______
19. 71–75	_______	_______	_______
20. 66–70	_______	_______	_______
21. 61–65	_______	_______	_______
22. 56–60	_______	_______	_______

57	91	76	89	82	59	72	88
76	84	67	59	77	66	56	76
77	84	85	79	69	88	75	58
85	65	67	66	93	83	69	81
80	64	78	76	72	90	79	90

2 Complete the following.

23. Use the information from Exercises 11–14 to make a histogram. Use the reference lines in Fig. 7–12 as guides.

24. Use the information from Exercises 15–22 to make a histogram.

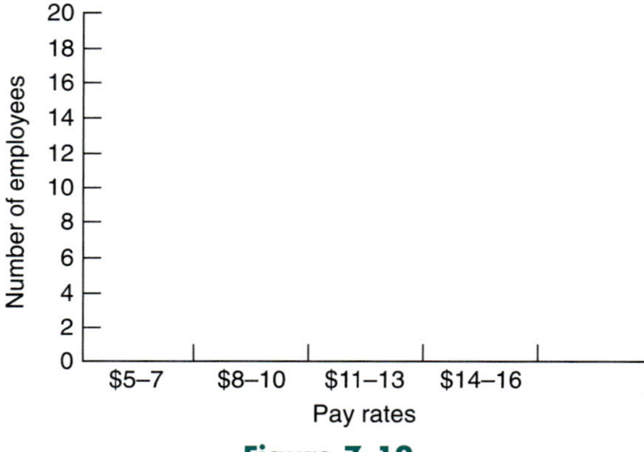

Figure 7–12

3

25. Use the information from Exercises 11–14 to make a frequency polygon. Use the reference lines in Fig. 7–12 as guides.

26. Use the information from Exercises 15–22 to make a frequency polygon.

7–3 FINDING STATISTICAL MEASURES

Learning Outcomes

1 Find the arithmetic mean or arithmetic average.

2 Find the median, the mode, and the range.

3 Find the standard deviation of a set of data.

In this age of information explosion and with the increasing use of computers, we have massive amounts of data available to us. To use these data effectively in the decision-making process, we need to examine the data and summarize key trends and characteristics. This summary is generally in the form of statistical measurements. A *statistical measurement* or *statistic* is a standardized, meaningful measure of a set of data that reveals a certain feature or characteristic of the data. Such statistics are called *descriptive statistics*.

1 Find the Arithmetic Mean or Arithmetic Average.

An *average* is an approximate number that is a central value or typical value of a group of data values. There are several different types of averages. The most common is the *arithmetic mean* or *arithmetic average*. This average is the sum of the quantities in the data set divided by the number of quantities.

A general practice in the study of statistics is to use symbols to write the procedures for statistical measures. To write the procedures for finding the mean symbolically, we need to introduce three new symbols: x_i, Σ, and $\bar{x}$. Using these symbols, we can write the formula for finding the mean as

$$\bar{x} = \frac{\Sigma x_i}{n}$$

This can be read as "the mean (or *x*-bar) is equal to the sum of each value of x (x sub i) divided by the number of values (n)." The Greek capital letter *sigma*, Σ, is a summation symbol and indicates the addition of a group of values. The notation x_i identifies each data value with a subscript: x_1 is the first value, x_2 is the second value, and x_n in the nth value.

To find the arithmetic mean or average:

Find the mean of 22, 31, and 37.
$$x_1 = 22, x_2 = 31, x_3 = 37$$

1. Add the quantities. $\quad \Sigma x_i$

$$\Sigma x_i = 22 + 31 + 37$$
$$\Sigma x_i = 90$$

2. Divide their sum by the number of quantities. $\quad \bar{x} = \dfrac{\Sigma x_i}{n}$

$$\bar{x} = \frac{\Sigma x_i}{n} = \frac{90}{3}$$
$$\bar{x} = 30$$

> **EXAMPLE** Find the mean of each group of quantities.
>
> (a) Pulse rates: 68, 84, 76, 72, 80
>   There are 5 pulse rates, so we find their sum and divide by 5:
>
> $$\bar{x} = \frac{\Sigma x_i}{n} = \frac{68 + 84 + 76 + 72 + 80}{5} = \frac{380}{5} = 76$$
>
> **The mean pulse rate is 76.**
>
> (b) Pounds: 21, 33, 12.5, 35.2 (to nearest hundredth)
>   There are 4 weights, so we find their sum and divide by 4.
>
> $$\bar{x} = \frac{\Sigma x_i}{n} = \frac{21 + 33 + 12.5 + 35.2}{4} = \frac{101.7}{4} = 25.425$$
>
> **The mean weight is 25.43 lb (to nearest hundredth).**

> **EXAMPLE** An automobile used 41 gallons of regular gasoline on a trip of 876 miles. What was the average miles per gallon (mpg) to the nearest tenth?
>
> The number of miles traveled (Σx_i) on the 41 gallons of gasoline is 876, and we divide by the number of gallons (n) used or 41.

$$\bar{x} = \frac{\Sigma x_i}{n} = \frac{876}{41} = 21.4 \qquad \text{To the nearest tenth.}$$

The car averaged 21.4 miles per gallon on the trip.

EXAMPLE   Table 7–6 lists the final grades of a horticulture student. Find the student's QPA (quality point average) to the nearest hundredth based on a 4-point system.

TABLE 7–6   Quality Points for Final Grades

Subject	Grade	Credit Hours (n)		Quality Points per Hour		Total Points (Σx_i)
Algebra 101	B	3	×	3	=	9
Spray chemicals 102	A	3	×	4	=	12
Landscape 301	A	3	×	4	=	12
English 101	C	4	×	2	=	8
		13 hours				41 points

To find the QPA for a term, we multiply the quality points for the letter grade of each course by the credit hours of each course to obtain the total quality points for each course. We divide this total by the total credit hours earned. The quality points awarded are A = 4, B = 3, and C = 2.

$$\bar{x} = \frac{\Sigma x_i}{n} = \frac{41}{13} = 3.153846154 \qquad \text{or} \qquad 3.15 \qquad \text{To the nearest hundredth.}$$

Thus, the student's QPA for the term is 3.15.

EXAMPLE   A community college student has the following grades in Physics 101: 73, 84, 80, 62, and 70. What grade is needed on the last test for the student to get a C on the final grade, or a 75 average?

One way to find out the needed grade is to assume that each grade is 75 for the 6 tests:

$$\frac{75 + 75 + 75 + 75 + 75 + 75}{6} = \frac{450}{6} = 75$$

Then, find the sum of the first five tests actually taken:

$$73 + 84 + 80 + 62 + 70 = 369$$

Subtract to find the difference between this sum and 450. The difference is the needed score.

$$450 - 369 = 81$$

The student must earn a score of 81. To check, use 81 for the last test and find the average.

$$\frac{73 + 84 + 80 + 62 + 70 + 81}{6} = \frac{450}{6} = 75$$

Thus, 81 is the score needed on the last test.

2   Find the Median, the Mode, and the Range.

Besides the mean or arithmetic average, we also use the *median* and the *mode* to describe groups of data.

Median.   The *median* is the middle value of a set of data values arranged in order of size.

> *To find the median:*
>
> 1. Arrange the values in order of size.
> 2. If the number of values is odd, the median is the middle value.
> 3. If the number of values is even, the median is the average of the two middle values.

EXAMPLE   A TPR chart shows a patient's *t*emperature, *p*ulse rate, and *r*espiration rate. The following pulse rates were recorded on a TPR chart: 68, 88, 76, 64, 72. What is the patient's median pulse rate?

In order of size:
```
88
76
72  ← median or middle value in odd number of values
68
64
```

The median pulse rate is 72.

EXAMPLE   The following temperatures were recorded: 56°, 48°, 66°, and 62°. What is the median temperature?

The number of temperatures is even, so we find the average of the two middle values.

In order of size:
$$48$$
$$\left.\begin{matrix} 56 \\ 62 \end{matrix}\right\} \frac{56 + 62}{2} = \frac{118}{2} = 59$$
$$66$$

Thus, the median temperature is 59°.

Mode.   The *mode* is the most frequently occurring value in the data set.

> *To find the mode:*
>
> 1. Identify the value that occurs most frequently as the mode.
> 2. If no value occurs more than another value, there is no mode for this data set.
> 3. If more than one value occurs with the same frequency that is the greatest frequency, the data set of values will have more than one mode.

EXAMPLE   The hourly pay rates at a local fast-food restaurant are as follows: cooks, $5.50; servers, $5.15; bussers, $5.15; dishwashers, $5.25; managers, $7.50. Find the mode.

The hourly pay rate of $5.15 occurs more than any other rate. It is the mode.

EXAMPLE   The daily work shifts at a mall clothing store are 4, 6, and 8 hours. Find the mode.

No shift occurs more frequently than another, so there is no mode.

Range.   The mean, the median, and the mode are *measures of central tendency*. Another group of statistical measures are *measures of variation or dispersion*. One of these measures is the *range*. The range is the difference between the highest value and the lowest value in a set of data.

To find the range:
1. Find the highest and lowest values.
2. Find the difference between the highest and lowest values.

EXAMPLE Find the range for the data described in the example for fast-food restaurant hourly pay rates.

The high value is $7.50. The low value is $5.15.

$$\text{Range} = \$7.50 - \$5.15 = \mathbf{\$2.35}$$

Another measure of variation or dispersion is *standard deviation*. We examine standard deviation in a Career Application later in this chapter.

Tip!	***Use More Than One Statistical Measure.***

A common mistake when making conclusions or inferences from statistical measures is to examine only one statistic, such as the mean. To obtain a complete picture of the data requires looking at more than one statistic.

3 | Find the Standard Deviation of a Set of Data.

Although the range gives us some information about dispersion, it does not tell us whether the highest or lowest values are typical values or extreme outliers. We can get a clearer picture of the data set by examining how much each data point *differs* or *deviates* from the mean.

The *deviation from the mean* of a data value is the difference between the value and the mean.

To find the deviations from the mean:

Data set: 38, 43, 45, 44.

1. Find the mean of a set of data.

$$\overline{x} = \frac{\text{Sum of data values}}{\text{Number of values}} = \frac{\Sigma x_i}{n}$$

$$\frac{38 + 43 + 45 + 44}{4} = \frac{170}{4} = 42.5$$

2. Find the amount that each data value deviates or is different from the mean.

$$\text{Data value} - \text{Mean} = x_i - \overline{x}$$

$$38 - 42.5 = -4.5 \text{ (below the mean)}$$
$$43 - 42.5 = 0.5 \text{ (above the mean)}$$
$$45 - 42.5 = 2.5 \text{ (above the mean)}$$
$$44 - 42.5 = 1.5 \text{ (above the mean)}$$

When the value is smaller than the mean, the difference is represented by a negative amount. This shows that the value is *below* or less than the mean. When the value is larger than the mean, the difference is represented by a positive amount. This shows that the value is *above* or greater than the mean. *The sum of the deviations below the mean should equal the sum of the deviations above the mean.* In the example in the box, only one value is below the mean, and its deviation is −4.5. Three values are above the mean, and the sum of these deviations is 0.5 + 2.5 + 1.5 = 4.5. We say that *the sum of all deviations from the mean is zero.* This is true for all sets of data.

We have not gained any statistical insight or new information by analyzing the sum of the deviations of the mean or even by analyzing the average of the deviations.

$$\text{Average deviation} = \frac{\text{Sum of deviations}}{\text{Number of values}} = \frac{0}{n} = 0$$

EXAMPLE Find the deviations from the mean for the set of data 45, 63, 87, and 91.

$$\bar{x} = \frac{45 + 63 + 87 + 91}{4} = \frac{286}{4} = 71.5 \qquad \text{Mean.}$$

Now let's define the deviation from the mean symbolically as $x_i - \bar{x}$. That is, we subtract the mean $\bar{x}$ from each value of x. We arrange these values in a table and find the sum of the deviations, $(x_i - \bar{x})$:

Values i	Mean x_i	Deviations $x_i - \bar{x}$
1	45	$45 - 71.5 = -26.5$
2	63	$63 - 71.5 = -8.5$
3	87	$87 - 71.5 = 15.5$
4	91	$91 - 71.5 = 19.5$
Total	286	0

As we might expect, the sum of the deviations equals zero because the sum of the negative deviations $(-26.5 + -8.5 = -35)$ equals the sum of the positive deviations $(15.5 + 19.5 = 35)$. To avoid this situation, mathematicians employ a statistical measure called the *standard deviation*, which uses the square of each deviation from the mean. The square of a negative value is always positive. The squared deviations are averaged (mean) and the result is called the *variance*. The square root is taken of the variance so that the result can be interpreted within the context of the problem. Other formulas exist for finding the standard deviation of a set of values, but we examine only one formula. This formula requires several calculations.

$$s = \sqrt{\frac{\Sigma(x_i - \bar{x})^2}{n - 1}}$$

To find the standard deviation of a set of data:

1. Find the mean, $\bar{x}$.
2. Find the deviation of each value from the mean. $(x_i - \bar{x})$
3. Square each deviation. $(x_i - \bar{x})^2$
4. Find the sum of the squared deviations. $\Sigma(x_i - \bar{x})^2$
5. Divide the sum of the squared deviations by *one less than* the number of values in the data set. This amount is called the *variance, v.* $\dfrac{\Sigma(x_i - \bar{x})^2}{n - 1}$
6. Find the square root of the variance. $\sqrt{\dfrac{\Sigma(x_i - \bar{x})^2}{n - 1}}$

To calculate the standard deviation manually, it helps to arrange our calculations in a table. To find the standard deviation, we average the squared deviations using one less than the number of values $(n - 1)$ and then find the square root of this average. The formula we are using deals with a sample of an entire population or a small set of data. In this formula instead of dividing by n, the number of data values in the sample, we divide by $n - 1$. In very large sets of data, there is very little difference in the calculations using n versus $n - 1$. The logical basis for using $n - 1$ assumes that one data value is *exactly* the mean (which may or may not be true) and there are $n - 1$ data values that deviate from the mean.

EXAMPLE Find the standard deviation for the values 45, 63, 87, and 91.

i	x_i	$x_i - \bar{x}$	$(x_i - \bar{x})^2$
1	45	$45 - 71.5 = -26.5$	$(-26.5)^2 = 702.25$
2	63	$63 - 71.5 = -8.5$	$(-8.5)^2 = 72.25$
3	87	$87 - 71.5 = 15.5$	$(15.5)^2 = 240.25$
4	91	$91 - 71.5 = 19.5$	$(19.5)^2 = 380.25$
Total	286	0	1,395

$$v = \frac{\Sigma(x_i - \bar{x})^2}{n - 1}, \ s = \sqrt{v}$$

$$s = \sqrt{\frac{1,395}{3}} = \sqrt{465} = 21.563859 \ or \ \textbf{21.6 standard deviation}$$

To interpret the meaning of standard deviation, note that a smaller value indicates that the mean is a typical value in the data set. A large standard deviation indicates that the mean is not typical, and other statistical measures should be examined to better understand the characteristics of the data set. In this example, the mean of 71.5 with a standard deviation of 21.6 indicates that the data values do not cluster closely around the mean.

SELF-STUDY EXERCISES 7–3

1 Find the mean of the given values. Round to hundredths if necessary.

1. 12, 14, 16, 18, 20

2. 13, 15, 17, 19, 21

3. 68, 54, 73, 69

4. 85, 68, 77, 65

5. 37.6, 29.8

6. 65.3, 67.9

7. 32°F, 41°F, 54°F

8. 10°C, 13°C, 15°C

9. $27, $32, $65, $29, $21

10. $32, $43, $22, $63, $36

11. 11 in., 17 in., 16 in.

12. 9 in., 7 in., 8 in.

13. Respiration rates: 16, 24, 20

14. Pulse rates: 68, 84, 76

Solve the problems.

15. A baseball player batted 276 home runs over a 16-year period. What was the average number of home runs per year to the nearest tenth?

16. Noel Womack scored 83, 96, 86, 92, and 93 in English 101. What must he score on the last test to earn an average score of 92?

17. An automobile used 32 gallons of regular gasoline on a 786-mi trip. What was the average miles per gallon to the nearest tenth?

18. A pickup truck used 25 gallons of regular gasoline on a 256-mi trip. What was the average miles per gallon to the nearest tenth?

2 Find the median for each data set.

19. 32, 56, 21, 44, 87

20. 78, 23, 56, 43, 38

21. 12, 21, 14, 18, 15, 16

22. 21, 33, 18, 32, 19, 44

23. $22, $35, $45, $30, $29

24. $66, $54, $76, $55, $69

25. The following hourly pay rates are used at fast-food restaurants: cooks, $5.15; servers, $5.25; bussers, $5.15; dishwashers, $5.25; managers, $8.25. Find the median pay rate.

26. The following hourly pay rates are used at a locally owned store: clerks, $5.45; bookkeepers, $6.25; operators, $5.15; assistant managers, $7.95. Find the median pay rate.

Find the mode for each data set.

27. 2, 4, 6, 2, 8, 2

28. 5, 12, 5, 5, 20

29. 21, 32, 67. 34, 23, 22

30. 32, 45, 41, 23, 56, 77

31. $56, $67, $32, $78, $67, $20, $67, $56

32. S32, $87, $67, $32, $32, $87, $77, $22

33. These weekend work shifts are in effect at a mall clothing store: 4 hours in A.M., 6 hours in P.M., 4 hours in P.M. Find the mode for the number of hours.

34. These special prices are in effect at a fast-food restaurant: $1.75, hamburgers; $1.97, hot ham sandwiches; $2.38, chicken fillet sandwiches; S1.97, roast beef sandwiches. Find the mode.

Find the range for each data set.

35. 22, 36, 41, 41, 17

36. 28, 33, 36, 13, 28

37. 10, 23, 12, 17, 13, 16

38. 23, 23, 18, 32, 29, 14

39. $25, $15, $25, $40, $19

40. $36, $44, $26. $52, $19

41. 23°F, 37°F, 29°F, 54°F, 46°F, 71°F, 67°F

3 Find the standard deviation for each data set.

42. 12, 14, 16, 18, 20

43. 58, 54, 73, 69

44. 32°F, 41°F, 54°F

45. $27, $32, $65, $29, $21

46. Respiration rates: 16, 24, 20

47. Pulse rates: 68, 84, 76

Teaching: Using Statistics to Examine Class Performance

Teachers use a variety of statistical measures to evaluate their students' understanding of a particular concept. These statistical measures are usually applied to one or more classes. Chloe Duke teaches two classes of technical math. She has given each class the same quiz on probability. When she examines the mean and range for each class, she discovers that they are exactly the same, but from a visual inspection of the scores it appears that the morning class has a better understanding of probability than the afternoon class. She decides to justify her speculation by calculating the standard deviation for each set of grades.

The quiz scores are as follows:

Morning class: 79, 80, 80, 59, 81, 80, 80, 35, 80, 85

Afternoon class: 66, 100, 64, 99, 50, 99, 69, 63, 67, 62

For each class, find the

1. mean score
2. range
3. median
4. mode
5. standard deviation
6. Explain why you agree or disagree with Ms. Duke's speculation.

Answers

1. Mean of morning class = 73.9; mean of afternoon class = 73.9.
2. Range of morning class = 50; range of afternoon class = 50.
3. Median of morning class = 80; median of afternoon class = 66.5
4. Mode for morning class = 80; mode for afternoon class = 99.
5. Standard deviation of morning class = 15.4; standard deviation of afternoon class = 18.3.
6. The high range for each class indicates that the scores have a wide dispersion. In examining the median for each set of scores, you see that the median for the morning class is significantly higher than that for the afternoon class.

The mode for the afternoon class is not at all descriptive of class scores as a whole. The standard deviation for the afternoon class is larger than that of the morning class, which indicates that the scores of the afternoon are not as closely clustered about the mean as those of the morning class. In looking at the total picture, you see that the median and the standard deviation give you the most useful information in agreeing with Ms. Duke that the morning class as a whole has a better understanding of probability than the afternoon class.

ASSIGNMENT EXERCISES

Section 7–1

Use Fig. 7–13 to answer Exercises 1–4.

1. In what year(s) did women use more sick days than men?
2. In what year(s) did men use about five sick days?
3. In what year(s) did men use more sick days than women?
4. What was the greatest number of sick days for men?

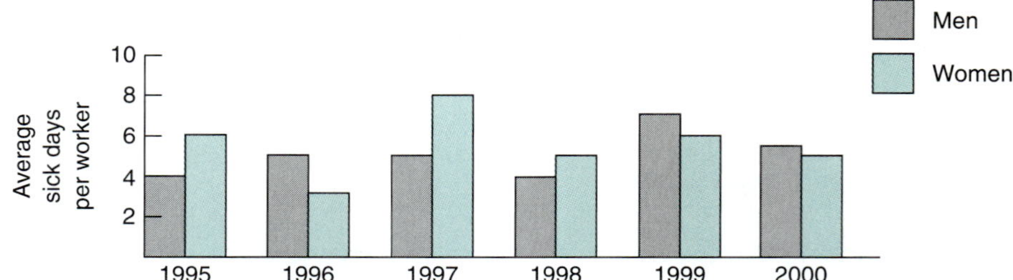

Figure 7–13 Comparison of sick days for men and women.

Use Fig. 7–14 to answer Exercises 5–7.

5. What percent of the total cost is the cost of the lot? Round to the nearest tenth.
6. What percent of the total cost is the cost of the house? Round to the nearest tenth.
7. The cost of the lot and landscaping represents what percent of the total cost? Round to the nearest tenth.

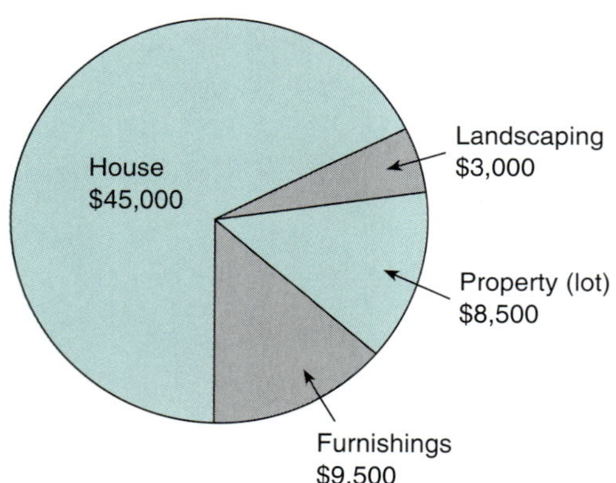

Figure 7–14 Distribution of costs for a $66,000 home.

Use Fig. 7–15 to answer Exercises 8–10.

8. What was the patient's highest pulse rate? The highest respiration rate?
9. On what date and time were the patient's pulse rate and respiration rate at their highest level?
10. On what days was the patient's respiration rate below 25?

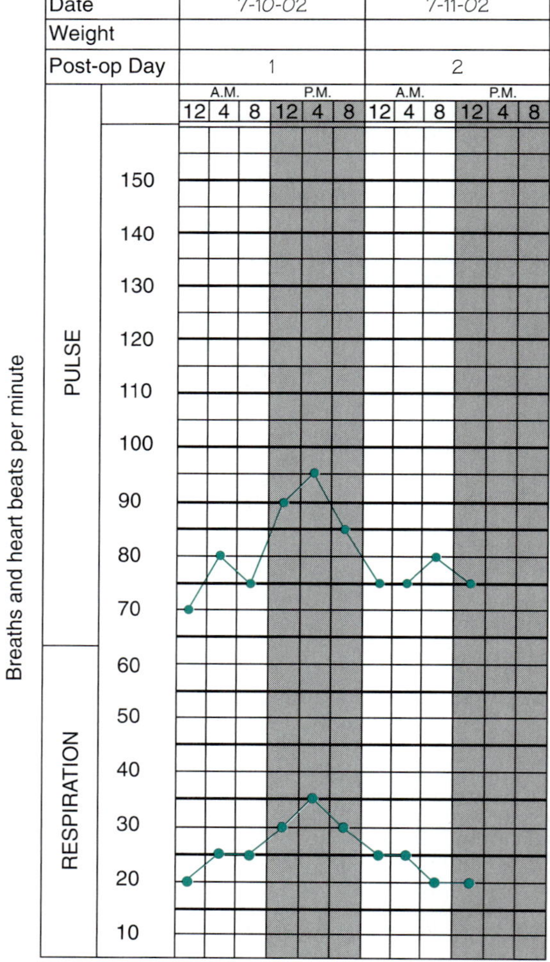

Figure 7–15 Graphic respiration/pulse chart.

    CHAPTER 7   Interpreting and Analyzing Data

Use the histogram in Fig. 7–16 to answer the questions about computer sales at three retail stores.

11. How many computers were sold at Randle?
12. How many computers were sold at Weppner?
13. How many computers were sold by all three stores?
14. How many computers were sold by Bassett and Weppner combined?
15. What is the ratio of computers sold at Weppner to those sold at Bassett?

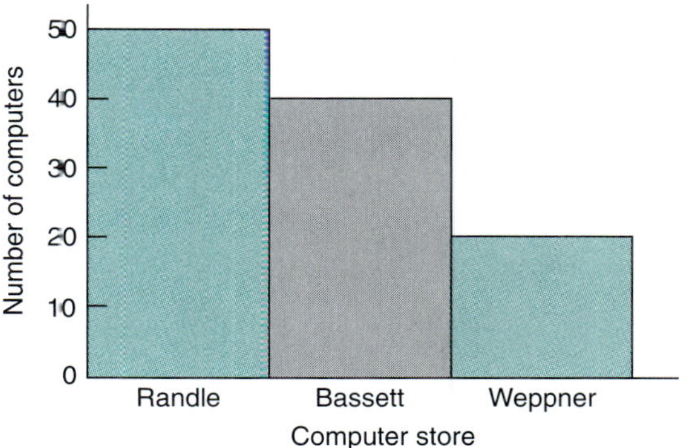

Figure 7–16 Computer sales at three retail stores.

A regional horticultural association is composed of 54 members of several clubs. Use the format shown in Table 7–7 and make a frequency distribution of the members' ages: 17, 18, 20, 21, 21, 24, 24, 29, 29, 29, 31, 31, 33, 33, 34, 35, 35, 38, 38, 38, 39, 41, 42, 43, 43, 43, 43, 45, 45, 47, 47, 48, 48, 48, 49, 50, 51, 51, 52, 56, 56, 58, 58, 60, 60, 62, 64, 64, 65, 66, 68, 70, 71, 71.

TABLE 7–7 Ages of Club Members of Regional Horticultural Association

	Class Interval	Midpoint	Tally	Class Frequency
16.	66–75	__________	__________	__________
17.	56–65	__________	__________	__________
18.	46–55	__________	__________	__________
19.	36–45	__________	__________	__________
20.	26–35	__________	__________	__________
21.	16–25	__________	__________	__________

22. Make a histogram of the members' ages from the previous frequency distribution. Use the reference lines in Fig. 7–17 as guides.

23. Make a frequency polygon from the histogram made in Exercise 22.

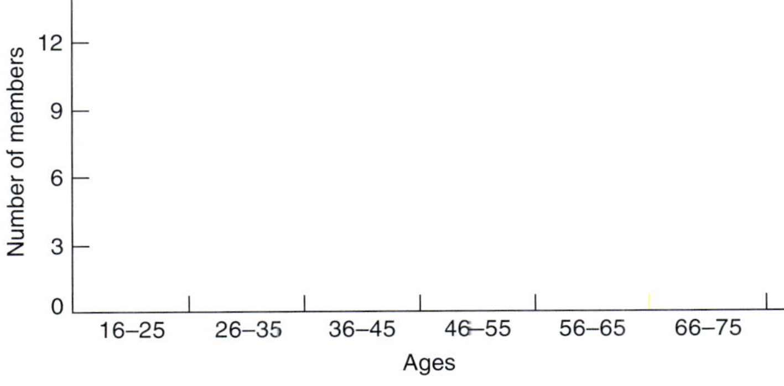

Figure 7–17 Ages of members of horticultural association.

Answer Exercises 24–31 based on the information in Exercises 22 and 23.

24. Which age group has the least number of members?
25. Which age group has the greatest number of members?
26. How many members are under age 36?
27. How many members are over age 55?

28. What is the ratio of the number of members age 66–75 to members age 16–25?

29. What is the ratio of the number of members age 66–75 to the number of members age 46–55?

30. What percent (to the nearest tenth of a percent) of the number of members age 46–55 are the members age 16–25?

31. What percent (to the nearest tenth of a percent) of the total number of members are the members age 46–55?

32. Make a frequency distribution of these scores on an English language test: 68, 70, 74, 77, 78, 82, 82, 84, 86, 86, 86, 88, 89, 90, 90, 93.

33. Make a frequency distribution of these mileages (miles per gallon, or mpg) reported one week by customers to an automotive rental company for six-cylinder cars: 22, 22, 23, 23, 23, 24, 24, 25, 26, 26, 27, 29, 29, 30, 30, 31, 31.

34. Make a histogram with the information from Exercise 33.

35. Make a frequency polygon based on the histogram you made for Exercise 34.

Section 7–3

Solve the problems.

36. Jim Smith made 176 baskets over a 16-game period. What was the average number of baskets per game to the nearest tenth?

37. Lee Vance sold 63 new cars over a 5-month period. What was the average number of cars sold per month to the nearest tenth?

38. An automobile used 22 gallons of regular gasoline on a trip of 358 miles. What was the average miles per gallon to the nearest tenth?

39. A delivery truck used 21 gallons of regular gasoline on a trip of 289 miles. What was the average miles per gallon to the nearest tenth?

40. These hourly pay rates are used at fast-food restaurants: cooks, $6.25; servers, $6.95; bussers, $5.20; dishwashers, $5.20; managers, $8.25. Find the median pay rate.

41. These hourly pay rates are used at a locally owned store: office assistants, $5.85; bookkeepers, $6.20; cashiers, $5.45; assistant managers, $7.90. Find the median pay rate.

42. These weekend work shifts are in effect in a mall clothing store: 3 hours in A.M., 6 hours in P.M., 3 hours in P.M. Find the mode.

43. These special prices are in effect at a fast-food restaurant: $1.85, hamburgers; $1.98, hot ham sandwiches; $2.28, chicken sandwiches; $1.85, roast beef sandwiches. Find the mode.

44. Cody Collier, a student at a technical college, earned the final grades listed in Table 7–8 for the past term. Find Cody's QPA (quality point average) to the nearest hundredth.

45. Tami Murphy earned the final grades listed in Table 7–9 for the past term. Find Tami's QPA (quality point average) to the nearest hundredth.

TABLE 7–8 Grade Distribution

Subject	Grade	Hours	Quality Points per Hour
Electronics 101	A	4	4
Circuits 201	A	4	4
Algebra 101	B	4	3

TABLE 7–9 Grade Distribution

Subject	Grade	Hours	Quality Points per Hour
English 201	A	3	4
History 202	B	3	3
Philosophy 101	A	3	4

46. Janice Van Dyke has these grades in Algebra 102: 98, 82, 87, 72, and 82. What minimum grade does she need on the last test to have a B or 85 average?

47. Sarah Smith has the following grades in American history: 79, 73, 71, 78, and 86. What grade does she need on the last test to get a C or 75 average?

48. Find the range, mean, median, and mode for these test scores: 67, 87, 76, 89, 70, 69, 82.

49. Find the range, mean, median and mode for Marcus Johnson's test scores of 92, 83, 39, 98, 88, 90.

50. Find the standard deviation of the test scores in Exercise 48.

51. Find the standard deviation of the test scores in Exercise 49.

52. Find the standard deviation of the hourly pay rates in Exercise 40.

53. Find the standard deviation of the hourly rates in Exercise 41.

One complete circle is made up of 360°. A measuring instrument called a *protractor* is used to draw angles of various sizes.

54. Use a protractor to make a circle graph of the sources of a construction company's yearly income of $150,000. Show 50% from small business construction, 25% from home remodeling, 15% from local government jobs, and 10% from miscellaneous work.

1. A _____ graph shows how different data values relate to each other.

2. A _____ graph shows how a whole quantity is related to its parts.

3. A _____ graph shows how an item or items change over time.

Answer Exercises 4 and 5 from the line graphs in Fig. 7–18.

4. How many degrees warmer was it indoors at midnight than outdoors?

5. What was the change in outdoor temperature between 11:00 A.M. and noon?

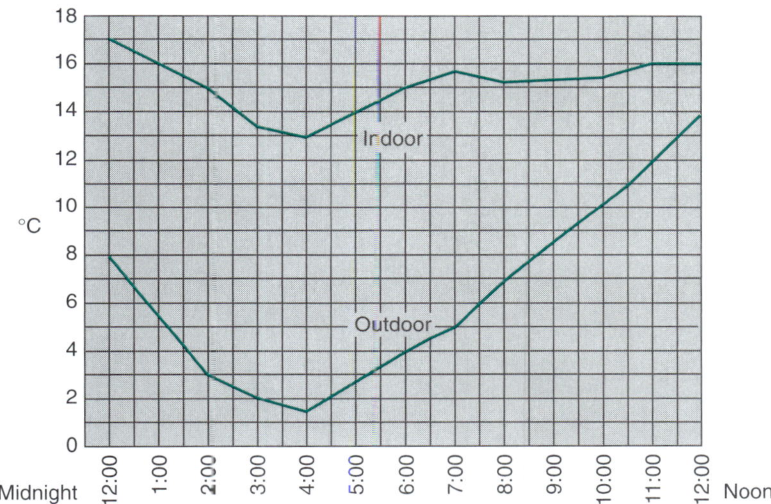

Figure 7–18 Indoor-outdoor temperature.

Answer the questions about the manufacturing costs for the electronic game as shown in the circle graph in Fig. 7–19.

6. What percent of the total cost is materials?
7. What would the profit be if there was no tariff on imported parts?
8. How much are overhead and materials together?
9. What percent of the total cost is the profit?
10. What is the ratio of labor to total cost?
11. What is the ratio of the tariff to total cost?

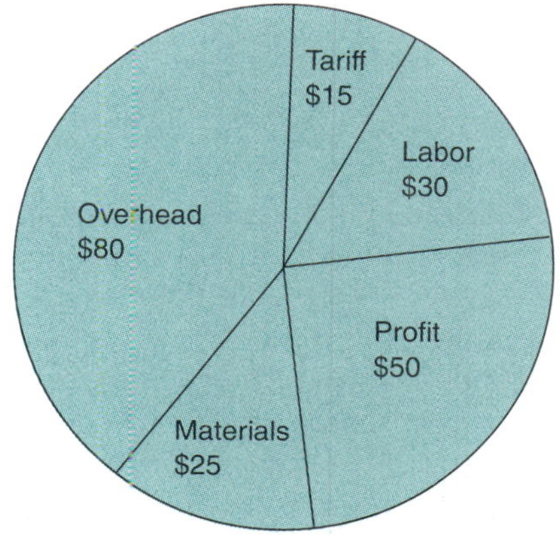

Figure 7–19 Manufacturing costs of a $200 electronic game.

Use the bar graph in Fig. 7–20 to answer the questions about the academic-year starting salaries of women and men college professors in various academic departments of a college.

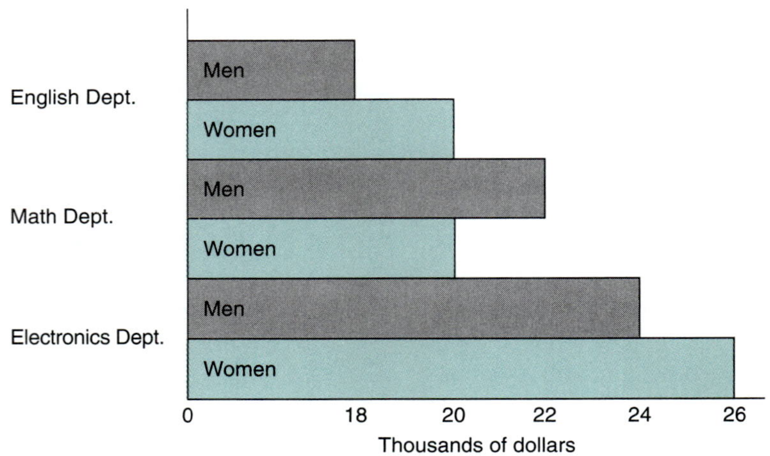

Figure 7–20   Salaries of women and men college professors.

12. In what departments do men make more than women?

13. In what departments do women make more than men?

14. What percent of women's salaries are men's salaries in the English department (to the nearest tenth of a percent)?

15. What percent of men's salaries are women's salaries in the electronics department (to the nearest tenth of a percent)?

16. What is the ratio of men's salaries to women's salaries in the math department?

17. What is the ratio of men's salaries to women's salaries in the English department?

18. Make a line graph to illustrate the following information about the average prices of two-, three-, and four-bedroom homes in a subdivision. Use the reference lines in Fig. 7–21 as guides.

    2-bedroom homes, average $80,000
    3-bedroom homes, average $90,000
    4-bedroom homes, average $100,000

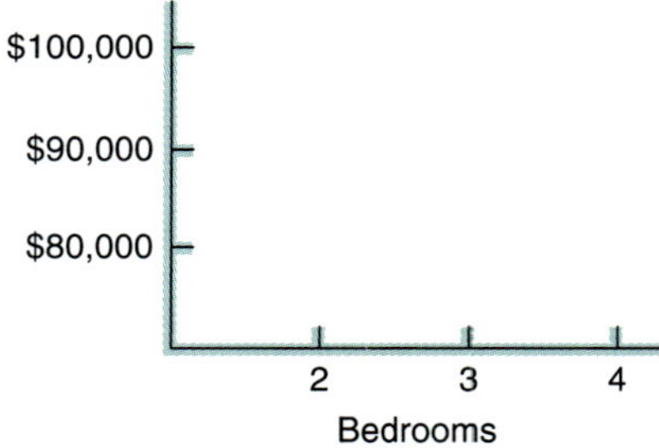

Figure 7–21   Prices of homes by number of bedrooms.

Use the frequency distribution shown in Table 7–10 to answer the questions that follow. The distribution shows the number of correct answers on a 30-question test in a science class.

19. How many students scored less than 16 correct?

20. How many students scored more than 25 correct?

21. What is the ratio of students scoring 6–10 correct to those scoring 21–25 correct?

22. What percent scored 21–30 correct?

TABLE 7–10   Frequency Distribution of Correct Answers

Class Interval	Midpoint	Tally	Class Frequency
26–30	28	/ / /	3
21–25	23	/ / / /	4
16–20	18	++++ / /	7
11–15	13	/ / /	3
6–10	8	/ /	2
1–5	3	/	1

Use the format shown in Table 7–12 and the ages of 24 children in a day-care center to complete a frequency distribution.

TABLE 7–12 Ages of 24 Day-Care Children

	Class Interval	Midpoint	Tally	Class Frequency
23.	4–6	__________	__________	__________
24.	1–3	__________	__________	__________

Ages of Children

1	1	1	1	1	1	1	2	2	2
3	3	3	3	3	4	4	4	4	5
5	5	5	6						

Use the histogram in Fig. 7–22 to answer Questions 25–27 about a company's software programs and the number of employees trained to use them.

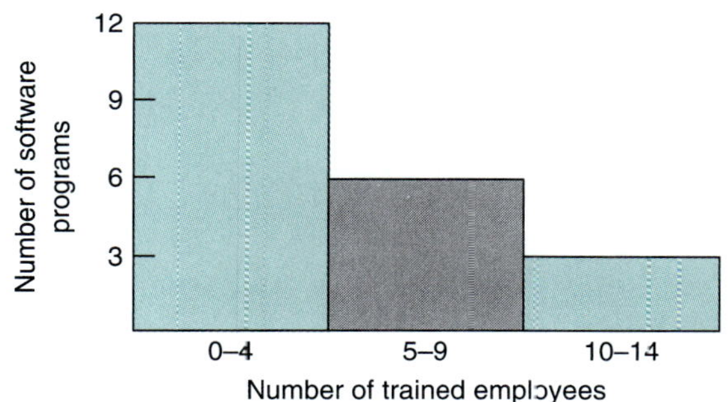

Figure 7–22   Company employees trained to use software.

25. How many employees at the most are able to use only three programs?

26. How many programs does the company have?

27. How many employees does the company have?

28. Find the mean, median, and mode for these test scores: 77, 87, 77, 89, 70, 69, 82.

29. Find the mean, median, and mode for these test scores: 81, 78, 69, 75, 81, 93, 68.

30. Find the standard deviation for the test scores in Exercise 29.

We use symbols in everyday life to overcome language barriers and to understand concepts. Pictures and symbols are easily associated with concepts. Even preschool children can distinguish between the symbols indicating male and female restroom facilities. A symbol is processed much more quickly than words. Our brains process 🚫 much faster than they process the words "no left turn."

In your teams use a flip chart or board to write the words and draw the nonmathematical symbols that any team member can recall. Have one team member be the recorder and another the reporter. Discuss the components of effective symbolic representation. Next, design signs that contain no words to distinguish a walking path from a bicycle path, to direct people to the information office on your campus, and to direct people to keep off the grass. In addition, identify one additional communication need on your campus and design a "no words" sign to meet this need.

Compare your team's signs with those of the other teams in your class. Share the "no words" sign you designed to meet the additional communication need on your campus and measure its effectiveness by asking classmates to interpret it. Extend this activity by researching international symbols for signs and investigate the history of this means of communication.

Linear Equations

8–1 Variable notation

1. Identify equations, terms, factors, constants, variables, and coefficients.
2. Write verbal interpretations of symbolic statements.
3. Translate statements into equations using variables.
4. Simplify variable expressions.

8–2 Solving basic linear equations

1. Solve basic equations.
2. Check the solutions of equations.

8–3 Isolating the variable in linear equations

1. Solve equations with like terms on the same side of the equation.
2. Solve equations with like terms on opposite sides of the equation.

8–4 Applying the distributive property in solving equations

1. Solve equations that contain parentheses.

8–5 Strategies for problem solving

1 Solve applied problems using equations.

8–6 Solving equations with fractions by clearing the denominators

1 Solve equations with fractions by clearing the denominators.

2 Solve applied problems involving rate, time, and work.

8–7 Solving equations with decimals

1 Solve equations with decimals.

2 Solve applied problems with decimal equations.

Symbols were introduced early in the development of our number system because symbols are effective in expressing numbers and mathematical concepts. Symbols allow mathematical expressions to be written in a concise form that overcomes many language and cultural barriers. For symbols to be effective, there must be a widespread understanding of the numbers and concepts represented by the symbols. We use common symbols to review many mathematical concepts. Later, we will expand on these concepts and introduce additional symbols.

8–1   VARIABLE NOTATION

Learning Outcomes

1 Identify equations, terms, factors, constants, variables, and coefficients.

2 Write verbal interpretations of symbolic statements.

3 Translate statements into equations using variables.

4 Simplify variable expressions.

An important use of symbols is in representing unknown values. Such notation allows us to develop rules of operations using symbols and thus find unknown values.

1 Identify Equations, Terms, Factors, Constants, Variables, and Coefficients.

Before we work with equations to solve problems, we need to understand the basic concepts and terminology of equations. In mathematics, we often use the symbol "=" to show that quantities are "equal to" each other. We write the statement "5 is equal to 2 plus 3" as $5 = 2 + 3$. This symbolic statement is called an *equation*.

■ **DEFINITION: Equation.** An *equation* is a symbolic statement that two expressions or quantities are equal in value.

An equation may be true or false. However, our purpose is to find values of the missing numbers that make the equation true.

EXAMPLE Verify that the statements are true equations.

(a) $3(8) = 24$ (b) $12 - 3 = 9$

To verify that these statements are true equations, we must *evaluate* or find the value of the expression on each side of the equal sign. To be a true equation, the value on the left side of the equation must equal the value on the right side of the equation.

(a) $3(8) = 24$ 3 times 8 is 24.
    $\mathbf{24 = 24}$ The equation is true.

(b) $12 - 3 = 9$ 12 minus 3 is 9.
    $\mathbf{9 = 9}$ The equation is true.

Some types of equations contain a *missing* or *unknown* number. We *solve* an equation by finding the missing number that makes the equation *true*. We use letters such as x, a, or z to represent the missing number in the equation. In each equation, the letter has a certain but unknown value. The same letter may be used in several equations, and its value in each equation depends on the other numbers and relationships in the equation. Because the value of these letters varies with each equation, we call them *variables*. The value of the letter is unknown until the equation is solved, so the letter is the *unknown* in the equation. The *solution*, also called the *root*, of an equation is the value of the variable that makes the equation true.

■ **DEFINITION: Variable or Unknown.** A letter that represents an unknown or missing number is called a *variable* or *unknown*.

■ **DEFINITION: Root or Solution.** The *root* or *solution* of an equation is the value of the variable that makes the equation true.

EXAMPLE Solve the equations.

(a) $x = 3 + 8$ (b) $\dfrac{18}{3} = n$ (c) $y = 3(4) - 5$

Because the value of one side of an equation must equal the value of the other side of the equation, we solve the equation by finding the missing number or variable. To do this, we evaluate the opposite side of the equation.

(a) $x = 3 + 8$ 8 added to 3 is 11.
    $\mathbf{x = 11}$

(b) $\dfrac{18}{3} = n$ 18 divided by 3 is 6.
    $\mathbf{6 = n}$

(c) $y = 3(4) - 5$ Remember, multiplication is worked first in the order of operations: 3 times 4 is 12.
    $y = 12 - 5$ Add 12 and -5. 12 plus $- 5 = 7$.
    $\mathbf{y = 7}$

Because both sides of an equation are equivalent, it does not matter which side comes first in the equation. In these examples, for instance, $x = 3 + 8$ means the same as $3 + 8 = x$; $\frac{18}{3} = n$ means the same as $n = \frac{18}{3}$; and $y = 3(4) - 5$ means the same as $3(4) - 5 = y$. Also, $n = 5$ is equivalent to $5 = n$, and so on. The unknown, or variable, may appear on either the left or right side of an equation.

Recall that numbers multiplied together are called *factors*. Division is the inverse operation of multiplication, so it can be rewritten as an equivalent product. Thus, $\frac{3}{y}$ can be rewritten as $3 \cdot \frac{1}{y}$, and division can be expressed as multiplication of factors. An algebraic expression, or quantity, can be written with a known factor multiplied by a variable, or unknown factor, such as $5x$. The number 5 is the known factor multiplied by some unknown number that we represent with the letter x.

An expression that contains a variable or unknown is an *algebraic expression*. If an algebraic expression contains only factors, then we refer to this expression as a *term*.

$2xy$ is a term containing the factors 2, x, and y.

$\dfrac{2x}{y}$ is a term containing the factors 2, x, and $\dfrac{1}{y}$.

The expression $2(x + 3)$ is a term containing the factors 2 and $(x + 3)$. Because $(x + 3)$ is grouped, it is one quantity and is considered a factor. However, if the distributive principle is applied, and each term within parentheses is multiplied by 2, then $2(x + 3)$ becomes $2x + 6$. In this expression, there are *two* terms, $2x$ and 6.

Tip! | **Multiplication Notation Conventions:**

When a term is the product of letters alone (such as ab or xyz) or the product of a number and one or more letters (such as $3x$ or $2ab$), parentheses or other symbols of multiplication are usually omitted. Thus, ab means a times b, $3x$ means 3 times x, and so on.

In algebra we carefully distinguish between terms and factors.

■ **DEFINITION: Factors.** Numbers or variables that are multiplied are called *factors*.

■ **DEFINITION: Terms.** Algebraic expressions that are single quantities or quantities that are added or subtracted are called *terms*.

A term can be a single letter; a single number; the product of numbers and letters; the product of numbers, letters, and/or groupings; or the quotient of numbers, letters, and/or groupings. The fraction line implies that the numerator and/or the denominator is grouped.

Tip! | **Identifying Terms:**

Plus and minus signs that are *not* within a grouping separate terms.

$$3a + b \qquad 3(a + b) \qquad \frac{3}{a + b}$$

$$\text{two terms} \qquad \text{one term} \qquad \text{one term}$$

EXAMPLE Identify the terms in each expression by drawing a box around each term.

(a) $a - 2$ (b) $3x + 5$ (c) $2x - 4(x + 7)$

(d) $2ab + 4a - b + 2(a + b)$ (e) $\dfrac{x}{2}$ (f) $\dfrac{2a + 1}{3}$ (g) $\dfrac{2a}{3} + 1$

(a) $\boxed{a} - \boxed{2}$

Terms are separated by "+" and "−" signs that are not within a grouping.

(b) $\boxed{3x} + \boxed{5}$

(c) $\boxed{2x} - \boxed{4(x + 7)}$

(d) $\boxed{2ab} + \boxed{4a} - \boxed{b} + \boxed{2(a + b)}$

(e) $\boxed{\dfrac{x}{2}}$

(f) $\boxed{\dfrac{2a + 1}{3}}$

The numerator or denominator of a fraction is a grouping, such as $2a + 1$.

(g) $\boxed{\dfrac{2a}{3}} + \boxed{1}$

Terms that contain only numbers are *number terms* or *constants,* and terms that contain only letters or that contain both numbers and letters are *letter terms* or *variable terms.*

■ **DEFINITION: Number Term** or **Constant.** A term that contains only numbers is a *number term* or *constant.*

■ **DEFINITION: Letter Term** or **Variable.** A term that contains only one letter, several letters used as factors, or a combination of letters and numbers used as factors is a *letter term* or *variable.*

If a term contains more than one factor, each factor is the *coefficient* of the other factors. Any factor(s) in a term containing more than one factor may be grouped as the coefficient of the remaining factor(s). In the term $2ab$,

$$
\begin{array}{rll}
2 & \text{is the coefficient of} & ab \\
a & \text{is the coefficient of} & 2b \\
b & \text{is the coefficient of} & 2a \\
2b & \text{is the coefficient of} & a \\
2a & \text{is the coefficient of} & b \\
ab & \text{is the coefficient of} & 2
\end{array}
$$

■ **DEFINITION: Coefficient.** Each factor of a term containing more than one factor is the *coefficient* of the remaining factor(s).

In algebraic expressions, we are usually interested in the *numerical coefficient.* In the term $2ab$, 2 is the *numerical coefficient* of ab.

Recall that mathematics makes use of signs, both positive and negative. If a term is positive, such as $2a$, there is no need to write the "+" sign before the term. But if the term is negative, such as $-5b$, then the "−" sign *must* be expressed. In the term $2a$, the numerical coefficient is 2, which is positive. In the term $-5b$, the numerical coefficient is negative, (-5).

If the term is of the type $-\frac{a}{3}$ or $\frac{2b}{7}$, the term is the product of a fractional coefficient and a letter; that is, the term $-\frac{a}{3} = -\frac{1}{3}\left(\frac{a}{1}\right)$ or $-\frac{1}{3}a$, and the term $\frac{2b}{7} = \frac{2}{7}\left(\frac{b}{1}\right)$ or $\frac{2}{7}b$. The *numerical coefficient* of a is $-\frac{1}{3}$. The *numerical coefficient* of b is $\frac{2}{7}$. Similarly, in the term $\frac{4n}{5}$ the *numerical coefficient* of n is $\frac{4}{5}$.

■ **DEFINITION: Numerical Coefficient.** The numerical factor of a term is the *numerical coefficient*. Unless otherwise specified, we will use *coefficient* to mean the *numerical coefficient*. The numerical factor should be written *in front* of the variable.

EXAMPLE Identify the numerical coefficient in each term.

(a) $2x$ (b) $-3ab$ (c) $4(x + 3)$ (d) $-\dfrac{n}{3}$ (e) $\dfrac{2b}{5}$

(a) The coefficient of x is **2.**
(b) The coefficient of ab is **-3.**
(c) The coefficient of $(x + 3)$ is **4.**
(d) $-\dfrac{n}{3}$ is the same as $-\dfrac{1}{3}n$, so the coefficient of n is $-\dfrac{1}{3}$.
(e) $\dfrac{2b}{5}$ is the same as $\dfrac{2}{5}\left(\dfrac{b}{1}\right)$ or $\dfrac{2}{5}b$, so the coefficient of b is $\dfrac{2}{5}$.

When a letter term has no written numerical coefficient, the coefficient is understood to be 1; that is, the coefficient of x is 1: $x = 1x$. Similarly, the coefficient of ab is 1: $ab = 1ab$. Likewise, the numerical coefficient of $-x$ is -1: $-x = -1x$. The coefficient of $-bc$ is -1: $-bc = -1bc$.

2 Write Verbal Interpretations of Symbolic Statements.

In early studies of mathematics, unknown values were represented by boxes, circles, underlining, or other symbols. That is, finding the missing value was written as $5 + \underline{\hspace{1cm}} = 8$. The underline identifies the missing value and the other numbers and symbols give the conditions of the problem.

We commonly use letters to represent the missing value. That is, we write $5 + x = 8$ or $5 + y = 8$ or $5 + a = 8$. The choice of letters is not significant. The position of the letter and the other conditions of the statement are important. This letter stands for the *variable* or *unknown.* We can phrase a mathematical statement several ways: $5 + x = 8$ can be stated

"What number added to 5 gives 8?"

or

"5 is added to a number and the result is 8. What is the missing number?"

In another convention in variable notation, we omit multiplication symbols when a number and a variable are multiplied. For instance, $6x$ means 6 times the value of x. If there is more than one missing number, different variables represent each different number. To write the multiplication of two unknowns in symbols, we use two different letters, xy, or the same letter with subscripts, x_1x_2. In both cases, the two unknowns are multiplied. For clarity, the raised dot or parentheses is used to indicate multiplication, even though it is not always necessary. We avoid the use of the multiplication sign ($\times$) so as not to confuse it with an unknown labeled x. A raised dot or parentheses can also be used with subscripted variables. For example, $x_1 \cdot x_2$ or $(x_1)(x_2)$ means that two variables are multiplied.

EXAMPLE State the equations in words.

(a) $x - 7 = 4$ (b) $\dfrac{x}{5} = 3$ (c) $2x + 3 = 15$ (d) $2(x + 3) = 14$

Each equation can be stated several ways. Here is one choice for each equation.

(a) **When 7 is subtracted from a number, the answer is 4.**

(b) **A number divided by 5 is 3.**

(c) **15 is the result when 3 is added to 2 times a number.**

(d) **If the sum of a number and 3 is doubled, the result is 14.**

3 Translate Statements into Equations Using Variables.

As we stated earlier, symbolic representations of written statements or real-life situations allow us to more systematically determine the value of missing amounts. We translate words into symbols just like we translate one language into another. To write words in symbols, we watch for key words or phrases that imply or suggest specific operations. The following are key words or phrases that distinguish between addition and subtraction.

Addition:	the sum of, plus, increased by, more than, added to, exceeds, longer, total, heavier, older, wider, taller, gain, greater than, more, expands
Subtraction:	less than, decreased by, subtracted from, the difference between, diminished by, take away, reduced by, less, minus, shrinks, younger, lower, shorter, narrower, slower, loss

EXAMPLE Translate the statements into symbols.

(a) The sum of 12, 23, and a third number is 52.

(b) When a number is subtracted from 45, the result is 17.

(a) The sum of 12, 23, and a third number is 52: $12 + 23 + x = 52.$

(b) When a number is subtracted from 45, the result is 17: $45 - x = 17.$

Key words or phrases that imply multiplication and division are

Multiplication:	times, multiply, of, the product of, multiplied by
Division:	divide, divided by, divided into, how big is each part, how many parts can be made from

Some words that imply multiplication or division may indicate a specific number in the multiplication or division. Examples include twice (2 times), double (2 times), triple (3 times), and half of (1/2 times or divided by 2).

EXAMPLE Translate the statements into symbols.

(a) If a value is tripled, the result is 63.

(b) How many shelves that are 3 feet long can be made from a board that is 12 feet long?

(a) If a value is tripled, the result is 63: $3x = 63.$

(b) How many shelves that are 3 feet long can be made from a board that is 12 feet long? $3x = 12$ or $\frac{12}{3} = x.$

Explain the difference between the two statements.

Four times the difference between a number and 8 is 24.
The difference between 4 times a number and 8 is 24.

Four times the difference between a number and 8 is 24. This statement is written symbolically as $4(x - 8) = 24$. It shows that the difference is taken first and the result is multiplied by 4.

The difference between 4 times a number and 8 is 24. This statement is written symbolically as $4x - 8 = 24$. A number is multiplied by 4 and then 8 is subtracted from the result.

4 Simplify Variable Expressions.

When we add and subtract fractions, we use like or common denominators. Similarly, when we add and subtract measures, we also use like units of measure. Thus, it should not be too surprising that when we add and subtract variables, we use like terms.

■ **DEFINITION: Like Terms.** Terms are *like terms* if they are numbers or if they are letter terms with exactly the same letter factors.

The terms $4y$ and $2y$ are like terms because both contain the same letter (y). Similarly, 3 and 1 are like terms because both are numbers. We *cannot* combine 3 and $4y$, because they are unlike terms (number and letter term), and we *cannot* combine $4y$ and $2x$, because they are also unlike terms (different letters).

To combine numbers:

1. Change subtraction to addition if appropriate.
2. Add the numbers using the appropriate rule for adding signed numbers.
3. The sum is a number.

To combine like variables:

1. Change subtraction to addition if appropriate.
2. Add the numerical coefficients of the variables using the appropriate rule for adding signed numbers.
3. The sum has the same letter factor or factors as the like terms being added.

EXAMPLE

Simplify the expressions by combining like terms.

(a) $5a + 2a - a$ (b) $3x + 5y + 8 - 2x + y - 12$

(a) $5a + 2a - a = 6a$ Add coefficients: $5 + 2 - 1 = 6$.

(b) $3x + 5y + 8 - 2x + y - 12$ Add like terms.

$\quad = x + 6y - 4$

$3x - 2x = x$
$5y + y = 6y$
$8 - 12 = -4$

The *distributive principle* can be extended to include multiplying numbers and variables. For now, we examine only multiplying numbers and variables by numbers.

EXAMPLE Simplify by applying the distributive principle.

(a) $3(5x - 2)$ (b) $-7(x + y - 3z)$ (c) $-(-3x + 4)$

$$\text{(a)}\ 3(5x - 2) = 3(5x) - 3(2)$$
$$= 15x - 6$$

Apply the distributive principle.

$$\text{(b)}\ -7(x + y - 3z) = -7(x) - 7(y) - 7(-3z)$$
$$= -7x - 7y + 21z$$

Apply the distributive principle.

$$\text{(c)}\ -(-3x + 4) = -1(-3x) - 1(+4)$$
$$= 3x - 4$$

Apply the distributive principle.

SELF-STUDY EXERCISES 8–1

1 Verify that the statements are true.

1. $5(-3) = -15$ **2.** $17 - 6 = 11$ **3.** $12 - 5(2) = -7 + 9$

4. $8 - 3(5) = 2(-1 - 3) + 1$

Solve the equations.

5. $n = 4 + 7$ **6.** $m = 8 - 2$ **7.** $5 - 9 = y$

8. $\dfrac{12}{2} = x$ **9.** $p = 2(5) - 1$ **10.** $b = 6 - 3(5)$

Identify the terms in each expression by drawing a box around each term.

11. $7 + c$ **12.** $4a - 7$ **13.** $3x - 2(x + 3)$

14. $\dfrac{a}{3}$ **15.** $7xy + 3x - 4 + 2(x + y)$ **16.** $14x + 3$

17. $\dfrac{7}{(a + 5)}$ **18.** $\dfrac{4x}{7} + 5$

19. Write an algebraic expression that contains three terms.

Identify the numerical coefficient of each term.

20. $5x$ **21.** $-4xy$ **22.** $\dfrac{n}{5}$ **23.** $\dfrac{2a}{7}$

24. $6(x + y)$ **25.** $\dfrac{-4}{5x}$

26. Write an expression that has one term and a numerical coefficient of -15.

2 State the equations in words.

27. $x + 4 = 7$ **28.** $x - 5 = 2$ **29.** $3x = 15$ **30.** $3x + 1 = 7$

3 Write the statements in symbols.

31. Five more than a number is 12.

32. A number divided by 6 is 9.

33. Four times the difference of a number and 3 is 12.

34. Three less than 4 times a number is 12.

35. The sum of 12, 7, and a third number is 17.

36. Seven more than twice a number is 21.

37. If 15° is added to a temperature, it will be 48°. What is the temperature?

38. If 15 mL of water is added to a medicine, there is 45 mL in all. What is the volume of the medicine?

39. How many 5-ft pieces of I-beam can be cut from a piece that is 45 ft long?

40. A piece of oak flooring that is 18 ft long has an unacceptable flaw that requires 3 ft to be trimmed from one end. How many 5-ft boards can be cut from the remaining length of flooring?

4

4   Simplify by combining like terms.

41. $3a - 7a + a$ **42.** $-8x + y - 3y$ **43.** $5x - 3y + 2x + y$
44. $-4a + b + 9 + a - b - 3$ **45.** $2x + 8 - x + 4 - x - 2$ **46.** $3a + 5b + 8c + 1 + b$

Simplify by applying the distributive principle and combining like terms if appropriate.

47. $5(2x - 4)$ **48.** $4 + 3(2x + 3)$ **49.** $3 - (4x - 2)$ **50.** $5 - 2(6a + 1)$

8-2   SOLVING BASIC LINEAR EQUATIONS

Learning Outcomes

1   Solve basic equations.

2   Check the solutions of equations.

Equations and formulas play an important role in the mathematics used in the workplace. We have found missing values in the percentage proportion, $\frac{R}{100} = \frac{P}{B}$, and in various area and perimeter formulas like $P = 2(l + w)$ or $A = \pi r^2$. Now we investigate how equations are used to model solutions to problems.

We will begin with linear equations in one variable. A *linear equation in one variable* is an equation in which the same letter is used in all variable terms, and the exponent of the letter is 1.

Algebraic equations help us solve many real-world problems. Suppose we know that five packages of copier paper cost \$15. Let's represent the cost of the five packages as $5p$, which is 5 times the unknown cost of each package (p). Since the five packages together cost a total of \$15, the $5p$ must equal \$15. We can write an equation to represent this relationship: $5p = 15$. This is known as a basic equation and is the subject of this section.

1   Solve Basic Equations.

■   **DEFINITION: Basic Equation.**   An equation that consists of only a variable and its coefficient on one side of the equal sign and only a number term on the other side is a *basic equation*.

Examples of basic equations are

$$5p = 15 \qquad -2 = 4b$$
$$2p = 11 \qquad -3y = 6$$
$$8 = 3a \qquad -15 = -3x$$

A basic equation states that a number times a variable equals a product. In the equation $5p = 15$, the number 5 times the variable p equals the product 15.

■ **Learning Strategy**   ***Use Simple Cases to Learn and Understand a Process or Procedure.***

Many basic equations can be solved intuitively, mentally, or by using basic arithmetic and our knowledge of the relationships among the operations of addition, subtraction, multiplication, and division. For example, in our example of a basic equation $5p = 15$, we can determine mentally that the missing number must be 3. Our goal is to learn a procedure for solving even the simplest equation so that we can apply the same procedure to more complicated equations.

An equation is *solved* when the letter is alone on one side of the equation; that is, the coefficient of the letter term is 1. The number on the side opposite the letter is called the *root* or *solution* of the equation.

In the equation $5p = 15$, the numerical coefficient of the letter term is 5. Since we would like the numerical coefficient of p to be 1, *we multiply both sides of the equation by the reciprocal of 5*, which is $\frac{1}{5}$. Recall that multiplication of a number by the reciprocal, or multiplicative inverse, of that number gives us 1.

$$\left(\frac{1}{5}\right)5p = 15\left(\frac{1}{5}\right)$$
Multiply by reciprocal of coefficient of letter term.

$$\left(\frac{1}{\cancel{5}}\right)\overset{1}{\cancel{5}}p = \overset{3}{\cancel{15}}\left(\frac{1}{\cancel{5}}\right)$$
Reduce where possible.

$$1p = 3$$
Work multiplication on each side.

$$p = 3$$

Before we lose sight of our original problem, let's interpret the solution within the context of the problem. *The cost of one package of paper is \$3.*

The *basic principle of equality* that underlies this procedure and many other procedures for solving equations is to preserve the equality of the two sides of the equation.

To preserve equality:

If we perform an operation on one side of an equation, we must perform the *same* operation on the other side.

For example, if we have a scale with 1 oz on one side and 1 oz on the other side, the scale is balanced. If we increase or decrease the weight on one side, we need to do the same on the other side or one side would be heavier than the other and the equality of both sides would be lost (see Fig. 8–1).

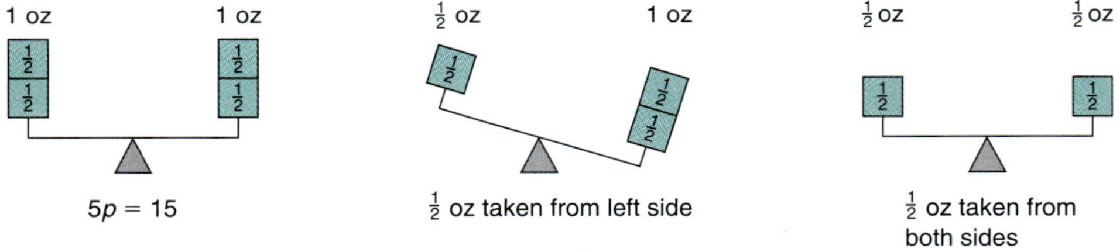

Figure 8–1

Let's look at the equation $5p = 15$ again. Notice that the numerical coefficient 5 is "attached" to the unknown, p, by multiplication. Since division is the inverse operation of multiplication, we can solve the equation by *dividing both sides of the equation by 5*. This procedure also gives us a numerical coefficient of 1 for p.

$$\frac{5p}{5} = \frac{15}{5}$$
Divide both sides by the coefficient of the letter term.

$$\frac{\overset{1}{\cancel{5}}p}{\underset{1}{\cancel{5}}} = \frac{\overset{3}{\cancel{15}}}{\underset{1}{\cancel{5}}}$$
Reduce within each fraction.

$$1p = 3$$

$$p = 3$$

In short, we can solve a basic equation by multiplication or by division, as indicated in the following rule.

To solve a basic equation:

Divide both sides of the equation by the numerical coefficient of the letter term;

or

Multiply both sides of the equation by the reciprocal of the coefficient of the letter term.

This rule is based on the *multiplication axiom.*

■ **DEFINITION: Multiplication Axiom.** Both sides of an equation may be multiplied by the same *nonzero* quantity without changing the equality of the two sides.

Symbolically,

$$\text{if } a = b \text{ and } c \neq 0, \text{ than } ac = bc.$$

Tip!	***Multiplication Axiom Applies to Both Multiplication and Division.***

Multiplication and division are inverse operations, so dividing by a number is the same as multiplying by its multiplicative inverse—its reciprocal.

$$\frac{1}{5}(5p) = 15\left(\frac{1}{5}\right) \qquad\qquad \frac{5p}{5} = \frac{15}{5}$$

Multiply by the reciprocal of the coefficient of the letter term.   **Divide by the coefficient of the letter term.**

$$p = 3 \qquad\qquad\qquad p = 3$$

Now let's solve several basic equations using multiplication or division.

EXAMPLE   Solve the equation $4x = 48$.

Using multiplication:

$$4x = 48$$
$$\left(\frac{1}{4}\right)4x = 48\left(\frac{1}{4}\right)$$

Multiply both sides of the equation by the reciprocal of the coefficient of the letter term.

$$x = 12$$

The same equation can be solved by dividing both sides of the equation by the coefficient of the letter term.

Using division:

$$4x = 48$$
$$\frac{4x}{4} = \frac{48}{4}$$

Divide both sides by the coefficient of the letter term.

$$1x = 12$$

This step is usually performed mentally and omitted from the written steps.

$$x = 12$$

EXAMPLE   Solve the equation $45 = 0.5x$

$$\frac{45}{0.5} = \frac{0.5x}{0.5}$$

Divide both sides by the coefficient of the letter term.

$$90 = x$$

<table>
<tr><td>Tip!</td><td>On Which Side Should the Letter Be?</td></tr>
</table>

Multiplication and division are effective no matter which side of the equation contains the letter term. Once the equation is solved, however, many people prefer to put the letter on the left ($x = 90$ instead of $90 = x$). Either way is correct.

EXAMPLE Solve the equation $\frac{1}{4}n = 7$.

Using multiplication:

$$\frac{1}{4}n = 7$$

$$\left(\frac{4}{1}\right)\frac{1}{4}n = 7\left(\frac{4}{1}\right) \qquad \text{The reciprocal of } \frac{1}{4} \text{ is } \frac{4}{1}.$$

$$n = 28$$

Using division:

$$\frac{1}{4}n = 7$$

$$\frac{\frac{1}{4}n}{\frac{1}{4}} = \frac{7}{\frac{1}{4}} \qquad 7 \div \frac{1}{4} = \frac{7}{1} \times \frac{4}{1} = 28$$

$$n = 28$$

<table>
<tr><td>Tip!</td><td>Dividing versus Multiplying by the Reciprocal.</td></tr>
</table>

When the equation contains only whole numbers or decimals, it is generally more convenient to divide by the coefficient of the variable or letter term than to multiply by the reciprocal of the coefficient of that variable. When the equation contains a fraction, it is usually preferable to multiply by the reciprocal of the coefficient of the variable.

EXAMPLE Solve the equation $-n = 25$.

The coefficient is -1, so the equation is not solved. The coefficient must be $+1$ for the equation to be solved.

$$\frac{-1n}{-1} = \frac{25}{-1} \qquad \text{Divide by coefficient of letter term.}$$

$$n = -25$$

<table>
<tr><td>Tip!</td><td>Don't Quit Too Soon.</td></tr>
</table>

An equation is *solved* when the coefficient of the letter term is $+1$. If the coefficient is -1, we apply the multiplication axiom to complete the solution.

2 Check the Solutions of Equations.

When an equation has been solved, the *solution*, or *root*, can be checked to avoid mistakes.

> ***To check or verify the solution or root of an equation:***
>
> **1.** Substitute the solution in place of the letter in the original equation.
> **2.** Perform all indicated operations.
> **3.** If the solution is correct, the value of the left side of the equation should equal the value of the right side.

EXAMPLE In a previous example the solution for the equation $4x = 48$ was found to be 12. Check or verify this root.

$$4x = 48 \qquad \text{Substitute 12 for } x.$$

$$4(12) = 48 \qquad \text{Perform the multiplication.}$$

$$\mathbf{48 = 48} \qquad \text{The root is verified if the final equality is true.}$$

SELF-STUDY EXERCISES 8–2

1 Solve the equations. Use multiplication or division. Show your work.

1. $7x = 56$	**2.** $2x = 18$	**3.** $15 = 3x$	**4.** $36 = 1.2n$
5. $-3a = 27$	**6.** $-18 = 9b$	**7.** $-7c = -49$	**8.** $-72 = 8x$
9. $5x = 32$	**10.** $36 = 8y$	**11.** $-6n = 15$	**12.** $-28 = -3b$
13. $-x = 7$	**14.** $-2 = -y$	**15.** $\dfrac{1}{3}x = 5$	**16.** $4 = \dfrac{1}{2}y$
17. $\dfrac{3}{5}x = 2$	**18.** $-\dfrac{3}{4}y = 5$	**19.** $-\dfrac{1}{2}x = -6$	**20.** $21 = \dfrac{3}{8}n$

2 **21.–30.** Check or verify each solution or root for the even-numbered Exercises 2–20.

8–3 ISOLATING THE VARIABLE IN LINEAR EQUATIONS

Learning Outcomes

1 Solve equations with like terms on the same side of the equation.

2 Solve equations with like terms on opposite sides of the equation.

Equations are not always basic equations. When they are more involved, we *simplify* them into the form of the basic equation so that they can be solved. In this section we show how to simplify equations that are not basic equations.

1 **Solve Equations with Like Terms on the Same Side of the Equation.**

Suppose a family of four is lunching at a fast-food restaurant. The mother has only $7 and the father has only $8, which they plan to combine to pay for lunch. At the last minute their daughter goes to a friend's house for lunch. An algebraic equation can be used to represent this scenario and find out how much each family member can spend on lunch after the daughter leaves.

Let x equal the amount of money each family member can spend. There are four members, so the total amount to be spent is 4 times x, or $4x$. Because the daughter left, we must subtract the amount (x) she would have spent. This gives us $4x - x$. The total amount of money is the mother's $7 plus the father's $8. These relationships can be expressed in equation form: $4x - x = 7 + 8$.

Recall that a basic equation has a single letter term for one side and a single number term for the opposite side. If an equation has more than one letter term on one side or more than one number term on the opposite side, we *combine* the terms on each side to form a basic equation.

We can combine only *like* terms. That means number terms can be added only to other number terms. Letter terms can be added only to other like letter terms. By "combine," we mean add like terms according to the appropriate signed number rule.

EXAMPLE Solve the equation $4x - x = 7 + 8$.

$$4x - x = 7 + 8 \qquad \text{The coefficient of } x \text{ is } -1.$$

$$4x - x = 7 + 8 \qquad \text{Think of } 4x \text{ plus } -1x = 3x. \text{ Then } 7 + 8 = 15.$$

$$3x = 15 \qquad \text{We now have a basic equation.}$$

We solve a basic equation as before, by multiplying or by dividing on both sides. Division is more convenient with whole numbers.

$$\frac{3x}{3} = \frac{15}{3} \qquad \text{Divide both sides by the numerical coefficient of } x \text{ to "undo" multiplication.}$$

$$x = 5$$

2 Solve Equations with Like Terms on Opposite Sides of the Equation.

In many equations, like terms do not appear on the same side of the equal sign. In such cases, the like terms must be manipulated so that they appear on the same side of the equal sign *before* they can be combined. The basic purpose is to *isolate* the variable or letter terms on one side of the equation and thereby obtain a basic equation to be solved.

Terms can be moved from one side of an equation to the other using the *addition axiom*.

■ **DEFINITION: Addition Axiom.** The same quantity can be added to both sides of an equation without changing the equality of the two sides.

Symbolically,

$$\text{if } a = b, \text{ then } a + c = b + c.$$

EXAMPLE Solve the equation $2x + 4 = 8$.

The like terms 4 and 8 are on opposite sides of the equal sign. To isolate the $2x$, we need to eliminate 4. Therefore, we use the addition axiom to move the number term 4 to the right side of the equation. Then, all the numbers are on one side and the letter term is isolated on the opposite side.

We move the 4 from the letter term side to the opposite side by adding the *opposite* of 4 (that is, -4) to *both* sides. Remember, whatever we do to one side, we must do to the other side.

$$2x + 4 = 8$$

$$2x + 4 - 4 = 8 - 4 \qquad \text{Add } -4, \text{ the opposite of 4, to both sides. (Addition axiom)}$$

$$2x + 0 = 4 \qquad 4 - 4 = 0, \; 8 - 4 = 4$$

$$2x = 4 \qquad 2x + 0 = 2x$$

Now we have a *basic equation* that we can solve by multiplying or by dividing on both sides. Let's use division.

$$\frac{2x}{2} = \frac{4}{2}$$ Divide both sides by the coefficient of the letter term.

$$x = 2$$

EXAMPLE Solve the equation $9 - 4x = 8x$.

In this equation the like terms $-4x$ and $8x$ are on opposite sides of the equation. We want to manipulate the terms in this equation so that both letter terms are isolated on one side of the equation and the number term is on the other side. Using the addition axiom, we eliminate $-4x$ from the left side of the equation by adding its opposite to both sides.

$$9 - 4x = 8x$$

$$9 - 4x + 4x = 8x + 4x$$ Add $4x$, the opposite of $-4x$, to both sides. (Addition axiom)

$$9 + 0 = 12x$$ $-4x - 4x = 0$, $8x + 4x = 12x$

$$9 = 12x$$ $9 + 0 = 9$

$$\frac{9}{12} = \frac{12x}{12}$$ Divide both sides by the coefficient of x. (Multiplication axiom)

$$\frac{9}{12} = x$$ Reduce to lowest terms.

$$\frac{3}{4} = x \qquad \text{or} \qquad x = \frac{3}{4}$$

EXAMPLE Solve the equation $7 - 5x = 12$.

The like terms are 7 and 12. We add -7 to both sides of the equation to isolate the variable term.

$$7 - 5x = 12$$

$$7 - 7 - 5x = 12 - 7$$ Addition axiom. $7 - 7 = 0$; $12 - 7 = 5$.

$$0 - 5x = 5$$ Combine like terms. $0 - 5x = -5x$.

$$-5x = 5$$

$$\frac{-5x}{-5} = \frac{5}{-5}$$ Note that both sides are divided by -5 because the coefficient is negative. (Multiplication axiom)

$$x = -1$$

Notice in the preceding examples that each time we add the opposite of a term to both sides we eliminate it from one side of the equation, and it appears as the *opposite* on the other side. Knowing this, we can *transpose* as a shortcut for adding opposites to both sides. *Transpose* means "move across as the opposite of."

Tip! | ***Sorting or Transposing Terms:***

Transposing is a shortcut for moving terms. To do this, we add the opposite of a term to both sides of an equation. In other words, we omit the term from one side of the equation and write its opposite on the other side. This process is better described as *sorting* the terms so that like terms are on the same side of the equation.

What we are doing is *mentally* adding the opposite term to both sides but showing only part of the procedure.

Let's rework the previous three examples transposing like terms to isolate the variable.

$$2x + 4 = 8 \qquad\qquad 9 - 4x = 8x \qquad\qquad 7 - 5x = 12$$

$$2x = 8 - 4 \qquad\qquad 9 = 8x + 4x \qquad\qquad -5x = 12 - 7$$

$$2x = 4 \qquad\qquad 9 = 12x \qquad\qquad -5x = 5$$

$$\frac{2x}{2} = \frac{4}{2} \qquad\qquad \frac{9}{12} = \frac{12x}{12} \qquad\qquad \frac{-5x}{-5} = \frac{5}{-5}$$

$$x = 2 \qquad\qquad \frac{3}{4} = x \qquad\qquad x = -1$$

Tip! | ***Align Equals Signs! One Equals Sign per Line!***

Organizing the steps of the solution to an equation reduces errors.

- Arrange steps under each other and align the equals signs in a vertical line. This helps you determine if a term has moved from one side to the other. As a term is eliminated from one side move it to the other side as its opposite.
- Have only one equals sign per line. Each line should be one complete equation or statement.

Good Form

$$x = \frac{2}{4}$$

$$x = \frac{1}{2}$$

Poor Form

$$x = \frac{2}{4} = \frac{1}{2}$$

EXAMPLE   Solve the equation $9x + 2 = 6x - 10$.

We first decide on which side of the equation to isolate the letter terms. Once we make that decision, we must collect the number terms on the opposite side. Let's solve this equation two ways: (1) by isolating the letter terms on the left and (2) by isolating the letter terms on the right. The main point is that, no matter which side we choose for the letter terms, the number terms must be on the opposite side.

Letter Terms Isolated on Left

$$9x + 2 = 6x - 10$$
$$9x - 6x = -10 - 2 \qquad \text{Sort or transpose.}$$
$$3x = -12 \qquad \text{Combine.}$$
$$\frac{3x}{3} = \frac{-12}{3} \qquad \text{Divide.}$$
$$x = -4$$

Letter Terms Isolated on Right

$$9x + 2 = 6x - 10$$
$$2 + 10 = 6x - 9x$$
$$12 = -3x$$
$$\frac{12}{-3} = \frac{-3x}{-3}$$
$$-4 = x$$

Tip! | ***Sequence of Strategies for Solving Equations.***

When an equation contains more terms, the steps for solving the equation fall into a pattern. No matter how many terms are involved or what types of coefficients are involved (whole numbers, integers, fractions, decimals), the process is the same.

1. Combine like terms on each side of the equation.
2. Sort or transpose terms to isolate the variable or letter terms on one side and the number terms on the other (addition axiom).
3. Combine like terms on each side to get a basic equation.
4. Solve the basic equation by multiplication or division (multiplication axiom).

Two visual representations of the sequence:

Combine like terms.	Combine like terms.
↓	↓
Use addition axiom.	Sort terms.
↓	↓
Combine like terms.	Combine like terms.
↓	↓
Use multiplication axiom.	Multiply or divide.

EXAMPLE Solve the equation $2x - 3 + 5x = 8 - 6x$.

$$2x - 3 + 5x = 8 - 6x$$

$$7x - 3 = 8 - 6x \qquad \text{Combine the like terms } 2x \text{ and } 5x \text{ on the left side.}$$

$$7x + 6x = 8 + 3 \qquad \text{Sort to collect like terms.}$$

$$13x = 11 \qquad \text{Combine the like terms to get a basic equation.}$$

$$\frac{13x}{13} = \frac{11}{13} \qquad \text{Divide by the coefficient of } x.$$

$$x = \frac{11}{13}$$

Number terms and coefficients in equations can also be fractions and decimals.

EXAMPLE Solve $9 + \frac{1}{4}b = \frac{3}{4}b$.

$$9 + \frac{1}{4}b = \frac{3}{4}b \qquad \text{Note that the letter terms are on opposite sides.}$$

$$9 = \frac{3}{4}b - \frac{1}{4}b \qquad \text{Sort. The denominators are the same.}$$

$$9 = \frac{2}{4}b \qquad \text{Combine the letter terms.}$$

$$9 = \frac{1}{2}b \qquad \text{Reduce.}$$

$$\left(\frac{2}{1}\right)9 = \left(\frac{2}{1}\right)\frac{1}{2}b \qquad \text{Multiply both sides by the reciprocal of } \frac{1}{2}.$$

$$18 = b$$

EXAMPLE Solve $1.1 = 3.4 + R$.

$$1.1 = 3.4 + R \qquad \text{Sort.}$$

$$1.1 - 3.4 = R \qquad \text{Combine terms.}$$

$$-2.3 = R$$

1 Solve the equations.

1. $3x + 7x = 60$
2. $42 = 8m - 2m$
3. $5a - 6a = 3$
4. $3m - 9m = 3$
5. $y + 3y = 32$
6. $0 = 2x - x$

2 Solve the following equations.

7. $b + 6 = 5$
8. $1 = x - 7$
9. $5t - 18 = 12$
10. $10 - 2x = 4$
11. $2y + 7 = 17$
12. $3x + 7 = x$
13. $3a - 8 = 7a$
14. $10x + 18 = 8x$
15. $4x + 7 = 8$
16. $4x = 5x + 8$
17. $2t + 6 = t + 13$
18. $12 + 5x = 6 - x$
19. $4y - 8 = 2y + 14$
20. $8 - 7y = y + 24$
21. $\dfrac{2x}{3} = 18$
22. $\dfrac{y + 1}{2} = 7$
23. $\dfrac{8 - R}{76} = 1$
24. $P = \dfrac{1}{2} + \dfrac{1}{3}$
25. $x + \dfrac{1}{7}x = 16$
26. $\dfrac{2}{5} - x = \dfrac{1}{2}x + \dfrac{4}{5}$
27. $\dfrac{3}{7}m - \dfrac{1}{2} = \dfrac{2}{3}$
28. $\dfrac{1}{4}s = \dfrac{1}{4} + \dfrac{1}{10} + \dfrac{1}{20}$
29. $m = 2 + \dfrac{1}{4}m$
30. $\dfrac{7}{9} + 3 = \dfrac{1}{2}T$
31. $2.3x = 4.6$
32. $0.8R = 0.6$ (round to nearest tenth)
33. $0.33x + 0.25x = 3.5$ (round to nearest hundredth)
34. $0.04x = 0.08 - x$ (round to nearest hundredth)
35. $0.47 = R + 0.4R$ (round to nearest hundredth)

| **8–4** | **APPLYING THE DISTRIBUTIVE PROPERTY IN SOLVING EQUATIONS** |

Learning Outcome

1 Solve equations that contain parentheses.

When an equation contains an addition or subtraction in parentheses and that quantity in parentheses is multiplied by another factor, we have an example of the *distributive property*. (See Chapter 1.)

1 Solve Equations That Contain Parentheses.

A restaurant has hired you to fill a vacated position. The manager tells you that you will be paid $2 more per hour than the previous worker. You forgot to ask the previous hourly rate, but in 3 hours you earn a gross pay of $24. You can use algebra to figure the pay rate of the previous worker. Just let x equal the previous worker's hourly pay rate and add $2 to it, which is $x + 2$. Then multiply that amount by 3 and make it equal to your gross pay of $24. The equation looks like this: $3(x + 2) = 24$.

If an equation contains a distributive multiplication, such as $3(x + 2)$, our first step in solving the equation is to apply the distributive property.

EXAMPLE Solve the equation $3(x + 2) = 24$ for x.

$$3(x + 2) = 24 \qquad \text{Multiply each term in parentheses by 3. (Distributive property)}$$

$$3x + 6 = 24$$

$$3x = 24 - 6 \qquad \text{Sort.}$$

$$3x = 18 \qquad \text{Combine.}$$

$$\frac{3x}{3} = \frac{18}{3} \qquad \text{Divide.}$$

$$x = 6$$

EXAMPLE Solve the equation $5x + 7 = 2(3 - x) - 13$ for x.

$$5x + 7 = 2(3 - x) - 13$$

$5x + 7 = 6 - 2x - 13 \qquad$ Multiply each term in parentheses by 2. (Distributive property)

$5x + 7 = -7 - 2x \qquad$ Combine. $6 - 13 = -7$

$5x + 2x = -7 - 7 \qquad$ Sort.

$7x = -14 \qquad$ Combine.

$\dfrac{7x}{7} = \dfrac{-14}{7} \qquad$ Divide.

$$x = -2$$

Recall from Section 8–2 that the solution or root of an equation can be checked by substituting the root for the letter in each place it appears in the equation and then evaluating both sides of the equation. If the root (solution) is correct, both sides of the equation will be the same number.

Let's check the root $x = -2$ for the previous example.

$$5x + 7 = 2(3 - x) - 13$$

$5(-2) + 7 = 2(3 - (-2)) - 13 \qquad$ Substitute -2 for x. Because the grouping $2(3 - (-2))$ has two negatives, be careful with the signs.

$5(-2) + 7 = 2(3 + (+2)) - 13 \qquad$ Evaluate the grouping first.

$5(-2) + 7 = 2(5) - 13 \qquad$ Perform all multiplications next.

$-10 + 7 = 10 - 13 \qquad$ Add last.

$$-3 = -3$$

Since $-3 = -3$, the solution checks.

Tip!	***The Sign of a Term.***

To solve any equation, it is very important to be careful with the sign of each term. The algebraic sign of any term is the sign that comes *before* the term.

EXAMPLE Solve the equation $6 - (x + 3) = 2x$ for x.

Since $(x + 3)$ is in parentheses, it is a grouping and we handle it as one term. The sign of the term is negative and the numerical coefficient is understood to be -1.

The first step in solving this equation is to apply the distributive property by multiplying $(x + 3)$ by its understood coefficient, -1.

$$6 - (x + 3) = 2x$$

$6 - 1(x + 3) = 2x \qquad$ -1 is the understood coefficient of $(x + 3)$.

$6 - x - 3 = 2x \qquad$ Distribute.

$6 - 3 = 2x + x \qquad$ Sort.

$3 = 3x \qquad$ Combine.

$$\frac{3}{3} = \frac{3x}{3} \qquad \text{Divide.}$$

$$1 = x$$

EXAMPLE Solve and check the equation $28 = 7x - 3(x - 4)$ for x.

$$28 = 7x - 3(x - 4) \qquad \text{Each term in parentheses is multiplied by } -3.$$

$$28 = 7x - 3x + 12 \qquad \text{Distribute.}$$

$$28 - 12 = 4x \qquad \text{Combine like terms on the right, then sort.}$$

$$16 = 4x \qquad \text{Combine.}$$

$$\frac{16}{4} = \frac{4x}{4} \qquad \text{Divide.}$$

$$4 = x$$

As equations get more involved, the importance of checking becomes more apparent. To check the root 4, we will substitute 4 for x in the equation.

$$28 = 7x - 3(x - 4)$$

$$28 = 7(4) - 3(4 - 4) \qquad \text{Substitute 4 for } x. \text{ Then simplify the grouping } 4 - 4 = 0.$$

$$28 = 7(4) - 3(0) \qquad \text{Multiply first.}$$

$$28 = 28 - 0 \qquad \text{Subtract.}$$

$$28 = 28$$

Here is a summary of some strategies for solving equations.

■ **Learning Strategy** *Write Procedures or Strategies in Your Own Words.*

We textbook authors carefully write procedures with lots of details to make sure you understand them. Try writing these procedures in your own words. Not only will you understand the procedure better, you will also remember it better.

Use key words as memory tools:

1. Distribute.
2. Combine.
3. Sort or transpose.
4. Combine.
5. Multiply or divide.

SELF-STUDY EXERCISES 8–4

 Solve the equations.

1. $3(7 + x) = 30$
2. $15 = 5(2 - y)$
3. $6(3x - 1) = 12$
4. $4t = 2(7 + 3t)$
5. $2x = 7 - (x + 6)$
6. $6x - 2(x - 3) = 30$
7. $4a + 5 = 3(2 + a) - 4$
8. $8x - (3x - 2) = 12$
9. $15 - 3(2x + 2) = 6$
10. $(2b + 1) = 5(b - 4) + 1$

8–5 STRATEGIES FOR PROBLEM SOLVING

Learning Outcome

 Solve applied problems using equations.

The four previous sections began with a problem that we then stated as an algebraic equation and solved. We showed how to solve specific kinds of linear equations. In this section, however, we show how our Six-Step Problem-Solving Plan helps you interpret and solve word problems using equations. To review this plan, let's reexamine the six steps.

1. **Unknown facts** means, What fact or facts am I trying to find? Usually you will choose a letter to represent the unknown fact or one of the unknown facts.
2. **Known facts** means, What relevant facts are known or given in the problem?
3. **Relationships** means, How are the known and unknown facts related? and What two quantities are equal? Express relationships first in words and then symbolically in an equation or formula.
4. **Estimation** means, What is a reasonable result or answer to the problem? Your estimation may sometimes be expressed as "more than . . ." or "less than . . ." or "between . . . and . . . ."
5. **Calculations** means to solve the equation or formula for the unknown facts and make appropriate calculations.
6. **Interpretation** means to express the answer in terms of the problem statement. For example, if the problem asks for the number of truckloads of gravel needed for a job, the answer is expressed in truckloads of gravel.

EXAMPLE Cher Hattier was taken to lunch by a friend. The order included a 12-in. turkey sub, a 6-in. turkey sub, and two iced teas. Cher didn't notice the price of the subs but did recall that the 12-in. sub cost $1.50 more than the smaller sub and that two teas cost $2.50. The lunch cost $10. How much did each sub cost?

Unknown facts The cost of each sub is requested.

Known facts The 12-in. sub was $1.50 more than the 6-in. sub. The two teas cost $2.50. The total cost of the meal was $10.

Relationships Let $x =$ the cost of the 6-in. sub. Then $x + \$1.50$ is the cost of the 12-in. sub. The cost of the two subs plus $2.50 for the iced teas equals $10; that is, $x + x + 1.50 + 2.50 = 10$.

Estimation

The subs together cost $7.50 because the teas cost $2.50 ($10 − $2.50 = $7.50). Thus, the subs cost roughly half of $7.50, or around $3.00 or $4.00 each.

Calculations

$$x + x + 1.50 + 2.50 = 10$$

$$2x + 4 = 10 \qquad \text{Combine. } x + x = 2x;\ 1.50 + 2.50 = 4.00,\ \text{or } 4$$

$$2x = 10 - 4 \qquad \text{Sort.}$$

$$2x = 6 \qquad \text{Combine.}$$

$$\frac{2x}{2} = \frac{6}{2} \qquad \text{Divide.}$$

$$x = 3 \qquad \text{Cost of 6-in. sub.}$$

Interpretation

The 6-in. sub cost $3.00, and the 12-in. sub cost $4.50 ($3 + $1.50).

Check:
$$x + x + 1.50 + 2.50 = 10$$
$$3 + 3 + 1.50 + 2.50 = 10$$
$$10 = 10$$

EXAMPLE

A student ordered a student stethoscope and two pairs of support hose. The total bill was $41.50, including a $2.50 shipping charge. If the student stethoscope costs twice as much as one pair of support hose, how much did she pay for the stethoscope and each pair of hose?

Unknown facts

The cost of the stethoscope and each pair of hose.

Known facts

The stethoscope cost twice what a pair of hose cost. The shipping was $2.50. The total bill was $41.50.

Relationships

Let x = the cost of a pair of hose. Then $2x$ = the cost of the stethoscope. Thus, the cost of two pair of hose ($x + x$) plus the cost of the stethoscope ($2x$) plus the shipping (2.50) equals $41.50. The equation is

$$x + x + 2x + \$2.50 = \$41.50$$

Estimation

The stethoscope cost twice what a pair of hose cost and there are two pairs of hose, so the stethoscope must cost around $20 ($\frac{1}{2}$ of $40), and each pair of hose must cost about $10 ($\frac{1}{2}$ of $20).

Calculations

$$x + x + 2x + 2.50 = 41.50 \qquad \text{Combine.}$$

$$4x + 2.50 = 41.50 \qquad \text{Sort.}$$

$$4x = 41.50 - 2.50 \qquad \text{Combine like terms.}$$

$$\frac{x}{4} = \frac{39}{4} \qquad \text{Divide.}$$

$$x = 9.75$$

Interpretation

One pair of hose cost $9.75. The stethoscope cost twice that amount ($2x$), so the stethoscope cost 2(9.75), or $19.50.

Check:
$$x + x + 2x + 2.50 = 41.50$$
$$9.75 + 9.75 + 2(9.75) + 2.50 = 41.50 \qquad \text{Substitute.}$$
$$41.50 = 41.50$$

EXAMPLE

A nursing student purchased a new pathology text, a used history text, and a notebook. When he checked his receipt for $132 plus tax, he discovered that the pathology text cost twice as much as the history text and notebook combined. If the notebook cost \$3.00, how much did he pay for the history text? For the pathology text?

Unknown facts

The cost of each text.

Known facts

The notebook cost \$3.00. The total cost was \$132.00. The pathology text cost twice as much as the history text and notebook combined.

Relationships

Let x = the cost of the history text. The notebook cost \$3. The pathology text cost 2 times the cost of the history text and the notebook, that is, $2(x + 3)$. The history text plus the pathology text plus the notebook equals \$132. Therefore, the equation is $x + 2(x + 3) + 3 = 132$.

Estimation

If we ignore the notebook, the pathology text cost twice the cost of the history text, so the ratio is 2 to 1. If these two books cost about \$132, a ratio of 2 to 1 would be \$88 to \$44 ($132/3 = 44$). So the pathology text was around \$88, and the history text was around \$44.

Calculations

$$x + 2(x + 3) + 3 = 132$$

$x + 2x + 6 + 3 = 132$	Distribute.
$3x + 9 = 132$	Combine like terms.
$3x = 132 - 9$	Sort.
$3x = 123$	Combine.
$\dfrac{3x}{3} = \dfrac{123}{3}$	Solve for x.
$x = 41$	

Interpretation

The history text cost \$41. The pathology text cost $2(x + 3)$. Thus, $2(41 + 3) = 82 + 6 = \$88$.

Check:
$$x + 2(x + 3) + 3 = 132$$
$$41 + 2(41 + 3) + 3 = 132$$
$$41 + 82 + 6 + 3 = 132$$
$$132 = 132$$

EXAMPLE A horticulturist marks off a 32-m-wide rectangular nursery plot with 158 m of fencing. Because the plants must be properly spaced, she needs to know the length of the plot. Find the length in meters.

Unknown facts

The length of the nursery plot in meters.

Known facts

The nursery plot is rectangular. The width is 32 m. The fencing gives us the perimeter of 158 m.

Relationships

Using the formula for the perimeter of a rectangle, we know that the perimeter is twice the sum of the length and width, or $p = 2(l + w)$.

Estimation

The length must be more than the width of 32 m and less than half the perimeter, that is, less than 79 m.

Calculations

$$p = 2(l + w)$$

$158 = 2(l + 32)$	Substitute in formula.
$158 = 2(l) + 2(32)$	Distribute.
$158 = 2l + 64$	
$158 - 64 = 2l$	Sort.

$$94 = 2l \qquad \text{Combine like terms.}$$

$$\frac{94}{2} = \frac{2l}{2} \qquad \text{Divide.}$$

$$47 = l$$

Interpretation   **Thus, the length of the nursery plot is 47 m.**

Check:

$$p = 2(l + w)$$
$$158 = 2(47 + 32)$$
$$158 = 94 + 64$$
$$158 = 158$$

EXAMPLE   A machine part weighs 2.7 kg and is to be shipped in a carton weighing x kg. If the total weight of three packaged machine parts is 9.3 kg, how much does each carton weigh? (See Fig. 8–2.)

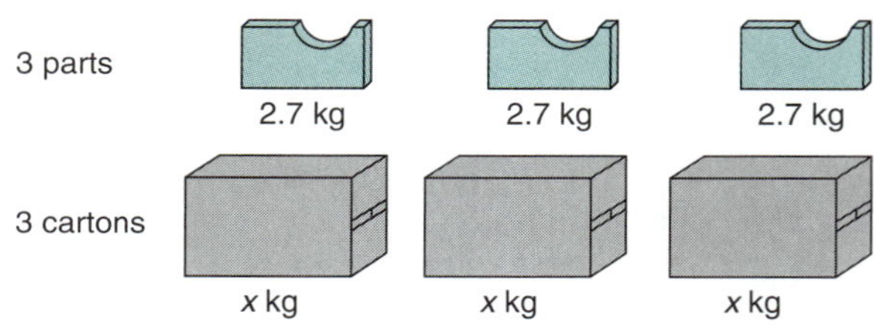

Figure 8–2

Unknown facts   The weight of each empty carton.

Known facts   There are three parts and three cartons. Each part weighs 2.7 kg. The three packaged parts (in cartons) weigh a total of 9.3 kg.

Relationships   Let x equal the weight of one empty carton. Each packaged part weighs the sum of the part (2.7 kg) plus the weight of a carton (x kg). The sum of the three packaged parts equals 9.3 kg; that is, $3(2.7 + x) = 9.3$ or $(2.7 + x) + (2.7 + x) + (2.7 + x) = 9.3$.

Estimation   Three packaged parts weigh 9.3 kg, and $\frac{9.3}{3} = 3.1$ kg per packaged carton. Since one part weighs 2.7 kg, then the carton weighs well under 1 kg.

Calculations

$$3(2.7 + x) = 9.3$$

$$3(2.7) + 3(x) = 9.3 \qquad \text{Distribute.}$$

$$8.1 + 3x = 9.3 \qquad \text{Make calculations.}$$

$$3x = 9.3 - 8.1 \qquad \text{Sort.}$$

$$3x = 1.2 \qquad \text{Combine.}$$

$$\frac{3x}{3} = \frac{1.2}{3} \qquad \text{Divide.}$$

$$x = 0.4$$

Interpretation   **Each carton weighs 0.4 kg.**

Check:

$$3(2.7 + x) = 9.3$$
$$3(2.7 + 0.4) = 9.3$$
$$9.3 = 9.3$$

- Read the problem carefully. Read it several times and phrase by phrase.
- Understand all the words in the problem.
- Analyze the problem:
    What are you asked to find?
    What facts are given?
    What facts are implied?
- Visualize the problem.
- State the conditions or relationships of the problem "symbolically."
- Examine the options.
- Develop a *plan* for solving the problem.
- Write your *plan* symbolically; that is, write an equation.
- Anticipate the characteristics of a reasonable solution.
- Solve the equation.
- Verify your answer with the conditions of the problem.

There are many different problem-solving plans. Additional strategies can be considered.

Spreadsheets.   Information is often displayed in a table of rows and columns called a *spreadsheet*. These rows and columns may show the results of calculations, such as totals or percents, in addition to the original data. Many software programs such as Excel, Lotus 123, and Quattro help you build an *electronic spreadsheet* and draw appropriate graphs using formulas and equations. The calculations are made automatically once the formulas are defined in the spreadsheet. Key information is placed in a specific location on the spreadsheet called an *address,* and formulas are developed using these addresses as the variables. Spreadsheet programs are useful because key information can be changed while retaining the basic formulas of the spreadsheet. This process allows one to see quickly how various changes in the key data affect certain results.

The 7th Inning Sports Memorabilia Shop is developing an annual operating budget. The budget categories and the projected amount of expense for each category are shown in the spreadsheet in Fig. 8–3. Additional information needed is the total operating budget for the year and the percent of total annual budget for each category in the projected annual budget. Formulas must be developed to calculate this information.

The spreadsheet program labels the columns with the letters A, B, and C, and the rows with numbers 1–14. We'll use row 1 for the title of the spreadsheet, row 3 to label the columns of data, and column A to label the rows of data. Each position on the spreadsheet is a *cell,* and the program identifies each cell by its column letter and row number. For example, the amount budgeted for taxes and insurance is in cell B9 and is $14,000.00.

Tip!	*Spreadsheet Addresses.*

A spreadsheet address is an individual cell that is the intersection or overlap of a column and a row. This address is generally given in two parts: column and row.

Address	Interpretation
C 8	Column C and Row 8
J 30	Column J and Row 30
AA 5	Column AA (follows Column Z) and Row 5

	A	B	C
1	The 7th Inning Budgeted Operating Expenses		
2			
3	Expense	Budget Amount	Percent of Total Budget
4			
5	Salaries	$42,000.00	
6	Rent	$36,000.00	
7	Depreciation	$9,000.00	
8	Utilities and Phone	$14,500.00	
9	Taxes and Insurance	$14,000.00	
10	Advertising	$3,500.00	
11	Purchases	$120,000.00	
12	Other	$500.00	
13			
14	Total		

Figure 8–3

Now we develop formulas to make the calculations. Each spreadsheet program gives various shortcuts for writing formulas and formats for giving instructions unique to that program; however, we will write the basic concepts used, and add program-specific conventions as appropriate.

EXAMPLE Write the formulas and make the calculations to complete the Fig. 8–3 spreadsheet.

To find the total to be placed in cell B14, we need to add the amounts in cells B5–B12. To do this, we give the addresses of the cells to be added in a formula: $B14 = B5 + B6 + B7 + B8 + B9 + B10 + B11 + B12$.

The percent of the total budget is calculated by dividing the specific amount by the total budget and then multiplying by 100. We will write a formula for each line of data. Most programs use an asterisk (∗) to show multiplication, and a forward slash (/) to show division.

$$C5 = B5/B14*100 \qquad C6 = B6/B14*100 \qquad C7 = B7/B14*100$$
$$C8 = B8/B14*100 \qquad C9 = B9/B14*100 \qquad C10 = B10/B14*100$$
$$C11 = B11/B14*100 \qquad C12 = B12/B14*100$$

There are two ways to determine the value for cell C14. If we use the percent method, the percentage and the base are the same amount, so the total percent is 100%. To build in a check against the spreadsheet formulas, however, it is advisable to find the total percent by adding the calculated percents. It is easy to make a typing error in the formulas or to place the formula in the wrong cell. The total should be 100% or extremely close. There may be a small discrepancy due to the effects of rounding. $C14 = C5 + C6 + C7 + C8 + C9 + C10 + C11 + C12$. The spreadsheet with the completed calculations is shown in Fig. 8–4.

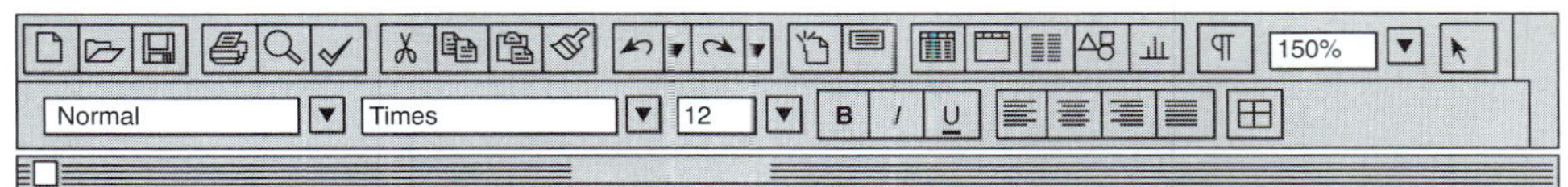

	A	B	C
1	The 7th Inning Budgeted Operating Expenses		
2			
3	Expense	Budget Amount	Percent of Total Budget
4			
5	Salaries	$42,000.00	17.5
6	Rent	$36,000.00	15.0
7	Depreciation	$9,000.00	3.8
8	Utilities and Phone	$14,500.00	6.1
9	Taxes and Insurance	$14,000.00	5.8
10	Advertising	$3,500.00	1.5
11	Purchases	$120,000.00	50.1
12	Other	$500.00	0.2
13			
14	Total	$239,500.00	100.0

Figure 8–4

SELF-STUDY EXERCISES 8–5

1 Write the statements as equations and solve.

1. The sum of x and 4 equals 12. Find x.

2. 4 less than 2 times a number is 6. Find the number.

3. Three parts totaling 27 lb are packaged for shipping. Two parts weigh the same. The third part weighs 3 lb less than each of the two equal parts. Find the weight of each part.

4. How many gallons of water must be added to 24 gal of pure cleaner to make 60 gal of diluted cleaner?

5. A wet casting weighing 4.03 kg weighs 3.97 kg after drying. Write and solve an algebraic equation to find the weight loss due to drying.

6. A plumber needs 3 times as much perforated pipe as solid pipe to lay a drain line 400 ft long. How much of each type of pipe is needed?

7. An engineering student purchased a new circuits text, a used Spanish text, and a graphing calculator. She remembered that the calculator cost twice as much as the Spanish book and that the circuits text cost $70.00. The total before tax was $235.00. What was the cost of the calculator and the Spanish text?

8. Mary Jefferson purchased a home on a square lot 150 feet on each side. She wants to enclose the entire lot with a cedar fence. One estimate for the job was for $14.00 per linear foot. How much will the fence cost?

9. Phil Chu, owner of Chu's Landscape Service, knows that one of his fertilizer tanks holds twice as many gallons of liquid fertilizer as a second tank. The two tanks together hold 325 gal. If both tanks are filled to capacity, how many gallons of fertilizer does each tank hold?

10. Carolina Villa hired a stone mason to build a triangular flower bed at one end of her patio. She will need enough mulch to cover 60 ft^2, the area of the flower bed. If the flower bed has an altitude of 10 ft, how long is the base?

11. Develop formulas to complete the spreadsheet in Fig. 8–5 to show data for the actual expenses for the 7th Inning Sports Memorabilia Shop.

12. Use the formulas to complete the spreadsheet for the 7th Inning Sports Memorabilia Shop.

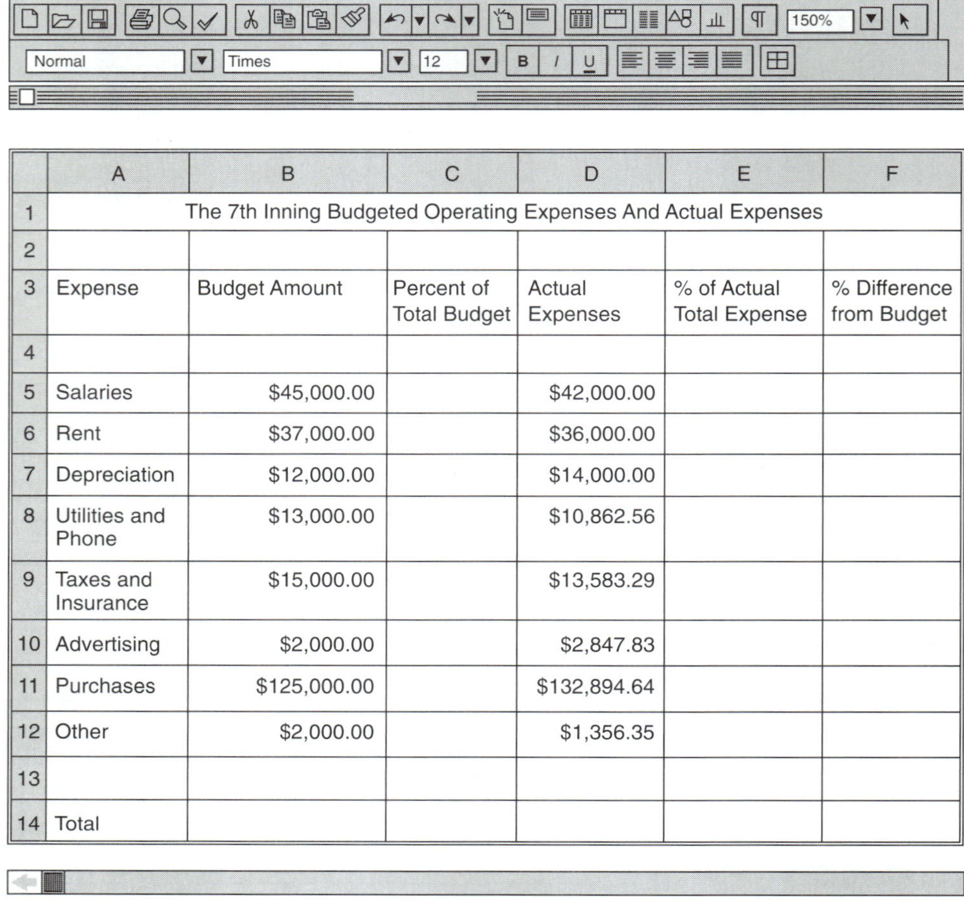

	A	B	C	D	E	F
1	The 7th Inning Budgeted Operating Expenses And Actual Expenses					
2						
3	Expense	Budget Amount	Percent of Total Budget	Actual Expenses	% of Actual Total Expense	% Difference from Budget
4						
5	Salaries	$45,000.00		$42,000.00		
6	Rent	$37,000.00		$36,000.00		
7	Depreciation	$12,000.00		$14,000.00		
8	Utilities and Phone	$13,000.00		$10,862.56		
9	Taxes and Insurance	$15,000.00		$13,583.29		
10	Advertising	$2,000.00		$2,847.83		
11	Purchases	$125,000.00		$132,894.64		
12	Other	$2,000.00		$1,356.35		
13						
14	Total					

Figure 8–5

8-6 SOLVING EQUATIONS WITH FRACTIONS BY CLEARING THE DENOMINATORS

Learning Outcomes

1. Solve equations with fractions by clearing the denominators.
2. Solve applied problems involving rate, time, and work.

Thus far, we have used mostly integers in equations. However, real-world situations require us to deal with various fractions and decimal quantities when we solve equations.

1 Solve Equations with Fractions by Clearing the Denominators.

The techniques we used earlier in the chapter will also help us solve equations with fractions and decimals. These techniques often require us to perform several calculations with fractions and decimals.

Other techniques minimize the calculations with fractions and decimals and allow more steps to be performed mentally. One of these techniques involves *clearing* the equation of all denominators in the first step. The resulting equation, which contains no fractions, is then solved using the procedures we learned earlier.

This process, also called *clearing the fractions,* is another application of the multiplication axiom. First, let's look at an example that contains only one fraction.

EXAMPLE   Solve $2x + \dfrac{3}{4} = 1$ by clearing the equation of fractions.

We *clear* an equation of a single fraction by multiplying the *entire* equation by the denominator of the fraction. *This means multiplying each term by the denominator.*

$$2x + \frac{3}{4} = 1$$

$$4(2x) + 4\left(\frac{3}{4}\right) = 4(1) \qquad \text{Multiply each term by the denominator 4.}$$

$$4(2x) + \overset{1}{4}\left(\frac{3}{\underset{1}{4}}\right) = 4(1) \qquad \text{Reduce and multiply.}$$

$$8x + 3 = 4 \qquad \text{Use procedures from Sections 8–2 to 8–4 to finish the solution.}$$

$$8x = 4 - 3 \qquad \text{Sort.}$$

$$8x = 1 \qquad \text{Combine.}$$

$$\frac{8x}{8} = \frac{1}{8} \qquad \text{Divide.}$$

$$x = \frac{1}{8}$$

$$\text{Check:} \qquad 2x + \frac{3}{4} = 1$$

$$\overset{1}{2}\left(\frac{1}{\underset{4}{8}}\right) + \frac{3}{4} = 1$$

$$\frac{1}{4} + \frac{3}{4} = 1$$

$$1 = 1$$

To clear an equation of a single fraction:

Apply the multiplication axiom by multiplying the entire equation by the fraction's denominator.

EXAMPLE   Solve $\dfrac{4d}{3} = 12$.

$$\frac{4d}{3} = 12$$

$$(\overset{1}{3})\frac{4d}{\underset{1}{3}} = (3)(12) \qquad \text{Multiply both sides by the denominator 3. Reduce where possible.}$$

$$4d = 36$$

$$\frac{4d}{4} = \frac{36}{4} \qquad \text{Divide.}$$

$$d = 9$$

$$\text{Check:} \quad \frac{4(\overset{3}{\cancel{9}})}{\underset{1}{\cancel{3}}} = 12$$

$$12 = 12$$

One advantage of clearing fractions is that it can be applied to equations in which the unknown is in the *denominator* of the fraction. Let's use this procedure to solve $\frac{10}{x} = 2$. Note that the term $\frac{10}{x}$ is *not* the product of 10 and x. Therefore, the coefficient of x is *not* 10. We solve the equation by clearing the fraction.

EXAMPLE   Solve $\dfrac{10}{x} = 2$.

$$\frac{10}{x} = 2$$

$$(\overset{1}{\cancel{x}})\frac{10}{\underset{1}{\cancel{x}}} = (x)2 \qquad \text{Multiply both sides by denominator } x. \text{ Reduce where possible.}$$

$$10 = 2x$$

$$\frac{10}{2} = \frac{2x}{2} \qquad \text{Divide.}$$

$$\mathbf{5 = x}$$

$$\text{Check:} \quad \frac{\overset{2}{\cancel{10}}}{\underset{1}{\cancel{5}}} = 2$$

$$2 = 2$$

Tip!	***Check for Extraneous Roots.***

When you solve equations with a variable in the fraction's denominator, you may get a "root" that does not make a true statement when substituted in the original equation. Solutions or roots that do not make a true statement in the original equation are called **extraneous roots. When solving an equation with the variable in the denominator, you *must* check the root to see if it makes a true statement in the original equation.**

One situation that produces an extraneous root is a value that causes the denominator of any fraction to be zero. Such values are called *excluded values*. The next example illustrates this situation.

To identify an excluded value that causes a denominator of zero:

1. Write an equation for each fraction that has a variable in the denominator by setting the denominator equal to zero.
2. Solve each resulting equation.
3. The solution for each equation from Step 1 is an excluded value.

EXAMPLE   Solve $\dfrac{0}{x} = 5$.

$$\frac{0}{x} = 5 \qquad \text{Multiply both sides by } x.$$

$$x\left(\frac{0}{x}\right) = 5(x)$$

$$\frac{0}{5} = \frac{5x}{5}$$

$$0 = x \qquad \text{Possible solution.}$$

Check: $\qquad \dfrac{0}{0} \neq 5 \qquad$ Does not check. Zero is an excluded value.

The equation $\frac{0}{x} = 5$ has no solution.

Equations in which the fraction contains more than one term in its numerator or denominator can also be cleared of fractions.

EXAMPLE Solve $\dfrac{12}{Q + 6} = 1$. Identify any excluded values.

The denominator is zero when $Q + 6 = 0$, or when $Q = -6$. Thus, -6 is an excluded value.

$$\frac{12}{Q + 6} = 1$$

$$(Q + 6)\frac{12}{Q + 6} = (Q + 6)\,1 \qquad$$ Multiply both sides of the equation by the denominator, the quantity $Q + 6$. Reduce where possible.

$$12 = Q + 6$$

$$12 - 6 = Q \qquad \text{Sort.}$$

$$6 = Q \qquad$$ Because the numerical coefficient of Q is already 1, the equation is solved.

Check: $\qquad \dfrac{12}{6 + 6} = 1 \qquad$ Substitute 6 for Q.

$$\frac{12}{12} = 1$$

$$1 = 1$$

The solution of the equation is $Q = 6$.

In the next example, we will look at an equation that has more than one fraction.

EXAMPLE Solve $-\dfrac{1}{4}x = 9 - \dfrac{2}{3}x$ by clearing all fractions first.

There are no excluded values, since there are no variables in the denominators. The denominators are 4 and 3. To clear *both* fractions at the same time, we multiply each term of the *entire* equation by both denominators or by the product of 4 and 3. This product is called a *common multiple* of 4 and 3 and is found the same way as a common denominator.

$$-\frac{1}{4}x = 9 - \frac{2}{3}x$$

$$(4)(3)\left(-\frac{1}{4}x\right) = (4)(3)(9) - (4)(3)\left(\frac{2}{3}x\right) \qquad$$ Multiply each term by both denominators.

$$\overset{1}{(\cancel{4})}(3)\left(-\frac{1}{\underset{1}{\cancel{4}}}x\right) = (4)(3)(9) - (4)\overset{1}{(\cancel{3})}\left(\frac{2}{\underset{1}{\cancel{3}}}x\right)$$ Reduce and multiply the remaining factors.

$$-3x = 108 - 8x$$ This equation contains no fractions.

$$-3x + 8x = 108$$ Sort.

$$5x = 108$$ Combine.

$$\frac{5x}{5} = \frac{108}{5}$$ Divide.

$$x = \frac{108}{5}$$

Check using a calculator:

$$-\frac{1}{4}x = 9 - \frac{2}{3}x$$

$$-\frac{1}{4}\left(\frac{108}{5}\right) = 9 - \frac{2}{3}\left(\frac{108}{5}\right)$$

Left Side: 1 $\boxed{a\ b/c}$ 4 $\boxed{+/-}$ $\boxed{\times}$ 108 $\boxed{a\ b/c}$ 5 $\boxed{=}$ $\Rightarrow$ $-5\frac{2}{5}$ or -5.4

Right Side: 9 $\boxed{-}$ 2 $\boxed{a\ b/c}$ 3 $\boxed{\times}$ 108 $\boxed{a\ b/c}$ 5 $\boxed{=}$ $\Rightarrow$ $-5\frac{2}{5}$ or -5.4

Tip!	***Improper Fractions versus Mixed Numbers.***

The solutions of many equations are not whole numbers. When this is the case, leave the solution as a *proper* or *improper fraction* in lowest terms. You can also change the solution to its decimal or mixed-number equivalent. When we solve applied problems, we will interpret the answer in an appropriate form. In the meantime, we leave the solutions as fractions in lowest terms.

For the fractions in the previous example, the common multiple of 12 (3 times 4) is also the *least common multiple* (LCM) of 3 and 4. In many instances, the LCM will be smaller than the product of the factors. For example, a common multiple for 4, 10, and 20 is 20.

To clear an equation of more than one fraction:

Multiply each term of the *entire* equation by the least common multiple (LCM) of the denominators of the equation.

Clearing fractions creates larger numbers in the new equation, but the numbers are integers instead of fractions.

EXAMPLE   Solve $\dfrac{1}{R} = \dfrac{1}{4} + \dfrac{1}{10} + \dfrac{1}{20}$ by clearing fractions first.

$R = 0$ is the excluded value.

$$\frac{1}{R} = \frac{1}{4} + \frac{1}{10} + \frac{1}{20}$$ The LCM is $20R$ because R, 4, 10, and 20 all divide evenly into $20R$.

$$\overset{}{20\cancel{R}}\left(\frac{1}{\cancel{R}}\right) = \overset{5}{\cancel{20}R}\left(\frac{1}{\underset{1}{\cancel{4}}}\right) + \overset{2}{\cancel{20}R}\left(\frac{1}{\underset{1}{\cancel{10}}}\right) + 20R\left(\frac{1}{\underset{1}{\cancel{20}}}\right)$$ Multiply each term in the *entire* equation by $20R$ and reduce.

$$20 = 5R + 2R + R \qquad \text{Combine like terms.}$$

$$\frac{20}{8} = \frac{8R}{8} \qquad \text{Divide by the coefficient of } R.$$

$$\frac{20}{8} = R$$

$$\frac{5}{2} = R \qquad \text{In lowest terms.}$$

$$\text{Check:} \quad \frac{1}{R} = \frac{1}{4} + \frac{1}{10} + \frac{1}{20} \qquad \text{Substitute } \frac{5}{2} \text{ for } R.$$

$$\frac{1}{\frac{5}{2}} = \frac{1}{4} + \frac{1}{10} + \frac{1}{20}$$

Left Side:

$$\frac{1}{\frac{5}{2}} = 1\left(\frac{2}{5}\right) = \frac{2}{5}$$

Right Side:

$$\frac{1}{4} + \frac{1}{10} + \frac{1}{20} =$$

$$\frac{5}{20} + \frac{2}{20} + \frac{1}{20} = \frac{8}{20} = \frac{2}{5}$$

The solution is $R = \dfrac{5}{2}$.

Tip!	*Fractions versus Decimals.*

Sometimes applied problems that require fractions in their equations require that their solutions be expressed as decimal numbers. In these cases, you will perform the division indicated by the fraction.

The equation in the preceding example is derived from the formula for finding total resistance in a parallel dc circuit with three branches rated at 4, 10, and 20 ohms. Ohms are expressed in decimal numbers, so in an application the solution should be converted to a decimal equivalent.

$$R = \frac{5}{2} \qquad \text{or} \qquad 2.5 \text{ ohms}$$

---

2 Solve Applied Problems Involving Rate, Time, and Work.

Our knowledge of fractional equations enables us to solve various types of applied problems. The first applied problem we will examine deals with the amount of work done, the rate of work, and the time worked.

Formula for amount of work:

Amount of work done = Rate of work × Time worked

As with percents and other rates we have studied, a rate measure involves a unit of time. If car A travels 50 miles in 1 hour, then car A's *rate of work* (travel) is 50 miles per 1 hour, or $\frac{50 \text{ mi}}{1 \text{ hr}}$ expressed as a fraction. If car A travels for 3 hours, then the product of the rate and time ($\frac{50\text{ mi}}{1\text{ hr}} \times 3$ hr) equals the total amount of work (travel) completed. Car A traveled (worked) 150 miles:

$$\text{Amount of work} = \text{rate} \times \text{time}$$

$$= \frac{50\text{ mi}}{1\text{ hr}} \times 3\text{ hr}$$

The unit of time in the rate measure must be compatible with the time measure.

$$= 150\text{ mi}$$

EXAMPLE A carpenter can install 1 door in 3 hours. How many doors can the carpenter install in 30 hours?

Known facts Rate of work = 1 door per 3 hours, $\frac{1}{3}$ door per hour, or $\frac{1\text{ door}}{3\text{ hr}}$
Time worked = 30 hours

Unknown facts W = Amount of work or number of doors installed

Relationships Amount of work = Rate of work $\times$ Time worked

Estimation Quite a few doors can be installed in 30 hr.

Calculations
$$W = \frac{1\text{ door}}{\overset{}{\underset{1}{3\text{ hr}}}} \times \overset{10}{30\text{ hr}}$$
$$W = 10\text{ doors}$$

Interpretation **Thus, 10 doors can be installed in 30 hours.**

If two workers or machines do a job together, we find the amount of work done by each worker or machine. Combined, the amounts equal 1 total job. Examine the basic formula for solving this type of work problem.

Formula for completing one job when A and B are working together:

$$\left(\begin{array}{c}\text{A's}\\\text{amount of}\\\text{work}\end{array}\right) + \left(\begin{array}{c}\text{B's}\\\text{amount of}\\\text{work}\end{array}\right) = 1\text{ completed job}$$

or

$$\left(\begin{array}{c}\text{A's}\\\text{rate of} \times \text{time}\\\text{work} \quad \text{worked}\end{array}\right) + \left(\begin{array}{c}\text{B's}\\\text{rate of} \times \text{time}\\\text{work} \quad \text{worked}\end{array}\right) = 1\text{ completed job}$$

EXAMPLE Pipe 1 fills a tank in 6 min and pipe 2 fills the same tank in 8 min (Fig. 8–6). How long does it take for both pipes together to fill the tank?

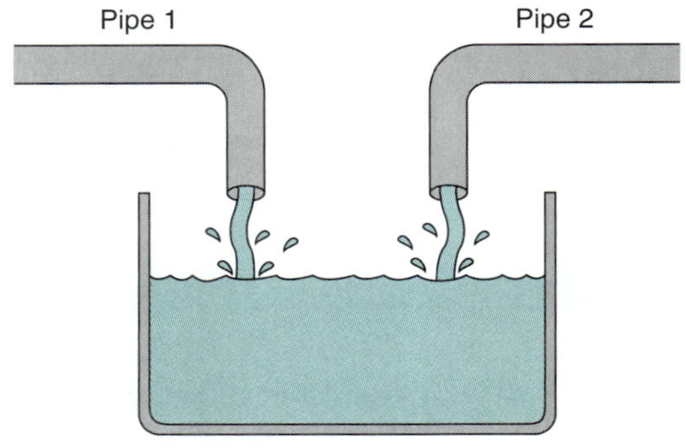

Figure 8–6

Known facts	Pipe 1 fills the tank at a rate of 1 tank per 6 min, $\frac{1}{6}$ tank per minute, or $\frac{1 \text{ tank}}{6 \text{ min}}$. Pipe 2 fills the tank at a rate of 1 tank per 8 min, $\frac{1}{8}$ tank per minute, or $\frac{1 \text{ tank}}{8 \text{ min}}$.
Unknown facts	T = time (in minutes) for both pipes together to fill the tank.
Relationships	Amount of work of pipe 1 = $\frac{1 \text{ tank}}{6 \text{ min}}(T)$. Amount of work of pipe 2 = $\frac{1 \text{ tank}}{8 \text{ min}}(T)$. Amount of work together = Pipe 1's work + Pipe 2's work.
Estimation	Both pipes together should fill the tank more quickly than the faster rate, or in less than 6 min.

Calculations

$$\frac{1 \text{ tank}}{6 \text{ min}}(T \text{ min}) + \frac{1 \text{ tank}}{8 \text{ min}}(T \text{ min}) = 1 \text{ tank} \qquad \text{The LCM is } 24.$$

$$(24)\left(\frac{1}{6}T\right) + (24)\left(\frac{1}{8}T\right) = (24)(1) \qquad \text{Clear fractions.}$$

$$\overset{4}{(24)}\left(\frac{1}{\underset{1}{6}}T\right) + \overset{3}{(24)}\left(\frac{1}{\underset{1}{8}}T\right) = (24)(1) \qquad \text{Reduce and multiply.}$$

$$4T + 3T = 24 \qquad \text{Combine.}$$

$$7T = 24$$

$$\frac{7T}{7} = \frac{24}{7} \qquad \text{Divide.}$$

$$T = \frac{24}{7}\left(\text{or } 3\frac{3}{7}\right)\text{min}$$

Interpretation Both pipes together fill the tank in $3\frac{3}{7}$ min.

Tip!	***Improper Fractions versus Mixed Numbers or Decimal Equivalents.***

In solving equations, a solution that is an improper fraction, like $\frac{24}{7}$, is left as an improper fraction. However, when you solve applied problems, it is desirable to change improper fractions like $\frac{24}{7}$ into mixed numbers or decimal equivalents.

The problem statement generally dictates the interpretation. In the preceding example, $\frac{24}{7}$ min is more appropriately interpreted as $3\frac{3}{7}$ min or 3.4 min (rounded). In some instances, you may need to change $\frac{3}{7}$ min to seconds ($\frac{3}{7}$ min $\times \frac{60 \text{ sec}}{1 \text{ min}}$ = 25.8 sec). Then, $3\frac{3}{7}$ min becomes 3 min 26 sec.

Now let's continue with work problems. Suppose two pipes, one a faucet and the other a drain, have opposite functions; that is, pipe 1 is filling the tank. Pipe 2 is a drain and is emptying the tank. In this case, we subtract the work done by the drain from the work done by the faucet. This combined action, if the faucet fills at a faster rate than the drain empties, will result in a full tank.

EXAMPLE A faucet fills a tank in 6 min. A drain empties the tank in 8 min (Fig. 8–7). If both the faucet and drain are open, in how many minutes will the tank start to overflow if the faucet is not turned off?

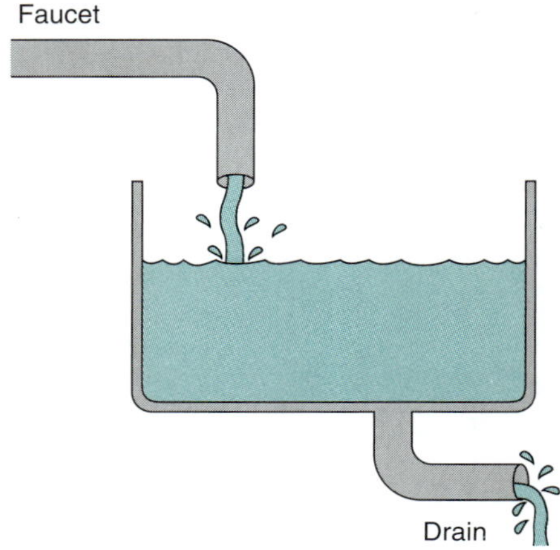

Figure 8–7

Known facts Faucet's rate of work $= 1$ tank filled per 6 min, $\frac{1}{6}$ tank per min, or $\frac{1 \text{ tank}}{6 \text{ min}}$.
Drain's rate of work $= 1$ tank emptied per 8 min, $\frac{1}{8}$ tank per min, or $\frac{1 \text{ tank}}{8 \text{ min}}$.

Unknown facts $T =$ time (min) until tank overflows with faucet and drain both working.

Relationships

$$\text{Amount of work of faucet} = \frac{1 \text{ tank}}{6 \text{ min}}(T)$$

$$\text{Amount of work of drain} = \frac{1 \text{ tank}}{8 \text{ min}}(T)$$

$$\begin{array}{c}\text{Amount of work when} \\ \text{both faucet and drain are open}\end{array} = \text{Faucet's work} - \text{Drain's work}$$

Estimation With both the faucet and drain open, it should take longer to fill the tank than if the faucet was open and the drain closed. It should take longer than 6 min.

Calculations

$$\frac{1 \text{ tank}}{6 \text{ min}}(T \text{ min}) - \frac{1 \text{ tank}}{8 \text{ min}}(T \text{ min}) = 1 \text{ tank}$$

$$\frac{1}{6}T - \frac{1}{8}T = 1$$

$$(24)\left(\frac{1}{6}T\right) - (24)\left(\frac{1}{8}T\right) = (24)(1)$$

$$(\overset{4}{\cancel{24}})\left(\frac{1}{\cancel{6}_{\,1}}T\right) - (\overset{3}{\cancel{24}})\left(\frac{1}{\cancel{8}_{\,1}}T\right) = (24)(1)$$

$$4T - 3T = 24$$

$$T = 24 \text{ min}$$

Interpretation With the faucet and drain both open, the tank will be full in 24 min.

SELF-STUDY EXERCISES 8–6

1 Solve the equations. Identify excluded values as appropriate.

1. $\dfrac{2}{7}x = 8$

2. $-7 = \dfrac{21}{33}p$

3. $\dfrac{1}{3}r = \dfrac{6}{7}$

4. $0 = -\dfrac{2}{5}c$

5. $-\dfrac{5}{3}m = 9$

6. $-9m = \dfrac{5}{3}$

7. $\dfrac{5}{8}t = 1$

8. $-\dfrac{5}{7}p = -\dfrac{11}{21}$

9. $10 = -\dfrac{1}{35}t$

10. $\dfrac{5}{12}z = 20$

11. $\dfrac{2x}{3} = 18$

12. $\dfrac{7}{Q} = 21$

13. $\dfrac{y + 1}{2} = 7$

14. $\dfrac{7}{p - 4} = -8$

15. $0 = \dfrac{x}{4}$

16. $-\dfrac{8}{P} = -72$

17. $\dfrac{P}{-8} = -72$

18. $-8 = \dfrac{4B}{B - 6}$

19. $\dfrac{3P}{7} = 12$

20. $\dfrac{8 - R}{76} = 1$

21. $\dfrac{2}{9}c + \dfrac{1}{3}c = \dfrac{3}{7}$

22. $-\dfrac{1}{4}x = 9 - \dfrac{2}{3}x$

23. $\dfrac{2}{7}y + \dfrac{3}{8} = \dfrac{1}{7}y + \dfrac{5}{3}$

24. $\dfrac{1}{3}x + \dfrac{1}{2}x = \dfrac{20}{3}$

25. $\dfrac{7}{R} - \dfrac{2}{R} = -1$

26. $S = \dfrac{1}{15} + \dfrac{1}{5} + \dfrac{1}{30}$

27. $18 - \dfrac{1}{4}x = \dfrac{1}{2}$

28. $\dfrac{1}{7}H - \dfrac{1}{3}H = 0$

29. $\dfrac{7}{16}h + \dfrac{1}{9} = \dfrac{1}{3}$

30. $x + \dfrac{1}{4}x = 8$

31. $3y + 9 = \dfrac{1}{4}y$

32. $18 = \dfrac{4}{3x} - \dfrac{3}{2x}$

33. $\dfrac{2}{7}p + 1 = \dfrac{1}{3}p$

34. $S = \dfrac{1}{10} + \dfrac{1}{25} + \dfrac{1}{50}$

35. $-x + \dfrac{1}{7} = \dfrac{1}{2}x$

36. $0 = 1 + \dfrac{2}{9}c - c$

37.–42. Check or verify the solutions of Exercises 12, 14, 16, 18, 25, and 32.

2 Set up an equation and solve.

43. A licensed electrician installs 8 light fixtures in 2 hr. How many light fixtures can she install in 20 hr?

44. A printing press produces 1 day's newspaper in 4 hr. A higher-speed press does 1 day's newspaper in 2 hr. How much time does it take both presses to produce 1 day's newspaper?

45. One machine packs 1 day's salmon catch in 8 hr. A second machine packs 1 day's catch in 5 hr. How much time does it require for 1 day's catch to be packed if both machines are used?

46. A painter can paint a house in 6 days. Another painter takes 8 days to paint the same house. If they work together, how much time will it take them to paint the house?

47. A tank has two pipes entering it and one leaving it. Pipe 1 fills the tank in 3 min. Pipe 2 takes 7 min to fill the same tank. Pipe 3, however, empties the tank in 21 min. How much time does it take to fill the tank with all three pipes operating at the same time?

Learning Outcomes

1. Solve equations with decimals.
2. Solve applied problems with decimal equations.

1 Solve Equations with Decimals.

As in equations with fractions, we can also clear an equation of decimals before starting to solve the equation. Like an equation cleared of fractions, an equation cleared of decimals contains only integers. It can then be solved using procedures previously studied.

To clear an equation of decimals:

Multiply each term of the *entire* equation by the least common multiple (LCM) of the fractional amounts following the decimal points.

EXAMPLE   Solve $1.1 = 3.4 + R$ by first clearing the equation of decimals.

In $1.1 = 3.4 + R$, the LCM is 10 because both decimals are in tenths (have understood denominators of 10).

$$1.1 = 3.4 + R$$

$$10(1.1) = 10(3.4) + 10R \qquad \text{Multiply entire equation by } 10.$$

$$11 = 34 + 10R \qquad \text{Sort.}$$

$$11 - 34 = 10R \qquad \text{Combine.}$$

$$-\frac{23}{10} = \frac{10R}{10} \qquad \text{Divide.}$$

$$-2.3 = R$$

Check: $1.1 = 3.4 + R$

$1.1 = 3.4 + (-2.3)$

$1.1 = 1.1$

Tip! | *LCM for Decimals.*

Digits to the right of the decimal point represent fractions whose denominators are determined by the place value. To find the LCM for all the decimal numbers in an equation, look at the decimal with the most digits after the decimal point. Use its denominator to clear the decimals.

This procedure allows you to avoid dividing by a decimal, which can be a common source of error when making calculations by hand.

EXAMPLE   Solve $0.38 + 1.1y = 0.6$ by first clearing the equation of decimals.

$$0.38 + 1.1y = 0.6 \qquad \text{The LCM is } 100.$$

$$100(0.38) + 100(1.1y) = 100(0.6) \qquad \text{Multiply by 100.}$$

$$38 + 110y = 60 \qquad \text{Sort.}$$

$$110y = 60 - 38 \qquad \text{Combine.}$$

$$\frac{110y}{110} = \frac{22}{110} \qquad \text{Divide.}$$

$$y = 0.2$$

2 Solve Applied Problems with Decimal Equations.

A number of applied problems are solved with decimal equations. One common problem involves interest on a loan or on an investment.

A basic tool for interest problems is the interest formula. It resembles the percentage formula, $P = RB$, but it includes the time period over which the money was borrowed or invested. If we know three of the four elements, we can find the fourth.

> **Formula for simple interest:**
>
> $$I = PRT$$
>
> where $I =$ **interest**, $P =$ **principal**, $R =$ **rate** or percent in decimal form, and $T =$ **time.**

For comparison with the percentage formula, interest is the part or percentage and the principal is the base.

EXAMPLE A \$1,000 investment is made for $2\frac{1}{2}$ years at 8.25%. Find the amount of interest.

When we use the interest formula, we change the percent to a decimal equivalent. When no time period is given for the rate, we assume the rate to be per year.

$I = PRT$ Substitute the values in the formula.

$$I = \underset{\text{principal}}{\$1{,}000} \times \underset{\text{rate}}{0.0825} \times \underset{\text{time}}{2.5 \text{ years}} \qquad 8.25\% = 0.0825;\ 2\tfrac{1}{2} \text{ years} = 2.5 \text{ years}$$

$$I = \$206.25$$

The interest for $2\frac{1}{2}$ years is \$206.25.

Tip!	***Hand-Clearing Decimals versus Using a Calculator:***

The preceding example illustrates that in some cases clearing decimals is not the most efficient way to solve an equation with decimals. The numbers produced may be extremely large and cumbersome, and the decimals may not all be cleared.

$$I = \$1{,}000 \times 0.0825 \times 2.5 \qquad \text{LCM is 10,000.}$$

$$10{,}000I = 10{,}000\left(\$1{,}000 \times \frac{825}{10{,}000} \times \frac{25}{10}\right) \qquad \text{Multiply by 10,000.}$$

$$10{,}000I = 100(825)(25)$$

$$10{,}000I = 2{,}062{,}500$$

$$I = \frac{2{,}062{,}500}{10{,}000}$$

$$I = \$206.25$$

A calculator gives the solution more quickly and more efficiently if decimals are used.

Let's use the interest formula to solve problems when facts other than the interest are missing.

EXAMPLE Aetna Photo Studio borrowed $3,500 for some darkroom equipment and has to pay $1,890 in interest over a 3-year period. What is the interest rate?

Known facts Principal = $3,500. Interest = $1,890. Time = 3 years.

Unknown facts R = Rate

Relationships $I = PRT$

Estimation One year's interest is about $\frac{1}{3}$ of the total interest, or about $600.

$$10\% \text{ of } \$3,500 = \$350$$

$$20\% \text{ of } \$3,500 = \$350 \times 2 = \$700$$

Therefore R is more than 10% and less than 20%.

Calculations
$$1,890 = 3,500 \times R \times 3$$

$$1,890 = 10,500R$$

$$\frac{1,890}{10,500} = \frac{10,500R}{10,500}$$

$$0.18 = R \qquad\qquad 0.18 = 18\%$$

Interpretation **The interest rate is 18%.**

Many other real-world situations can be solved with formulas that use decimal numbers. Let's examine a few.

EXAMPLE The distance formula is distance = rate × time. If a car was driven 82.5 mi at 55 mi per hr, how long did the trip take?

Known facts Distance = 82.5 mi. Rate = $\dfrac{55 \text{ mi}}{\text{hr}}$.

Unknown facts T = Time in hours

Relationships Distance = Rate × Time

Estimation At 55 mi per hr, a car would travel 110 mi in 2 hr. It would take less than 2 hr to travel 82.5 mi.

Calculations
$$82.5 = 55 \times T$$

$$82.5 = 55T$$

$$\frac{82.5}{55} = \frac{55T}{55}$$

$$1.5 = T$$

Interpretation **The trip took 1.5 or $1\frac{1}{2}$ hr.**

EXAMPLE The formula for voltage V is wattage W divided by amperage A: $V = \frac{W}{A}$. Find the voltage to the nearest hundredth needed for a circuit of 1,280 W with a current of 12.23 A.

Estimation $1,200 \div 12 = 100$

$$V = \frac{W}{A} \qquad\qquad \text{Substitute values.}$$

$$V = \frac{1{,}280}{12.23}$$

Divide.

$$V = 104.6606705 \quad\text{or}\quad 104.66 \text{ V}$$

Interpretation   The voltage is **104.66 V**.

EXAMPLE   Juan earned \$41.85 working 6.75 hr at a fast-food restaurant. What was his hourly wage?

Known facts   Pay = \$41.85. Hours = 6.75.

Unknown facts   Wage = W

Relationships   Hourly wage $\times$ Hours worked = Amount of pay

Estimation   7 hr at \$5 = \$35. 7 hr at \$6 = \$42. Hourly wage is close to \$6.

Calculations

$$W \times 6.75 = 41.85$$
$$6.75W = 41.85$$
$$\frac{6.75W}{6.75} = \frac{41.85}{6.75}$$
$$W = 6.2 \quad\text{or}\quad \$6.20 \text{ per hr}$$

Interpretation   **Juan's hourly wage is \$6.20 per hour.**

SELF-STUDY EXERCISES 8–7

1   Solve the equations.

1. $2.3x = 4.6$

2. $0.8R = 0.6$ (round to nearest tenth)

3. $0.33x + 0.25x = 3.5$ (round to nearest hundredth)

4. $0.3a = 4.8$

5. $1.5p = 7$ (round to nearest tenth)

6. $0.04x = 0.08 - x$ (round to nearest hundredth)

7. $0.4p = 0.014$

8. $0.47 = R + 0.4R$ (round to nearest hundredth)

9. $2.3 = 5.6 + y$

10. $4.3 = 0.3x - 7.34$

11. $2x + 3.7 = 10.3$

12. $0.16 + 2.3x = -0.3$

13. $1.5x + 2.1 = 3$

14. $3.82 - 2.5y = 1$

15. $0.15p = 2.4$

2   Solve the problems using decimal equations.

16. If the formula for force is force = pressure $\times$ area, how many pounds of force are produced by a pressure of 35 pounds per square inch (psi) on a piston whose surface area is 2.5 in.^{2}? Express the pounds of force as a decimal number.

17. The circumference of a circle equals π times the diameter. If a steel rod has a diameter of 1.5 in., what is the circumference of the rod to the nearest hundredth? (Circumference is the distance around a circle.)

18. The distance formula is distance = rate $\times$ time. If a trucker drove 682.5 mi at 55 miles per hour (mi/hr), how long did she drive? (Answer to the nearest whole number.)

19. The distance formula is distance = rate $\times$ time. If a tractor-trailer rig is driven 422.5 mi at 65 miles per hour (mph) on interstate highways, how long does the trip take?

20. Electrical resistance in ohms (Ω) is voltage V divided by amperes A. Find the resistance to the nearest tenth for a motor with a voltage of 12.4 V requiring 1.5 A.

21. Rita earned \$39.75 working 7.5 hr at a college bookstore. What was her hourly wage?

22. The formula for electrical power is $V = \frac{W}{A}$: voltage (V) equals wattage (W) divided by amperage (A). Find the voltage to the nearest hundredth needed for a circuit of 500 W with a current of 3.2 A.

23. Find the interest paid on a loan of $2,400 for 1 year at an interest rate of 11%.

24. Find the interest paid on a loan of $800 at $8\frac{1}{2}\%$ interest for 2 years.

25. Find the total amount of money (maturity value) that the borrower will pay back on a loan of $1,400 at $12\frac{1}{2}\%$ simple interest for 3 years.

26. Find the rate of interest on an investment of $2,500 made by Nurse Honda for a period of 2 years if she received $612.50 in interest.

27. Maddy Brown needed start-up money for her landscape service. She borrowed $12,000 for 30 months and paid $360 interest on the loan. What interest rate did she pay?

28. Raul Fletes needs money to buy lawn equipment. He borrows $500 for 7 months and pays $53.96 in interest. What is his rate of interest?

29. Linda Davis agrees to lend money to Alex Luciano at a special interest rate of 9%, on the condition that he borrow enough that he will pay her $500 in interest over a 2-year period. What is the minimum amount Alex can borrow?

30. Rob Thweatt needs money for medical school. He borrows $6,000 at 12% interest. If he pays $360 interest, what is the duration of the loan?

Electronics: Kirchhoff's Laws

Kirchhoff's current law (KCL) and Kirchhoff's voltage law (KVL) form the basis of all electronics. The only difficult thing about the two laws is spelling Kirchhoff—notice that there are two h's and two f's. The rest is easy.

Kirchhoff's current law (KCL) says that *the sum of all currents at a node equals zero*. A node is like an intersection near your house. At 3 A.M. there are no cars in the intersection. If each entering car counts $+1$, and each exiting car counts -1, then the sum of all the cars will equal zero. KCL works the same way, only with current, which is measured in amperes (A). Because an ampere is a large measuring unit, measurements are often in milliamperes (mA), or thousandths of an ampere. Figure 8–8 illustrates a node.

Arrows are often used to indicate current. *Let each entering arrow have a $+$ sign and each exiting arrow have a $-$ sign.* Assume that there is a node with two currents entering the node ($I_1 = 6$ mA, $I_2 = 5$ mA), an unknown current I_3, and three currents exiting the node ($I_4 = 2$ mA, $I_5 = 4$ mA, and $I_6 = 3$ mA). See Fig. 8–8, circuit 1. As an equation this is

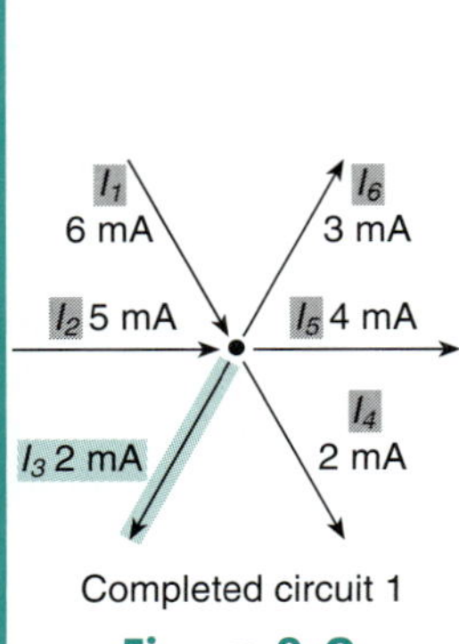

Figure 8–8

$$I_1 \quad + \quad I_2 \quad + I_3 \quad + \quad I_4 \quad + \quad I_5 \quad + \quad I_6 \quad = 0$$

$$6\,\text{mA} + 5\,\text{mA} + I_3 - 2\,\text{mA} - 4\,\text{mA} - 3\,\text{mA} = 0 \qquad \text{Insert appropriate numbers and signs.}$$

$$I_3 = 0 - 6\,\text{mA} - 5\,\text{mA} + 2\,\text{mA} + 4\,\text{mA} + 3\,\text{mA} \qquad \text{Solve for } I_3.$$

$$I_3 = -2\,\text{mA} \qquad \text{Combine like terms.}$$

This means that I_3 *is exiting the node and equals 2 mA*, as shown in Fig. 8–9. To verify this, substitute -2 mA into the original equation.

$$I_1 \quad + \quad I_2 \quad + \quad I_3 \quad + \quad I_4 \quad + \quad I_5 \quad + \quad I_6 \quad = 0$$

$$6\,\text{mA} + 5\,\text{mA} - 2\,\text{mA} - 2\,\text{mA} - 4\,\text{mA} - 3\,\text{mA} = 0 \qquad \text{It works!}$$

Kirchhoff's voltage law (KVL) says that *the sum of all voltages in any loop equals zero*. Picture a bug walking around any loop in a circuit. Voltage is always measured as a drop in potential across something, so there is a plus sign at one end

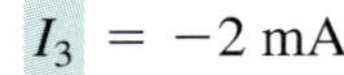

Figure 8–9

of the something and a minus at the other end. As the bug walks around the circuit, *the first sign encountered is the one used.* The equations for KVL look like those for KCL. The only difference is that a closed loop, rather than a node, is used.

Look at circuit 2 in Fig. 8–10, which has two voltage sources ($V_1 = 8$ V, $V_2 = 7$ V), an unknown voltage called V_3, and three voltage drops ($V_4 = 3$ V, $V_5 = 4$ V, $V_6 = 2$ V). Let's start at the top of the circuit and move down the left side to the bottom and then up the right side.

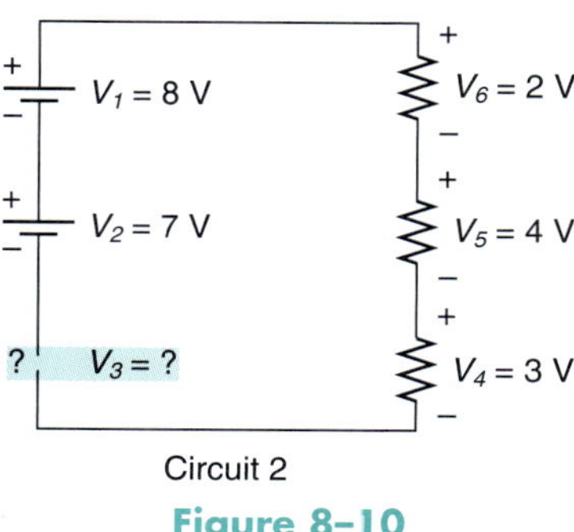

Circuit 2

Figure 8–10

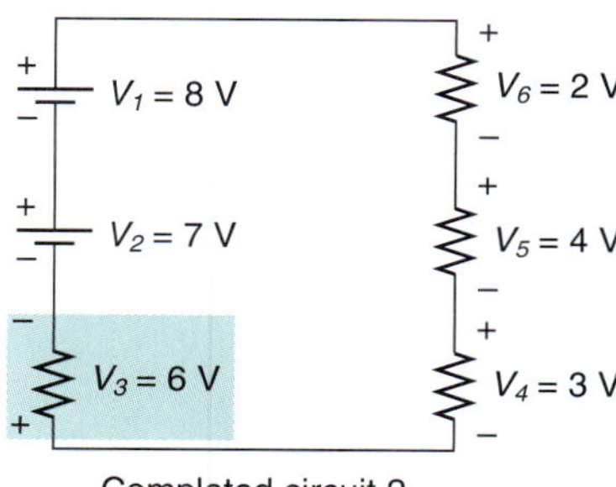

Completed circuit 2

Figure 8–11

As an equation,

$$V_1 + V_2 + V_3 + V_4 + V_5 - V_6 = 0$$

8 V $+ 7$ V $+ V_3 - 3$ V $- 4$ V $- 2$ V $= 0$ Insert appropriate numbers and signs.

$V_3 = 0 - 8$ V $- 7$ V $+ 3$ V $+ 4$ V $+ 2$ V   Solve for V_3.

$V_3 = -6$ V

This means that V_3 *is a voltage drop of 6 V* as shown in Fig. 8–11. To verify this, substitute -6 V into the original equation.

$$V_1 + V_2 + V_3 + V_4 + V_5 + V_5 = 0$$

8 V $+ 7$ V $- 6$ V $- 3$ V $- 4$ V $- 2$ V $= 0$ It works!

Exercises

Redraw each circuit in Figs. 8–12 to 8–17. Find the missing currents and voltages in all the circuits and draw the correct component on the circuit. For each circuit, write a complete equation using all the values to show that the algebraic sum of the currents at a node or of the voltages in a loop equals zero.

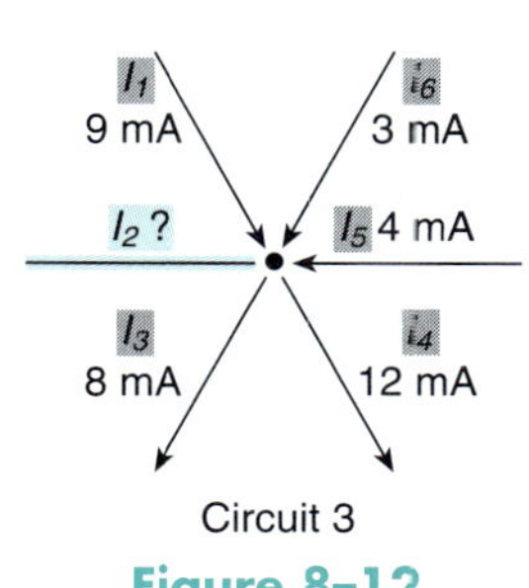

Circuit 3

Figure 8–12

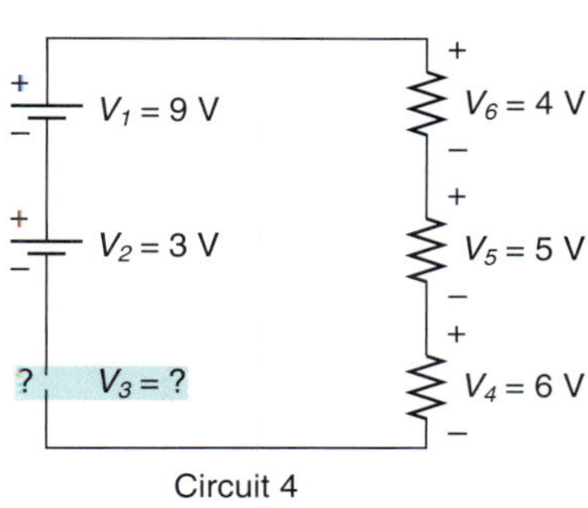

Circuit 4

Figure 8–13

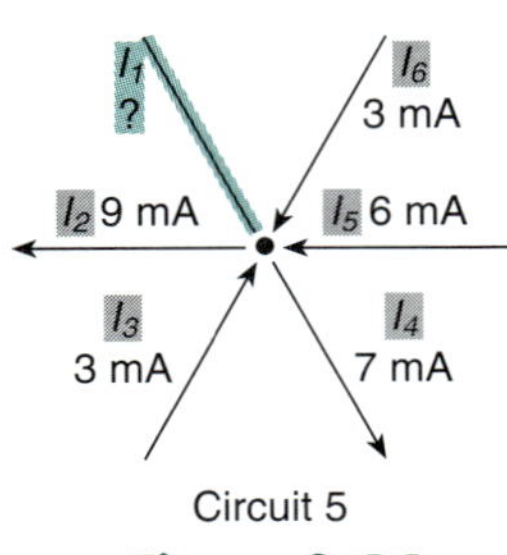

Circuit 5

Figure 8–14

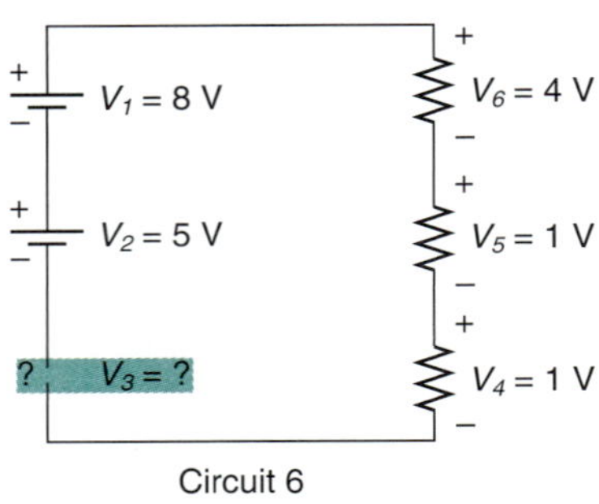

Circuit 6

Figure 8–15

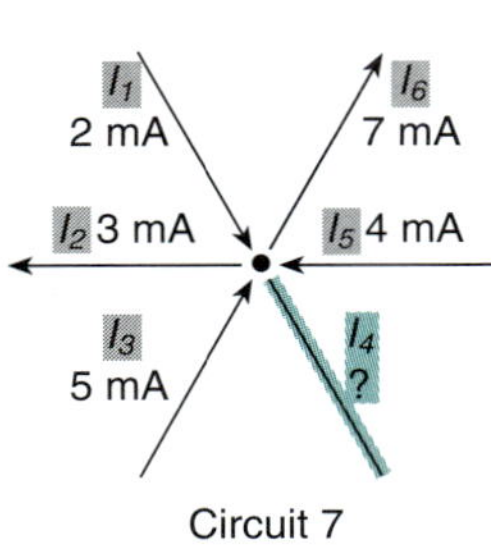

Circuit 7

Figure 8–16

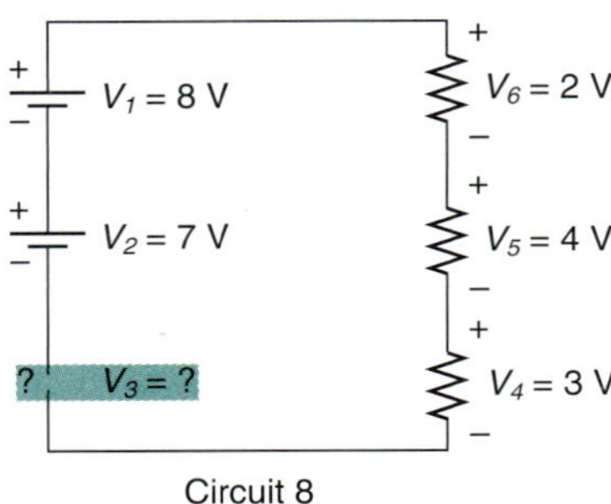

Circuit 8

Figure 8–17

Answers for Exercises

Circuit 3: $\quad I_1 \;+\; I_2 \;+\; I_3 \;+\; I_4 \;+\; I_5 \;+\; I_6 \;=\; 0$

$\qquad 9\text{ mA} + 4\text{ mA} - 8\text{ mA} - 12\text{ mA} + 4\text{ mA} + 3\text{ mA} = 0$

Circuit 4: $\qquad V_1 + V_2 + V_3 + V_4 + V_5 + V_6 = 0$

$\qquad 9\text{ V} + 3\text{ V} + 3\text{ V} - 6\text{ V} - 5\text{ V} - 4\text{ V} = 0$

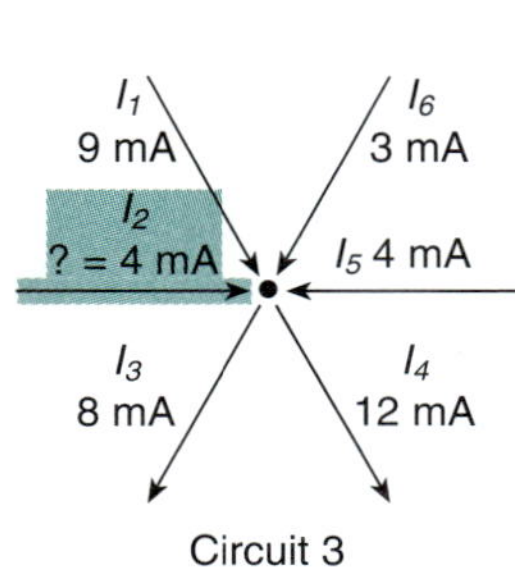

Circuit 3

Figure 8–18

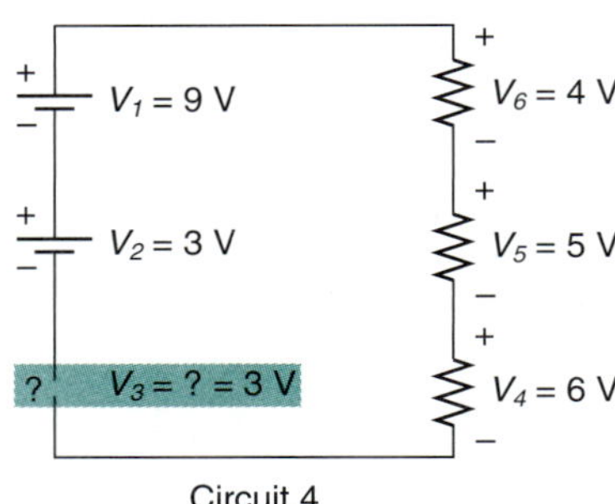

Circuit 4

Figure 8–19

Circuit 5: $\quad I_1 \;+\; I_2 \;+\; I_3 \;+\; I_4 \;+\; I_5 \;+\; I_6 \;=\; 0$

$\qquad 4\text{ mA} - 9\text{ mA} + 3\text{ mA} - 7\text{ mA} + 6\text{ mA} + 3\text{ mA} = 0$

Circuit 6: $\qquad V_1 + V_2 + V_3 + V_4 + V_5 + V_6 = 0$

$\qquad 8\text{ V} + 5\text{ V} - 7\text{ V} - 1\text{ V} - 1\text{ V} - 4\text{ V} = 0$

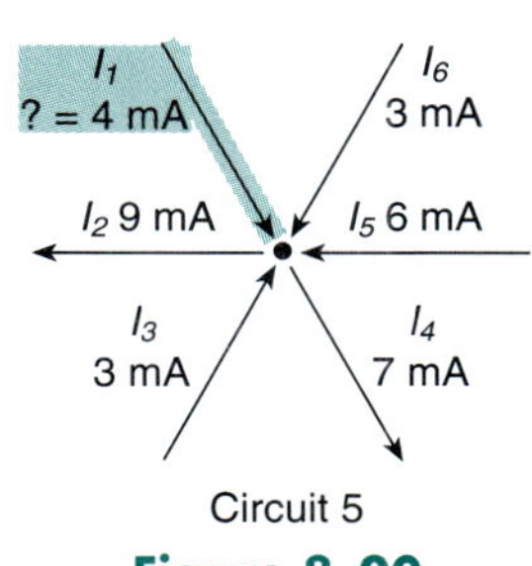

Figure 8–20

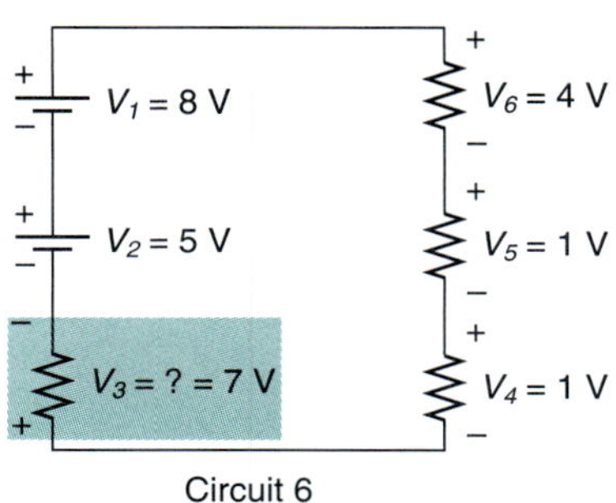

Figure 8–21

Circuit 7:
$$I_1 + I_2 + I_3 + I_4 + I_5 + I_6 = 0$$
$$2\text{ mA} - 3\text{ mA} + 5\text{ mA} - 1\text{ mA} + 4\text{ mA} - 7\text{ mA} = 0$$

Circuit 8:
$$V_1 + V_2 + V_3 + V_4 + V_5 + V_6 = 0$$
$$8\text{ V} + 7\text{ V} - 6\text{ V} - 3\text{ V} - 4\text{ V} - 2\text{ V} = 0$$

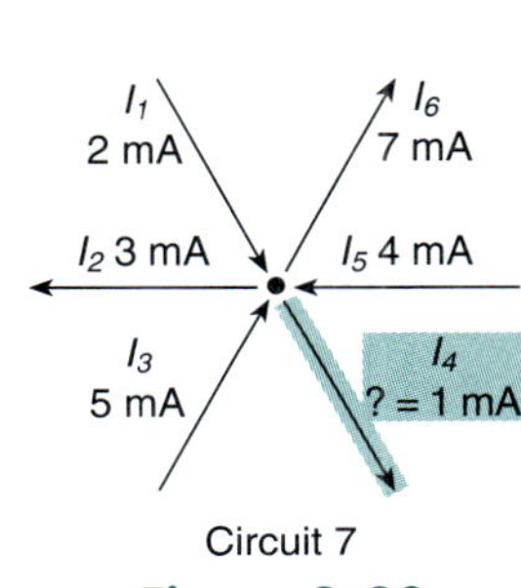

Figure 8–22

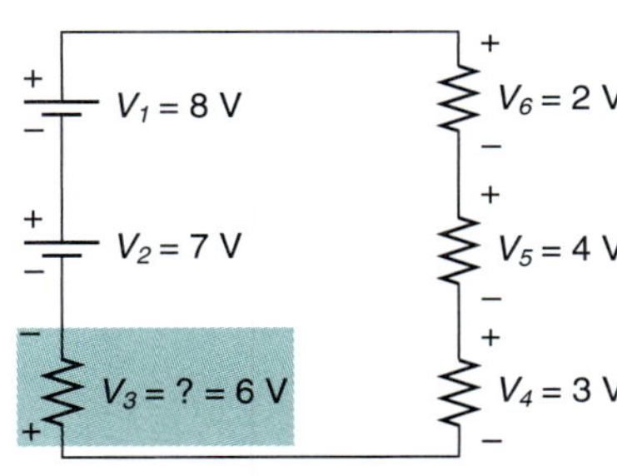

Figure 8–23

ASSIGNMENT EXERCISES

Section 8–1

Identify the terms in the expressions by drawing a box around each term.

1. $15x - \dfrac{3a}{7} + \dfrac{(x-7)}{5}$

2. $5x - 8 + \dfrac{3}{y}$

State the equations in words.

3. $x + 5 = 2$

4. $x - 7 = 11$

5. $\dfrac{x}{8} = 7$

6. $3x + 7 = -3$

7. $3(x + 7) = -3$

Write the following statements in symbols.

8. A number decreased by 5 is 8. What is the number?

9. Seven more than twice a number is 11. Find the number.

10. A certain stock listed on the New York Stock Exchange closed at $42\frac{3}{8}$, a decrease of $3\frac{1}{8}$ points from the opening price. What was the opening price?

11. Twice the sum of a number and 8 is 40. What is the number?

12. The print shop used 31 cases of copy paper during one month. End-of-the-month inventory indicated 172 cases on hand. How many cases of paper were on hand at the beginning of the month? Write an equation using addition to solve the problem. Then write an equation using subtraction that will also solve the problem.

Section 8–2

Solve the equations using multiplication or division. Check your answers.

13. $3x = 21$

14. $4x = -28$

15. $-15 = 2b$

16. $-5y = 30$

17. $-7y = -49$

18. $-5 = -m$

19. $8 = -x$

20. $0.6 = -a$

21. $3 = \dfrac{1}{5}x$

22. $-\dfrac{2}{7}x = 8$

23. $-\dfrac{3}{8}x = -24$

24. $\dfrac{1}{2}x = -5$

25. $42 = -\dfrac{6}{7}x$

26. $-\dfrac{5}{8}x = -10$

27. $\dfrac{1}{7}x = 12$

Section 8–3

Solve the equations.

28. $5y - 7y = 14$

29. $4x + x = 25$

30. $36 = 9a - 5a$

31. $2b - 7b = 10$

32. $0 = 4t - t$

33. $21 = x + 2x$

34. $8x - 3x = 6 + 9$

35. $20 - 4 = 2x - 6x$

36. $13 - 27 = 3x - 10x$

37. $x + 7 = 10$

38. $x - 5 = 3$

39. $x - 3 = -4$

40. $3 + y = -5$

41. $1 = a - 4$

42. $t + 7 = 12$

43. $3x + 4 = 19$

44. $4x - 3 = 9$

45. $15 - 3x = -6$

46. $5 = 3x - 7$

47. $-7 = 6x - 31$

48. $4 = 7 - 4x$

49. $-12 = -8 - 2x$

50. $2x + 6 = x$

51. $5x - 12 = 9x$

52. $12x + 27 = 3x$

53. $3x + 9 = 10$

54. $7x = 8x + 4$

55. $4y + 8 = 3y - 4$

56. $10 + 4x = 5 - x$

57. $7 - 4y = y + 22$

58. $7x - 1 = 4x + 17$

59. $4x - 3 = 2x + 6$

60. $y - 5 = 6y + 30$

61. $8 - 2y = 15 - 3y$

62. $6 - 7x = 15 - x$

63. $5x - 12 = 2x + 15$

64. $18x - 21 = 15x + 33$

65. $7x - 5 + 2x = 3 - 4x + 12$

66. $2x - 3 + 15 = 7x - 8 - 6x$

67. $3x - 5x + 2 = 6x - 5 + 12x$

68. $\dfrac{R}{7} - 6 = -R$

69. $0 = \dfrac{8}{9}c + \dfrac{1}{4}$

70. $\dfrac{2}{15}P - P = 4$

71. $6.7y - y = 8.4$ (round to tenths)

72. $0.9R = 0.3$ (round to tenths)

73. $0.86 = R + 0.4R$ (round to hundredths)

74. $0.04y = 0.02 - y$ (round to hundredths)

Section 8–4

Solve the equations.

75. $18 = 6(2 - y)$

76. $4(6 + x) = 36$

77. $3x = 3(9 + 2x)$

78. $4(5x - 1) = 16$

79. $7x - 3(x - 8) = 28$

80. $4a = 8 - (a + 7)$

81. $5x = 7 + (x + 5)$

82. $3(x + 2) - 5 = 2x + 7$

83. $4(3 - x) = 2x$

84. $3(2x - 4) = 4x - 6$

85. $-2(4 - 2x) = -16$

86. $3(2x + 1) = -3$

87. $5(3 - 2x) = -5$

88. $-16 = -2(-2x + 4)$

89. $8 = 6 - 2(3x - 1)$

90. $4x - (x + 3) = 3$

91. $3(x - 1) = 18 - 2(x + 3)$

92. $-(x - 1) = 2(x + 7)$

93. $-(2x + 1) = -7$

94. $2 + 3(x - 4) = 2x - 5$

95. $7 = 3 + 4(x + 2)$

96. $7(x + 2) = -6 + 2x$

97. $3(4x + 3) = 3 - 4(x - 1)$

98. $3(2 - x) - 1 = 4(3 - x)$

Section 8–5

Write the statements as equations and solve.

99. The difference between x and 6 is 8. Find x.

100. Twice a number increased by 5 is 17. Find the number.

101. Five times the sum of x and 6 is 42 more than x. Find x.

102. How many gallons of water must be added to 46 gal of pure alcohol to make 100 gal of alcohol solution?

103. If one technician works 3 hours less than another and their total hours worked are 51, how many hours has each technician worked?

104. The shorter side of an L-shaped carpenter's square is 6 in. shorter than the longer side. If the total length of the carpenter's square is 24 in., what is the measure of each side?

105. A 6-in. square of sheet metal is formed into a circular tube. Will the tube fit into a round hole 1.5 in. in diameter?

106. LaShaundria worked 14 hours in one week and then received a $0.50 per hour pay raise. She then worked 21 hours the remainder of the week at the new pay rate. She earned $220.50 for the week. What was her old pay rate? Her new pay rate?

107. Ms. Galendez's back yard is a rectangle whose length is twice the width. If the perimeter of the yard is 720 ft, what are the dimensions of her yard?

108. Develop spreadsheet formulas for the Samaritan Home Health Company payroll sheet shown in Table 8–1. The company pays time and a half for any hours in excess of 40 during any work week.

TABLE 8–1

Employee	Date	Hourly Rate	Hours Worked for Week	Regular Pay	Overtime Pay	Total Gross Pay
Gayden, Bertha	8/19	$ 9.25	40			
Harrover, Roy	8/19	$13.60	42			
Harshman, Carol	8/19	$ 7.80	38			
Kearney, Claude	8/19	$16.50	48			
Morgan, Relbue	8/19	$12.75	30			
Oswalt, Alisa	8/19	$ 6.45	40			
Stapleton, Iven	8/19	$ 9.15	45			
Total Gross Payroll						

109. Make the calculations to complete the spreadsheet for Samaritan Home Health Company for the given week.

Make up a word problem or situation that would use the following equations in the solution.

110. $2x + 7 = 47$

111. $3(x + 5) + 2 = 80$

Section 8–6

Solve the equations.

112. $M + \dfrac{1}{4} = \dfrac{3}{4}$

113. $P = \dfrac{1}{2} + \dfrac{1}{3}$

114. $x + \dfrac{1}{7}x = 16$

115. $\dfrac{2}{5} - x = \dfrac{1}{2}x + \dfrac{4}{5}$

116. $\dfrac{R}{7} - 6 = -R$

117. $\dfrac{3}{7}m - \dfrac{1}{2} = \dfrac{2}{3}$

118. $0 = \dfrac{8}{9}c + \dfrac{1}{4}$

119. $\dfrac{2}{15}P - P = 4$

120. $\dfrac{1}{4}S = \dfrac{1}{4} + \dfrac{1}{10} + \dfrac{1}{20}$

121. $m = 2 + \dfrac{1}{4}m$

122. $\dfrac{7}{9} + 3 = \dfrac{1}{2}T$

123.–127. Check or verify Exercises 114, 116, 118, 120, and 122.

Solve the equations.

128. $6x + \dfrac{1}{4} = 5$

129. $2x + 4 = \dfrac{1}{2}$

130. $\dfrac{2}{5}x + 6 = \dfrac{2}{3}x$

131. $\dfrac{3}{10}x = \dfrac{1}{8}x + \dfrac{2}{5}$

132. $\dfrac{2}{5} + 3x = \dfrac{1}{10} - x$

133. $\dfrac{1}{x} = \dfrac{2}{5} + \dfrac{3}{10}$

134. $\dfrac{1}{R} = \dfrac{1}{10} + \dfrac{1}{3} + \dfrac{1}{6}$

135. $\dfrac{2}{P} = \dfrac{1}{2} + \dfrac{1}{4} - \dfrac{5}{12}$

136. $\dfrac{3}{x} + 4 = \dfrac{1}{5} - 7$

137. $\dfrac{5}{12}x - \dfrac{3}{4} = \dfrac{1}{9} - \dfrac{2}{3}x$

Set up an equation with fractions and solve.

138. A plastic tube fills a container in 10 min. A drain in the container, however, empties the container in 30 min. If the drain is open, how long does it take to fill the container?

139. Melissa can complete a landscape project in 3 hr. Henry can complete the same project in 7 hr. How long would it take Melissa and Henry working together to complete the landscape project?

140. One optical scanner reads a stack of sheets in 20 min. A second scanner reads the same stack in 12 min. How long does it take for both scanners together to process the one stack of sheets?

141. An apprentice electrician can install 5 light fixtures in 2 hr. How many light fixtures can be installed in 10 hr?

142. A brick mason can erect a retaining wall in 6 hr. The brick mason's apprentice can do the same job in 10 hr. How much time does it take both of them working together to erect a retaining wall?

Section 8–7

Solve the equations.

143. $3.4 = 1.5 + T$

144. $2y + 2.9 = 11.7$

145. $2.3x - 4.1 = 0.5$

146. $0.22 + 1.6x = -0.9$

147. $6.8 = 0.2y - 8.64$

148. $1.4x - 7.2 = 3.5x - 4.3$

149. $0.3x - 2.15 = 0.8x + 3.75$

150. $1.3x + 2 = 8.6x - 3.24$

151. $2.7 - x = 5 + 2x$

152. $4x - 3.2 + x = 3.3 - 2.4x$

153. $6.7y - y = 8.4$ (round to tenths)

154. $0.9R = 0.3$ (round to tenths)

155. $\dfrac{4x}{0.7} = \dfrac{3}{1.2}$ (round to hundredths)

156. $0.86 = R + 0.4R$ (round to hundredths)

157. $0.04y = 0.02 - y$ (round to hundredths)

158. $\dfrac{x}{6} = \dfrac{1.8}{3}$

159. $\dfrac{2.1}{x} = \dfrac{4.3}{7}$ (round to tenths)

160. The distance formula is distance = rate × time. If a portable MRI unit traveled 350.8 mi to and from a rural hospital at 50 mph, how long to the nearest hour did the trip to and from the hospital take?

161. Electrical resistance in ohms (Ω) is voltage (V) divided by amperes (A). Find the resistance of a small motor with a voltage of 8.5 V requiring 0.5 A.

162. Lester earned $29.69 working $4\frac{3}{4}$ hr for a landscape service. What was his hourly wage?

163. The formula for voltage is $V = \dfrac{W}{A}$: voltage (V) equals wattage (W) divided by amperage (A). Find the voltage to the nearest hundredth needed for a circuit of 385 W with a current of 3.5 A.

164. If the formula for simple interest is interest = principal × rate × time, find the interest to the nearest cent on a principal of $1,000 at a 19.5% rate for 1.5 years.

165. If the formula for force is force = pressure × area, how many pounds of force are produced by a pressure of 30 psi on a piston whose surface area is 12.5 in.^{2}?

Identify the terms in each expression by drawing a box around each.

1. $5x^2$

2. $7(3x - 1) - 5x - 3$

Write an equation for each statement.

3. Five added to a number gives a result of 35.

4. Seven times the difference of 8 and a number is 11.

Write the equations in words.

5. $x + 8 = 21$

6. $3x - 8 = 14$

Solve the equations.

7. $5x = 30$

8. $-\dfrac{3}{5}m = -6$

9. $y - 2y = 15$

10. $2x = 3x - 7$

11. $5 - 2x = 3x - 10$

12. $5x + 3 - 7x = 2x + 4x - 11$

13. $3(x + 4) = 18$

14. $2(x - 7) = -10$

15. $7 - (x - 3) = 4x$

16. 5 added to x equals 16. Find x.

17. Twice the sum of a number and 3 is 10. Find the number.

18. One tank holds four times as many gallons as a smaller tank. If the tanks hold 2,580 gal together, how much does each tank hold?

19. The Ross Prairie Wildlife Management Area (WMA) is located in Marion County, Florida, and the Potts Wildlife Management Area is located in Citrus County, Florida. Together the two WMAs have 11,034 acres. Potts WMA has 4,980 acres more than Ross Prairie. How many acres are in each WMA? Write an equation to express the relationship.

20. Acme Medical Corporation operates two hospitals in one small county. The chief executive officer reported for one month that an increased use of the outpatient center at one hospital was twice that of the other hospital's outpatient center. If the number of outpatients served by both facilities for the month was 2,550, how many outpatients used each facility for the month?

Solve the equations.

21. $\dfrac{3}{8}y = 6$

22. $4 = \dfrac{1}{3}x + 2$

23. $\dfrac{8}{y + 2} = -7$

24. $\dfrac{4}{5}z + z = 8$

25. $\dfrac{2}{7}x = \dfrac{1}{2}x + 4$

26. $-\dfrac{2}{3}x = \dfrac{1}{4}x - 11$

27. $5x + \dfrac{3}{5} = 2$

28. $\dfrac{1}{x} = \dfrac{1}{3} + \dfrac{5}{6}$

Solve the equations. Round to hundredths when necessary.

29. $1.3x = 8.02$

30. $4.5y + 1.1 = 3.6$

31. $0.18x = 300 - x$

32. $4.3 = 7.6 + x$

Solve the problems involving fractions, decimal numbers, and proportions.

33. Using the formula pressure $= \dfrac{\text{force}}{\text{area}}$, determine how much pressure a force of 32.75 lb exerts on a surface of 24.65 in.^{2} in a hydraulic system. Answer in pounds per square inch (psi) rounded to hundredths.

34. A pipe fills 1 tank in 4 hr. If a second pipe empties 1 tank in 6 hr, how long does it take for the tank to fill with both pipes operating?

35. Resistance in a parallel dc circuit equals voltage divided by amperes. What is the resistance (in ohms) if the voltage is 40 V and amperage is 3.5 A? Express the answer as a decimal rounded to thousandths.

36. One employee can wallpaper a room in 2 hr. Another employee can wallpaper the same room in 3 hr. How long will it take both employees to wallpaper the room when working together?

Ratio and Proportion

GOOD DECISIONS THROUGH TEAMWORK

The proportionality of the sides of similar triangles gives rise to the classic problem of finding the unknown height of an object, such as a building, if the building casts a shadow and a nearby object whose height is known also casts a shadow.

Your team is to find the height of a measurable object, such as a post holding a stop sign, using the similar triangles formed by a team member and his or her shadow and by the object and its shadow. Measure the length of the team member's shadow or use a yardstick or other rule and compare to the length of the object's shadow whose height is unknown.

Solve the similar triangles. What did you find for the height of the object? Now measure the object to verify your findings. How close is your computed height to the actual height? Is there a difference? Why or why not? If there is a difference, what is the percent of difference?

Repeat the process with two other objects that are easily measured. Find the percent of difference between the calculated and measured heights. Next, find the height of an object that you cannot measure directly. Establish a reasonable range of values that your team expects the actual height to be.

Present your team's findings to the class. Include a demonstration of the method you used and the solutions of the similar right triangles. Make a chart or transparency to aid in the presentation. Defend your estimate of the unmeasurable object.

9–1 Ratios and proportions

1. Verify that two ratios are proportional.
2. Solve fractional equations that are proportions.

9–2 Using proportions to solve problems

1. Solve problems with direct proportions.
2. Solve problems with inverse proportions.
3. Solve problems that involve similar triangles.

One of the most significant uses of fractions in problem solving is with the proportion. We have examined proportions in working with fractions, converting from one measuring unit to another, converting monetary units, and developing conversion factors. Again, we will build on our knowledge of proportions to develop some additional problem-solving techniques.

9-1   RATIOS AND PROPORTIONS

Learning Outcomes

1. Verify that two ratios are proportional.
2. Solve fractional equations that are proportions.

1   Verify That Two Ratios Are Proportional.

As we learned before, a ratio is the comparison of two quantities, and the most common notation for expressing ratios is the fraction. When we have two ratios that are equal they form a proportion. We can use the properties of proportions to verify whether two fractions are proportional.

In a proportion, *each side* of the equation is a fraction or ratio. (See Chapter 4 for an introduction to ratios and proportions.)

The relationship of the terms of the ratios or fractions in a proportion such as

$$\frac{6}{9} = \frac{2}{3}$$

can be stated as

"6 is to 9 as 2 is to 3"

Still another way of representing a proportion is

$6:9 = 2:3,$ which is also read   "6 is to 9 as 2 is to 3"

As noted in Chapter 4, the *cross products* are equal in a proportion. A cross product is the product of the numerator of one ratio and the denominator of the other. Let's state this property symbolically.

Property of proportions:

$$\text{If} \quad \frac{a}{b} = \frac{c}{d}, \quad \text{then} \quad ad = bc, \quad b, d \neq 0$$

Remember, properties do not apply when denominators are zero.

Applying this property, we get

$$\text{if} \quad \frac{6}{9} = \frac{2}{3}$$

$$\text{then} \quad 6(3) = 9(2)$$

$$\text{or} \quad 18 = 18$$

Another way to state this is the product of the *extremes* (end factors a and d) equals the product of the *means* (middle factors b and c).

To verify that two ratios form a proportion:

$$\text{Do } \frac{4}{12} \text{ and } \frac{6}{18} \text{ form a proportion?}$$

1. Find the two cross products.

2. Compare the two cross products.

3. If the cross products are equal, the two ratios form a proportion.

$4 \times 18 = 72 \qquad 6 \times 12 = 72$
Cross products are equal.
The ratios are proportional.

EXAMPLE Select two ratios that form a proportion from each set of ratios.

(a) $\dfrac{1}{2}, \dfrac{3}{8}, \dfrac{6}{12}, \dfrac{5}{15}$ (b) $\dfrac{3}{7}, \dfrac{3}{4}, \dfrac{7}{12}, \dfrac{15}{20}$

(a) $\dfrac{1}{2}, \dfrac{3}{8}$ $1 \times 8 = 8, \ 2 \times 3 = 6$ no

$\dfrac{1}{2}, \dfrac{6}{12}$ $1 \times 12 = 12, \ 2 \times 6 = 12$ yes

(b) $\dfrac{3}{7}, \dfrac{3}{4}$ $3 \times 4 = 12, \ 7 \times 3 = 21$ no

$\dfrac{3}{7}, \dfrac{7}{12}$ $3 \times 12 = 36, \ 7 \times 7 = 49$ no

$\dfrac{3}{7}, \dfrac{15}{20}$ $3 \times 20 = 60, \ 7 \times 15 = 105$ no

$\dfrac{3}{4}, \dfrac{7}{12}$ $3 \times 12 = 36, \ 4 \times 7 = 28$ no

$\dfrac{3}{4}, \dfrac{15}{20}$ $3 \times 20 = 60, \ 4 \times 15 = 60$ yes

EXAMPLE Verify that $\dfrac{6}{9}$ and $\dfrac{8}{12}$ are proportional.

$6 \times 12 = 72$, and $9 \times 8 = 72$.   Therefore, the ratios are proportional.

2 Solve Fractional Equations That Are Proportions.

The concept of proportions is most useful in problem solving when only three of the four terms of a proportion are known. We can use our knowledge of solving equations to find the missing term.

EXAMPLE Solve the proportions.

(a) $\dfrac{x}{4} = \dfrac{9}{6}$ (b) $\dfrac{4x}{5} = \dfrac{17}{20}$ (c) $\dfrac{x-2}{x+8} = \dfrac{3}{5}$ (d) $\dfrac{3}{x} = 7$

(a) $\dfrac{x}{4} = \dfrac{9}{6}$

$6x = 36$ Cross multiply. $x(6) = 6x$; $4(9) = 36$.

$\dfrac{6x}{6} = \dfrac{36}{6}$ Solve for x.

$x = 6$

(b) $\dfrac{4x}{5} = \dfrac{17}{20}$

$80x = 85$ Cross multiply. $4x(20) = 80x$; $5(17) = 85$.

$\dfrac{80x}{80} = \dfrac{85}{80}$ Solve for x.

$x = \dfrac{85}{80}$ Reduce.

$x = \dfrac{17}{16}$

(c) $\dfrac{x - 2}{x + 8} = \dfrac{3}{5}$

$5(x - 2) = 3(x + 8)$ Cross multiply.

$5x - 10 = 3x + 24$ Distribute.

$5x - 3x = 24 + 10$ Sort.

$2x = 34$ Combine terms.

$\dfrac{2x}{2} = \dfrac{34}{2}$ Solve for x.

$x = 17$

(d) $\dfrac{3}{x} = 7$ $7 = \dfrac{7}{1}$

$\dfrac{3}{x} = \dfrac{7}{1}$

$3 = 7x$ Cross multiply.

$\dfrac{3}{7} = \dfrac{7x}{7}$ Solve for x.

$\dfrac{3}{7} = x$

SELF-STUDY EXERCISES 9–1

1   Select two ratios that form a proportion.

1. $\dfrac{1}{3}, \dfrac{3}{4}, \dfrac{4}{12}, \dfrac{5}{10}$ **2.** $\dfrac{2}{5}, \dfrac{7}{8}, \dfrac{5}{10}, \dfrac{4}{10}$

3. $\dfrac{4}{8}, \dfrac{5}{10}, \dfrac{7}{12}, \dfrac{3}{5}$ **4.** $\dfrac{18}{24}, \dfrac{6}{12}, \dfrac{6}{18}, \dfrac{9}{12}$

Verify that the two ratios form a proportion.

5. $\dfrac{7}{8} = \dfrac{21}{24}$ **6.** $\dfrac{3}{5} = \dfrac{12}{20}$ **7.** $\dfrac{3}{8} = \dfrac{6}{24}$ **8.** $\dfrac{4}{5} = \dfrac{8}{10}$

2   Solve the proportions.

9. $\dfrac{x}{5} = \dfrac{9}{15}$ **10.** $\dfrac{3x}{16} = \dfrac{3}{8}$ **11.** $\dfrac{x - 1}{x + 6} = \dfrac{4}{5}$

12. $\dfrac{5}{x} = 8$

13. $\dfrac{2}{7} = \dfrac{x - 4}{x + 3}$

14. $\dfrac{2x + 1}{8} = \dfrac{3}{7}$

15. $\dfrac{3x - 2}{3} = \dfrac{2x + 1}{3}$

16. $\dfrac{5}{2x - 2} = \dfrac{1}{8}$

17. $\dfrac{8}{3x + 2} = \dfrac{8}{14}$

18. $\dfrac{2x}{8} = \dfrac{3x + 1}{7}$

9–2 USING PROPORTIONS TO SOLVE PROBLEMS

Learning Outcomes

1 Solve problems with direct proportions.
2 Solve problems with inverse proportions.
3 Solve problems that involve similar triangles.

1 Solve Problems with Direct Proportions.

Many problems in the workplace can be solved using proportion. These problems have a common characteristic: the details of the problem can be grouped into two pairs of data. These two pairs of data can be *directly* related or *inversely* related.

■ **DEFINITION: Direct Proportion.** A *direct proportion* is one in which the quantities being compared are directly related, so that as one quantity increases (or decreases), the other quantity also increases (or decreases). This relationship is also called a *direct variation.*

Apples	Cost
4	$1
8	$2
12	$3
16	$4

Data are often arranged in data tables so that these relationships can be examined. If 4 apples cost $1, let's set up a data table to examine related costs of other amounts of apples.

We can also use this relationship to find the cost of a given number of apples or to determine how many apples can be purchased with a given amount of money.

> *To set up a direct proportion:*
>
> **1.** Establish two pairs of related data.
> **2.** Write one pair of data in the numerators of the two ratios.
> **3.** Write the other pair of data in the denominators of the two ratios.
> **4.** Form a proportion using the two ratios.

EXAMPLE (a) Find the cost of 10 apples if 4 apples cost $1. (b) How many apples can be purchased for $10?

(a) Pair 1: 4 apples cost $1.

Pair 2: 10 apples cost c dollars.

Estimation 8 apples would cost $2. 10 apples will cost more than $2.

$$\frac{4 \text{ apples}}{10 \text{ apples}} = \frac{\$1}{\$c}$$
Pair 1 is the numerator of each ratio.
Pair 2 is the denominator of each ratio.

$$4c = 10$$
Cross multiply.

$$\frac{4c}{4} = \frac{10}{4}$$
Divide.

$$c = 2.50 \qquad \text{To nearest cent.}$$

Interpretation

Ten apples cost \$2.50.

(b) Pair 1: 4 apples cost \$1.

   Pair 2: a apples cost \$10.

Estimation

If 10 apples cost \$2.50, 4 times as many apples can be bought for \$10.
40 apples can be bought.

$$\frac{4 \text{ apples}}{a \text{ apples}} = \frac{\$1}{\$10} \qquad \begin{array}{l} \text{Pair 1} \\[4pt] \text{Pair 2} \end{array}$$

$$40 = a \qquad \text{Cross multiply.}$$

Interpretation

Forty apples can be bought for \$10.

Directly related data pairs can also be set up by making each pair a ratio. In the preceding example, we would write the ratio as:

$$\frac{4 \text{ apples}}{\$1} = \frac{10 \text{ apples}}{\$c} \qquad \begin{array}{l} \text{Pair 1} \\[4pt] \text{Pair 2} \end{array}$$

The solution is the same as before after cross multiplying.

   A *third way* to use directly related data is to identify a data pair in which both values are known and then use that relationship to find a conversion factor. For example, if 4 apples cost \$1, how much does 1 apple cost?

$$\$1 \div 4 \text{ apples} = \$0.25 \text{ per apple}$$

A conversion factor is used to multiply by the number of items. If each apple costs \$0.25, then 10 apples cost $10 \times \$0.25$, or \$2.50. This conversion factor is also called a *constant of direct variation*. The direct variation formula is $y = kx$, where x is the independent variable, y is the dependent variable and k is the constant of direct variation. Another way to express this is $k = \frac{y}{x}$.

To find the constant of direct variation:

1. Identify a pair of related data in which both values are known.
2. Write the data pair as a ratio. The units in the denominator of the ratio should match the units of the independent variable.
3. Leave the ratio as a fraction or change it to a decimal equivalent.

Any of these three methods work with directly related data. However, we will continue with the first method.

EXAMPLE A truck travels 102 mi on 6 gal of gasoline. How far will it travel on 30 gal of gasoline?

Known facts Pair 1: 102 mi uses 6 gal of gasoline.

Unknown facts Pair 2: m mi uses 30 gal of gasoline.

Estimation 30 gal $\div$ 6 gal = 5. Then, approximately 5 times 100 miles can be driven. More than 500 miles can be traveled on 30 gal.

Dimension Analysis

Calculations

$$\frac{102 \text{ mi}}{m \text{ mi}} = \frac{6 \text{ gal}}{30 \text{ gal}} \qquad \frac{\text{distance}_1}{\text{distance}_2} = \frac{\text{gasoline}_1}{\text{gasoline}_2}$$

$$\frac{102}{m} = \frac{6}{30} \qquad\qquad \frac{\text{mi}}{\text{mi}} = \frac{\text{gal}}{\text{gal}}$$

$$102(30) = 6m \qquad\qquad \text{Cross multiply. mi(gal)} = \text{gal(mi)}$$

$$\frac{3{,}060}{6} = \frac{6m}{6} \qquad\qquad \text{Divide by gal. Reduce. } \frac{\text{mi(gal)}}{\text{gal}} = \frac{\text{gal(mi)}}{\text{gal}}$$

$$510 = m \qquad\qquad m \text{ is expressed in miles.}$$

Interpretation   **The truck will travel 510 mi on 30 gal of gasoline.**

Tip!	***Always Analyze Dimensions.***

Even though we often remove the written dimensions from an equation, we must always analyze the dimensions to be sure we use the correct units in the solution.

EXAMPLE   If a metal rod tapers 1 in. for every 24 in. of length, what is the amount of taper of a 30-in. piece of rod? (see Fig. 9–1.)

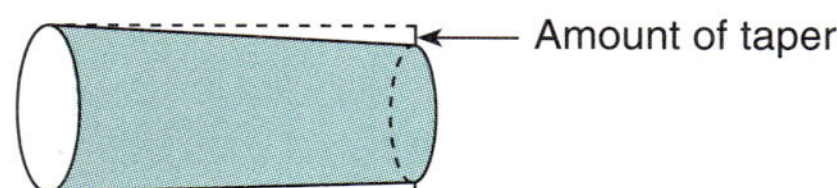

Figure 9–1

Known facts   Pair 1:   1-in. taper for 24-in. length

Unknown facts   Pair 2: x-in. taper for 30-in. length

Estimation   A 30-in. rod will taper more than 1 inch.

Calculations

$$\frac{1\text{-in. taper}}{x\text{-in. taper}} = \frac{24\text{-in. length}}{30\text{-in. length}} \qquad \begin{array}{l}\text{Pair 1}\\[4pt]\text{Pair 2}\end{array}$$

$$\frac{1}{x} = \frac{24}{30}$$

$$1(30) = 24x \qquad\qquad \text{Cross multiply.}$$

$$30 = 24x$$

$$\frac{30}{24} = \frac{24x}{24} \qquad\qquad \text{Divide.}$$

$$\frac{5}{4} = x \qquad \text{or} \qquad 1\frac{1}{4} = x$$

Interpretation   **The amount of taper for a 30-in. length of rod is $1\frac{1}{4}$ in.**

EXAMPLE   The ratio of chicory to coffee in a New Orleans coffee mixture is $1:8$. If the coffee company uses 75 lb of chicory for a batch of the coffee mixture, how many pounds of coffee are needed?

Known facts   Pair 1:   1 lb chicory for 8 lb coffee

Unknown facts   Pair 2:   75 lb chicory for x lb coffee

Estimation   50 lb of chicory would require 50×8 or 400 lb of coffee. 100 lb of chicory would require 100×8 or 800 lb of coffee. Between 400 lb and 800 lb of coffee is needed.

$$\frac{1 \text{ lb chicory}}{75 \text{ lb chicory}} = \frac{8 \text{ lb coffee}}{x \text{ lb coffee}}$$

Pair 1

Pair 2

$$\frac{1}{75} = \frac{8}{x}$$

$$1x = 75(8) \qquad \text{Cross multiply.}$$

$$x = 600$$

Thus, 600 lb of coffee is needed for one batch of the coffee mixture.

2 Solve Problems with Inverse Proportions.

Now that we have seen how to set up a direct proportion, let's turn our attention to relationships in which two quantities are *inversely proportional.*

■ **DEFINITION: Inverse Proportion.** An *inverse proportion* is one in which the quantities being compared are inversely related. That is, as one quantity increases, the other decreases, or as one quantity decreases, the other increases. This relationship is also called *inverse variation.*

For example, as we *increase* pressure on materials such as foam rubber, gases, loose dirt, and so on, the materials compress and so *decrease* in size or volume. Or, as we *decrease* pressure on these same materials, they *increase* in size or volume as they expand. This relationship is the opposite, or inverse, of direct proportion.

Here is another example: 3 workers frame a house in 2 weeks. If the contractor *increases* the number of workers to 6, the framing time *decreases* to 1 week, assuming the workers work at the same rate. In other words, the framing time is *inversely proportional* to the number of workers on the job.

Unlike the directly related ratios in a proportion, inversely related ratios do not allow us the flexibility we had in setting up ratios of unlike measures. As we proceed we will analyze dimensions to show the problems that arise when using ratios with unlike measures.

To set up an inverse proportion:

1. Establish two pairs of related data.
2. Arrange one pair as the numerator of one ratio and the denominator of the other.
3. Arrange the other pair so that each ratio contains like measures.
4. Form a proportion using the two ratios.

If the intensity of a light illuminating a wall is 30 candelas when the light source is 10 ft from the wall, to what level does the intensity in candelas decrease if the light source is 15 ft from the wall and if the intensity is inversely proportional to the square of the distance from its source?

The relationship here is one of *inverse variation.* As the distance between the wall and light source *increases,* the intensity *decreases.*

Pair 1:   30 candelas at 10 ft

Pair 2:   x candelas at 15 ft

The light intensity will be less than 30 candelas because the distance from the light source is greater.

Calculations

Dimension Analysis

$$\frac{30 \text{ candelas at } 10 \text{ ft}}{x \text{ candelas at } 15 \text{ ft}} = \frac{(15 \text{ ft})^2}{(10 \text{ ft})^2} \qquad \frac{\text{intensity}_1}{\text{intensity}_2} = \frac{(\text{distance}_2)^2}{(\text{distance}_1)^2}$$

$$\frac{30}{x} = \frac{15^2}{10^2} \qquad \frac{\text{candelas}}{\text{candelas}} = \frac{\text{ft}^2}{\text{ft}^2}$$

$$30(100) = 225x \qquad \text{Cross multiply. candelas(ft}^2) = \text{ft}^2(\text{candelas})$$

$$3{,}000 = 225x$$

$$\frac{3{,}000}{225} = x \qquad \text{Divide by ft}^2: \frac{\text{candelas(ft}^2)}{\text{ft}^2} = \text{candelas}$$

Interpretation

$$x = 13.33 \textbf{ candelas at 15 ft}$$

Tip!	***Why Is Estimation So Important?***

Suppose we arrange the data so that each pair forms a ratio and then we invert one of the ratios. Why can't we form ratios of unlike measures in our problem? Look at the value to see if the answer conforms to what we expect the answer to be.

Pair 1   Reciprocal of Pair 2

$$\frac{30 \text{ candelas at } 10 \text{ ft}}{(10 \text{ ft})^2} = \frac{(15 \text{ ft})^2}{x \text{ candelas at } 15 \text{ ft}}$$

$$30x = 22{,}500$$

$$x = 750$$

Did we expect the intensity to increase? No. We expected our answer to be less than 30 candelas. Analyze the dimensions.

$$\frac{\text{candelas}}{\text{ft}^2} = \frac{\text{ft}^2}{\text{candelas}}$$

$$\text{candelas(candelas)} = \text{ft}^2(\text{ft}^2)$$

Is this a true statement? No. These measures are not equal. Estimation catches proportions that are set up incorrectly!

EXAMPLE If 5 machines take 12 days to complete a job, how long will it take for 8 machines to do the job?

As the number of machines *increases,* the amount of time required to do the job *decreases.* Thus, the quantities are *inversely proportional.*

Pair 1: 5 machines finish in 12 days .

Pair 2: 8 machines finish in x days.

Estimation We expect more machines to do the job in less than 12 days.

$$\frac{5 \text{ machines}}{8 \text{ machines}} = \frac{x \text{ days for 8 machines}}{12 \text{ days for 5 machines}}$$

Each ratio uses like measures. The pairs are arranged inversely.

$$\frac{5}{8} = \frac{x}{12}$$

$\dfrac{\text{machines}}{\text{machines}} = \dfrac{\text{days}}{\text{days}}$

$$5(12) = 8x$$

Cross multiply: machines(days) = machines(days)

$$60 = 8x$$

$$\frac{60}{8} = x$$

Divide by machines.

$$\frac{\text{machines (days)}}{\text{machines}} = \text{days}$$

$$x = \frac{15}{2} \quad \text{or} \quad 7\frac{1}{2} \text{ days}$$

Interpretation   **It will take $7\frac{1}{2}$ days for 8 machines to do the job.**

Inverse variation has a conversion factor similar to the constant of direct variation. The inverse variation formula is $y = \frac{k}{x}$, where x is the independent variable, y is the dependent variable and k is the constant of indirect variation. This formula can also be written as $k = xy$.

To find the constant of indirect variation:

1. Identify a pair of related data in which both values are known.
2. Write the related pair as a product.

The constant of indirect variation for the preceding example is found using the data pair 5 machines times 12 days. We finish the example using the inverse variation formula:

$$y = \frac{k}{x}$$

$$y = \frac{(5 \text{ machines})(12 \text{ days})}{8 \text{ machines}}$$

$$y = \frac{60 \text{ days}}{8}$$

$$y = 7\frac{1}{2} \text{ days}$$

Gears and pulleys involve *inverse* relationships. Suppose that a large gear and a small gear are in mesh, or a large pulley is connected by a belt to a smaller pulley. The larger gear or pulley has the *slower* speed, and the smaller gear or pulley has the *faster* speed. A bicycle is an example. The larger gear or pulley is turned by the pedals, and this motion is transmitted to the smaller gear or pulley turning the rear wheel.

If the diameter (greatest distance across a circular object) of the larger gear or pulley is increased, the larger size means that the larger gear or pulley will take even longer to make a complete revolution. However, it will turn the smaller gear or pulley even faster. Thus, the larger gear or pulley will revolve *more slowly*, whereas the smaller one will revolve *more quickly*.

EXAMPLE   A 10-in.-diameter gear is in mesh with a 5-in.-diameter gear (Fig. 9–2). If the larger gear has a speed of 25 revolutions per minute (rpm), at how many rpm does the smaller gear turn?

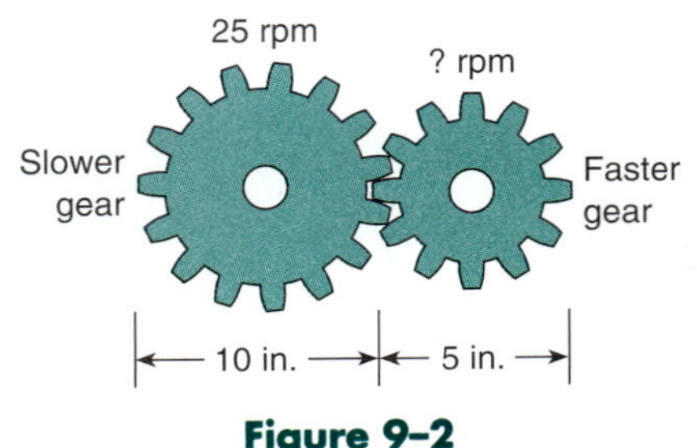

Figure 9–2

Because gears in mesh are *inversely* related, we set up an inverse proportion. Each ratio uses like measures and the ratios are in inverse order.

Pair 1:   25 rpm of larger gear for 10-in. size of larger gear.

Pair 2: x rpm of smaller gear for 5-in. size of smaller gear.

Estimation

We expect the speed of the smaller gear to be faster than the 25-rpm speed of the larger gear.

$$\frac{25 \text{ rpm of larger gear}}{x \text{ rpm of smaller gear}} = \frac{5 \text{ in. of smaller gear}}{10 \text{ in. of larger gear}} \quad \begin{array}{l}\text{Pair 2}\\\text{Pair 1}\end{array}$$

$$\frac{25}{x} = \frac{5}{10}$$

$$25(10) = 5x \qquad \text{Cross multiply.}$$

$$250 = 5x$$

$$\frac{250}{5} = x$$

$$50 = x$$

Interpretation   **Thus, the smaller gear turns at the faster speed of 50 rpm.**

Tip!	***How to Distinguish between Direct and Inverse Variation.***

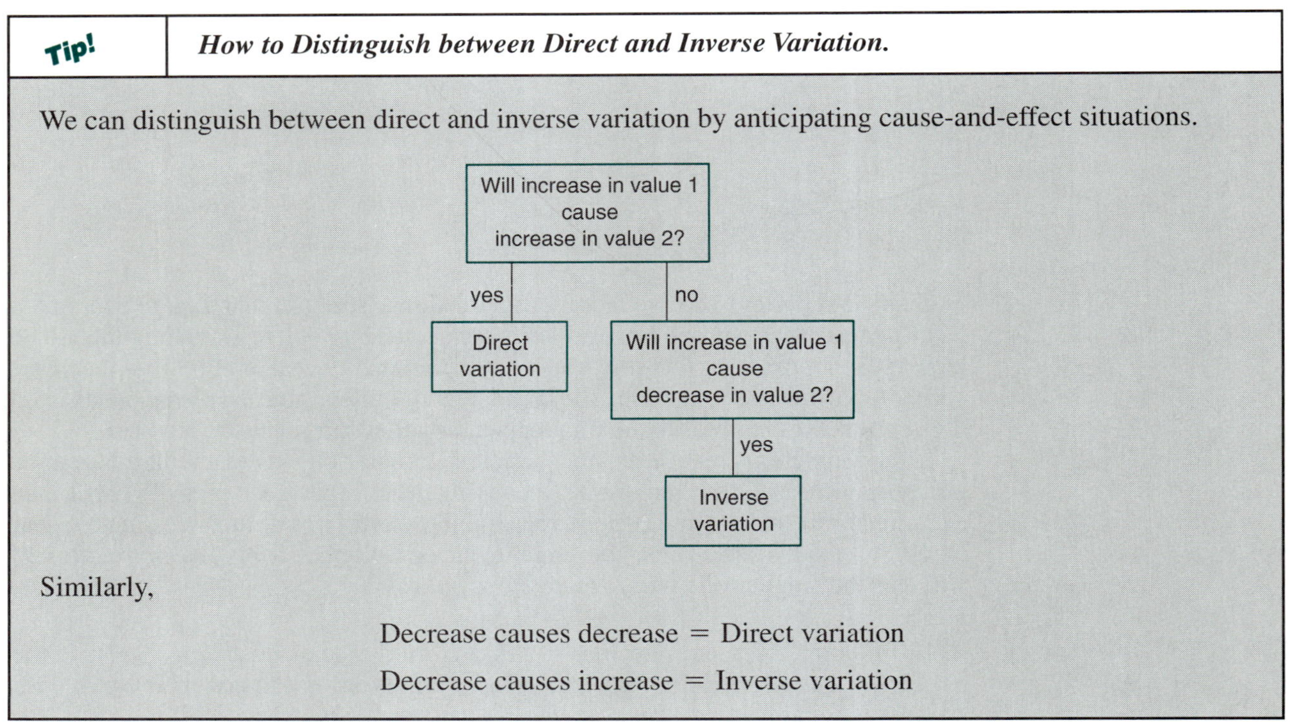

We can distinguish between direct and inverse variation by anticipating cause-and-effect situations.

Similarly,

Decrease causes decrease = Direct variation

Decrease causes increase = Inverse variation

 Solve Problems That Involve Similar Triangles.

Many applications involving proportions are based on the properties of similar triangles. Before we solve applied problems using similar triangles, let's examine the definitions of similar and congruent triangles. *Congruent triangles* have the same size and shape. *Similar triangles* have the same shape but not the same size.

Every triangle has six parts: three angles and three sides. Each angle or side of one similar or congruent triangle has a corresponding angle or side in the other similar or congruent triangle. The symbol for showing congruency is ≅ and is read "is congruent to." The symbol for a triangle is △.

Examine the triangles in Fig. 9–3.

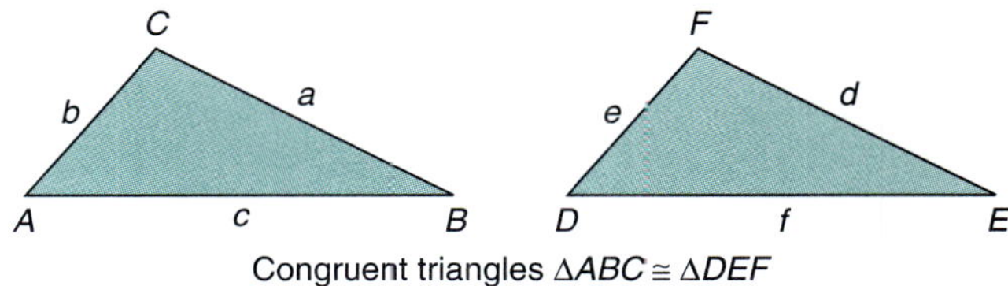
Congruent triangles △*ABC* ≅ △*DEF*

Figure 9–3

Corresponding angles of congruent triangles are the same size; that is, they are equal in measure.

Angle *A* = angle *D*
Angle *B* = angle *E*
Angle *C* = angle *F*

Corresponding sides of congruent triangles are the same size; that is, they are equal in measure.

Side *a* = side *d*
Side *b* = side *e*
Side *c* = side *f*

The symbol for showing similarity is ~ and is read "is similar to." Examine Fig. 9–4, which shows two similar triangles.

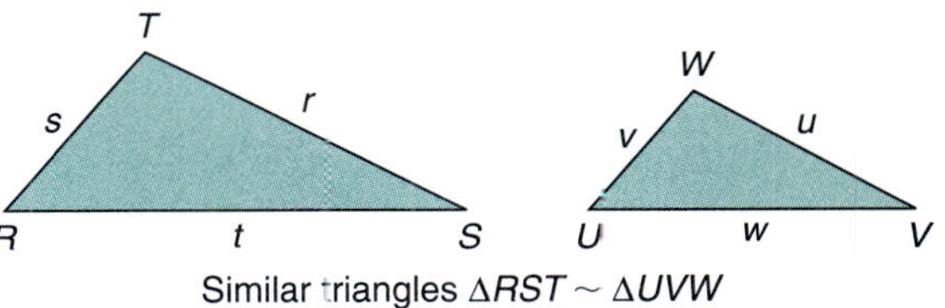
Similar triangles △*RST* ~ △*UVW*

Figure 9–4

Corresponding angles of similar triangles are equal in size.

Angle *R* = angle *U*
Angle *S* = angle *V*
Angle *T* = angle *W*

Corresponding sides of similar triangles are directly proportional.

Side *r* corresponds to side *u*
Side *s* corresponds to side *v*
Side *t* corresponds to side *w*

■ **DEFINITION: Congruent Triangles.** Two triangles are *congruent* if they have the same size and shape. Each angle of one triangle is equal to its corresponding angle in the other triangle. Each side of one triangle is equal to its corresponding side in the other triangle.

■ **DEFINITION: Similar Triangles.** Two triangles are *similar* if they have the same shape but different size. Each angle of one triangle is equal to its corresponding angle in the other triangle. Each side of one triangle is directly proportional to its corresponding side in the other triangle.

In Fig. 9–5, △*ABC* is similar to △*MNQ*. The corresponding angles are ∠*A* and ∠*M*, ∠*B* and ∠*N*, ∠*C* and ∠*Q*. (The symbol for angle is ∠.) The angles correspond because each pair is equal. The triangles are similar, so their corresponding sides are proportional. Corresponding sides are opposite or across from equal angles. The corresponding sides in Fig. 9–5 are *BC* and *NQ*, *AC* and *MQ*, *AB* and *MN*. We commonly label the angles of a triangle with capital letters and the sides opposite these angles with the corresponding lowercase letters. The corresponding sides are proportional, so their ratios are equal. For example, we can write the proportion

$$\frac{BC}{NQ} = \frac{AC}{MQ} = \frac{AB}{MN} \qquad \text{or} \qquad \frac{a}{m} = \frac{b}{n} = \frac{c}{q}$$

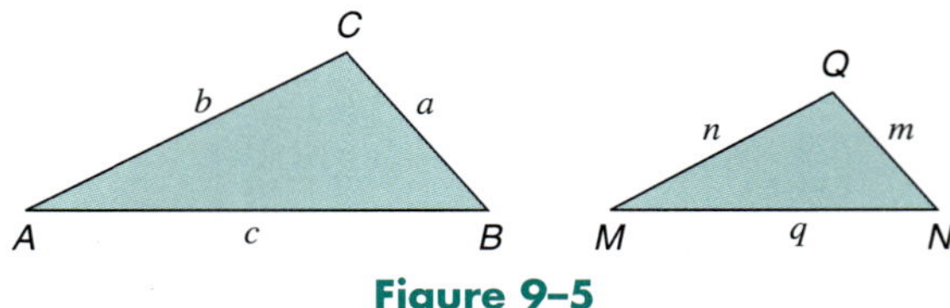

Figure 9–5

Similar triangles:

If two triangles are *similar,* the ratios of their corresponding sides are equal. That is, similar triangles are directly proportional.

We use this property of similar triangles to solve many problems.

EXAMPLE Find the missing side in Fig. 9–6 if $\triangle ABC \sim \triangle DEF$.

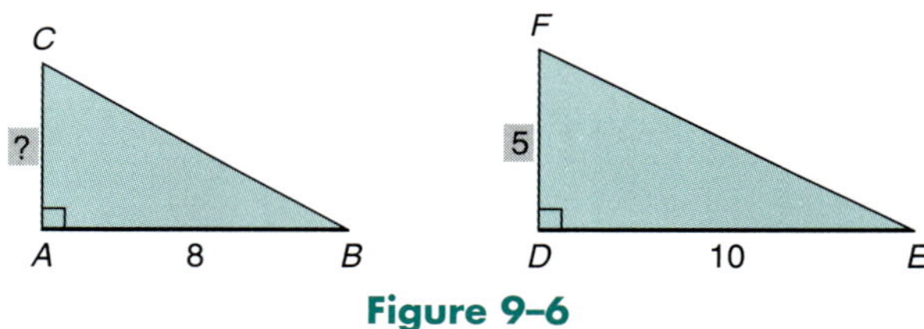

Figure 9–6

Write ratios of corresponding sides in a proportion. For convenience, single lowercase letters may be used to identify sides in similar triangles. Side *a* is opposite $\angle A$, side *b* is opposite $\angle B$, and so on.

Pair 1: 8 from $\triangle ABC$ corresponds and is proportional to 10 from $\triangle DEF$.

Pair 2: *b* from $\triangle ABC$ corresponds and is proportional to 5 from $\triangle DEF$.

Estimation Side *b* should be less than 5.

Pair 1
Pair 2
$$\frac{8}{b} = \frac{10}{5} \qquad \text{Substitute values and find missing side } b.$$

$$10b = 8(5) \qquad \text{Cross multiply.}$$

$$10b = 40$$

$$b = 4$$

Interpretation **Side *b* is 4 units long.**

A tree surgeon must know the height of a tree to determine which way to fell it so that it does not endanger lives, traffic, or property. A 6-ft pole casts a 4-ft shadow when the tree casts a 20-ft shadow (Fig. 9–7). What is the height of the tree?

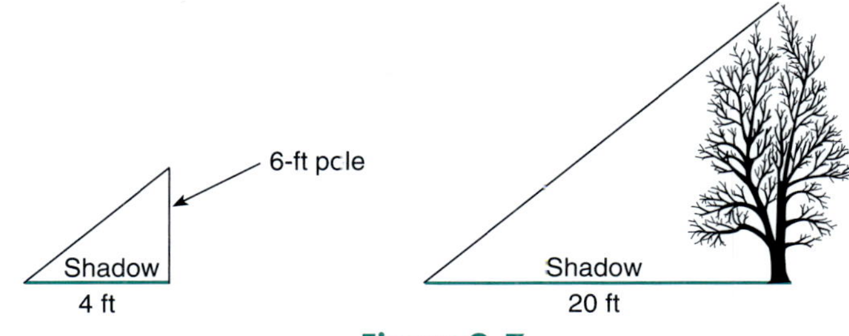

Figure 9–7

The triangles formed are similar, so we use a proportion to solve the problem. Let x = the height of the tree.

Pair 1: 6-ft pole casts a 4-ft shadow.

Pair 2: x-ft tree casts a 20-ft shadow.

Estimation Tree is more than 6 ft tall.

$$\frac{6 \text{ (height of pole)}}{x \text{ (height of tree)}} = \frac{4 \text{ (shadow of pole)}}{20 \text{ (shadow of tree)}}$$

$$6(20) = 4x \qquad \text{Cross multiply.}$$

$$120 = 4x$$

$$30 = x$$

Interpretation **The tree is 30 ft tall.**

A building lies between points A and B, so the distance between these points cannot be measured directly by a surveyor. Find the distance using the similar triangles shown in Fig. 9–8.

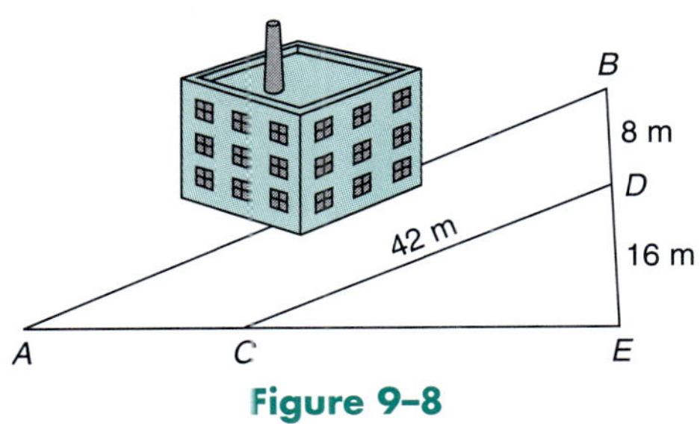

Figure 9–8

$\triangle ABE \sim \triangle CDE$. Note that CD must be made parallel to AB for the triangles to be similar. Parallel means the lines are the same distance apart from end to end.

Visualize the triangles as separate triangles (Fig. 9–9). Using the lowercase letter e for the missing AB measure is confusing because in $\triangle CDE$, side CD can also be thought of as side e.

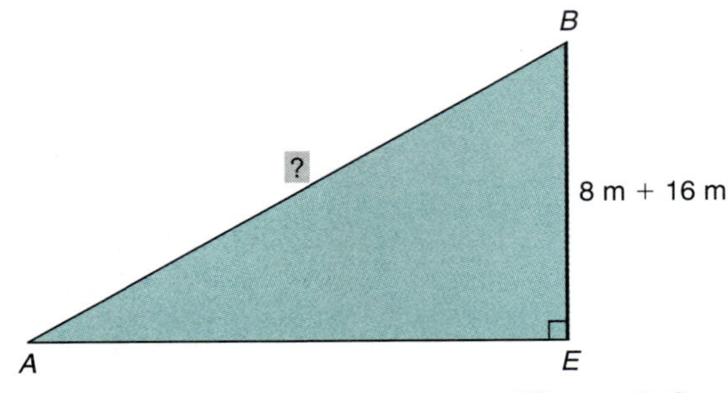

Figure 9–9

Pair 1: *AB* from △ *ABE* corresponds to *CD* from △*CDE*.

Pair 2: *BE* from △ *ABE* corresponds to *DE* from △*CDE*.

Estimation   *AB* is more than 42 m.

$$\frac{AB}{BE} = \frac{CD}{DE}$$

$BE = BD + DE = 8 + 16 = 24$

$$\frac{AB}{24} = \frac{42}{16}$$

$$16AB = (24)42$$

$$16AB = 1,008$$

$$AB = 63 \text{ m}$$

Interpretation   **Thus, the distance from *A* to *B* is 63 m.**

SELF-STUDY EXERCISES 9–2

1 Solve using proportions.

1. If 7 cans of dog food sell for $4.13, how much will 10 cans sell for?

2. If 6 cans of coffee sell for $22.24, how much will 20 cans sell for?

3. A mechanic took 7 hr to tune up 9 fuel-injected engines. At this rate, how many fuel-injected engines can be tuned up in 37.5 hr? Round to the nearest whole number.

4. A costume maker took 9 hr to make 4 headpieces for a Mardi Gras ball. At this rate, how many complete headpieces can be made in 35 hr?

5. How far can a family travel in 5 days if it travels at the rate of 855 mi in 3 days?

6. How far can a tractor-trailer rig travel in 8 days if it travels at the rate of 1,680 mi in 4 days?

7. How much crystallized insecticide does 275 acres of farmland need if the insecticide treats 50 acres per 100 lb?

8. How much fertilizer does 2,625 ft^2 of lawn need if the fertilizer treats 1,575 ft^2 per gallon? Express the answer to the nearest tenth of a gallon.

2 Solve using proportions.

9. The fan pulley and alternator pulley are connected by a fan belt on an automobile engine. The fan pulley is 225 cm in diameter and the alternator pulley is 125 cm in diameter. If the fan pulley turns at 500 rpm, at how many revolutions per minute does the faster alternator pulley turn?

10. The volume of a certain gas is inversely proportional to the pressure on it. If the gas has a volume of 160 in.^{3} under a pressure of 20 pounds per square inch (psi), what is the volume if the pressure is decreased to 16 psi?

11. A pulley that measures 15 in. across (diameter) turns at 1,600 rpm and drives a larger pulley at the rate of 1,200 rpm. What is the diameter of the larger pulley in this inverse relationship?

12. Six painters can trim the exterior of all the new brick homes in a subdivision in 9 weeks. The contractor wants to have the homes ready in just 3 weeks and so needs more painters. How many painters are needed for the job? Assume that the painters all work at the same rate.

13. Two groundskeepers take 25 hr to prepare a golf course for a tournament. How long would it take 5 groundskeepers to prepare the golf course?

14. A gear measures 5 in. across. It turns another gear 2.5 in. across. If the larger gear has a speed of 25 revolutions per minute (rpm), what is the rpm of the smaller gear?

15. A small pulley 3 in. in diameter turns 250 rpm and drives a larger pulley at 150 rpm. What is the diameter (distance across) of the larger pulley?

16. Two painters working at the same speed can paint 800 ft^2 of wall space in 6 hr. If a third painter paints at the same speed, how long will it take the three of them to paint the same wall space?

17. Three machines complete a printing project in 5 hr. How many machines are needed to finish the same project in 3 hr?

18. A gear measures 6 in. across. It is in mesh with another gear with a diameter of 3 in. If the larger gear has a speed of 60 revolutions per minute (rpm), what is the rpm of the smaller gear?

19. Nurse Lee prepares dosages for her patients in 30 min. If she gets help from assistants, who also work at her rate, and together they complete the preparation in 6 min, how many helpers did she get?

20. A 4.5-in. pulley turning at 1,000 revolutions per minute (rpm) is belted to a larger pulley turning at 500 rpm. What is the size of the larger pulley?

3 Write the corresponding parts not given for the congruent triangles in Exercises 21–23 (Figs. 9–10 to 9–12).

21. $AB = FE$
$BC = DE$
$AC = DF$

22. $\angle R$ and $\angle U = 90°$
$QR = TU$
$RP = SU$

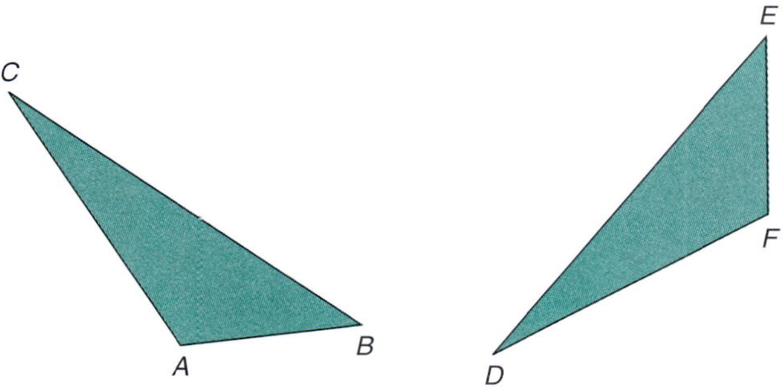

Figure 9–10

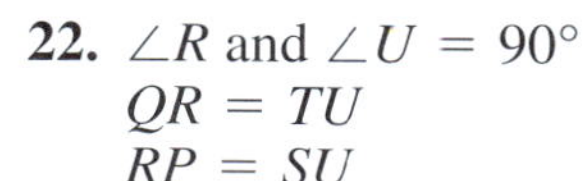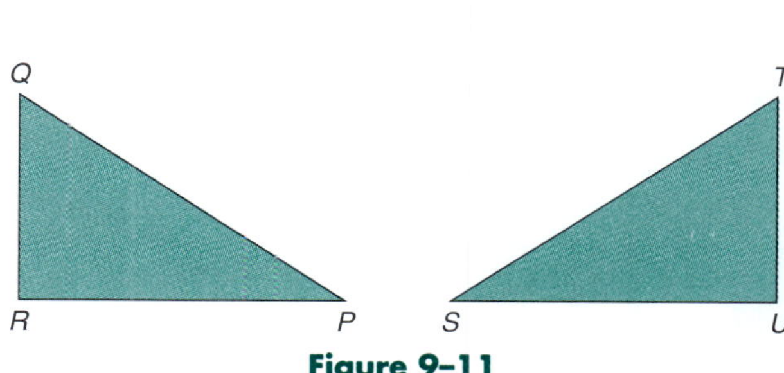

Figure 9–11

23. $JK = NM$
$\angle J = \angle M$
$\angle K = \angle N$

24. $\triangle ABC \sim \triangle EDF$. $\angle A = \angle E$, $\angle C = \angle F$. Find a and d. Use Fig. 9–13.

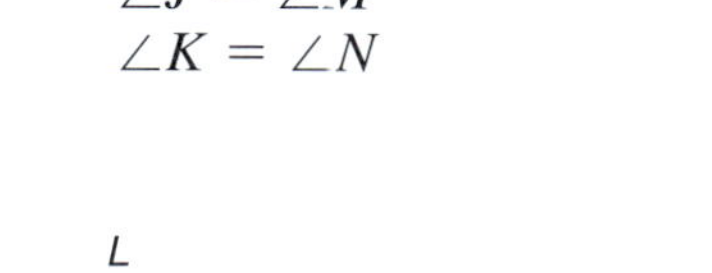

Figure 9–12

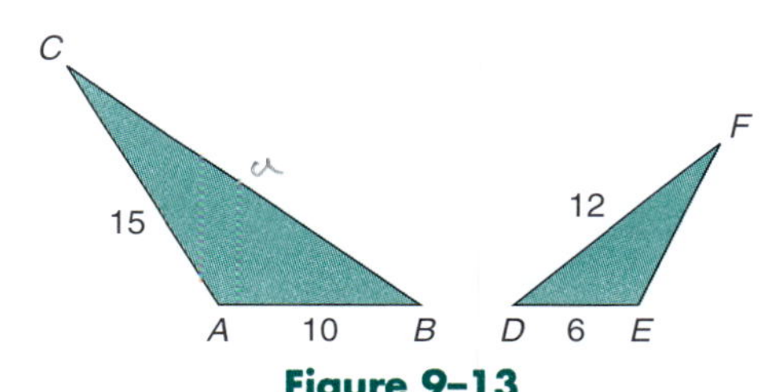

Figure 9–13

25. $\triangle ABC \sim \triangle DEC$. $\angle 1$ and $\angle 2$ have the same measure. Find DC and DE. (*Hint:* Let $DC = x$ and $AC = x + 3$. Use Fig. 9–14.)

26. Find the height of a tree that casts a 30-ft shadow when a 6-ft 6-in. pole casts a shadow of 3 ft.

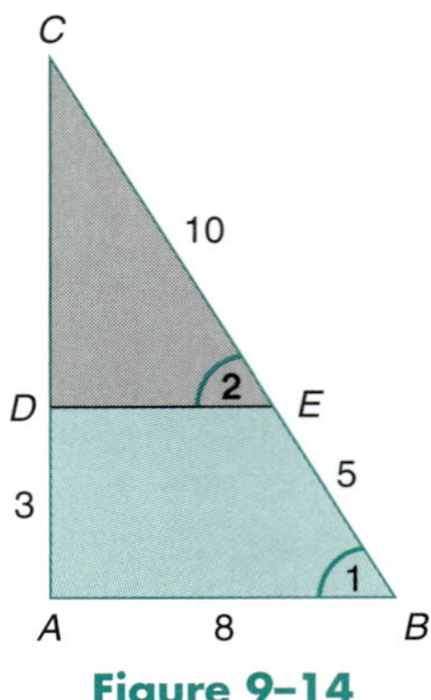

Figure 9–14

CAREER APPLICATION

Analyzing Nutrition Labels

Outcome: Use proportions to analyze nutrition labels.

In 1990 Congress passed the Nutrition Labeling and Education Act. This act requires that nutritional labels following the new standards appear on all processed food products. Nutrition information for fresh produce, meat, poultry, and fish must appear in the areas of the store where they are displayed.

To interpret the information given on the label it is important to determine the *base* for the percents given on the label. Regulations also require that *standard* daily nutritional value be used as the basis for all labels. The column heading "% Daily Value" has an asterisk that guides us to a footnote on the label that reads, "Percent Daily Values are based on a 2,000 calorie diet. Your daily values may be higher or lower depending on your calorie needs." The label also includes the recommended daily intakes that are used as a basis for a 2,000- and 2,500-calorie diet.

Nutrition Facts
Serving Size 1/2 cup (114g)
Servings per Container 4

Amount Per Serving
Calories 260   Calories from Fat 120

	% Daily Value*
Total Fat 13g	20%
Saturated Fat 5g	25%
Cholesterol 30mg	10%
Sodium 660mg	28%
Total Carbohydrate 31g	11%
Dietary Fiber 0g	0%
Sugar 5g	
Protein 5g	

Vitamin A 4%	Vitamin C 2%
Calcium 15%	Iron 4%

*Percent Daily Values are based on a 2,000 calorie diet.
Your daily values may be higher or lower depending on your calorie needs:

	Calories:	2,000	2,500
Total Fat	Less than	65 g	80 g
Sat Fat	Less than	20 g	25 g
Cholesterol	Less than	300 mg	300 mg
Sodium	Less than	2,400 mg	2,400 mg
Total Carbohydrate		300 g	375 g
Dietary Fiber		25 g	30 g

Are the recommended daily intakes proportional to the daily calories?

First, examine the recommended daily intake values for total fat. If the values are proportional,

$$\frac{65 \text{ g}}{80 \text{ g}} = \frac{2,000 \text{ calories}}{2,500 \text{ calories}}$$

Before we continue we need to realize that the recommended values have been rounded, so if cross products are *reasonably* close, we will say the relationships are proportional. To verify that the total fat recommended is or is not proportional to the number of calories per day, check the cross products.

$$65 \times 2,500 = 2,000 \times 80$$

$$162,500 = 160,000$$

If we round the values to the nearest ten thousand, we can say that the recommended daily intake of total fat is proportional.

If total fat is proportional to calories, then we can determine the recommended total fat intake for diets based on any number of calories.

EXAMPLE Find the recommended intake value of total fat for a 1,600-calorie diet. Round the value to the nearest multiple of 5.

$$\frac{x}{1,600} = \frac{65 \text{ g}}{2,000}$$ Find cross products.

$$2,000x = 1,600 \times 65$$ Multiply 1,600 × 65.

$$2,000x = 140,000$$

$$x = \frac{140,000}{2,000}$$ Divide by 2,000.

$$x = 52 \text{ g}$$

Rounded to the nearest 5 grams, the recommended total fat for a 1,600-calorie diet is 50 grams.

Exercises

1. Determine if the following recommended daily intake values are proportional (based on rounded Self-Study amounts).

   Saturated Fat _________   Cholesterol _________
   Sodium _________   Total Carbohydrate _________
   Dietary Fiber _________

2. Find the recommended daily intake values for a 1,600-calorie diet for the other recommended values that are proportional.
3. Find the recommended daily intake values for a 1,200-calorie diet for the recommended values that are proportional.
4. Find the recommended daily intake values for a 1,000-calorie diet for the recommended values that are proportional.
5. Calculate the percent daily value for each proportional category of one serving of the food item for the given nutrition label based on a 2,500-calorie diet.

Answers

1. Saturated fat—proportional
   Cholesterol—not proportional
   Sodium—not proportional
   Total carbohydrate—proportional

Dietary fiber—proportional
2. Saturated fat—16 grams
Total carbohydrate—240 grams
Dietary fiber—20 grams
3. Total fat—39 grams
Saturated fat—12 grams
Total carbohydrate—180 grams
Dietary fiber—15 grams
4. Total fat—33 grams
Saturated fat—10 grams
Total carbohydrate—150 grams
Dietary fiber—13 grams
5. Total fat—16%
Saturated fat—20%
Cholesterol—10%
Sodium—28%
Total carbohydrate—8%
Dietary fiber—0%

ASSIGNMENT EXERCISES

Section 9–1

Select two ratios that form a proportion.

1. $\dfrac{5}{8}, \dfrac{3}{4}, \dfrac{15}{24}, \dfrac{8}{16}$

2. $\dfrac{6}{8}, \dfrac{7}{12}, \dfrac{3}{5}, \dfrac{9}{12}$

Verify that the two ratios form a proportion.

3. $\dfrac{7}{9} = \dfrac{49}{63}$

4. $\dfrac{9}{10} = \dfrac{6}{8}$

5. $\dfrac{4}{10} = \dfrac{6}{15}$

Solve the proportions.

6. $\dfrac{x}{6} = \dfrac{5}{3}$

7. $\dfrac{3x}{8} = \dfrac{3}{4}$

8. $\dfrac{x-3}{x+6} = \dfrac{2}{5}$

9. $\dfrac{7}{x} = 6$

10. $\dfrac{2}{3} = \dfrac{x+3}{x-7}$

11. $\dfrac{4x+3}{15} = \dfrac{1}{3}$

12. $\dfrac{3x-2}{3} = \dfrac{2x+1}{4}$

13. $\dfrac{5}{4x-3} = \dfrac{3}{8}$

14. $\dfrac{8}{3x-2} = \dfrac{2}{3}$

15. $\dfrac{4x}{7} = \dfrac{2x+3}{3}$

16. $\dfrac{2x}{3x-2} = \dfrac{5}{8}$

17. $\dfrac{5x}{3} = \dfrac{2x+1}{4}$

18. $\dfrac{5}{9} = \dfrac{x}{2x-1}$

19. $\dfrac{7}{x} = \dfrac{5}{4x+3}$

20. $\dfrac{3}{5} = \dfrac{2x-3}{7x+4}$

Section 9–2

Solve the problems using proportions.

21. If 15 machines can complete a job in 6 weeks, how many machines are needed to complete the job in 4 weeks?

22. In preparing a banquet for 30 people, Cedric Henderson uses 9 lb of potatoes. How many pounds of potatoes will be needed for a banquet for 175 people?

23. A car with a speed-control device travels 100 mi at 50 mph. The trip takes 2 hr. If the car traveled at 40 mph, how much time would the driver need to reach the same destination?

24. A 6-ft landscape engineer casts a 5-ft shadow on the ground. How tall is a nearby tree that casts a 30-ft shadow?

25. There are 25 women in a class of 35 students. If this is typical of all classes in the college, how many women are enrolled in the college if it has 6,300 students altogether?

26. It takes five people 7 days to clear an acre of land of debris left by a tornado; inversely, more people can do the job in less time. How long would it take seven people all working at the same rate?

27. A gear with a diameter of 45 cm is in mesh with a gear that has a diameter of 30 cm. If the larger gear turns at 1,000 rpm, how many revolutions per minute does the smaller gear make?

28. A certain cloth sells at a rate of 3 yd for $7.00. How many yards can Clemetee Whaley buy for $35.00?

29. Three workers take 5 days to assemble a shipment of microwave ovens; inversely, more workers can do the job in less time. How long will it take five workers to do the same job?

30. An architect's drawing is scaled at $\frac{3}{4}$ in. = 6 ft. What is the actual height of a door that measures $\frac{7}{8}$ in. on the drawing?

31. A CAD program scales a blueprint so that $\frac{5}{8}$ in. = 2 ft. What is the actual measure of a wall that is shown as $1\frac{5}{16}$ in. on the blueprint?

32. A 10-in. pulley makes 900 revolutions every minute. It drives a larger pulley at 500 rpm. What is the diameter of the larger pulley in this inverse relationship?

33. On recent trips to rural health centers, a portable mammography unit used 81.2 gal of unleaded gasoline. If the travel involved a total of 845 mi, how many gallons of gasoline would be used for travel of 1,350 mi? Round to tenths.

34. A pulley whose diameter is 3.5 in. is belted to a pulley whose diameter is 8.5 in. In this inverse relationship, if the smaller, faster pulley turns at the rate of 1,200 rpm, what is the rpm of the slower pulley? Round to the nearest whole number.

35. A contractor estimates that a painter can paint 300 ft^2 of wall space in 3.5 hr. How many hours should the contractor estimate for the painter to paint 425 ft^2? Express your answer to the nearest tenth.

36. A coffee company mixes 1.6 lb of chicory with every 3.5 lb of coffee. At this ratio, how many pounds of chicory are needed to mix with 2,500 lb of coffee? Round to the nearest whole number.

37. A wire 825 ft long has a resistance of 1.983 ohms (Ω). How long is a wire of the same diameter if the resistance is 3.247 Ω? Round to the nearest whole foot.

38. The ceramic tile for a job weighs 510 lb. If the average weight of the ceramic tile is 4.25 lb per square foot, how many square feet are estimated for this job?

39. A gear is 4 in. across. It turns a smaller gear 2 in. across. If the larger gear has a speed of 30 rpm, what is the rpm of the smaller gear?

40. A small pulley with a 6-in. diameter turns 350 rpm and drives a larger pulley at 150 rpm. What is the diameter (distance across) of the larger pulley?

41. Two machines can complete a printing project in 6 hr. How many machines would be needed to finish the same project in 4 hr?

42. Waylon can install a hard drive in 10 PCs in 6 hr. He gets help from assistants who work at his rate and together they complete the installation in 2 hr. How many helpers did he get?

43. Write proportions for the similar triangles in Fig. 9–15.

44. $\triangle LMN \sim \triangle XYZ$, $\angle L = \angle X$, $\angle M = \angle Y$. Find m and x (see Fig. 9–16).

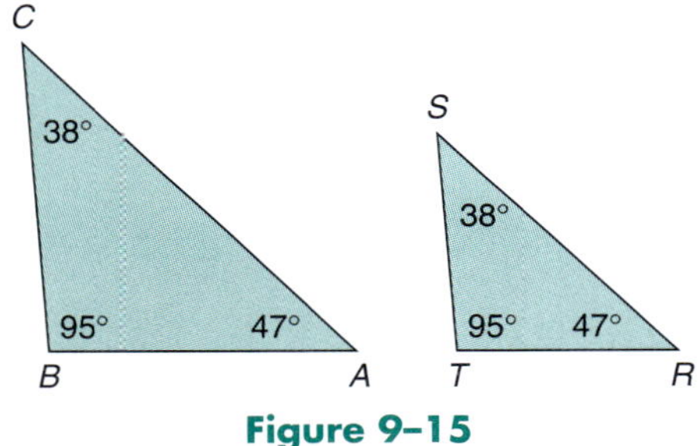

Figure 9–15

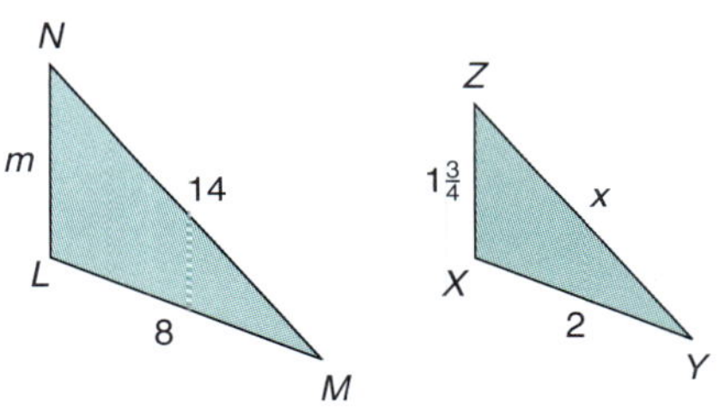

Figure 9–16

45. Find *AB* if $\triangle DEC \sim \triangle AEB$ (see Fig. 9–17).

Figure 9–17

CHALLENGE PROBLEMS

46. Make up a word problem that can be solved with a direct proportion. Include a lawn mower, tanks of gasoline, and acres to be mowed.

48. Make up a word problem that can be solved with an inverse proportion. Include a belt, pulleys, rpm's, and diameters of the pulleys.

47. A gear turning at 130 rpm has 50 teeth. It is in mesh with another gear that turns at 65 rpm. How many teeth does the other gear have?

49. The ratio of water to antifreeze in a mixture of radiator solution is 2 to 5. If the radiator is filled with 10 gal of liquid, how much is water and how much is antifreeze?

CHAPTER TRIAL TEST

1. Select two fractions that are proportional from $\dfrac{3}{4}, \dfrac{4}{5}, \dfrac{5}{6},$ and $\dfrac{6}{8}$.

Solve the equations.

2. $\dfrac{R}{7} = \dfrac{2}{5}$

3. $\dfrac{3 + Q}{1} = \dfrac{4}{5}$

4. $\dfrac{3}{y + 2} = \dfrac{2}{3}$

5. $\dfrac{8}{y + 2} = -7$

6. $\dfrac{1}{x} = \dfrac{7}{6}$

Solve the equations. Round to hundredths when necessary.

7. $\dfrac{1.2}{x} = 4.05$

8. $\dfrac{3.8}{6} = \dfrac{0.05}{R}$

9. A 9-in. gear is in mesh with a 4-in. gear. If the larger gear makes 75 rpm, how many revolutions per minute does the smaller gear make in this inverse relationship?

10. If three workers take 8 days to complete a job, how many workers would be needed to finish the same job in only 6 days if each worked at the same rate? (More workers take fewer days.)

11. If a compact car used 62.5 L of unleaded gasoline to travel 400 mi, how many liters of gasoline would the driver use to travel 350 mi? Round to tenths.

12. The ratio of men to women in technical and trade occupations is estimated to be 3 to 1, that is, $\frac{3}{1}$. If 56,250 men are employed in such occupations in a certain city, how many employees are women?

13. If an ice maker produces 75 lb of ice in $3\frac{1}{2}$ hr, how many pounds of ice would it produce in 5 hr? Round to the nearest whole number.

14. Find *HI* if $\triangle ABC \sim \triangle GHI$ (see Fig. 9–18).

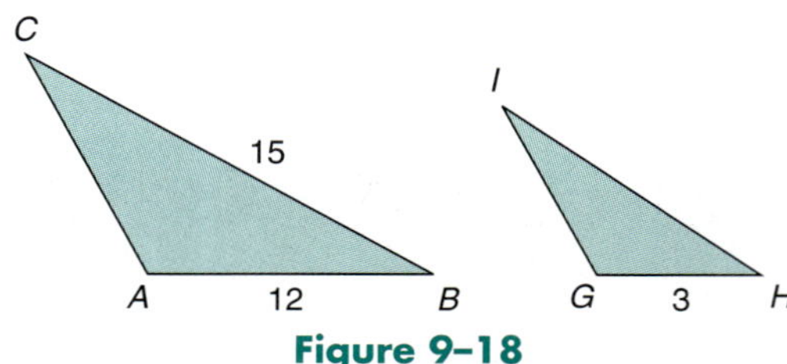

Figure 9–18

15. Find *DB* if $\triangle ABE \sim \triangle CDE$, $CD = 9$, $AB = 12$, $DE = 15$ (see Fig. 9–19).

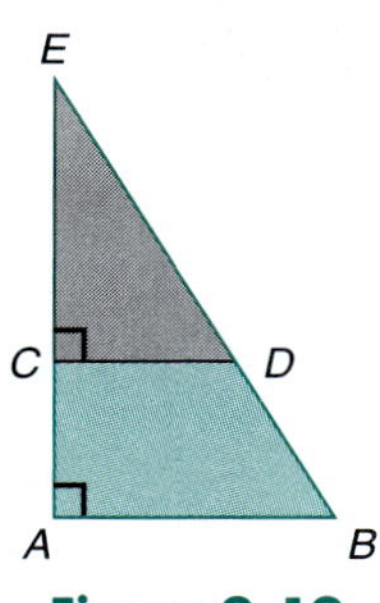

Figure 9–19

16. A model of a triangular part is shown in Fig. 9–20. Find side *x* if the part is to be similar to the model.

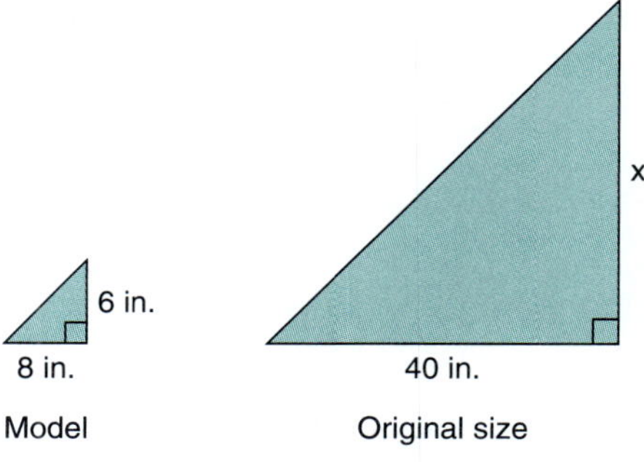

Figure 9–20

CHAPTER 10

Powers and Polynomials

GOOD DECISIONS THROUGH TEAMWORK

One of the most important advantages of working in a team is the opportunity to benefit from the work of fellow team members. In our everyday lives, time is one of our most important resources. For example, if you are interested in gathering information about one or more topics, the tasks can be divided among team members and your team can have all the information in a fraction of the time. This team project involves gathering information on roots, irrational numbers, and imaginary numbers.

The concept of taking roots introduces a new type of number called an irrational number. For this project, refer to historical records to answer the following questions. When was this type of number first recorded and what purpose did it serve? When was a radical symbol first used for indicating a root of a number? When did it become standard practice for fractional or rational exponents to be used for indicating a root of a number? Do you think this notation for roots will become more or less common in the future? Why or why not? Find examples of formulas that use radicals or rational exponents. Explain what each formula can be used to find.

When planning team strategies, each team member should work on some portion of the project, and the team as a whole should discuss some resources for finding the desired information. A decision must be made on how much detail each team member should record for his or her findings and how the team will share and compile the information.

Prepare a written report using the information the team has compiled. This project could be extended to include a similar investigation of imaginary and complex numbers.

10–1 Laws of exponents

1. Multiply powers with like bases.
2. Divide powers with like bases.
3. Find a power of a power.

10–2 Basic operations with algebraic expressions containing powers

1. Add or subtract like terms.
2. Multiply or divide algebraic expressions.

10–3 Powers of 10 and scientific notation

1. Multiply and divide by powers of 10.
2. Change a number from scientific notation to ordinary notation.
3. Change a number from ordinary notation to scientific notation.
4. Multiply and divide numbers in scientific notation.

10–4 Polynomials

1. Identify polynomials, monomials, binomials, and trinomials.
2. Identify the degree of terms and polynomials.
3. Arrange polynomials in descending order.

10–5 Roots and notation conventions

1. Write powers and roots using rational exponent and radical notation.
2. Use laws of exponents to simplify and evaluate expressions with rational exponents.

10–6 Complex and imaginary numbers

1. Write imaginary numbers using the letter i.
2. Raise imaginary numbers to powers.
3. Write real and imaginary numbers in complex form, $a + bi$.
4. Combine complex numbers.

10–7 Equations with squares and square roots

1. Solve equations with squared letter terms.
2. Solve equations with square-root radical terms.

In Chapter 8, we saw that repeated multiplication forms *powers* whenever values have a positive integral exponent. In this chapter, we extend our study of powers to include variable powers and logarithms. Powers are found in numerous applications that use scientific notation and occur frequently in formulas used to solve career-related problems.

10–1 LAWS OF EXPONENTS

Learning Outcomes

1. Multiply powers with like bases.
2. Divide powers with like bases.
3. Find a power of a power.

We have seen in various equations and formulas that a variable can represent all types of numbers: natural numbers, whole numbers, integers, fractions, decimals, and signed numbers. When raising a variable to a power, we must consider all the types of numbers the variable can be. As we study other types of numbers, we will look at their powers. Examine the chart of values for x^2, x^3, and x^4 in Table 10–1.

TABLE 10–1 Powers of x.

x	x^2	x^3	x^4
5	$(5)(5) = 25$	$(5)(5)(5) = 125$	$(5)(5)(5)(5) = 625$
-3	$(-3)(-3) = 9$	$(-3)(-3)(-3) = -27$	$(-3)(-3)(-3)(-3) = 81$
$\dfrac{1}{3}$	$\left(\dfrac{1}{3}\right)\left(\dfrac{1}{3}\right) = \dfrac{1}{9}$	$\left(\dfrac{1}{3}\right)\left(\dfrac{1}{3}\right)\left(\dfrac{1}{3}\right) = \dfrac{1}{27}$	$\left(\dfrac{1}{3}\right)\left(\dfrac{1}{3}\right)\left(\dfrac{1}{3}\right)\left(\dfrac{1}{3}\right) = \dfrac{1}{81}$
0.7	$(0.7)(0.7) = 0.49$	$(0.7)(0.7)(0.7) = 0.343$	$(0.7)(0.7)(0.7)(0.7) = 0.2401$
-2.1	$(-2.1)(-2.1) = 4.41$	$(-2.1)(-2.1)(-2.1) = -9.261$	$(-2.1)(-2.1)(-2.1)(-2.1) = 19.4481$

As we continue our study of powers we will first review expressions with natural-number exponents. Then we will include exponents that are whole numbers, integers, fractions, decimals, and other types of numbers.

Several laws of exponents apply to terms that have *like* bases. The *laws of exponents* are presented in this section.

1 Multiply Powers with Like Bases.

To multiply 2^3 by 2^2 using repeated multiplication, we have $2(2)(2)$ times $2(2)$ or $2(2)(2)(2)(2)$. Another way of writing this is $2^3 \times 2^2 = 2^5$. We can multiply the terms x^4 and x^2. Even though the value of x is not known, $x^4(x^2)$ is $x(x)(x)(x)$ times $x(x)$ or $x(x)(x)(x)(x)(x)$.

$$x^4(x^2) = x^6$$

The product contains $4 + 2$ or 6 factors of x. The following rule is a shortcut for using repeated multiplication.

To multiply powers that have like bases:

The exponent of the product is the *sum* of the exponents of the powers, and the base of the product remains unchanged. This rule can be stated symbolically as

$$a^m(a^n) = a^{m+n}$$

where a, m, and n are various types of numbers.

EXAMPLE Write the products.

(a) $y^4(y^3)$ (b) $a(a^2)$ (c) $b(b)$ (d) $x(x^3)(x^2)$ (e) $x^2(y^3)$

(a) $y^4(y^3) = y^{4+3} = \mathbf{y^7}$ Like bases, add exponents.

(b) $a(a^2) = a^{1+2} = \mathbf{a^3}$ When there is no written exponent, the exponent is 1.

(c) $b(b) = b^{1+1} = \mathbf{b^2}$

(d) $x(x^3)(x^2) = x^{1+3+2} = \mathbf{x^6}$

(e) $x^2(y^3) = \mathbf{x^2 y^3}$ Bases are unlike.

Tip!   ***Do You Always Add Exponents When Multiplying?***

The previous rule applies *only* to expressions with *like* bases. Thus, $x^2(y^3)$ can be written only as $x^2 y^3$. No other simplification can be made.

---

2 Divide Powers with Like Bases.

When reducing fractions, we can cancel or reduce to a factor of 1 any factors common to both the numerator and denominator. We use this concept when dividing powers that have like bases.

EXAMPLE Reduce or simplify the fractions; that is, perform the division indicated by each fraction.

(a) $\dfrac{x^5}{x^2}$ (b) $\dfrac{a^2}{a}$ (c) $\dfrac{b^7}{c^5}$ (d) $\dfrac{m^3}{m^3}$ (e) $\dfrac{y^3}{y^4}$

CHAPTER 10   Powers and Polynomials

(a) $\dfrac{x^5}{x^2} = \dfrac{x(x)(x)(\cancel{x})(\cancel{x})}{\cancel{x}(\cancel{x})} = x^3$

The quotient contains $5 - 2 = 3$ factors of x, or x^3.

(b) $\dfrac{a^2}{a} = \dfrac{a(\cancel{a})}{\cancel{a}} = a$ (or a^1)

The quotient contains $2 - 1 = 1$ factor of a.

(c) $\dfrac{b^7}{c^5} = \dfrac{b^7}{c^5}$

Bases are unlike, so no simplification can be made.

(d) $\dfrac{m^3}{m^3} = \dfrac{(\cancel{m})(\cancel{m})(\cancel{m})}{(\cancel{m})(\cancel{m})(\cancel{m})} = 1$

All factors of m reduce to 1.

(e) $\dfrac{y^3}{y^4} = \dfrac{(\cancel{y})(\cancel{y})(\cancel{y})}{(y)(\cancel{y})(\cancel{y})(\cancel{y})} = \dfrac{1}{y}$

The denominator has more factors of y than the numerator.

Examine the exponents in the division and the exponent of the quotient. In parts a and b, the exponent of the quotient is the difference of the exponents of the dividend and the divisor.

(a) $\dfrac{x^5}{x^2} = x^{5-2} = x^3$ (b) $\dfrac{a^2}{a} = a^{2-1} = a^1 = a$

Are the results in parts c and d consistent with this interpretation and our previous knowledge of exponents?

(c) $\dfrac{m^3}{m^3} = m^{3-3} = m^0 = 1$

Any nonzero base raised to the zero power equals 1.

(d) $\dfrac{y^3}{y^4} = y^{3-4} = y^{-1} = \dfrac{1}{y^1} = \dfrac{1}{y}$

An expression with a negative exponent equals the reciprocal of the expression written with a positive exponent with the same absolute value.

To divide powers that have like bases:

The exponent of the quotient is the *difference* between the exponents of the dividend and divisor, and the base of the quotient remains unchanged. This rule can be stated symbolically as

$$\dfrac{a^m}{a^n} = a^{m-n}$$

where a, m, and n are various types of numbers except that $a \neq 0$.

Because expressions with positive integral exponents are evaluated by using repeated multiplication, it is preferable to rewrite expressions with negative exponents as equivalent expressions with positive exponents. This manipulation is accomplished by applying the definition of negative exponents and exponents of zero.

EXAMPLE Write the quotients using positive exponents.

(a) $\dfrac{x^5}{x^8}$ (b) $\dfrac{a}{a^4}$ (c) $\dfrac{b}{b^2}$ (d) $\dfrac{y^{-3}}{y^2}$ (e) $\dfrac{x^3}{x^{-5}}$

(a) $\dfrac{x^5}{x^8} = x^{5-8} = x^{-3} = \dfrac{1}{x^3}$

(d) $\dfrac{y^{-3}}{y^2} = y^{-3-2} = y^{-5} = \dfrac{1}{y^5}$

(b) $\dfrac{a}{a^4} = a^{1-4} = a^{-3} = \dfrac{1}{a^3}$

(e) $\dfrac{x^3}{x^{-5}} = x^{3-(-5)} = x^{3+5} = x^8$

(c) $\dfrac{b}{b^2} = b^{1-2} = b^{-1} = \dfrac{1}{b}$

<table>
<tr><td>**Tip!**</td><td>*Manipulating Negative and Positive Exponents.*</td></tr>
</table>

The inverse relationship of multiplication and division allows us flexibility in applying the laws of exponents.

Look at the previous part a.

$$\frac{x^5}{x^8} \quad \text{is} \quad x^5 \div x^8 \quad \text{or} \quad x^5 \cdot \frac{1}{x^8} \quad \text{or} \quad x^5 \cdot x^{-8}$$

The multiplication law of exponents can be used.

$$x^5 \cdot x^{-8} = x^{5+(-8)} = x^{-3}$$

A factor can be moved from a numerator to a denominator (or vice versa) by changing the sign of its exponent.

$$x^{-2} = \frac{x^{-2}}{1} = \frac{1}{x^2}, \qquad \frac{1}{x^{-3}} = \frac{x^3}{1} = x^3$$

This property *does not apply* to a term that is part of a numerator or denominator that has two or more terms.

$$\frac{x^{-2} + 1}{3} \qquad \text{does not equal} \qquad \frac{1}{3x^2}.$$

In other words, *only factors* of the entire numerator or denominator can be moved. If more than one term is in the numerator, each term is divided by the denominator.

$$\frac{x^{-2} + 1}{3} = \frac{x^{-2}}{3} + \frac{1}{3} = \frac{1}{3x^2} + \frac{1}{3}$$

The most important point to remember with exponents is that the laws of exponents for multiplying and dividing powers apply only to factors that have like bases. Let's look at some examples that emphasize these points.

EXAMPLE Simplify the expressions and make all exponents positive.

$$\text{(a)} \ \frac{a^2 b^{-3}}{ab^{-1}} \qquad \text{(b)} \ \frac{xy^{-1}}{xy^2} \qquad \text{(c)} \ \frac{x^3 + y^2}{xy}$$

There are several ways to approach the expressions.

(a)

Option 1

$$\frac{a^2 b^{-3}}{ab^{-1}} = a^{2-1}b^{-3-(-1)}$$

Apply the division law of exponents. $2 - 1 = 1$; $-3 - (-1) = -3 + 1 = -2$

$$= ab^{-2}$$

Make all exponents positive.

$$= \frac{a}{b^2}$$

Option 2

$$\frac{a^2 b^{-3}}{ab^{-1}} = \frac{a^2 a^{-1} b^{-3} b^1}{1}$$

Move all exponents to the numerator.

$$= ab^{-2} = \frac{a}{b^2}$$

Apply multiplication law of exponents.

$$\text{(b)} \ \frac{xy^{-1}}{xy^2} = x^{1-1}y^{-1-(2)}$$

Apply the division law of exponents.

$$= y^{-3}$$

Make exponents positive.

$$= \frac{1}{y^3}$$

(c) $\dfrac{x^3 + y^2}{xy} = \dfrac{x^3}{xy} + \dfrac{y^2}{xy}$

Separate into two terms.

$$= \frac{x^{3-1}}{y} + \frac{y^{2-1}}{x}$$

Reduce each term.

$$= \frac{x^2}{y} + \frac{y}{x}$$

3 Find a Power of a Power.

In the term $(2^3)^2$, we have a power raised to a power. 2^3 is the base of the expression and 2 is the exponent.

$$(2^3)^2 \quad \text{is} \quad (2^3)(2^3) = 2^{3+3} = 2^6 \quad \text{or} \quad 64$$

Let's look at some examples of raising numerical and variable powers to a power.

EXAMPLE Find the powers of powers by first expressing each expression as repeated multiplication and then applying the multiplication law of exponents.

(a) $(3^2)^3$ (b) $(x^3)^4$ (c) $(a^2)^2$ (d) $(n^3)^5$

(a) $(3^2)^3 = (3^2)(3^2)(3^2) = 3^{2+2+2} = 3^6 = \mathbf{729}$

(b) $(x^3)^4 = (x^3)(x^3)(x^3)(x^3) = x^{3+3+3+3} = \boldsymbol{x^{12}}$

(c) $(a^2)^2 = (a^2)(a^2) = a^{2+2} = \boldsymbol{a^4}$

(d) $(n^3)^5 = (n^3)(n^3)(n^3)(n^3)(n^3) = n^{3+3+3+3+3} = \boldsymbol{n^{15}}$

From the preceding example we can see a pattern developing. In each case, if we multiply the exponent, we get the exponent of the new power. This pattern leads us to the following rule.

To raise a power to a power:

Multiply exponents and keep the same base.

$$(a^m)^n = a^{mn}$$

Applying this rule to the problems in the preceding example, we have

(a) $(3^2)^3 = 3^{2(3)} = 3^6 = 729$

(b) $(x^3)^4 = x^{3(4)} = x^{12}$

(c) $(a^2)^2 = a^{2(2)} = a^4$

(d) $(n^3)^5 = n^{3(5)} = n^{15}$

There are other laws of exponents that are extensions or combinations of the three laws we have already examined. Although these additional laws are not necessary, they are useful tools for simplifying expressions containing exponents.

To raise a fraction or quotient to a power:

Raise both the numerator and denominator to that power.

$$\left(\frac{a}{b}\right)^n = \frac{a^n}{b^n} \qquad b \neq 0$$

EXAMPLE Raise the fractions to the indicated power.

(a) $\left(\dfrac{2}{3}\right)^2$ (b) $\left(\dfrac{-3}{4}\right)^2$ (c) $\left(\dfrac{-1}{3}\right)^3$ (d) $\left(\dfrac{x}{y^3}\right)^3$ (e) $\left(\dfrac{x^2}{y^3}\right)^4$

(a) $\left(\dfrac{2}{3}\right)^2 = \dfrac{2^2}{3^2} = \dfrac{4}{9}$

(b) $\left(\dfrac{-3}{4}\right)^2 = \dfrac{(-3)^2}{4^2} = \dfrac{9}{16}$ $\qquad (-3)(-3) = +9$

(c) $\left(\dfrac{-1}{3}\right)^3 = \dfrac{(-1)^3}{3^3} = \dfrac{-1}{27}$ $\qquad (-1)(-1)(-1) = -1$

(d) $\left(\dfrac{x}{y^3}\right)^3 = \dfrac{x^3}{(y^3)^3} = \dfrac{x^3}{y^9}$

(e) $\left(\dfrac{x^2}{y^3}\right)^4 = \dfrac{(x^2)^4}{(y^3)^4} = \dfrac{x^8}{y^{12}}$

To raise a product to a power:

Raise each factor to the indicated power.

$$(ab)^n = a^n b^n$$

EXAMPLE Raise the products to the indicated powers.

(a) $(ab)^2$ (b) $(a^2 b)^3$ (c) $(xy^2)^2$ (d) $(3x)^2$ (e) $(2x^2 y)^3$ (f) $(-5xy)^2$

(a) $(ab)^2 = a^{1(2)} b^{1(2)} = a^2 b^2$

(b) $(a^2 b)^3 = a^{2(3)} b^{1(3)} = a^6 b^3$

(c) $(xy^2)^2 = x^{1(2)} y^{2(2)} = x^2 y^4$

(d) $(3x)^2 = 3^{1(2)} x^{1(2)} = 3^2 x^2 = 9x^2$

(e) $(2x^2 y)^3 = 2^{1(3)} x^{2(3)} y^{1(3)} = 2^3 x^6 y^3 = 8x^6 y^3$

(f) $(-5xy)^2 = (-5)^{1(2)} x^{1(2)} y^{1(2)} = (-5)^2 x^2 y^2 = 25x^2 y^2$

In applying the laws of exponents, it is very important to recognize differences in expressions.

Tip!	*Limitations of the Laws of Exponents.*

It is very important to understand what the laws of exponents *do not* include.

- The product-raised-to-power law applies to factors, not terms.

$$(a + b)^3 \qquad \textbf{does not equal} \qquad a^3 + b^3$$

- The multiplication-of-powers law applies to like bases.

$$a^2(b^3) \quad \textbf{does not equal} \quad ab^5 \quad \text{or} \quad (ab)^5$$

- An exponent affects only the one factor or grouping immediately to the left.

$$3x^2 \quad \text{and} \quad (3x)^2 \quad \textbf{are not equal}$$

In the term $3x^2$, the numerical coefficient 3 is multiplied times the square of x. In the term $(3x)^2$, $3x$ is squared. Thus, $(3x)^2 = 3^2x^2 = 9x^2$.

- A negative coefficient of a base is not affected by the exponent.

$$-x^3 \quad \textbf{means} \quad -(x)(x)(x)$$

If $x = 4$, $-x^3 = -(4)^3$ or $-(64) = -64$. If $x = -4$, $-x^3 = -(-4)^3$ or $-(-64) = 64$.

SELF-STUDY EXERCISES 10–1

1 Write the products.

1. $x^3(x^4)$ **2.** $m(m^3)$ **3.** $a(a)$ **4.** $x^2(x^3)(x^5)$

5. $y(y^2)(y^3)$ **6.** $a^2(b)$ **7.** $a^5(b^7)$ **8.** $x^3(x^8)$

2 Write the quotients. Write the answers with positive exponents.

9. $\dfrac{y^7}{y^2}$ **10.** $\dfrac{x^5}{x}$ **11.** $\dfrac{a^3}{a^4}$ **12.** $\dfrac{b^6}{b^5}$

13. $\dfrac{m^2}{m^2}$ **14.** $\dfrac{x}{x^3}$ **15.** $\dfrac{y^5}{y}$ **16.** $\dfrac{n^2}{n^{-5}}$

17. $\dfrac{x^{-4}}{x^6}$ **18.** $\dfrac{n^2}{n^3}$ **19.** $\dfrac{x^7}{x^0}$ **20.** $\dfrac{x^{-3}}{x^2}$

21. $\dfrac{x^2y^{-2}}{xy^4}$ **22.** $\dfrac{ab^2}{a^{-2}b^4}$ **23.** $\dfrac{x^2 + y^4}{xy^2}$ **24.** $\dfrac{a^3 - b^2}{a^2b^3}$

3 Simplify the expressions.

25. $(4^2)^3$ **26.** $(x^4)^2$ **27.** $(y^7)^0$ **28.** $(x^{10})^4$

29. $(a^7)^3$ **30.** $\left(-\dfrac{1}{2}\right)^3$ **31.** $\left(-\dfrac{2}{7}\right)^2$ **32.** $\left(\dfrac{a}{b}\right)^4$

33. $(2m^2n)^3$ **34.** $\left(\dfrac{x^2}{y}\right)^3$ **35.** $(x^2y^4)^3$ **36.** $(-2a)^2$

37. $(x^2y)^3$ **38.** $(-3ab^2)^3$ **39.** $(-7x^2)^5$ **40.** $(4x^2y^8)^2$

10–2 BASIC OPERATIONS WITH ALGEBRAIC EXPRESSIONS CONTAINING POWERS

Learning Outcomes

1 Add or subtract like terms.

2 Multiply or divide algebraic expressions.

1 Add or Subtract Like Terms.

Now that variables with exponents have been introduced, we can broaden our concept of *like terms*. For letter terms to be like terms, the letters as well as the exponents of the letters must be exactly the same; that is, $2x^2$ and $-4x^2$ are like terms because the x's are both squared, but $2x^2$ and $4x$ are not like terms because the x's do not have the same exponents.

EXAMPLE Tell whether the pairs of terms are like terms.

(a) $3a^2b$ and $-\frac{2}{3}a^2b$ (b) $-8xy^2$ and $7x^2y$ (c) $10ab^2$ and $(-2ab)^2$
(d) $5x^4y^3$ and $-2y^3x^4$ (e) $2x^2$ and $3x^3$

(a) **$3a^2b$ and $-\frac{2}{3}a^2b$ are like terms.** All letters and their exponents are the same.
(b) **$-8xy^2$ and $7x^2y$ are *not* like terms.** In $-8xy^2$ the exponent of x is 1, and in $7x^2y$ the exponent of x is 2. Also, the exponents of y are not the same.
(c) **$10ab^2$ and $(-2ab)^2$ are *not* like terms.** In the term $10ab^2$, only the b is squared. In $(-2ab)^2$, the entire term is squared, and the result of the squaring is $4a^2b^2$.
(d) **$5x^4y^3$ and $-2y^3x^4$ are like terms.** Both x factors have an exponent of 4, and both y factors have an exponent of 3. Because multiplication is commutative, the order of factors does not matter.
(e) **$2x^2$ and $3x^3$ are *not* like terms.** The exponents of x are not the same.

When an algebraic expression contains several terms, we simplify the expression as much as possible by combining like terms.

To combine like terms:

1. Combine the coefficients using the rules for adding or subtracting signed numbers.
2. The letter factors and exponents do not change.

Tip!   *Is Combining Like Terms the Same as Adding? and, What Does **Simplify** Mean?*

- Combining versus adding

With the introduction of integers into our number system, we broaden our concept of addition to include both addition and subtraction. That is, when adding integers with like signs, we add absolute values, and when adding integers with unlike signs, we subtract absolute values. Also, we developed a strategy for interpreting any subtraction as an equivalent addition by changing the subtrahend to its opposite.
It is common to use the word *combine* when referring to addition or to subtraction of signed numbers.

- Simplifying

The instructions to "simplify the expression" are vague but often used in mathematics exercises. In general, to simplify an expression means to write the expression using fewer terms or reduced or with lower coefficients or exponents. When the instructions say "simplify," the intent is for you to examine the expression and see which laws or operations allow you to rewrite the expression in a simpler form.

EXAMPLE Simplify the algebraic expressions by combining like terms.

(a) $5x^3 + 2x^3$ (b) $x^5 - 4x^5$ (c) $a^3 + 4a^2 + 3a^3 - 6a^2$
(d) $3m + 5n - m$ (e) $y + y - 5y^2 + y^3$

(a) $5x^3 + 2x^3 = (5 + 2)x^3 = 7x^3$

$5x^3$ and $2x^3$ are like terms; therefore, add the coefficients 5 and 2. The answer has the same letter factor and exponent as the like terms.

(b) $x^5 - 4x^5 = (1 - 4)x^5 = -3x^5$

x^5 and $-4x^5$ are like terms. $1 - 4 = -3$. The answer has the same letter factor and exponent as the like terms.

(c) $a^3 + 4a^2 + 3a^3 - 6a^2 = 4a^3 - 2a^2$

a^3 and $3a^3$ are like terms. $4a^2$ and $-6a^2$ are like terms. Combine coefficients mentally.

(d) $3m + 5n - m = \mathbf{2m} + \mathbf{5n}$ $3m$ and $-m$ are like terms.

(e) $y + y - 5y^2 + y^3 = \mathbf{2y} - \mathbf{5y^2} + \mathbf{y^3}$ y and y are like terms. Coefficients are 1.

When an algebraic expression contains a grouping preceded immediately by a minus sign, we subtract the entire grouping. The grouping is multiplied by -1, the implied coefficient of the grouping. This causes *each* sign within the grouping to be changed to its opposite and at the same time removes the parentheses. If the grouping is preceded by a plus sign or an unexpressed positive sign, parentheses are removed without changing signs, as if each term in the grouping were multiplied by $+1$, the implied coefficient.

EXAMPLE   Simplify the expressions.

(a) $y^2 + 2y - (3y^2 + 5y)$
(b) $(m^3 - 3m^2 - 5m + 4) - (4m^3 - 2m^2 - 5m + 2)$

(a) $y^2 + 2y - (3y^2 + 5y) =$

$y^2 + 2y - 1(3y^2 + 5y) =$ Distribute the implied coefficient of -1.

$y^2 + 2y - 3y^2 - 5y =$ Combine like terms.

$-2y^2 - 3y$

(b) $(m^3 - 3m^2 - 5m + 4) - (4m^3 - 2m^2 - 5m + 2) =$

$(m^3 - 3m^2 - 5m + 4) - 1(4m^3 - 2m^2 - 5m + 2) =$ Distribute the implied coefficient of -1.

$m^3 - 3m^2 - 5m + 4 - 4m^3 + 2m^2 + 5m - 2 =$ Combine like terms.

$-3m^3 - m^2 + 2$

2 Multiply or Divide Algebraic Expressions.

Simplifying algebraic expressions may also involve multiplication or division.

To multiply or divide algebraic expressions:

1. Multiply or divide the coefficients using the rules for signed numbers.
2. Multiply or divide the letter factors using the laws of exponents for factors with like bases.

EXAMPLE   Multiply or divide. Express answers with positive exponents.

(a) $(4x)(3x^2)$ (b) $(-6y^2)(2y^3)$ (c) $(-a)(-3a)$ (d) $\dfrac{2x^4}{x}$

(e) $\dfrac{-6y^5}{2y^3}$ (f) $\dfrac{-4x}{4x^2}$ (g) $\dfrac{-5x^4}{15x^2}$ (h) $\dfrac{3x}{12x^3}$

(a) $(4x)(3x^2) = 4(3)(x^{1+2}) = \mathbf{12\,x^3}$ Multiply coefficients. Add exponents.

(b) $(-6y^2)(2y^3) = -6(2)(y^{2+3}) = \mathbf{-12\,y^5}$ Multiply coefficients. Add exponents.

(c) $(-a)(-3a) = -1(-3)(a^{1+1}) = \mathbf{3\,a^2}$ Multiply coefficients. Add exponents. The coefficient of $-a$ is -1. The exponent of $-a$ is 1.

(d) $\dfrac{2x^4}{x} = \dfrac{2}{1}(x^{4-1}) = 2x^3$

Divide coefficients. Subtract exponents. The coefficient of x in the denominator is 1.

(e) $\dfrac{-6y^5}{2y^3} = \dfrac{-6}{2}(y^{5-3}) = -3y^2$

Divide coefficients. Subtract exponents.

(f) $\dfrac{-4x}{4x^2} = \dfrac{-4}{4}(x^{1-2}) = -1x^{-1}$ or $-\dfrac{1}{x}$

Divide coefficients. Subtract exponents. Write factors with negative exponents as equivalent positive exponents.

This result can also be written as $\dfrac{1}{-x}$ or $\dfrac{-1}{x}$.

(g) $\dfrac{-5x^4}{15x^2} = \dfrac{-5}{15}(x^{4-2}) = -\dfrac{1}{3}x^2$ or $-\dfrac{x^2}{3}$

Reduce coefficients.

$-\dfrac{1}{3}x^2$ is the same as $-\dfrac{1}{3}\left(\dfrac{x^2}{1}\right)$ or $-\dfrac{x^2}{3}$.

(h) $\dfrac{3x}{12x^3} = \dfrac{3}{12}(x^{1-3}) = \dfrac{1}{4}x^{-2} = \dfrac{1}{4x^2}$

Reduce coefficients.
Make exponent positive.

Since $\dfrac{1}{4}x^{-2} = \dfrac{1}{4}\left(\dfrac{1}{x^2}\right)$, this can be written as $\dfrac{1}{4x^2}$.

If factors are to be multiplied times more than one term, apply the distributive property.

EXAMPLE   Perform the multiplications.

(a) $2x(x^2 - 4x)$ (b) $-2y^3(2y^2 + 5y - 6)$ (c) $4a(3a^3 - 2a^2 - a)$

(a) $2x(x^2 - 4x) = 2x^3 - 8x^2$ Distribute.

(b) $-2y^3(2y^2 + 5y - 6) = -4y^5 - 10y^4 + 12y^3$ Distribute.

(c) $4a(3a^3 - 2a^2 - a) = 12a^4 - 8a^3 - 4a^2$ Distribute.

When more than one term is divided by a single term, we must divide *each* term in the dividend (numerator) by the divisor (denominator). Actually, each term in the numerator represents the numerator of a separate fraction with the given denominator. This is also an application of the distributive property.

EXAMPLE   Perform the divisions.

(a) $\dfrac{18a^4 + 15a^3 - 9a^2 - 12a}{3a}$ (b) $\dfrac{3x^3 - x^2}{x^3}$ (c) $\dfrac{6x^3 + 2x^2}{2x^2}$

(a) $\dfrac{18a^4 + 15a^3 - 9a^2 - 12a}{3a} = \dfrac{18a^4}{3a} + \dfrac{15a^3}{3a} - \dfrac{9a^2}{3a} - \dfrac{12a}{3a}$

Write as separate terms.

$= 6a^3 + 5a^2 - 3a - 4$

Simplify each term.

(b) $\dfrac{3x^3 - x^2}{x^3} = \dfrac{3x^3}{x^3} - \dfrac{x^2}{x^3} = 3 - x^{-1}$ or $3 - \dfrac{1}{x}$

Write as separate terms and simplify each term.

(c) $\dfrac{6x^3 + 2x^2}{2x^2} = \dfrac{6x^3}{2x^2} + \dfrac{2x^2}{2x^2} = 3x + 1$

<table>
<tr><td>**Tip!**</td><td>**Why Can't We Cancel Terms?**</td></tr>
</table>

We have said before but we want to say again: **You reduce or cancel *factors*, not *terms*.** Look at part c of the previous example, $\dfrac{6x^3 + 2x^2}{2x^2}$.

A common *mistake* is to cancel the terms.

$$\frac{6x^3 + 2\cancel{x^2}}{2\cancel{x^2}} = 6x^3 \qquad \textbf{Incorrect!}$$

Why is this not correct? To check a division, multiply the quotient by the divisor (denominator). The result will be the dividend (numerator).

Does $6x^3(2x^2) = 6x^3 + 2x^2$? NO!

Now, let's simplify expressions that have a variety of operations.

EXAMPLE Simplify the expressions and write all exponents as positive exponents.

(a) $\dfrac{5x^5}{10x^2} + 3x(2x^2)$ (b) $\dfrac{3ab^2 - a^2b + 4}{ab}$ (c) $3x(4xy^2)^2$

(a) $\dfrac{5x^5}{10x^2} + 3x(2x^2) =$ Simplify each term.

$\dfrac{1}{2}x^3 + 6x^3 =$ Combine terms.

$\mathbf{6.5x^3}$ **or** $\dfrac{\mathbf{13}}{\mathbf{2}}\mathbf{x^3}$ **or** $\dfrac{\mathbf{13x^3}}{\mathbf{2}}$ $\quad \dfrac{1}{2} + 6 = 0.5 + 6 = 6.5 \quad$ or $\quad \dfrac{1}{2} + \dfrac{12}{2} = \dfrac{13}{2}$

(b) $\dfrac{3ab^2 - a^2b + 4}{ab} =$ Write or mentally visualize as separate terms.

$\dfrac{3ab^2}{ab} - \dfrac{a^2b}{ab} + \dfrac{4}{ab} =$ Simplify each term.

$\mathbf{3b - a + \dfrac{4}{ab}}$

(c) $3x(4xy^2)^2 =$ Follow the order of operations. Raise to a power first.

$3x(16x^2y^4) =$ Multiply.

$\mathbf{48x^3y^4}$

SELF-STUDY EXERCISES 10–2

1 Simplify.

1. $3a^2 + 4a^2$

2. $5x^3 - 2x^3$

3. $b^2 + 3a^2 + 2b^2 - 5a^2$

4. $3a - 2b - a$

5. $x - 3x - 2x^2 - 3x^2$

6. $3a^2 - 2a^2 + 4a^2$

7. $x^2 + 3y - (2x^2 + 5y)$

8. $4m^2 - 2n^2 - (2m^2 - 3n^2)$

9. $7a + 3b + 8c + 2a - (b - 2c)$

10. $5x + 3y - (7x - 2z)$

11. $7x(2x^2)$ **12.** $(-2m)(-m^2)$ **13.** $(-3m)(7m)$ **14.** $(-y^3)(2y^3)$

15. $\dfrac{6x^4}{3x^2}$ **16.** $\dfrac{-5a^2}{10a}$ **17.** $\dfrac{-7x}{-14x^3}$ **18.** $\dfrac{-9x^5}{12x^2}$

19. $3x(x - 6)$ **20.** $4x(3x^2 - 7x + 8)$ **21.** $-4x(2x - 3)$ **22.** $2x^2(5 + 2x)$

23. $\dfrac{6x^2 - 4x}{2x}$ **24.** $\dfrac{12x^5 - 6x^3 - 3x^2}{3x^2}$ **25.** $\dfrac{7x^4 - x^2}{x^3}$ **26.** $\dfrac{8x^4 + 6x^3}{2x^2}$

27. $\dfrac{6x^4}{18x^2} + 5x(3x^3)$ **28.** $\dfrac{5a^2b^3 - 3ab^2 - 7}{ab}$ **29.** $4a(3a^2x^3)^2$ **30.** $6xy(2x^3y^2)^4$

10-3 POWERS OF 10 AND SCIENTIFIC NOTATION

Learning Outcomes

1 Multiply and divide by powers of 10.
2 Change a number from scientific notation to ordinary notation.
3 Change a number from ordinary notation to scientific notation.
4 Multiply and divide numbers in scientific notation.

1 Multiply and Divide by Powers of 10.

We learned that our number system is based on the number 10; that is, each place value is a *power of 10*. Look at the place-value chart in Fig. 10–1 to see how our number system relates to powers of 10.

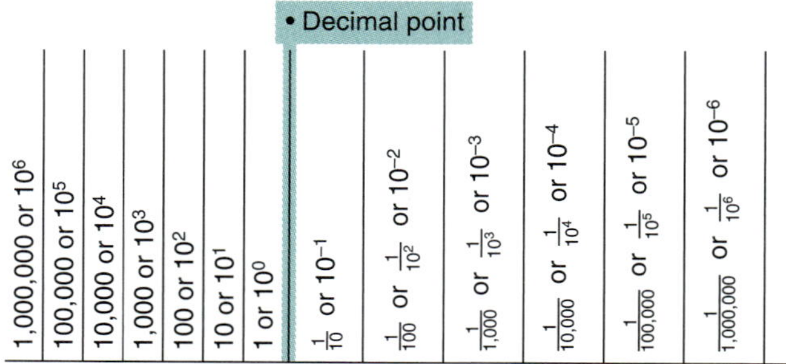

Figure 10–1 Base-ten place-value chart.

The ones place is 10^0. This is consistent with the definition of zero exponents. The tenths place is $\frac{1}{10}$ or 10^{-1}, which is consistent with the definition of negative exponents.

Notice the relationship between the exponent of 10 and the number of zeros in the following list of ordinary numbers.

Whole-number part		**Fractional or decimal part**	
Millions	$1,000,000 = 10^6$	$\dfrac{1}{10} = 10^{-1}$	Tenths
Hundred-thousands	$100,000 = 10^5$	$\dfrac{1}{100} = 10^{-2}$	Hundredths
Ten-thousands	$10,000 = 10^4$	$\dfrac{1}{1,000} = 10^{-3}$	Thousandths

Thousands	$1{,}000 = 10^3$	$\dfrac{1}{10{,}000} = 10^{-4}$	Ten-thousandths
Hundreds	$100 = 10^2$	$\dfrac{1}{100{,}000} = 10^{-5}$	Hundred-thousandths
Tens	$10 = 10^1$	$\dfrac{1}{1{,}000{,}000} = 10^{-6}$	Millionths
Ones	$1 = 10^0$		

Tip! | ***What Does the Exponent in a Power of 10 Tell Us?***

The absolute value of the exponent equals the *number of zeros* in the ordinary number in whole-number or fractional form. The absolute value of the exponent equals the *number of places to the right of the decimal* in the ordinary number in decimal form.

We learned in arithmetic that we can quickly multiply or divide by 10, 100, 1,000, and so on, by shifting the decimal point. When we are multiplying or dividing by a power of 10, **the exponent of 10 tells how many places the decimal is to be shifted and in what direction.**

To multiply by a power of 10:

1. If the exponent is positive, shift the decimal point to the *right* the number of places indicated by the *positive* exponent. Attach zeros as necessary.
2. If the exponent is negative, shift the decimal point to the *left* the number of places indicated by the *negative* exponent. Insert zeros as necessary.

To divide by a power of 10:

1. Change the division to an equivalent multiplication.
2. Use the rule for multiplying by a power of 10.

EXAMPLE Perform the multiplications or divisions using powers of 10.

(a) 275×10 (b) 0.18×100 (c) $2.4 \times 1{,}000$ (d) 43×0.1
(e) $3.14 \div 10$ (f) $0.48 \div 100$ (g) $20.1 \div 1{,}000$

(a) $275 \times 10 = 275 \times 10^1 = \mathbf{2{,}750}$

    The exponent is $+1$, so move the decimal one place to the right. Attach one zero.

(b) $0.18 \times 100 = 0.18 \times 10^2 = \mathbf{18}$

    The exponent is $+2$, so move the decimal two places to the right.

(c) $2.4 \times 1{,}000 = 2.4 \times 10^3 = \mathbf{2{,}400}$

    The exponent is $+3$, so move the decimal three places to the right. Attach two zeros.

(d) $43 \times 0.1 = 43 \times 10^{-1} = \mathbf{4.3}$

    The exponent is -1, so move the decimal one place to the left.

(e) $3.14 \div 10 = 3.14 \times \frac{1}{10}$

    Change the division to an equivalent multiplication.

$\qquad = 3.14 \times 10^{-1}$

    Express the fraction $\frac{1}{10}$ as a power of 10.

$\qquad = \mathbf{0.314}$

    Because the exponent of 10 is -1, move the decimal one place to the *left*.

(f) $0.48 \div 100 = 0.48 \times \frac{1}{100} = 0.48 \times 10^{-2} = \mathbf{0.0048}$

    Move decimal two places to the left. Insert two zeros.

(g) $20.1 \div 1{,}000 = 20.1 \times \frac{1}{1{,}000} = 20.1 \times 10^{-3} = \mathbf{0.0201}$ Move the decimal three places to the left. Insert one zero.

We can multiply or divide a power of 10 by another power of 10 by following the laws of exponents.

EXAMPLE   Multiply or divide using the laws of exponents.

(a) $10^5(10^2)$ (b) $10^{-1}(10^2)$ (c) $10^0(10^3)$ (d) $\dfrac{10}{10^3}$

(e) $\dfrac{10^5}{10^4}$ (f) $\dfrac{10^2}{10^2}$ (g) $\dfrac{10^{-2}}{10^3}$

(a) $10^5(10^2) = 10^{5+2} = \mathbf{10^7}$ Like bases, add exponents.

(b) $10^{-1}(10^2) = 10^{(-1+2)} = \mathbf{10^1}$ or $\mathbf{10}$ Like bases, add exponents.

(c) $10^0(10^3) = 10^{0+3} = \mathbf{10^3}$ Like bases, add exponents.

(d) $\dfrac{10}{10^3} = 10^{1-3} = \mathbf{10^{-2}}$ or $\dfrac{\mathbf{1}}{\mathbf{10^2}}$ Like bases, subtract exponents.

(e) $\dfrac{10^5}{10^4} = 10^{5-4} = \mathbf{10^1}$ or $\mathbf{10}$ Like bases, subtract exponents.

(f) $\dfrac{10^2}{10^2} = 10^{2-2} = \mathbf{10^0}$ or $\mathbf{1}$ Like bases, subtract exponents.

(g) $\dfrac{10^{-2}}{10^3} = 10^{-2-3} = \mathbf{10^{-5}}$ or $\dfrac{\mathbf{1}}{\mathbf{10^5}}$ Like bases, subtract exponents.

2 Change a Number from Scientific Notation to Ordinary Notation.

Powers of 10 are used in many applications in a special form that saves time when operations are performed with very large or very small numbers. This special form for writing such numbers is called *scientific notation*.

To write a number in scientific notation, shift the decimal so the number has a numerical value of at least 1 but less than 10; that is, the whole-number part of the value is a single, nonzero digit. Then, multiply that value by the appropriate power of 10 so that the entire expression has the same value as the original number.

■ **DEFINITION: Scientific Notation.** A number is expressed in *scientific notation* if it is the product of two factors. The absolute value of the first factor is a number greater than or equal to 1 but less than 10. The second factor is a power of 10.

■ **DEFINITION: Ordinary Notation.** A number that is written strictly according to place value is an *ordinary number.*

Tip!	***Characteristics of Scientific Notation.***

- Numbers between 0 and 1 and between -1 and 0 require negative exponents when written in scientific notation.
- The first factor in scientific notation will always have only one nonzero digit to the left of the decimal.
- The use of the times sign ($\times$) for multiplication is the most common representation for scientific notation.

EXAMPLE Which of the terms is expressed in scientific notation?

(a) 4.7×10^2 (b) 0.2×10^{-1} (c) -3.4×5^2 (d) 2.7×10^0
(e) 34×10^4 (f) 8×10^{-6} (g) $-2.8 \div 10^3$

(a) 4.7 is more than 1 but less than 10. 10^2 is a power of 10. Multiplication is indicated. **Thus, the term is in scientific notation.**
(b) Even though 10^{-1} is a power of 10, **this term is not in scientific notation** because the first factor (0.2) is less than 1.
(c) The absolute value of -3.4 is more than 1 and less than 10, but 5^2 is not a power of 10. **Thus, this term is not in scientific notation.**
(d) 2.7 is more than 1 and less than 10. 10^0 is a power of 10. **Thus, this term is in scientific notation.**
(e) 34 is greater than 10. Even though 10^4 is a power of 10, **this term is not in scientific notation** because the first factor (34) is 10 or more.
(f) 8 is more than 1 and less than 10. 10^{-6} is a power of 10. **Thus, this term is in scientific notation.**
(g) The absolute value of -2.8 is more than 1 and less than 10, but division rather than multiplication is the indicated operation. **This term is not in scientific notation.**

To change from a number written in scientific notation to an ordinary number:

1. Perform the indicated multiplication by moving the decimal point in the first factor the appropriate number of places. Affix or insert zeros as necessary.
2. Omit the power-of-10 factor.

Remember, when we are multiplying by a power of 10, the exponent of 10 tells us how many places and in which direction to move the decimal.

EXAMPLE Change to ordinary numbers.

(a) 3.6×10^4 (b) 2.8×10^{-2} (c) 1.1×10^0 (d) 6.9×10^{-5}
(e) 9.7×10^6

(a) $3.6 \times 10^4 = 36000. = \mathbf{36,000}$ Move the decimal four places to the right.

(b) $2.8 \times 10^{-2} = .028 = \mathbf{0.028}$ Move the decimal two places to the left.

(c) $1.1 \times 10^0 = \mathbf{1.1}$ Move the decimal no (zero) places.

(d) $6.9 \times 10^{-5} = .000069 = \mathbf{0.000069}$ Move the decimal five places to the left.

(e) $9.7 \times 10^6 = 9700000. = \mathbf{9,700,000}$ Move the decimal six places to the right.

3 Change a Number from Ordinary Notation to Scientific Notation.

If we want to express an ordinary number in scientific notation, we are reversing the procedures we used before. Shifting the decimal point changes the value of a number. The power-of-10 factor is used to offset or balance this change. The original value of the number must be maintained.

To change from a number written in ordinary notation to scientific notation:

1. Indicate where the decimal should be positioned in the ordinary number so that the absolute value of the number is valued at 1 or between 1 and 10 by inserting a caret ($\wedge$) in the proper place.
2. Determine how many places and in which direction the decimal shifts *from* the new position (caret) *to* the old position (decimal point). This number is the exponent of the power of 10.

Tip! | ***A Balancing Act: Why Count from the New to the Old?***

Moving the decimal in the ordinary number changes the value of the number unless you balance the effect of the move in the power-of-10 factor. When a decimal is moved in the first factor, the value is changed. To offset this change, an opposite change must be made in the power-of-10 factor. From the new to the old position indicates the proper number of places and the direction (positive or negative) for balancing with the power-of-10 factor.

Remember the word *NO*. Count from *New* to *Old*.

$$3{,}800 = 3.8 \times 10^3 \qquad 3{_\wedge}800. \times 10^3 \qquad N \to O = +3$$

$$0.0045 = 4.5 \times 10^{-3} \qquad (0.004{_\wedge}5 \times 10^{-3} \qquad N \to O = -3)$$

EXAMPLE Express in scientific notation.

   (a) 285   (b) 0.007   (c) 9.1   (d) 85,000   (e) 0.00074

(a) $285 \to 2{_\wedge}85 = \mathbf{2.85 \times 10^2}$
The unwritten decimal is after the 5. Place the caret between 2 and 8 so the number 2.85 is between 1 and 10. Count *from* the caret *to* the decimal to determine the exponent of 10. A move two places to the right represents the exponent $+2$.

(b) $0.007 \to 0.007{_\wedge} = \mathbf{7 \times 10^{-3}}$
7 is between 1 and 10. Count *from* the caret *to* the decimal. A move three places to the left represents the exponent -3.

(c) $9.1 = \mathbf{9.1 \times 10^0}$
9.1 is already between 1 and 10, so the decimal does not move; that is, the decimal moves zero places.

(d) $85{,}000 \to 8{_\wedge}5000 = \mathbf{8.5 \times 10^4}$
From the caret *to* the decimal is four places to the right.

(e) $0.00074 \to 0.0007{_\wedge}4 = \mathbf{7.4 \times 10^{-4}}$
From the caret *to* the decimal is four places to the left.

Occasionally a number is in power-of-10 notation but not in scientific notation because the first factor is not equal to 1 or is not between 1 and 10. When this is the case, shift the decimal to the proper place and adjust the original power-of-10 factor appropriately.

Tip! | ***Between 1 and 10: One Nonzero Digit to the Left of the Decimal.***

The expression "between 1 and 10" means any number that is more than 1 but less than 10. The first factor in scientific notation must be equal to 1 *or* between 1 and 10. That means **there will be one and only one nonzero digit to the left of the decimal point.**

$$\text{\small\textbf{EXAMPLE}}\quad \text{Express in scientific notation.}$$

EXAMPLE Express in scientific notation.

(a) 37×10^5 (b) 0.03×10^3

(a) $3{\scriptstyle\wedge}7 \times 10^5 = 3.7 \times 10^1 \times 10^5$ $N \rightarrow O = +1$

$\qquad\qquad\qquad\qquad = 3.7 \times 10^6$

(b) $0.03{\scriptstyle\wedge} \times 10^3 = 3. \times 10^{-2} \times 10^3$ $N \rightarrow O = -2$

$\qquad\qquad\qquad\qquad = 3 \times 10^1$

4 Multiply and Divide Numbers in Scientific Notation.

We can multiply or divide numbers expressed in scientific notation without first having to convert them to ordinary numbers.

To multiply numbers in scientific notation:

1. Multiply the first factors using the rules of signed numbers.
2. Multiply the power-of-10 factors using the laws of exponents.
3. Examine the first factor of the product (Step 1) to see if its value is equal to 1 or between 1 and 10.
   (a) If so, write the results of Steps 1 and 2.
   (b) If not, shift the decimal so the first factor is equal to 1 or is between 1 and 10, and adjust the exponent of the power-of-10 factor accordingly.

EXAMPLE Multiply.

(a) $(4 \times 10^2)(2 \times 10^3)$ (b) $(3.7 \times 10^3)(2.5 \times 10^{-1})$
(c) $(8.4 \times 10^{-2})(5.2 \times 10^{-3})$

(a) $(4 \times 10^2)(2 \times 10^3) = 8 \times 10^5$
Because 8 is between 1 and 10, we do not make any adjustments.

(b) $(3.7 \times 10^3)(2.5 \times 10^{-1}) = 9.25 \times 10^2$
Because 9.25 is between 1 and 10, we do not make any adjustments.

(c) $(8.4 \times 10^{-2})(5.2 \times 10^{-3}) = 43.68 \times 10^{-5}$

43.68 is not between 1 and 10. We must make adjustments.

$$43.68 \rightarrow 4{\scriptstyle\wedge}3.68 \qquad \text{or} \qquad 4.368 \times 10^1$$

Now, multiply 4.368×10^1 times 10^{-5}.

$$4.368 \times 10^1 \times 10^{-5} = 4.368 \times 10^{1-5}$$
$$= 4.368 \times 10^{-4}$$

Division involving numbers in scientific notation follows a similar procedure.

To divide numbers in scientific notation:

1. Divide the first factors using the rules of signed numbers.
2. Divide the power-of-10 factors using the laws of exponents.
3. Examine the first factor of the quotient (Step 1) to see if its value is equal to 1 or between 1 and 10.
   (a) If so, write the results of Steps 1 and 2.
   (b) If not, shift the decimal so that the first factor is equal to 1 or is between 1 and 10, and adjust the exponent of the power-of-10 factor accordingly.

(a) $\dfrac{3 \times 10^5}{2 \times 10^2}$ (b) $\dfrac{1.44 \times 10^{-3}}{6 \times 10^{-5}}$ (c) $\dfrac{9.6 \times 10^{29}}{3.2 \times 10^{111}}$ (d) $\dfrac{1.25 \times 10^3}{5}$

(a) $\dfrac{3 \times 10^5}{2 \times 10^2} = \dfrac{3}{2} \times 10^{5-2} = \mathbf{1.5 \times 10^3}$

Because 1.5 is between 1 and 10, no adjustments are necessary. The first factor is usually written in decimal notation.

(b) $\dfrac{1.44 \times 10^{-3}}{6 \times 10^{-5}} = \dfrac{1.44}{6} \times 10^{-3-(-5)} = 0.24 \times 10^{-3+5} = 0.24 \times 10^2$

0.24 is less than 1, so adjustments are necessary.

$$0.24 \to 0.2_\wedge 4 = 2.4 \times 10^{-1}$$

Then $2.4 \times 10^{-1} \times 10^2 = \mathbf{2.4 \times 10^1}$.

(c) $\dfrac{9.6 \times 10^{29}}{3.2 \times 10^{111}} = \dfrac{9.6}{3.2} \times 10^{29-111} = \mathbf{3 \times 10^{-82}}$

(d) $\dfrac{1.25 \times 10^3}{5}$ 5 is the same as 5×10^0.

$\dfrac{1.25 \times 10^3}{5 \times 10^0} = \dfrac{1.25}{5} \times 10^{3-0} = 0.25 \times 10^3$ 0.25 is less than 1.

$$0.25 \to 0.2_\wedge 5 = 2.5 \times 10^{-1}$$

Then, $2.5 \times 10^{-1} \times 10^3 = 2.5 \times 10^{-1+3} = \mathbf{2.5 \times 10^2}$

Tip!   *Negative Exponents and Mental Adjustment of Exponents.*

In scientific notation, we **do** leave exponents as negative numbers when appropriate.

Once we understand the concept of balancing the effect of moving a decimal by adjusting the exponent of the power-of-10 factor, we can perform this adjustment mentally.

$43.68 \times 10^{-5} = 4_\wedge 3.68 \times 10^{-5}$ $N \to O = +1$

$\phantom{43.68 \times 10^{-5}} = 4.368 \times 10^{-5+1}$ Adjust mentally.

$\phantom{43.68 \times 10^{-5}} = 4.368 \times 10^{-4}$

$0.25 \times 10^3 = 0.2_\wedge 5 \times 10^3$ $N \to O = -1$

$ = 2.5 \times 10^{3-1}$ Adjust mentally.

$ = 2.5 \times 10^2$

Tip!   *Scientific Notation and the Calculator.*

Power-of-10 Key

The power-of-10 key, labeled $\boxed{\text{EXP}}$ or $\boxed{\text{EE}}$ on most calculators, is a shortcut key for entering the following keys:

$$\boxed{\times}\ 10\ \boxed{x^y}$$

The shortcut key is used only for power-of-10 factors, and only the *exponent* of 10 is entered. If you enter $\boxed{\times}$ 10, then $\boxed{\text{EXP}}$, your answer will have one extra factor of 10.

Look at $\dfrac{(3 \times 10^5)}{(2 \times 10^2)}$ on the calculator.

$$3 \boxed{\text{EXP}}\, 5 \boxed{\div}\, 2 \boxed{\text{EXP}}\, 2 \boxed{=} \Rightarrow 1500$$

This result may be expressed in scientific notation if desired: $1{,}500 = 1.5 \times 10^3$.

The internal program of a calculator has predetermined how the output of a calculator is displayed. For example, even if you would like an answer displayed in scientific notation, it may fall within the guidelines for display as an ordinary number. You must make the conversion to scientific notation yourself. The reverse may also be true.

There are times in real-life applications that numbers are expressed in scientific notation.

EXAMPLE A star is 4.2 light-years from Earth. If 1 light-year is 5.87×10^{12} miles, how many miles from Earth is the star?

To solve this problem, express the given information in a direct proportion of two fractions relating light-years to miles.

Pair 1: 1 light-year $= 5.87 \times 10^{12}$ mi
Pair 2: 4.2 light-years $= x$ mi

Estimation The star will be more than 5.87×10^{12} miles away.

$$\frac{1 \text{ light-year}}{4.2 \text{ light-years}} = \frac{5.87 \times 10^{12} \text{ mi}}{x \text{ mi}}$$

Pair 1 becomes the two numerators.
Pair 2 becomes the two denominators.
(Direct proportion)

$$\frac{1}{4.2} = \frac{5.87 \times 10^{12}}{x}$$

Cross multiply.

$$x = (5.87 \times 10^{12})(4.2)$$

$4.2 = 4.2 \times 10^0$

$$x = 24.654 \times 10^{12}$$

Perform scientific notation adjustment.

$$x = 2.4654 \times 10^{12+1}$$

$N \to O = +1$

$$x = 2.4654 \times 10^{13}$$

or $\quad 2.5 \times 10^{13}$

Round the first factor to tenths.

Interpretation **The star is 2.5×10^{13} miles from Earth.**

EXAMPLE An angstrom (Å) is 1×10^{-7} mm. What is the length in millimeters of 14.82 Å?

We will set up a direct proportion of fractions relating angstrom units to millimeters.

Pair 1: 1 Å $= 1 \times 10^{-7}$ mm
Pair 2: 14.82 Å $= x$ mm

Estimation 14.82 angstrom units are longer than 1×10^{-7} mm.

$$\frac{1 \text{ Å}}{14.82 \text{ Å}} = \frac{1 \times 10^{-7} \text{ mm}}{x \text{ mm}}$$

Pair 1 becomes the two numerators.
Pair 2 becomes the two denominators.
(Direct proportion)

$$\frac{1}{14.82} = \frac{1 \times 10^{-7}}{x}$$

Cross multiply.

$$x = 14.82 \times 10^{-7}$$

$$x = 1.482 \times 10^{1} \times 10^{-7}$$

Adjust.

$$x = 1.482 \times 10^{-6} \text{ mm}$$

Interpretation **The length of 14.82 Å is 1.482×10^{-6} mm.**

EXAMPLE   One coulomb (C) is approximately 6.28×10^{18} electrons. How many coulombs do 2.512×10^{21} electrons represent?

Pair 1: 1 C = 6.28×10^{18} electrons
Pair 2: x C = 2.512×10^{21} electrons

Estimation   Since 2.512×10^{21} is larger than 6.28×10^{18}, there is more than one coulomb.

$$\frac{1 \text{ C}}{x \text{ C}} = \frac{6.28 \times 10^{18} \text{ electrons}}{2.512 \times 10^{21} \text{ electrons}}$$

Pair 1 becomes the two numerators.
Pair 2 becomes the two denominators.
(Direct proportion)

$$\frac{1}{x} = \frac{6.28 \times 10^{18}}{2.512 \times 10^{21}}$$

Cross multiply.

$$(6.28 \times 10^{18})(x) = 2.512 \times 10^{21}$$

Divide by the coefficient of x.

$$x = \frac{2.512 \times 10^{21}}{6.28 \times 10^{18}}$$

Divide coefficients, subtract exponents.

$$x = 0.4 \times 10^{3}$$

Adjust.

$$x = 4 \times 10^{3-1}$$

$N \to O = -1$

$$x = 4 \times 10^{2}$$

Write as an ordinary number.

$$x = 400 \text{ C}$$

Interpretation 2.512×10^{21} **electrons represent 400 C.**

SELF-STUDY EXERCISES 10–3

1   Multiply or divide as indicated.

1. 0.37×10^{2}
2. 1.82×10^{3}
3. 5.6×10^{-1}
4. 142×10^{-2}
5. 78×10^{4}
6. 62×10^{0}
7. $4.6 \div 10^{4}$
8. $6.1 \div 10$
9. $7.2 \div 10^{1}$
10. $42 \div 10^{0}$
11. $10^{4}(10^{6})$
12. $10^{-3}(10^{-4})$
13. $10^{0}(10^{-3})$
14. $10^{-3}(10^{4})$
15. $10(10^{2})$
16. $\dfrac{10^{4}}{10^{2}}$
17. $\dfrac{10}{10^{4}}$
18. $\dfrac{10^{4}}{10^{4}}$
19. $\dfrac{10^{-2}}{10^{3}}$
20. $\dfrac{10^{0}}{10^{1}}$

2   Write as ordinary numbers.

21. 4.3×10^{2}
22. 6.5×10^{-3}
23. 2.2×10^{0}
24. 7.3×10
25. 9.3×10^{-2}
26. 8.3×10^{4}
27. 5.8×10^{-3}
28. 8×10^{4}
29. 6.732×10^{0}
30. 5.89×10^{-3}

3   Express in scientific notation.

31. 392
32. 0.02
33. 7.03
34. 42,000
35. 0.081
36. 0.0021
37. 23.92
38. 0.101
39. 1.002
40. 721

Write in scientific notation.

41. 42×10^{4}
42. 32.6×10^{3}
43. 0.213×10^{2}
44. 0.0062×10^{-3}
45. $56,000 \times 10^{-3}$

4   Perform the indicated operations. Express answers in scientific notation.

46. $(6.7 \times 10^{4})(3.2 \times 10^{2})$
47. $(1.6 \times 10^{-1})(3.5 \times 10^{4})$
48. $(5.0 \times 10^{-3})(4.72 \times 10^{0})$
49. $(8.6 \times 10^{-3})(5.5 \times 10^{-1})$
50. $\dfrac{3.15 \times 10^{5}}{4.5 \times 10^{2}}$
51. $\dfrac{4.68 \times 10^{3}}{7.2 \times 10^{7}}$
52. $\dfrac{4.55 \times 10^{-1}}{6.5 \times 10^{-4}}$
53. $\dfrac{7.84 \times 10^{-2}}{9.8 \times 10^{0}}$

54. A star is 5.5 light-years from Earth. If one light-year is 5.87×10^{12} miles, how many miles from Earth is the star?

55. An angstrom (Å) is 1×10^{-7} mm. How many angstroms are in 4.2×10^{-5} mm?

Learning Outcomes

1. Identify polynomials, monomials, binomials, and trinomials.
2. Identify the degree of terms and polynomials.
3. Arrange polynomials in descending order.

1 Identify Polynomials, Monomials, Binomials, and Trinomials.

As we continue to manipulate and simplify algebraic expressions we look at some new terms that are useful in identifying types of expressions and terms. A *polynomial* is a special type of algebraic expression that contains variables and exponents.

■ **DEFINITION: Polynomial.** A *polynomial* is an algebraic expression in which the exponents of the variables are nonnegative integers.

EXAMPLE Identify which expressions are polynomials. If an expression is not a polynomial, explain why.

(a) $5x^2 + 3x + 2$ (b) $5x - \dfrac{3}{x}$ (c) 9 (d) $-\dfrac{1}{2}x + 3x^{-2}$

(a) $5x^2 + 3x + 2$ **is a polynomial.**
(b) $5x - \frac{3}{x}$ **is not a polynomial** because the term $-\frac{3}{x}$ is equivalent to $-3x^{-1}$. A polynomial cannot have a variable with a negative exponent.
(c) **9 is a polynomial** because it is equivalent to $9x^0$ and the exponent, zero, is a nonnegative integer.
(d) $-\frac{1}{2}x + 3x^{-2}$ **is not a polynomial** because $3x^{-2}$ has a negative exponent.

Some polynomials have special names, depending on the number of terms contained in the polynomial.

■ **DEFINITION: Monomial.** A *monomial* is a polynomial containing one term. A term may have more than one factor.

$$3, \quad -2x, \quad 5ab, \quad 7xy^2, \quad \frac{3a^2}{4} \quad \text{are monomials.}$$

■ **DEFINITION: Binomial.** A *binomial* is a polynomial containing two terms.

$$x + 3, \quad 2x^2 - 5x, \quad x + \frac{y}{4}, \quad 3(x - 1) + 2 \quad \text{are binomials.}$$

■ **DEFINITION: Trinomial.** A *trinomial* is a polynomial containing three terms.

$$a + b + c, \quad x^2 - 3x + 4, \quad x + \frac{2a}{7} - 5 \quad \text{are trinomials.}$$

EXAMPLE Identify which of the expressions are polynomials, then state whether each polynomial is a monomial, binomial, or trinomial.

(a) $x^2y - 1$ (b) $4(x - 2)$

(c) $\dfrac{2x + 5}{2y}$ (d) $3x^2 - x + 1$

(e) $3x^2 - (x + 1)$

(a) $x^2y - 1$ **Binomial**

(b) $4(x - 2)$ **Monomial.** If the distributive property is applied, the expression will become a binomial: $4(x - 2) = 4x - 8$.

(c) $\dfrac{2x + 5}{2y}$ This one-term expression is **not a monomial because it is not a polynomial** (the exponent of y would be -1 if it were in the numerator).

(d) $3x^2 - x + 1$ **Trinomial**

(e) $3x^2 - (x + 1)$ **Binomial** $(x + 1)$ is a grouping and counts as one term. If expanded to be $3x^2 - x - 1$, then the expression is a trinomial.

Tip!	**Counting Terms:**

Terms in an algebraic expression are separated by plus or minus signs that are not in a grouping.

$$5(x + 3) - 4 \qquad \text{2 terms}$$

$$(3x - 4)(x + 1) \qquad \text{1 term}$$

2 Identify the Degree of Terms and Polynomials.

The *degree of a term* that has only one variable with a nonnegative exponent is the same as the exponent of the variable.

$$3x^4, \quad \text{fourth degree} \qquad -5x, \quad \text{first degree}$$

Number terms other than zero have a degree of 0. A variable to the zero power is implied.

$$5 = 5x^0, \quad \text{degree zero}$$

If a term has more than one variable, the degree of the term is the *sum* of the exponents of all the variable factors.

$$2xy = 2x^1y^1, \quad \text{second degree} \qquad -2ab^2 = -2a^1b^2, \quad \text{third degree}$$

EXAMPLE Identify the degree of each term in the polynomial $5x^3 + 2x^2 - 3x + 3$.

$5x^3$, degree 3 (or third degree)
$2x^2$, degree 2 (or second degree)
$-3x$, degree 1 (or first degree)
3, degree 0

Special names are associated with terms of degree 0, 1, 2, and 3. A *constant term* has degree 0. A *linear term* has degree 1. A *quadratic term* has degree 2. A *cubic term* has degree 3.

■ **DEFINITION: Degree of a Polynomial.**   The *degree of a polynomial* that has only one variable and only positive integral exponents is the degree of the term with the largest exponent.

A *linear polynomial* has degree 1. A *quadratic polynomial* has degree 2. A *cubic polynomial* has degree 3.

$$5x^3 - 2 \text{ has a degree of 3 and is a cubic polynomial.}$$

$$x + 7 \text{ has a degree of 1 and is a linear polynomial.}$$

$$7x^2 - 4x + 5 \text{ has a degree of 2 and is a quadratic polynomial.}$$

EXAMPLE Identify the degree of the polynomials.

(a) $5x^4 + 2x - 1$ (b) $3x^3 - 4x^2 + x - 5$ (c) 7 (d) $x - \dfrac{1}{2}$

(a) $5x^4 + 2x - 1$ has a **degree of 4.**
(b) $3x^3 - 4x^2 + x - 5$ has a **degree of 3 and is a cubic polynomial.**
(c) 7 has a **degree of 0 and is a constant.**
(d) $x - \frac{1}{2}$ has a **degree of 1 and is a linear polynomial.**

3 Arrange Polynomials in Descending Order.

The terms of a polynomial are customarily arranged in order based on the degree of each term of the polynomial. The terms can be arranged beginning with the term with the highest degree (*descending order*) or beginning with the term with the lowest degree (*ascending order*).

Tip!	***Most Common Arrangement of Polynomials:***

Polynomials are most often arranged in **descending order** so that the degree of the polynomial is the degree of the first term.

The first term of a polynomial arranged in descending order is called the *leading term* of the polynomial. The coefficient of the leading term of a polynomial is called the *leading coefficient.*

EXAMPLE Arrange each polynomial in descending order and identify the degree, the leading term, and the leading coefficient of the polynomial.

(a) $5x + 3x^3 - 7 + 6x^2$ (b) $x^4 - 2x + 3$
(c) $4 + x$ (d) $x^2 + 5$

(a) $5x + 3x^3 - 7 + 6x^2 = 3x^3 + 6x^2 + 5x - 7.$
   Third degree, leading term is $3x^3$, leading coefficient is 3.

(b) $x^4 - 2x + 3$ is already in descending order.
   Fourth degree, leading term is x^4, leading coefficient is 1.

(c) $4 + x = x + 4.$
   First degree or linear polynomial, leading term is x, leading coefficient is 1.

(d) $x^2 + 5$ is already in descending order.
   Second degree or quadratic polynomial, leading term is x^2, leading coefficient is 1.

Tip!	***Coefficients of Zero:***

A polynomial arranged in descending order can be written with every successive degree represented. Missing terms have a coefficient of 0.
$x^4 - 2x + 3$ is the same as $x^4 + 0x^3 + 0x^2 - 2x^1 + 3x^0$.

Terminology sometimes seems overwhelming. To learn several new terms, look at the similarities and differences in terms.

- Examine prefixes and word stems.

Prefix	Stem
poly-	-nomial
mo- (one)	-nomial
bi- (two)	-nomial
tri- (three)	-nomial

In the mathematical context, polynomial is the global term, and monomials, binomials, and trinomials are specific types of polynomials.

- Group new words with similar mathematical concepts, and relate differences numerically if possible.

Constant	degree 0
Linear	degree 1
Quadratic	degree 2
Cubic	degree 3

SELF-STUDY EXERCISES 10–4

1   Identify each of the following expressions as a monomial, binomial, or trinomial.

1. $2x^3y - 7x$ **2.** $5xy^2 + 8y$ **3.** $3xy$ **4.** $7ab$

5. $5(3x - y)$ **6.** $(x - 5)(2x + 4)$ **7.** $5x^2 - 8x + 3$ **8.** $7y^2 + 5y - 1$

9. $\dfrac{4x - 1}{5}$ **10.** $\dfrac{x}{6} - 5$ **11.** $4(x^2 - 2) + x^3$ **12.** $3x^2 - 7(x - 8)$

2   Identify the degree of each term in each polynomial.

13. $6x$ **14.** $8x^2$ **15.** $6x^2 - 8x + 12$ **16.** $7x^3 - 8x + 12$

17. $x - 12$ **18.** $3x^2 - 8$ **19.** 15 **20.** 21

21. $2x - \dfrac{1}{4}$ **22.** $8x^2 + \dfrac{5}{6}$

Identify the degree of the following polynomials.

23. $5x^2 + 8x - 14$ **24.** $x^3 - 8x^2 + 5$ **25.** $9 - x^3 + x^6$ **26.** $12 - 15x^2 - 7x^5$

27. $2x - \dfrac{4}{5}x^2$ **28.** $\dfrac{7}{8} - x$

3   Arrange each polynomial in descending order, and identify the degree, leading term, and leading coefficient of the polynomial.

29. $5x - 3x^2$ **30.** $7 - x^3$ **31.** $4x - 8 + 9x^2$

32. $5x^2 + 8 - 3x$ **33.** $7x^3 - x + 8x^2 - 12$ **34.** $7 - 15x^4 + 12x$

35. $-7x + 8x^6 - 7x^3$ **36.** $15 - 14x^8 + x$

Learning Outcomes

 Write powers and roots using rational exponent and radical notations.

 Use laws of exponents to simplify and evaluate expressions with rational exponents.

Previously, we examined the symbolic representation for irrational numbers that resulted from taking roots of numbers that are not perfect powers. We extend our study of roots to include roots of variables.

As we saw before, two common notations represent roots: fractional or rational exponents and radical notation. We will look at both types of notation and the conventions used with each.

The calculator can be used to illustrate that the relationship between a number raised to the one-half power and the square root of a number are the same.

$$8^{1/2} = 2.828427125 \qquad \sqrt{8} = 2.828427125$$

Similarly, a number raised to the one-third power and the cube root of a number are the same.

$$8^{1/3} = 2 \qquad \sqrt[3]{8} = 2$$

Thus, we determined that the index or order of a root is indicated by the number in the denominator of a rational exponent or the number in the "$\sqrt{}$" portion of the radical symbol. By combining our knowledge of roots, the laws of exponents, and the arithmetic of rational numbers, we can extend our study of powers and roots to include variables.

1 Write Powers and Roots Using Rational Exponent and Radical Notations.

For our discussion of powers and roots of variables, unless otherwise specified, *we will consider variables to represent positive values.* Suppose that $\sqrt{x}$ is raised to the fourth power. In rational exponent notation, $\sqrt{x}$ is written as $x^{1/2}$.

$$(x^{1/2})^4 = x^{4/2} = x^2 \qquad \text{Multiply exponents and reduce.}$$

We apply the laws of exponents and arithmetic properties of fractions.

What happens if $\sqrt{x}$ is raised to the third power (cubed)?

$$(x^{1/2})^3 = x^{3/2}$$

This exponent can be left in improper fraction form, which brings up the need for further interpretation of the meaning of a rational or fractional exponent.

To convert between rational exponent notation and radical notation:

The numerator of a rational exponent represents the power, and the denominator of the rational exponent represents the index or order of the root.

$$x^{\text{power/root}} = \sqrt[\text{root}]{x^{\text{power}}} \qquad \text{or} \qquad \left(\sqrt[\text{root}]{x}\right)^{\text{power}}$$

Then, is $\left(\sqrt{x}\right)^2$ the same as $\sqrt{x^2}$? *Yes,* for nonnegative values of x or $x \geq 0$.

$$(x^{1/2})^2 = x^{2/2} = x^1 = x, \qquad (x^2)^{1/2} = x^{2/2} = x^1 = x$$

That means, in the order of operations, powers and roots (Exponents in "**Please Excuse My Dear Aunt Sally**" from Section 1–7) have the same priority.

Let's expand this concept to show the relationships between powers and roots.

Powers and roots—inverse operations:

Raising to powers and extracting roots are inverse operations for nonnegative values. The order in which the operations are performed does not matter.

Rational Exponent Notation	Radical Notation
$(x^{1/n})^n = x$	$(\sqrt[n]{x})^n = x$
$(x^n)^{1/n} = x$	$\sqrt[n]{x^n} = x$

for $x \geq 0$.

EXAMPLE Write both the rational exponent and radical notations.

(a) The fourth root of the square of x.
(b) The square of the fourth root of x.
(c) The square root of the cube of x.
(d) The cube of the square root of x.

	Rational Exponent Notation	Radical Notation
(a)	$(x^2)^{1/4}$	$\sqrt[4]{x^2}$
(b)	$(x^{1/4})^2$	$(\sqrt[4]{x})^2$
(c)	$(x^3)^{1/2}$	$\sqrt{x^3}$
(d)	$(x^{1/2})^3$	$(\sqrt{x})^3$

Even though the calculator makes the pencil-and-paper method of taking roots obsolete, it is still helpful to perform some mental manipulations on expressions containing powers and roots before evaluating these expressions using the calculator. One common manipulation is to simplify rational exponents and radical expressions by applying the laws of exponents.

To simplify radicals using rational exponents and the laws of exponents:

1. Convert the radicals to equivalent expressions using rational exponents.
2. Apply the laws of exponents and the arithmetic of fractions.
3. Convert simplified expressions back to radical notation if desired.

EXAMPLE Convert the radical expressions to equivalent expressions using rational exponents and simplify if appropriate.

(a) $\sqrt[3]{x}$ (b) $\sqrt[5]{2y}$ (c) $(\sqrt{ab})^3$ (d) $\sqrt[4]{16b^8}$ (e) $(\sqrt[3]{27xy^5})^4$

(a) $\sqrt[3]{x} = x^{1/3}$ (b) $\sqrt[5]{2y} = (2y)^{1/5}$ or $2^{1/5}y^{1/5}$

(c) $(\sqrt{ab})^3 = (ab)^{3/2}$ or $a^{3/2}b^{3/2}$

(d) $\sqrt[4]{16b^8} = (2^4b^8)^{1/4} = (2^4)^{1/4}(b^8)^{1/4} = 2b^2$

The fourth root of 16 is the number used as a factor 4 times to equal 16. By inspection (or by using a calculator), $\sqrt[4]{16} = 2$, and $2^4 = 16$.

(e) $(\sqrt[3]{27xy^5})^4 = (3^3x^1y^5)^{4/3} = (3^3)^{4/3}x^{4/3}(y^5)^{4/3} = 3^4x^{4/3}y^{20/3} = 81x^{4/3}y^{20/3}$

27 is a perfect cube; $3^3 = 27$. Therefore, $27^{4/3} = (3^3)^{4/3} = (3)^4 = 81$.

Tip!	**Simplifying Coefficients.**

When coefficients of variable terms have rational exponents, the numerical equivalent can be determined if desired. Usually, if the coefficient is a perfect power of the indicated root, we will evaluate the coefficient. Otherwise, we may leave the coefficient with the rational exponent.

$$8^{2/3} = (8^{1/3})^2 = 2^2 = 4$$

2 Use Laws of Exponents to Simplify and Evaluate Expressions with Rational Exponents.

Other laws of exponents are applied to rational exponents. Remember, the laws of exponents apply to factors having *like bases*. The following example illustrates other laws of exponents applied to rational exponents.

EXAMPLE Perform the following operations and simplify. Express answers with positive exponents in lowest terms.

(a) $(x^{3/2})(x^{1/2})$ (b) $(3a^{1/2}b^3)^2$ (c) $\dfrac{x^{1/2}}{x^{1/3}}$ (d) $\dfrac{10a^3}{2a^{1/2}}$

(a) $(x^{3/2})(x^{1/2}) = x^{3/2+1/2} = x^{4/2} = \boldsymbol{x^2}$ Add exponents.

(b) $(3a^{1/2}b^3)^2 = 3^2ab^6 = \boldsymbol{9ab^6}$ Multiply exponents, $\dfrac{1}{2} \cdot 2 = 1; 3 \cdot 2 = 6$

(c) $\dfrac{x^{1/2}}{x^{1/3}} = x^{1/2-1/3} = \boldsymbol{x^{1/6}}$ Subtract exponents. $\dfrac{1}{2} - \dfrac{1}{3} = \dfrac{3}{6} - \dfrac{2}{6} = \dfrac{1}{6}$

(d) $\dfrac{10a^3}{2a^{1/2}} = 5a^{3-1/2} = \boldsymbol{5a^{5/2}}$ Reduce coefficients. Subtract exponents. $3 - \dfrac{1}{2} = \dfrac{3}{1} - \dfrac{1}{2} = \dfrac{6}{2} - \dfrac{1}{2} = \dfrac{5}{2}$

Let's evaluate some expressions both before and after we simplify.

EXAMPLE Evaluate the expressions from the preceding example for $x = 2$, $a = 3$, and $b = 4$ both before and after simplifying.

Before	**After**

(a) $(x^{3/2})(x^{1/2}) = (2^{3/2})(2^{1/2})$ $x^2 = 2^2$

$2 \boxed{x^y} 3 \boxed{a\,b/c} 2 \boxed{\times} 2 \boxed{x^y} 1$ $2 \boxed{x^2} \Rightarrow \boldsymbol{4}$

$\boxed{a\,b/c} 2 \boxed{=} \Rightarrow \boldsymbol{4}$

(b) $(3a^{1/2}b^3)^2 = (3 \cdot 3^{1/2} \cdot 4^3)^2$ $9ab^6 = 9 \cdot 3 \cdot 4^6$

$\boxed{(} 3 \boxed{\times} 3 \boxed{x^y} \boxed{(} 1 \boxed{a\,b/c} 2 \boxed{)} \boxed{\times} 4$ $9 \boxed{\times} 3 \boxed{\times} 4 \boxed{x^y} 6 \boxed{=} \Rightarrow \boldsymbol{110{,}592}$

$\boxed{x^y} 3 \boxed{)} \boxed{x^y} 2 \boxed{=} \Rightarrow \boldsymbol{110{,}592}$

(c) $\dfrac{x^{1/2}}{x^{1/3}} = \dfrac{2^{1/2}}{2^{1/3}}$ $x^{1/6} = 2^{1/6}$

$2 \boxed{x^y} \boxed{(} 1 \boxed{a\,b/c} 2 \boxed{)} \boxed{\div} 2 \boxed{x^y} \boxed{(} 1$ $2 \boxed{x^y} \boxed{(} 1 \boxed{a\,b/c} 6 \boxed{)} \boxed{=} \Rightarrow$

$\boxed{a\,b/c} 3 \boxed{)} \boxed{=} \Rightarrow \boldsymbol{1.122462048}$ $\boldsymbol{1.122462048}$

(d) $\dfrac{10a^3}{2a^{1/2}} = \dfrac{10 \cdot 3^3}{2 \cdot 3^{1/2}}$

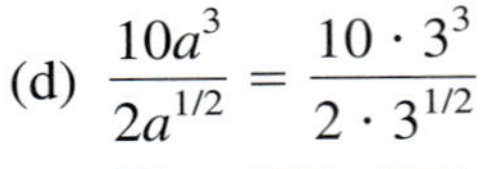

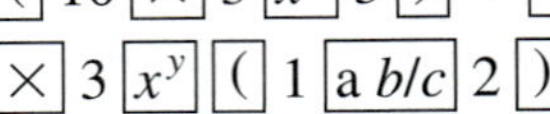

$= \Rightarrow 77.94228634$

$5a^{5/2} = 5 \cdot 3^{5/2}$

77.94228634

Is it really worth the effort to simplify an expression before evaluating? Yes, in most cases. Long sequences of calculator steps are very tedious to enter.

Decimal exponents can also be used to show roots: $\sqrt{x} = x^{1/2} = x^{0.5}$.

EXAMPLE   Perform the operations. Express the answers with positive exponents.

(a) $a^{2.3}(a^3)$ (b) $(5a^{3.5}b^{0.5})^2$

(a) $a^{2.3}(a^3) = a^{2.3+3} = a^{5.3}$

What does $a^{5.3}$ mean? If written as an improper fraction, $a^{5.3} = a^{5\,3/10} = a^{53/10}$.

This means we take the tenth root of a to the 53rd power or $\sqrt[10]{a^{53}}$.

(b) $(5a^{3.5}b^{0.5})^2 = 5^2 a^{3.5(2)} b^{0.5(2)}$
$= 25a^7b$

SELF-STUDY EXERCISES 10–5

1   Write both the rational exponent and radical notations.

1. The fifth root of the cube of x.
2. The cube of the fifth root of x.
3. The square root of the fourth power of x.
4. The square of the sixth root of x.
5. The fourth root of the cube of x.
6. The fourth power of the square root of x.

Convert the radical expressions to equivalent expressions using rational exponents and simplify.

7. $\sqrt[4]{a}$ **8.** $\sqrt[3]{m}$ **9.** $\sqrt[5]{3x}$ **10.** $\sqrt[4]{7a}$ **11.** $\left(\sqrt{xy}\right)^4$

12. $\sqrt[3]{27a^6}$ **13.** $\sqrt[5]{32x^7}$ **14.** $\left(\sqrt{36x^3y^4}\right)^4$ **15.** $\left(\sqrt[3]{16a^4b^7}\right)^2$

2   Perform the operations and simplify. Express the answers with positive exponents in lowest terms.

16. $(a^{5/2})(a^{1/2})$ **17.** $(b^{3/5})(b^{1/5})$ **18.** $(4x^{1/3}y^4)^3$ **19.** $(2m^{1/5}n^2)^5$

20. $\dfrac{a^{3/5}}{a^{1/5}}$ **21.** $\dfrac{b^{1/5}}{b^{3/5}}$ **22.** $\dfrac{m^{7/8}}{m^{3/4}}$ **23.** $\dfrac{x^{1/3}}{x^{5/6}}$

24. $\dfrac{a^2}{a^{1/3}}$ **25.** $\dfrac{14a^4}{7a^{3/5}}$ **26.** $\dfrac{8a^{2/3}}{24a^5}$ **27.** $\dfrac{6a^{7/8}}{27a^2}$

Evaluate the expressions for $a = 2$, $b = 4$, and $c = 5$ both before and after simplifying. Verify that the results are the same.

28. $(a^{1/3})(a^{4/3})$ **29.** $(4a^{1/3}b^2)^3$ **30.** $\dfrac{c^{3/5}}{c^{1/4}}$ **31.** $\dfrac{5b^2}{10b^{1/3}}$

Perform the operations and express the answers with positive exponents.

32. $(a^{3.2})(a^4)$ **33.** $\dfrac{a^{5.6}}{a^7}$ **34.** $(3a^{1.2}b^2)^3$ **35.** $(4c^{1.3}d^{0.3})^{10}$

Learning Outcomes

1. Write imaginary numbers using the letter i.
2. Raise imaginary numbers to powers.
3. Write real and imaginary numbers in complex form, $a + bi$.
4. Combine complex numbers.

1 Write Imaginary Numbers Using the Letter i.

Taking the square root of a negative number introduces a new type of number. This new type of number is called an *imaginary number.*

■ **DEFINITION: Imaginary Number.** An *imaginary number* has a factor of $\sqrt{-1}$, which is represented by i $\left(i = \sqrt{-1}\right)$.

Imaginary numbers and real numbers combine to form the set of complex numbers (Fig. 10–2).

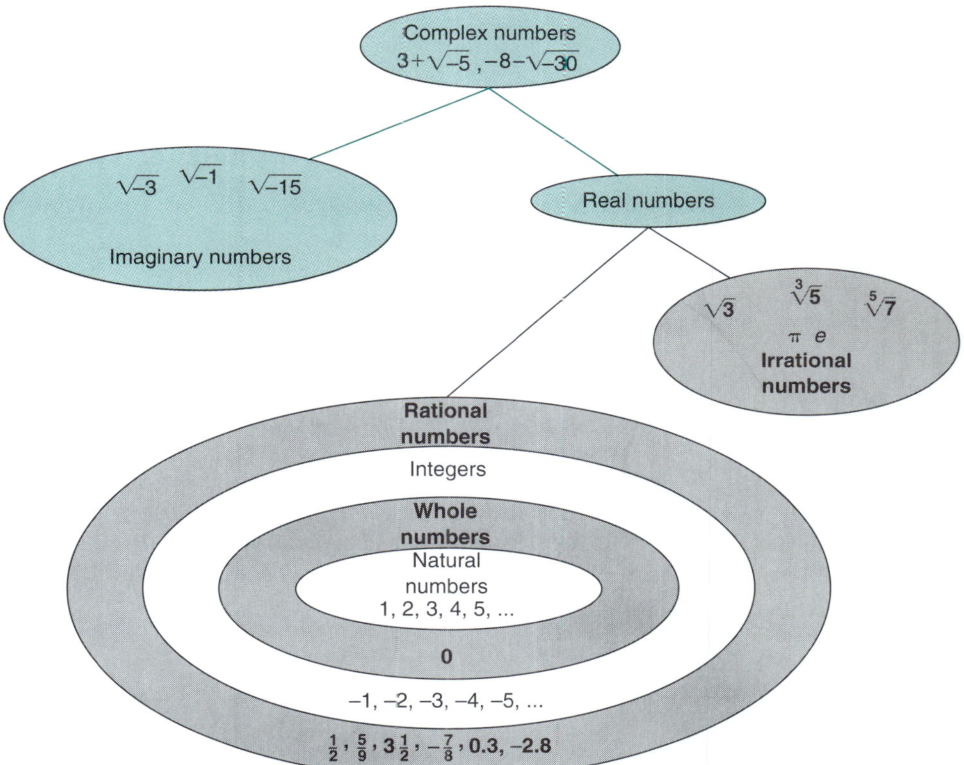

Figure 10–2 Complex-number system.

■ **DEFINITION: Complex Number.** A *complex number* is a number that can be written in the form $a + bi$, where a and b are real numbers and i is $\sqrt{-1}$.

The square root of a negative number, say, $\sqrt{-16}$, can be simplified as $\sqrt{-1 \cdot 16}$ or $4\sqrt{-1}$. Because the square root of -1 is an *imaginary* factor and 4 is a real factor, we will use the letter i to represent $\sqrt{-1}$ and rewrite $4\sqrt{-1}$ as $4i$. Similarly, $\sqrt{-4} = \sqrt{-1 \cdot 4} = 2\sqrt{-1} = 2i$.

Electronics and other applications of imaginary numbers may use j rather than i to represent $\sqrt{-1}$. Thus, $3 + 4i$ and $3 + 4j$ represent the same amount.

EXAMPLE   Rewrite the imaginary numbers using the letter i for $\sqrt{-1}$.

(a) $\sqrt{-9}$ (b) $\sqrt{-25}$ (c) $\sqrt{-7}$

(a) $\sqrt{-9} = \sqrt{-1 \cdot 9} = 3\sqrt{-1} = 3i$

(b) $\sqrt{-25} = \sqrt{-1 \cdot 25} = 5\sqrt{-1} = 5i$

(c) $\sqrt{-7} = \sqrt{-1 \cdot 7} = \sqrt{7} \cdot \sqrt{-1} = \sqrt{7}i$ **or** $i\sqrt{7}$.

$\sqrt{7}i$ and $\sqrt{7i}$ do not represent the same amount. In the first term, $\sqrt{7}i$, i is not under the radical symbol. In the second term, $\sqrt{7i}$, i is under the radical symbol. Because it is easy to confuse the two terms, when the coefficient of an imaginary number is an irrational number, we write the i factor first: $\sqrt{7}i = i\sqrt{7}$

2   Raise Imaginary Numbers to Powers.

Some powers of imaginary numbers are real numbers. Examine the pattern that develops with powers of i.

$$i = \sqrt{-1} = i \qquad i^5 = i^4 \cdot i^1 = 1(i) = i \qquad i^9 = i^4 \cdot i^4 \cdot i^1 = 1(1)(i) = i$$

$$i^2 = \left(\sqrt{-1}\right)^2 = -1 \qquad i^6 = i^4 \cdot i^2 = 1(-1) = -1 \qquad i^{10} = i^4 \cdot i^4 \cdot i^2 = 1(1)(-1) = -1$$

$$i^3 = i^2 \cdot i^1 = -1i = -i \qquad i^7 = i^4 \cdot i^3 = 1(-i) = -i \qquad i^{11} = i^4 \cdot i^4 \cdot i^3 = 1(1)(-i) = -i$$

$$i^4 = i^2 \cdot i^2 = -1(-1) = 1 \qquad i^8 = i^4 \cdot i^4 = 1(1) = 1 \qquad i^{12} = i^4 \cdot i^4 \cdot i^4 = 1(1)(1) = 1$$

- All powers of i simplify to either i, -1, $-i$, or 1.
- If the exponent of i is a multiple of 4, the result is 1.
- If the exponent of i is even but not a multiple of 4, the result is -1.
- All even powers of i are real numbers.

These and other patterns allow us to develop a shortcut process for simplifying powers of i.

To simplify a power of i:

1. Divide the exponent by 4 and examine the **remainder.**
2. The power of i simplifies as follows, based on the remainder in Step 1.

Remainder of $0 \Rightarrow i^0 = 1$	Remainder of $2 \Rightarrow i^2 = -1$
Remainder of $1 \Rightarrow i^1 = i$	Remainder of $3 \Rightarrow i^3 = -i$

EXAMPLE   Simplify the powers of i.

(a) i^{15} (b) i^{20} (c) i^{33} (d) i^{18}

(a) $i^{15} = -i$ $15 \div 4 = 3$ R3, remainder of $3 \Rightarrow i^3 = -i$

(b) $i^{20} = 1$ $20 \div 4 = 5$,   remainder $0 \Rightarrow i^0 = 1$

(c) $i^{33} = i$ $33 \div 4 = 8$ R1, remainder $1 \Rightarrow i^1 = i$

(d) $i^{18} = -1$ $18 \div 4 = 4$ R2, remainder $2 \Rightarrow i^2 = -1$

3   Write Real and Imaginary Numbers in Complex Form, $a + bi$.

A complex number has two parts, a *real part* and an *imaginary part*. In $a + bi$, if $a = 0$, then $a + bi$ is the same as $0 + bi$ or bi, which is an imaginary number. If $b = 0$, then $a + bi$ is the same as $a + 0 \cdot i$ or a, which is a real number. Thus, real numbers and imaginary numbers are also complex numbers.

EXAMPLE   Rewrite in the form $a + bi$.

(a) 5   (b) $\sqrt{-36}$ (c) $-3i^2$ (d) $6 - \sqrt{-5}$

(a) $5 = 5 + 0i$ $a = 5,\ b = 0$

(b) $\sqrt{-36} = 6i = 0 + 6i$ $a = 0,\ b = 6$

(c) $-3i^2 = -3(-1) = 3 = 3 + 0i$ $a = 3,\ b = 0$

(d) $6 - \sqrt{-5} = 6 - \sqrt{(5)(-1)} = 6 - i\sqrt{5}$ $a = 6,\ b = -\sqrt{5}$

4   Combine Complex Numbers.

Complex numbers are combined by adding *like* parts. Real parts are added together and imaginary parts are added together.

> **To combine complex numbers:**
>
> **1.** Add real parts for the real part of the answer.
> **2.** Add imaginary parts for the imaginary part of the answer.
>
> $$(a + bi) + (c + di) = (a + c) + (b + d)i$$

EXAMPLE   Combine the complex numbers.

(a) $(3 + 5i) + (8 - 2i)$ (b) $(-3 + i) - (7 - 4i)$ (c) $-2 + (5 + 6i)$

(a) $(3 + 5i) + (8 - 2i) = 11 + 3i$ $3 + 8 = 11;\ 5i - 2i = 3i$

(b) $(-3 + i) - (7 - 4i) =$ Distribute -1.

     $(-3 + i) + (-7 + 4i) = -10 + 5i$ $-3 - 7 = -10;\ i + 4i = 5i$

(c) $-2 + (5 + 6i) = 3 + 6i$ $-2 + 5 = 3$

SELF-STUDY EXERCISES 10–6

1   Write the imaginary numbers using the letter i. Simplify if possible.

1. $\sqrt{-25}$ **2.** $\sqrt{-36}$ **3.** $\sqrt{-64x^2}$ **4.** $\sqrt{-32y^5}$

2   Simplify the powers of i.

5. i^{17} **6.** i^{20} **7.** i^{24} **8.** i^{32}

3   Write the real and imaginary numbers in complex form.

9. 15   **10.** $33i$ **11.** $5 + \sqrt{-4}$ **12.** $8 + \sqrt{-32}$

13. $7 - \sqrt{-3}$ **14.** $-7i^3$ **15.** $4i^6$

16. $(4 + 3i) + (7 + 2i)$

17. $\sqrt{12} - 5\sqrt{-3} + \left(\sqrt{8} - \sqrt{-27}\right)$

18. $12 - 3i - (8 - 4i)$

19. $15 + 8i - (3 - 12i)$

20. $(4 + 7i) - (3 - 2i)$

10-7 EQUATIONS WITH SQUARES AND SQUARE ROOTS

Learning Outcomes

1 Solve equations with squared letter terms.

2 Solve equations with square-root radical terms.

1 Solve Equations with Squared Letter Terms.

In solving equations with squared letter terms, we use the same procedures for solving equations that we used in Chapter 8. When squared letter terms are involved, we need to take one more step. After simplifying the equation to a basic equation with a coefficient of $+1$, we take the square root of both sides of the equation to find the roots. In doing so, we express the solutions with $\pm$ to show both the negative and the positive square-root values. Equations that we examine in this section are *quadratic equations* because they contain a quadratic term. In Chapter 13, we define quadratic equations and examine several types of quadratic equations.

To solve an equation containing only squared variable terms and number terms:

1. Perform all normal steps to isolate the squared variable and to obtain a numerical coefficient of 1.
2. Take the square root of both sides.
3. Identify both roots of the equation.

EXAMPLE Solve the equations:

(a) $x^2 = 4$ (b) $x^2 + 9 = 25$ (c) $5x^2 + 10 = 30$

(d) $\dfrac{3}{25} = \dfrac{x^2}{48}$ (e) $x^2 + 16 = 0$

(a) $x^2 = 4$ — Take the square root of both sides.

$x = \pm 2$ — Identify both roots.

(b) $x^2 + 9 = 25$ — Isolate the squared variable term.

$x^2 = 25 - 9$ — Combine like terms.

$x^2 = 16$ — Take the square root of both sides.

$x = \pm 4$ — Identify both roots.

(c) $5x^2 + 10 = 30$ — Isolate the squared variable term.

$5x^2 = 30 - 10$ — Combine like terms.

$5x^2 = 20$ — Divide by the coefficient of x^2.

$\dfrac{5x^2}{5} = \dfrac{20}{5}$

$x^2 = 4$ — Take the square root of both sides.

$x = \pm 2$ — Identify both roots.

(d) $\dfrac{3}{25} = \dfrac{x^2}{48}$ — To solve a proportion, cross multiply.

$3(48) = 25x^2$

$144 = 25x^2$ — Divide by the coefficient of x^2.

$$\frac{144}{25} = \frac{25x^2}{25}$$

$$\frac{144}{25} = x^2 \qquad \text{Take the square root of both sides.}$$

$$\pm\frac{12}{5} = x \qquad \text{Identify both roots.}$$

(e) $x^2 + 16 = 0$ Isolate the squared variable term.

$\quad\quad x^2 = -16$ Take the square root of both sides.

$\quad\quad x = \pm\sqrt{-16}$ Express the roots as imaginary numbers using i.

$\quad\quad x = \pm 4i$ Identify both roots.

> **Tip!**
>
> ***Imaginary Roots Versus No Real Roots.***
>
> Sometimes within the context of an applied problem, imaginary or complex roots are unacceptable. Often a problem will specify that only real roots are acceptable.

2 Solve Equations with Square-Root Radical Terms.

In this discussion we consider only equations containing a radical isolated on either or both sides of the equation. Whenever this is the case, we solve the equation by first squaring both sides of the equation. This rids the equation of radicals. Then, we use previously learned procedures to finish solving the equation.

> ***To solve an equation containing square-root radicals that are isolated on either or both sides of the equation:***
>
> 1. Square both sides to eliminate any radicals.
> 2. Perform all normal steps to isolate the variable and solve the equation.

EXAMPLE Solve the following equations.

(a) $\sqrt{x} = \dfrac{3}{4}$ (b) $\sqrt{x - 2} = 5$ (c) $\sqrt{5x^2 - 36} = 8$

(d) $\sqrt{1.7x^2} = \sqrt{6.8}$

(a) $\sqrt{x} = \dfrac{3}{4}$ Square both sides of the equation.

$\quad\quad \left(\sqrt{x}\right)^2 = \left(\dfrac{3}{4}\right)^2$ $(\sqrt{x})^2 = x$

$\quad\quad x = \dfrac{9}{16}$

(b) $\sqrt{x - 2} = 5$ Square both sides of the equation.

$\quad\quad \left(\sqrt{x - 2}\right)^2 = 5^2$ Any square-root radical squared equals the radicand.

$\quad\quad x - 2 = 25$ Isolate the variable.

$\quad\quad x = 25 + 2$ Combine like terms.

$\quad\quad x = 27$

(c) $\sqrt{5x^2 - 36} = 8$

$\quad\quad \left(\sqrt{5x^2 - 36}\right)^2 = 8^2$ Do this step mentally.

$\quad\quad 5x^2 - 36 = 64$ Isolate the variable term.

$\quad\quad 5x^2 = 64 + 36$ Combine like terms.

$$
\begin{aligned}
5x^2 &= 100 &&\text{Divide by the coefficient of } x^2. \\
\frac{5x^2}{5} &= \frac{100}{5} &&\text{Do this step mentally.} \\
x^2 &= 20 &&\text{Take the square root of both sides.} \\
x &= \pm\sqrt{20} &&\text{Identify both roots.} \\
\mathbf{x} &= \mathbf{\pm 2\sqrt{5}} &&\text{Exact roots} \\
\mathbf{x} &\approx \mathbf{\pm 4.472} &&\text{Approximate roots}
\end{aligned}
$$

$$
\begin{aligned}
\text{(d)} \quad \sqrt{1.7x^2} &= \sqrt{6.8} &&\text{Square both sides to eliminate radicals.} \\
1.7x^2 &= 6.8 &&\text{Divide by the coefficient of } x^2. \\
x^2 &= \frac{6.8}{1.7} &&\text{Simplify.} \\
x^2 &= 4 &&\text{Take the square root of both sides.} \\
\mathbf{x} &= \mathbf{\pm 2} &&\text{Identify both roots.}
\end{aligned}
$$

Tip!	***When Do I Take the Square Root of Both Sides? When Do I Square Both Sides?***

These are important questions.

- If an equation has a squared letter term, take the square root of both sides at the *end* of the solution. That is, when the squared letter has been isolated with a coefficient of $+1$, take the square root of both sides.
- If an equation has one or two radical terms but each radical is isolated on one side of the equal sign, square both sides of the equation at the *beginning* of the solution. Then, solve the remaining equation that contains no radicals.

There are other situations in which an equation may have one or more radical terms that are not isolated. We will not discuss those situations at this time.

EXAMPLE Shirley Riddle needs to prepare a flower bed for planting. She needs a square bed with an area of 128 ft^2. How long should each side of the bed be?

Estimation

$11^2 = 121$

$12^2 = 144$

The bed should be between 11 and 12 ft on a side.

$$
\begin{aligned}
A &= s^2 &&\text{Formula for the area of a square.} \\
128 &= s^2 &&\text{Substitute 128 for } A. \\
\sqrt{128} &= s &&\text{Take square root of both sides.} \\
\mathbf{11.3} &\approx s &&\text{Approximate root}
\end{aligned}
$$

Interpretation **Each side of the bed should be 11.3 ft.**

Tip!	***Examining Roots.***

Exact Roots versus Approximate Roots

A root is an exact amount when no rounding has been done. If an exact root is desired, give the root in radical notation.

If the root has been rounded, it becomes an approximate root. Approximate roots are often more meaningful than exact roots, especially in applied problems.

One Root versus Two Roots

Look again at the four problems in the example starting at the bottom of page 431. Parts a and b have variables to the first power after the radical is removed. Therefore, there will be *at most* one solution. Parts c and d have variables that are squared when the radical is removed. These equations will have *at most* two roots or solutions.

Extraneous Roots

The techniques of squaring both sides of an equation and taking the square root of both sides of an equation are equivalent to multiplying or dividing by a variable, thus introducing the possibilities of gaining an extraneous root or losing a root. Be sure to check your roots.

SELF-STUDY EXERCISES 10–7

1 Solve the equations. Use your calculator to evaluate to the nearest thousandth when necessary.

1. $y^2 = 25$ **2.** $3x^2 = 75$ **3.** $3x^2 = 36$ **4.** $\dfrac{3}{81} = \dfrac{x^2}{12}$

5. $3 + R^2 = 12$ **6.** $\dfrac{1}{2}P^2 = \dfrac{3}{7}$ **7.** $\dfrac{4}{9} = \dfrac{x^2}{3}$ **8.** $16 - T^2 = 0$

9. $5x^2 - 10 = 20$ **10.** $x^2 + 25 = 0$ **11.** $\dfrac{x^2}{6} = \dfrac{12}{7}$ **12.** $-9 + 3x^2 = 18$

2 Solve the equations. Use your calculator if necessary. Evaluate to the nearest thousandth.

13. $\sqrt{x} = 9$ **14.** $\sqrt{\dfrac{y}{3}} = 8$ **15.** $8 = \sqrt{y - 2}$ **16.** $\sqrt{3x^2 - 3} = 9$

17. $\sqrt{\dfrac{3}{x + 2}} = 12$ **18.** $\sqrt{x^2} = 2$ **19.** $10 = \sqrt{2x}$ **20.** $\sqrt{\dfrac{2}{3}x^2} = 4$

21. $\sqrt{1.3y^2} = 2.4$ **22.** $\sqrt{y^2} = 2.9$ **23.** $\sqrt{2x^2 - 1} = 5$ **24.** $\sqrt{x - 5} = 2$

CAREER APPLICATION

Horticulture: Insect Mating

Thomasville, Georgia, is known as the Rose Capital of the World for the abundant native and imported roses that bloom almost year-round due to its mild climate and excellent iron-rich soil. In preparation for its annual Rose Festival each April, local horticulturists try to control as many problems as they can to maximize the quality and quantity of April roses.

One big problem is aphids, the tiny sucking insects that secrete a sweet juice ants love to eat. Ants will even carry around and nurse aphids just to get their honeydew. While the aphids eat the new growth on leaves, stems, and buds the ants loosen soil around roots, causing rose plants to wilt and die. Aphids and ants can destroy a rose garden in a few days.

Horticulture researchers study aphid mating in the field and in the laboratory. When equal numbers of male and female aphids are in close proximity, the number of their matings per hour depends on the temperature according to the following formula:

$$M = 5t^{3/2} - 3t^{1/2}$$

where M is the number of group matings per hour and t is the temperature in degrees Celsius (°C). This formula approximates the matings for temperatures between 4° and 25°C, inclusive.

Knowing the temperatures at the maximum and minimum number of matings helps horticulturists control aphids and ants before they affect the quality and quantity of rose production. Aphids can be eradicated with natural predators (ladybugs, which eat aphids), manual labor (washing them off with soap), or insecticide. The type of treatment depends on the degree of infestation, how much time is available for eradication, treatment cost, and one's attitudes toward environmental protection.

Exercises

Use the preceding information to answer the following. Round answers to the nearest unit.

1. Approximate the number of matings at the following Celsius temperatures: (a) 9° (b) 16° (c) 6° (d) 20.5° (e) 36°
2. Convert the exponential mating formula to radical form. Now, use the square-root calculator key to perform the same calculations in Exercise 1. If your answers differ, explain why. Which form uses fewer calculator key strokes?
3. Plot your results from Exercise 1 on a rectangular coordinate system.
4. Use your answers from Exercise 1 to state the trend of mating frequency as temperature increases. Do you think this trend will continue for 36°C? Explain why or why not.
5. At what temperatures do the maximum and minimum numbers of matings occur? State your answers in degrees Celsius and degrees Fahrenheit. What are the maximum and minimum numbers of matings for this temperature range?
6. If Thomasville's average February temperature of 51.5°F is predicted to be 60°F this year, use the formula to calculate the expected percentage increase or decrease in matings this year compared with matings in typical years. How should horticulturists plan to alter their aphid-control measures?
7. If Thomasville's average October temperature of 66.0°F is predicted to be 57°F this year, use the formula to calculate the expected percentage increase or decrease in matings this year compared with matings in typical years. How should horticulturists plan to alter their aphid-control measures?

Answers

1. (a) 126 matings (b) 308 matings (c) 66 matings (d) 451 matings (e) Unknown because 36° is outside the temperature range.
2. The answers should be the same. The radical form uses fewer keystrokes.
3. See Fig. 10–3.

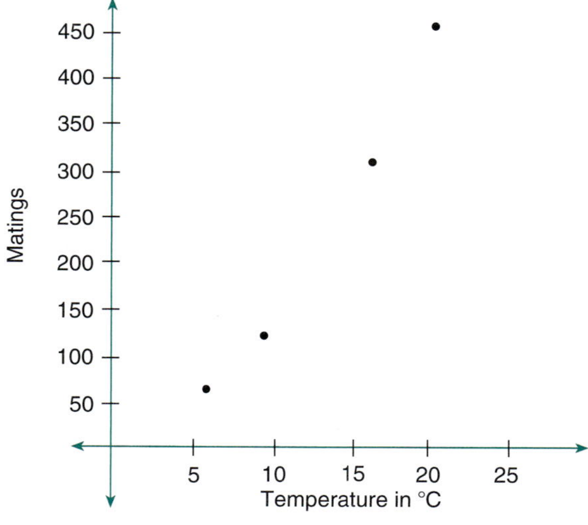

Figure 10–3

CHAPTER 10   Powers and Polynomials

4. As temperature increases, the number of matings increases. The trend is hard to predict. At 36°C (97°F), the matings will probably decrease or stop completely (especially if the humidity is also high). Matings might actually increase, however, in a frantic effort to reproduce before death.

5. The maximum number of matings is about 610, which occurs at 25°C (77°F). The minimum number of matings is about 34, which occurs at 4°C (39°F).

6. At 51.5°F (10.8°C), 168 matings are predicted. At the higher temperature of 60°F (15.6°C), 296 matings are expected. This 76% increase in aphid mating means that horticulturists should be prepared to increase their aphid-control measures.

7. At 66°F (19°C), 401 matings are predicted. At the lower temperature of 57°F (14°C), 251 matings are expected. This 37% decrease in aphid mating means that horticulturists should be prepared to decrease their aphid-control measures.

CAREER APPLICATION

Electronics: Prefixes Used in Technical Fields

Technicians work with numbers that are very large or very small. It is difficult to read and to write numbers that have many decimal places, both to the left and to the right of the decimal point. To make things easier, these numbers are usually changed to power-of-10 form, that is, the number times 10 to some power. Then the power-of-10 factor can be replaced with a letter that stands for that value.

$$9.876 \times 10^3 = 9.876 \times 1,000 = 9,876$$

The 3 is the exponent and is read as $+3$, which means that the decimal point moves three places to the right from its original location in 9.876; so you end up with 9,876. The 10^3 can be replaced with k for *kilo*, which means "thousand." If the exponent is $+6$, then the 10^6 can be replaced with M for *mega* (million), and the decimal point moves 6 places to the right. If the exponent is $+9$, then the 10^9 can be replaced with G for *giga* (billion), and the decimal point moves 9 places to the right.

$$9.876 \times 10^3 = 9.876 \text{ k} = 9.876 \times 1,000 = 9,876$$

$$9.876 \times 10^6 = 9.876 \text{ M} = 9.876 \times 1,000,000 = 9,876,000$$

$$9.876 \times 10^9 = 9.876 \text{ G} = 9.876 \times 1,000,000,000 = 9,876,000,000$$

$$9.876 \times 10^{12} = 9.876 \text{ T} = 9.876 \times 1,000,000,000,000$$

$$= 9,876,000,000,000$$

$$9.876 \times 10^{15} = 9.876 \text{ P} = 9.876 \times 1,000,000,000,000,000$$

$$= 9,876,000,000,000,000$$

When the exponent for the power-of-10 factor is a multiple of 3 and this factor is multiplied times a number equal to or more than 1 and less than 1,000, this notation is called *engineering notation*. The first factor in engineering notation can have one, two, or three digits to the left of the decimal.

Values in engineering notation can be written using prefixes and the appropriate unit of measure.

$$400 \times 10^6 \text{ hertz (hz)} = 400 \text{ Mhz (megahertz)}$$

$$2.8 \times 10^9 \text{ bytes} = 2.8 \text{ GB (gigabytes)}$$

Prefixes commonly used in engineering notation are

$$10^{15} = P = peta = quadrillion \qquad 10^{-3} = m = milli = thousandth$$

$$10^{12} = T = tera = trillion \qquad 10^{-6} = \mu = micro = millionth$$

$$10^{9} = G = giga = billion \qquad 10^{-9} = n = nano = billionth$$

$$10^{6} = M = mega = million \qquad 10^{-12} = p = pico = trillionth$$

$$10^{3} = k = kilo = thousand \qquad 10^{-15} = f = femto = quadrillionth$$

$$10^{0} = (unit = no\ prefix)$$

If the exponent is negative, the decimal point must move to the left. The prefix that replaces 10^{-3} is m for milli, and the prefix that replaces 10^{-6} is μ for micro. (The Greek letter mu, μ, was used to get an "m" sound for the prefix micro.)

$$9.876 \times 10^{-3} = 9.876\ m = 9.876 \times 0.001 = 0.009876$$

$$9.876 \times 10^{-6} = 9.876\ \mu = 9.876 \times 0.000001 = 0.000009876$$

$$9.876 \times 10^{-9} = 9.876\ n = 9.876 \times 0.000000001 = 0.000000009876$$

$$9.876 \times 10^{-12} = 9.876\ p = 9.876 \times 0.000000000001$$

$$= 0.000000000009876$$

$$9.876 \times 10^{-15} = 9.876\ f = 9.876 \times 0.000000000000001$$

$$= 0.000000000000009876$$

Exercises on Prefixes

$$P = 10^{15} \qquad T = 10^{12} \qquad G = 10^{9} \qquad M = 10^{6} \qquad k = 10^{3}$$
$$m = 10^{-3} \qquad \mu = 10^{-6} \qquad n = 10^{-9} \qquad p = 10^{-12} \qquad f = 10^{-15}$$

Convert each of the following numbers into a basic unit with no prefix. Follow the examples. The letter that follows the prefix is a unit that you keep. These are units commonly used in electronics.

1. $9{,}753$ kW = __9,753,000__ W
2. $15.68\ \mu$A = __0.00001568__ A
3. 4.567 mA = __________ A
4. 3.471 MW = __________ W
5. $65.42\ \mu$S = __________ S
6. $4{,}892$ pF = __________ F
7. 15.89 km = __________ m

8. 3.51 GV = __________ V
9. 53.2 kΩ = __________ Ω
10. 1.30 fS = __________ S
11. 5.69 nH = __________ H
12. 75.6 pF = __________ F
13. 24.56 mm = __________ m
14. $876.54\ \mu$S = __________ S

Answers for Exercises on Prefixes

3. 4.567 mA = __0.004567__ A
4. 3.471 MW = __3,471,000__ W
5. $65.42\ \mu$S = __0.00006542__ S
6. $4{,}892$ pF = __0.000000004892__ F
7. 15.89 km = __15,890__ m
8. 3.51 GV = __3,510,000,000__ V

9. 53.2 kΩ = __53,200__ Ω
10. 1.30 fS = __0.0000000000000013__ S
11. 5.69 nH = __0.00000000569__ H
12. 75.6 pF = __0.0000000000756__ F
13. 24.56 mm = __0.02456__ m
14. $876.54\ \mu$S = __0.00087654__ S

CHAPTER 10 Powers and Polynomials

Section 10–1
Perform the indicated operations. Write the answers with positive exponents.

1. $x^5 \cdot x^5$

2. $x^2(x^4)$

3. $\dfrac{x^8}{x^5}$

4. $\dfrac{x^3}{x^5}$

5. $\dfrac{x^3 y^{-1}}{x^2 y^2}$

6. $\dfrac{x^3 + y^2}{x^2 y}$

7. $(x^3)^4$

8. $(-x^3)^3$

9. $(x^{-3})^{-5}$

10. $(x^2 y^5)^2$

11. $(-3x^2)^3$

12. $\dfrac{a^2 + b^3}{ab^2}$

13. $\dfrac{x - y^3}{xy}$

Section 10–2
Simplify the following.

14. $4x^3 + 7x - 3x^3 - 5x$

15. $8x - 2x^4 - 3x^3 + 5x - x^3$

16. $5x - 3x + 7x^2 - 8x$

17. $4x^2 - 3y^2 + 7x^2 - 8y^2$

18. $4x^3(-3x^4)$

19. $-7x^8(-3x^{-2})$

20. $\dfrac{12x^5}{6x^3}$

21. $\dfrac{12x^7}{-18x^4}$

22. $\dfrac{11x^4}{22x^7}$

23. $5x(2x^2 + 3x - 4)$

24. $8x^3(2x - 6)$

25. $\dfrac{6x^3 - 12x^2 + 21x}{3x}$

26. $\dfrac{25y^5 - 85y^3 + 70y^2}{-5y}$

27. $\dfrac{4x^5}{8x^2} - 3x^2(2x^4)$

28. $2x(3x^2 y)^3$

Section 10–3
Perform the indicated operations. Express as ordinary numbers.

29. $10^5 \cdot 10^7$

30. $10^{-2} \cdot 10^8$

31. $10^7 \cdot 10^{-10}$

32. 4.2×10^5

33. $8.73 \div 10^{-3}$

34. $5.6 \div 10^{-2}$

Write in scientific notation.

35. $52{,}000$

36. 160

37. 0.00017

Perform the indicated operation and write the result in scientific notation.

38. $(4.2 \times 10^5)(3.9 \times 10^{-2})$

39. $(7.8 \times 10^{53})(5.6 \times 10^{72})$

40. $\dfrac{5.2 \times 10^8}{6.1 \times 10^5}$

41. $\dfrac{1.25 \times 10^3}{3.7 \times 10^{-8}}$

Section 10–4
42. Describe a polynomial and give an example.

43. Is the expression $5x^3 - 3x^{-2}$ a polynomial? Why or why not?

44. Arrange the following polynomials in descending order and identify the degree of each polynomial, the leading term, and the leading coefficient.
(a) $5x + 3x^3 - 8 + x^2$
(b) $3y^5 - 7y - 8y^4 + 12$

Section 10–5
Write in both rational exponent and radical notations.

45. The square root of the seventh power of x

46. The cube root of the fourth power of x

47. The square of the cube root of x

48. The fifth power of the fifth root of x

Write using rational exponents.

49. $\sqrt{x}$

50. $\sqrt[3]{x^5}$

51. $\sqrt[5]{x^4}$

52. $\sqrt[4]{9x}$

53. $\left(\sqrt[3]{xy}\right)^4$

54. $\sqrt[3]{64x^{10}}$

55. $\sqrt{7}$

Write in radical format.

56. $x^{5/8}$

57. $y^{3/5}$

58. $a^{1/4}$

Convert the radicals to equivalent expressions using rational exponents.

59. $\sqrt[3]{x}$

60. $\sqrt[4]{p}$

61. $\sqrt[5]{4y}$

62. $\left(\sqrt{ab}\right)^6$

63. $\sqrt[3]{8b^{12}}$

64. $\left(\sqrt{49x^2y^3}\right)^4$

Perform the operations. Express the answers with positive exponents in lowest terms.

65. $(a^{1/2})(a^{3/2})$

66. $(x^{4/3})(x^{2/3})$

67. $y^{3/4} \cdot y^{1/4}$

68. $y^{5/8} \cdot y^{1/8}$

69. $(3x^{1/4}y^2)^3$

70. $(2x^{3/4}y)^2$

71. $(4ax^{1/2})^3$

72. $(x^{1/2})^{1/3}$

73. $\dfrac{x^{3/4}}{x^{1/4}}$

74. $\dfrac{x^{1/6}}{x^{5/6}}$

75. $\dfrac{a^{5/6}}{a^{-1/3}}$

76. $\dfrac{a^{7/10}}{a^{2/5}}$

77. $\dfrac{x^{5/8}}{x^{3/4}}$

78. $\dfrac{y^{1/3}}{y^{5/6}}$

79. $\dfrac{a^3}{a^{1/3}}$

Evaluate the expressions in Exercises 80 to 86 for $a = 2$, $b = 1$, and $x = 3$ both before and after simplifying.

80. $\dfrac{a^2}{a^{3/5}}$

81. $\dfrac{12a^4}{6a^{1/2}}$

82. $\dfrac{27x^3}{9x^{2/3}}$

83. $\dfrac{15a^{3/5}}{10a^5}$

84. $\dfrac{14a^{5/6}}{24a^2}$

85. $a^{2.3}(a^4)$

86. $(3a^{1.2}b^2)^3$

Section 10–6

Write the numbers using the letter i. Simplify if possible.

87. $\sqrt{-100}$

88. $-\sqrt{-16x^2}$

89. $\pm\sqrt{-24y^7}$

Simplify the powers of i.

90. i^5

91. i^{14}

92. i^{98}

93. i^{77}

Write as complex numbers in simplified form.

94. 5

95. $15i$

96. $3 + \sqrt{-9}$

97. $-12i^5$

98. $-6i^{11}$

Simplify.

99. $(5 + 3i) + (2 - 7i)$

100. $(4 - i) = (3 - 2i)$

101. $\left(7 - \sqrt{-9}\right) + \left(4 + \sqrt{-16}\right)$

Section 10–7

Solve the equations containing squares and radicals. Use your calculator if needed. Evaluate to the nearest thousandth.

102. $q^2 = 81$

103. $x^2 - 36 = 0$

104. $3x^2 - 2 = 7$

105. $x^2 - 4 = 0$

106. $7 + P^2 = 107$

107. $x^2 + 4 = 0$

108. $x^2 + 81 = 0$

109. $18 = 2x^2$

110. $\sqrt{P + 2} = 12$

111. $\sqrt{\dfrac{27}{2}} = x$

112. $\sqrt{3 + x} = 14$

113. $\sqrt{q + 3} = 7$

114. $\sqrt{\dfrac{1}{4x}} = 2$

115. $\sqrt{1.3x^2} = 11.7$

116. $\sqrt{x^2 + 1} = 5$

117. $\sqrt{x^2 + 2} = 9$

118. $\sqrt{3 + y^2} = 10$

119. $\sqrt{Q^2 - 1} = 0$

120. $0 = \sqrt{z^2 - 4}$

121. $\sqrt{2 + y^2} = 8$

CHALLENGE PROBLEMS

Perform the indicated operation and express in simplest form.

122. $i(4 + 2i)$

123. $5i(3 - 4i)$

124. $12(7 + 2i)$

Perform the indicated operations. Write the answers with positive exponents.

1. $(x^4)(x)$

2. $\dfrac{x^0}{x^2}$

3. $\left(\dfrac{4}{7}\right)^2$

4. $(6a^2b)^2$

5. $\left(\dfrac{x^2}{y}\right)^2$

6. $\dfrac{12x^2}{4x^3}$

7. $4a(3a^2 - 2a + 5)$

8. $\dfrac{60x^3 - 45x^2 - 5x}{5x}$

9. $(10^3)^2$

10. $\dfrac{10^{-5}}{10^3}$

Write as ordinary numbers.

11. 42×10^3

12. 0.83×10^2

Write in scientific notation.

13. 240

14. 5.2301

15. 783×10^{-5}

Perform the indicated operations. Express the answers in scientific notation.

16. $(5.9 \times 10^5)(3.1 \times 10^4)$

17. $\dfrac{5.25 \times 10^4}{1.5 \times 10^2}$

18. A star is 3.4 light-years from Earth. If 1 light-year is 5.87×10^{12} mi, how many miles from Earth is the star?

19. The total resistance (in ohms) of a dc series circuit equals the total voltage divided by the total amperage. If the total voltage is 3×10^3 V and the total amperage is 2×10^{-3} A, find the total resistance (in ohms) expressed as an ordinary number.

Convert the radical expressions to equivalent expressions using rational exponents and simplify.

20. $\sqrt[6]{x}$

21. $\sqrt[3]{27x^{15}}$

22. $\left(\sqrt{5x^4}\right)^6$

23. $\sqrt[5]{x^{10}y^{15}z^{30}}$

Perform the operations. Express answers with positive exponents in lowest terms.

24. $a^{4/5} \cdot a^{1/5}$

25. $(125x^{1/2}y^6)^{1/3}$

26. $\dfrac{b^{3/4}}{b^{1/4}}$

27. $\dfrac{12x^{3/5}}{6x^{-2/5}}$

28. $\dfrac{r^{-1/5}s^{1/3}}{r^{3/5}s^{-5/3}}$

Simplify.

29. i^{23}

30. i^{88}

Add or subtract.

31. $(5 + 3i) - (8 - 2i)$

32. $(7 - i) + (4 - 3i)$

Solve the equations containing square-root radicals or squared terms.
Evaluate to the nearest thousandth.

33. $x^2 = 81$

34. $x^2 = 144$

35. $\sqrt{5 - x^2} = 2$

36. $8 = y^2 - 1$

37. $7 = \sqrt{Q}$

38. $\sqrt{\dfrac{x^2}{2}} = 6$

39. $\sqrt{\dfrac{6}{y}} = \sqrt{\dfrac{2}{3}}$

40. $\sqrt{4x} = 20$

41. $x^2 + 49 = 0$

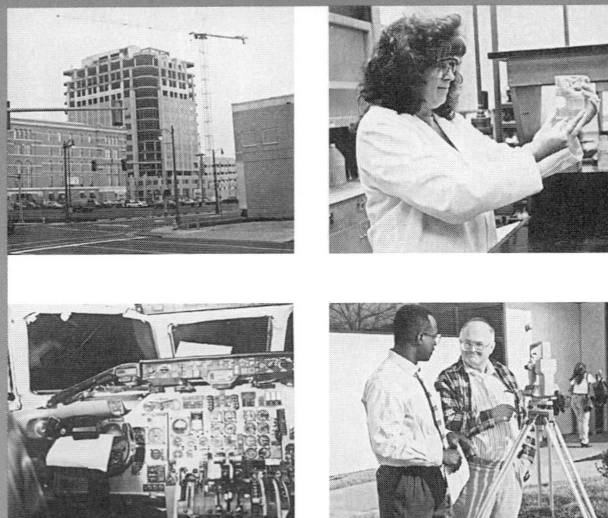

Formulas and Applications

GOOD DECISIONS THROUGH TEAMWORK

Formulas are equations we use so frequently in certain applications that they are now standard in those applications. Examples are the simple interest formula, $I = PRT$, and Ohm's law in electronics, $E = IR$. Many fields, like ballistics, accounting, manufacturing, photography, real estate, physics, mechanics, construction estimation, statistics, medicine, and engineering, use formulas to make calculations quickly and conveniently.

Your team project is to have team members investigate their major field or some other field of interest to determine what formulas are used, if any. Collect three to five formulas for each field, including the meaning of the symbols, what the formulas are used for, variations of the formulas, and how to solve the formulas or variations. You might interview instructors, recognized experts, and other professionals for this purpose.

Discuss with your team members the advantages of using the formulas, difficulties in using them, and whether computers are used with them. Discuss also why some fields may not use formulas. Then present your team's findings to your class.

6 Find the volume of prisms and cylinders.

7 Find the surface area and volume of a sphere.

8 Find the surface area and volume of a cone.

11–5 Miscellaneous technical formulas

1 Evaluate miscellaneous technical formulas.

We solved a number of applied problems with percent, area, perimeter, and interest *formulas,* although we did not always call them formulas. The procedure for work problems, rate $\times$ time = amount of work, was a formula.

Formulas are really procedures that have been used so frequently to solve certain types of problems that they have become the accepted means of solving these problems. Most formulas are expressed with one or more letter terms rather than words, and these procedures are written in an abbreviated, symbolic equation, such as $A = \pi r^2$ and $P = 2(l + w)$, the formulas for the area of a circle and the perimeter of a rectangle, respectively. Electronic spreadsheets and calculator and computer programs are developed through formulas.

11–1 FORMULA EVALUATION

Learning Outcome

1 Evaluate formulas for a given variable.

The most common use of formulas is for finding missing values. If we know values for all but one variable of a formula, we can find the unknown value.

1 Evaluate Formulas for a Given Variable.

To *evaluate* a formula is to substitute known values for some variables and perform the indicated operations to find the value of the variable in question. In so doing, we may need to use any or all of the steps and procedures for solving equations.

Many formulas involve more than one operation. When evaluating such formulas, we follow the rules for the order of operations.

Order of operations:

Perform operations in the following order as they appear from left to right.

1. Parentheses used as groupings and other grouping symbols
2. Exponents (powers and roots)
3. Multiplications and divisions
4. Additions and subtractions

This section illustrates several formula evaluations to help develop a concept of formula evaluation. Subsequent sections focus on more specific formulas and their evaluations.

The percentage proportion studied previously (Chapter 4) is actually a formula and shows how any formula written as a proportion may be evaluated. This is a good place to begin formula evaluation.

EXAMPLE   Evaluate the formula $\frac{R}{100} = \frac{P}{B}$ if $R = 20$ and $P = 10$.

$$\frac{R}{100} = \frac{P}{B}$$

$$\frac{20}{100} = \frac{10}{B} \qquad \text{Substitute values.}$$

$$20(B) = 10(100) \qquad \text{Cross multiply.}$$

$$20B = 1{,}000 \qquad \text{Divide.}$$

$$\frac{20B}{20} = \frac{1{,}000}{20}$$

$$B = 50 \qquad \text{Interpret solution.}$$

The base is 50.

The formulas in the next three examples are variations of the percentage proportion. In the first formula, the rate is isolated.

EXAMPLE   Evaluate the formula $R = \frac{100P}{B}$ if $P = \$75$ and $B = \$125$.

$$R = \frac{100P}{B} \qquad \text{Rate} = 100 \text{ times the percentage divided by the base.}$$

$$R = \frac{100(75)}{125} \qquad \text{Substitute values.}$$

$$R = \frac{7{,}500}{125} \qquad \text{Perform calculations.}$$

$$R = 60 \qquad \text{Interpret solution.}$$

The rate is 60%.

Sometimes the variable we are asked to solve for is not isolated. In this case, we substitute the given values and isolate the variable as in any equation. The next example shows the variable in the numerator of a fraction.

EXAMPLE   Solve the formula $P = \frac{RB}{100}$ for B if $P = 45$ and $R = 30$.

$$P = \frac{RB}{100} \qquad \text{Percentage} = \text{rate times base divided by 100.}$$

$$45 = \frac{30B}{100} \qquad \text{Substitute values.}$$

$$45(100) = \frac{30B}{100}(100) \qquad \text{Multiply to eliminate denominator.}$$

$$4{,}500 = \frac{30B}{\overset{1}{\underset{1}{100}}}(\cancel{100}) \qquad \text{Reduce whenever possible.}$$

$$\frac{4{,}500}{30} = \frac{30B}{30} \qquad \text{Divide.}$$

$$150 = B \qquad \text{Interpret solution.}$$

The base is 150.

The variable may also be in the denominator of a fraction.

EXAMPLE Evaluate the formula $B = \frac{100P}{R}$ if $P = 60$ m and $B = 200$ m.

$$B = \frac{100P}{R}$$

Base = 100 times percentage divided by the rate.

$$200 = \frac{100(60)}{R}$$

Substitute values.

$$200 = \frac{6,000}{R}$$

Multiply in numerator.

$$200(R) = \frac{6,000}{R}(R)$$

Multiply to eliminate denominator.

$$200R = \frac{6,000}{\cancel{R}}(\cancel{R})$$

Reduce.

$$200R = 6,000$$

$$\frac{200R}{200} = \frac{6,000}{200}$$

Divide.

$$R = 30$$

Interpret solution.

The rate is 30%.

To evaluate a formula:

1. Write the formula.
2. Rewrite the formula substituting known values for variables of the formula.
3. Solve the equation from Step 2 for the missing variable.
4. Interpret the solution within the context of the formula.

Let's look at the formula for the perimeter of a rectangle (Chapter 6). We solve for a variable that is not isolated. In this example, the variable is within parentheses and must be removed from the parentheses.

EXAMPLE Solve the formula $P = 2(l + w)$ for w if $P = 12$ ft and $l = 4$ ft.

$$P = 2(l + w)$$

Perimeter of a rectangle = two times the sum of the length and the width.

$$12 = 2(4 + w)$$

Substitute values.

$$12 = 8 + 2w$$

Apply the distributive property.

$$12 - 8 = 2w$$

Isolate the term with the variable.

$$4 = 2w$$

Combine like terms.

$$\frac{4}{2} = \frac{2w}{2}$$

Divide.

$$2 = w$$

Interpret solution.

The width is 2 ft.

The formula for the area of a circle (Chapter 6) contains a squared variable. In the next example, we solve for the squared variable.

EXAMPLE Evaluate the formula $A = \pi r^2$ for r if $A = 144$ in.^{2}

$$A = \pi r^2 \qquad \text{Area of a circle} = \pi \text{ times the radius squared.}$$

$$144 = \pi r^2 \qquad \text{Substitute values.}$$

$$\frac{144}{\pi} = \frac{\pi r^2}{\pi} \qquad \text{Isolate the variable by dividing both sides by } \pi.$$

$$\frac{144}{\pi} = r^2 \qquad \text{Take the square root of both sides.}$$

$$r = \sqrt{\frac{144}{\pi}} \qquad \text{For convenience, rewrite with } r \text{ on the left side. Use the calculator value for } \pi.$$

$$r = \sqrt{45.83662361} \qquad \text{Only the positive square root is appropriate with measures.}$$

$$r = 6.770275003 \qquad \text{Interpret solution.}$$

The radius is 6.77 in. (rounded).

A variation of the same formula for the area of a circle allows us to solve for the variable under a square-root radical.

EXAMPLE Evaluate the formula $r = \sqrt{\frac{A}{\pi}}$ for A if $r = 6.25$ cm.

$$r = \sqrt{\frac{A}{\pi}} \qquad \text{Radius} = \text{square root of the area of the circle divided by } \pi.$$

$$6.25 = \sqrt{\frac{A}{\pi}} \qquad \text{Substitute values.}$$

$$(6.25)^2 = \left(\sqrt{\frac{A}{\pi}}\right)^2 \qquad \text{Square both sides.}$$

$$39.0625 = \frac{A}{\pi} \qquad \text{Perform calculations.}$$

$$39.0625(\pi) = \frac{A}{\cancel{\pi}}\,\frac{1}{\cancel{1}}(\cancel{\pi}) \qquad \text{Eliminate denominator.}$$

$$39.0625\pi = A \qquad \text{Evaluate using the calculator value for } \pi.$$

$$122.718463 = A \qquad \text{Interpret solution.}$$

The area is 122.72 cm^2 (rounded).

The interest formula may be used to illustrate a variable that is one of several factors.

EXAMPLE Evaluate the formula $I = PRT$ for principal (P) if interest $(I) = \$94.50$, rate $(R) = 21\%$, and time $(T) = \frac{1}{2}$ year.

For convenience in using a calculator, convert $\frac{1}{2}$ year to 0.5 year. In this formula, the rate is expressed as a decimal equivalent, $21\% = 0.21$.

$$I = PRT$$

$$94.50 = P(0.21)(0.5) \qquad \text{Substitute values.}$$

$$94.50 = 0.105\,P \qquad \text{Multiply 0.21 and 0.5.}$$

$$\frac{94.50}{0.105} = \frac{0.105\,P}{0.105} \qquad \text{Divide.}$$

Chapter 11   Formulas and Applications

$$900 = P \qquad \text{Interpret solution.}$$

The principal is \$900.

SELF-STUDY EXERCISES 11–1

1 Evaluate the formulas. Use $A = 5$, $B = 6$, and $C = 8$.

1. $X = B + (3C - A)$

2. $X = A^2 - (2C - B)$

3. $X = C(B^2 + 4AC) - 3$

4. $X = \dfrac{3(C - A)}{3} + B$

5. $X = \dfrac{2(C - A)^2}{A - 1}$

Evaluate the interest formula $I = PRT$ using the following values.

6. Find the interest if $P = \$800$, $R = 15.5\%$ and $T = 2\frac{1}{2}$ years.

7. Find the rate if $I = \$427.50$, $P = \$1,500$, and $T = 2$ years.

8. Find the time if $I = \$236.25$, $P = \$750$, and $R = 10.5\%$.

9. Find the principal if $I = \$838.50$, $R = 21\frac{1}{2}\%$, and $T = 1\frac{1}{2}$ years.

Evaluate using the percentage formula $P = \frac{RB}{100}$.

10. Find the percentage if $R = 15\%$ and $B = 600$ lb.

11. Find the rate if $P = 24$ kg and $B = 300$ kg.

12. Find the base if $P = \$250$ and $R = 7.4\%$. Round to hundredths.

13. Evaluate the rate formula $R = \frac{100P}{B}$ for P if $R = 16\%$ and $B = 85$.

14. Evaluate the base formula $B = \frac{100P}{R}$ for R if $B = \$2,200$ and $P = \$374$.

15. Find the length of a rectangular work area if the perimeter is 180 in. and the width is 24 in. Use the formula $P = 2(l + w)$.

16. Find the radius of a circle whose area is 254.5 cm^2 using the formula $r = \sqrt{\frac{A}{\pi}}$. Round to the nearest tenth.

17. Find the cost (C) if the markup (M) on an item is \$5.25 and the selling price (S) is \$15.75. Use the formula $M = S - C$.

18. Using the formula for the side of a square, $s = \sqrt{A}$, find the length of a side of a square field whose area is $\frac{1}{16}$ mi^2.

19. Evaluate the formula for the area of a circle, $A = \pi r^2$, if $r = 7$ in. Round to the nearest tenth.

20. Evaluate the formula for the area of a square, $A = s^2$, if $s = 2.5$ km.

21. Using the markup formula $M = S - C$, find the selling price (S) if the markup (M) is \$12.75 and the cost (C) is \$36.

11–2   FORMULA REARRANGEMENT

Learning Outcome

1 Rearrange formulas to solve for a given variable.

We saw in Section 11–1 that we can solve a formula for any missing variable, wherever the missing variable is located in the formula. In developing formulas for spreadsheets or in programming calculators or computers, the missing values must be indicated by a variable that is isolated on the left side of the equation. This may require that the formula be rearranged.

1 Rearrange Formulas to Solve for a Given Variable.

Formula rearrangement generally refers to isolating a letter term other than the one already isolated in the formula. Solving formulas in this manner shortens our work when doing repeated formula evaluations. Since we do not evaluate the formulas after we rearrange them, we do not discuss what each formula represents. Here we are interested only in the techniques used in the rearrangement of the formulas. After we solve the formulas for the desired variable, we rewrite the formula with the variable on the left side for convenience.

The following formulas require applying the addition axiom (transposing or sorting) and careful handling of signs.

EXAMPLE Solve the formula $M = S - C$ for S.

$$M = S - C \qquad \text{Transpose to isolate } S.$$

$$M + C = S \qquad \text{The coefficient of } S \text{ is positive, so the formula is solved.}$$

$$S = M + C \qquad \text{Rewrite with } S \text{ on the left for convenience.}$$

Tip!	***Where Is the Variable or Unknown in a Formula?***

It may help to think of the letter term we are solving for as the unknown or variable and to think of the other letter terms *as if* they were the number terms in an ordinary equation.

EXAMPLE Solve the formula $M = S - C$ for C.

$$M = S - C \qquad \text{Transpose to isolate } C.$$

$$M - S = -C$$

$$\frac{M - S}{-1} = \frac{-C}{-1} \qquad \text{The coefficient of } C \text{ is negative, so divide both sides by } -1.$$

$$-M + S = C \qquad \text{Note effect on signs after division by } -1.$$

$$S - M = C \qquad \text{Write the positive term first.}$$

$$C = S - M \qquad \text{Rewrite with } C \text{ on the left for convenience.}$$

When a formula contains a term of several factors and we need to solve for one of those factors, we treat the factor being solved for as the variable and the other factors as its coefficient.

To rearrange a formula:

1. Determine which variable of the formula will be isolated (solved for).
2. Highlight or mentally locate all instances of the variable to be isolated.
3. Treat all other variables of the formula as you would numbers in an equation and perform normal techniques for solving an equation.
4. If the isolated variable is on the right side of the equation, interchange the sides so that it appears on the left side.

EXAMPLE Solve for R in the formula $I = PRT$.

$$I = PRT \qquad \text{Since we are solving for } R, PT \text{ is its coefficient.}$$

$$\frac{I}{PT} = \frac{PRT}{PT} \qquad \text{Divide both sides by the coefficient of the variable.}$$

$$\frac{I}{PT} = R \qquad \text{Rewrite with } R \text{ on the left.}$$

$$R = \frac{I}{PT}$$

Sometimes formulas contain addition (or subtraction) and multiplication in which we must use the distributive property to solve for a particular letter. In the formula $A = P(1 + ni)$, we must use the distributive property to solve for either n or i because each appears inside the grouping.

EXAMPLE Solve for n in $A = P(1 + ni)$.

$$A = P(1 + ni)$$ Identify the variable to be isolated.

$$A = P + Pni$$ Use the distributive property to remove n from parentheses.

$$A - P = Pni$$ Transpose P to isolate the term with n.

$$\frac{A - P}{Pi} = \frac{Pni}{Pi}$$ Divide both sides by the coefficient of n.

$$\frac{A - P}{Pi} = n$$ Rewrite.

$$n = \frac{A - P}{Pi}$$

Sometimes we use the distributive property *in reverse* to rearrange formulas. If we wish to solve the formula $S = FC + VC$ for C, we must first rewrite the formula by using the distributive property because the C appears in two terms. This is the same as expressing $FC + VC$ as the product of C and $(F + V)$.

EXAMPLE Solve for C in $S = FC + VC$.

$$S = FC + VC$$ Since C appears in two terms, use the distributive property to express the terms as the product of C and $(F + V)$.

$$S = C(F + V)$$

$$\frac{S}{F + V} = \frac{C(F + V)}{F + V}$$ Since we are solving for C, divide both sides by the coefficient of C, which is $(F + V)$.

$$\frac{S}{F + V} = C$$ Rewrite.

$$C = \frac{S}{F + V}$$

When the formula contains division, we eliminate the denominator before taking further steps.

EXAMPLE Solve the formula $R = \frac{V}{I}$ for V.

$$R = \frac{V}{I}$$ Identify the variable to be isolated.

$$(I)R = \frac{V}{I}(I)$$ Multiply both sides by the denominator I to eliminate it.

$$IR = V$$ Rewrite.

$$V = IR$$

When we wish to rearrange formulas written in the form of a proportion, we use the property of proportions that allows cross multiplication. Once we have cross multiplied, we use methods previously discussed to solve for the appropriate letter.

EXAMPLE Solve for T_1 in the formula $\dfrac{T_1}{T_2} = \dfrac{V_1}{V_2}$.

$$\frac{T_1}{T_2} = \frac{V_1}{V_2} \qquad \text{Identify the variable to be isolated. Watch the subscripts.}$$

$$T_1 V_2 = T_2 V_1 \qquad \text{Cross multiply.}$$

$$\frac{T_1 V_2}{V_2} = \frac{T_2 V_1}{V_2} \qquad \text{Divide both sides by the coefficient of } T_1 \text{, which is } V_2.$$

$$T_1 = \frac{T_2 V_1}{V_2}$$

To solve a formula for a letter that is squared, we first solve for the squared letter and then take the square root of both sides of the formula.

EXAMPLE Solve for a in the formula $c^2 = a^2 + b^2$.

$$c^2 = a^2 + b^2 \qquad \text{Identify the variable to be isolated.}$$

$$c^2 - b^2 = a^2 \qquad \text{Transpose } b^2 \text{ to isolate the letter term being solved for.}$$

$$\sqrt{c^2 - b^2} = \sqrt{a^2} \qquad \text{Take the square root of both sides.}$$

$$\sqrt{c^2 - b^2} = a \qquad \text{Rewrite.}$$

$$a = \pm \sqrt{c^2 - b^2}$$

To rearrange a formula with a single term on each side, when one side is a square root, we *begin* by squaring both sides of the formula. Then, we solve for the desired letter by using the same techniques used earlier to rearrange formulas.

EXAMPLE Solve for A in the formula $s = \sqrt{A}$.

$$s = \sqrt{A} \qquad \text{Identify the variable to be isolated.}$$

$$s^2 = (\sqrt{A})^2$$

Square both sides to eliminate the $\sqrt{\ }$.
Notice that $(\sqrt{A})^2 = \sqrt{A} \cdot \sqrt{A} = A$, for $A > 0$.

$$s^2 = A$$

$$A = s^2$$

SELF-STUDY EXERCISES 11–2

1 Rearrange the formulas.

1. Solve $E = X + I$ for X.

2. Solve $A = M - K$ for K.

3. Solve $S = 2\pi rh$ for r.

4. Solve $m = bx + by$ for b.

5. Solve $A = m(x + 2y)$ for y.

6. Solve $\dfrac{T_1}{T_2} = \dfrac{V_1}{V_2}$ for T_2.

7. Solve $V = \pi r^2 h$ for r.

8. Solve $m = \sqrt{XY}$ for X.

9. Solve $\dfrac{R}{100} = \dfrac{P}{B}$ for R.

10. Solve $P = 2(b + s)$ for b.

11. Solve $C = 2\pi r$ for r.

12. Solve $A = lw$ for l.

13. Solve $R = AC - BC$ for C.

14. Solve $A = s^2$ for s.

15. Solve $D = RT$ for R.

16. Solve $S = P - D$ for D.

17. Solve $A = \pi r^2$ for r.

18. Solve $a^2 + b^2 = c^2$ for c.

19. The formula for finding the amount of a repayment on a loan is $A = I + P$, where A is the amount of the repayment, I is the interest, and P is the principal. Solve the formula for interest.

20. The formula for finding interest is $I = PRT$, where I represents interest, P represents principal, R represents rate, and T represents time. Rearrange the formula to find the time.

11–3 TEMPERATURE FORMULAS

Learning Outcomes

1 Convert Fahrenheit to Celsius temperatures.

2 Convert Celsius to Fahrenheit temperatures.

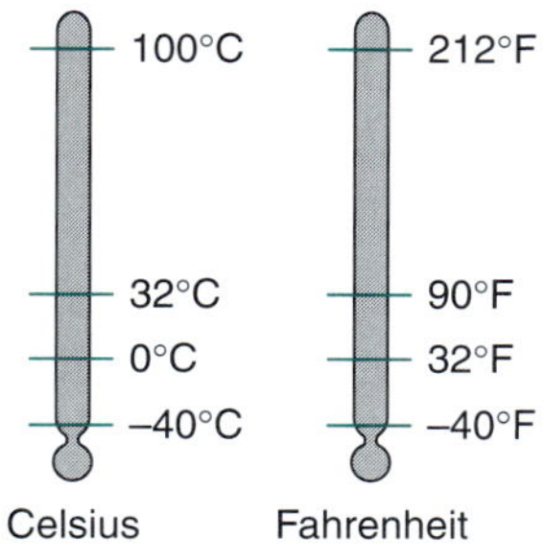

Figure 11–1

Many technical applications require conversions between different temperature scales. We are concerned with temperature conversions involving the Celsius and Fahrenheit scales.

The U.S. customary temperature scale is the *Fahrenheit* (abbreviated °F) scale, which places the freezing point of water at 32°. The metric scale used to measure temperature is the *Celsius* scale (abbreviated °C), which has 0° as its freezing point of water (Fig. 11–1).

1 **Convert Fahrenheit to Celsius Temperatures.**

The Celsius and Fahrenheit scales are the most common temperature scales used for reporting air and body temperatures. Since we still use both scales, we need to be able to convert Fahrenheit temperatures to Celsius and Celsius temperatures to Fahrenheit.

> ***To convert Fahrenheit degrees to Celsius degrees:***
>
> $$°C = \frac{5}{9}(°F - 32)$$

If we are given a temperature reading on the Fahrenheit scale and need to change it to the Celsius scale, we use the formula $°C = \frac{5}{9}(°F - 32)$.

Change 212°F (the boiling point of water) to degrees Celsius.

$$^\circ C = \frac{5}{9}(^\circ F - 32)$$

Write 212 in place of the °F in the formula.

$$^\circ C = \frac{5}{9}(212 - 32)$$

Work within groupings; subtract 212 − 32.

Recall that a number, $\frac{5}{9}$, written in front of a grouping ($^\circ F - 32$) means *multiply* the number $\frac{5}{9}$ by the value of the grouping.

$$^\circ C = \frac{5}{9}(180)$$

$\frac{5}{9}(180)$ means $\frac{5}{9} \times 180$

$$^\circ C = \frac{5}{9}\left(\frac{180}{1}\right)$$

Multiply or divide. $\frac{5}{9} \times \frac{\overset{20}{180}}{1} = 100$

$$^\circ C = 100$$

212°F = 100°C.

The order of these operations can be reviewed in Chapter 1 and Section 11–1.

According to Dr. Shotwell, an antifungal powder containing tolnaftate may be stored at a room temperature of 77°F. Change 77°F to degrees Celsius.

$$^\circ C = \frac{5}{9}(^\circ F - 32)$$

Estimation

In Figure 11–1 we see that the Celsius temperature is a smaller value than the Fahrenheit temperature for values near 77°F.

$$^\circ C = \frac{5}{9}(77 - 32)$$

Write 77 in place of °F.

$$^\circ C = \frac{5}{9}(45)$$

Work grouping.

$$^\circ C = \frac{5}{\overset{}{\underset{1}{9}}} \times \frac{\overset{5}{45}}{1}$$

Multiply. Reduce if possible.

$$^\circ C = 25$$

Interpretation

77°F = 25°C.

2 Convert Celsius to Fahrenheit Temperatures.

When a temperature is expressed in Celsius and needs to be expressed in Fahrenheit, we use the formula $^\circ F = \frac{9}{5}{}^\circ C + 32$ and follow the rules of the order of operations. When no groupings are present, multiplication is done before addition.

To convert Celsius degrees to Fahrenheit degrees:

$$^\circ F = \frac{9}{5}{}^\circ C + 32$$

EXAMPLE   Change 100°C to degrees Fahrenheit.

$$°F = \frac{9}{5}°C + 32$$

A number, $\frac{9}{5}$, written in front of a letter, °C, indicates multiplication.

$$°F = \frac{9}{5}(100) + 32$$

Write 100 in place of C.

$$°F = \frac{9}{\cancel{5}}\left(\frac{\overset{20}{\cancel{100}}}{1}\right) + 32$$

Multiply first, reducing if possible.

$$°F = 180 + 32$$

Add.

$$°F = 212$$

100°C = 212°F.

EXAMPLE   The label on a dropper bottle of ofloxacin ophthalmic solution warns that the medicine must not be stored at a temperature above 25°C. Change 25°C to degrees Fahrenheit.

$$°F = \frac{9}{5}°C + 32$$

Estimation   In Figure 11–1 we see that the Fahrenheit temperature is a larger value than the Celsius temperature for values near 25°C. In most cases, it is more than twice the Celsius temperature.

$$°F = \frac{9}{5}(25) + 32$$

Write 25 in place of C.

$$°F = \frac{9}{\cancel{5}}\left(\frac{\overset{5}{\cancel{25}}}{1}\right) + 32$$

Multiply first, reducing if possible.

$$°F = 45 + 32$$

Add.

$$°F = 77$$

Interpretation   **25°C = 77°F.**

Tip!	***Unnecessary Duplication of Formulas?***

In this section, two formulas are given instead of one. Each formula could have been derived by rearranging the other.

$$°C = \frac{5}{9}(°F - 32) \qquad °F = \frac{9}{5}°C + 32$$

Argument for Unnecessary Formulas

If formulas are easily accessible and in a format that does not cause you to lose valuable time *searching* for the appropriate formula, then all variations of a formula are handy. You can save calculation time and use calculators and computers most efficiently by selecting the formula with the variable that matches your missing information on the left.

Argument against Unnecessary Formulas

If formulas are not easily accessible and if you need to commit formulas to memory, memorize the fewest number of formulas possible.

1 Change the Fahrenheit temperatures to Celsius.

1. 95°F **2.** 32°F **3.** 113°F **4.** 41°F **5.** 59°F
6. 50°F **7.** 149°F **8.** 122°F **9.** 176°F **10.** 248°F

2 Change the Celsius temperatures to Fahrenheit.

11. 70°C **12.** 15°C **13.** 45°C **14.** 50°C **15.** 20°C
16. 215°C **17.** 310°C **18.** 410°C **19.** 185°C **20.** 0°C

11–4 GEOMETRIC FORMULAS

Learning Outcomes

1 Find the perimeter and area of a trapezoid.
2 Find the perimeter and area of a triangle.
3 Find the area of a triangle using Heron's formula.
4 Use the Pythagorean theorem to find the missing side of a right triangle.
5 Find the surface area of prisms and cylinders.
6 Find the volume of prisms and cylinders.
7 Find the surface area and volume of a sphere.
8 Find the surface area and volume of a cone.

Geometry involves the study and measurement of shapes. We have already seen in Chapter 6 four geometric figures: the square, the rectangle, the parallelogram, and the circle. We used formulas to find their perimeters and areas. These formulas are listed in Table 11–1.

TABLE 11–1 Formulas from Chapter 6

Square		
Perimeter: $P = 4s$	Area: $A = s^2$	s is the measure of one side
Rectangle		
Perimeter: $P = 2(l + w)$	Area: $A = lw$	l is length and w is width
Parallelogram		
Perimeter: $P = 2(b + s)$	Area: $A = bh$	b is base, h is height, and s is adjacent side
Circle		
Circumference: $C = \pi d$ or $2\pi r$	Area: $A = \pi r^2$	π is an irrational constant, r is the radius, and d is the diameter

Geometric layouts and figures used in work-related applications are not limited to squares, rectangles, and parallelograms. Other common figures are *trapezoids* and *triangles*.

■ Learning Strategy ***Be Realistic about Memorizing Formulas.***

Do you normally remember details or general information? Most of us remember the context of a particular conversation with a friend, but not the exact words that were spoken. The formulas you use often are generally easy to memorize, but formulas you use occasionally are easily forgotten.

Instead of memorizing formulas, focus on

- The purpose for which the formula is used
- The interpretation of each letter of that formula
- Terminology that is useful in locating a specific formula when needed
- Memorizing only formulas that are required for a specific objective (a test or exam, formulas used often on the job, and so on)
- Relating formulas to other formulas you have studied (area of parallelogram to area of rectangle, perimeter of square to perimeter of rectangle, and so on)

Formulas that are memorized but not used regularly are generally only in your short-term memory.

1 Find the Perimeter and Area of a Trapezoid.

Figure 11–2

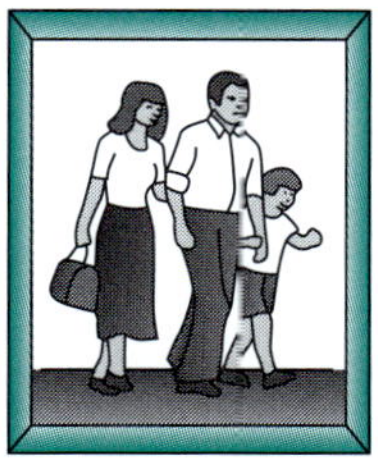

Trapezoids are used in the construction of many objects such as each side of a picture frame (illustrated in Fig. 11–2), table tops, windows, and roofs.

■ **DEFINITION: Trapezoid.** A *trapezoid* is a four-sided polygon having only two parallel sides.

Unlike a parallelogram, the trapezoid's four sides may all be of unequal size, as illustrated in Fig. 11–3.

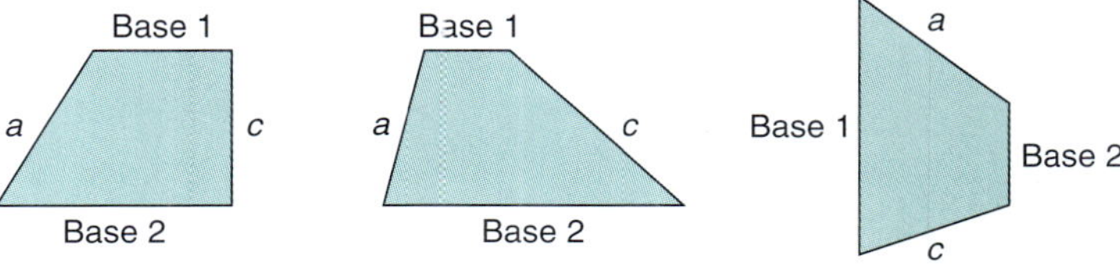

Figure 11–3

Note that the *parallel* sides are called *bases* (b_1 and b_2) with the subscripts 1 and 2 used to distinguish them. The nonparallel sides are designated side a and side c. The perimeter is the sum of the lengths of all four sides, or $b_1 + b_2 + a + c$. Since all four sides may be unequal, neither the bases nor the nonparallel sides may be combined to simplify the formula.

Perimeter of a trapezoid:

$$P = b_1 + b_2 + a + c$$

where b_1, b_2, a, and c are sides of the trapezoid.

EXAMPLE The rear window of a pickup truck is a trapezoid whose shorter base is 40 in., whose longer base is 48 in., and whose nonparallel sides are each 16 in. How many inches of rubber gasket are needed to surround the window?

$P = b_1 + b_2 + a + c$

Estimation By rounding, the perimeter is 40 + 50 + 20 + 20 or 130 in.

$$P = 40 + 48 + 16 + 16 \qquad \text{Substitute in formula.}$$

$$P = 120 \text{ in.}$$

Interpretation

The rubber gasket to surround the rear window must be 120 in. long.

The area of a trapezoid can be represented by dividing it into squares (Fig. 11–4). However, if we recognize that by placing two congruent trapezoids end to end we form a parallelogram, then the area of one trapezoid is half the area of the larger parallelogram. Since the area of a parallelogram is the height times the base, the area of a trapezoid is *one-half* the height times the "base" of the parallelogram formed by joining two equal trapezoids end to end.

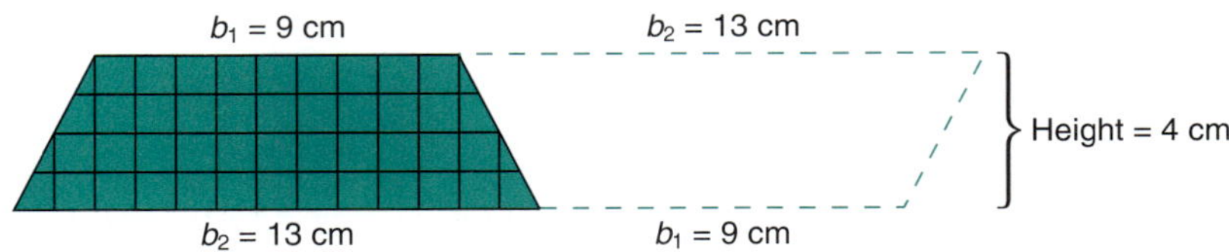

Figure 11–4

Area of a trapezoid:

$$A = \frac{1}{2}h(b_1 + b_2)$$

where b_1 is one of two parallel sides, b_2 is the other parallel side, and h is the perpendicular distance between the bases.

EXAMPLE A trapezoidal table top for special use in a reading classroom measures 60 cm on the shorter base and 90 cm on the longer base. If the height (perpendicular distance between the bases) is 50 cm, how much Formica is needed to re-cover the top?

$$A = \frac{1}{2}h(b_1 + b_2)$$

Estimation

A rectangle whose length equals the shorter base (60 cm) and height equals 50 cm has an area of (60)(50) or 3,000 cm². A rectangle whose length equals the longer base has an area of (90)(50) or 4,500 cm². The area of the trapezoid is between 3,000 cm² and 4,500 cm².

$$A = \frac{1}{2}(50)(60 + 90) \qquad \text{Substitute in formula.}$$

$$A = \frac{1}{2}(\overset{25}{\cancel{50}})(150) \qquad \text{Reduce where possible.}$$

$$A = 3{,}750 \text{ cm}^2$$

Interpretation

3,750 cm² of Formica is needed.

2   Find the Perimeter and Area of a Triangle.

■ **DEFINITION: Triangle.**   A *triangle* is a polygon that has three sides.

A triangle also has three vertices. The sides of a triangle form three angles. The *degree* is the unit we use to measure angles. The degree used as an angle measure is discussed more thoroughly in Chapter 16.

<table><tr><td>***Angle Measures of Any Triangle.***</td></tr></table>

In any triangle, no matter what its shape or size, the sum of the measures of its three angles is 180°.

Before we examine the area and perimeter of triangles, let's look at some terminology that identifies specific properties of a triangle.

Classifying Triangles by Angles

There are two primary classifications of triangles according to angles: *right triangles* and *oblique triangles*. These are triangles that have one right (90°) angle and triangles that do not have a right angle, respectively. For this discussion, see Fig. 11–5.

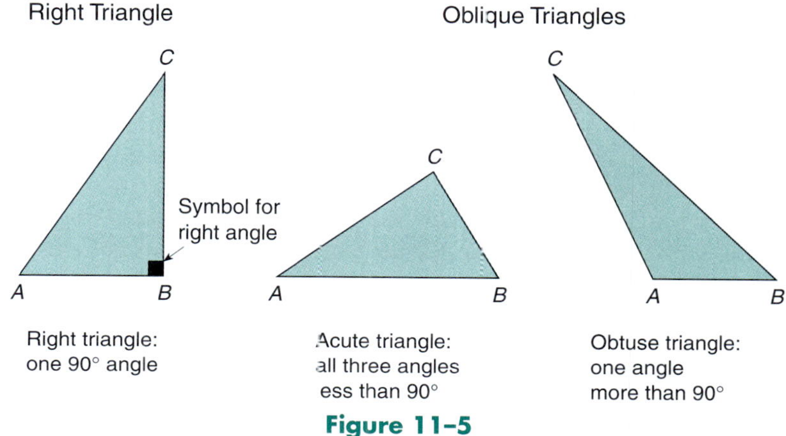

Figure 11–5

■ **DEFINITION: Right Triangle.** A *right triangle* has one right (90°) angle. The sum of the other two angles is 90°.

One side of a right triangle is used quite frequently in mathematics and deserves special attention. This side is the *hypotenuse*.

■ **DEFINITION: Hypotenuse.** The side of a right triangle that is opposite the right angle is called the *hypotenuse*.

The other sides are called *legs,* or the vertical side may be called the *altitude* and the horizontal side the *base*. The right angle is usually marked ⌐ or ⌐.

Right triangles are used in many technologies. Surveyors, carpenters, and navigators are but a few who frequently use right triangles to solve problems involving relationships between lines and angles.

■ **DEFINITION: Oblique Triangle.** An *oblique triangle* is any triangle that does not contain a right angle.

There are two kinds of oblique triangles: *acute* and *obtuse*. All the angles of an acute triangle are each less than 90°. One angle of an obtuse triangle is more than 90°. For a more thorough discussion of acute and obtuse angles and triangles, see Chapter 16.

The perimeter of a triangle is the sum of the lengths of the three sides (Fig. 11–6).

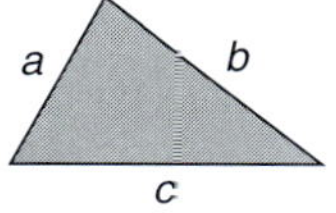

Figure 11–6

EXAMPLE The triangular gables of an apartment unit under construction are outlined in contrasting trim. If each gable is 32 ft wide and 18 ft on each side, how many feet of contrasting wood trim are needed for each gable? Disregard overlap at the corners (Fig. 11–7).

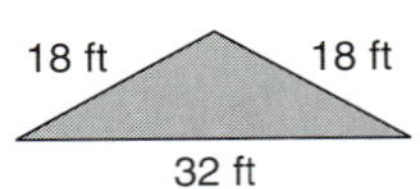

Figure 11–7

$P = a + b + c$

$P = 18 + 32 + 18$ Substitute in formula.

$P = 68$ ft

Each gable will require 68 linear feet of trim.

The area of a triangle, like the area of a trapezoid, can be found by dividing it into square units (Fig. 11–8). Notice that one side is designated as the base and that the height is the perpendicular distance from the base to the top. Like the trapezoid, the triangle forms a parallelogram if an identical triangle is placed beside the original triangle. A triangle, then, is half a parallelogram, and its area can be expressed as half the area of a parallelogram, that is, one-half the base times the height. Any side can serve as the base of a triangle. However, the height of a triangle is the perpendicular distance from the selected base to the vertex of the angle opposite the base. This vertex opposite the base is the *apex* of the triangle.

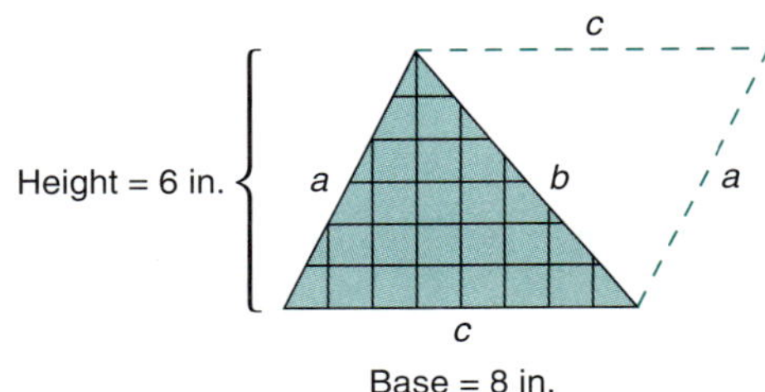

Figure 11–8

EXAMPLE A gable on a house has a rise of 7 ft 6 in. and a span of 30 ft 6 in. What is its area? The rise is the height of the gable, and the span is the base of the gable (Fig. 11–9).

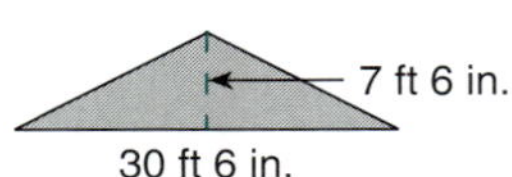

Figure 11–9

$A = \dfrac{1}{2}bh$ Substitute in formula after converting to one common unit of measure.
7 ft 6 in. = 7.5 ft, 30 ft 6 in = 30.5 ft because 1 ft = 12 in.

$A = \dfrac{1}{2}(30.5)(7.5)$

$A = 0.5(30.5)(7.5)$

$A = 114.375$ ft^2

The area of the gable is 114.375 ft^2.

In a right triangle, the sides adjacent to the hypotenuse or the sides that form the right angle are perpendicular to each other. One is the height and the other is the base. If a triangle is known to be a right triangle, the height and the base can be identified even if they are not labeled as such.

EXAMPLE Find the area of a piece of sheet metal cut as a right triangle. One side is 12 in., one is 16 in., and the third is 20 in. (Fig. 11–10).

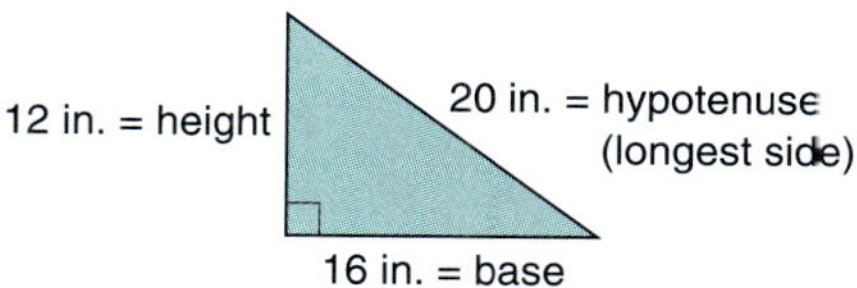

Figure 11–10

Designate one side as the base and another as the height. The longest side is the hypotenuse.

$$A = \frac{1}{2}bh$$

$$A = \frac{1}{2}(\overset{8}{\cancel{16}})(12) \qquad \text{Substitute in formula.}$$

$$A = 96 \text{ in.}^2$$

The area of the sheet metal is 96 in.²

In some triangles, the height is measured along an imaginary line outside the triangle from the base to the highest point of the triangle. The triangles illustrated in Fig. 11–11 are examples of triangles with the height measured outside the triangle itself. The area is calculated using the standard formula. Let's figure the area of $\triangle DEF$. ($\triangle$ means triangle.)

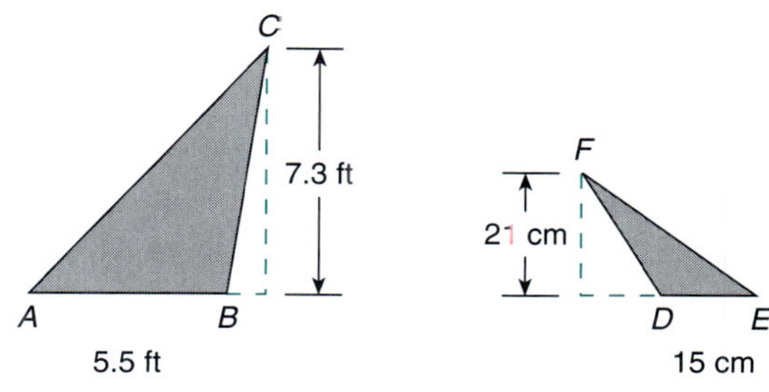

Figure 11–11

EXAMPLE The base of $\triangle DEF$ in Fig. 11–11 is 15 cm and the height is 21 cm. Find the area.

$$A = \frac{1}{2}bh$$

$$A = \frac{1}{2}(15)(21) \qquad \text{Substitute in formula.}$$

$$A = \frac{1}{2}(315)$$

$$A = 157.5 \text{ cm}^2$$

The area of $\triangle DEF$ is 157.5 cm².

3 Find the Area of a Triangle Using Heron's Formula.

When the three sides of a triangle are known and the height is not known, the area of a triangle can be found using a different formula. The formula is known as *Heron's formula.*

> **Heron's formula for the area of a triangle:**
>
> $$\text{Area} = \sqrt{s(s - a)(s - b)(s - c)}$$
>
> where $s = \frac{1}{2}(a + b + c)$ and a, b, and c are the lengths of the three sides.

EXAMPLE Find the area of the triangle in Fig. 11–12 by using Heron's formula.

Find s.

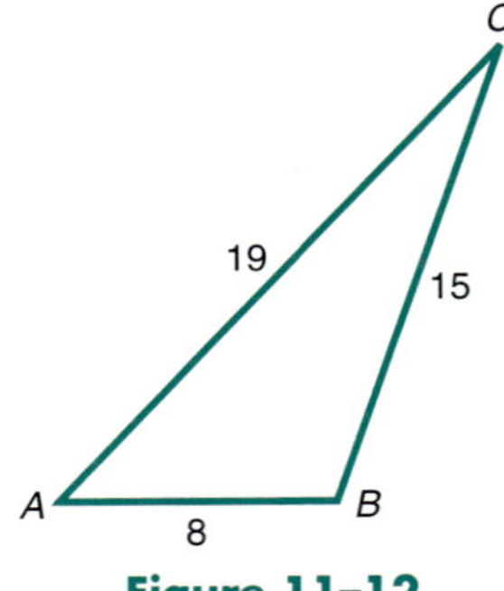

Figure 11–12

$$s = \frac{1}{2}(a + b + c) \qquad \text{Substitute.}$$

$$s = \frac{1}{2}(19 + 15 + 8)$$

$$s = 21$$

$$\text{Area} = \sqrt{s(s - a)(s - b)(s - c)} \qquad \text{Heron's formula.}$$

$$\text{Area} = \sqrt{21(21 - 19)(21 - 15)(21 - 8)} \qquad \text{Substitute.}$$

$$\text{Area} = \sqrt{21(2)(6)(13)}$$

$$\text{Area} = \sqrt{3{,}276}$$

$$\text{Area} = 57.23635209$$

Area = 57.24 square units Four significant digits.

We now have two tools or formulas that we can use to find the area of any triangle. We select the appropriate tool based on what we know and what we need to find about the triangle. If we know the base and height, we can use the familiar formula, $A = \frac{1}{2}\,bh$. If we know all three sides we can use Heron's formula, $A = \sqrt{s(s - a)(s - b)(s - c)}$.

EXAMPLE A triangular piece of property has sides that measure 120 ft long, 150 ft long, and 100 ft long. Find the area of the lot.

Find s.

$$s = \frac{1}{2}(120 + 150 + 100) = \frac{370}{2} = 185$$

$$\text{Area} = \sqrt{s(s - a)(s - b)(s - c)}$$

$$\text{Area} = \sqrt{185(185 - 120)(185 - 150)(185 - 100)} \qquad \text{Heron's formula.}$$

$$\text{Area} = \sqrt{185(65)(35)(85)}$$

$$\text{Area} = \sqrt{35{,}774{,}375}$$

$$\text{Area} = 5{,}981.168364$$

Area = 5,981 ft^2 Four significant digits.

4 **Use the Pythagorean Theorem to Find the Missing Side of a Right Triangle.**

One of the most famous and useful theorems in mathematics is the Pythagorean theorem. It is named for the Greek mathematician Pythagoras.

■ **DEFINITION: Pythagorean Theorem.** The *Pythagorean theorem* states that the square of the hypotenuse of a right triangle is equal to the sum of the squares of the two legs of the triangle.

The symbolic representation of the Pythagorean theorem follows.

Formula for Pythagorean theorem:

$$c^2 = a^2 + b^2$$

where c is the hypotenuse of a right triangle; a and b are the legs (Fig. 11–13).

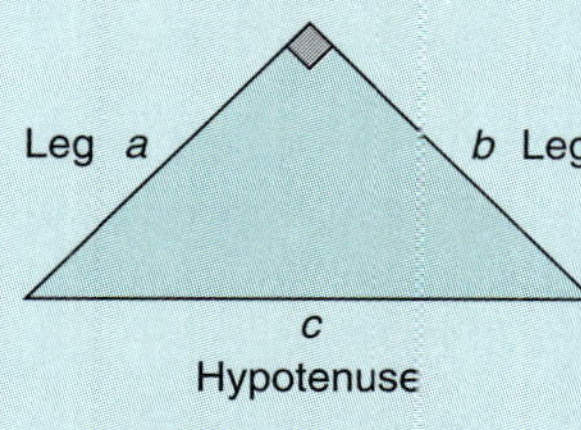

Figure 11–13

This theorem has numerous applications in technical fields. When working with the Pythagorean theorem, we will know two sides of a right triangle and can expect to find the remaining side.

Figure 11–14

EXAMPLE   In $\triangle ABC$ (Fig. 11–14), if $AB = 4$ ft and $AC = 3$ ft, find BC.

$(BC)^2 = (AC)^2 + (AB)^2$ State theorem symbolically.

$(BC)^2 = 3^2 + 4^2$ Substitute values.

$(BC)^2 = 9 + 16$

$(BC)^2 = 25$

$BC = \sqrt{25}$ Take the square root of both sides.

$BC = 5$ **ft**   Use only the positive square root.

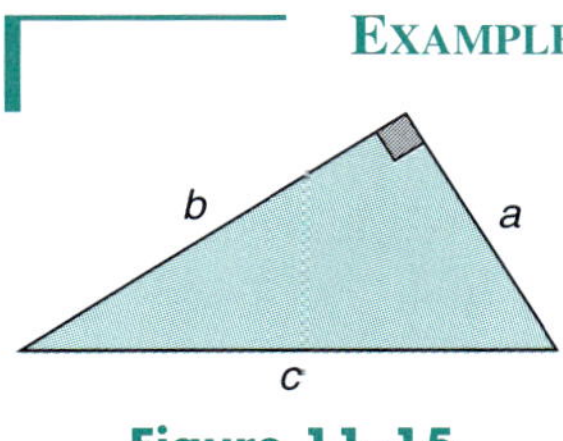

Figure 11–15

EXAMPLE   If $a = 8$ mm and $c = 17$ mm, find b (Fig. 11–15).

$c^2 = a^2 + b^2$ State theorem symbolically.

$17^2 = 8^2 + b^2$ Substitute values.

$289 = 64 + b^2$

$289 - 64 = b^2$ Transpose to isolate.

$225 = b^2$

$\sqrt{225} = b$ Take the square root of both sides.

15 mm $= b$

Many times we are given problems that contain "hidden" triangles. In these cases, we need to visualize the triangle or triangles in the problems. Drawing one or more of the sides of the "hidden" triangle helps solve the problem.

EXAMPLE Find the center-to-center distance between pulleys A and C (Fig. 11–16).

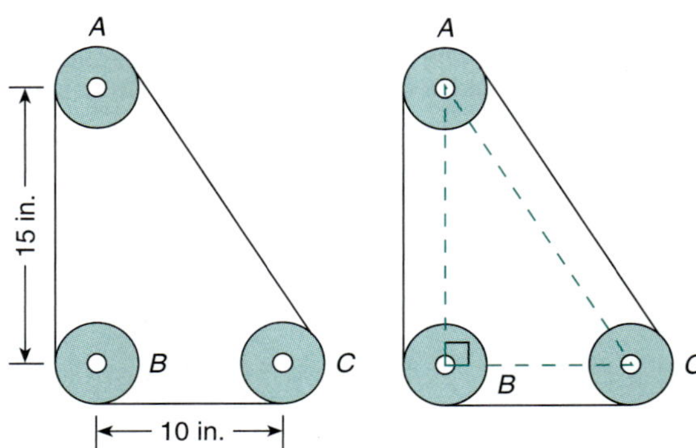

Figure 11–16

Connect the center points of the three pulleys to form a right triangle. The hypotenuse is the distance between pulleys A and C. Use the Pythagorean theorem and substitute the given values for the two known sides.

$(AC)^2 = (AB)^2 + (BC)^2$ — State theorem symbolically.

$(AC)^2 = 15^2 + 10^2$ — Substitute values.

$(AC)^2 = 225 + 100$

$(AC)^2 = 325$ — Take the square root of both sides.

$AC = \sqrt{325}$

$AC = 18.0277564$

The distance from pulley A to pulley C is 18.0 in. (to the nearest tenth).

EXAMPLE The head of a bolt is a square 0.5 in. on a side (distance across flats). What is the distance from corner to corner (distance across corners)? (See Fig. 11–17.)

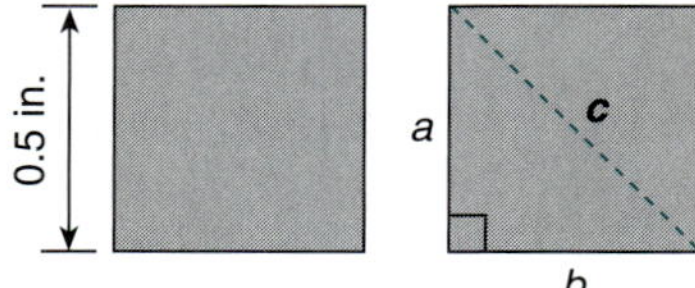

Figure 11–17

The sides of the bolt head form the legs of a triangle if a diagonal line is drawn from one corner to the opposite corner. This diagonal forms the hypotenuse of the right triangle. Because the legs of the triangle are known to be 0.5 in. each, we can substitute in the Pythagorean theorem to find the diagonal line, the hypotenuse, which is the distance across corners.

$c^2 = a^2 + b^2$ — State theorem symbolically.

$c^2 = 0.5^2 + 0.5^2$ — Substitute values.

$c^2 = 0.25 + 0.25$

$$c^2 = 0.5 \qquad \text{Take the square root of both sides.}$$

$$c = \sqrt{0.5}$$

$$c = 0.707106781$$

The distance across corners is 0.71 in. (to the nearest hundredth).

Sometimes right triangles are used in technical fields to represent certain relationships, such as forces acting on an object at right angles or electrical and electronic phenomena related in the way the three sides of a right triangle are related. Let's look at an example.

EXAMPLE Forces A and B come together at a right angle to produce force C (Fig. 11–18). If force A is 74.8 lb and the resulting force C is 91.5 lb, what is force B?

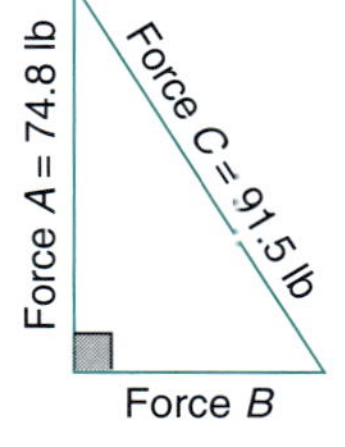

Figure 11–18

Since the forces are related in the way the sides of a right triangle are related, we may use the Pythagorean theorem to find the missing force B, a leg of the triangle.

$$A^2 + B^2 = C^2 \qquad \text{State theorem symbolically.}$$

$$74.8^2 + B^2 = 91.5^2 \qquad \text{Substitute values.}$$

$$5{,}595.04 + B^2 = 8{,}372.25 \qquad \text{Transpose to isolate.}$$

$$B^2 = 8{,}372.25 - 5{,}595.04$$

$$B^2 = 2{,}777.21 \qquad \text{Take the square root of both sides.}$$

$$B = \sqrt{2{,}777.21}$$

$$B = 52.7 \qquad \text{To the nearest tenth.}$$

Force B is 52.7 lb.

5 Find the Surface Area of Prisms and Cylinders.

Common household items like ice cubes, cardboard storage boxes, and toy building blocks are examples of three-dimensional geometric figures classified generally as *prisms*. Cans and pipes are examples of *cylinders*.

■ **DEFINITION: Prism.** A *prism* is a three-dimensional figure with bases (ends) that are parallel, congruent polygons and faces (sides) that are parallelograms, rectangles, or squares. In a *right prism*, the faces are perpendicular to the bases.

■ **DEFINITION: Cylinder.** A right circular *cylinder* is a three-dimensional figure with a curved surface and two circular bases such that the height is perpendicular to the bases.

■ **DEFINITION: Height.** The *height* of a three-dimensional figure with two bases is the shortest distance between the two bases.

In right circular cylinders and in right prisms, the height is the same as the length of a side or face. However, in oblique prisms and cylinders the height is the perpendicular distance between the bases and is different from the length of a side or face (Fig. 11–19).

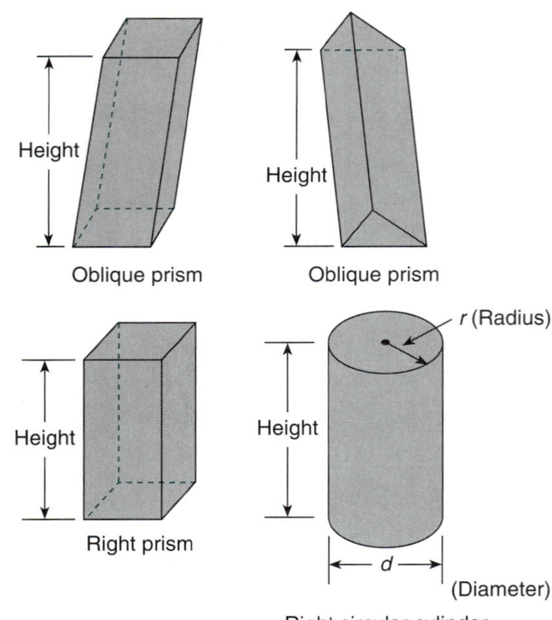

Figure 11–19

Tip!	***Three-Dimensional Figure Versus Solid.***

Sometimes three-dimensional figures are referred to as solids.

When we refer to a figure as a solid, it may not be solid at all, but it has length, width, and height. A plane figure, on the other hand, is two-dimensional.

Some jobs require us to find the area of a three-dimensional figure. The area of a three-dimensional figure can refer to just the area of the *sides* of the figure. Or area can refer to the overall area, including the bases along with the sides.

■ **DEFINITION: Lateral Surface Area.**   The *lateral surface area* (LSA) of a three-dimensional figure is the area of its sides only.

■ **DEFINITION: Total Surface Area.**   The *total surface area* (TSA) of a three-dimensional figure is the area of the sides plus the area of its base or bases.

To find the lateral surface area, we find the sum of the areas of each side using the formulas from Chapter 6. However, we can also find the lateral surface area (LSA) of a prism or cylinder by multiplying the perimeter of the base times the height of the solid. In the formulas and examples that follow, only *right* prisms and *right* circular cylinders are considered.

Lateral surface area of a right prism or cylinder:

$$\text{LSA} = ph$$

where *p* is the perimeter of the base, and *h* is the height of the three-dimensional figure.

To get the total surface area (TSA), add the areas of the two bases to the lateral surface area.

> **Total surface area of a right prism or cylinder:**
>
> $$\text{TSA} = ph + 2B$$
>
> where p is the perimeter of base, h is the height of the three-dimensional figure, and B is the area of the base.

EXAMPLE Find the lateral surface area of a rectangular shipping carton measuring 24 in. in length, 12 in. in width, and 20 in. in height (Fig. 11–20).

$\text{LSA} = ph$

$\text{LSA} = (2l + 2w)h$ — Since the base is a rectangle, its perimeter is $2l + 2w$.

$\text{LSA} = [2(24) + 2(12)]20$ — Substitute numerical values.

$\text{LSA} = [72]20$

$\text{LSA} = 1{,}440 \text{ in.}^2$

The lateral surface area of the carton is 1,440 in.^{2}

20 in.
12 in.
24 in.

Figure 11–20

EXAMPLE How many square centimeters of sheet metal are required to manufacture a can that has a radius of 4.5 cm and height of 9 cm? Assume no waste or overlap.

$\text{TSA} = ph + 2B$ — Total surface area is needed.

$\text{TSA} = 2\pi rh + 2\pi r^2$ — $p = 2\pi r, \ B = \pi r^2$.

$\text{TSA} = 2(\pi)(4.5)(9) + 2(\pi)(4.5)^2$ — Substitute values.

$\text{TSA} = 254.4690049 + 127.2345025$

$\text{TSA} = 381.70 \text{ cm}^2$ — Rounded.

The can requires 381.70 cm^2 of sheet metal.

EXAMPLE Find the total surface area of the triangular prism shown in Fig. 11–21.

$\text{TSA} = ph + 2B$

$\text{TSA} = 9(15) + (2)\left(\dfrac{1}{2}\right)(3)(2.6)$ — Perimeter of triangular base is $3 + 3 + 3 = 9$ cm. Area of triangular base is $\frac{1}{2}bh$, or $\frac{1}{2}(3)(2.6)$.

$\text{TSA} = 135 + 7.8$

$\text{TSA} = 142.8 \text{ cm}^2$

The total surface area of the triangular prism is 142.8 cm^2.

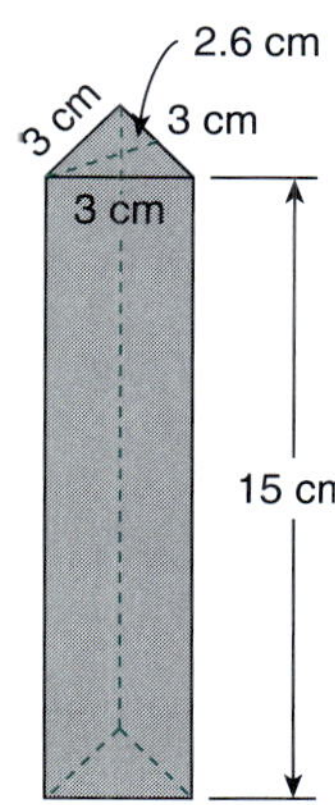

Figure 11–21

6 Find the Volume of Prisms and Cylinders.

At times we need to know the *volume* of an object, such as a container, to estimate, for example, how many of these containers can be loaded into a given size storage area or shipped in a tractor-trailer rig of certain dimensions.

■ **DEFINITION:** **Volume.** The *volume* of a three-dimensional geometric figure is the amount of space it occupies, measured in terms of three dimensions (length, width, and height).

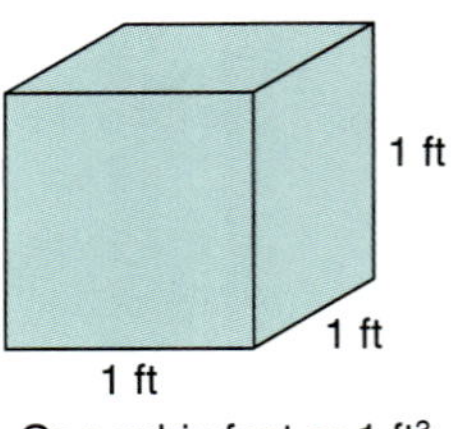

1 ft

1 ft

1 ft

One cubic foot or 1 ft³

Figure 11–22

If we have a rectangular box measuring 1 ft long, 1 ft wide, and 1 ft high (Fig. 11–22), it will be a cube representing 1 cubic foot (ft^3). Its volume is calculated by multiplying length $\times$ width $\times$ height, or 1 ft $\times$ 1 ft $\times$ 1 ft = 1 ft^3. We indicate a cubic measure with an exponent 3 after the unit of measure, meaning that the measure is *"cubed."*

From this concept of 1 ft^3 or 1 cubic foot comes the formula for the volume of a rectangular box as length $\times$ width $\times$ height or $V = lwh$. Note that $l \times w$ is the formula for the area of the rectangle (or square) that forms the base of the rectangular box. If the base of the prism is a triangle, pentagon, hexagon, or other polygon, we use the appropriate formula for its area. In the case of a cylinder, we use the formula for the area of a circle because its base is a circle. The general formula for the volume of *any* right prism or cylinder follows.

Volume of right prism or cylinder:

$$V = Bh$$

where B is the area of the base, and h is the height of the prism or cylinder.

EXAMPLE Find the volume of the triangular prism in the preceding example, if the height is 15 cm and the bases are triangles 3 cm on a side and 2.6 cm in height.

$V = Bh_2$ Substitute the formula for the area of the triangular base for B.

$V = \left(\dfrac{1}{2}bh_1\right)h_2$ h_1 = 2.6 cm (height of prism base), h_2 = 15 cm (height of prism)

$V = \left[\dfrac{1}{2}(3)(2.6)\right]15$ Substitute values.

$V = 58.5 \text{ cm}^3$

The volume of the prism is 58.5 cm³.

EXAMPLE What is the cubic-inch displacement (space occupied) of a cylinder whose diameter is 5 in. and whose height is 4 in.?

$V = Bh$ Substitute the formula for the area of the circular base for B.

$V = \pi r^2 h$ Substitute values; $r = \frac{1}{2}$ diameter, or 2.5.

$V = \pi(2.5)^2(4)$

$V = 78.5 \text{ in.}^3$ Rounded.

The cylinder displacement is 78.5 in.³.

7   ## Find the Surface Area and Volume of a Sphere.

Soccer balls, golf balls, tennis balls, baseballs, and ball bearings are *spheres*. Spheres are also used as tanks to store gas and water because spheres hold the greatest volume for a specified amount of surface area.

■   **DEFINITION: Sphere.**   A *sphere* is a solid formed by a curved surface whose points are all equidistant from a point inside called the *center* (Fig. 11–23).

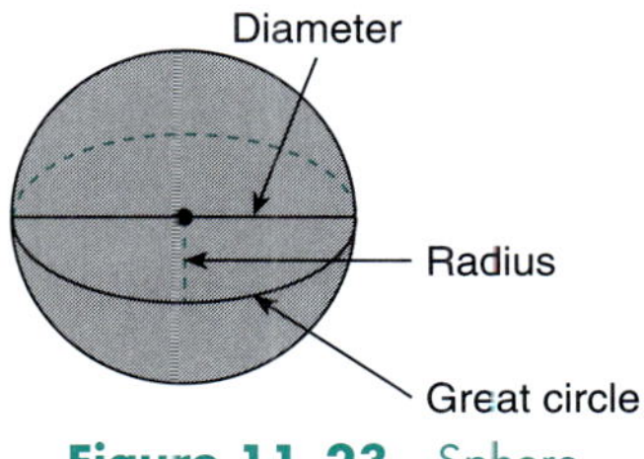

Figure 11–23   Sphere

■   **DEFINITION: Great Circle.**   The *great circle* divides the sphere in half and is formed by a plane through the center of the sphere.

A sphere does not have bases like prisms and cylinders. The surface area of a sphere includes *all* the surface so there is only one formula. Because of the relationship of the sphere to the circle, the formula includes elements of the formula for the area of a circle. The total surface area of the sphere is 4 times the area of a circle with the same radius.

Total surface area of a sphere:

$$\text{TSA} = 4\pi r^2$$

where r is the radius.

The formula for the volume of a sphere also contains elements found in formulas for a circle.

Volume of a sphere:

$$V = \frac{4\pi r^3}{3}$$

where r is the radius.

Note that the radius is *cubed,* or raised to the power of 3, indicating volume.

EXAMPLE   Find the surface area and volume of a sphere whose diameter is 90 cm.

$\text{TSA} = 4\pi r^2$

$\text{TSA} = 4(\pi)(45)^2$ Substitute values. $\frac{1}{2}d = r$, so $r = 45$.

$\textbf{TSA} = \textbf{25,447 cm}^2$ Rounded

$V = \dfrac{4\pi r^3}{3}$

$$V = \frac{4(\pi)(45)^3}{3}$$ Substitute values.

$$V = 381{,}704 \text{ cm}^3$$ Rounded

8 Find the Surface Area and Volume of a Cone.

One example of a *cone* is the funnel or the circular rain cap placed on top of stove vent pipes extending through the roof of some homes.

■ **DEFINITION: Cone.** A right *cone* is a solid whose base is a circle and whose side surface tapers to a point, called the *vertex* or *apex,* and whose height is a perpendicular line between the base and apex (Fig. 11–24).

Figure 11–24 shows the perpendicular height and the *slant height* of a cone.

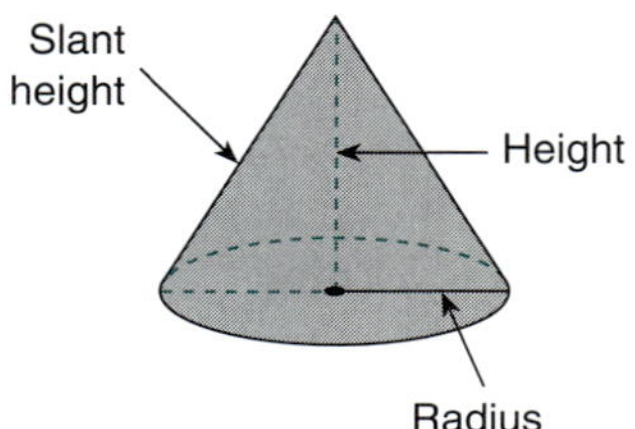

Figure 11–24 Circular cone

■ **DEFINITION: Slant Height.** The *slant height* of a cone is the distance along the side from the base to the apex.

The lateral surface area of a cone equals the circumference of the base times $\frac{1}{2}$ the slant height, or $\text{LSA} = 2\pi r \frac{s}{2}$, which simplifies as πrs.

Lateral surface area of a cone:

$$\text{LSA} = \pi rs$$

where r is the radius, and s is the slant height.

The total surface area, then, is the lateral surface area plus the area of the base.

Total surface area of a cone:

$$\text{TSA} = \pi rs + \pi r^2$$

where r is the radius of the circular base, s is the slant height, and πr^2 is the area of the base.

The volume of a cone equals one-third the area of the base times the *height of the cone* (*not* the slant height).

Volume of a cone:

$$V = \frac{\pi r^2 h}{3}$$

where πr^2 is the area of the circular base, and h is the height of the cone.

Find the lateral surface area, total surface area, and volume of a cone whose diameter is 8 cm, height is 6 cm, and slant height is 7 cm. Round to hundredths.

$$\text{LSA} = \pi rs$$

$$\text{LSA} = (\pi)(4)(7)$$ Substitute values: $r = \tfrac{1}{2}d$, or 4.

$$\textbf{LSA} = \textbf{87.96 cm}^2$$ Rounded from 87.9645943.

$$\text{TSA} = \pi rs + \pi r^2$$

$$\text{TSA} = 87.9645943 + (\pi)(4)^2$$ Substitute values.

$$\textbf{TSA} = \textbf{138.23 cm}^2$$ Rounded.

$$V = \frac{\pi r^2 h}{3}$$

$$V = \frac{(\pi)(4)^2(6)}{3}$$ Substitute values.

$$\textbf{\emph{V}} = \textbf{100.53 cm}^3$$ Rounded.

EXAMPLE Find the weight of the cast-iron solid shown in Fig. 11–25 if cast iron weighs 0.26 lb per cubic inch. Round to the nearest whole pound.

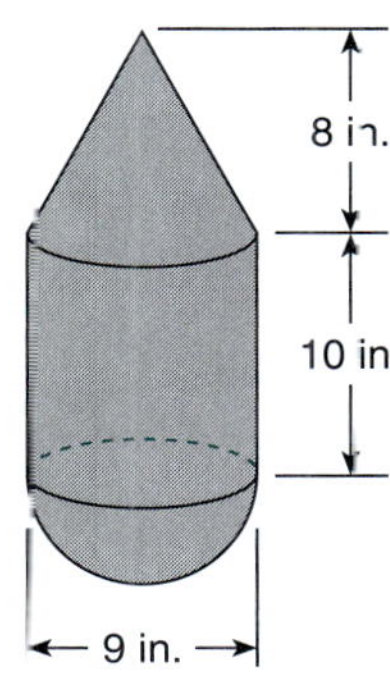

Figure 11–25

The solution requires finding the volume of the cone that forms the top of the solid, the volume of the cylinder that forms the middle portion of the solid, and the volume of the *hemisphere* (half sphere) that forms the bottom of the solid.

$$V_{\text{cone}} = \frac{\pi r^2 h}{3} \qquad V_{\text{cylinder}} = \pi r^2 h \qquad V_{\text{hemisphere}} = \frac{1}{2}\left[\frac{4\pi r^3}{3}\right]$$

$$V_{\text{cone}} = \frac{(\pi)(4.5)^2(8)}{3} \qquad V_{\text{cylinder}} = (\pi)(4.5)^2(10) \qquad V_{\text{hemisphere}} = \frac{1}{2}\left[\frac{4(\pi)(4.5)^3}{3}\right]$$

$$V_{\text{cone}} = 169.6460033 \text{ in.}^3 \qquad V_{\text{cylinder}} = 636.1725124 \text{ in.}^3 \qquad V_{\text{hemisphere}} = 190.8517537 \text{ in.}^3$$

$$\text{Total volume} = V_{\text{cone}} + V_{\text{cylinder}} + V_{\text{hemisphere}}$$

$$\text{Total volume} = 996.6702694 \text{ in.}^3$$

Convert to pounds:

$$\frac{996.6702694 \text{ in.}^3}{1} \times \frac{0.26 \text{ lb}}{1 \text{ in.}^3} = 259 \text{ lb} \qquad \text{Rounded}$$

To the nearest whole pound, the solid cast-iron figure weighs 259 lb.

1 Find the perimeter and the area of Figs. 11–26 and 11–27. Round to hundredths.

1.

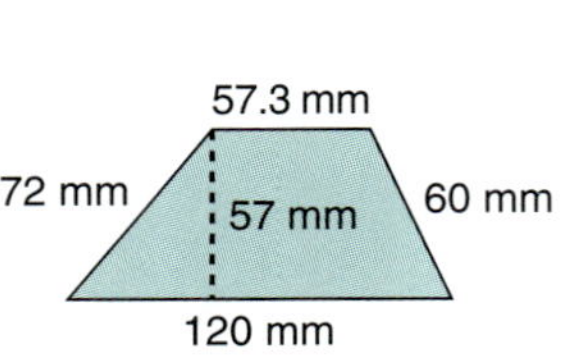

Figure 11–26

2.

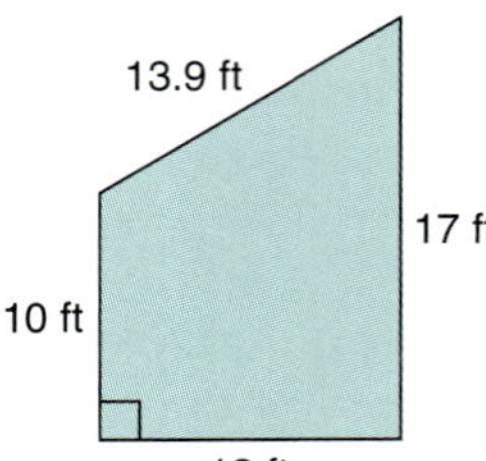

Figure 11–27

Solve these problems involving perimeter and area.

3. Each of six glass panes in a kitchen light fixture measures $4\frac{1}{2}$ in. along the top and 10 in. along the bottom. The top and bottom are parallel. The height of each pane is 8 in. What is the combined area of the six trapezoidal panes?

4. A section of a hip roof is a trapezoid measuring 38 ft at the bottom, 14 ft at the top, and 10 ft high. Find the area of this section of the roof in square feet.

5. A lot in an urban area is 60 ft wide. The sides are 120 ft and 154 ft, and they are perpendicular to the width of the lot. Find the area of the trapezoidal property.

6. A swimming pool is fashioned in a trapezoidal design. The parallel sides are $18\frac{1}{2}$ ft and 31 ft. The other sides are $24\frac{1}{2}$ ft and 24 ft. What is the perimeter of the pool?

2 Find the perimeter and area of the triangles in Figs. 11–28 and 11–29. Round to hundredths.

7.

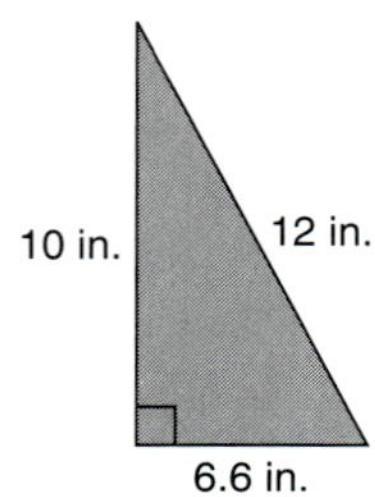

Figure 11–28

8.

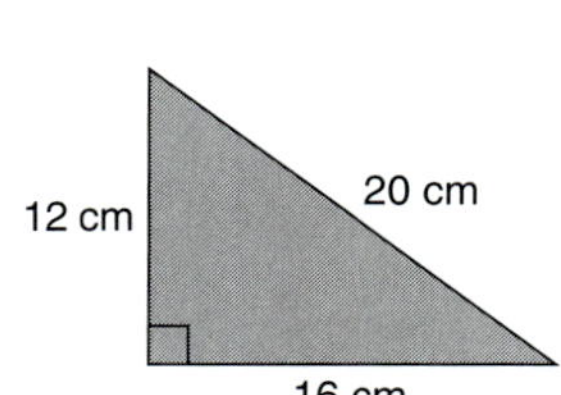

Figure 11–29

9. Find the area of the triangle in Fig. 11–30.

Figure 11–30

Solve these problems involving perimeter and area.

10. Dave Long is planning a patio that will adjoin the sides of his L-shaped home. One side of the home is 24 ft and the other is 18 ft. The shape of the patio is triangular. Draw a representation of the patio and find the perimeter if its hypotenuse is 30 ft. Find the number of square feet of surface area to be covered with concrete.

11. A louver, or triangular vent, for a gable roof measures 6 ft 6 in. wide and stands 3 ft high. What is the area in square feet of the vented portion of the gable?

12. A mason lays tile in the form of a triangle. The height of the triangle is 12 ft and the base is 5 ft. What is the area?

13. A metal worker cuts a triangular plate with a base of 11 in. and a height of $4\frac{1}{2}$ in. from a piece of metal. What is the area of the plate?

14. If aluminum siding costs \$6.75 a square yard installed, how much does it cost to put the siding on the two triangular gable ends of a roof under construction? Each gable has a span (base) of 30 ft 6 in. and a rise (height) of 7 ft 6 in. Any portion of a square yard is rounded to the next highest square yard.

15. Find the height of a triangle if its base is 15 ft 6 in. and its area is 62 ft^2.

16. Find the perimeter of a triangle that has sides measuring 2 ft 6 in., 1 yd 8 in., and 4 ft 6 in.

3 Find the area of the triangles shown in Figs. 11–31 to 11–34. Use Heron's formula. Round to four significant digits.

17.

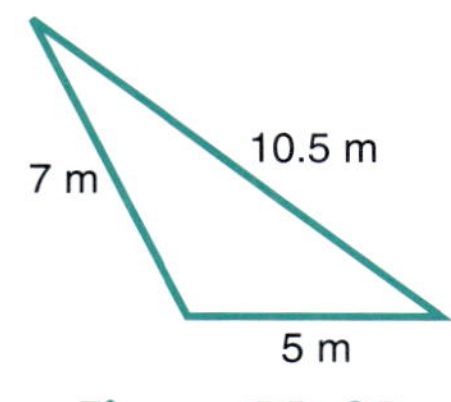

Figure 11–31

18.

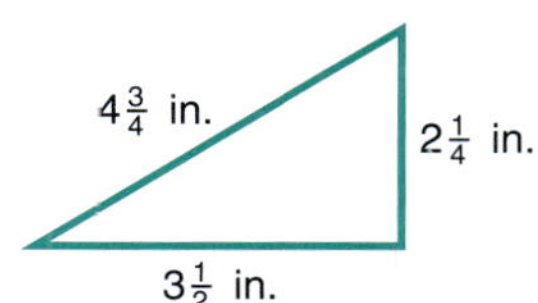

Figure 11–32

19.

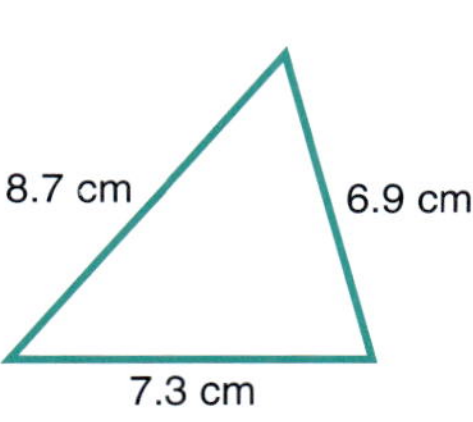

Figure 11–33

20.

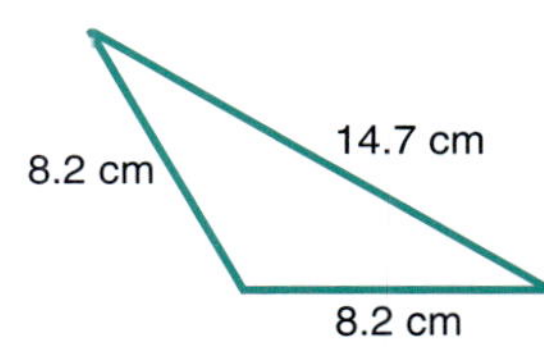

Figure 11–34

4 Find the missing side of the right triangle. Round the final answers to the nearest thousandth. Use Fig. 11–35 for Exercises 21–26.

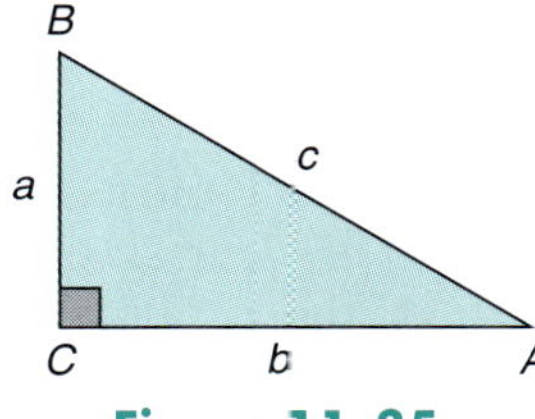

Figure 11–35

21. $AC = ?$
$BC = 7$ cm
$AB = 25$ cm

22. $AC = 24$ mm
$BC = ?$
$AB = 26$ mm

23. $AC = 15$ yd
$BC = 8$ yd
$AB = ?$

24. $a = 5$ cm
$b = 4$ cm
$c = ?$

25. $a = ?$
$b = 9$ m
$c = 11$ m

26. $a = 9$ ft
$b = ?$
$c = 15$ ft

Solve the problems. Round final answers to the nearest thousandth if necessary.

27. A light pole will be braced with a wire that is to be tied to a stake in the ground 18 ft from the base of the pole, which extends 26 ft above the ground. If the wire is attached to the pole 2 ft from the top, how much wire must be used to brace the pole? (Fig. 11–36.)

28. Find the center-to-center distance between holes A and C in a sheet metal plate if the distance between the centers of A and B is 16.5 cm and the distance between the centers of B and C is 36.2 cm (Fig. 11–37).

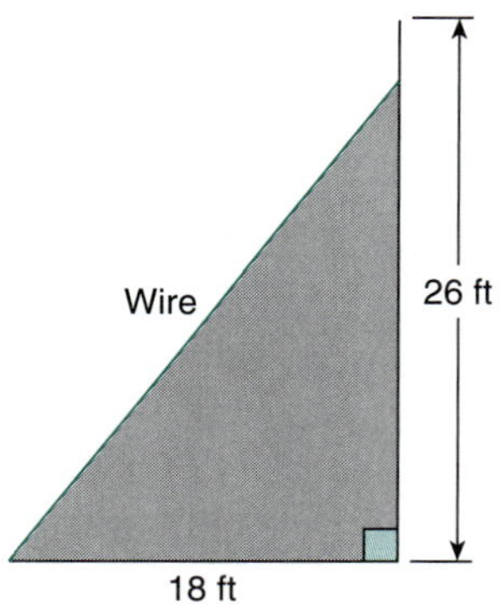

Figure 11–36

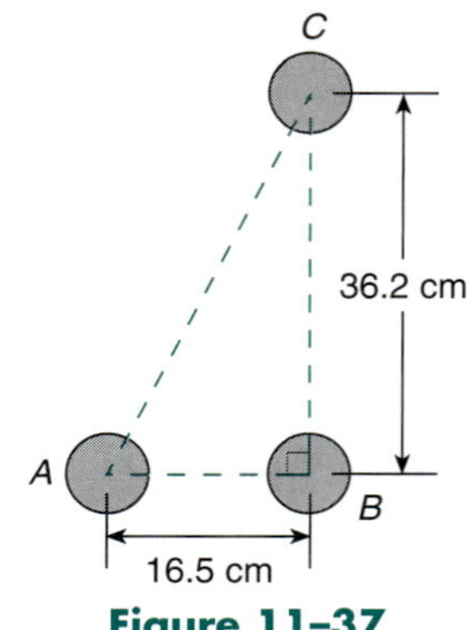

Figure 11–37

29. Find the length of a rafter that has a 10-in. overhang if the rise of the roof is 10 ft and the joists are 48 ft long (Fig. 11–38).

30. A stair stringer is 8 ft high and extends 10 ft from the wall (Fig. 11–39). How long will the stair stringer be?

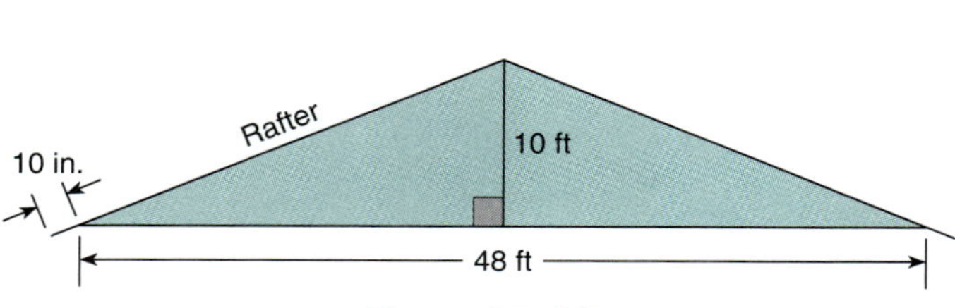

Figure 11–38

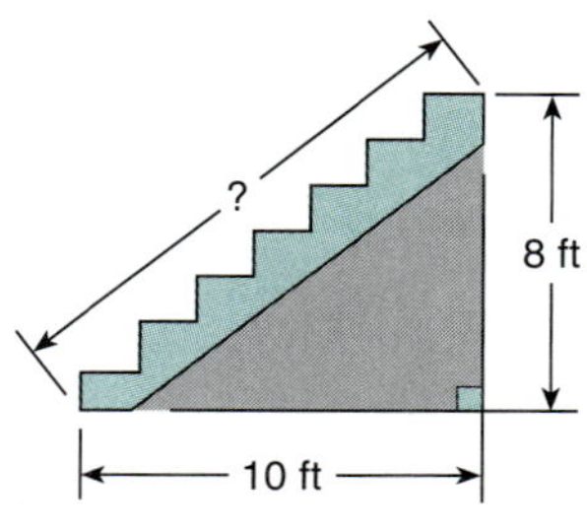

Figure 11–39

31. A machinist wishes to strengthen an L bracket that is 5 cm by 12 cm by welding a brace to each end of the bracket (Fig. 11–40). How much metal rod is needed for the brace?

32. To make a rectangular table more stable, a diagonal brace is attached to the underside of the table surface. If the table is 27 dm by 36 dm, how long is the brace?

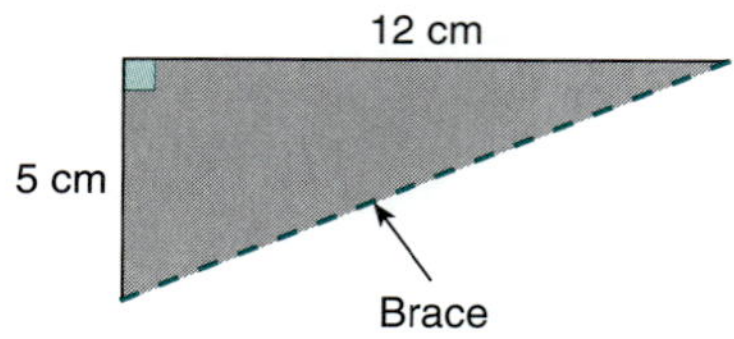

Figure 11–40

33. Find the length of the side of the largest square nut that can be milled from a piece of round stock whose diameter is 15 mm (Fig. 11–41).

34. A rigid length of electrical conduit must be shaped as shown in Fig. 11–42 to clear an obstruction. What total length of the conduit is needed? (*Hint:* Don't forget to include *AB* and *CD* in the total length.)

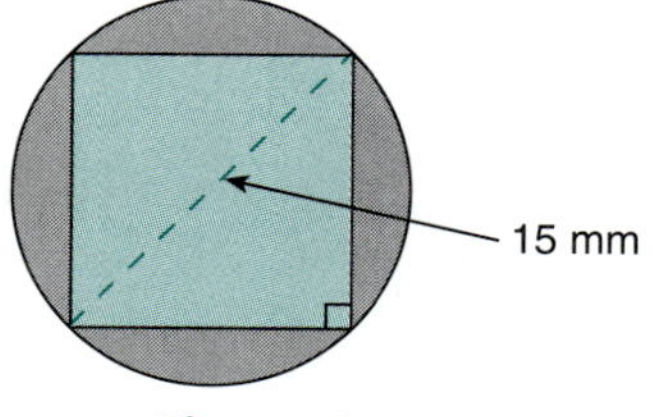

Figure 11–41

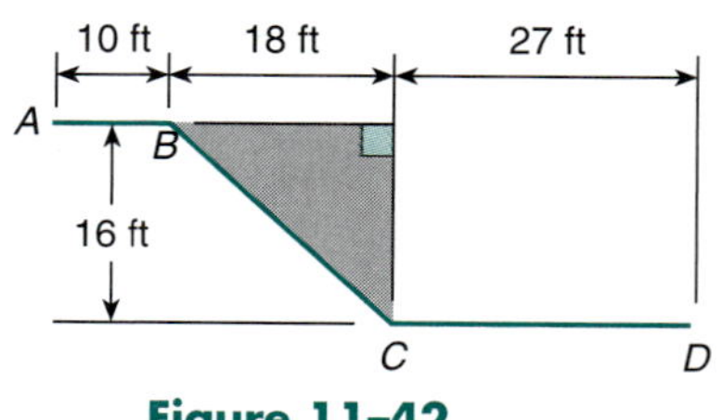

Figure 11–42

35. The vector diagram in Fig. 11–43 is used in electrical applications. Find the voltage of E_a.

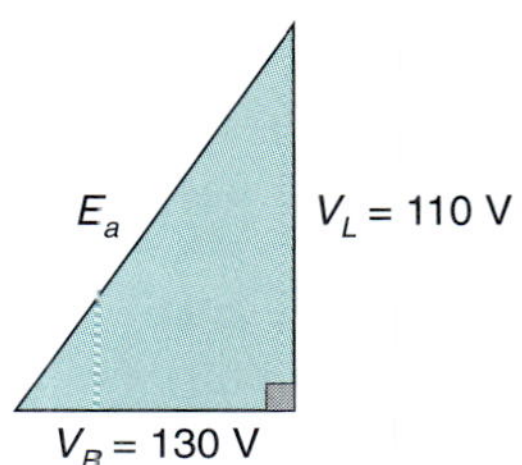

Figure 11–43

5 Find the lateral surface area and total surface area of the three-dimensional figures in Figs. 11–44 and 11–45. Round to the nearest hundredth if necessary.

36.

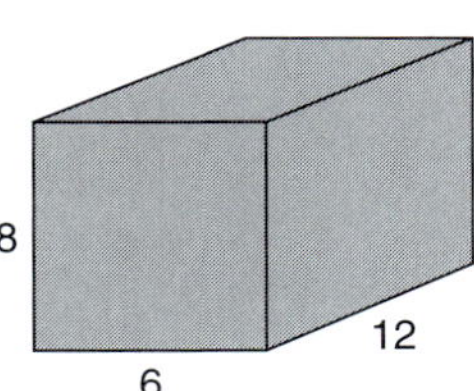

Figure 11–44

37.

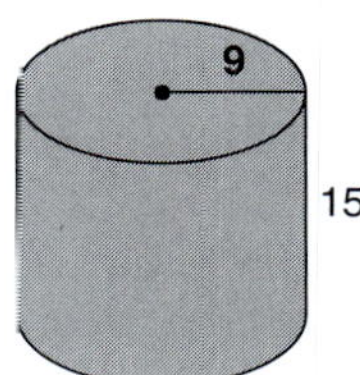

Figure 11–45

Solve.

38. What is the total surface area of a cylindrical oil storage tank 35 ft in diameter and 12 ft tall? Round to the nearest whole number.

39. What is the total surface area of a triangular prism with a height of 6 in. and triangular base of 2 in. on each side and a height of 1.7 in.?

40. A solid chocolate candy bar is molded in the form of a prism. The manufacturer wants to wrap the sides of the bar with paper indicating the company name, ingredients, and weight of the bar. What is the lateral area to be wrapped if the bar measures as shown in Fig. 11–46?

41. Find the number of square inches needed to make a paper label for an aluminum can $2\frac{1}{2}$ in. in diameter and $4\frac{3}{4}$ in. tall. Assume no waste or overlap. Round to tenths.

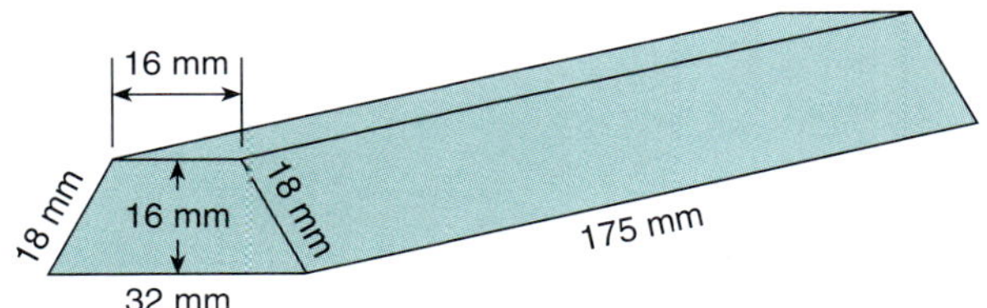

Figure 11–46

42. A wall-mounted three-way speaker is covered with wood-grained vinyl. What lateral area is to be covered if the speaker is 19 in. high, 11 in. wide, and 10 in. deep? (*Hint:* Assume that the 11×19 rectangle is the base.)

43. If the side of a 1-lb coffee can is imprinted to show the manufacturer and contents, how many square inches of lateral surface are imprinted if the can is 4 in. across and $5\frac{1}{2}$ in. high? Round to tenths.

6 Find the volume in Figs. 11–47 and 11–48. Round to the nearest hundredth if necessary.

44.

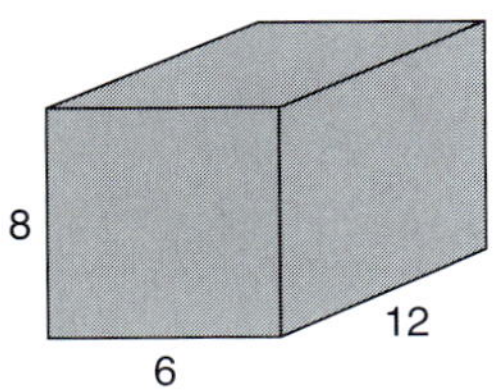

Figure 11–47

45.

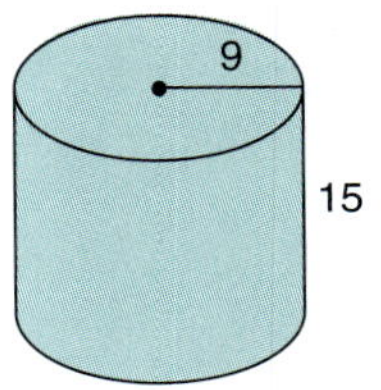

Figure 11–48

Solve.

46. How many cubic inches are in an aluminum can with a $2\frac{1}{2}$ in. diameter and $4\frac{3}{4}$ in. height? Round to tenths.

47. What is the volume of a cylindrical oil storage tank that has a 40 ft diameter and 15 ft height? Round to the nearest whole number.

48. A right pentagonal prism is 10 cm high. If the area of each pentagonal base is 32 cm^2, what is the volume of the prism?

49. What is the volume of a triangular prism that has a height of 8 in., a triangular base that measures 4 in. on each side, and a height of 3.46 in.? Round to hundredths.

50. A cylindrical water well is 1,200 ft deep and 6 in. across. How much soil and other material was removed? Round to the nearest cubic foot. (*Hint:* Convert measures to a common unit.)

 Solve. Round to tenths.

51. Find the surface area of a sphere with a radius of 5 cm.

52. Find the volume of a sphere with a radius of 6 in.

53. How many square feet of steel are needed to manufacture a spherical water tank with a diameter of 45 ft?

54. If 1 ft^3 = 7.48 gal, how many gallons can the water tank in Exercise 53 hold?

55. A spherical propane tank has a diameter of 4 ft. How many square feet of surface area need to be painted?

56. If a propane tank is filled to 90% of its capacity, how many gallons of propane does the tank in Exercise 55 hold? (1 ft^3 = 7.48 gal.)

 Solve. Round to tenths.

57. Find the lateral surface area, total surface area, and volume of the cone in Fig. 11–49.

58. How many cubic feet are in a conical pile of sand that is 30 ft in diameter and is 20 ft high?

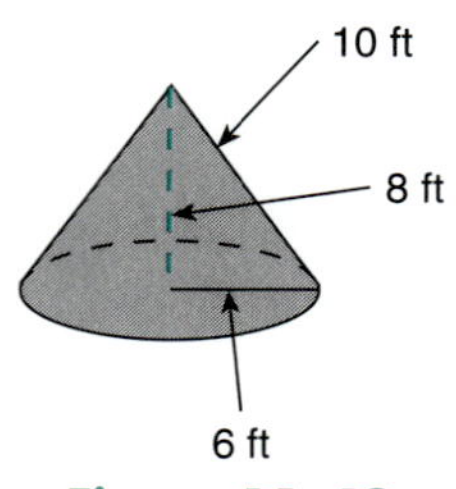

Figure 11–49

59. How many square centimeters of sheet metal are needed to form a conical rain cap 25 cm in diameter if the slant height is 15 cm?

60. A cone-shaped storage container holds a photographic chemical. If the container is 80 cm wide and 30 cm high, how many liters of the chemical does it hold if 1 L = 1,000 cm^3?

61. Find the total surface area of a conical tank that has a radius of 15 ft and a slant height of 20 ft.

62. Find the height of a conical tank with a volume of 261.67 ft^3 and a radius of 5 ft. (*Hint:* Rearrange the volume formula to find the height.)

63. A cylindrical water tower with a conical top and hemispheric bottom (see Fig. 11–50) needs to be painted. If the cost is \$2.19 per square foot, how much does it cost (to the nearest dollar) to paint the tank?

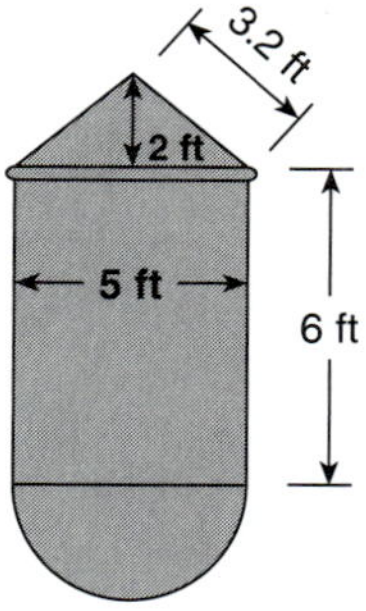

Figure 11–50

MISCELLANEOUS TECHNICAL FORMULAS

Learning Outcome

1 Evaluate miscellaneous technical formulas.

In addition to temperature conversion formulas and geometric formulas, several other formulas are used in technical applications. In this section, a miscellaneous assortment of technical formulas is presented.

1 Evaluate Miscellaneous Technical Formulas.

EXAMPLE Evaluate the formula $E = \frac{I - P}{I}$ if $I = 24{,}000$ calories (cal) and $P = 8{,}600$ cal.

$$E = \frac{I - P}{I}$$
Engine efficiency = difference between heat input and output divided by heat input.

$$E = \frac{24{,}000 - 8{,}600}{24{,}000}$$
Substitute given values. Perform calculations in numerator grouping.

$$E = \frac{15{,}400}{24{,}000}$$
Divide.

$$E = 0.642$$
Nearest thousandth.

The engine efficiency is 0.642 (64.2% efficient).

EXAMPLE Evaluate the formula $R_T = \frac{R_1 R_2}{R_1 + R_2}$ if $R_1 = 10\ \Omega$ and $R_2 = 6\ \Omega$.

$$R_T = \frac{R_1 R_2}{R_1 + R_2}$$
Total resistance = product of first resistance and second resistance divided by sum of first and second resistances.

$$R_T = \frac{10(6)}{10 + 6}$$
Substitute given values. Perform calculations in numerator and denominator groupings.

$$R_T = \frac{60}{16}$$
Divide.

$$R_T = 3.75$$

The total resistance in the circuit is 3.75 ohms.

In the next formula, don't forget to square the D.

EXAMPLE Evaluate the formula $H = \frac{D^2 N}{2.5}$ if $D = 4$ and $N = 8$.

$$H = \frac{D^2 N}{2.5}$$
Horsepower = diameter of the cylinder squared times the number of cylinders divided by 2.5.

$$H = \frac{4^2(8)}{2.5}$$
Perform power operation.

$$H = \frac{16(8)}{2.5}$$
Perform multiplication.

$$H = \frac{128}{2.5}$$
Divide.

$$H = 51.2$$

The engine is rated at 51.2 hp.

Sometimes we are asked to solve for a letter term that is not isolated. Let's look again at the horsepower formula and solve for a letter term other than horsepower.

EXAMPLE Solve the formula for horsepower $H = \frac{D^2N}{2.5}$ for the diameter of the piston D (in inches) if $H = 45$ and $N = 6$.

$$H = \frac{D^2N}{2.5}$$

$$45 = \frac{D^2(6)}{2.5} \qquad \text{Substitute values.}$$

$$(2.5)45 = \frac{D^2(6)}{2.5}(2.5) \qquad \text{Multiply to eliminate denominator.}$$

$$112.5 = 6D^2 \qquad \text{Simplify.}$$

$$\frac{112.5}{6} = D^2 \qquad \text{Divide by the coefficient.}$$

$$18.75 = D^2 \qquad \text{Take the square root of both sides.}$$

$$D = 4.330 \qquad \text{Rounded.}$$

The diameter of the piston is 4.330 in.

Tip!	***Interpret the Solution within the Context of the Formula.***

In practical problems where a negative square root is unrealistic, we use *only* the positive square root. In the previous example, for instance, we cannot have a negative diameter of a piston.

Caution: The next formula includes both powers and roots. Because the process for solving these equations involves multiplying and dividing by variables, the solution should be checked for extraneous roots.

EXAMPLE In the formula $Z = \sqrt{R^2 + X^2}$, solve for R (in ohms) if $Z = 12.4$ ohms and $X = 12$ ohms.

$$Z = \sqrt{R^2 + X^2} \qquad \text{Impedance} = \text{square root of the sum of the resistance squared plus the reactance squared.}$$

$$12.4 = \sqrt{R^2 + (12)^2} \qquad \text{Substitute values.}$$

$$(12.4)^2 = \left(\sqrt{R^2 + 144}\right)^2 \qquad \text{Square both sides to eliminate radical.}$$

$$153.76 = R^2 + 144 \qquad \text{Isolate } R^2 \text{ (sort terms).}$$

$$153.76 - 144 = R^2 \qquad \text{Subtract.}$$

$$9.76 = R^2 \qquad \text{Take the square root of both sides.}$$

$$R = 3.124 \qquad \text{Rounded.}$$

The resistance is 3.1 ohms (rounded to tenths).

Note the many steps in the following formula evaluation and the importance of observing the order of operations. The formula is for finding the length of a belt connecting two pulleys.

EXAMPLE Evaluate $L = 2C + 1.57(D + d) + \frac{D + d}{4C}$ if $C = 24$ in., $D = 16$ in., and $d = 4$ in. (Fig. 11–51).

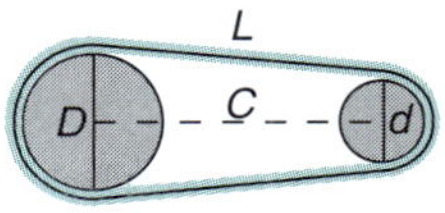

Figure 11–51

L = length of belt joining two pulleys
C = distance between centers of pulleys
D = diameter of large pulley
d = diameter of small pulley

$$L = 2C + 1.57(D + d) + \frac{D + d}{4C}$$

$$L = 2(24) + 1.57(16 + 4) + \frac{16 + 4}{4(24)}$$

Work groupings in parentheses, numerator, and denominator.

$$L = 2(24) + 1.57(20) + \frac{20}{96}$$

Work multiplication and division.

$$L = 48 + 31.4 + 0.20833333$$

Add.

$$L = 79.61$$

Rounded.

The pulley belt is 79.61 in. long.

SELF-STUDY EXERCISES 11–5

1 Evaluate.

1. Use the formula $R_T = \frac{R_1 R_2}{R_1 + R_2}$ to find the total resistance (R_T) if one resistance (R_1) is 12 ohms (Ω) and the second resistance (R_2) is 8 ohms (Ω).

2. What is the percent efficiency (E) of an engine if the input (I) is 25,000 calories and the output (P) is 9,600 calories? Use the formula $E = \frac{I - P}{I}$.

3. Distance is rate times time, or $D = RT$. Find the rate if the distance traveled is 140 mi and the time traveled is 4 hr.

4. The formula for voltage (Ohm's law) is $E = IR$. Find the amperes of current (I) if the voltage (E) is 120 V and the resistance (R) is 80 ohms (Ω).

5. According to Boyle's law, if temperature is constant, the volume of a gas is inversely proportional to the pressure on it. Find the final volume (V_2) of a gas using the formula $\frac{V_1}{V_2} = \frac{P_2}{P_1}$ if the original volume (V_1) is 15 ft^3, the original pressure (P_1) is 60 lb per square inch (psi), and the final pressure (P_2) is 150 psi.

6. The formula for power (P) in watts (W) is $P = I^2 R$. Find the current (I) in amperes if a device draws 63 W and the resistance (R) is 7 ohms (Ω).

7. Use the formula $H = \frac{D^2 N}{2.5}$ to find the number of cylinders (N) required in an engine of 3.2 hp (H) if the cylinder diameter (D) is 2 in.

8. The formula for the speed (s) of a driven pulley in revolutions per minute (rpm) is $s = \frac{DS}{d}$. Find the speed of a driven pulley with a diameter (d) of 5 in. if the diameter (D) of the driving pulley is 10 in. and its speed (S) is 800 rpm.

9. If the distance (C) between the centers of the pulleys in Exercise 8 is 24 in., find the length (L) of the belt connecting them using the formula

$$L = 2C + 1.57(D + d) + \frac{D + d}{4C}$$

Round to hundredths.

10. Find the reactance (X) in ohms using the formula $Z = \sqrt{R^2 + X^2}$ if the impedance (Z) is 10 ohms and the resistance (R) is 9 ohms. Round to tenths.

Electronics: Ohm's Law

Ohm's law is derived from Kirchhoff's laws to form the basis for much of the work done in electronics. In its simplest forms, Ohm's law is represented by three equations:

$$\text{Ohm's law:} \qquad E = IR, \qquad P = IE, \qquad G = \frac{1}{R}$$

where E = potential difference, in volts (V); I = current, in amperes (A); R = resistance, in ohms (Ω); P = power, in watts (W); G = conductance, in siemens (S).

First, consider $E = IR$. The R can be replaced by $\frac{1}{G}$, so $E = I\left(\frac{1}{G}\right)$ or $E = \frac{I}{G}$. This equation solved for I is $I = GE$, and solved for G is $G = \frac{I}{E}$.

The equation $P = IE$ can be rearranged as $I = \frac{P}{E}$ and $E = \frac{P}{I}$. Also, IR can be substituted for E in the equation $P = IE$ to get $P = I(IR) = I^2R$, and $\frac{1}{G}$ can be substituted for R to get $P = \frac{I^2}{G}$.

Assume that you start with $P = \frac{I^2}{G}$ and wish to solve for I or G. Eliminate denominators (multiply both sides of equation by G) to get $PG = I^2$. Solved for I you have $I = \sqrt{GP}$. Solved for G you have $G = \frac{I^2}{P}$.

Suppose that $P = 4$ mW and $G = 7$ μS. Then

$$I = \sqrt{(4 \times 10^{-3})(7 \times 10^{-6})}$$

$$= \sqrt{28 \times 10^{-9}}$$

$$= \sqrt{280 \times 10^{-10}}$$

$$= 16.7 \times 10^{-5}$$

$$= 167 \times 10^{-6}$$

$$= 167 \text{ μA} \qquad\qquad \text{Rounded.}$$

$$\textit{Proof:} \quad P = \frac{I^2}{G} = \frac{(167 \text{ μA})^2}{7 \text{ μS}} = 3{,}984 \text{ W or 4 mW} \qquad \text{Rounded.}$$

Exercises

Solve each equation for the indicated letter. Then, evaluate the new equation for the given values. Verify the numerical answer obtained by substituting back into the original equation. Be sure to use the given prefixes.

1. Solve for I in $P = I^2R$. What is I if $P = 8$ W and $R = 2$ Ω?
2. Solve for G in $E = \frac{I}{G}$. What is G if $E = 12$ V and $I = 4$ mA?
3. Solve for P in $G = \frac{I^2}{P}$. What is P if $G = 8$ mS and $I = 6$ mA?
4. Solve for R in $P = I^2R$. What is R if $I = 4$ mA and $P = 8$ mW?
5. Solve for E in $P = \frac{E^2}{R}$. What is E if $P = 9$ W and $R = 63$ Ω?
6. Solve for E in $P = E^2G$. What is E if $P = 8$ W and $G = 2$ mS?
7. Solve for R_3 in $R_t = R_1 + R_2 + R_3$. What is R_3 if $R_t = 1{,}410$ Ω, $R_1 = 420$ Ω, and $R_2 = 440$ Ω?
8. Solve for G_3 in $G_t = G_1 + G_2 + G_3$. What is G_3 if $G_t = 92$ mS, $G_1 = 22$ mS, and $G_2 = 24$ mS?
9. Solve for R_1 in $V_1 = \frac{R_1 V_t}{R_1 + R_2}$. What is R_1 if $V_1 = 8$ V, $V_t = 12$ V, and $R_2 = 41\Omega$?
10. Solve for G_2 in $I_2 = \frac{G_2 I_t}{G_1 + G_2}$. What is G_2 if $I_2 = 5$ mA, $I_t = 8$ mA, and $G_1 = 6$ mS?

Answers for Exercises

1. Solve for I in $P = I^2R$. What is I if $P = 8$ W and $R = 2$ Ω?

$$I = \sqrt{\frac{P}{R}} = \sqrt{\frac{8}{2}} = 2 \text{ A} \qquad Proof: \quad 2^2 \times 2 = 8 \text{ W}$$

2. Solve for G in $E = \frac{I}{G}$. What is G if $E = 12$ V and $I = 4$ mA?

$$G = \frac{I}{E} = \frac{4 \text{ mA}}{12 \text{ V}} = 0.333 \text{ mS} = 333 \text{ } \mu\text{S} \qquad Proof: \quad \frac{4 \text{ mA}}{0.333 \text{ mS}} = 12 \text{ V}$$

3. Solve for P in $G = \frac{I^2}{P}$. What is P if $G = 8$ mS and $I = 6$ mA?

$$P = \frac{I^2}{G} = \frac{(6 \text{ mA})^2}{8 \text{ mS}} = 4.5 \text{ mW} \qquad Proof: \quad \frac{(6 \text{ mA})^2}{4.5 \text{ mW}} = 8 \text{ mS}$$

4. Solve for R in $P = I^2R$. What is R if $I = 4$ mA and $P = 8$ mW?

$$R = \frac{P}{I^2} = \frac{8 \text{ mW}}{(4 \text{ mA})^2} = 0.5 \text{ k}\Omega = 500 \text{ }\Omega \qquad Proof: \quad (4 \text{ mA})^2(500 \text{ }\Omega) = 8 \text{ mW}$$

5. Solve for E in $P = \frac{E^2}{R}$. What is E if $P = 9$ W and $R = 63$ Ω?

$$E = \sqrt{PR} = \sqrt{9 \times 63} = 23.8 \text{ V} \qquad Proof: \quad \frac{(23.8 \text{ V})^2}{63 \text{ }\Omega} = 9 \text{ W}$$

6. Solve for E in $P = E^2G$. What is E if $P = 8$ W and $G = 2$ mS?

$$E = \sqrt{\frac{P}{G}} = \sqrt{\frac{8 \text{ W}}{2 \text{ mS}}} = \sqrt{\frac{8 \text{ W}}{0.002 \text{ S}}} = 63.2 \text{ V}$$

$$Proof: \quad (63.2 \text{ V})^2(2 \text{ mS}) = 7,988.48 \text{ mW} = 8 \text{ W}$$

7. Solve for R_3 in $R_t = R_1 + R_2 + R_3$. What is R_3 if $R_t = 1{,}410$ Ω, $R_1 = 420$ Ω, and $R_2 = 440$ Ω?

$$R_3 = R_t - R_1 - R_2 = 1{,}410 \text{ }\Omega - 420 \text{ }\Omega - 440 \text{ }\Omega = 550 \text{ }\Omega$$
$$Proof: \quad 420 \text{ }\Omega + 440 \text{ }\Omega + 550 \text{ }\Omega = 1{,}410 \text{ }\Omega$$

8. Solve for G_3 in $G_t = G_1 + G_2 + G_3$. What is G_3 if $G_t = 92$ mS, $G_1 = 22$ mS, and $G_2 = 24$ mS?

$$G_3 = G_t - G_1 - G_2 = 92 \text{ mS} - 22 \text{ mS} - 24 \text{ mS} = 46 \text{ mS}$$
$$Proof: \quad 22 \text{ mS} + 24 \text{ mS} + 46 \text{ mS} = 92 \text{ mS}$$

9. Solve for R_1 in $V_1 = \frac{R_1 V_t}{R_1 + R_2}$. What is R_1 if $V_1 = 8$ V, $V_t = 12$ V, and $R_2 = 41$ Ω?

$$R_1 = \frac{V_1 R_2}{V_t - V_1} = \frac{8 \text{ V} (41 \text{ }\Omega)}{12 \text{ V} - 8 \text{ V}} = 82 \text{ }\Omega$$

$$Proof: \quad \frac{(82 \text{ }\Omega)(12 \text{ V})}{82 \text{ }\Omega + 41 \text{ }\Omega} = 8 \text{ V}$$

10. Solve for G_2 in $I_2 = \frac{G_2 I_t}{G_1 + G_2}$. What is G_2 if $I_2 = 5$ mA, $I_t = 8$ mA, and $G_1 = 6$ mS?

$$G_2 = \frac{I_2 G_1}{I_t - I_2} = \frac{5 \text{ mA} (6 \text{ mS})}{8 \text{ mA} - 5 \text{ mA}} = 10 \text{ mS}$$

$$Proof: \quad \frac{(10 \text{ mS})(8 \text{ mA})}{6 \text{ mS} + 10 \text{ mS}} = 5 \text{ mA}$$

Section 11–1

Evaluate the formulas. Use $D = 2.5$, $E = 3$, and $F = 4$.

1. $Y = E + (3F - D)$

2. $Y = F(E^2 + 4DF) - 3$

3. $Y = \dfrac{2(F - D)^2}{D - 1}$

4. $Y = \dfrac{3(F - D)}{3} + E$

5. $Y = D^2 - (2F - E)$

Evaluate the percentage formula $P = \dfrac{RB}{100}$ using the given values.

6. Find the percentage if $R = 10\%$ and $B = 300$ lb.

7. Find the rate if $P = 12$ kg and $B = 125$ kg.

8. Find the base if $P = \$28.05$ and $R = 8.5\%$.

9. Evaluate the rate formula $R = \dfrac{100P}{B}$ for P if $R = 12\%$ and $B = 90$.

10. Evaluate the base formula $B = \dfrac{100P}{R}$ for R if $B = \$5,000$ and $P = \$450$.

Evaluate the interest formula $I = PRT$ using the given values.

11. Find the interest if $P = \$440$, $R = 16\%$, and $T = 2\frac{3}{4}$ years.

12. Find the rate if $I = \$2,484$, $P = \$4,600$, and $T = 3$ years.

13. Find the time if $I = \$387.50$, $P = \$1,550$, and $R = 12.5\%$.

14. Find the principal if $I = \$1,665$, $R = 18\frac{1}{2}\%$ per yr, and $T = 1\frac{1}{2}$ yr.

15. Find the length of a rectangular work area if the perimeter is 160 in. and the width is 30 in. Use the formula $P = 2(l + w)$.

16. Find the radius of a circle whose area is 132.7 mm^2 using the formula $r = \sqrt{\frac{A}{\pi}}$. Round to the nearest tenth.

17. Find the cost (C) if the markup (M) on an item is \$25.75 and the selling price (S) is \$115.25. Use the formula $M = S - C$.

18. Using the formula for the side of a square, $s^2 = A$, find the length of a side of a square field that has an area of $\frac{1}{4}$ mi^2.

19. Evaluate the formula for the area of a circle, $A = \pi r^2$, if $r = 5.5$ in. Round to the nearest tenth.

20. Evaluate the formula for the area of a square, $A = s^2$, if $s = 3.25$ km.

Section 11–2

Solve the formulas for the indicated variable.

21. $D = F - (m + n)$ for F

22. $H = C - S$ for S

23. $V = lwh$ for w

24. $R = h(p + 3q)$ for q

25. $c = ah + ab$ for a

26. $K = \dfrac{m + n}{p}$ for m

27. $B = cr^2x$ for r

28. $r = \sqrt{s^2 - t^2}$ for t

29. $PB = A$ for B

30. $V = lwh$ for h

31. $I = Prt$ for r

32. $s = c + m$ for c

33. $s = r - d$ for r

34. $s = r - d$ for d

35. $v = v_0 - 32t$ for t

36. $P = 2(l + w)$ for w

37. $A = P(1 + rt)$ for t

38. $V = \frac{1}{3}Bh$ for h

39. The formula for finding the sale price on an item is $S = P - D$, where S is the sale price, P is the original price, and D is the discount. Solve the formula for the original price.

40. The formula for finding tax is $T = RM$, where T represents tax, R represents the tax rate, and M represents the marked price. Rearrange the formula to find the marked price.

Section 11–3

Make the temperature conversions.

41. $95°C = $ _____ $°F$

42. $86°F = $ _____ $°C$

43. The label on a container of the acid-reducer famotidine states that storage above 40°C should be avoided. What is 40°C on the Fahrenheit scale?

44. The freezing point of benzene is 5°C. What is this temperature on the Fahrenheit scale?

45. Candy that should reach a cooking temperature of 365°F should reach what temperature using the Celsius scale?

46. Summer road surface temperatures of 122°F give what reading on the Celsius scale?

Find the perimeter and the area of Figs. 11–52 to 11–55.

47.

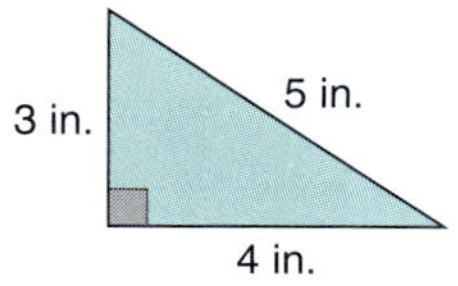

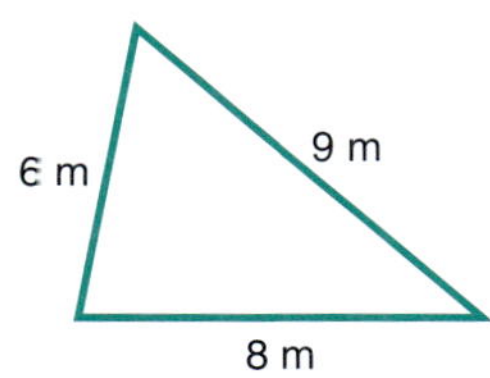

3 in. 5 in.

4 in.

Figure 11–52

48.

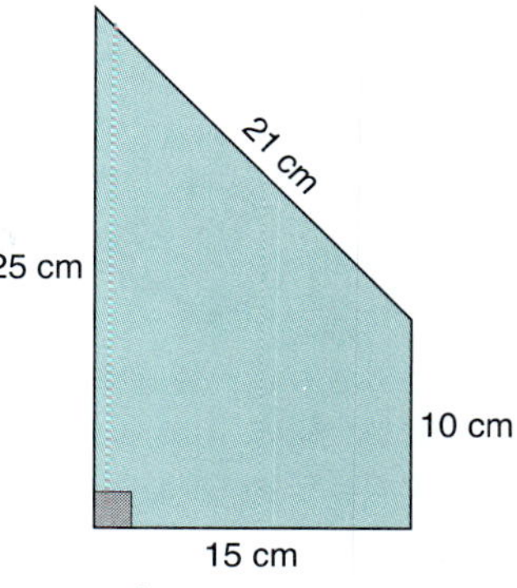

Figure 11–53

49.

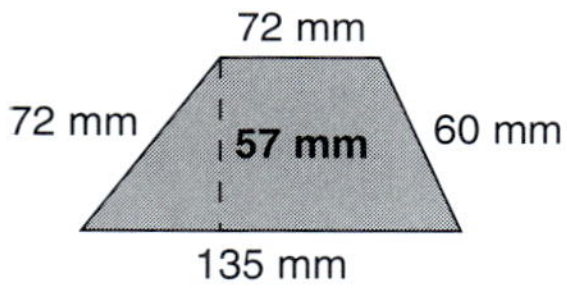

72 mm

72 mm   **57 mm**   60 mm

135 mm

Figure 11–54

50.

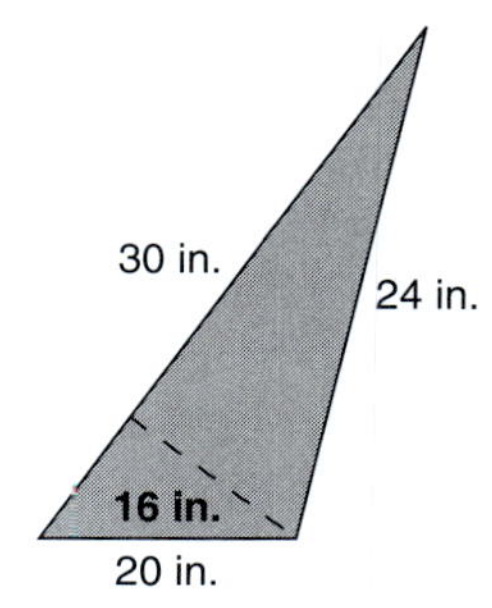

Figure 11–55

51. Find the area of the triangle in Fig. 11–56.

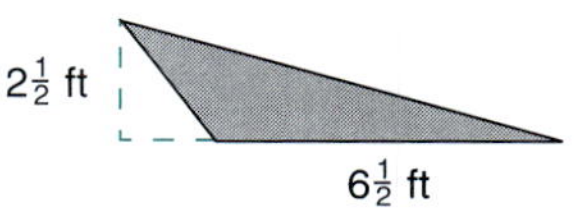

Figure 11–56

Find the area of the triangles shown in Figs. 11–57 to 11–60. Use Heron's formula. Round the answers to four significant digits.

52.

6 m   9 m

8 m

Figure 11–57

53.

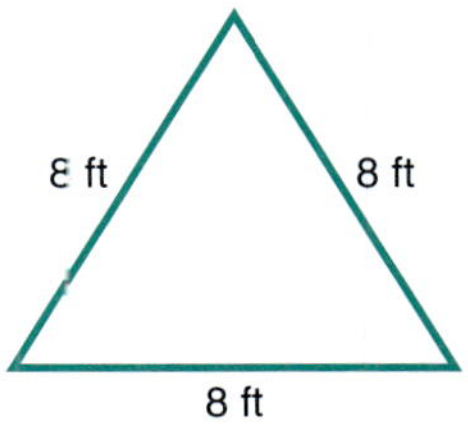

8 ft   8 ft

8 ft

Figure 11–58

54.

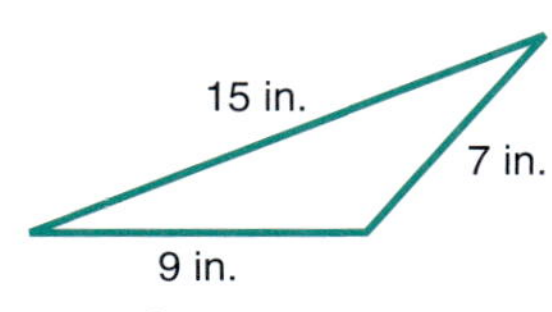

Figure 11–59

55.

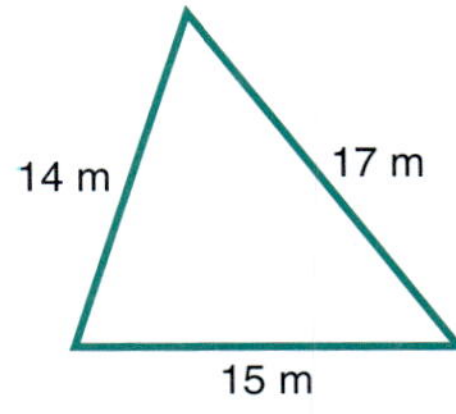

Figure 11–60

56. What is the area of a triangular plot of ground with sides that measure 146 ft, 85 ft, and 195 ft?

57. Find the area of a triangular tabletop that has sides measuring 46 in., 37 in., and 40 in.

Solve.

58. A trapezoidal wall section of a contemporary home has a height of 9 ft, a top length of 7 ft, and a bottom length of 14 ft. Find the area of the wall section.

59. The sides of a triangle are $2\frac{1}{2}$ ft, $3\frac{3}{4}$ ft, and 5 ft. If the triangle is to be made into an advertising sign for a restaurant, how many feet of contrasting trim will be needed to outline the sign?

60. Using your knowledge of formula rearrangement and triangles, find the base of a triangle that has a height of 26.25 ft and an area of 236.25 ft^2.

61. If a triangular louver that has a height of 3 ft and a base of 6 ft 6 in. is installed in each gable end of a house, how many square feet of ventilation do the two louvers provide?

62. One trapezoidal side of a hip roof measures $21\frac{1}{2}$ ft along the ridge line (upper base) and $45\frac{1}{2}$ ft at the bottom (lower base). The height of the side is 16 ft. How many bundles of roofing shingles are necessary to cover the side at a 5-in. exposure if 3.2 bundles cover each square (100 ft^2)?

63. A stairway to the upstairs portion of a house has a run (base) of 10 ft and a rise (height) of $7\frac{1}{2}$ ft. How many sheets of 4-ft $\times$ 8-ft paneling are needed to finish the exposed side? Disregard waste.

64. A gable end of a house has a span (base) of 30 ft and a rise (height) of $5\frac{1}{2}$ ft. If the gable is to be covered with 8-in. wide bevel siding with a $6\frac{1}{2}$ in. exposure to the weather, how many square feet of siding are needed if we assume a 23% loss for lap and waste? (Round any portion of a square foot to the next highest square foot.)

65. If the three sides of the gable end of a roof are each 16 ft, how many feet of trim would be needed to surround the gable end?

66. A trapezoidal flower bed has its longer base along the entire length of a 45-ft outside wall of a building. If the shorter base is 25 ft and the adjacent sides are each 11.5 ft, how many feet of garden edging material are needed to surround the flower bed if no edging is used against the building wall?

67. If the flower bed in Exercise 66 projects 5.7 ft from the building wall, how many square feet of surface are available for flowering plants?

Use Fig. 11–61 to solve the following exercises. Round the final answers to the nearest thousandth if necessary.

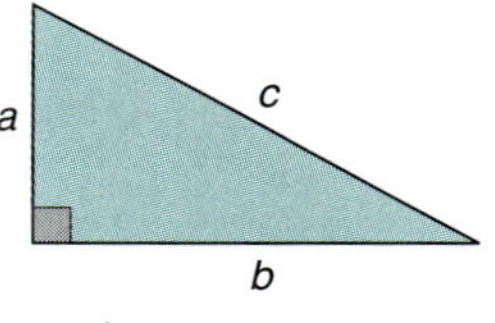

Figure 11–61

68. $a = 3$ m
$b = ?$
$c = 5$ m

69. $a = 9$ in.
$b = 12$ in.
$c = ?$

70. $a = 8$ cm
$b = 15$ cm
$c = ?$

71. $a = 7$ ft
$b = ?$
$c = 10$ ft

72. $a = 8$ mm
$b = ?$
$c = 17$ mm

73. $a = ?$
$b = 15$ yd
$c = 17$ yd

74. $a = ?$
$b = 12$ km
$c = 15$ km

75. $a = 11$ mi
$b = 17$ mi
$c = ?$

76. $a = 10$ in.
$b = 24$ in.
$c = ?$

77. $a = ?$
$b = 40$ cm
$c = 50$ cm

Solve the following problems. Round the final answers to the nearest thousandth if necessary.

78. If the base of a ladder is placed on the ground 4 ft from a house, how tall must the ladder be (to the nearest foot) to reach the chimney top that extends $18\frac{1}{2}$ ft above the ground?

79. In an automobile, three pulleys are connected by one belt. The center-to-center distance between the pulleys farthest apart cannot be measured conveniently. The other center-to-center distances are 12 in. and 18 in. (Fig. 11–62). Find the distance between the pulleys farthest apart.

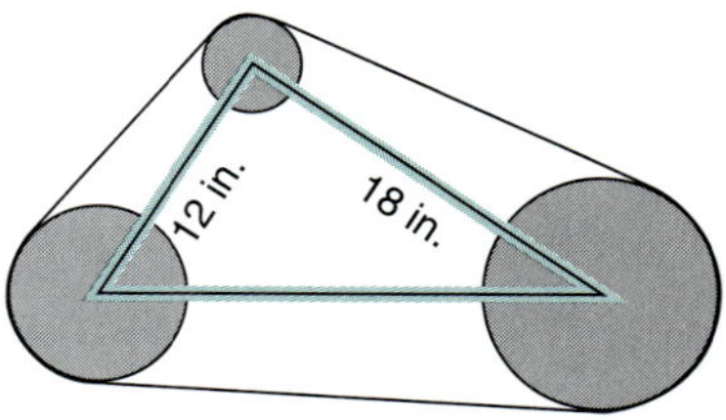

Figure 11–62

80. A central vacuum outlet is installed in one corner of a rectangular room that measures $9' \times 12'$. How long must the nonelastic hose be to reach all parts of the room?

82. Find the distance across the corners of a square nut that is 7.9 mm on a side (Fig. 11–63).

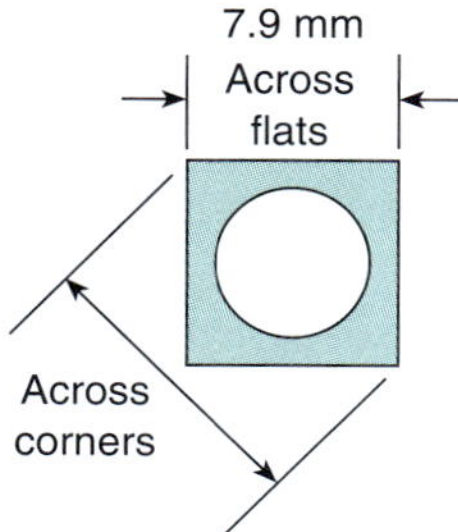

Figure 11–63

81. Find the diameter of a piece of round steel from which a 3-in. square nut can be milled.

83. Find the volume of the triangular prism in Fig. 11–64.

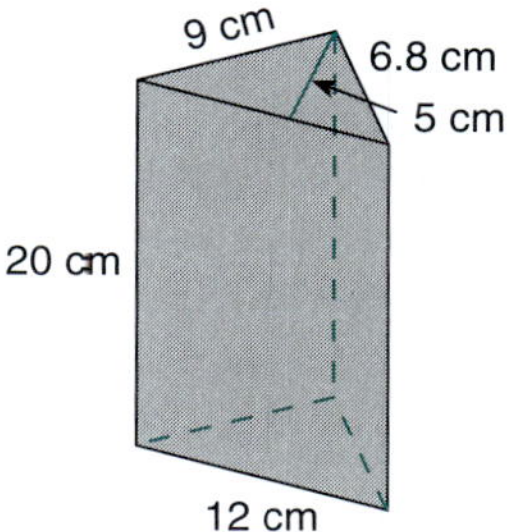

Figure 11–64

84. Find the lateral surface area of the triangular prism in Fig. 11–64.

86. Find the lateral surface area of the cylinder in Fig. 11–65.

85. Find the total surface area of the triangular prism in Fig. 11–64.

87. Find the total surface area of the cylinder in Fig. 11–65.

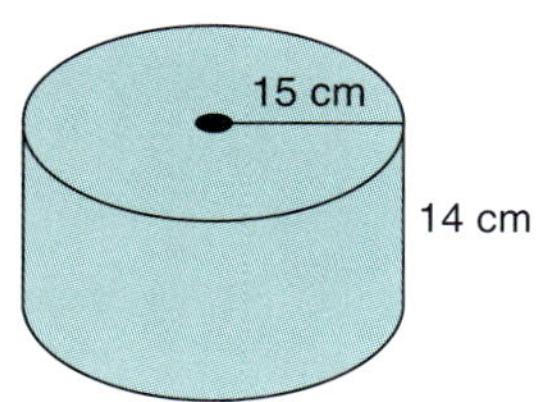

Figure 11–65

88. Find the volume of the cylinder of Fig. 11–65.

90. If concrete weighs 160 lb per cubic foot, what is the weight of a concrete circular slab 4 in. thick and 15 ft across?

92. An interstate highway is repaired in one section 48 ft across, 25 ft long, and 8 in. deep. If concrete costs $25.50 per cubic yard, what is the cost of the concrete needed to repair the highway rounded to the nearest dollar? ($27 \text{ ft}^3 = 1 \text{ yd}^3$.)

94. How many barrels of oil will a tank hold if its height is $65\frac{1}{2}$ ft and its radius is 20 ft? Round to the nearest whole barrel. (31.5 gal = 1 barrel and $1 \text{ ft}^3 = 7.48$ gal.)

89. How many cubic yards of top soil are needed to cover an 85-ft by 65-ft area for landscaping if the topsoil is 6 in. deep? Round to the nearest whole number. ($27 \text{ ft}^3 = 1 \text{ yd}^3$)

91. What is the lateral surface area of a hexagonal column 20 ft high that measures 6 in. on a side?

93. A pipeline to carry oil between two towns 5 mi apart has an inside diameter of 18 in. If 1 mi = 5,280 ft and $1 \text{ ft}^3 = 7.48$ gal, how many gallons of oil will the pipeline hold (to the nearest gallon)?

Solve. Round the final answer to the nearest tenth unless otherwise specified.

95. Find the total surface area of a sphere with a radius of 9 m.

97. Find the volume of a sphere that has a radius of 12 ft.

99. Find the lateral surface area of a cone with a radius of 6 cm and a slant height of 9 cm.

96. Find the total surface area of a sphere if its diameter is 20 cm.

98. Find the volume of a sphere that has a diameter of 30 cm.

100. Find the total surface area of a cone that has a radius of 4 m and a slant height of 8 m.

101. Find the volume of a cone with a radius of 6 in. and a height of 10 in.

102. If 1 gal $= 231$ in.^{3}, find the number of gallons that a conical oil container 18 in. high and 23 in. across holds.

103. The entire exterior surface of a conical tank with a slant height of 12 ft and a diameter of 18 ft is being painted. If the paint covers at a rate of 350 ft^2 per gallon, how many gallons of paint are needed for the job? Round any fraction of a gallon to the next whole gallon.

104. A hopper deposits sand in a cone-shaped pile with a diameter of 9′6″ and a height of 8′3″. To the nearest cubic foot, how much sand is deposited?

105. How many square feet of plastic material are needed to devise a wind-tunnel cone that has a base of 6 ft across and height of 12 ft (Fig. 11–66). (*Hint:* To find the slant height, consider it to be the hypotenuse of a right triangle.)

106. To the nearest square foot, how much nylon material is needed to construct a conical tent-like pavilion 30 ft in diameter and 5 ft from the apex to the base? (*Hint:* See Exercise 103.)

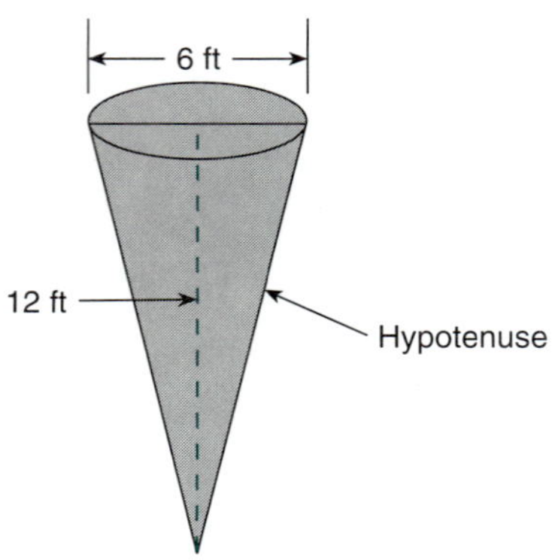

Figure 11–66

107. A No. 5 soccer ball has a diameter of 8.8 in. How much leather is needed to cover the surface?

108. How much does a 4-in. lead ball weigh if lead weighs 1 lb per 2.4 in.^{3}?

109. A spherical tank is anchored halfway in the ground. How many cubic feet of earth are excavated to install the tank? The tank measures 20 ft across. Round to the nearest whole cubic foot.

Section 11-5

110. Use the formula $R_t = \frac{R_1 R_2}{R_1 + R_2}$ to find the total resistance (R_t) if one resistance (R_1) is 10 ohms (Ω) and the second resistance (R_2) is 9 ohms (Ω). Round to tenths.

111. Ohm's law is $E = IR$. Find the amperes of current (I) if the voltage (E) is 220 V and the resistance (R) is 80 ohms (Ω).

112. Distance is rate times time, or $D = RT$. Find the rate if the distance traveled is 260 mi and the time traveled is 4 hr.

113. According to Boyle's law, if temperature is constant, the volume of a gas is inversely proportional to the pressure on it. Find the final volume (V_2) of a gas using the formula $\frac{V_1}{V_2} = \frac{P_2}{P_1}$ if the original volume (V_1) is 30 ft^3, the original pressure (P_1) is 75 psi (pounds per square inch), and the final pressure (P_2) is 225 psi.

114. Use the formula $E = \frac{I - P}{I}$ to find the percent efficiency (E) of an engine if the input (I) is 22,600 calories and the output (P) is 5,600 calories. Round to the nearest tenth of a percent.

115. The formula for power (P) in watts (W) is $P = I^2 R$. Find the current (I) in amperes if a device draws 80 W and the resistance (R) is 8 Ω. Round to tenths.

116. Find the impedance (Z) in ohms using the formula $Z = \sqrt{R^2 + X^2}$ if the resistance (R) is 4 Ω and the reactance (X) is 7 Ω. Round to tenths.

117. The formula for the speed (s) of a driven pulley in revolutions per minute (rpm) is $s = \frac{DS}{d}$. Find the speed of a driven pulley with diameter (d) of 3 in. if the diameter (D) of the driving pulley is 7 in. and its speed (S) is 600 rpm.

118. If the distance (C) between the centers of the pulleys in Exercise 117 is 30 in., find the length (L) of the belt connecting them using the formula

$$L = 2C + 1.57(D + d) + \frac{D + d}{4C}$$

Round to hundredths.

119. Use the formula $H = \frac{D^2 N}{2.5}$ to find the number of cylinders (N) required in an engine of 5 hp (H) if the cylinder diameter (D) is 2.5 in.

<hr>

CHALLENGE PROBLEMS

120. Devise your own formulas for the following relationships.

   (a) An electrical power company computes the monthly charges by multiplying the kilowatts of power used times the cost per kilowatt and adds to that a fixed monthly fee.

   (b) A store figures ending balance on a charge account by multiplying the interest rate times the previous unpaid balance and then adding the previous balance and purchases and subtracting payments.

   (c) Profit on the sale of a particular item is the product of the number of items sold and the difference between the selling price of the item and its cost to the seller.

121. Explain the usefulness of formula rearrangement in solving applied problems. Use at least one formula to illustrate.

<hr>

CHAPTER TRIAL TEST

1. The electrical resistance of a wire is found from the formula $R = \frac{PL}{A}$. Rearrange the formula to find the length L of the wire.

2. The formula for the volume (V) of a solid rectangular figure is $V = lwh$ (length $\times$ width $\times$ height). If the volume of a mailing container is 7.5 cm^3, its length is 1.5 cm, and its width is 0.5 cm, what is its height?

3. Engine displacement d is found using the formula $d = \pi r^2 sn$. Solve to find r (the radius of the bore).

4. Using $d = 351$ in.^{3}, $s = 3.5$ in. stroke, and $n = 8$ cylinders, calculate to the nearest tenth the radius of the bore with the rearranged formula in Exercise 3.

Use the temperature formulas for Exercises 5–6:

5. $88°C = $ _____ $°F$

6. $104°F = $ _____ $°C$

7. Find the area of triangle RST in Fig. 11–67. Round to four significant digits.

8. What is the area of a triangular cast if the sides measure 31 mm, 42 mm, and 27 mm?

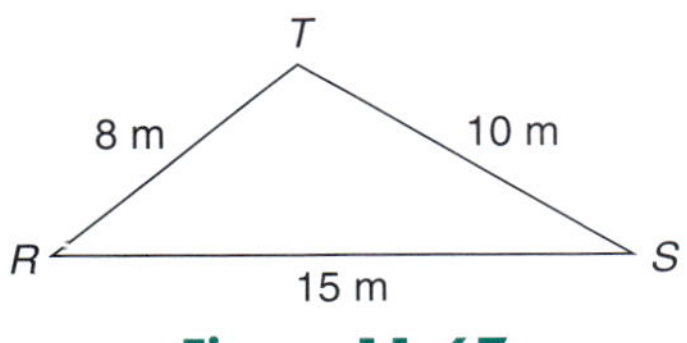

Figure 11–67

9. A section of a hip roof is a trapezoid measuring 35 ft at the bottom, 15 ft at the top, and 10 ft high. Find the area of this section of the roof in square feet.

10. The gable end of a roof is a triangle that has a rise (height) of 7 ft and a span (base) of 24 ft. The gable end contains a window 24 in. $\times$ 36 in. How much of the gable area needs to be painted?

11. A metal rod is welded to a metal support (Fig. 11–68). If the two sides of the metal support are 8 dm and 6 dm, find the length of the metal rod needed to make the brace.

12. Find the perimeter of the polygons shown in Fig. 11–69.

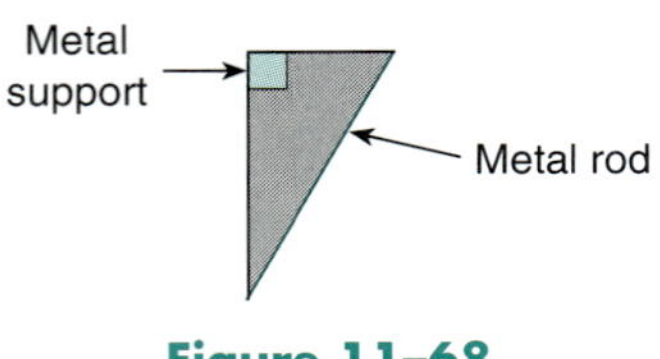

Figure 11–68

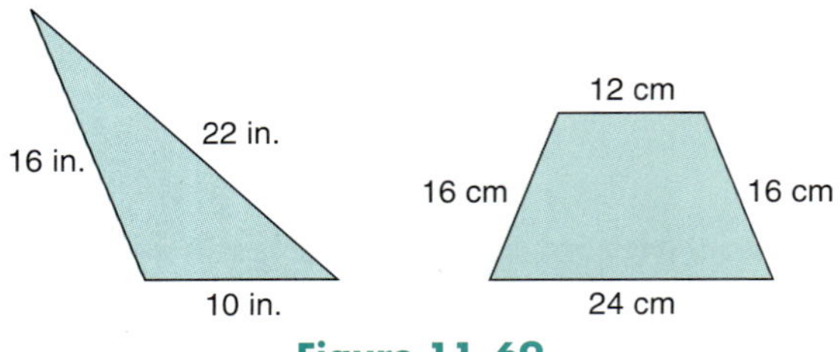

Figure 11–69

Solve.

13. A right pentagonal prism (5 sides) measures 1 in. on each side of its base and has a height of 10 in. What is its lateral surface area?

14. How much space is required to store 30 micro-computers crated in boxes that measure 62 cm by 70 cm by 50 cm?

15. A spherical tank 12 ft in diameter can hold how many gallons of fluid if $1 \text{ ft}^3 = 7.48$ gal? Answer to the nearest whole gallon.

16. The bases of a brass prism are equilateral triangles with sides of 3 in. and heights of 2.6 in. If the prism's height is 8 in., what is the volume of the prism?

17. Hard coal broken into small pieces is dumped into a cone-shaped pile. The base is 35 ft across and the pile stands 12 ft tall. How many cubic feet of coal are in the pile? Round to tenths.

18. A spherical gas storage tank 2.5 m wide needs to be sandblasted, primed, and refinished. The owner received an estimate of $8.50 per square meter. How much should the job cost to the nearest dollar?

19. A sheet metal worker wants to make a tin cone. If the base of the cone is 30 cm across and its height is 25 cm, what is the total surface area required to the nearest tenth? (*Hint:* Use your knowledge of a right triangle to find the slant height.)

20. A cold-water pipe with an outside diameter of $\frac{7}{8}$ in. is 23 ft long. How many whole rolls of insulating wrap are needed if one roll covers $5\frac{1}{2}$ ft^2 and no allowance is made for overlap? (*Hint:* $\frac{7}{8}$ in. is $\frac{7}{8} \times \frac{1}{12} = 0.0729$ ft.)

21. A steel rod has a diameter of 20 in. Find the volume of steel in a 5-ft length of the rod measured in cubic feet to the nearest tenth.

22. Find the total surface area of a cylindrical storage tank if it is 30 ft tall and has a diameter of 12 ft. Round to hundredths.

23. Using the formula $P = \frac{1.27F}{D^2}$, calculate force F (in pounds) if the pressure P is 180 psi and the piston diameter is 3.25 in. Round to the nearest hundredth.

24. Use the formula $R_t = \frac{R_1 R_2}{R_1 + R_2}$ to find the total resistance (R_t) if one resistance (R_1) is 9 Ω and the second resistance (R_2) is 8 Ω. Round to tenths.

25. If the efficiency (E) of an engine is 70% and the input (I) is 40,000 calories, find the output (P) in calories. Use the formula $E = \frac{I - P}{I}$.

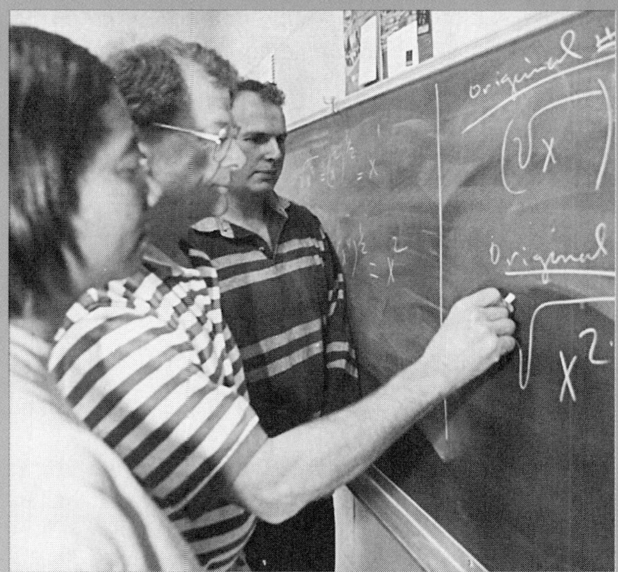

12

Products and Factors

GOOD DECISIONS THROUGH TEAMWORK

In this chapter, we will study methods of factoring, like removing common factors and grouping, and even factoring of special products, like the difference of two perfect squares and perfect-square trinomials. When faced with one or more problems to solve, however, there are always the same questions: Do I have to factor? If I do, where do I start? Is there a way to select the most appropriate method of factoring for a given problem?

Your team should answer these questions. Some of your team members might use the library to examine several mathematics textbooks and, if possible, solutions manuals to discover the approaches of different authors. Other team members might discuss these questions with mathematics instructors at your school, at nearby colleges, and in local high schools. Then, as a team, discuss your findings and use them to formulate answers to the three questions. In some cases, there may be more than one right answer.

Make an oral presentation to your class related to the questions, how you researched the answers to the questions, and the answers themselves. You may want to experiment with presentation software for this presentation. Be prepared to answer questions on your methods and solutions.

12–1 The distributive property and common factors

1. Factor an expression containing a common factor.

12–2 Multiplying polynomials

1. Use the FOIL method to multiply two binomials.
2. Multiply polynomials to obtain two special products.

12–3 Factoring special products

1. Recognize and factor the difference of two perfect squares.
2. Recognize and factor a perfect-square trinomial.

12–4 Factoring general trinomials

1. Factor general trinomials whose squared term has a coefficient of 1.
2. Remove common factors after grouping an expression.
3. Factor a general trinomial by grouping.
4. Factor a general trinomial by trial and error.
5. Factor any binomial or trinomial that is not prime.

Throughout our study of mathematics, we have examined products and factors. To reduce fractions, we looked for factors common to both the numerator and denominator. In examining perfect squares, we looked for two identical factors to find the square root of a value. In this chapter, we again find it useful to examine products and factors.

<table>
<tr><td>12-1</td><td></td></tr>
</table>

12-1 THE DISTRIBUTIVE PROPERTY AND COMMON FACTORS

1 Factor an expression containing a common factor.

We applied the distributive property and finding common factors earlier in the text and applied them in different contexts. In this section, rather than use the distributive property to multiply and obtain a product, we will start with a product and re-generate the factors that produce the product. In other words, we want to undo the multiplication. Factoring resembles division, which is the inverse operation of multiplication.

1 Factor an Expression Containing a Common Factor.

The multiplication problem $7a(3a + 2)$ is the indicated product of $7a$ and the grouped quantity $3a + 2$. This is the *factored* form of the expression. After the expression is multiplied, we have two terms written as the indicated sum $21a^2 + 14a$. This is the *expanded* form. To rewrite the expression $21a^2 + 14a$ as the indicated product $7a(3a + 2)$ is to *factor* it.

Let's look at a general example of the distributive property:

$$a(x + y) = ax + ay$$

Notice that a appears as a factor in both terms on the right side of the equals sign. When a factor appears in each of several terms, it is called a *common factor* of the terms. The distributive property in reverse can be used to write the addition as a multiplication. In other words, we can *factor* the expression.

EXAMPLE Write $3a + 3b$ in factored form.

Because 3 is the common factor in both terms, we can use the distributive property to factor the expression.

$$3a + 3b = 3(a + b)$$

The distributive property also applies if we have more than two terms.

EXAMPLE Write $3ab + 9a + 12b$ in factored form.

Because 3 is the only factor that appears in all three terms of the expression, it is the common factor.

$$3ab + 9a + 12b = 3(ab + 3a + 4b)$$

$$3 \cdot 3 \qquad 3 \cdot 4$$

> **To factor an expression containing a common factor:**
>
> **1.** Find the *greatest* factor common to *each* term of the expression.
> **2.** Divide each term by the common factor.
> **3.** Rewrite the expression as the indicated product of the greatest common factor (GCF) and the quotients in Step 2.

When looking for a common factor, we always look for *all* common factors.

EXAMPLE   Factor $10a^2 + 6a$ completely.

Because 2 and a are common factors, the greatest common factor is $2a$.

$$10a^2 + 6a = 2a(5a + 3) \qquad \frac{10a^2}{2a} = 5a, \frac{6a}{2a} = 3$$

We check our factoring by multiplying.

$$2a(5a + 3) = 10a^2 + 6a$$

This process of checking can be done mentally.

EXAMPLE   Factor $2x^2 + 4x^3$ completely.

Because 2 and x^2 are common factors, the greatest common factor is $2x^2$.

$$2x^2 + 4x^3 = 2x^2(1 + 2x) \qquad \frac{2x^2}{2x^2} = 1, \frac{4x^3}{2x^2} = 2x$$

Note that a 1 remains in the grouping because $2x^2$ is the product of $2x^2$ and 1.

Tip!	**When Is It Necessary to Write a 1?**

We have found that it is not always necessary to write the number 1. When is this the case? When 1 is a *factor* (multiplicative property of 1, $1 \cdot n = n$), the 1 does not have to be written. When the exponent of a factor is 1 (definition of the exponent of 1, $a^1 = a$), the 1 does not have to be written. On the other hand, **when 1 is a term** instead of a factor **it must be written.** Look at the previous example again.

To check factoring, multiply the factors. $2x^2(1 + 2x) = 2x^2 + 4x^3$. The product $2x^2(2x)$ would not regenerate the original polynomial.

Examine the factor $(a + b)$ in the example $3a + 3b = 3(a + b)$. This factor results from the division of $\frac{3a}{3}$ and $\frac{3b}{3}$. In each case, the quotient is $1a$ and $1b$, respectively; but the 1's are factors of a term instead of separate terms. Thus, the factors of 1 do not have to be written.

Now, let's answer the question, When is it necessary to write a 1?

Term of 1	Factor of 1, Exponent of 1
Necessary	**Optional**

SELF-STUDY EXERCISES 12–1

1   Factor completely. Check.

1. $7a + 7b$

2. $m^2 + 2m$

3. $6x^2 + 3x$

4. $5ab + 10a + 20b$

5. $4ax^2 + 6a^2x$

6. $5a - 7ab$

7. $12a^2 - 15a + 6$

8. $3x^3 - 9x^2 - 6x$

9. $8a^2b + 14ab^3$

10. $3m^2 - 6m^3$

Learning Outcomes

1   Use the FOIL method to multiply two binomials.
2   Multiply polynomials to obtain two special products.

In this section, we multiply an expression by another expression rather than by a group of factors. The basic concept involves using the distributive property more than once. Procedures that encourage doing some of the steps mentally are introduced.

1   Use the FOIL Method to Multiply Two Binomials.

As we expand our experiences with multiplication, we need to use the appropriate terminology associated with polynomials. Recall:

- A *polynomial* is an expression with constants and variables with whole-number exponents and contains one or more terms with at least one variable term.

Monomials, binomials, and trinomials are polynomials.

- A *monomial* contains one term, such as $5x^2$.
- A *binomial* contains two terms, such as $3a + 4$.
- A *trinomial* contains three terms, such as $4x^2 + x - 2$.

We multiply two binomials by using the distributive property. According to the distributive property, each term of the first factor is multiplied by each term of the second factor. This means we use the distributive property more than once.

EXAMPLE   Multiply $(x + 4)(x + 2)$.

$$(x + 4)(x + 2) = x(x + 2) + 4(x + 2) \qquad \text{Distributive property.}$$
$$= x^2 + 2x + 4x + 8 \qquad \text{Distribute.}$$
$$= x^2 + 6x + 8 \qquad \text{Combine like terms.}$$

To guide you through the applications of the distributive property, we use the word **FOIL.** Each letter stands for a step in the process.

To multiply two binomials by the FOIL method:

First   Last   <u>F</u>irst   <u>O</u>uter   <u>I</u>nner   <u>L</u>ast

$$(a + b)(c + d) = ac + ad + bc + bd$$

Inner

Outer

In this systematic process using the word FOIL, F is the product of the *first* terms in each factor, O is the product of the two *outer* terms, I is the product of the *inner* terms, and L is the product of the *last* terms in each factor. If the inner and outer products are like terms, they should be combined.

EXAMPLE   Use the FOIL method to multiply $(2x - 3)(x + 1)$.

$$\overset{\text{F}}{} \quad \overset{\text{O}}{} \quad \overset{\text{I}}{} \quad \overset{\text{L}}{}$$
$$(2x - 3)(x + 1) = 2x^2 + 2x - 3x - 3$$
$$= 2x^2 - x - 3$$

2 Multiply Polynomials to Obtain Two Special Products.

When the factors being multiplied have certain characteristics or conditions, we can anticipate the product without going through all the steps. Identifying and using these special characteristics allows us to do many multiplications mentally.

When we multiply the sum of two terms by the difference of the same two terms, we make some special observations about the product.

$$\overset{\mathbf{F}\quad\mathbf{O}\quad\mathbf{I}\quad\mathbf{L}}{(x + 3)(x - 3) = x^2 - 3x + 3x - 9 = x^2 - 9}$$

$$\underbrace{}_{0}$$

Examine the final product. Notice that the product has only *two* terms and is a *difference*. The sum of the outer and inner products is zero ($-3x + 3x = 0$). The terms of the product are *perfect squares*. Now compare the product with its factors. The first term of the product is the *square of the first term* in either of its factors. The second term in the product is the *square of the second term* in either of its factors.

We call pairs of factors like $(a + b)(a - b)$ *the sum and difference of the same two terms* or *conjugate pairs*. The product, $a^2 - b^2$, is called *the difference of two perfect squares*.

To mentally multiply the sum and difference of the same two terms:

1. Square the first term.
2. Insert a minus sign.
3. Square the second term.

EXAMPLE Find the products mentally.

(a) $(x + 2)(x - 2)$ (b) $(a - 4)(a + 4)$
(c) $(2m + 3)(2m - 3)$ (d) $(2a + y)(2a - y)$

Since each of these is the sum and difference of the same two terms, the product is the difference of two perfect squares.

(a) $(x + 2)(x - 2) = x^2 - 4$

(b) $(a - 4)(a + 4) = a^2 - 16$

(c) $(2m + 3)(2m - 3) = 4m^2 - 9$

(d) $(2a + y)(2a - y) = 4a^2 - y^2$

When we multiply the same two binomials, we also make some special observations. Because $x \cdot x = x^2$, $(x + 3)(x + 3) = (x + 3)^2$. The quantity $(x + 3)^2$ is called a *binomial square* or the *square of a binomial*.

EXAMPLE Find the products of the binomial squares.

(a) $(x + 2)^2$ (b) $(x - 3)^2$ (c) $(2x - 5)^2$ (d) $(3a - 2b)^2$

To find the products, write each binomial square as two factors; then use the FOIL method to multiply.

(a) $(x + 2)^2 = (x + 2)(x + 2) = x^2 + 4x + 4$ Outer + Inner: $2x + 2x = 4x$

(b) $(x - 3)^2 = (x - 3)(x - 3) = x^2 - 6x + 9$ Outer + Inner: $-3x - 3x = -6x$

(c) $(2x - 5)^2 = (2x - 5)(2x - 5)$

$\qquad = 4x^2 - 20x + 25$

Outer + Inner: $-10x - 10x = -20x$

(d) $(3a - 2b)^2 = (3a - 2b)(3a - 2b)$

$\qquad = 9a^2 - 12ab + 4b^2$

Outer + Inner: $-6ab - 6ab = -12ab$

Notice that each product is a trinomial (three terms). These are special trinomials called *perfect-square trinomials.* In each, the first and last terms are positive and perfect squares, and result from squaring the terms of the binomial. To get the middle term of the trinomial, we combine the outer and inner products. Because the outer and inner products are the same, the middle term can be found mentally by doubling the product of the two terms of the binomial. Let's apply these observations to an example.

EXAMPLE Find the product of $(2x + 5)^2$.

$(2x + 5)^2 = 4x^2 + \underline{\quad} + 25$ Square binomial terms.

$(2x + 5)^2 = 4x^2 + 20x + 25$ Double product of binomial terms.
$2x(5) = 10x$; $2(10x) = 20x$.

To mentally square a binomial:

1. Square the first term.
2. Double the product of the two terms.
3. Square the second term.

EXAMPLE Square the binomials mentally.

(a) $(x + 2)^2$ (b) $(2x - 3)^2$ (c) $(5x + 1)^2$ (d) $(7x + 2)^2$

	Square first term	**Double product of terms**	**Square second term**
(a) $(x + 2)^2 =$	$(x)^2$ x^2	$2 \cdot (2x)$ $+ \ 4x$	$(2)^2$ $+ \ 4$
(b) $(2x - 3)^2 =$	$(2x)^2$ $4x^2$	$2 \cdot (-6x)$ $- \ 12x$	$(-3)^2$ $+ \ 9$
(c) $(5x + 1)^2 =$	$(5x)^2$ $25x^2$	$2 \cdot (5x)$ $+ \ 10x$	$(1)^2$ $+ \ 1$
(d) $(7x + 2)^2 =$	$(7x)^2$ $49x^2$	$2 \cdot (14x)$ $+ \ 28x$	$(2)^2$ $+ \ 4$

SELF-STUDY EXERCISES 12–2

1 Use the FOIL method to find the products. Practice combining the outer and inner products mentally.

1. $(a + 3)(a + 8)$
2. $(x - 4)(x + 5)$
3. $(y - 7)(y - 3)$
4. $(2a - 3)(a + 4)$
5. $(3a - 2b)(a - 2b)$
6. $(5x - y)(c - 5y)$
7. $(3x - 4)(2x - 3)$
8. $(a - b)(2a - 5b)$
9. $(7 - m)(3 - 7m)$
10. $(5 - 2x)(8 - x)$
11. $(x + 7)(x + 4)$
12. $(y - 7)(y - 5)$
13. $(m + 3)(m - 7)$
14. $(3b - 2)(x + 6)$
15. $(4r - 5)(3r + 2)$
16. $(5 - x)(7 - 3x)$
17. $(4 - 2m)(1 - 3m)$
18. $(2 + 3x)(3 + 2x)$
19. $(x + 3)(2x - 5)$
20. $(5x - 7y)(4x + 3y)$
21. $(2a + 3b)(7a - b)$
22. $(5a + 2b)(6a - 5b)$
23. $(9x - 2y)(3x + 4y)$
24. $(5x - 8y)(4x - 3y)$
25. $(7m - 2n)(3m + 5n)$

2 Find the special products mentally.

26. $(a + 3)(a - 3)$ **27.** $(2x + 3)(2x - 3)$ **28.** $(a - y)(a + y)$
29. $(4r + 5)(4r - 5)$ **30.** $(5x + 2)(5x - 2)$ **31.** $(7 + m)(7 - m)$
32. $(2 - 3x)^2$ **33.** $(3x + 4)^2$ **34.** $(Q + L)^2$
35. $(a^2 + 1)^2$ **36.** $(2d - 5)^2$ **37.** $(3a + 2x)^2$
38. $(3x - 7)(3x + 7)$ **39.** $(6 + Q)^2$ **40.** $(y - 5x)(y + 5x)$
41. $(4 - 3j)^2$ **42.** $(3m - 2p)(3m + 2p)$ **43.** $(m^2 + p^2)^2$
44. $(2a - 7c)(2a + 7c)$ **45.** $(9 - 13a)^2$

12–3   FACTORING SPECIAL PRODUCTS

Learning Outcomes

1 Recognize and factor the difference of two perfect squares.
2 Recognize and factor a perfect-square trinomial.

To factor any of the special products of the preceding section, we apply the inverse of the process. That is, we start with the product and "work back" to the factors that produce these special products.

To rewrite a special product in factored form, we must recognize the product as a pattern. Once we identify the special product, then we must know the pattern of the product in factored form.

1 Recognize and Factor the Difference of Two Perfect Squares.

The difference of two perfect squares, such as $x^2 - a^2$, factors into the product of two binomials, $(x + a)(x - a)$. These factors are the sum and difference of the same two terms. The terms of each binomial are the square roots of the terms of the expression to be factored. Before we can factor such a special product, we must be able to recognize an expression as the difference of two perfect squares.

EXAMPLE Identify the special products that are the difference of two perfect squares.

(a) $x^2 - 9$ (b) $a^2 + 49$ (c) $m^2 - 27$ (d) $3y^2 - 25$
(e) $9x^2 - 4$ (f) $4x^2 - 4x + 1$

(a) Difference of two perfect squares.
(b) Not the difference of two perfect squares; this is a *sum,* not a difference.
(c) Not the difference of two perfect squares; 27 is not a perfect square.
(d) Not the difference of two perfect squares; in $3y^2$, 3 is not a perfect square.
(e) Difference of two perfect squares.
(f) Not the difference of two perfect squares; this is not a binomial.

To factor the difference of two perfect squares:

1. Take the square root of the first term.
2. Take the square root of the second term.
3. Write one factor as the *sum* of the square roots found in Steps 1 and 2, and write the other as the *difference* of the square roots found in Steps 1 and 2.

EXAMPLE Factor the special products, which are the differences of two perfect squares.

(a) $a^2 - 9$ (b) $x^2 - 36$ (c) $4x^2 - 1$ (d) $16m^2 - 49$

(a) $a^2 - 9 = (a + 3)(a - 3)$ (b) $x^2 - 36 = (x + 6)(x - 6)$
(c) $4x^2 - 1 = (2x + 1)(2x - 1)$ (d) $16m^2 - 49 = (4m + 7)(4m - 7)$

<table>
<tr><td>Tip!</td><td>Order of Factors.</td></tr>
</table>

Because multiplication is commutative, the factors given as answers in the preceding example may be expressed in any order, such as $(a + 3)(a - 3)$ or $(a - 3)(a + 3)$, $(x + 6)(x - 6)$ or $(x - 6)(x + 6)$, and so on.

2 Recognize and Factor a Perfect-Square Trinomial.

A trinomial is a *perfect-square trinomial* if the first and last terms are positive perfect squares and the absolute value of the middle term is *twice* the product of the square roots of the first and last terms. Again, we need to be able to distinguish these special products from other expressions before we factor them.

EXAMPLE Tell why the trinomials are *not* perfect-square trinomials.

(a) $x^2 + 2x - 1$ (b) $4x^2 + 6x + 9$
(c) $x^2 - 5x + 4$ (d) $-4x^2 - 4x + 1$

(a) The last term, -1, is negative. This term must be positive in a perfect-square trinomial.
(b) The middle term, $6x$, is not twice the product of $2x$ and 3.
(c) The middle term, $-5x$, is not twice the product of x and 2.
(d) The first term, $-4x^2$, is negative. It should be positive.

EXAMPLE Verify that the trinomials are perfect-square trinomials.

(a) $x^2 + 14x + 49$ (b) $4m^2 - 12m + 9$ (c) $9x^2 + 24xy + 16y^2$

(a) The first and last terms, x^2 and 49, are positive perfect squares. The middle term, $14x$, has an absolute value of twice the product of the square roots of x^2 and 49. And, $2(7x) = 14x$.
(b) The first and last terms, $4m^2$ and 9, are positive perfect squares. The middle term, $-12m$, has an absolute value of twice the product of the square roots of $4m^2$ and 9. And, $2(2m \cdot 3) = 12m$.
(c) The first and last terms, $9x^2$ and $16y^2$, are positive perfect squares. The middle term, $24xy$, has an absolute value of twice the product of the square roots of $9x^2$ and $16y^2$. And, $2(3x \cdot 4y) = 24xy$.

To factor a perfect-square trinomial:

1. Write the square root of the first term.
2. Write the sign of the middle term.
3. Write the square root of the last term.
4. Indicate the square of this binomial quantity.

EXAMPLE Factor the perfect-square trinomials in the preceding example.

(a) $x^2 + 14x + 49$

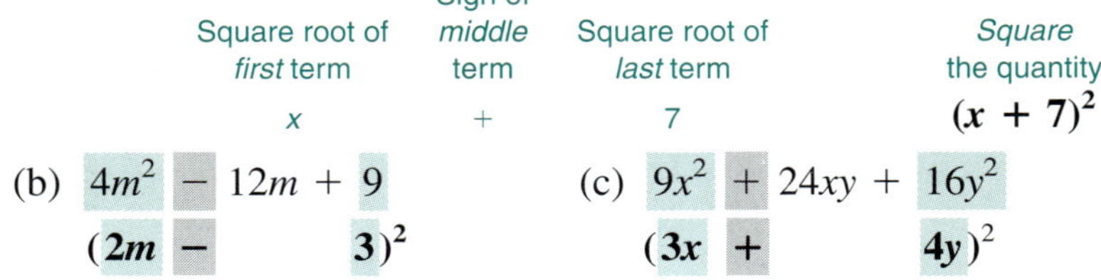

1 Identify the special products that are the difference of two perfect squares.

1. $r^2 - s^2$

2. $d^2 - 4d + 10$

3. $4y^2 - 16$

4. $36m^2 - 9n^2$

5. $9 + 4a^2$

6. $64p^2 - q^2$

Factor the following special products.

7. $y^2 - 49$

8. $16x^2 - 1$

9. $9a^2 - 100$

10. $4m^2 - 81n^2$

11. $9x^2 - 64y^2$

12. $25x^2 - 64$

13. $100 - 49x^2$

14. $4x^2 - 49y^2$

15. $121m^2 - 49n^2$

16. $81x^2 - 169$

2 Verify which of the trinomials are perfect-square trinomials.

17. $4y^2 + 2y + 16$

18. $9m^2 - 24mn + 16n^2$

19. $16a^2 + 8a - 1$

20. $-9r^2 + 12r + 4$

21. $y^2 - 14y + 49$

22. $p^2 + 10p + 25$

Factor the special products.

23. $x^2 + 6x + 9$

24. $x^2 + 14x + 49$

25. $x^2 - 12x + 36$

26. $x^2 - 16x + 64$

27. $4a^2 + 4a + 1$

28. $25x^2 - 10x + 1$

29. $9m^2 - 48m + 64$

30. $4x^2 - 36x + 81$

31. $x^2 - 12xy + 36y^2$

32. $4a^2 - 20ab + 25b^2$

12–4 FACTORING GENERAL TRINOMIALS

Learning Outcomes

1 Factor general trinomials whose squared term has a coefficient of 1.

2 Remove common factors after grouping an expression.

3 Factor a general trinomial by grouping.

4 Factor a general trinomial by trial and error.

5 Factor any binomial or trinomial that is not prime.

We often have expressions that do not fit the special cases described in Section 12–3. We now look at trinomials that are not special products. There are many instances in the study of mathematics when writing polynomials in factored form facilitates the solving of certain types of problems. We use factoring as one method for solving quadratic equations (Chapter 13).

1 **Factor General Trinomials Whose Squared Term Has a Coefficient of 1.**

Recall the FOIL method of multiplying two binomials, $(x + 3)(x + 2)$. x^2 is the product of the First terms of each binomial. $2x + 3x = 5x$ is the sum of the products of the Outer and Inner terms. $+6$ is the product of the Last terms of each binomial.

The product is a trinomial, $x^2 + 5x + 6$. If we wish to factor this trinomial, we begin by indicating that it factors into two binomials and by looking at characteristics of the trinomial.

$$(\quad)(\quad)$$

If the sign of the *third* term is positive, the signs of the two binomial factors are alike. The sign of the middle term tells us which sign to use. In the trinomial $x^2 + 5x + 6$, the sign of the 6 is positive, so the binomial signs are alike. Because the sign of the $5x$ is positive, the signs of both binomial factors are positive.

$$(\quad + \quad)(\quad + \quad)$$

Next, we write the factors of the *first* term in the *first* position of each binomial.

$$(x +)(x +)$$

The factors of x^2 are x and x.

Then, we write the factors of the *last* term in the *last* position of each binomial. We have more than one pair of factors whose product is $+6$ ($+3$ and $+2$; -3 and -2; $+6$ and $+1$; -6 and -1). *We choose the pair that will give us the correct middle term, including its sign when we multiply the two binomials together.* The positive sign on the third term indicates that we need two factors whose *sum* is $+5$, so we choose $+3$ and $+2$.

$$(x + 3)(x + 2) = x^2 + 5x + 6 \qquad \text{Only this choice gives the correct middle term, } +5x.$$

$$\left.\begin{array}{l} (x - 3)(x - 2) = x^2 - 5x + 6 \\ (x + 6)(x + 1) = x^2 + 7x + 6 \\ (x - 6)(x - 1) = x^2 - 7x + 6 \end{array}\right\} \quad \text{These choices give the wrong sign and/or the wrong value for the middle term.}$$

Therefore, the factors of the trinomial $x^2 + 5x + 6$ are $(x + 3)(x + 2)$. Because multiplication is commutative, the two factors may be written in the opposite order, $(x + 2)(x + 3)$.

Tip!	***Start with the Signs.***

In the examples that follow, pay particular attention to the signs of the middle and last terms of the trinomials and their effect on the signs of the binomial factors.

One procedure for determining the factors that produce a trinomial product is to *factor by inspection.* This means that after examining the problem, we logically deduce the appropriate solution.

EXAMPLE Factor $x^2 + 6x + 5$ by inspection.

$$()() \qquad \text{Write parentheses for two binomial factors.}$$

$$(x +)(x +) \qquad \text{The sign on } +5 \text{ indicates like signs, and the sign on the } +6x \text{ indicates both signs are positive. Write the factors of } x^2 \text{ as the first term of each binomial.}$$

$$(x + 5)(x + 1) \qquad \text{Write the factors of } +5, (+5 \text{ and } +1), \text{ as the last term of each binomial. } Note: +5 \text{ of the trinomial has two sets of factors: } +5, +1 \text{ and } -5, -1. \text{ We choose } +5 \text{ and } +1 \text{ because the sign of the middle term is positive.}$$

$$(x + 5)(x + 1) = x^2 + 6x + 5 \qquad \text{Multiply the binomials to check factoring.}$$

The correct factoring of $x^2 + 6x + 5$ is $(x + 5)(x + 1)$.

EXAMPLE Factor $x^2 - 4x + 3$ by inspection.

$$()() \qquad \text{Indicate binomial factors.}$$

$$(x -)(x -) \qquad \text{The sign on } +3 \text{ indicates that both signs are the same; the sign on } -4x \text{ tells us to use negative signs. Write the factors of } x^2 \text{ as the first term of each binomial.}$$

$$(x - 1)(x - 3) \qquad \text{Write the factors of } +3, (-1 \text{ and } -3), \text{ as the last terms. } Note: +3 \text{ of the trinomial has two sets of factors: } -1, -3 \text{ and } +1, +3. \text{ We choose } -1 \text{ and } -3 \text{ because the sign of the middle term is negative.}$$

$$(x - 1)(x - 3) = x^2 - 4x + 3 \qquad \text{Check by multiplying.}$$

The correct factoring of $x^2 - 4x + 3$ is $(x - 1)(x - 3)$.

Factor $x^2 - 2x - 8$ by inspection.

$$(\quad)(\quad) \qquad \text{Indicate binomial factors.}$$

$$(x - \quad)(x + \quad) \qquad \text{Write the factors of } x^2. \text{ The sign of } -8 \text{ indicates that the signs will be different.}$$

$$(x - 2)(x + 4)$$

$$(x - 4)(x + 2)$$

To find the factors of -8 that give a middle term of $-2x$, find the two factors of 8 that have a difference of 2. Then, use the negative sign with the larger factor.

$$(x - 8)(x + 1)$$

$$(x - 1)(x + 8)$$

$$(x - 4)(x + 2) = x^2 - 2x - 8 \qquad \text{Check by multiplying.}$$

The correct factoring of $x^2 - 2x - 8$ is $(x - 4)(x + 2)$.

When trinomials have terms whose coefficients have several pairs of factors, such as in the trinomial $12a^2 + 28a + 15$, more cases have to be considered. Two methods of factoring general trinomials are common. The one method is systematic, while the other requires inspection or trial and error. Before we use the systematic method, we need to learn to factor by grouping.

2 Remove Common Factors after Grouping an Expression.

Algebraic expressions that have no factors common to every term in the expression may have common factors for some groups of terms. In these cases, the expression is written in factored form. In the expression $2x^2 - 2xb + ax - ab$, there are four terms and no factor is common to each term. To factor, write the expression as two terms by *grouping* the first two terms and the last two terms; then look for common factors in each grouping.

$$(2x^2 - 2xb) + (ax - ab)$$

Tip!	*Groupings Are Not Necessarily Factors.*

The groupings $(2x^2 - 2xb)$ and $(ax - ab)$ are not factors. Notice that they are groupings that are added. Additional factoring techniques are used to write the expression in factored form.

The first grouping, $(2x^2 - 2xb)$, has a common factor of $2x$ and can be written in factored form as $2x(x - b)$. The second grouping, $(ax - ab)$, has a common factor of a and can be written in factored form as $a(x - b)$. If we look at the entire expression $2x(x - b) + a(x - b)$, we see the common binomial factor $(x - b)$. When we factor this common factor from both terms, we have $(x - b)(2x + a)$. The expression is now one term that is written as a product of two factors. We can check the factoring by using the distributive property or the FOIL method of multiplying.

$$(x - b)(2x + a) = 2x^2 + ax - 2xb - ab$$

This result is the same as the original expression.

EXAMPLE   Write the expressions in factored form by using grouping.

(a) $mx + 2m - 4x - 8$ (b) $y^2 + 2xy + 3y + 6x$
(c) $3x^2 - 9x - 7x + 21$ (d) $2x^2 + 8x + 5y - 15$

(a) $mx + 2m - 4x - 8 =$ Group the four-termed expression into two terms.

$(mx + 2m) + (-4x - 8) =$ Factor out any common factor in each of the two terms.

$m(x + 2) + -4(x + 2) =$ Convert double signs to an equivalent single sign.

$m(x + 2) - 4(x + 2) =$ Factor out the common factor $(x + 2)$.

$(x + 2)(m - 4)$ Check the result by using the FOIL method.

$(x + 2)(m - 4) = xm - 4x + 2m - 8$
$\qquad\qquad\quad = mx + 2m - 4x - 8$ Rearrange terms and factors (commutative properties of multiplication and addition).

(b) $y^2 + 2xy + 3y + 6x =$ Group into two terms.

$(y^2 + 2xy) + (3y + 6x) =$ Factor out the common factor in each term.
$y(y + 2x) + 3(y + 2x) =$ Factor into one term by factoring out the common binomial factor.

$(y + 2x)(y + 3)$

(c) $3x^2 - 9x - 7x + 21 =$ Group into two terms.

$(3x^2 - 9x) + (-7x + 21) =$ Factor each term.

$3x(x - 3) + -7(x - 3) =$ Convert double signs to a single sign.

$3x(x - 3) - 7(x - 3) =$ Factor into one term by factoring out the common binomial factor.

$(x - 3)(3x - 7)$

(d) $2x^2 + 8x + 5y - 15 =$ Group into two terms.

$(2x^2 + 8x) + (5y - 15) =$ Factor each term.

$2x(x + 4) + 5(y - 3)$

In this example, the two terms do not have a common factor. Thus, the expression cannot be factored into one term. Even if we rearrange the terms, we will not be able to write the expression as a single term.

$$2x^2 + 8x + 5y - 15 \text{ is prime}$$

Tip! | ***Make Leading Coefficient of Binomial Positive.***

When factoring by grouping, manipulate the signs of the common factor so that the leading coefficient of the binomial is positive.

$$-3x + 9$$

Factor as $-3(x - 3)$, *not* $3(-x + 3)$. Parts a and c of the preceding example illustrate this tip.

---

3   **Factor a General Trinomial by Grouping.**

We can now use a systematic method to factor general trinomials. First, we factor trinomials of the type $ax^2 + bx + c$.

> **To factor a general trinomial of the form $ax^2 + bx + c$ by grouping:**
>
> **1.** Multiply the coefficient of the first term by the coefficient of the last term.
> **2.** Factor the product from Step 1 into two factors:
>   **(a)** whose *sum* is the coefficient of the middle term if the sign of the last term is positive, or
>   **(b)** whose *difference* is the coefficient of the middle term if the sign of the last term is negative.
> **3.** Rewrite the trinomial so that the middle term is the sum or difference from Step 2.
> **4.** Divide the trinomial from Step 3 into two groups with a common factor in each.
> **5.** Factor out the common factor.

We learned earlier that trinomials with a positive third term factor into two binomials with like signs.

EXAMPLE Factor $6x^2 + 19x + 10$ by grouping.

$$6(10) = 60$$
Multiply the coefficients of the first and third terms.

$$60 = (1)60$$
List all factor pairs of 60.

$$(2)30$$

$$(3)20$$

$$(4)15$$
Identify the pair that *adds* to 19, since the sign of the last term is positive.

$$(5)12$$

$$(6)10$$

$$6x^2 + 19x + 10 =$$
Separate $+19x$ into two terms using the coefficients 4 and 15.

$$6x^2 + 4x + 15x + 10 =$$
Group into two terms.

$$(6x^2 + 4x) + (15x + 10) =$$
Factor each term.

$$2x(3x + 2) + 5(3x + 2) =$$
Factor the common binomial.

$$(3x + 2)(2x + 5)$$
Check using the FOIL method.

EXAMPLE Factor $20x^2 - 23x + 6$ by grouping.

$$20(6) = 120$$
Find the product of 20 and 6.

List the factor pairs of 120 and select the pair that has a sum of 23.

$$120 = (1)120$$
List all factor pairs of 120.

$$(2)60$$

$$(3)40$$

$$(4)30$$

$$(5)24$$

$$(6)20$$

$$(8)15$$
Identify the pair that adds to 23, since the sign of the last term is positive.

$$(10)12$$

$$20x^2 - 23x + 6 =$$
Separate $-23x$ into two terms using the coefficients -8 and -15.

$$20x^2 - 8x - 15x + 6 =$$
Group into two terms.

$$(20x^2 - 8x) + (-15x + 6) =$$
Factor each term.

$$4x(5x - 2) - 3(5x - 2) =$$ Factor the common binomial.

$$(5x - 2)(4x - 3)$$ Check using the FOIL method.

If the third term of the trinomial has a negative sign, we use the same procedure but look for two factors whose *difference* is the coefficient of the middle term.

EXAMPLE   Factor $10x^2 + 19x - 15$ by grouping.

$10(15) = 150$ Find the product of 10 and 15.

Systematically factor 150 to find two factors whose difference is 19.

$150 = (1)150$ List all factor pairs of 150.
$(2)75$
$(3)50$
$(5)30$
$(6)25$ Identify the pair that subtracts to 19, since the last term is negative.
$(10)15$

The factors 25 and 6 have a difference of 19. When you rewrite the trinomial, write $19x$ as $25x - 6x$.

$$10x^2 + 19x - 15 =$$ Separate $+19x$ into two terms using the coefficients -6 and 25. The signs will be different and the larger factor will be positive.

$$10x^2 + 25x - 6x - 15 =$$ Group into two terms.

$$(10x^2 + 25x) + (-6x - 15) =$$ Factor each term.

$$5x(2x + 5) - 3(2x + 5) =$$ Factor the common binomial.

$$(2x + 5)(5x - 3)$$ Check using the FOIL method.

Now, let's factor by grouping a trinomial with a middle term and end term that are negative. Because the last term is negative, we look for two factors with a *difference* that is the coefficient of the middle term.

EXAMPLE   Factor $x^2 - 2x - 8$ by grouping.

$1(8) = 8$ Find the product of 1 and 8, the coefficients of the first and last terms.

Factor 8 so that the difference of the absolute values of the factors is 2, the coefficient of the middle term.

$8 = (1)8$ List all factor pairs of 8.
$(2)4$ Identify the pair that subtracts to 2.

$$x^2 - 2x - 8 =$$ Separate $-2x$ into two terms using the coefficients 2 and 4. The signs will be different. The larger factor will be negative.

$$x^2 + 2x - 4x - 8 =$$ Group into two terms.

$$(x^2 - 4x) + (2x - 8) =$$ Factor each term.

$$x(x - 4) + 2(x - 4) =$$ Factor the common binomial.

$$(x - 4)(x + 2)$$ Check using the FOIL method.

<table>
<tr><td>Tip!</td><td>Shortening the Process for Factoring by Grouping.</td></tr>
</table>

The following is a variation of the factor-by-grouping method. The variation is mathematically sound and employs a strategy using the property of 1 that is often overlooked.

Factor $6x^2 + 7x - 20$.

$6x^2 + 7x - 20$		Multiply the coefficients of the first and third terms: $6(-20) = -120$.
$120 =$	1 120	A negative product means the factor pair has unlike signs. List all factor pairs of 120.
	2 60	
	3 40	
	4 30	
	5 24	
	6 20	
	8 15	Identify the factor pair with a *difference* of $+7$. The larger factor will be positive. $-8 + 15 = 7$
	10 12	

Variation in procedure begins here.

$$\dfrac{(6x \quad)(6x \quad)}{6}$$ Use the coefficient of the first term as the coefficient of the first term in *each* binomial. This gives you an extra factor of 6. Compensate by *dividing* the *expression* by 6.

$$\dfrac{(6x - 8)(6x + 15)}{6} =$$ Use the factors of -120 that have an algebraic sum of 7 as the second term of each binomial.

$$\dfrac{2(3x - 4)(3)(2x + 5)}{6} =$$ Factor the common factors from each binomial.

$$\dfrac{6(3x - 4)(2x + 5)}{6} =$$ The factors of the trinomial will be the two binomial factors. The factor 6 reduces.

$$(3x - 4)(2x + 5)$$ Check using the FOIL method.

Why does this variation work? The coefficient of the first term in the original trinomial, 6, was used in each binomial. Thus, the resulting product would have an extra factor of 6. This extra factor of 6 must be removed to obtain the correct factors. **Now, let's shorten the process again.**

Once you have selected the pair of factors whose sum or difference matches the middle term, *make two fractions using the factor pair as the denominators*. The *leading coefficient of the trinomial will be the numerator of each fraction*.

$$6x^2 + 7x - 20 \qquad 6(-20) = -120$$

The factor pair of -120 that has a sum of $+7$ is -8 and $+15$. Make fractions and reduce each fraction.

$$\frac{6}{-8} = \frac{3}{-4} \qquad \frac{5}{+15} = \frac{2}{+5}$$

Each fraction gives the coefficients of one of the binomial factors. 3, -4 and 2, $+5$.

$$(3x - 4)(2x + 5)$$

If the trinomial has a common factor, you must factor the common factor before finding the factor pair and writing the fractions. The common factor will be part of the final answer.

Factor $30x^2 + 8 - 32x$.

$30x^2 - 32x + 8$	Arrange in descending powers of *x*.
$2(15x^2 - 16x + 4)$	Factor any common factors.

(continued)

$$15(4) = 60$$

1	60
2	30
3	20
4	15
5	12
6	10

Factor pairs of 60.

Identify the pair that has a *sum* of −16.

−6 + (−10) = −16

Make fractions of the factor pair −6 and −10 and the leading coefficient of the trinomial, 15. Reduce each fraction.

$$\frac{15}{-6} = \frac{5}{-2} \qquad \frac{15}{-10} = \frac{3}{-2}$$

Write the factors using the coefficients 5, −2 and 3, −2. Don't forget the common factor.

$$2(5x - 2)(3x - 2)$$

4 Factor a General Trinomial by Trial and Error.

Another method of factoring is commonly referred to as *trial and error,* but this name is a bit misleading. The procedure involves mentally using the FOIL method for all possible binomial pairs that give the correct first and last terms until we find the one that gives the correct middle term. The binomial pairs that do not produce the correct middle term are not actually errors. They are logical choices that did not produce the desired result. With experience using this method, the correct choice is often found by testing choices mentally and only writing the correct choice.

To factor any general trinomial by trial and error:

1. Indicate binomial factors with double parentheses (   )(   ).
2. List all possible factors of the first and last terms of the trinomial.
3. List all possible binomial factors using factors from Step 2.
4. Multiply binomial factors from Step 3 to find the pair that gives the correct middle term of the trinomial.
5. Check all work to be sure that the signs are correct.

To factor the trinomial $2a^2 + 5a + 3$ by trial and error, we place the factors of the first term, $2a^2$, in the first position of each binomial factor. There is only one choice.

$$(2a \quad)(a \quad)$$

Place the factors of the last term, 3, in the second position of each binomial factor. There is one factor pair but two possible arrangements.

$$(2a \quad 3)(a \quad 1)$$

or

$$(2a \quad 1)(a \quad 3)$$

The signs of 3 and 1 are both positive because all signs in the trinomial are positive.

$$(2a + 3)(a + 1)$$

or

$$(2a + 1)(a + 3)$$

To determine the correct factoring, multiply each product to see which yields $2a^2 + 5a + 3$.

$$(2a + 3)(a + 1) = 2a^2 + 2a + 3a + 3$$
$$= 2a^2 + 5a + 3$$
$$(2a + 1)(a + 3) = 2a^2 + 6a + a + 3$$
$$= 2a^2 + 7a + 3$$

The first factoring, $(2a + 3)(a + 1)$, is correct.

Tip!	***Shortening the Trial-and-Error Method.***

To make your work easier and less time-consuming, obtain the outer and inner products and add them *mentally.*

EXAMPLE Factor $6x^2 + 7x - 5$ by the trial-and-error method.

$6x$ and x; $3x$ and $2x$ Choices for the first terms of the binomials.

-5 and 1; 5 and -1 Choices for the second terms of the binomials.

Possible binomial factors	**Middle term**	**Observations**
$(6x + 5)(x - 1)$	$-x$	Since the sign of the last term, -5, is negative, we must use unlike signs for the factors.
$(6x - 5)(x + 1)$	$+x$	
$(6x + 1)(x - 5)$	$-29x$	
$(6x - 1)(x + 5)$	$+29x$	
$(3x + 5)(2x - 1)$	$+7x$	
$(3x - 5)(2x + 1)$	$-7x$	
$(3x + 1)(2x - 5)$	$-13x$	
$(3x - 1)(2x + 5)$	$+13x$	

The fifth pair of factors, $(3x + 5)(2x - 1)$, gives the correct middle term, $+7x$. **Thus, $(3x + 5)(2x - 1)$ is the correct factoring of $6x^2 + 7x - 5$.**

■ **Learning Strategy** *Practice Moves You from Systematic Processes to Intuitive Processes.*

Systematic processes help you understand the concepts and build confidence. They also help you develop your mathematical senses like your number sense, your spatial sense, and so on. The more you practice, the more you develop your intuitive senses.

Many students instinctively and automatically move to the trial-and-error process as these senses develop. The old adage "practice makes perfect" is true.

5 Factor Any Binomial or Trinomial That Is Not Prime.

Now let's develop a strategy for factoring any binomial or trinomial that can be factored. This strategy allows us to say with confidence that a particular binomial or trinomial will not factor or is prime. We used the word *prime* to identify positive numbers that have no factors except themselves and 1. When the word *prime* refers to algebraic expressions, it describes algebraic expressions that have only themselves and 1 as factors.

> ***To factor any binomial or trinomial:***
>
> Perform the following steps in order.
>
> **1.** Factor out the greatest common factor (if any).
> **2.** Check the binomial or trinomial to see if it is a special product.
>   **(a)** If it is the difference of two perfect squares, use the pattern.
>   **(b)** If it is a perfect-square trinomial, use the pattern.
> **3.** If there is no special product, factor by grouping or by trial and error.
> **4.** Examine each factor to see that it cannot be factored further.
> **5.** Check factoring by multiplying.

The steps for factoring any binomial or trinomial are presented visually in the flow chart (Fig. 12–1).

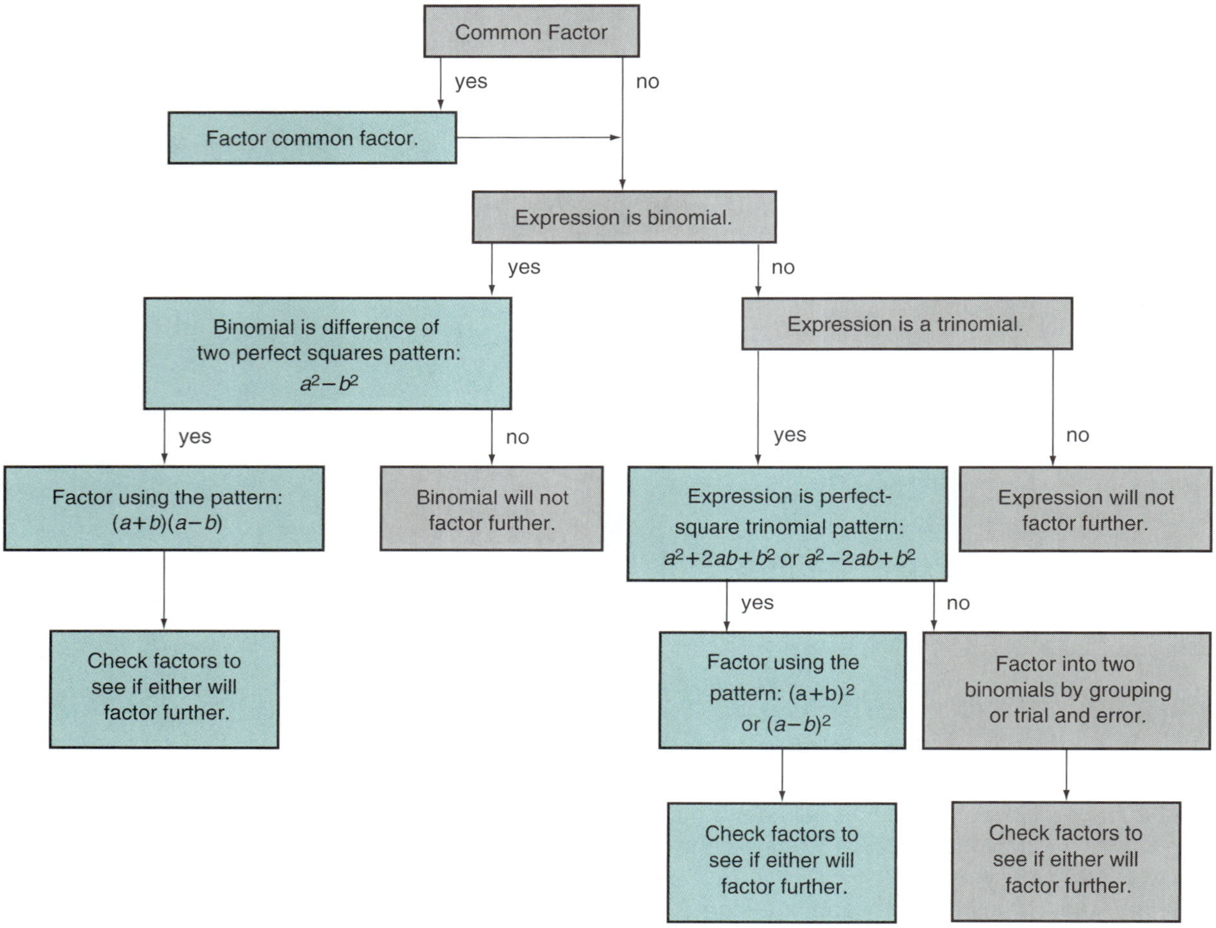

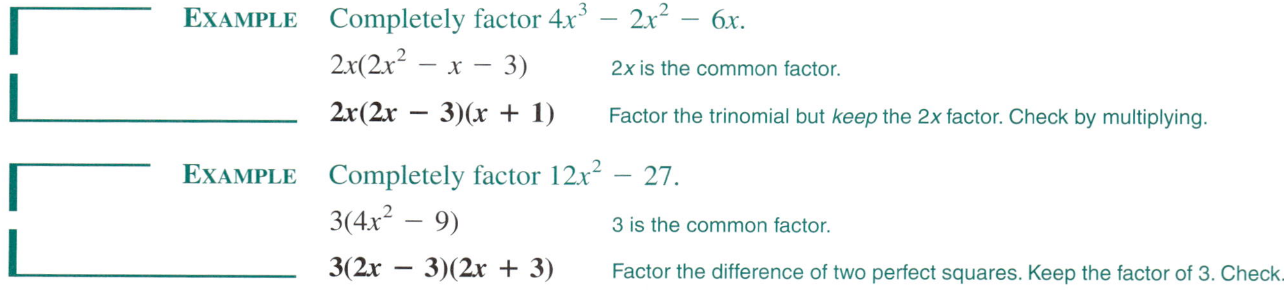

Figure 12-1   Factoring flow chart

EXAMPLE   Completely factor $4x^3 - 2x^2 - 6x$.

$2x(2x^2 - x - 3)$ $2x$ is the common factor.

$\mathbf{2x(2x - 3)(x + 1)}$ Factor the trinomial but *keep* the $2x$ factor. Check by multiplying.

EXAMPLE   Completely factor $12x^2 - 27$.

$3(4x^2 - 9)$ 3 is the common factor.

$\mathbf{3(2x - 3)(2x + 3)}$ Factor the difference of two perfect squares. Keep the factor of 3. Check.

EXAMPLE Completely factor $-18x^3 + 24x^2 - 8x$.

$$-2x(9x^2 - 12x + 4)$$

Look for a common factor: $-2x$. We often factor a negative when there are more negative terms than positive terms or when the term with the highest degree is negative. It is preferred that the leading coefficient in the binomial or trinomial be positive.

$$-2x(3x - 2)^2$$

Factor the perfect-square trinomial.

Tip!	**Why Do We Factor?**

There is general disagreement on the importance of factoring polynomials. Although the issue may never be resolved, examine the arguments for and against so that you can better understand the controversy.

For factoring:

- Many problem-solving techniques that use factoring can be done more quickly than with the formula-based or systematic technique.
- Properties and patterns are generally easier to recognize in factored form.
- Computers (and even humans) can often evaluate factored expressions more quickly.

Evaluate the following when $x = 4$ and $y = 3$.

Factoring technique:

$4x(2x + y)(x - 2y)$

$4 \cdot 4(2 \cdot 4 + 3)(4 - 2 \cdot 3)$

$16(11)(-2)$

-352

Formula-based technique:

$8x^3 - 12x^2y - 8xy^2$

$8 \cdot 4^3 - 12 \cdot 4^2 \cdot 3 - 8 \cdot 4 \cdot 3^2$

$8 \cdot 64 - 12 \cdot 16 \cdot 3 - 8 \cdot 4 \cdot 9$

$512 - 576 - 288$

-352

Against factoring:

- Real-world applications rarely have numbers and expressions that factor.
- Factoring techniques apply only to special cases. Formula-based techniques apply in general.
- With the accessibility of calculators and inexpensive computer software, formula-based techniques are more practical.

SELF-STUDY EXERCISES 12–4

1 Factor.

1. $x^2 + 5x + 6$
2. $x^2 - 11x + 28$
3. $x^2 + 8x + 12$
4. $x^2 - 4x + 3$
5. $x^2 + 8x + 7$
6. $x^2 + 7x + 10$
7. $x^2 - x - 12$
8. $y^2 - 3y - 10$
9. $y^2 - y - 6$
10. $a^2 + a - 12$
11. $b^2 + 2b - 3$
12. $14 - 5b - b^2$
13. $x^2 - 7x + 12$
14. $x^2 - x - 30$
15. $x^2 + 11x + 18$
16. $x^2 - 9x + 18$
17. $x^2 - 7x - 18$
18. $x^2 + 17x - 18$
19. $x^2 + 9x + 20$
20. $x^2 - 12x + 20$
21. $x^2 - 10x + 16$
22. $x^2 - 17x + 16$
23. $x^2 - 13x - 14$
24. $x^2 - 5x - 14$

2 Factor the polynomials by removing the common factors after grouping.

25. $x^2 + xy + 4x + 4y$
26. $6x^2 + 4x - 3xy - 2y$
27. $3mx + 5m - 6nx - 10n$
28. $30xy - 35y - 36x + 42$
29. $x^2 - 2x + 8x - 16$
30. $6x^2 - 2x - 21x + 7$
31. $x^2 - 4x + x - 4$
32. $8x^2 - 4x + 6x - 3$
33. $x^2 - 5x + 4x - 20$
34. $3x^2 - 6x + 5x - 10$

3   Factor the trinomials by grouping.

35. $3x^2 + 7x + 2$	**36.** $3x^2 + 14x + 8$	**37.** $6x^2 + 13x + 6$
38. $x^2 - 18x + 32$	**39.** $6x^2 - 17x + 12$	**40.** $2x^2 - 9x + 10$
41. $6x^2 - 13x + 5$	**42.** $p^2 + 5p - 36$	**43.** $6x^2 - 11x + 5$
44. $8x^2 + 26x + 15$	**45.** $15x^2 - 22x - 5$	**46.** $y^2 - 15y + 36$
47. $2x^2 - 5x - 7$	**48.** $12x^2 + 8x - 15$	**49.** $10x^2 + x - 3$
50. $Q^2 - 7Q - 44$		

4   Factor the trinomials by trial and error.

51. $6x^2 + 7xy - 10y^2$	**52.** $6a^2 - 17ab - 14b^2$	**53.** $18x^2 - 3x - 10$
54. $20x^2 - xy - 12y^2$	**55.** $x^2 + x - 56$	**56.** $a^2 + 21a + 38$

5   Factor the polynomials completely. Be sure to look for the common factors first and then look for special cases.

57. $4x - 4$	**58.** $x^2 + x - 6$	**59.** $2x^2 + x - 3$
60. $x^2 - 9$	**61.** $4x^2 - 16$	**62.** $m^2 + 2m - 15$
63. $2a^2 + 6a + 4$	**64.** $b^2 + 6b + 9$	**65.** $16m^2 - 8m + 1$
66. $x^2 + 8x + 7$	**67.** $2m^2 + 5m + 2$	**68.** $2m^2 - 5m - 3$
69. $2a^2 - 3a - 5$	**70.** $3x^2 + 10x - 8$	**71.** $6x^2 + x - 15$
72. $8x^2 + 10x - 3$		

CAREER APPLICATION

Information Systems: Security Coding

One of the most important areas of information systems is security for computer programs, files, and data bases. Governments, companies, and individuals use sophisticated methods of coding passwords and log-on procedures. In one type of advanced code, a branch of mathematics called *matrix theory* encodes and decodes. A rectangular array of numbers, called a *matrix,* is multiplied by the password matrix after the letters of the password have been assigned numeric values (A=1, B=2, for example). The inverse of the encoding matrix is then used to decode the password. It is desirable that the encoding matrix not have a determinant that is a prime number because prime numbers make it much easier for hackers to learn the encoding matrix (and hence the password).

About 250 B.C., a Greek scholar named Eratosthenes devised a method, called a *sieve,* for finding all prime numbers up to any given number *N*. Use his method to find all prime numbers up to 200 (which would be eliminated as choices for encoding matrix determinants). Continue the process shown in the Tip box on page 116 for the numbers 1–200. The best choices for encoding matrix determinants are composite numbers with numerous factors. These numbers would have been crossed out repeatedly.

Exercises

Use the sieve that you constructed above to answer the following.

1. Count and list the prime numbers less than 200. Remember that 1 is neither prime nor composite because it has exactly one factor; the number 1. (Prime numbers have exactly two factors, and composite numbers have more than two factors.)
2. Which was the first prime number to have no multiples crossed out? Eratosthenes proved that it is never necessary to go higher than the square root of *N*. Did your sieve comply with this rule?
3. Select five good choices for encoding matrix determinants, which would be composite numbers less than or equal to 200 with many factors.

4. Write a computer program to construct the same sieve, and compare results between the computer method and the manual method. If they differ, explain why.
5. Use the program from Exercise 4 to find the primes less than 400. List and count them. Are there double the number of primes less than 200? Why do you think this is so?
6. Alter your program to list the composites with 10 or more factors.

Answers

1. 46 primes: 2, 3, 5, 7, 11, 13, 17, 19, 23, 29, 31, 37, 41, 43, 47, 53, 59, 61, 67, 71, 73, 79, 83, 89, 97, 101, 103, 107, 109, 113, 127, 131, 137, 139, 149, 151, 157, 163, 167, 173, 179, 181, 191, 193, 197, and 199.
2. 17 was the first prime number to have no multiples crossed out. Note that the square root of 200 is about 14.2, and the next largest prime is 17.
3. Good choices, 48, 60, 72, 90, 96, 120, 144, 168, 180, and 192.
4. The results should be the same, but a common error is to forget to eliminate the number 1 from the list of primes.
5. 78 primes: 2, 3, 5, 7, 11, 13, 17, 19, 23, 29, 31, 37, 41, 43, 47, 53, 59, 61, 67, 71, 73, 79, 83, 89, 97, 101, 103, 107, 109, 113, 127, 131, 137, 139, 149, 151, 157, 163, 167, 173, 179, 181, 191, 193, 197, 199, 211, 223, 227, 229, 233, 239, 241, 251, 257, 263, 269, 271, 277, 281, 283, 293, 307, 311, 313, 317, 331, 337, 347, 349, 353, 359, 367, 373, 379, 383, 389, and 397. There are fewer and fewer primes as the numbers increase because larger numbers are more likely than smaller numbers to have additional factors. An interesting question in mathematics is, Do primes continue into infinity or stop at a certain point?
6. Good choices: 240, 252, 264, 270, 288, 300, 336, 360, 384, and 400.

ASSIGNMENT EXERCISES

Section 12–1

Factor completely. Check.

1. $5x + 5y$
2. $2x + 5x^2$
3. $12m^2 - 8n^2$
4. $25x^2y - 10xy^3 + 5xy$
5. $2a^3 - 14a^2 - 2a$
6. $30a^3 - 18a^2 - 12a$
7. $15x^3 - 5x^2 - 20x$
8. $9a^2b^3 - 6a^3b^2$
9. $18a^3 + 12a^2$
10. $a^2b + ab^2$

Section 12–2

Use the FOIL method to find the products. Combine the outer and inner products mentally.

11. $(x + 7)(x + 4)$
12. $(y - 7)(y - 5)$
13. $(m + 3)(m - 7)$
14. $(3b - 2)(x + 6)$
15. $(4r - 5)(3r + 2)$
16. $(5 - x)(7 - 3x)$
17. $(4 - 2m)(1 - 3m)$
18. $(2 + 3x)(3 + 2x)$
19. $(x + 3)(2x - 5)$
20. $(5x - 7y)(4x + 3y)$
21. $(2a + 3b)(7a - b)$
22. $(5a + 2b)(6a - 5b)$
23. $(9x - 2y)(3x + 4y)$
24. $(5x - 8y)(4x - 3y)$
25. $(7m - 2n)(3m + 5n)$

Find the special products mentally.

26. $(y - 4)(y + 4)$
27. $(6x - 5)(6x + 5)$
28. $(3m + 4)(3m - 4)$
29. $(7y + 11)(7y - 11)$
30. $(5x - 2y)(5x + 2y)$
31. $(8a - 5b)(8a + 5b)$
32. $(12r - 7s)(12r + 7s)$
33. $(8x + y)(8x - y)$
34. $(5m + 11n)(5m - 11n)$
35. $(3y - 4z)(3y + 4z)$
36. $(x + 8)^2$
37. $(x + 9)^2$
38. $(x - 7)^2$
39. $(x - 3)^2$
40. $(2x - 3)^2$
41. $(4x - 15)^2$
42. $(5 + 3m)^2$
43. $(8 + 7m)^2$
44. $(5x - 13)^2$
45. $(4x - 11)^2$

Identify the special products that are the difference of two perfect squares.

46. $25m^2 - 4n^2$

47. $64 + 4a^2$

48. $4f^2 - 9g^2$

49. $H^2 - G^2$

50. $s^2 - 4s + 10$

51. $64b^2 - 49$

Verify which of the following trinomials are perfect-square trinomials.

52. $16c^2 + 8c - 1$

53. $-9x^2 + 12x + 4$

54. $4d^2 + 2d + 16$

55. $9t^2 - 24tp + 16p^2$

56. $a^2 - 14a + 49$

57. $j^2 + 10j + 25$

Factor the expressions that are special products. Explain why the other expressions are not special products.

58. $x^2 - 81$

59. $25y^2 - 4$

60. $100a^2 - 8ab^2$

61. $a^2b^2 + 49$

62. $121 - 9m^2$

63. $a^2 + 2a + 1$

64. $4x^2 + 12x + 9$

65. $16c^2 - 24bc + 9b^2$

66. $9y^2 - 12y - 4$

67. $n^2 + 169 - 26n$
(*Hint:* Rearrange.)

68. $16d^2 - 20d + 25$

69. $36a^2 + 84ab + 49b^2$

70. $4x^2 - 25y^2$

71. $49 - 14x + x^2$

72. $9x^2y^2 - 49z^2$

73. $64 + 25x^2$

74. $4x^2 + 12xy + 9y^2$

75. $16x^2 + 24x + 9y^2$

76. $36 - x^2$

77. $49 - 81y^2$

78. $64x^2 - 25y^2$

79. $9x^2 - 100y^2$

80. $a^2 - 10a + 25$

81. $9x^2 - 6xy + y^2$

82. $25 - 16a^2b^2$

83. $9x^2y^2 - z^2$

84. $x^2 - y^2$

85. $x^2 + 4x + 4$

86. $\frac{1}{4}x^2 - \frac{1}{9}y^2$

87. $\frac{4}{25}x^2 - \frac{1}{16}y^2$

Factor the trinomials.

88. $x^2 + 10x + 21$

89. $x^2 + 11x + 24$

90. $x^2 + 29x + 28$

91. $x^2 + 13x + 30$

92. $x^2 - 13x + 40$

93. $x^2 - 9x + 8$

94. $x^2 - 17x - 18$

95. $x^2 - 11x - 26$

96. $x^2 + x - 30$

97. $x^2 + 5x - 24$

98. $6x^2 + 25x + 14$

99. $6x^2 + 25x + 4$

100. $4x^2 - 23x + 15$

101. $5x^2 - 34x + 24$

102. $3x^2 - x - 14$

103. $6x^2 - x - 35$

104. $3x^2 + 11x - 4$

105. $7x^2 - 13x - 24$

Factor the polynomials. Look for common factors and special cases.

106. $5mn - 25m$

107. $9a^2 - 100$

108. $a^2 - b^2$

109. $2x^2 - 3x - 2$

110. $b^3 - 8b^2 - b$

111. $a^2 - 81$

112. $x^2 - 14x + 13$

113. $y^2 - 14y + 49$

114. $m^2 - 3m + 2$

115. $b^2 + 8b + 15$

116. $x^2 - 13x + 30$

117. $169 - m^2$

118. $5x^2 + 13x + 6$

119. $x^2 - 4x - 32$

120. $x^2 + 9x + 14$

121. $x^2 + 19x - 20$

122. $x^2 + 8x - 20$

123. $2x^2 - 4x - 16$

124. $5x^2 - 20$

125. $2x^3 - 10x^2 - 12x$

CHALLENGE PROBLEMS

126. Explain the value of factoring a general trinomial by the grouping method as opposed to the trial-and-error method.

127. What is the value of being able to recognize the special products when we factor algebraic expressions?

A standard-sized rectangular swimming pool is 25 ft long and 15 ft wide. This gives a water-surface area of 375 ft^2. A customer wants to examine some options for varying the size of the pool. If x is the amount of adjustment to the length and y is the amount of adjustment to the width, find the following.

128. Write a formula in both factored and expanded form for finding the water-surface area of a pool that is adjusted x feet in length and y feet in width.

129. Write a formula using one variable in both factored and expanded form for the water-surface area of a pool when the length and width are increased the same amount.

130. Will the formulas in Problems 128 and 129 work for both increasing and decreasing the size of the pool?

131. Illustrate your answer in Problem 130 with numerical examples.

CHAPTER TRIAL TEST

Find the products.

1. $3(x + 2y)$

2. $7x(x - 5)$

3. $7x(2x^2 + 3x - 5)$

4. $2x^2(2x + 3)$

5. $(m - 7)(m + 7)$

6. $(3x - 2)(3x + 2)$

7. $(a + 3)^2$

8. $(2x - 7)^2$

9. $(x - 3)(2x - 5)$

10. $(7x - 3)(2x + 1)$

Factor completely.

11. $7x^2 + 8x$

12. $6ax + 15bx$

13. $7a^2b - 14ab$

14. $9x^2 - 42x + 49$

15. $x^2 + 9x + 8$

16. $x^2 - 5x - 36$

17. $x^2 + 5x - 24$

18. $6x^2 - 5x - 6$

19. $3x^2 + 23x + 30$

20. $12y^2 - 27$

21. $a^2 + 16ab + 64b^2$

22. $y^2 - y - 6$

23. $b^2 - 3b - 10$

24. $x^2 - 7x + 10$

25. $3x^2 + x - 4$

26. $a^2 - 4a + 3$

27. $3m^2 - 5m + 2$

28. $4c^2 - 28c + 49$

29. $3x^2 + 11xy + 3y^2$

Solving Quadratic Equations

GOOD DECISIONS THROUGH TEAMWORK

Today's law enforcement officers are Sherlock Holmes in disguise when they investigate car accidents. If the drivers die in the accident or survive and remember the details incorrectly, the officers are missing key information or must resolve conflicting information as they try to piece together what happened. In addition, eyewitnesses may not be available or they may give conflicting reports. Therefore, physical evidence in the form of skid marks, distance, and the percentage of braking before stopping, as well as the computed drag factor, are used in mathematical equations to estimate the speed of the cars involved. Officers can then state with confidence that the driver was driving at a speed of at least a certain number of miles per hour.

The drag factor, or coefficient of friction, is a number between 0.03 (poor stopping conditions) and 1.3 (ideal stopping conditions). The drag factor varies according to road surface type (asphalt, concrete, gravel, dirt, or grass), age of road, weather conditions (precipitation, wind, air temperature, air pressure, relative humidity), road surface conditions (wet, dry, icy, oily, sandy, covered with dirt, gravel, or debris), grade of road, and so on. Officers can conduct drag factor tests under similar conditions to determine the drag factor.

Each test consists of braking to a stop from a specified speed (usually 25–35 mph), denoted by s, then measuring the distance of the skid marks (D). At least two tests are conducted, using the longest skid distance (if they are within 5% of each other) in the following formula to calculate the drag factor (f):

13–1 Quadratic equations

1 Write quadratic equations in standard form.

2 Identify the coefficients of the quadratic, linear, and constant terms of a quadratic equation.

13–2 Solving quadratic equations using the quadratic formula

1 Solve quadratic equations using the quadratic formula.

2 Solve applied problems that use quadratic equations.

13–3 Solving pure quadratic equations: $ax^2 + c = 0$

1 Solve pure quadratic equations by the square-root method $(ax^2 + c + 0)$.

13–4 Solving incomplete quadratic equations: $ax^2 + bx = 0$

1 Solve incomplete quadratic equations by factoring $(ax^2 + bx = 0)$.

$$f = \frac{s^2}{30D}$$

These tests are especially important if a fatality is involved. The weight of the vehicle is usually negligible in determining speed, but weight can be a factor under extremely poor stopping conditions such as black ice, oil slick, or oil mixed with water (which happens at the first rainstorm after a long dry period).

Under most stopping conditions, the *distance* of the skid marks, D, is related to the *speed* of the car, s; the *drag factor f*; and the *percentage* of braking before stopping, n (written as a decimal), by the formula:

$$D = \frac{s^2}{30fn}$$

Contact the police departments of a large metropolitan city (population of 1 million or more) and a small town (population of less than 25,000). Divide responsibilities among team members to gather additional information. Ask each officer to describe the process used to determine a car's speed when given the length of the skid marks and the road conditions. Find out how the calculations are done (by hand, with a calculator, with tables or charts, or with computer software). Describe the advantages and disadvantages of each method.

Ask a law enforcement officer to describe how accident conditions are simulated accurately during drag factor tests. What happens if their calculated speed is disputed?

13–5 Solving complete quadratic equations by factoring

1 Solve complete quadratic equations by factoring ($ax^2 + bx + c = 0$).

13–6 Selecting an appropriate method for solving quadratic equations

1 Determine the nature of the roots of a quadratic equation by examining the discriminant.

Most of the equations we have been solving so far are called linear equations. In such equations, the variable appears only to the first power, such as x or y. In Chapter 10, we looked briefly at one type of quadratic equation. In this chapter we continue to examine quadratic equations.

13–1 QUADRATIC EQUATIONS

Learning Outcomes

1 Write quadratic equations in standard form.
2 Identify the coefficients of the quadratic, linear, and constant terms of a quadratic equation.

1 Write Quadratic Equations in Standard Form.

There are several methods for solving quadratic equations. Each method has strengths and weaknesses depending on the characteristics of the equation being solved. First, let's examine the basic characteristics of quadratic equations, starting with quadratic equations having only one variable. We shall then examine quadratic equations written as functions, which involve both independent and dependent variables.

■ **DEFINITION: Quadratic Equation.** A *quadratic equation* is an equation in which at least one letter term is raised to the second power, and no letter term has a power higher than 2 or less than 0. The *standard form* for a quadratic equation is $ax^2 + bx + c = 0$, where a, b, and c are real numbers and $a > 0$.

EXAMPLE Arrange the quadratic equations in standard form.

(a) $3x^2 - 3 + 5x = 0$ (b) $7x - 2x^2 - 8 = 0$
(c) $12 = 7x^2 - 4x$ (d) $7x^2 = 3$

(a) $3x^2 - 3 + 5x = 0$
$\mathbf{3x^2 + 5x - 3 = 0}$ Arrange the terms in descending powers of x.

(b) $7x - 2x^2 - 8 = 0$
       $-2x^2 + 7x - 8 = 0$ Arrange the terms in descending powers of x.
$-1(-2x^2 + 7x - 8) = -1(0)$ Multiply both sides by -1 to make the coefficient of the x^2 term positive.

       $\mathbf{2x^2 - 7x + 8 = 0}$ Standard form.

(c) $12 = 7x^2 - 4x$
        $0 = 7x^2 - 4x - 12$ Rearrange the terms so that one side of the equation equals zero.
  $\mathbf{7x^2 - 4x - 12 = 0}$ Interchange sides of equation.

(d) $7x^2 = 3$
   $\mathbf{7x^2 - 3 = 0}$ Rearrange so that one side equals zero.

A quadratic equation is a second-degree equation, which means that it can have as many as two real solutions. As we examine the process for solving quadratic equations we shall see quadratic equations that have two, one, or no solutions.

2 Identify the Coefficients of the Quadratic, Linear, and Constant Terms of a Quadratic Equation.

Some methods for solving quadratic equations can be used for any type of quadratic equation but are very time-consuming. Other methods are quicker but apply only to certain types of quadratic equations. In choosing an appropriate method it is helpful to be able to recognize similar and different characteristics of quadratic equations.

The standard form of quadratic equations, $ax^2 + bx + c = 0$, has three types of terms.

1. ax^2 is a *quadratic term;* that is, the degree of the term is 2. In standard form, this term is the leading term and a is the leading coefficient.
2. bx is a *linear term;* that is, the degree of the term is 1 and the coefficient is b.
3. c is a *number* term or *constant term;* that is, the degree of the term is 0. The coefficient is c ($c = cx^0$).

A quadratic equation that has all three types of terms is sometimes referred to as a *complete quadratic equation.*

$ax^2 + bx + c = 0$ Complete quadratic equation.

A quadratic equation that has a quadratic term (ax^2) and a linear term (bx) but no constant term (c) is sometimes referred to as an *incomplete quadratic equation.*

$ax^2 + bx = 0$ Incomplete quadratic equation.

CHAPTER 13   Solving Quadratic Equations

A quadratic equation that has a quadratic term (ax^2) and a constant term (c) but no linear term (bx) is sometimes referred to as a *pure quadratic equation*, the type that we worked with in Chapter 10.

$$ax^2 + c = 0 \qquad \text{Pure quadratic equation.}$$

It is helpful in planning our strategy for solving a quadratic equation to be able to identify the types of terms in a quadratic equation and to identify the coefficients a, b, and c.

EXAMPLE Write each equation in standard form and identify a, b, and c.

(a) $3x^2 + 5x - 2 = 0$ (b) $5x^2 = 2x - 3$
(c) $3x = 5x^2$ (d) $-6x^2 + 2x = 0$
(e) $5x^2 - 4 = 0$ (f) $x^2 = 9$
(g) $x^2 + 3x = 5x$ (h) $7 - x^2 = 3 + x$

(a) $3x^2 + 5x - 2 = 0$ In standard form.
$a = 3, \quad b = 5, \quad c = -2$

(b) $5x^2 = 2x - 3$
$5x^2 - 2x + 3 = 0$ Write in standard form.
$a = 5, \quad b = -2, \quad c = 3$

(c) $3x = 5x^2$
$0 = 5x^2 - 3x$ Write in standard form.
$5x^2 - 3x = 0$ Interchange sides of equation.
$a = 5, \quad b = -3, \quad c = 0$ No constant term means $c = 0$.

(d) $-6x^2 + 2x = 0$
$-1(-6x^2 + 2x = 0)$ Multiply by -1.
$6x^2 - 2x = 0$ Standard form.
$a = 6, \quad b = -2, \quad c = 0$ No constant term.

(e) $5x^2 - 4 = 0$ Standard form.
$a = 5, \quad b = 0, \quad c = -4$ No linear term means $b = 0$.

(f) $x^2 = 9$
$x^2 - 9 = 0$ Write in standard form.
$a = 1, \quad b = 0, \quad c = -9$ No linear term.

(g) $x^2 + 3x = 5x$
$x^2 + 3x - 5x = 0$ Rearrange.
$x^2 - 2x = 0$ Combine like terms.
$a = 1, \quad b = -2, \quad c = 0$ No constant term.

(h) $7 - x^2 = 3 + x$
$-x^2 - x + 7 - 3 = 0$ Rearrange.
$-x^2 - x + 4 = 0$ Combine like terms.
$-1(-x^2 - x + 4 = 0)$ Multiply by -1.
$x^2 + x - 4 = 0$ Standard form.
$a = 1, \quad b = 1, \quad c = -4$

SELF-STUDY EXERCISES 13–1

1 Arrange the quadratic equations in standard form.

1. $5 - 4x + 7x^2 = 0$
2. $8x^2 - 3 = 6x$
3. $7x^2 = 5$
4. $x^2 = 6x - 8$
5. $x^2 - 3x = 6x - 8$
6. $8 - x^2 + 6x = 2x$
7. $3x^2 + 5 - 6x = 0$
8. $5 = x^2 - 6x$
9. $x^2 - 16 = 0$
10. $8x^2 - 7x = 8$
11. $8x^2 + 8x - 2 = 8$
12. $0.3x^2 - 0.4x = 3$

2 Write each equation in standard form and identify a, b, and c.

13. $5x = x^2$
14. $7x - 3x^2 = 5$
15. $7x^2 - 4x = 0$
16. $8 + 3x^2 = 5x$
17. $5x - 6 = x^2$
18. $11x^2 = 8x$
19. $x = x^2$
20. $9x^2 - 7x = 12$
21. $5 = x^2$
22. $-x^2 - 6x + 3 = 0$
23. $0.2x - 5x^2 = 1.4$
24. $\dfrac{2}{3}x^2 - \dfrac{5}{6}x = \dfrac{1}{2}$
25. $1.3x^2 - 8 = 0$
26. $\sqrt{3}x^2 + \sqrt{5}x - 2 = 0$

Write the equations in standard form.

27. The coefficient of the quadratic term is 8; the coefficient of the linear term is -2; the constant term is -3.

28. The constant is 0, the coefficient of the linear term is 3, and the coefficient of the quadratic term is 1.

29. $a = 5$, $b = 2$, $c = -7$

30. $a = 2.5$, $c = 0.8$

13-2 SOLVING QUADRATIC EQUATIONS USING THE QUADRATIC FORMULA

Learning Outcomes

1 Solve quadratic equations using the quadratic formula.

2 Solve applied problems that use quadratic equations.

The first method for solving quadratic equations examined here works for all types of quadratic equations. It is time-consuming and requires cumbersome calculations. Using a programmable calculator or a computer program makes this a popular method for problem solving.

1 Solve Quadratic Equations Using the Quadratic Formula.

One method for solving quadratic equations is to use the *quadratic formula*. This formula results from solving the standard quadratic equation, $ax^2 + bx + c = 0$, by a method called *completing the square*.

Quadratic formula:

$$x = \frac{-b \pm \sqrt{b^2 - 4ac}}{2a}$$

where a, b, and c are coefficients of a quadratic equation in the form $ax^2 + bx + c = 0$.

Because the formula is derived from the standard quadratic equation $ax^2 + bx + c = 0$, the a, b, and c in the formula are the same a, b, and c in the standard quadratic equation. To evaluate the formula, we identify a, b, and c, then we evaluate the formula to solve equations. Even though we have one formula, we can have *two* values for x because the formula requires us to *add* or to *subtract* the radical. This means that we evaluate the formula two times: once with the plus sign and again with the minus sign.

$$x = \frac{-b + \sqrt{b^2 - 4ac}}{2a} \qquad x = \frac{-b - \sqrt{b^2 - 4ac}}{2a}$$

<table>
<tr><td>**Tip!**</td><td>*Common Cause for Errors.*</td></tr>
<tr><td colspan="2">When you use the quadratic formula to solve problems, begin by writing the formula to help you remember it. When you write the formula, be sure to extend the fraction bar beneath the *entire* numerator. This omission is a common cause for errors.</td></tr>
</table>

EXAMPLE Use the quadratic formula to solve $x^2 + 5x + 6 = 0$ for x.

$$a = 1, \qquad b = 5, \qquad c = 6 \qquad \text{Identify } a, b, \text{ and } c.$$

$$x = \frac{-b \pm \sqrt{b^2 - 4ac}}{2a} \qquad \text{Quadratic formula. Substitute for } a, b, \text{ and } c.$$

$$x = \frac{-5 \pm \sqrt{5^2 - 4 \cdot 1 \cdot 6}}{2 \cdot 1} \qquad \text{Multiply in groupings.}$$

$$x = \frac{-5 \pm \sqrt{25 - 24}}{2} \qquad \text{Combine terms under radical.}$$

$$x = \frac{-5 \pm \sqrt{1}}{2} \qquad \text{Evaluate radical.}$$

$$x = \frac{-5 \pm 1}{2}$$

At this point, we separate the formula into two parts, one using the $+1$ and the other using the -1.

$$x = \frac{-5 + 1}{2} \qquad\qquad\qquad x = \frac{-5 - 1}{2}$$

$$x = \frac{-4}{2} \qquad\qquad\qquad x = \frac{-6}{2}$$

$$\mathbf{x = -2} \qquad\qquad\qquad \mathbf{x = -3}$$

$$\text{Check } x = -2 \qquad\qquad \text{Check } x = -3$$

$$x^2 + 5x + 6 = 0 \qquad\qquad x^2 + 5x + 6 = 0$$

$$(-2)^2 + 5(-2) + 6 = 0 \qquad\qquad (-3)^2 + 5(-3) + 6 = 0$$

$$4 - 10 + 6 = 0 \qquad\qquad 9 - 15 + 6 = 0$$

$$-6 + 6 = 0 \qquad\qquad -6 + 6 = 0$$

$$0 = 0 \qquad\qquad 0 = 0$$

<table>
<tr><td>**Tip!**</td><td>*Completing the Square and the Quadratic Formula.*</td></tr>
<tr><td colspan="2">The completing-the-square method for solving quadratic equations also works for all types of quadratic equations, but it requires several tedious manipulations. We use this method to show how the quadratic formula was derived, but we won't use it in the examples in this chapter. In Chapter 10, we saw that we could take the square root of both sides of an equation and maintain equality. That principle is applied in this method.</td></tr>
</table>

$$ax^2 + bx + c = 0 \qquad a, b, \text{ and } c \text{ are real numbers and } a > 0.$$

$$ax^2 + bx = -c \qquad \text{Isolate the terms containing variables.}$$

$$x^2 + \frac{bx}{a} = -\frac{c}{a} \qquad \text{Divide the entire equation by } a \text{ so that the coefficient of } x^2 \text{ is 1.}$$

$$x^2 + \frac{bx}{a} + \frac{b^2}{4a^2} = \frac{b^2}{4a^2} - \frac{c}{a} \qquad \text{Add a constant to each side so that the left side becomes a perfect-square trinomial. The appropriate constant will be } \left(\frac{1}{2}\frac{b}{a}\right)^2 \text{ or } \left(\frac{1}{4}\frac{b^2}{a^2}\right) \text{ or } \frac{b^2}{4a^2}$$

$$\left(x + \frac{b}{2a}\right)^2 = \frac{b^2}{4a^2} - \frac{c}{a} \qquad \text{Factor the left side.}$$

$$\sqrt{\left(x + \frac{b}{2a}\right)^2} = \sqrt{\frac{b^2}{4a^2} - \frac{c}{a}} \qquad \text{Take the square root of both sides.}$$

$$x + \frac{b}{2a} = \pm\sqrt{\frac{b^2}{4a^2} - \frac{c}{a}} \qquad \text{Get a common denominator for the terms under the radical.}$$
$$-\frac{c}{a} \cdot \frac{4a}{4a} = -\frac{4ac}{4a^2}.$$

$$x + \frac{b}{2a} = \pm\sqrt{\frac{b^2}{4a^2} - \frac{4ac}{4a^2}} \qquad \text{Simplify the radicand.}$$

$$x + \frac{b}{2a} = \frac{\pm\sqrt{b^2 - 4ac}}{2a} \qquad \text{Solve for } x.$$

$$x = -\frac{b}{2a} \pm \frac{\sqrt{b^2 - 4ac}}{2a} \qquad \text{Combine like fractions.}$$

$$x = \frac{-b \pm \sqrt{b^2 - 4ac}}{2a} \qquad \text{Quadratic formula.}$$

EXAMPLE Use the quadratic formula to solve $3x^2 - 3x - 7 = 0$ for x. Round answers to the nearest hundredth.

$$a = 3, \qquad b = -3, \qquad c = -7 \qquad \text{Identify } a, b, \text{ and } c.$$

$$x = \frac{-b \pm \sqrt{b^2 - 4ac}}{2a} \qquad \text{Quadratic formula.}$$

$$x = \frac{+3 \pm \sqrt{(-3)^2 - 4(3)(-7)}}{2 \cdot 3} \qquad \text{Substitute } -(b) = -(-3) = +3.$$

$$x = \frac{+3 \pm \sqrt{9 + 84}}{6} \qquad \text{Perform calculations under the radical.}$$

$$x = \frac{+3 \pm \sqrt{93}}{6} \qquad \text{Evaluate the radical.}$$

$$x = \frac{+3 \pm 9.643650761}{6} \qquad \text{Separate two solutions.}$$

$$x = \frac{3 + 9.643650761}{6} \qquad\qquad x = \frac{3 - 9.643650761}{6}$$

$$x = \frac{12.643650761}{6} \qquad\qquad x = \frac{-6.643650761}{6}$$

$$x = 2.11 \quad \text{Rounded.} \qquad\qquad x = -1.11 \quad \text{Rounded.}$$

Notice that we round to hundredths *after* making the final calculation.

Check $x = 2.11$	Check $x = -1.11$
$3x^2 - 3x - 7 = 0$	$3x^2 - 3x - 7 = 0$
$3(2.11)^2 - 3(2.11) - 7 = 0$	$3(-1.11)^2 - 3(-1.11) - 7 = 0$
$3(4.4521) - 6.33 - 7 = 0$	$3(1.2321) + 3.33 - 7 = 0$
$13.3563 - 6.33 - 7 = 0$	$3.6963 + 3.33 - 7 = 0$
$0.0263 \doteq 0$	$0.0263 \doteq 0$

Tip!	***Rounding Discrepancies.***

The symbol $\doteq$ means "is approximately equal to." Another symbol for approximations is $\approx$. If we had checked *before* rounding, the checks would have been "closer" to zero.

Tip!	***Check with Calculator.***

With a formula as complex as the quadratic formula, it is important to check your solutions.

Use the full calculator value to check. As we mentioned in the previous tip, the full calculator value improves the accuracy of the check. Calculate the first root in the previous example.

$$3 \boxed{+} \boxed{\sqrt{}} \; 93 \boxed{=} \boxed{\div} 6 \boxed{=} \Rightarrow 2.107275127$$

Check:

If your calculator allows you to insert the answer of your last calculation, use this function in checking. Otherwise, you can store an answer in memory and recall as appropriate.

$$3 \boxed{\text{ANS}} \boxed{x^2} \boxed{-} 3 \boxed{\text{ANS}} \boxed{-} 7 \boxed{=} \Rightarrow -1.6 \times 10^{-9} \text{ or } -0.0000000016 \qquad \textbf{Very close to zero.}$$

Calculate the second root.

$$3 \boxed{-} \boxed{\sqrt{}} \; 93 \boxed{=} \boxed{\div} 6 \boxed{=} \Rightarrow -1.107275127$$

Check:

$$3 \boxed{\text{ANS}} \boxed{x^2} \boxed{-} 3 \boxed{\text{ANS}} \boxed{-} 7 \boxed{=} \Rightarrow 8 \times 10^{-12} \text{ or } 0.000000000008 \qquad \textbf{Very close to zero.}$$

In the next example, the equation contains no squared term at first, but the squared term appears *after* the multiplication to eliminate the denominator.

EXAMPLE   Solve $\dfrac{3}{5 - x} = 2x$ for x. Round answers to the nearest hundredth.

$$\dfrac{3}{5 - x} = 2x \qquad \text{Eliminate the denominator and then write the equation in standard form.}$$

$$(5 - x)\dfrac{3}{5 - x} = 2x(5 - x) \qquad \text{Multiply both sides by the denominator to eliminate it.}$$

$$3 = 2x(5 - x) \qquad \text{Distribute.}$$

$$3 = 10x - 2x^2$$

$$2x^2 - 10x + 3 = 0 \qquad \text{Write the equation in standard form.}$$

$$a = 2, \qquad b = -10, \qquad c = 3 \qquad \text{Identify } a, b, \text{ and } c.$$

$$x = \frac{-b \pm \sqrt{b^2 - 4ac}}{2a}$$

Write the formula.

$$x = \frac{+10 \pm \sqrt{(-10)^2 - 4(2)(3)}}{2(2)}$$

Substitute and simplify radicand.

$$x = \frac{10 \pm \sqrt{100 - 24}}{4}$$

$$x = \frac{10 \pm \sqrt{76}}{4}$$

$$x = \frac{10 \pm 8.717797887}{4}$$

Separate into two solutions.

$$x = \frac{10 + 8.717797887}{4} \qquad x = \frac{10 - 8.717797887}{4}$$

$$x = \frac{18.717797887}{4} \qquad x = \frac{1.282202113}{4}$$

$$x = 4.68 \qquad\qquad x = 0.32 \qquad \text{Round after final calculation.}$$

Check $x = 4.68$ Check $x = 0.32$

$$\frac{3}{5 - x} = 2x \qquad\qquad \frac{3}{5 - x} = 2x$$

$$\frac{3}{5 - (4.68)} = 2(4.68) \qquad\qquad \frac{3}{5 - (0.32)} = 2(0.32)$$

$$\frac{3}{0.32} = 9.36 \qquad\qquad \frac{3}{4.68} = 0.64$$

$$9.375 \doteq 9.36 \qquad\qquad 0.641025641 \doteq 0.64$$

Tip!	***Extraneous Roots.***

When an equation is multiplied by a factor that contains a variable, as in the preceding example, an *extraneous root* can be introduced. An extraneous root is a root that will *not* check when substituted into the original equation. Thus, it is very important to check each root.

2 Solve Applied Problems That Use Quadratic Equations.

Many applications require quadratic equations. Even though quadratic equations have two roots, a root may be disregarded in an applied problem because it is not appropriate within the context of the problem.

EXAMPLE   Find the length and width of a rectangular table if the length is 8 inches more than the width and the area is 260 in^2.

Unknown facts   Length and width of rectangle

Known facts   Area = 260 in.^{2}, length = 8 in. more than width

Relationships   Let x = number of inches in the width

$x + 8$ = number of inches in the length

Area = length times width, or $A = lw$

Estimation The square root of 260 is between 16 and 17; the width should be less than 16 and the length more than 16.

Calculations

$$260 = (x + 8)(x) \qquad \text{Substitute into area formula.}$$

$$260 = x^2 + 8x \qquad \text{Distribute.}$$

$$x^2 + 8x - 260 = 0 \qquad \text{Write in standard form.}$$

$$a = 1, \qquad b = 8, \qquad c = -260$$

$$x = \frac{-b \pm \sqrt{b^2 - 4ac}}{2a} \qquad \text{Quadratic formula.}$$

$$x = \frac{-8 \pm \sqrt{(8)^2 - 4(1)(-260)}}{2(1)} \qquad \text{Substitute.}$$

$$x = \frac{-8 \pm \sqrt{64 + 1{,}040}}{2}$$

$$x = \frac{-8 \pm \sqrt{1{,}104}}{2} \qquad \text{Combine terms in radicand.}$$

$$x = \frac{-8 \pm 33.22649545}{2} \qquad \text{Separate into two solutions.}$$

$$x = \frac{-8 + 33.22649545}{2} \qquad\qquad x = \frac{-8 - 33.22649545}{2}$$

$$x = \frac{25.22649545}{2} \qquad\qquad x = \frac{-41.22649545}{2}$$

$$x = 12.6 \qquad\qquad x = -20.6$$

$$x + 8 = 20.6$$

Interpretation Measurements are positive, so disregard the negative solution. **The width is 12.6 in. and the length is 20.6 in.**

SELF-STUDY EXERCISES 13–2

1 Solve the quadratic equations by using the quadratic formula.

1. $3x^2 - 7x - 6 = 0$
2. $x^2 + x - 12 = 0$
3. $5x^2 - 6x = 11$
4. $x^2 - 6x + 9 = 0$
5. $x^2 - x - 6 = 0$
6. $8x^2 - 2x = 3$

Solve the quadratic equations by using the quadratic formula. Round each final answer to the nearest hundredth.

7. $x^2 = -9x + 20$
8. $3x^2 + 6x + 1 = 0$
9. $\dfrac{2}{x} + \dfrac{5}{2} = x$

10. $\dfrac{2}{x} + 3x = 8$
11. $2x^2 + 3x + 3 = 0$
12. $x^2 - x + 2 = 0$

2 Solve the applied problems.

13. A rectangular table top is 3 cm longer than it is wide. Find the length and width to the nearest hundredth if the area is 47.5 cm². (Area = length × width, or $A = lw$.)

14. A rectangular instrument case has an area of 40 in². If the length is 6 in. more than the width, find the dimensions (length and width) of the instrument case.

15. A bricklayer plans to build an arch with a span (s) of 8 m and a radius (r) of 4 m. How high (h) is the arch? (Use the formula $h^2 - 2hr + \frac{s^2}{4} = 0$.)

16. A rectangular patio slab is 180 ft^2. If the length is 1.5 times the width, find the width and length of the slab to the nearest whole number.

17. A farmer normally plants a field 80 ft by 120 ft. This year the government requires the planting area to be decreased by 20%, and the farmer chooses to decrease the width and length of the field by an equal amount. Draw both the original field and the reduced-size field. If x is the amount by which the length and width are decreased, find the length and width of the new field.

18. The recommended dosage of a certain type of medicine is determined by the patient's weight. The formula to determine the dosage is given by $D = 0.1w^2 + 5w$, where D is the dosage in milligrams (mg) and w is the patient's body weight in kilograms. A doctor ordered a dosage of 1,800 mg. This dosage is to be administered to what weight patient?

13–3   SOLVING PURE QUADRATIC EQUATIONS: $ax^2 + c = 0$

Learning Outcome

1   Solve pure quadratic equations by the square-root method ($ax^2 + c = 0$).

The quadratic formula can be used to solve any quadratic equation; however, other methods are less cumbersome for equations with certain characteristics. Alternative methods for solving pure, incomplete, and complete quadratic equations are examined.

1   Solve Pure Quadratic Equations by the Square-Root Method ($ax^2 + c = 0$).

To solve pure quadratic equations, solve for the squared letter, then take the square root of both sides of the equation. This procedure is the same one we used in Chapter 10. All roots or solutions should continue to be checked.

EXAMPLE   Solve $3y^2 = 27$.

$3y^2 = 27$ Divide both sides by the coefficient of the quadratic term.

$y^2 = 9$ Take the square root of both sides.

$y = \pm 3$

Check $y = +3$ Check $y = -3$

$3y^2 = 27$ $3y^2 = 27$

$3(3)^2 = 27$ $3(-3)^2 = 27$

$3(9) = 27$ $3(9) = 27$

$27 = 27$ $27 = 27$

To solve a pure quadratic equation ($ax^2 + c = 0$):

1. Rearrange the equation, if necessary, so that quadratic or squared-letter terms are on one side and constants or number terms are on the other side of the equation.
2. Combine like terms, if appropriate.
3. Solve for the squared letter, so that the coefficient is $+1$.
4. Apply the square-root principle of equality by taking the square root of both sides.
5. Check both answers.

EXAMPLE Solve $4x^2 - 9 = 0$.

$$4x^2 - 9 = 0 \qquad \text{Transpose or sort terms.}$$

$$4x^2 = 9 \qquad \text{Divide by 4.}$$

$$x^2 = \frac{9}{4} \qquad \text{Take the square root of the numerator and denominator if they are perfect-square whole numbers.}$$

$$x = \pm\frac{3}{2} \qquad \text{Divide if a decimal answer is desired.}$$

$$x = \pm 1.5$$

Check $x = +1.5$	Check $x = -1.5$
$4x^2 - 9 = 0$	$4x^2 - 9 = 0$
$4(1.5)^2 - 9 = 0$	$4(-1.5)^2 - 9 = 0$
$4(2.25) - 9 = 0$	$4(2.25) - 9 = 0$
$9 - 9 = 0$	$9 - 9 = 0$
$0 = 0$	$0 = 0$

In some applications, the fractional answer may be more convenient. In other applications the decimal answer may be more convenient.

Not all pure quadratic equations have *real solutions*. The next example illustrates such an equation.

EXAMPLE Solve $3m^2 + 48 = 0$.

$$3m^2 + 48 = 0$$

$$3m^2 = -48 \qquad \text{Isolate the variable.}$$

$$m^2 = \frac{-48}{3} \qquad \text{Solve for } m^2.$$

$$m^2 = -16$$

$$m = \pm\sqrt{-16} \qquad \text{Apply the square-root principle.}$$

There are **no real solutions** to this equation. **If imaginary roots are acceptable, the roots are $\pm 4i$.**

Tip! | ***Roots of Pure Quadratic Equations.***

In a pure quadratic equation, the value of b is zero: $ax^2 + 0x + c = 0$.

$$x = \frac{-b \pm \sqrt{b^2 - 4ac}}{2a}$$

$$x = \frac{0 \pm \sqrt{0^2 - 4ac}}{2a}$$

$$x = \frac{\pm\sqrt{-4ac}}{2a}$$

- The equation has real solutions only if c is negative. Remember, standard form requires that a be positive.
- The two solutions have the same absolute value.

SELF-STUDY EXERCISES 13–3

1 Solve the equations. Round to thousandths when necessary.

1. $x^2 = 9$
2. $x^2 - 49 = 0$
3. $9x^2 = 64$
4. $16x^2 - 49 = 0$
5. $0.09y^2 = 0.81$
6. $0.04x^2 = 0.81$
7. $2x^2 = 10$
8. $3x^2 + 4 = 7$
9. $5x^2 - 8 = 12$
10. $7x^2 - 2 = 19$

11. The label on a new tarpaulin shows that it is a square and contains 529 ft^2. What is the length of each side of the tarpaulin?

12. A farmer has a field designed in the shape of a circle for efficiency in irrigation. The area of the field is known to be 21,000 yd^2. What length of irrigation line is required to provide water to the entire field? (*Note:* The line moves from the center of the field in a circle around the field.)

13. Describe the process for solving a pure quadratic equation.

14. What is the relationship between the roots of a pure quadratic equation?

13–4 SOLVING INCOMPLETE QUADRATIC EQUATIONS: $ax^2 + bx = 0$

Learning Outcome

1 Solve incomplete quadratic equations by factoring ($ax^2 + bx = 0$).

Incomplete quadratic equations have no number terms or constants. Thus, in standard form, $c = 0$. The two remaining terms both have variable factors. This characteristic leads to another method for solving quadratic equations.

1 Solve Incomplete Quadratic Equations by Factoring ($ax^2 + bx = 0$).

Incomplete quadratic equations like $x^2 + 2x = 0$ can be solved by factoring a common factor from $x^2 + 2x$. An incomplete quadratic equation always has a variable common factor, so we can factor the expression to $x(x + 2) = 0$. We now have two factors that have a product of 0. If we multiply two factors and get a product of 0, one or both factors *must* be 0 because $0 \cdot 0 = 0$ as well as $0 \cdot a = 0$ or $a \cdot 0 = 0$, where a is any number. This is called the *zero-product property*.

Zero-product property:

If $ab = 0$, then a or b or both equal zero.

If $x = 0$, we can check by substituting 0 for x wherever x occurs in the original equation.

$$\text{For } x = 0, \ x^2 + 2x = 0$$
$$0^2 + 2(0) = 0$$
$$0 + 0 = 0$$
$$0 = 0$$

If $x + 2 = 0$, we first solve for x.

$$x + 2 = 0$$
$$x = -2$$

Because $x = -2$, we can check by substituting -2 for x wherever x occurs in the original equation.

For $x = -2$, $x^2 + 2x = 0$

$$(-2)^2 + 2(-2) = 0 \qquad (-2)^2 = +4$$
$$4 + (-4) = 0$$
$$0 = 0$$

Therefore, the original equation checks with either value of x. This means that x may be 0 or -2 in this incomplete quadratic equation.

To solve an incomplete quadratic equation ($ax^2 + bx = 0$) by factoring:

1. If necessary, sort or transpose so that all terms are on one side of the equation in standard form, with zero on the other side.
2. Factor the common factor and set each factor equal to zero.
3. Solve for the variable in each equation formed in Step 2.

EXAMPLE Solve $2x^2 = 5x$ for x.

$$2x^2 = 5x \qquad \text{Put in standard form.}$$
$$2x^2 - 5x = 0$$
$$x(2x - 5) = 0 \qquad \text{Factor common factor.}$$
$$x = 0 \quad 2x - 5 = 0 \qquad \text{Set each factor equal to zero.}$$
$$x = 0 \qquad 2x = 5 \qquad \text{Solve each equation.}$$
$$x = \frac{5}{2}$$

$$\text{Check } x = 0 \qquad\qquad \text{Check } x = \frac{5}{2}$$

$$2x^2 = 5x \qquad\qquad 2x^2 = 5x$$

$$2(0)^2 = 5(0) \qquad 2\left(\frac{5}{2}\right)^2 = 5\left(\frac{5}{2}\right)$$

$$2(0) = 5(0) \qquad \overset{1}{\cancel{2}}\left(\frac{25}{\underset{2}{\cancel{4}}}\right) = \frac{25}{2}$$

$$0 = 0 \qquad\qquad \frac{25}{2} = \frac{25}{2}$$

EXAMPLE Solve $4x^2 - 8x = 0$ for x.

$$4x^2 - 8x = 0 \qquad \text{Standard form.}$$
$$4x(x - 2) = 0 \qquad \text{Factor common factors.}$$

$$4x = 0 \quad x - 2 = 0 \qquad \text{Set each factor equal to zero.}$$

$$x = \frac{0}{4} \qquad x = 2 \qquad \text{Solve each equation.}$$

$$x = 0$$

$$\text{Check } x = 0 \qquad\qquad \text{Check } x = 2$$
$$4(0)^2 - 8(0) = 0 \qquad 4(2)^2 - 8(2) = 0$$
$$4(0) - 8(0) = 0 \qquad 4(4) - 8(2) = 0$$
$$0 - 0 = 0 \qquad 16 - 16 = 0$$
$$0 = 0 \qquad\qquad 0 = 0$$

Tip!	***Roots of Incomplete Quadratic Equations.***

In an incomplete quadratic equation, the value of $c = 0$: $ax^2 + bx + 0 = 0$.

$$x = \frac{-b \pm \sqrt{b^2 - 4a(c)}}{2a}$$

$$x = \frac{-b \pm \sqrt{b^2}}{2a} \qquad\qquad \mathbf{4ac = 4a(0) = 0.}$$

$$x = \frac{-b + b}{2a} \qquad x = \frac{-b - b}{2a}$$

$$x = \frac{0}{2a} \qquad x = \frac{-2b}{2a}$$

$$x = 0 \qquad x = \frac{-b}{a}$$

- One root is always zero.
- The other root is $\frac{-b}{a}$.

SELF-STUDY EXERCISES 13–4

1 Solve.

1. $x^2 - 3x = 0$
2. $3x^2 = 7x$
3. $5x^2 - 10x = 0$
4. $2x^2 + x = 0$
5. $8x^2 - 4x = 0$
6. $5x^2 = 15x$
7. $x^2 - 6x = 0$
8. $y^2 = -4y$
9. $3x^2 + 2x = 0$
10. $9x^2 - 12x = 0$

11. The square of a number is 8 times the number. Find the number.

12. A square rug has an area 15 times the length of one of the sides. What is the length of a side?

13. How does an incomplete quadratic equation differ from a pure quadratic equation?

14. Will one of the two roots of an incomplete quadratic equation always be zero? Justify your answer.

13–5 SOLVING COMPLETE QUADRATIC EQUATIONS BY FACTORING

Learning Outcome

1 Solve complete quadratic equations by factoring ($ax^2 + bx + c = 0$).

Complete quadratic equations have all three types of terms. If the expression on the left will factor into two binomials, we can apply the zero-product property.

1 **Solve Complete Quadratic Equations by Factoring ($ax^2 + bx + c = 0$).**

In this section, we use knowledge of factoring trinomials to solve quadratic equations by factoring. If the trinomial will not factor, we must use another method for solving quadratic equations.

> *To solve a complete quadratic equation by factoring:*
>
> 1. If necessary, transpose and arrange terms so that all terms are on one side of the equation in standard form, with zero on the other side.
> 2. Factor the trinomial, if possible, into the product of two binomials, and set each factor equal to zero.
> 3. Solve for the variable in each equation formed in Step 2.

EXAMPLE Solve $x^2 + 6x + 5 = 0$ for x.

$$x^2 + 6x + 5 = 0 \qquad \text{Standard form.}$$

$$(x + 5)(x + 1) = 0 \qquad \text{Factor into the product of two binomials.}$$

$$x + 5 = 0 \qquad x + 1 = 0 \qquad \text{Set each factor equal to 0.}$$

$$x = -5 \qquad x = -1 \qquad \text{Solve for } x \text{ in each equation.}$$

$$\text{Check } x = -5 \qquad\qquad \text{Check } x = -1$$

$$x^2 + 6x + 5 = 0 \qquad\qquad x^2 + 6x + 5 = 0$$

$$(-5)^2 + 6(-5) + 5 = 0 \qquad (-1)^2 + 6(-1) + 5 = 0$$

$$25 + (-30) + 5 = 0 \qquad 1 + (-6) + 5 = 0$$

$$-5 + 5 = 0 \qquad\qquad -5 + 5 = 0$$

$$0 = 0 \qquad\qquad 0 = 0$$

EXAMPLE Solve $6x^2 + 4 = 11x$ for x.

$$6x^2 + 4 = 11x$$

$$6x^2 - 11x + 4 = 0 \qquad \text{Transpose into standard form.}$$

$$6x^2 - 3x - 8x + 4 = 0 \qquad \text{Write } -11x \text{ as } -3x - 8x.$$

$$(6x^2 - 3x) + (-8x + 4) = 0 \qquad \text{Group.}$$

$$3x(2x - 1) - 4(2x - 1) = 0 \qquad \text{Factor common factors.}$$

$$(3x - 4)(2x - 1) = 0 \qquad \text{Factor the common grouping.}$$

$$3x - 4 = 0 \qquad 2x - 1 = 0 \qquad \text{Set each factor equal to 0.}$$

$$3x = 4 \qquad 2x = 1 \qquad \text{Solve each equation.}$$

$$x = \frac{4}{3} \qquad x = \frac{1}{2}$$

$$\text{Check } x = \frac{4}{3} \qquad\qquad \text{Check } x = \frac{1}{2}$$

$$6x^2 + 4 = 11x \qquad\qquad 6x^2 + 4 = 11x$$

$$6\left(\frac{4}{3}\right)^2 + 4 = 11\left(\frac{4}{3}\right) \qquad 6\left(\frac{1}{2}\right)^2 + 4 = 11\left(\frac{1}{2}\right)$$

$$\overset{2}{\cancel{6}}\left(\frac{16}{\underset{3}{\cancel{9}}}\right) + 4 = \frac{44}{3} \qquad\qquad \overset{3}{\cancel{6}}\left(\frac{1}{\underset{2}{\cancel{4}}}\right) + 4 = \frac{11}{2}$$

$$\left(4 = \tfrac{12}{3}\right) \qquad \frac{32}{3} + \frac{12}{3} = \frac{44}{3} \qquad\qquad \frac{3}{2} + \frac{8}{2} = \frac{11}{2} \qquad \left(4 = \tfrac{8}{2}\right)$$

$$\frac{44}{3} = \frac{44}{3} \qquad\qquad\qquad \frac{11}{2} = \frac{11}{2}$$

Tip!	***Roots of Complete Quadratic Equations.***

Because all three types of terms are included in complete quadratic equations, we see that the roots do *not* have the same characteristics as the roots of pure or incomplete quadratic equations.

- Zero is not a root.
- The two roots do not have the same absolute value.

SELF-STUDY EXERCISES 13–5

1   Solve the equations by factoring.

1. $x^2 + 5x + 6 = 0$
2. $x^2 - 6x + 9 = 0$
3. $x^2 - 5x - 14 = 0$
4. $x^2 + 3x - 18 = 0$
5. $x^2 + 7x + 12 = 0$
6. $y^2 - 8y = -15$
7. $a^2 - 13a - 14 = 0$
8. $b^2 - 9b = -18$
9. $2x^2 - 7x + 3 = 0$
10. $3x^2 + 13x + 4 = 0$
11. $10x^2 - x = 3$
12. $6x^2 + 11x + 3 = 0$
13. $2x^2 + 13x + 15 = 0$
14. $3x^2 - 10x + 8 = 0$
15. $6x^2 + 17x = 3$

16. A rectangular hallway is 6 ft longer than its width. The area is 55 ft². What are the length and width of the hallway?

17. A rectangular metal plate covering a spare tire well that has an opening of 378 in.² is broken and must be reconstructed. The width is 3 in. less than the length. What dimensions should the metalsmith use when making the replacement part?

13–6   SELECTING AN APPROPRIATE METHOD FOR SOLVING QUADRATIC EQUATIONS

Learning Outcome

1   Determine the nature of the roots of a quadratic equation by examining the discriminant.

In the previous sections, we made various observations about the roots of the different types of quadratic equations. Now, we make additional observations about the roots of quadratic equations.

1   Determine the Nature of the Roots of a Quadratic Equation by Examining the Discriminant.

Which method for solving quadratic equations is best? Since any methods that are mathematically sound and produce consistently correct solutions are good, a "best" method may be evaluated differently. If you have mastered factoring, the factoring methods are generally quicker, especially when a calculator or computer is unavailable.

In general, follow these suggestions for solving quadratic equations.

1. Write the equation in standard form: $ax^2 + bx + c = 0$.
2. Identify the type of quadratic equation.
3. Solve pure quadratic equations by the square-root method.

4. Solve incomplete quadratic equations by finding common factors.
5. Solve complete quadratic equations by factoring into two binomials, if possible.
6. If factoring is not possible or is difficult, use the quadratic formula to solve complete quadratic equations.

All types of quadratic equations can be solved using the completing-the-square method (as presented on pages 513–514) or the quadratic formula. Only certain types of quadratic equations can be solved by factoring or applying the square-root property. The radicand of the radical portion of the quadratic formula, $b^2 - 4ac$, is called the *discriminant*. The general characteristics of the roots of a quadratic equation can be determined by examining the discriminant.

Properties of the discriminant, $b^2 - 4ac$:

1. If $b^2 - 4ac \geq 0$, the equation has real-number roots.
   a. If $b^2 - 4ac$ is a perfect square, there are two rational roots.
   b. If $b^2 - 4ac = 0$, there is one rational root (sometimes called a *double root*).
   c. If $b^2 - 4ac$ is not a perfect square, there are two irrational roots.
2. If $b^2 - 4ac < 0$, the equation has no real-number roots. The roots are imaginary or complex.

EXAMPLE Examine the discriminant of each equation and determine the nature of the roots. Then solve the equation.

(a) $5x^2 + 3x - 1 = 0$ (b) $3x^2 + 5x = 2$ (c) $4x^2 + 2x + 3 = 0$

(a) $5x^2 + 3x - 1 = 0$ $a = 5, b = 3, c = -1$
$$b^2 - 4ac = 3^2 - 4(5)(-1)$$ Examine the discriminant.
$$= 9 + 20$$
$$= \boxed{29}$$ There will be two irrational roots.

$$x = \frac{-b \pm \sqrt{b^2 - 4ac}}{2a}$$ Use the quadratic formula.

$$x = \frac{-3 \pm \sqrt{29}}{2(5)}$$ Substitute 29 for the discriminant.

$$x = \frac{-3 + \sqrt{29}}{10} \quad \text{or} \quad x = \frac{-3 - \sqrt{29}}{10}$$ Exact irrational roots.

(b)
$$3x^2 + 5x = 2$$
$$3x^2 + 5x - 2 = 0$$ Standard form: $a = 3, b = 5, c = -2$
$$b^2 - 4ac = 5^2 - 4(3)(-2)$$ Examine the discriminant.
$$= 25 + 24$$
$$= \boxed{49}$$ There are two rational roots and the trinomial will factor.
$$(3x - 1)(x + 2) = 0$$ Factor.
$$3x - 1 = 0 \quad x + 2 = 0$$
$$3x = 1$$
$$x = \frac{1}{3} \qquad x = -2$$ Exact rational roots.

(c) $4x^2 + 2x + 3 = 0$ $a = 4, b = 2\ c = 3$
$$b^2 - 4ac = 2^2 - 4(4)(3)$$ Examine the discriminant.
$$= 4 - 48$$
$$= \boxed{-44}$$ There are no real roots. The roots are imaginary or complex.

$$x = \frac{-2 \pm \sqrt{-44}}{2 \cdot 4}$$ Quadratic equation.

$$x = \frac{-2 \pm 2i\sqrt{11}}{8} \qquad \text{Simplify the radical and reduce.}$$

$$x = \frac{2(-1 \pm i\sqrt{11})}{8} \qquad \text{Factor and reduce.}$$

$$x = \frac{-1 \pm i\sqrt{11}}{4} \qquad \text{The roots are complex.}$$

> ■ **Learning Strategy**   *Understand the Concepts.*
>
> When calculators and computers can so easily find solutions to all types of equations, why do we spend so much time with paper-and-pencil techniques? For our knowledge of algebra and mathematics to be useful in real-world situations, we must understand the concepts and know when to use them, and we must understand what the solutions represent.
>
> In developing this understanding of the concepts, we must examine a wide range of situations that might be encountered with the concept. In practice, technological tools can be used to solve equations once we have established appropriate equations through a critical examination of the situation.

SELF-STUDY EXERCISES 13–6

1   Examine the discriminant of each equation and find the roots. Round to hundredths if necessary.

1. $3x^2 + x - 2 = 0$ **2.** $x^2 - 3x = -1$ **3.** $2x^2 + x = 2$

4. $3x^2 - 2x + 1 = 0$ **5.** $x^2 - 3x - 7 = 0$ **6.** $3x^2 + 5x - 6 = 0$

7. Use the discriminant to write a quadratic equation that has real roots.

8. Use the discriminant to write a quadratic equation that has real, rational, and unequal roots.

CAREER APPLICATION

Civil Engineering: Freeway Supports

Safe, dependable, durable freeway-support design is extremely important because highways carry an ever-increasing number of heavy vehicles (such as tandem tractor-trailer trucks) and heavier weight loads. Civil engineers try to design supports that hold up during temperature extremes, earthquakes, and high wind conditions (for example, tornadoes and hurricanes). When designing a rectangular box support, researchers found that support is optimal when the depth is at least one-half of the width, and the volume is at least 10 ft^3 per ft of support height.

Civil engineers lobby a state government to allot construction money for a freeway passing through an earthquake fault zone at a higher, safer volume rate of 12.5 ft^3 per ft of support height. At intersections, rectangular box supports 16 ft high are required.

Exercises

Use the information for the higher, safer volume rate to answer the questions. Round answers to the nearest tenth.

1. Draw a diagram of a rectangular box support with a 16-ft height. Label the sides W for width and D for depth.

2. What is the required volume of this safer 16-ft-high support?

3. Use W to write an expression for the depth if it is to be half the width.

4. Use the volume formula $V = LWD$ to write an equation to find the width and depth of the support. Solve the equation, and state your answers rounded to the nearest tenth of a foot. What is the volume of this support?

   CHAPTER 13   Solving Quadratic Equations

5. If a support has a depth three-fourths the size of the width, find its dimensions and volume.
6. This freeway will be in an earthquake fault zone, so state civil engineers decide to make the depth equal to the width of the support to withstand maximum vibration stress. Find the dimensions and volume of each support.
7. Why do the dimensions you found in Exercise 4 have a volume of exactly 200 ft^3, while the dimensions you found in Exercises 5 and 6 have volumes over and under the prescribed 200 ft^3?
8. A recent computer modeling simulation found that supports with the smallest cross-sectional perimeter per volume are best able to endure vibration stress, which means that supports in earthquake zones should be in the shape of a cylinder (a cylinder has a circular cross-sectional perimeter). Use $V = \pi r^2 h$ to find the radius of the support with a height of 16 ft and a volume of 200 ft^3.
9. Find the cross-sectional perimeters of the supports designed in Exercises 4, 5, 6, and 8. Does the circular cross section have the smallest perimeter?

Answers

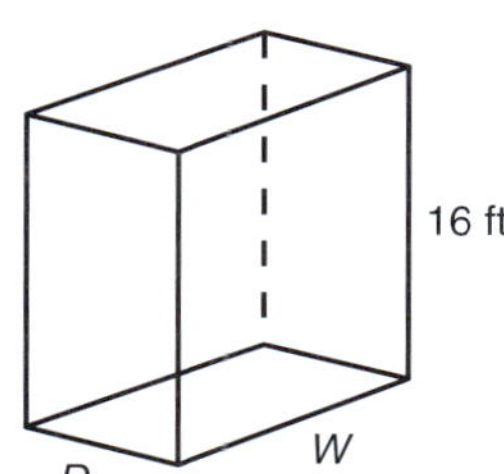

1. Support diagram:
2. At 12.5 ft^3 per ft of height, 200 ft^3 is allotted.
3. $D = \left(\frac{1}{2}\right)W$ or $D = \frac{W}{2}$ or $D = 0.5W$.
4. The width should be 5.0 ft and the depth 2.5 ft, with a volume of 200.0 ft^3.
5. From the equation $16(0.75W)(W) = 200$, the width should be 4.1 ft, the depth should be 3.1 ft, and the height should be 16 ft. These dimensions have a volume of 203.4 ft^3.
6. With $D = W$, both the width and depth should be 3.5 ft, and the height should be 16 ft. These dimensions have a volume of 200.5 ft^3.
7. When the pure quadratic equation of Exercise 4 was solved by the square-root method, an exact square root was found. In Exercises 5 and 6, square roots were approximated to the nearest tenth of a foot, which resulted in round-off error propagation in the volume formula.
8. The radius should be 2.0 ft, which produces a volume of 201.1 ft^3.
9. The perimeter of the rectangular cross section is 15 ft in Exercise 4 and 14.4 ft in Exercise 5. The square cross section of Exercise 6 is 14.2 ft, while the perimeter of the circular cross section of Exercise 8 is 12.6 ft. The perimeter of the circular cross section has the smallest perimeter per volume.

ASSIGNMENT EXERCISES

Section 13–1
Identify the quadratic equations as pure, incomplete, or complete.

1. $x^2 = 49$
2. $x^2 - 5x = 0$
3. $5x^2 - 45 = 0$
4. $3x^2 + 2x - 1 = 0$
5. $8x^2 + 6x = 0$
6. $5x^2 + 2x + 1 = 0$
7. $x^2 - 32 = 0$
8. $x^2 + x = 0$
9. $3x^2 + 6x + 1 = 0$

Section 13–2
Indicate the values for a, b, and c in the quadratic equations.

10. $5x^2 + x + 6 = 0$
11. $x^2 - 2x = 8$
12. $x^2 - 7x + 12 = 0$
13. $x^2 + 3x = 4$
14. $3x^2 = 2x + 7$
15. $x^2 - 3x = -2$

Solve the quadratic equations by using the quadratic formula.

16. $x^2 - 9x + 20 = 0$
17. $x^2 - 8x - 9 = 0$
18. $x^2 - 5x = -6$
19. $x^2 + 2x = 8$
20. $x^2 - x - 12 = 0$
21. $2x^2 - 3x - 2 = 0$

Solve the quadratic equations by using the quadratic formula. Round each final answer to the nearest hundredth.

22. $3x^2 + 6x + 2 = 0$

23. $2x^2 - 3x - 1 = 0$

24. $5x^2 + 4x - 8 = 0$

25. $3x^2 + 5x + 1 = 0$

26. A bricklayer plans to build an arch with a span (s) of 10 m and a radius (r) of 5 m. How high (h) is the arch? (Use the formula $h^2 - 2hr + \frac{s^2}{4} = 0$.)

27. A rectangular kitchen contains 240 ft^2. If the length is two times the width, find the length and width of the room to the nearest whole number. (Area = length × width, or $A = lw$.)

28. What are the dimensions of a rectangular tool storage room if the area is 45.5 m^2 and the room is 0.5 m longer than it is wide? Round to the nearest tenth.

29. Find the length and width of a piece of fiberglass if its length is three times the width and the area is 591 in.^{2}. Round to the nearest inch.

Section 13–3

Solve the equations. Round to thousandths when necessary.

30. $x^2 = 121$

31. $x^2 = 100$

32. $x^2 - 64 = 0$

33. $4x^2 = 9$

34. $64x^2 - 49 = 0$

35. $0.36y^2 = 1.09$

36. $0.16x^2 = 0.64$

37. $5x^2 = 40$

38. $2x^2 - 5 = 3$

39. $6x^2 + 4 = 34$

40. $5x^2 - 6 = 19$

41. $3x^2 = 12$

42. $3x^2 - 4 = 8$

43. $2x^2 = 34$

44. $5x^2 - 9 = 30$

45. $3y^2 - 36 = -8$

46. $2x^2 + 3 = 51$

47. $\frac{1}{2}x^2 = 8$

48. $\frac{2}{3}x^2 = 24$

49. $\frac{1}{4}x^2 - 1 = 15$

50. $\frac{2}{5}x^2 + 2 = 8$

51. A circle has an area of 845 cm^2. What is the radius of the circle?

52. An oil painting that is a square is known to be 9,072 cm^2. What are the inside dimensions of its picture frame?

Section 13–4

Solve the equations.

53. $x^2 - 5x = 0$

54. $4x^2 = 8x$

55. $6x^2 - 12x = 0$

56. $3x^2 + x = 0$

57. $10x^2 + 5x = 0$

58. $3y^2 = 12y$

59. $y^2 - 5y = 0$

60. $x^2 = 16x$

61. $12x^2 + 8x = 0$

62. $8x^2 - 12x = 0$

63. $x^2 + 3x = 0$

64. $4x^2 - 28x = 0$

65. $5x^2 = 45x$

66. $7x^2 = 28x$

67. $y^2 + 8y = 0$

68. $z^2 - 6z = 0$

69. $3m^2 - 5m = 0$

70. $4n^2 - 3n = 0$

71. $2x^2 = x$

72. $5y^2 = y$

73. Three times the square of a number is the same as 12 times the number. What is the number?

74. Describe the steps for solving an incomplete quadratic equation. Include a clear description of the nature of the roots.

Section 13–5

Solve the equations by factoring.

75. $x^2 - 4x + 3 = 0$

76. $x^2 + 7x + 12 = 0$

77. $x^2 + 3x = 10$

78. $x^2 - 7x + 12 = 0$

79. $x^2 + 7x = -6$

80. $x^2 + 3 = -4x$

81. $x^2 - 6x + 8 = 0$

82. $6y + 7 = y^2$

83. $6y^2 - 5y - 6 = 0$

84. $5y^2 + 23y = 10$

85. $10y^2 - 21y - 10 = 0$

86. $6x^2 - 16x + 8 = 0$

87. $4x^2 + 7x + 3 = 0$

88. $3x^2 = -7x + 6$

89. $12y^2 - 5y - 3 = 0$

90. $x^2 - 3x = 18$

91. $x^2 + 19x = 42$

92. $3x^2 + x - 2 = 0$

93. $3y^2 + y - 2 = 0$

94. $2x^2 - 4x - 6 = 0$

95. $2x^2 - 10x + 12 = 0$

96. $y^2 + 18y + 45 = 0$

97. $x^2 - 3x - 18 = 0$

98. $3x^2 - 9x - 30 = 0$

99. $2y^2 + 22y + 60 = 0$

100. An office building in the shape of a rectangle is known to have 47,500 ft^2 of space on the ground floor. The tenant wants to landscape the two longer sides of the building. The tenant also knows that the building is about 60 ft longer than it is wide. How many feet of land along the building need to be landscaped?

101. Jerri Amour is an architect who is designing a hospital. She knows that a kidney dialysis machine needs a space that is 7 ft longer than it is wide. The total area needed is 228 ft^2 of space. Advise Jerri about the number of feet of space she should include in the length and width of the planned space.

Use the discriminant of the quadratic formula to describe the roots of the equations.

102. $x^2 - 3x + 2 = 0$

103. $x^2 + 8x + 16 = 0$

104. $2x^2 - 3x - 5 = 0$

105. $5x^2 - 100 = 0$

106. $3x^2 - 2x + 4 = 0$

107. $2x = 5x^2 - 3$

CHALLENGE PROBLEM

108. Many objects are designed with dimensions according to the ratio of the *Golden Ratio*. Objects that have measurements according to this ratio are said to be most pleasing to the eye. The *Golden Rectangle* has dimensions of length (l) and height (h) that satisfy the formula

$$\frac{l + h}{l} = \frac{l}{h}$$

You can cross-multiply to obtain a quadratic equation for the Golden Rectangle.

You have been commissioned to construct a wall hanging for the lobby of a new office building. The wall is 32 ft long and the ceiling is 20 ft high. The owners want the wall hanging to be at least 2 ft from the ceiling, the floor, and each of the side corners. They also want the wall hanging to have dimensions according to the Golden Ratio.

Using the given information, determine the largest-size wall hanging that can be placed in the lobby that also has the dimensions of the Golden Rectangle.

CHAPTER TRIAL TEST

Identify the quadratic equations as pure, incomplete, or complete.

1. $3x^2 = 42$

2. $7x^2 - 3x + 2 = 0$

3. $5x^2 = 7x$

4. $4x^2 - 1 = 0$

Solve the quadratic equations. Find the exact solutions. Also, give approximate solutions to the nearest hundredth when appropriate.

5. $x^2 = 81$

6. $x^2 - 32 = 0$

7. $9x^2 = 16$

8. $81x^2 - 64 = 0$

9. $0.09x^2 = 0.49$

10. $2x^2 + 3x + 1 = 0$

11. $3x^2 - 6x = 0$

12. $3x^2 - 6x - 1 = 0$

13. $x^2 - 5x + 6 = 0$

14. $x^2 - 3x - 4 = 0$

15. $2x^2 + 12 = 11x$

16. $3x^2 - 5x + 4 = 0$

17. Find the diameter (d) in mils to the nearest hundredth of a copper wire conductor whose resistance (R) is 1.314 Ω and whose length (L) is 3,642.5 ft. (Formula: $R = \frac{KL}{d^2}$, where K is 10.4 for copper wire.)

18. Find the radius (r) of a circle whose area (A) is 35.15 cm^2. Round the answer to the nearest hundredth centimeter. (Formula: $A = \pi r^2$.)

19. What is the current in amps (I) to the nearest hundredth if the resistance (R) of the circuit is 52.29 Ω and the watts (W) used are 205? (Formula: $R = \frac{W}{I^2}$.)

20. A square parcel of land is 156.25 m^2 in area. What is the length of a side to the nearest hundredth? (Use the formula $A = s^2$, where A is the area and s is the length of a side.)

21. In the formula $E = 0.5\,mv^2$, solve for v if $E = 180$ and $m = 10$.

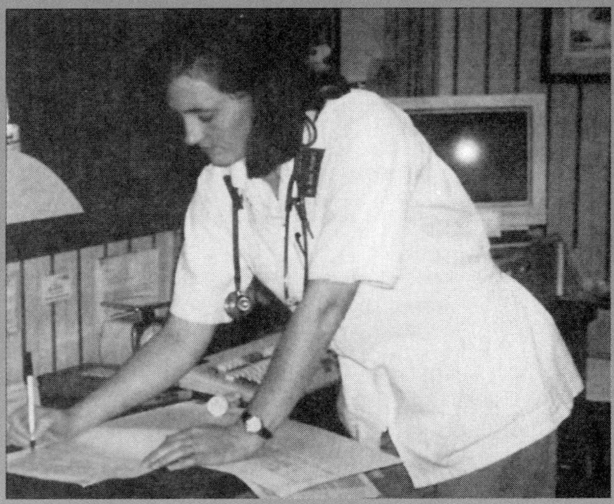

14

Graphing

GOOD DECISIONS THROUGH TEAMWORK

A technical college graduate with a degree in computer engineering has started a company called Health Sciences Analysis, Inc., which designs software. It allows physicians to enter hospital chart data electronically at each patient's bedside so that nurses don't have to transcribe this same data. Her first year of operation in 1999 yielded a gross income (revenue before expenses) of about $185,000. In 2002, her gross income was $326,250.

As financial analysts, you and your team have been hired to track company growth for this company and also to derive an equation to predict future growth. You are to assume that growth will follow the same linear trend. Your team will use the 1999 and 2002 gross income figures to form ordered pairs for writing a linear equation that relates the year to gross income. Let $x = 1$ represent 1999 and $x = 4$ represent 2002. Let y represent the corresponding yearly gross income. Find the y-intercept and give its real-world meaning, and graph the given information.

From the graph, predict gross income for the years 2003, 2004, and 2005. If this trend continues, in what year will gross income top $500,000? Verify your predictions using the equation of the graph.

As financial analysts, you and your team members must explain to your client how you arrived at your model and what assumptions you used. But you must also discuss any limitations of your model. Prepare a presentation for the company owner.

14–1 Graphical representation of equations

1. Represent the solutions of a function in a table of values and on a graph.

14–2 Linear equations in two variables and function notation

1. Check the solution of an equation with two variables.
2. Find a specific solution of an equation with two variables when given the value of one variable.

14–3 Graphing linear equations with two variables

1. Graph linear equations using intercepts.
2. Graph linear equations using the slope-intercept method.
3. Graph linear equations using a graphing calculator.

14–4 Slope

1. Calculate the slope of a line, given two points on the line.
2. Determine the slope of a horizontal or vertical line.
3. Find the equation of a line, given the slope and one point.

4. Find the equation of a line, given two points on the line.

5. Find the slope and y-intercept of a line, given the equation of the line.

6. Find the equation of a line, given the slope and y-intercept.

14–5 Parallel and perpendicular lines

1. Find the equation of a line, given a point on the line and the equation of a line parallel to the line.

2. Find the equation of a line, given a point on the line and the equation of a line perpendicular to the line.

14–6 Graphing quadratic equations

1. Identify nonlinear equations.

2. Graph quadratic equations using the table-of-solutions method or by examining properties.

3. Graph quadratic equations using a graphing calculator.

In previous chapters, we solved various types of equations. The solution was usually one or two values that made a true statement when substituted into the equation. In this chapter, we look at the graphical representation of some of these types of equations.

14–1   GRAPHICAL REPRESENTATION OF EQUATIONS

Learning Outcome

1. Represent the solutions of a function in a table of values and on a graph.

Sometimes the solution of an equation answers a specific question. How much can we pay for a new copier with a 3-year service contract, if the service contract is $100 and the total amount we are budgeted to spend is $900? We can represent this situation with the equation $x + \$100 = \$900$, where x represents the cost of the copier.

In other situations an equation serves as a *model to show the effect of change*. These models can be written as functions and can be represented graphically.

1. Represent the Solutions of a Function in a Table of Values and on a Graph.

The monthly cost of a cellular phone is $15 plus $0.25 for each minute of use. An equation to represent this situation must have two variables. One variable represents the number of minutes of phone use (m), and the other variable represents the monthly cost (c). One variable is dependent on the other. For instance, the monthly cost is dependent on the number of minutes of use: $c = \$15 + \$0.25m$. We can also write equations with two variables in function notation: $C(m) = \$15 + \$0.25m$ or $f(x) = 15 + 0.25x$. Function notation is a notation showing the relationship between an independent variable (input) and a dependent variable (output). This type of function is often referred to as a *cost function*.

EXAMPLE Make a table of values and graph the function $f(x) = 15 + 0.25m$ for five values of x in 50-min intervals.

x	$f(x)$
0	$15.00
50	$27.50
100	$40.00
150	$52.50
200	$65.00

$$f(0) = 15 + 0.25(0) \qquad f(50) = 15 + 0.25(50) \qquad f(100) = 15 + 0.25(100)$$

$$f(0) = 15 + 0 \qquad f(50) = 15 + 12.5 \qquad f(100) = 15 + 25$$

$$\mathbf{f(0) = 15, \ or \ \$15.00} \qquad \mathbf{f(50) = 27.5, \ or \ \$27.50} \qquad \mathbf{f(100) = 40, \ or \ \$40.00}$$

$$f(150) = 15 + 0.25(150) \qquad f(200) = 15 + 0.25(200)$$

$$f(150) = 15 + 37.5 \qquad f(200) = 15 + 50$$

$$\mathbf{f(150) = 52.5, \ or \ \$52.50} \qquad \mathbf{f(200) = 65, \ or \ \$65.00}$$

To represent this model graphically, use the x- or horizontal axis to represent the values of the independent variable. Use the y- or vertical axis to represent the values of the dependent variable (Fig. 14–1).

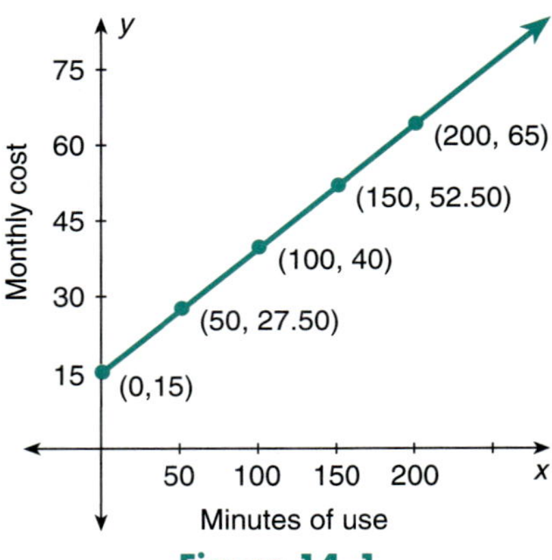

Figure 14–1

Let's examine the graphical solution. What does the point (0, $15) represent? This is a *fixed cost*. Even if no minutes are used, the monthly cost will be $15. What if the monthly phone use is more than 200 minutes? When the table of values was represented on the graph, a pattern was established. We can use this pattern or model to project other values of x.

EXAMPLE Find the monthly cost from the graph of $f(x) = 15 + 0.25x$ (Fig. 14–2) for 250 minutes and 300 minutes.

From the graph find the corresponding values for $f(x)$ when $x = 250$ and when $x = 300$.

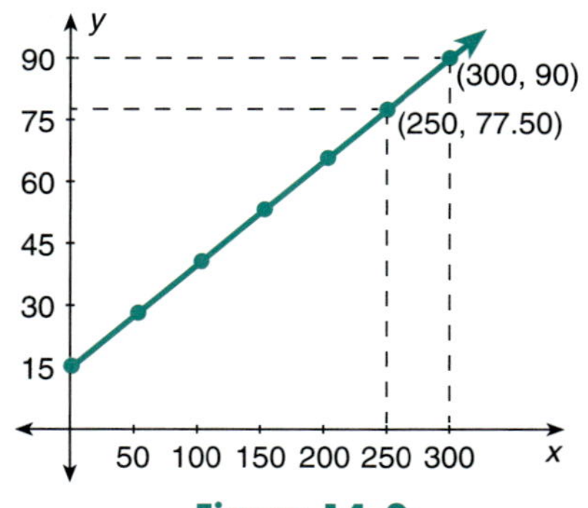

For $x = 250$, $f(x) =$ slightly more than $75.
For $x = 300$, $f(x) =$ $90.00.

Figure 14–2

Exact values are difficult and sometimes impossible to determine from a graph. To find the exact cost for 250 min of phone use we evaluate the function $f(x) = 15 + 0.25x$ for $x = 250$. $f(250) = 15 + 0.25(250)$; $f(250) = 15 + 62.50$; $f(250) = \$77.50$.

CHAPTER 14 Graphing

In the preceding two examples, we were looking at a solution that restricted the values of x and $f(x)$ to positive values. In many cases, we want to view the pattern established by considering all types of numbers. After viewing all values, we may still disregard some of these values because they are impractical or impossible within the context of the situation.

> **To graph a function using a table of values:**
>
> 1. Prepare a table of values by evaluating the function at different values (at least three) of the independent variable.
> 2. Plot the points in the table of values on the rectangular coordinate system, with the independent variable on the x-axis and the dependent variable on the y-axis.
> 3. Connect the points with a straight line or a smooth, continuous curve, as appropriate.

The *table-of-values* or *table-of-solutions* procedure for graphing an equation works for *all* types of equations. To emphasize the breadth of this procedure, we give examples of quadratic, logarithmic, and exponential equations. Other methods for graphing equations focus on specific properties of each type of equation; these methods are generally less time-consuming and more efficient than the table-of-values procedure. However, the table-of-values method can be used as a backup or check if uncertainty arises with the other methods.

EXAMPLE Prepare a table of values and graph the function $f(x) = x^2 + x - 6$ using integral values of x between -3 and 3, inclusive.

x	$f(x)$
-3	0
-2	-4
-1	-6
0	-6
1	-4
2	0
3	6

$$f(-3) = (-3)^2 + (-3) - 6$$
$$f(-3) = 9 - 3 - 6$$
$$\boldsymbol{f(-3) = 0}$$

$$f(-2) = (-2)^2 + (-2) - 6$$
$$f(-2) = 4 - 2 - 6$$
$$\boldsymbol{f(-2) = -4}$$

$$f(-1) = (-1)^2 + (-1) - 6$$
$$f(-1) = 1 - 1 - 6$$
$$\boldsymbol{f(-1) = -6}$$

$$f(0) = 0^2 + 0 - 6$$
$$f(0) = 0 + 0 - 6$$
$$\boldsymbol{f(0) = -6}$$

$$f(1) = 1^2 + 1 - 6$$
$$f(1) = 1 + 1 - 6$$
$$\boldsymbol{f(1) = -4}$$

$$f(2) = 2^2 + 2 - 6$$
$$f(2) = 4 + 2 - 6$$
$$\boldsymbol{f(2) = 0}$$

$$f(3) = 3^2 + 3 - 6$$
$$f(3) = 9 + 3 - 6$$
$$\boldsymbol{f(3) = 6}$$

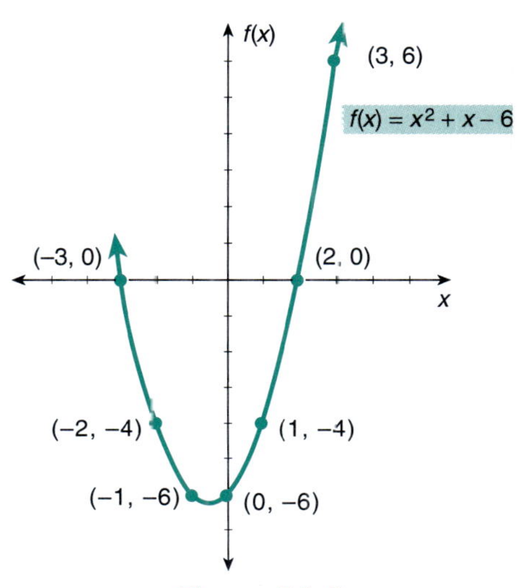

Figure 14–3

Next, plot the points indicated in the table of values and connect the points with a smooth, continuous curve (Fig. 14–3).

EXAMPLE Prepare a table of values and graph the function $f(x) = 2^x$ using integral values of x between -3 and 3, inclusive.

x	$f(x)$
-3	0.125
-2	0.25
-1	0.5
0	1
1	2
2	4
3	8

$$f(-3) = 2^{-3}$$
$$f(-3) = \frac{1}{2^3}$$
$$f(-3) = \tfrac{1}{8}, \text{ or } 0.125$$

$$f(-2) = 2^{-2}$$
$$f(-2) = \frac{1}{2^2}$$
$$f(-2) = \tfrac{1}{4}, \text{ or } 0.25$$

$$f(-1) = 2^{-1}$$
$$f(-1) = \frac{1}{2^1}$$
$$f(-1) = \tfrac{1}{2}, \text{ or } 0.5$$

$$f(0) = 2^0$$
$$f(0) = 1$$

$$f(1) = 2^1$$
$$f(1) = 2$$

$$f(2) = 2^2$$
$$f(2) = 4$$

$$f(3) = 2^3$$
$$f(3) = 8$$

Next, plot the points indicated in the table of values and connect the points with a smooth, continuous curve (Fig. 14–4).

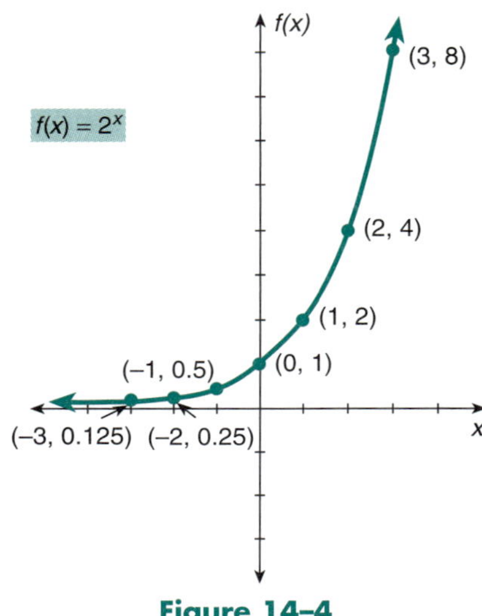

Figure 14–4

As you can see from these examples, all functions can be expressed graphically by using a table of values.

■ **Learning Strategy** *Always Have a Backup or Spot-Check Plan.*

You will see in later sections and in future courses that graphing is done most efficiently by examining and applying various properties of equations. Also, with the easy availability of graphing calculators and computer graphing software, if your work requires graphing on a regular basis, you will most likely use the technological tools available.

However, sometimes you will need a backup or spot-check plan to verify a solution that you are unsure about or to get started if you experience a memory block for a specific situation. The table-of-values procedure works well as a backup or spot-check plan.

The table-of-values method of graphing and the graphing method of solving an equation are tedious. These methods may be used when a computer algebra system and/or a graphing calculator is available. It is helpful when you can visualize a problem and solution. However, other methods for graphing are less time-consuming because they use the properties of various types of equations. *In the sections that follow, we will examine several different types of equations and observe properties that allow us to sketch a graph before we use the computer or calculator. Sketching is a form of estimating.*

1 Make a table of values for each function for five values of x and graph the function.

1. $f(x) = 3x$

2. $f(x) = -3x$

3. $f(x) = 2x - 3$

4. $f(x) = -2x - 3$

5. $f(x) = \dfrac{1}{2}x + 2$

6. $f(x) = 3x - 7$

7. $f(x) = -\dfrac{1}{3}x + 1$

8. Rental cars cost $39 for a day plus $0.10 per mile. Write a function that represents the cost of the car for a single day if x represents the number of miles and $f(x)$ represents the car rental cost.

9. Make a table of values for the function in Exercise 8. Graph the function.

10. From the graph prepared in Exercise 9, determine the cost of driving the rental car for 400 miles in a single day.

Prepare a table of values and graph each of the following functions.

11. $f(x) = x^2 - 4x + 5$

12. $f(x) = 3^x$

14-2   **LINEAR EQUATIONS IN TWO VARIABLES AND FUNCTION NOTATION**

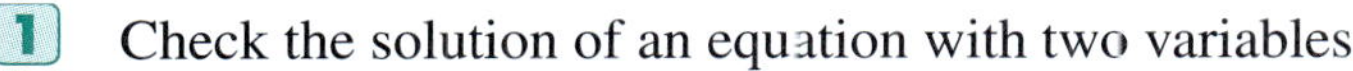

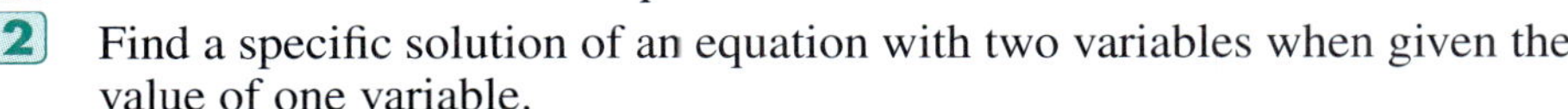

Learning Outcomes

1 Check the solution of an equation with two variables.

2 Find a specific solution of an equation with two variables when given the value of one variable.

Most of the equations we have studied in this text have had only one variable, such as x or y, and took the form $3x - 2 = 7$ or $5y - 3 = 9$. However, equations may have two variables, both x *and* y, such as $2x + 3y = 12$. Both types of equations are *linear equations*.

■   **DEFINITION: Linear Equation.** A *linear equation* is an equation for which the graph is a straight line. It can be written in the standard form

$$ax + by = c$$

where a, b, and c are real numbers and a and b are not both zero. The degree of any term in the equation is 1 or zero.

Before we examine properties of linear equations, let's be sure we can verify that a possible solution is correct or that we can find a specific solution if we know the value of one variable of the solution.

1   **Check the Solution of an Equation with Two Variables.**

The solutions of a linear equation in two variables are *ordered pairs* of numbers. There is one number for each variable. The numbers are generally ordered in the alphabetical order of the two variables rather than in the order the letters appear in the equation. Thus, in the equation $2y + x = 8$, a solution of $(4, 2)$ indicates that $x = 4$ and $y = 2$. An ordered pair is written in point notation (enclosed in parentheses and the numbers are separated by a comma).

If the equation is in function notation, the ordered pair gives the value for the independent variable first and the dependent variable second. In point notation, a solution for a function of x would be written as $(x, f(x))$.

Solutions of equations in two variables are ordered pairs of numbers that make a true statement when used to evaluate the original equation. An equation in two variables has many solutions. We can check each solution of an equation in two variables by substituting the values for the respective variables and performing the operations indicated.

EXAMPLE Check the solution $(-2, 16)$ for the equation $2x + y = 12$.

$$2x + y = 12$$
$$2(-2) + 16 = 12 \qquad \text{Substitute } -2 \text{ for } x \text{ and } 16 \text{ for } y.$$
$$-4 + 16 = 12$$
$$12 = 12 \qquad \text{True.}$$

The ordered pair makes the equation true, so $(-2, 16)$ is a solution.

EXAMPLE Check the ordered pair $(3, -2)$ for the equation $y = 2x - 5$.

$$y = 2x - 5$$
$$-2 = 2(3) - 5 \qquad \text{Substitute 3 for } x \text{ and } -2 \text{ for } y.$$
$$-2 = 6 - 5$$
$$-2 = 1 \qquad \text{False.}$$

The ordered pair does not make the equation true, so $(3, -2)$ is not a solution.

2 Find a Specific Solution of an Equation with Two Variables When Given the Value of One Variable.

If we are given the value of one variable for an equation with two variables, we can substitute the given value in the equation and solve the equation for the other variable.

EXAMPLE Solve the equation $2x - 4y = 14$, if $x = 3$.

$$2x - 4y = 14$$
$$2(3) - 4y = 14 \qquad \text{Substitute 3 for } x.$$
$$6 - 4y = 14$$
$$-4y = 14 - 6 \qquad \text{Sort.}$$
$$-4y = 8$$
$$\frac{-4y}{-4} = \frac{8}{-4} \qquad \text{Divide both sides by } -4.$$
$$y = -2$$

The solution is $(3, -2)$; $x = 3$ and $y = -2$.

The standard form of *a linear equation in two variables, $ax + by = c$,* is such that both a or b are not zero. However, either a or b may be zero.

When a or b is zero, the variable for which zero is the coefficient is eliminated from the equation. Let's take the equation $5x + 3y = 15$ and replace the coefficient 3 with zero:

$$5x + 3y = 15$$

CHAPTER 14 Graphing

$$5x + 0(y) = 15 \qquad \text{Replace 3 with 0 to get a new equation in one variable.}$$

$$5x + 0 = 15 \qquad 0(y) = 0 \text{ since 0 times any number is 0.}$$

$$5x = 15$$

$$\frac{5x}{5} = \frac{15}{5} \qquad \text{Divide both sides by 5.}$$

$$x = 3$$

Linear equations may also be equations in one variable. Thus, an equation like $x = 3$ is a linear equation and we think of the coefficient of the missing variable term as zero:

$$x = 3 \qquad \text{is considered as} \qquad x + 0y = 3$$

This means that, for any value of y, x is 3.

EXAMPLE For the equation $x = 6$, complete the given ordered pairs: (, 2); (, −3); (, 1).

Equation   **Ordered pairs**

$x = 6$ (, 2)   (, −3)   (, 1)

(6, 2)   **(6, −3)**   **(6, 1)** x is 6 for any value of y.

Numerous application problems can be solved with equations in two variables.

EXAMPLE A small business photocopied a report that included 12 black-and-white pages and 25 color pages. The cost was \$23.70. Letting $b = $ the cost for a black-and-white page and $c = $ the cost for a color page, set up an equation with two variables.

Cost of black-and-white copies: $12b$ (12 times cost per page)

Cost of color copies: $25c$ (25 times cost per page)

Total cost: black-and-white plus color = \$23.70

Equation: $12b + 25c = 23.70$

An equation containing two variables shows the relationship between the two unknown amounts. If a value is known for either amount, the other amount can be found by solving the equation for the missing amount.

EXAMPLE Solve the equation from the preceding example to find the cost for each black-and-white copy if color copies cost \$0.90 each.

$$12b + 25c = 23.70$$

$$12b + 25(0.90) = 23.70 \qquad \text{Substitute 0.90 for } c.$$

$$12b + 22.50 = 23.70$$

$$12b = 23.70 - 22.50 \qquad \text{Transpose.}$$

$$12b = 1.20$$

$$\frac{12b}{12} = \frac{1.20}{12} \qquad \text{Divide both sides by 12.}$$

$$b = 0.10$$

Each black-and-white copy costs \$0.10.

1 Determine which of the ordered pairs are solutions for the equation $2x - 5y = 9$.

1. $(4, 5)$ **2.** $(17, 5)$ **3.** $(12, 3)$
4. $(6, 1)$ **5.** $(7, 1)$ **6.** $(4.5, 0)$

Determine which of the ordered pairs are solutions for the equation $3y = 2 - 4x$.

7. $(2, -2)$ **8.** $(-2, 2)$ **9.** $(5, -6)$
10. $(-6, 5)$ **11.** $(0, \frac{2}{3})$ **12.** $(6, -7\frac{1}{3})$

Determine which of the ordered pairs are solutions for the equation $4y - x = 7$.

13. $(1, 2)$ **14.** $(-2, -12)$ **15.** $(-2, -15)$
16. $(3, 5)$ **17.** $(5, 3)$ **18.** $(7, 0)$

2 Write the specific solution for each equation in ordered-pair form.

19. $x - y = 3$, if $x = 8$ **20.** $x + y = 7$, if $x = 2$
21. $x + 2y = -4$, if $y = 1$ **22.** $x - 3y = 12$, if $y = 3$
23. $5x - y = 10$, if $x = 5$ **24.** $4x + y = 8$, if $x = 2$
25. $3x + 4y = 16$, if $y = -2$ **26.** $5x + 2y = 9$, if $y = -3$

27. $3x - y = 16$, if $x = 6$ **28.** $x + 3y = 12$, if $x = 9$ **29.** $\frac{2}{3}x - y = 6$, if $y = 2$

30. $x - \frac{3}{4}y = 9$, if $x = 0$ **31.** $6 = x + 3y$, if $y = -4$ **32.** $8 = 2x + y$, if $x = -2$

Complete the ordered pairs for each equation.

33. $x = 1$ $(\ , 3); (\ , -2); (\ , 0); (\ , 1)$ **34.** $y = 4$ $(-1, \ ); (3, \ ); (0, \ ); (2, \ )$
35. $y = -7$ $(2, \ ); (-3, \ ); (0, \ ); (5, \ )$ **36.** $x = 9$ $(\ , -2); (\ , 3); (\ , 0); (\ , 2)$

Solve using equations with two variables.

37. Diane purchased three shirts at one price and five belts at another price. Let s = the price of each shirt and b = the price of each belt. Each shirt cost \$22. How much did each belt cost if the total purchase price was \$126?

38. Paulette and Terry Fink paid \$13 for four spark plugs and five quarts of motor oil. Let p = the cost of each plug and q = the cost of each quart of oil. If the oil cost \$1 per quart, how much did each spark plug cost?

39. Paul and Donna were charged \$160 for four pairs of pants and two shirts. If the shirts cost \$12 each, how much did all four pairs of pants cost? Let p = the cost of each pair of pants and s = the cost of each shirt.

40. David paid \$15,000 for an automobile. The purchase price included three extended warranty payments of \$400 each. How much did the car cost without the extended warranty? Let c = the cost of the car and w = the cost of each warranty payment.

14–3   GRAPHING LINEAR EQUATIONS WITH TWO VARIABLES

Learning Outcomes

1 Graph linear equations using intercepts.
2 Graph linear equations using the slope-intercept method.
3 Graph linear equations using a graphing calculator.

We have graphed linear equations using a table of values. Now let's examine other procedures and specific properties of linear equations.

1   Graph Linear Equations Using Intercepts.

■ **DEFINITION:** **x-intercept.** The *x-intercept* is the point on the *x*-axis through which the line of the equation passes; that is, the *y*-value is zero $(x, 0)$.

■ **DEFINITION:** *y-intercept.* The *y-intercept* is the point on the *y*-axis through which the line of the equation passes; that is, the *x*-value is zero $(0, y)$.

The following rule may help.

> ### To find the intercepts of a linear equation:
>
> **1.** To find the *x*-intercept, let $y = 0$ and solve for *x*.
> **2.** To find the *y*-intercept, let $x = 0$ and solve for *y*.

EXAMPLE Find the intercepts of the equation $3x - y = 5$.

$$3x - y = 5 \qquad \text{For the } x\text{-intercept, let } y = 0.$$

$$3x - 0 = 5$$

$$3x = 5$$

$$x = \frac{5}{3} \qquad \text{A mixed number or decimal is easier to plot than an improper fraction.}$$

$$x = 1\frac{2}{3} \quad \text{or} \quad 1.67$$

Thus, the *x*-intercept is $(1\frac{2}{3}, 0)$.

$$3x - y = 5 \qquad \text{For the } y\text{-intercept, let } x = 0.$$

$$3(0) - y = 5$$

$$-y = 5$$

$$y = -5$$

Thus, the *y*-intercept is $(0, -5)$.

> ### To graph linear equations by the intercepts method:
>
> **1.** Find the *x*- and *y*-intercepts.
> **2.** Plot the intercepts on a rectangular coordinate system.
> **3.** Draw the line through the two points and extend it beyond each point.
> **4.** Check by examining one additional solution of the equation.

EXAMPLE Graph the equation $3x - y = 5$ by using the intercepts of each axis.

Plot the two intercepts found in the preceding example, $(1\frac{2}{3}, 0)$ and $(0, -5)$. Draw the line connecting these points and extending beyond (Fig. 14–5). We can check by finding one other point in the usual way and plotting it. If it is on the line, the graph is correct. Check for $x = 2$:

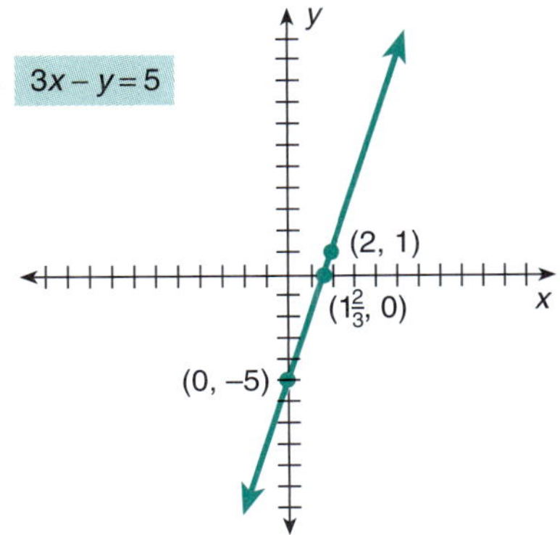

$$3x - y = 5$$

$$3(2) - y = 5 \qquad \text{Substitute 2 for } x.$$

$$6 - 5 = y \qquad \text{Solve for } y.$$

$$1 = y \qquad \text{The point (2, 1) is on the graph.}$$

Figure 14–5

	Graphing Equations When Both Intercepts Are (0, 0).

If both intercepts are (0, 0), they coincide on the origin, and form only *one point*. An additional point must be found by the table-of-values method so that you will have two distinct points for drawing the line. A third point is still useful to check your work.

$$y = 7x$$

If $x = 0$, then $y = 7(0)$ or $y = 0$. Both the x- and y-intercepts are (0, 0). So let $x = 1$, then $y = 7(1)$ or $y = 7$. Plot the points (1, 7) and (0, 0) and draw the graph.

EXAMPLE   Graph $y = 7x$ using the intercepts method.

Plot the points (1, 7) and (0, 0) as shown in the Tip feature. Draw the line through the two points and extend it beyond each point, as shown in Fig. 14–6.

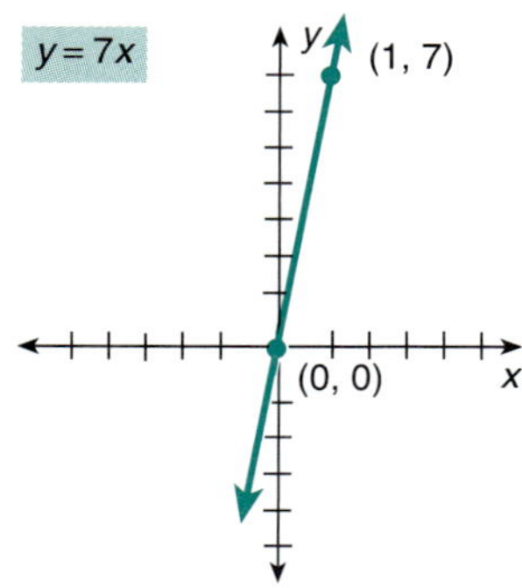

Figure 14–6

2 Graph Linear Equations Using the Slope-Intercept Method.

We can identify characteristics of the graph of an equation by inspection. First, let's examine the characteristics identified by the constant. We will let the coefficient of x equal 1 and look at the graphs of the equations in Fig. 14–7.

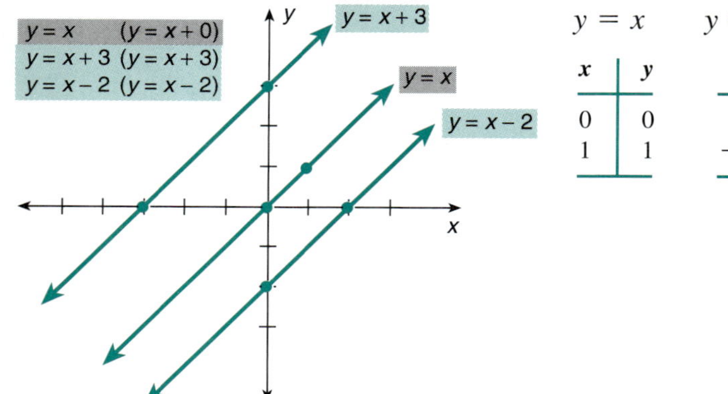

$y = x$		$y = x + 3$		$y = x - 2$	
x	y	x	y	x	y
0	0	0	3	0	−2
1	1	−3	0	2	0

Figure 14–7

The three graphs have the same slope (steepness or slant), but they have different y-intercepts.

$$y = x + 0 \quad \text{crosses the } y\text{-axis at } 0; \ y\text{-intercept} = (0, 0)$$

$$y = x + 3 \quad \text{crosses the } y\text{-axis at } +3; \ y\text{-intercept} = (0, 3)$$

$$y = x - 2 \quad \text{crosses the } y\text{-axis at } -2; \ y\text{-intercept} = (0, -2)$$

Notice that the point where the graph crosses the y-axis is the same as the constant in the equation. The constant (b) in an equation in the form $y = mx + b$ is the y-coordinate of the y-intercept.

Now, let's examine equations that have graphs with a different slant or steepness but the same y-intercept. The x- and y-intercepts in each of the three equations are $(0, 0)$. Thus, we will find one additional point to graph (Fig. 14–8).

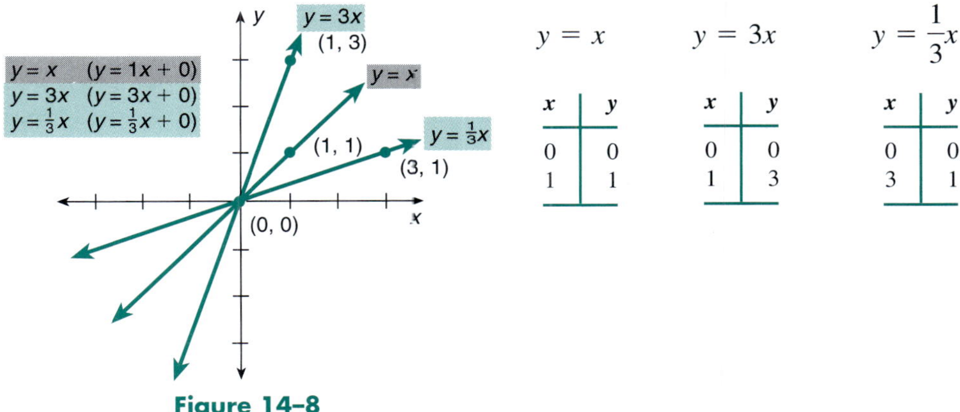

Figure 14-8

The three graphs have the same y-intercepts, but they have different slopes (steepness or slants). We define this slope or steepness so that a numerical value identifies it. The *slope* of a line is the ratio of the vertical change to the horizontal change.

A slope of 1 (written as $\frac{1}{1}$) means a change of 1 vertical unit for every 1 horizontal unit of change. A slope of 3 (written as the ratio $\frac{3}{1}$) means a change of 3 vertical units for every 1 horizontal unit. A slope of $\frac{1}{3}$ means a change of 1 vertical unit for every 3 horizontal units (Fig. 14–9). This slope can be determined from any two points on the graph.

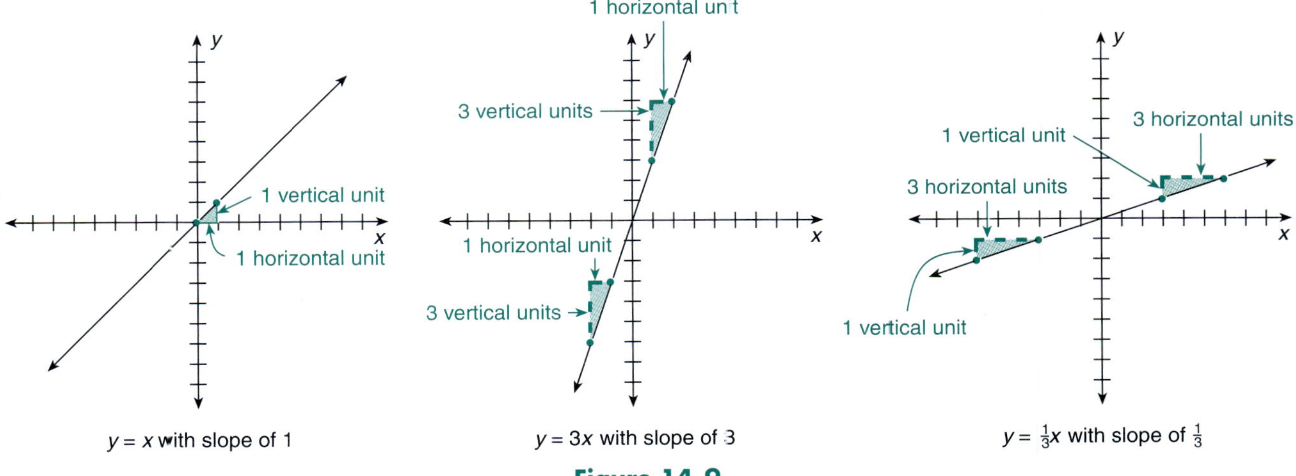

Figure 14-9

Notice that the slope of the line is the same as the coefficient of the x-term in the equation. We must make one final observation. Look again at the equations in Figs. 14–7 and 14–8.

From Fig. 14–7

$$y = x$$

$$y = x + 3$$

$$y = x - 2$$

From Fig. 14–8

$$y = x$$

$$y = 3x$$

$$y = \frac{1}{3}x$$

In each case, the equation is solved for y. This form of equation is called the *slope-intercept form* of a linear equation, or $y = mx + b$, where m is the slope and b is the y-coordinate of the y-intercept. When equations are written in this form, the slope and y-intercept can be identified by inspection.

Slope-intercept form of an equation:

$$y = mx + b$$

where m is the slope, and b is the y-coordinate of the y-intercept

---

EXAMPLE Write the equations in slope-intercept form and identify the slope and y-intercept.

(a) $2x + y = 4$ (b) $5x - y = -2$ (c) $3x + 4y = -12$

(a) $2x + y = 4$ Solve for y.

$$y = -2x + 4$$

$$\text{slope} = -2 \text{ or } -\frac{2}{1} \text{ or } \frac{2}{-1}$$ Coefficient of x.

$$y\text{-intercept} = 4 \text{ or } (0, 4)$$ Constant.

(b) $5x - y = -2$ Solve for y.

$$\frac{-y}{-1} = \frac{-5x - 2}{-1}$$ Divide by -1.

$$y = 5x + 2$$

$$\text{slope} = 5 \text{ or } \frac{5}{1}$$ Coefficient of x.

$$y\text{-intercept} = 2 \text{ or } (0, 2)$$ Constant.

(c) $3x + 4y = -12$ Solve for y. Sort terms to isolate $4y$. Divide by 4.

$$\frac{4y}{4} = \frac{-3x - 12}{4}$$ Write the right side as separate terms and reduce.

$$y = -\frac{3}{4}x - 3$$

$$\text{slope} = -\frac{3}{4}$$ Coefficient of x.

$$y\text{-intercept} = -3 \text{ or } (0, -3)$$ Constant.

An equation in the slope-intercept form can be graphed using just the slope and y-intercept.

To graph a linear equation in the form $y = mx + b$ using the slope and y-intercept:

1. Locate the y-intercept on the y-axis.
2. Using the slope, determine the amount of vertical and horizontal movement indicated.
3. From the y-intercept, locate additional points on the graph of the equation by counting the indicated vertical and horizontal movement.
4. Draw the line connecting the points, and extend it beyond the points.

EXAMPLE Graph the equations using the slope and y-intercept.

(a) $2x + y = 4$ (b) $5x - y = -2$ (c) $3x + 4y = -12$

(a) $2x + y = 4$

$$y = -2x + 4$$ Solve for y.

y-intercept $= (0, 4)$ Locate this point on the y-axis.

$$\text{slope} = \frac{-2}{1} \text{ or } \frac{2}{-1}$$

$\frac{-2}{1}$ indicates vertical movement of -2 and horizontal movement of $+1$ from the y-intercept. $\frac{2}{-1}$ indicates vertical movement of $+2$ and horizontal movement of -1 (Fig. 14–10).

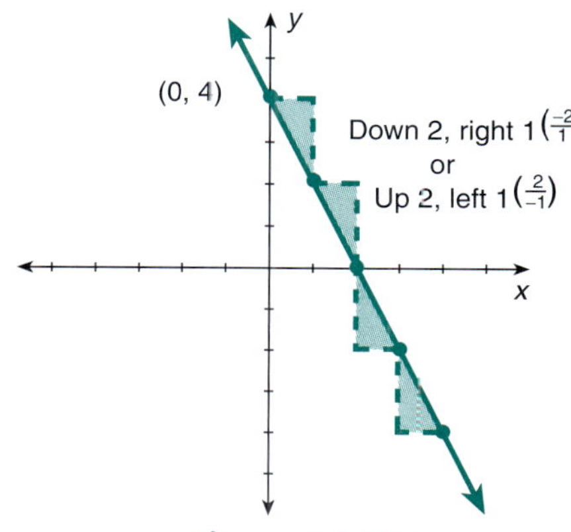

Figure 14–10

(b) $5x - y = -2$ Solve for y.

$$y = 5x + 2$$

y-intercept $= (0, 2)$ Locate this point on the y-axis.

$$\text{slope} = \frac{5}{1} \text{ or } \frac{-5}{-1}$$

$\frac{5}{1}$ indicates vertical movement of $+5$ and horizontal movement of $+1$ from the y-intercept. $\frac{-5}{-1}$ indicates vertical movement of -5 and horizontal movement of -1 (Fig. 14–11).

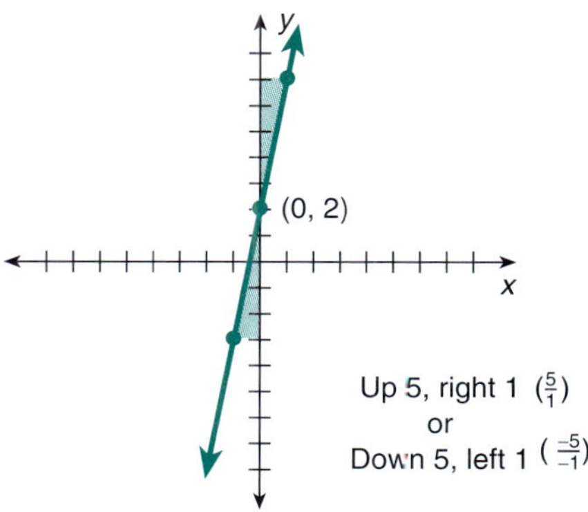

Figure 14–11

(c) $3x + 4y = -12$ Solve for y.

$$y = -\frac{3}{4}x - \frac{12}{4}$$

$$y = -\frac{3}{4}x - 3$$

y-intercept $= (0, -3)$ Locate this point on the y-axis.

$$\text{slope} = \frac{-3}{4} \text{ or } \frac{3}{-4}$$

$\frac{-3}{4}$ indicates vertical movement of -3 and horizontal movement of $+4$ from the y-intercept. $\frac{3}{-4}$ indicates vertical movement of $+3$ and horizontal movement of -4 (Fig. 14–12).

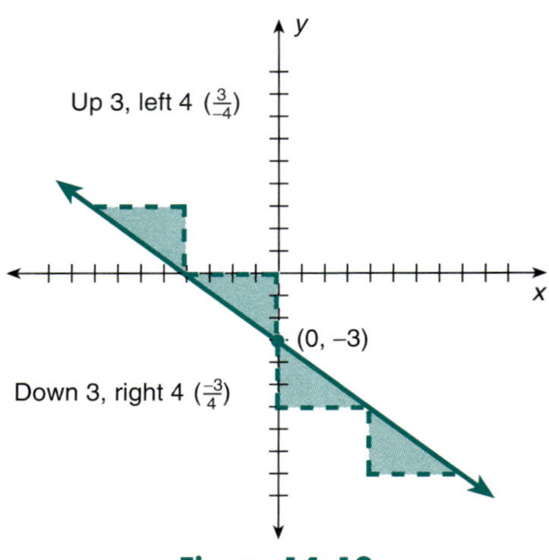

Figure 14–12

3 Graph Linear Equations Using a Graphing Calculator.

A graphing calculator can be used to graph linear equations written in the form $y = mx + b$. Also, computer software can be used to graph linear equations. These tools free us from the tedious task of graphing equations and allow us to focus on the patterns and properties of graphs.

Every model of graphing calculator and every brand of computer software may have a different set of keystrokes for graphing equations. You will need to refer to the owner's manual to adapt to a particular model or brand. However, we will illustrate the usefulness of these tools by showing a few features common to most calculators.

Some of the basic features that you will need to examine on any graphing calculator or software include

Feature	**Purpose**
Range	Setting the range (high and low values of each axis).
Graph/Text	Toggling between graph screen and text screen
x, Θ, t	Entering variables
Zoom	Zooming to get a closer or wider view
Trace	Tracing to locate specific points and determine their coordinates

EXAMPLE   Graph $y = 5x + 2$ using a graphing calculator.

Perform the appropriate function on the graphing calculator of your choice.

Clear screen.	Erases previous graphs.
Set or initialize range.	Define the viewing range of the graph window. One option is to "initialize" the range. That means the range will be reset at a predetermined, factory-set range.
Enter equation.	Equation must be solved for y. Enter the equation using the appropriate key for the variable.
View graph.	Determine a different viewing range by resetting the range or defining a box.

| Determine key points on the graph. | Using the trace function, move the cursor so that the y-value is as close to zero as possible. For example, two choices of y for a particular range may be |

$$y = -0.021276 \qquad \text{Closer to 0.}$$

$$y = 0.0851063$$

Then, the corresponding x-value (rounded) is -0.4, and the coordinates of the x-intercept are $(-0.4, 0)$. To find the y-intercept, find the value of x closest to 0. Choices may be

$$x = -4 \times 10^{-14} \qquad \text{Closer to 0.}$$

$$x = 0.0212765$$

The corresponding y-value (rounded) is 2. The coordinates of the y-intercept are $(0, 2)$.

SELF-STUDY EXERCISES 14–3

1 Graph the equations using the intercepts procedure.

1. $x + y = 5$　　　　**2.** $x + 3y = 5$　　　　**3.** $\dfrac{1}{2}y = 4 + x$

4. $y = 3x - 1$　　　　**5.** $5x = y + 2$　　　　**6.** $x = -2y + 3$

2 Graph the equations using the slope-intercept procedure.

7. $y = 2x - 3$　　　　**8.** $y = -\dfrac{1}{2}x - 2$　　　　**9.** $y = -\dfrac{3}{5}x$

10. $x - 2y = 3$　　　　**11.** $2x + y = 1$

3 Verify the graphs for Exercises 1–11 using a graphing calculator.

14–4　SLOPE

Learning Outcomes

1 Calculate the slope of a line, given two points on the line.
2 Determine the slope of a horizontal or vertical line.
3 Find the equation of a line, given the slope and one point.
4 Find the equation of a line, given two points on the line.
5 Find the slope and y-intercept of a line, given the equation of the line.
6 Find the equation of a line, given the slope and y-intercept.

The concept of slope is important in real-world applications. We use slope when we consider the slant of a roof or the grade of a roadway, and we also use slope in applications dealing with change, like changes in temperature, in the quality of a product, or in sales. In Section 14–3, we learned to graph linear equations using slope. Now we extend our knowledge of slope and examine procedures for calculating it.

1　Calculate the Slope of a Line, Given Two Points on the Line.

When we know the model or the equation of a line, we find the slope by rewriting the equation in the form $y = mx + b$. If we do not know the equation of a line, we can calculate the slope using a formula.

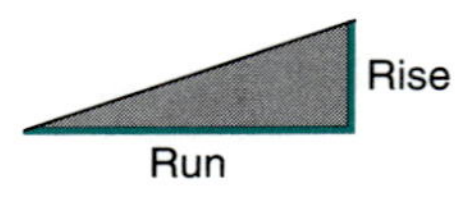

Figure 14–13

■ **DEFINITION: Slope.** The *slope* of a line is the ratio of the vertical rise of a line to the horizontal run of a line (Fig. 14–13).

The definition of slope may be expressed as a formula.

Find the slope from two given points:

$$\text{Slope} = \frac{\text{Rise}}{\text{Run}}$$

$$\text{Slope} = \frac{\Delta y}{\Delta x} = \frac{y_2 - y_1}{x_2 - x_1}$$

where Δy is vertical change and Δx is horizontal change.

$$P_1 = (x_1, y_1), \qquad P_2 = (x_2, y_2)$$

P_1 and P_2 are any two points on the line.

The formula gives us a way to find the slope from the coordinates of two points on a line. We calculate the change (*difference*) in

- y-coordinates to find the vertical *rise*
- x-coordinates to find the horizontal *run*

Then we make a ratio of the rise to the run and reduce the ratio to lowest terms.

EXAMPLE Find the slope of a line if the points $(2, -1)$ and $(5, 3)$ are on the line (Fig. 14–14).

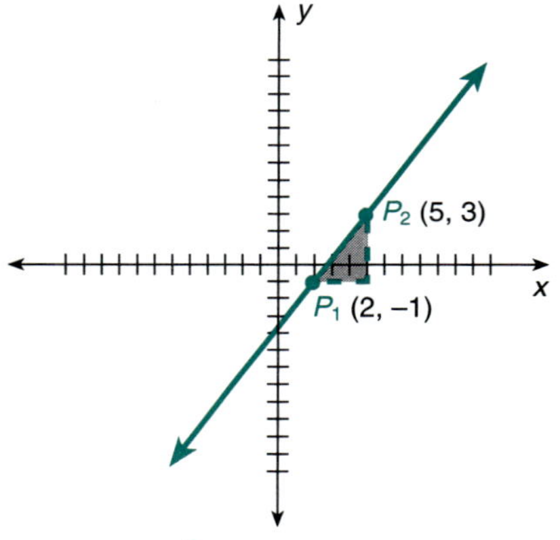

Figure 14–14

First, label $(2, -1)$ as point 1 (P_1) and $(5, 3)$ as point 2 (P_2). The coordinates of P_1 are (x_1, y_1) and the coordinates of P_2 are (x_2, y_2). We use the Greek capital letter delta (Δ) to indicate a change and to write the slope definition symbolically.

$$\text{Change in } y = \Delta y = \text{rise} = y_2 - y_1 = 3 - (-1) = 4$$

$$\text{Change in } x = \Delta x = \text{run} = x_2 - x_1 = 5 - 2 = 3$$

$$\text{Slope} = \frac{\Delta y}{\Delta x} = \frac{\text{rise}}{\text{run}} = \frac{y_2 - y_1}{x_2 - x_1} = \frac{3 - (-1)}{5 - 2} = \frac{4}{3}$$

The slope of the line through points $(2, -1)$ and $(5, 3)$ is $\frac{4}{3}$.

Any line has an infinite number of points. Any two points on the line can be used to find the slope. Also, the designation of P_1 and P_2 is not critical. Let's find the slope of the line in the preceding example by designating P_1 as $(5, 3)$ and P_2 as $(2, -1)$.

$$\Delta y = \text{Rise} = y_2 - y_1 = -1 - 3 = -4$$

$$\Delta y = \text{Run} = x_2 - x_1 = 2 - 5 = -3$$

$$\frac{\Delta y}{\Delta x} = \frac{\text{Rise}}{\text{Run}} = \frac{-4}{-3} = \frac{4}{3}$$

Notice in the preceding example that when either point is designated as point 1, the slope is positive. When the slope of a line is positive, the line *rises* from left to right. When the slope of a line is negative, the line *falls* from left to right. Figure 14–15 shows a line with a slope of $\frac{4}{3}$ and a line with a slope of $-\frac{4}{3}$.

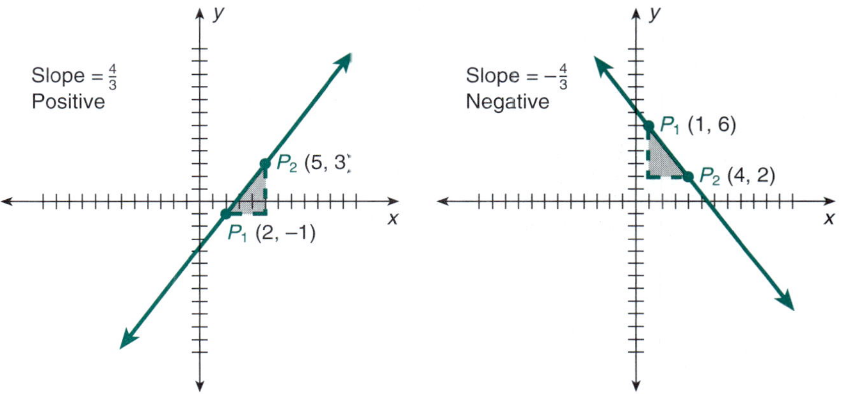

Figure 14–15

EXAMPLE Find the slope of the line passing through the pairs of points.

(a) $(1, 6)$ and $(4, 2)$ (b) $(-5, -1)$ and $(3, -2)$ (c) $(0, 0)$ and $(-5, 3)$

(a) Either point can be designated as P_1. The other point will be P_2. If $P_1 = (x_1, y_1)$ and $P_2 = (x_2, y_2)$, then the slope $= \frac{\Delta y}{\Delta x} = \frac{y_2 - y_1}{x_2 - x_1}$. Let $P_1 = (1, 6)$ and $P_2 = (4, 2)$. Then

$$\text{Slope} = \frac{\Delta y}{\Delta x} = \frac{2 - 6}{4 - 1} = \frac{-4}{3} = -\frac{4}{3}$$

The slope is $-\frac{4}{3}$.

(b) Let $P_1 = (-5, -1)$ and $P_2 = (3, -2)$.

$$\text{Slope} = \frac{\Delta y}{\Delta x} = \frac{-2 - (-1)}{3 - (-5)} = \frac{-2 + 1}{3 + 5} = \frac{-1}{8} = -\frac{1}{8}$$

The slope is $-\frac{1}{8}$.

(c) Let $P_1 = (0, 0)$ and $P_2 = (-5, 3)$.

$$\text{Slope} = \frac{\Delta y}{\Delta x} = \frac{3 - 0}{-5 - 0} = \frac{3}{-5} = -\frac{3}{5}$$

The slope is $-\frac{3}{5}$.

2 Determine the Slope of a Horizontal or Vertical Line.

Let's examine the slopes of a horizontal line and a vertical line (Figs. 14–16 and 14–17). First look at the horizontal line that passes through the points (3, 2) and (−3, 2).

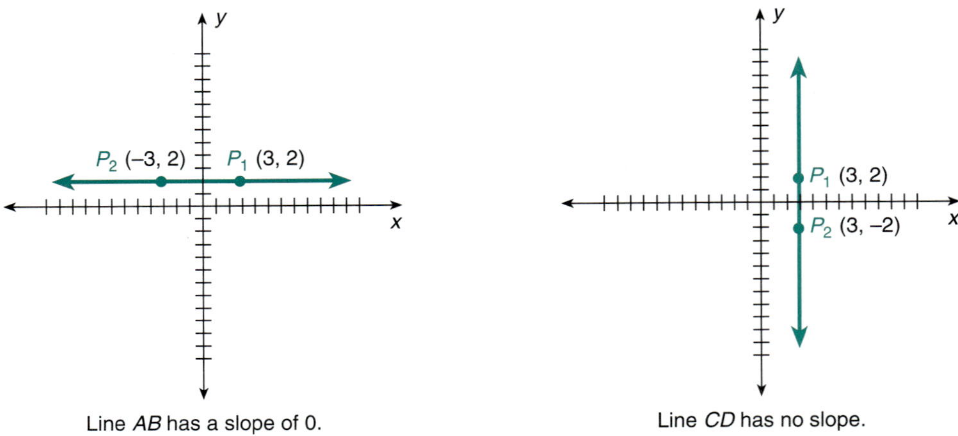

Line *AB* has a slope of 0.

Figure 14–16

Line *CD* has no slope.

Figure 14–17

EXAMPLE Find the slope of the horizontal line that passes through (3, 2) and (−3, 2).

Let $P_1 = (3, 2)$ and $P_2 = (-3, 2)$.

$$\text{Slope} = \frac{\Delta y}{\Delta x} = \frac{y_2 - y_1}{x_2 - x_1} = \frac{2 - 2}{-3 - 3} = \frac{0}{-6} = 0$$

Since any horizontal line, as in Fig. 14–16, has no change in *y*-values, or no rise, *the slope of any horizontal line is zero.*

Slope of horizontal lines:

- The slope of a horizontal line is zero.
- Two points are on the same horizontal line if their *y*-coordinates are equal.

Next, let's look at the vertical line that passes through the points (3, 2) and (3, −2).

EXAMPLE Find the slope of the vertical line that passes through the points (3, 2) and (3, −2).

Let $P_1 = (3, 2)$ and $P_2 = (3, -2)$.

$$\text{Slope} = \frac{\Delta y}{\Delta x} = \frac{y_2 - y_1}{x_2 - x_1} = \frac{-2 - 2}{3 - 3} = \frac{-4}{0}$$

Recall that division by zero is impossible; **therefore, the slope of the vertical line passing through (3, 2) and (3, −2) is undefined.**

Since any vertical line has no change in *x*-values or no run, *the slope of any vertical line is undefined.* We say that a vertical line has *no slope.*

Slope of vertical lines:

- A vertical line has no slope; that is, the slope of a vertical line is undefined.
- Two points are on the same vertical line if their x-coordinates are equal.

Tip! | ***Slope of Zero Versus No Slope:***

A slope of zero and no slope are *not* the same. Zero is a real number. A slope of zero is real. When zero is the denominator of a fraction, the number is undefined. "No slope" is the same as an undefined slope. As a memory aid, think of the number zero as a ball. It will balance on a horizontal line. It will not balance on a vertical line.

3 Find the Equation of a Line, Given the Slope and One Point.

The slope of a line can be determined from any two points on that line. However, it is often desirable to know the equation of the line. The equation of the line can serve as a model for solving various applications.

To write the equation of a line, we need to know certain facts about the line. From the basic definition of slope, we can determine the minimum amount of information needed. To help in writing equations, we rewrite the slope definition using the letter m to represent slope.

$$\text{Slope} = \frac{y_2 - y_1}{x_2 - x_1} \qquad m = \frac{y_2 - y_1}{x_2 - x_1}$$

We used this definition in the last section to find the slope when two points are known. Now, the definition is used to find the equation of a line when the slope and one point are known.

Point-slope form of an equation of a straight line:

$$y - y_1 = m(x - x_1) \qquad \text{or} \qquad m = \frac{y - y_1}{x - x_1}$$

where y_1 and x_1 are the coordinates of the given point and m is the slope of the line passing through that point. The x and y represent the variables of the equation being sought.

EXAMPLE Find the equation of the line passing through the point $(3, -2)$ with a slope of $\frac{2}{3}$.

$$y - y_1 = m(x - x_1) \qquad \text{Substitute into the point-slope form.}$$

$$y - (-2) = \frac{2}{3}(x - 3) \qquad m = \frac{2}{3}; \, x_1 = 3; \, y_1 = -2$$

$$y + 2 = \frac{2}{3}(x - 3) \qquad \text{Distribute. } \frac{2}{\cancel{3}} \cdot \frac{\overset{-1}{\cancel{-3}}}{1} = -2$$

$$y + 2 = \frac{2}{3}x - 2 \qquad \text{Transpose or sort terms.}$$

$$y = \frac{2}{3}x - 2 - 2 \qquad \text{Combine like terms.}$$

$$y = \frac{2}{3}x - 4 \qquad \text{Solve for } y.$$

Another method for finding the equation uses a variation of the slope formula. The coordinates of the known point are (x_1, y_1). The coordinates of the unknown point are (x, y).

$$m = \frac{y - y_1}{x - x_1}$$

Substitute into variation of point-slope form.

$$\frac{2}{3} = \frac{y - (-2)}{x - 3}$$

$m = \frac{2}{3}; x_1 = 3; y_1 = -2$

$$\frac{2}{3} = \frac{y + 2}{x - 3}$$

Solve the equation for y.

$$3(y + 2) = 2(x - 3)$$

Cross multiply.

$$3y + 6 = 2x - 6$$

Distribute.

$$3y = 2x - 6 - 6$$

Transpose.

$$3y = 2x - 12$$

$$\frac{3y}{3} = \frac{2x - 12}{3}$$

Divide.

$$y = \frac{2x - 12}{3} \quad \text{or} \quad y = \frac{2x}{3} - \frac{12}{3}$$

$$y = \frac{2x}{3} - 4 \quad \text{or} \quad y = \frac{2}{3}x - 4$$

Tip!	**Which Point-Slope Formula Is Better?**

If you surveyed numerous math students and teachers to determine the preferred version of the point-slope formula, you would probably get an inconclusive result. It is strictly a matter of personal preference. Some may argue that they prefer one version if the slope is a fraction and another if the slope is an integer.

4 Find the Equation of a Line, Given Two Points on the Line.

Many times the facts that we are given do not match the facts that are required in a particular formula. In these cases, we need to use more than one formula. To find the equation of a line when two points on the line are known, we apply slope and point-slope formulas.

EXAMPLE   Find the equation of the line that passes through the points $(0, 8)$ and $(5, 0)$.

We need to find the slope of the line passing through the points $(0, 8)$ and $(5, 0)$. Let $P_1 = (0, 8)$ and $P_2 = (5, 0)$.

$$\text{Slope} = \frac{\Delta y}{\Delta x} = \frac{y_2 - y_1}{x_2 - x_1} = \frac{0 - 8}{5 - 0} = -\frac{8}{5}$$

Slope formula.

Now we use the slope and one of the points and the point-slope form of the equation to write the desired equation. Using P_1 we have

$$y - y_1 = m(x - x_1)$$

Point-slope form.

$$y - 8 = -\frac{8}{5}(x - 0)$$

$m = -\frac{8}{5}; x_1 = 0; y_1 = 8$

$$y - 8 = -\frac{8}{5}x \qquad\qquad \text{Solve for } y.$$

$$y = -\frac{8}{5}x + 8$$

Suppose we use P_2 (5, 0) instead of P_1 in the point-slope form of the equation. Because only one line passes through (0, 8) and (5, 0), we have the same equation if either point is used. Let's find the equation of the line passing through (0, 8) and (5, 0) using the point (5, 0) and the slope $-\frac{8}{5}$.

$$y - y_1 = m(x - x_1)$$

$$y - 0 = -\frac{8}{5}(x - 5) \qquad\qquad m = -\frac{8}{5}; \ x_1 = 5; \ y_1 = 0$$

$$y = -\frac{8}{5}x - \frac{8}{5}(-5) \qquad\qquad \text{Distribute.}$$

$$y = -\frac{8}{5}x + 8 \qquad\qquad \text{Same equation as before.}$$

EXAMPLE   Find the equation of the line that passes through the points $(-3, 4)$ and $(7, 4)$.

We let $P_1 = (-3, 4)$ and $P_2 = (7, 4)$.

$$\text{Slope} = \frac{\Delta y}{\Delta x} = \frac{y_2 - y_1}{x_2 - x_1} = \frac{4 - 4}{7 - (-3)} = \frac{0}{7 + 3} = \frac{0}{10} = 0$$

Use the point-slope form of an equation and P_1.

$$y - y_1 = m(x - x_1)$$

$$y - 4 = 0[x - (-3)]$$

$$y - 4 = 0$$

$$y = 4$$

The situation in the preceding example is generalized in the equation of a horizontal line.

Equation of a horizontal line:

The equation of a line with a slope of zero (a horizontal line) is

$$y = k$$

where k is the common y-coordinate for all points on the line.

There is one special situation in which the point-slope form of an equation *cannot* be used to determine the equation of a line, namely, when the slope is undefined. Remember, *any vertical line has an undefined slope.* We can readily identify such a line if the coordinates of the two points have the same x-coordinate. The equation of the line is $x = k$, where k is the common x-coordinate.

Equation of a vertical line:

The equation of a line with a slope that is undefined (a vertical line) is

$$x = k$$

where k is the common x-coordinate for all points on the line.

EXAMPLE Find the equation of the line that passes through the points $(5, 3)$ and $(5, 7)$.

Since the x-coordinate is the same in both points, the slope of the line is undefined, and the equation of the line is $x = k$, where k is the common x-coordinate. In this example, $k = 5$. **The equation of the line through (5, 3) and (5, 7) is $x = 5$.**

5 Find the Slope and y-Intercept of a Line, Given the Equation of the Line.

In discussing equations of straight lines, we have always preferred the final form of the equation be solved for y. When a linear equation is solved for y, we say it is the *slope-intercept form* of an equation of a straight line.

Slope-intercept form of an equation of a straight line:

$$y = mx + b$$

where m is the slope and b is the y-coordinate of the y-intercept, $(0, b)$.

When an equation is written in the form $y = mx + b$, the slope of the line and the y-intercept can be determined by inspection.

EXAMPLE Find the slope and y-intercept of the lines with the given equations.

(a) $y = 4x + 3$ (b) $y = -5x - 3$

(c) $y = -\dfrac{3}{5}x + \dfrac{1}{2}$ (d) $y = 2.3x - 4.7$

(a) $y = 4x + 3$
Slope, $m = 4$ (coefficient of x); $b = 3$, y-intercept $= (0, 3)$

(b) $y = -5x - 3$
Slope, $m = -5$; $b = -3$, y-intercept $= (0, -3)$

(c) $y = -\dfrac{3}{5}x + \dfrac{1}{2}$

Slope, $m = -\dfrac{3}{5}$; $b = \dfrac{1}{2}$, y-intercept $= \left(0, \dfrac{1}{2}\right)$

(d) $y = 2.3x - 4.7$
Slope, $m = 2.3$; $b = -4.7$, y-intercept $= (0, -4.7)$

Tip! | ***Why Do We Say b Equals the y-Intercept?***

Even though b is the y-coordinate of the y-intercept, we commonly refer to b as the y-intercept. This is an acceptable shortcut.

EXAMPLE Rewrite the following equations in the slope-intercept form. Then, find the slope and y-intercept.

(a) $x = y + 5$ (b) $2x - 3y = 9$ (c) $5y + 2x = 0$
(d) $1.5x - 3y = 4.5$ (e) $2y = 4$

To rewrite an equation in slope-intercept form, solve the equation for y. Then, the coefficient of x is the slope and the number term or constant is the y-intercept.

(a)

$$x = y + 5 \qquad \text{Transpose.}$$
$$x - 5 = y \qquad \text{Interchange sides.}$$
$$y = x - 5$$

Slope, $m = 1$; y-intercept, $b = -5$

(b) $2x - 3y = 9 \qquad \text{Transpose.}$
$$2x - 9 = 3y \qquad \text{Interchange sides.}$$
$$3y = 2x - 9 \qquad \text{Divide.}$$

$$\frac{3y}{3} = \frac{2x - 9}{3}$$

$$y = \frac{2x}{3} - \frac{9}{3} \qquad \text{Reduce.}$$

$$y = \frac{2}{3}x - 3$$

Slope, $m = \frac{2}{3}$; y-intercept, $b = -3$

(c) $5y + 2x = 0 \qquad \text{Transpose}$
$$5y = -2x \qquad \text{Divide.}$$
$$\frac{5y}{5} = -\frac{2x}{5}$$

$$y = -\frac{2}{5}x \qquad \text{Because there is no number term, the value of } b \text{ is zero.}$$

Slope, $m = -\frac{2}{5}$; y-intercept, $b = 0$

(d) $1.5x - 3y = 4.5 \qquad \text{Transpose.}$
$$1.5x - 4.5 = 3y \qquad \text{Interchange sides.}$$
$$3y = 1.5x - 4.5 \qquad \text{Divide.}$$
$$\frac{3y}{3} = \frac{1.5x - 4.5}{3}$$
$$y = 0.5x - 1.5$$

Slope, $m = 0.5$; y-intercept, $b = -1.5$

(e) $2y = 4 \qquad \text{Divide.}$
$$\frac{2y}{2} = \frac{4}{2}$$
$$y = 2 \qquad \text{Because there is no } x\text{-term, the coefficient of } x, \text{ or the slope, is zero.}$$

Slope, $m = 0$; y-intercept, $b = 2$

Tip! | ***Vertical Lines and the Slope-Intercept Form.***

Vertical lines are exceptions when we use the slope-intercept form of an equation. In the equation of a vertical line such as $x = 4$, there is no y-term, so the equation cannot be solved for y. Recall that the graph is a vertical line that crosses the x-axis at $(4, 0)$. However, it does not cross the y-axis, so it does not have a y-intercept. Recall that the slope of a vertical line is undefined. Thus, we say it has no slope.

So far we have written the equation of a line when we know either two points on the line or the slope and at least one point on the line. If the one point is the y-intercept $(0, b)$, then we use the slope-intercept form of an equation, $y = mx + b$, and we can write the equation by inspection.

EXAMPLE   Write the equation for a line with a slope of -3 and a y-intercept of 5.

Slope $= m = -3$;   y-intercept $= b = 5$

$y = mx + b$ Substitute values.

$y = -3x + 5$

The necessary facts for writing the equation of a line are often obtained from the graph of the equation.

EXAMPLE   Figure 14–18 shows the cost of producing picture frames.

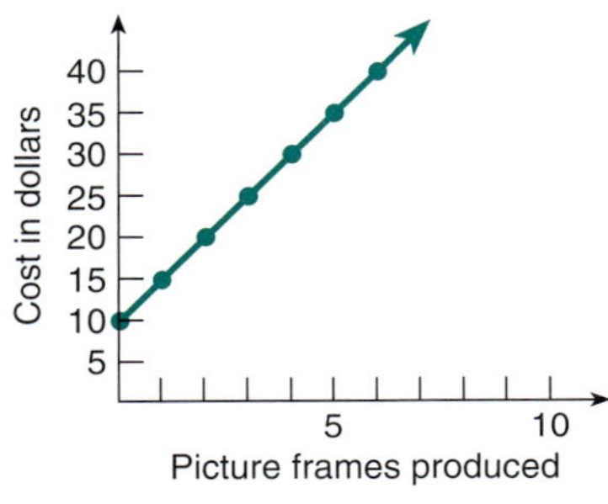

Figure 14–18

(a) Write an equation that represents the graph.
(b) Using the equation, find the cost of producing 20 picture frames.

(a) The y-intercept represents the fixed cost of producing picture frames. An example of a fixed cost is the cost of the necessary tools. From the graph, $b = \$10$. The slope m is 5 units vertical change for every 1 unit horizontal change, which is $\frac{5}{1} = 5$.

$y = 5x + 10$ Equation of a line.

This type of equation is typically referred to in the business world as a *cost function*. The function notation used for the relationship described is $C(x) = 5x + 10$.

(b) Use the equation of the line found in part (a) to find y when $x = 20$.

$$y = 5x + 10 \qquad \text{or} \qquad C(x) = 5x + 10$$
$$y = 5(20) + 10 \qquad\qquad C(20) = 5(20) + 10$$
$$y = 100 + 10 \qquad\qquad C(20) = 100 + 10$$
$$y = \$110 \qquad\qquad C(20) = \$110$$

The cost of producing 20 frames is \$110.

1 Find the slope of the line passing through the pairs of points.

1. $(-3, 3)$ and $(1, 5)$ **2.** $(4, -1)$ and $(1, 4)$ **3.** $(3, 1)$ and $(5, 7)$

4. $(-1, -1)$ and $(3, 3)$ **5.** $(4, 5)$ and $(-4, -1)$ **6.** $(7, 2)$ and $(-3, 2)$

7. $(4, -5)$ and $(0, 0)$ **8.** $(2, -2)$ and $(5, -5)$ **9.** $(-5, 2)$ and $(-5, -4)$

10. $(4, -4)$ and $(1, 5)$

2 **11.–20.** Identify the horizontal and vertical lines in Exercises 1–10 and explain why you identified each as horizontal or vertical.

3 Find the equation of a line passing through the given point with the given slope. Solve the equation for y when necessary.

21. $(-8, 3), m = \dfrac{2}{3}$ **22.** $(4, 1), m = -\dfrac{1}{2}$ **23.** $(-3, -5), m = 2$ **24.** $(0, -1), m = 1$

4 Find the equation of a line passing through the given pairs of points. Solve the equation for y.

25. $(4, 6)$ and $(7, 1)$ **26.** $(-1, 6)$ and $(-1, 4)$ **27.** $(-1, -3)$ and $(3, -3)$

28. $(-4, 4)$ and $(-4, -2)$ **29.** $(-2, -4)$ and $(5, 10)$ **30.** $(-4, 0)$ and $(6, 0)$

5 Determine the slope and y-intercept by inspection.

31. $y = 4x + 3$ **32.** $y = -5x + 6$ **33.** $y = -\dfrac{7}{8}x - 3$

34. $y = 3$

Rewrite the following equations in slope-intercept form and determine the slope and y-intercept.

35. $4x - 2y = 10$ **36.** $2y = 5$ **37.** $\dfrac{1}{2}y + x = 3$

38. $2.1y - 4.2x = 10.5$ **39.** $2x = 8$ **40.** $x - 5 = 4$

6 Write the equations of lines with the given slopes and y-intercepts.

41. $m = \dfrac{1}{4}, b = 7$ **42.** $m = -8, b = -4$

Write the equations of the lines that are graphed in Figs. 14–19 through 14–21.

43.
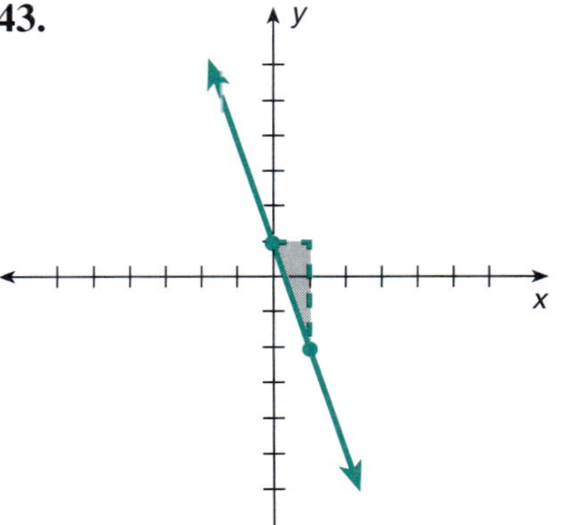

Figure 14–19

44.
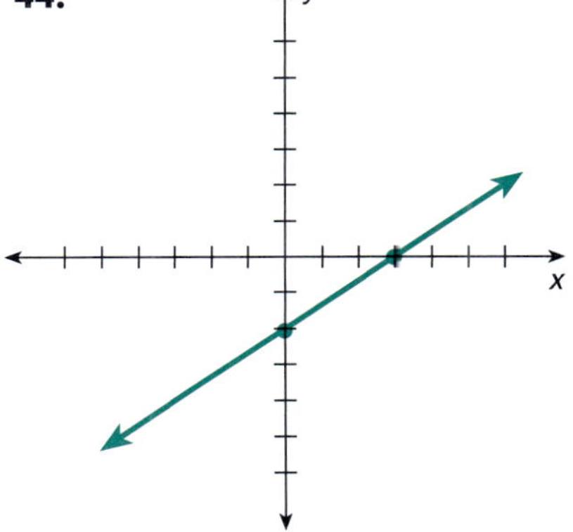

Figure 14–20

45.
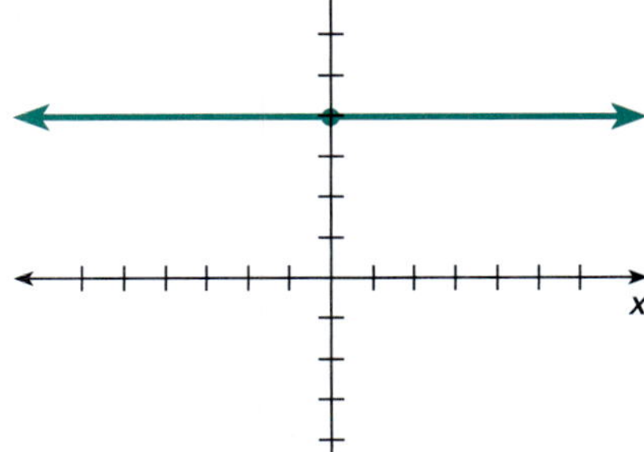

Figure 14–21

Solve Exercises 46 and 47 using the information given in Fig. 14–22, which shows
the cost in dollars of producing widgets.

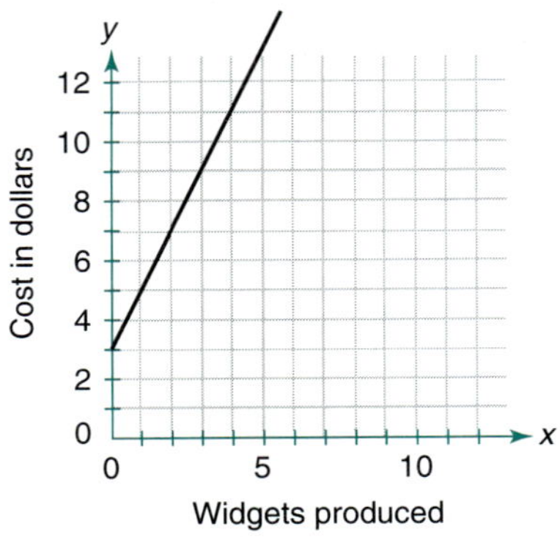

Figure 14–22

46. Write an equation that represents the cost of producing widgets shown in the graph.

47. Use the equation to find the cost of producing 10 widgets and verify by locating the data on the graph.

14–5   PARALLEL AND PERPENDICULAR LINES

Learning Outcomes

1 Find the equation of a line, given a point on the line and the equation of a line parallel to the line.

2 Find the equation of a line, given a point on the line and the equation of a line perpendicular to the line.

1 Find the Equation of a Line, Given a Point on the Line and the Equation of a Line Parallel to the Line.

Parallel lines are lines that are the same distance apart. No matter how far the lines are extended, the lines do not get any closer together or farther apart. Since slope represents the slant or steepness of a line, parallel lines have the same slope. Any line can have several lines parallel to it. Figure 14–23 shows two lines that are parallel to the line $x + y = 5$.

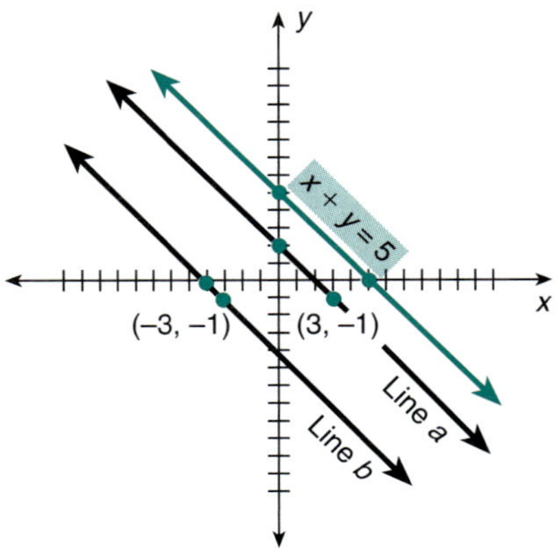

Figure 14–23

■ **DEFINITION: Parallel Lines.** *Parallel lines* are two or more lines that are the same distance apart everywhere. They have no points in common.

> ***Slope of parallel lines:***
>
> The slopes of parallel lines are equal.

If we write $x + y = 5$ in slope-intercept form, we see that $y = -x + 5$, and the slope is -1. *Any* equation that has a slope of -1 is either parallel to or coincides with the line formed by the equation $x + y = 5$. *Coincides* means one line lies on top of the other; that is, they are the same line.

We examined the similarities of parallel lines, but what are the differences? Parallel lines have different x- and y-intercepts.

EXAMPLE   Write the equations for lines a and b in Figure 14–23.

Line a has a y-intercept of 2 and is parallel to $x + y = 5$. Solved for y, the equation $x + y = 5$ is $y = -x + 5$, with a slope of -1.

Line a: $y = -x + 2$ $m = -1, b = 2$

Line b: $y = -x - 4$ $m = -1, b = -4$

The *standard form* of an equation of a line has the variables on the left, with the x-term written first, and the constant on the right. The equation of line a in Figure 14–23 is $y = -1x + 2$, or $y = -x + 2$, or $x + y = 2$. The equation $x + y = 2$ is written in *standard form*.

> ***Standard form of an equation of a straight line:***
>
> $$ax + by = c$$
>
> where a, b, and c are integers, and $a \geq 0$.

The equation of a parallel line can be found if any point, not just the y-intercept, is known. In line a of Figure 14–23, one point is $(3, -1)$. Substitute $x_1 = 3$, $y_1 = -1$, and $m = -1$ into the point-slope form of an equation.

$$y - y_1 = m(x - x_1)$$
$$y - (-1) = -1(x - 3)$$
$$y + 1 = -x + 3$$
$$y = -x + 3 - 1$$
$$y = -x + 2$$

Any point we choose on line a gives the same equation, $y = -x + 2$ or $x + y = 2$.

Let's write the equation for line b using the slope and a known point other than the y-intercept. To do so, we use the procedure outlined below. Line b in Fig. 14–23 is parallel to the line $x + y = 5$ and passes through the point $(-3, -1)$.

> ***To find the equation of a line that passes through a given point and is parallel to a given line:***
>
> **1.** Determine the slope (m) of the *given* line. The slope of the parallel line is the same as the slope of the given line.
> **2.** Write the equation for the parallel line by substituting the values for m, x_1, and y_1 into the point-slope form of the equation, $y - y_1 = m(x - x_1)$.
> **3.** Write the equation in slope-intercept or standard form.

EXAMPLE   Find the equation of line b in Fig. 14–23.

The slope of the line $x + y = 5$ is -1 because $x + y = 5$ is equal to $y = -x + 5$. Substitute -1 for m, -3 for x_1, and -1 for y_1 into the point-slope form of the equation.

$$y - y_1 = m(x - x_1)$$

$$y - (-1) = -1[x - (-3)] \qquad m = -1;\ x_1 = -3;\ y_1 = -1$$

$$y + 1 = -1(x + 3) \qquad \text{Distribute.}$$

$$y + 1 = -x - 3 \qquad \text{Solve for } y.$$

$$y = -x - 3 - 1 \qquad \text{Combine like terms.}$$

$$y = -x - 4 \qquad \text{Slope-intercept form.}$$

$$x + y = -4 \qquad \text{Standard form.}$$

EXAMPLE   Find the equation of a line that is parallel to $2y = 3x + 8$ and passes through the point $(2, 1)$.

Find the slope of the given line, $2y = 3x + 8$.

$$2y = 3x + 8 \qquad \text{Solve for } y.$$

$$\frac{2y}{2} = \frac{3x + 8}{2}$$

$$y = \frac{3}{2}x + 4 \qquad \text{Identify the slope.}$$

The slope of $2y = 3x + 8$ is $\dfrac{3}{2}$.

Next, find the equation of a line passing through the point $(2, 1)$ that is parallel to $2y = 3x + 8$.

$$y - y_1 = m(x - x_1) \qquad \text{Substitute. The slope of a parallel line is equal to the slope of the given line.}$$

$$y - 1 = \frac{3}{2}(x - 2) \qquad m = \frac{3}{2};\ x_1 = 2;\ y_1 = 1$$

$$y - 1 = \frac{3}{2}x - 3 \qquad \text{Distribute.}$$

$$y = \frac{3}{2}x - 3 + 1 \qquad \text{Transpose and combine like terms.}$$

$$y = \frac{3}{2}x - 2 \qquad \text{Slope-intercept form.}$$

To write the equation in standard form, start by clearing fractions.

$$2y = 2\left(\frac{3}{2}x - 2\right) \qquad \text{Clear fractions.}$$

$$2y = 3x - 4 \qquad \text{Transpose variable terms to the left.}$$

$$-3x + 2y = -4 \qquad \text{Multiply each term by } -1 \text{ so that the coefficient of } x \text{ is positive.}$$

$$3x - 2y = 4 \qquad \text{Standard form.}$$

2   Find the Equation of a Line, Given a Point on the Line and the Equation of a Line Perpendicular to the Line.

Perpendicular lines are lines that intersect and make a square corner. Any line can have several lines perpendicular to it, but only one line perpendicular to a *given* line passes through a *given* point. Figure 14–24 shows a line that is perpendicular to the line $2x + y = 7$ and passes through the point $(6, 8)$. In some cases, the given point may lie on the given line.

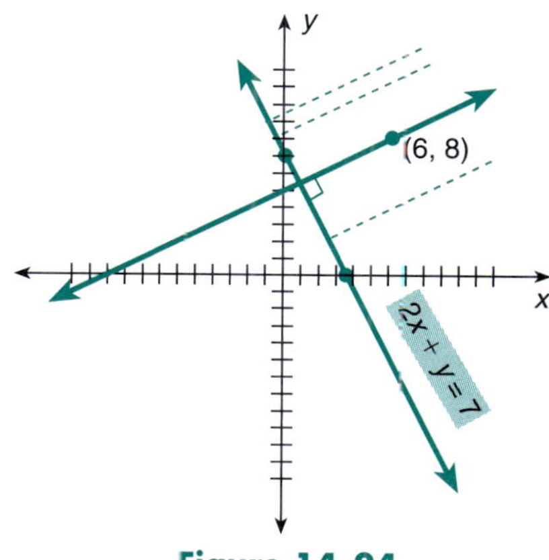

Figure 14–24

■   **DEFINITION: Perpendicular Lines.** *Perpendicular lines* are two lines that intersect to form right angles ($90°$ angles). Another term for *perpendicular* is *normal*.

In Fig. 14–24, the perpendicular (or normal) line that passes through the given point is indicated by the symbol ⌐, which means that a right or $90°$ angle is formed by the lines.

The slope of the equation $2x + y = 7$ is -2, which we can see by inspection if the equation is written in the slope-intercept form as $y = -2x + 7$. We can determine the equation of a line that passes through $(6, 8)$ and is perpendicular to $2x + y = 7$ using an important property of perpendicular lines.

Slope of perpendicular lines:

The slope of *any* line perpendicular to a given line is the *negative reciprocal* of the slope of the given line.

If the slope of $2x + y = 7$ or $y = -2x + 7$ is -2, then the slope of *any* line perpendicular to that line is $+\frac{1}{2}$ (the negative reciprocal of -2, or $-\frac{2}{1}$).

Now, we know the slope $(+\frac{1}{2})$ and one point $(6, 8)$ through which the perpendicular line passes. We can find the equation of the perpendicular line by using the point-slope form of the equation.

$$y - y_1 = m(x - x_1) \qquad m = \frac{1}{2}, P_1 = (6, 8)$$

$$y - 8 = \frac{1}{2}(x - 6) \qquad \text{Substitute.}$$

$$y - 8 = \frac{1}{2}x - 3 \qquad \text{Distribute.}$$

$$y = \frac{1}{2}x - 3 + 8 \qquad \text{Transpose and combine like terms.}$$

$$y = \frac{1}{2}x + 5$$

The equation of the perpendicular line is

$$y = \frac{1}{2}x + 5 \qquad \text{Slope-intercept form.}$$

or

$$x - 2y = -10 \qquad \text{Standard form (Fig. 14–25).}$$

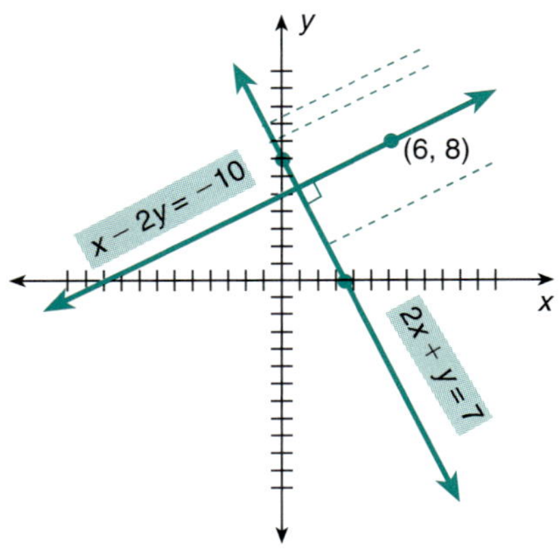

Figure 14–25

Tip!	*Negative Reciprocals.*

The negative reciprocal of a number is not necessarily a negative value. It has the opposite sign.
   To find the negative reciprocal:

1. Interchange the numerator and denominator.
2. Give the reciprocal the opposite sign.

The negative reciprocal of -5 is $+\frac{1}{5}$. The negative reciprocal of $-\frac{3}{4}$ is $+\frac{4}{3}$.
The negative reciprocal of $\frac{4}{5}$ is $-\frac{5}{4}$. The negative reciprocal of 3 is $-\frac{1}{3}$.

To further examine the relationship of the slopes of perpendicular lines, examine the graphs in Fig. 14–26.

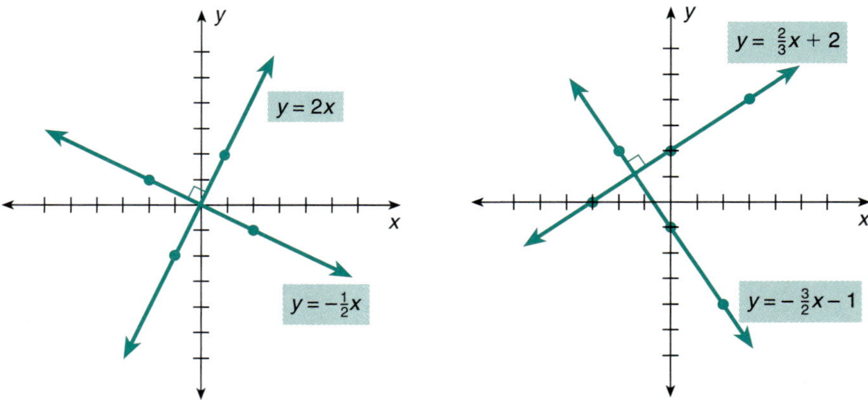

Figure 14–26

A summary of the procedures for finding the equation of a line perpendicular to a given line follows:

To find the equation of a line that is perpendicular (normal) to a given line and passes through a given point:

1. Determine the slope of the *given* line.
2. Find the *negative reciprocal* of this slope. The negative reciprocal is the slope of the perpendicular line.
3. Write the equation for the perpendicular line by substituting the coordinates of the *given* point for x_1 and y_1 and the slope of the *perpendicular* line for m into the point-slope form of the equation $y - y_1 = m(x - x_1)$.
4. Write the equation in slope-intercept or standard form.

EXAMPLE Find the equation of a line that is perpendicular to the line $y = \frac{2}{3}x + 8$ and passes through the point $(3, 10)$.

First, find the slope of the given line, $y = \frac{2}{3}x + 8$. The negative reciprocal of $\frac{2}{3}$ is $-\frac{3}{2}$.

Write the equation.

$$y - y_1 = m(x - x_1)$$
Point-slope form.

$$y - 10 = -\frac{3}{2}(x - 3)$$
Substitute values: $m = -\frac{3}{2}$; $x_1 = 3$; $y_1 = 10$.

$$y - 10 = -\frac{3}{2}x + \frac{9}{2}$$
Distribute.

$$y = -\frac{3}{2}x + \frac{9}{2} + 10$$
Solve for y.

$$y = -\frac{3}{2}x + \frac{9}{2} + \frac{20}{2}$$
Combine like terms.

$$y = -\frac{3}{2}x + \frac{29}{2}$$
Slope-intercept form.

or

$$2y = -3x + 29$$
Clear fractions.

$$3x + 2y = 29$$
Standard form.

Tip! | **Slope-Intercept Form Versus Standard Form.**

The variations of the equation in the preceding example represent the same line. The slope-intercept form is useful for determining properties of a graph by inspection. The standard form is desirable because it contains no fractions.

Tip! | **Words as Subscripts.**

In the next example, we clarify the notation with subscripts. The phrase *slope of a given line* is notated as "$\text{slope}_{\text{given}}$." The slope of the perpendicular line is notated as "$\text{slope}_{\text{perpendicular}}$."

EXAMPLE Find the equation of the line normal to $4x + y = -3$ and passing through $(0, -3)$.

$$4x + y = -3 \qquad \text{Given equation.}$$

$$y = -4x - 3 \qquad \text{Solve for } y.$$

$$\text{Slope}_{\text{given}} = -4$$

$$\text{Slope}_{\text{perpendicular}} = +\frac{1}{4} \qquad \text{Negative reciprocal of } -4.$$

$$y - y_1 = m(x - x_1) \qquad \text{Point-slope form.}$$

$$y - (-3) = \frac{1}{4}(x - 0) \qquad m = \frac{1}{4}, P_1 = (0, -3)$$

$$y + 3 = \frac{1}{4}x \qquad \text{Solve for } y.$$

$$y = \frac{1}{4}x - 3 \qquad \text{Slope-intercept form.}$$

or

$$4y = x - 12 \qquad \text{Clear fractions and rearrange.}$$

$$x - 4y = 12 \qquad \text{Standard form.}$$

EXAMPLE Which of the equations represents a normal of the line $2x - 3y = 1$ passing through $(2, -1)$?

(a) $y = \frac{2}{3}x - \frac{1}{3}$ (b) $y = \frac{2}{3}x - \frac{7}{3}$

(c) $y = -\frac{3}{2}x - \frac{1}{3}$ (d) $y = -\frac{3}{2}x + 2$

$$2x - 3y = 1 \qquad \text{Rewrite in slope-intercept form.}$$

$$-3y = -2x + 1$$

$$\frac{-3y}{-3} = \frac{-2x}{-3} + \frac{1}{-3}$$

$$y = \frac{2}{3}x - \frac{1}{3} \qquad \text{Slope-intercept form.}$$

$$\text{Slope}_{\text{given}} = \frac{2}{3} \qquad \text{From } y = \frac{2}{3}x - \frac{1}{3}$$

$$\text{Slope}_{\text{perpendicular}} = -\frac{3}{2} \qquad \text{Negative reciprocal.}$$

Choices are now limited to (c) or (d).

$$y - y_1 = m(x - x_1) \qquad \text{Substitute } m = -\frac{3}{2}, x_1 = 2, y_1 = -1.$$

$$y - (-1) = -\frac{3}{2}(x - 2)$$

$$y + 1 = -\frac{3}{2}x + 3$$

$$y = -\frac{3}{2}x + 3 - 1$$

$$y = -\frac{3}{2}x + 2$$

The correct equation is $y = -\frac{3}{2}x + 2$, or choice (d).

1 Write the equations in standard form. Verify your answers using a graphing calculator.

1. Find the equation of the line that is parallel to the line $x + y = 6$ and passes through the point $(2, -3)$.

2. Find the equation of the line that is parallel to the line $2x + y = 5$ and passes through the point $(1, 7)$.

3. Find the equation of the line that is parallel to the line $3y = x - 2$ and passes through the point $(4, 0)$.

4. Find the equation of the line that is parallel to the line $3x - y = -2$ and passes through the point $(-3, -2)$.

5. Find the equation of the line that is parallel to the line $x + 3y = 7$ and passes through the point $(4, 1)$.

6. Find the equation of the line that is parallel to the line $3x - y = 4$ and passes through the point $(0, 3)$.

7. Find the equation of the line that is parallel to the line $2x + 3y = 5$ and passes through the point $(1, 1)$.

8. Find the equation of the line that is parallel to the line $3x + 2y = 1$ and passes through the point $(2, 0)$.

9. Find the equation of the line that is parallel to the line $2x - 5y = 0$ and passes through the point $(3, -1)$.

10. Find the equation of the line that is parallel to the line $-3x + 4y = -1$ and passes through the point $(\frac{1}{2}, 0)$.

2 Write the equation for each exercise in standard form. Verify your results using a graphing calculator. Be sure the range is set so the vertical and horizontal increments are equal.

11. Find the equation of the line that is perpendicular to the line $x + y = 6$ and passes through the point $(2, 3)$.

12. Find the equation of the line that is normal to the line $2x + y = 5$ and passes through the point $(1, 7)$.

13. Find the equation of the line that is perpendicular to the line $3y = x - 2$ and passes through the point $(4, 0)$.

14. Find the equation of the line that is normal to the line $3x - y = 2$ and passes through the point $(-3, -2)$.

15. Find the equation of the line that is normal to the line $x + 3y = 7$ and passes through the point $(4, 1)$.

16. Find the equation of the line that is perpendicular to the line $x + 2y = 7$ and passes through the point $(-2, 3)$.

17. Find the equation of the line that is perpendicular to the line $2x + 3y = 4$ and passes through the point $(3, -1)$.

18. Find the equation of the line that is normal to the line $4x + y = 1$ and passes through the point $(0, 0)$.

19. Find the equation of the line that is perpendicular to the line $2x + 2y = 3$ and passes through the point $(\frac{1}{2}, 2)$.

20. Find the equation of the line that is perpendicular to the line $5x - y = 6$ and passes through the point $(5, -\frac{1}{5})$.

14–6 GRAPHING QUADRATIC EQUATIONS

Learning Outcomes

1 Identify nonlinear equations.

2 Graph quadratic equations using the table-of-solutions method or by examining properties.

3 Graph quadratic equations using a graphing calculator.

1 Identify Nonlinear Equations.

In previous sections, we graphed linear equations and inequalities. Linear equations have two basic characteristics.

1. No term has a letter with an exponent greater than 1.
2. No term has two or more different letter factors.

In other words, at least one term of the equation has a degree of 1, and no term has a degree higher than 1.

Quadratic equations also have two basic characteristics.

1. No term has a degree greater than 2. Quadratic equations may have terms with two different letter factors as long as the degree of the term does not exceed 2.
2. At least one term of the equation must have degree 2.

EXAMPLE Identify the equations as linear, quadratic, or other nonlinear equations.

(a) $3x + 4y = 12$ (b) $3x + 4 = 8$
(c) $3x + 4xy + 2y = 0$ (d) $y = x^2 + 4$
(e) $y = 3x^3 - 2x^2 + x$

(a) $3x + 4y = 12$
This is a linear equation. $3x$ and $4y$ are terms of degree 1. No terms have a degree other than 1 or 0.
(b) $3x + 4 = 8$
This is a linear equation. $3x$ has a degree of 1 and no higher degree is present.
(c) $3x + 4xy + 2y = 0$
This is *not* a linear equation. The term $4xy$ has two different letter factors and a degree of 2. **This is a quadratic equation.**
(d) $y = x^2 + 4$
This is *not* a linear equation. The term x^2 has a degree of 2. **This is a quadratic equation.**
(e) $y = 3x^3 - 2x^2 + x$
This equation is neither linear nor quadratic. It has a term that has degree 3. **This is a *cubic equation*.**

2 Graph Quadratic Equations Using the Table-of-Solutions Method or by Examining Properties.

The graphs of linear equations are *straight* lines. We will examine the graphs of some equations that have a degree higher than 1, which are not linear equations.

The graph of an equation of a degree higher than 1 is a *curved* line. The curved line can be a parabola, hyperbola, circle, ellipse, or an irregular curved line. We do not define these terms at this time, but Fig. 14–27 illustrates them. The equation for a parabola that opens up or down is distinguished from other quadratic equations. The y variable has degree 1. The x variable must have one term with degree 2. Such a quadratic equation can be written in function notation: $f(x) = ax^2 + bx + c$.

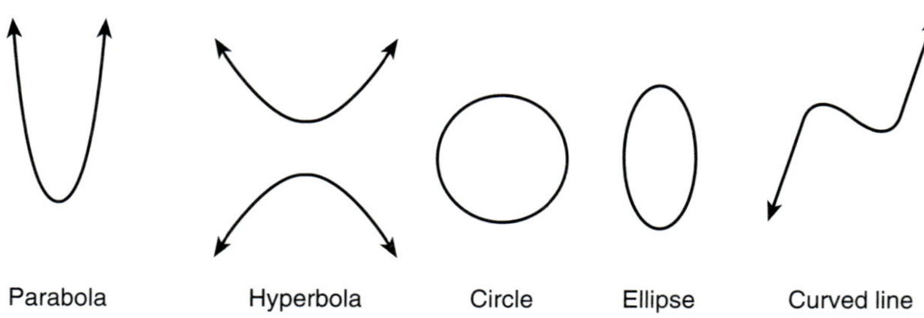

Figure 14–27

One method of graphing nonlinear equations is to form a table of values or solutions and plot the points as we did in Section 14–1. This method requires *more* points than we customarily use in graphing linear equations or inequalities. Other methods of graphing nonlinear equations involve examining characteristics or properties of the equation.

A parabola is symmetrical; that is, it can be folded in half and the two halves match. The fold line is called the *axis of symmetry*. For a parabola in the form $y = ax^2 + bx + c$, the equation of the axis of symmetry is $x = -\frac{b}{2a}$.

The point of the graph that crosses the axis of symmetry is the *vertex* of the parabola. Thus, the x-coordinate of the vertex of the parabola is $-\frac{b}{2a}$.

To graph quadratic equations in the form $y = ax^2 + bx + c$:

1. Find the axis of symmetry: $x = -\frac{b}{2a}$.
2. Find the vertex: x-coordinate of vertex $= -\frac{b}{2a}$. To find the y-coordinate of the vertex, substitute the x-coordinate into the original equation and solve for y.
3. Find one or two additional points that are to the right of the axis of symmetry.
4. Apply the property of symmetry to find additional points to the left of the axis of symmetry.
5. Connect the plotted points with a smooth, continuous curved line.

EXAMPLE Graph the equation $y = x^2$ by finding the vertex, the axis of symmetry, and some additional points.

$$y = x^2$$

$$y = x^2 + 0x + 0 \qquad \text{Standard form: } a = 1,\ b = 0,\ c = 0.$$

$$x = -\frac{b}{2a}$$

$$x = -\frac{0}{2(1)}$$

$$x = 0 \qquad \text{Axis of symmetry and } x\text{-coordinate of vertex.}$$

$$y = 0^2 \qquad \text{Substitute 0 for } x.$$

$$y = 0 \qquad y\text{-coordinate of vertex.}$$

Vertex: $(0, 0)$
Axis of symmetry: $x = 0$

Find y for $x = 1$: $y = 1^2$
$$y = 1$$

Find y for $x = 2$: $y = 2^2$
$$y = 4$$

Find y for $x = 3$: $y = 3^2$
$$y = 9$$

x	y
-3	9
-2	4
-1	1
Vertex 0	0
1	1
2	4
3	9

Apply the principle of symmetry to complete the table. Plot the points and connect them with a smooth, continuous curve (Fig. 14–28).

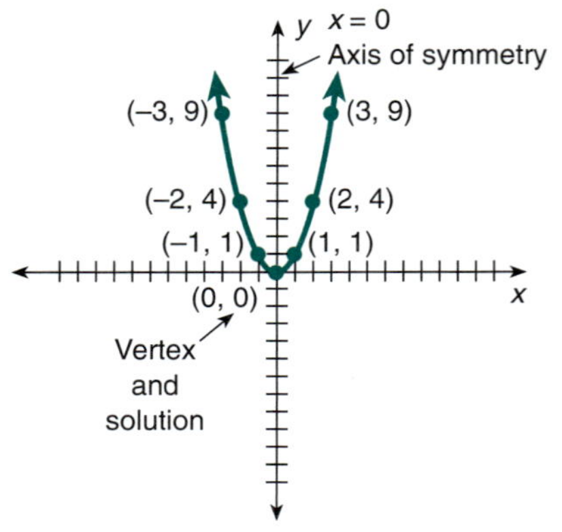

Figure 14–28

If the equation $y = x^2$ is written in function notation as $f(x) = x^2$, *the solutions of the equation are the values or value of x that make the function zero.* If $0 = x^2$ or $x^2 = 0$, then $x = 0$ is the only solution of the equation. Also, the solutions of the equation are found where the graph crosses or meets the x-axis. When we graph a quadratic equation, we can test the solutions as points on the graph. *If the graph does not cross or meet the x-axis, the equation does not have any real-number solutions.*

EXAMPLE Graph the equation $y = x^2 + x - 6$ by using the axis of symmetry, the vertex, and the solutions.

For the parabola $y = x^2 + x - 6$, the equation of the axis of symmetry is

Axis of symmetry: $\quad x = -\dfrac{b}{2a}$

$$x = -\dfrac{1}{2(1)} \qquad (a = 1, b = 1)$$

$$x = -\dfrac{1}{2} \qquad \text{Also, } x\text{-coordinate of vertex.}$$

Vertex: $\left(-\dfrac{1}{2}, y\right) \qquad$ Use the x-coordinate, $-\dfrac{1}{2}$, to find the y-coordinate of the vertex.

$$y = x^2 + x - 6 \quad \text{for} \quad x = -\dfrac{1}{2}$$

$$y = \left(-\dfrac{1}{2}\right)^2 + \left(-\dfrac{1}{2}\right) - 6$$

$$y = \dfrac{1}{4} - \dfrac{1}{2} - 6$$

$$y = \dfrac{1}{4} - \dfrac{2}{4} - \dfrac{24}{4}$$

$\left(-\dfrac{1}{2}, -6\dfrac{1}{4}\right) \quad y = -\dfrac{25}{4} \quad \text{or} \quad -6\dfrac{1}{4}$

Solutions: $\qquad 0 = x^2 + x - 6$

$$0 = (x + 3)(x - 2)$$

$$x + 3 = 0 \qquad x - 2 = 0$$

$$x = -3 \qquad\quad x = 2$$

$$(-3, 0); (2, 0)$$

Using the axis of symmetry, vertex, and two solutions, we can get a general idea of the shape of the graph (Fig. 14–29).

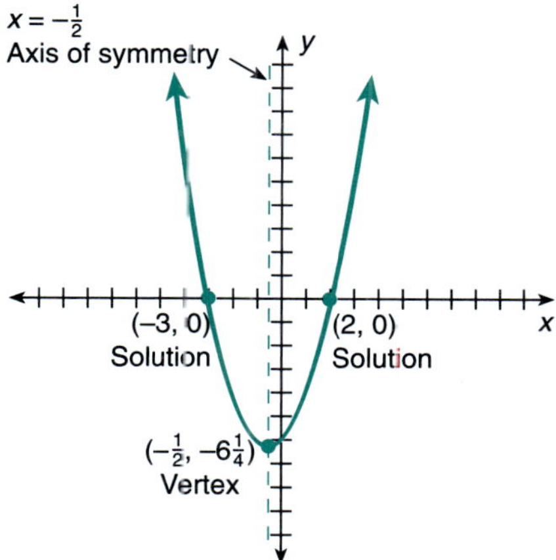

Figure 14–29

EXAMPLE Graph the equation $y = -x^2 + 3$ by using the axis of symmetry, vertex, and the solutions.

Axis of symmetry: $\quad x = -\dfrac{b}{2a} \qquad\qquad y = -x^2 + 0x + 3; a = -1, b = 0$

$$x = -\dfrac{0}{2(-1)}$$

$$\mathbf{x = 0}$$

Vertex: $(0, y)$

$$y = -0^2 + 0 + 3$$

$$y = 3$$

$$\mathbf{(0, 3)}$$

Solutions: $\qquad 0 = -x^2 + 3$

$$x^2 = 3$$

$$x = \pm\sqrt{3}$$

$$x \approx \pm 1.7$$

$$\mathbf{(1.7, 0); (-1.7, 0)}$$

Plot the points and connect them with a smooth, continuous curve. Two additional points are plotted to give a more complete view of the parabola (Fig. 14–30).

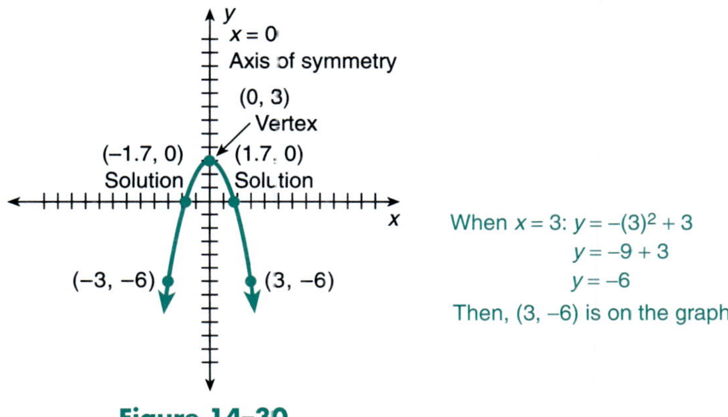

Figure 14–30

Tips for Sketching Curves.

When the coefficient of the squared letter term is negative as in the preceding example, the graph of the parabola opens downward. When the coefficient of the squared letter term is positive, the graph of the parabola opens upward. Think of the parabola as a cup or glass: positive holds water; negative spills water.

Figure 14–31

3 **Graph Quadratic Equations Using a Graphing Calculator.**

Quadratic equations are graphed on graphing calculators using the same procedures as for linear equations. The equation must first be solved for y.

EXAMPLE Graph the equation $y = 2x^2 - 5x - 3$ using a graphing calculator.

$$y = 2x^2 - 5x - 3$$

1. Clear previous graphs.
2. Initialize range.
3. Enter equation.
4. Graph (Fig. 14–32).

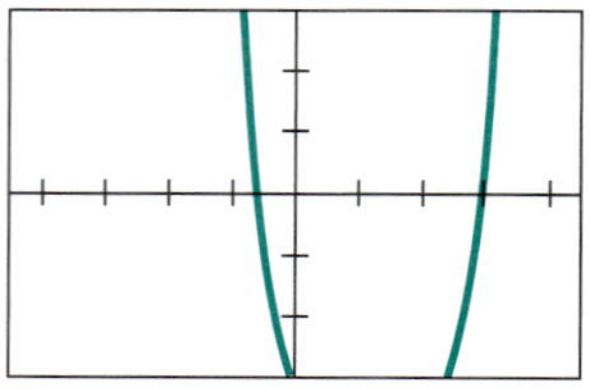

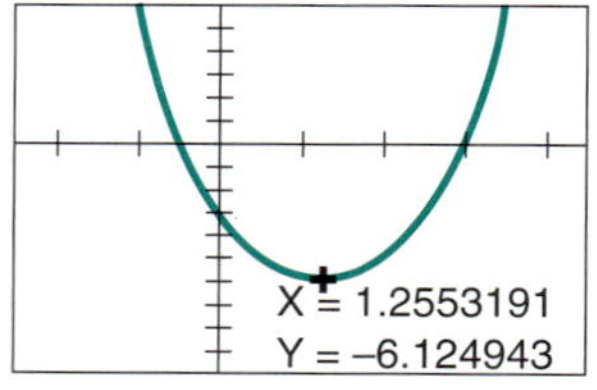

Figure 14–32

To view the portion of the parabola containing the vertex, change the range to show the x-values from -2 to $+4$ and the y-values from -10 to 5. Leave the scale at 1.

Graph again (Fig. 14–33).

Different views can be obtained by changing the range. More accurate values for x and y at critical points such as x- and y-intercepts and the vertex can be displayed by using the trace function with closer views of particular points.

Figure 14–33

Features of the Graphing Calculator.

Other features of graphing calculators, such as zoom and box, allow you to view different portions of the graph. A smaller range allows a closer examination of critical points. A larger range allows a more complete overview of the graph.

On most calculators, the initial settings for the range give a view with x and y intervals equally spaced, thus yielding a minimum of distortion. Multiplying (or dividing) the x- and y-values by the same amount allows different views with a minimum of distortion. Many calculators allow you to change the viewing box by using a "Factor" feature.

1 Identify the equations as linear, quadratic, or other.

1. $y = 3x - 7$

2. $y = x^2 - 4x + 2$

3. $x^2 + y = 5x$

4. $x^2 + y^2 - 3 = 2x$

5. $2x + 3 = 8$

6. $\dfrac{x}{y} = 3$

2 Use a table of solutions to graph the quadratic equations.

7. $y = x^2$

8. $y = 3x^2$

9. $y = \dfrac{1}{3}x^2$

10. $y = -4x^2$

11. $y = -\dfrac{1}{4}x^2$

12. $y = x^2 - 4$

13. $y = x^2 + 4$

14. $y = x^2 - 6x + 9$

15. $y = -x^2 + 6x - 9$

3 Graph using a graphing calculator. Reset the range if necessary to show the vertex of the parabola.

16. $y = 3x^2 + 5x - 2$

17. $y = (2x - 3)(x - 1)$

18. $y = 2x^2 - 9x - 5$

19. $y = -2x^2 + 9x + 5$

CAREER APPLICATION

Statistics: Predicting College Class Size

A statistics professor at a major university noticed a relationship between the number of hours of homework assigned and the number of students continuing in the class. She found that as the number of homework hours increased in her statistics class that met 5 hr per week, the number of students decreased. In a class beginning with 225 students, she started with no required homework and then systematically increased the homework requirement three times during the semester, recording the class enrollment one week after each increase. She used these data to derive the following function:

$$f(x) = -6x + 225$$

where x represents the approximate number of hours of homework required weekly, and $f(x)$ represents the approximate class enrollment.

Exercises

Use the preceding function to answer the following in complete sentences. Show all computations. Round to the nearest number of people or hours.

1. Approximate the statistics class enrollment when the following number of hours of weekly homework was required: (a) none, (b) 5, (c) 10, (d) 15.
2. Graph the function. Increment by 1's along the x axis, and by 25's along the $f(x)$ axis. Recall that $f(x) = y$ for graphing purposes.
3. Approximate the weekly homework requirement if the enrollment was the following: (a) 201 students, (b) 177 students, (c) 190 students, (d) 100 students.
4. Explain the real-life meaning of the graph's slope: $m = -\dfrac{6}{1}$.
5. Use the graph to estimate the homework required for all students to drop the class. Now find this answer exactly by using the function definition. Explain any difference in these two answers.
6. From prior experience, the professor knows that students learn statistics much more effectively when they spend about 12 hours per week (12.5 hr) on homework. How many students can be expected to stay in the class?
7. Do you think this function is valid for all nonnegative values of x? If you do not, state the highest value of x you think is appropriate, and give your reason(s).

8. Do you think a homework requirement is the only factor influencing class size? If not, give other appropriate factors.
9. Which factor do you think most influences the class retention rate, and why?
10. What would you say to this professor regarding her use of this function to predict class size?

Answers

1. (a) 225, (b) 195, (c) 165, (d) 135
2. See Fig. 14–34.
3. (a) 4, (b) 8, (c) 6, (d) 21

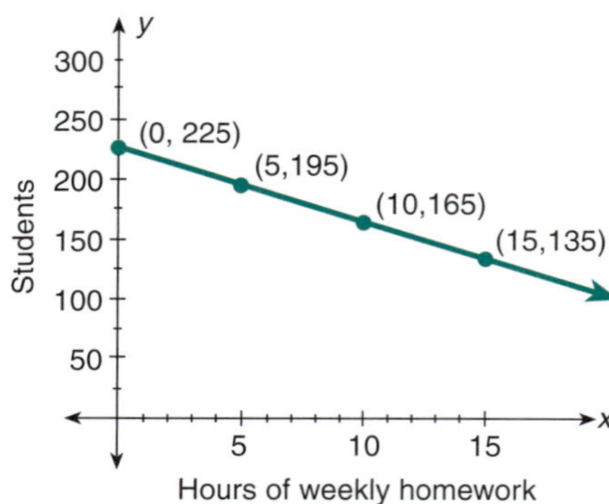

Figure 14–34

4. The slope of the graph in this application means that for every hour of homework required 6 students withdraw from the class.
5. Graph estimate: ~35 hr
   Exact answer: 37.5, or ~38, hr
6. Approximately 150 students.
7. Clearly x cannot be greater than 37.5 hr because a negative number of students in the class would result! Also, a "rule of thumb" for a weekly homework requirement is two or three times the number of class hours per week. The professor should require approximately 10 to 15 hr per week.
8. Other factors that can influence class size include the instructor's teaching style, length and difficulty of exams, changes in students' work schedules, students' prerequisite mathematical skills, students' efforts in the class, and amount of time available in students' lives to devote to the class.
9. Answers will vary.
10. Answers will vary.

Electronics: Using the Slope to Find Resistance

To find the slope of the line that passes through the two points (x_1, y_1) and (x_2, y_2), use the formulas

$$\text{Slope} = \frac{\text{rise}}{\text{run}} = \frac{y_2 - y_1}{x_2 - x_1} \qquad \text{or} \qquad \text{Slope} = \frac{y_1 - y_2}{x_1 - x_2}$$

Assume that you have a graph with current on the horizontal axis and voltage on the vertical axis (Fig. 14–35). Find the slope of the line through the points (1 mA, 3 V) and (4 mA, 6 V).

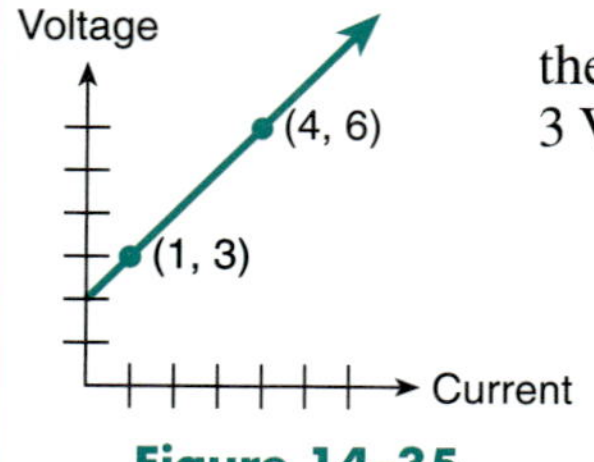

Figure 14–35

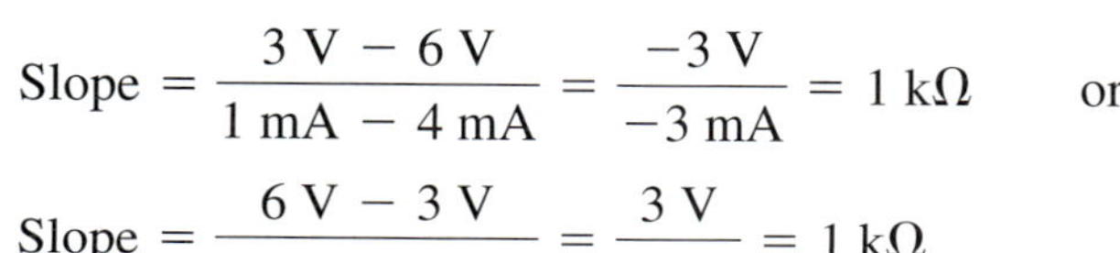

$$\text{Slope} = \frac{3 \text{ V} - 6 \text{ V}}{1 \text{ mA} - 4 \text{ mA}} = \frac{-3 \text{ V}}{-3 \text{ mA}} = 1 \text{ k}\Omega \qquad \text{or}$$

$$\text{Slope} = \frac{6 \text{ V} - 3 \text{ V}}{4 \text{ mA} - 1 \text{ mA}} = \frac{3 \text{ V}}{3 \text{ mA}} = 1 \text{ k}\Omega$$

CHAPTER 14 Graphing

To analyze the dimensions, use the following relationships: volts divided by amperes gives ohms, $\frac{V}{A} = \Omega$. $m = \frac{1}{1,000}$; $1 \div \frac{1}{1,000} = 1,000$; $1,000 = k$; that is, the reciprocal of the unit m (milli-) is the unit k (kilo-).

Find the slope of the line through (4 V, 5 mA) and (6 V, 4 mA).

$$\text{Slope} = \frac{5\text{ mA} - 4\text{ mA}}{4\text{ V} - 6\text{ V}} = \frac{1\text{ mA}}{-2\text{ V}} = -0.5\text{ mS}$$

To analyze the dimensions, amperes divided by volts gives siemens, $\frac{A}{V} = S$.

A graph like the one in Fig. 14–36 that compares current and voltage for a **series circuit** is usually drawn with the current or the independent variable on the horizontal axis, because the current is the reference. The voltage or dependent variable is on the vertical axis. A line drawn on this graph that goes through the origin (vertical or y-intercept $= 0$) has

$$\text{Slope} = \frac{\text{Rise}}{\text{Run}} = \frac{\Delta\text{ Voltage }(E)}{\Delta\text{ Current }(I)} = \text{Resistance }(R \text{ or } Z)$$

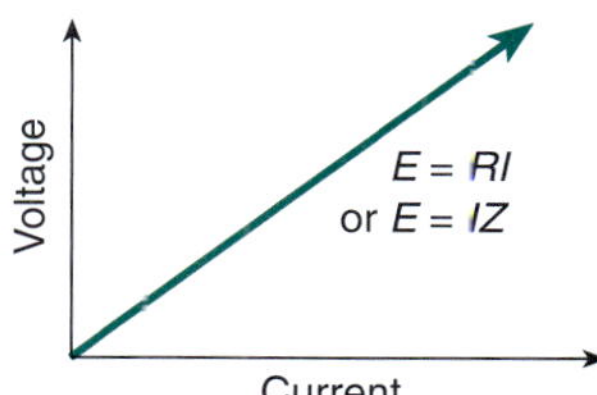

Figure 14–36   Series circuit.

The equation of the line for a **series circuit** is

$$E = RI \text{ (for dc circuits)} \qquad \text{or} \qquad E = IZ \text{ (for ac circuits)}$$

A graph like the one in Fig. 14–37 that compares current and voltage for a **parallel circuit** is usually drawn with the voltage on the horizontal axis because the voltage is the reference or independent variable, and the current is on the vertical axis as the dependent variable. A line drawn on this graph that passes through the origin has

$$\text{Slope} = \frac{\text{Rise}}{\text{Run}} = \frac{\Delta\text{ Current }(I)}{\Delta\text{ Voltage }(E \text{ or } V)} = \text{Conductance }(G \text{ or } Y)$$

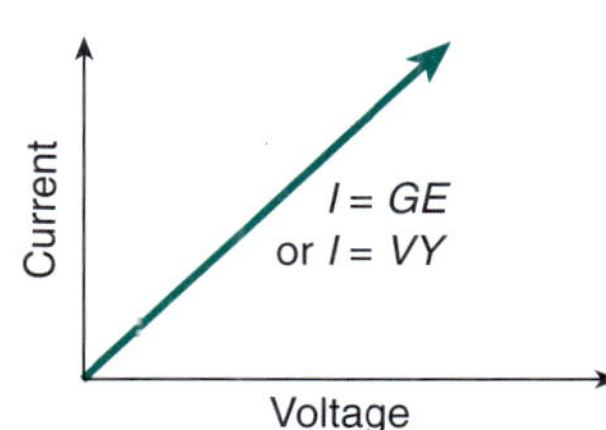

Figure 14–37   Parallel circuit.

The equations of the line for a parallel circuit are

$$I = GE \text{ (for dc circuits)} \qquad \text{or} \qquad I = VY \text{ (for ac circuits)}$$

Exercises

Find the slope of the line that passes through each of the given pairs of points. Be sure to include sign, prefix, and units as well as the number in the answer.

1. (4 mA, 12 V) and (7 mA, 15 V)
2. (4 mA, 12 V) and (9 mA, 8 V)
3. (15 V, 7 mA) and (20 V, 9 mA)

   4. (25 V, 4 mA) and (37 V, 1 mA)
   5. (0.65 V, 1 mA) and (0.75 V, 2 mA)
   6. (9 μS, 11 V) and (12 μS, 17 V)
   7. (9 μS, 11 V) and (12 μS, 5 V)
   8. (4 hr, 200 mi) and (5 hr, 250 mi)
   9. (7 hr, 700 km) and (9 hr, 900 km)
   10. (2 class-hours, 4 hours of homework) and (3 class-hours, 6 hours of homework)

Answers for Exercises

1. $\dfrac{12\ \text{V} - 15\ \text{V}}{4\ \text{mA} - 7\ \text{mA}} = \dfrac{-3\ \text{V}}{-3\ \text{mA}} = +1\ \text{k}\Omega$

2. $\dfrac{12\ \text{V} - 8\ \text{V}}{4\ \text{mA} - 9\ \text{mA}} = \dfrac{4\ \text{V}}{-5\ \text{mA}} = -0.8\ \text{k}\Omega \text{ or } -800\ \Omega$

3. $\dfrac{7\ \text{mA} - 9\ \text{mA}}{15\ \text{V} - 20\ \text{V}} = \dfrac{-2\ \text{mA}}{-5\ \text{V}} = 0.4\ \text{mS} \text{ or } 400\ \mu\text{S}$

4. $\dfrac{4\ \text{mA} - 1\ \text{mA}}{25\ \text{V} - 37\ \text{V}} = \dfrac{3\ \text{mA}}{-12\ \text{V}} = -0.25\ \text{mS} \text{ or } -250\ \mu\text{S}$

5. $\dfrac{1\ \text{mA} - 2\ \text{mA}}{0.65\ \text{V} - 0.75\ \text{V}} = \dfrac{-1\ \text{mA}}{-0.10\ \text{V}} = 10\ \text{mS}$

6. $\dfrac{11\ \text{V} - 17\ \text{V}}{9\ \mu\text{S} - 12\ \mu\text{S}} = \dfrac{-6\ \text{V}}{-3\ \mu\text{S}} = 2\ \text{M}\Omega \text{ or } 2{,}000{,}000\ \Omega$

7. $\dfrac{11\ \text{V} - 5\ \text{V}}{9\ \mu\text{S} - 12\ \mu\text{S}} = \dfrac{+6\ \text{V}}{-3\ \mu\text{S}} = -2\ \text{M}\Omega \text{ or } -2{,}000{,}000\ \Omega$

8. $\dfrac{200\ \text{mi} - 250\ \text{mi}}{4\ \text{hr} - 5\ \text{hr}} = 50\ \text{mph or } 50\ \text{mi/hr}$

9. $\dfrac{700\ \text{km} - 900\ \text{km}}{7\ \text{hr} - 9\ \text{hr}} = 100\ \text{km/hr}$

10. $\dfrac{4\ \text{hours of homework} - 6\ \text{hours of homework}}{2\ \text{class-hours} - 3\ \text{class-hours}} = \dfrac{2\ \text{hours of homework}}{\text{per class-hour}}$

ASSIGNMENT EXERCISES

Section 14–1

Represent the solutions of the functions in a table of values and on a graph.

1. $f(x) = 2x - 3$ **2.** $f(x) = -4x + 1$ **3.** $f(x) = 3x$

4. $f(x) = 4x$ **5.** $f(x) = -3x$ **6.** $f(x) = -4x$

7. $f(x) = 2x + 1$ **8.** $f(x) = 2x + 5$ **9.** $f(x) = x^2 + 2x - 8$

10. $f(x) = x^2 - 3$ **11.** $f(x) = 2x^2 - 5x - 3$ **12.** $f(x) = x^3$

13. $f(x) = x - 6$ **14.** $f(x) = 3x - 8$ **15.** $f(x) = 3x + 6$

16. $f(x) = 2x$ **17.** $f(x) = 2x - 2$ **18.** $f(x) = 5x - 7$

Section 14–2

Which of the coordinate pairs are solutions for the equation $2x - 3y = 12$?

19. $(-2, -3)$ **20.** $(1, -3)$ **21.** $(3, -2)$

22. $(0, 4)$ **23.** $(6, 0)$

24. In the equation $2x + y = 8$, find x when $y = 10$.

25. Find y in the equation $x - 3y = 5$ when $x = 8$.

26. Find y if $x = 7$ in the equation $3x - 4 = -2$.

Graph the equations using the intercepts procedure.

27. $x = -4y - 1$ **28.** $x + y = -4$ **29.** $3x - y = 1$
30. $x = -4y$

Graph using the slope-intercept procedure.

31. $y = 5x - 2$ **32.** $y = -x$ **33.** $y = -3x - 1$

34. $y = \dfrac{1}{2}x + 3$ **35.** $x - y = 4$ **36.** $2y + 4 = -3$

37. $x - 2y = -1$ **38.** Verify Exercises 27–37 with a graphing calculator.

Find the slope of the line passing through the given pairs of points.

39. $(-2, 2)$ and $(1, 3)$ **40.** $(3, -1)$ and $(1, 3)$ **41.** $(3, 2)$ and $(5, 6)$
42. $(-1, -1)$ and $(2, 2)$ **43.** $(4, 3)$ and $(-4, -2)$ **44.** $(6, 2)$ and $(-3, 2)$
45. $(3, -4)$ and $(0, 0)$ **46.** $(1, -1)$ and $(5, -5)$ **47.** $(-4, 1)$ and $(-4, 3)$
48. $(4, -4)$ and $(1, 3)$ **49.** $(5, 0)$ and $(-2, 4)$ **50.** $(-2, 1)$ and $(0, 3)$
51. $(-4, -8)$ and $(-2, -1)$ **52.** $(3, 3)$ and $(3, 0)$ **53.** $(5, -3)$ and $(-1, -3)$
54. $(-5, -1)$ and $(-7, -3)$ **55.** $(-7, 0)$ and $(-7, 5)$ **56.** $(3, 5)$ and $(2, 5)$
57. $(5, 9)$ and $(7, 11)$ **58.** $(3, 5)$ and $(5, 3)$

59. Write the coordinates of two points that lie on the same horizontal line.

60. Write the coordinates of two points that lie on the same vertical line.

Find the equation of a line passing through the given point with the given slope. Solve the equation for y if necessary.

61. $(-6, 2), m = \dfrac{1}{3}$ **62.** $(3, 2), m = -\dfrac{2}{5}$ **63.** $(4, 0), m = \dfrac{3}{4}$

64. $(0, -2), m = 2$ **65.** $(2, 3), m = 4$ **66.** $(6, 0), m = -1$

67. $(5, -4), m = -\dfrac{2}{3}$ **68.** $(-1, -5), m = -3$

Find the equation of a line passing through the given pairs of points. Solve the equation for y if necessary.

69. $(-5, 2)$ and $(6, 1)$ **70.** $(1, 4)$ and $(-1, 3)$ **71.** $(-1, -3)$ and $(3, 4)$
72. $(-3, 0)$ and $(4, 0)$ **73.** $(-2, -3)$ and $(3, 6)$ **74.** $(2, -4)$ and $(3, -4)$
75. $(5, 2)$ and $(6, 3)$ **76.** $(4, 6)$ and $(1, -1)$ **77.** $(-1, -2)$ and $(-3, -4)$
78. $(4, 0)$ and $(4, -3)$ **79.** $(5, -2)$ and $(3, -2)$ **80.** $(5, 4)$ and $(0, 4)$

Determine the slope and y-intercept of the given equations by inspection.

81. $y = 3x + \dfrac{1}{4}$ **82.** $y = \dfrac{2}{3}x - \dfrac{3}{5}$ **83.** $y = -5x + 4$ **84.** $y = 7$

85. $x = 8$ **86.** $y = \dfrac{1}{3}x - \dfrac{5}{8}$ **87.** $y = \dfrac{x}{8} - 5$ **88.** $y = -\dfrac{x}{5} + 2$

Write the given equations in slope-intercept form and determine the slope and y-intercept.

89. $2x + y = 8$ **90.** $4x + y = 5$ **91.** $3x - 2y = 6$ **92.** $5x - 3y = 15$

93. $\dfrac{3}{5}x - y = 4$ **94.** $2.2y - 6.6x = 4.4$ **95.** $3y = 5$ **96.** $3x - 6y = 12$

Write the equations using the given slope and y-intercept.

97. $m = 3, b = -2$ **98.** Slope $= \dfrac{3}{5}$; y-intercept $= -7$

Write the equations using information from the graphs in Figs. 14–38 and 14–39.

99.

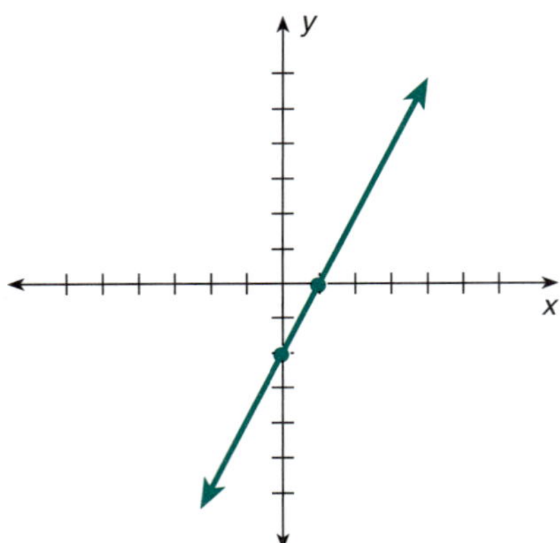

Figure 14–38

100.

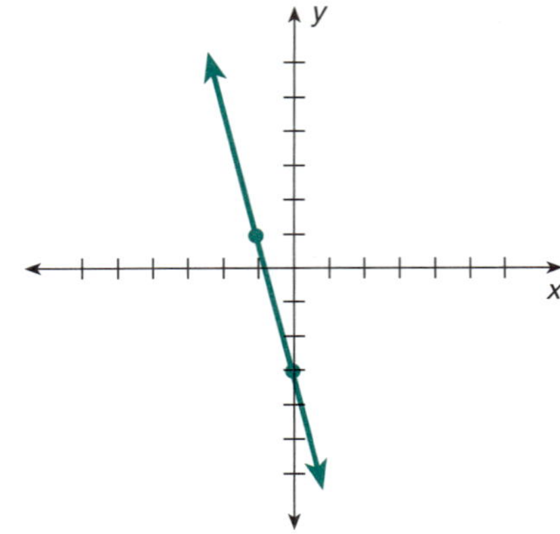

Figure 14–39

Section 14–5

Write equations in standard form. Graph both equations for each exercise on a graphing calculator to verify parallelism.

101. Find the equation of the line that is parallel to the line $x + y = 4$ and passes through the point $(2, 5)$.

102. Find the equation of the line that is parallel to the line $3x + y = 6$ and passes through the point $(1, 0)$.

103. Find the equation of the line that is parallel to the line $2y = x - 3$ and passes through the point $(2, -3)$.

104. Find the equation of the line that is parallel to the line $4x - y = -1$ and passes through the point $(0, -2)$.

105. Find the equation of the line that is parallel to the line $x - 3y = 5$ and passes through the point $(5, -5)$.

106. Find the equation of the line that is parallel to the line $3x - 2y = 2$ and passes through the point $(0, -3)$.

107. Find the equation of the line that is parallel to the line $x + 3y = 6$ and passes through the point $(-4, -2)$.

108. Find the equation of the line that is parallel to the line $4x + 3y = 1$ and passes through the point $(3, -\frac{1}{2})$.

109. Find the equation of the line that is parallel to the line $3x - 4y = 0$ and passes through the point $(\frac{1}{3}, 2)$.

110. Find the equation of the line that is parallel to the line $-2x + 3y = 2$ and passes through the point $(-1, -1)$.

Write the equations in standard form. Use a graphing calculator to verify that the two lines in each exercise are perpendicular.

111. Find the equation of the line that is perpendicular to the line $x + y = 4$ and passes through the point $(-3, 1)$.

112. Find the equation of the line that is normal to the line $3x + y = 6$ and passes through the point $(1, 4)$.

113. Find the equation of the line that is perpendicular to the line $x + 2y = 5$ and passes through the point $(-2, 0)$.

114. Find the equation of the line that is normal to the line $2x + 2y = 4$ and passes through the point $(0, 0)$.

115. Find the equation of the line that is normal to the line $5x + y = 8$ and passes through the point $(-1, 2)$.

116. Find the equation of the line that is perpendicular to the line $3y = x - 4$ and passes through the point $(2, 3)$.

117. Find the equation of the line that is perpendicular to the line $5x - y = 10$ and passes through the point $(\frac{1}{2}, 3)$.

118. Find the equation of the line that is normal to the line $x - 3y = 6$ and passes through the point $(-2, 4)$.

119. Find the equation of the line that is perpendicular to the line $4x - y = 8$ and passes through the point $(4, -\frac{1}{2})$.

120. Find the equation of the line that is perpendicular to the line $4y = 2x + 1$ and passes through the point $(3, -1)$.

Section 14–6

Identify the equations as linear, quadratic, or other nonlinear equations.

121. $y = 3x + 7$

122. $4x - 7y = 8$

123. $y = 5x + 3x^2$

124. $x^2 + y = x + 2$

125. $xy = 5x - 5y$

Graph the quadratic equations by examining their properties, and verify with a graphing calculator.

126. $y = x^2 - 1$ **127.** $y = -x^2 - 1$ **128.** $y = x^2 + 3x - 10$
129. $y = x^2 - 6x + 8$ **130.** $y = x^2 - 2x + 1$ **131.** $y = -x^2 + 2x - 1$
132. $y = x^2 - 4x + 4$ **133.** $y = -x^2 + 4x - 4$ **134.** $y = 2x^2 + 1$

CHALLENGE PROBLEMS

135. A 1-gallon can of indoor house paint is advertised to cover 400 ft^2 of wall surface.

  (b) Graph these data.

  (d) The paint being used for this job can be purchased for $19.95 a gallon. Find the cost of the paint for the 5,500 ft^2 of wall surface.

  (a) Make a table to show the amount of wall surface area that can be covered by 1, 2, 3, . . . , or 10 gallons of paint.

  (c) Use the graph to decide how many 1-gal cans of paint it would take to cover 5,500 ft^2 of wall surface.

  (e) The sales tax rate is 8.25%. Calculate the total cost of the paint.

136. Use column 1 of the following table for the horizontal scale and one of the other columns for the vertical scale.

  (a) Which graph in Fig. 14–40 illustrates the function pattern for the column you selected?

Payment Number	Payment Amount	Applied to Interest	Applied to Principal	Balance Owed	Total Interest
1	611.09	580.00	31.09	69,568.91	580.00
2	611.09	579.74	31.35	69,537.56	1,159.74
3	611.09	579.48	31.61	69,505.95	1,739.22
4	611.09	579.22	31.87	69,474.08	2,318.44
5	611.09	578.95	32.14	69,441.94	2,897.39
6	611.09	578.68	32.41	69,409.53	3,476.07
7	611.09	578.41	32.68	69,376.85	4,054.48
8	611.09	578.14	32.95	69,343.90	4,632.62
9	611.09	577.87	33.22	69,310.68	5,210.49
10	611.09	577.59	33.50	69,277.18	5,788.08

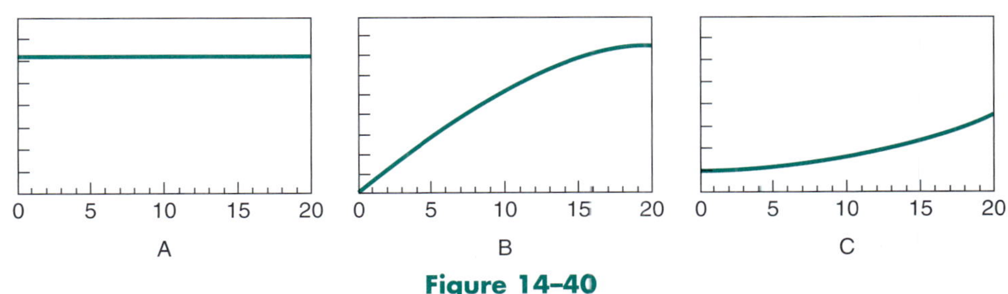

Figure 14–40

  (b) Sketch graphs to describe the pattern shown by the numbers in the two columns not illustrated by graphs A, B, and C.

Christy Hunsucker is introducing a new lipstick in Brightglow's product line. As product development manager, she estimates the cost of the new lipstick to be $4.53 per item, plus an additional cost of $5,000.00 for product development.

137. Make a table of the estimated cost for producing 0 lipsticks, 100 lipsticks, 1,000 lipsticks, and 2,000 lipsticks.

138. Represent these costs as ordered pairs (number of lipsticks, cost).

139. Plot the ordered pairs and graph a line that fits the ordered pairs.

140. Write a formula (using function notation) that represents the cost of the new lipstick as a function of the number of lipsticks produced.

141. Christy Hunsucker projects the selling price of the new lipstick to be $8.99. How many lipsticks must be sold for the company to recover its cost of producing the new item?

CHAPTER TRIAL TEST

Make a table of values to represent the solutions to the following equations and show the solutions on a graph.

1. $f(x) = \dfrac{1}{2}x$

2. $f(x) = \dfrac{1}{2}x + 1$

3. $f(x) = 2x - 4$

4. $f(x) = 5 - x$

Write the specific solution for each equation in ordered-pair form.

5. $x + y = 7$, if $x = 2$

6. $2x - y = 1$, if $y = -3$

7. Find the x- and y-intercepts of the equation $x - 4y = 2$.

8. Find the x- and y-intercepts of the equation $y = -8 + x$.

9. Plot the graph for the relationship between horsepower (hp) and revolutions per minute (rpm) in a test engine. Use the same intervals as those in the data table.

x-axis (rpm)	y-axis (hp)
500	30
1,000	45
1,500	60
2,000	75

Graph using the intercepts method.

10. $2x - 3y = 6$

11. $x + 2y = 8$

Graph using the slope-intercept method.

12. $y = -3x + 1$

13. $2x + y = -3$

14. $x + 2y = 1$

15. $y = x - 5$

16. $y = 4x$

Find the slope of the line passing through the given pairs of points.

17. $(-3, 6)$ and $(3, 2)$

18. $(0, 4)$ and $(-1, 6)$

19. $(1, -5)$ and $(3, 0)$

20. $(5, 3)$ and $(-2, 3)$

21. $(-1, -1)$ and $(2, 2)$

22. $(-1, 5)$ and $(-1, 7)$

Find the equation of the line passing through the given point with the given slope. Solve the equation for y.

23. $(3, -5)$, $m = \dfrac{2}{3}$

24. $(5, 1)$, $m = -2$

Find the equation of the line passing through the given pairs of points. Solve the equation for y.

25. $(1, 3)$ and $(4, 5)$

26. $(-1, 1)$ and $(4, -4)$

27. $(5, 2)$ and $(-1, 2)$

28. $(7, 4)$ and $(-3, -1)$

29. What is the slope and y-intercept of a line whose equation is $y = 3x - 22$?

Write the equations in slope-intercept form and determine the slope and y-intercept.

30. $x - y = 4$

31. $x = 4y$

32. $2y - x = 3$

33. $\dfrac{1}{3}y + 2x = 1$

Write equations in slope-intercept form using the given slope and y-intercept or the graph in Fig. 14–41.

34. slope $= -2$, y-intercept $= -3$

35.

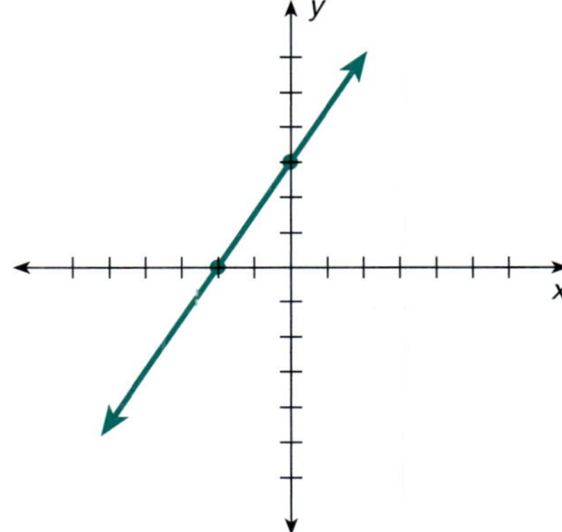

Figure 14–41

Write the equations in standard form.

36. Find the equation of the line that is parallel to $y - 2x = 3$ and passes through the point (2, 5).

37. Find the equation of the line that is parallel to $2x + y = 4$ and passes through the point (4, −3).

38. Find the equation of the line perpendicular to $y - 2x = 3$ and passing through the point (2, 5).

39. Find the equation of the line perpendicular to $2x + y = 4$ and passing through the point (4, −3).

Graph the quadratic equations by examining properties. Show the axis of symmetry, vertex, and solutions.

40. $y = x^2 + 2$

41. $y = x^2 + 2x + 1$

42. $y = x^2 + 4x + 4$

Systems of Equations

GOOD DECISIONS THROUGH TEAMWORK

While going to school at night to get a degree in horticulture, Mario Menendez decides to give up his full-time, minimum-wage job selling clothes for a job that requires less time, pays the same or more money, is in his field (horticulture), and allows for flexible working hours. He's going to capitalize on his hobby of several years—making holiday table centerpieces out of greenery, pine cones, dried flowers, and candles. He already has a large stockpile of greenery, cones, and flowers he's collected. He will have to buy candles, ribbon, and foam bases, but he can buy these in bulk at low prices. He has a workshop set up in his garage, so he can work at home.

He knows from experience that most customers want either a 12-in. or a 16-in. base, and that he can sell these to local florist shops for $20 and $25, respectively. (The florist shops resell them for $40 and $50). The 12-in. base requires a half hour of production time and costs $3 for materials. The 16-in. base requires the same amount of production time but costs $4.50 for materials.

You and your team are helping Mario with his economic forecasting. Let x equal the number of 12-in. centerpieces and y equal the number of 16-in. centerpieces he can make per week in the home business while attending college full-time. Derive the following forecasting formulas for different production levels: Write a linear equation in x and y to show the revenue received for any number of 12-in. and 16-in. centerpieces. Write a linear equation in x and y to show the cost of materials for making any number of 12-in. and 16-in. centerpieces. Finally, write a linear equation in x and y to show the profit made on any number of 12-in. and 16-in. centerpieces.

15–1 Solving systems of equations graphically

1 Solve a system of equations by graphing.

15–2 Solving systems of equations using the addition method

1 Use the addition method to solve a system of equations that contains opposite variable terms.

2 Use the addition method to solve a system of equations that does not contain opposite variable terms.

3 Apply the addition method to a system of equations with no solution or with many solutions.

15–3 Solving systems of equations using the substitution method

1 Use the substitution method to solve a system of equations.

Graph your cost and revenue equations. Shade the area representing profit and the area representing loss, and identify the breakeven point. Determine how many hours per week Mario must work to make more than minimum wage for 40 hours per week. In your report, offer your advice on this business venture.

15–4 Problem solving using systems of equations

1. Use a system of equations to solve application problems.

Many real-world situations involve problems in which several conditions or constraints have to be considered. These conditions can be written in separate equations that form a system of equations. The solution of the system will be the value or values that satisfy all conditions.

15–1 SOLVING SYSTEMS OF EQUATIONS GRAPHICALLY

Learning Outcome

1. Solve a system of equations by graphing.

In Chapter 14, we learned to graph equations with two variables. Because an equation with two variables has many ordered pairs of solutions, a graph gives an overall pictorial view of these solutions. In this section we look at solutions that two or more equations have in common.

1 Solve a System of Equations by Graphing.

A *system of two linear equations,* having two variables each, is solved when we find the one ordered pair of solutions that satisfies *both* equations. One method of solving systems of two equations is to graph the ordered pairs of solutions of each equation and find the intersection of these graphs. The point where the two graphs intersect represents the ordered pair of solutions that the two graphs have in common.

To solve a system of two equations with two variables by graphing:

1. Graph each equation on the same pair of axes.
2. The solution will be the common point or points.

Now let's look at a system of two equations with two variables.

EXAMPLE A board is 20 ft long. It needs to be cut so that one piece is 2 ft longer than the other. What should be the length of each piece?

First, we write equations to describe all the conditions of the problem. Since the board is not cut into equal pieces, we let the letter l represent the *longer* piece and the letter s represent the *shorter* piece.

The first condition of the problem states that the total length of the board is 20 ft. Thus, the two pieces (l and s) total 20 ft: $l + s = 20$ (condition 1).

The second condition is that one piece is 2 ft longer than the other. Thus, the shorter piece plus 2 ft equals the longer piece: $s + 2 = l$ (condition 2).

The two equations become a *system of equations.*

$l + s = 20$ Condition 1.

$s + 2 = l$ Condition 2.

Graph each equation on the same set of axes and examine the intersection of the graphs. To do this, we make a table of solutions for each equation.

Condition 1   **Condition 2**

$l + s = 20$ $s + 2 = l$

or   or

$l = 20 - s$ $l = s + 2$

s	l		s	l
9	11		9	11
10	10		10	12
11	9		11	13

Values for s like 1 and 2 would require 19 and 18 units for l on the graph. Therefore, to fit the values easily on a smaller graph, we select s-values near 10. Now we graph the s-values on the x-axis and the l-values on the y-axis. We then write each equation along its line graph for easy identification (Fig. 15–1).

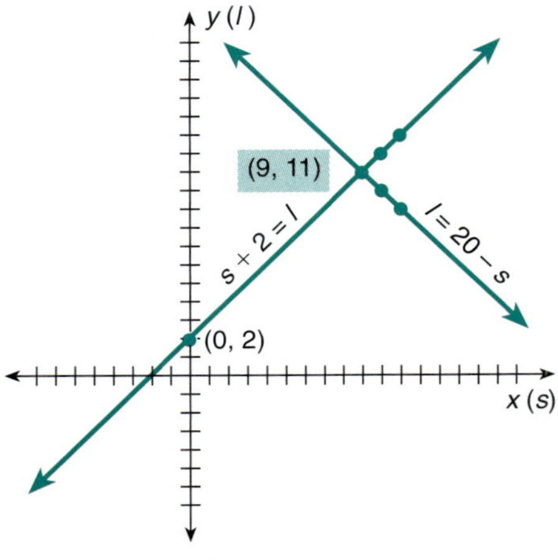

Figure 15–1

The point of intersection is (9, 11), which means that $s = 9$, $l = 11$. **The short length is 9 ft and the longer length is 11 ft.**
We can check to see if the solution ($s = 9$, $l = 11$) satisfies both equations.

$l + s = 20$ $s + 2 = l$

$11 + 9 = 20$ $9 + 2 = 11$

$20 = 20$ $11 = 11$

The ordered pair checks in both equations.

Line relationships:

When two straight lines are graphed, three possibilities can occur.

1. The two lines intersect in **just one point.** This means the system is *independent.*
2. The two lines **do not intersect** at all. This means that the system is *inconsistent,* and no pair of values satisfies both equations.
3. The two lines **coincide** or fall exactly in the same place. This means that the system is *dependent* and the equations are identical or are multiples, and any pair of values that satisfies one equation satisfies both equations.

The last two possibilities rarely occur in practical applications.

SELF-STUDY EXERCISES 15–1

1 Solve the systems of equations by graphing.

1. $x + y = 12$
$x - y = 2$

2. $2x + y = 9$
$3x - y = 6$

3. $2x - y = 5$
$4x - 2y = 8$

4. $x - y = 9$
$3x - 3y = 27$

5. $3x + 2y = 8$
$x + y = 2$

15–2 SOLVING SYSTEMS OF EQUATIONS USING THE ADDITION METHOD

Learning Outcomes

1 Use the addition method to solve a system of equations that contains opposite variable terms.

2 Use the addition method to solve a system of equations that does not contain opposite variable terms.

3 Apply the addition method to a system of equations with no solution or with many solutions.

We can see that solving systems of equations graphically is tedious and time-consuming. Also, many solutions to systems of equations are not whole numbers. Graphically, it is difficult to plot fractions or to read fractional intersection points. Therefore, we need a convenient algebraic procedure for solving systems of equations. We examine the two methods most commonly used to solve a system of equations with two variables, but there are other methods. The first method we introduce is the *addition method*.

1 Use the Addition Method to Solve a System of Equations That Contains Opposite Variable Terms.

The addition method incorporates the concept that equals added to equals give equals. Thus, when we add two equations, the result is still an equation. However, the addition method is effective *only* if one variable is *eliminated* in the addition process. This method is also called the *elimination method*. The following examples contain two opposite terms that add to zero when the equations are added.

EXAMPLE Add the following pairs of equations.

(a) $x + y = 7$
$x - y = 5$

(b) $3x + 8y = 7$
$-3x - 3y = 8$

(c) $x - 2y = 4$
$3x + 2y = 8$

(a)
$$\begin{array}{r} x + y = 7 \\ x - y = 5 \\ \hline 2x + 0y = 12 \\ 2x = 12 \end{array}$$

(b)
$$\begin{array}{r} 3x + 8y = 7 \\ -3x - 3y = 8 \\ \hline 0x + 5y = 15 \\ 5y = 15 \end{array}$$

(c)
$$\begin{array}{r} x - 2y = 4 \\ 3x + 2y = 8 \\ \hline 4x + 0y = 12 \\ 4x = 12 \end{array}$$

In each problem in the preceding example, one unknown was eliminated by the addition method. Now the remaining equation contains only one variable and it can be solved to find the value of that variable in the system. The other variable can then be found by substitution. Let's finish solving each problem in the preceding example.

EXAMPLE Solve the systems of equations in the preceding example by the addition method.

(a) $x + y = 7$ Equation 1.

$\underline{x - y = 5}$ Equation 2.

$2x = 12$

$x = 6$ Substitute 6 in place of x in *either* of the given equations.

$x + y = 7$ Equation 1.

$6 + y = 7$

$y = 7 - 6$

$y = 1$

The solution is $x = 6$ and $y = 1$, or (6, 1).
   Check in the *other* equation:

$x - y = 5$ Equation 2.

$6 - 1 = 5$

$5 = 5$

Checking in the *other* equation helps ensure that no errors have been made in the addition process or in the substitution process.

(b) $3x + 8y = 7$ Equation 1.

$\underline{-3x - 3y = 8}$ Equation 2.

$5y = 15$

$y = 3$ Substitute 3 for y in Equation 1.

$3x + 8y = 7$ Equation 1.

$3x + 8(3) = 7$

$3x + 24 = 7$

$3x = 7 - 24$

$3x = -17$

$x = \dfrac{-17}{3}$

The solution is $(-\frac{17}{3}, 3)$.
   Check in the other equation:

$-3x - 3y = 8$ Equation 2.

$-3\left(\dfrac{-17}{3}\right) - 3(3) = 8$

$17 - 9 = 8$

$8 = 8$

(c)
$$x - 2y = 4 \qquad \text{Equation 1.}$$
$$\underline{3x + 2y = 8} \qquad \text{Equation 2.}$$
$$4x \quad\;\; = 12$$
$$x = 3 \qquad \text{Substitute 3 for } x \text{ in Equation 1.}$$
$$x - 2y = 4 \qquad \text{Equation 1.}$$
$$3 - 2y = 4$$
$$-2y = 4 - 3$$
$$-2y = 1$$
$$y = -\frac{1}{2}$$

The solution is $(3, -\frac{1}{2})$.

Check:
$$3x + 2y = 8 \qquad \text{Equation 2.}$$
$$3(3) + 2\left(-\frac{1}{2}\right) = 8$$
$$9 + (-1) = 8$$
$$8 = 8$$

2 Use the Addition Method to Solve a System of Equations That Does Not Contain Opposite Variable Terms.

Remember, in the addition method one unknown must be eliminated; that is, the terms must add to 0. We can also use the property that both sides of an equation may be multiplied by the same number so that the equality is preserved and the value of the variable remains the same. Thus, if neither pair of variable terms in a system of equations are opposite terms that will add to 0, we multiply one or both of the equations by numbers that *cause* the terms of one variable to add to 0.

To solve a system of equations using the addition method (elimination):

1. If necessary, multiply one or both equations by numbers that cause the terms of one variable to add to zero.
2. Add the two equations to eliminate a variable.
3. Solve the equation from Step 2.
4. Substitute the solution from Step 3 in either equation and solve for the remaining variable.
5. Check in one or both original equations.

EXAMPLE Solve the system of equations: $2x + y = 7$ and $x + y = 3$.

In this system, neither pair of like variable terms adds to 0. We can use several different strategies to make this happen: If the x in the second equation has a coefficient of -2, the x-terms add to 0; or if the y-term in either the first or second equation has a coefficient of -1, the y-terms add to 0. Thus, we have at least three choices: multiply the first equation by -1, multiply the second equation by -1, or multiply the second equation by -2.

Choice 1.

$$-2x - y = -7 \quad \text{Multiply the first equation by } -1.$$
$$\underline{x + y = \quad 3} \quad \text{Add the equations.}$$
$$-x \quad\quad = -4$$
$$\boxed{x = \quad 4}$$

Choice 2.

$$2x + y = \quad 7$$
$$\underline{-x - y = -3} \quad \text{Multiply the second equation by } -1$$
$$\boxed{x \quad\quad = \quad 4} \quad \text{and add the equations.}$$

Choice 3.

$$2x + \quad y = \quad 7$$
$$\underline{-2x - 2y = -6} \quad \text{Multiply the second equation by } -2$$
$$-y = \quad 1 \quad \text{and add the equations.}$$
$$\boxed{y = -1}$$

In choices 1 and 2, substitute $x = 4$.

$$-2x - y = -7$$
$$-2(4) - y = -7$$
$$-8 - y = -7$$
$$-y = -7 + 8$$
$$-y = 1$$
$$\boxed{y = -1}$$

In choice 3, substitute $y = -1$.

$$2x + y = 7$$
$$2x + (-1) = 7$$
$$2x - 1 = 7$$
$$2x = 7 + 1$$
$$2x = 8$$
$$\boxed{x = 4}$$

The solution is $(4, -1)$.

Check the solution in the first original equation.

$$2x + y = 7$$
$$2(4) + (-1) = 7$$
$$8 + (-1) = 7$$
$$7 = 7$$

Check the solution in the second original equation.

$$x + y = 3$$
$$4 + (-1) = 3$$
$$3 = 3$$

Tip!	***Importance of Checking Solutions.***

When altering an original equation, it is very important to check your solution in *both* original equations. This enables you to identify mistakes such as forgetting to multiply *each* term in the equation by a number.

EXAMPLE Solve the system of equations: $x = 3y + 7$ and $2x + 3y = 2$.

In this system, the first equation does not have both letter terms on one side of the equation. Our standard form for solving systems of equations using the addition method requires that the letter terms be on one side and the number term on the other. Also, the letter terms in both equations need to be in the same order so that like terms will be in columns. Thus, we rearrange the first equation.

$$x = 3y + 7 \rightarrow x - 3y = 7$$

Now, we continue to solve the system.
Add to eliminate the y-terms.

584   CHAPTER 15   Systems of Equations

$$x - 3y = 7 \qquad \text{Equation 1.}$$
$$\underline{2x + 3y = 2} \qquad \text{Equation 2.}$$
$$3x = 9$$
$$x = 3$$
$$x - 3y = 7 \qquad \text{Substitute } x = 3.$$
$$3 - 3y = 7$$
$$-3y = 7 - 3$$
$$-3y = 4$$
$$y = -\frac{4}{3}$$

The solution is $(3, -\frac{4}{3})$.

Check the solution in the original equations:

Equation 1:

$$x = 3y + 7$$
$$3 = 3\left(-\frac{4}{3}\right) + 7$$
$$3 = -4 + 7$$
$$3 = 3$$

Equation 2:

$$2x + 3y = 2$$
$$2(3) + 3\left(-\frac{4}{3}\right) = 2$$
$$6 + (-4) = 2$$
$$2 = 2$$

EXAMPLE Solve the system of equations: $2x + 3y = 1$ and $3x + 4y = 2$.

In this system, multiplying just one equation by a number will not eliminate a letter. Therefore, we need to multiply each equation by some number. There are several possibilities; we examine just two.

Choice 1. We multiply the first equation by -3 so that the coefficient of x is -6. Then we multiply the second equation by $+2$ so that the coefficient of x is $+6$. We then have

$$-6x - 9y = -3 \qquad -3(\text{Equation 1}).$$
$$\underline{6x + 8y = 4} \qquad 2(\text{Equation 2}).$$
$$-y = 1$$
$$y = -1$$
$$2x + 3y = 1 \qquad \text{Substitute } y = -1.$$
$$2x + 3(-1) = 1$$
$$2x - 3 = 1$$
$$2x = 1 + 3$$
$$2x = 4$$
$$x = 2$$

Choice 2. To eliminate y, we could multiply the first equation by 4 and the second equation by -3. There is no real preference between choices 1 and 2.

$$8x + 12y = 4 \qquad 4(\text{Equation 1}).$$
$$\underline{-9x - 12y = -6} \qquad -3(\text{Equation 2}).$$
$$- x = -2$$

$$x = 2$$

$$2x + 3y = 1 \qquad \text{Substitute } x = 2.$$

$$2(2) + 3y = 1$$

$$3y = 1 - 4$$

$$3y = -3$$

$$y = -1$$

The solution is $(2, -1)$.

Check the solution in the original equations:

$$2x + 3y = 1 \qquad\qquad 3x + 4y = 2$$

$$2(2) + 3(-1) = 1 \qquad 3(2) + 4(-1) = 2$$

$$4 + (-3) = 1 \qquad\qquad 6 + (-4) = 2$$

$$1 = 1 \qquad\qquad\qquad 2 = 2$$

3 Apply the Addition Method to a System of Equations with No Solution or with Many Solutions.

In Section 15–1, we noted that sometimes there is no solution to a system of equations. In one situation, the graphs of two equations do not intersect, and in the other, the graphs of the two equations coincide. Let's look at examples of these situations and attempt to solve them using the addition method.

EXAMPLE Solve the system $x + y = 7$ and $x + y = 5$.

To eliminate the x-terms, let's multiply the second equation by -1 and add the equations.

$$x + y = 7 \qquad \text{Equation 1.}$$

$$\underline{-x - y = -5} \qquad -1(\text{Equation 2}).$$

$$0 = 2$$

Notice that both variables are eliminated and the resulting equation, $0 = 2$, is *false*. **Thus, there are no solutions to this system.**

Tip!	*Inconsistent Equations—No Solution.*

When solving a system of equations, if both variables are eliminated and the resulting statement is false, the equations are *inconsistent* and have no solution. The graphs of the equations are parallel lines.

EXAMPLE Solve the system $2x - y = 7$ and $4x - 2y = 14$.

We can eliminate the y's by multiplying the first equation by -2 and adding the equations.

$$-4x + 2y = -14 \qquad -2(\text{Equation 1}).$$

$$\underline{4x - 2y = 14} \qquad \text{Equation 2.}$$

$$0 = 0$$

Again, both variables are eliminated; however, this time the result is a *true* statement $(0 = 0)$. In this situation, **all solutions of one equation are also solutions of the other equation.**

CHAPTER 15 Systems of Equations

SELF-STUDY EXERCISES 15–2

1   Solve the systems of equations using the addition method.

1. $a - 2b = 7$
   $3a + 2b = 13$

2. $3m + 4n = 8$
   $2m - 4n = 12$

3. $x - 4y = 5$
   $-x - 3y = 2$

4. $a - b = 6$
   $2a + b = 3$

5. $x + 2y = 5$
   $3x - 2y = 3$

2   Solve the systems of equations using the addition method.

6. $3x + y = 9$
   $x + y = 3$

7. $7x + 2y = 17$
   $y = 3x + 2$

8. $a + 6b = 18$
   $4a - 3b = 0$

9. $3x + y = -1$
   $4x - 2y = -8$

10. $a = 6y$
   $2a - y = 11$

3   Solve the systems of equations using the addition method.

11. $x + y = 8$
   $x + y = 3$

12. $x + 2y = 9$
   $x + y = 3$

13. $3x - 2y = 6$
   $9x - 6y = 18$

14. $2a + 4b = 10$
   $a + 2b = 5$

15. $3a - b = 14$
   $a - 3b = 2$

15–3   SOLVING SYSTEMS OF EQUATIONS USING THE SUBSTITUTION METHOD

Learning Outcome

1   Use the substitution method to solve a system of equations.

1   Use the Substitution Method to Solve a System of Equations.

Another method for solving systems of equations is by substitution. Recall that in formula rearrangement (Section 11–2), whenever more than one variable is used in an equation or formula, we can rearrange the equation or solve for a particular variable. In the *substitution method* for solving systems of equations, we solve one equation for one variable and then substitute the equivalent expression in place of the variable in the other equation.

> *To solve a system of equations by substitution:*
>
> **1.** Rearrange either equation to isolate one variable.
> **2.** Substitute the equivalent expression from Step 1 into the *other* equation and solve for the remaining variable.
> **3.** Substitute the solution from Step 2 into the equation from Step 1 to find the value of the substituted variable.
> **4.** Check both original equations.

EXAMPLE Solve the system of equations using the substitution method.

$$x + y = 15 \qquad \text{Equation 1.}$$
$$y = 2x \qquad \text{Equation 2.}$$

$$x + y = 15$$
$$y = 2x \qquad \text{Equation 2 is already solved for } y \text{ (Step 1).}$$
$$x + y = 15 \qquad \text{Substitute } 2x \text{ for } y \text{ in Equation 1 and solve (Step 2).}$$
$$x + 2x = 15$$
$$3x = 15$$
$$x = 5$$
$$y = 2x \qquad \text{Substitute the solution for } x \text{ in Equation 2 to find } y \text{ (Step 3).}$$
$$y = 2(5)$$
$$y = 10$$

The solution is (5, 10).
Check:

$$x + y = 15 \qquad y = 2x \qquad \text{Step 4.}$$
$$5 + 10 = 15 \qquad 10 = 2(5)$$
$$15 = 15 \qquad 10 = 10 \qquad \text{The solution checks in both equations.}$$

EXAMPLE Solve the following system of equations using the substitution method.

$$2x - 3y = -14$$
$$x + 5y = 19$$

When using the substitution method, you can solve either equation for either unknown. In this example, the x-term in the second equation has a coefficient of 1, so the simplest choice would be to solve the second equation for x.

Step 1	**Step 2**	**Step 3**
$x + 5y = 19$	$2x - 3y = -14$	$x = 19 - 5y$
$x = 19 - 5y$	$2(19 - 5y) - 3y = -14$	$x = 19 - 5(4)$
	$38 - 10y - 3y = -14$	$x = 19 - 20$
	$38 - 13y = -14$	$x = -1$
	$-13y = -14 - 38$	
	$-13y = -52$	
	$\dfrac{-13y}{-13} = \dfrac{-52}{-13}$	
	$y = 4$	

The solution is $(-1, 4)$.
Check the roots $x = -1$, $y = 4$ in both original equations.

Step 4

$$2x - 3y = -14 \qquad x + 5y = 19$$
$$2(-1) - 3(4) = -14 \qquad -1 + 5(4) = 19$$
$$-2 - 12 = -14 \qquad -1 + 20 = 19$$
$$-14 = -14 \qquad 19 = 19$$

Sometimes we let ourselves become overwhelmed by the mere length of a problem. Look at the previous example. Each step of the solution involves skills that we have previously used many times. Here are some tips to help you manage longer problems.

- Get a global or overall understanding of the problem you are solving.
- Make a prediction or estimate of the solution.
- Get a global or overall understanding of the process you are using to solve the problem.
- List in your own words (as briefly as possible) the steps of the process.
- Focus on one step at a time.
- Examine the solution to see if it matches your prediction or estimate.

SELF-STUDY EXERCISES 15–3

1 Solve the systems of equations using the substitution method.

1. $2a + 2b = 60$
  $a = 10 + b$

2. $7r + c = 42$
  $3r - 8 = c$

3. $x - 35 = -2y$
  $3x - 2y = 17$

4. $x + y = 12$
  $x = 2 + y$

5. $2p + 3k = 2$
  $2p - 3k = 0$

6. $x + 2y = 5$
  $x = 3y$

7. $x - 3 = 2y$
  $x = 3y - 2$

8. $a = 3x - 1$
  $x = a + 5$

15–4 PROBLEM SOLVING USING SYSTEMS OF EQUATIONS

Learning Outcome

1 Use a system of equations to solve application problems.

1 Use a System of Equations to Solve Application Problems.

Many job-related problems can be solved by setting up and solving systems of equations.

EXAMPLE A television repair person purchased 10 pairs of rabbit-ear antennas and four power-supply lines for $48. Soon after, the repair person purchased another three pairs of rabbit-ear antennas and five power-supply lines for $22. How much did each antenna and each power-supply line cost?

Known facts 10 pairs of antennas and 4 power-supply lines cost $48
3 antennas and 5 power-supply lines cost $22

Unknown facts Cost of one antenna (a)
Cost of one power-supply line (l)

Relationships Total cost = Number of items × Cost of each item

$10a + 4l = \$48$ Equation 1.

$3a + 5l = \$22$ Equation 2.

Estimation In Equation 1, 14 items cost $48, which averages to less than $4 per item.
In Equation 2, 8 items cost $22, which also averages to less than $4 per item.
The more expensive item should be approximately $4 and the other item should be less than $4.

$$10a + 4l = \$48 \qquad \text{Eliminate } a \text{ by multiplying Equation 1 by 3 and Equation 2 by } -10.$$

$$3a + 5l = \$22$$

$$30a + 12l = 144 \qquad \text{Add equations and solve.}$$

$$-30a - 50l = -220$$

$$-38l = -76$$

$$l = \frac{-76}{-38}$$

$$l = 2$$

A power-supply line costs \$2.00. Substituting into equation 1, we get

$$10a + 4l = 48$$

$$10a + 4(2) = 48$$

$$10a + 8 = 48$$

$$10a = 48 - 8$$

$$10a = 40$$

$$a = \frac{40}{10}$$

$$a = 4$$

Interpretation

Rabbit-ear antennas cost \$4.00 each and a supply line costs \$2.00.

EXAMPLE A restaurant ordered three hampers of blue crabs and one hamper of shrimp that together weighed 89 lb. In another order, two hampers of shrimp and five hampers of blue crabs weighed 160 lb. How much did one hamper of crabs weigh and how much did one hamper of shrimp weigh?

Known facts

3 hampers of blue crabs and 1 hamper of shrimp weigh 89 lb
5 hampers of blue crabs and 2 hampers of shrimp weigh 160 lb

Unknown facts

What is the weight of one hamper of blue crabs (c)?
What is the weight of one hamper of shrimp (s)?

Relationships

Total weight = Number of containers $\times$ Weight of each container

$$3c + 1s = 89 \qquad \text{Equation 1.}$$

$$5c + 2s = 160 \qquad \text{Equation 2.}$$

Estimation

In Equation 1, 4 hampers weigh 89 lbs, which averages to more than 20 lb per hamper. In Equation 2, 7 hampers weigh 160 lb, which also averages to more than 20 lb per hamper. One type of seafood is expected to weigh more than 20 lb and the other type is expected to weigh 20 lb or less.

Calculations

$$3c + 1s = 89 \qquad \text{Eliminate } s \text{ by multiplying Equation 1 by } -2.$$

$$5c + 2s = 160$$

$$-6c - 2s = -178$$

$$5c + 2s = 160$$

$$-c = -18$$

$$c = 18$$

$$5c + 2s = 160 \qquad \text{Substitute for } c \text{ in Equation 2.}$$

$$5(18) + 2s = 160$$
$$90 + 2s = 160$$
$$2s = 160 - 90$$
$$2s = 70$$
$$s = \frac{70}{2}$$
$$s = 35$$

Interpretation

One hamper of blue crabs weighed 18 lb; one hamper of shrimp weighed 35 lb.

EXAMPLE Two dry cells connected in series have a total internal resistance of 0.09 ohm. The difference between the internal resistances of the two dry cells is 0.03 ohm. How much is each internal resistance?

Known facts The total internal resistance of the two dry cells is 0.09 ohm
The difference in internal resistances of the two dry cells is 0.03 ohm

Unknown facts What is the resistance of dry cell 1 (r_1)?
What is the resistance of dry cell 2 (r_2)?

Relationships $r_1 + r_2 = 0.09$ Equation 1.

$r_1 - r_2 = 0.03$ Equation 2.

Estimation If resistances were the same, they would each be 0.045 ohm. Because they are not the same, one will be more than 0.045 and one will be less than 0.045.

Calculations
$$r_1 + r_2 = 0.09$$
$$\underline{r_1 - r_2 = 0.03} \qquad \text{Add and solve.}$$
$$2r_1 = 0.12$$
$$r_1 = \frac{0.12}{2}$$
$$r_1 = 0.06$$
$$r_1 + r_2 = 0.09 \qquad \text{Substitute for } r_1 \text{ in Equation 1.}$$
$$0.06 + r_2 = 0.09$$
$$r_2 = 0.09 - 0.06$$
$$r_2 = 0.03$$

Interpretation

Thus, the larger internal resistance is 0.06 ohm, and the smaller internal resistance is 0.03 ohm.

EXAMPLE A tank holds a solution that is 10% herbicide. Another tank holds a solution that is 50% herbicide. If a farmer wants to mix the two solutions to get 200 gallons of a solution that is 25% herbicide, how many gallons of each solution should be mixed?

Known facts There are two strengths of herbicide, 10% and 50%.
200 gallons of 25% herbicide are needed.

Unknown facts How many gallons of 10% herbicide (h) are needed?
How many gallons of 50% herbicide (H) are needed?

Relationships 200 gallons of the new herbicide are needed: $h + H = 200$
Amount of pure herbicide in h gallons of 10% herbicide: $0.10h$

Amount of pure herbicide in H gallons of 50% herbicide: $0.50H$
Amount of pure herbicide in 200 gallons of 25% herbicide: $0.25(200)$

$$h + H = 200 \qquad \text{Equation 1 (total gallons).}$$

$$0.10h + 0.50H = 0.25(200) \qquad \text{Equation 2 (gallons of pure herbicide).}$$

Estimation

If equal amounts of herbicide were needed, we would need 100 gallons of each solution. However, since the desired solution strength is not exactly halfway between the two original herbicide strengths, we will need unequal amounts of herbicide. One amount will be less than 100 gallons and the other will be more than 100 gallons.

Calculations

Solve by the substitution method.

$$h + H = 200 \qquad \text{Solve Equation 1 for } h.$$

$$h = 200 - H \qquad \text{Substitute in Equation 2.}$$

$$0.10h + 0.50H = 0.25(200)$$

$$0.10(200 - H) + 0.50H = 0.25(200) \qquad \text{Substitute in Equation 2: } h = 200 - H.$$

$$20 - 0.10H + 0.50H = 50$$

$$20 + 0.40H = 50$$

$$0.40H = 50 - 20$$

$$0.40H = 30$$

$$H = \frac{30}{0.40}$$

$$H = 75 \text{ gal}$$

$$h + H = 200 \qquad \text{Substitute 75 for } H \text{ in Equation 1.}$$

$$h + 75 = 200$$

$$h = 200 - 75$$

$$h = 125 \text{ gal}$$

Interpretation

The farmer must mix 75 gallons of the 50% solution and 125 gallons of the 10% solution to make 200 gallons of a 25% solution.

EXAMPLE

Rosita has $5,500 to invest and for tax purposes wants to earn exactly $500 interest for 1 year. She wants to invest part at 10% and the remainder at 5%. How much must she invest at each interest rate to earn exactly $500 interest in 1 year?

Let $x =$ the amount invested at 10%. Let $y =$ the amount invested at 5%. Interest for one year $=$ rate $\times$ amount invested. Remember to convert percents to decimals. Using these relationships, we derive a system of equations.

Known facts

Total of $5,500 to be invested
$500 interest to be earned in one year

Unknown facts

How much should be invested at 10%?
How much should be invested at 5%?

Relationships

Amount invested at 10%: x
Interest earned at 10%: $0.1x$
Amount invested at 5%: y
Interest earned at 5%: $0.05y$

$$x + y = 5,500 \qquad \text{Equation 1 (total investment).}$$

$$0.1x + 0.05y = 500 \qquad \text{Equation 2 (total interest in one year).}$$

CHAPTER 15   Systems of Equations

Estimation

If the total amount was invested at 10%, the interest (in one year) would be $550 (0.1 × $5,500). Since we want $500 in interest, most of the money will need to be invested at 10%.

Calculations

Solve by the substitution method.

$$x + y = 5,500 \qquad \text{Solve Equation 1 for } x.$$
$$x = 5,500 - y \qquad \text{Substitute into Equation 2.}$$
$$0.1x + 0.05y = 500$$
$$0.1(5,500 - y) + 0.05y = 500 \qquad \text{Substitute } 5,500 - y \text{ for } x.$$
$$550 - 0.1y + 0.05y = 500$$
$$550 - 0.05y = 500 \qquad -0.10 + 0.05 = -0.05$$
$$-0.05y = 500 - 550$$
$$-0.05y = -50$$
$$y = \frac{-50}{-0.05}$$
$$y = \$1,000 \text{ at } 5\%$$
$$x + y = \$5,500 \qquad \text{Substitute } \$1,000 \text{ for } y \text{ in Equation 1.}$$
$$x + 1,000 = 5,500$$
$$x = 5,500 - 1,000$$
$$x = \$4,500 \text{ at } 10\%$$

Interpretation

Rosita must invest $4,500 at 10% and $1,000 at 5% for 1 year to earn $500 interest.

SELF-STUDY EXERCISES 15–4

1 Solve the problems using systems of equations with two unknowns.

1. Two boards together are 48 in. If one board is 17 in. shorter than the other, find the length of each board.

2. A broker invested $35,000 in two different stocks. One earned dividends at 4% and the other at 5%. If a $1,570 dividend was earned on both stocks together, how much was invested in each? (*Reminder:* Change 4% to 0.04 and 5% to 0.05.)

3. A department store buyer ordered 12 shirts and 8 hats for $180 one month and 24 shirts and 10 hats for $324 the following month. What was the cost of each shirt and each hat?

4. A mechanic makes $15 on each 8-cylinder engine tune-up and $10 on each 4-cylinder engine tune-up. If the mechanic did 10 tune-ups and made a total of $135, how many 8-cylinder jobs and how many 4-cylinder jobs were completed?

5. Thirty resistors and 15 capacitors cost $12. Ten resistors and 20 capacitors cost $8.50. How much does each capacitor and resistor cost?

6. A private airplane flew 420 miles in 3 hr with the wind. The return trip against the wind took 3.5 hr. Find the rate of the plane in calm air and the rate of the wind.

7. In 1 year, John Kirk earned $660 in interest on two investments totaling $8,000. If he received 7% and 9% rates of return, how much did he invest at each rate?

8. A motorboat went 40 mi with the current in 3 hr. The return trip against the current took 4 hr. How fast was the current? What would have been the speed of the boat in calm water?

9. A visitor to south Louisiana purchased 3 lb of dark-roast pure coffee and 4 lb of coffee with chicory for $15.10 in a local supermarket. Another visitor at the same store purchased 2 lb of coffee with chicory and 5 lb of dark-roast pure coffee for $16.30. How much did each coffee cost per pound?

10. A lawn-care technician wants to spread a 200-lb seed mixture that is 50% bluegrass. If the technician has on hand a mixture that is 75% bluegrass and a mixture that is 10% bluegrass, how many pounds of each mixture are needed to make 200 lb of the 50% mixture? Round to the nearest whole lb.

11. A college bookstore received a partial shipment of 50 scientific calculators and 25 graphing calculators at a total cost of $1,300. Later the bookstore received the balance of the calculators: 25 scientific and 50 graphing at a cost of $1,775. Find the cost of each calculator.

12. A consumer received two 1-yr loans totaling $10,000 at interest rates of 10% and 15%. If the consumer paid $1,300 interest, how much money was borrowed at each rate?

13. For the first performance at the Overton Park Shell, 40 reserved seats and 80 general admission seats were sold for $2,000. For the second performance, 50 reserved seats and 90 general admission seats were sold for $2,350. What was the cost for a reserved seat and for a general admission seat?

14. A photographer has a container with a solution of 75% developer and a container with a solution of 25% developer. If she wants to mix the solutions to get 8 pt of solution with 50% developer, how many pints of each solution does she need to mix?

Electronics: Mesh Currents

Circuits are frequently solved using mesh currents to determine individual currents in a circuit. For instance, the analysis of a circuit with three currents might yield the following system of equations for I_1, I_2, and I_3.

$$I_1 - I_2 - I_3 = 0 \qquad \text{Equation 1.}$$

$$6I_1 + 4I_3 = 12 \text{ A} \qquad \text{Equation 2.}$$

$$3I_2 - 4I_3 = -3 \text{ A} \qquad \text{Equation 3.}$$

There are several ways to solve this system. One way is to solve the first equation for I_1 in terms of I_2 and I_3 and substitute that into the second equation. Then the second and third equations form a pair of equations in two unknowns, which are easy to solve using several methods. The proof is to put all the final values back into the original equation to see if they work.

$$I_1 - I_2 - I_3 = 0 \qquad \text{Rearrange Equation 1 to give } I_1 = I_2 + I_3.$$

Substituting that expression for I_1 into the second equation gives

$$6I_1 + 4I_3 = 12 \qquad \text{Equation 2.}$$

$$6(I_2 + I_3) + 4I_3 = 12 \qquad \text{Distribute and combine like terms.}$$

$$6I_2 + 10I_3 = 12 \qquad \text{Divide each term by 2.}$$

$$3I_2 + 5I_3 = 6$$

This gives the following pair from the modified Equation 2 and the original Equation 3. (Illustrated is the addition method of solving a pair of equations.)

			Substituting into Equation 3:
New Equation 2:	$3I_2 + 5I_3 = 6$ A	$3I_2 + 5I_3 = 6$	$3I_2 - 4I_3 = -3$ A
Equation 3:	$3I_2 - 4I_3 = -3$ A	$\underline{-3I_2 + 4I_3 = 3}$	$3I_2 - 4(1\text{ A}) = -3$ A
		$9I_3 = 9$	$3I_2 = -3$ A $+ 4$ A
		$I_3 = 1$ A	$3I_2 = 1$ A
			$I_2 = 0.333333$ A

Then we go back to rearranged Equation 1 and substitute values for I_2 and I_3.

$$I_1 = I_2 + I_3 = 0.333333 \text{ A} + 1 \text{ A} = +1.333333 \text{ A}$$

Thus, the solution is $I_1 = 1.333333$ A, $I_2 = 0.333333$ A, and $I_3 = 1$ A.

Proof: **Equation 1**

$$I_1 - I_2 - I_3 = 0$$
$$1.333333 \text{ A} - 0.333333 \text{ A} - 1 \text{ A} = 0$$
$$0 = 0$$

Equation 2

$$6I_1 + 4I_3 = 12 \text{ A}$$
$$6(1.333333 \text{ A}) + 4(1 \text{ A}) = 12 \text{ A}$$
$$12 \text{ A} = 12 \text{ A}$$

Equation 3

$$3I_2 - 4I_3 = -3 \text{ A}$$
$$3(0.333333 \text{ A}) - 4(1 \text{ A}) = -3 \text{ A}$$
$$-3 \text{ A} = -3 \text{ A}$$

Exercises

Solve each systems of equations, which are derived from actual circuits. Prove your answers by going back to the original equations.

1. $I_1 - 2I_2 + 3I_3 = 4$ A
$2I_1 + I_2 - 4I_3 = 3$ A
$I_1 + 2I_3 = 8$ A

2. $I_1 + I_2 + I_3 = -4$ A
$2I_1 + 3I_2 + 4I_3 = 0$
$-I_1 - I_2 + 2I_3 = 8$ A

3. $I_1 + I_2 + I_3 = 4$ A
$2I_1 + 3I_2 + 4I_3 = 0$
$-I_1 - I_2 + 2I_3 = 8$ A

4. $3I_1 + 2I_2 + 2I_3 = 3$ A
$2I_1 + 6I_2 + 3I_3 = 0$
$I_1 + 2I_2 + I_3 = 1$ A

5. $I_1 - I_2 + 3I_3 = 2$ A
$-I_1 + I_2 = 7$ A
$-I_1 + 2I_2 + 6I_3 = 4$ A

	System	Solutions	Proof
1.	$I_1 - 2I_2 + 3I_3 = 4$ A $2I_1 + I_2 - 4I_3 = 3$ A $I_1 + 2I_3 = 8$ A	$I_1 = 4$ A $I_2 = 3$ A $I_3 = 2$ A	$4 - 6 + 6 = 4$ $8 + 3 - 8 = 3$ $4 + 4 = 8$
2.	$I_1 + I_2 + I_3 = -4$ A $2I_1 + 3I_2 + 4I_3 = 0$ $-I_1 - I_2 + 2I_3 = 8$ A	$I_1 = -10.667$ A $I_2 = 5.333$ A $I_3 = 1.333$ A	$-10.667 + 5.333 + 1.333 = -4$ $-21.333 + 16 + 5.333 = 0$ $10.667 - 5.333 + 2.667 = 8$
3.	$I_1 + I_2 + I_3 = 4$ A $2I_1 + 3I_2 + 4I_3 = 0$ $-I_1 - I_2 + 2I_3 = 8$ A	$I_1 = 16$ A $I_2 = -16$ A $I_3 = 4$ A	$16 - 16 + 4 = 4$ $32 - 48 + 16 = 0$ $-16 + 16 + 8 = 8$
4.	$3I_1 + 2I_2 + 2I_3 = 3$ A $2I_1 + 6I_2 + 3I_3 = 0$ $I_1 + 2I_2 + I_3 = 1$ A	$I_1 = 3$ A $I_2 = 1$ A $I_3 = -4$ A	$9 + 2 - 8 = 3$ $6 + 6 - 12 = 0$ $3 + 2 - 4 = 1$
5.	$I_1 - I_2 + 3I_3 = 2$ A $-I_1 + I_2 = 7$ A $-I_1 + 2I_2 + 6I_3 = 4$ A	$I_1 = -28$ A $I_2 = -21$ A $I_3 = 3$ A	$-28 + 21 + 9 = 2$ $28 - 21 = 7$ $28 - 42 + 18 = 4$

ASSIGNMENT EXERCISES

Section 15–1

Solve the systems of equations by graphing.

1. $x + y = 8$
$x - y = 2$

2. $3x + 2y = 13$
$x - 2y = 7$

3. $2x + 2y = 10$
$3x + 3y = 15$

4. $x + y = 1$
$3x - 4y = 10$

5. $2x - y = 5$
$4x - 2y = 2$

6. $2x + y = 5$
$x = 3$

Section 15–2

Solve the systems of equations using the addition method.

7. $3x + y = 9$
$2x - y = 6$

8. $2a + 3b = 8$
$a - b = 4$

9. $Q = 2P + 8$
$2Q + 3P = 2$

10. $4j + k = 3$
$8j + 2k = 6$

11. $r = 2y + 6$
$2r + y = 2$

12. $3a + 3b = 3$
$2a - 2b = 6$

13. $c = 2y$
$2c + 3y = 21$

14. $2x + 4y = 9$
$x + 2y = 3$

15. $3R - 2S = 7$
$-14 = -6R + 4S$

16. $x - 3 = -y$
$2y = 9 - x$

17. $c = 2 + 3d$
$3c - 14 = d$

18. $Q - 10 = T$
$T = 2 - 2Q$

19. $x - 18 = -6y$
$4x - 0 = 3y$

20. $R + S = 3$
$S - 9 = -3R$

21. $3a - 2b = 6$
$6a - 12 = b$

22. $a - b = 2$
$a + b = 12$

23. $x + 2y = 7$
$x - y = 1$

24. $2c + 3b = 2$
$2c - 3b = 0$

25. $x + 2r = 5.5$
$2x = 1.5r$

26. $7x + 2y = 6$
$4y = 12 - x$

Section 15–3

Solve the systems of equations using either the addition or the substitution method.

27. $a + 7b = 32$
$3a - b = 8$

28. $x + y = 1$
$4x + 3y = 0$

29. $c - d = 2$
$c = 12 - d$

30. $3a + 4b = 0$
$a + 3b = 5$

31. $7x - 4 = -4y$
$3x + y = 6$

32. $5Q - 4R = -1$
$R + 3Q = -38$

33. $a = 2b + 11$
$3a + 11 = -5b$

34. $y = 5 - 2x$
$3x - 2y = 4$

35. $c = 2q$
$2c + q = 2$

36. $3x + 2y = 10$
$y = 6 - x$

37. $4x - 2.5y = 2$
$2x - 1.5y = -10$

38. $2a - c = 4$
$a = 2 + c$

39. $4d - 7 = -c$
$3c - 6 = -6d$

40. $x = 10 - y$
$5x + 2y = 11$

41. $3.5a + 2b = 2$
$0.5b = 3 - 1.5a$

42. $c + d = 12$
$c - d = 2$

43. $x + 4y = 20$
$4x + 5y = 58$

44. $a + 5y = 7$
$a + 4y = 8$

45. $3a + 1 = -2b$
$4b + 23 = 15a$

46. $6y + 0 = -5p$
$4y - 3p = 38$

Section 15–4

Solve the problems using systems of equations with two unknowns.

47. Three electricians and four apprentices earned a total of \$365 on one job. At the same rate of pay, one electrician and two apprentices earned a total of \$145. How much pay did each apprentice and electrician receive?

48. Six bushels of bran and 2 bushels of corn weigh 182 lb. If 2 bushels of bran and 4 bushels of corn weigh 154 lb, how much do 1 bushel of bran and 1 bushel of corn weigh each?

49. A painter paid \$22.50 for 2 qt of white shellac and 5 qt of thinner. If 3 qt of shellac and 2 qt of thinner cost the painter's helper \$14.50, what is the cost of each qt of shellac and thinner?

50. A main current of electricity is the sum of two smaller currents whose difference is 0.8 A. What are the two smaller currents if the main current is 10 A?

51. The sum of two angles is 175°. Their difference is 63°. What is the measure of each angle?

52. A johnboat traveled 20 mi in 2 hr with the current. The return trip took 3 hr. Find the rate of the boat in calm water and the rate of the current.

53. In 1 year, Sholanda Brown earned \$560 on two investments totaling \$5,000. If she received 10% and 12% rates of return, how much did she invest at each rate?

54. A plane flew 300 km against the wind in 4 hr. The return trip with the wind took 3 hr. How fast was the wind? What would have been the speed of the plane in calm air?

55. A restaurant purchased 30 lb of Colombian coffee and 10 lb of blended coffee for \$190. In a second purchase, the same restaurant paid \$120 for 20 lb of Colombian coffee and 5 lb of blended coffee. How much did each coffee cost per lb?

56. A rancher wants to spread a 300-lb grass seed mixture that is 50% tall fescue. If the rancher has on hand a seed mixture that is 80% tall fescue and a mixture that is 20% tall fescue, how many pounds of each mixture are needed to make 300 lb of the 50% mixture?

57. An automotive service station purchased 25 maps of Ohio and 8 maps of Alaska at a total cost of \$65.55. Later the station purchased 20 maps of Ohio and 5 maps of Alaska for \$49.50. Find the cost of each map.

58. Bev Witonski made two 1-yr investments totaling \$7,000 at interest rates of 8% and 12%. If she received \$760 in return, how much money was invested at each rate?

59. At the first of the month, store buyer Kathy Miller placed a \$12,525 order for 20 name-brand suits and 35 suits with generic labels. At the end of the month, she placed a \$15,725 order for 30 name-brand suits and 35 generic-label suits. How much did she pay for each type of suit?

60. A taxidermist has a container with a solution of 10% tanning chemical and a container with a solution of 50% tanning chemical. If the taxidermist wants to mix the solutions to get 10 gal of solution with 25% tanning chemical, how many gal of each solution should be mixed?

61. Jorge makes 5% commission on telephone sales and 6% commission on showroom sales. If his sales totaled \$40,000 and his commission was \$2,250, how much did he sell by telephone? How much did he sell on the showroom floor?

62. Sing-Fong has 60 coins in nickels and quarters. The total value of the coins is \$12. How many coins of each type does she have?

63. How many gal of 75% fertilizer and 25% fertilizer are needed to make a mixture of 8 gal of 50% fertilizer?

Examine the problems to see whether they are worked correctly. If there are errors, make corrections.

64. Solve by addition: $a - b = 6$ and $a + b = 2$.

$$
\begin{array}{lll}
a - b = 6 & 4 - b = 6 & \text{Check:} \quad a - b = 6 \\
a + b = 2 & -b = 6 - 4 & 4 - (-2) = 6 \\
\hline
2a \quad\;\; = 8 & -b = 2 & 4 + 2 = 6 \\
a = 4 & b = -2 & 6 = 6
\end{array}
$$

65. Solve by addition: $2c + 3d = 9$ and $3c + d = 10$.

$$
\begin{array}{ll}
2c + 3d = 9 & 2c + 3d = 9 \\
3c + d = 10 \quad \text{Multiply by } -3 & 2(5.25) + 3d = 9 \\
2c + 3d = 9 & 10.5 + 3d = 9 \\
-6c - 3d = -30 & 3d = 9 - 10.5 \\
\hline
-4c \quad\quad = -21 & 3d = -1.5 \\
\dfrac{-4c}{-4} = \dfrac{-21}{-4} & \dfrac{3d}{3} = -\dfrac{1.5}{3} \\
c = 5.25 & d = -0.5
\end{array}
$$

<hr>

CHALLENGE PROBLEMS

66. Write an applied mixture problem that can be solved with the given system of equations with two unknowns. Solve the system and check the results.

$$0.5x + 0.3y = 8$$
$$x + y = 20$$

67. Write an applied problem that can be solved with the given system of equations with two unknowns. Solve the system and check the results.

$$3x + 5y = 28$$
$$x - y = 4$$

68. Solve Exercise 51 by writing one equation with one variable.

<hr>

CHAPTER TRIAL TEST

Solve the systems of equations graphically.

1. $2a + b = 10$
$\quad a - b = 5$

2. $x + y = 8$
$\quad 3x + 2y = 12$

3. $3x + 4y = 6$
$\quad x + y = 5$

4. $a - 3b = 7$
$\quad a - 5 = b$

5. $2c - 3d = 6$
$\quad c - 12 = 3d$

Solve the systems of equations using the addition method.

6. $x + y = 6$
$\quad x - y = 2$

7. $p + 2m = 0$
$\quad 2p = -m$

8. $6p + 5t = -16$
$\quad 3p - 3 = 3t$

9. $3x + y = 5$
$\quad 2x - y = 0$

10. $7c - 2b = -2$
$\quad c - 4b = -4$

Solve the systems of equations with two unknowns using the substitution method.

11. $4x + 3y = 14$
$\quad x - y = 0$

12. $a + 2y = 6$
$\quad a + 3y = 3$

13. $7p + r = -6$
$\quad 3p + r = 6$

Solve the systems of equations with two unknowns using either the addition or the substitution method.

14. $4x + 4 = -4y$
$6 + y = -6x$

15. $38 + d = -3a$
$5a + 1 = 4d$

Solve the problems using systems of equations with two unknowns.

16. Two lengths of stereo speaker wire total 32.5 ft. One length is 2.9 ft longer than the other. How long is each length of speaker wire?

17. Two currents add to 35 A and their difference is 5 A. How many amperes are in each current?

18. Six packages of common nails and four packages of finishing nails weigh 6.5 lb. If two packages of common nails and three packages of finishing nails weigh 3.0 lb, how much does one package of each kind of nail weigh?

19. The length of a piece of sheet metal is $1\frac{1}{2}$ times the width. The difference between the length and the width is 17 in. Find the length and the width.

20. A mixture of fieldstone is needed for a construction job and will cost $378 for the 27 tons of stone. The stone is of two types, one costing $18 per ton and one costing $12 per ton. How many tons of each are required?

21. A broker invested $25,000 in two different stocks. One stock earned dividends at 11% and the other at 12.5%. If a dividend of $3,050 was earned on both stocks together, how much was invested in each stock?

22. A mason purchased 2-in. cold-rolled channels and $\frac{3}{4}$-in. cold-rolled channels whose total weight was 820 lb. The difference in weight between the heavier 2-in. and lighter $\frac{3}{4}$-in. channels was 280 lb. How many pounds of each type of channel did the mason purchase?

23. A total capacitance in parallel is the sum of two capacitances. If the capacitance totals 0.00027 farad (F) and the difference between the two capacitances is 0.00016 F, what is the value of each capacitance in the system?

16

Selected Concepts of Geometry

GOOD DECISIONS THROUGH TEAMWORK

What's the best pizza deal in town? How do you shop comparatively when different pizza stores have different size pans? Are prices for some large pizzas for a particular store proportional to the amount of pizza for each size? Does any combination of two pan sizes give a better buy than a larger pan size? Is pizza available in shapes besides circles? If so, what are the advantages and disadvantages of these shapes from a marketing viewpoint? From a consumer viewpoint?

To answer these and other questions, your team will collect statistics on pan sizes, prices, topping options, and other information affecting a decision to purchase pizza. As a class, before you begin your research, rate the pizza available in your area based on quality or taste and on your initial impressions of which pizza store has the best deal in town.

To begin your research, compare cost per square inch of pizza for each pan size offered by a restaurant. Then compare cost per square inch of pizza with customer-chosen toppings versus store-established combinations of toppings. Compare cost per square inch of pizza for small, medium, and large pizzas among various restaurants.

After the information is collected, discuss your findings among team members to determine how it will be organized for reporting team results. Charts, tables, graphs, lists, and narrative are some choices you may want to consider for your presentation.

16–1 Basic terminology and notation

1. Use various notations to represent points, lines, line segments, rays, and planes.
2. Distinguish among lines that intersect, that coincide, and that are parallel.
3. Use various notations to represent angles.
4. Classify angles according to size.

16–2 Angle calculations

1. Add and subtract angle measures.
2. Change minutes and seconds to a decimal part of a degree.
3. Change the decimal part of a degree to minutes and seconds.
4. Multiply and divide angle measures.

16–3 Triangles

1. Classify triangles by sides.
2. Relate the sides and angles of a triangle.
3. Determine if two triangles are congruent using inductive and deductive reasoning.
4. Use the properties of a 45°, 45°, 90° triangle to find missing parts and to solve applied problems.
5. Use the properties of a 30°, 60°, 90° triangle to find missing parts and to solve applied problems.

16–4 Polygons

1. Find the missing dimensions of composite figures.
2. Find the perimeter and area of composite figures.
3. Find the number of degrees in each angle of a regular polygon.

16–5 Sectors and segments of a circle

1. Find the area of a sector.
2. Find the arc length of a sector.
3. Find the area of a segment.

16–6 Inscribed and circumscribed regular polygons and circles

1. Use the properties of inscribed and circumscribed equilateral triangles to find missing amounts and to solve applied problems.
2. Use the properties of inscribed and circumscribed squares to find missing amounts and to solve applied problems.
3. Use the properties of inscribed and circumscribed hexagons to find missing amounts and to solve applied problems.

Geometry is one of the oldest and most useful of the mathematical sciences. *Geometry* involves the study and measurement of shapes according to their sizes, volumes, and positions. A knowledge of geometry is necessary in many careers. Notice how the meanings of common words you are already familiar with change when they are precisely defined for geometry.

16–1   BASIC TERMINOLOGY AND NOTATION

Learning Outcomes

1. Use various notations to represent points, lines, line segments, rays, and planes.
2. Distinguish among lines that intersect, that coincide, and that are parallel.
3. Use various notations to represent angles.
4. Classify angles according to size.

1   Use Various Notations to Represent Points, Lines, Line Segments, Rays, and Planes.

Geometry is the study of size, shape, position, and other properties of the objects around us. The basic terms used in geometry are *point, line,* and *plane.* Generally, these terms are not defined. Instead, they are only described. Once described, they are used in definitions of other terms and concepts.

A *point* is a location or position that has no size or dimension. A dot is used to represent a point, and a capital letter is usually used to label the point (Fig. 16–1).

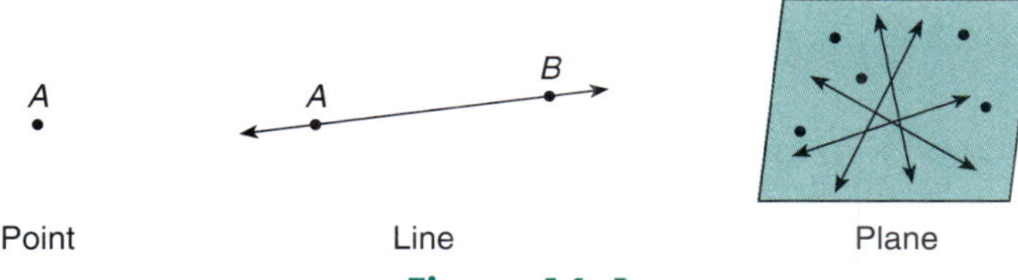

Figure 16–1

A *line* extends indefinitely in both directions and contains an infinite number of points. It has length but no width. In our discussions, the word *line* always refers to a straight line unless otherwise specified. In Fig. 16–1, the line can be identified by naming any two points on the line (such as *A* and *B*).

A *plane* is a flat, smooth surface that extends indefinitely in all directions. A plane contains an infinite number of points and lines (Fig. 16–1).

Since a line extends indefinitely in both directions, most geometric applications deal with parts of lines. A part of a line is called a *line segment* or *segment*. A line segment starts and stops at distinct points that we call *end points*. A line segment is defined as follows:

Figure 16–2

■ **DEFINITION: Line Segment** or **Segment.** A *line segment*, or *segment*, consists of all points on the line between and including two points that are called end points (Fig. 16–2).

The notation for a line that extends through points A and B is $\overleftrightarrow{AB}$ (read "line AB"). The notation for the line segment including points A and B and all the points between is $\overline{AB}$ (read "line segment AB").

Another term used in connection with parts of a line is *ray*. Before we give the definition of a ray, consider the beam of light from a flashlight. The beam is like a ray. It seems to continue indefinitely in only one direction.

Figure 16–3

■ **DEFINITION: Ray.** A *ray* consists of a point on a line and all points of the line on one side of the point (Fig. 16–3).

The point from which the ray originates is called the *end point,* and all other points on the ray are called *interior points* of the ray. A ray is named by its end point and any interior point on the ray. In Fig. 16–3, we use the notation $\overrightarrow{RS}$ to denote the ray whose end point is R and that passes through S.

To see the contrast in the notation used for a line, line segment, and ray, look at Fig. 16–4.

To illustrate appropriate notations for lines, segments, and rays, consider a line with several points designated on the line (Fig. 16–5).

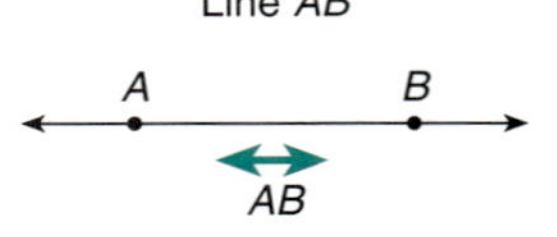

Line *AB*

Line segment *AB*

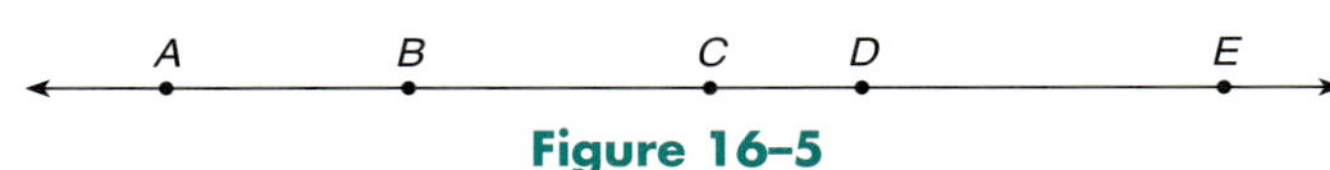

Figure 16–5

Ray *AB*

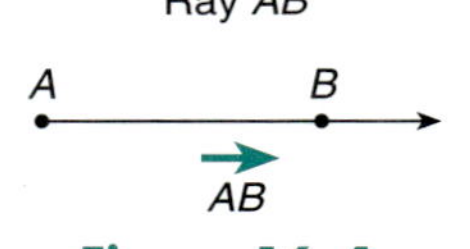

Figure 16–4

Any two points can be used to name the line in Fig. 16–5. For example, $\overleftrightarrow{AB}$, $\overleftrightarrow{AC}$, $\overleftrightarrow{CE}$, $\overleftrightarrow{BD}$, and $\overleftrightarrow{BC}$ are some of the possible ways to name the line. However, a segment is named *only* by its end points. Thus, in Fig. 16–5, $\overline{AB}$ is not the same segment as $\overline{AC}$, but $\overleftrightarrow{AB}$ and $\overleftrightarrow{AC}$ represent the same line.

In Fig. 16–5, $\overrightarrow{BC}$ and $\overrightarrow{BD}$ represent the same ray, but $\overline{BC}$ and $\overline{BD}$ do not represent the same segment.

■ **Learning Strategy**   *Using Notation as Clues.*

Notation is important in geometry because it allows us to shorten the amount of writing needed to describe a situation, and it allows us to recognize or recall certain properties at a glance.

Use geometric notations as visual clues for understanding the notation.

$\overleftrightarrow{AB}$ line continues indefinitely in both directions.

$\overrightarrow{AB}$ ray continues indefinitely in one direction.

$\overline{AB}$ line segment starts and stops at specific points.

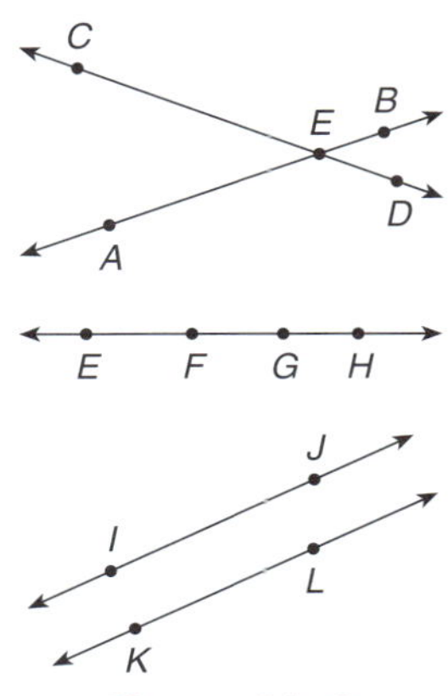

Figure 16–6

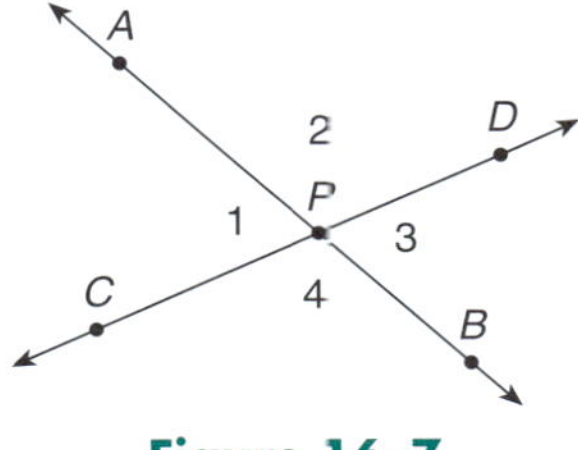

Figure 16–7

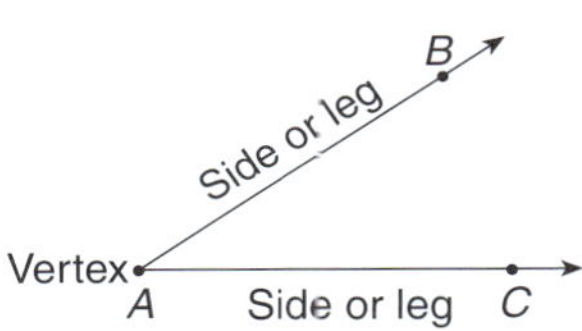

Figure 16–8

2 **Distinguish Among Lines That Intersect, That Coincide, and That Are Parallel.**

A line can be extended indefinitely in either direction. If two lines are drawn in the same plane, three things can happen:

1. The two lines *intersect* in *one and only one point*. In Fig. 16–6, $\overleftrightarrow{AB}$ and $\overleftrightarrow{CD}$ intersect at point E.
2. The two lines *coincide;* that is, one line fits exactly on the other. In Fig. 16–6, $\overleftrightarrow{EF}$ and $\overleftrightarrow{GH}$ coincide.
3. The two lines never intersect. In Fig. 16–6, $\overleftrightarrow{IJ}$ and $\overleftrightarrow{KL}$ are the same distance from each other along their entire lengths and so never touch.

The relationship described in the third situation has a special name, *parallel lines*. The symbol ‖ is used for parallel lines. (See Section 16–4.)

3 **Use Various Notations to Represent Angles.**

This outcome deals with a special intersection of straight lines—the angle. Most of us are familiar with angles in everyday life. The corners of a soccer field are angles. The lines between bricks and tiles form angles, and so on. This section and the ones that follow introduce formally the study of angles for technical applications.

When two lines intersect in a point, four *angles* are formed, as shown in Fig. 16–7.

■ **DEFINITION: Angle.** An *angle* is a geometric figure formed by two rays that intersect in a point, and the point of intersection is the end point of each ray.

In Fig. 16–8, rays $\overrightarrow{AB}$ and $\overrightarrow{AC}$ intersect at point A. Point A is the end point of $\overrightarrow{AB}$ and $\overrightarrow{AC}$. $\overrightarrow{AB}$ and $\overrightarrow{AC}$ are called the *sides* or *legs* of the angle. Point A is called the *vertex* of the angle.

Angles can be named in several ways. An angle can be named using a number or lowercase letter, by the capital letter that names the vertex point, or by using three capital letters. If three capital letters are used, two of the letters name interior points of each of the two rays, and the middle letter names the vertex point of the angle. Using the symbol $\angle$ for angle, the angle in Fig. 16–9 can be named $\angle 1$, $\angle KLM$, $\angle MLK$, or $\angle L$. If one capital letter is used to name an angle, the letter is always the vertex letter. One capital letter is used only when it is perfectly clear which angle is designated by the letter. If three letters are used, the vertex letter is the center letter. To name the angle in Fig. 16–10 with three letters, we write $\angle XZY$ or $\angle YZX$. This angle can also be named $\angle a$ or $\angle Z$.

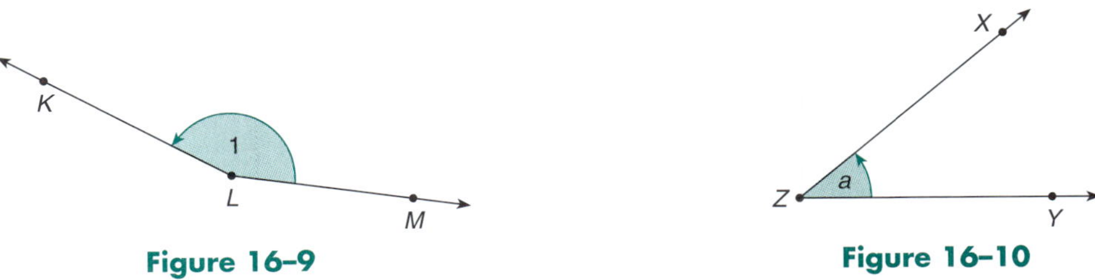

Figure 16–9

Figure 16–10

Figure 16–11 illustrates how the intersection of two rays actually forms two angles. In this text, we refer to the smaller of the two angles formed by two rays unless the other angle is specifically indicated. Arcs (curved lines) and arrows are often used to clarify which angle we are considering.

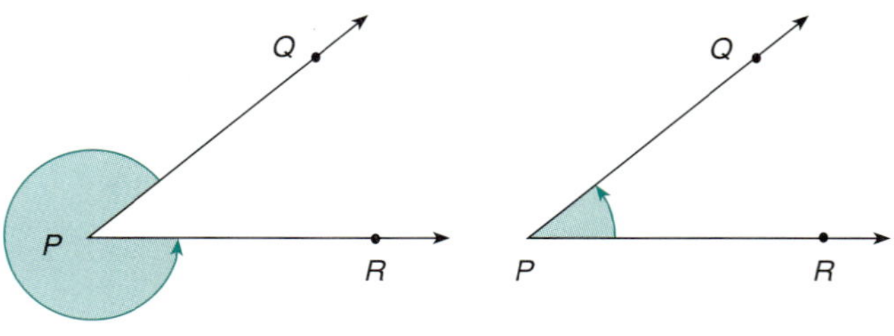

Figure 16–11

4 Classify Angles According to Size.

The measure of an angle is determined by the amount of opening between the two sides of the angle. The length of the sides does not affect the angle measure. Two units are commonly used to measure angles, *degrees* and *radians*. In this chapter, only degrees are used to measure angles. Radians are discussed in Chapter 17.

Consider the hands of a clock as the sides of an angle. When the two hands both point to the same number, the measure of the angle formed is 0 degrees (0°). An angle of 0 degrees is used in trigonometry but is seldom used in geometric applications. During 1 hour, the minute hand makes one complete revolution. If we ignore the movement of the hour hand, this revolution of the minute hand contains 360 degrees (360°). Figure 16–12 shows a revolution or rotation of 360°. Note that A is kept as a fixed point and B rotates around point A then back to its original position. This rotation can be either clockwise or counterclockwise. A complete rotation counterclockwise is +360°. A complete rotation clockwise is −360°.

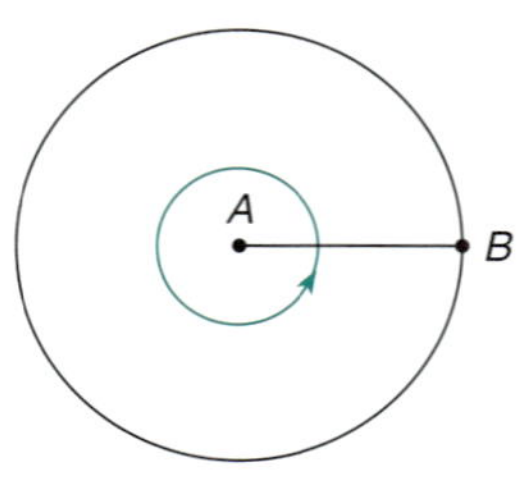

Figure 16–12

Tip!	*Clockwise and Counterclockwise.*

The customary rotation around a point progresses *opposite* to the normal rotation of the hands of a clock. The direction is called *counterclockwise* (Fig. 16–13).

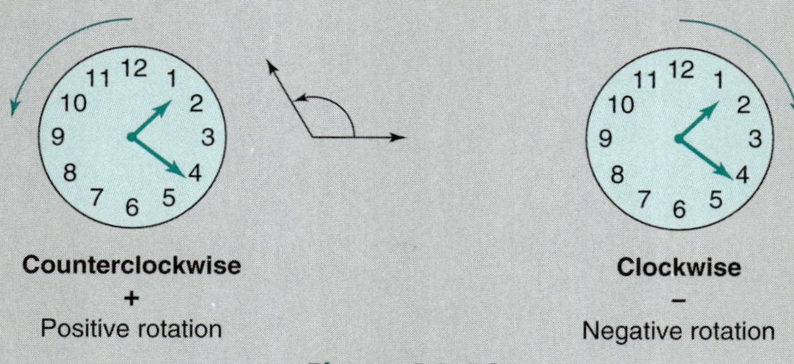

Figure 16–13

Sometimes, traditional notations or conventions are not always logical.

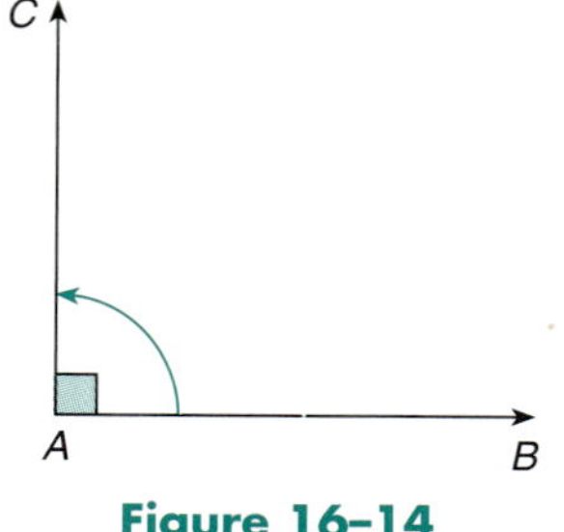

Figure 16–14

■ **Definition:** **Degree.** A *degree* is a unit for measuring angles. It represents $\frac{1}{360}$ of a complete rotation about the vertex.

Suppose, in Fig. 16–14, that $\overrightarrow{AC}$ rotates from $\overrightarrow{AB}$ through one-fourth of a circle. Then $\overrightarrow{AC}$ and $\overrightarrow{AB}$ form a 90° angle ($\frac{1}{4}$ of 360 = 90). This angle is a *right angle*. The symbol for a right angle is ⌐.

If two lines intersect so that right angles are formed (90° angles), the lines are *perpendicular* to each other. (See Section 14–5.) The symbol for "perpendicular" is ⊥. However, the right-angle symbol also implies the lines forming the angle are perpendicular.

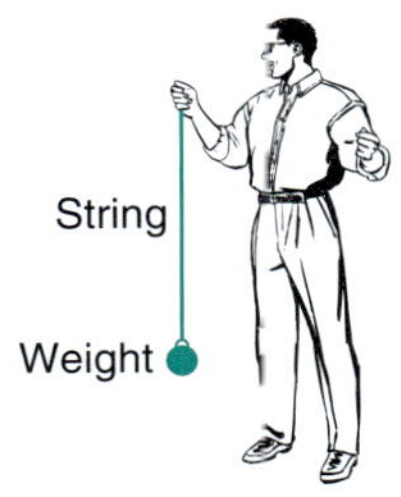

Figure 16–15

If a string is suspended at one end and weighted at the other (Fig. 16–15), the line it forms is a *vertical line*. A line that is perpendicular to the vertical line is a *horizontal line*. In Fig. 16–16, $\overleftrightarrow{AB}$ is a vertical line, and $\overleftrightarrow{AB}$ and $\overleftrightarrow{CD}$ form right angles. Thus, $\overleftrightarrow{AB} \perp \overleftrightarrow{CD}$, and $\overleftrightarrow{CD}$ is a horizontal line.

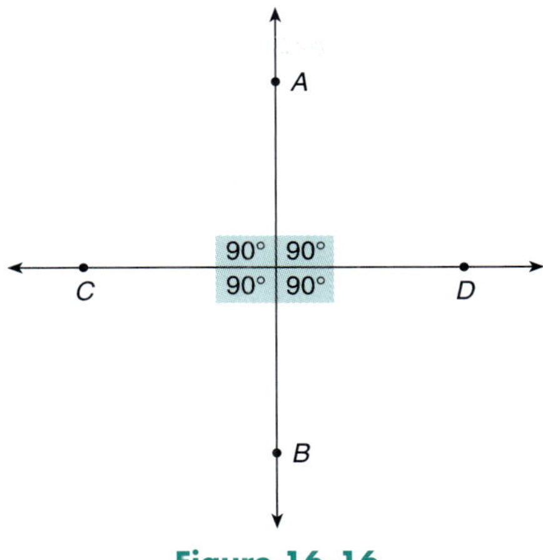

Figure 16–16

Now look at Fig. 16–17. When $\overrightarrow{ML}$ rotates half a circle from $\overrightarrow{MN}$, an angle of 180° is formed ($\frac{1}{2}$ of 360 = 180). This angle is a *straight angle*.

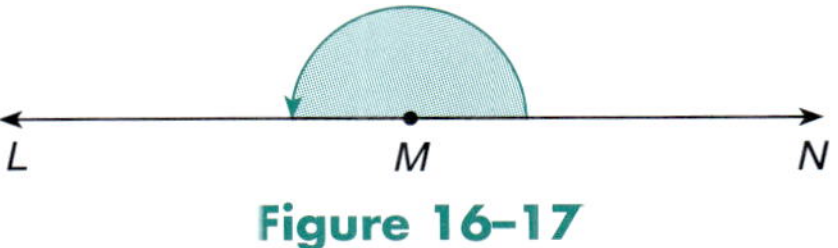

Figure 16–17

We use these two special angles, the right angle (90°) and the straight angle (180°), to define two sets of angles that occur frequently in geometry. An angle that is less than 90° but more than 0° is an *acute angle*. An angle that is more than 90° but less than 180° is an *obtuse angle* (Fig. 16–18).

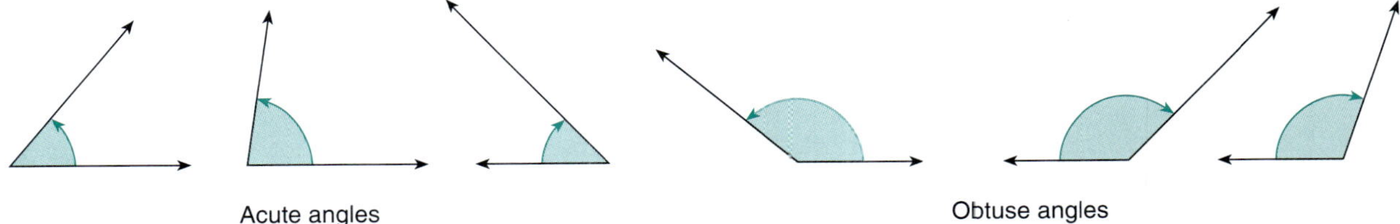

Figure 16–18

The definitions below summarize the classification of angles according to size.

■ **DEFINITION:** **Right Angle.** A *right angle* (90°) represents one-fourth of a circle or one-fourth of a complete rotation.

■ **DEFINITION:** **Straight Angle.** A *straight angle* (180°) represents half of a circle or half of a complete rotation.

■ **DEFINITION:** **Acute angle.** An *acute angle* is less than 90° but more than 0°.

■ **DEFINITION:** **Obtuse Angle.** An *obtuse angle* is more than 90° but less than 180°.

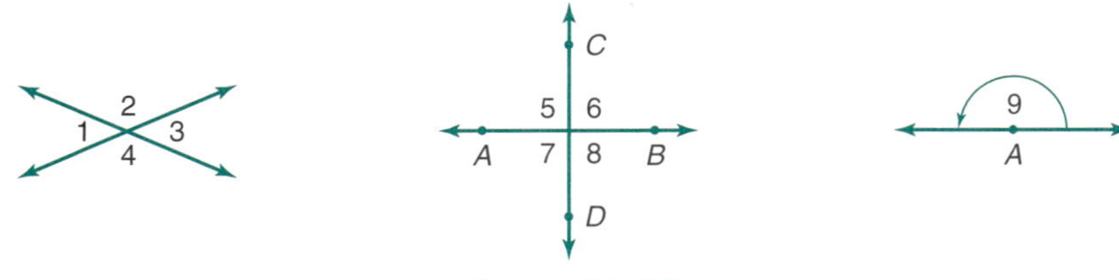

Figure 16–19

∠**1 is acute.** $\overleftrightarrow{AB} \perp \overleftrightarrow{CD}$ ∠**9 is a straight angle.**
∠**2 is obtuse.**   ∠**5, ∠6, ∠7, and ∠8**
∠**3 is acute.**   **are all right angles.**
∠**4 is obtuse.**

The angle classifications used so far—right, straight, acute, and obtuse—deal with one angle at a time. If two angles together form a right angle or if their measures total 90°, they are *complementary angles.* If two angles together form a straight angle or if their measures total 180°, they are *supplementary angles.*

■ **DEFINITION: Complementary Angles.** *Complementary angles* are two angles that have a sum equivalent to one right angle or 90° (Fig. 16–20).

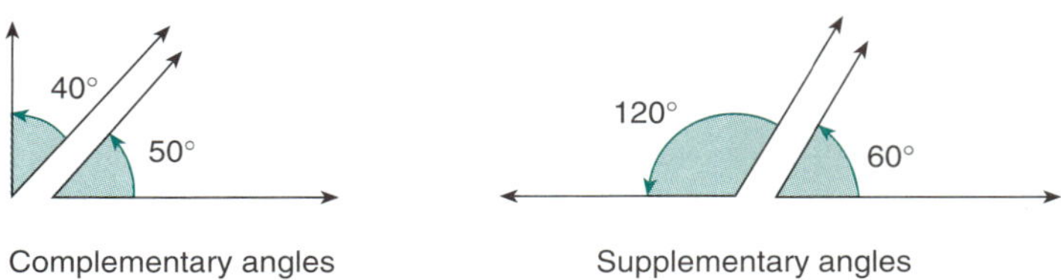

Complementary angles   Supplementary angles

Figure 16–20

■ **DEFINITION: Supplementary Angles.** *Supplementary angles* are two angles that have a sum equivalent to one straight angle or 180° (Fig. 16–20).

Tip!	***Equal Versus Congruent: Precise Terminology and Casual Conventions!***

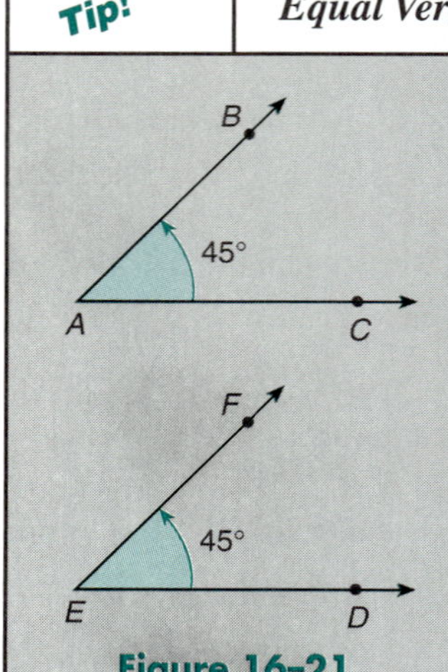

Figure 16–21

Equal Versus Congruent. When the word *equal* is used to describe the relationship between two angles, it implies the measures of the angles are equal. Another word often used in geometry is *congruent.* When geometric figures are congruent, one figure can be placed on top of the other, and the two figures match perfectly. If two angles are congruent, the measures of the angles are equal. Also, if the measures of two angles are equal, the angles are congruent. The symbol for congruence is ≅. For instance, in Fig. 16–21, the two angles have equal measures, both 45°. Because they have the same measures, they are congruent; that is, ∠*BAC* ≅ ∠*FED*.

Traditionally, we indicate the measures of angles by placing the letter *m* before the angle symbol. Thus, $m\angle BAC = m\angle FED$ expresses that the measures of angles *BAC* and *FED* are equal. Casually, we interpret ∠*BAC* = ∠*FED* and $m\angle BAC = m\angle FED$ as giving us the same information.

If you know the measure of an angle you can find the *complement* or *supplement* of the angle by subtracting from 90° or 180°, respectively.

EXAMPLE Find the complement and supplement of an angle that measures 57°.

Complement

$90° − 57° = 33°$

The complement of 57° is 33°.

Supplement

$180° − 57° = 123°$

The supplement of 57° is 123°.

SELF-STUDY EXERCISES 16–1

1 Use Fig. 16–22 for Exercises 1–3.

1. Name the line in three different ways.
2. Name in two different ways the ray with end point P and with interior points Q and R.
3. Name the segment with end points Q and R.

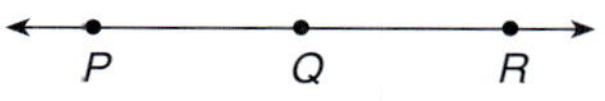

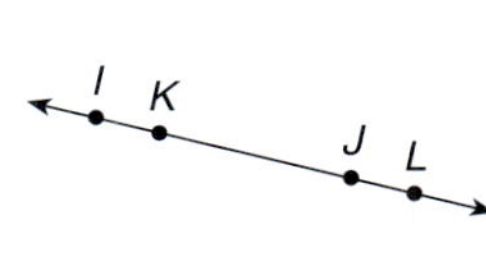
Figure 16–22

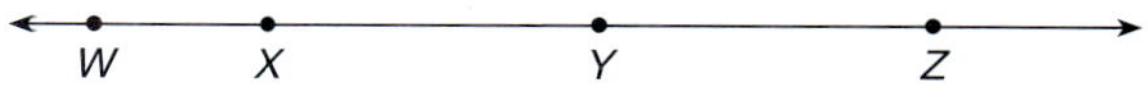

Use Fig. 16–23 for Exercises 4–10.

Figure 16–23

4. Does $\overleftrightarrow{XY}$ represent the same line as $\overleftrightarrow{YZ}$?

5. Does $\overrightarrow{XY}$ represent the same ray as $\overrightarrow{YZ}$?

6. Does $\overline{XY}$ represent the same segment as $\overline{YZ}$?

7. Is $\overleftrightarrow{WX}$ the same as $\overleftrightarrow{WY}$?

8. Is $\overrightarrow{XY}$ the same as $\overrightarrow{XZ}$?

9. Is $\overline{XY}$ the same as $\overline{YX}$?

10. Is $\overrightarrow{XW}$ the same as $\overrightarrow{XY}$?

2 Use Fig. 16–24 for Exercises 11–15.

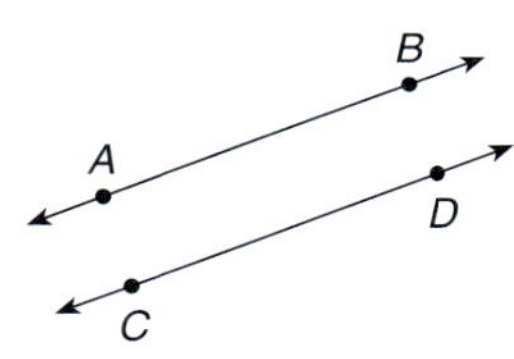

Figure 16–24

11. $\overleftrightarrow{AB}$ and $\overleftrightarrow{CD}$ are _____ lines.

12. $\overleftrightarrow{EF}$ and $\overleftrightarrow{GH}$ _____ at point O.

13. $\overleftrightarrow{IJ}$ and $\overleftrightarrow{KL}$ _____.

14. $\overleftrightarrow{AB}$ and $\overleftrightarrow{CD}$ will never _____.

15. Name two lines that intersect in exactly one point.

3 Use Fig. 16–25 for Exercises 16–18.

16. Name the angle in two different ways using three capital letters.
17. Name the angle using one capital letter.
18. Name the angle using a number.

Use Fig. 16–26 for Exercises 19–21.

19. Name the angle using one lowercase letter.
20. Name the angle using three capital letters.
21. Name the angle using one capital letter.

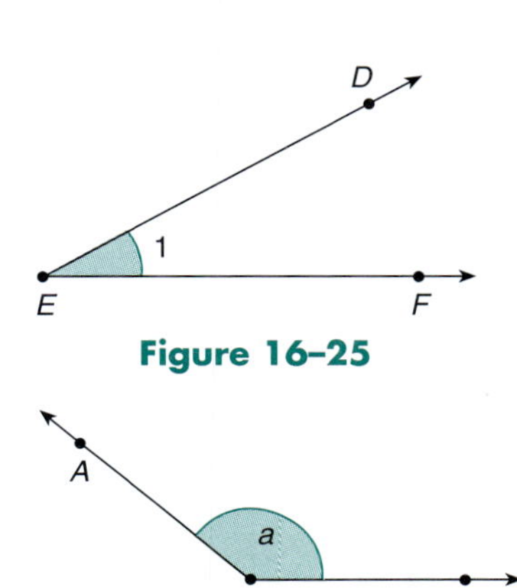

Figure 16–25

Figure 16–26

4 Fill in the blanks.

22. One complete rotation is _____°.
23. Half a complete rotation is _____°.
24. One-fourth a complete rotation is _____°.

Classify the angle measures using the terms *right, straight, acute,* or *obtuse.*

25. 38° **26.** 95° **27.** 90° **28.** 153°
29. 10° **30.** 180° **31.** 60° **32.** 120°

Tell whether the angle pairs are complementary.

33. 17°, 73° **34.** 38°, 142° **35.** 52°, 48°

Tell whether the angle pairs are supplementary.

36. 60°, 30° **37.** 110°, 70° **38.** 42°, 138° **39.** 23°, 67°
40. Two perpendicular lines form a _____° angle. **41.** Two 35° angles are said to be _____ angles.

16-2 ANGLE CALCULATIONS

Learning Outcomes

1 Add and subtract angle measures.
2 Change minutes and seconds to a decimal part of a degree.
3 Change the decimal part of a degree to minutes and seconds.
4 Multiply and divide angle measures.

A device used to measure angles is called a *protractor.* The most common protractor is a semicircle with two scales from 0° to 180°. An *index mark* is in the middle of the straight edge of the protractor (Fig. 16–27).

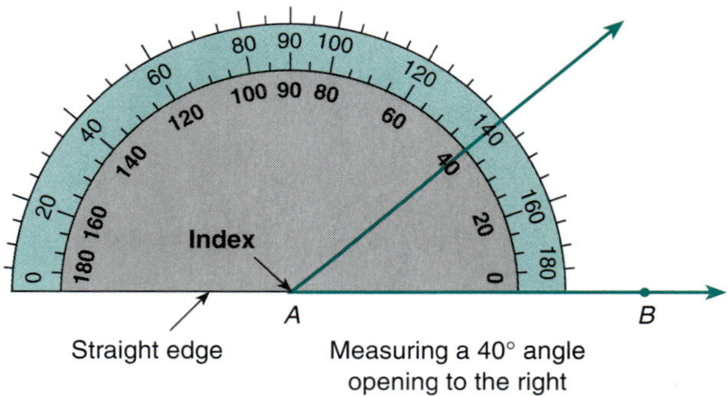

Figure 16–27

To measure an angle that opens to the right, read the degree measure from the lower scale. Notice in Fig. 16–27 that the lower scale starts with 0 at the right. The index of the protractor is aligned with the vertex of the angle, and the straight edge lies along the lower side of the angle.

To measure an angle that opens to the left, read the degree measure from the upper scale. Notice in Fig. 16–28 the upper scale starts with 0 at the left. The index of the protractor is aligned with the vertex of the angle, and the straight edge lies along the lower side of the angle.

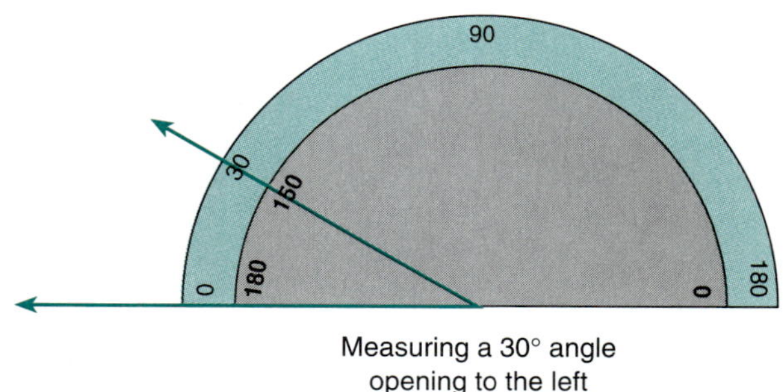

Figure 16–28

Degrees are divided into 60 equal parts. Each part is called a *minute*.

$$1 \text{ degree } (1°) = 60 \text{ minutes } (60')$$

The symbol ′ is used for minutes. Similarly, one minute is divided into 60 equal parts called *seconds*.

$$1 \text{ minute } (1') = 60 \text{ seconds } (60'')$$

The symbol ″ is used for seconds. Thus,

$$1° = 60' = 3{,}600''$$

1 Add and Subtract Angle Measures.

Angle measures can be added or subtracted. Keep in mind that only like measures can be added or subtracted.

EXAMPLE Add $12°15'54''$ and $82°28'19''$.

$$12°15'54''$$
$$+\ 82°28'19''$$

Arrange the measures in columns of like measures and add.

$$94°43'\quad 73''$$

Write in standard notation.

$$+1'\ -\ 60''$$
$$94°44'\quad 13''$$

Because $73'' = 1'13''$, simplify by adding $1'$ to the minutes column and subtracting $60''$ from the seconds column. **Thus, $94°43'73'' = 94°44'13''$.**

EXAMPLE Add $71°14'$ and $82°12''$.

$$71°14'$$
$$+\ 82°\quad\quad 12''$$

Arrange in columns of like measures and add.

$$\mathbf{153°14'12''}$$

EXAMPLE Subtract $15°32'$ from $37°15'$.

$$37°15' = 36°75'$$

Borrow $1°$ from $37°$; $1° = 60'$ and $60' + 15' = 75'$.

$$-\ 15°32' = 15°32'$$
$$\mathbf{21°43'}$$

EXAMPLE Subtract $3°12'30''$ from $15°$.

$$15° = 14°59'60''$$

Borrow $1°$ from $15°$; $1° = 60'$. Borrow $1'$ from $60'$; $1' = 60''$, then subtract.

$$-\ 3°12'30'' = 3°12'30''$$
$$\mathbf{11°47'30''}$$

EXAMPLE Find the complement of an angle of $35°25'40''$.

To find the complement of an angle, subtract the given angle measure from $90°$.

$$90° = 89°59'60''$$
$$-\ 35°25'40'' = 35°25'40''$$
$$\mathbf{54°34'20''}$$

2 **Change Minutes and Seconds
to a Decimal Part of a Degree.**

With the increased popularity of the calculator, it is sometimes necessary to
change minutes or seconds to decimal equivalents. We first change minutes or seconds to a fractional part of a degree, then we change the fraction to its decimal
equivalent by dividing the numerator by the denominator. Remember, $1' = \frac{1}{60}$ of a
degree, and $1'' = \frac{1}{3,600}$ of a degree.

To change minutes to a decimal part of a degree:

Divide minutes by 60.

$$\frac{x \text{ min}}{1} \times \frac{1°}{60 \text{ min}} = \frac{x°}{60}$$

To change seconds to a decimal part of a degree:

Divide seconds by 3,600.

$$\frac{x \text{ sec}}{1} \times \frac{1 \text{ min}}{60 \text{ sec}} \times \frac{1°}{60 \text{ min}} = \frac{x°}{3,600}$$

EXAMPLE   Change 15′ to its decimal degree equivalent.

$$\frac{15}{60} = \textbf{0.25°}$$

EXAMPLE   Change 37″ to its decimal degree equivalent.

$$\frac{37}{3,600} = \textbf{0.0103°} \qquad \text{To the nearest ten-thousandth.}$$

EXAMPLE   Change 22′35″ to its decimal degree equivalent.

$$22' = \frac{22}{60} = 0.3667° \qquad \text{To the nearest ten-thousandth.}$$

$$35'' = \frac{35}{3,600} = 0.0097° \qquad \text{To the nearest ten-thousandth.}$$

$$0.3667° + 0.0097° = \textbf{0.3764°}$$

Tip!	***Working with Degrees, Minutes, and Seconds Using a Calculator***

Many calculators have a key or menu choice that automatically converts between degrees, minutes, and
seconds and decimal degrees. The key is labeled $\boxed{° \,' \,''}$. This function key is also called a sexagesimal
function key because of its relationship with the number 60. The same key is also used for hours, minutes, and seconds.

3   Change the Decimal Part of a Degree to Minutes and Seconds.

We can change a decimal part of a degree to minutes and seconds by reversing the procedure for changing minutes and seconds to degrees. To change a decimal part of a degree to minutes, we multiply by 60. Similarly, to change the decimal part of a minute to seconds, we multiply by 60.

> **To change a decimal part of a degree to minutes:**
>
> Multiply the decimal part of a degree by 60.
>
> $$\frac{x°}{1} \times \frac{60 \text{ min}}{1°} = 60x \text{ min}$$

> **To change a decimal part of a minute to seconds:**
>
> Multiply the decimal part of a minute by 60.
>
> $$\frac{x \text{ min}}{1} \times \frac{60 \text{ sec}}{1 \text{ min}} = 60x \text{ sec}$$

If we are changing a decimal part of a *degree* to *seconds,* we can multiply by 3,600.

$$\frac{x°}{1} \times \frac{60 \text{ min}}{1°} \times \frac{60 \text{ sec}}{1 \text{ min}} = 3{,}600 \text{ sec}$$

EXAMPLE   Change 0.75° to minutes.

$$\mathbf{0.75 \times 60 = 45'}$$

EXAMPLE   Change 0.43° to minutes and seconds.

$0.43 \times 60 = 25.8'$	Degrees to minutes.
$0.8 \times 60 = 48''$	Decimal part of minute to seconds.

Thus, 0.43° = 25′48″.

4   Multiply and Divide Angle Measures.

Angle calculations involving multiplication and division are sometimes necessary. We may multiply or divide angle measures by a number as we do with U.S. customary measures and write in standard notation.

EXAMPLE   An angle of 42°27′32″ needs to be increased to 4 times its size. Find the measure of the new angle in degrees, minutes, and seconds.

$$
\begin{array}{rrr l}
42° & 27' & 32'' & \text{Multiply each part of the measure by 4.}\\
\times & & 4 & \\
\hline
168° & 108' & 128'' & \\
& +2' & -120'' & \text{Two minutes equal 120 seconds.}\\
\hline
168° & 110' & 8'' & \\
+1° & -60' & & \\
\hline
\mathbf{169°} & \mathbf{50'} & \mathbf{8''} & \text{Write in standard notation.}
\end{array}
$$

EXAMPLE An angle of 70°15′16″ is divided into three equal angles. Find the measure of each angle in degrees, minutes, and seconds.

$$23° \quad 25′ \quad 5\frac{1}{3}″ \quad \text{or} \quad \mathbf{23°25′5″} \qquad \text{To the nearest second.}$$

$$
\begin{array}{r}
3\,\overline{)70°\quad 15′16″} \\
69° \\
\hline
1° = 60′ \\
75′ \\
75′ \\
\hline
16″ \\
15″ \\
\hline
1″
\end{array}
$$

EXAMPLE An angle of 57°42′17″ is divided into four equal parts. Change the measure to its decimal equivalent and find the measure of each part in decimal degrees.

$$57°42′17″ = 57 + \frac{42}{60} + \frac{17}{3,600}$$

$$= 57 + 0.7 + 0.0047 = 57.7047° \qquad \text{Rounded.}$$

$$\frac{57.7047}{4} = \mathbf{14.4262°} \qquad \text{Rounded.}$$

SELF-STUDY EXERCISES 16–2

1 Add or subtract as indicated. Write in standard notation.

1. 15°47′18″
+ 38°12′42″

2. 83°19′54″
− 37°11′36″

3. 152°28′19″
− 114°35′23″

4. 45°15′38″
+ 28°47′34″

5. 47°30′
− 30°30′15″

6. 90°
− 35°15′48″

7. Find the supplement of an angle whose measure is 115°35′14″.

2 Change to decimal degree equivalents. Express the decimals to the nearest ten-thousandth.

8. 47′ **9.** 36″ **10.** 5′14″ **11.** 10′15″

12. An angle of 59°24′ is divided into two equal angles. Change the measure to its decimal equivalent and express the measure of the two equal angles in decimal degrees.

3 Change to equivalent minutes and seconds. Round to the nearest second when necessary.

13. 0.35° **14.** 0.20° **15.** 0.12° **16.** 0.213° **17.** 0.3149°

4 In these exercises, round angle measures to the nearest second when necessary.

18. An angle of 90° is divided into 12 equal parts. Find the measure of each angle in degrees and minutes.

19. An angle of 80° is divided into seven equal parts. Find the measure of each angle in degrees, minutes, and seconds.

20. An angle of 12°43′49″ needs to be three times as large. Find the measure of the new angle in degrees to the nearest ten-thousandth.

21. An angle of 25°17′39″ is divided into two equal angles. Find the measure of each angle in degrees, minutes, and seconds.

Learning Outcomes

1. Classify triangles by sides.
2. Relate the sides and angles of a triangle.
3. Determine if two triangles are congruent using inductive and deductive reasoning.
4. Use the properties of a 45°, 45°, 90° triangle to find missing parts and to solve applied problems.
5. Use the properties of a 30°, 60°, 90° triangle to find missing parts and to solve applied problems.

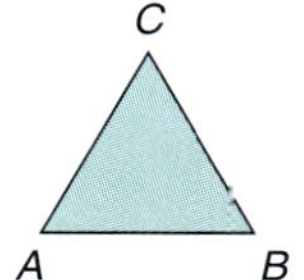

Equilateral △*ABC*
AB = *BC* = *AC*
∠*A* = ∠*B* = ∠*C*

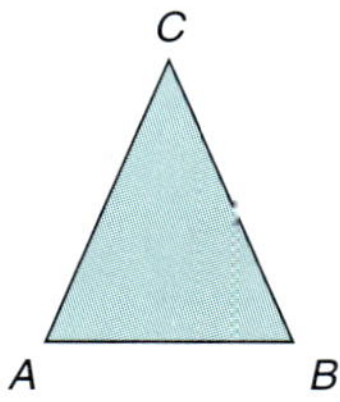

Isosceles △*ABC*
AC = *BC*
∠*A* = ∠*B*

Figure 16–29

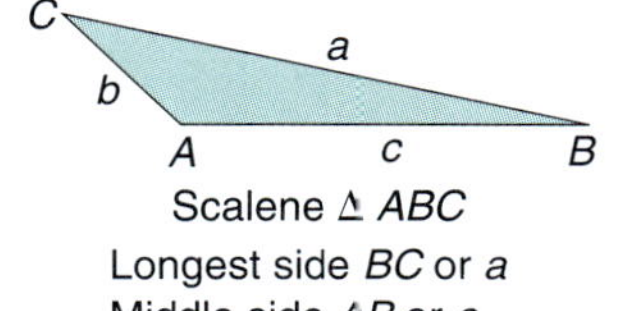

Scalene △ *ABC*
Longest side *BC* or *a*
Middle side *AB* or *c*
Shortest side *AC* or *b*
Largest angle ∠*A*
Middle angle ∠*C*
Smallest angle ∠*B*

Figure 16–30

We have examined various relationships among lines and angles in the previous sections. In this section, we study one of the most useful mathematical figures for any technician who uses geometry—the *triangle*. The relationships among the sides and angles in the triangle enable us to obtain much information that is implied but not always expressed in certain applications.

Triangles can be classified according to the relationship of their sides or their angles. We classified triangles by angles in Chapter 11. In this section, we classify triangles by their sides.

1 Classify Triangles by Sides.

Three relationships are possible among the three sides of a triangle, and result in special names for each type of triangle.

■ **DEFINITION: Equilateral Triangle.** An *equilateral triangle* is a triangle with three equal sides. The three angles of an equilateral triangle are also equal. Each angle measures 60° (Fig. 16–29, top).

■ **DEFINITION: Isosceles Triangle.** An *isosceles triangle* is a triangle with *exactly* two equal sides. The angles opposite these equal sides are also equal (Fig. 16–29, bottom).

■ **DEFINITION: Scalene Triangle.** A *scalene triangle* is a triangle with *all* three sides unequal (Fig. 16–30).

2 Relate the Sides and Angles of a Triangle.

An equilateral (three sides equal) triangle also has three equal angles. The two equal sides of an isosceles triangle have opposite angles that are equal (Fig. 16–29).

In a scalene triangle, where no sides are equal, we can state an important relationship between the sides and their opposite angles (Fig. 16–30).

To determine the longest and shortest sides of a triangle:

If the three sides of a triangle are unequal, the *largest* angle is opposite the *longest* side, and the *smallest* angle is opposite the *shortest* side.

EXAMPLE   Identify the longest and shortest sides of the triangle in Fig. 16–31.

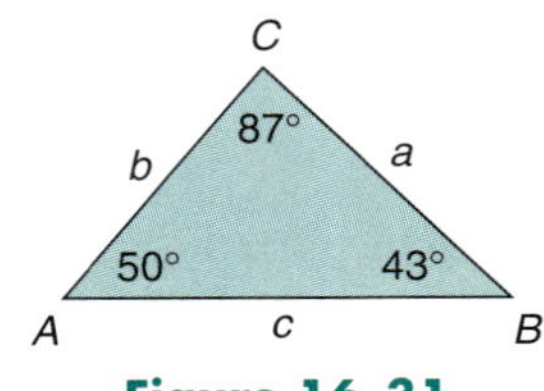

Figure 16–31

The longest side is *AB* **or** *c* (87° is the largest angle). **The shortest side is** *AC* **or** *b* (43° is the smallest angle).

EXAMPLE   Identify the largest and smallest angles in the triangle in Fig. 16–32.

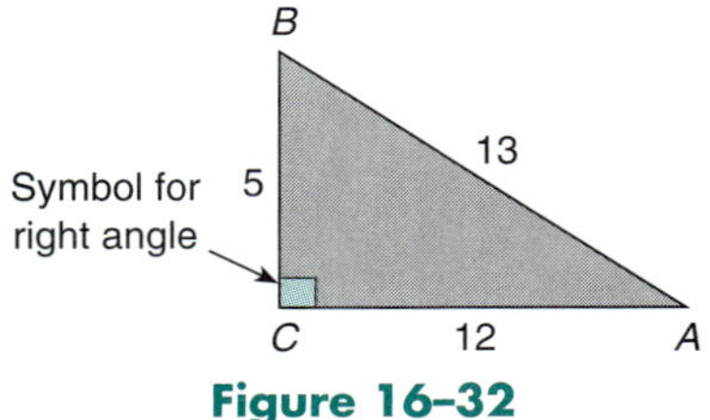

Figure 16–32

The largest angle is $\angle C$ (13 is the longest side). **The smallest angle is** $\angle A$ (5 is the shortest side).

When applying the rule to determine the longest side of a triangle, we see that the hypotenuse of a right triangle (the side opposite the 90° angle) is always the longest side. Figure 16–32 shows the hypotenuse (side *AB*) to be the longest side.

3   Determine If Two Triangles Are Congruent Using Inductive and Deductive Reasoning.

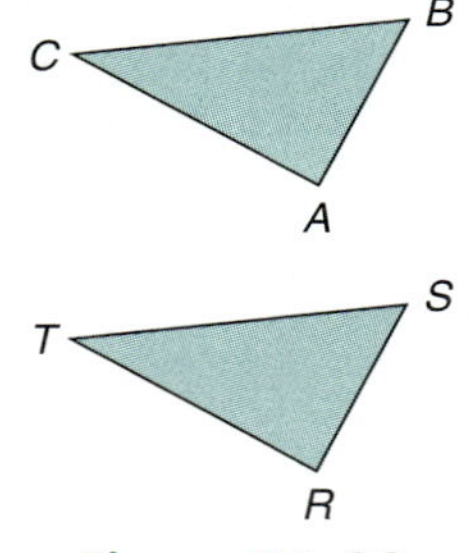

Figure 16–33

Triangles that have the same size and shape are *congruent* triangles. These triangles fit exactly on top of each other.

In Fig. 16–33, $\triangle ABC$ fits exactly over $\triangle RST$. They are congruent triangles. The symbol $\cong$ means congruent. That is, $\triangle ABC \cong \triangle RST$. Each angle in $\triangle ABC$ has an angle in $\triangle RST$ that is its equal. We say that these pairs of angles *correspond*. In Fig. 16–33, $\angle A$ corresponds to $\angle R$ because they are equal. Also, $\angle C$ corresponds to $\angle T$, and $\angle B$ corresponds to $\angle S$. The equal sides also correspond. $\overline{AB}$ corresponds to $\overline{RS}$, $\overline{CB}$ corresponds to $\overline{TS}$, and $\overline{AC}$ corresponds to $\overline{RT}$.

■ **DEFINITION: Congruent Triangles.**   *Congruent triangles* are triangles in which the corresponding sides and angles are equal.

We can establish congruent triangles even when we know that only certain angles and sides are equal.

> **Use three sides to determine congruent triangles:**
>
> If the three sides of one triangle are equal to the corresponding three sides of another triangle, the triangles are congruent (side-side-side or SSS).

EXAMPLE Write the corresponding sides of the two triangles in Fig. 16–34.

$\triangle ABC \cong \triangle FDE$

$\overline{AB}$ corresponds to $\overline{FD}$

$\overline{BC}$ corresponds to $\overline{DE}$

$\overline{AC}$ corresponds to $\overline{FE}$

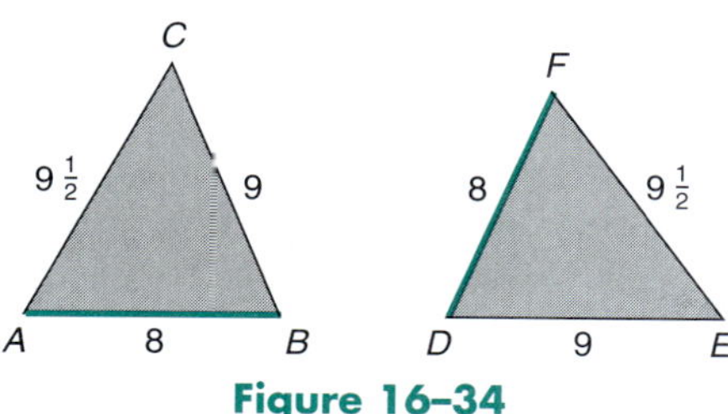

Figure 16–34

Use two sides and the included angle to determine congruent triangles:

If two sides and the *included angle* of one triangle are equal to two sides and the included angle of another triangle, the triangles are congruent (side-angle-side or SAS).

EXAMPLE List the corresponding and equal sides and angles of the two triangles in Fig. 16–35.

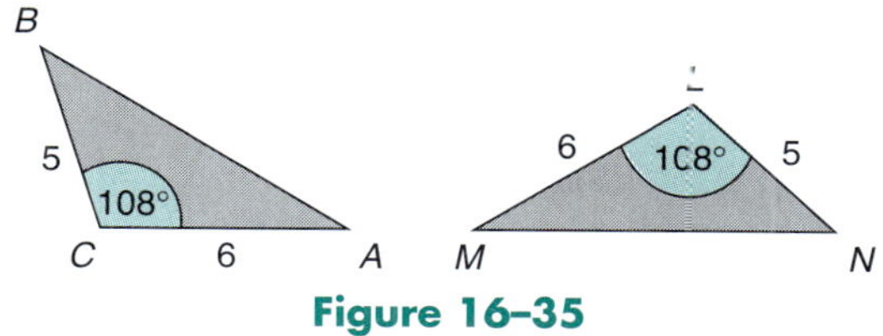

$\overline{BC} = \overline{NL}.$
$\overline{AC} = \overline{ML}.$
$\angle C = \angle L.$

Figure 16–35

The triangles are congruent because two sides and the included angle of one triangle are equal to two sides and the included angle of the other (SAS).

$\overline{AB} = \overline{MN}.$ They are opposite equal angles.
$\angle A = \angle M.$ They are opposite equal sides.
$\angle B = \angle N.$ They are opposite equal sides.

Use two angles and the included side to determine congruent triangles:

If two angles and the common side of one triangle are equal to two angles and the common side of another triangle, the triangles are congruent (angle-side-angle or ASA).

EXAMPLE List the corresponding and equal angles and sides of the two triangles in Fig. 16–36.

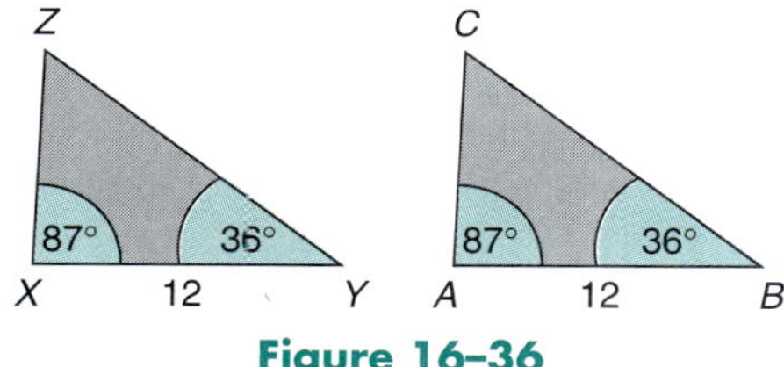

$\angle X = \angle A.$
$\angle Y = \angle B.$
$\angle Z = \angle C.$

The triangles are congruent because of the ASA relationship.

$\overline{XY} = \overline{AB}.$
$\overline{XZ} = \overline{AC}.$
$\overline{YZ} = \overline{BC}.$

Figure 16–36

The study of geometry applies two types of reasoning, *inductive* and *deductive* reasoning. *Inductive reasoning* starts with investigation and experimentation. Early mathematicians discovered the properties of geometry through inductive reasoning. After extensive investigation, general conclusions are drawn from the results of specific cases.

Let's use inductive reasoning to examine the results of cutting a square into four parts by cutting along the two diagonals (Fig. 16–37). Compare the resulting four

Figure 16–37

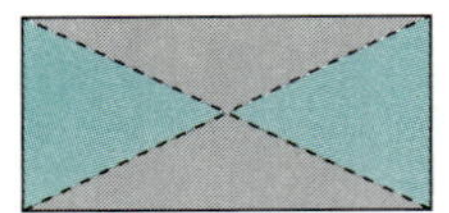

Figure 16–38

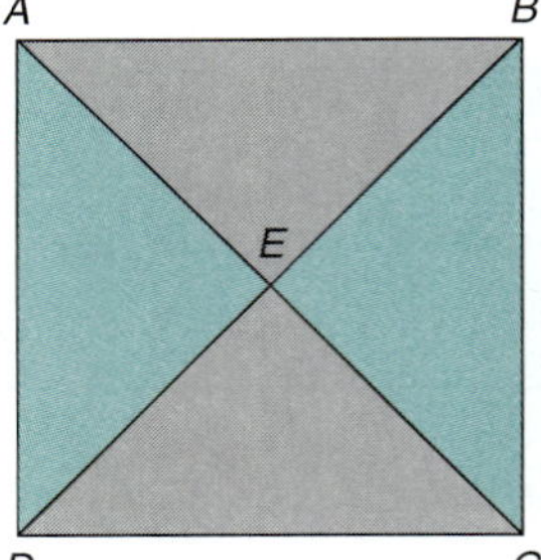

Figure 16–39

triangles. Are they congruent triangles? Make a square of a different size and repeat the exercise. Continued repetitions lead you to conclude that the diagonals of a square divide the square into four congruent triangles.

Now, cut a rectangle along the two diagonals (Fig. 16–38). Are the resulting four triangles congruent? No. What conclusions appear to result from this experiment? Repeat the experiment with different sizes of rectangles. What properties seem apparent? There are two pairs of congruent triangles.

Deductive reasoning starts with accepted principles, then additional properties are concluded from these principles. Accepting the congruent triangle properties and an additional property that the diagonals of a square bisect (cut in half) the angles of the square, we can use deductive reasoning to show that the two diagonals of a square form four congruent triangles.

Square $ABCD$ forms the four triangles $\triangle AED$, $\triangle AEB$, $\triangle BEC$, and $\triangle DEC$ (Fig. 16–39). The four sides of a square are equal in measure, so sides AD and AB are equal in measure. Because the diagonals of a square bisect the angles, $\angle DAE$, $\angle EAB$, $\angle ABE$, $\angle EBC$, $\angle BCE$, $\angle ECD$, $\angle CDE$, and $\angle EDA$ are all 45° angles. Then, in $\triangle AED$ and $\triangle AEB$ we have two angles and a common side of one triangle equal to two angles and a common side of the other. By applying the congruent triangle property for two angles and a common side, we can say that $\triangle AED$ and $\triangle AEB$ are congruent. Similarly, we can continue the argument to show that all four triangles are congruent.

In a systematic or axiomatic study of geometric concepts, all properties are developed or proved from a limited number of basic principles. In our study of geometric concepts, we use an informal approach and apply both inductive and deductive reasoning in our arguments for solving problems. We do not introduce all the concepts necessary to make formal proofs of geometric properties.

4 Use the Properties of a 45°, 45°, 90° Triangle to Find Missing Parts and to Solve Applied Problems.

An isosceles triangle is a triangle with two equal sides. An *isosceles right triangle* is a right triangle that has equal legs. The angles opposite the equal legs are *base angles*. The two base angles are also equal because they are the angles opposite the equal sides (Fig. 16–40).

Because the sum of the angles of the triangle is 180°, the sum of the two base angles of a right triangle is 180° − 90° or 90°. If both angles are equal, as in the isosceles right triangle, then each angle is $\frac{1}{2}(90°)$ or 45°. Thus, we get the name 45°, 45°, 90° triangle.

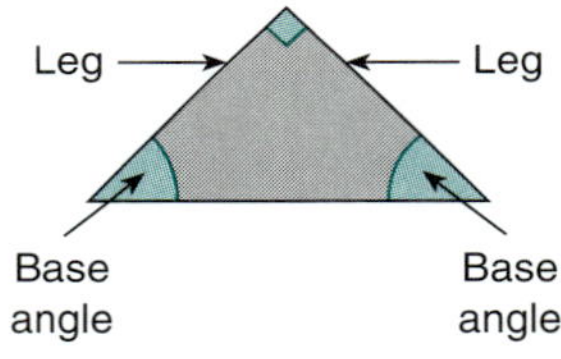

Figure 16–40

■ **DEFINITION: 45°, 45°, 90° Triangle.** A *45°, 45°, 90° triangle* is an isosceles right triangle in which the two base angles are 45° each.

This triangle is frequently used in applications of the Pythagorean theorem.

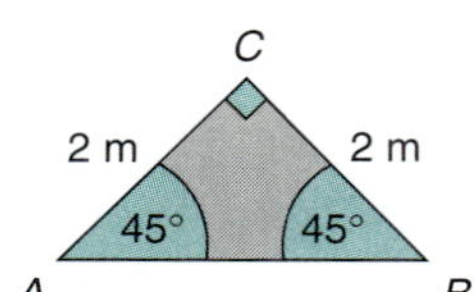

Figure 16–41

EXAMPLE Find the hypotenuse of an isosceles right triangle that has equal sides of 2 m (Fig. 16–41).

Using the Pythagorean theorem $c^2 = a^2 + b^2$, we have

$$(AB)^2 = 2^2 + 2^2$$

$$(AB)^2 = 4 + 4$$

$$(AB)^2 = 8$$

$$AB = \sqrt{8}$$

$$AB = 2.828 \qquad \text{Rounded.}$$

The hypotenuse is 2.828 m.

We can also calculate $AB = 2.828$ by simplifying $\sqrt{8}$.

$$AB = \sqrt{8} = \sqrt{4 \cdot 2} = 2\sqrt{2} \qquad \text{The hypotenuse is the product of a leg and } \sqrt{2}.$$

$$AB = 2(1.414213562)$$

$$AB = 2.828 \qquad \text{Rounded.}$$

To find the hypotenuse of a 45°, 45°, 90° triangle:

Multiply the measure of a leg by $\sqrt{2}$, or 1.414213562 (Fig. 16–43); that is,

$$\text{Hypotenuse} = \text{Leg}\sqrt{2}$$

Remember this relationship and it will save you time on your job.

If we need to find a leg of a 45°, 45°, 90° triangle, we divide the hypotenuse by $\sqrt{2}$.

EXAMPLE   Find AC and BC if $AB = 5$ cm (Fig. 16–42).

The hypotenuse is equal to the product of a leg and $\sqrt{2}$.

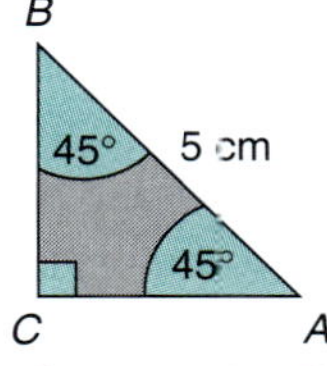

Figure 16–42

$$5 = AC\sqrt{2}$$

$$\frac{5}{\sqrt{2}} = \frac{AC\sqrt{2}}{\sqrt{2}} \qquad \text{Solve for } AC, \text{ a leg.}$$

$$\frac{5}{\sqrt{2}} = AC$$

$$\frac{5}{\sqrt{2}} \cdot \frac{\sqrt{2}}{\sqrt{2}} = AC \qquad \text{Rationalize.}$$

$$\frac{5\sqrt{2}}{2} = AC$$

$$3.536 \text{ cm} = AC \qquad \text{Rounded.}$$

Using a calculator, we can find the decimal value of AC without first rationalizing the denominator.

$$AC = \frac{5}{\sqrt{2}}$$

$AC = \textbf{3.536 cm}$, from $\boxed{5}\ \boxed{\div}\ \boxed{\sqrt{}}\ \boxed{2}\ \boxed{=}\ \Rightarrow\ 3.536$ Rounded.

Since $AC = BC$ then $BC = 3.536$ cm.

To find a leg of a 45°, 45°, 90° triangle:

Divide the product of the hypotenuse and $\sqrt{2}$ by 2 (Fig. 16–43); that is,

$$\text{Leg} = \frac{\text{Hypotenuse } \sqrt{2}}{2} \quad \text{or} \quad \frac{\text{Hypoteneuse}}{\sqrt{2}}$$

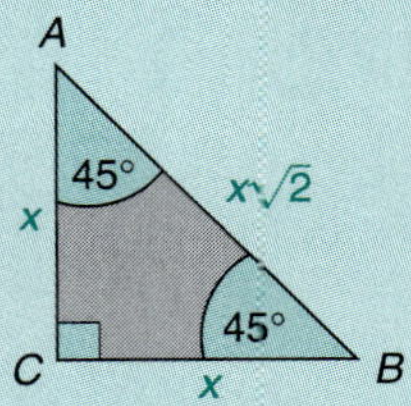

Figure 16–43

If the sides opposite the 45° angles are x units, the sides of a 45°, 45°, 90° triangle are x, x, $x\sqrt{2}$, respectively.

5 Use the Properties of a 30°, 60°, 90° Triangle to Find Missing Parts and to Solve Applied Problems.

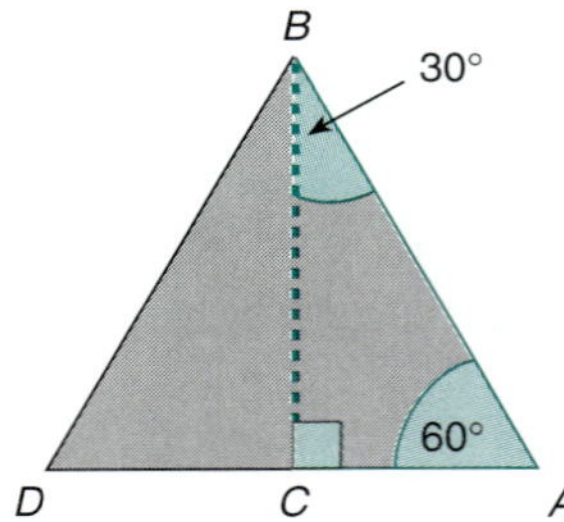
Figure 16–44

Another special case of the Pythagorean theorem is the *30°, 60°, 90° triangle,* which arises often in applications. If we draw the altitude of an *equilateral* triangle, we form two 30°, 60°, 90° triangles (Fig. 16–44).

■ **DEFINITION: Altitude of Equilateral Triangle.** The *altitude of an equilateral triangle* is a line drawn from the midpoint of the base to the opposite vertex, dividing the triangle into two congruent right triangles.

Because the altitude of an equilateral triangle bisects (divides in half) the vertex angle and the base, two 30° angles are formed at B and $AC = CD$. We can also say that AC is $\frac{1}{2}AD$ or half any side of the original equilateral triangle. That means $AC = \frac{1}{2}AB$, or the side opposite the 30° angle is half the hypotenuse.

EXAMPLE If $AC = 5$ cm, find AB and BC (Fig. 16–45).

Side AB (the hypotenuse) is twice AC because the altitude divides AD in half, and in an equilateral triangle all sides are equal. Therefore, **$AB = 2(5) = 10$ cm.** Now that we know two sides of the right triangle, we can use the Pythagorean theorem to find the third side.

$$(AB)^2 = (AC)^2 + (BC)^2$$
$$10^2 = 5^2 + (BC)^2$$
$$100 = 25 + (BC)^2$$
$$75 = (BC)^2$$
$$\sqrt{75} = BC$$
$$8.660 = BC \qquad \text{Rounded.}$$

Thus, BC is 8.660 cm.
We can also calculate $BC = 8.660$ by simplifying $\sqrt{75}$.

$$BC = \sqrt{75} = \sqrt{25 \cdot 3} = 5\sqrt{3}$$
$$BC = 5(1.732050808)$$
$$BC = 8.660 \qquad \text{Rounded.}$$

Then BC (the side opposite the 60° angle) is the product of AC (the side opposite the 30° angle) and $\sqrt{3}$.

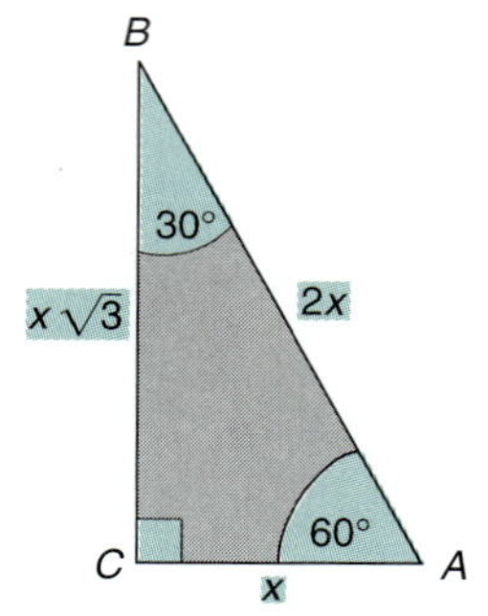
Figure 16–45

Once we have learned the relationships of the three sides, we can find two sides of any 30°, 60°, 90° triangle if we know only one side (Fig. 16–46).

Figure 16–46

> **To find the hypotenuse of a 30°, 60°, 90° triangle:**
>
> Multiply the side opposite the 30° angle by 2.

> **To find the side opposite the 30° angle in a 30°, 60°, 90° triangle:**
>
> Divide the hypotenuse by 2.

> **To find the side opposite the 60° angle in a 30°, 60°, 90° triangle:**
>
> Multiply the side opposite the 30° angle by $\sqrt{3}$.

Tip! | *Summary of 30°, 60°, 90° Triangle.*

To summarize the relationships in Fig. 16–46:

$$AB = 2(AC) \text{ or } \frac{2\sqrt{3}BC}{3} \qquad AC = \frac{AB}{2} \text{ or } \frac{BC\sqrt{3}}{3} \qquad BC = AC\sqrt{3} \text{ or } \frac{AB\sqrt{3}}{2}$$

If the side opposite the 30° angle is x units, the sides of a 30°, 60°, 90° triangle are x, $x\sqrt{3}$, $2x$, respectively.

EXAMPLE Find AC and AB if $BC = 8$ cm (Fig. 16–47).

$$BC = x\sqrt{3} = 8 \qquad x = AC$$

Solving for x, we have $\dfrac{x\sqrt{3}}{\sqrt{3}} = \dfrac{8}{\sqrt{3}}$. Rationalizing, we have

$$x = \frac{8}{\sqrt{3}} \cdot \frac{\sqrt{3}}{\sqrt{3}}$$

$$x = \frac{8\sqrt{3}}{3}$$

$$x = 4.619 \text{ cm} \qquad \text{Rounded.}$$

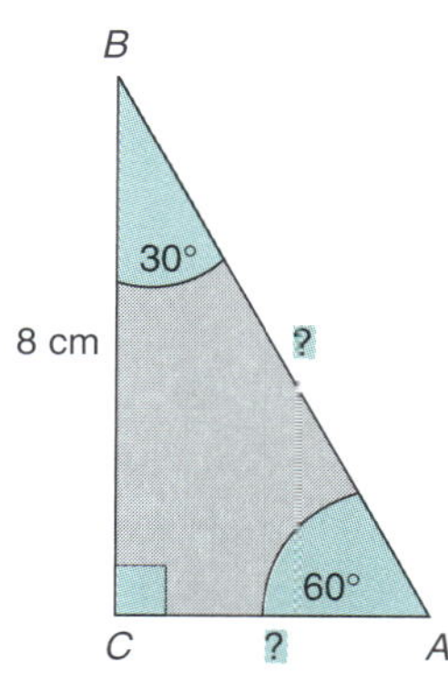

Figure 16–47

$$AB = 2x = 2\left(\frac{8\sqrt{3}}{3}\right) = \frac{16\sqrt{3}}{3} = 9.238 \text{ cm} \qquad \text{Rounded.}$$

Using a scientific calculator, we can find the decimal value of AC without rationalizing.

$$AC = \frac{8}{\sqrt{3}}$$

$$AC = 4.619 \text{ cm} \qquad \text{From } \boxed{8}\,\boxed{\div}\,\boxed{\sqrt{\ }}\,\boxed{3}\,\boxed{=} \Rightarrow 4.618802154.$$

$AB = 9.238$ cm and $AC = 4.619$ cm.

1   Fill in the blanks.

1. A triangle with no equal sides is called a(n) _____ triangle.

2. A triangle with three equal sides is called a(n) _____ triangle.

3. A triangle with only two equal sides is called a(n) _____ triangle.

2   Identify the longest and shortest sides in Figs. 16–48 and 16–49.

4.

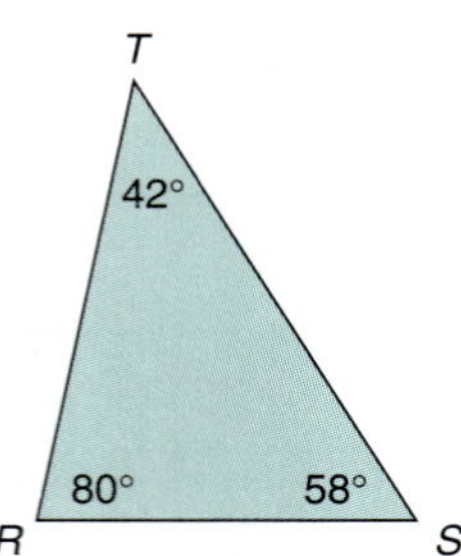

Figure 16–48

5.

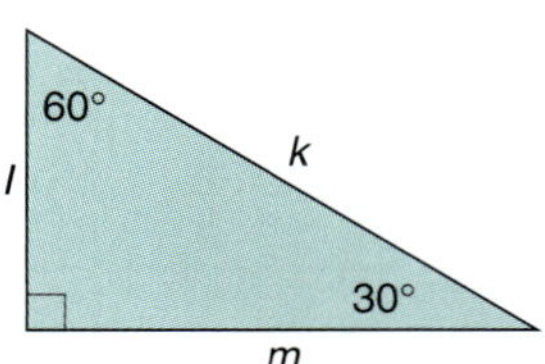

Figure 16–49

List the angles in order of size from largest to smallest in Figs. 16–50 and 16–51.

6.

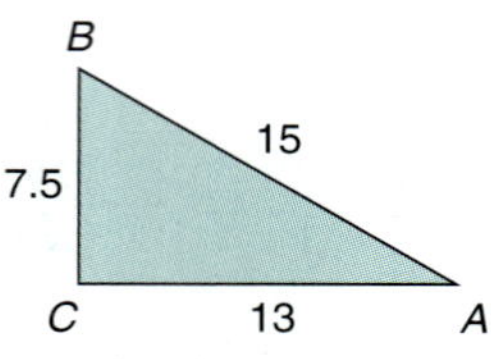

Figure 16–50

7.

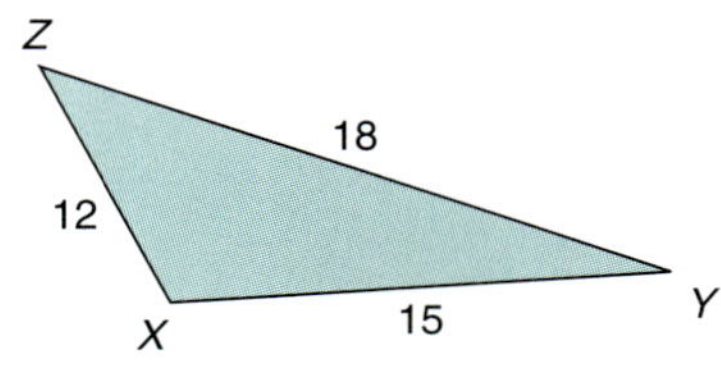

Figure 16–51

3   Write the corresponding parts not given for the congruent triangles in Figs. 16–52 to 16–54.

8.

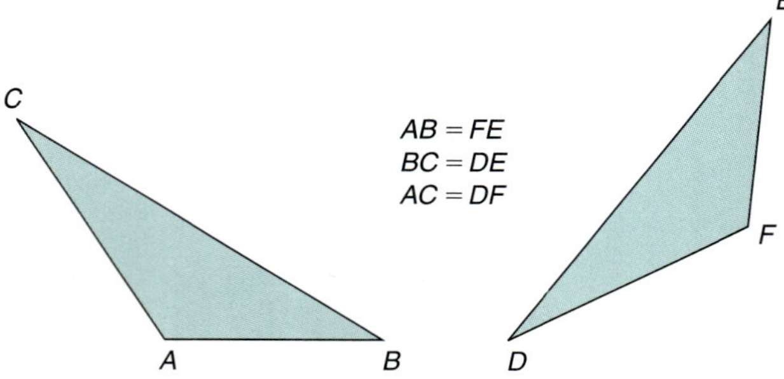

Figure 16–52

9.

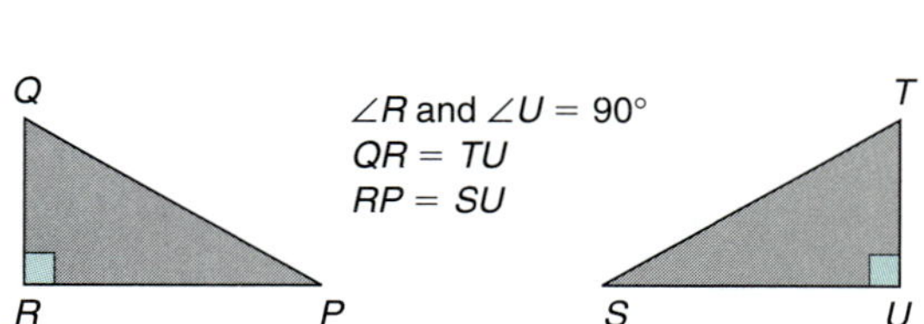

Figure 16–53

10.

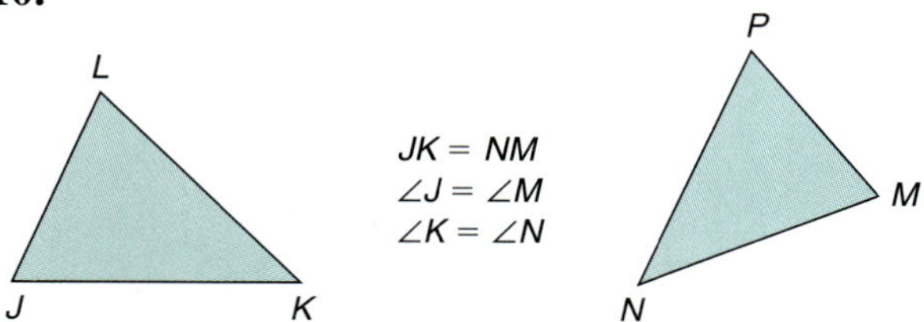

Figure 16–54

 Use Fig. 16–55 to solve. Round the final answers to the nearest thousandth if necessary.

11. $AC = 12$ cm; find BC and AB.
12. $AB = 10$ m; find AC and BC.
13. $BC = 7\sqrt{2}$ m; find AC and AB.
14. $AB = 8\sqrt{2}$ m; find AC and BC.
15. $AB = 12\sqrt{3}$ mm; find AC and BC.

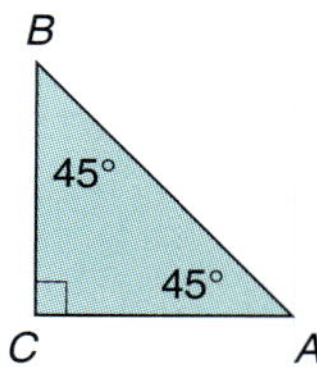

Figure 16–55

16. A rafter 19.5 ft long makes a 45° angle with a joist. If the rafter has an 18-in. overhang, find the length of the joist to the nearest inch (Fig. 16–56).

17. An elevated tank is connected to a pipe. The connecting pipe is 53.5 ft long and forms a 45° angle where it connects to the tank (Fig. 16–57). At what horizontal distance from the tank should the pipe stop to make the connection?

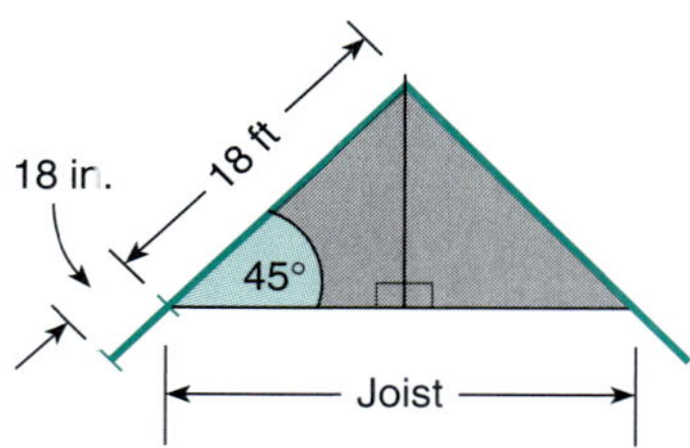

Figure 16–56

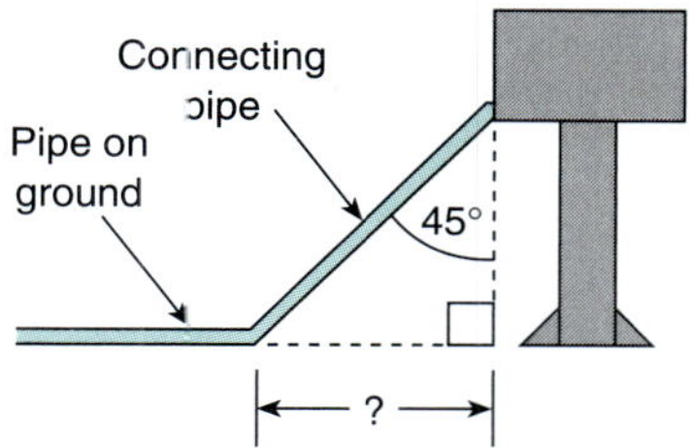

Figure 16–57

 Use Fig. 16–58 to solve. Round the final answers to the nearest thousandth if necessary.

18. $AC = 6$ cm; find AB and BC.
19. $AB = 18$ mm; find AC and BC.
20. $BC = 8$ in.; find AC and AB.
21. $BC = 7\sqrt{2}$ cm; find AC and AB.
22. $AC = 2$ ft 9 in.; find AB and BC to the nearest inch.

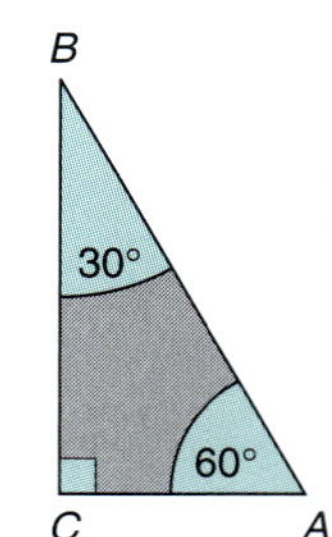

Figure 16–58

Solve. Round the final answers to the nearest thousandth if necessary.

23. Find the length of the conduit $ABCD$ (Fig. 16–59) if $AB = 18$ ft, $CK = 6$ ft, $CD = 5$ ft, and $\angle KCB = 60°$.

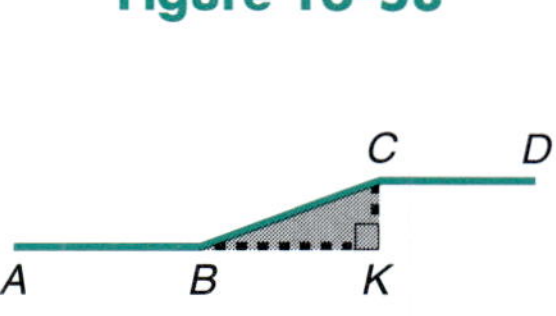

Figure 16–59

24. A rafter makes a 30° angle with the horizontal. If the rise is 9 ft, find the rafter length and the run to the nearest inch (Fig. 16–60).

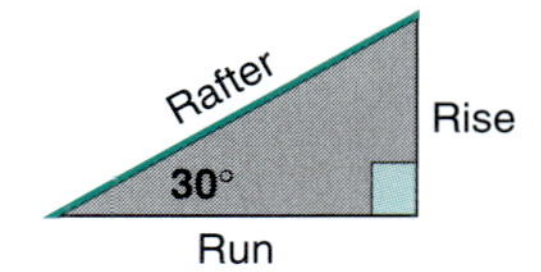

Figure 16–60

25. Find the depth of a V-slot in the form of an equilateral triangle if the cross-sectional opening is 5.2 cm across (Fig. 16–61).

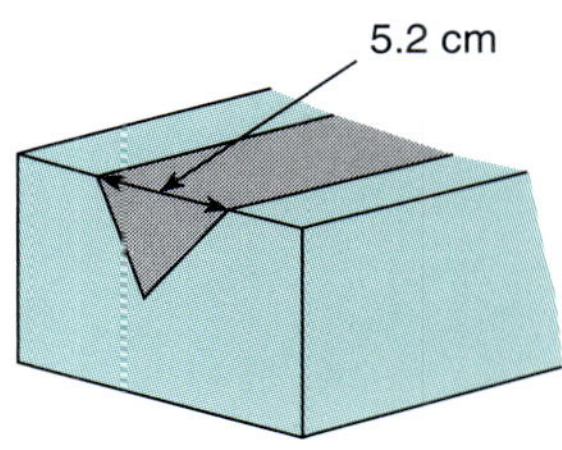

Figure 16–61

Learning Outcomes

1 Find the missing dimensions of composite figures.
2 Find the perimeter and area of composite figures.
3 Find the number of degrees in each angle of a regular polygon.

As we look about us we see polygons that are not squares, rectangles, parallelograms, trapezoids, or triangles. We can use what we already know to find both perimeter and area. These figures are called *composites*.

> ■ **DEFINITION: Composite figure.** A *composite figure* is a geometric figure made up of two or more geometric figures.

For instance, an L-shaped slab foundation for a building is a composite of two rectangles. As suggested in Fig. 16–62, the two rectangles forming the composite figure may be considered in more than one way. Layout I is partitioned vertically, whereas layout II is partitioned horizontally.

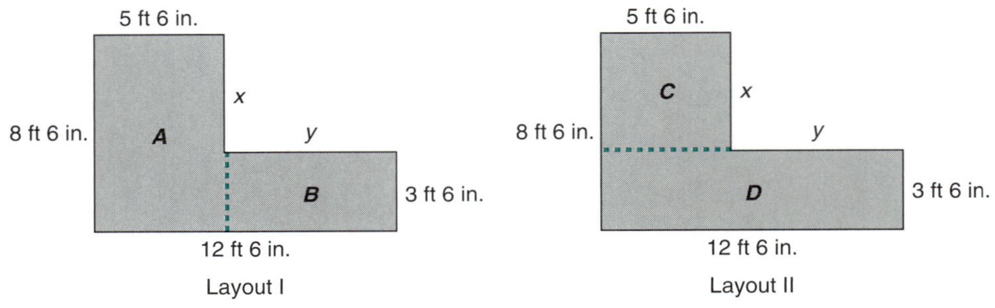

Figure 16–62

1 **Find the Missing Dimensions
of Composite Figures.**

Layouts don't always have all their dimensions indicated. However, we can infer or calculate the missing dimensions from our knowledge of polygons and the dimensions that are noted.

EXAMPLE Find the missing dimensions x and y on the slab layout in Fig. 16–62.

In layout I, the side of B opposite its $3'6''$ side is also $3'6''$ because opposite sides of a rectangle are equal. The side of A opposite its $8'6''$ side is, for the same reason, $8'6''$. Dimension x is the difference between $8'6''$ and $3'6''$.

$$x = 8'6'' - 3'6''$$

$$x = 5'$$

Think of layout II as two horizontal rectangles and find dimension y. The side opposite the $5'6''$ side of rectangle C is $5'6''$. The side of D opposite the $12'6''$ side is $12'6''$. Dimension y is the difference between $12'6''$ and $5'6''$.

$$y = 12'6'' - 5'6''$$

$$y = 7'$$

The missing dimensions are $x = 5'$ and $y = 7'$.

2 ## Find the Perimeter and Area of Composite Figures.

Perimeter is the sum of the lengths of the sides of a figure or layout. The number and the lengths of the sides vary from one composite figure to another, so no specific formula covers the variety of composite figures that exist, but we can use a general formula.

> ***Perimeter of a composite figure:***
>
> $$P = a + b + c + \cdots$$
>
> That is, the perimeter is the sum of the measures of the sides of the figure.

EXAMPLE Find the number of feet of 4-in. stock needed for the base plates of a room that has the layout shown in Fig. 16–63. Make no allowances for openings when estimating the linear footage of the base plates that form the perimeter.

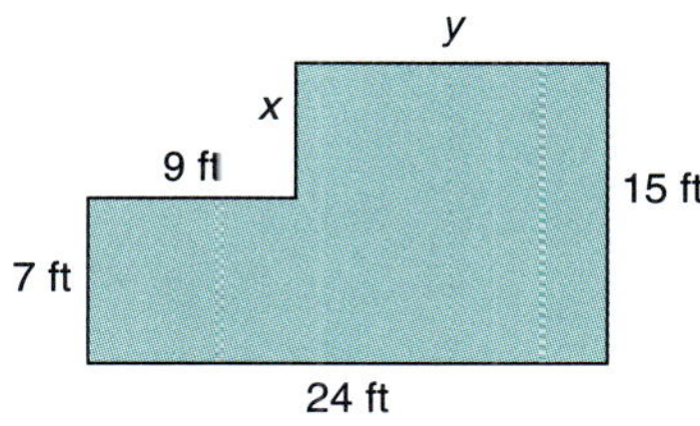

Figure 16–63

1. Find the missing dimensions.

$$x = 15 \text{ ft} - 7 \text{ ft} \qquad y = 24 \text{ ft} - 9 \text{ ft}$$
$$x = 8 \text{ ft} \qquad\qquad y = 15 \text{ ft}$$

2. Apply the general formula for the perimeter of a polygon.

$$P = a + b + c + \cdots$$
$$P = 7 \text{ ft} + 24 \text{ ft} + 15 \text{ ft} + 15 \text{ ft} + 8 \text{ ft} + 9 \text{ ft} = 78 \text{ ft}$$

3. Count the number of sides on the layout to check that each has been substituted into the formula.

The room needs 78 ft of 4-in. stock for the base plates.

> ***Area of a composite figure:***
>
> $$A_{\text{total}} = A + B + C + \cdots$$
>
> That is, the total area is the sum of the areas of the figures that make up the composite.

EXAMPLE Find the number of square yards of carpeting required for the room in the previous example ($9 \text{ ft}^2 = 1 \text{ yd}^2$).

1. Divide the composite figure into two polygons that have areas we can compute (Fig. 16–64). In this case, A is a rectangle and B is a square.

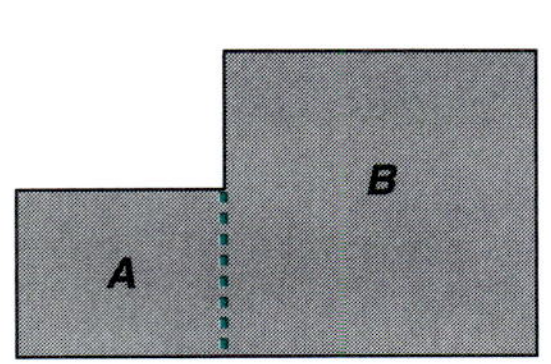

Figure 16–64

2. Find the area of each smaller polygon and add.

Rectangle A	**Square B**
$A_1 = lw$	$A_2 = s^2$
$A_1 = 9 \times 7$	$A_2 = 15^2$
$A_1 = 63 \text{ ft}^2$	$A_2 = 225 \text{ ft}^2$

$$A_1 + A_2 = \text{Total area}$$
$$63 + 225 = 288 \text{ ft}^2$$

3. Convert square feet to square yards using a unity ratio.

$$\frac{\overset{32}{\cancel{288} \text{ ft}^2}}{1} \times \frac{1 \text{ yd}^2}{\underset{1}{\cancel{9} \text{ ft}^2}} = 32 \text{ yd}^2$$

The room requires 32 yd^2 of carpeting.

EXAMPLE   Find the area of a gable end of a gambrel roof that has dimensions as shown in Fig. 16–65.

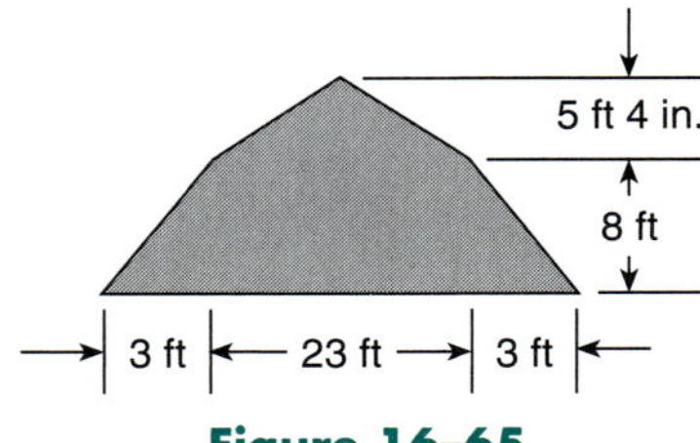

Figure 16–65

1. Divide the gable end of the gambrel roof into polygons with areas that can be calculated (Fig. 16–66).

2. Calculate the areas of the three triangles and the rectangle. Find the sum of the areas.

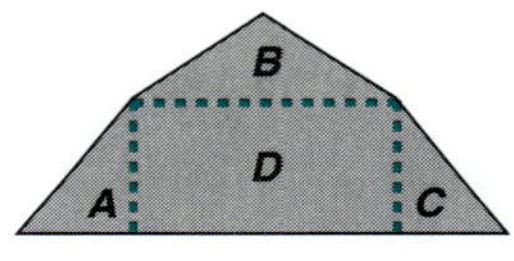

Figure 16–66

Triangle B	**Triangle A or C**	**Rectangle D**
Base $= 23'$	Base $= 3'$	Length $= 23'$
Height $= 5'4'' = 5\frac{1}{3}'$	Height $= 8'$	Width $= 8'$
$A_1 = \dfrac{1}{2} bh$	$A_2 = \dfrac{1}{2} bh$	$A_3 = lw$
$A_1 = \dfrac{1}{2}(23)\left(5\dfrac{1}{3}\right)$	$A_2 = \dfrac{1}{\underset{1}{\cancel{2}}}(3)(\overset{4}{\cancel{8}})$	$A_3 = 23(8)$
$A_1 = \dfrac{1}{\underset{1}{\cancel{2}}}(23)\left(\dfrac{\overset{8}{\cancel{16}}}{3}\right)$	$A_2 = 12 \text{ ft}^2$	$A_3 = 184 \text{ ft}^2$
$A_1 = \dfrac{184}{3} = 61\dfrac{1}{3} \text{ ft}^2$		

Doubled to include $\triangle A$ and $\triangle C$

$$A_1 + 2(A_2) + A_3 = \text{Area of gable end}$$

$$61\frac{1}{3} + 2(12) + 184 = 61\frac{1}{3} + 24 + 184 = 269\frac{1}{3} \text{ ft}^2$$

The area of the gable end of the gambrel roof is $269\frac{1}{3}$ ft^2.

Sometimes when we figure area, we find it more convenient to calculate an overall area and subtract a smaller area. We saw this for rectangles when we figured the area of a room and then subtracted the area of the fireplace hearth that projected into the room. We needed to find the area to be covered with carpet, so we had to exclude the area of the fireplace.

The following example is a similar case that involves composite figures.

EXAMPLE Find the area of the flat metal piece shown in Fig. 16–67.

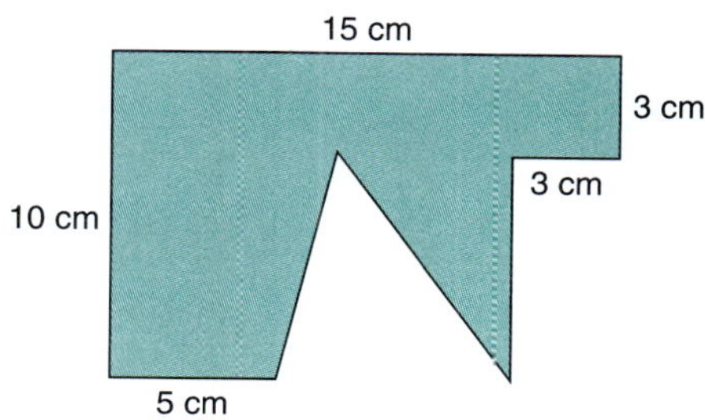

Figure 16–67

1. Divide the figure into polygons for which we can calculate areas (Fig. 16–68).

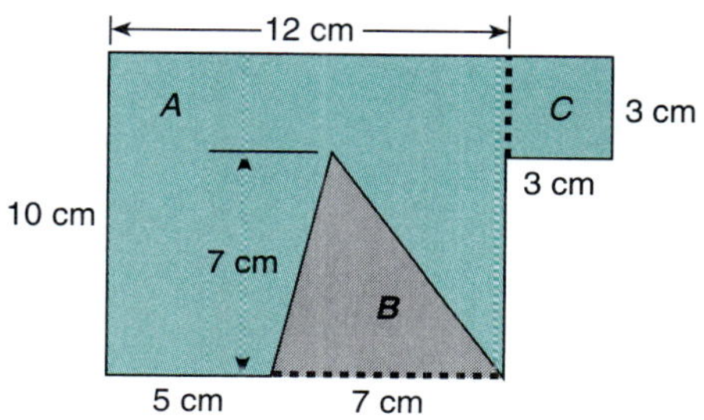

Figure 16–68

2. Find the missing dimensions.
3. Find the area of the square C (A_1), rectangle A (A_2), and triangle B (A_3). The area of the piece of metal is $A_1 + A_2 - A_3$.

Square	Rectangle	Triangle
$A_1 = s^2$	$A_2 = lw$	$A_3 = \dfrac{1}{2}bh$
$A_1 = 3^2$	$A_2 = 12(10)$	$A_3 = \dfrac{1}{2}(7)(7)$
$A_1 = 9 \text{ cm}^2$	$A_2 = 120 \text{ cm}^2$	$A_3 = \dfrac{1}{2}(49)$
		$A_3 = 24.5 \text{ cm}^2$

$$\text{Area of piece} = A_1 + A_2 - A_3 = 9 + 120 - 24.5 = 104.5 \text{ cm}^2$$

The area of the flat piece of metal is 104.5 cm².

3 Find the Number of Degrees in Each Angle of a Regular Polygon.

Polygons with equal sides and angles, such as squares, are called *regular polygons*. Several other regular polygons with definite shapes and specific names are treated as composite figures when calculating their areas. Let's examine regular polygons more closely.

■ **DEFINITION: Regular Polygon.** A *regular polygon* is a polygon with equal sides and equal angles.

To find the number of degrees in each angle of a regular polygon:

Multiply the number of sides less 2 (which is the number of triangles) by 180° and divide by the number of sides or angles.

$$\text{Degrees per angle} = \frac{180° \, (\text{Number of sides} - 2)}{\text{Number of sides}}$$

EXAMPLE Find the number of degrees in each angle of the regular polygons.

| Triangle | (3 sides) | Pentagon | (5 sides) | Octagon | (8 sides) |
| Quadrilateral | (4 sides) | Hexagon | (6 sides) | | |

Triangle:
$$\frac{180° \, (3 - 2)}{3} = \frac{180° \, (1)}{3} = 60°$$

Quadrilateral:
$$\frac{180° \, (4 - 2)}{4} = \frac{180° \, (2)}{4} = \frac{360°}{4} = 90°$$

Pentagon:
$$\frac{180° \, (5 - 2)}{5} = \frac{180° \, (3)}{5} = \frac{540°}{5} = 108°$$

Hexagon:
$$\frac{180° \, (6 - 2)}{6} = \frac{180° \, (4)}{6} = \frac{720°}{6} = 120°$$

Octagon:
$$\frac{180° \, (8 - 2)}{8} = \frac{180° \, (6)}{8} = \frac{1,080°}{8} = 135°$$

To form congruent triangles in a regular polygon:

Draw lines from the center of the regular polygon to each vertex. Congruent triangles are formed.

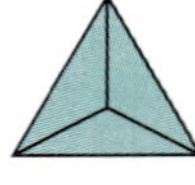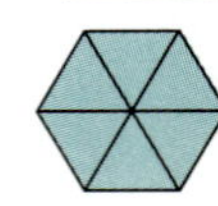

Figure 16–69 shows how this property applies to regular polygons: the equilateral triangle, square, regular pentagon, and regular hexagon. To accept this property using inductive reasoning, construct various regular polygons, draw lines from the vertex to the center, and compare the resulting triangles.

Figure 16–69

EXAMPLE Find the floor area of a recreational building at a park if it forms a regular hexagon and each side is 20 ft long. The perpendicular distance from one side to the center of the building is 17.3 ft (Fig. 16–70).

Divide the regular hexagon into congruent triangles by connecting the center of the hexagon with each vertex of the hexagon.

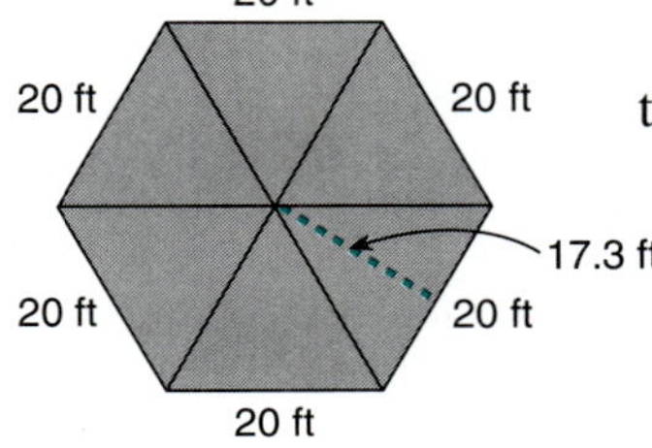

Figure 16–70

$$A = \frac{1}{2}hb \qquad \text{Find the area of one triangle.}$$

$$A = \frac{1}{2}(17.3)(\overset{10}{\underset{1}{20}}) \qquad \text{Use the known distance from the side of the hexagon to the center for the height of the triangle.}$$

$$A = 173 \text{ ft}^2$$

$$173 \times 6 = 1{,}038 \text{ ft}^2 \quad$$ Multiply the area of the one triangle by 6 because the hexagon has been divided into six congruent triangles.

The floor area of the recreational building is 1,038 ft².

SELF-STUDY EXERCISES 16–4

1 Find the missing dimensions x and y of Figs. 16–71 and 16–72.

1.

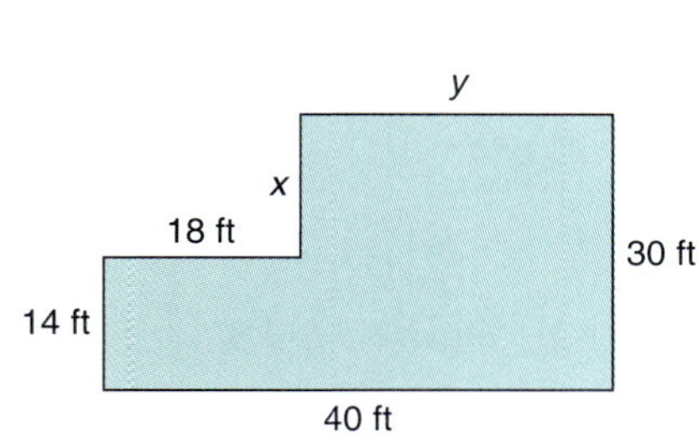

Figure 16–71

2.

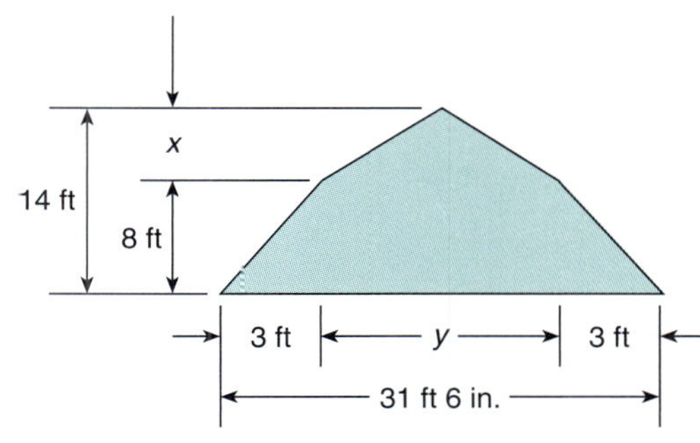

Figure 16–72

2 Find the perimeter and the area of Figs. 16–73 and 16–74.

3.

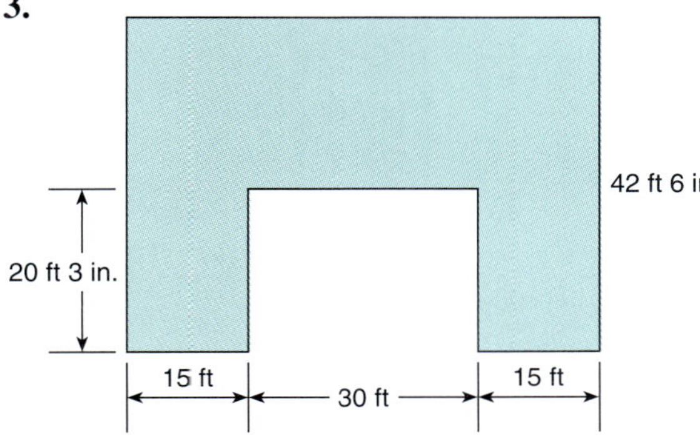

Figure 16–73

4.

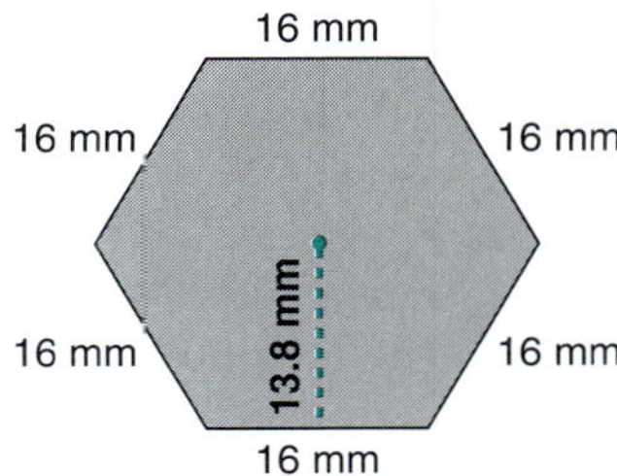

Figure 16–74

Solve the problems involving perimeter and area of composite figures.

5. A house features a den with the layout shown in Fig. 16–75. Find the number of feet of 4-in. stock needed for the base plates around the perimeter. Make no allowances for doorways or other openings.

6. If the den in Exercise 5 is covered with roll vinyl that costs \$16.50 per yd² for materials and labor, how much does it cost to install the vinyl? (*Note:* Round to the next whole square yard before figuring cost.)

7. The metal piece in Fig. 16–76 has the dimensions indicated. If the 18-gauge steel weighs 2 lb per square foot, how much does the piece weigh to the nearest pound? ($144 \text{ in}^2 = 1 \text{ ft}^2$)

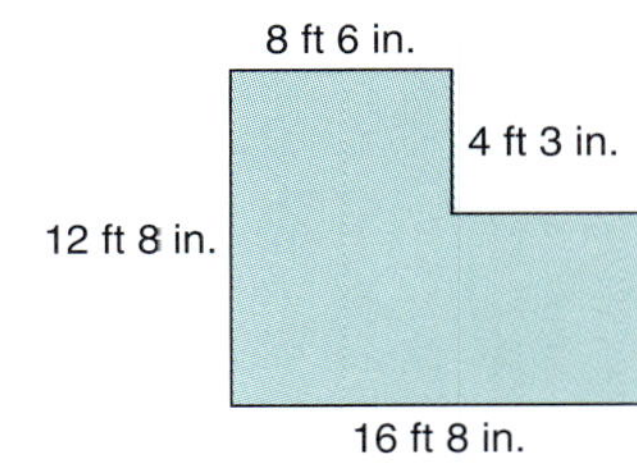

Figure 16–75

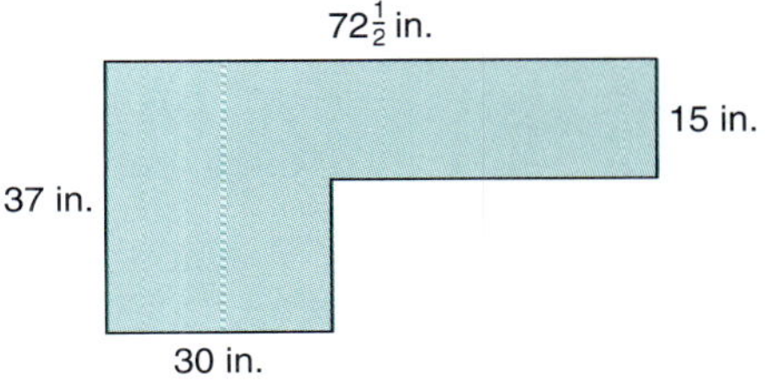

Figure 16–76

8. A swimming pool has an octagon shape 15 ft on each side. If the distance from the center of the pool to the midpoint of a side is 18.1 ft, what is the area of the pool? The pool is protected by an octagonal cover. What is the measure of each angle of the cover?

10. Find the area of the layout shown in Fig. 16–77.

9. A No. 5 soccer ball covering is made of 12 colored pentagons 1.75 in. on a side and 20 white hexagons 3.19 in. on a side. If the distance from the midpoint of a side of a colored pentagon to the center of the pentagon is 1.20 in., how many square inches of the covering are colored?

11. Find the area of the layout shown in Fig. 16–78.

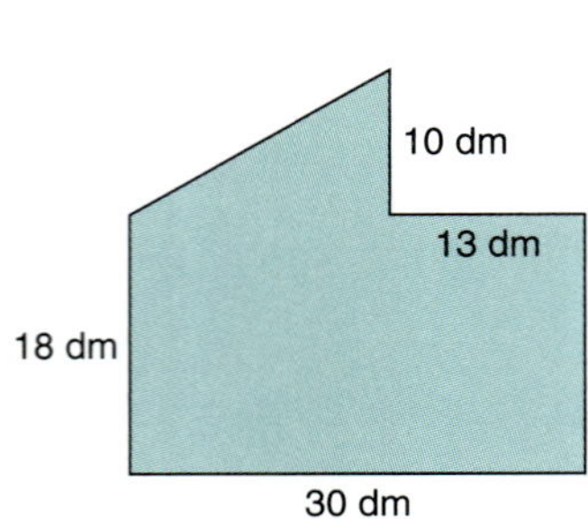

Figure 16–77

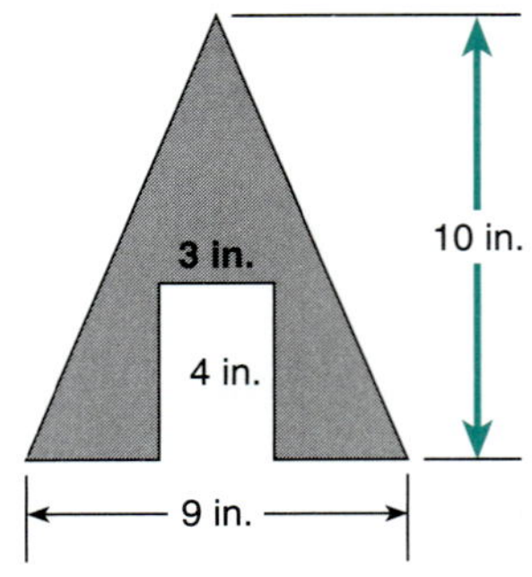

Figure 16–78

3 Give the specific name of each shape in Figs. 16–79 to 16–81 and the number of degrees in each angle.

12.

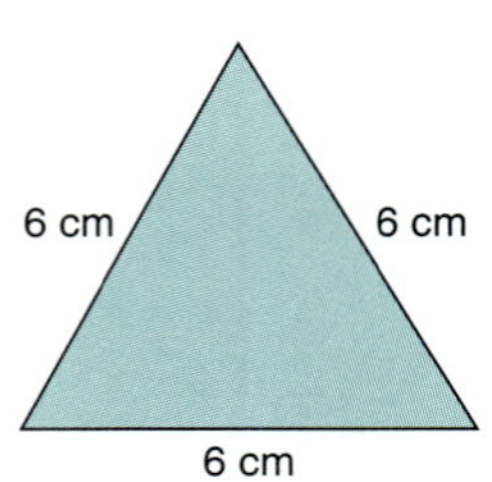

Figure 16–79

13.

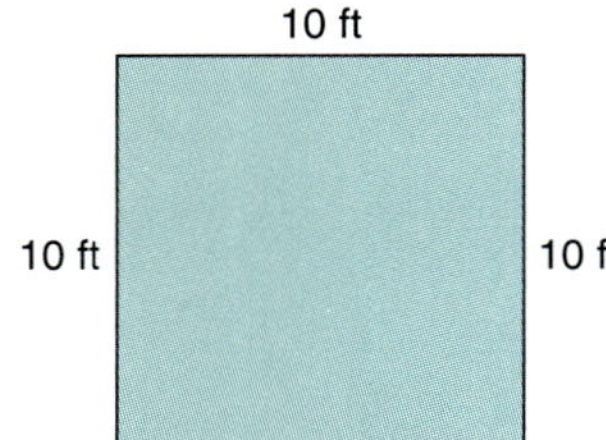

Figure 16–80

14.

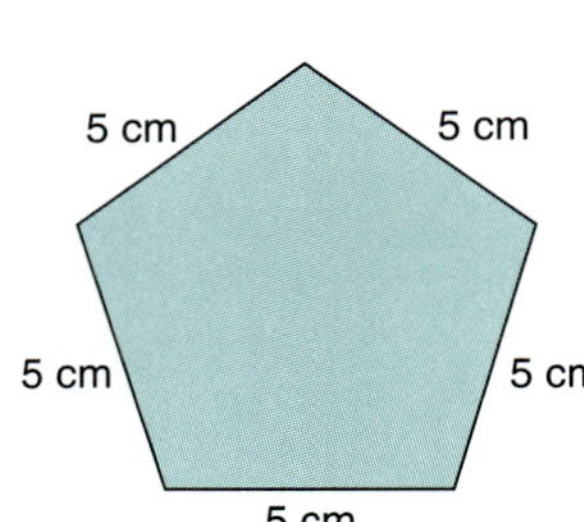

Figure 16–81

15. Identify the figure in Exercise 4 by giving its specific name and the number of degrees in each angle.

<h2>16–5   SECTORS AND SEGMENTS OF A CIRCLE</h2>

Learning Outcomes

1 Find the area of a sector.
2 Find the arc length of a sector.
3 Find the area of a segment.

We often work with figures that are less than a whole circle. For example, earlier we worked with the semicircle in several composite figures. *Sectors* and *segments* are both parts of a circle.

1 **Find the Area of a Sector.**

■ **DEFINITION: Sector.** A *sector* of a circle is the portion of the area of a circle cut off by two radii.

To find the area of a sector, calculate the portion of the circle taken up by the sector. In Fig. 16–82, the sector takes up 45° of the 360° of the whole circle. Represent this fractional part of the circle as $\frac{45}{360}$. The area of the sector is $\frac{45}{360}$ of the area of the circle.

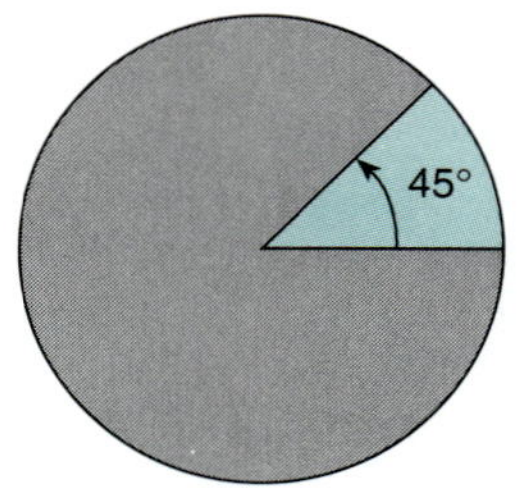

Figure 16–82

We use the Greek letter *theta* (θ) to represent the number of degrees in the angle of the sector, and express the area of a sector as shown in the following formula.

Area of a sector:

$$A = \frac{\theta}{360}\,\pi r^2$$

where $\frac{\theta}{360}$ is a fractional part of the circle, πr^2 is the area of the circle, and θ is a central angle measured in degrees.

EXAMPLE Find the area of a sector with a central angle of $45°$ in a circle with a radius of 10 in. Round to hundredths.

$$A = \frac{\theta}{360}\,\pi r^2$$

$$A = \frac{45}{360}\,(\pi)(10)^2 \qquad \text{Substitute for } \theta \text{ and } r.$$

$$A = 0.125\,(\pi)(100) \qquad \text{Perform the indicated operations.}$$

$$A = 39.26990817 \text{ in}^2$$

The area of the sector is 39.27 in.^{2}.

EXAMPLE A cone is made from sheet metal. To form a cone, a sector with a central angle of $40°20'$ is cut from a metal circle whose diameter is 20 in. Find the area of the stretchout (portion of the circle) formed into the cone to the nearest hundredth (Fig. 16–83).

Area used for cone = Area of circle − Area of sector

Circle

$$A_1 = \pi r^2$$

$$A_1 = \pi(10)^2$$

$$A_1 = \pi(100)$$

$$A_1 = 314.1592654 \text{ in.}^2$$

Sector

$$A_2 = \frac{\theta}{360}\,\pi r^2$$

$$A_2 = \frac{40.3\overline{3}}{360}\,(314.1592654)$$

$$A_2 = 0.112037037(314.1592654)$$

$$A_2 = 35.19747324 \text{ in.}^2$$

Convert $40°20'$ to $40.333333°$. πr^2 is figured as $314.1592654 \text{ in.}^2$.

Area used for cone = $A_1 - A_2$

$$A_3 = 314.1592654 - 35.19747324$$

$$A_3 = 278.96 \text{ in.}^2 \qquad \text{Rounded.}$$

The area of the metal sector used to form the cone is 278.96 in.^{2}.

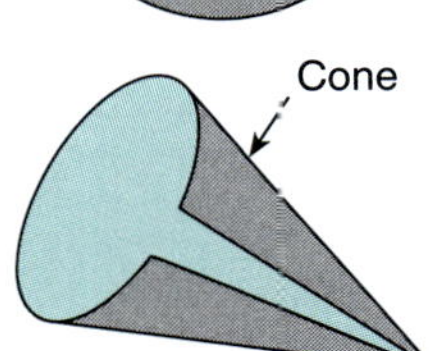

Figure 16–83

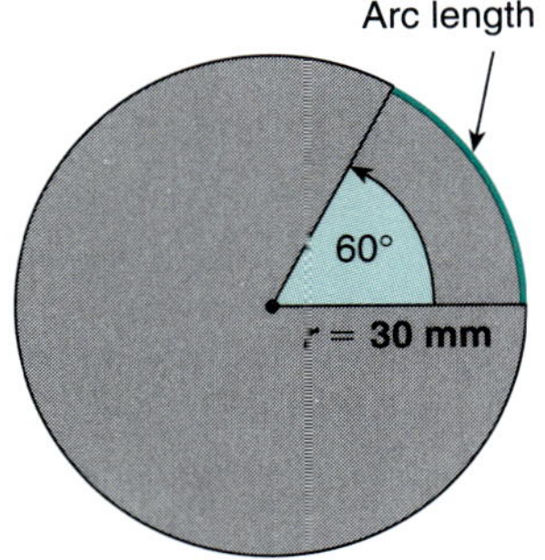

Figure 16–84

2   **Find the Arc Length of a Sector.**

We often need to find the arc length of a sector. The *arc length* is the portion of the circumference intercepted by the sides of the sector (see Fig. 16–84). A formula for finding *arc* is given in the following box.

Formula for arc length:

$$s = \frac{\theta}{360}(2\pi r) \qquad \text{or} \qquad s = \frac{\theta}{360}(\pi d)$$

where θ is a central angle measured in degrees.

EXAMPLE Find the arc length of the sector formed by a 60° central angle if the radius is 30 mm (Fig. 16–84).

$$s = \frac{\theta}{360}(2\pi r)$$

$$s = \frac{60}{360}(2)(\pi)(30) \qquad \text{Substitute values.}$$

$$s = 31.41592654$$

$$\mathbf{s = 31.42\ mm} \qquad \text{Rounded.}$$

3 Find the Area of a Segment.

If a line segment (called a *chord*) joins the end points of the radii that form a sector, the sector is divided into two figures, a triangle and a *segment* (Fig. 16–85).

Figure 16–85

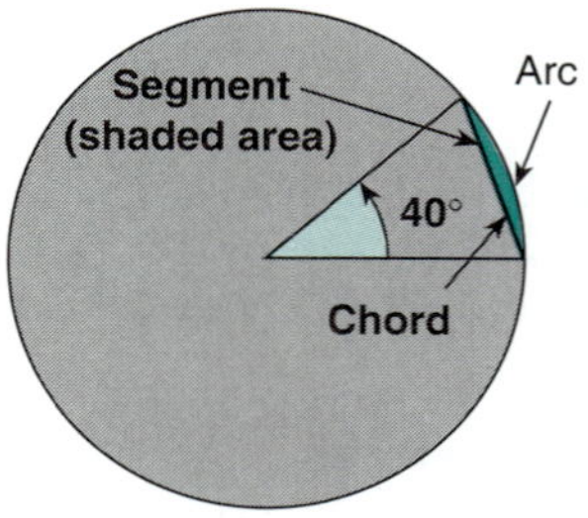

■ **DEFINITION: Chord.** A *chord* is a line segment joining two points on the circumference of a circle.

■ **DEFINITION: Arc.** The portion of the circumference cut off by a chord is an *arc*.

■ **DEFINITION: Segment.** A *segment* is the portion of the area of a circle bounded by a chord and an arc.

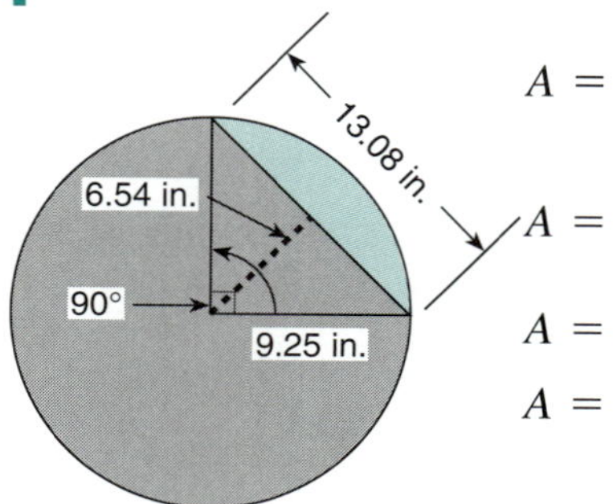

Because the chord divides the sector into a triangle and a segment, we can calculate the area of the segment by subtracting the area of the triangle from the area of the sector. Expressed symbolically, we have the formula for the area of a segment.

Area of a segment:

$$A = \tfrac{\theta}{360}\pi r^2 - \tfrac{1}{2}bh$$

$\tfrac{\theta}{360}\pi r^2 = \text{Area of sector} \qquad \tfrac{1}{2}bh = \text{Area of triangle}$

EXAMPLE Find the area of the segment in circle A of Fig. 16–86 to the nearest hundredth.

Figure 16–86

$$A = \frac{\theta}{360}\pi r^2 - \frac{1}{2}bh$$

$$A = \frac{90}{360}(\pi)(9.25)^2 - \frac{1}{2}(13.08)(6.54) \qquad \text{Substitute in formula.}$$

$$A = 67.20063036 - 42.7716$$

$$A = 24.43\ \text{in}^2 \qquad \text{Rounded.}$$

The area of the segment is 24.43 in.^{2}.

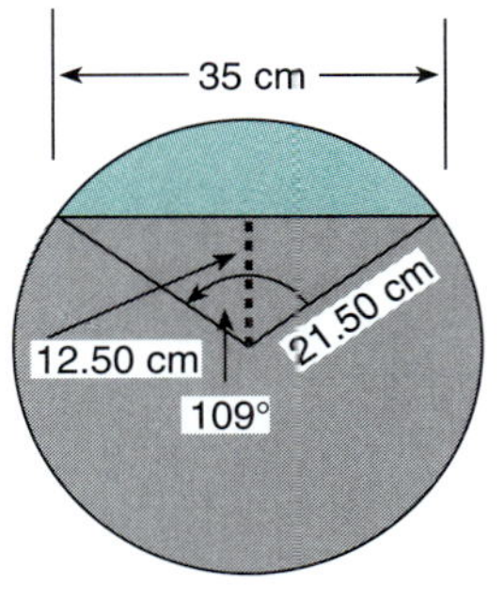

Figure 16–87

EXAMPLE A segment of circle B in Fig. 16–87 is removed so that a template for a cam is made from the rest of the circle. What is the area of the template? Give the answer to the nearest hundredth.

Area of template = Area of circle − Area of segment

Circle

$$A_1 = \pi r^2$$

$$A_1 = \pi(21.50)^2$$

$$A_1 = 1,452.201204 \text{ cm}^2$$

Segment

$$A_2 = \frac{\theta}{360}\pi r^2 - \frac{1}{2}bh$$

$$A_2 = \frac{109}{360}(1,452.201204) - \frac{1}{2}(35)(12.50)$$

$$A^2 = 220.9442535 \text{ cm}^2$$

Area$_3$ (template) $= A_1 - A_2$

$$A_3 = 1,452.201204 - 220.9442535$$

$$A_3 = 1,231.26 \text{ cm}^2 \qquad \text{Rounded.}$$

The area of the template is 1,231.26 cm^2.

SELF-STUDY EXERCISES 16–5

1 Find the area of the sectors of a circle using Fig. 16–88. Round to hundredths.

1. $\angle = 54°$
   $r = 16$ cm

2. $\angle = 25°16'$
   $r = 30$ mm

3. $\angle = 120°30'$
   $r = 1.52$ ft

4. $\angle = 65°$
   $r = 5$ in.

5. $\angle = 150°$
   $r = 1.45$ m

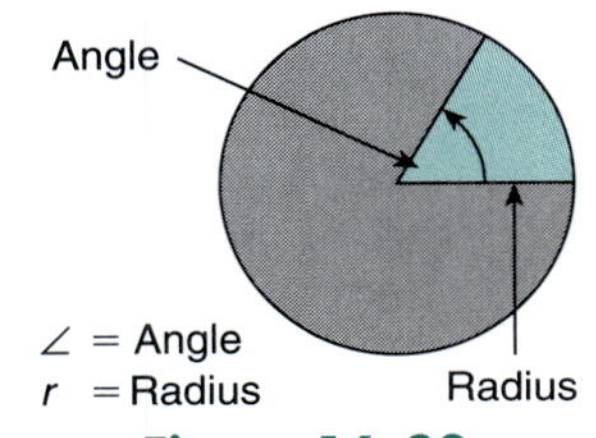

Figure 16–88

6. The library of a contemporary elementary school is circular (Fig. 16–89). The floor plan includes sectors reserved for science materials, literary materials, reference materials, and so on. Find the area of the reference section excluding its storage area.

7. A mason lays a tile mosaic featuring a four-sector design (color portion of Fig. 16–90). What is the area of the design to the nearest hundredth?

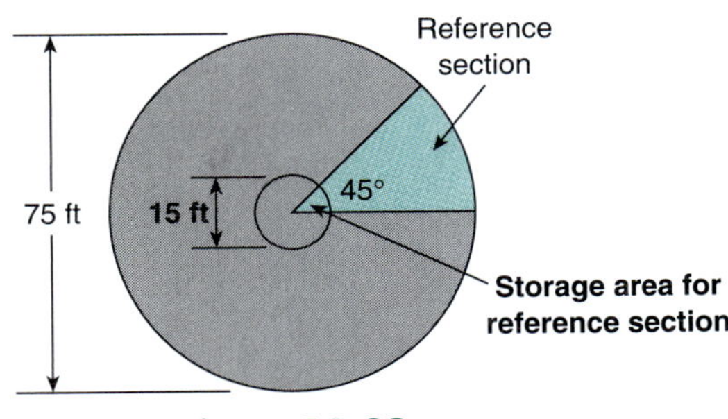

Figure 16–89

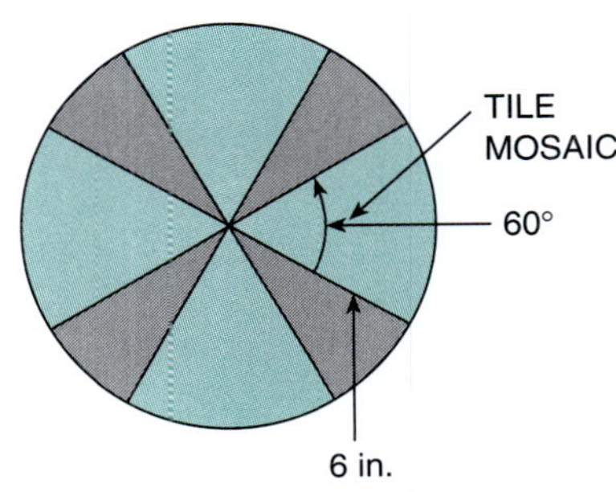

Figure 16–90

8. If the angle formed by a pendulum swing is 19°30′ and the pendulum is 10 in. long (Fig. 16–91), what area does the pendulum move across as it swings? Express your answer to the nearest tenth. (60′ = 1°.)

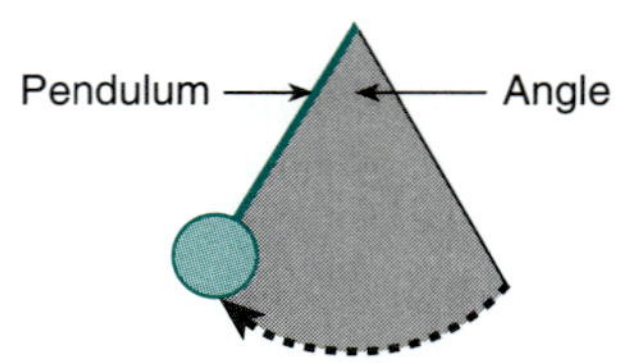

Figure 16–91

2 Find the arc length of the sectors of a circle. Round to hundredths.

9. ∠ = 54°
  r = 30 mm

10. ∠ = 150°
  r = 5 in.

11. ∠ = 120°30′
  r = 1.52 ft

12. ∠ = 25°16′
  r = 16 cm

13. ∠ = 65°
  r = 1.45 m

14. ∠ = 90°
  r = 10 in.

3 Find the area of the segments of a circle using Fig. 16–92. Round to hundredths.

15. ∠ = 60°
  r = 13.3 cm
  h = 11.52 cm
  b = 13.3 cm

16. ∠ = 110°
  r = 10 in.
  h = 5.74 in.
  b = 16.38 in.

17. ∠ = 105°
  r = 11.25 in.
  h = 6.85 in.
  b = 17.85 in.

18. ∠ = 60°
  r = 24 cm
  h = 20.8 cm
  b = 24 cm

19. ∠ = 108°
  r = 14 in.
  h = 8.25 in.
  b = 22.65 in.

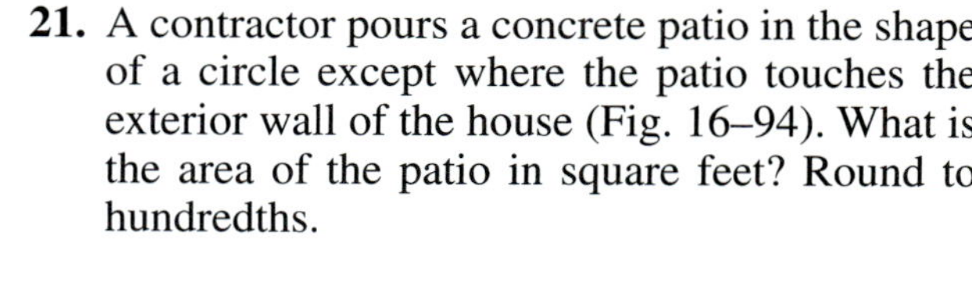

Figure 16–92

20. A motor shaft has milled on it a flat for a setscrew to rest so that it can hold a pulley on the shaft (Fig. 16–93). What is the cross-sectional area of the shaft after being milled? Round to hundredths.

21. A contractor pours a concrete patio in the shape of a circle except where the patio touches the exterior wall of the house (Fig. 16–94). What is the area of the patio in square feet? Round to hundredths.

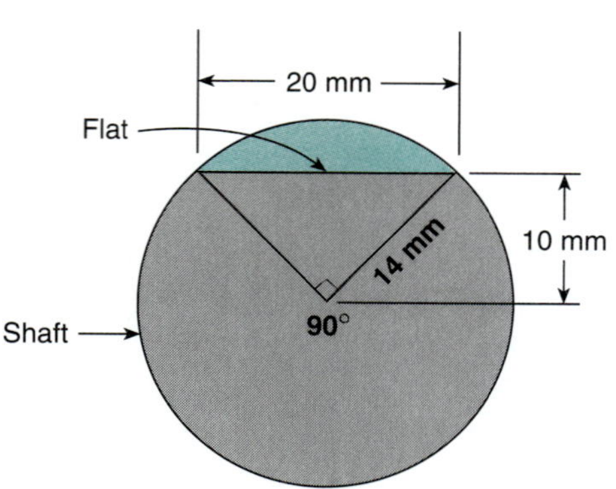

Figure 16–93

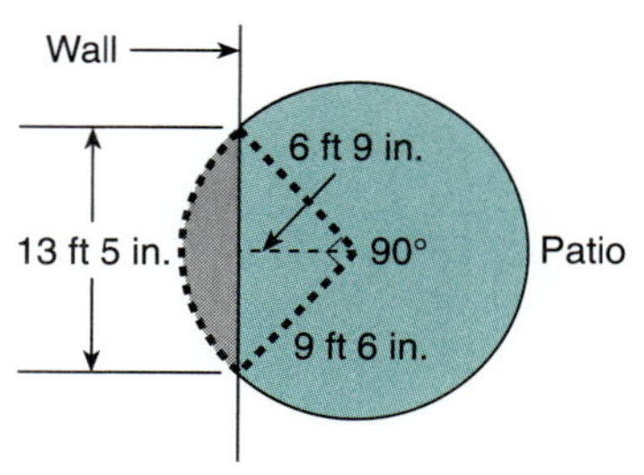

Figure 16–94

Learning Outcomes

1. Use the properties of inscribed and circumscribed equilateral triangles to find missing amounts and to solve applied problems.

2. Use the properties of inscribed and circumscribed squares to find missing amounts and to solve applied problems.

3. Use the properties of inscribed and circumscribed hexagons to find missing amounts and to solve applied problems.

Previously, we studied polygons and their areas and perimeters. In this section, we study regular polygons *inscribed* in a circle and *circumscribed* about a circle.

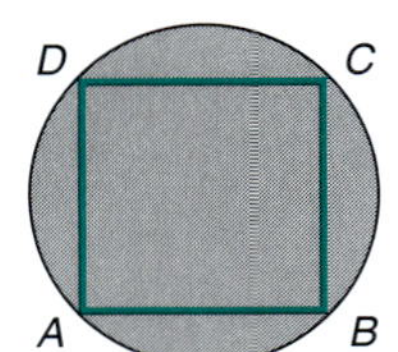

Inscribed square:
Vertices lie on circle
at points *A*, *B*, *C*, and *D*.

■   **DEFINITION: Inscribed Polygon.** A polygon is *inscribed in a circle* when it is inside the circle and all its *vertices* (points where sides of each angle meet) are on the circle. The circle is said to be *circumscribed about the polygon* (Fig. 16–95).

■   **DEFINITION: Circumscribed Polygon.** A polygon is *circumscribed about a circle* when it is outside the circle and all its sides are *tangent* to (intersecting in exactly one point) the circumference. The circle is said to be *inscribed in the polygon* (Fig. 16–95).

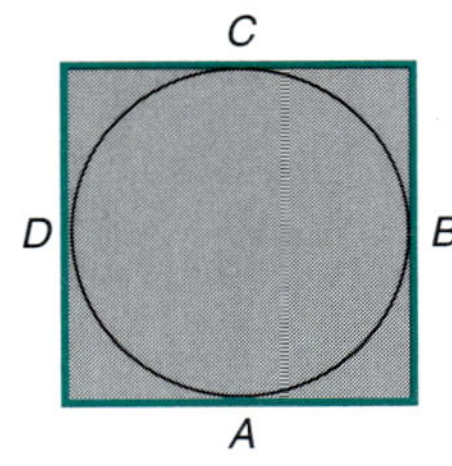

Circumscribed square:
Sides of square are tangent
at points *A*, *B*, *C*, and *D*.

Figure 16–95

In this section, we examine the three most common polygons inscribed in or circumscribed about a circle: the triangle, the square, and the hexagon.

1. **Use the Properties of Inscribed and Circumscribed Equilateral Triangles to Find Missing Amounts and to Solve Applied Problems.**

An equilateral triangle can be inscribed in a circle or circumscribed about a circle. If the height (or altitude) is drawn, two congruent right triangles are formed, as shown in Fig. 16–96. Each congruent triangle formed in this way is a 30°, 60°, 90° triangle in which the hypotenuse is twice the shortest side. The special 30°, 60°, 90° triangle and its properties may be reviewed in Section 16–3.

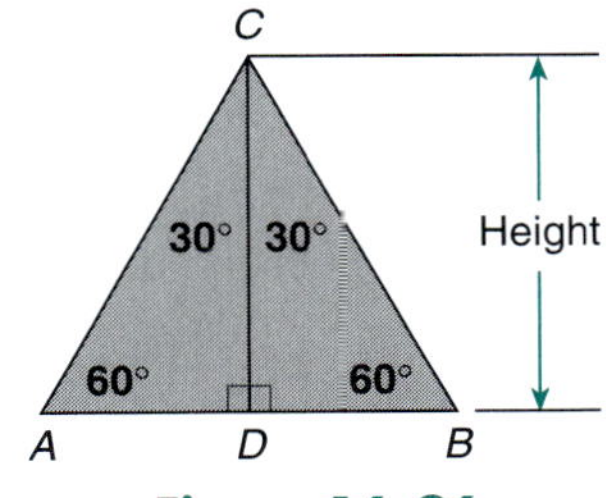

Figure 16–96

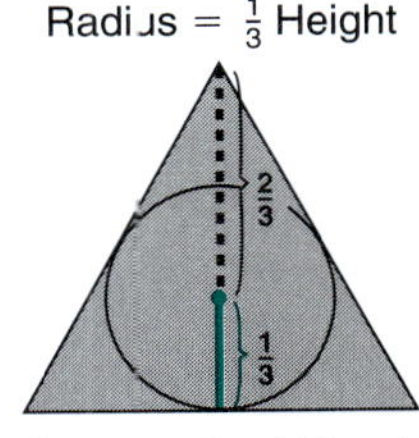

Figure 16–97

Examine Fig. 16–97. When the equilateral triangle is inscribed in a circle, the radius (from the vertex of the triangle to the center of the circle) is two-thirds the height of the triangle. When the equilateral triangle is circumscribed about a circle, the radius (from the center of the circle to the base of the triangle) is one-third the height of the triangle.

Height of an equilateral triangle:

The height of an equilateral triangle forms two congruent 30°, 60°, 90° right triangles.

Radius of a circle circumscribed about an equilateral triangle:

The radius of a circle circumscribed about an equilateral triangle is two-thirds the height of the triangle or twice the distance from the center of the circle to the base of the triangle.

Radius of a circle inscribed in an equilateral triangle:

The radius of a circle inscribed in an equilateral triangle is one-third the height of the triangle or half the distance from the vertex of the triangle to the center of the circle.

EXAMPLE Find the dimensions for the inscribed equilateral triangle in Fig. 16–98, where $CD = 13$ mm and $AO = 15$ mm.

(a) Height of $\triangle ABC$ (b) BC (c) BD (d) $\angle CAD$
(e) $\angle ACD$ (f) Area of $\triangle ABC$

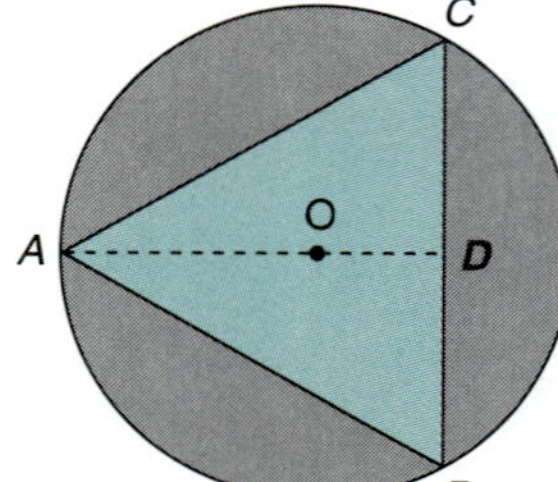

Figure 16–98

(a) $\qquad AO = \dfrac{2}{3}AD \qquad\qquad$ $AO = \frac{2}{3}$ height

$\qquad\qquad 15 = \dfrac{2}{3}AD \qquad\qquad$ Substitute.

$\qquad \dfrac{3}{2}(15) = \left(\dfrac{3}{2}\right)\dfrac{2}{3}AD \qquad$ Solve for AD.

$\qquad\qquad \dfrac{45}{2} = AD$

$\qquad$ **22.5 mm $= AD$**

Or use the formula for finding a leg of a 30°, 60°, 90° triangle (Section 16–3).

$AD = CD\sqrt{3}$

$AD = 13\sqrt{3}$

$\mathbf{AD = 22.5}$ **mm**   Rounded.

(b) $BC = 2CD$ 30°, 60°, 90° $\triangle$

$\quad BC = 2(13)$

$\quad \mathbf{BC = 26\ mm}$

(c) $\mathbf{BD = 13\ mm}$ $\triangle ADC \cong \triangle ADB$; $BD = CD$

(d) $\mathbf{\angle CAD = 30°}$

(e) $\mathbf{\angle ACD = 60°}$

(f) $A = \dfrac{1}{2}bh$

$\quad A = \dfrac{1}{2}(BC)(AD)$

$$A = \frac{1}{2}(26)(22.5)$$

$$A = 292.5 \text{ mm}^2$$

EXAMPLE One end of a shaft 35 mm in diameter is milled as shown in Fig. 16–99. Find the radius of the smaller circle.

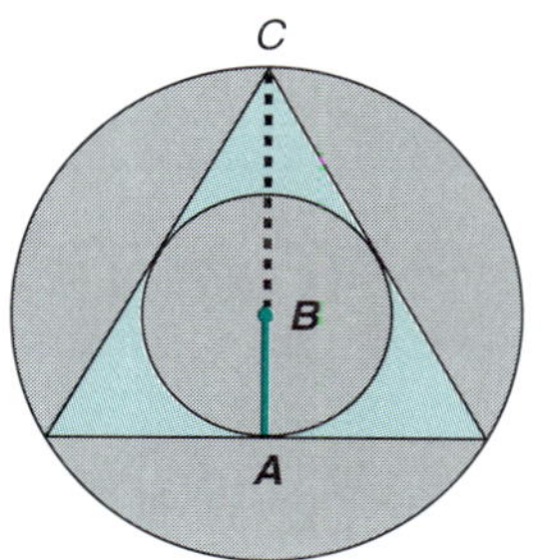

Figure 16–99

$$BC = \frac{1}{2}(\text{Diameter of large circle})$$

$$BC = \frac{1}{2}(35)$$

$$BC = 17.5 \text{ mm}$$

$$AB = \frac{1}{2}BC$$

$$AB = \frac{1}{2}(17.5)$$

$$AB = 8.75 \text{ mm}$$

The radius of the smaller circle is 8.75 mm.

EXAMPLE Find the area of the larger circle in Fig. 16–99. Round to hundredths.

$$A = \pi r^2 = \pi(17.5)^2 = 962.11 \text{ mm}^2 \qquad \text{Rounded.}$$

The area of the larger circle is 962.11 mm^2.

2 Use the Properties of Inscribed and Circumscribed Squares to Find Missing Amounts and to Solve Applied Problems.

A square has four equal sides and four equal angles, each 90°. If a diagonal is drawn, the result is two congruent right triangles that have angles of 45°, 45°, 90°. A second diagonal divides the square into four 45°, 45°, 90° congruent right triangles, as shown in Fig. 16–100. The special 45°, 45°, 90° triangle and its properties may be reviewed in Section 16–3. Diagonals AC and BD are equal in length and bisect each other. Point O, where the diagonals intersect, is the center of the square and the center of any inscribed or circumscribed circle (Fig. 16–101).

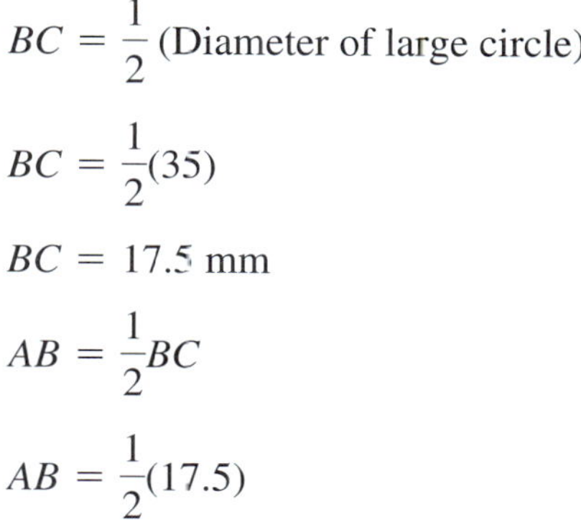

Figure 16–100

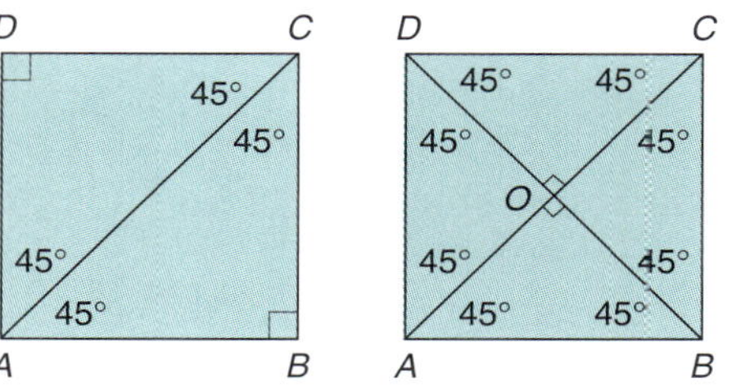

Figure 16–101

When the circle is inscribed in the square, the diameter (such as EF) equals the length of a side. The radius (such as FO) equals the height of a 45°, 45°, 90° triangle formed (such as $\triangle AOB$) by the two diagonals.

When the circle is circumscribed about the square, the diameter (such as LJ) equals the diagonal of the square. The radius (such as KO) is the height of a 45°, 45°, 90° right triangle (such as $\triangle JKL$) formed by one diagonal.

> **Diagonals of a square:**
>
> The diagonals of a square form congruent 45°, 45°, 90° triangles.

> **Diameter of a circle inscribed in a square:**
>
> The diameter of a circle inscribed in a square equals a side of the square.

> **Diameter of a circle circumscribed about a square:**
>
> The diameter of a circle circumscribed about a square equals a diagonal of the square.

> **Radius of a circle inscribed in a square:**
>
> The radius of a circle inscribed in a square equals the height of a 45°, 45°, 90° triangle formed by two diagonals.

> **Radius of a circle circumscribed about a square:**
>
> The radius of a circle circumscribed about a square equals the height of a 45°, 45°, 90° triangle formed by one diagonal.

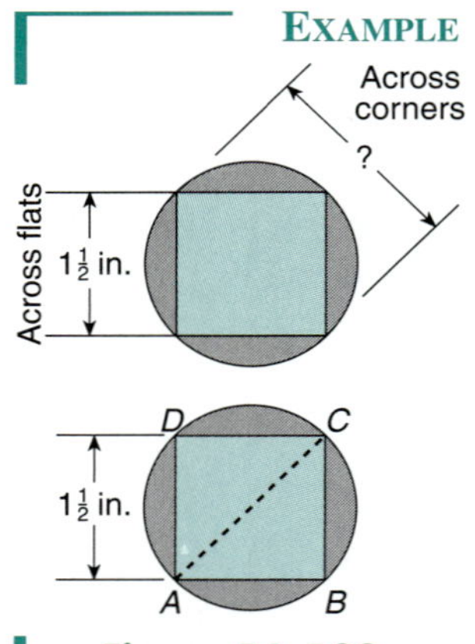

EXAMPLE To mill a square bolt head $1\frac{1}{2}$ in. across flats, round stock of what diameter is needed? Answer to the nearest hundredth (Fig. 16–102).

The distance across corners is the diameter of the circle and the diagonal of the square. It is also the hypotenuse of the two 45°, 45°, 90° triangles formed by one diagonal. Use the property of 45°, 45°, 90° triangles to find the hypotenuse.

$$\text{Hypotenuse} = \text{Leg } \sqrt{2}$$
$$\text{Hypotenuse} = 1.5\sqrt{2} \qquad 1\tfrac{1}{2}\text{ in.} = 1.5\text{ in.}$$
$$\text{Hypotenuse} = 2.12 \text{ in.} \qquad \text{Rounded.}$$

Figure 16–102

The diameter of the round stock is 2.12 in.

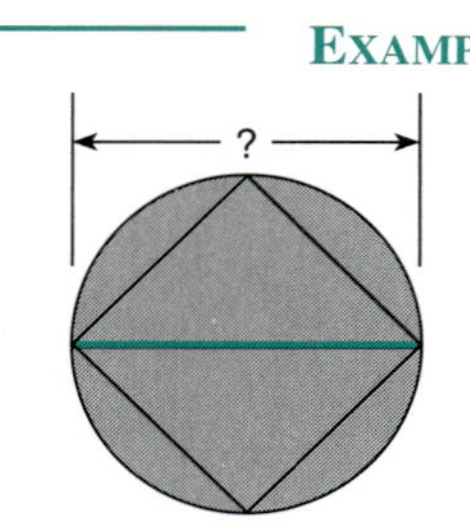

EXAMPLE Find the diagonal of a square if the area of the circumscribed circle is 22 in.^{2}. Round to hundredths (Fig. 16–103).

$$A = \pi r^2$$
$$22 = \pi r^2 \qquad \text{Substitute.}$$
$$\frac{22}{\pi} = r^2$$
$$\sqrt{\frac{22}{\pi}} = r$$

Figure 16–103

$$2.646283714 = r$$

Diagonal = Diameter = 2 × Radius = 2(2.646283714) = 5.29 in.   Rounded.

The diagonal of the inscribed square is 5.29 in.

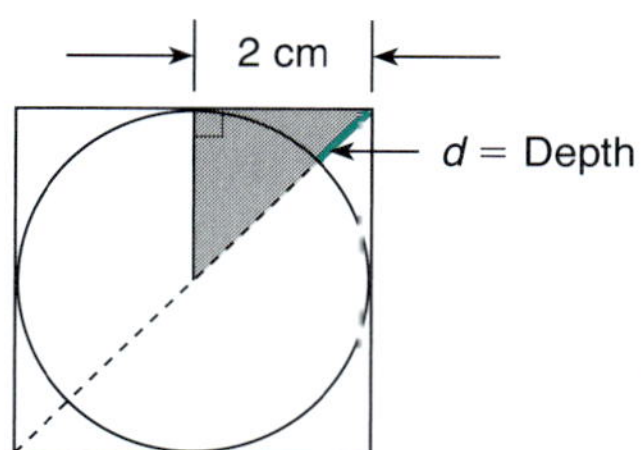

Figure 16–104

EXAMPLE   What is the minimum depth of cut required to mill a circle on the end of a square piece of stock with a side measuring 4 cm? Give the answer to hundredths (Fig. 16–104).

The side of the square is 4 cm, so the diameter of the circle is 4 cm, and the radius is 2 cm. If the radius is drawn perpendicular to the top side, a right triangle is formed as indicated (shading).

Let d = depth of milling, which equals the hypotenuse of the shaded triangle minus the radius of the circle. Use the Pythagorean theorem, where c = hypotenuse.

$$c^2 = a^2 + b^2$$
$$c^2 = 2^2 + 2^2$$
$$c^2 = 4 + 4$$
$$c^2 = 8$$
$$c = \sqrt{8}$$
$$c = 2.828427125 \text{ cm}$$

Hypotenuse − Radius = Depth of milling.

2.828427125 − 2 = 0.83 cm   Rounded.

The minimum depth of milling is 0.83 cm.

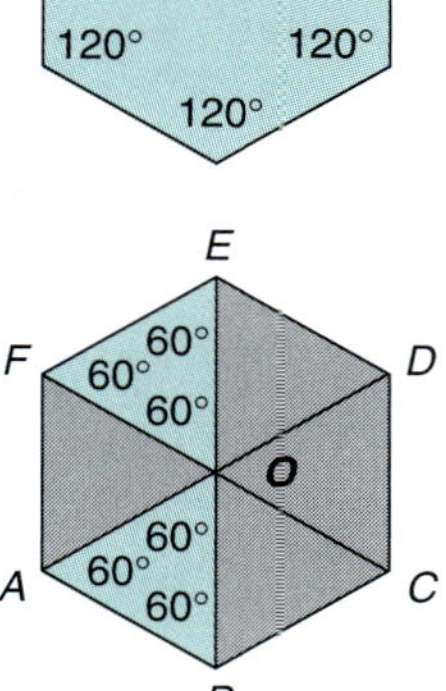

Figure 16–105

3 Use the Properties of Inscribed and Circumscribed Hexagons to Find Missing Amounts and to Solve Applied Problems.

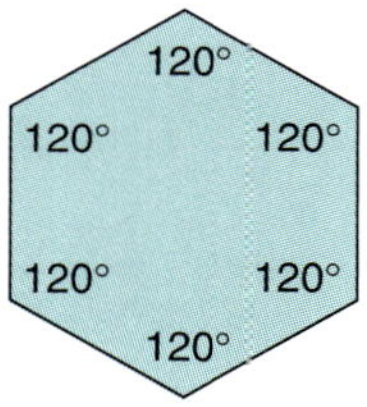

A regular hexagon is a figure with six equal sides and six equal angles of 120° each. Three diagonals joining pairs of opposite vertices divide the hexagon into six congruent equilateral triangles (all angles 60°) (see Fig. 16–105).

In Fig. 16–106, for the inscribed circle, the height of an equilateral triangle (GO) is the radius, and for the circumscribed circle, any side of the equilateral triangles (AO) equals the radius. The height divides any of the six triangles into two congruent 30°, 60°, 90° right triangles.

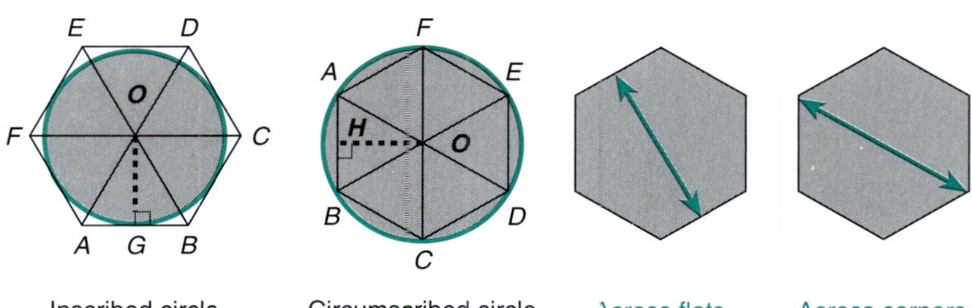

Inscribed circle   Circumscribed circle   Across flats   Across corners

Figure 16–106

Diagonals join opposite vertices of a regular hexagon:

The three diagonals that join opposite vertices of a regular hexagon form six congruent equilateral triangles.

EXAMPLE Find the indicated dimensions. When appropriate, round to hundredths. Use Fig. 16–107 for (a)–(f) and Fig. 16–108 for (g)–(j).

(a) $\angle EOD$ (b) $\angle COH$ (c) $\angle BHO$ (d) Radius (e) Distance across corners
(f) FO (g) Distance across flats (h) EO (i) Diagonal (j) $\angle FED$

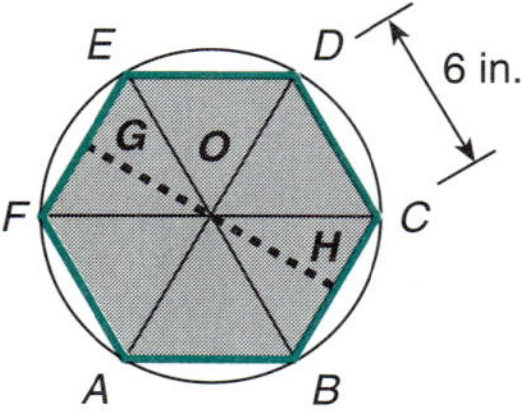

Figure 16–107

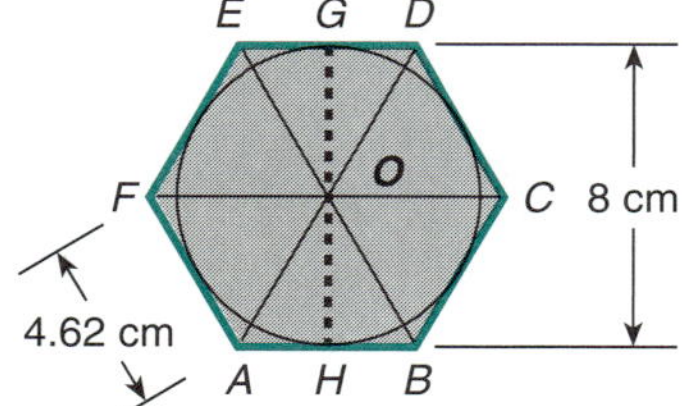

Figure 16–108

(a) $\angle EOD = 60°$ This is an angle of an equilateral triangle.

(b) $\angle COH = 30°$ The height bisects the 60° angle to form a 30°, 60°, 90° triangle.

(c) $\angle BHO = 90°$ The height forms a right angle with the base.

(d) Radius = 6 in.   The radius is a side of the equilateral triangle.

(e) Distance across corners = 12 in.   The distance across corners is the diameter.

(f) $FO = 6$ in.   This is a side of an equilateral triangle.

(g) Distance across flats $= 8$ cm   This is the diameter of the inscribed circle.

(h) $EO = 4.62$ cm   An equilateral triangle has equal sides.

(i) Diagonal $= 9.24$ cm   The diagonal is twice the side of the equilateral triangle.

(j) $\angle FED = 120°$ This is the angle at the vertex of a hexagon.

EXAMPLE If a regular hexagon is cut from a circle with a radius of 6.45 in., how much of the circle is not used? Round to hundredths (Fig. 16–109).

To solve, we find the area of the circle and the area of the hexagon and subtract to find the difference (portion not used).

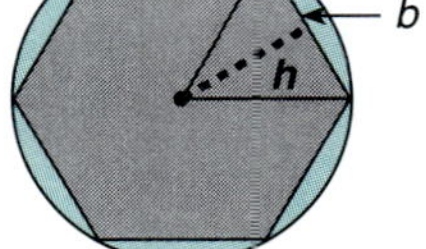

$$A_1 = \pi r^2 \qquad\qquad A_1 = \text{area of circle.}$$

$$A_1 = \pi(6.45)^2$$

$$A_1 = \pi(41.6025)$$

$$A_1 = 130.6981084 \text{ in.}^2$$

$$A_2 = 6\left(\frac{1}{2}bh\right)$$

$A_2 = $ area of six triangles.
Because the height of an equilateral triangle forms a 30°, 60°, 90° triangle, $h = \frac{b}{2}\sqrt{3} = 3.225\sqrt{3} = 5.585863854$ in.

Figure 16–109

$$A_2 = 6\left[\frac{1}{2}(6.45)(5.585863854)\right]$$

$$A_2 = 108.0864656 \text{ in.}^2$$

$$\text{Waste} = A_1 - A_2$$

$$\text{Waste} = 22.61 \text{ in.}^2 \qquad \text{Rounded.}$$

There are 22.61 in.^{2} of waste from the circle.

EXAMPLE Use Fig. 16–110 to answer the questions. Round answers to hundredths.

(a) What is the smallest-diameter round stock from which a hex-bolt head $\frac{1}{4}$ in. on a side can be milled?
(b) What is the distance across the corners of the hex-bolt head?
(c) What is the distance across the flats?

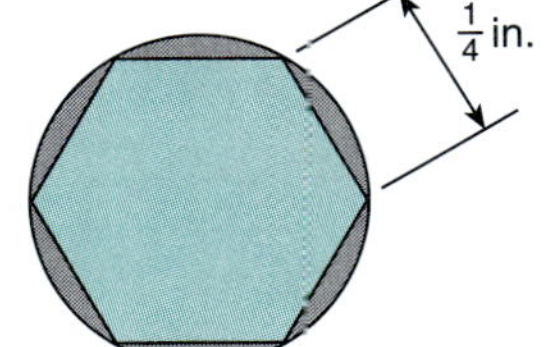

(a) Diagonal $= 2 \times$ Side $=$ Diameter $2 \times \frac{1}{4} = \frac{1}{2}$

Smallest-diameter round stock $= \frac{1}{2}$ in.

(b) Diagonal $=$ Distance across corners

Distance across corners $= \frac{1}{2}$ in.

(c) Use the properties of a 30°, 60°, 90° right triangle.

$$a = b\sqrt{3} \qquad\qquad \begin{aligned} a &= \text{Height of 30°, 60°, 90° triangle.} \\ b &= \text{Base of 30°, 60°, 90° triangle.} \end{aligned}$$

$$a = \frac{1}{8}\sqrt{3}$$

Figure 16–110

$$a = 0.21650635 \text{ in.}$$

Distance across flats $= 2 \times$ Height $= 2 \times 0.21650635 = 0.43$ in.   To nearest hundredth.

The distance across the flats is 0.43 in.

SELF-STUDY EXERCISES 16–6

1 Use Fig. 16–111 to find the dimensions for an equilateral triangle inscribed in a circle. Round to hundredths.

1. $\angle CAB$	**2.** $\angle BCD$	**3.** $\angle ADC$
4. CB	**5.** AC	**6.** Diameter
7. Area of $\triangle ABC$	**8.** AD	**9.** Radius
10. Circumference		

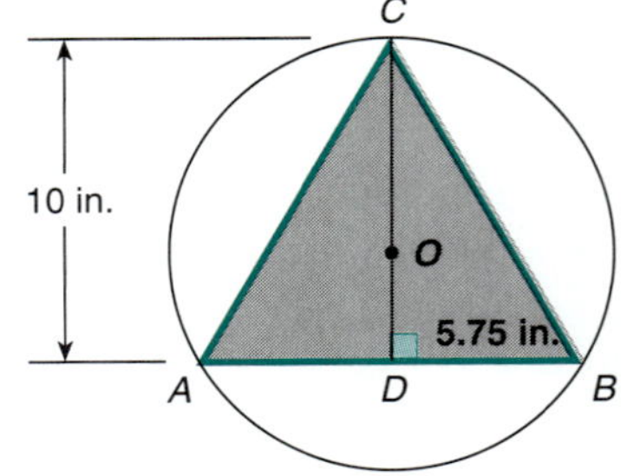

Figure 16–111

Solve the problems involving inscribed or circumscribed equilateral triangles. Round to hundredths.

11. The area of an equilateral triangle with a base of 10 cm is 43.5 cm². The triangle is circumscribed about a circle. What is the radius of the inscribed circle?

12. One end of a circular steel rod is milled into an equilateral triangle whose height is $\frac{3}{4}$ in. What is the diameter of the rod if the smallest diameter possible was used for the job?

2 Find the measures of the inscribed square in Fig. 16–112.

13. $\angle DOC$	**14.** $\angle BCO$	**15.** BO
16. $\angle BOE$	**17.** CE	

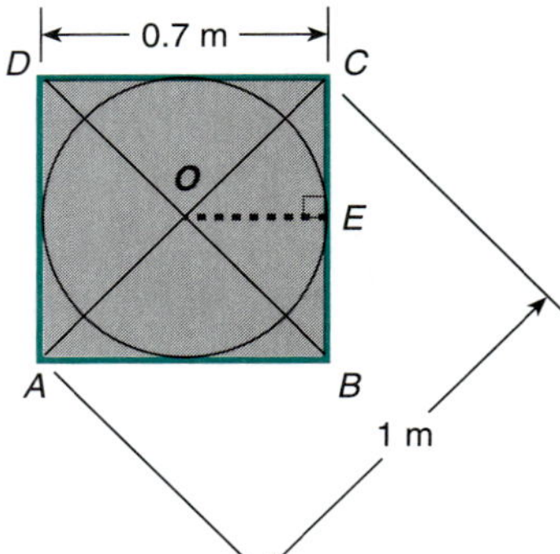

Figure 16–112

Find the measures of the circumscribed square in Fig. 16–113.

18. DO	**19.** Radius of circle
20. $\angle AOB$	**21.** BE

Figure 16–113

Solve the problems involving squares and circles. Round the answers to hundredths.

22. To what depth must a 2-in. shaft (Fig. 16–114) be milled on an end to form a square 1.414 in. on a side?

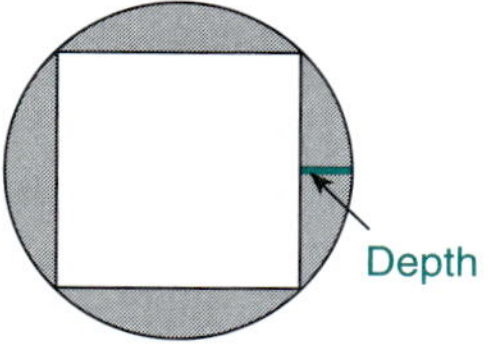

Figure 16–114

23. What is the smallest-diameter round stock (Fig. 16–115) needed to mill a square $\frac{3}{4}$ in. on a side on the end of the stock?

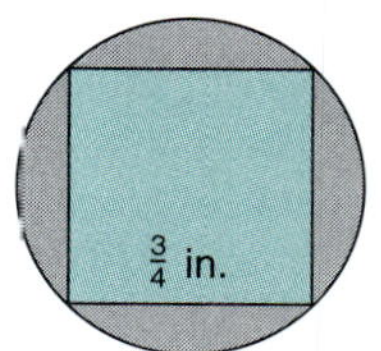

Figure 16–115

24. What is the area of the largest circle inscribed in a square with a perimeter of 36 dkm?

3 Find the dimensions for Fig. 16–116 if $GO = 17.32$ mm and $FO = 20$ mm.

25. Distance across flats
27. $\angle EFO$
29. $\angle FGO$
31. AB

26. $\angle CDE$
28. Distance across corners
30. GA

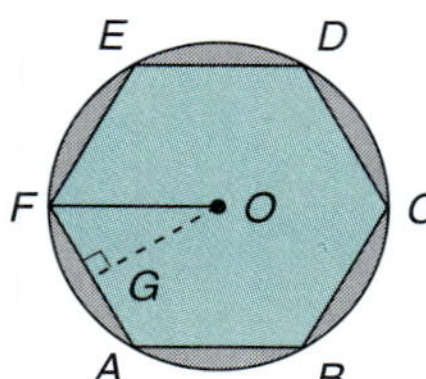

Figure 16–116

Find the dimension for Fig. 16–117 if $FO = 3$ in.

32. $\angle ODC$
33. $\angle EOG$
34. OC

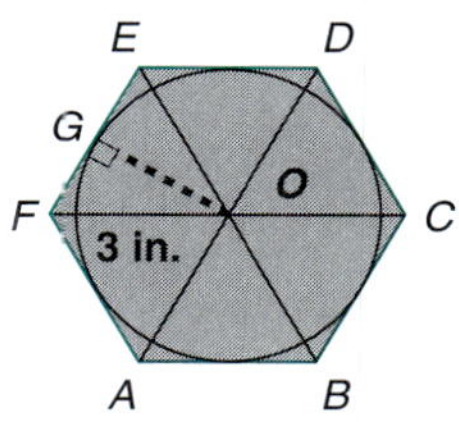

Figure 16–117

35. What is the distance across the flats of a hexagon cut from a circle blank with a circumference of 145 mm (Fig. 16–118)? Round to hundredths.

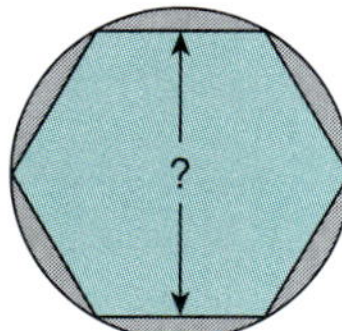

Figure 16–118

36. The end of a hexagonal rod is milled into the largest circle possible (Fig. 16–119). What is the distance across the corners of the hexagon if the area of the circle is 4.75 in.^{2}? Round to hundredths.

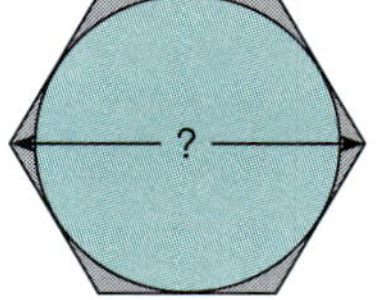

Figure 16–119

Electronics: Phase Angle

The phase angle is very important in electronics. We use it to compare voltage and current.

An inductive circuit contains an inductor or coil, usually denoted by the symbol L, and the inductance across the coil is measured in henrys (H). If an alternating current (ac) is applied, the voltage in that circuit always leads the current. This can be measured on an oscilloscope. We also say that the current lags the voltage.

Assume that the sine waves that represent the voltage and the current are shown on the oscilloscope. The two sine waves are slightly offset from each other. The distance of separation is constant and is called the *phase angle*.

Suppose that one full cycle (from zero up to the maximum and then back to zero and to the negative minimum and then back to zero) measures 8 cm on the oscilloscope, and the amount of offset is 1 cm. The two sine waves are $\frac{1}{8}$ of a cycle apart. One cm of offset can be converted to degrees by taking $360° \times \frac{1}{8} = 45° = $ the phase angle. If the offset is 0.8 cm, then the phase angle $= 360° \times \frac{0.8}{8} = 36°$.

If the inductive circuit is a simple series circuit with only one inductor and only one resistor, and an alternating current is applied, then the voltage measured across the inductor (called V_L) leads the voltage across the resistor (called V_R) by 90°. We can state this result by saying that the voltage across the resistor lags the voltage across the inductor by 90°. This can be measured with an oscilloscope.

If the circuit is capacitive (rather than inductive) and is a simple series circuit with only one capacitor and only one resistor, and an alternating current is applied, the voltage measured across the capacitor (called V_C) lags the voltage across the resistor (called V_R) by 90°. We can state this result by saying that the voltage across the capacitor lags the voltage across the resistor by 90°. This can be measured with an oscilloscope.

When current and voltage are measured for a capacitive circuit, the current always leads the voltage, or the voltage always lags the current. When current and voltage are measured for an inductive circuit, the current always lags the voltage, or the voltage always leads the current. We can remember this by using

ELI the ICE man

where E = electromotive force (voltage), I = current, L = inductive circuit, and C = capacitive circuit.

Exercises

Complete the following chart, which shows readings and interpretations from two sine waves on an oscilloscope. For each pair of sine waves, is the circuit inductive or capacitive?

	Sine Wave 1	Sine Wave 2	Length of One Cycle (cm)	Amount of Offset (cm)	Phase Angle (°)	Inductive or Capacitive?
1.	E	I	8	3		
2.	E	I	6	2		
3.	E	I	5	1		
4.	I	E	4	0.5		
5.	I	E	8	0.5		
6.	I	E	6	1		
7.	V_L	V_R	5			
8.	V_R	V_L		1		
9.	V_C	V_R		2		
10.	V_R	V_C		1.5		

CHAPTER 16   Selected Concepts of Geometry

Answers for Exercises

	Sine Wave 1	Sine Wave 2	Length of One Cycle (cm)	Amount of Offset (cm)	Phase Angle (°)	Inductive or Capacitive?
1.	E	I	8	3	135	Inductive
2.	E	I	6	2	120	Inductive
3.	E	I	5	1	72	Inductive
4.	I	E	4	0.5	45	Capacitive
5.	I	E	8	0.5	22.5	Capacitive
6.	I	E	6	1	60	Capacitive
7.	V_L	V_R	5	1.25	90	Inductive
8.	V_R	V_L	4	1	90	Inductive
9.	V_C	V_R	8	2	90	Capacitive
10.	V_R	V_C	6	1.5	90	Capacitive

ASSIGNMENT EXERCISES

Section 16–1

Give the proper notation.

1. Line AB

2. Segment AB

3. Ray AB

Use Fig. 16–120 for Exercises 4–10.

4. Name the line in two different ways.

5. Name the ray with end point N and with an interior point O.

6. Name the ray with end point M and with an interior point L.

7. Is $\overline{MN}$ the same as $\overline{MO}$?

8. Is $\overleftrightarrow{NO}$ the same as $\overleftrightarrow{NM}$?

9. Is $\overrightarrow{NO}$ the same as $\overrightarrow{NM}$?

10. Is $\overrightarrow{MN}$ the same as $\overrightarrow{MO}$?

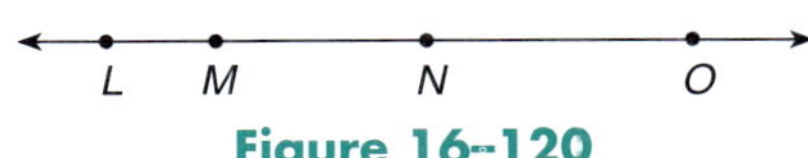

Figure 16–120

Use Fig. 16–121 for Exercises 11–15.

11. Which lines are parallel?
12. Which lines coincide?
13. Which lines intersect?

14. When will $\overleftrightarrow{AB}$ and $\overleftrightarrow{CD}$ meet?

15. Is $\overleftrightarrow{IJ}$ the same as $\overleftrightarrow{KJ}$?

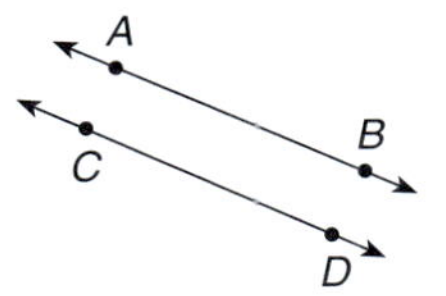

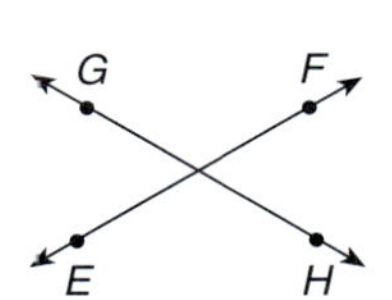

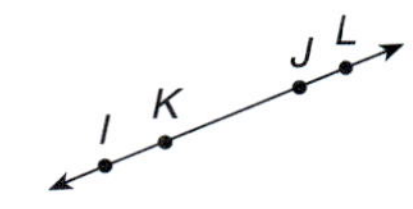

Figure 16–121

Use Fig. 16–122 for Exercises 16–21.

16. Name $\angle a$ using three capital letters.
18. Name $\angle c$ using three capital letters.
20. Name $\angle b$ using one capital letter.

17. Name $\angle b$ using three capital letters.
19. Name $\angle a$ using one capital letter.
21. Name $\angle c$ using one capital letter.

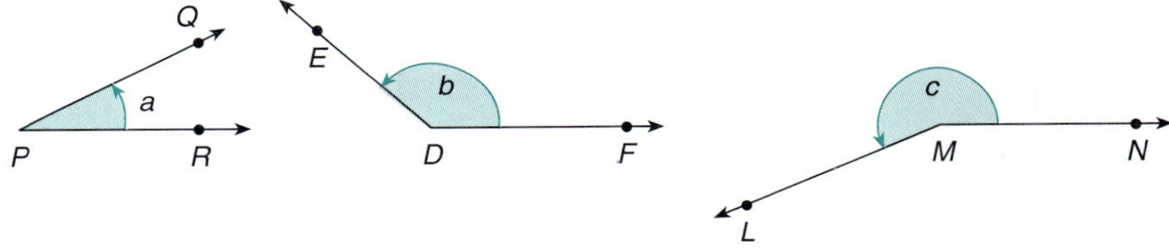

Figure 16–122

Classify the angle measures using the terms *right, straight, acute,* or *obtuse.*

22. 50° **23.** 90° **24.** 120° **25.** 18° **26.** 75°
27. 180° **28.** 89° **29.** 179° **30.** 30°

State whether the angle pairs are complementary, supplementary, or neither.

31. 63°, 37° **32.** 98°, 62° **33.** 135°, 45°
34. 45°, 35° **35.** 21°, 79° **36.** 90°, 90°
37. Congruent angles have ____ measures. **38.–39.** Perpendicular lines are formed when a ____ line and a ____ line intersect.

Section 16–2

Add or subtract as indicated. Write in standard notation.

40. $34°23'41''$
$+\ 18°37'50''$

41. $115°34'29''$
$-\ 84°26'18''$

42. $34°29'35''$
$+\ 19°30'25''$

43. $64°15'37''$
$-\ 29°37'41''$

44. $80°$
$-\ 28°14'28''$

45. $74°$
$-\ 13°19'42''$

46. Find the complement of an angle whose measure is $35°29'14''$.

Change to decimal degree equivalents. Express the decimals to the nearest ten-thousandth.

47. $29'$ **48.** $47''$ **49.** $7'34''$
50. An angle of $34°36'48''$ is divided into two equal angles. Express the measure of each angle in decimal degrees. Round to the nearest ten-thousandth.

Change to equivalent minutes and seconds. Round to the nearest second when necessary.

51. $0.75°$ **52.** $0.46°$ **53.** $0.2176°$

In the exercises, round angle measures to the nearest second when necessary.

54. A right angle is divided into eight equal parts. Find the measure of each angle in degrees and minutes.

55. An angle of $140°$ is divided into six equal parts. Find the measure of each angle in degrees and minutes.

56. An angle of $18°52'48''$ needs to be three times as large. Find the measure of the new angle in decimal degrees.

57. An angle of $37°19'41''$ is divided into two equal angles. Find the measure of each equal angle in degrees, minutes, and seconds.

Section 16–3

Classify the triangles in Figs. 16–123 to 16–125 according to their sides.

58. 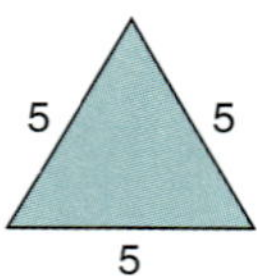

Figure 16–123

59. 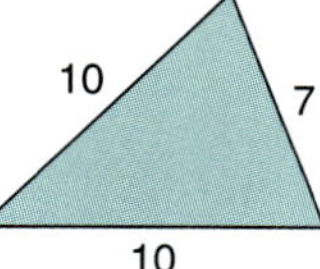

Figure 16–124

60. 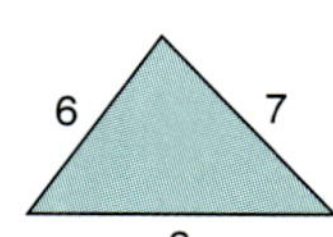

Figure 16–125

Identify the longest and shortest sides in Figs. 16–126 and 16–127.

61. 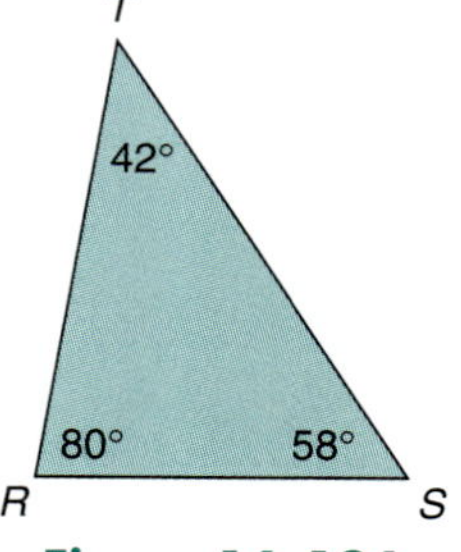

Figure 16–126

62. 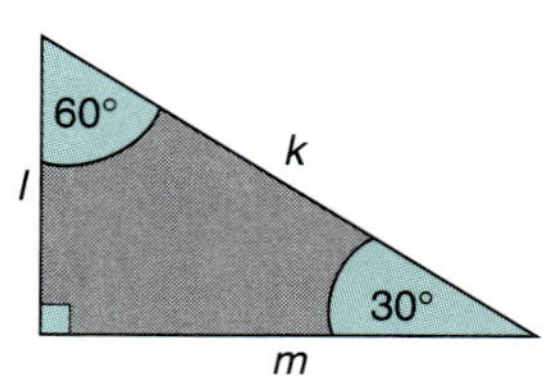

Figure 16–127

List the angles in order of size from largest to smallest in Figs. 16–128 and 16–129.

63.

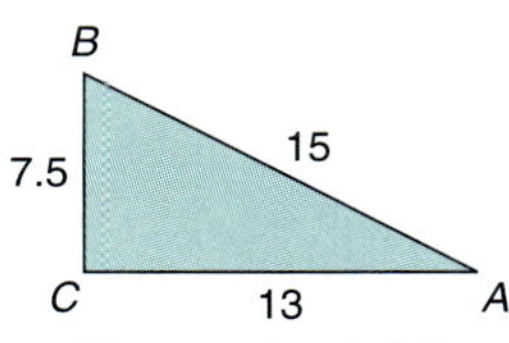

Figure 16–128

64.

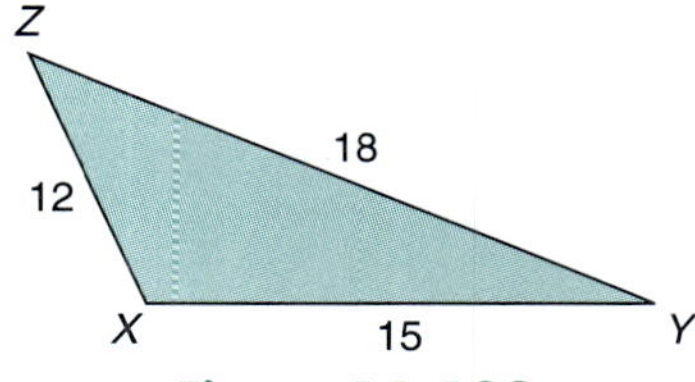

Figure 16–129

Write the corresponding parts not given for the congruent triangles in Figs. 16–130 through 16–132.

65.

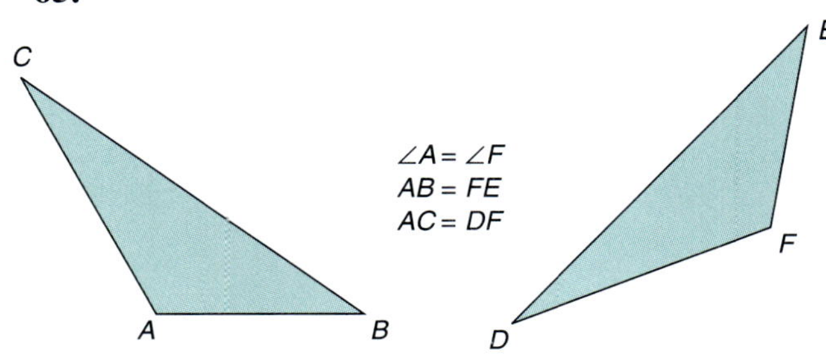

Figure 16–130

66.

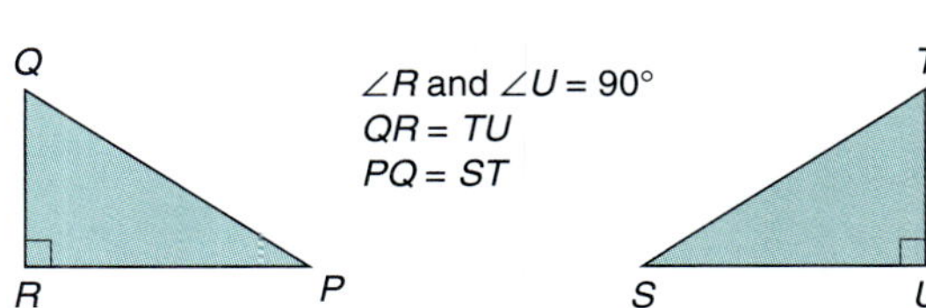

Figure 16–131

67.

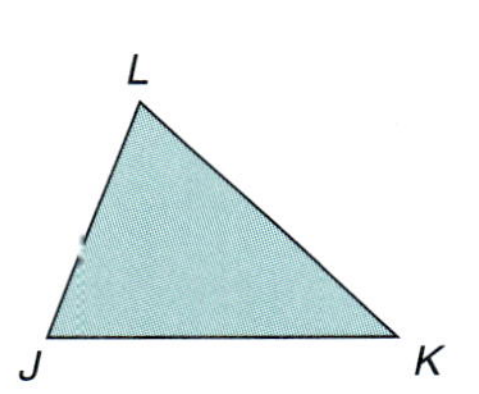

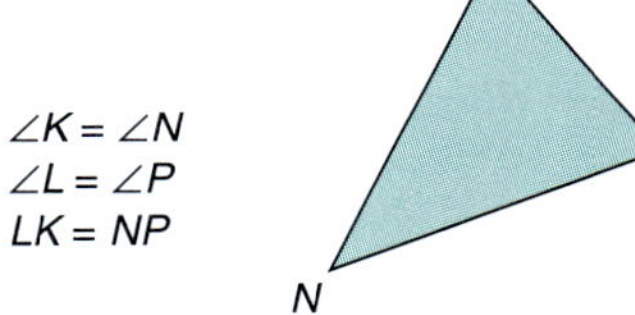

Figure 16–132

Use Fig. 16–133 to solve the following exercises. Round your final answers to the nearest thousandth if necessary.

68. $RS = 8$ mm; find ST and RT.
69. $RT = 15$ cm; find RS and ST.
70. $ST = 7.2$ ft; find RS and RT.
71. $RT = 9\sqrt{2}$ hm; find RS and ST.
72. $RT = 8\sqrt{5}$ dkm; find RS and ST.

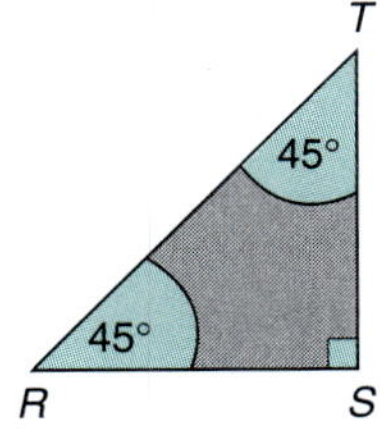

Figure 16–133

Use Fig. 16–134 to solve the following exercises. Round your final answers to the nearest thousandth if necessary.

73. $AC = 12$ dm; find AB and BC.
74. $AB = 15$ km; find AC and BC.
75. $BC = 10$ in.; find AC and AB.
76. $BC = 8\sqrt{2}$ hm; find AC and AB.
77. $AC = 40$ ft 7 in.; find AB and BC to the nearest inch.

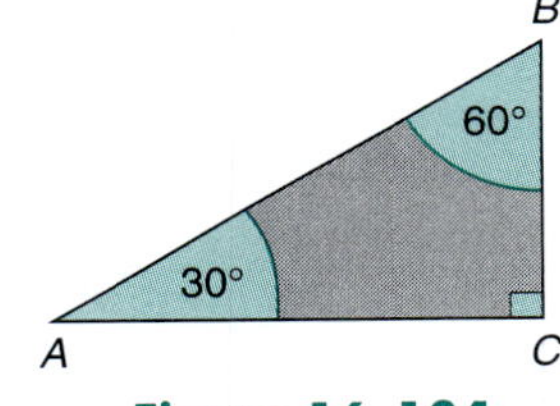

Figure 16–134

Solve. Round your final answers to the nearest thousandth if necessary.

78. The sides of an equilateral triangle are 4 dm in length. Find the altitude of the triangle. Round the answer to the nearest thousandth of a decimeter.

79. Find the total length to the nearest inch of the conduit *ABCD* if *XY* = 8 ft, *CE* = 14 in., and angle *CBE* = 30° (Fig. 16–135).

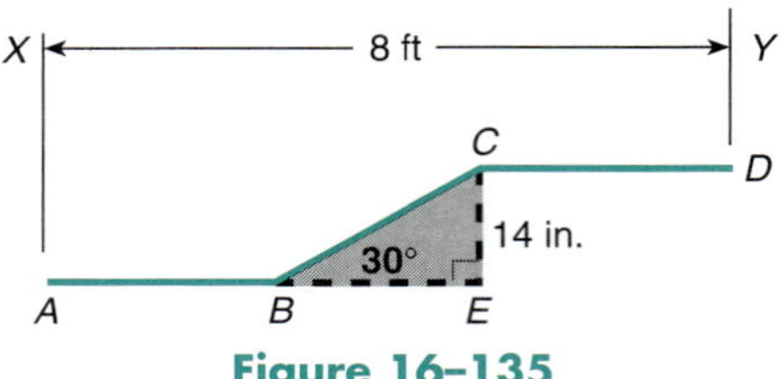

Figure 16–135

80. A piece of round steel is milled to a point at one end. The angle formed by the point is the angle of taper. If the steel has a 16-mm diameter and a 60° angle of taper, find the length *c* of the taper (Fig. 16–136).

81. A V-slot forms a triangle whose angle at the vertex is 60°. The depth of the slot is 17 mm (Fig. 16–137). Find the width of the V-slot.

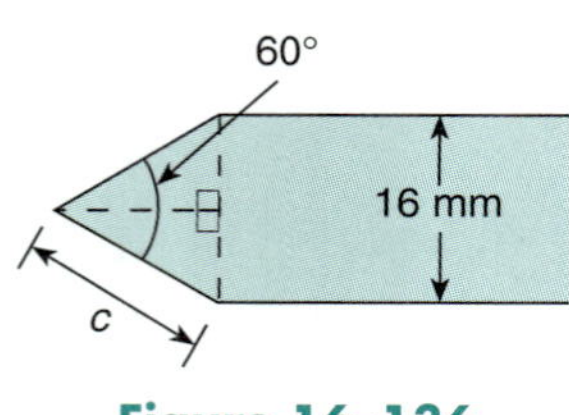

Figure 16–136

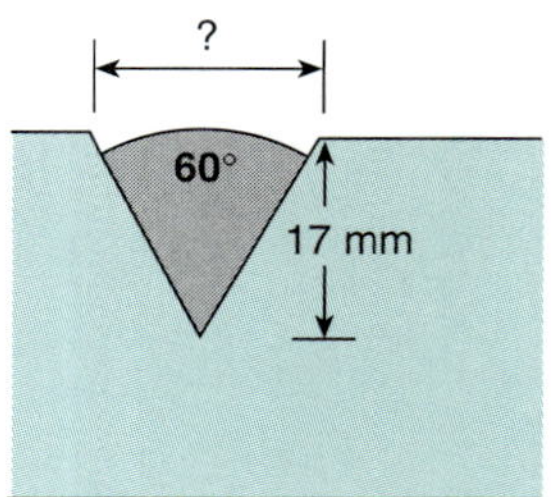

Figure 16–137

82. A manufacturer recommends attaching a guy wire to its 30-ft antenna at a 45° angle. If the antenna is installed on a flat surface, how long must the guy wire be (to the nearest foot) if it is attached to the antenna 4 ft from the top?

Section 16–4

Find the perimeter and the area of Figs. 16–138 and 16–139.

83.

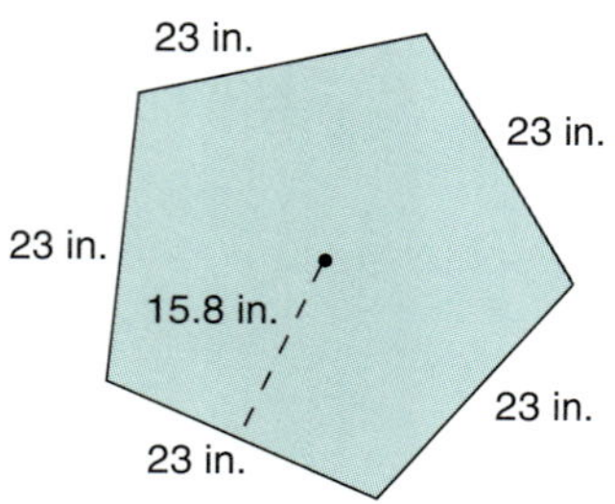

Figure 16–138

84.

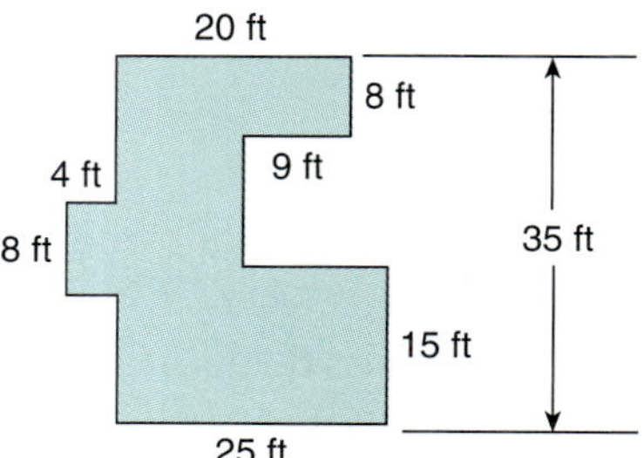

Figure 16–139

85. Identify the figure in Exercise 83 by giving its specific name and the number of degrees in each angle.

Solve the problems involving the perimeter and area of composite figures.

86. How many sections of 2 ft × 4 ft ceiling tiles will be required for the layout in Fig. 16–140?

87. How many feet of baseboard are needed to install around the room in Fig. 16–140 if we make no allowance for openings or doorways?

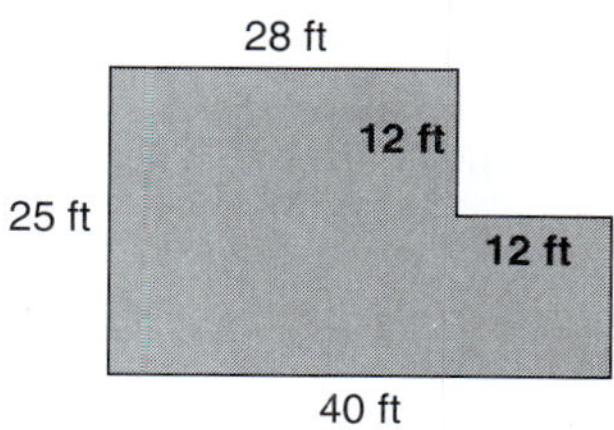

Figure 16–140

88. A gable end of a gambrel roof (Fig. 16–141) is to be covered with 12-in. bevel siding at an $8\frac{1}{2}$-in. exposure to the weather. Estimate the square footage of siding needed if 18% is allotted for lap and waste. Answer to the nearest foot.

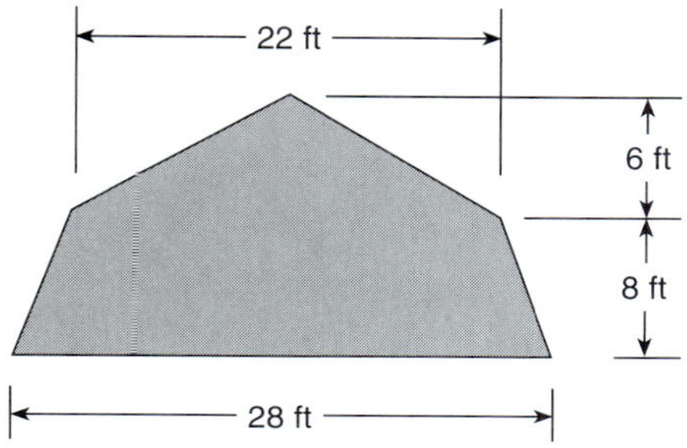

89. Find the square feet of floor space inside the walls of the plan of Fig. 16–142. Answer to the nearest square foot.

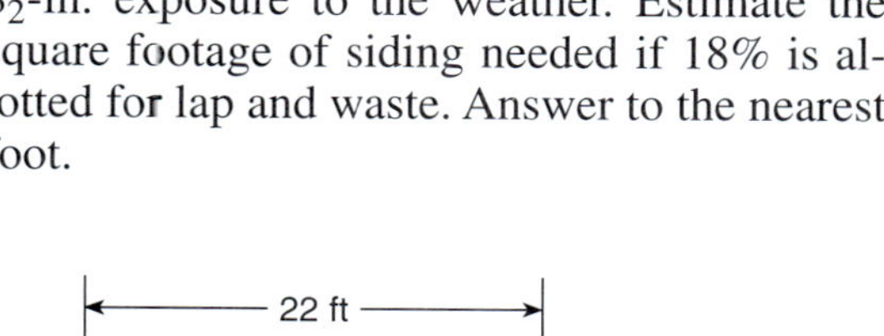

Figure 16–141

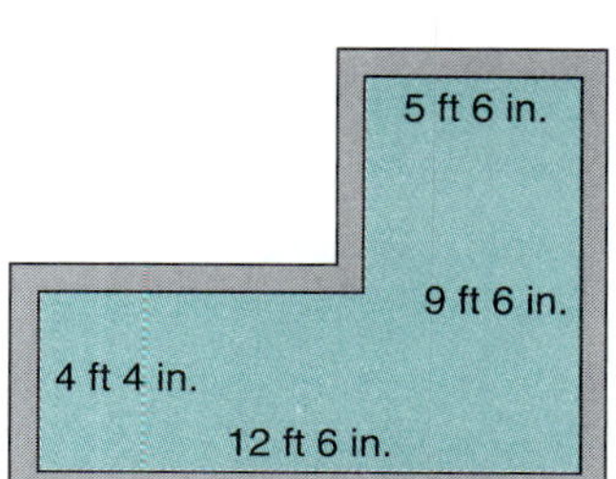

Figure 16–142

90. If one 2 in. × 4 in. stud is estimated for every linear foot of a wall when studs are 16 in. on center, how many studs are needed for the outside walls of the floor plan in Fig. 16–143?

91. A home has concrete steps in the rear (Fig. 16–144). The two sides are covered with a brick facing. How many bricks are needed if $\frac{1}{4}$-in. mortar joints are used? (Six bricks cover 1 ft^2.) Round any part of a square foot to the next-highest square foot. (144 in^2 = 1 ft^2.)

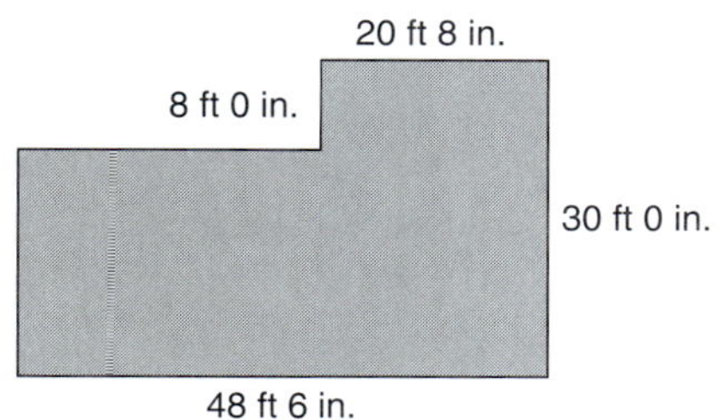

Figure 16–143

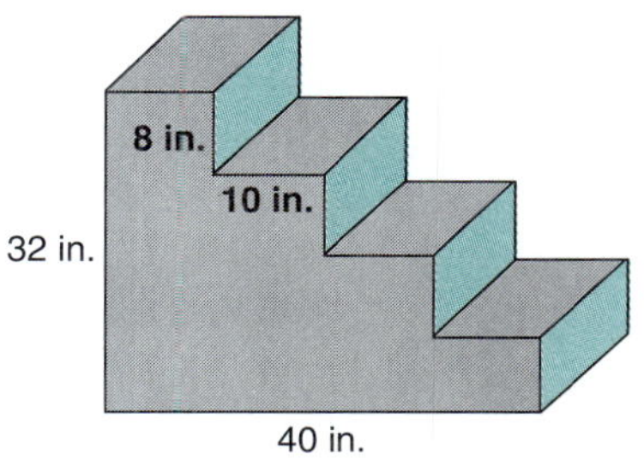

Figure 16–144

92. A mason lays tile to cover an area in the form of a regular octagon with eight sides, each 18 ft. If the distance from the midpoint of a side to the center of the octagon is 21.7 ft, how many tiles are required if $7\frac{1}{2}$ tiles will cover 1 square foot, and 36 tiles are added for waste?

Find the area of the sectors of a circle using Fig. 16–145. Round to hundredths.

93. $\angle = 45°9'$
   $r = 2.58$ cm

94. $\angle = 165°$
   $r = 15$ in.

95. $\angle = 15°15'$
   $r = 110$ mm

96. $\angle = 75°$
   $r = 1$ m

97. $\angle = 40°$
   $r = 2\frac{1}{2}$ ft

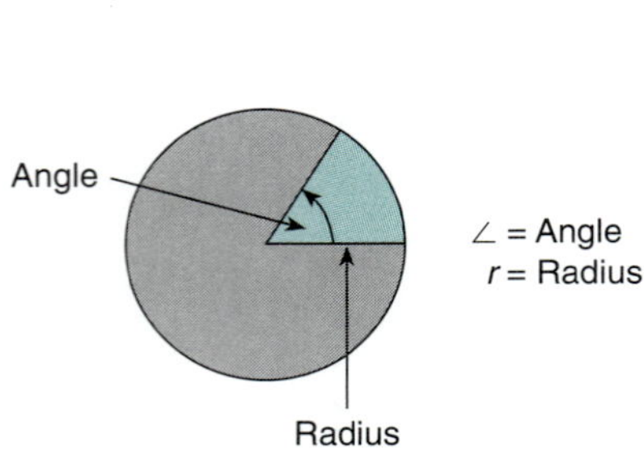

Figure 16–145

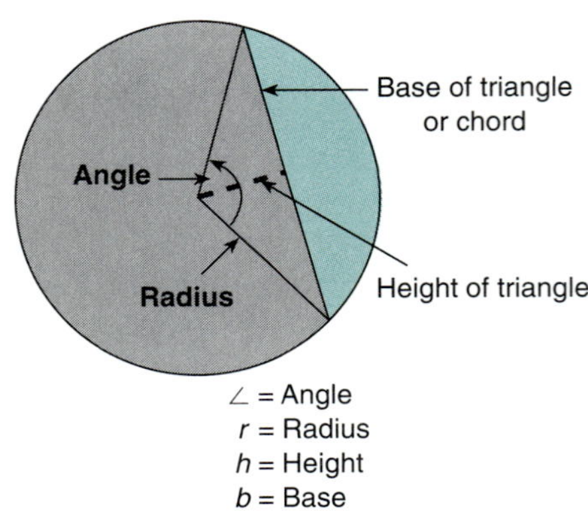

Figure 16–146

Find the area of the segments of a circle using Fig. 16–146. Round to hundredths.

98. $\angle = 52°30'$
   $r = 11.25$ in.
   $h = 10.09$ in.
   $b = 10$ in.

99. $\angle = 45°$
   $r = 14$ mm
   $h = 12.9$ mm
   $b = 10.9$ mm

100. $\angle = 55°$
   $r = 10''$
   $h = 8.9''$
   $b = 9.2''$

101. $\angle = 30°$
   $r = 24$ cm
   $h = 23.2$ cm
   $b = 12.4$ cm

102. $\angle = 60°$
   $r = 13.3$ cm
   $h = 11.5$ cm
   $b = 13.3$ cm

Solve.

103. What is the area (to the nearest tenth) of the flat-topped bifocal lens shown in Fig. 16–147?

104. The minute hand on a grandfather clock makes an angle of 30° as it moves from 12 to 1 on the clock face. If the minute hand is 27.5 cm long, what is the area of the clock face over which the minute hand moves from 12 to 1? Round to tenths.

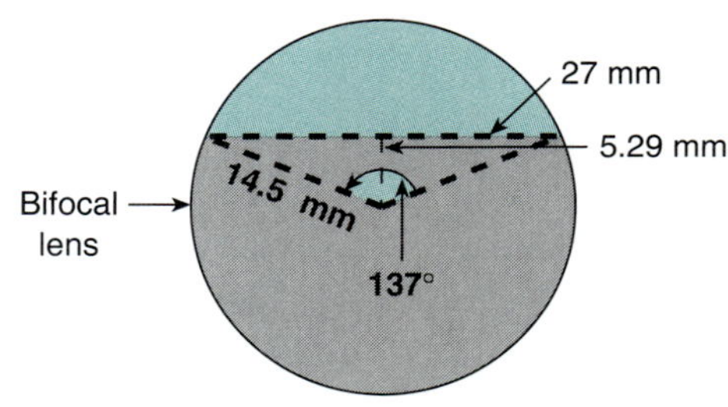

Figure 16–147

105. What is the area of a patio (Fig. 16–148) that has one rounded corner? Round to hundredths.

106. A machine cuts a 12-in.-diameter frozen pizza into slices with sides that form 72° angles at the center of the pizza. What is the surface area of each slice to the nearest square inch?

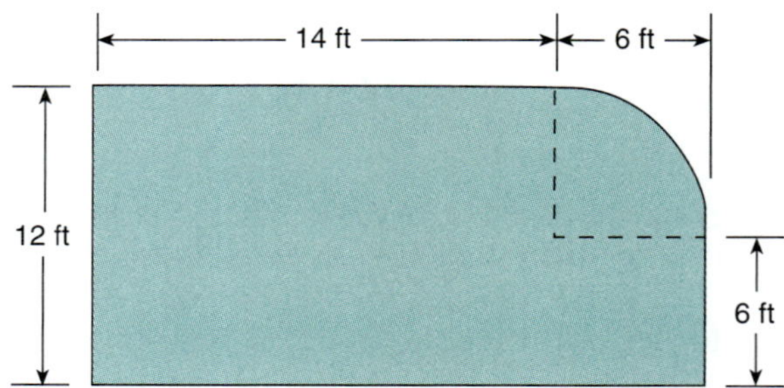

Figure 16–148

107. A drain pipe with a 20-in. diameter has 4 in. of water in it (Fig. 16–149). What is the cross-sectional area of the water in the pipe to the nearest hundredth?

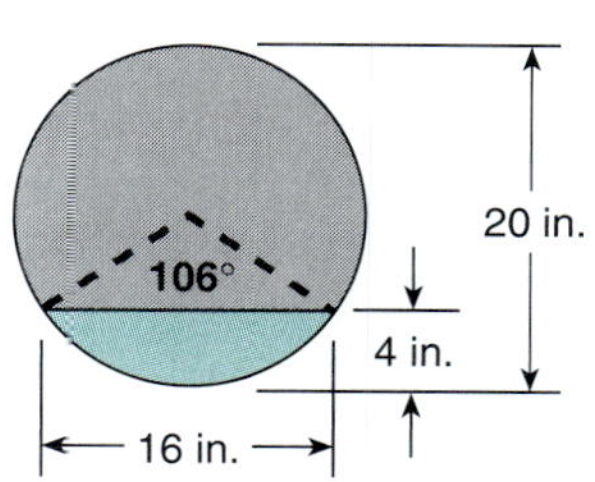

Figure 16–149

Find the arc length of the sectors of a circle using Fig. 16–145. Round to hundredths.

108. $\angle = 40^c$
   $r = 1.45$ ft

109. $\angle = 180°$
   $r = 10$ in.

110. $\angle = 30°15'$
   $r = 20$ cm

111. $\angle = 70°10'$
   $r = 30$ mm

Section 16–6

For each figure, supply the missing dimensions. For an inscribed triangle, see Fig. 16–150.

112. If $FB = 9$, $FO = $ _______
113. If $AB = 10$, $AE = $ _______
114. $\angle ABF = $ _______

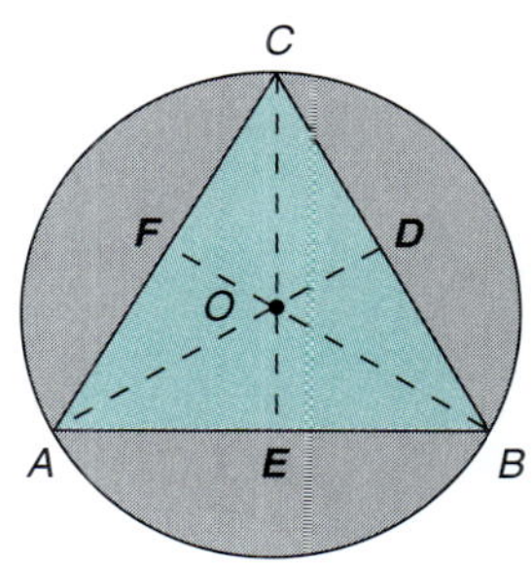

Figure 16–150

For a circumscribed square, see Fig. 16–151.

115. $\angle GJO = $ _______
116. If $GO = 7$, $HO = $ _______
117. If $KO = 10$, $IJ = $ _______

For an inscribed hexagon, see Fig. 16–152.

118. If $QO = 15$, $MN = $ _______
119. $\angle MOP = $ _______
120. $\angle MNO = $ _______

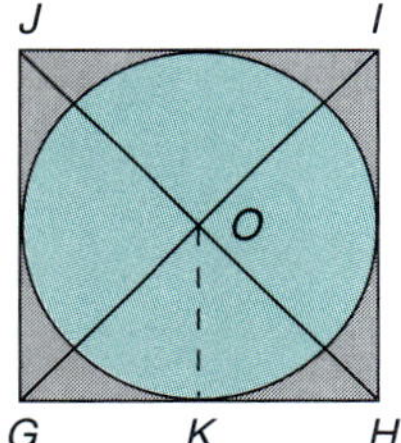

Figure 16–151

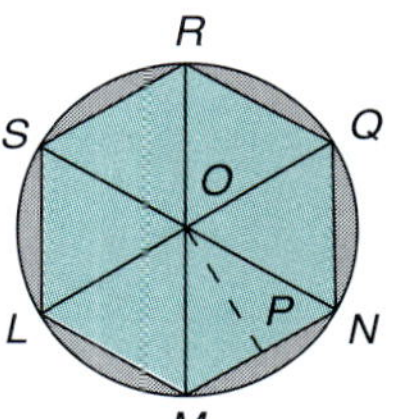

Figure 16–152

Solve the problems involving polygons and circles. Round the answers to hundredths.

121. A shaft with a 20-mm diameter is milled at one end, as illustrated in Fig. 16–153. What is the radius of the inscribed circle?

122. A hex nut 0.87 in. on a side is milled from the smallest-diameter round stock possible (Fig. 16–154). What is the diameter of the stock?

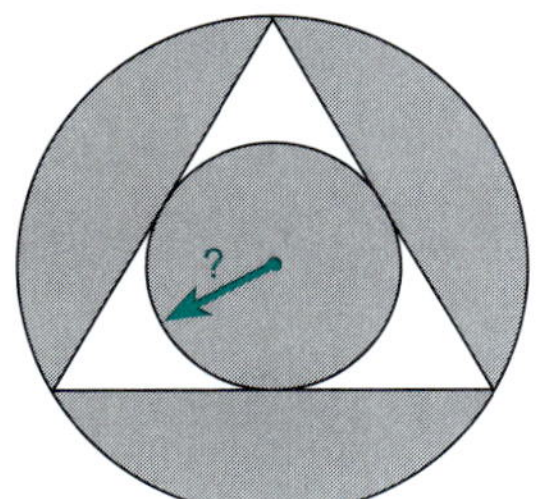

Figure 16–153

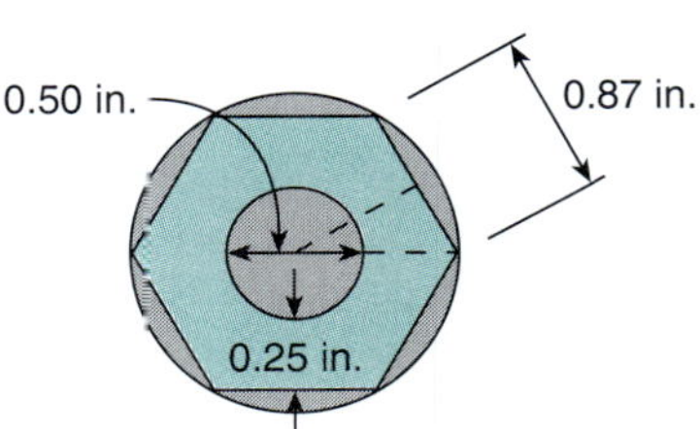

Figure 16–154

123. A square is milled on the end of a 5-cm shaft (Fig. 16–155). If the square is the largest that can be milled on the 5-cm shaft, what is the length of a side to hundredths?

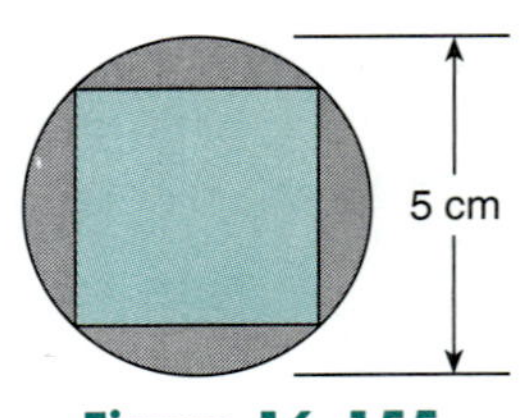

Figure 16–155

<hr>

CHALLENGE PROBLEMS

124. Compare the area of a circle with a radius of 5 cm to the area of an inscribed equilateral triangle. Compare the area of a circle with a radius of 5 cm to the area of an inscribed square. Compare the area of a circle with a radius of 5 cm to the area of the inscribed polygon in Fig. 16–156.

125. Estimate the area of the inscribed regular pentagon in Fig. 16–157 by giving two values the area is between. Present a convincing argument for your estimate. Similarly, estimate the area of an inscribed regular octagon and give an argument for your estimate.

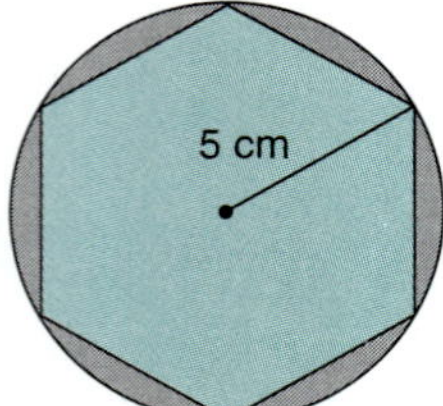

Figure 16–156

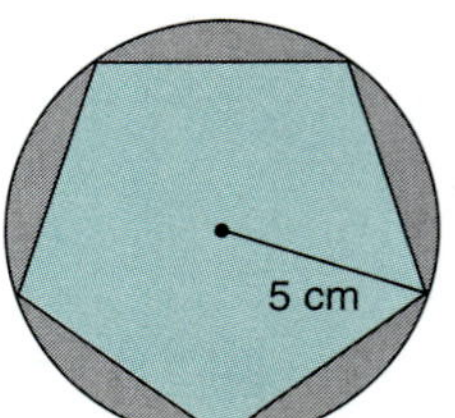

Figure 16–157

<hr>

CHAPTER TRIAL TEST

1. Show the proper notation for naming the line segment from B to C in Fig. 16–158.

2. Name the angle in Fig. 16–159 in four different ways.

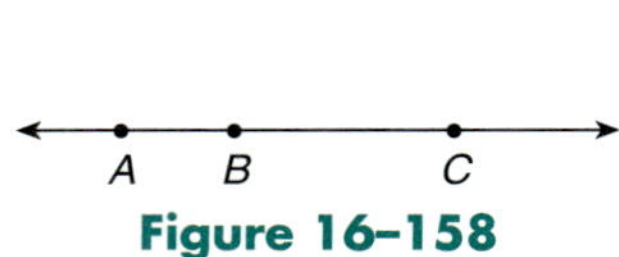

Figure 16–158

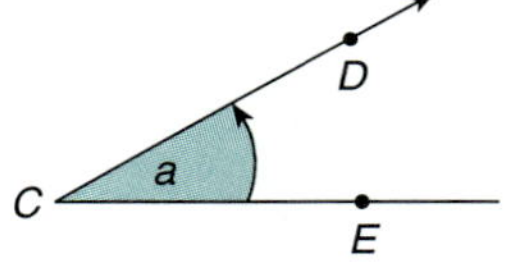

Figure 16–159

Classify the following angles as right, straight, acute, or obtuse.

3. 42°

4. 158°

Identify the lines in Figs. 16–160 to 16–162 as intersecting, parallel, or perpendicular.

5.

Figure 16–160

6.

Figure 16–161

7.

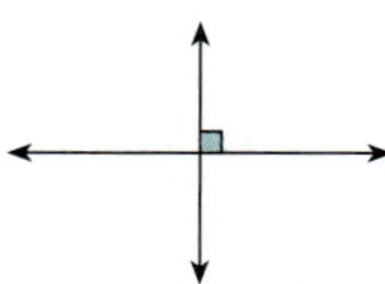

Figure 16–162

8. Subtract 40°37′26″ from 75°.

9. An angle of 47°16′28″ is divided into three equal angles. Find the measure of each angle in degrees, minutes, and seconds (to the nearest second).

10. Change 15′32″ to a decimal degree to the nearest ten-thousandth.

11. Change 0.3125° to minutes and seconds.

12. Name the following triangles: (a) all sides unequal, (b) exactly two sides equal, and (c) all sides equal.

13. Identify the largest and the smallest angles in Fig. 16–163.

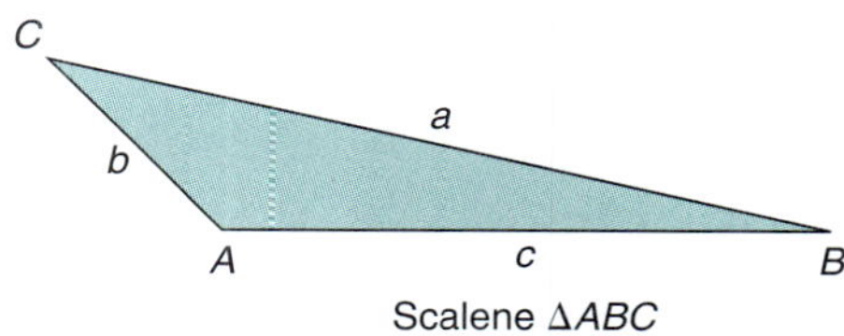

Figure 16–163

14. In Fig. 16–164, identify (a) the angle corresponding to ∠DEF and (b) the side corresponding to HI.

15. A conduit ABCD is made so that ∠CBE is 45° (Fig. 16–165). If BE = 4 cm and AK = 12 cm, find the length of the conduit to the nearest thousandth centimeter.

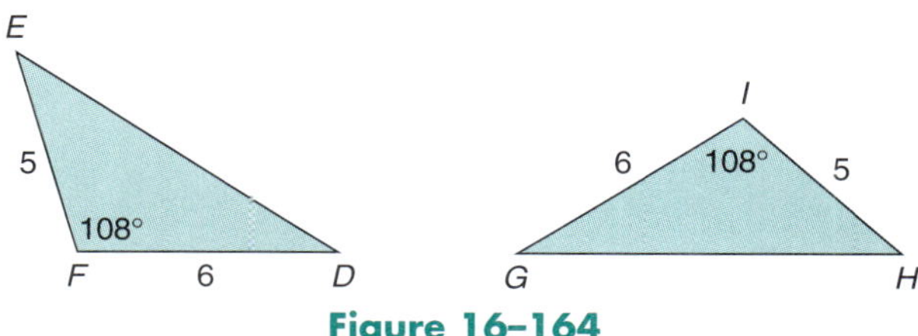

Figure 16–164

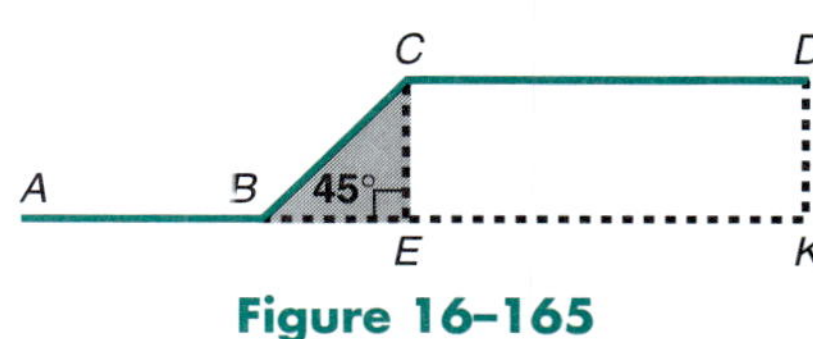

Figure 16–165

16. The rafters of a house make a 30° angle with the joists (Fig. 16–166). If the rafters have an 18-in. overhang and the center of the roof is 10 ft above the joists, how long must the rafters be cut?

17. Find the perimeter of the composite figure in Fig. 16–167.

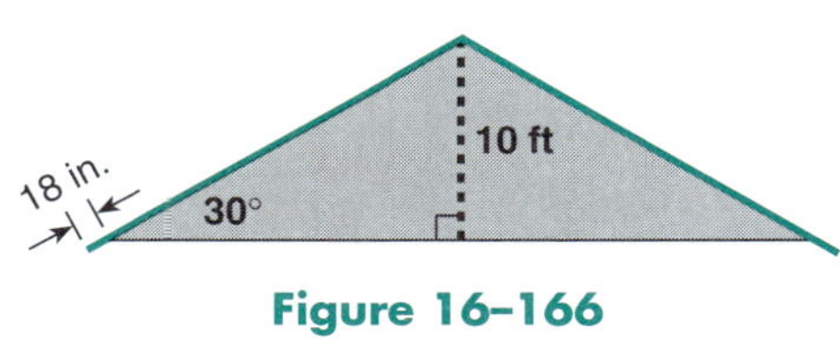

Figure 16–166

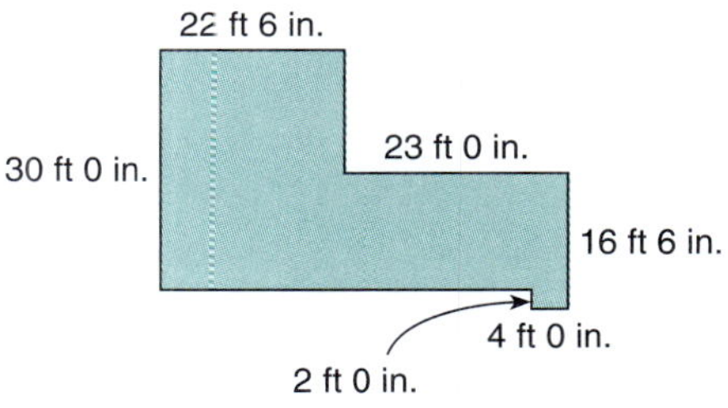

Figure 16–167

18. A section of a hip roof is a trapezoid measuring 35 ft at the bottom, 15 ft at the top, and 10 ft high. Find the area of this section of the roof in square feet.

19. Find the number of degrees in each angle of a regular octagon (eight sides).

20. Find the area of the sector in Fig. 16–168 if the central angle is 55° and the radius is 45 mm. Round to hundredths.

21. Find the arc length cut off by the sector in Fig. 16–168. Round to hundredths.

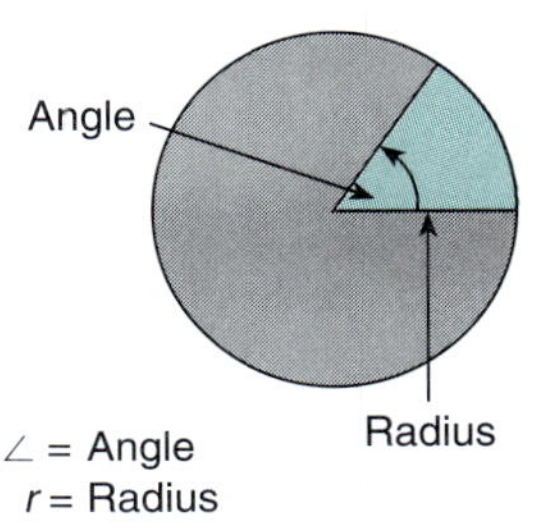

Figure 16–168

22. A segment is removed from a flat metal circle so that the piece rests on a horizontal base (Fig. 16–169). Find the area of the segment that is removed.

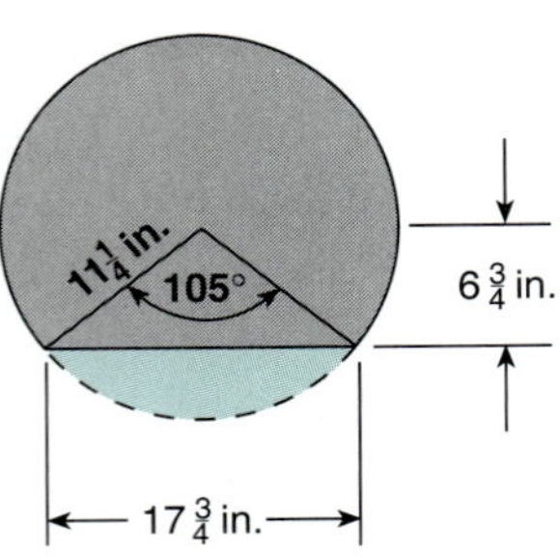

Figure 16–169

23. Find the circumference of a circle inscribed in an equilateral triangle if the height of the triangle is 12 in. (Fig. 16–170).

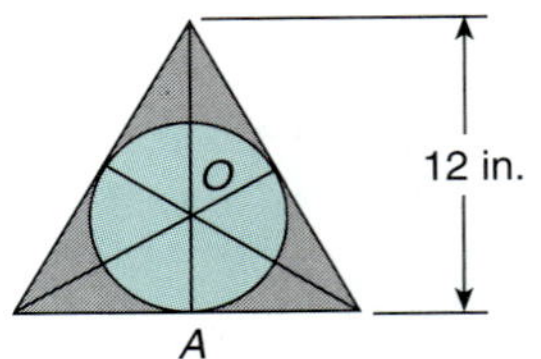

Figure 16–170

24. Taylor Fink milled a square metal rod so that a circle is formed at the end (Fig. 16–171). If the square cross section is 1.8 in. across the corners, what is the diameter of the circle?

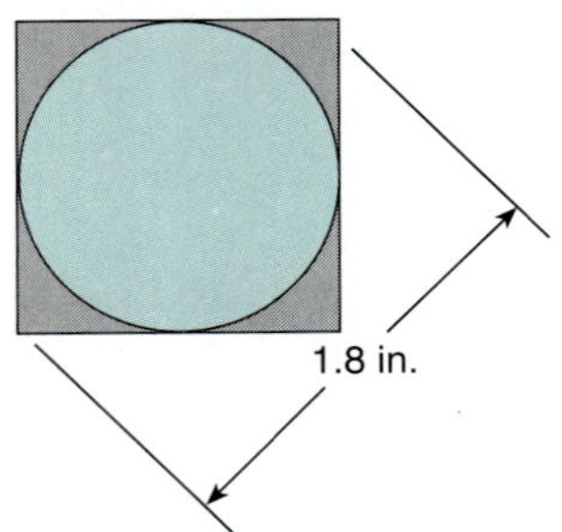

Figure 16–171

25. Haley Fink milled a circle on the cross-sectional end of a hexagonal rod (Fig. 16–172). If the distance across the flats of the cross section is $\frac{1}{2}$ in., what is the circumference of the circle?

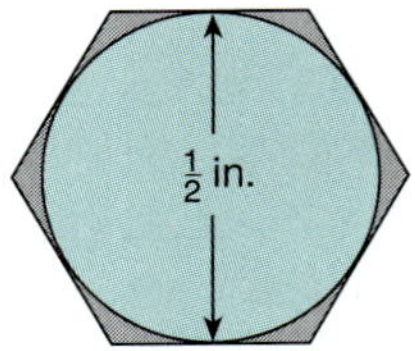

Figure 16–172

17

Right-Triangle Trigonometry

GOOD DECISIONS THROUGH TEAMWORK

Trigonometry allows us to find lengths, angles, or other values, such as the length of the arc formed when a pendulum swings or the force of an electric current, that cannot be measured directly. Your team assignment is to find some of the major practical uses of trigonometry, the types of professionals who use trigonometry in the workplace, and the names of some of these workplaces.

To complete this project, your team may investigate other mathematics books and library resources, and they can interview professionals like surveyors, manufacturers, architects, engineers, and college professors in technical and scientific fields. Team members may also want to visit workplaces suggested in the interviews, such as manufacturing plants, architectural firms that design buildings and roads, and airports.

Collect the information gathered by team members and organize it into categories, such as "Users," "Uses," and "Places." Prepare a written report and an oral presentation to share your findings.

17–1 Radians and degrees

1. Find arc length, given a central angle in radians.
2. Find the area of a sector, given a central angle in radians.
3. Convert angle measures from degrees to radians.
4. Convert angle measures from radians to degrees.

17–2 Trigonometric functions

1. Find the sine, cosine, and tangent of angles of right triangles, given the measures of at least two sides.
2. Find the cosecant, secant, and cotangent of angles of right triangles, given the measures of at least two sides.

17–3 Using a calculator to find trigonometric values

1. Find trigonometric values for sine, cosine, and tangent using a calculator.
2. Find the angle measure, given a trigonometric value.
3. Find the trigonometric values for cosecant, secant, and cotangent using the reciprocal relationship.

17–4 Sine, cosine, and tangent functions for right triangles

1. Find the missing parts of a right triangle using the sine function.
2. Find the missing parts of a right triangle using the cosine function.
3. Find the missing parts of a right triangle using the tangent function.

17–5 Applied problems using right-triangle trigonometry

1. Select the most direct method for solving right triangles.
2. Solve applied problems using right-triangle trigonometry.

One of the most important uses of *trigonometry* is to find by indirect measurement lengths or distances that are difficult or impossible to measure directly. Although trigonometry, which means "triangle measurement," involves more than a study of just the triangle, we limit our study to trigonometry of the triangle.

In our study of trigonometry, we build on our knowledge of angles and triangles from geometry. With trigonometry certain calculations are made more quickly and in fewer steps than they are made using geometry alone. Also, certain calculations that cannot be made using geometry alone can be made using trigonometry.

17–1 RADIANS AND DEGREES

Learning Outcomes

1. Find arc length, given a central angle in radians.
2. Find the area of a sector, given a central angle in radians.
3. Convert angle measures from degrees to radians.
4. Convert angle measures from radians to degrees.

In this section, we study the angles formed by radii of a circle and learn how such angles are measured. We also use these angles to find additional information about the circle.

In geometry, we learned that angles can be measured in units called *degrees*. Angles can also be measured in *radians*.

■ **DEFINITION: Radian.** A *radian* is the measure of a central angle of a circle whose intercepted arc is equal in length to the radius of the circle (Fig. 17–1). The abbreviation for radian is *rad*.

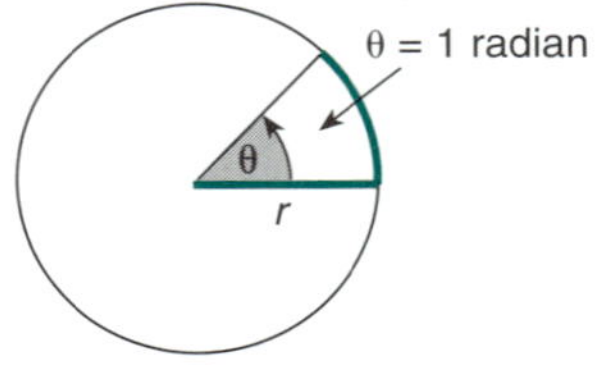

Figure 17–1

The circumference of a circle is related to the radius by the formula $C = 2\pi r$. Thus, the ratio of the circumference to the radius of any circle is $\frac{C}{r} = 2\pi$; that is, the radius could be measured off 2π times (about 6.28 times) along the circumference. A complete rotation is 2π radians (Fig. 17–2).

How are radians and degrees related? A central angle measuring 1 radian makes an arc length equal to the radius, so we use the arc-length formula from Chapter 16 to find the equivalent degree measure for 1 radian.

$$\text{Arc length} = \frac{\theta}{360}\, 2\pi r; \ \theta \text{ in degrees}$$

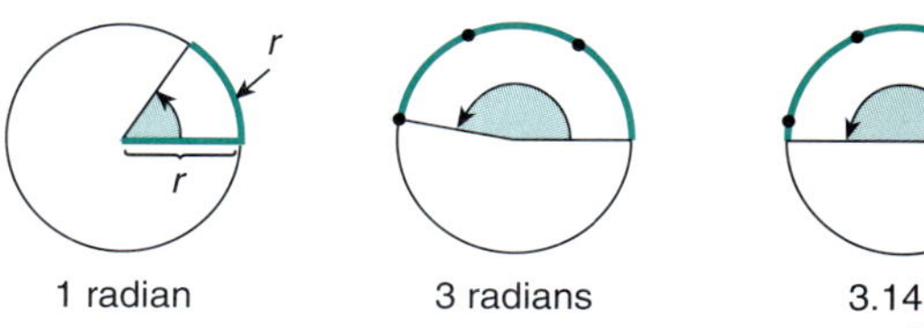

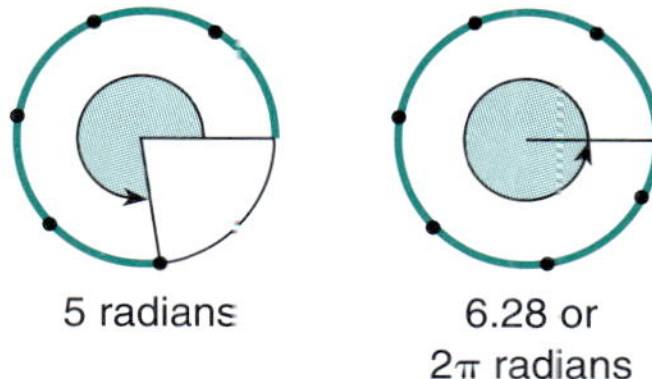

Figure 17–2

If the radius and arc length are equal to 1,

$$1 = \frac{\theta}{360}\, 2\pi(1)$$

$$360 = \theta(2\pi) \qquad \text{Multiply both sides by 360.}$$

$$\frac{360}{2\pi} = \frac{\theta(2\pi)}{2\pi} \qquad \text{Divide both sides by } 2\pi.$$

$$57.29577951 = \theta$$

One radian is approximately 57.3^c.

Tip!	**_Degree and Radian Notation._**

Angles measured in degrees always require the word *degree* or the degree symbol to be written. No comparable symbol exists for the radian. The abbreviation *rad,* or no unit at all indicates that radian is the unit.

1   Find Arc Length, Given a Central Angle in Radians.

Radians are used frequently in physics and engineering mechanics. One application that uses radians is the determination of *arc length,* the length of the arc intercepted by a central angle.

The circumference of a circle is $2\pi r$, where r is the radius of the circle. 2π is the radian measure of a complete rotation. To find the arc length, s, we can multiply the radian measure of the central angle (formed by the end points of the arc and the center of the circle) times the radius of the circle.

$$\text{Arc length} = (\text{Angle measure in radians})(\text{Radius})$$

We use the Greek lowercase letter theta (θ) to represent the central angle measured in radians in the formula for arc length.

Arc length:

$$s = \theta r$$

where θ is the central angle measured in radians, and r is the radius of the circle.

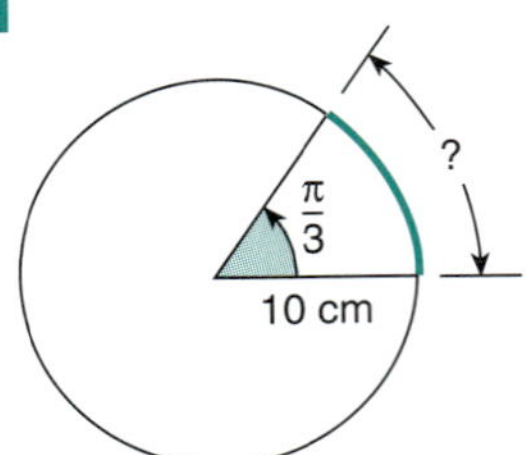

Figure 17–3

EXAMPLE Find the arc length intercepted on the circumference of the circle in Fig. 17–3 by a central angle of $\frac{\pi}{3}$ radians (rad) if the radius of the circle is 10 cm.

$$s = \theta r$$
Substitute $\frac{\pi}{3}$ for θ and 10 cm for r.

$$s = \frac{\pi}{3}(10\ \text{cm})$$

$$s = \frac{\pi(10\ \text{cm})}{3}$$

$$s = 10.47\ \text{cm}$$
To the nearest hundredth.

Thus, the arc length of the intercepted arc is 10.47 cm.

Tip!	*Analyzing Arc Length Dimensions*

In the preceding example, $(\frac{\pi}{3}\ \text{rad})(10\ \text{cm}) = 10.47\ \text{cm}$. What happened to the radians? In the definition of a radian, we relate the measure of an arc connecting the end points of a central angle to the measure of the radius of the circle. Therefore, the arc length will have the same measuring unit as the radius.

EXAMPLE Find the radians of an angle at the center of a circle of radius 5 m. The angle intercepts an arc length of 12.5 m (Fig. 17–4).

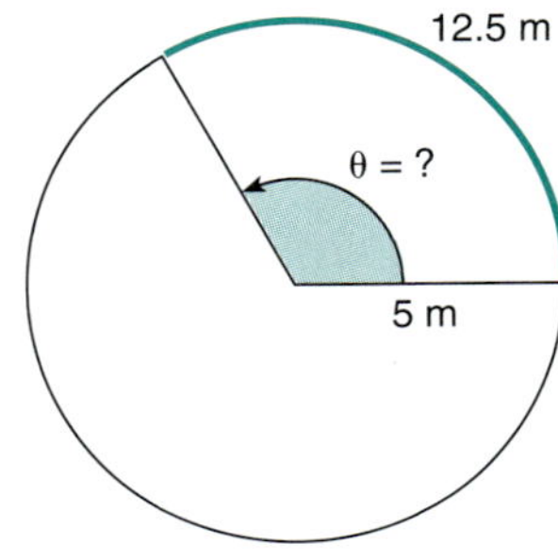

Figure 17–4

Because $s = \theta r$,

$$\theta = \frac{s}{r}$$
Rearrange the formula for θ.

$$\theta = \frac{12.5\ \text{m}}{5\ \text{m}}$$
Substitute 12.5 m for s and 5 m for r. Arc length and radius measuring units are compatible.

$$\theta = 2.5\ \text{rad}$$

Thus, the angle is 2.5 rad.

EXAMPLE Find the radius of an arc if the length of the arc is 8.22 cm and the intercepted central angle is 3 rad (Fig. 17–5).

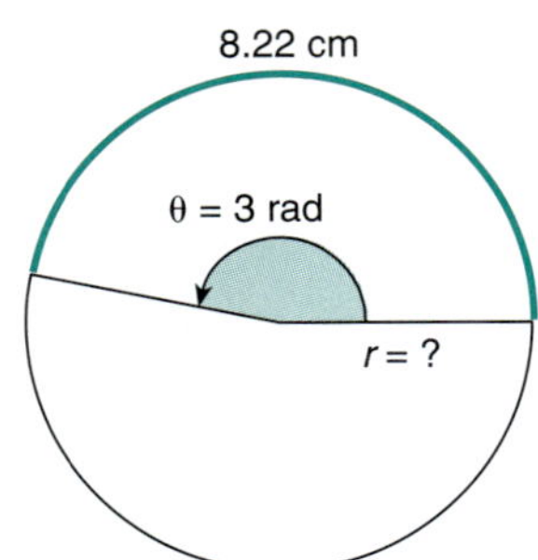

Figure 17–5

CHAPTER 17   Right-Triangle Trigonometry

Because $s = \theta r$,

$$r = \frac{s}{\theta}$$ Rearrange the formula for *r*.

$$r = \frac{8.22 \text{ cm}}{3}$$ Substitute 8.22 cm for *s* and 3 rad for θ.

$$r = 2.74 \text{ cm}$$ Same measuring unit as arc length.

Therefore, the radius of the arc is 2.74 cm.

2   Find the Area of a Sector, Given a Central Angle in Radians.

The formula for the area of a circle is $A = \pi r^2$. A sector is a part of a circle. In this formula, we examine how the area of a complete rotation (circle) relates to the angle measure in radians of a complete rotation (2π). First, we rewrite the area formula so that we can see the factors 2π. Because $\frac{1}{2}(2) = 1$ and multiplying by 1 does not change the value of an expression, we rewrite $A = \pi r^2$ as $A = \frac{1}{2}(2)\pi r^2$. Then, $A = \frac{1}{2}(2\pi)r^2$, or $A = \frac{1}{2}$(angle measure in radians of a complete rotation)r^2. In Section 16–5, we defined a *sector* as the part of a circle determined by two radii of the circle and the included arc. The area of the sector was a fraction of the area of a complete circle, $A = \frac{\theta}{360}\pi r^2$, where θ is the central angle measure of the sector in degrees. Similarly, the area of a sector can be written as $A = \frac{1}{2}\theta r^2$, where θ is the central angle of the sector in radians.

Area of sector:

$$A = \frac{1}{2}\theta r^2$$

where θ is the central angle measured in radians, and *r* is the radius of the circle.

EXAMPLE   Find the area of a sector that has a central angle of 5 rad and a radius of 7.2 in.

$$A = \frac{1}{2}\theta r^2$$

$$A = \frac{1}{2}(5)(7.2 \text{ in.})^2$$

$$A = 129.6 \text{ in}^2$$

The area of the sector is 129.6 in².

3   Convert Angle Measures from Degrees to Radians.

Many job-related problems require us to convert radians to degrees for certain applications and to convert degrees to radians for others. A complete rotation is 360°, or 2π rad. To convert from one unit of angle measure to another, we multiply by a *unity ratio* that relates degrees and radians. Because $360° = 2\pi$ rad, we can simplify the relationship to $180° = \pi$ rad $\left(\frac{360°}{2} = \frac{2\pi}{2} \text{ rad}\right)$.

To convert degrees to radians:

Multiply degrees by the unity ratio $\frac{\pi \text{ rad}}{180°}$.

EXAMPLE Convert the angle measures to radians. Use the calculator value for π to change to a decimal equivalent rounded to the nearest hundredth.

(a) 20°   (b) 50°   (c) 175°   (d) 270°   (e) 67°

(a) Because $360° = 2\pi$ rad or $180° = \pi$ rad, we use this relationship in a unity ratio to convert degrees to radians. Like the procedure we used to convert units in Chapter 5, we arrange the unity ratio so the original units cancel.

$$20° \times \frac{\pi \text{ rad}}{180°} = \frac{20(\pi)}{180} = \textbf{0.35 rad}$$

(b) $50° \times \dfrac{\pi \text{ rad}}{180°} = \dfrac{50(\pi)}{180} = \textbf{0.87 rad}$

(c) $175° \times \dfrac{\pi \text{ rad}}{180°} = \dfrac{175(\pi)}{180} = \textbf{3.05 rad}$

(d) $270° \times \dfrac{\pi \text{ rad}}{180°} = \dfrac{270(\pi)}{180} = \textbf{4.71 rad}$

(e) $67° \times \dfrac{\pi \text{ rad}}{180°} = \dfrac{67(\pi)}{180} = \textbf{1.17 rad}$

When degrees are expressed in decimals, we use the same procedure as before to convert to radians. When angle measures are expressed in degrees, minutes, and seconds, we first convert the minutes and seconds to a decimal degree before converting to radians. This procedure was discussed in detail in Section 16–2.

EXAMPLE Convert the following measures to radians. Round the degrees to the nearest ten-thousandth; then round the radians to the nearest hundredth.

(a) 15.25°   (b) 25°20′   (c) 110°25′30″

(a) $15.25° \times \dfrac{\pi \text{ rad}}{180°} = \dfrac{15.25(\pi)}{180} = \textbf{0.27 rad}$ Rounded.

(b) $25°20′ = 25.3333°$

Convert minutes to a decimal degree: $20′ \times \dfrac{1°}{60′} = \dfrac{20°}{60} = 0.3333°$

$$25.3333° \times \frac{\pi \text{ rad}}{180°} = \frac{25.3333(\pi)}{180} = \textbf{0.44 rad}$$ Rounded.

(c) $110°25′30″ = 110.425°$

Convert minutes and seconds to a decimal degree:

$110 + \dfrac{25}{60} + \dfrac{30}{3{,}600} = 110.425$

$$110.425° \times \frac{\pi \text{ rad}}{180°} = \frac{110.425(\pi)}{180} = \textbf{1.93 rad}$$ Rounded.

4 **Convert Angle Measures from Radians to Degrees.**

When converting from radians to degrees, we multiply by a unity ratio so that the radians cancel and are replaced by degrees.

To convert radians to degrees:

Multiply radians by the unity ratio $\dfrac{180°}{\pi \text{ rad}}$.

EXAMPLE Convert to degrees. Round to the nearest ten-thousandth of a degree.

(a) 2 rad   (b) $\dfrac{\pi}{2}$ rad   (c) 0.25 rad

(a) $2 \text{ rad} \times \dfrac{180°}{\pi \text{ rad}} = \dfrac{360°}{\pi} = \mathbf{114.5916°}$ Substitute calculator value of π and round.

(b) $\dfrac{\overset{1}{\cancel{\pi}}}{2} \text{ rad} \times \dfrac{180°}{\underset{1}{\cancel{\pi}} \text{ rad}} = \dfrac{180°}{2} = \mathbf{90°}$ Because π canceled, we do not substitute for π.

(c) $0.25 \text{ rad} \times \dfrac{180°}{\pi \text{ rad}} = \dfrac{(0.25)(180°)}{\pi} = \mathbf{14.3239°}$ Use calculator value for π and round.

EXAMPLE Convert to degrees, minutes, and seconds. Round to the nearest second.

(a) 1 rad   (b) 3.2 rad

(a) $1 \text{ rad} \times \dfrac{180°}{\pi \text{ rad}} = \dfrac{180°}{\pi} = 57.29577951°$ Continue with the decimal part of the degree measure.

$0.29577951° \times \dfrac{60'}{1°} = 17.74677077'$ Continue with the decimal part of the minute measure.

$0.74677077' \times \dfrac{60''}{1'} = 45''$ Convert the decimal part of minute to seconds, to the nearest second.

Thus, 1 rad $= 57°17'45''$.

(b) $3.2 \text{ rad} \times \dfrac{180°}{\pi \text{ rad}} = \dfrac{3.2(180°)}{\pi} = 183.3464944°$

$0.3464944° \times \dfrac{60}{1°} = 20.7896664'$ Convert the decimal part of degree to minutes.

$0.7896664' \times \dfrac{60''}{1'} = 47''$ Convert the decimal part of minute to seconds, to the nearest second.

Thus, 3.2 rad $= 183°20'47''$.

Tip!	***Using Only the Decimal Part of a Calculator Result.***

When a procedure requires us to continue a calculation with only the decimal part of a result, *first subtract the whole-number part.* Look at the calculator sequence for the preceding example, part a:

$\boxed{180}\ \boxed{\div}\ \boxed{\pi}\ \boxed{=}\ \boxed{-}\ 57\ \boxed{=}\ \boxed{\times}\ 60\ \boxed{=}\ \Rightarrow\ 17.74677077$

17.74677077 (in display) $\boxed{-}\ 17\ \boxed{=}\ \boxed{\times}\ 60\ \boxed{=}\ \Rightarrow\ 44.8062432 = 45$ (rounded)

Whenever the central angle measure is expressed in degrees, we may use either

$$A = \dfrac{\theta}{360}\,\pi r^2 \ (\theta \text{ in degrees}) \qquad \text{or} \qquad A = \dfrac{1}{2}\,\theta r^2 \ (\theta \text{ in radians})$$

To use $A = \dfrac{1}{2}\,\theta r^2$, we must convert from degrees to radians. In the following example, we find the area using both formulas.

 Find the area of a sector whose central angle is 135° and whose radius is 2.7 cm.

Using $A = \frac{\theta}{360}\pi r^2$, where θ is given in degrees:

$$A = \left(\frac{135}{360}\right)(\pi)(2.7 \text{ cm})^2 = 8.59 \text{ cm}^2 \qquad \text{To nearest hundredth.}$$

Using $A = \frac{1}{2}\theta r^2$, where θ is given in radians, we first change 135° to radians.

$$135° = 135° \times \frac{\pi \text{ rad}}{180°} = \frac{135(\pi)}{180} = 2.35619449 \text{ rad}$$

$$A = \frac{1}{2}(2.35619449)(2.7)^2 = 8.59 \text{ cm}^2 \qquad \text{To nearest hundredth.}$$

Therefore, the area of the sector is approximately 8.59 cm^2.

SELF-STUDY EXERCISES 17–1

1 Solve the following problems. Round answers to hundredths if necessary.

1. Find the arc length intercepted on the circumference of a circle by a central angle of 2.15 rad if the radius of the circle is 3 in.

2. Find the arc length intercepted on the circumference of a circle by a central angle of 4 rad if the radius of the circle is 3.5 cm.

3. Use radians to find an angle at the center of a circle of radius 2 in. if the angle intercepts an arc length of 8.5 in.

4. Use radians to find an angle at the center of a circle of radius 4.3 cm if the angle intercepts an arc length of 15 cm.

5. Find the radius of an arc if the length of the arc is 14.7 cm and the intercepted central angle is 2.1 rad.

6. Find the radius of an arc if the length of the arc is 12.375 in. and the intercepting central angle is 2.75 rad.

2

7. Find the area of a sector whose central angle is 2.14 rad and whose radius is 4 in.

8. Find the area of a sector that has a central angle of 6 rad and a radius of 1.2 cm.

9. Find the radius of a sector if the area of the sector is 7.5 cm^2 and the central angle is 3 rad.

10. How many radians does the central angle of a sector measure if its area is 1.7 in.^{2} and its radius is 2 in.?

3 Convert the measures to radians rounded to the nearest hundredth.

11. 45°   **12.** 56°   **13.** 78°   **14.** 140°

Convert the measures to degrees rounded to the nearest ten-thousandth. Then convert to radians to the nearest hundredth.

15. 21°45′   **16.** 177°33′   **17.** 44°54′12″   **18.** 10°31′15″

4 Convert the measures to degrees. Round to the nearest ten-thousandth of a degree.

19. $\frac{\pi}{4}$ rad   **20.** $\frac{\pi}{6}$ rad   **21.** 2.5 rad   **22.** 1.4 rad

Convert the measures to degrees, minutes, and seconds. Round to the nearest second.

23. 0.5 rad   **24.** $\frac{\pi}{8}$ rad   **25.** 0.75 rad   **26.** 1.1 rad

27. Find the arc length of an arc whose intercepting angle is 38° and whose radius is 2.3 cm. Round to hundredths.

28. Find the number of degrees in a central angle whose arc length is 3.2 in. and whose radius is 3 in. Round to the nearest whole degree.

29. Find the area of a sector whose central angle is 105° and whose radius is 7.2 cm. Round to hundredths.

30. Find the number of degrees to the nearest ten-thousandth of a central angle of a sector whose area is 5.6 cm^2 and whose radius is 4 cm.

Learning Outcomes

1 Find the sine, cosine, and tangent of angles of right triangles, given the measures of at least two sides.

2 Find the cosecant, secant, and cotangent of angles of right triangles, given the measures of at least two sides.

1   Find the Sine, Cosine, and Tangent of Angles of Right Triangles, Given the Measures of at Least Two Sides.

In geometry, we studied the basic properties of similar triangles and right triangles that allowed us to find missing measures of the sides of the triangle. Using trigonometry, we can determine the measure of either acute angle of a right triangle if we know the measure of at least two sides of the right triangle. We will define several functions that are ratios of various sides of a triangle, and these ratios will be used later to determine the angles of a triangle.

Figure 17–6 shows a right triangle, *ABC*, with the sides of the triangle labeled according to their relationship to angle *A*. The *hypotenuse* is the side opposite the right angle of the triangle, and the hypotenuse forms one side of angle *A*. The other side that forms angle *A* is the *adjacent side* of angle *A*. The third side of the triangle is the *opposite side* of angle *A*.

In Fig. 17–7, the sides of the right triangle *ABC* are labeled according to their relationship to angle *B*. The hypotenuse forms one side of angle *B*, the other side that forms angle *B* is the adjacent side of angle *B*, and the third side is the opposite side of angle *B*. In Section 16–3, we used the term *hypotenuse*, but the terms *adjacent side* and *opposite side* are also very important in understanding trigonometric functions.

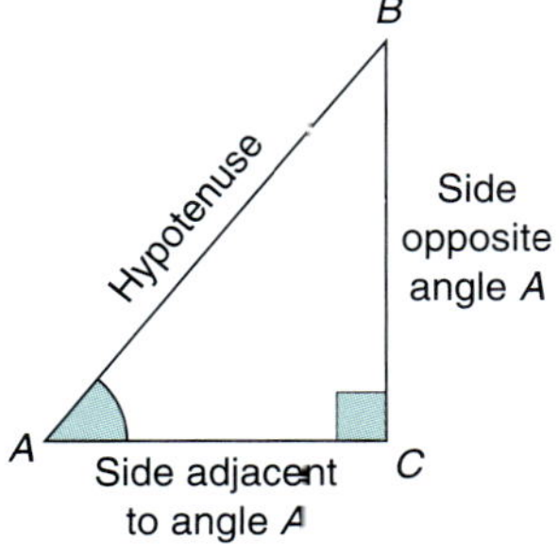

Figure 17–6

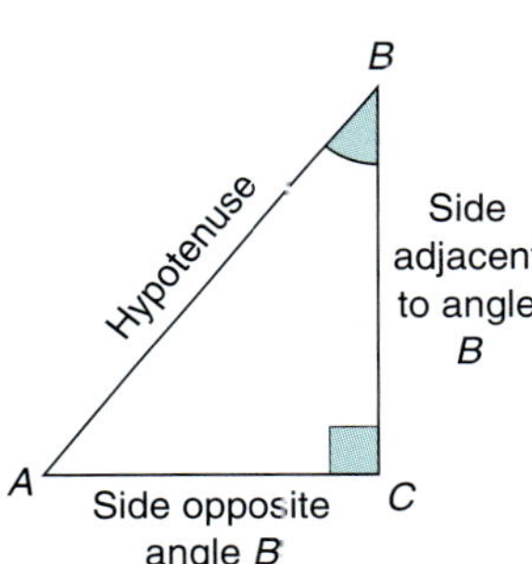

Figure 17–7

■   **DEFINITION: Adjacent Side.**   The *adjacent side* of an acute angle of a right triangle is the side that forms the angle with the hypotenuse.

■   **DEFINITION: Opposite Side.**   The *opposite side* of an acute angle of a right triangle is the side that does not form the given angle.

The three most commonly used trigonometric functions are the *sine, cosine,* and *tangent*. The sine, cosine, and tangent of angle *A* in Fig. 17–8 are defined as ratios of the sides of the right triangle.

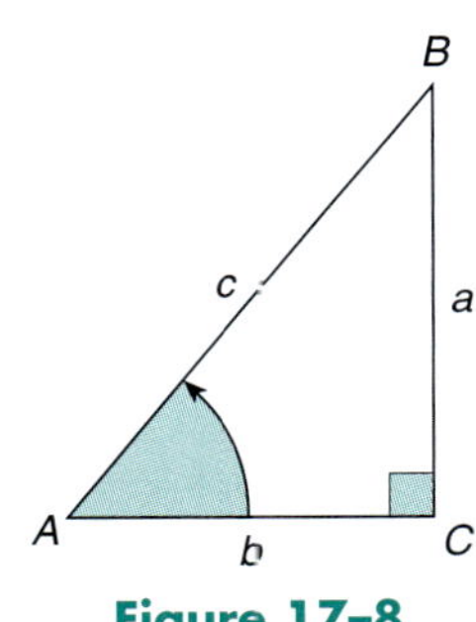

Figure 17–8

$$\text{sine of angle } A = \frac{\text{side opposite angle } A}{\text{hypotenuse}}$$

$$\text{cosine of angle } A = \frac{\text{side adjacent to angle } A}{\text{hypotenuse}}$$

$$\text{tangent of angle } A = \frac{\text{side opposite angle } A}{\text{side adjacent to angle } A}$$

For convenience, the sine, cosine, and tangent functions are abbreviated as *sin, cos,* and *tan,* respectively. Furthermore, we will use letters to designate the sides of the *standard right triangle* in Fig. 17–8. In the standard right triangle, side *a* is opposite angle *A*, side *b* is opposite angle *B*, and side *c* (hypotenuse) is opposite angle *C* (right angle). Thus, we can identify the functions of angle *A* as follows.

Trigonometric functions of angle A in a standard right triangle:

$$\sin A = \frac{\text{side opposite } \angle A}{\text{hypotenuse}} = \frac{a}{c}$$

$$\cos A = \frac{\text{side adjacent to } \angle A}{\text{hypotenuse}} = \frac{b}{c}$$

$$\tan A = \frac{\text{side opposite } \angle A}{\text{side adjacent to } \angle A} = \frac{a}{b}$$

Similarly, we can identify the sine, cosine, and tangent of the other acute angle, angle *B*.

Trigonometric functions of angle B in a standard right triangle:

$$\sin B = \frac{\text{side opposite } \angle B}{\text{hypotenuse}} = \frac{b}{c}$$

$$\cos B = \frac{\text{side adjacent to } \angle B}{\text{hypotenuse}} = \frac{a}{c}$$

$$\tan B = \frac{\text{side opposite } \angle B}{\text{side adjacent to } \angle B} = \frac{b}{a}$$

■ **Learning Strategy**   *Words Are Easier to Remember Than Letters.*

We often substitute letters for words in relationships because it shortens the amount of information we need to write. However, letters sometimes are arbitrary and meaningless and words are easier to understand and remember. Here are two common tips for remembering the trigonometric relationships of sin, cos, and tan.

- "Oscar had a a heap of apples."

$$\sin = \frac{\text{opposite}}{\text{hypotenuse}} = \frac{\text{Oscar}}{\text{had}}$$

$$\cos = \frac{\text{adjacent}}{\text{hypotenuse}} = \frac{\text{a}}{\text{heap}}$$

$$\tan = \frac{\text{opposite}}{\text{adjacent}} = \frac{\text{of}}{\text{apples}}$$

- "Chief Soh-Cah-Toa."

$$\sin = \frac{\text{opposite}}{\text{hypotenuse}} \qquad \cos = \frac{\text{adjacent}}{\text{hypotenuse}} \qquad \tan = \frac{\text{opposite}}{\text{adjacent}}$$

We can now determine the sine, cosine, and tangent of a right triangle by writing the appropriate ratio of the sides of the triangle and expressing the ratio in lowest terms or as a decimal equivalent.

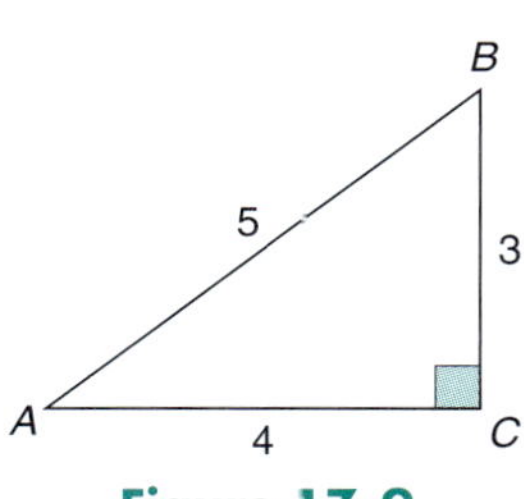

Figure 17–9

EXAMPLE Find the sine, cosine, and tangent of angles A and B in Fig. 17–9. Express your answers as fractions in lowest terms.

$$\sin A = \frac{\text{opposite}}{\text{hypotenuse}} = \frac{3}{5}$$

$$\cos A = \frac{\text{adjacent}}{\text{hypotenuse}} = \frac{4}{5}$$

$$\tan A = \frac{\text{opposite}}{\text{adjacent}} = \frac{3}{4}$$

$$\sin B = \frac{\text{opposite}}{\text{hypotenuse}} = \frac{4}{5}$$

$$\cos B = \frac{\text{adjacent}}{\text{hypotenuse}} = \frac{3}{5}$$

$$\tan B = \frac{\text{opposite}}{\text{adjacent}} = \frac{4}{3}$$

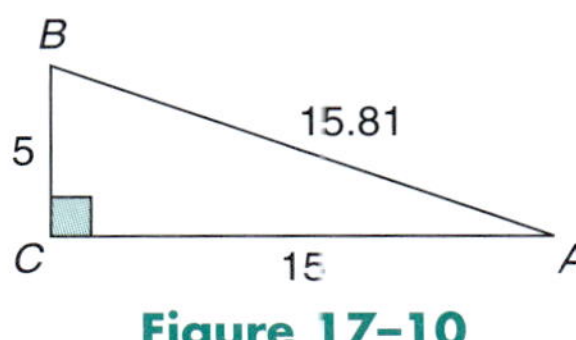

Figure 17–10

EXAMPLE Find the sine, cosine, and tangent of angles A and B in Fig. 17–10. Express your answers as decimals. Round to the nearest ten-thousandth.

$$\sin A = \frac{\text{opposite}}{\text{hypotenuse}} = \frac{5}{15.81} = \mathbf{0.3163}$$

$$\cos A = \frac{\text{adjacent}}{\text{hypotenuse}} = \frac{15}{15.81} = \mathbf{0.9488}$$

$$\tan A = \frac{\text{opposite}}{\text{adjacent}} = \frac{5}{15} = \mathbf{0.3333}$$

$$\sin B = \frac{\text{opposite}}{\text{hypotenuse}} = \frac{15}{15.81} = \mathbf{0.9488}$$

$$\cos B = \frac{\text{adjacent}}{\text{hypotenuse}} = \frac{5}{15.81} = \mathbf{0.3163}$$

$$\tan B = \frac{\text{opposite}}{\text{adjacent}} = \frac{15}{5} = \mathbf{3}$$

Tip!	***Is a Trigonometric Ratio Expressed as a Unit of Measure?***

When both terms of a ratio are expressed in the same unit of measure, that is, *like* units, then the ratio itself is unitless because the common units cancel, for example, $\frac{3 \text{ in.}}{5 \text{ in.}} = \frac{3}{5}$. Thus, trigonometric ratios are numerical values with no unit of measure.

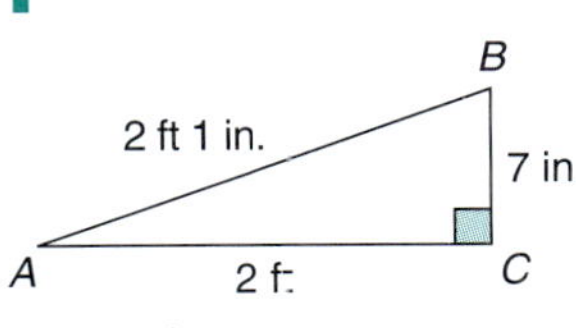

Figure 17–11

EXAMPLE Find the sine, cosine, and tangent of angles A and B in Fig. 17–11. Leave the answers as fractions in lowest terms.

$$\sin A = \frac{\text{opposite}}{\text{hypotenuse}} = \frac{7 \text{ in.}}{2 \text{ ft 1 in.}} = \frac{7 \text{ in.}}{25 \text{ in.}} = \frac{7}{25}$$

$$\cos A = \frac{\text{adjacent}}{\text{hypotenuse}} = \frac{2 \text{ ft}}{2 \text{ ft 1 in.}} = \frac{24 \text{ in.}}{25 \text{ in.}} = \frac{24}{25}$$

$$\tan A = \frac{\text{opposite}}{\text{adjacent}} = \frac{7 \text{ in.}}{2 \text{ ft}} = \frac{7 \text{ in.}}{24 \text{ in.}} = \frac{7}{24}$$

$$\sin B = \frac{\text{opposite}}{\text{hypotenuse}} = \frac{2 \text{ ft}}{2 \text{ ft } 1 \text{ in.}} = \frac{24 \text{ in.}}{25 \text{ in.}} = \frac{24}{25}$$

$$\cos B = \frac{\text{adjacent}}{\text{hypotenuse}} = \frac{7 \text{ in.}}{2 \text{ ft } 1 \text{ in.}} = \frac{7 \text{ in.}}{25 \text{ in.}} = \frac{7}{25}$$

$$\tan B = \frac{\text{opposite}}{\text{adjacent}} = \frac{2 \text{ ft}}{7 \text{ in.}} = \frac{24 \text{ in.}}{7 \text{ in.}} = \frac{24}{7}$$

2 Find the Cosecant, Secant, and Cotangent of Angles of Right Triangles, Given the Measures of at Least Two Sides.

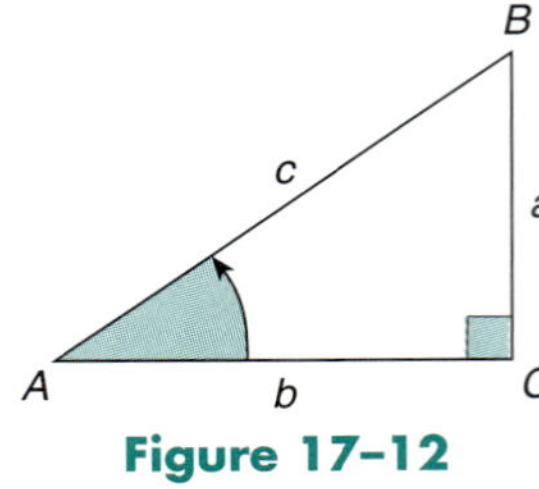

Figure 17–12

Each basic trigonometric function (sine, cosine, and tangent) has a *reciprocal function*. The *cosecant* (*csc*) function is the reciprocal of the sine function, the *secant* (*sec*) function is the reciprocal of the cosine function, and the *cotangent* (*cot*) function is the reciprocal of the tangent function.

Using Fig. 17–12, we can write the reciprocal trigonometric functions of angle A.

Reciprocal trigonometric functions of angle A in a standard right triangle:

$$\csc A = \frac{\text{hypotenuse}}{\text{opposite}} = \frac{c}{a}$$

$$\sec A = \frac{\text{hypotenuse}}{\text{adjacent}} = \frac{c}{b}$$

$$\cot A = \frac{\text{adjacent}}{\text{opposite}} = \frac{b}{a}$$

Tip! *Reciprocal Trigonometric Functions.*

$$\sin A = \frac{1}{\csc A} \qquad \cos A = \frac{1}{\sec A} \qquad \tan A = \frac{1}{\cot A}$$

$$\csc A = \frac{1}{\sin A} \qquad \sec A = \frac{1}{\cos A} \qquad \cot A = \frac{1}{\tan A}$$

For example, if $\sin A = \frac{a}{c}$, then

$$\frac{1}{\frac{a}{c}} = 1 \div \frac{a}{c} = \frac{c}{a}$$

Thus, $\csc A = \frac{c}{a}$.

Similarly, we can write the reciprocal trigonometric functions of angle B in Fig. 17–12.

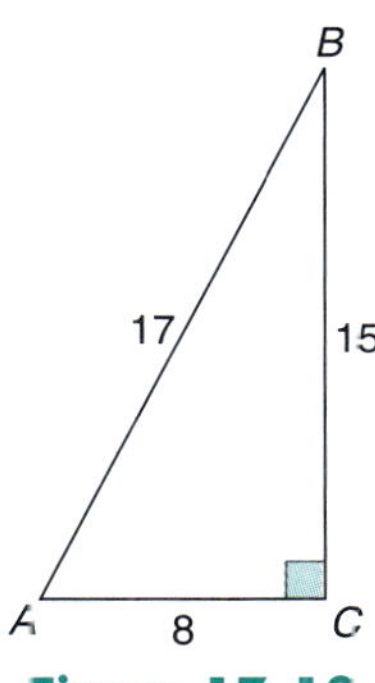

Figure 17–13

EXAMPLE   Write the six trigonometric ratios for angles A and B in Fig. 17–13. For convenience, *opposite, hypotenuse,* and *adjacent* are now abbreviated.

Angle A

$$\sin A = \frac{\text{opp}}{\text{hyp}} = \frac{15}{17} \qquad \csc A = \frac{\text{hyp}}{\text{opp}} = \frac{17}{15}$$

$$\cos A = \frac{\text{adj}}{\text{hyp}} = \frac{8}{17} \qquad \sec A = \frac{\text{hyp}}{\text{adj}} = \frac{17}{8}$$

$$\tan A = \frac{\text{opp}}{\text{adj}} = \frac{15}{8} \qquad \cot A = \frac{\text{adj}}{\text{opp}} = \frac{8}{15}$$

Angle B

$$\cos B = \frac{\text{adj}}{\text{hyp}} = \frac{15}{17} \qquad \sec B = \frac{\text{hyp}}{\text{adj}} = \frac{17}{15}$$

$$\sin B = \frac{\text{opp}}{\text{hyp}} = \frac{8}{17} \qquad \csc B = \frac{\text{hyp}}{\text{opp}} = \frac{17}{8}$$

$$\tan B = \frac{\text{opp}}{\text{adj}} = \frac{8}{15} \qquad \cot B = \frac{\text{adj}}{\text{opp}} = \frac{15}{8}$$

EXAMPLE   Find the six trigonometric functions for angles A and B in Fig. 17–14. Express ratios to the nearest ten-thousandth.

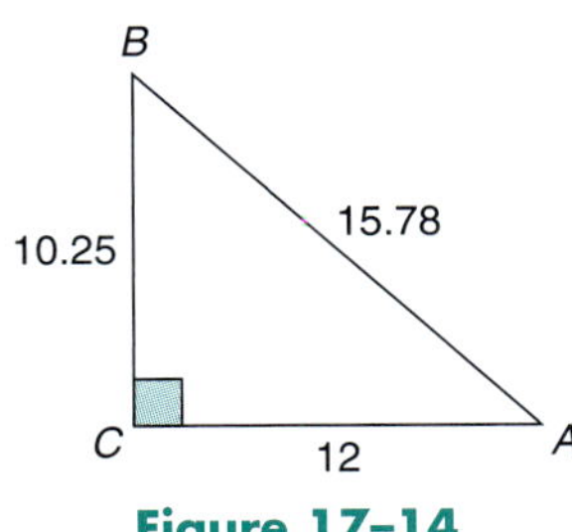

Figure 17–14

$$\sin A = \frac{10.25}{15.78} = 0.6496 \qquad \sin B = \frac{12}{15.78} = 0.7605$$

$$\cos A = \frac{12}{15.78} = 0.7605 \qquad \cos B = \frac{10.25}{15.78} = 0.6496$$

$$\tan A = \frac{10.25}{12} = 0.8542 \qquad \tan B = \frac{12}{10.25} = 1.1707$$

$$\cot A = \frac{12}{10.25} = 1.1707 \qquad \cot B = \frac{10.25}{12} = 0.8542$$

$$\csc A = \frac{15.78}{10.25} = 1.5395 \qquad \csc B = \frac{15.78}{12} = 1.3150$$

$$\sec A = \frac{15.78}{12} = 1.3150 \qquad \sec B = \frac{15.78}{10.25} = 1.5395$$

<table>
<tr><td>Tip!</td><td>Patterns and Relationships.</td></tr>
</table>

Examine the boxes in the two previous examples. The ratios are arranged and highlighted in the first example to illustrate *reciprocal relationships*. The arrangement and highlighting in the second example illustrates the relationships of ratios for angle A and angle B.

$$\sin A = \cos B \qquad \csc A = \sec B \qquad \cos A = \sin B \qquad \sec A = \csc B \qquad \tan A = \cot B \qquad \cot A = \tan B$$

The relationships are referred to as *cofunctions*.

SELF-STUDY EXERCISES 17–2

1 Use Fig. 17–15 to find the indicated trigonometric ratios.

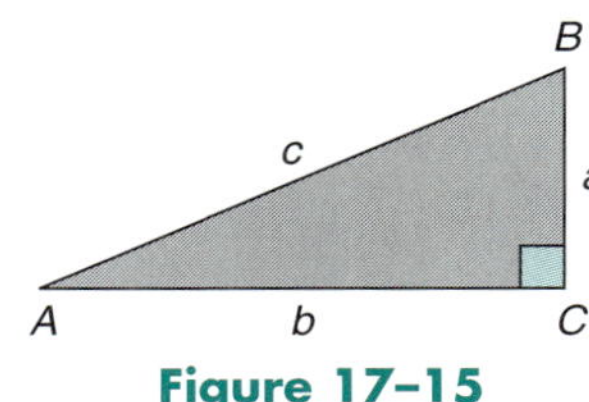

Figure 17–15

Use $a = 5$, $b = 12$, and $c = 13$ in Exercises 1–6 and express the ratios as fractions in lowest terms.

1. $\sin A$ **2.** $\cos A$ **3.** $\tan A$ **4.** $\sin B$ **5.** $\cos B$ **6.** $\tan B$

Use $a = 9$, $b = 12$, and $c = 15$ in Exercises 7–12 and express the ratios as fractions in lowest terms.

7. $\sin A$ **8.** $\cos A$ **9.** $\tan A$ **10.** $\sin B$ **11.** $\cos B$ **12.** $\tan B$

Use $a = 16$, $b = 30$, and $c = 34$ in Exercises 13–18 and express the ratios as fractions in lowest terms.

13. $\sin A$ **14.** $\cos A$ **15.** $\tan A$ **16.** $\sin B$ **17.** $\cos B$ **18.** $\tan B$

Use $a = 9$, $b = 14$, and $c = 16.64$ in Exercises 19–24 and express the ratios as decimals to the nearest ten-thousandth.

19. $\sin A$ **20.** $\cos A$ **21.** $\tan A$ **22.** $\sin B$ **23.** $\cos B$ **24.** $\tan B$

2 Find the six trigonometric ratios for angles A and B in Fig. 17–15 using the values given in Exercises 25 and 26.

25. $a = 12$, $b = 16$, $c = 20$

26. $a = 8.15$, $b = 5.32$, $c = 9.73$. Express the ratios as decimals to the nearest ten-thousandth.

17–3 USING A CALCULATOR TO FIND TRIGONOMETRIC VALUES

Learning Outcomes

1 Find trigonometric values for sine, cosine, and tangent using a calculator.

2 Find the angle measure, given a trigonometric value.

3 Find the trigonometric values for cosecant, secant, and cotangent using the reciprocal relationship.

In studying similar triangles, we found that corresponding angles of similar triangles are equal and that corresponding sides are proportional. These properties of similar triangles lead us to a very important property of trigonometric functions.

For centuries, trigonometric values were recorded in tables and were readily available in mathematics textbooks to facilitate calculations. Today, scientific and graphing calculators and software packages have these values of the trigonometric functions already programmed. These calculator values are expressed to more decimal places than even the most accurate of written tables, thus producing more accurate results.

1 Find Trigonometric Values for Sine, Cosine, and Tangent Using a Calculator.

We recommend that you use your calculator to find trigonometric values. To find trigonometric values on your calculator, the keys $\boxed{\text{SIN}}$, $\boxed{\text{COS}}$, and $\boxed{\text{TAN}}$ are used most often. The angle measure can be entered in degrees or radians, depending on the selected mode.

Tip!	*Finding Trigonometric Values Using the Calculator.*

To find trigonometric values for sine, cosine, and tangent using most scientific and graphing calculators, follow this procedure:

1. Set your calculator to the desired mode of angle measure. Calculators generally have three angle modes: degrees, radians, and gradients.
2. Press the appropriate function key (sin, cos, tan) and then enter the angle measure.
3. Display the result by pressing the $\boxed{=}$ or $\boxed{\text{ENTER}}$ key.

To find trigonometric values for sine, cosine, and tangent in certain scientific calculators, reverse the sequence in the preceding steps. Enter the angle measure, then press the appropriate function key. The result is displayed immediately after pressing the trigonometric function key.

EXAMPLE Using a calculator, find the trigonometric values.

(a) $\sin 27°$ (b) $\cos 52°$ (c) $\tan 85°$ (d) $\sin 20°30'$

(e) $\cos 1.34$ (f) $\tan\dfrac{\pi}{4}$

(a) Be sure your calculator is in degree mode.

$\boxed{\text{SIN}}\ 27\ \boxed{=} \Rightarrow 0.4539904997$ Keystrokes may vary.

(b) Be sure your calculator is in degree mode.

$\boxed{\text{COS}}\ 52\ \boxed{=} \Rightarrow 0.6156614753$

(c) Be sure your calculator is in degree mode.

$\boxed{\text{TAN}}\ 85\ \boxed{=} \Rightarrow 11.4300523$

(d) Be sure your calculator is in degree mode.

$\boxed{\text{SIN}}\ 20.5\ \boxed{=} \Rightarrow 0.3502073813$

With some scientific calculators angle measures can be entered using degrees, minutes, and seconds *or* using the decimal equivalent.

(e) Reset your calculator to radian mode.

$\boxed{\text{COS}}$ 1.34 $\boxed{=}$ $\Rightarrow$ **0.2287528078**

(f) Be sure your calculator is in radian mode.

$\boxed{\text{TAN}}$ $\boxed{(}$ $\boxed{\pi}$ $\boxed{\div}$ 4 $\boxed{)}$ $\boxed{=}$ $\Rightarrow$ **1**

2 Find the Angle Measure, Given a Trigonometric Value.

When the measures of two sides of a right triangle are known, we can find the angle measure from a trigonometric value.

Tip!	***Finding Angle Measures Using the Calculator.***

Finding one of the acute angle measures of a right triangle when given the trigonometric value is the *inverse operation* of finding the trigonometric value when given the angle measure. The notation most commonly used is $\sin^{-1}$, $\cos^{-1}$, or $\tan^{-1}$. Another notation is arcsin, arccos, or arctan.

Most Scientific and Graphing Calculators
1. Set your calculator to the desired mode of angle measure (degrees or radians).
2. Select the appropriate inverse trigonometric function key or menu option ($\boxed{\text{SIN}^{-1}}$, $\boxed{\text{COS}^{-1}}$, or $\boxed{\text{TAN}^{-1}}$). Enter the trigonometric value and $\boxed{=}$ or $\boxed{\text{ENTER}}$.

Some calculators use an inverse key $\boxed{\text{INV}}$ or shift key $\boxed{\text{SHIFT}}$ to find the inverse of a function.

Other Calculators
1. Set your calculator to the desired mode of angle measure (degrees or radians).
2. Enter the trigonometric value and press the *inverse* function key $\boxed{\text{INV}}$ or shift function key, then the appropriate trigonometric function key ($\boxed{\text{SIN}}$, $\boxed{\text{COS}}$, $\boxed{\text{TAN}}$). Some calculators have $\boxed{\text{SIN}^{-1}}$, $\boxed{\text{COS}^{-1}}$, and $\boxed{\text{TAN}^{-1}}$ as inverse function keys.

Test your calculator with a known value. For example, in the previous example we found sin 27° = 0.4539904997. $\sin^{-1}$ 0.4539904997 should be 27°.

EXAMPLE Find the angle in degrees given the trigonometric values in parts a and b. θ represents the unknown angle measure. Round to the nearest tenth of a degree. For part c, find the radians to the nearest thousandth.

(a) $\sin \theta = 0.6561$ (b) $\cos \theta = 0.4226$ (c) $\tan \theta = 2.825$

(a) Be sure your calculator is in degree mode.

$\boxed{\text{SIN}^{-1}}$ .6561 $\boxed{=}$ $\Rightarrow$ 41.0031105 $\approx$ **41.0°**   Rounded.

(b) Be sure your calculator is in degree mode.

$\boxed{\text{COS}^{-1}}$ .4226 $\boxed{=}$ $\Rightarrow$ 65.00115448 $\approx$ **65.0°**   Rounded.

(c) Be sure your calculator is in radian mode.

$\boxed{\text{TAN}^{-1}}$ 2.825 $\boxed{=}$ $\Rightarrow$ 1.230578215 $\approx$ **1.231 rad**   Rounded.

3 Find the Trigonometric Values for Cosecant, Secant, and Cotangent Using the Reciprocal Relationship.

The reciprocal key ($\boxed{x^{-1}}$ or $\boxed{1/x}$) on calculators can be used to find the value of *reciprocal trigonometric functions*. For instance, after finding the sine of an angle, we can determine the cosecant by pressing the reciprocal key.

The following statements indicate the reciprocal relationships of the tangent and cotangent, the sine and cosecant, and the cosine and secant functions.

$$\cot \theta = \frac{1}{\tan \theta} \qquad \csc \theta = \frac{1}{\sin \theta} \qquad \sec \theta = \frac{1}{\cos \theta}$$

A relationship statement such as $\cot \theta = \dfrac{1}{\tan \theta}$ is a true statement or fact. A relationship statement such as this is a *trigonometric identity*.

EXAMPLE Determine the value of the trigonometric functions.

(a) csc 18.5°   (b) sec 1.0821

(a) $\csc 18.5° = \dfrac{1}{\sin 18.5°} = \dfrac{1}{0.317304656} = \mathbf{3.1515}$ To the nearest ten-thousandth.

In degree mode:

$\boxed{\text{SIN}}\ 18.5\ \boxed{=}\ \boxed{x^{-1}}\ \boxed{=}\ \Rightarrow\ 3.151545305$

(b) $\sec 1.0821 = \dfrac{1}{\cos 1.0821} = \dfrac{1}{0.469475214} = \mathbf{2.1300}$ To the nearest ten-thousandth.

In radian mode:

$\boxed{\text{COS}}\ 1.0821\ \boxed{=}\ \boxed{x^{-1}}\ \boxed{=}\ \Rightarrow\ 2.130037898$

EXAMPLE Find an angle in degrees whose trigonometric value is given. Express the measure to the nearest tenth of a degree.

(a) sec θ = 2.7320   (b) csc θ = 5.9137

Find the radians to the nearest hundredth radian.

(c) cot θ = 0.2167

First, find the reciprocal of the given value. Then, find the inverse of the reciprocal function.

(a) If sec θ = 2.7320, then $\cos \theta = \frac{1}{2.732}$. In degree mode:

2.732 $\boxed{x^{-1}}$ $\boxed{=}$ $\boxed{\text{COS}^{-1}}$ $\boxed{\text{ANS}}$ $\boxed{=}$ $\boxed{\text{ANS}}$ enters the previous answer.

θ = 68.5°

(b) In degree mode:

5.9137 $\boxed{x^{-1}}$ $\boxed{=}$ $\boxed{\text{SIN}^{-1}}$ $\boxed{\text{ANS}}$ $\boxed{=}$

θ = 9.7°

(c) In radian mode:

.2167 $\boxed{x^{-1}}$ $\boxed{=}$ $\boxed{\text{TAN}^{-1}}$ $\boxed{\text{ANS}}$ $\boxed{=}$

θ = 1.36 rad

SELF-STUDY EXERCISES 17–3

1 Use your calculator to find the trigonometric values. Express your answers in ten-thousandths.

1. sin 21°	**2.** cos 3.5°	**3.** tan 47°	**4.** cos 52.5°
5. sin 0.5498	**6.** cos 21°30′	**7.** cos 1.1519	**8.** cos 0.3665
9. sin 53°30′	**10.** tan 42.5°	**11.** tan 47.7°	**12.** sin 62°10′
13. cos 12°40′	**14.** cos 1.0530	**15.** tan 73°14′	**16.** sin 1.2363
17. cos 46.8°	**18.** cos 0.3549	**19.** cos 1.1636	**20.** tan 12.4°

2 Find the angles of the trigonometric values in degrees. θ represents the unknown measure. Express each answer to the nearest tenth of a degree.

21. $\sin \theta = 0.3420$ **22.** $\cos \theta = 0.9239$ **23.** $\tan \theta = 2.356$
24. $\cos \theta = 0.4617$ **25.** $\cos \theta = 0.540$ **26.** $\sin \theta = 0.5712$
27. $\tan \theta = 1.265$ **28.** $\cos \theta = 0.137$ **29.** $\sin \theta = 0.6298$
30. $\cos \theta = 0.9325$

Find the angles of the trigonometric values in radians. Express each answer to the nearest ten-thousandth.

31. $\sin \theta = 0.5299$ **32.** $\tan \theta = 0.8098$ **33.** $\cos \theta = 0.6947$
34. $\cos \theta = 0.3907$ **35.** $\cos \theta = 0.968$ **36.** $\sin \theta = 0.9959$
37. $\tan \theta = 0.3160$ **38.** $\tan \theta = 2.430$ **39.** $\cos \theta = 0.9610$
40. $\cos \theta = 0.4210$

3 Find the indicated trigonometric values using the reciprocal relationship. Round to the nearest ten-thousandth.

41. cot 24.5° **42.** csc 42° **43.** sec 0.2443 **44.** csc 1.3788
45. sec 1.2165 **46.** cot 28.6° **47.** cot 87° **48.** sec 42°30′
49. csc 0.2136 **50.** sec 1.0372

Find θ to the nearest tenth of a degree.

51. $\cot \theta = 0.4238$ **52.** $\sec \theta = 1.8291$ **53.** $\csc \theta = 3.7129$
54. $\sec \theta = 8.2156$ **55.** $\cot \theta = 1.7318$

17–4 SINE, COSINE, AND TANGENT FUNCTIONS FOR RIGHT TRIANGLES

Learning Outcomes

1 Find the missing parts of a right triangle using the sine function.
2 Find the missing parts of a right triangle using the cosine function.
3 Find the missing parts of a right triangle using the tangent function.

We use right triangles extensively in real-world applications. Thus, a working knowledge of solving right triangles, that is, finding the measures of all sides and angles, is important. We also need to practice using right triangles to solve career-related applications. By using the trigonometric relationships we learned in the preceding sections, we can find all the angles and sides of a right triangle if we know the measure of one side and any other part.

We often use the sine, cosine, and tangent functions to find parts of a right triangle. To do this, we manipulate formulas and use other algebraic principles depending on what information is given and what information needs to be found.

1 Find the Missing Parts of a Right Triangle Using the Sine Function.

We abbreviate the sine function to read $\sin \theta = \frac{\text{opp}}{\text{hyp}}$. Using this relationship, we can find parts of right triangles when we know any two parts that involve the sine function: one acute angle, the side opposite the known acute angle, and the hypotenuse.

Tip!	***Does It Matter Which Acute Angle Is Known?***

Not really. The acute angles of a right triangle are complementary. Therefore, if we know either acute angle (a), we can find the other one ($90° - a$). Thus, we can use the sine function if we know *either* angle and any side.

EXAMPLE Find angle A if $a = 7$ and $c = 21$ (Fig. 17–16).

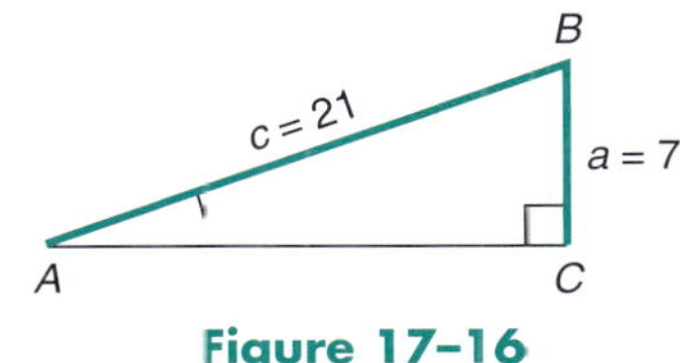

Figure 17–16

The two known values are the side opposite angle A and the hypotenuse, so we use the sine function for the acute angle A.

$$\sin A = \frac{\text{opp}}{\text{hyp}}$$

Substitute the known values: opp = 7, hyp = 21.

$$\sin A = \frac{7}{21}$$

Convert the ratio to a decimal equivalent.

$$\sin A = 0.3333333333$$

Find $\boxed{\sin^{-1}}$ of 0.3333333333.

$$A = 19.47122063° \text{ or } 19.5°$$

Round to the nearest tenth of a degree.

Tip!	***How Do I Round My Answers?***

Rounding practices are generally dictated by the context of the problem or industry standards; however, **for consistency, we round all trigonometric ratios to four significant digits and all angle values to the nearest 0.1° throughout Chapters 17 and 18 unless otherwise indicated.**
   The *significant digits* of a whole number or integer are the digits beginning with the first nonzero digit on the left and ending with the last nonzero digit on the right. The significant digits of a decimal number are the digits beginning with the first nonzero digit on the left and ending with the last digit on the *right of the decimal point.*

5,000	1 significant digit	5 is first *and* last nonzero digit.
250	2 significant digits	2 is first and 5 is last nonzero digit.
205	3 significant digits	2 is first and 5 is last nonzero digit.
0.004	1 significant digit	4 is first nonzero *and* last digit.
3.05	3 significant digits	3 is first nonzero and 5 is last digit.
2.070	4 significant digits	2 is first nonzero digit and 0 is last digit.

(continued)

To round a number to a certain number of significant digits:

1. From the left, count the number of significant digits desired. Notice the last of these significant digits.
2. If the next digit to the right is less than 5, do not change the last significant digit. If the next digit to the right is 5 or greater, add 1 to the last significant digit.
3. If the last significant digit precedes the decimal point, replace all digits after the last significant digit up to the decimal point with zeros and drop all digits after the decimal point. If the last significant digit follows the decimal point, drop all digits after the last significant digit.

In the first example, the angle measure is rounded to the nearest tenth of a degree. In the next example, the side measure will be rounded to four significant digits.

EXAMPLE Find side a in triangle ABC (Fig. 17–17).

We are given $\angle A$ and the hypotenuse and are asked to find the measure of the side opposite $\angle A$, so we use the sine function.

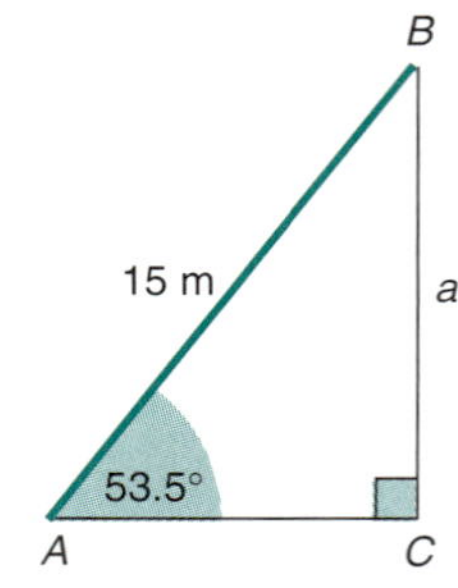

Figure 17–17

$$\sin A = \frac{\text{opp}}{\text{hyp}}$$

Substitute known values: $\angle A = 53.5°$, hyp $= 15$ m.

$$\sin 53.5° = \frac{a}{15}$$

$\sin 53.5° = 0.80385686$

$$0.80385686 = \frac{a}{15}$$

$$15(0.80385686) = a$$

$$a = 12.05785291 \qquad \text{or} \qquad 12.06 \text{ m}$$

Rounded to four significant digits.

Tip! | **Rearrange the Formula Before You Calculate.**

In the preceding example and in those to follow, we can visualize a continuous sequence of calculator steps if we rearrange the formula for the missing part *before* we make any calculations.

$$\sin 53.5° = \frac{a}{15}$$

$$15(\sin 53.5°) = a$$

Then we make a continuous series of calculations. In degree mode:

$$15 \; \boxed{\times} \; \boxed{\text{SIN}} \; 53.5 \; \boxed{=} \; \Rightarrow 12.05785291$$

EXAMPLE Find the hypotenuse in triangle RST (Fig. 17–18).

We are given an acute angle and the side opposite the angle and are asked to find the hypotenuse, so we use the sine function.

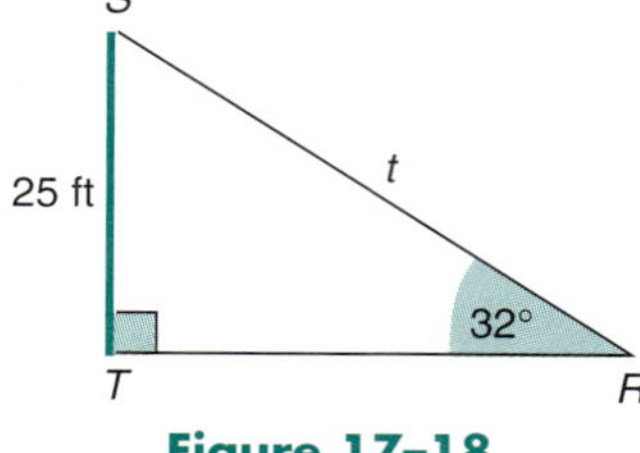

Figure 17–18

$$\sin R = \frac{\text{opp}}{\text{hyp}}$$

Substitute known values: $\angle R = 32°$, opp $= 25$ ft.

$$\sin 32° = \frac{25}{t}$$

Rearrange for t.

$$t(\sin 32°) = 25$$

CHAPTER 17   Right-Triangle Trigonometry

$$t = \frac{25}{\sin 32°}$$

$$t = 47.17699787$$

$$\mathbf{t = 47.18 \ ft} \qquad \text{Rounded to four significant digits.}$$

<table>
<tr><td>Tip!</td><td>Choose Given Values over Calculated Values Whenever Possible:</td></tr>
</table>

It is best to use given values rather than calculated values when finding missing parts of a triangle. Because the rounded value for one missing part is sometimes used to find other missing parts, final answers may vary slightly due to rounding discrepancies. For instance, the angles of a triangle may add to as little as 179° or as much as 181°, or the length of a side may be slightly different in the last significant digit. If the full calculator value of a side or angle is used to find other missing parts, the rounding discrepancy is reduced.

To *solve* a triangle means to find the measures of all sides and all angles. A right triangle can be solved if we know one side and any other part besides the right angle. In the next example, we find values of all sides and angles of the triangle using the sine function.

EXAMPLE Solve triangle *DEF* (Fig. 17–19). One side and one other part besides the right angle are known.

Because we have one acute angle and a side that is not opposite the known angle, we must find the other angle of the triangle. Angles *D* and *E* are complementary, so

$$\angle E = 90° - 32.5° = 57.5°$$

Now we use the sine function to find the hypotenuse because we have an acute angle, *E*, and its opposite side, *e*.

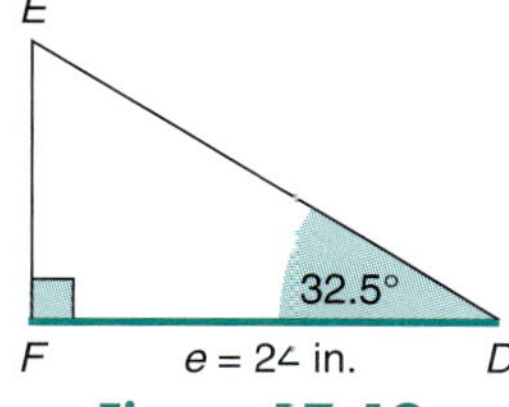

Figure 17–19

$$\sin E = \frac{\text{opp}}{\text{hyp}} \qquad \text{Substitute.}$$

$$\sin 57.5° = \frac{24}{f} \qquad \text{Solve for } f.$$

$$f(\sin 57.5°) = 24$$

$$f = \frac{24}{\sin 57.5°}$$

$$f = 28.45653714$$

$$\mathbf{f = 28.46 \ in.} \qquad \text{Rounded to four significant digits.}$$

To find side *d*, we have

$$\sin D = \frac{\text{opp}}{\text{hyp}} \qquad \text{Substitute; use full calculator value for } f \text{ to get the most accurate result.}$$

$$\sin 32.5° = \frac{d}{28.45653714} \qquad \text{Solve for } d.$$

$$28.45653714(\sin 32.5°) = d$$

$$15.28968626 = d$$

$$\mathbf{15.29 \ in. = d} \qquad \text{Rounded to four significant digits.}$$

The solved triangle is shown in Fig. 17–20.

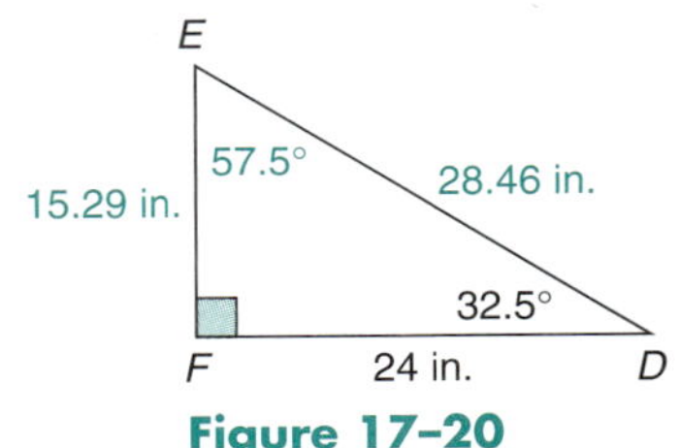

Figure 17–20

| Tip! | ***Check Computations Using the Pythagorean Theorem.*** |

In this and other problems involving right triangles, you can check your computations using the Pythagorean theorem. Let's check the solution to the preceding example.

$$(\text{hyp})^2 = (\text{leg})^2 + (\text{leg})^2$$
$$(28.45653714)^2 = (15.28968626)^2 + (24)^2$$
$$809.774506 = 233.7745059 + 576$$
$$809.774506 = 809.7745059 \qquad \text{Difference due to rounding.}$$

Rounding discrepancies are minimized when more significant digits are used.

■ **Learning Strategy** *See the BIG Picture, Then Focus on the Little Parts.*

In solving a right triangle, we are generally given the values of three parts and are asked to find the values of the three missing parts. If you find the problem overwhelming, look at the big picture first. In the preceding example, we know the value of two angles ($\angle D$ and $\angle F$) and one side (the leg DF). We are asked to find $\angle E$, leg EF, and hypotenuse DE.

Next, plan a strategy for finding each missing part.

To find $\angle E$, use the property that the two acute angles of a right triangle are complementary.

To find the hypotenuse DE, use the sine function and $\angle E$. Now, relax and focus on one part at a time. Remember, *LONG* doesn't have to mean *HARD*. Long can just be long.

Use the Pythagorean theorem and the property that the three angles of a triangle add to $180°$ to check.

2 Find the Missing Parts of a Right Triangle Using the Cosine Function.

In some of the previous problems, when we found a side not opposite the given angle, we had to find the other angle first by subtracting the given acute angle from $90°$. If we use the cosine function, however, we can find the same side by using the given angle rather than its complement. The abbreviated cosine ratio is $\cos \theta = \frac{\text{adj}}{\text{hyp}}$.

Keep in mind that angle θ is made up of two sides of the right triangle. One side is the hypotenuse and the other side is the side *adjacent* to angle θ.

Use the cosine function to find unknown parts of a right triangle:

1. Two of these three parts of a right triangle must be known:
   a. One acute angle
   b. The side adjacent to the known acute angle
   c. The hypotenuse
2. Substitute two known values in the ratio $\cos \theta = \frac{\text{adj}}{\text{hyp}}$.
3. Solve for the missing part.

Figure 17–21

EXAMPLE Find angle A of Fig. 17–21.

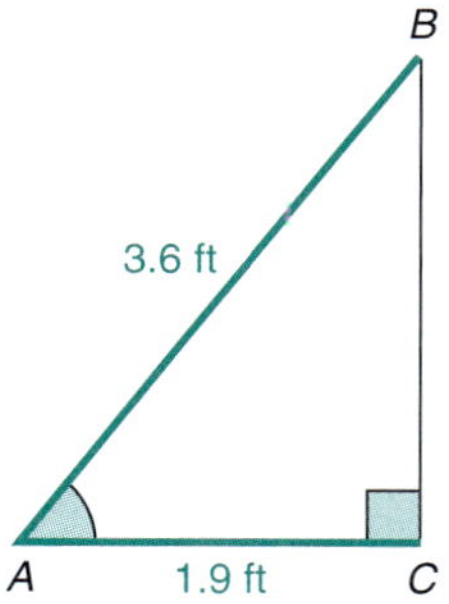

We are asked to find an angle and we are given the hypotenuse and the side adjacent to the angle, so we use the cosine function.

$$\cos A = \frac{\text{adj}}{\text{hyp}}$$ Substitute known values.

$$\cos A = \frac{1.9}{3.6}$$ Or $\cos^{-1}\frac{1.9}{3.6} = A$.

$$A = 58.14456918$$

$$\boldsymbol{A = 58.1°}$$ Rounded to nearest 0.1°.

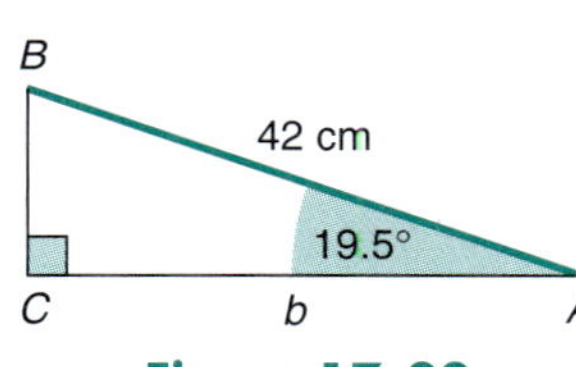

Figure 17–22

EXAMPLE Find side b of Fig. 17–22.

We can use either the sine or the cosine function because we are given the hypotenuse and an angle. However, we do not have to find the complement of the given angle if we use the cosine function.

$$\cos A = \frac{\text{adj}}{\text{hyp}}$$ Substitute.

$$\cos 19.5° = \frac{b}{42}$$ Solve for b.

$$42(\cos 19.5°) = b$$

$$39.59094263 = b$$

$$\boldsymbol{39.59 \text{ cm} = b}$$ Four significant digits.

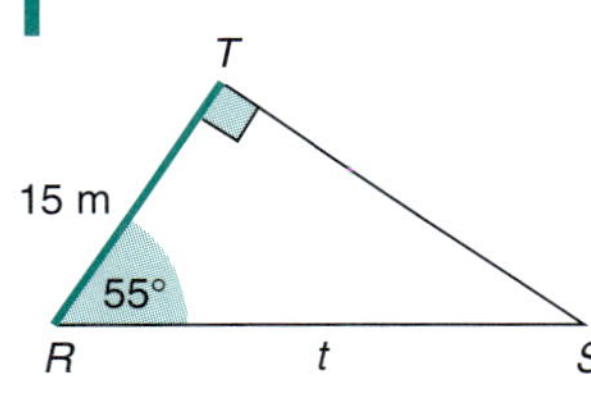

Figure 17–23

EXAMPLE Find side t of Fig. 17–23.

The cosine function is the most efficient function to use because we are finding the hypotenuse and we are given an acute angle and its adjacent side.

$$\cos R = \frac{\text{adj}}{\text{hyp}}$$ Substitute known values.

$$\cos 55° = \frac{15}{t}$$ Solve for t.

$$t(\cos 55°) = 15$$

$$t = \frac{15}{\cos 55°}$$

$$t = 26.15170193$$

$$\boldsymbol{t = 26.15 \text{ m}}$$ Four significant digits.

3 Find the Missing Parts of a Right Triangle Using the Tangent Function.

If we have a right triangle in which we know only the length of the two legs, we cannot use the sine or cosine function to solve the triangle unless we use the Pythagorean theorem to find the length of the hypotenuse. However, we can use the tangent function directly: $\tan \theta = \frac{\text{opp}}{\text{adj}}$.

Whenever we wish to check our calculations of the sides, we can use the Pythagorean theorem, as we did earlier in the section.

EXAMPLE Find angle A of Fig. 17–24.

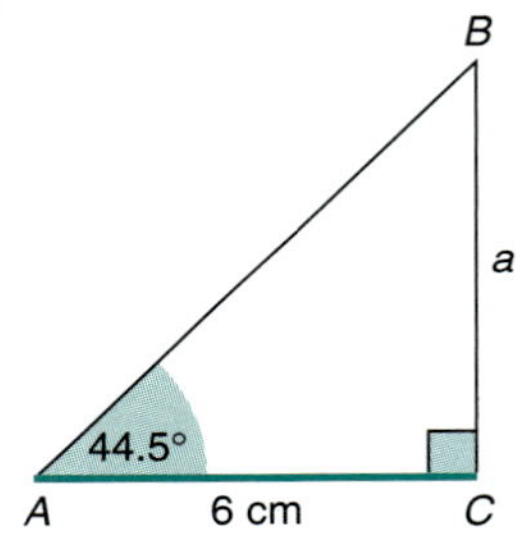

Figure 17–24

We are looking for an angle and given the side opposite and the side adjacent to the angle, so we use the tangent function.

$$\tan A = \frac{\text{opp}}{\text{adj}} \qquad \text{Substitute known values.}$$

$$\tan A = \frac{14}{16} \qquad \text{Or } \tan^{-1}\frac{14}{16} = A.$$

$$A = 41.18592517°$$

$$\mathbf{A = 41.2°} \qquad \text{Rounded to nearest } 0.1°.$$

EXAMPLE Use the tangent function to find a of Fig. 17–25.

Figure 17–25

We are given an acute angle and the side adjacent to the acute angle. We are looking for the opposite side so use the tangent function.

$$\tan A = \frac{\text{opp}}{\text{adj}} \qquad \text{Substitute known values.}$$

$$\tan 44.5° = \frac{a}{6} \qquad \text{Solve for } a.$$

$$6(\tan 44.5°) = a$$

$$5.896183579 = a$$

$$\mathbf{5.896 \text{ cm} = a} \qquad \text{Four significant digits}$$

EXAMPLE Find side b of Fig 17–26.

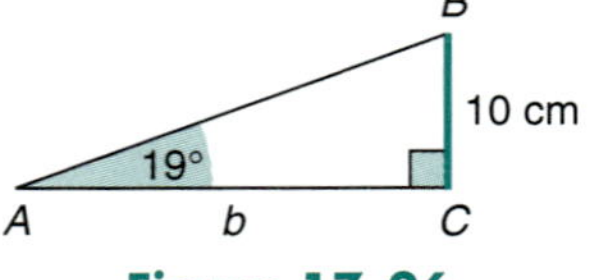

Figure 17–26

We know an acute angle and its opposite side. We are looking for the side adjacent to the acute angle.

$$\tan A = \frac{\text{opp}}{\text{adj}} \qquad \text{Substitute known values.}$$

$$\tan 19° = \frac{10}{b} \qquad \text{Solve for } b.$$

$$b(\tan 19°) = 10$$

$$b = \frac{10}{\tan 19°}$$

$$b = 29.04210878$$

$$\mathbf{b = 29.04 \text{ cm}} \qquad \text{Four significant digits.}$$

1 Use the sine function to find the indicated parts of the triangle *LMN* in Fig. 17–27. Round lengths of sides to four significant digits and angles to the nearest 0.1°.

1. Find M if $n = 15$ m and $m = 7$ m.
2. Find l if $n = 13$ in. and $L = 32°$.
3. Find m if $l = 15$ m and $L = 28°$.
4. Find n if $l = 12$ ft and $M = 42°$.
5. Find M if $m = 13$ cm and $n = 19$ cm.
6. Find M if $n = 3.7$ in. and $l = 2.4$ in.

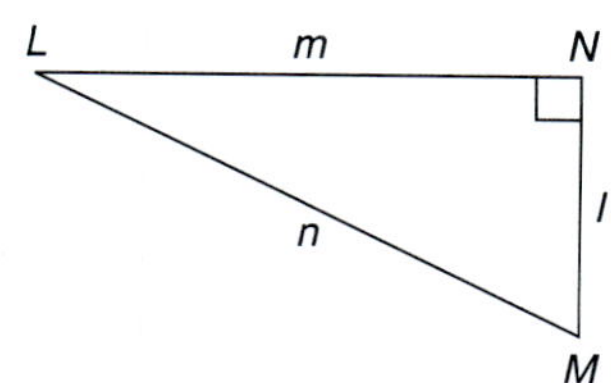

Figure 17–27

Solve triangle *STU* of Fig. 17–28 using the sine function. Check the measures of the sides by using the Pythagorean theorem. Round as above.

7. Solve if $t = 18$ yd and $s = 14$ yd.
8. Solve if $U = 45°$ and $u = 4.7$ m.
9. Solve if $S = 34.5°$ and $t = 8.5$ mm.
10. Solve if $S = 16°$ and $s = 14$ m.

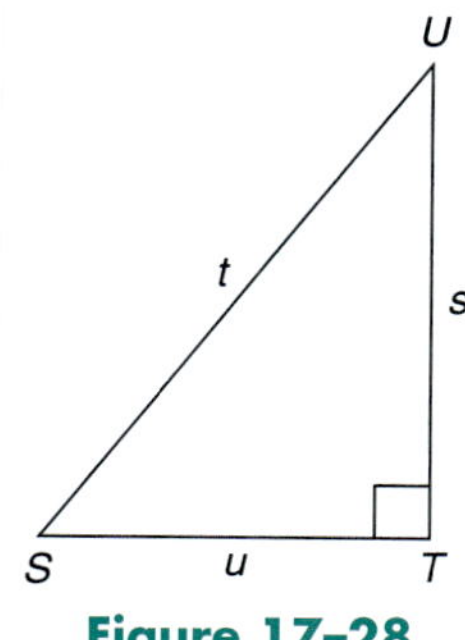

Figure 17–28

2 Use the cosine function to find the indicated parts of triangle *KLM* in Fig. 17–29. Round sides to four significant digits and angles to the nearest 0.1°.

11. Find M if $k = 13$ m and $l = 16$ m.
12. Find k if $l = 11$ cm and $M = 24°$.
13. Find l if $M = 31°$ and $k = 27$ ft.
14. Find l if $m = 15$ dm and $M = 25°$.
15. Find k if $K = 72°$ and $l = 16.7$ mm.
16. Find l if $K = 67°$ and $k = 13$ yd.

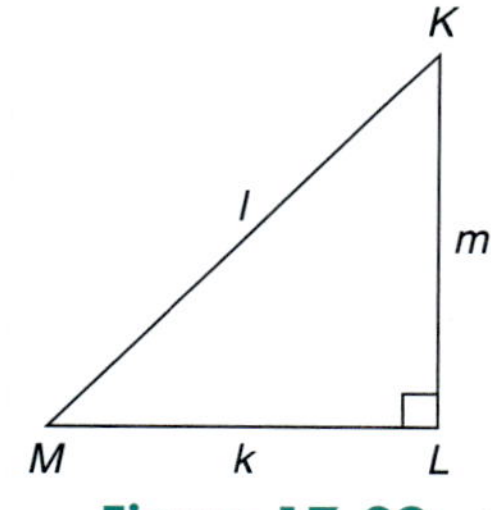

Figure 17–29

Use the sine or cosine function to solve triangle *QRS* of Fig. 17–30. Round as above.

17. Solve if $s = 23$ ft and $q = 16$ ft.
18. Solve if $s = 17$ cm and $R = 46°$.
19. Solve if $q = 14$ dkm and $Q = 73.5°$.
20. Solve if $R = 59.5°$ and $q = 8$ m.

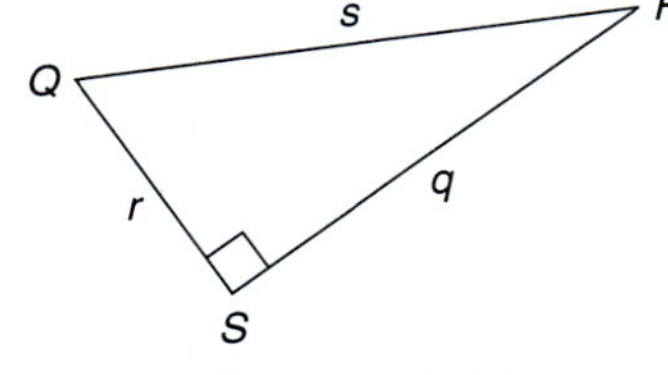

Figure 17–30

3 Use the tangent function to find the indicated parts of triangle *ABC* of Fig. 17–31. Round sides to four significant digits and angles to the nearest 0.1°.

21. Find A if $b = 11$ cm and $a = 6$ cm.
22. Find b if $a = 1.9$ m and $A = 25°$.
23. Find a if $A = 40.5°$ and $b = 7$ ft.
24. Find A if $b = 10.8$ m and $a = 4.7$ m.
25. Find a if $A = 43°$ and $b = 0.05$ cm.
26. Find a if $B = 68°$ and $b = 0.03$ m.

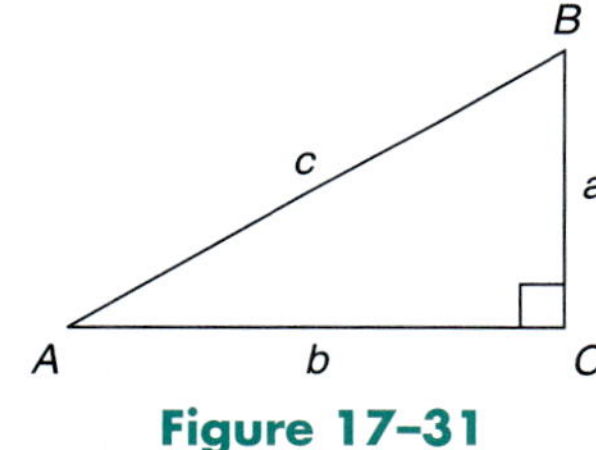

Figure 17–31

Solve triangle *DEF* of Fig. 17–32. Round sides to four significant digits and angles to the nearest 0.1°.

27. Solve if $e = 4.6$ m and $d = 3.2$ m.
28. Solve if $D = 42°$ and $e = 7$ ft.
29. Solve if $E = 73.5°$ and $e = 20.13$ in.
30. Solve if $d = 11$ ft and $e = 8$ ft.

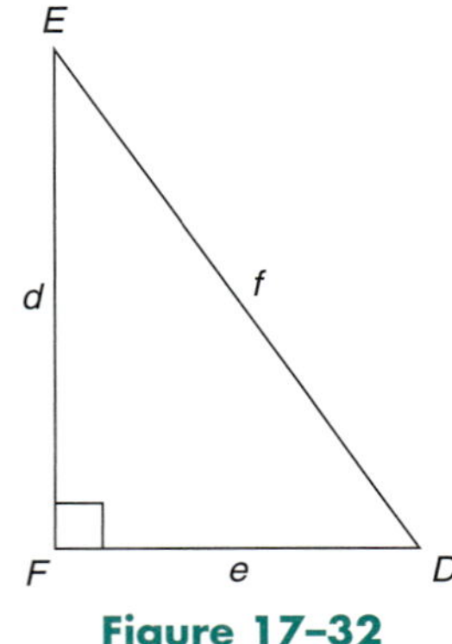
Figure 17–32

17–5 APPLIED PROBLEMS USING RIGHT-TRIANGLE TRIGONOMETRY

Learning Outcomes

1. Select the most direct method for solving right triangles.
2. Solve applied problems using right-triangle trigonometry.

In this section, we learn to select the function (theorem, property, or formula) that minimizes the number of steps needed to solve a problem or that uses the fewest calculations. This selection may improve both efficiency and accuracy.

1 Select the Most Direct Method for Solving Right Triangles.

Our first task in solving any problem involving right triangles is to *select the most convenient and efficient function*. Following are two basic guidelines.

> **To select the most direct method for solving a right triangle:**
>
> 1. Where possible, choose the function that uses given parts rather than unknown parts that must be calculated.
> 2. Where possible, choose the function that gives the desired part directly, that is, without having to find other parts first.

EXAMPLE In $\triangle ABC$ of Fig. 17–33, find c.

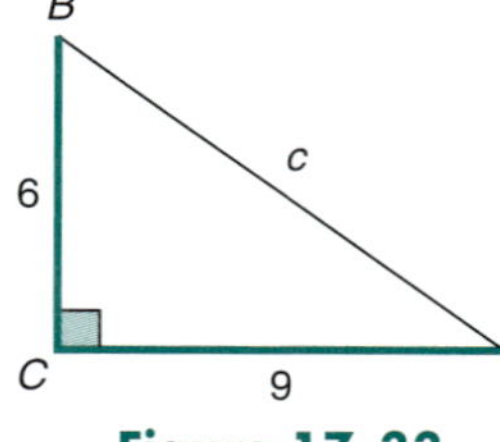
Figure 17–33

The most direct way of finding side c is to use the *Pythagorean theorem*. The two legs of a right triangle are given.

$$c^2 = a^2 + b^2 \qquad \text{Substitute.}$$
$$c^2 = 6^2 + 9^2$$
$$c^2 = 36 + 81$$
$$c^2 = 117 \qquad \text{Take the square root of both sides.}$$
$$c = 10.81665383$$
$$\mathbf{c = 10.82} \qquad \text{Four significant digits.}$$

Now let's see how trigonometric functions can be used to find c.

$$\tan A = \frac{\text{opp}}{\text{adj}} \qquad \text{Opposite and adjacent sides are given for this function.}$$

$$\tan A = \frac{6}{9} \qquad \text{Solve for } A.$$

$$A = \tan^{-1}\frac{6}{9}$$

$$A = 33.69006753$$

$$A = 33.7°$$ Rounded to the nearest 0.1°

$$\sin A = \frac{\text{opp}}{\text{hyp}}$$ Substitute 33.69006753 for A and 6 for the opposite side.

$$\sin 33.69006753 = \frac{6}{c}$$ Solve for the hypotenuse.

$$c = \frac{6}{\sin 33.69006753}$$

$$c = 10.81665383$$

$$\mathbf{c = 10.82}$$ Four significant digits.

EXAMPLE   In $\triangle ABC$ of Fig. 17–34, find angle B.

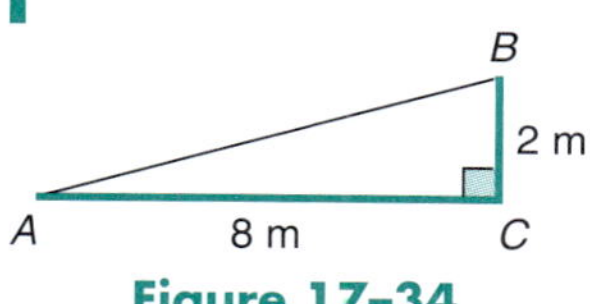

Figure 17–34

The Pythagorean theorem cannot be used to find any angle measure. In this problem, we are given the sides adjacent to and opposite angle B, so we should use the *tangent function* for a quick, direct solution.

$$\tan \theta = \frac{\text{opp}}{\text{adj}}$$ Substitute.

$$\tan B = \frac{8}{2}$$ Reduce.

$$\tan B = 4$$ Or $\tan^{-1} 4 = B$.

$$B = 75.96375653$$

$$\mathbf{B = 76.0°}$$ Round to the nearest 0.1°.

2 Solve Applied Problems Using Right-Triangle Trigonometry.

Many technical applications can be solved using right triangles. The following examples show some career-related applications. In solving technical problems, it's a good idea to draw diagrams or pictures to visualize the various relationships.

EXAMPLE   A jet takes off at a 30° angle (Fig. 17–35). If the runway (from takeoff) is 875 ft long, find the altitude of the airplane as it flies over the end of the runway.

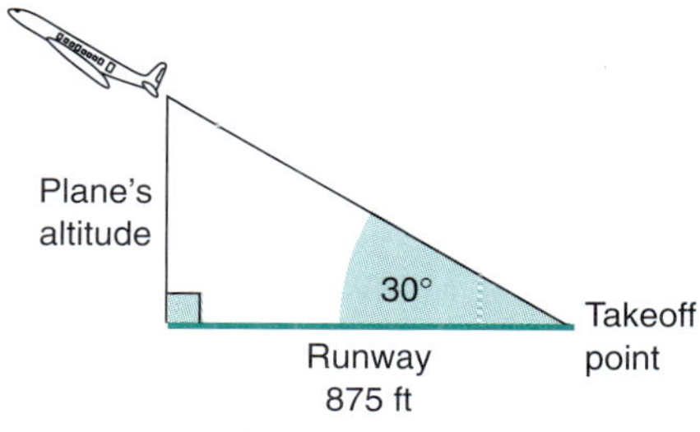

Figure 17–35

Known facts   One acute angle = 30°; adjacent side = 875 ft

Unknown fact   Plane's altitude at end of runway or opposite side

<table>
<tr><td>

Relationship

</td><td>

$$\tan \theta = \frac{\text{opp}}{\text{adj}}$$

</td></tr>
<tr><td>

Estimation

</td><td>

In a 45° 45° 90° right triangle, the legs are equal. Because 30° is less than 45°, the side opposite the 30° angle should be less than 875 ft.

</td></tr>
<tr><td>

Calculations

</td><td>

$$\tan \theta = \frac{\text{opp}}{\text{adj}}$$

$$\tan 30° = \frac{a}{875}$$

$$875(\tan 30°) = a$$

$$505.1814855 = a$$

</td></tr>
<tr><td>

Interpretation

</td><td>

505.2 ft = plane's altitude   Four significant digits.

</td></tr>
</table>

Many right–triangle applications use the terminology *angle of elevation* and *angle of depression* (Fig. 17–36). The angle of elevation is generally used when we are looking *up* at an object. We use the angle of depression to describe the location of an objective *below* our eye level. *Both* angles are formed by a line of sight and a horizontal line from the point of sight.

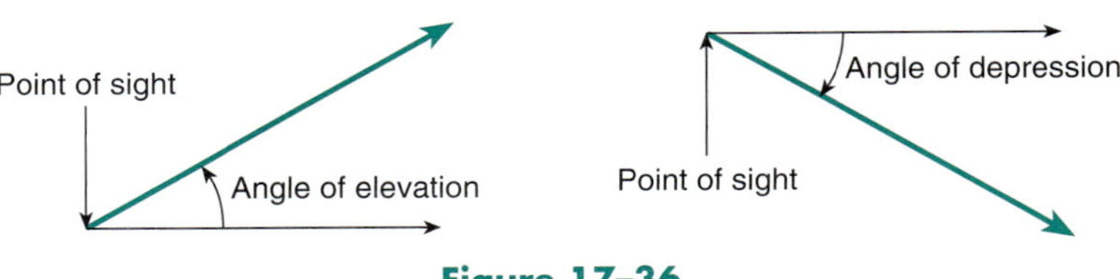

Figure 17–36

EXAMPLE A stretch of roadway drops 30 ft for every 300 ft of road (Fig. 17–37). Find the *angle of declination* of the road.

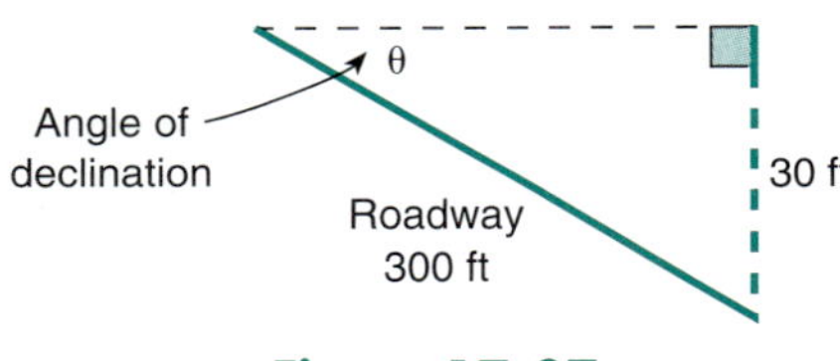

Figure 17–37

The *angle of declination* is the angle of depression. The opposite side and the hypotenuse are given.

$$\sin \theta = \frac{\text{opp}}{\text{hyp}}$$ Substitute.

$$\sin \theta = \frac{30}{300}$$

$$\sin \theta = 0.1$$ Or $\sin^{-1} 0.1 = \theta$.

$$\theta = 5.739170477°$$

The angle of declination of the road is 5.7° rounded to the nearest 0.1°.

EXAMPLE A surveyor locates two points on a steel column so that it can be set plumb (perpendicular to the horizon). If the angle of elevation is 15° and the surveyor's transit is 175 ft from the column (Fig. 17–38), find the distance from the transit to the upper point (point B) on the column. (A *transit* is a surveying instrument used for measuring angles.)

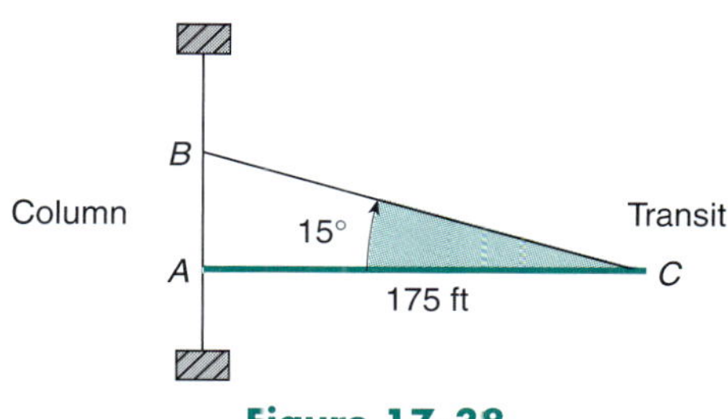

Figure 17–38

An acute angle and the adjacent side are given. To find the distance from the transit to point B on the column (hypotenuse), we use the *cosine function.*

$$\cos \theta = \frac{\text{adj}}{\text{hyp}} \qquad \text{Substitute.}$$

$$\cos 15° = \frac{175}{\text{hyp}} \qquad \text{Solve for the hypotenuse.}$$

$$\text{hyp}(\cos 15°) = 175$$

$$\text{hyp} = \frac{175}{\cos 15°}$$

$$\text{hyp} = 181.1733316$$

$$\text{hyp} = 181.2 \text{ ft} \qquad \text{Four significant digits.}$$

Point B is 181.2 ft from the transit.

EXAMPLE Find the angle a rafter makes with a joist of a house if the rise is 12 ft and the span is 30 ft (Fig. 17–39). Also, find the length of the rafter.

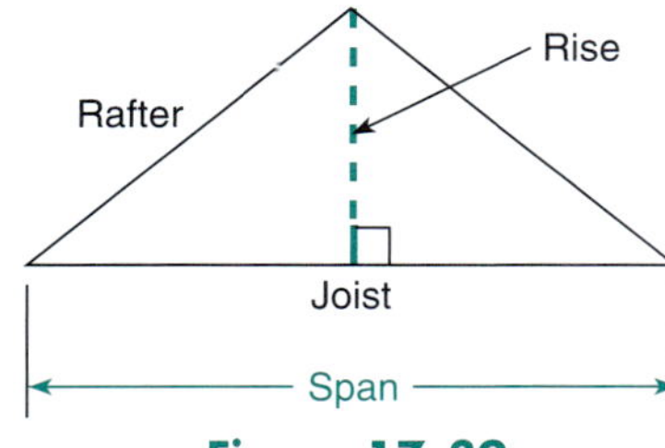

Figure 17–39

The span is twice the distance from the outside stud to the center point of the joist. Therefore, to solve the right triangle for the desired angle, we draw the triangle shown in Fig. 17–40. Because the legs of a right triangle are given, we use the tangent function.

$$\tan \theta = \frac{\text{opp}}{\text{adj}} \qquad \text{Substitute.}$$

$$\tan \theta = \frac{12}{15} \qquad \operatorname{Tan}^{-1}\frac{12}{15} = \theta$$

$$\theta = 38.65980825$$

$$\theta = 38.7° \qquad \text{Round to the nearest } 0.1°.$$

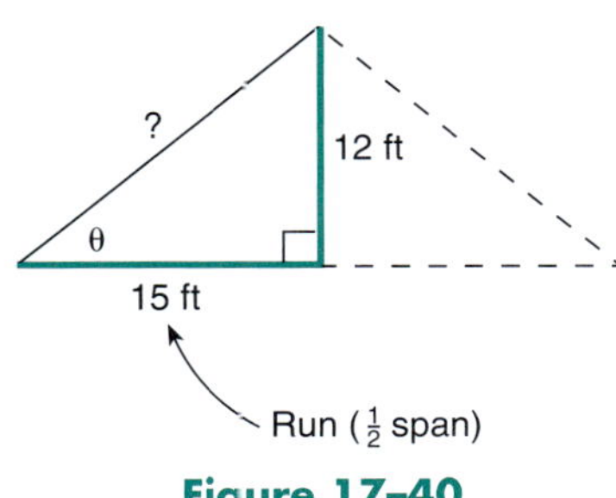

Figure 17–40

We use the *Pythagorean theorem* to find the length of the rafter directly.

$$(\text{hyp})^2 = 12^2 + 15^2$$

$$(\text{hyp})^2 = 144 + 225$$

$$(\text{hyp})^2 = 369$$

$$\text{hyp} = 19.20937271$$

$$\text{hyp} = 19.21 \text{ ft} \qquad \text{Four significant digits.}$$

The angle the rafter makes with the joist is 38.7°, and the rafter is 19.21 ft long.

EXAMPLE Find the angle formed by the connecting rod in the mechanical assembly shown in Fig. 17–41.

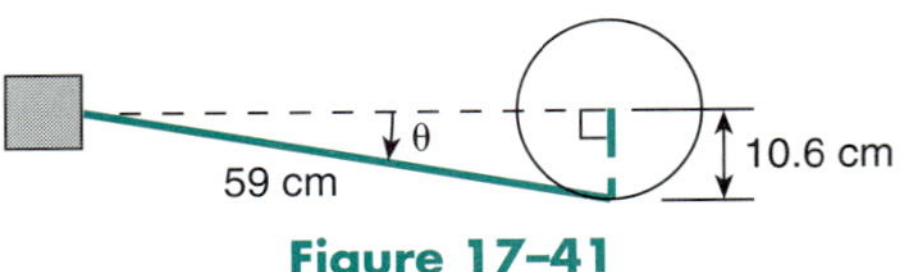

Figure 17–41

Given the hypotenuse and the side opposite the desired angle, we use the *sine function.*

$$\sin \theta = \frac{\text{opp}}{\text{hyp}} \qquad \text{Substitute.}$$

$$\sin \theta = \frac{10.6}{59} \qquad \operatorname{Sin}^{-1} \frac{10.6}{59} = \theta$$

$$\theta = 10.35001563$$

$$\theta = 10.4° \qquad \text{Rounded.}$$

The angle formed by the connecting rod is 10.4°.

EXAMPLE Find the impedance Z of a circuit with 20 Ω of reactance X_L represented by the vector diagram in Fig. 17–42.

Because we are looking for the hypotenuse Z and are given an acute angle and the opposite side, we use the *sine function.*

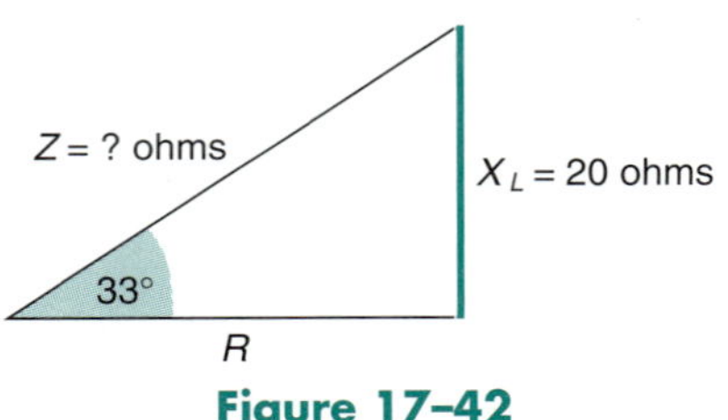

Figure 17–42

$$\sin \theta = \frac{\text{opp}}{\text{hyp}} \qquad \text{Substitute.}$$

$$\sin 33° = \frac{20}{Z}$$

$$Z (\sin 33°) = 20$$

$$Z = \frac{20}{\sin 33°}$$

$$Z = 36.72156918$$

$$Z = 36.72 \ \Omega$$

The impedance, Z, is 36.72 Ω.

Chapter 17   Right-Triangle Trigonometry

1   Find the indicated part of the right triangles in Figs. 17–43 to 17–47 by the most direct method.

1.

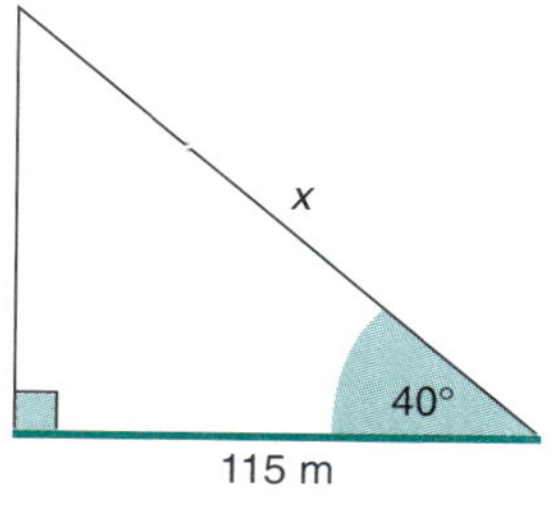

Figure 17–43

2.

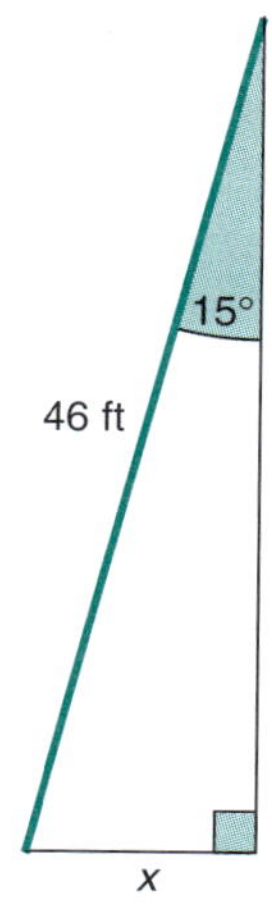

Figure 17–44

3.

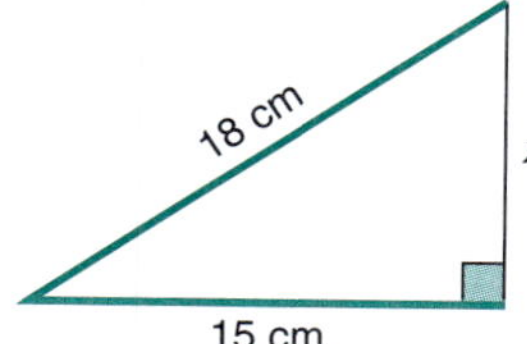

Figure 17–45

4.

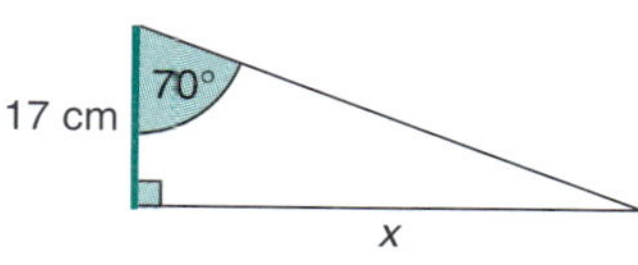

Figure 17–46

5.

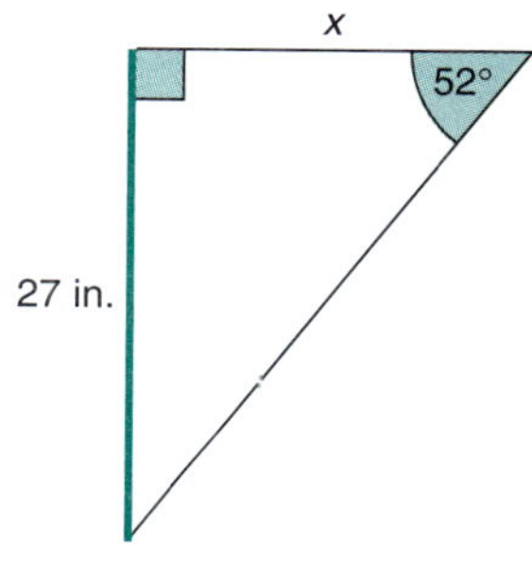

Figure 17–47

2   Use the trigonometric functions to solve the technical problems. Round side lengths to four significant digits and angles to 0.1°.

6. A sign is attached to a building by a triangular brace. If the horizontal length of the brace is 48 in. and the angle at the sign is 25° (Fig. 17–48), what is the length of the wall support piece?

7. A surveyor uses right triangles to measure inaccessible property lines. To measure a property line that crosses a pond, a surveyor sights to a point across the pond, then makes a right angle, measures 50 ft, and sights the point across the pond with a 47° angle (Fig. 17–49). Find the distance across the pond from the initial point.

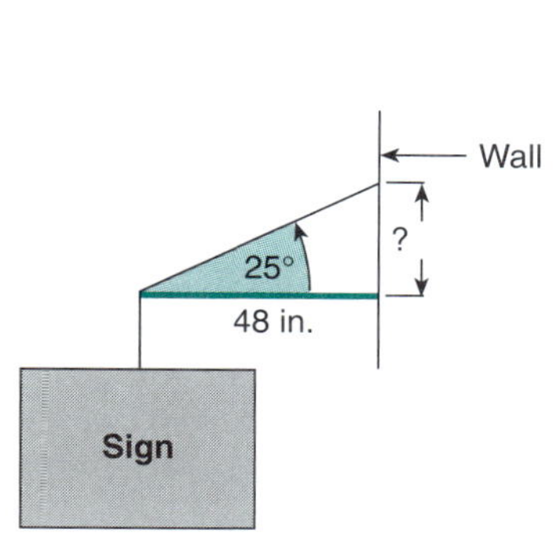

Figure 17–48

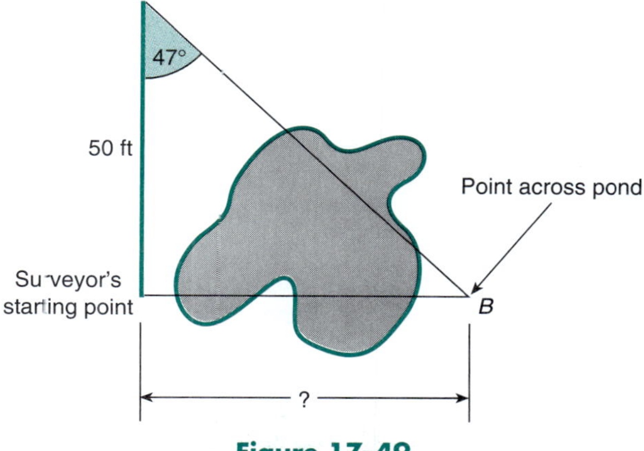

Figure 17–49

8. At what angle must a jet descend if it is 900 ft above the end of the runway and must touch down 1,500 ft from the runway's end?

9. A 50-ft wire is used to brace a utility pole. If the wire is attached 4 ft from the top of the 35-ft pole, how far from the base of the pole will the wire be attached to the ground?

10. A roadway rises 4 ft for every 15 ft along the road. What is the angle of inclination of the roadway?

11. A shadow cast by a tree is 32 ft long when the angle of inclination of the sun is 36°. How tall is the tree?

12. The vector diagram of the circuit in Fig. 17–50 has a known impedance Z. Find the reactance X_L. All units are in ohms.

13. Using Fig. 17–50, find resistance R.

14. A piston assembly at the midpoint of its stroke forms a right triangle (Fig. 17–51). Find the length of rod R.

15. Find the angle a rafter makes with a joist of a house if the rise is 18 ft and the span is 50 ft. Refer to Fig. 17–39 on page 681.

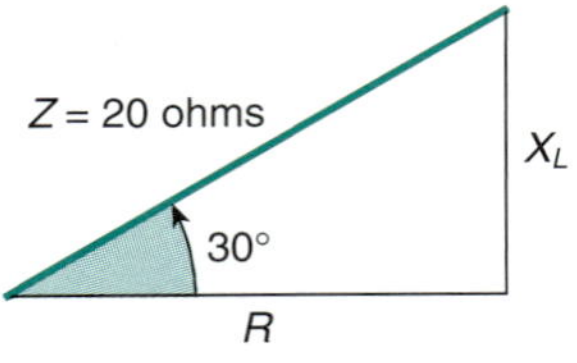

Figure 17–50

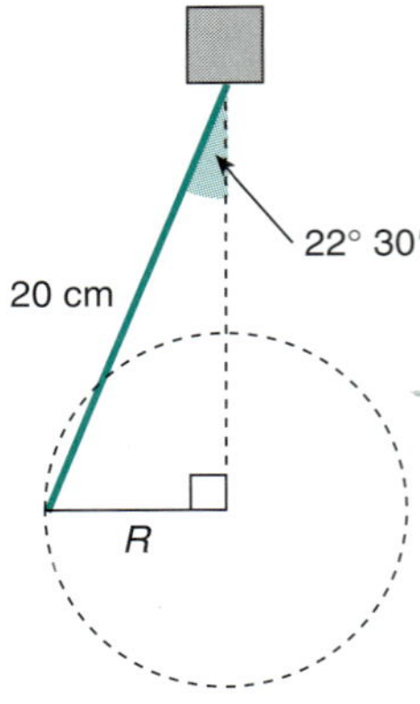

Figure 17–51

Electronics: Series Circuits and Parallel Circuits

Electronic technicians use many right triangles, but the two most common ones are the ohms triangle and the siemens triangle. Ohms and siemens are reciprocals of each other. Series circuits use the ohms triangle (Fig. 17–52) and parallel circuits use the siemens triangle (Fig. 17–53).

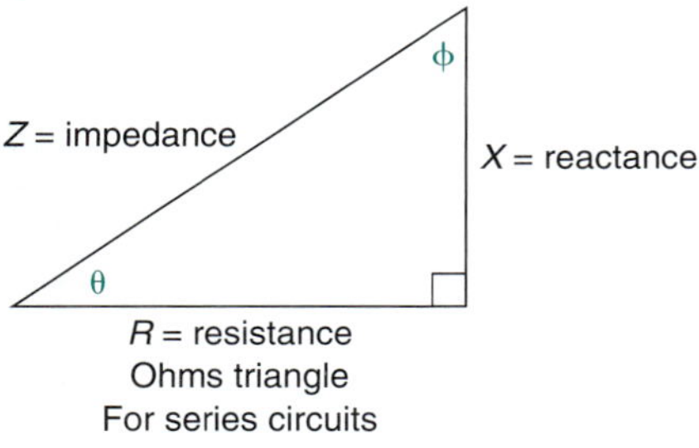

Figure 17–52

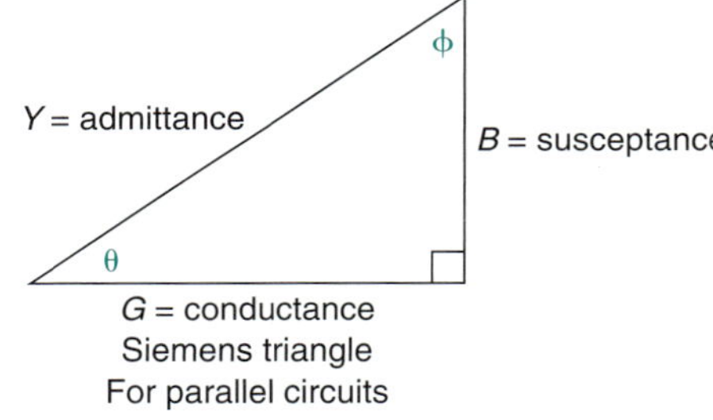

Figure 17–53

For each triangle, the units are the same on all three legs. The hypotenuse and the angle of rotation combine to give the polar notation, and the horizontal side and the vertical side combine to give the coordinates of a point in rectangular notation. A polar number shows the location of a point differently from the rectangular coordinate system. The coordinates of a polar number are the angle of rotation and the length of the hypotenuse.

Most analyses are done with rectangular numbers. All measurements are done with polar numbers. It is important to understand the terminology of this application. All the triangles are handled in a similar way.

Some basic facts about circuits will help us in interpreting a problem.

- All three sides in the siemens triangle are measured in siemens, the reciprocal of ohms.
- If a component has a high resistance, it has a low conductance, and vice versa.
- R and G are used for both dc (direct current) and ac (alternating current) analyses.
- X, Z, B, and Y are used only for ac analyses.

The Pythagorean theorem applies to any right triangle. The formula can be rearranged to solve for any missing side if the other two sides are known. The following equations give a variation of the Pythagorean theorem for each side in both the ohms and siemens triangles.

Ohms Triangle for Series Circuits   **Siemens Triangle for Parallel Circuits**

$$Z = \sqrt{R^2 + X^2} \qquad\qquad Y = \sqrt{G^2 + B^2}$$
$$R = \sqrt{Z^2 - X^2} \qquad\qquad G = \sqrt{Y^2 - B^2}$$
$$X = \sqrt{Z^2 - R^2} \qquad\qquad B = \sqrt{Y^2 - G^2}$$

The following are some additional basic facts in electronics:

- The sides for R and G are always positive and rest on the horizontal axis.
- X and B are always perpendicular to the horizontal axis, and they may be positive or negative depending on whether the circuit is inductive or capacitive.
- The hypotenuse, Z or Y, is considered positive because it is a measure of magnitude only rather than magnitude *and* direction found in the vertical and horizontal measures.
- The angle θ is also called the *phase angle* and is either positive or negative, depending on the slope of the hypotenuse.
- The angle θ is the angle between the hypotenuse and the vertical side.

The angles θ and ϕ can be calculated in several different ways, depending on which sides or angles you were given originally. Another standard notation for $\sin^{-1}$, $\cos^{-1}$, and $\tan^{-1}$ is arcsin, arccos, and arctan, respectively.

Ohms

$$\theta = \arctan\left(\frac{\text{opp}}{\text{adj}}\right) \qquad \theta = \arctan\left(\frac{X}{R}\right) \qquad \theta = \tan^{-1}\left(\frac{X}{R}\right)$$

$$\theta = \text{arcsin}\left(\frac{\text{opp}}{\text{hyp}}\right) \qquad \theta = \text{arcsin}\left(\frac{X}{Z}\right) \qquad \theta = \sin^{-1}\left(\frac{X}{Z}\right)$$

$$\theta = \arccos\left(\frac{\text{adj}}{\text{hyp}}\right) \qquad \theta = \arccos\left(\frac{R}{Z}\right) \qquad \theta = \cos^{-1}\left(\frac{R}{Z}\right)$$

Siemens

$$\theta = \arctan\left(\frac{\text{opp}}{\text{adj}}\right) \qquad \theta = \arctan\left(\frac{B}{G}\right) \qquad \theta = \tan^{-1}\left(\frac{B}{G}\right)$$

$$\theta = \arcsin\left(\frac{\text{opp}}{\text{hyp}}\right) \qquad \theta = \arcsin\left(\frac{B}{Y}\right) \qquad \theta = \sin^{-1}\left(\frac{B}{Y}\right)$$

$$\theta = \arccos\left(\frac{\text{adj}}{\text{hyp}}\right) \qquad \theta = \arccos\left(\frac{G}{Y}\right) \qquad \theta = \cos^{-1}\left(\frac{G}{Y}\right)$$

Finally, θ and ϕ are complementary angles.

$$\phi = 90° - \theta \qquad \text{and} \qquad \theta = 90° - \phi$$

The calculator sequence for finding an arcfunction is the same as the inverse function.

Find all indicated missing sides and angles in Figs. 17–54 and 17–55.

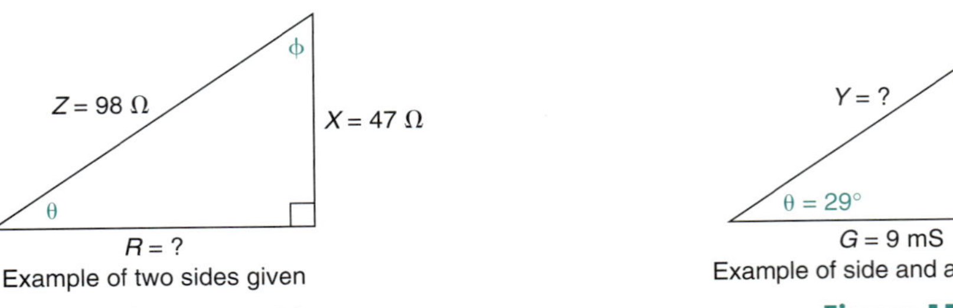

Figure 17–54

Figure 17–55

$$Z^2 = R^2 + X^2 \qquad \text{Solve for } R.$$

$$R^2 = Z^2 - X^2$$

$$R = \sqrt{Z^2 - X^2}$$

$$R = \sqrt{98^2 - 47^2}$$

$$\mathbf{R = 85.99\ \Omega}$$

$$\sin \theta = \frac{\text{opp}}{\text{hyp}} = \frac{X}{Z}$$

$$\theta = \arcsin\left(\frac{X}{Z}\right) = \sin^{-1}\left(\frac{X}{Z}\right)$$

$$\theta = \arcsin\left(\frac{47}{98}\right) = \sin^{-1}\left(\frac{47}{98}\right)$$

$$\boldsymbol{\theta = 28.7°}$$

$$\boldsymbol{\phi = 90° - 28.7° = 61.3°}$$

Proof: $\quad \tan 28.7° \overset{?}{=} \dfrac{47}{85.99}$ True, except for rounding discrepancy.

$$0.547484008 \approx 0.546575183$$

$$\cos \theta = \frac{G}{Y} \qquad\qquad \tan \theta = \frac{B}{G}$$

$$Y = \frac{G}{\cos \theta} \qquad\qquad B = G \tan \theta$$

$$Y = \frac{9}{\cos 29°} \qquad\qquad B = 9 \tan 29°$$

$$\mathbf{Y = 10.29\ mS} \qquad \mathbf{B = 4.99\ mS}$$

$$\boldsymbol{\phi = 90° - 29° = 61°}$$

Proof: $\quad Y^2 = G^2 + B^2$

$$10.29^2 \overset{?}{=} 9^2 + 4.99^2$$

$$105.8841 \overset{?}{=} 81 + 24.9001$$

$$105.8841 \overset{?}{=} 105.9001 \qquad \text{True, except for rounding discrepancy.}$$

$$\phi + \theta = 90°$$

$$61 + 29 \overset{?}{=} 90° \qquad \text{True.}$$

Fill in all answers on the triangles in Figs. 17–56 and 17–57 with the correct units. Always give a proof.

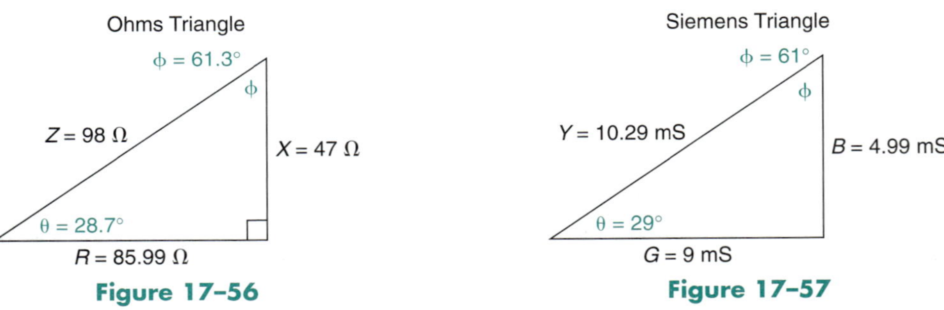

Figure 17–56

Figure 17–57

CHAPTER 17　Right-Triangle Trigonometry

Exercises

Redraw the triangles in Figs. 17–58 to 17–65. Fill in all missing sides and angles. Show your proof. Include correct measuring units. Round angle measures to the nearest 0.1° and side measures to hundreths.

1.

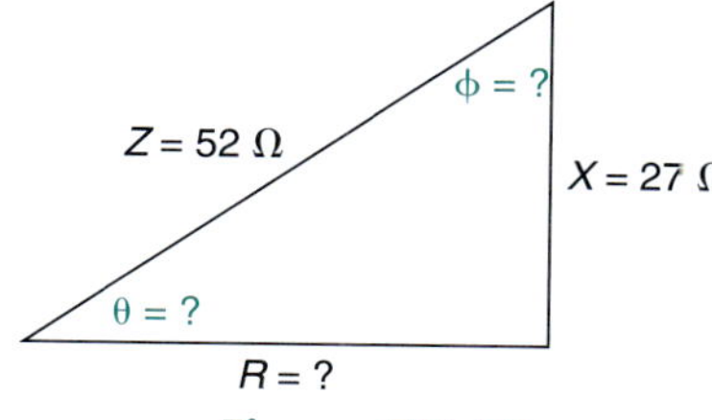

Figure 17–58

2.

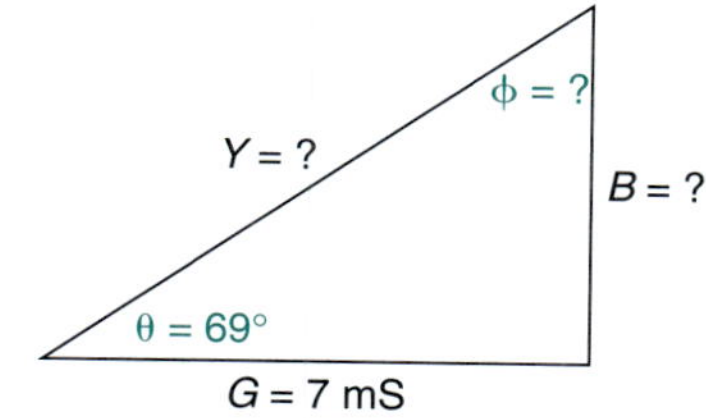

Figure 17–59

3.

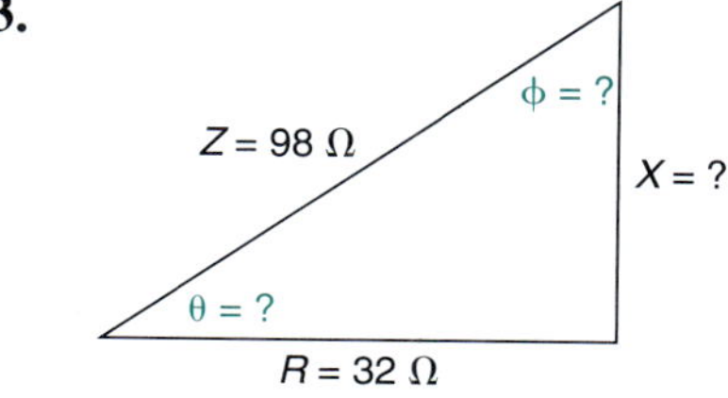

Figure 17–60

4.

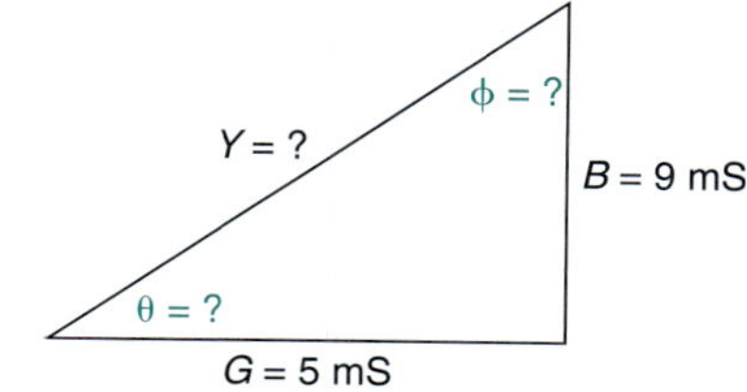

Figure 17–61

5.

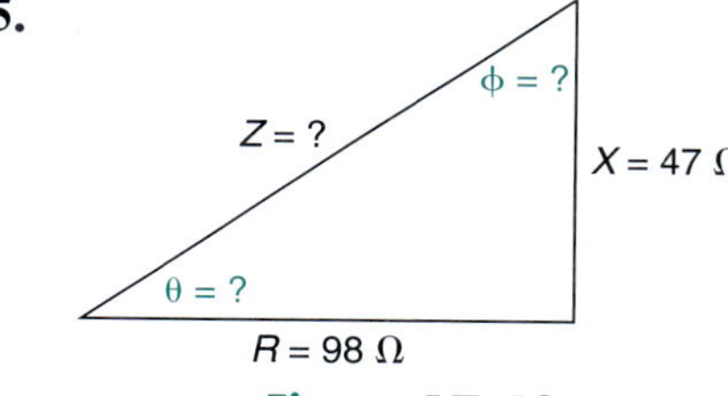

Figure 17–62

6.

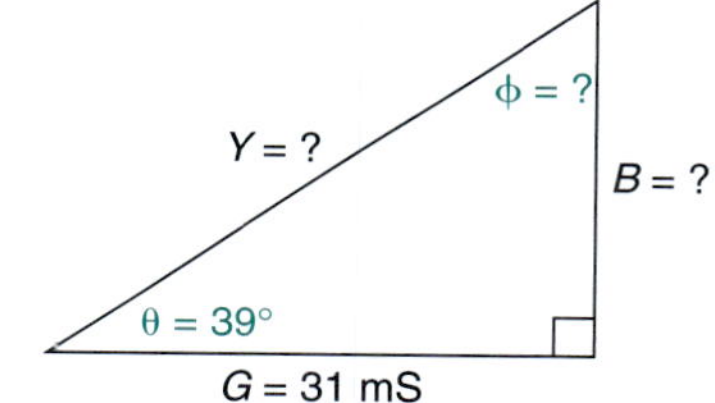

Figure 17–63

7.

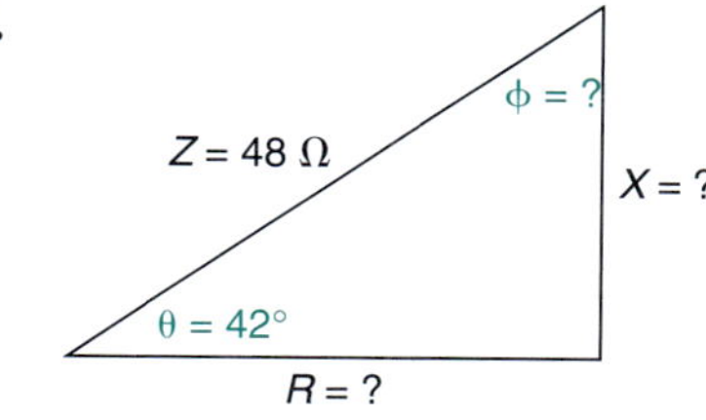

Figure 17–64

8.

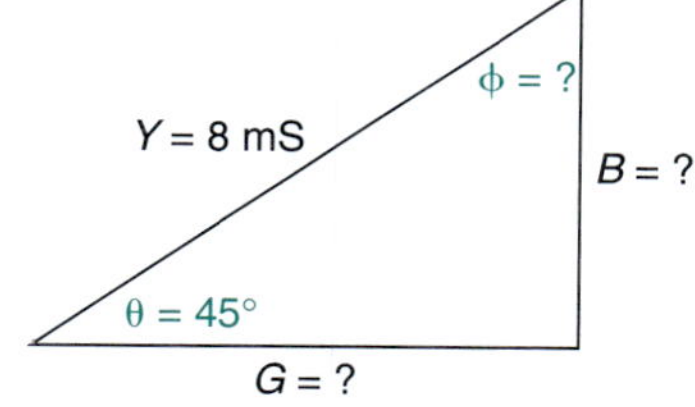

Figure 17–65

Answers for Exercises

1.

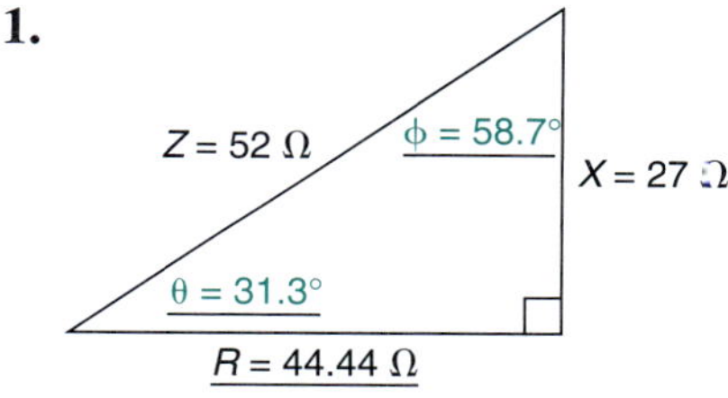

Figure 17–66

2.

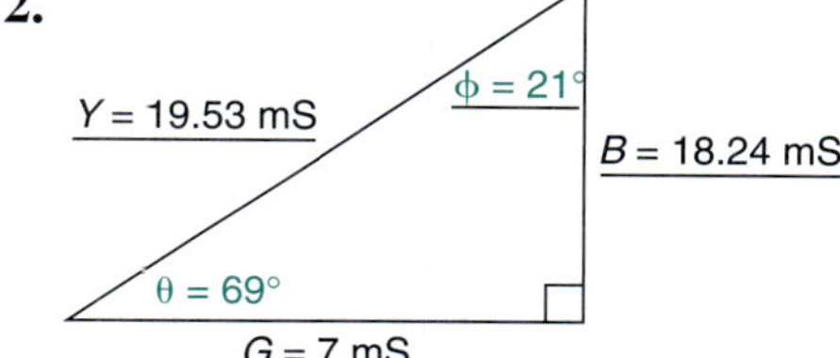

Figure 17–67

3.

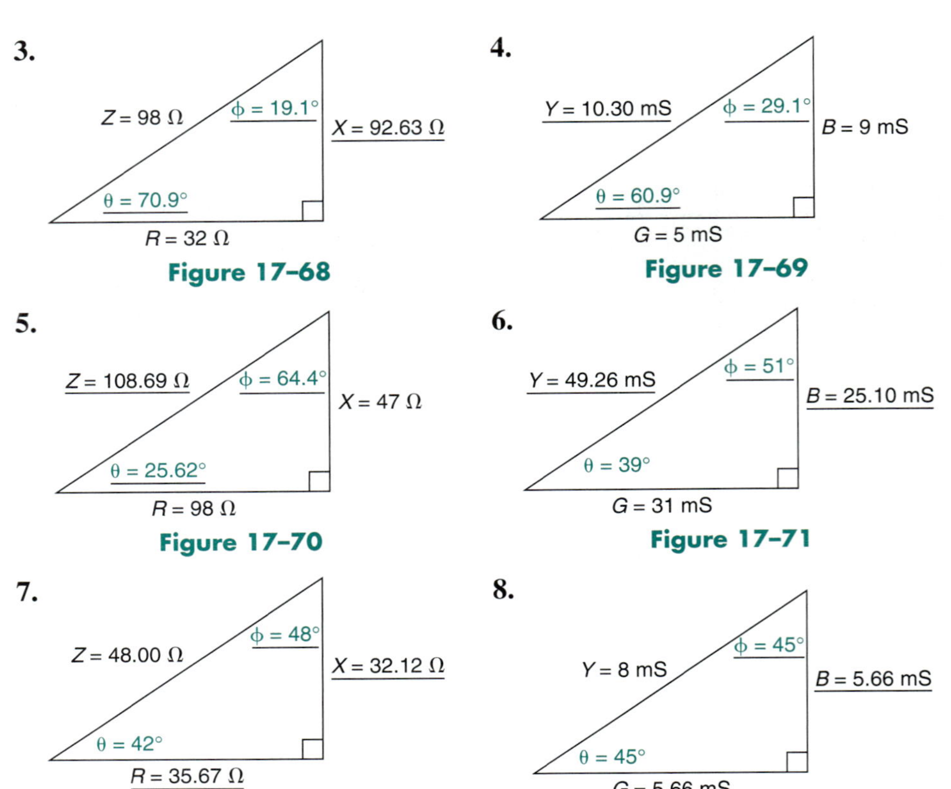

$Z = 98\ \Omega$ $\phi = 19.1°$ $X = 92.63\ \Omega$

$\theta = 70.9°$

$R = 32\ \Omega$

Figure 17–68

4.

$Y = 10.30\ \text{mS}$ $\phi = 29.1°$ $B = 9\ \text{mS}$

$\theta = 60.9°$

$G = 5\ \text{mS}$

Figure 17–69

5.

$Z = 108.69\ \Omega$ $\phi = 64.4°$ $X = 47\ \Omega$

$\theta = 25.62°$

$R = 98\ \Omega$

Figure 17–70

6.

$Y = 49.26\ \text{mS}$ $\phi = 51°$ $B = 25.10\ \text{mS}$

$\theta = 39°$

$G = 31\ \text{mS}$

Figure 17–71

7.

$Z = 48.00\ \Omega$ $\phi = 48°$ $X = 32.12\ \Omega$

$\theta = 42°$

$R = 35.67\ \Omega$

Figure 17–72

8.

$Y = 8\ \text{mS}$ $\phi = 45°$ $B = 5.66\ \text{mS}$

$\theta = 45°$

$G = 5.66\ \text{mS}$

Figure 17–73

CAREER APPLICATION

Tornado-Chasing Expeditions

Since the hit movie *Twister* travel agents in Oklahoma City have done a brisk business booking adventure expeditions with local storm-chasers. Most tornadoes occur during hot weather, so bookings follow a periodic yearly pattern, as shown in the graph in Fig. 17–74. The following formula is used to plot the values on the graph.

$$B = 80 \sin\left[\left(\frac{\pi}{26}\right)w\right] + 80$$

where B is the number of individual bookings per week, and w is the number of weeks since the week of April 1 (the first week of tornado season).

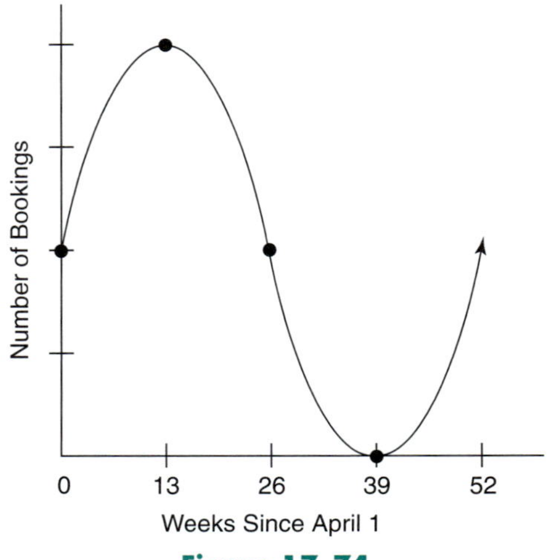

Figure 17–74

CHAPTER 17   Right-Triangle Trigonometry

Exercises

Use the preceding information to answer the following. Round answers to the nearest unit.

1. The numbers on the vertical axis of the graph have been omitted. Use the formula and your calculator to fill in the numbers.
2. Under the five numbers on the horizontal axis, write the name of the corresponding month of the year.
3. Compare this graph with the standard graph of $y = \sin\theta$, for θ between 0 and 2π rad. Discuss amplitude (maximum high and low values), period (length of a complete cycle), and phase shift (shift vertically or horizontally).
4. Use the graph to estimate two dates at which the number of weekly bookings is about 50.
5. Now substitute your guesses from Exercise 4 into the formula to see how close you are.
6. Use the formula to find the number of bookings for the following weeks after the start of tornado season: (a) 3, (b) 14, (c) 20, (d) 45
7. Use the graph to verify your answers in Exercise 6. If your answers differ, explain why.
8. During which months do bookings increase? During which months do they decrease?

Answers

1. The vertical axis is incremented by 40. The missing numbers are 40, 80, 120, and 160.
2. Because w = weeks since April 1, 0 = April, 13 = July, 26 = October, 39 = January, and 52 = April.
3. The graphs are very similar; both have the same shape, the same orientation, and the same period. In the bookings graph, the amplitude of the standard sine graph has been multiplied by 30, and the whole standard sine graph has been shifted vertically +80 units.
4. Around weeks 29 and 49.
5. Answers will vary. About 51 weeks.
6. (a) About 108, (b) about 159, (c) about 133, (d) about 20
7. Answers will vary, but they should be close.
8. Bookings increase from January to July, and they decrease from July to January.

ASSIGNMENT EXERCISES

Section 17–1

Convert the measures to radians rounded to the nearest hundredth.

1. $60°$
2. $212°$
3. $300°$

Convert the measures to degrees rounded to the nearest ten-thousandth, then convert to radians to the nearest hundredth.

4. $25°30'$
5. $99°45'$
6. $120°20'40''$

Convert the measures to degrees. Round to the nearest ten-thousandth of a degree.

7. $\dfrac{5\pi}{6}$ rad
8. 2.4 rad
9. 1.7 rad

Convert the measures to degrees, minutes, and seconds. Round to the nearest second.

10. 0.9 rad
11. $\dfrac{3\pi}{8}$ rad
12. 1.2 rad

Find the arc length, radius, or central angle (in radians) for each of the following.
Round to hundredths when necessary.

13. Find s if $\theta = 0.7$ rad and $r = 2.3$ cm.
14. Find θ if $s = 6.2$ cm and $r = 5$ cm.
15. Find r if $\theta = 2.1$ rad and $s = 3.6$ ft.

Find the area of the sector, the radius, and the central angle (in radians) of the sector for each of the following. Round to hundredths when necessary.

16. Find A if $\theta = 0.88$ rad and $r = 1.5$ m.
17. Find r if $\theta = 4.2$ rad and $A = 24$ in^2.
18. Find θ if $r = 4$ cm and $A = 12$ cm^2.
19. A pendulum 6 in. long swings through an angle of 20°. Find the arc length the pendulum swings over from one extreme position to the other. Round to hundredths.
20. A movable part on a machine swings through an angle of 40° with an arc length of 8 in. What is the length of the part? Round to hundredths.
21. Find the area of a sector with a central angle of 85° and whose radius is 4.6 cm. $\left(A = \frac{1}{2}\theta r^2,\right.$ where θ is in radians.) Round to hundredths.
22. Find the number of degrees to the nearest ten-thousandth of a central angle of a sector that has an area of 8.4 cm^2 and a radius of 6 cm.

Section 17–2
Find the trigonometric ratios using Fig. 17–75.

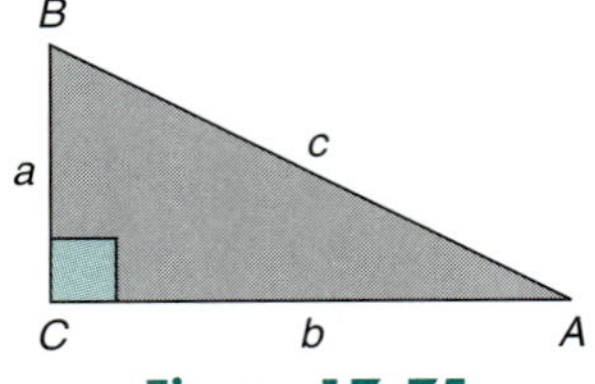

Figure 17–75

Use $a = 15$, $b = 20$, $c = 25$ in Exercises 23–34 and express the ratios as fractions in lowest terms.

23. $\sin A$ **24.** $\cos A$ **25.** $\tan A$ **26.** $\csc A$ **27.** $\sec A$ **28.** $\cot A$
29. $\sin B$ **30.** $\cos B$ **31.** $\tan B$ **32.** $\csc B$ **33.** $\sec B$ **34.** $\cot B$

Use $a = 2$ ft, $b = 10$ in., $c = 2$ ft 2 in. in Exercises 35–40 and express the ratios as fractions in lowest terms.

35. $\sin A$ **36.** $\sec B$ **37.** $\cot B$ **38.** $\csc A$ **39.** $\tan B$ **40.** $\cos A$

Use $a = 7$, $b = 10.5$, $c = 12.62$ in Exercises 41–47 and express the ratios in decimals to the nearest ten-thousandth.

41. $\cos B$ **42.** $\csc A$ **43.** $\tan A$ **44.** $\sin B$ **45.** $\sec A$ **46.** $\sin A$
47. $\cot A$

Section 17–3
Use your calculator to find the trigonometric values. Round to the nearest ten-thousandth.

48. $\cos 42.5°$ **49.** $\sin 0.4712$ **50.** $\sin 65.5°$ **51.** $\cot 73°$
52. $\tan 1.0210$ **53.** $\tan 47°$ **54.** $\tan 15.6°$ **55.** $\sin 0.8610$
56. $\tan 25°40'$ **57.** $\cos 32°50'$ **58.** $\cot 0.7510$ **59.** $\cos 80°10'$

Find an angle (in degrees) having the trigonometric values. θ represents the unknown measure. Express answers to the nearest tenth of a degree.

60. $\sin \theta = 0.5446$ **61.** $\cos \theta = 0.6088$ **62.** $\tan \theta = 0.8720$
63. $\cot \theta = 0.9884$ **64.** $\cos \theta = 0.8897$ **65.** $\cot \theta = 3.340$

Find an angle (in radians) having the trigonometric values. Round to the nearest ten-thousandth of a radian.

66. $\sin \theta = 0.9205$ **67.** $\tan \theta = 2.723$ **68.** $\cos \theta = 0.9450$
69. $\cot \theta = 0.3772$ **70.** $\sin \theta = 0.2896$ **71.** $\tan \theta = 0.3440$

Find the trigonometric values. Round to the nearest ten-thousandth.

72. $\sec 15.5°$ **73.** $\csc 71°$ **74.** $\sec 0.4363$
75. $\csc 1.082$ **76.** $\sec 1.2886$

Solve the triangles in Figs. 17–76 to 17–85. Round sides to four significant digits and round angles to the nearest 0.1°.

77.

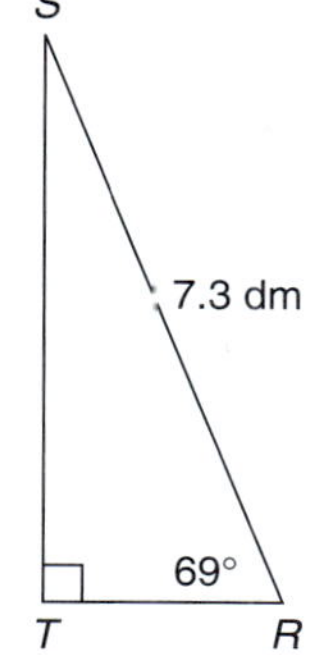

Figure 17–76

78.

Figure 17–77

79.

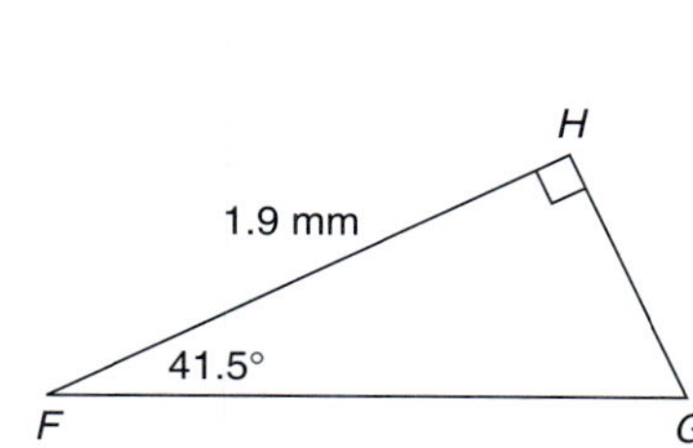

Figure 17–78

80.

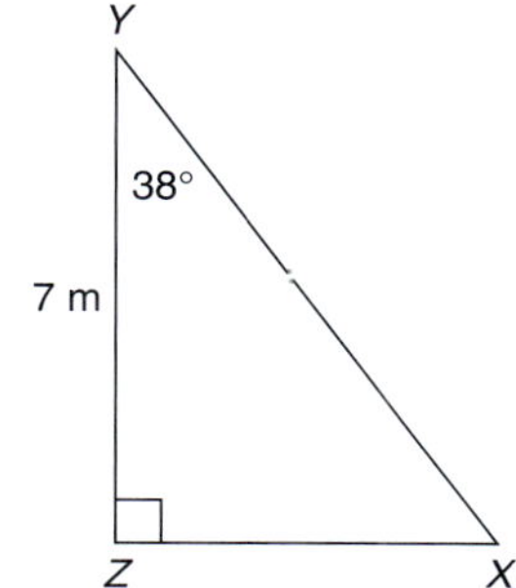

Figure 17–79

81.

Figure 17–80

82.

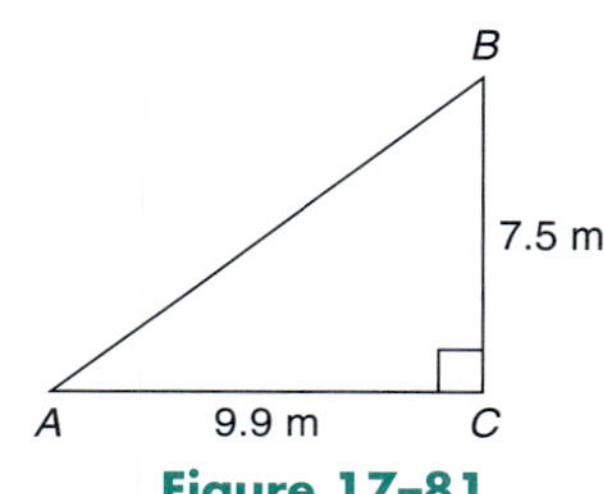

Figure 17–81

83.

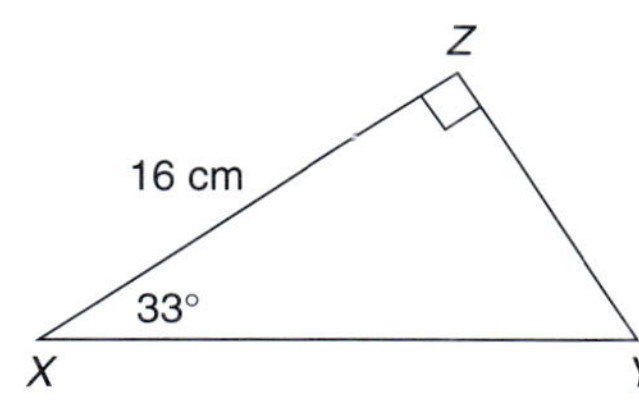

Figure 17–82

84.

Figure 17–83

85.

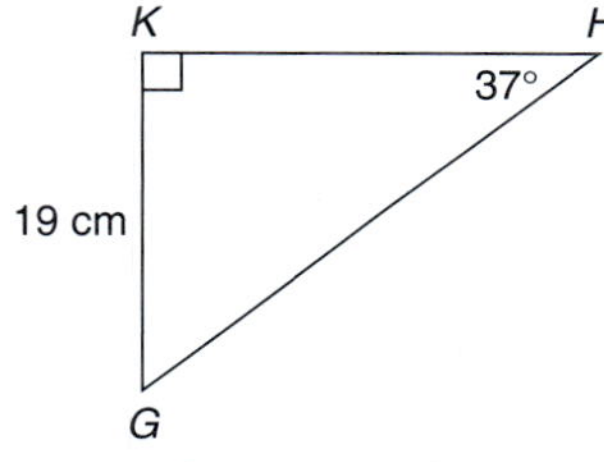

Figure 17–84

86.

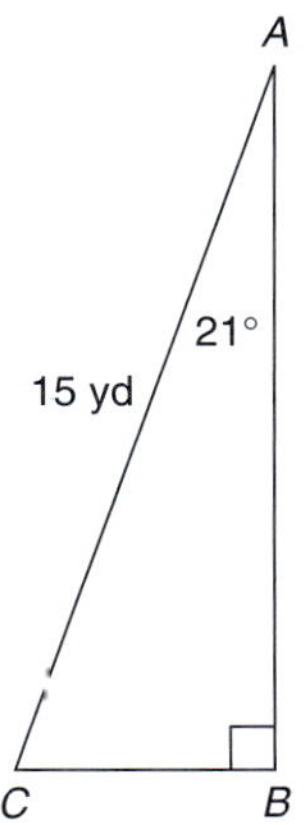

Figure 17–85

Find the indicated parts of the triangles in Figs. 17–86 to 17–88. Round lengths of sides to four significant digits and angle measures to the nearest 0.1°.

87. Find A.

88. Find A.

89. Find a.

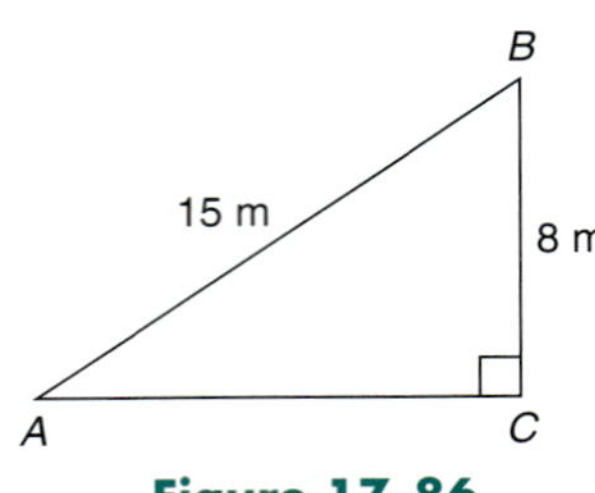

Figure 17–86

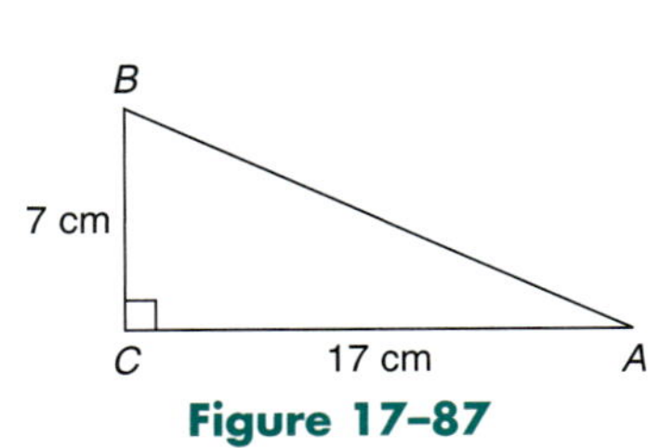

Figure 17–87

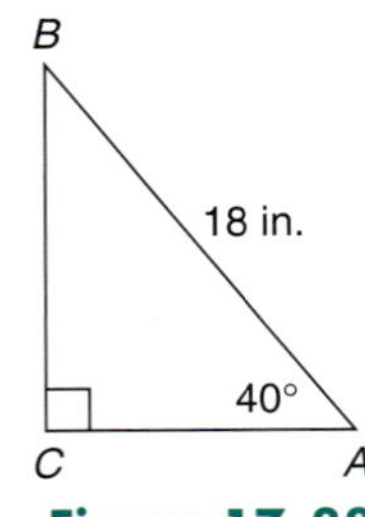

Figure 17–88

Solve the triangles in Figs. 17–89 and 17–90.

90.

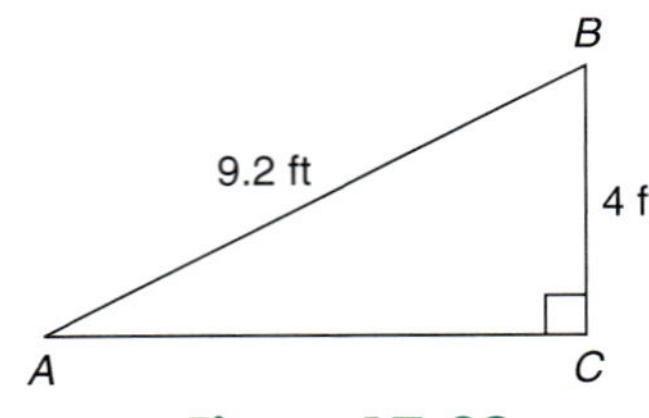

Figure 17–89

91.

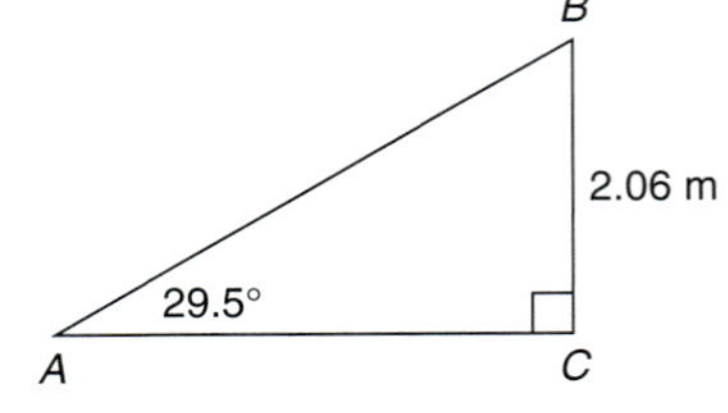

Figure 17–90

92. A railway inclines 14°. How many feet of track must be laid if the hill is 15 ft high?

93. A corner shelf is cut so that the sides placed on the wall are 37 in. and 42 in. What are the measures of the acute angles?

94. From a point 5 ft above the ground and 20 ft from the base of a building, a surveyor uses a transit to sight an angle of 38° to the top of the building. Find the height of the building.

95. A surveyor makes the measures indicated in Fig. 17–91. Solve the triangle. All parts of the triangle should be included in a report.

96. What length of rafter is needed for a roof if the rafters form an angle of 35.5° with a joist, and the rise is 8 ft?

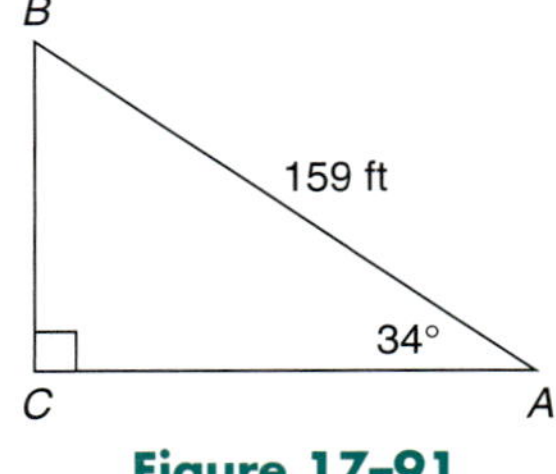

Figure 17–91

CHALLENGE PROBLEMS

97. Is there an angle between 0° and 360° for which the sine and cosine of the angle are equal? Justify your answer by making a table of values for the sine and cosine of angles between 0° and 360° and by graphing $\sin x$ and $\cos x$ for values between 0° and 360°.

98. Investigate the relationship described in Exercise 97 for the sine and tangent functions and justify your answer.

99. You are designing a stairway for a home and need to determine how many steps the stairway will have. If the stairway is 8 ft high and the floorspace allotted is 14 ft, determine how many steps are needed and determine the optimum measure (rise and run) for each step.

100. Draw triangle ABC so that angle C is 90° and angle A is 70°. Point E is on side $\overline{AB}$ and is 10 cm from A. Point D is on $\overline{BC}$, and $\overline{DE}$ is parallel to $\overline{CA}$. If $\overline{CA}$ is 24 cm, find the length of $\overline{BD}$.

Convert the measures to radians rounded to the nearest hundredth.

1. 35°   **2.** 122°   **3.** 315°   **4.** 240°

Convert the measures to degrees rounded to the nearest ten-thousandth, then convert to radians to the nearest hundredth.

5. 15°25′   **6.** 142°32′15″   **7.** 16°12′   **8.** 32°18′37″

Convert the measures to degrees. Round to the nearest ten-thousandth of a degree.

9. $\dfrac{5\pi}{8}$ rad   **10.** 3.1 rad

Convert the measures to degrees, minutes, and seconds. Round to the nearest second.

11. 1.2 rad   **12.** $\dfrac{\pi}{6}$ rad

Use the relationships of arc length, area, central angle, and radius to solve the problems relating to sectors. Round to hundredths if necessary.

13. Find s if $\theta = 0.5$ and $r = 2$ in.   **14.** Find θ if $s = 5.3$ m and $r = 7$ m.
15. Find r if $\theta = 1.7$ and $s = 2.9$ m.   **16.** Find A if $\theta = 35°$ and $r = 7.3$ cm.

Write the ratios as fractions in lowest terms for the following trigonometric functions using triangle ABC in Fig. 17–92.
$a = 10, b = 24, c = 26$.

17. $\sin A$
18. $\tan B$
19. $\csc A$

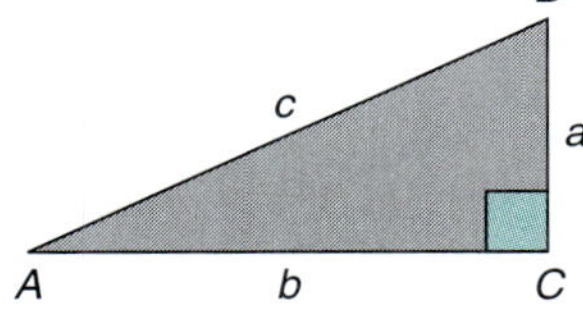

Figure 17–92

Express the trigonometric values in decimals to the nearest ten-thousandth. Refer to Fig. 17–92 and use $a = 5, b = 11.5, c = 12.54$.

20. $\cot A$ **21.** $\cos A$ **22.** $\sin B$

Use your calculator to find the trigonometric values.

23. $\sin 53°$ **24.** $\sin 61°10′$ **25.** $\sin 1.1519$ **26.** $\cos 1.0297$

Find an angle (in degrees) having the trigonometric values; θ represents the unknown angle measure. Express your answers to the nearest tenth of a degree.

27. $\sin \theta = 0.2756$ **28.** $\tan \theta = 1.280$
29. $\cos \theta = 0.9426$ **30.** $\cot \theta = 1.540$

Find an angle (in radians) having the trigonometric values. Round to the nearest ten-thousandth of a degree.

31. $\sin \theta = 0.7660$ **32.** $\cos \theta = 0.8387$ **33.** $\tan \theta = 0.3259$

Find the indicated trigonometric values. Round to the nearest ten-thousandth.

34. $\sec 25.5°$ **35.** $\csc 47°$ **36.** $\cot 0.3316$

The measures are indicated for a standard right triangle, ABC, where C is the right angle. Draw the figures and use the sine, cosine, or tangent function to find the indicated parts of the triangle. Round side lengths to four significant digits and angles to the nearest 0.1°.

37. $a = 16$ m, $b = 14$ m, find A.   **38.** $a = 7$ in., $A = 33°$, find c.
39. $c = 17$ ft, $B = 25°$, find a.   **40.** $a = 21$ m, $A = 48.5°$, find b.
41. $a = 32$ cm, $c = 47$ cm, find A.   **42.** $c = 12$ m, $A = 35°$, find a.
43. $b = 21$ cm, $A = 17°$, find a.   **44.** $b = 1$ cm, $A = 87°$, find c.
45. $b = 3.1$ m, $c = 6.8$ m, find A.   **46.** $a = 0.15$ m, $c = 0.46$ m, find A.

47. Solve triangle *ABC* in Fig. 17–93.

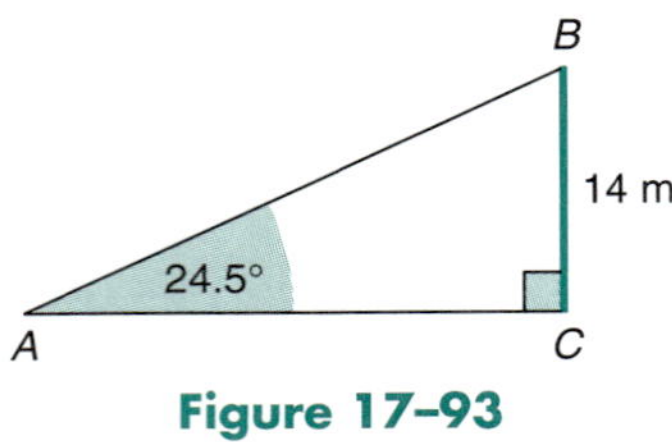

Figure 17–93

48. Solve triangle *DEF* in Fig. 17–94. Round as above.

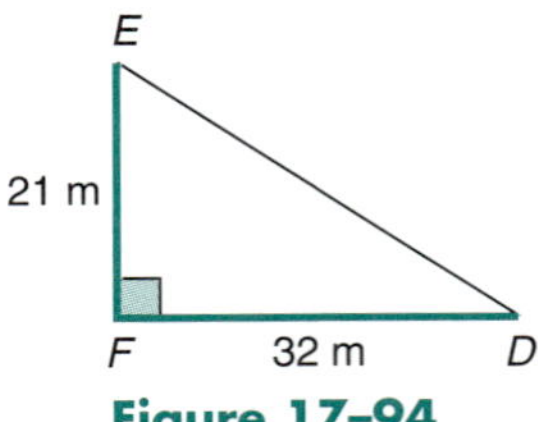

Figure 17–94

49. A stair has a rise of 4 in. for every 5 in. of run. What is the angle of inclination of the stair?

50. A surveyor uses indirect measurement to find the width of a river at a certain point. The surveyor marks off 50 ft along the river bank at right angles with the river and then sights an angle of 43° to point *A* across the river (Fig. 17–95). Find the width of the river.

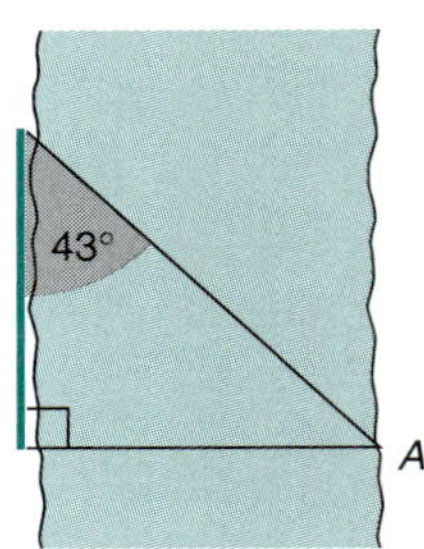

Figure 17–95

51. Steel girders are reinforced by placing steel supports between two runners so that right triangles are formed (Fig. 17–96). If the runners are 24 in. apart and a 30° angle is desired between the support and a runner, find the length of the support to be placed at a 30° angle.

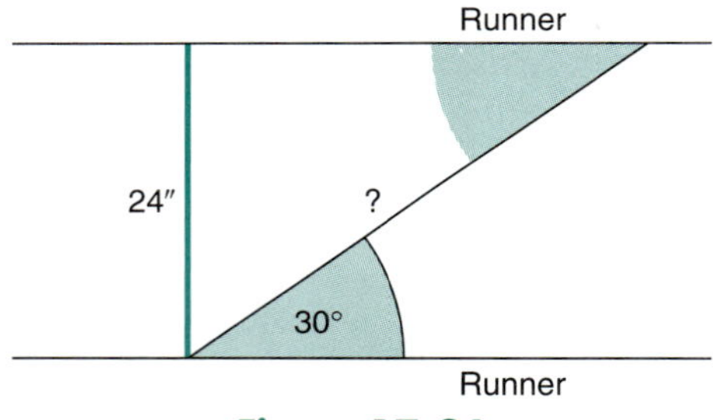

Figure 17–96

52. In the circuit represented by the diagram of Fig. 17–97, find the total current I_t. All units are in amps.

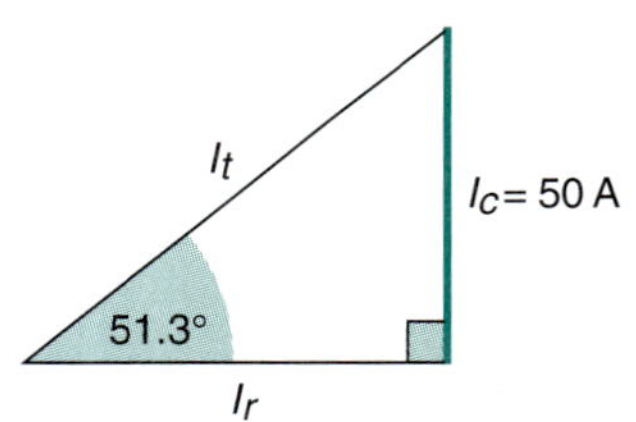

Figure 17–97

53. Find the current in the resistance branch I_r of the circuit in Exercise 52.

54. The minimum clearances for the installation of a metal chimney pipe are shown in Fig. 17–98. How far from the ridge should the hole be cut for the pipe to pass through the roof?

55. Refer to Fig. 17–98. What angle is formed by the chimney pipe and the roof where the pipe passes through the roof?

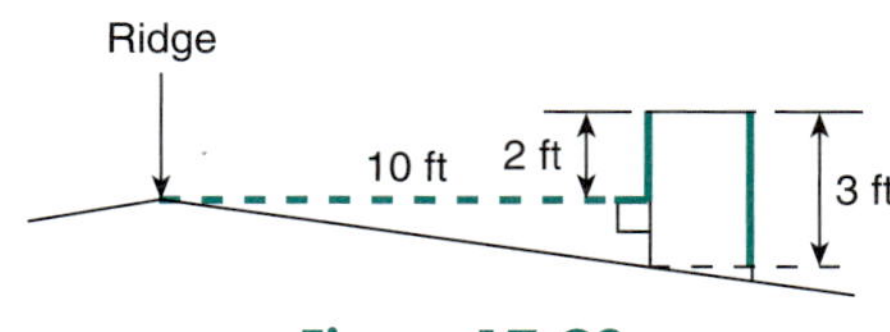

Figure 17–98

56. Elbows are used to form bends in rigid pipe and are measured in degrees. What is the angle of bend of the elbow in the installation shown in Fig. 17–99 to the nearest whole degree?

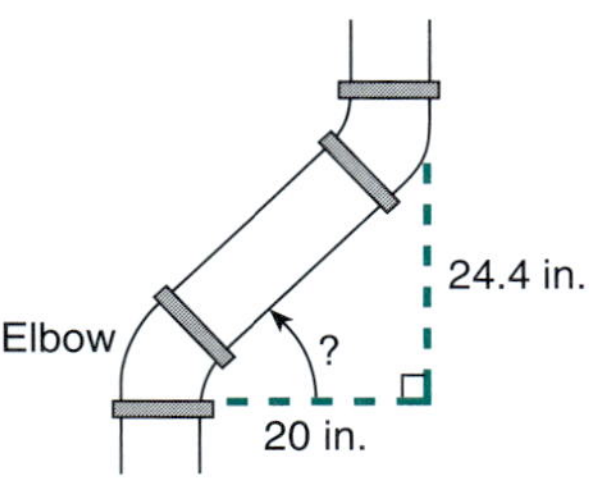

Figure 17–99

57. Find the angle formed by the connecting rod and the horizontal in the mechanical assembly shown in Fig. 17–100.

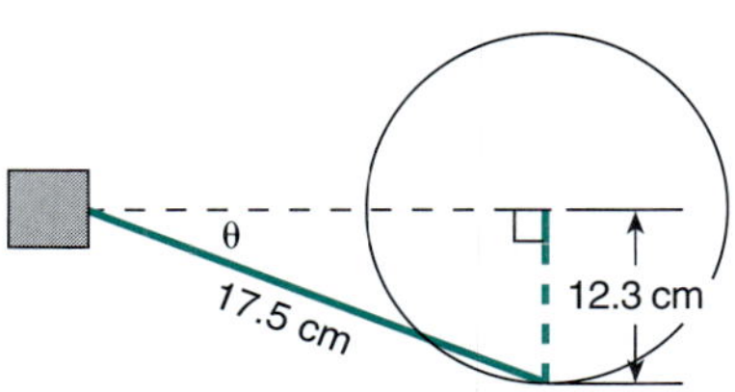

Figure 17–100

58. A utility pole is 40 ft above ground level. A guy wire must be attached to the pole 3 ft from the top to give it support. If the guy wire forms a 20° angle with the ground, how long is the guy wire? Disregard the length needed for attaching the wire to the pole or ground. Round to hundredths.

59. Solve for reactance X_L and resistance R in Fig. 17–101. All units are in ohms. Round to hundredths.

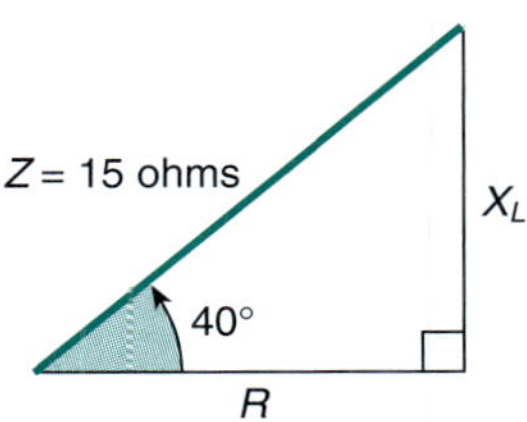

Figure 17–101

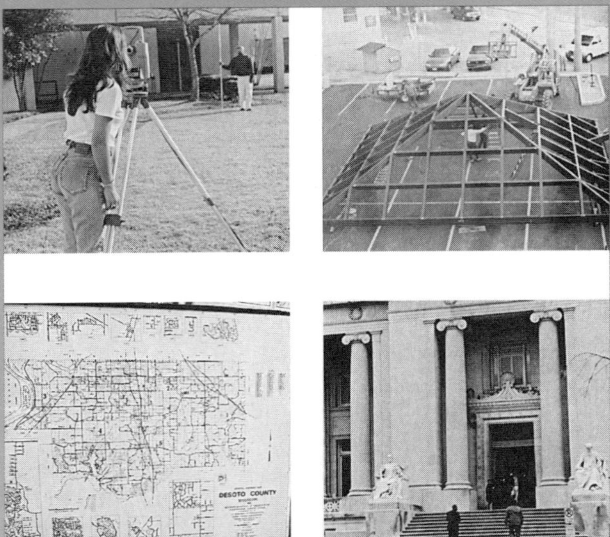

18

Oblique Triangles

GOOD DECISIONS THROUGH TEAMWORK

Survey reports describe property lines determined in part through the use of triangles, including oblique triangles, particularly for irregularly shaped properties. For this project, your team will diagram two irregularly shaped properties described in survey reports. To begin, draw a scaled diagram for the property described below, including angle measures for each vertex. Calculate the acreage to verify that the acreage shown on the survey report is correct. The report reads: "Beginning at the northwest corner of said forty; thence east along the north line of said forty 600 feet; thence south 313 feet to the center of County Road as now located; thence in a northwesterly direction along center of said county road 613 feet to the point of beginning (3 acres)."

Next, your team should locate another survey report of an irregularly shaped property. You might check with real estate offices, abstract companies, or court house records for suitable survey reports. Then draw a diagram of the selected property as you did above.

Keep a record of the calculations your teammates had to make and any landmarks or other points used by the surveyors as reference points. After the diagrams are completed, report to your class on the tools you used to make the calculations and draw the diagrams, the reference points used in the survey reports, the calculations your team made, and the diagrams of the properties.

18–1 Vectors

1. Find the magnitude of a vector in standard position, given the coordinates of the end point.
2. Find the direction of a vector in standard position, given the coordinates of the end point.
3. Find the sum of vectors.

18–2 Trigonometric functions for any angle

1. Find related acute angles for angles or vectors in quadrants II, III, and IV.
2. Determine the signs of trigonometric values of angles of more than 90°.
3. Find the trigonometric values of angles of more than 90° using a calculator.

18–3 Law of sines

1. Find the missing parts of an oblique triangle, given two angles and an opposite side.
2. Find the missing parts of an oblique triangle, given two sides and an angle opposite one of them.
3. Solve applied problems using the law of sines.

18-4 Law of cosines

1. Find the missing parts of an oblique triangle, given three sides of the triangle.

2. Find the missing parts of an oblique triangle, given two sides and the included angle of the triangle.

3. Solve applied problems using the law of cosines and the law of sines.

Although right triangles are perhaps the most frequently used triangles, we often work with other kinds of triangles. In this chapter, we apply the laws of sines and cosines to oblique triangles, find the area of oblique triangles when only selected parts are given, and use *vectors* for solving triangles that have an angle greater than 90°.

18-1   VECTORS

Learning Outcomes

1. Find the magnitude of a vector in standard position, given the coordinates of the end point.

2. Find the direction of a vector in standard position, given the coordinates of the end point.

3. Find the sum of vectors.

1   Find the Magnitude of a Vector in Standard Position, Given the Coordinates of the End Point.

Quantities that we have discussed so far in this text have been described by specifying their size or magnitude. Quantities such as area, volume, length, and temperature, which are characterized by magnitude only, are called *scalars*. There are many other quantities such as electrical current, force, velocity, and acceleration that are called *vectors*.

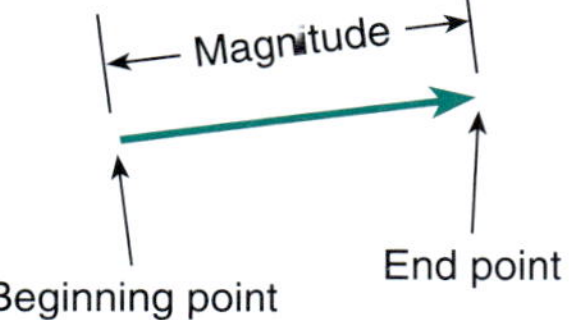

Figure 18–1

■ **DEFINITION: Vector.** A quantity described by magnitude (or length) and direction is a *vector.* A vector represents a shift from one point to another.

When we consider the speed of a plane to be 500 mph, we are considering a scalar quantity. However, when we consider the speed of a plane *traveling northeast from a given location,* we are concerned with both the distance traveled and the direction. This is a vector quantity. We have seen such quantities in the vector diagrams used to solve electronic problems by means of right triangles in Chapter 17 and elsewhere.

Vectors are represented by straight arrows. The length of the arrow represents the *magnitude* of the vector (Fig. 18–1). The curved arrow shows the counterclockwise *direction* of the vector (Fig. 18–2), with the end of the arrow being the beginning point and the arrowhead being the end point.

In relating trigonometric functions to vector quantities, we will place the vectors in *standard position.* Standard position places the beginning point of the vector at the origin of a rectangular coordinate system. *The direction of the vector is the counterclockwise angle measured from the positive x-axis (horizontal axis).*

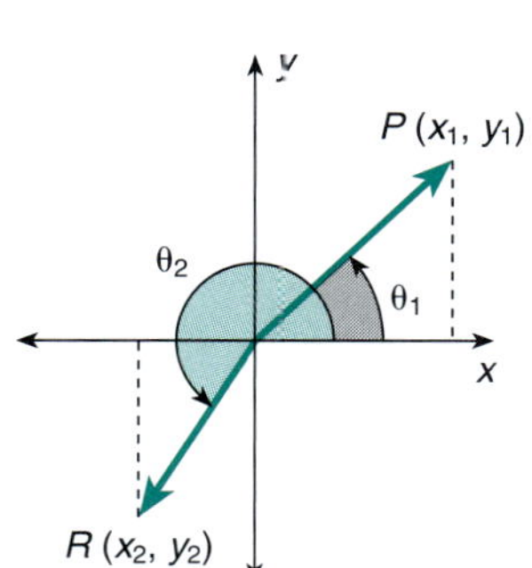

Figure 18–2

In Fig. 18–2, vector P is a first-quadrant vector that has a direction of θ_1 and an end point at the point P. The magnitude of a vector can be determined if the vector is in standard position and the x- and y-coordinates of the end point are known. Vector R is a third-quadrant vector that has a direction of θ_2 and an end point at point R. Its magnitude may be determined similarly.

Magnitude of a vector in standard position:

The *magnitude* of a vector in standard position is the length of the hypotenuse of the right triangle formed by the vector, the x-axis, and the vertical line from the end point of the vector to the x-axis.

To find the magnitude of a vector in standard position:

1. Substitute into the Pythagorean theorem the coordinates of the end points of the vector as the legs of a right triangle.
2. Solve for the hypotenuse.

EXAMPLE Find the magnitude of vector P in Fig. 18–2 if the coordinates of P are $(4, 3)$.

From P, we can draw a vertical line to the x-axis, forming a right triangle with the x-axis and the vector P. The length of the side of the triangle along the x-axis is the x-coordinate of the point P, or 4. The length of the vertical side of the triangle is the y-coordinate of the point P, or 3. The length of the hypotenuse of the triangle is the magnitude of the vector. We will use the Pythagorean theorem to determine the magnitude of vector P.

$$p^2 = x^2 + y^2 \qquad \text{Substitute } x = 4 \text{ and } y = 3.$$

$$p^2 = 4^2 + 3^2 \qquad \text{Solve for } p.$$

$$p^2 = 16 + 9$$

$$p^2 = 25$$

$$p = \pm\sqrt{25}$$

$$p = \pm 5 \qquad \text{Or } +5 \text{ because length or magnitude is positive.}$$

The magnitude of vector P is 5.

EXAMPLE Find the magnitude of vector r in Fig. 18–2 if the coordinates of point r are $(-2, -5)$.

The magnitude of a vector is always positive; however, the x- and y-coordinates can be negative.

$$r^2 = x^2 + y^2 \qquad \text{Substitute } x = -2 \text{ and } y = -5.$$

$$r^2 = (-2)^2 + (-5)^2 \qquad \text{Solve for } r.$$

$$r^2 = 4 + 25$$

$$r^2 = 29$$

$$r = \sqrt{29}$$

$$r = 5.385 \qquad \text{Rounded.}$$

The magnitude of vector r is 5.385.

2 Find the Direction of a Vector in Standard Position, Given the Coordinates of the End Point.

If the vector in standard position falls in quadrant I, we can use right-triangle trigonometry to find the direction of the vector. For a vector in standard position with an end point in quadrant I, the tangent function can be used to find the angle or direction of the vector.

CHAPTER 18   Oblique Triangles

To find the direction of a quadrant I vector in standard position:

1. Identify the coordinates of the end point of the vector.
2. Substitute the coordinates of the end point into the tangent function: $\tan \theta = \frac{\text{opp}}{\text{adj}}$ or $\tan \theta = \frac{y}{x}$.
3. Solve for θ.

EXAMPLE Find the direction of vector P in Fig. 18–2 if the coordinates of P are (4, 3).

The right triangle is formed by the vector, the x-axis, and a line from the end point of the vector to the x-axis. The angle at the origin, which is in the direction of the vector P, has an opposite-side value of 3 and an adjacent-side value of 4.

$$\tan \theta = \frac{y}{x} \qquad \text{Substitute } x = 4 \text{ and } y = 3.$$

$$\tan \theta = \frac{3}{4} \qquad \text{Solve for } \theta.$$

$$\tan \theta = 0.75 \qquad \text{Or } \tan^{-1} 0.75 = \theta$$

$$\theta = 36.9° \qquad \text{Nearest } 0.1°.$$

The direction of vector P is 36.9°.

3 Find the Sum of Vectors.

Two vectors with the same direction can be added by aligning the beginning point of one vector with the end point of the other. The vector represented by the sum is called the *resultant* vector, or resultant. The resultant has the same direction as the vectors being added and a magnitude that is the sum of the two magnitudes.

To add vectors of the same direction:

1. Add the magnitudes of each vector.
2. The resultant vector will have the same direction as the original vectors.

EXAMPLE Add two vectors with a direction of 35° if the magnitudes of the vectors are 4 and 5, respectively (Fig. 18–3).

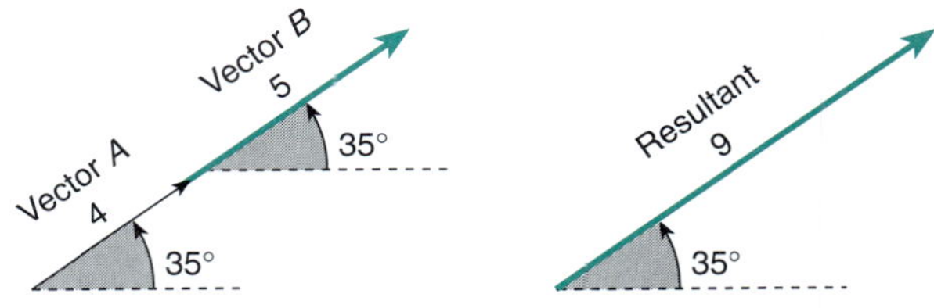

Figure 18–3

The resultant vector has a magnitude of 9 and a direction of 35°.

Two vectors have opposite directions if the directions of the vectors differ by 180°. Two vectors with opposite directions can be added by aligning the beginning point of the vector with the smaller magnitude, with the end point of the vector with the larger magnitude. The resultant has the same direction as the vector with the larger magnitude and a magnitude that is the difference of the two magnitudes.

EXAMPLE Add a vector with a direction of 60° and a magnitude of 6 to a vector with a direction of 240° and a magnitude of 4 (Fig. 18–4).

Subtract the magnitudes: $6 - 4 = 2$.

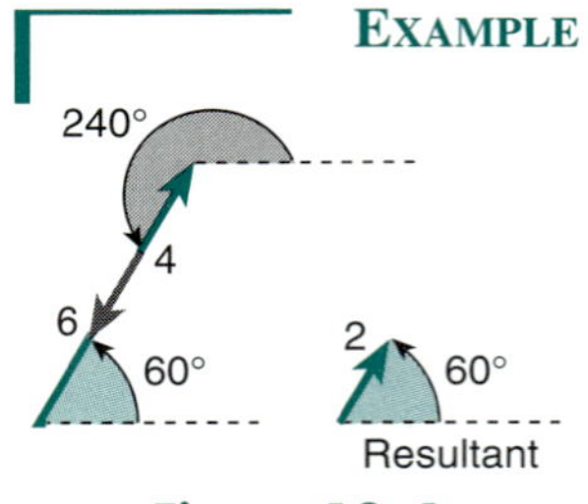

Figure 18–4

The resultant has a direction of 60° and a magnitude of 2.

Adding any two vectors that have different directions is accomplished by performing the shifts of each vector in succession.

EXAMPLE Show a graphical representation of the sum of a 45° vector with a magnitude of 5 (vector A) and a 60° vector with a magnitude of 6 (vector B) (Fig. 18–5).

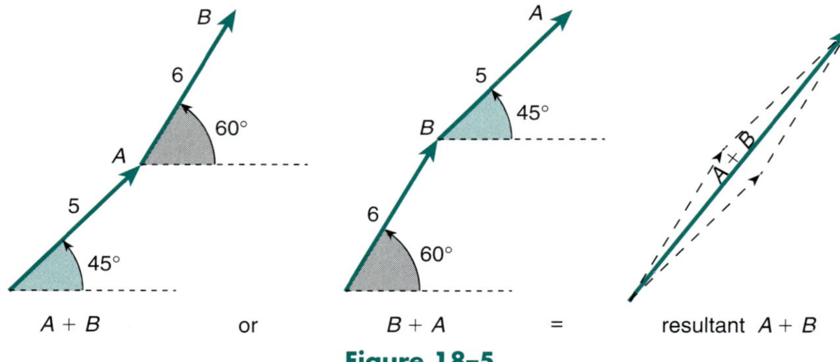

Figure 18–5

Complex numbers in the form $a + bi$ are sometimes used to represent vectors. The x-coordinate of a vector in standard position is represented as the real component a. The y-coordinate of a vector in standard position is represented as the imaginary component bi. Vector A can be written in complex notation by first finding the x- and y-coordinates of the end point.

<table><tr><td>**Tip!**</td><td>***Vectors Using Complex Notation.***</td></tr></table>

Vector in complex form = (x-coordinate of end point) + (y-coordinate of end point)i

To find the x-coordinate, use the relationship

$$\cos \theta = \frac{\text{adj}}{\text{hyp}} = \frac{x}{r}, \text{ and solve for } x. \qquad r = \text{magnitude}$$

To find the y-coordinate, use the relationship

$$\sin \theta = \frac{\text{opp}}{\text{hyp}} = \frac{y}{r}, \text{ and solve for } y. \qquad r = \text{magnitude}$$

EXAMPLE Write the vectors A, B, and $A + B$ from Fig. 18–5 in complex notation.

Because the magnitude and direction are known for both vectors A and B, we can find the x-coordinates by using the cosine function, $\cos \theta = \frac{\text{adj}}{\text{hyp}} = \frac{x}{r}$, where r is the magnitude. We can find the y-coordinates by using the sine function, $\sin \theta = \frac{\text{opp}}{\text{hyp}} = \frac{y}{r}$.

Vector A: *x*-coordinate *y*-coordinate

$$\cos 45° = \frac{x}{5} \qquad \sin 45° = \frac{y}{5}$$

$$5(\cos 45°) = x \qquad 5(\sin 45°) = y$$

$$3.535533906 = x \qquad 3.535533906 = y$$

Vector A in complex form: 3.535533906 + 3.535533906i

Vector B: *x*-coordinate *y*-coordinate

$$\cos 60° = \frac{x}{6} \qquad \sin 60° = \frac{y}{6}$$

$$6(\cos 60°) = x \qquad 6(\sin 60°) = y$$

$$3 = x \qquad 5.196152423 = y$$

Vector B in complex form: 3 + 5.196152423i

Resultant $A + B = (3.535533906 + 3.535533906i) + (3 + 5.196152423i)$

Resultant $A + B$ = 6.535533906 + 8.731686329i or   6.536 + 8.732i

EXAMPLE Find the magnitude and direction of the resultant $A + B$ in the preceding example.

The resultant in complex form is $6.535533906 + 8.731686329i$; thus, the x-coordinate of the end point is 6.535533906 and the y-coordinate is 8.731686329.

$$\text{Magnitude} = \sqrt{(6.535533906)^2 + (8.731686329)^2}$$

$$\text{Magnitude} = \sqrt{118.9555496}$$

Magnitude = 10.91

$$\tan \theta = \frac{8.731686329}{6.535533906}$$

$$\tan \theta = 1.336032596 \qquad \text{Or } \tan^{-1} 1.336032596 = \theta.$$

$$\theta = 53.18570658$$

Direction: θ = 53.2° Nearest 0.1°.

Vectors written in complex form are often used in electronics. The imaginary part is referred to as the *j*-factor.

Resultant vector $A + B$ from the first example on the preceding page would be written as $6.536 + j8.732$. In customary notation *j* is followed by the coefficient.

SELF-STUDY EXERCISES 18–1

1 Find the magnitude of the vectors in standard position with end points at the indicated points. Round to the nearest thousandth.

1. $(5, 12)$ **2.** $(-12, 9)$ **3.** $(2, -7)$ **4.** $(-8, -3)$ **5.** $(1.5, 2.3)$

2 Find the direction of the vectors in standard position with end points at the indicated points. Round to the nearest hundredth.

6. $(5, 12)$ **7.** $(6, 8)$ **8.** $(8, 3)$ **9.** $(2, 5)$ **10.** $(1, 4)$

3

11. Find the resultant vector of two vectors that have a direction of 42° and magnitudes of 7 and 12, respectively.

12. Two vectors have a direction of 72°. Find the sum of the vectors if their magnitudes are 1 and 7, respectively.

13. Find the sum of two vectors if one has a direction of 45° and a magnitude of 7 and the other has a direction of 225° and a magnitude of 8.

14. Find the sum of two vectors if one has a direction of 75° and a magnitude of 15 and the other has a direction of 255° and a magnitude of 9.

Find the magnitude and direction of the resultant of the sum of the two given vectors in complex notation.

15. $5 + 3i$ and $7 + 2i$

16. $1 + 2i$ and $5 + 2i$

18–2   TRIGONOMETRIC FUNCTIONS FOR ANY ANGLE

Learning Outcomes

1 Find related acute angles for angles or vectors in quadrants II, III, and IV.
2 Determine the signs of trigonometric values of angles of more than 90°.
3 Find the trigonometric values of angles of more than 90° using a calculator.

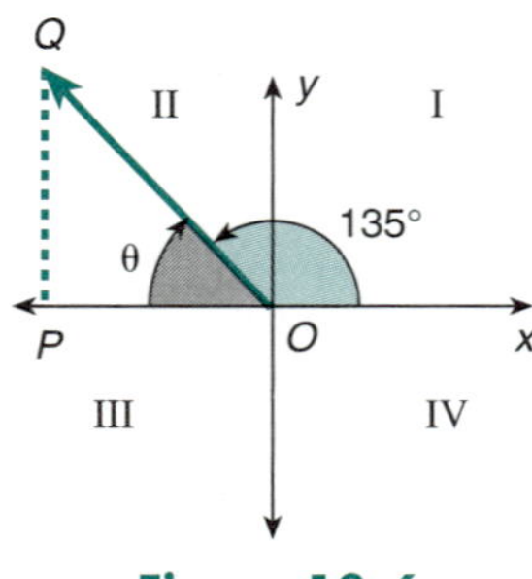

Figure 18–6

1 Find Related Acute Angles for Angles or Vectors in Quadrants II, III, and IV.

We can determine the direction of any vector in standard position, if we know the coordinates of the end point, by applying our knowledge of trigonometric functions. We can form a right triangle that we will refer to as our *reference triangle* by drawing a vertical line from the end point of the vector to the *x*-axis. See the example of a reference triangle in Fig. 18–6. The angle θ is called the *related angle*.

■ **DEFINITION: Related Angle.** The *related angle* is the acute angle formed by the *x*-axis and the vector.

In quadrant I, the related angle is the same as the direction of the vector. Therefore, the direction of the vector in quadrant I is always less than 90°. For vectors in quadrants II, III, and IV, see Figs. 18–6, 18–7, and 18–8.

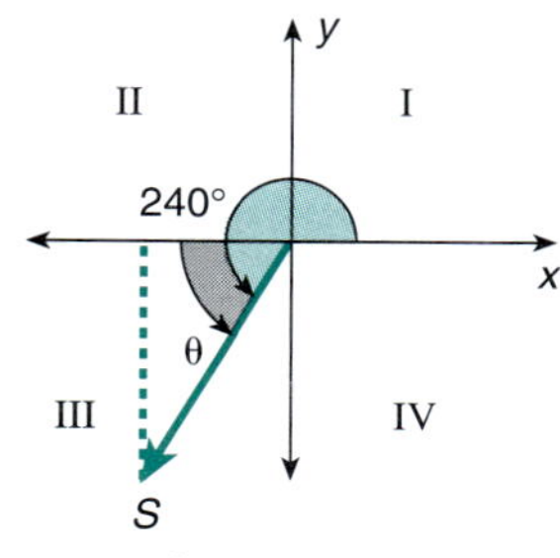

Figure 18–7

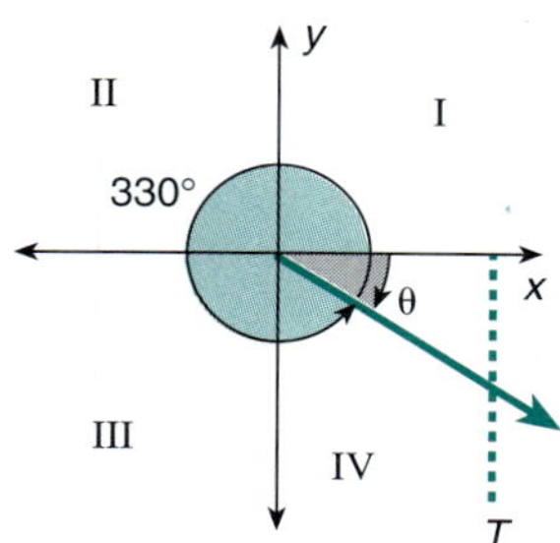

Figure 18–8

In Fig. 18–6, $\overline{PQ}$, the x-axis, and vector Q form a right triangle. The direction of the vector is 135°; therefore, the related angle is 45° (180° − 135°). 135° is a second-quadrant angle. Second-quadrant angles are more than 90° and less than 180°, or more than $\frac{\pi}{2}$ radians (1.57) and less than π radians (3.14). The related angle for any second-quadrant vector can be found by subtracting the direction of the vector from 180° (or π radians).

Third-quadrant angles are more than 180° and less than 270° or more than π radians (3.14) and less than $\frac{3\pi}{2}$ radians (4.71). In Fig. 18–7, vector S is a third-quadrant vector. The related angle θ is 60° (240° − 180°). The related angle for any third-quadrant vector is found by subtracting 180° (or π radians) from the direction of the vector.

Vector T in Fig. 18–8 is a fourth-quadrant vector. Fourth-quadrant angles are more than 270° and less than 360°, or more than $\frac{3\pi}{2}$ radians (4.71) and less than 2π radians (6.28). The related angle θ is 30° (360° − 330°). The related angle for any fourth-quadrant vector is found by subtracting the direction of the vector from 360° (or 2π radians).

To find related angles for vectors that are more than 90°:

The related angle for quadrant I angles and vectors is equal to the direction (angle) of the vector. The related angle for angles and vectors more than 90° can be found as follows:

Quadrant II angle:	$180° - \theta_2$	or	$\pi - \theta_2$
Quadrant III angle:	$\theta_3 - 180°$	or	$\theta_3 - \pi$
Quadrant IV angle:	$360° - \theta_4$	or	$2\pi - \theta_4$

Related angles are alway angles less than 90° or $\frac{\pi}{2}$ radians.

EXAMPLE Find the related angle for the following angles.

(a) 210°   (b) 1.93 rad

(a) 210° is between 180° and 270° and is a quadrant III angle. Then 210° − 180° = 30°. **The related angle is 30°.**

(b) 1.93 rad is between $\frac{\pi}{2}$ and π rad and is a quadrant II angle. Then $\pi - 1.93 = 1.211592654$ rad. **The related angle is 1.21 rad to the nearest hundredth.**

2 Determine the Signs of Trigonometric Values of Angles of More Than 90°.

To determine the appropriate sign of trigonometric functions for angles more than 90°, we examine the trigonometric functions in each quadrant. The sign of the function indicates the direction or quadrant of the vector.

The vector in Fig. 18–9 has a magnitude of r and direction of θ. To relate our trigonometric functions to vectors, the magnitude of the vector is the hypotenuse, the side opposite θ is y, and the side adjacent to θ is x. The magnitude of a vector (r) is always positive. In quadrant I, the x-value and the y-value are both positive.

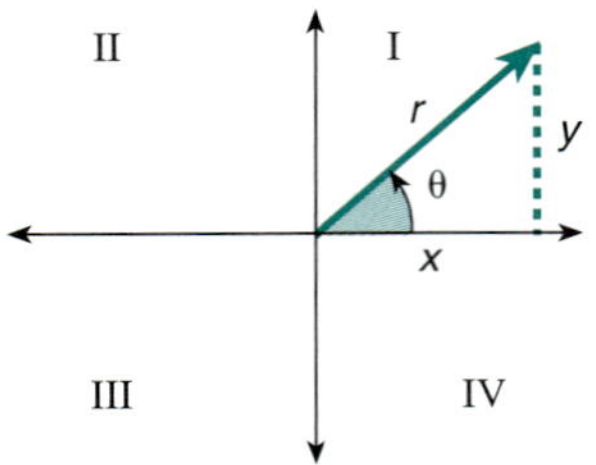

Figure 18–9

■ **DEFINITION:** **Trigonometric Functions in Quadrant I:**

$$\sin\theta_1 = \frac{y}{r} \qquad \cos\theta_1 = \frac{x}{r} \qquad \tan\theta_1 = \frac{y}{x}$$

$$\csc\theta_1 = \frac{r}{y} \qquad \sec\theta_1 = \frac{r}{x} \qquad \cot\theta_1 = \frac{x}{y}$$

In quadrant I, the sign of all six trigonometric functions is positive.

For quadrant II vectors (Fig. 18–10), again the magnitude (r) is positive and the y value is positive, but the x value is negative.

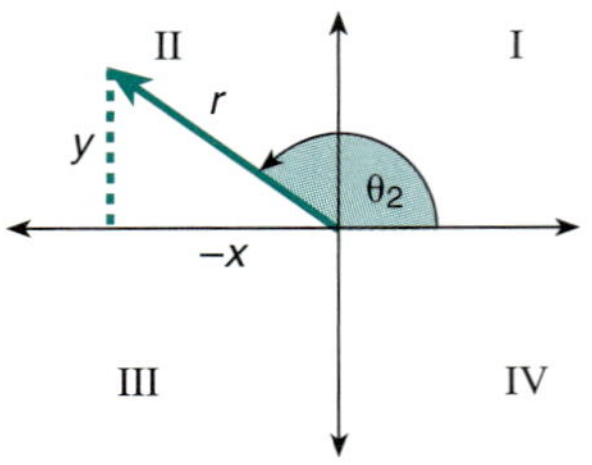

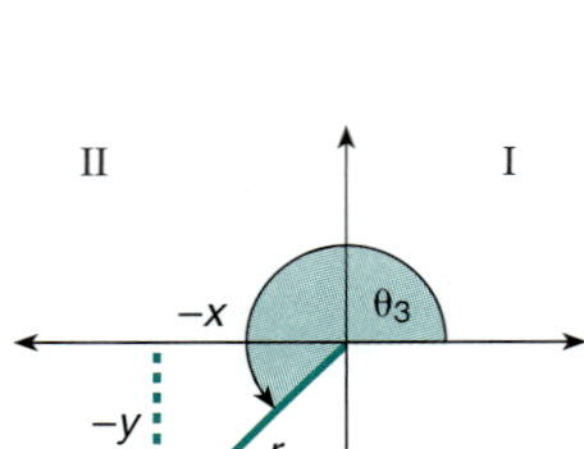

Figure 18–10

■ **DEFINITION:** **Trigonometric Functions in Quadrant II:**

$$\sin\theta_2 = \frac{y}{r} \qquad \cos\theta_2 = \frac{-x}{r} = -\frac{x}{r} \qquad \tan\theta_2 = \frac{y}{-x} = -\frac{y}{x}$$

$$\csc\theta_2 = \frac{r}{y} \qquad \sec\theta_2 = -\frac{r}{x} \qquad \cot\theta_2 = -\frac{x}{y}$$

Thus, in quadrant II the sine and cosecant functions are positive, and the remaining functions are negative.

Quadrant III vectors (Fig. 18–11) have a positive magnitude (r), negative x value, and negative y value.

■ **DEFINITION:** **Trigonometric Functions in Quadrant III:**

$$\sin\theta_3 = \frac{-y}{r} = -\frac{y}{r} \qquad \cos\theta_3 = \frac{-x}{r} = -\frac{x}{r} \qquad \tan\theta_3 = \frac{-y}{-x} = \frac{y}{x}$$

$$\csc\theta_3 = -\frac{r}{y} \qquad \sec\theta_3 = -\frac{r}{x} \qquad \cot\theta_3 = \frac{x}{y}$$

Figure 18–11

Thus, in quadrant III the tangent and cotangent functions are positive, and the remaining functions are negative.

Quadrant IV vectors (Fig. 18–12) have a positive magnitude (r), positive x value, and negative y value.

■ **DEFINITION:** **Trigonometric Functions in Quadrant IV:**

$$\sin\theta_4 = \frac{-y}{r} = -\frac{y}{r} \qquad \cos\theta_4 = \frac{x}{r} \qquad \tan\theta_4 = \frac{-y}{x} = -\frac{y}{x}$$

$$\csc\theta_4 = -\frac{r}{y} \qquad \sec\theta_4 = \frac{r}{x} \qquad \cot\theta_4 = -\frac{x}{y}$$

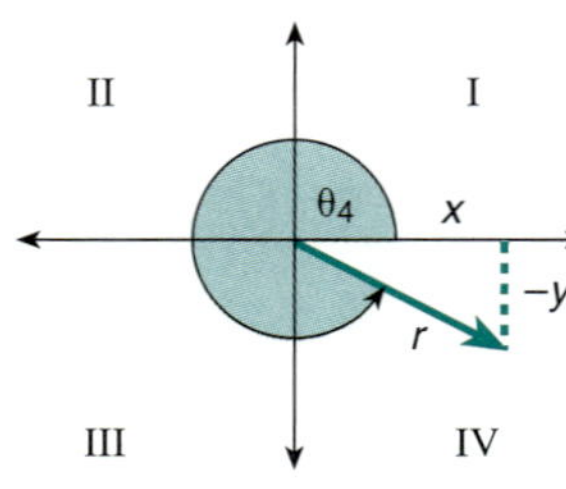

Figure 18–12

Thus, in quadrant IV the cosine and secant functions are positive and the remaining functions are negative.

■ **Learning Strategy** **Understanding Beats Memorizing.**

We have said it before but it is worth saying again, *Understand rather than memorize.*

  CHAPTER 18 Oblique Triangles

To understand the signs of the trigonometric functions in each quadrant, we fit together information we already know:

> r or the hypotenuse is always positive.
> Two negatives in division make a positive quotient.
> One negative and one positive make a negative quotient.

Use these three tools along with the fractional relationships for sine, cosine, and tangent to determine the sign of the sine, cosine, and tangent for each of the four quadrants.

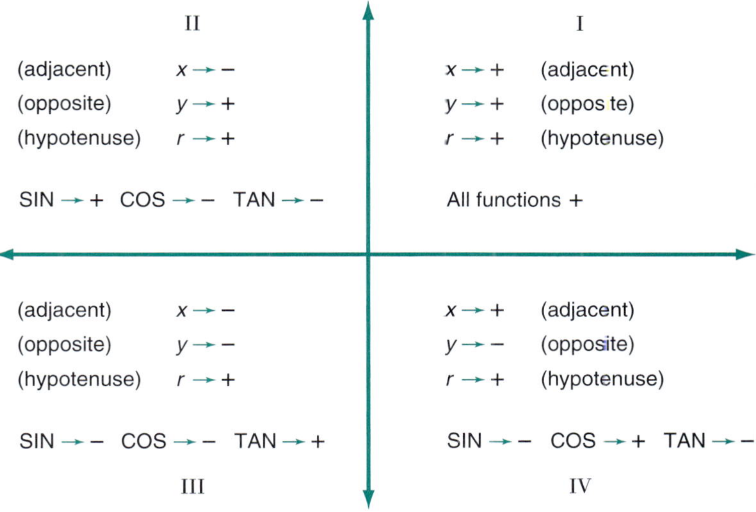

Trigonometric functions, just like algebraic functions, can be represented graphically. This graphical representation can help you visualize the sign patterns of the various functions and draws attention to other properties of trigonometric functions. To graph these functions, we make a table of values from 0° to 360°, or from 0 radians to 2π radians (Figs. 18–13 to 18–15).

		Quadrant I				Quadrant II				Quadrant III				Quadrant IV			
$\sin\theta$	0	0.5	0.71	0.87	1	0.87	0.71	0.5	0	-0.5	-0.71	-0.87	-1	-0.87	-0.71	-0.5	0
θ	0	$\dfrac{\pi}{6}$	$\dfrac{\pi}{4}$	$\dfrac{\pi}{3}$	$\dfrac{\pi}{2}$	$\dfrac{2\pi}{3}$	$\dfrac{3\pi}{4}$	$\dfrac{5\pi}{6}$	π	$\dfrac{7\pi}{6}$	$\dfrac{5\pi}{4}$	$\dfrac{4\pi}{3}$	$\dfrac{3\pi}{2}$	$\dfrac{5\pi}{3}$	$\dfrac{7\pi}{4}$	$\dfrac{11\pi}{6}$	2π
$\theta°$	0	30	45	60	90	120	135	150	180	210	225	240	270	300	315	330	360

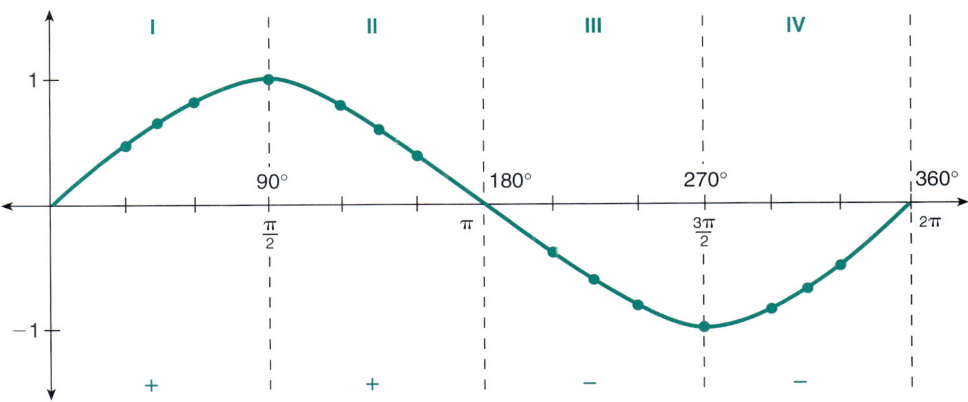

Figure 18–13 Graph of Sine θ

cos θ	1	0.87	0.71	0.5	0	−0.5	−0.71	−0.87	−1	−0.87	−0.71	−0.5	0	0.5	0.71	0.87	1
		Quadrant I				**Quadrant II**				**Quadrant III**				**Quadrant IV**			
θ	0	$\frac{\pi}{6}$	$\frac{\pi}{4}$	$\frac{\pi}{3}$	$\frac{\pi}{2}$	$\frac{2\pi}{3}$	$\frac{3\pi}{4}$	$\frac{5\pi}{6}$	π	$\frac{7\pi}{6}$	$\frac{5\pi}{4}$	$\frac{4\pi}{3}$	$\frac{3\pi}{2}$	$\frac{5\pi}{3}$	$\frac{7\pi}{4}$	$\frac{11\pi}{6}$	2π
θ°	0	30	45	60	90	120	135	150	180	210	225	240	270	300	315	330	360

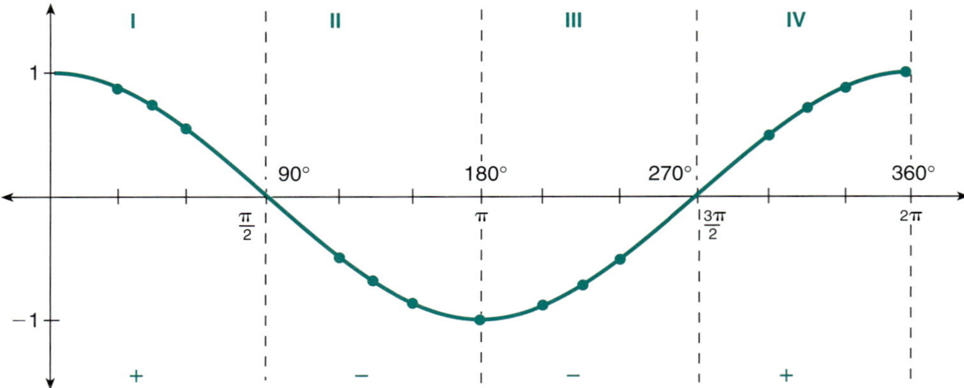

Figure 18–14 Graph of Cosine θ

tan θ	0	0.58	1	1.7	∞	−1.7	−1	−0.58	0	0.58	1	1.7	∞	−1.7	−1	−0.58	0
		Quadrant I				**Quadrant II**				**Quadrant III**				**Quadrant IV**			
θ	0	$\frac{\pi}{6}$	$\frac{\pi}{4}$	$\frac{\pi}{3}$	$\frac{\pi}{2}$	$\frac{2\pi}{3}$	$\frac{3\pi}{4}$	$\frac{5\pi}{6}$	π	$\frac{7\pi}{6}$	$\frac{5\pi}{4}$	$\frac{4\pi}{3}$	$\frac{3\pi}{2}$	$\frac{5\pi}{3}$	$\frac{7\pi}{4}$	$\frac{11\pi}{6}$	2π
θ°	0	30	45	60	90	120	135	150	180	210	225	240	270	300	315	330	360

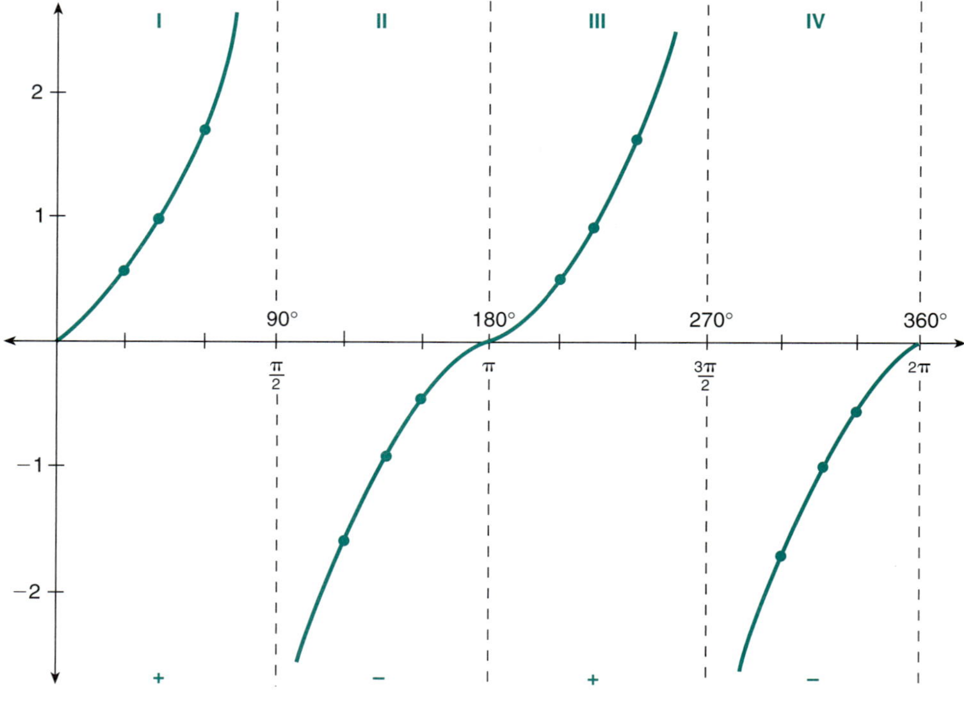

Figure 18–15 Graph of Tangent θ

<table>
<tr><td>Tip!</td><td>Signs of Trigonometric Functions.</td></tr>
</table>

To help remember the signs of the trigonometric functions in the various quadrants, you may find the following reminder helpful:

ALL-SIN-TAN-COS

This reminder gives the positive functions in the quadrants I to IV, respectively. Also, the functions cosecant, cotangent, and secant have the same signs as their respective reciprocal functions.

3 Find the Trigonometric Values of Angles of More Than 90° Using a Calculator.

<table>
<tr><td>Tip!</td><td>General Tips for Using the Calculator.</td></tr>
</table>

To find the value of the trigonometric function of an angle that is more than 90° using a calculator:

1. Set the calculator to degree or radian mode as desired.
2. Enter the trigonometric function.
3. Enter the angle measure in degrees or radians.
4. Press $=$ or ENTER.

To find the related angle, given the end point of a vector:

1. Set the calculator to degree or radian mode as desired.
2. Press the inverse tangent TAN⁻¹ or ARCTAN function.
3. Enter the coordinates of the end point as appropriate in the tangent ratio.
4. Press $=$ or ENTER.

EXAMPLE Using a calculator and the π key, find the values of the trigonometric functions. Round the final answer to four significant digits.

(a) $\sin 155°$ (b) $\cos 3$ (c) $\tan 208°$ (d) $\sin 4.2$

(e) $\cos 304.5°$ (f) $\tan\dfrac{5\pi}{3}$

(a) $\sin 155° = \mathbf{0.4226}$ (b) $\cos 3 = \mathbf{-0.9900}$
(c) $\tan 208° = \mathbf{0.5317}$ (d) $\sin 4.2 = \mathbf{-0.8716}$
(e) $\cos 304.5° = \mathbf{0.5664}$ (f) $\tan\dfrac{5\pi}{3} = \mathbf{-1.732}$

EXAMPLE Find the direction in degrees of a vector in standard position if the coordinates of its end point are $(6, -8)$.

From the coordinates of the end point, we can determine that the vector is a quadrant IV vector. Also, from the coordinates of the end point we know that the x-coordinate is 6, and the y-coordinate is -8. Then, $\tan\theta_4 = \dfrac{-8}{6} = -1.333333333$. From the calculator, we see that $\theta = -53.1°$ to the nearest 0.1°. The related angle for a quadrant IV angle of $-53.1°$ is $53.1°$. If $360° - \theta_4 = 53.1°$, then $\theta_4 = 306.9°$.

Thus, the direction of the vector is 306.9°.

When Do We Use the Cosecant, Secant, and Cotangent Functions?

As you have probably noticed, we don't often use the cosecant, secant, and cotangent functions.

What causes a fraction or ratio to be undefined? A denominator of zero. One use of reciprocal functions would be to avoid undefined terms. For example, if $x = 0$ and $y = 3$, the tangent function is undefined: $\tan \theta = \dfrac{y}{x} = \dfrac{3}{0} = \infty$. However, the cotangent function is defined: $\cot \theta = \dfrac{x}{y} = \dfrac{0}{3} = 0$.

SELF-STUDY EXERCISES 18–2

1　Find the related angle for the angles. (Use the calculator value for π and round to hundredths.)

1. 120°　　**2.** 195°　　**3.** 290°　　**4.** 345°　　**5.** 148°

6. 250°　　**7.** 212°　　**8.** 118°　　**9.** 2.18 rad　　**10.** 5.84 rad

2　Give the signs of all six trigonometric functions of vectors with the indicated end points.

11. (3, 5)　　**12.** (−2, 6)　　**13.** (−4, −2)　　**14.** (5, −3)　　**15.** (5, 0)

3　Using a calculator and the π key, find the value of the functions. Round to four significant digits.

16. sin 210°　　**17.** tan 140°　　**18.** cos 2.5　　**19.** cos 4　　**20.** sin 300°

21. tan 6　　**22.** cos 100°　　**23.** $\sin \dfrac{5\pi}{6}$

24. Find the direction in degrees of a vector in standard position if the coordinates of its end point are $(-3, 2)$. Round to the nearest 0.1°.

25. Find the direction in radians of a vector in standard position if the coordinates of its end point are $(-2, -1)$. Round to the nearest hundredth.

18–3　LAW OF SINES

Learning Outcomes

1　Find the missing parts of an oblique triangle, given two angles and an opposite side.

2　Find the missing parts of an oblique triangle, given two sides and an angle opposite one of them.

3　Solve applied problems using the law of sines.

In Chapter 17, we defined an oblique triangle as a triangle that does not contain a right angle. Because these triangles do not have right angles, we cannot use the trigonometric functions directly as we did previously to find sides and angles of right triangles. However, two formulas based on the trigonometric functions of right triangles can be used to solve oblique triangles. In this section we study one of these formulas, the *law of sines*.

Law of sines:

The **law of sines** states that the ratios of the sides of a triangle to the sines of the angles opposite these respective sides are equal (Fig. 18–16).

$$\frac{a}{\sin A} = \frac{b}{\sin B} = \frac{c}{\sin C}$$

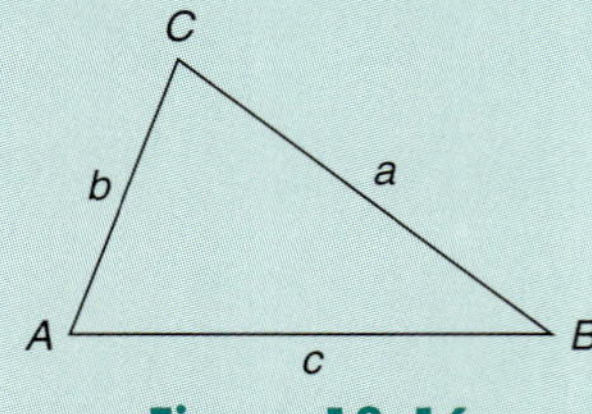

Figure 18–16

Conditions for Using the Law of Sines.

The law of sines is used to solve triangles when either of the conditions exists:

1. Two angles and the side opposite one of them are known.
2. Two sides and the angle opposite one of them are known.

If we know two angles of a triangle, we can always find the third angle. Therefore, in Step 1, if we know two angles and any side, we can solve the triangle.

1 Find the Missing Parts of an Oblique Triangle, Given Two Angles and an Opposite Side.

Application problems often require solving triangles when we know two angles and a side of a triangle. In the examples that follow, all digits of the calculated value that show in the calculator display will be given. Rounding should be done after the last calculation has been made.

EXAMPLE Solve triangle ABC if $A = 50°$, $B = 75°$, and $b = 12$ ft.

The first step in solving the triangle is to sketch it and label the parts, as shown in Fig. 18–17. To solve the triangle, we need to find angle C, side a, and side c. We find angle C by applying the property that the sum of the angles of a triangle equals 180°.

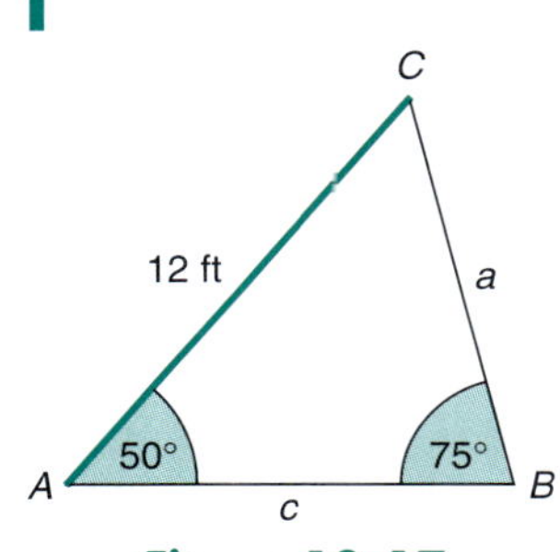

Figure 18–17

$$A + B + C = 180° \qquad \text{Substitute } A = 50° \text{ and } B = 75°.$$

$$50° + 75° + C = 180° \qquad \text{Solve for } C.$$

$$C = 180° - 50° - 75°$$

$$C = 55°$$

To find a and c, we use the law of sines. First, we find a. To find a, we choose the proportion that contains a (the unknown quantity) and three known quantities.

$$\frac{a}{\sin A} = \frac{b}{\sin B} \qquad \text{Substitute } b = 12, A = 50°, B = 75°.$$

$$\frac{a}{\sin 50°} = \frac{12}{\sin 75°} \qquad \text{Solve for } a.$$

$$a \sin 75° = 12 \sin 50°$$

$$a = \frac{12 \sin 50°}{\sin 75°} \qquad \text{Perform the calculations.}$$

$$a = 9.516810781$$

$$a = 9.517 \text{ ft} \qquad \text{Four significant digits.}$$

To find c, we choose a proportion that contains the unknown quantity c and three known quantities.

$$\frac{b}{\sin B} = \frac{c}{\sin C} \qquad \text{Substitute } b = 12, B = 75°, \text{ and } C = 55°.$$

$$\frac{12}{\sin 75°} = \frac{c}{\sin 55°} \qquad \text{Solve for } c.$$

$$12 \sin 55° = c \sin 75°$$

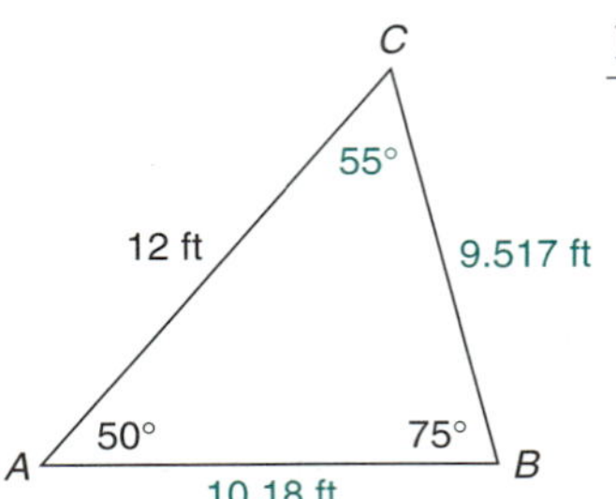

Figure 18–18

$$\frac{12\sin 55°}{\sin 75°} = c \qquad \text{Perform the calculations.}$$

$$c = 10.1765832$$

$$c = \mathbf{10.18 \ ft} \qquad \text{Four significant digits.}$$

As a check for our work, the longest side should be opposite the largest angle, and the shortest side should be opposite the smallest angle (Fig. 18–18).

EXAMPLE Solve triangle ABC if $A = 35°$, $a = 7$ cm, and $B = 40°$.

We first sketch the triangle and label its parts (Fig. 18–19). To solve the triangle, we need to find angle C, side b, and side c.

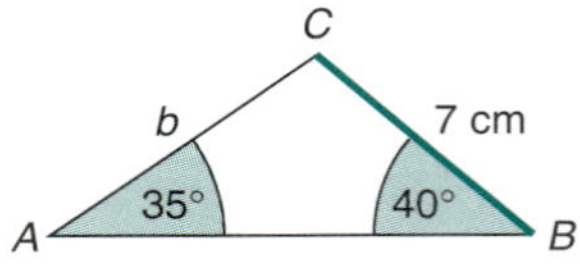

Figure 18–19

$$A + B + C = 180° \qquad \text{Substitute } A = 35° \text{ and } B = 40°.$$

$$35° + 40° + C = 180° \qquad \text{Solve for } C.$$

$$C = 180° - 35° - 40°$$

$$C = \mathbf{105°}$$

To find b and c, we use the law of sines. First, we find b.

$$\frac{a}{\sin A} = \frac{b}{\sin B} \qquad \text{Substitute } a = 7, A = 35°, \text{ and } B = 40°.$$

When choosing the proportion, such as the one above, be sure to choose one in which substitutes can be made for three terms.

$$\frac{7}{\sin 35°} = \frac{b}{\sin 40°} \qquad \text{Solve for } b.$$

$$7 \sin 40° = b \sin 35°$$

$$\frac{7 \sin 40°}{\sin 35°} = b \qquad \text{Evaluate.}$$

$$b = 7.844661989$$

$$b = \mathbf{7.845 \ cm} \qquad \text{Four significant digits.}$$

To find c, we choose the proportion that allows substitution for as many of the *originally* given sides and angles as possible.

$$\frac{a}{\sin A} = \frac{c}{\sin C} \qquad \text{Substitute } a = 7, A = 35°, \text{ and } C = 105°.$$

$$\frac{7}{\sin 35°} = \frac{c}{\sin 105°} \qquad \text{Solve for } c.$$

$$7 \sin 105° = c \sin 35°$$

$$\frac{7 \sin 105°}{\sin 35°} = c \qquad \text{Evaluate.}$$

$$c = 11.78828201$$

$$c = \mathbf{11.79 \ cm} \qquad \text{Four significant digits.}$$

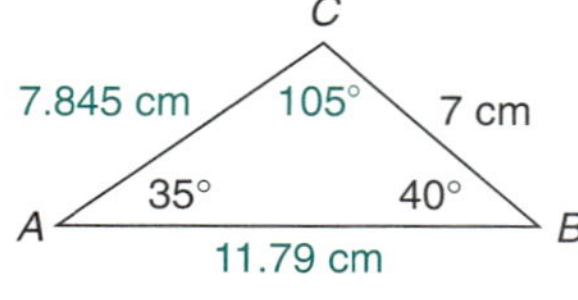

Figure 18–20

As a check for our work, the longest side should be opposite the largest angle, and the shortest side should be opposite the smallest angle (Fig. 18–20).

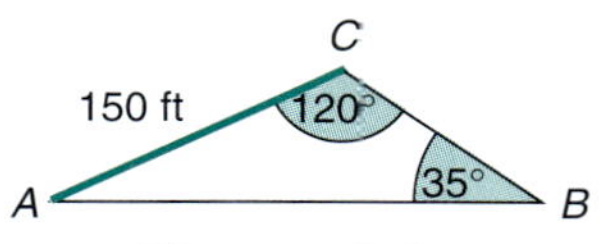

Figure 18–21

EXAMPLE ◆ Determine the unknown angles and sides of a piece of land described by the triangle in Fig. 18–21.

Find A.

$$A + B + C = 180°$$
Substitute $B = 35°$ and $C = 120°$.

$$A + 35° + 120° = 180°$$

$$A = 180° - 35° - 120°$$

$$\boldsymbol{A = 25°}$$

Next, we find side a.

$$\frac{b}{\sin B} = \frac{a}{\sin A}$$
Substitute $A = 25°$, $b = 150$, and $B = 35°$.

$$\frac{150}{\sin 35°} = \frac{a}{\sin 25°}$$
Solve for a.

$$150 \sin 25° = a \sin 35°$$

$$\frac{150 \sin 25°}{\sin 35°} = a$$
Evaluate.

$$a = 110.5218681$$

$$\boldsymbol{a = 110.5 \text{ ft}}$$
Four significant digits.

Next, we find c.

$$\frac{b}{\sin B} = \frac{c}{\sin C}$$
Substitute $B = 35°$, $b = 150$, and $C = 120°$.

$$\frac{150}{\sin 35°} = \frac{c}{\sin 120°}$$
Solve for c.

$$150 \sin 120° = c \sin 35°$$

$$\frac{150 \sin 120°}{\sin 35°} = c$$
Evaluate.

$$c = 226.4803823$$

$$\boldsymbol{c = 226.5 \text{ ft}}$$
Four significant digits.

2 **Find the Missing Parts of an Oblique Triangle, Given Two Sides and an Angle Opposite One of Them.**

When two sides of a triangle and an angle opposite one of them are known, we do not always have a single triangle. If the given sides are a and b and the given angle is B, three possibilities may exist (Fig. 18–22).

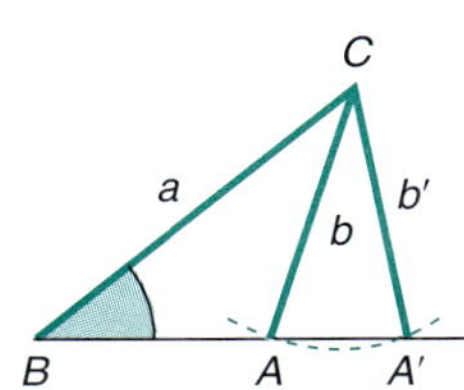

Case 1 ($b < a$)

Two solutions for A, C, and c since b can meet side c in

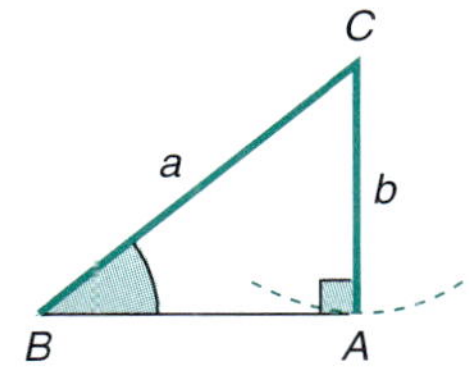

Case 2 ($A = 90°$)

One solution since b meets side c in exactly one point.

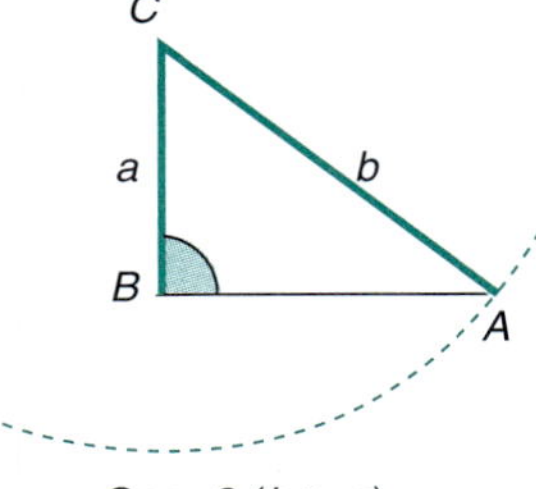

Case 3 ($b \geq a$)

One solution since b meets side c in only one point.

Figure 18–22

- If $b < a$ and $\angle A \neq 90°$, we have two possible solutions (case 1).
- If $b < a$ and $\angle A = 90°$, we have one solution (case 2).
- If $b \geq a$, and $\angle A \neq 90°$, we have one solution (case 3).

Note that side b above could be any side of the triangle that is opposite the given angle. Side a is then the other given side. Because of the lack of clarity when two sides and an angle opposite one of them are given, the situation is called the *ambiguous case*. (*Ambiguous* means that the given information can be interpreted in more than one way.)

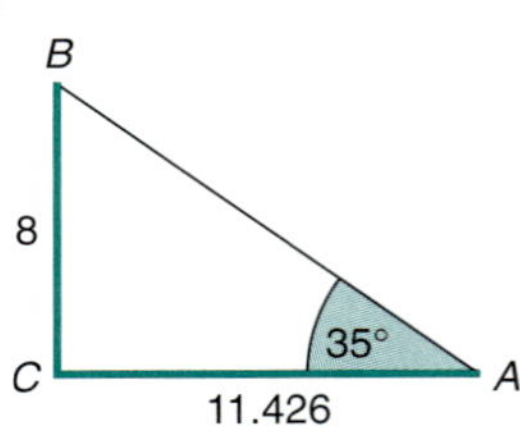

Figure 18–23

EXAMPLE Solve triangle ABC if $A = 35°$, $a = 8$, and $b = 11.426$ (Fig. 18–23).

$$\frac{a}{\sin A} = \frac{b}{\sin B}$$

Substitute $A = 35°$, $a = 8$, and $b = 11.426$.

$$\frac{8}{\sin 35°} = \frac{11.426}{\sin B}$$

Solve for B.

$$8 \sin B = 11.426 \sin 35°$$

$$\sin B = \frac{11.426 \sin 35°}{8}$$

Evaluate.

$$\sin B = 0.8192105452 \qquad \sin^{-1} 0.8192105452 = B.$$

Possible solution 1: $\sin^{-1} 0.8192105452 = 55.00584421°$

$$B = 55.0° \qquad \text{Nearest } 0.1°$$

$$A + B + C = 180°$$

$$35° + 55° + C = 180°$$

$$C = 180° - 55° - 35°$$

$$C = 90° \qquad ABC \text{ is a right triangle.}$$

This is a right triangle, but we still have two possible solutions. The angle opposite the second given side is not the 90° angle. To find the third side of this right triangle, we can use the Pythagorean theorem or the law of sines.

Using the law of sines, we have

$$\frac{a}{\sin A} = \frac{c}{\sin C}$$

Substitute $a = 8$, $A = 35°$, and $C = 90°$.

$$\frac{8}{\sin 35°} = \frac{c}{\sin 90°}$$

Solve for c.

$$8 \sin 90° = c \sin 35°$$

$$\frac{8 \sin 90°}{\sin 35°} = c$$

Evaluate.

$$c = 13.94757436$$

$$c = 13.95 \qquad \text{Four significant digits.}$$

The solved triangle for this possibility is shown in Fig. 18–24.
The second possibility for angle B is

$$180° - 55.00584421° = 124.9941558°$$

To obtain angle C, we have

$$C = 180° - (35° + 124.9941558°) = 20.0058442°$$

$$C = 20.0°$$

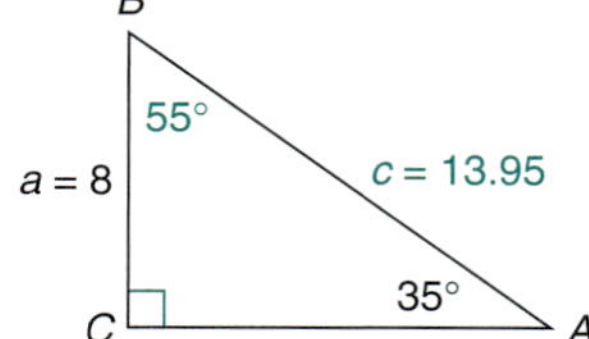

Figure 18–24

CHAPTER 18   Oblique Triangles

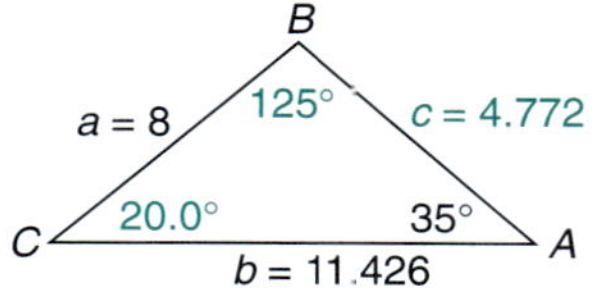

Figure 18–25

We can find side c using the law of sines.

$$\frac{a}{\sin A} = \frac{c}{\sin C}$$

$$\frac{8}{\sin 35^\circ} = \frac{c}{\sin 20.0058442^\circ}$$

$$c \sin 35^\circ = 8 \sin 20.0058442^\circ$$

$$c = \frac{8 \sin 20.0058442^\circ}{\sin 35^\circ}$$

$$c = 4.771688222 \quad \text{or} \quad c = 4.772$$

The solved triangle for this possibility is shown in the triangle Fig. 18–25.

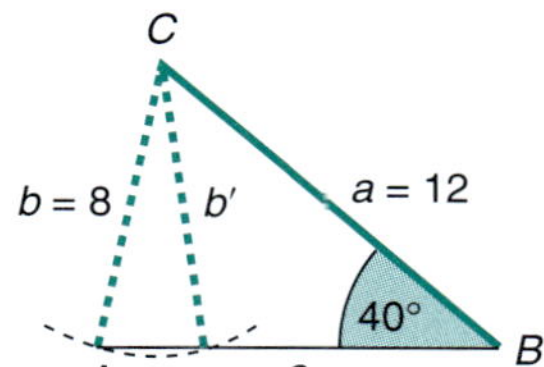

Figure 18–26

EXAMPLE Solve triangle ABC if $a = 12$, $b = 8$, and $B = 40^\circ$ (Fig. 18–26).

To find A, we have

$$\frac{a}{\sin A} = \frac{b}{\sin B} \qquad \text{Substitute } a = 12, b = 8, \text{ and } B = 40^\circ.$$

$$\frac{12}{\sin A} = \frac{8}{\sin 40^\circ} \qquad \text{Solve for } A.$$

$$12 \sin 40^\circ = 8 \sin A$$

$$\frac{12 \sin 40^\circ}{8} = \sin A \qquad \text{Evaluate.}$$

$$\sin A = 0.9641814145 \qquad \sin^{-1} 0.9641814145 = A$$

$$\mathbf{A = 74.61856831^\circ} \qquad \textbf{or} \qquad \mathbf{105.3814317^\circ}$$

There are two angles less than 180° that have a sine of approximately 0.9641814145. These angles are 74.61856831° and 105.3814317°. The angle 74.61856831° is in quadrant I. The other angle is in quadrant II and has 74.61856831° as its related angle. Therefore, $180^\circ - 74.61856831^\circ = 105.3814317^\circ$ is the measure of the other angle. You can verify this by comparing the sine of the two angles. This situation occurs when the side opposite the given angle is less than the other given side. As a result, we have *two* possible solutions.

Possible solution 1: A = 74.61856831° (Fig. 18–27).

$$C = 180^\circ - 74.61856831^\circ - 40^\circ$$

$$C = 65.38143169^\circ$$

$$\mathbf{C = 65.4^\circ} \qquad \text{Nearest } 0.1^\circ.$$

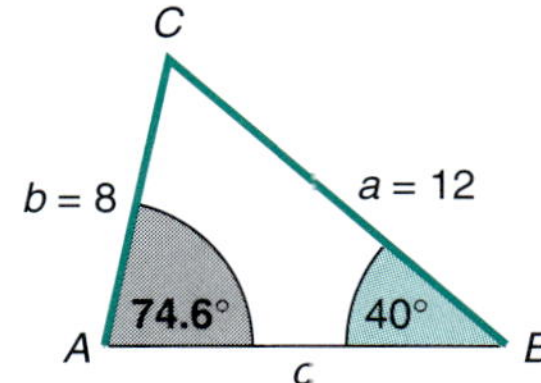

Figure 18–27

Now we find side c.

$$\frac{b}{\sin B} = \frac{c}{\sin C} \qquad \text{Substitute } b = 8, B = 40^\circ, C = 65.38143169^\circ. \text{ Use the full calculator value for } C \text{ to maximize accuracy.}$$

$$\frac{8}{\sin 40^\circ} = \frac{c}{\sin 65.38143169^\circ} \qquad \text{Solve for } c.$$

$$8 \sin 65.38143169^\circ = c \sin 40^\circ$$

$$\frac{8 \sin 65.38143169^\circ}{\sin 40^\circ} = c \qquad \text{Evaluate.}$$

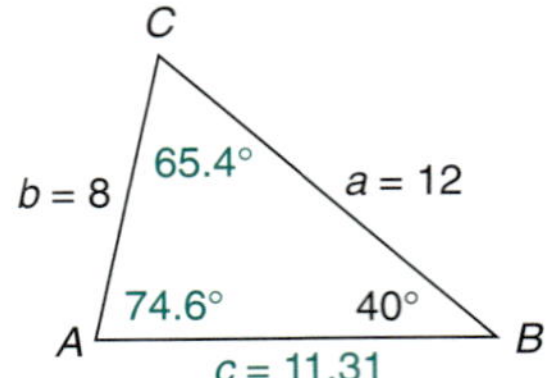

Figure 18–28

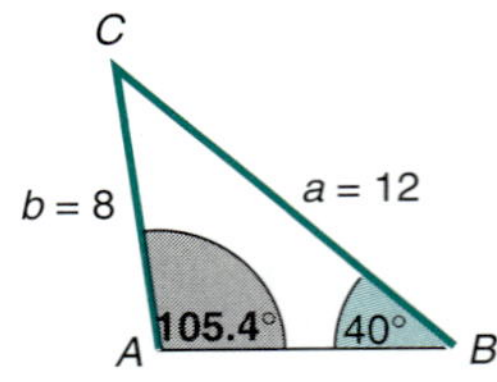

Figure 18–29

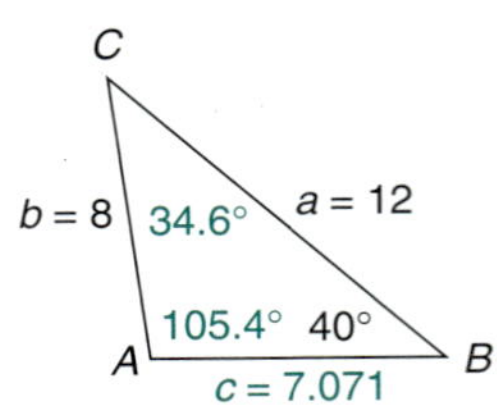

Figure 18–30

$$c = 11.31448261$$

$$\mathbf{c = 11.31}$$ Four significant digits.

The solved triangle is shown in Fig. 18–28.

Possible solution 2: A $= 105.3814317°$ (Fig. 18–29).

$$C = 180° - 105.3814317° - 40°$$

$$C = 34.6185683°$$

$$\mathbf{C = 34.6°}$$ Nearest 0.1°.

Now we find side c.

$$\frac{b}{\sin B} = \frac{c}{\sin C}$$ Substitute. Use the full calculator value for *C* to maximize accuracy.

$$\frac{8}{\sin 40°} = \frac{c}{\sin 34.6185683°}$$ Solve for *c*.

$$8 \sin 34.6185683° = c \sin 40°$$

$$\frac{8 \sin 34.6185683°}{\sin 40°} = c$$

$$c = 7.070584025$$

$$\mathbf{c = 7.071}$$ Four significant digits.

The solved triangle is shown in Fig. 18–30.

EXAMPLE Solve triangle ABC if $b = 10$, $c = 12$, and $C = 40°$ (Fig. 18–31).

We solve for B first. Because c, the side opposite the given angle C, is *longer* than the other side, b, we expect *one* solution.

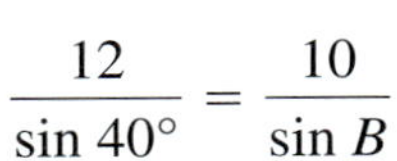

Figure 18–31

$$\frac{c}{\sin C} = \frac{b}{\sin B}$$ Substitute.

$$\frac{12}{\sin 40°} = \frac{10}{\sin B}$$ Solve for *B*.

$$12 \sin B = 10 \sin 40°$$

$$\sin B = \frac{10 \sin 40°}{12}$$ Evaluate.

$$\sin B = 0.5356563414$$ $\sin^{-1} 0.5356563414 = B$

$$B = 32.38843382°$$

$$\mathbf{B = 32.4°}$$ Nearest 0.1°.

Both $32.38843382°$ and $147.6115662°$ have a sine of approximately 0.5356563414; however, we cannot have an angle of $147.6115662°$ in this triangle because the triangle already has a 40° angle. $147.6115662° + 40° = 187.6115662°$. The *three* angles of a triangle total only 180°; therefore, we have only *one* solution. Finding A, we have

$$A = 180° - 40° - 32.38843382°$$ Use the full calculator value for *B*.

$$A = 107.6115662°$$

$$\mathbf{A = 107.6°}$$

Next, we find a.

$$\frac{a}{\sin A} = \frac{c}{\sin C}$$
Substitute full calculator values.

$$\frac{a}{\sin 107.6115662°} = \frac{12}{\sin 40°}$$
Solve for a.

$$a \sin 40° = 12 \sin 107.6115662°$$

$$a = \frac{12 \sin 107.6115662°}{\sin 40°}$$
Evaluate.

$$a = 17.79367732$$

$$\boldsymbol{a = 17.79}$$
Four significant digits.

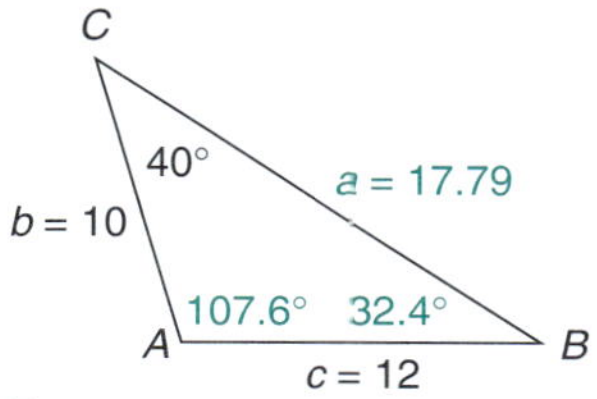

Figure 18–32

The solved triangle is shown in Fig. 18–32.

3 Solve Applied Problems Using the Law of Sines.

Because real-world applications often do not involve right triangles, the law of sines is very useful in solving applied problems.

EXAMPLE A technician checking a surveyor's report is given the information shown in Fig. 18–33. Calculate the missing information.

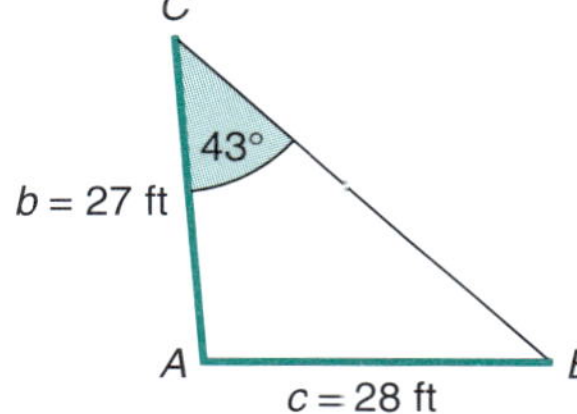

Figure 18–33

Find B.

$$\frac{c}{\sin C} = \frac{b}{\sin B}$$
Substitute.

$$\frac{28}{\sin 43°} = \frac{27}{\sin B}$$
Solve for B.

$$28 \sin B = 27 \sin 43°$$

$$\sin B = \frac{27 \sin 43°}{28}$$

$$\sin B = 0.6576412758$$
$\sin^{-1} 0.6576412758 = B$

$$B = 41.12023021°$$

$$\boldsymbol{B = 41.1°}$$
Nearest 0.1°.

There is only one case here because we are given the triangle. Also, the other angle whose sine is 0.6576412758 is 138.8797698°, which we exclude because 138.8797698° + 43° = 181.8797698°.

To find angle A, we have

$$A = 180° - (43° + 41.12023021°)$$

$$A = 95.87976979°$$

$$\boldsymbol{A = 95.9°}$$
Nearest 0.1°

$$\frac{a}{\sin A} = \frac{c}{\sin C}$$
Substitute.

$$\frac{a}{\sin 95.87976979°} = \frac{28}{\sin 43°}$$

$$a \sin 43° = 28 \sin 95.87976979°$$

$$a = \frac{28 \sin 95.87976979°}{\sin 43°}$$
Evaluate.

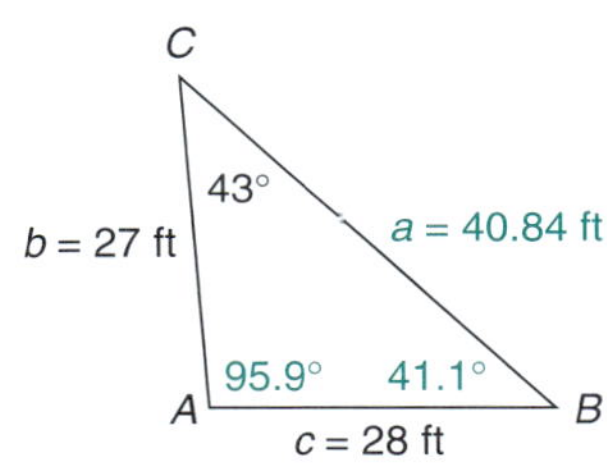

Figure 18–34

$$a = 40.83982458$$

$$a = \mathbf{40.84\ ft} \qquad \text{Four significant digits.}$$

The completed survey should have the measures shown in Fig. 18–34.

SELF-STUDY EXERCISES 18–3

 Solve each of the oblique triangles using the law of sines. Round the final answer for sides to four significant digits and angles to the nearest 0.1°.

1. $a = 46$, $A = 65°$, $B = 52°$
2. $b = 7.2$, $B = 58°$, $C = 72°$
3. $a = 65$, $B = 60°$, $A = 87°$
4. $c = 3.2$, $A = 120°$, $C = 30°$
5. $b = 12$, $A = 95°$, $B = 35°$

2 Use the law of sines to solve the triangles. If a triangle has two possibilities, find both solutions. Round sides to four significant digits and angles to the nearest 0.1°.

6. $a = 42$, $b = 24$, $A = 40°$
7. $b = 15$, $c = 3$, $B = 70°$
8. $a = 18$, $c = 9$, $C = 20°$
9. $a = 8$, $b = 4$, $A = 30°$

3

10. Find the missing angle and sides of the plot of land described by Fig. 18–35.

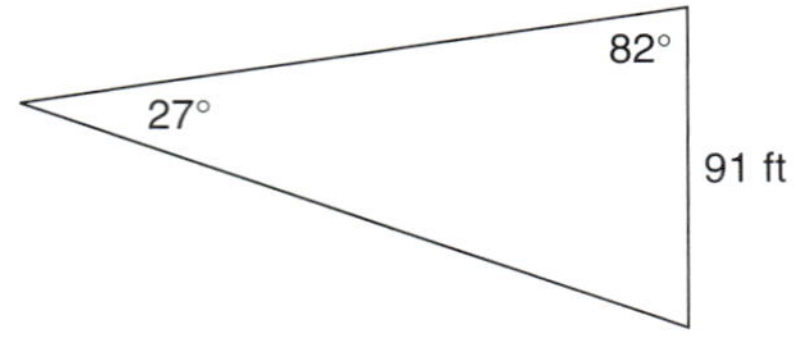

Figure 18–35

11. Find the distance from A to B on the surveyed plot shown in Fig. 18–36.

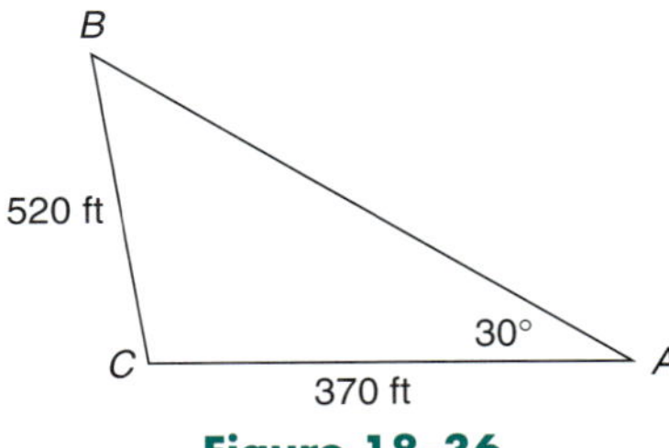

Figure 18–36

18–4   LAW OF COSINES

Learning Outcomes

1 Find the missing parts of an oblique triangle, given three sides of the triangle.

2 Find the missing parts of an oblique triangle, given two sides and the included angle of the triangle.

3 Solve applied problems using the law of cosines and the law of sines.

In some cases, our given information does not allow us to use the law of sines. For example, if we know all three sides of a triangle, we cannot use the law of sines to find the angles. In a case such as this, however, we can use the *law of cosines*. This law is based on the trigonometric functions just as the law of sines is.

Law of cosines:

The **law of cosines** states that the square of any side of a triangle equals the sum of the squares of the other sides minus twice the product of the other two sides and the cosine of the angle opposite the first side. For triangle *ABC*:

$$a^2 = b^2 + c^2 - 2bc \cos A$$

$$b^2 = a^2 + c^2 - 2ac \cos B$$

$$c^2 = a^2 + b^2 - 2ab \cos C$$

When Do You Use the Law of Cosines?

You can use the law of cosines efficiently when you are given

1. Three sides of a triangle

or

2. Two sides and the included angle of a triangle.

You can use *either* the law of sines or the law of cosines to solve certain triangles; however, whenever possible, the law of sines is generally preferred because it involves fewer calculations.

1   Find the Missing Parts of an Oblique Triangle, Given Three Sides of the Triangle.

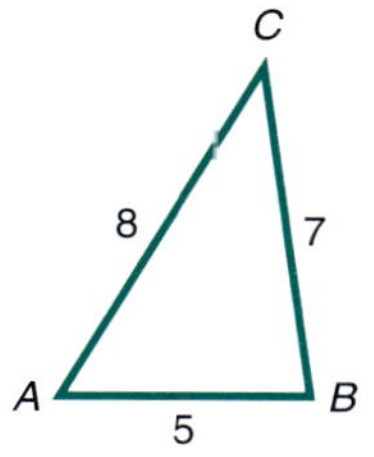

EXAMPLE   Find the angles in triangle ABC (Fig. 18–37).

Figure 18–37

We may find any one of the angles first. If we choose to find angle A first, we must use the formula that contains $\cos A$: $a^2 = b^2 + c^2 - 2bc \cos A$. We rearrange the formula to solve for $\cos A$.

$$a^2 - b^2 - c^2 = -2bc \cos A$$

Multiply each term on both sides by -1 to reduce the number of negative signs.

$$-a^2 + b^2 + c^2 = 2bc \cos A$$

Solve for $\cos A$.

$$\frac{-a^2 + b^2 + c^2}{2bc} = \cos A$$

Substitute.

$$\frac{-(7)^2 + 8^2 + 5^2}{2(8)(5)} = \cos A$$

Solve for A.

$$\frac{-49 + 64 + 25}{80} = \cos A$$

$$\frac{40}{80} = \cos A$$

$$0.5 = \cos A \qquad \cos^{-1} 0.5 = A$$

$$\mathbf{60° = A}$$

To find angle B, we use the law of cosines and only given values. We could use the law of sines, but that would involve using a calculated value.

$$b^2 = a^2 + c^2 - 2ac \cos B$$

Solve for $\cos B$.

$$2ac \cos B = a^2 + c^2 - b^2$$

$$\cos B = \frac{a^2 + c^2 - b^2}{2ac}$$

Substitute.

$$\cos B = \frac{7^2 + 5^2 - (8)^2}{2(7)(5)}$$

$$\cos B = \frac{49 + 25 - 64}{70}$$

$$\cos B = \frac{10}{70}$$

$$\cos B = 0.1428571429 \qquad \cos^{-1} 0.1428571429 = B$$

$$B = 81.7867893°$$

$$\boldsymbol{B = 81.8°} \qquad \text{Nearest } 0.1°.$$

We can use the law of cosines or the law of sines to find the third angle; however, the quickest way to find this angle is to subtract the sum of A and B from $180°$.

$$C = 180° - (A + B)$$

$$C = 180° - 60° - 81.7867893°$$

$$C = 180° - 141.7867893°$$

$$C = 38.2132107°$$

$$\boldsymbol{C = 38.2°} \qquad \text{Nearest } 0.1°$$

The solved triangle is shown in Fig. 18–38.

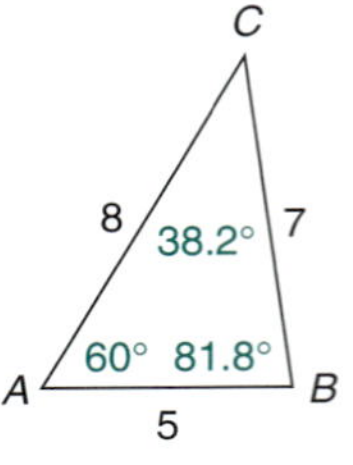

Figure 18–38

EXAMPLE Find the angles in triangle ABC (Fig. 18–39).

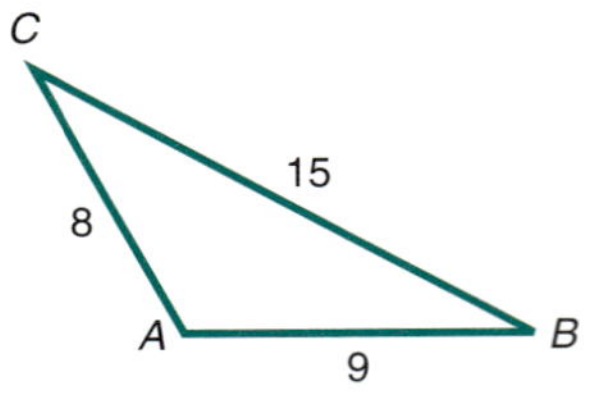

We will find A first.

$$a^2 = b^2 + c^2 - 2bc \cos A \qquad \text{Solve for } \cos A.$$

$$\cos A = \frac{b^2 + c^2 - a^2}{2bc} \qquad \text{Substitute.}$$

$$\cos A = \frac{8^2 + 9^2 - 15^2}{2(8)(9)} \qquad \text{Evaluate.}$$

$$\cos A = \frac{64 + 81 - 225}{144}$$

$$\cos A = \frac{-80}{144}$$

$$\cos A = -0.5555555556 \qquad \cos^{-1} -0.5555555556 = A$$

$$A = 123.7489886°$$

$$\boldsymbol{A = 123.7°} \qquad \text{Nearest } 0.1°.$$

Figure 18–39

Recall from Section 18–2 that the cosine is *negative* in the second and third quadrants. Because A is either acute ($<90°$) or obtuse ($>90°$ but $<180°$) and because its cosine is negative (Fig. 18–40), it must be in the second quadrant. If we use a calculator and enter -0.5555555556 the display will show $123.7489886°$. We can use the law of sines to find the second angle, but keep in mind the risk involved in continuing with a calculated value instead of a given value.

Figure 18–40

$$\frac{a}{\sin A} = \frac{b}{\sin B} \qquad \text{Substitute.}$$

$$\frac{15}{\sin 123.7489886°} = \frac{8}{\sin B} \qquad \text{Solve for } B.$$

$$15 \sin B = 8 \sin 123.7489886°$$

$$\sin B = \frac{8 \sin 123.7489886°}{15}$$

$$\sin B = 0.4434556903 \qquad \sin^{-1} 0.4434556903 = B$$

$$B = 26.32457654°$$

$$\boldsymbol{B = 26.3°} \qquad \text{Rounded to } 0.1°$$

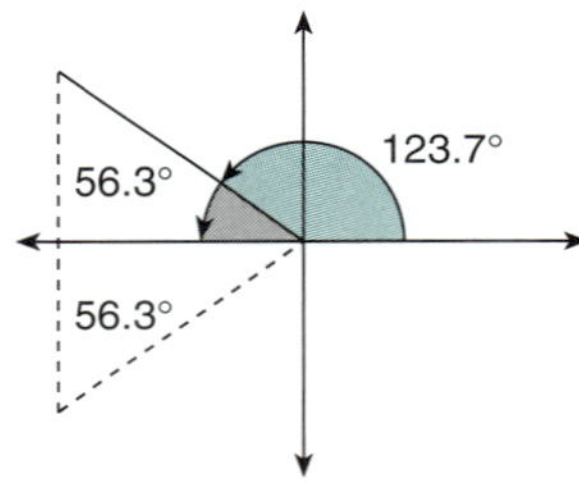

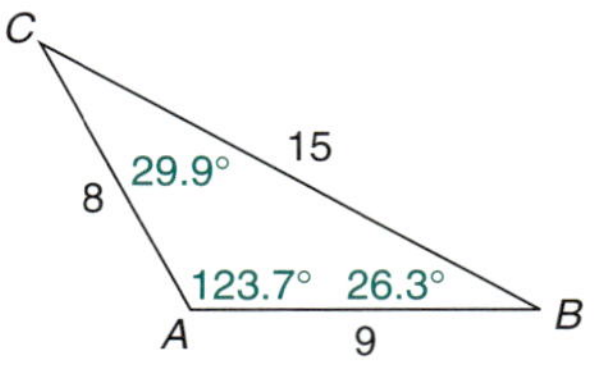

Figure 18–41

   CHAPTER 18   Oblique Triangles

$$C = 180° - A - B$$
$$C = 180° - 123.7489886° - 26.32457654°$$
$$\mathbf{C = 29.9°}$$

The solved triangle is given in Fig. 18–41.

2 Find the Missing Parts of an Oblique Triangle, Given Two Sides and the Included Angle of the Triangle.

The law of cosines is needed to solve oblique triangles if two sides and the included angle are given.

EXAMPLE Solve triangle ABC in Fig. 18–42.

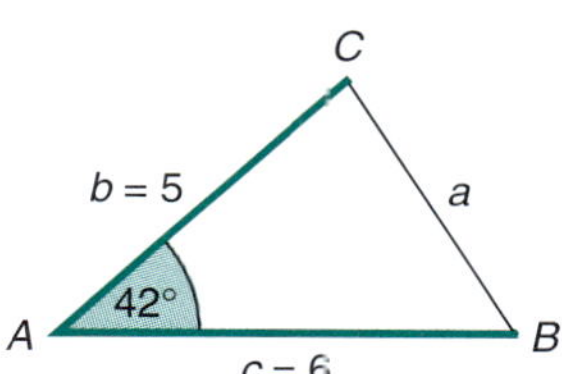

Figure 18–42

We can find a by using the law of cosines.

$$a^2 = b^2 + c^2 - 2bc \cos A \qquad \text{Substitute.}$$
$$a^2 = 5^2 + 6^2 - 2(5)(6) \cos 42°$$
$$a^2 = 25 + 36 - 60 \cos 42° \qquad \cos 42° = 0.7431448255$$
$$a^2 = 25 + 36 - 44.58868953$$
$$a^2 = 16.41131047$$
$$a = 4.051087566$$
$$\mathbf{a = 4.051} \qquad \text{Four significant digits.}$$

Use the law of sines to find each of the two missing angles.

Find B.

$$\frac{a}{\sin A} = \frac{b}{\sin B} \qquad \text{Substitute.}$$
$$\frac{4.051087566}{\sin 42°} = \frac{5}{\sin B} \qquad \text{Solve for } B.$$
$$4.051087566 \sin B = 5 \sin 42°$$
$$\sin B = \frac{5 \sin 42°}{4.051087566}$$
$$\sin B = 0.8258653947$$
$$B = 55.67632739°$$
$$\mathbf{B = 55.7°} \qquad \text{Nearest 0.1°.}$$

Find C.

$$\frac{a}{\sin A} = \frac{c}{\sin C} \qquad \text{Substitute.}$$
$$\frac{4.051087566}{\sin 42°} = \frac{6}{\sin C} \qquad \text{Solve for } C.$$
$$4.051087566 \sin C = 6 \sin 42°$$
$$\sin C = \frac{6 \sin 42°}{4.051087566}$$
$$\sin C = 0.9910384737 \qquad \sin^{-1} 0.9910384737 = C$$

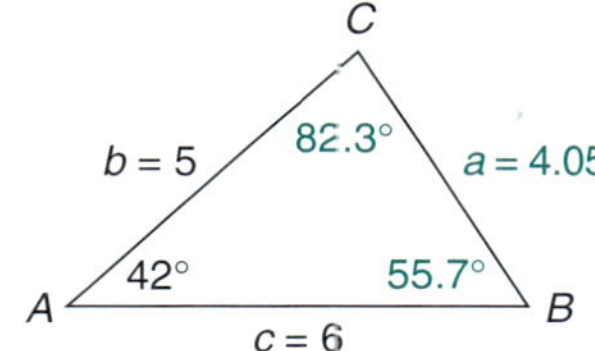

Figure 18–43

$$C = 82.32367267°$$

$$C = 82.3°$$ Nearest 0.1°.

The solved triangle is given in Fig. 18–43.

To check, verify that the sum of the angles adds to 180°, the longest side is opposite the largest angle, and the shortest side is opposite the smallest angle.

3 Solve Applied Problems Using the Law of Cosines and the Law of Sines.

EXAMPLE A vertical 45-ft pole is placed on a hill that is inclined 17° to the horizontal (Fig. 18–44). How long a guy wire is needed if the guy wire is placed 5 ft from the top of the pole and attached to the ground at a point 32 ft uphill from the base of the pole?

Point A is where the pole enters the ground. Point B is where the wire is attached to the ground. Point C is where the wire is attached to the pole (5 ft from the top of the pole). Side b is the length of the pole from the wire to the ground $(45 - 5 = 40$ ft$)$. The length of the wire is side a in the triangle. Because AC makes a 90° angle with the horizontal, we know that angle A is $90° - 17°$, or 73°. Using the law of cosines, we have

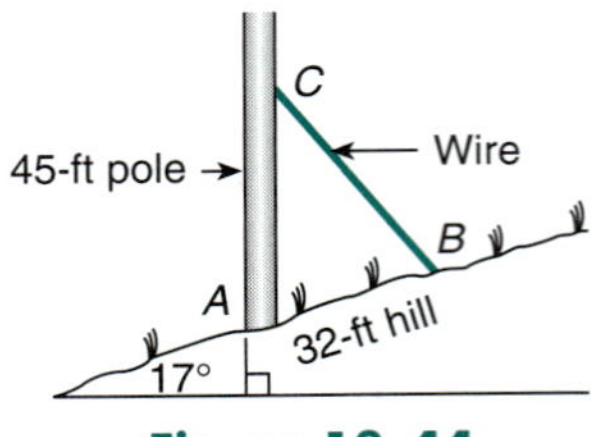

Figure 18–44

$$a^2 = b^2 + c^2 - 2bc \cos A$$ Substitute.

$$a^2 = 40^2 + 32^2 - 2(40)(32) \cos 73°$$

$$a^2 = 1{,}600 + 1{,}024 - 2{,}560(0.2923717047)$$

$$a^2 = 1{,}600 + 1{,}024 - 748.471564$$

$$a^2 = 1{,}875.528436$$

$$a = 43.30737161$$

$$a = 43.31 \text{ ft}$$ Four significant digits.

The length of the wire is 43.31 ft.

SELF-STUDY EXERCISES 18–4

1 Solve the triangles in Figs. 18–45 and 18–46. Round side lengths to four significant digits and angles to the nearest 0.1°.

1.

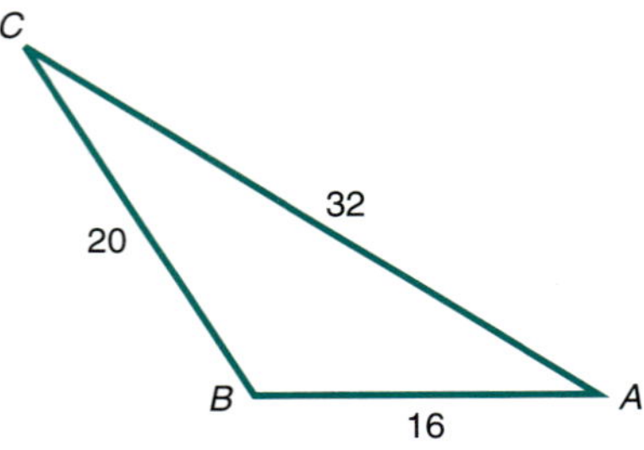

Figure 18–45

2.

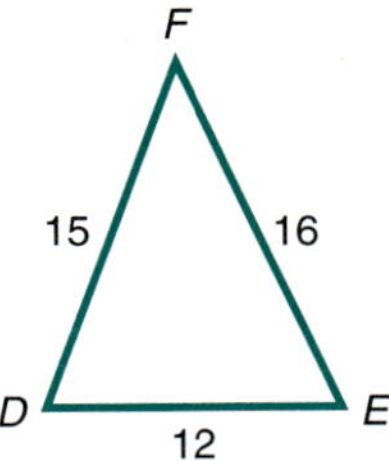

Figure 18–46

CHAPTER 18 Oblique Triangles

2 Solve the triangles in Figs. 18–47 and 18–48.

3.

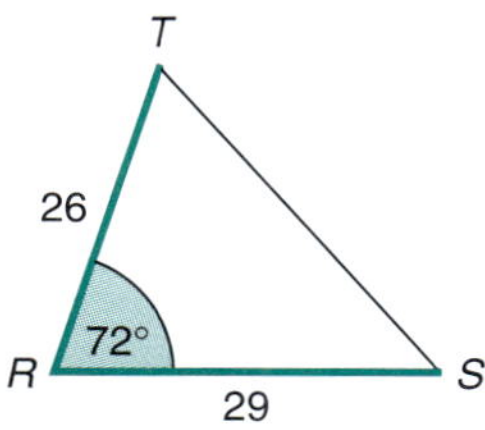

Figure 18–47

4.

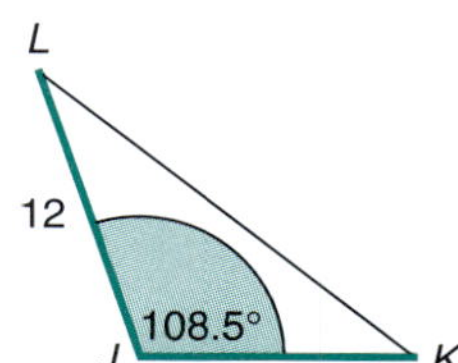

Figure 18–48

3 Solve the problems. Round as above.

5. A hill is inclined 20° to the horizontal. A pole stands vertically on the side of the hill with 35 ft above the ground. How much wire will it take to reach from a point 2 ft from the top of the pole to a point on the ground 27 ft downhill from the base of the pole?

6. A triangular tabletop is to be 8.4 ft by 6.7 ft by 9.3 ft. What angles must be cut?

Law Enforcement: Vehicle Vaults

Today's traffic accident investigators often act as traffic reconstructionists because drivers may lie about their driving speed, be seriously injured (and so be unable to think clearly and answer questions), or die in the accident. When a vehicle becomes airborne with an uphill takeoff angle of at least 6°, investigators use trigonometry to estimate the vehicle's speed at takeoff with the following vault formula based on an oblique triangle:

$$S = \frac{2.73 \cdot D}{\sqrt{D \cdot \cos \theta \cdot \sin \theta \pm [H \cdot (\cos \theta)^2]}}$$

where S = speed in mph, D = horizontal distance from takeoff point to first contact, H = vertical distance from takeoff point to first contact, and θ = takeoff angle in degrees. Use + in the formula when the vehicle lands at a point lower than takeoff, and − when it lands at a point higher than takeoff.

The takeoff angle is usually the same as the road grade. The grade of a road is the slope of the road written as a percent, or

$$\text{Grade} = \frac{\text{Vertical rise}}{\text{Horizontal run}} \cdot 100$$

The takeoff angle in degrees is $\tan^{-1}$ of the grade (in decimal form). In the case of a motorcycle driver vaulting off his vehicle, the takeoff angle is assumed to be 45°, and the driver's vertical drop on level ground is assumed to be 3 ft.

Exercises

Use the preceding information to estimate the speed of the vehicle in each of the following vault accidents. Round answers to the nearest unit.

1. The driver of an empty tractor-trailer truck falls asleep at the wheel, "straightens the curve" on an uphill 7% grade road, and vaults off a cliff. The truck's first contact point is 40 vertical ft and 55 horizontal ft below its takeoff point. The driver of the car behind the truck said the truck was going between 20 and 25 mph when it left the road. Is this correct?

2. On a level street, a driver runs a red light at a T intersection and is hit by a motorcycle whose driver vaults over the top of the car. The motorcycle driver hits the street 57 horizontal feet from the takeoff point. How fast was the motorcycle traveling when it hit the car?

3. A car traveling on a divided highway blows a tire, crosses the median ditch (which dips 5 vertical feet, then rises sharply), and vaults into the oncoming traffic lanes. It makes first contact at a point 12 ft higher than the takeoff point and 162 ft horizontally from the takeoff point. Combining the 6% uphill road grade with the ditch angle gives a takeoff angle of 32°. Find the car's speed.

4. A drunk driver in a high-rise sport utility vehicle with big wheels drifts into the median strip of a divided highway at an overpass. He hits a large mound of dirt piled in the median, and the truck is launched upward at a 50° angle. The truck vaults over the oncoming traffic lane, and falls 30 ft to the ground below. Find the takeoff speed if the horizontal distance is 153 ft.

5. Do you think the takeoff speed of the vehicle from Exercise 4 is the same speed the truck was going just before it hit the median strip? Explain.

Answers

1. The driver of the car estimated the truck's speed correctly at about 22 mph.
2. The motorcycle was traveling at least 25 mph.
3. The car's speed was at least 55 mph.
4. When the truck left the top of the dirt mound, its speed was at least 44 mph.
5. Climbing the dirt mound would have slowed the vehicle considerably, so the truck's speed when it left the road was greater.

ASSIGNMENT EXERCISES

Section 18–1

Find the direction and magnitude of vectors in standard position with the indicated end points. Round lengths to four significant digits and angles to the nearest tenth of a degree.

1. (4, 6)
2. (2, 8)
3. (6, 1)

Find the magnitude and direction of the resultant of the sum of the two given vectors in complex notation. Round as above.

4. $3 + 5i$ and $2 + 3i$
5. $1 + 4i$ and $6 + 8i$

Section 18–2

Find the related angle for the angles.

6. 115°
7. 3.04 rad
8. 4.75 rad
9. 221°
10. 305°
11. 5.4 rad
12. 138.5°
13. 212°15′10″

Using a calculator, find the values of the functions. Round to four significant digits.

14. $\cos 250°$
15. $\sin 2.1$
16. $\tan 175°$
17. $\sin 340°$
18. $\tan 4.5$
19. $\cos 290°$
20. $\sin \dfrac{3\pi}{4}$
21. $\tan \dfrac{5\pi}{4}$

22. Find the magnitude and direction of a vector of an electrical current in standard position if the coordinates of the end point of the vector are $(2, -3)$. Round the magnitude (in amps) to the nearest hundredth. Express the direction in degrees to the nearest 0.1°.

23. Find the direction and magnitude of a vector in standard position if the coordinates of the end point of the vector are $(-2, 2)$. Express its direction in radians and round its direction and its magnitude to the nearest hundredth.

Section 18–3

Solve the triangles. If a triangle has two possibilities, find both solutions. Round sides to tenths and angles to the nearest 0.1°.

24. $A = 60°, B = 40°, b = 20$

25. $B = 120°, C = 20°, a = 8$

26. $A = 60°, B = 60°, a = 10$

27. $a = 5, c = 7, C = 45°$

28. $b = 10, c = 8, B = 52°$

29. $a = 9.2, b = 6.8, B = 28°$

30. A surveyor needs the measure of *JK* in Fig. 18–49. Find *JK* to the nearest foot.

31. In Fig. 18–50, find *RS*.

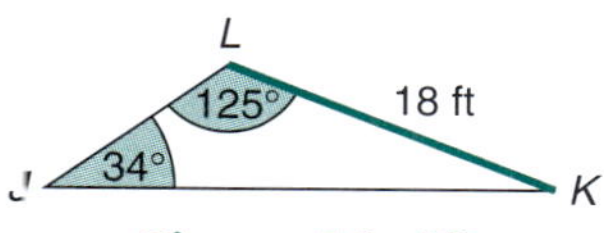

Figure 18–49

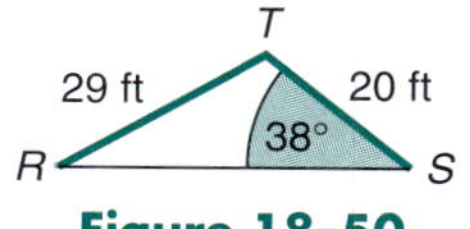

Figure 18–50

Section 18–4

Solve the triangles in Figs. 18–51 to 18–56. Round sides to four significant digits and angles to the nearest 0.1°.

32.

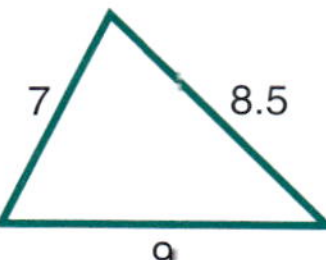

Figure 18–51

33.

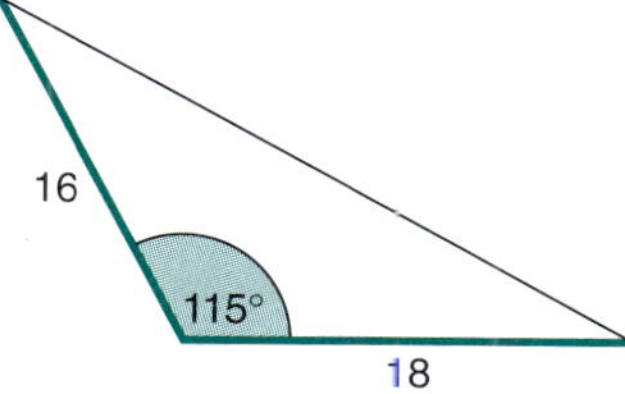

Figure 18–52

34.

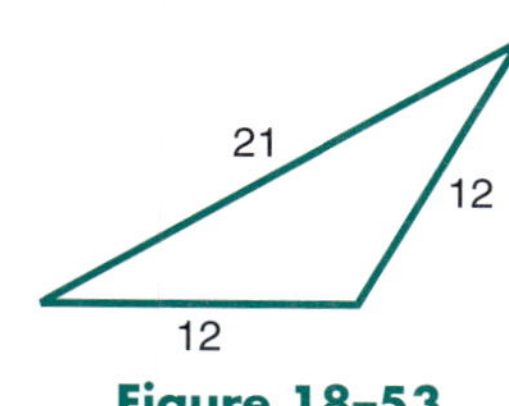

Figure 18–53

35.

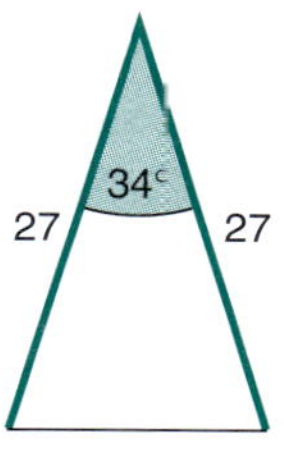

Figure 18–54

36.

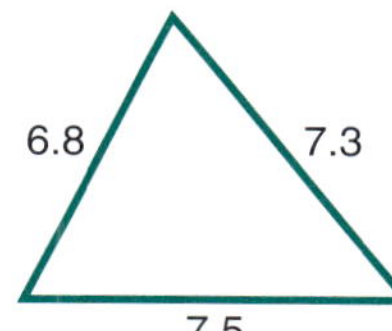

Figure 18–55

37.

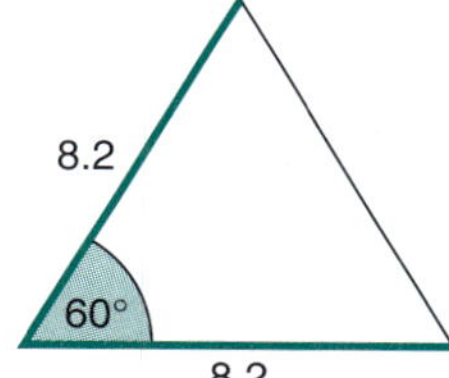

Figure 18–56

Solve the problems. Round sides to four significant digits and angles to the nearest 0.1°.

38. A triangular lot has sides 180 ft long, 160 ft long, and 123.5 ft long. Find the angles of the lot.

39. A vertical 50-ft pole stands on top of a hill inclined 18° to the horizontal. What length of wire is needed to reach from a point 6 ft from the pole's top to a point 75 ft downhill from the base of the pole?

40. A ship sails from a harbor 35 nautical miles east, then 42 nautical miles in a direction 32° south of east (Fig. 18–57). How far is the ship from the harbor?

41. A hill with a 35° grade (inclined to the horizontal) is cut down for a roadbed to a 10° grade. If the distance from the base to the top of the original hill is 800 ft (Fig. 18–58), how many vertical feet will be removed from the top of the hill, and what is the distance from the bottom to the top of the hill for the roadbed?

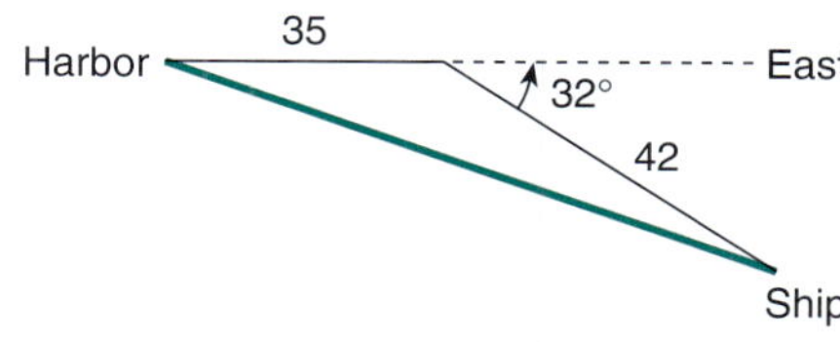

Figure 18–57

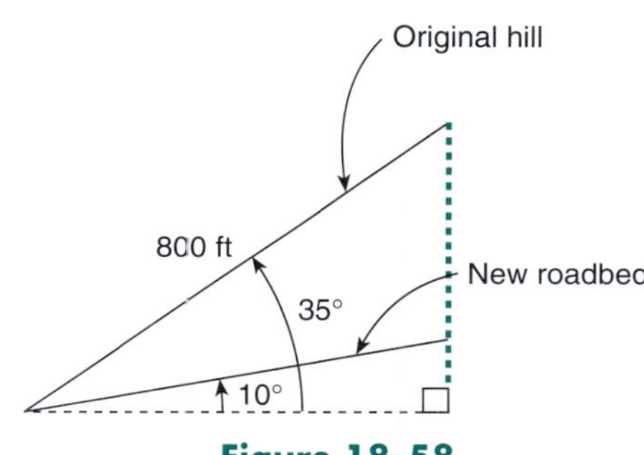

Figure 18–58

Find the values. Round to four significant digits.

1. sin 125°
2. tan 140°
3. cos 160°
4. Find the length of the vector that has end point coordinates of (8, 15).
5. Find the angle of the vector whose end point coordinates are $(-8, 8)$.

Use the law of sines or the law of cosines to find the side or angle indicated in Figs. 18–59 to 18–66. Round sides to four significant digits and angles to the nearest 0.1°.

6.

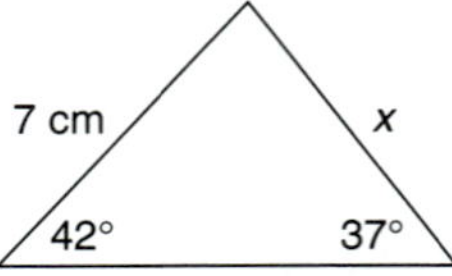

Figure 18–59

7.

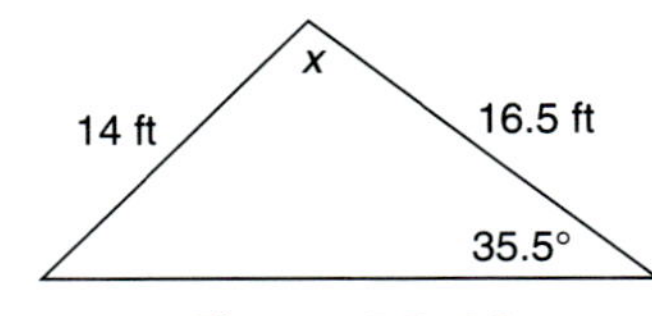

Figure 18–60

8.

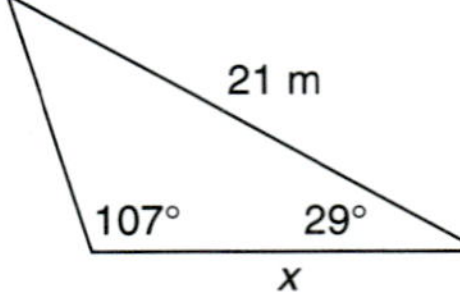

Figure 18–61

9.

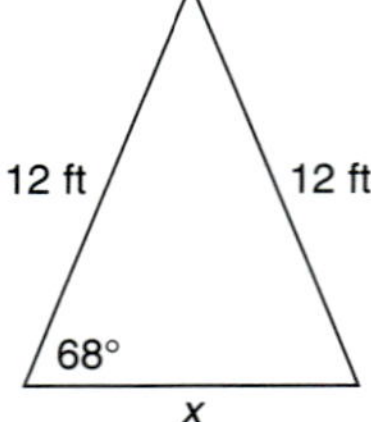

Figure 18–62

10.

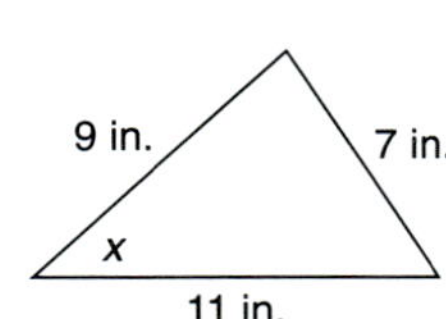

Figure 18–63

11.

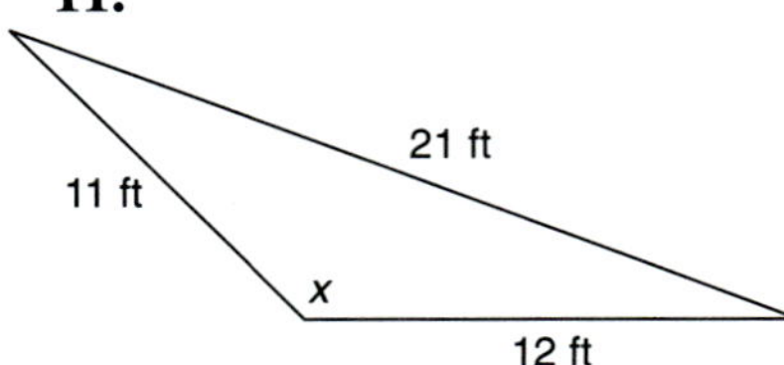

Figure 18–64

12.

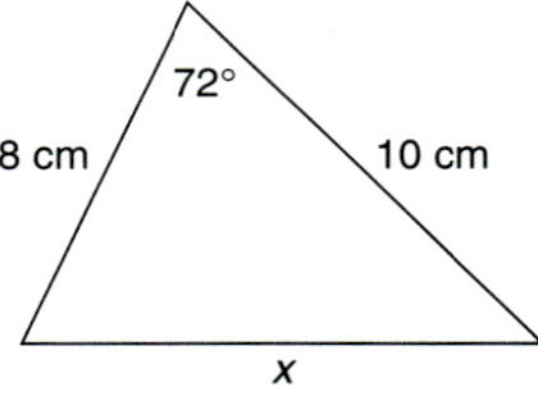

Figure 18–65

13.

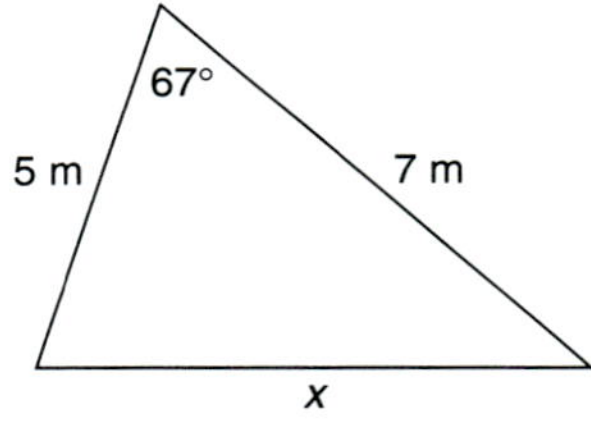

Figure 18–66

14. Find the direction in degrees (to the nearest 0.1°) of the vector I_t (total current) in Fig. 18–67.

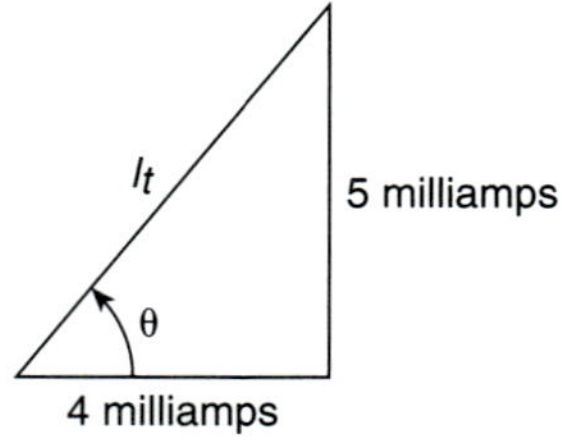

Figure 18–67

15. A surveyor measures two sides of a triangular lot and the angle formed by these two sides. What is the length of the third side if the two measured sides are 76 ft and 110 ft and the angle formed by these two sides is 107°?

16. What is the area of a triangular cast if the sides measure 31 mm, 42 mm, and 27 mm?

18. A plane is flown from an airport due east for 32 mi, and then turns 15° north of east and travels 72 mi. How far is the plane from the airport?

20. A connecting rod 30 cm long joins a crank 20 cm long to form a triangle with a third, imaginary line (Fig. 18–68). If the crank and the imaginary line form an angle of 150°, find the angle formed by the connecting rod and the imaginary line.

17. Find the area of a triangular flower bed if two sides measure 12 ft and 15 ft, and the included angle measures 48°.

19. Find the magnitude of the vector I_t in Exercise 14 to the nearest tenth.

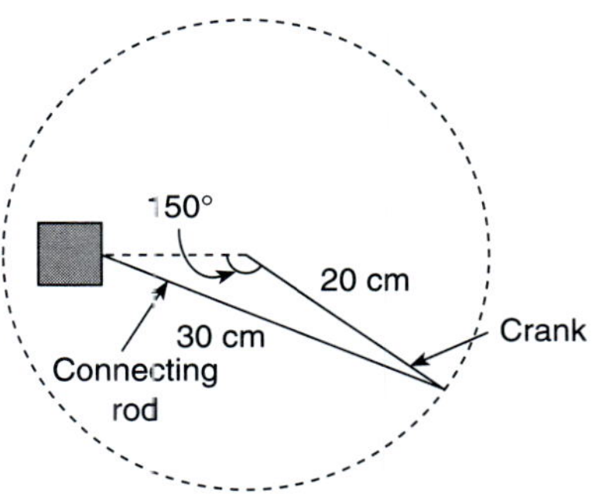

Figure 18–68

21. Find the magnitude of vector Z in Fig. 18–69 if vectors R and X_L form a right angle. All units are in ohms.

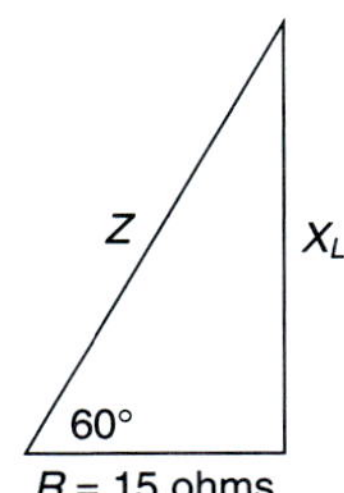

Figure 18–69

Selected Answers to Student Exercise Material

Answers to all Self-Study Exercises and answers to odd-numbered Assignment Exercises and odd-numbered problems in the Chapter Trial Tests are included here. Answers to all even-numbered exercises and problems and the concepts analysis questions may be found in the Instructor's Resource Manual.

Self-Study Exercises 1–1

1   **1.** hundred millions   **2.** ten thousands   **3.** hundreds   **4.** ten millions   **5.** billions
6. hundred thousands   **7.** tens   **8.** ten millions   **9.** billions   **10.** ten thousands

2   **11.** six thousand, seven hundred, four   **12.** eighty-nine thousand, twenty-one   **13.** six hundred sixty-two million, nine hundred thousand, seven hundred fourteen   **14.** three million, one hundred one
15. fifteen billion, four hundred seven million, two hundred ninety-four thousand, three hundred seventy-six.   **16.** one hundred fifty   **17.** 7,000,000,400   **18.** 1,627,106   **19.** 58,201   **20.** $1,006
21. seven hundred eighteen   **22.** forty two   **23.** one thousand, nine hundred eighty three   **24.** eight thousand, twenty one   **25.** fifty two thousand, ten   **26.** seven hundred   **27.** 537   **28.** 9,008
29. 65   **30.** 40,217

3   **31.** thousandths   **32.** ones   **33.** hundred-thousandths   **34.** millionths   **35.** ten-thousandths
36. 4   **37.** 0   **38.** 6   **39.** 3   **40.** 7

4   **41.** twenty-one and three hundred eighty-seven thousandths   **42.** four hundred twenty and fifty-nine thousandths   **43.** eighty-nine hundredths   **44.** five hundred sixty-eight ten-thousandths   **45.** thirty and two thousand three hundred seventy-nine hundred-thousandths   **46.** twenty-one and two hundred five thousand eighty-five millionths   **47.** 3.42   **48.** 78.195   **49.** 500.0005   **50.** 0.75034

5   **51.** 0.5   **52.** 0.23   **53.** 0.07   **54.** 6.83   **55.** 0.079   **56.** 0.468   **57.** 5.87   **58.** 0.108
59. 6.03   **60.** 4.00

6   **61.** 3.72   **62.** 7.08   **63.** 0.3   **64.** 0.56   **65.** 2.75   **66.** 0.2   **67.** 8.88   **68.** 0.25
69. 0.913   **70.** 0.76   **71.** 5.983   **72.** 1.972   **73.** 0.179, 0.23, 0.314   **74.** 1.87, 1.9, 1.92
75. 72.07, 72.1, 73   **76.** 0.837 in.   **77.** the reading   **78.** yes   **79.** 0.04 in.   **80.** No. 10 wire
81. 42 > 38; 38 < 42   **82.** 168 > 160; 160 < 168   **83.** 709 < 721; 721 > 709   **84.** 5,297 > 984; 984 < 5,297   **85.** 2,742 < 27,420; 27,420 > 2,742   **86.** 150,000 > 134,812; 134,812 < 150,000
87. 119 > 62; 62 < 119; The distance from Cleveland to Toledo is greater than the distance from Detroit to Toledo.   **88.** 96° > 82°; 82° < 96°; The average temperature in July at Laredo is higher than the average temperature at El Paso.   **89.** 26° < 70°; 70° > 26°   **90.** 57° < 71°; 71° > 57°   **91.** San Francisco 1906, 8.3   **92.** 1.14 in 1980 > 1.02 in 1990; falling growth rate   **93.** 1.7 in Asia < 2.0 in dev. world; Asia growing slower than dev. world   **94.** 0.139 Rolling 5 < 0.167 Rolling 6; 6
95. 2.4 kW > 1.21 kW; 100 watt bulb   **96.** 0.394 < 0.621; 0.621

7   **97.** 500   **98.** 6,200   **99.** 8,000   **100.** 430,000   **101.** 40,000   **102.** 40,000
103. 285,000,000   **104.** 470   **105.** 83,000,000,000   **106.** 300,000,000   **107.** Corpus Christi 40, Jacksonville 20; tens   **108.** Denver 5,300, Salt Lake City 4,200; hundreds   **109.** Indianapolis 700, Salt Lake City 4,200; hundreds   **110.** Answers may vary. Compare the number of digits for the altitude in the pair of cities. Round to the same place when comparing each pair.   **111.** 43   **112.** 367   **113.** 8
114. 103   **115.** 3   **116.** 8.1   **117.** 12.9   **118.** 42.6   **119.** 83.2   **120.** 6.0   **121.** 7.04

122. 42.07   **123.** 0.79   **124.** 3.20   **125.** 7.77   **126.** 0.217   **127.** 0.020   **128.** 1.509
129. 4.238   **130.** 7.004   **131.** $219   **132.** $83   **133.** $507   **134.** $3   **135.** $6   **136.** $8.24
137. $0.29   **138.** $0.53   **139.** $5.80   **140.** $238.92   **141.** 0.784   **142.** 3.82 in.   **143.** 2.8 A

8   **144.** 500   **145.** 8   **146.** 60   **147.** 0.5   **148.** 0.009   **149.** 0.1   **150.** 3   **151.** 50
152. 80   **153.** 50   **154.** $3   **155.** 20   **156.** 0.4, 2, 0.5, 1, 0.07, 2   **157.** 0.03, 0.03, 0.02, 0.03, 0.04

Self-Study Exercises 1–2

1   **1.** 15   **2.** 27   **3.** 22   **4.** 36   **5.** 37 bolts   **6.** 9,192   **7.** 16,956   **8.** 106,285   **9.** 2,310
10. 97,614   **11.** $456   **12.** 24,349 lb   **13.** 298 screws   **14.** 4,553 bricks   **15.** 50 gal

2   **16.** 15.7   **17.** 34.18   **18.** 8.13   **19.** 87.4   **20.** 129.97   **21.** 26.03   **22.** 78.2   **23.** 45.7
24. 261.335   **25.** 356.612   **26.** 6.984   **27.** 9.525   **28.** 1,126.6   **29.** 413.6   **30.** 18.1
31. 18.8   **32.** 0.81805   **33.** 1.17642   **34.** 55.513   **35.** 85.411   **36.** 69.987   **37.** 3.077 in.
38. 15.503 A   **39.** $181.25   **40.** 391.4 ft   **41.** $19.93   **42.** $5.04   **43.** 100.9   **44.** $28,334.01
45. $224.2800   **46.** 3.8 million   **47.** $2,189.45   **48.** 36,901.9   **49.** $16.26   **50.** $304.97
51. 30,900; 30,986.41   **52.** 4,700; 4,740.80   **53.** 13,100; 13,157.42   **54.** 70; 72.3

3   **55.** 17,100; 17,018.21   **56.** 84,200; 84,213   **57.** 402,300; 402,199   **58.** $3,440; $3,443.60
59. $2,300; $2,260   **60.** 8,900; 9,239   **61.** 600; 624.18   **62.** 9,000; 8,449   **63.** 800; 801.39
64. 60,000; 52,801   **65.** 50,000; 49,241   **66.** 159 lb   **67.** Yes, total capacity available is 115 gal.
68. Yes, 479 pages needed   **69.** 350 ft

4   **70.** 190,000; 190,786   **71.** 1,900,000; 1,859,867   **72.** 50,000; 54,359.65   **73.** 51,000,000; 47,520,014   **74.** $17,039.04

Self-Study Exercises 1–3

1   **1.** 5   **2.** 1   **3.** 7   **4.** 2   **5.** 3   **6.** 7   **7.** 5   **8.** 2   **9.** 4   **10.** 1   **11.** 24   **12.** 401
13. 1,020   **14.** 115   **15.** 53,036   **16.** 22 bags   **17.** 341 boxes   **18.** 321 ft

2   **19.** 6.93   **20.** 15.834   **21.** 803.693   **22.** 56.14   **23.** 4.094   **24.** 3.9   **25.** 291.82
26. 7.5   **27.** 310.8   **28.** 4.4   **29.** 12.7   **30.** 5°   **31.** 15.1 lb   **32.** 12.08 in., 12.10 in.
33. 59.83 cm   **34.** 4.189 in., 4.201 in.   **35.** 0.22 dm   **36.** 11.55 A   **37.** $8.75   **38.** $2.25
39. $4.75

3   **40.** 45   **41.** 608   **42.** 1,573   **43.** 22,205   **44.** 100,000; 91,034   **45.** 2,000; 2,088.884
46. 0; 174   **47.** 200; 599.15

4   **48.** 4,000; 4,397   **49.** 1,000; 1,187.251   **50.** 500,000; 508,275   **51.** 17 L   **52.** 186 bricks
53. 213.8 in.   **54.** 125.5 in.   **55.** 45 in.

Self-Study Exercises 1–4

1   **1.** (a) 15   (b) 56   (c) 63   (d) 24   **2.** $42   **3.** commutative property of multiplication
4. Numbers may be grouped in any way for multiplication.
   $3[(5)(2)] = [3(5)]2$
    $3(10) = (15)2$
     $30 = 30$ Answers may vary.
5. 0   **6.** 378   **7.** 84   **8.** 0   **9.** 224   **10.** 581   **11.** 630   **12.** 102,612   **13.** 864 pieces
14. 4,096 washers   **15.** 84 books   **16.** 700 tickets   **17.** $288   **18.** 249 students

2   **19.** 15.486   **20.** 56.55   **21.** 3.2445   **22.** 0.05805   **23.** 0.08672   **24.** 7.141   **25.** 0.0834
26. 170.12283   **27.** 0.38381871   **28.** 4.9386274   **29.** 30.66   **30.** 596.97   **31.** 50.7357
32. 38.6232   **33.** 339.04   **34.** 540.27   **35.** 254   **36.** 184.2   **37.** 0.9307   **38.** 0.7602
39. 0.58635   **40.** 0.73265   **41.** 9.21702   **42.** 11.42356   **43.** 0.0176   **44.** 0.027045

45. 0.915371   **46.** 0.390483   **47.** 0.00015   **48.** 0.00056   **49.** 3.957 in.   **50.** 5.25 in.
51. $27.48   **52.** 151.2 in.   **53.** $64.20   **54.** $10,728   **55.** 0.375 in.   **56.** $1,960.50   **57.** 15 A
58. $78

3   **59.** 55   **60.** 60   **61.** 196   **62.** 12   **63.** 15.6   **64.** 5.6   **65.** $P = 40$ ft   **66.** $P = 61$ in.
67. 56 ft   **68.** 170 ft

4   **69.** $600; $768   **70.** 123.54 ft^2 **71.** 1,164 cm^2 **72.** 1,200 ft^2; 1,374.75 ft^2 **73.** 96 ft^2

5   **74.** 3,578,040.664   **75.** 23,379,045   **76.** 561,500,160   **77.** 0   **78.** $36,180   **79.** $22,006.40
80. 147,000,000,000   **81.** 16,755,200,000   **82.** 420,000,000

Self-Study Exercises 1–5

1   **1.** $8 \div 4; 4\overline{)8}; \dfrac{8}{4}$ **2.** $9 \div 3; 3\overline{)9}; \dfrac{9}{3}$ **3.** $24 \div 6; 6\overline{)24}; \dfrac{24}{6}$ **4.** $30 \div 7; 7\overline{)30}; \dfrac{30}{7}$ **5.** $6 \div 2; 2\overline{)6}; \dfrac{6}{2}$

6. $8 \div 4; 4\overline{)8}; \dfrac{8}{4}$

2   **7.** 75   **8.** 23   **9.** 20   **10.** 3   **11.** 16   **12.** 12   **13.** 43   **14.** 47   **15.** 12   **16.** 23R1
17. 124R30   **18.** 56R80   **19.** 32 ft   **20.** $1,245   **21.** 6 in.

3   **22.** 1.26   **23.** 3.09   **24.** 0.063   **25.** 285   **26.** 5.9   **27.** 45   **28.** 10.9   **29.** 0.19
30. 1.06   **31.** 0.33   **32.** 25   **33.** 20,700   **34.** 90,200   **35.** 10,700   **36.** 1.8   **37.** 6 lb
38. 23 rolls   **39.** 600 revolutions   **40.** 0.7 ft   **41.** 93.75 volts   **42.** A measure divided by a number
equals a measure.   **43.** $19.84\dfrac{8}{13}$ **44.** $1.82\dfrac{1}{12}$ **45.** $2,601.16\dfrac{2}{3}$ **46.** $1.98\dfrac{4}{7}$ **47.** $0.07\dfrac{27}{29}$
48. 169.3   **49.** 0.16   **50.** 9   **51.** $0.96   **52.** $1   **53.** $575   **54.** $1.98   **55.** 0.0812 in.
56. 0.2 in.

4   **57.** 100; 125.6   **58.** 4,000; 3,151R90   **59.** 4,000; 3,731R53   **60.** 8; 8.43   **61.** 128 loads
62. 127 cords

5   **63.** 80.4°   **64.** 85   **65.** 454 lb   **66.** $765.32   **67.** 1.69 in.   **68.** 3.5 A

6   **69.** 54   **70.** 301   **71.** 40   **72.** 89,500   **73.** 40,200   **74.** 6,201   **75.** 1,070 bricks
76. $10,500   **77.** $104.75   **78.** $1,950   **79.** $2,465   **80.** $4.45   **81.** 24 parts   **82.** 10 lengths
83. $78,505   **84.** $59.71   **85.** 0.7 in.

Self-Study Exercises 1–6

1   **1.** 4; 3   **2.** 9; 4   **3.** 2.7; 9   **4.** 3.375   **5.** 49   **6.** 1,000   **7.** 16   **8.** 11.56   **9.** 15
10. 8   **11.** 1   **12.** 8^1 **13.** 14.5^1 **14.** 12^1 **15.** 23^1 **16.** leaves base unchanged   **17.** changes
value to 1; 0^0 is undefined

2   **18.** 64   **19.** 4   **20.** 12.25   **21.** 1.96   **22.** 169   **23.** 1   **24.** 10,000   **25.** 64   **26.** 81
27. 289   **28.** 324   **29.** 10,201   **30.** 484   **31.** 5   **32.** 7   **33.** 9   **34.** 14   **35.** Use it as a
factor two times.   **36.** Find a number that is used as a factor two times to give the desired number.

3   **37.** $10^3 = 1,000$ **38.** 32,000   **39.** 30,000   **40.** 20,000   **41.** 10,200   **42.** 22,000   **43.** 25
44. 21   **45.** 3   **46.** 9   **47.** 250   **48.** 1.2   **49.** Shift the decimal to the right in the number being
multiplied by the power of ten as indicated by the exponent.   **50.** Shift the decimal to the left in the
number being multiplied by the power of ten as indicated by the exponent.

4   **51.** 225   **52.** 343   **53.** 78,125   **54.** 18   **55.** 28   **56.** 33

Self-Study Exercises 1–7

1   **1.** 26   **2.** 18   **3.** 9   **4.** 12.5   **5.** 24   **6.** 32   **7.** 30   **8.** 15   **9.** 156   **10.** 109
11. 23   **12.** 145   **13.** 15.6   **14.** 25.04   **15.** 9   **16.** 5   **17.** 7   **18.** 15   **19.** 160
20. 1,458

Assignment Exercises, Chapter 1

1. (a) 0.3 (b) 0.15 (c) 0.04   **3.** thousandths   **5.** tenths   **7.** (a) tens (b) ten-thousands
(c) hundreds; millions   **9.** fifty-six million, one hundred nine thousand, one hundred ten
11. 1,265,401   **13.** six and eight hundred three thousandths   **15.** 0.625   **17.** (a) 40 (b) 70 (c) 24
(d) $43 (e) 80 (f) $8.94 (g) $1.00 (h) 0.0970   **19.** (a) 320 (b) 7,000 (c) 500 (d) 50,000
(e) 27,000,000,000 (f) 41.4 (g) 6.90 (h) 23.4610   **21.** 4.79   **23.** 0.02; 0.021; 0.0216   **25.** $\dfrac{7}{8}$

27. (a) 23 (b) 18 (c) 28 (d) 25   **29.** 34.9 kW   **31.** (a) 29,000; 29,092.09 (b) 36,000; 36,048
33. (a) 4.61 (b) 3.127 (c) 2 (d) 5 (e) 0 (f) 204.899 (g) 144 (h) 12,140   **35.** 8.291 in.;
8.301 in.   **37.** 200 miles, 190 miles   **39.** 0.430   **41.** 8.930 in.; 8.940 in.   **43.** 84   **45.** 1,143
47. 13,725   **49.** 3,349,890   **51.** 394,254,080   **53.** $1,407   **55.** $43,920   **57.** 20   **59.** 60
61. 80   **63.** $328   **65.** $14,800, $13,140   **67.** 1,140,000 ft^2; 1,204,010.28 ft^2 **69.** 0.12096 in.

71. $5 \div 3, 3\overline{)5}, \dfrac{5}{3}$ **73.** 1   **75.** not defined   **77.** 5.26875   **79.** 13   **81.** 2,008.4   **83.** 23 R11

85. 5.8375   **87.** 5; 5.52   **89.** $10; $10.63   **91.** 50; 48 79 ft   **93.** (a) 7, 3, 343 (b) 2.3, 4, 27.9841
(c) 8, 4, 4,096   **95.** (a) 0.9 (b) 35 (c) 1   **97.** (a) 1 (b) 1 (c) 1 (d) 1   **99.** (a) 1
(b) 15,625 (c) 31.36 (d) 441   **101.** (a) 10^1 (b) 10^3 (c) 10^4 (d) 10^5 **103.** (a) 7
(b) 0.04056 (c) .605 (d) 2.3079 (e) 44.582   **105.** 75   **107.** 3   **109.** 13.6   **111.** 113.608
113. 49 boxes   **115.** 16,309 people   **117.** $12,880   **119.** 45 yd × 48 yd   **121.** (a) 0.8 (b) 0.3
(c) 0.03 (d) 0.25 (e) 0.004 (f) 0.63245 . . . or 0.6325 (rounded)

Trial Test, Chapter 1

1. five million, thirty thousand, one hundred two   **3.** 7.027   **5.** 2,700   **7.** 5.09   **9.** 48.3
11. 1,007   **13.** $9,271,314   **15.** 134   **17.** 106   **19.** 0.0086   **21.** $310, $310   **23.** $10, $14.00
25. Commutative means we can add numbers in any order. 2 + 4 = 6; 4 + 2 = 6. Associative means we
can group numbers differently. (2 + 1) + 3 = 6; 2 + (1 + 3) = 6.   **27.** 42,730   **29.** 11.6   **31.** 83
33. 0.6   **35.** 70 or between 70 and 80   **37.** $17,500

Self-Study Exercises 2–1

1 **1.** right   **2.** left   **3.** infinity   **4.** 1   **5.** neither   **6.** −3, −2, −1, 0, 1, 2, 3   **7.** to the right
8. to the left   **9.**   **10.**

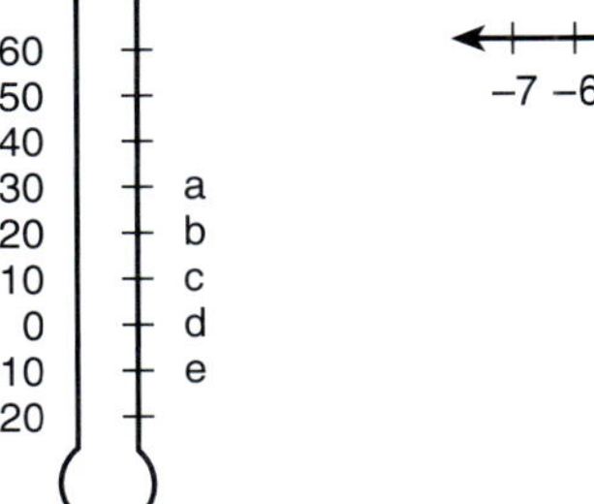

2 **11.** <   **12.** >   **13.** <   **14.** <   **15.** >   **16.** <   **17.** >   **18.** <   **19.** $x < y$
20. $a > b$ **21.** 3 + 5 > 6   **22.** 9 < 18 − 6   **23.** $k > t$ **24.** $r < s$

3 **25.** 23   **26.** 0   **27.** 10   **28.** −17   **29.** 13   **30.** 345   **31.** 67   **32.** 61

4 **33.** (−7, 0, 7)   **34.** (−8, 0, 8)   **35.** (−4, 0, 4)
36. (−12, 0, 12)   **37.** 0   **38.** a negative integer; a positive integer

5 **39.** *R*: horizontal, 4; vertical, 2   **40.** *S*: horizontal, -4; vertical, 3   **41.** *T*: horizontal, 5; vertical, 3
42. *U*: horizontal, -3; vertical, 5   **43.** *V*: horizontal, 5; vertical, 0   **44.** *W*: horizontal, 0; vertical, 5
45. *X*: horizontal, 0; vertical, 0   **46.** *Y*: horizontal, -3; vertical, 1
47. $A = (4, 2)$ $B = (-3, 2)$ $C = (-2, -1)$ $D = (3, -2)$ **48.–53.**
54. *y*-value $= 0$ **55.** *x*-value $= 0$ **56.** $(-x, +y)$ **57.** $(+x, -y)$

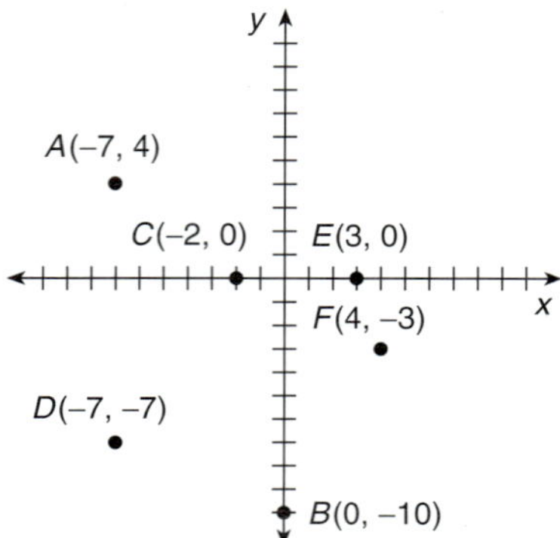

Self-Study Exercises 2–2

1 **1.** 17   **2.** -13 **3.** 99   **4.** -59 **5.** -48 **6.** -161

2 **7.** -6 **8.** 4   **9.** -2 **10.** 2   **11.** -10 **12.** -40 **13.** 7   **14.** 15   **15.** -8 **16.** -33
17. 9   **18.** -11 **19.** 13   **20.** 9   **21.** 3   **22.** 0   **23.** 0   **24.** 0   **25.** 0   **26.** 0   **27.** 9
28. -3 **29.** -7 **30.** 18   **31.** -8 **32.** -28 **33.** $5 + (-5) = 0$ or other example
34. 0 (zero)   **35.** Adding numbers of like signs increases the absolute value whether positive or negative.
36. \$36   **37.** \$66   **38.** $+2{:}00$ **39.** $-8{:}00$ **40.** $-23°$

Self-Study Exercises 2–3

1 **1.** -12 **2.** 6   **3.** -6 **4.** 4   **5.** -25 **6.** -3 **7.** 8   **8.** -3 **9.** -9 **10.** 13
11. -1 **12.** -8

2 **13.** 15   **14.** -8 **15.** -12 **16.** 8   **17.** 7   **18.** 10   **19.** 56   **20.** -92 **21.** 14
22. -36

3 **23.** -4 **24.** 1   **25.** 14   **26.** -3 **27.** -9 **28.** -5 **29.** 13   **30.** 2   **31.** -1
32. $107°$ **33.** $107°$ **34.** \$115,054   **35.** Subtracting zero from a number results in the same number
with the same sign. Subtracting a number from zero results in the opposite of the number.   **36.** $+3{:}00$

Self-Study Exercises 2–4

1 **1.** 40   **2.** 12   **3.** 35   **4.** 21   **5.** 24   **6.** 6   **7.** -15 **8.** -10 **9.** -32 **10.** -12
11. -56 **12.** -24 **13.** $4 \cdot (-28) = -\$112$ **14.** $(7)(-40) = -\$280$ **15.** When multiplying
signed numbers, the absolute values are always multiplied. When numbers with like signs are multiplied, the
result is positive. When numbers with unlike signs are multiplied, the result is negative.

2 **16.** -60 **17.** 36   **18.** 0   **19.** 90   **20.** 0   **21.** -42 **22.** 54   **23.** -210 **24.** 2,268
25. $-20,160$

3 **26.** 0   **27.** 0   **28.** 0   **29.** 0   **30.** 0   **31.** 0   **32.** 0   **33.** 0   **34.** 0   **35.** 0   **36.** -30
37. 168   **38.** The multiplicative inverse of a number is the number that, when multiplied by the original
number, results in 1, the multiplicative identity.

4 **39.** 9   **40.** -8 **41.** 25   **42.** -25 **43.** -8 **44.** 1   **45.** 625   **46.** 81   **47.** 1   **48.** 1
49. 604,800   **50.** 1,209,600   **51.** No, we don't normally use 000 as the first 3 digits.   **52.** 17,576,000
53. 10,000   **54.** $45°$

Self-Study Exercises 2–5

1 **1.** 2   **2.** -3 **3.** -4 **4.** -4 **5.** -4 **6.** 8   **7.** 5   **8.** -6 **9.** 8   **10.** 5
11. $\dfrac{-1,800}{6} = -\$300.00$

2 **12.** not defined **13.** not defined **14.** 0 **15.** 0 **16.** 0 **17.** not defined **18.** not defined
19. not defined **20.** 0 **21.** not defined **22.** multiplication and division **23.** $-10°$
24. $3,472.50

Self-Study Exercises 2–6

1 **1.** -4 **2.** -30 **3.** -9 **4.** 17 **5.** 20 **6.** 24 **7.** 8 **8.** 24 **9.** -5 **10.** -7
11. -3 **12.** 14

2 **13.** -12 **14.** $-1,260$ **15.** 27 **16.** -14 **17.** 28 **18.** -35 **19.** -29 **20.** -132
21. $-E-$; indicating that division by zero is not possible. **22.** Numbers can be added in any order.
23. When the order of subtraction is changed, the result is not the same. **24.** Numbers can be multiplied
in any order. **25.** The order of division of two numbers cannot be changed.

Assignment Exercises, Chapter 2

1. G **3.** F **5.** E **7.** A **9.** L **11.** M
13. 5, 21, 987 (Answers may vary.) **15.** 0
17. 1 **19.** $>$ **21.** $>$ **23.** $<$ **25.** 5
27. 3 **29.** 7 **31.** 11 **33.** 12 **35.** -15
37. 2 **39.** 42 **41.** -87
49. $(0, 0)$ **51.** $A(3, 0)$ $B(2, 2)$ $C(2, -5)$
$D(-4, -1)$ $E(-3, 1)$ **53.** 17 **55.** -7 **57.** -13
59. 0 **61.** -8 **63.** -6 **65.** run up the middle, 2 yd for
first down **67.** $569 **69.** The additive identity, 0, if added to any
number, will result in the same number. Numbers that are additive
inverses, when added, result in zero. **71.** -13 **73.** 14 **75.** -15
77. -5 **79.** 2 **81.** 36 **83.** $70°$ **85.** $-5.4°$ **87.** 40
89. -14 **91.** -12 **93.** 0 **95.** 0 **97.** -168
99. 343 **101.** -16 **103.** $-10°$ **105.** higher, 20, $-4 \times 5 = -20$ **107.** 4 **109.** 4 **111.** -4
113. undefined **115.** $15°$ **117.** 56 **119.** 1 **121.** 36 **123.** 73 **125.** -2 **127.** 391,464
129.

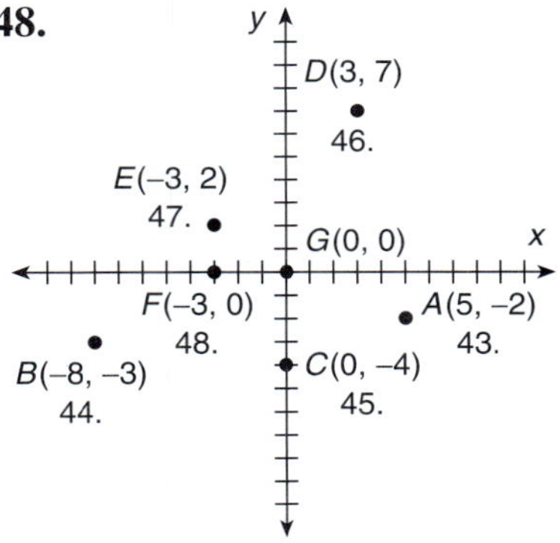

9:00 A.M.: $-1°$C	10:00 A.M.: $0°$C	11:00 A.M.: $0°$C	12:00 P.M.: $1°$C
1:00 P.M.: $1°$C	2:00 P.M.: $4°$C	3:00 P.M.: $0°$C	4:00 P.M.: $-7°$C
5:00 P.M.: $-15°$C	6:00 P.M.: $-27°$C		

43.–48.

Trial Test, Chapter 2

1. $<$ **3.** $<$ **5.** -8 **7.** 4 **9.** -48 **11.** 3 **13.** 8 **15.** 0 **17.** -7 **19.** not defined
21. -3 **23.** -2 **25.** 1 **27.** 33 **29.** 22 **31.** -5 **33.** If a number is multiplied by the
multiplicative identity, or if a number is added to the additive identity, the result is the original number.
$5 \times 1 = 5$ or $3 + 0 = 3$ **35.** $175°$

Self-Study Exercises 3–1

1 **1.** $\dfrac{5}{6}$ **2.** $\dfrac{3}{4}$ **3.** $\dfrac{3}{8}$ **4.** $\dfrac{7}{12}$ **5.** $\dfrac{1}{16}$ **6.** $\dfrac{3}{3}$ **7.** $\dfrac{4}{6}$ **8.** (a) 6 (b) 5 (c) 5 (d) 6 (e) 5; 6
(f) 6 (g) 5 (h) proper (i) 5; 6 (j) less than 1 **9.** (a) 8 (b) 8 (c) 8 (d) 8 (e) 8; 8 (f) 8
(g) 8 (h) improper (i) 8; 8 (j) equal to 1 **10.** (a) 5 (b) 11 (c) 11 (d) 5 (e) 11; 5 (f) 5
(g) 11 (h) improper (i) 11; 5 (j) greater than 1 **11.** (a) 3 (b) 12 (c) 12 (d) 3 (e) 12; 3
(f) 3 (g) 12 (h) improper (i) 12; 3 (j) greater than 1 **12.** b **13.** a **14.** f **15.** b **16.** d
17. d **18.** a **19.** e **20.** c **21.** c **22.** c **23.** $6\dfrac{27}{100}$ **24.** $\dfrac{3}{10}$ **25.** $3\dfrac{7}{10}$ **26.** $1\dfrac{53}{100}$

Self-Study Exercises 3–2

1 **1.** $5 = 5 \times 1, 10 = 5 \times 2, 15 = 5 \times 3, 20 = 5 \times 4, 25 = 5 \times 5, 30 = 5 \times 6$ **2.** $6 = 6 \times 1,$
$12 = 6 \times 2, 18 = 6 \times 3, 24 = 6 \times 4, 30 = 6 \times 5, 36 = 6 \times 6$ **3.** $8 = 8 \times 1, 16 = 8 \times 2,$

$24 = 8 \times 3, 32 = 8 \times 4, 40 = 8 \times 5, 48 = 8 \times 6$ **4.** $9 = 9 \times 1, 18 = 9 \times 2, 27 = 9 \times 3, 36 = 9 \times 4,$ $45 = 9 \times 5, 54 = 9 \times 6$ **5.** $10 = 10 \times 1, 20 = 10 \times 2, 30 = 10 \times 3, 40 = 10 \times 4, 50 = 10 \times 5,$ $60 = 10 \times 6$ **6.** $30 = 30 \times 1, 60 = 30 \times 2, 90 = 30 \times 3, 120 = 30 \times 4, 150 = 30 \times 5, 180 = 30 \times 6$ **7.** $5 \times 1 = 5, 5 \times 2 = 10, 5 \times 3 = 15, 5 \times 4 = 20, 5 \times 5 = 25$ **8.** $12 \times 1 = 12, 12 \times 2 = 24,$ $12 \times 3 = 36, 12 \times 4 = 48, 12 \times 5 = 60$ **9.** $7 \times 1 = 7, 7 \times 2 = 14, 7 \times 3 = 21, 7 \times 4 = 28,$ $7 \times 5 = 35$ **10.** $3 \times 1 = 3, 3 \times 2 = 6, 3 \times 3 = 9, 3 \times 4 = 12, 3 \times 5 = 15$ **11.** $50 \times 1 = 50,$ $50 \times 2 = 100, 50 \times 3 = 150, 50 \times 4 = 200, 50 \times 5 = 250$ **12.** $4 \times 1 = 4, 4 \times 2 = 8, 4 \times 3 = 12,$ $4 \times 4 = 16, 4 \times 5 = 20$

2 **13.** no; 405 remainder 4 **14.** yes; $230 \div 5 = 46$ **15.** no, 608 remainder 2 **16.** yes; $1{,}221 \div 3 = 407$ **17.** yes; $756 \div 7 = 108$ **18.** yes; $920 \div 8 = 115$ **19.** yes; $621 \div 3 = 207$ **20.** yes; $426 \div 6 = 71$ **21.** yes; $1{,}232 \div 2 = 616$

3 **22.** $1 \cdot 24, 2 \cdot 12, 3 \cdot 8, 4 \cdot 6$ **23.** $1 \cdot 36, 2 \cdot 18, 3 \cdot 12, 4 \cdot 9, 6 \cdot 6$ **24.** $1 \cdot 45, 3 \cdot 15, 5 \cdot 9$ **25.** $1 \cdot 32, 2 \cdot 16, 4 \cdot 8$ **26.** $1 \cdot 16, 2 \cdot 8, 4 \cdot 4$ **27.** $1 \cdot 27, 3 \cdot 9$ **28.** $1 \cdot 20, 2 \cdot 10, 4 \cdot 5$ **29.** $1 \cdot 30, 2 \cdot 15, 3 \cdot 10, 5 \cdot 6$ **30.** $1 \cdot 12, 2 \cdot 6, 3 \cdot 4$ **31.** $1 \cdot 8, 2 \cdot 4$ **32.** $1 \cdot 4, 2 \cdot 2$ **33.** $1 \cdot 15, 3 \cdot 5$ **34.** $1 \cdot 81, 3 \cdot 27, 9 \cdot 9$ **35.** $1 \cdot 64, 2 \cdot 32, 4 \cdot 16, 8 \cdot 8$ **36.** $1 \cdot 38, 2 \cdot 19$ **37.** $1 \cdot 46, 2 \cdot 23$ **38.** $1 \cdot 51, 3 \cdot 17$ **39.** $1 \cdot 18, 2 \cdot 9, 3 \cdot 6$ **40.** $1 \cdot 72, 2 \cdot 36, 3 \cdot 24, 4 \cdot 18,$ $6 \cdot 12, 8 \cdot 9$

Self-Study Exercises 3–3

1 **1.** $1 \cdot 14, 2 \cdot 7; 1, 2, 7, 14$ **2.** $1 \cdot 22, 2 \cdot 11; 1, 2, 11, 22$ **3.** $1 \cdot 11; 1, 11$ **4.** $1 \cdot 17; 1, 17$ **5.** $1 \cdot 18, 2 \cdot 9, 3 \cdot 6; 1, 2, 3, 6, 9, 18$ **6.** $1 \cdot 24, 2 \cdot 12, 3 \cdot 8, 4 \cdot 6; 1, 2, 3, 4, 6, 8, 12, 24$

2 **7.** $2 \cdot 5 \cdot 5$ **8.** $2 \cdot 2 \cdot 13$ **9.** $3 \cdot 3 \cdot 5 \cdot 5$ **10.** $5 \cdot 5 \cdot 5$ **11.** $2 \cdot 2 \cdot 5 \cdot 5$ **12.** $2 \cdot 2 \cdot 2 \cdot 5 \cdot 5$ **13.** $5 \cdot 13$ **14.** $3 \cdot 5 \cdot 5$ **15.** $11 \cdot 11$ **16.** $2 \cdot 2 \cdot 2 \cdot 2 \cdot 3 \cdot 3$ **17.** $2^3 \cdot 71$ **18.** $2^4 \cdot 7$ **19.** $2^2 \cdot 31$ **20.** $2^2 \cdot 41$ **21.** $2^3 \cdot 3^2$ **22.** $2^2 \cdot 3^2 \cdot 5^2$

Self-Study Exercises 3–4

1 **1.** 6 **2.** 30 **3.** 56 **4.** 12 **5.** 180 **6.** 60 **7.** 24 **8.** 18 **9.** 24 **10.** 700 **11.** 27 **12.** 16 **13.** 90 **14.** 120 **15.** 180 **16.** 60 **17.** 96 **18.** 72 **19.** 60 **20.** 300 **21.** 66 **22.** 312

2 **23.** 6 **24.** 5 **25.** 1 **26.** 1 **27.** 2 **28.** 2 **29.** 15 **30.** 5 **31.** 12 **32.** 6

Self-Study Exercises 3–5

1 **1.** $\dfrac{4}{5}, \dfrac{8}{10}, \dfrac{12}{15}, \dfrac{16}{20}, \dfrac{20}{25}, \dfrac{24}{30}$ **2.** $\dfrac{7}{10}, \dfrac{14}{20}, \dfrac{21}{30}, \dfrac{28}{40}, \dfrac{35}{50}, \dfrac{42}{60}$ **3.** $\dfrac{3}{4} = \dfrac{18}{24}$ **4.** 6 **5.** 12 **6.** 36 **7.** 5 **8.** 20 **9.** 21 **10.** 12

2 **11.** $\dfrac{1}{2}$ **12.** $\dfrac{3}{5}$ **13.** $\dfrac{3}{4}$ **14.** $\dfrac{5}{16}$ **15.** $\dfrac{1}{2}$ **16.** $\dfrac{7}{8}$ **17.** $\dfrac{5}{16}$ **18.** $\dfrac{1}{4}$ **19.** $\dfrac{1}{4}$ **20.** $\dfrac{6}{25}$ **21.** $\dfrac{5}{8}$ **22.** $\dfrac{1}{4}$ **23.** $\dfrac{3}{4}$ **24.** $\dfrac{3}{16}$ **25.** $\dfrac{7}{32}$

3 **26.** $\dfrac{1}{2}$ **27.** $\dfrac{1}{10}$ **28.** $\dfrac{1}{5}$ **29.** $\dfrac{7}{10}$ **30.** $\dfrac{1}{4}$ **31.** $\dfrac{1}{40}$ **32.** $3\dfrac{9}{10}$ **33.** $4\dfrac{4}{5}$ **34.** $\dfrac{189}{500}$ **35.** $\dfrac{7}{8}$ **36.** $\dfrac{3}{8}$ **37.** $\dfrac{5}{8}$ **38.** $\dfrac{3}{4}$ **39.** $\dfrac{3}{16}$ **40.** $2\dfrac{3}{8}$ **41.** $\dfrac{5}{6}$ **42.** $\dfrac{5}{16}$ **43.** $3\dfrac{1}{8}$

4 **44.** 0.4 **45.** 0.3 **46.** 0.875 **47.** 0.625 **48.** 0.45 **49.** 0.98 **50.** 0.21 **51.** 3.875 **52.** 1.4375 **53.** 4.5625 **54.** $0.\overline{6}$ or $0.66\ldots$ **55.** $0.\overline{27}$ or $0.2727\ldots$ **56.** $0.\overline{7}$ or $0.77\ldots$ **57.** $0.\overline{384615}$ or $.384615\ldots$ **58.** $0.83\dfrac{1}{3}$ **59.** $0.58\dfrac{1}{3}$ **60.** 2.046875 in. **61.** 4.5% **62.** 0.125 in.

Self-Study Exercises 3–6

1   **1.** $2\frac{2}{5}$ **2.** $1\frac{3}{7}$ **3.** 1   **4.** $4\frac{4}{7}$ **5.** 4   **6.** $2\frac{1}{7}$ **7.** $2\frac{5}{9}$ **8.** $9\frac{2}{5}$ **9.** $9\frac{5}{9}$ **10.** $1\frac{17}{21}$

11. $3\frac{4}{5}$ **12.** 16   **13.** $7\frac{1}{5}$ **14.** $9\frac{1}{2}$ **15.** 9

2   **16.** $\frac{7}{3}$ **17.** $\frac{25}{8}$ **18.** $\frac{15}{8}$ **19.** $\frac{77}{12}$ **20.** $\frac{77}{8}$ **21.** $\frac{31}{8}$ **22.** $\frac{89}{12}$ **23.** $\frac{103}{16}$ **24.** $\frac{257}{32}$

25. $\frac{69}{64}$ **26.** $\frac{73}{10}$ **27.** $\frac{26}{3}$ **28.** $\frac{100}{3}$ **29.** $\frac{200}{3}$ **30.** $\frac{25}{2}$ **31.** 15   **32.** 18   **33.** 56   **34.** 32

35. 48

Self-Study Exercises 3–7

1   **1.** 72   **2.** 30   **3.** 50   **4.** 48   **5.** 24

2   **6.** $\frac{2}{3}$ **7.** $\frac{7}{16}$ **8.** $\frac{8}{9}$ **9.** $\frac{11}{16}$ **10.** $\frac{15}{32}$ **11.** $\frac{7}{12}$ **12.** $\frac{4}{5}$ **13.** $\frac{9}{10}$ **14.** $\frac{4}{15}$ **15.** $\frac{1}{2}$

16. No, $\frac{15}{64}$ is greater   **17.** Yes, $\frac{3}{8}$ is greater   **18.** Yes, $\frac{7}{16}$ is greater   **19.** Yes   **20.** $\frac{5}{8}$

21. No   **22.** No   **23.** Yes   **24.** No   **25.** too small   **26.** 0.03   **27.** 0.392   **28.** 5.38

29. 4.71   **30.** 98.6°   **31.** 0.0256 cm   **32.** 0.973 mills   **33.** Answers may vary. To compare fractions, find a common denominator and compare numerators. To compare decimals, compare the whole numbers. If they are equal, compare decimal digits by each place value.

Self-Study Exercises 3–8

1   **1.** $\frac{3}{8}$ **2.** $1\frac{3}{8}$ **3.** $\frac{5}{8}$ **4.** $\frac{25}{32}$ **5.** $\frac{9}{16}$ **6.** $1\frac{7}{16}$ **7.** $\frac{11}{64}$ **8.** $1\frac{19}{40}$ **9.** $1\frac{23}{36}$ **10.** $1\frac{1}{12}$

11. $\frac{15}{16}$ in.   **12.** $1\frac{27}{32}$ in.   **13.** $1\frac{15}{16}$ in.   **14.** $1\frac{7}{8}$ in.   **15.** $1\frac{1}{4}$ in.

2   **16.** $6\frac{4}{5}$ **17.** $4\frac{1}{8}$ **18.** $16\frac{13}{16}$ **19.** $1\frac{11}{18}$ **20.** $4\frac{13}{16}$ **21.** $5\frac{17}{32}$ **22.** $4\frac{11}{16}$ **23.** $13\frac{1}{2}$

24. $4\frac{7}{15}$ **25.** $12\frac{11}{16}$ **26.** $8\frac{13}{16}$ in.   **27.** $3\frac{15}{16}$ in.   **28.** $11\frac{5}{8}$ gal   **29.** $24\frac{5}{32}$ in.   **30.** $22\frac{7}{8}$ in.

Self-Study Exercises 3–9

1   **1.** $\frac{1}{4}$ **2.** $\frac{3}{16}$ **3.** $\frac{1}{16}$ **4.** $\frac{1}{8}$ **5.** $\frac{9}{64}$ **6.** $\frac{1}{8}$

2   **7.** $4\frac{11}{16}$ **8.** $17\frac{3}{4}$ **9.** $4\frac{15}{16}$ **10.** $5\frac{21}{32}$ **11.** $8\frac{13}{16}$ in.   **12.** $10\frac{1}{8}$ in.   **13.** $3\frac{1}{10}$ lb   **14.** $\frac{3}{16}$ in.

15. $\frac{29}{32}$

Self-Study Exercises 3–10

1   **1.** $\frac{3}{32}$ **2.** $\frac{7}{32}$ **3.** $\frac{7}{16}$ **4.** $\frac{7}{12}$ **5.** $\frac{1}{3}$ **6.** $\frac{5}{32}$ **7.** $\frac{1}{15}$ **8.** $\frac{11}{25}$ **9.** $\frac{2}{25}$ **10.** $\frac{21}{32}$

2   **11.** $21\frac{7}{8}$ **12.** 75   **13.** $4\frac{1}{8}$ **14.** $36\frac{1}{10}$ **15.** $1\frac{21}{40}$ **16.** $2\frac{1}{6}$ **17.** $18\frac{3}{4}$ L   **18.** 90 in.

19. $15\frac{5}{8}$ in.   **20.** 264 kg copper, 84 kg tin, 36 kg zinc

Self-Study Exercises 3–11

1   **1.** $\frac{7}{3}$ **2.** $\frac{1}{8}$ **3.** $\frac{5}{11}$ **4.** $\frac{5}{1}$ **5.** $\frac{1}{7}$

2   **6.** $\frac{9}{10}$ **7.** $\frac{11}{12}$ **8.** 2   **9.** $1\frac{1}{20}$ **10.** $2\frac{1}{12}$ **11.** $4\frac{2}{3}$ **12.** $\frac{2}{5}$ **13.** 8

3   **14.** $13\frac{1}{3}$ **15.** 32   **16.** $\frac{5}{8}$ **17.** $14\frac{3}{4}$ **18.** 7   **19.** 18 2 × 4's   **20.** 10 lengths

21. 12 shovels   **22.** $23\frac{11}{12}$ in.   **23.** 20 ft × 15 ft   **24.** 4 whole pieces   **25.** 7 strips

26. 23 straws, $3\frac{3}{4}$ in. left   **27.** 8 pieces

4   **28.** $\frac{1}{10}$ **29.** 3   **30.** $5\frac{3}{5}$ **31.** $\frac{12}{25}$ **32.** $\frac{1}{3}$ **33.** 5   **34.** 28   **35.** $\frac{2}{9}$ **36.** 3   **37.** $\frac{1}{3}$

38. $\frac{1}{6}$ **39.** $\frac{5}{6}$ **40.** $\frac{2}{3}$

Self-Study Exercises 3–12

1   **1.** $-\frac{-5}{8}, -\frac{5}{-8}, \frac{-5}{-8}$ **2.** $-\frac{-3}{-4}, \frac{-3}{4}, \frac{3}{-4}$ **3.** $-\frac{2}{-5}, -\frac{-2}{5}, \frac{2}{5}$ **4.** $-\frac{7}{8}, \frac{-7}{8}, \frac{7}{-8}$

5. $-\frac{-7}{8}, -\frac{7}{-8}, \frac{-7}{-8}$

2   **6.** $-\frac{2}{8}$ or $-\frac{1}{4}$ **7.** $-1\frac{1}{10}$ **8.** $1\frac{1}{10}$ **9.** $\frac{6}{11}$ **10.** $-\frac{25}{32}$

3   **11.** -26.297 **12.** -1.11 **13.** -91.44 **14.** -110.72 **15.** -59.04 **16.** 340.71
17. $-27.7\overline{3}$ **18.** 0.413   **19.** -26.6 **20.** 7.73

Self-Study Exercises 3–13

1   **1.** $1\frac{5}{24}$ **2.** $-2\frac{16}{35}$ **3.** $14\frac{59}{96}$ **4.** $1\frac{3}{4}$ **5.** $\frac{4}{5}$ **6.** $\frac{1}{2}$ **7.** $-\frac{1}{24}$ **8.** $282\frac{1}{10}$ **9.** $11\frac{7}{40}$

10. $-1\frac{5}{16}$

Self-Study Exercises 3–14

1   **1.** 40%   **2.** 70%   **3.** $62\frac{1}{2}\%$ **4.** $77\frac{7}{9}\%$ **5.** $\frac{7}{10}\%$ **6.** $\frac{2}{7}\%$ **7.** 20%   **8.** 14%   **9.** 0.7%

10. 1.25%   **11.** 500%   **12.** 800%   **13.** $133\frac{1}{3}\%$ **14.** 350%   **15.** 430%   **16.** 220%

17. 305%   **18.** 720%   **19.** 1,510%   **20.** 3,625%

2   **21.** $\frac{9}{25}$, 0.36   **22.** $\frac{9}{20}$, 0.45   **23.** $\frac{1}{5}$, 0.20   **24.** $\frac{3}{4}$, 0.75   **25.** $\frac{1}{16}$, 0.0625   **26.** $\frac{5}{8}$, 0.625

27. $\frac{2}{3}$, $0.66\frac{2}{3}$ **28.** $\frac{3}{500}$, 0.006 **29.** $\frac{1}{500}$, 0.002 **30.** $\frac{1}{2,000}$, 0.0005 **31.** $\frac{1}{12}$, $0.08\frac{1}{3}$

32. $\frac{3}{16}$, 0.1875 **33.** 8   **34.** 4   **35.** $2\frac{1}{5}$, 2.5 **36.** $4\frac{1}{4}$, 4.25 **37.** $1\frac{19}{25}$, 1.76 **38.** $3\frac{4}{5}$, 3.8

39. $1\frac{3}{8}$, 1.375 **40.** $3\frac{7}{8}$, 3.875 **41.** $1\frac{2}{3}$ **42.** $3\frac{1}{6}$ **43.** 1.153 **44.** 2.125 **45.** 1.0625

46. $\frac{1}{10}$, 0.1 **47.** 25%, 0.25 **48.** 20%, $\frac{1}{5}$ **49.** $33\frac{1}{3}$%, $0.33\frac{1}{3}$ **50.** $\frac{1}{2}$, 0.5 **51.** 80%, 0.8

52. 75%, $\frac{3}{4}$ **53.** $\frac{2}{3}$, $0.66\frac{2}{3}$ **54.** 100%, $\frac{1}{1}$ **55.** 30%, 0.3 **56.** $\frac{2}{5}$, 0.4 **57.** 70%, $\frac{7}{10}$

58. 90%, 0.9 **59.** $\frac{3}{5}$, 0.6 **60.** 100%, 1   **61.** 25%, $\frac{1}{4}$ **62.** $66\frac{2}{3}$%, $0.66\frac{2}{3}$ **63.** $\frac{7}{10}$, 0.7

64. 50%, $\frac{1}{2}$ **65.** 60%, 0.6 **66.** 10%, $\frac{1}{10}$ **67.** $\frac{1}{5}$, 0.2 **68.** $33\frac{1}{3}$%, $\frac{1}{3}$ **69.** $66\frac{2}{3}$%, $\frac{2}{3}$

70. 75%, 0.75 **71.** $\frac{3}{10}$, 0.3 **72.** 90%, $\frac{9}{10}$ **73.** 20%, 0.2 **74.** 60%, $\frac{3}{5}$ **75.** $\frac{1}{1}$, 1   **76.** 60%

77. 40%   **78.** 70%   **79.** 63%   **80.** 45%

Assignment Exercises, Chapter 3

1. $\frac{3}{8}$ **3.** $\frac{1}{3}$ **5.** $\frac{13}{24}$ **7.** (a) 9   (b) 4   (c) 4   (d) 9   (e) 9, 4   (f) 4   (g) 9   (h) improper

(i) >1 (j) 9, 4   **9.** f   **11.** d   **13.** e   **15.** d   **17.** $\frac{87}{100}$ **19.** $2\frac{3}{100}$ **21.** 4, 6, 8, 10, 12

23. 42, 63, 84, 105, 126   **25.** 14, 21, 28, 35, 42   **27.** 16, 24, 32, 40, 48   **29.** yes; $153 \div 3 = 51$
31. no, remainder of 2   **33.** $1 \times 48, 2 \times 24, 3 \times 16, 4 \times 12, 6 \times 8$; 1, 2, 3, 4, 6, 8, 12, 16, 24, 48
35. $1 \times 51, 3 \times 17$; 1, 3, 17, 51   **37.** $2 \cdot 2 \cdot 11$ or $2^2 \cdot 11$ **39.** $2 \cdot 2 \cdot 2 \cdot 3 \cdot 3 \cdot 3$ or $2^3 \cdot 3^3$ **41.** 360

43. 180   **45.** 2   **47.** 6   **49.** $\frac{15}{24}$ **51.** $\frac{25}{60}$ **53.** $\frac{10}{15}$ **55.** $\frac{24}{32}$ **57.** $\frac{11}{55}$ **59.** $\frac{1}{2}$ **61.** $\frac{1}{8}$

63. $\frac{1}{4}$ **65.** $\frac{17}{32}$ **67.** $\frac{3}{8}$ **69.** $\frac{3}{4}$ **71.** $\frac{7}{10}$ **73.** $\frac{19}{20}$ **75.** $\frac{109}{125}$ **77.** $\frac{1}{50}$ **79.** 0.2 **81.** 0.625

83. $0.\overline{81}$ or $0.8181\ldots$ **85.** $3\frac{3}{5}$ **87.** $4\frac{7}{8}$ **89.** $5\frac{3}{8}$ **91.** $87\frac{1}{2}$ **93.** $\frac{8}{1}$ **95.** $\frac{57}{8}$ **97.** $\frac{147}{16}$

99. $\frac{23}{5}$ **101.** $\frac{12}{1}$ **103.** 20   **105.** 33   **107.** 60   **109.** 16   **111.** 60   **113.** greater than

115. no   **117.** smaller   **119.** $\frac{3}{8}$ **121.** $\frac{3}{16}$ **123.** $\frac{27}{32}$ **125.** $\frac{9}{19}$ **127.** $\frac{21}{64}$ **129.** $1\frac{13}{30}$

131. $8\frac{19}{32}$ **133.** $10\frac{9}{32}$ **135.** $18\frac{1}{16}$ in.   **137.** $12\frac{19}{32}$ in.   **139.** $15\frac{25}{32}$ in.   **141.** $4\frac{3}{4}$ in.   **143.** $\frac{7}{8}$ in.

145. $\frac{1}{3}$ **147.** $1\frac{5}{8}$ **149.** $5\frac{31}{32}$ **151.** $7\frac{11}{16}$ **153.** $\frac{7}{16}$ in.   **155.** $1\frac{29}{64}$ in.   **157.** $\frac{7}{24}$ **159.** $\frac{7}{24}$

161. $\frac{1}{2}$ **163.** $7\frac{7}{8}$ **165.** $1\frac{1}{5}$ **167.** $100\frac{1}{2}$ in.   **169.** $2\frac{3}{4}$ cups   **171.** $1\frac{1}{6}$ **173.** $9\frac{1}{3}$ **175.** 24

177. 2   **179.** $\frac{3}{5}$ **181.** $2\frac{55}{64}$ in.   **183.** $\frac{3}{16}$ yd   **185.** $\frac{1}{18}$ **187.** $5\frac{1}{3}$ **189.** $\frac{1}{4}$ **191.** $\frac{1}{8}$

193. $-\frac{3}{-8}, +\frac{-3}{8}, \frac{3}{-8}$ **195.** $\frac{7}{8}, -\frac{-7}{8}, \frac{-7}{-8}$ **197.** $\frac{62}{63}$ **199.** $1\frac{5}{7}$ **201.** 0.0351 **203.** -62

205. $-1\frac{7}{24}$ **207.** $-11\frac{13}{16}$ **209.** 70%   **211.** 12,500%   **213.** 1,730%   **215.** 0.72 or $\frac{18}{25}$

217. 0.125 or $\frac{1}{8}$ **219.** $0.0066\frac{2}{3}$ or $\frac{1}{150}$ **221.** 2.75 or $2\frac{3}{4}$ **223.** 1.125 or $1\frac{1}{8}$ **225.** 2.272

227. 0.09275   **229.** 3.4

Trial Test, Chapter 3

1. $\dfrac{3}{4}$ **3.** 3   **5.** $\dfrac{34}{7}$ **7.** $2^5 \cdot 3$ **9.** $\dfrac{1}{4}$ **11.** $3\dfrac{8}{9}$ **13.** $13\dfrac{1}{2}$ **15.** $1\dfrac{5}{12}$ **17.** $7\dfrac{13}{14}$ **19.** $\dfrac{1}{9}$

21. $5\dfrac{1}{10}$ **23.** $\dfrac{5}{16}$ **25.** 60%   **27.** $\dfrac{2}{7}$ **29.** $3\dfrac{5}{6}$ cups   **31.** 8 yd

Self-Study Exercises 4–1

1   **1.** $x = 5$ **2.** $b = 4$ **3.** $a = 9$ **4.** $c = 10$ **5.** $x = 4$ **6.** $a = 8$ **7.** $x = 6$ **8.** $b = 18$
9. $d = 1.5$ **10.** $x = 3.33$ **11.** $c = .8889$ **12.** $b = 0.75$

2   **13.** R missing; $B = 10$; $P = 2$ **14.** $P = 2$; $R = 20\%$; B missing   **15.** $R = 20\%$; $B = 10$;
P missing   **16.** $P = 3$; R missing; $B = 4$ **17.** $R = 15\%$; B missing; $P = \$9$ **18.** R missing; $B = 25$;
$P = 5$ **19.** $P = 6$; $B = 15$; R missing   **20.** P missing; $R = 20\%$; $B = 15$ **21.** $R = 35\%$; B missing;
$P = 70$ **22.** R missing; $B = \$45$; $P = \$3.15$

3   **23.** 75   **24.** 63   **25.** 206   **26.** 115.92   **27.** 0.675   **28.** 0.94   **29.** 154.1   **30.** 231

31. 924   **32.** 345   **33.** 25%   **34.** 30%   **35.** $33\dfrac{1}{3}\%$ **36.** 32.9%   **37.** 15.75%   **38.** 16%

39. 0.8%   **40.** $0.66\dfrac{2}{3}\%$ **41.** 500%   **42.** 111.25%   **43.** 72   **44.** 50   **45.** 344   **46.** 46

47. 360   **48.** 275   **49.** 75   **50.** 250   **51.** 18.4   **52.** 261   **53.** 3.75   **54.** 0.625   **55.** $66\dfrac{2}{3}\%$

56. $37\dfrac{1}{2}\%$ **57.** 14.25   **58.** 350   **59.** 9.375   **60.** 220   **61.** 20%   **62.** 200

4   **63.** 1.0625 lb   **64.** 3%   **65.** 200 hp   **66.** 0.021 lb   **67.** 5%   **68.** \$575   **69.** 1,200 lb
70. 11%   **71.** 9,625 parts   **72.** 100,880 welds

Self-Study Exercises 4–2

1   **1.** 108   **2.** 109.2   **3.** 10.2   **4.** 33.75

2   **5.** 27 in.   **6.** 1,815 board feet   **7.** 2,393 board feet   **8.** 25,908 bricks   **9.** \$16,802.50
10. 1,366.4 yd^3

3   **11.** 7%   **12.** 45%   **13.** 20%   **14.** 5%   **15.** 350 hp   **16.** 10.5 yd^3 **17.** 3,019 board feet
18. 28.5 lb   **19.** 40 in.   **20.** 20%   **21.** 150 hp

Assignment Exercises, Chapter 4

1. 4.5 or $4\dfrac{1}{2}$ **3.** 48   **5.** 1.33 or $1\dfrac{1}{3}$ **7.** 6   **9.** 2.5 or $2\dfrac{1}{2}$ **11.** 0.67 or $\dfrac{2}{3}$ **13.** R missing; $B = 25$;
$P = 5$ **15.** $R = 5\%$; $B = 180$; P missing   **17.** $R = 45\%$; B missing; $P = \$36$ **19.** $P = 6$; R missing;
$B = 25$ **21.** $R = 18\%$; $B = 150$; P missing   **23.** 24   **25.** 0.4375   **27.** 60%   **29.** 250%   **31.** 83
33. 152   **35.** 15.3%   **37.** 500%   **39.** 84   **41.** 37.5%   **43.** 1.14%   **45.** \$266   **47.** 4%
49. 200 students   **51.** 10%   **53.** 7%   **55.** 14%   **57.** 142.1 kg   **59.** 62 cm, 63 cm   **61.** \$1.69
63. 12.5%   **65.** 57.5 lb   **67.** $33\dfrac{1}{3}\%$ **69.** 4.8%   **71.** 289 hp   **73.** 26.6%   **75.** 28.6%

Trial Test, Chapter 4

1. 8   **3.** 12   **5.** $R = 40\%$; $B = 10$; P missing   **7.** $P = 9$; R missing; $B = 27$ **9.** $R = 12\%$; $B = 50$;
P missing   **11.** 9   **13.** 305   **15.** 115   **17.** 67.10   **19.** \$2,500   **21.** 21.14%   **23.** 13.17%
25. 5%   **27.** \$104.87

Self-Study Exercises 5–1

1   **1.** lb   **2.** oz   **3.** lb or oz   **4.** gal   **5.** qt   **6.** T   **7.** lb   **8.** qt   **9.** ft and in.   **10.** c
11. in.   **12.** yd   **13.** mi   **14.** oz   **15.** oz   **16.** mi   **17.** lb or oz   **18.** lb   **19.** lb and oz
20. oz

2   **21.** $\dfrac{\text{pint}}{1}\left(\dfrac{1\ \text{quart}}{2\ \text{pints}}\right)$ **22.** $\dfrac{\text{feet}}{1}\left(\dfrac{1\ \text{mile}}{5{,}280\ \text{feet}}\right)$ **23.** $\dfrac{\text{inches}}{1}\left(\dfrac{1\ \text{foot}}{12\ \text{inches}}\right)$ **24.** $\dfrac{\text{feet}}{1}\left(\dfrac{1\ \text{yard}}{3\ \text{feet}}\right)$
25. $\dfrac{\text{days}}{1}\left(\dfrac{1\ \text{week}}{7\ \text{days}}\right)$ **26.** $\dfrac{\text{quarts}}{1}\left(\dfrac{1\ \text{gallon}}{4\ \text{quarts}}\right)$

3   **27.** 48 in.   **28.** 21 ft   **29.** 4,400 yd   **30.** $9\frac{1}{3}$ yd   **31.** 2 mi   **32.** 48 oz   **33.** $2\frac{3}{10}$ or 2.3 lb
34. 732.8 oz   **35.** 19.2 oz   **36.** 408 oz, 25.5 lb   **37.** 20 qt   **38.** 13 pt   **39.** $1\frac{1}{2}$ gal   **40.** 60 pt
41. 9 gal   **42.** 256 oz   **43.** $1\frac{1}{2}$ or 1.5 yd   **44.** 108 in.   **45.** 7,040 ft   **46.** 24 c

4   **47.** feet to inches = 12; inches to feet = $\frac{1}{12}$ **48.** quarts to gallons = $\frac{1}{4}$; gallons to quarts = 4
49. quarts to pints = 2; pints to quarts = $\frac{1}{2}$ or 0.5   **50.** ounces to cups = $\frac{1}{8}$ or 0.125; cups to ounces = 8
51. 28.75 pints   **52.** 92 pints   **53.** 16.67 yards   **54.** 25 gallon   **55.** 36.25 lb   **56.** 153 inches
57. 720 inches   **58.** 4.67 feet   **59.** 7,920 feet   **60.** 28 quarts

5   **61.** 3 ft 8 in.   **62.** 2 mi 1,095 ft   **63.** 3 lb $3\frac{1}{2}$ oz   **64.** 2 gal 1 qt   **65.** 2 gal   **66.** 2 T 500 lb
67. 2 yd 2 ft 11 in.   **68.** 2 qt 2 c 2 oz   **69.** 2 ft 10 in.   **70.** 6 lb 9 oz   **71.** 4 gal 2 qt 16 oz or
4 gal 2 qt 1 pt   **72.** 6 qt 20 oz or 1 gal 2 qt 1 pt 4 oz   **73.** 2 mi 5 yd 2 ft 1 in.   **74.** 5 ft 4 in. or
1 yd 2 ft 4 in.   **75.** 3 T 600 lb 15 oz   **76.** 4 lb 5 oz   **77.** 2 lb 5 oz   **78.** 2 yd 2 ft 4 in.
79. 2 yd 2 ft 11 in.   **80.** 1 mi 875 ft 6 in.

Self-Study Exercises 5–2

1   **1.** 2 lb 5 oz   **2.** 4 ft 7 in.   **3.** 15 lb 11 oz   **4.** 18 ft 3 in.   **5.** 8 qt $\frac{1}{2}$ pt or 2 gal 1 c
6. 14 gal 1 qt   **7.** 9 yd 1 ft   **8.** 5 c 1 oz or 2 pt 1 c 1 oz   **9.** 5 ft 4 in. or 1 yd 2 ft 4 in.
10. 7 yd 1 ft 5 in.   **11.** 2 ft 7 in.   **12.** 11 ft or 3 yd 2 ft   **13.** 5 lb 15 oz   **14.** 12 lb

2   **15.** 6 in.   **16.** 1 pt   **17.** 2 ft   **18.** 4 lb 14 oz   **19.** 1 lb 6 oz   **20.** 6 lb 11 oz   **21.** 11 in.
22. 10 lb 13 oz   **23.** 2 in.   **24.** 3 gal 3 qt $1\frac{1}{2}$ pt or 3 gal 3 qt 1 pt 1 c   **25.** 4 ft 2 in.   **26.** 2 ft 3 in.
27. 45 sec   **28.** 39 sec   **29.** 69 lb 7 oz   **30.** 16 lb 14 oz

Self-Study Exercises 5–3

1   **1.** 60 mi   **2.** 108 gal   **3.** 252 lb   **4.** 14 qt or 3 gal 2 qt   **5.** 57 lb 8 oz   **6.** 38 gal 2 qt
7. 43 gal 3 qt   **8.** 58 ft or 19 yd 1 ft   **9.** 36 lb   **10.** 7 qt 1 pt or 1 gal 3 qt 1 pt

2   **11.** 35 in.^{2} **12.** 108 ft^2 **13.** 180 yd^2 **14.** 108 mi^2 **15.** 378 tiles   **16.** 47 gal 2 oz
17. 7 qt 1 pt or 1 gal 3 qt 1 pt   **18.** 6 lb 4 oz   **19.** 7 lb 8 oz

3   **20.** 6 gal   **21.** 1 day 15 hr   **22.** 10 yd 1 ft 3 in.   **23.** 1 yd 1 ft 7 in.   **24.** 3 qt $\frac{1}{2}$ pt or 3 qt 1 c
25. $6\frac{2}{3}$ gal   **26.** 1 hr 40 min   **27.** $5\frac{1}{4}$ qt or 5 qt 1 c   **28.** 7 ft 6 in.   **29.** 3 gal 2 qt   **30.** 9 pieces
31. 5 ft   **32.** 3 gal 1 qt 5 oz   **33.** 24 lb 3 oz

4 **34.** 3 **35.** 17 **36.** 8 **37.** 18 **38.** 3 **39.** 8 pieces **40.** 9 boxes **41.** 24 cans
42. 9 tickets **43.** 9 pieces

5 **44.** $15\dfrac{\text{gal}}{\text{sec}}$ **45.** $\dfrac{3}{4}\dfrac{\text{lb}}{\text{min}}$ **46.** $15{,}840\dfrac{\text{ft}}{\text{hr}}$ **47.** $2{,}304\dfrac{\text{oz}}{\text{min}}$ **48.** $2\dfrac{\text{qt}}{\text{sec}}$ **49.** $20\dfrac{\text{oz}}{\text{hr}}$ **50.** $40\dfrac{\text{ft}}{\text{min}}$

51. $440\dfrac{\text{ft}}{\text{sec}}$ **52.** $44\dfrac{\text{ft}}{\text{sec}}$ **53.** $3\dfrac{\text{qt}}{\text{min}}$ **54.** $53\dfrac{1}{3}\dfrac{\text{lb}}{\text{min}}$ **55.** $\dfrac{5}{6}\dfrac{\text{gal}}{\text{sec}}$

Self-Study Exercises 5–4

1 **1.** (a) 1,000 meters (b) 10 liters (c) $\dfrac{1}{10}$ of a gram (d) $\dfrac{1}{1{,}000}$ of a meter (e) 100 grams

(f) $\dfrac{1}{100}$ of a liter **2.** b **3.** b **4.** c **5.** b **6.** a **7.** d **8.** b **9.** c **10.** a **11.** b
12. c **13.** b **14.** a **15.** b **16.** b

2 **17.** 40 **18.** 70 **19.** 580 **20.** 80 **21.** 2.5 **22.** 210 **23.** 85 **24.** 142 **25.** 153 mL
26. 460 m **27.** 75 dkg **28.** 160 mm **29.** 400 **30.** 8,000 **31.** 58,000 **32.** 800 **33.** 250
34. 2,100 **35.** 102,500 **36.** 8,330 **37.** 2,000,000 **38.** 70 **39.** 236 L **40.** 467 cm
41. 38,000 dg **42.** 13,000 cm **43.** 2.8 **44.** 23.8 **45.** 10.1 **46.** 6 **47.** 2.9 **48.** 19.25
49. 1.7 **50.** 438.9 dm **51.** 4.7 g **52.** 0.225 dL **53.** 2.743 **54.** 0.385 **55.** 0.15 **56.** 0.08
57. 2,964.84 **58.** 0.2983 **59.** 0.0003 **60.** 0.004 **61.** 0.002857 **62.** 15.285 **63.** 0.0297 hm
64. 0.00003 L

3 **65.** 11 m **66.** 12 hL **67.** 6 cg **68.** 2.4 dm or 24 cm **69.** 5.9 cL or 59 mL **70.** cannot
add **71.** 10.1 kL or 101 hL **72.** 0.55 g or 55 cg **73.** cannot subtract **74.** 7.002 km or 7,002 m
75. 1,000 mL **76.** 1.47 kL or 147 dkL **77.** 516 m **78.** 40.8 m **79.** 150.96 dm **80.** 969.5 m
81. 13 m **82.** 9 cL **83.** 163 g **84.** 0.4 m or 4 dm **85.** 5 **86.** 30 **87.** 16 prescriptions
88. 80 containers **89.** 5.74 dL or 57.4 cL **90.** 2.3 dkm or 23 m **91.** 165.7 hm or 16.57 km
92. 9.1 kL or 91 hL **93.** 1.25 cL **94.** 70 mm **95.** 7.5 dm **96.** 50 mL **97.** 21,250 containers
98. 19 vials

Self-Study Exercises 5–5

1 **1.** 354.33 in. **2.** 130.8 yd **3.** 26.04 mi **4.** 6.36 liq qt **5.** 11 L **6.** 59.4 lb **7.** 22.5 kg
8. 17.78 cm **9.** 5.4 m **10.** 1.32288 oz **11.** 21.85 L **12.** 30 m **13.** 27 kg **14.** 241.5 km
15. 32.7 yd **16.** 235.6 mi **17.** 7.44 mi **18.** 13.2 L **19.** 328 ft

Self-Study Exercises 5–6

1 **1.** 1.5 days **2.** 9.67 9 min − 40 sec **3.** 150 min **4.** 318 sec **5.** 210 min
6. 3 min 2 sec **7.** 432 min **8.** 1 min 28 sec, 1.467 min

2 **9.** 4,320 lb/hr **10.** 2,400 gal/hr **11.** 10,950 lb/yr **12.** 1,680 vehicles/hr **13.** 0.0167 mi/sec
14. 2.0833 gal/min **15.** 60 lb/min **16.** 1.6 gal/sec

3 **17.** 6:00 P.M. **18.** New York 9:00 A.M.–11:00 A.M.; St. Louis 8:00 A.M.–10.00 A.M.;
London 2:00 P.M.–4:00 P.M. **19.** Paris 4:00 P.M.; Stanford 8:00 A.M. **20.** +1 hr **21.** 11:00 P.M.
22. 10 hours

Self-Study Exercises 5–7

1 **1.** 3% **2.** 6.25% **3.** 1.2%

2 **4.** 3 **5.** 4 **6.** 2 **7.** 4 **8.** 5 **9.** 3 **10.** $\dfrac{1}{32}$ in. **11.** 0.05 mm **12.** $\dfrac{1}{8}$ in. **13.** 0.5 oz
14. 0.5 L

 Selected Answers to Student Exercise Material

3   **15.** $4\frac{9}{16}$ in.   **16.** $4\frac{1}{16}$ in.   **17.** $3\frac{13}{16}$ in.   **18.** $3\frac{3}{8}$ in.   **19.** $2\frac{1}{4}$ in.   **20.** 2 in.   **21.** $1\frac{3}{4}$ in.

22. $1\frac{3}{16}$ in.   **23.** $\frac{3}{4}$ in.   **24.** $\frac{3}{8}$ in.   **25.** 0.44 in.   **26.** 0.85 in.   **27.** 1.20 in.   **28.** 1.70 in.

4   **29.** 115 mm or 11.5 cm   **30.** 102 mm or 10.2 cm   **31.** 96 mm or 9.6 cm   **32.** 85 mm or 8.5 cm
33. 57 mm or 5.7 cm   **34.** 50 mm or 5 cm   **35.** 44 mm or 4.4 cm   **36.** 30 mm or 3 cm   **37.** 19 mm
or 1.9 cm   **38.** 10 mm or 1 cm

5   **39.** 115.90 mm; 4.560 in.   **40.** 8.05 mm; 0.312 in.   **41.** 26.15 mm; 1.069 in.   **42.** 45.20 mm;
1.777 in.

6   **43.** 0.150 in.   **44.** 4.225 in.   **45.** 2.275 in.   **46.** 4.359 in.   **47.** 1.242 in.   **48.** 0.2507 in.
49. 2.2505 in.   **50.** 1.2510 in.   **51.** 4.2655 in.   **52.** 2.2550 in.

7   **53.** absolute error = 0.02   **54.** absolute error = 0.2
       relative error = 0.00038   relative error = 0.0038
       percent error = 0.038%   percent error = 0.38%

8   **55.** 419,850 ft^3 **56.** 787,516 ft^3 **57.** 145,556 ft^3 **58.** 328,153 ft^3 **59.** 138,444 ft^3
60. 287,747 ft^3 **61.** 7,555 kWh   **62.** 3,421 kWh   **63.** 0.616 mm   **64.** 0.636 mm   **65.** 0.317 mm
66. 0.382 mm   **67.** 135 volts   **68.** 210 volts   **69.** 4 ohms   **70.** 5.5 ohms   **71.** Answers will vary
72. 1.0
    0.03
    0.004
    0.0005
    0.00006
    1.03456

Assignment Exercises, Chapter 5

1. lb or oz   **3.** qt   **5.** lb or oz   **7.** T   **9.** qt   **11.** c   **13.** yd   **15.** $\frac{4\,c}{1\,qt}, \frac{1\,qt}{4\,c}$

17. $\frac{2,000\,lb}{1\,T}, \frac{1\,T}{2,000\,lb}$ **19.** 4 yd   **21.** 6,336 ft   **23.** 80 oz   **25.** $42\frac{1}{2}$ lb   **27.** 304 oz or 19 lb

29. 15 pt   **31.** 24 pt   **33.** 6,600 ft   **35.** 7 ft 5 in.   **37.** 13 lb $1\frac{1}{2}$ oz   **39.** 2 gal 1 pt   **41.** 4 yd 4 in.

43. 4 ft 1 in. or 1 yd 1 ft 1 in.   **45.** 2 gal 16 oz or 2 gal 1 pt   **47.** 3 ft 11 in.   **49.** 10 ft 11 in. or
3 yd 1 ft 11 in.   **51.** 8 gal 2 qt   **53.** 14 oz   **55.** 1 ft 9 in.   **57.** 1 ft 5 in.   **59.** 504 ft^2 **61.** 63 in.^{2}

63. 10 yd 1 ft 3 in.   **65.** $5\frac{5}{12}$ ft   **67.** 4 lb 8 oz   **69.** 3.5   **71.** $4\frac{4}{9}$ **73.** $300\frac{mi}{hr}$ **75.** $60\frac{mi}{hr}$

77. $1.25\frac{gal}{min}$ **79.** $1\frac{1}{2}$ qt   **81.** kilo-   **83.** milli-   **85.** centi-   **87.** 10 times   **89.** $\frac{1}{1,000}$ of

91. 1,000 times   **93.** a   **95.** a   **97.** c   **99.** b   **101.** 6.71 dkm   **103.** 2,300 mm
105. 12,300 mm   **107.** 230,000 mm   **109.** 413.27 km   **111.** 3.945 hg   **113.** 30.00974 kg
115. Cannot add unlike measures.   **117.** 748 cg or 7.48 g   **119.** 61.47 cg   **121.** 15   **123.** 8.5 hL
125. 18.9 m   **127.** 245 mL or 24.5 cL   **129.** 6 m   **131.** 100 servings   **133.** 40 shirts
135. 234.35 yd   **137.** 15.9 liq qt   **139.** 70.4 lb   **141.** 22.86 cm   **143.** 156.88 qt   **145.** 60 m
147. 281.75 km   **149.** 6 days   **151.** 2 hours, 38 minutes   **153.** 8 days   **155.** 3 years, 3 months
157. 1%   **159.** 3 significant digits   **161.** 0.05 cg   **163.** $5\frac{1}{4}$ in.   **165.** $4\frac{7}{16}$ in.   **167.** $3\frac{15}{16}$ in.

169. $3\frac{9}{16}$ in.   **171.** $2\frac{3}{4}$ in.   **173.** 11.8 cm or 118 mm   **175.** 99 mm or 9.9 cm   **177.** 60 mm or 6 cm
179. 45 mm or 4.5 cm   **181.** 20 mm or 2 cm   **183.** 1.367 in.   **185.** 2.550 in.   **187.** 0.500 in.
189. 4.3801 in.   **191.** absolute error = 0.48; relative error = 0.04; percent error = 4.03%
193. 441,822 cubic feet   **195.** 4,968 kWh   **197.** 5.15 mm   **199.** 25 volts   **201.** 300 ohms

Trial Test, Chapter 5

1. 36 in.   **3.** 8 gal   **5.** $4\frac{qt}{sec}$ **7.** 165 yd   **9.** $80\frac{2}{3}\frac{ft}{sec}$ **11.** $1\frac{1}{4}$ in.   **13.** 2 gal 1 qt 4 oz
15. 1 hr   **17.** deci-   **19.** 0.298 km   **21.** 9.48 L or 94.8 dL   **23.** 120.75 km   **25.** 8.48 pt
27. $735\frac{m}{sec}$ **29.** 3.912 in.   **31.** 9.27 mm   **33.** Answers will vary

Self-Study Exercises 6–1

1   **1.** $P = 12$ cm   **2.** square   **3.** 193 ft^2 **4.** 590 ft   **5.** 900 yd^2 **6.** 16 mi^2 **7.** 5,268 ft
    $A = 9$ cm^2
8. 81 tiles   **9.** 36 tiles   **10.** perimeter = 80 in.
           area = 400 in.^{2}

2   **11.** $P = 10$ ft   **12.** rectangle   **13.** 42,500 ft^2 **14.** 180 ft^2 **15.** 241 ft^2
    $A = 6$ ft^2
16. 59 ft   **17.** 142.5 board feet   **18.** 156 in. or 13 ft   **19.** 17 ft   **20.** 130 in.

3   **21.** $P = 38$ in.   **22.** parallelogram   **23.** 156 in. or 13 ft   **24.** 9,000 ft^2 **25.** 124 in.
    $A = 72$ in.^{2}
26. 18 tiles   **27.** 1,111 parallelograms   **28.** 160 in.   **29.** 108 in.   **30.** 9 signs

Self-Study Exercises 6–2

1   **1.** 25.1 cm   **2.** 18.8 in.   **3.** 9.4 ft   **4.** 17.3 m

2   **5.** 50.3 cm^2 **6.** 28.3 in.^{2} **7.** 7.1 ft^2 **8.** 23.8 m^2

3   **9.** $y = 22$ ft   **10.** $y = 25$ ft $-$ 6 in
    $x = 16$ ft $x = 6$ ft
11. 0.6 m^2 **12.** 1.2 ft^2 **13.** 14.6 cm^2 **14.** 1.0 in.^{2} **15.** 191.5 ft   **16.** 1,942.5 ft^2
17. 58 ft 8 in.   **18.** \$330.00   **19.** 24.3 lb   **20.** \$69.50   **21.** 592.7 ft^2 **22.** 268.27 in.
23. 1,178.1 mm^2 **24.** 168.5 mm^2 **25.** Yes, the cross-section area of the third pipe is larger than the
combined area of the other two pipes.   **26.** Yes, the combined cross-sectional area of the two pipes is
25.1 in.^{2} which is greater than 20 in.^{2}—the area of the large pipe.   **27.** 2,827.4 ft/min   **28.** 1,570.8 ft/min
29. 78.5 ft^2 **30.** 0.448 in.   **31.** 4.4 m

Assignment Exercises, Chapter 6

1. $P = 57$ cm   **3.** $P = 210$ mm   **5.** 836 ft   **7.** 33 yd^2 **9.** \$14.25   **11.** 3 rolls   **13.** 235 ft^2
    $A = 162$ cm^2 $A = 2,450$ mm^2
15. $C = 25.13$ m   **17.** $C = 17.42$ cm   **19.** 5.8 m^2 **21.** 2.4 in.   **23.** 33 ft/min   **25.** 15.0 in.
    $A = 50.27$ m^2 $A = 12$ cm^2
27. 1.4 yd^2 **29.** 59.1 in.   **31.** 66 in. diameter

Trial Test, Chapter 6

1. $P = 91$ ft   **3.** $P = 12.2$ in.   **5.** \$52   **7.** 282 ft^2 **9.** center   **11.** radius   **13.** 37.70 in.
    $A = 514.5$ ft^2 $A = 6.08$ in.^{2}
15. 385.62 ft   **17.** 0.29 in.^{2} **19.** 0.15 in.^{2}

Self-Study Exercises 7–1

1   **1.** 8.6%   **2.** 35.5%   **3.** 56.3%

2   **4.** Debt retirement   **5.** Misc. expenses and general government   **6.** Social projects and education
costs

3   **7.** 5 Amps   **8.** 50 volts   **9.** 35 volts   **10.** 25 ohms

1 **1.** 10 **2.** 2 **3.** $\dfrac{1}{3}$ **4.** $\dfrac{1}{7}$ **5.** 12% **6.** 28% **7.** 35–37 and 38–40 **8.** 20–22 and 23–25

9. 7 **10.** 16

	Midpoint	Tally	Class Frequency
11.	15		2
12.	12		4
13.	9		7
14.	6		20

	Midpoint	Tally	Class Frequency
15.	93		2
16.	88		5
17.	83		8
18.	78		9
19.	73		3
20.	68		6
21.	63		2
22.	58		5

2 **23.**

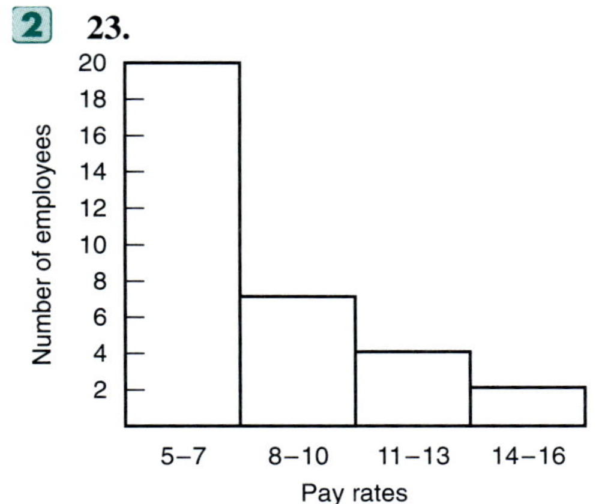

24.

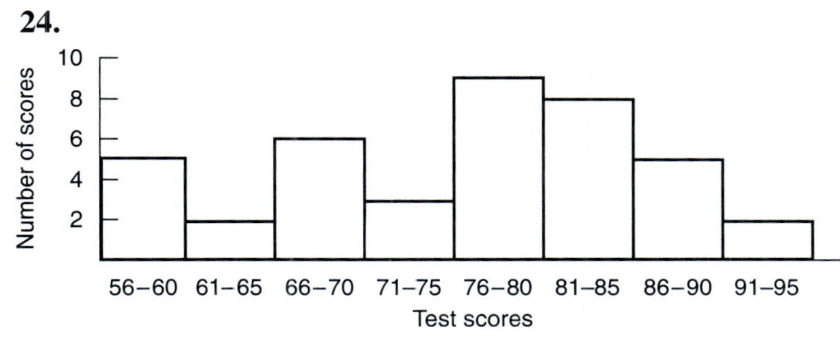

3 **25.**

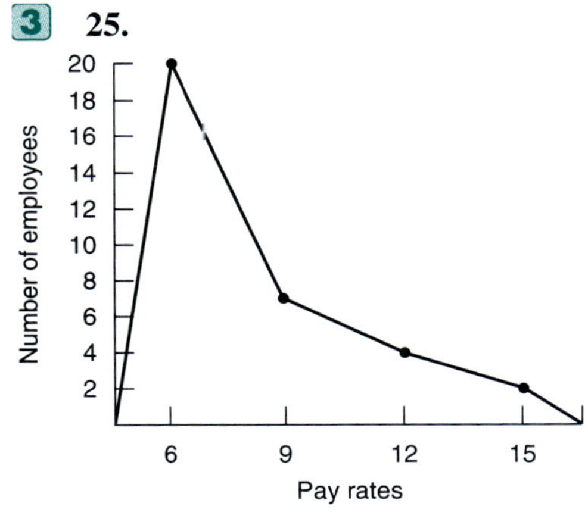

26.

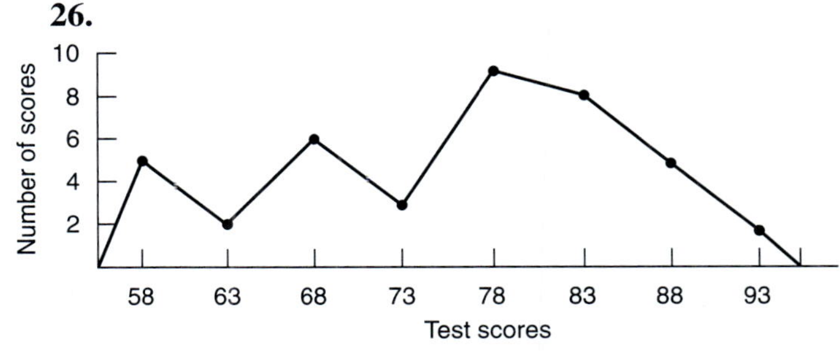

1 **1.** 16 **2.** 17 **3.** 66 **4.** 73.75 **5.** 33.7 **6.** 66.6 **7.** 42.33°F **8.** 12.67°C **9.** $34.80
10. $39.20 **11.** 14.67 in. **12.** 8 in. **13.** 20 **14.** 76 **15.** 17.3 runs **16.** 78 cars
17. 24.6 mpg **18.** 10.2 mpg

Assignment Exercises, Chapter 7

1. 1995, 1997, 1998   **3.** 1996, 1999, 2000   **5.** 12.9%   **7.** 17.4%   **9.** 7-10-02 @ 4:00 P.M.

11. 50   **13.** 110   **15.** $\dfrac{1}{2}$ **23.**

	Midpoint	Tally	Class Frequency
17.	60.5	ℍ ℍ	10
19.	40.5	ℍ ℍ ‖	12
21.	20.5	ℍ ‖	7

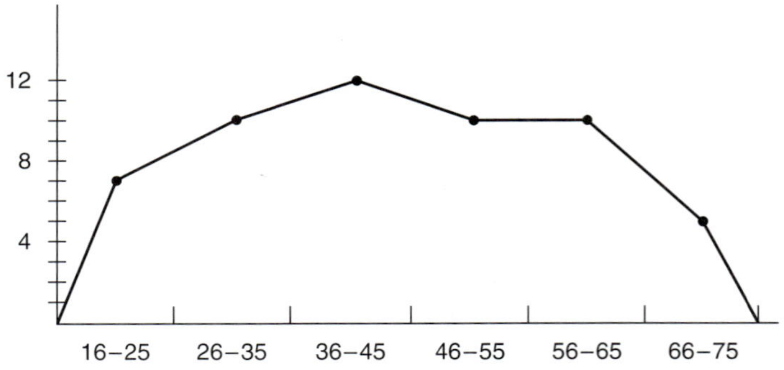

25. 36–45   **27.** 15   **29.** $\dfrac{1}{2}$ **31.** $\dfrac{10}{54} = 18.5\%$

33.

miles per gallon	Midpoint	Tally	MPG Frequency	(Answers may vary)
20–24	22	ℍ ‖	7	
25–29	27	ℍ ∣	6	
30–34	32	‖‖	4	

35.

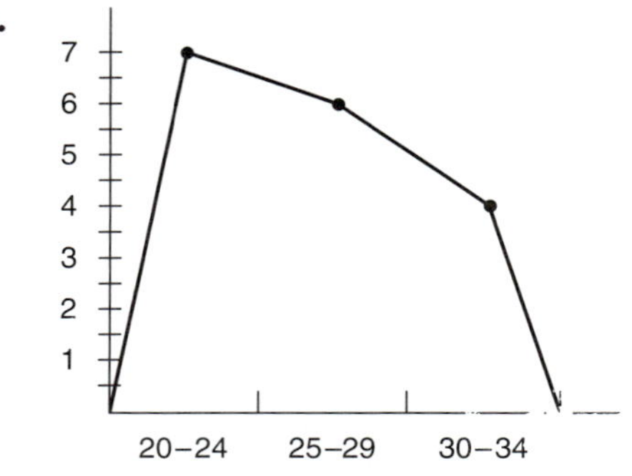

37. 12.6 cars   **39.** 13.8 mpg   **41.** 3   **43.** $1.85   **45.** 3.67   **47.** 63   **49.** mean = 81.67
Median = 89
Mode = none

51. 21.47 (rounded)   **53.** 1.078 (rounded)

Trial Test, Chapter 7

1. bar   **3.** line   **5.** 2 degrees   **7.** $65   **9.** 25%   **11.** $\dfrac{15}{200} = \dfrac{3}{40}$ **13.** English dept.
and Electronics dept.   **15.** 108.3%   **17.** $\dfrac{18}{20} = \dfrac{9}{10}$ **19.** 6   **21.** $\dfrac{1}{2}$

	Midpoint	Tally	Class Frequency
23.	5	ℍ ‖‖	9

25. 14   **27.** cannot be determined   **29.** mean: 77.9
median: 78
mode: 81

Self-Study Exercises 8–1

1 1. $-15 = -15$ 2. $11 = 11$ 3. $2 = 2$ 4. $-7 = -7$ 5. $n = 11$ 6. $m = 6$ 7. $y = -4$
8. $x = 6$ 9. $p = 9$ 10. $b = -9$ 11. $\boxed{7} + \boxed{c}$ 12. $\boxed{4a} - \boxed{7}$ 13. $\boxed{3x} - \boxed{2(x + 3)}$

14. $\boxed{\dfrac{a}{3}}$ 15. $\boxed{7xy} + \boxed{3x} - \boxed{4} + \boxed{2(x + y)}$ 16. $\boxed{14x} + \boxed{3}$ 17. $\boxed{\dfrac{7}{(a + 5)}}$ 18. $\boxed{\dfrac{4x}{7}} + \boxed{5}$

19. $\boxed{11x} - \boxed{5y} + \boxed{15xy}$ 20. 5   21. -4 22. $\dfrac{1}{5}$ 23. $\dfrac{2}{7}$ 24. 6   25. $-\dfrac{4}{5}$

26. $-15c$ (Answers will vary.)

2 27. Four more than a number is seven. (Answers may vary.)   28. Five less than a number is two.
(Answers may vary.)   29. Three times a number is fifteen. (Answers may vary.)   30. One more than
three times a number is seven. (Answers may vary.)

3 31. $x + 5 = 12$ 32. $\dfrac{x}{6} = 9$ 33. $4(x - 3) = 12$ 34. $4x - 3 = 12$ 35. $12 + 7 + x = 17$

36. $2x + 7 = 21$ 37. $x + 15° = 48°$ 38. $x + 15 \text{ mL} = 45 \text{ mL}$ 39. $x \cdot 5 = 45$ 40. $\dfrac{18 - 3}{5} = x$

4 41. $-3a$ 42. $-8x - 2y$ 43. $7x - 2y$ 44. $-3a + 6$ 45. 10   46. $3a + 6b + 8c + 1$
47. $10x - 20$ 48. $6x + 13$ 49. $-4x + 5$ 50. $-12a + 3$

Self-Study Exercises 8–2

1 1. $x = 8$ 2. $x = 9$ 3. $x = 5$ 4. $n = 30$ 5. $a = -9$ 6. $b = -2$ 7. $c = 7$ 8. $x = -9$
9. $x = \dfrac{32}{5}$ 10. $y = \dfrac{9}{2}$ 11. $n = -\dfrac{5}{2}$ 12. $b = \dfrac{28}{3}$ 13. $x = -7$ 14. $y = 2$ 15. $x = 15$

16. $y = 8$ 17. $\dfrac{10}{3}$ 18. $-\dfrac{20}{3}$ 19. $x = 12$ 20. $n = 56$

2 21.–30. Each solution should check.

Self-Study Exercises 8–3

1 1. $x = 6$ 2. $m = 7$ 3. $a = -3$ 4. $m = -\dfrac{1}{2}$ 5. $y = 8$ 6. $x = 0$

2 7. $b = -1$ 8. $x = 8$ 9. $t = 6$ 10. $x = 3$ 11. $y = 5$ 12. $x = -\dfrac{7}{2}$ 13. $a = -2$

14. $x = -9$ 15. $x = \dfrac{1}{4}$ 16. $x = -8$ 17. $t = 7$ 18. $x = -1$ 19. $y = 11$ 20. $y = -2$

21. $x = 27$ 22. $y = 13$ 23. $R = -68$ 24. $P = \dfrac{5}{6}$ 25. $x = 14$ 26. $x = -\dfrac{4}{15}$ 27. $m = \dfrac{49}{18}$

28. $s = \dfrac{8}{5}$ 29. $m = \dfrac{8}{3}$ 30. $T = \dfrac{68}{9}$ 31. $x = 2$ 32. $R = 0.8$ 33. $x = 6.03$ 34. $x = 0.08$
35. $R = 0.34$

Self-Study Exercises 8–4

1 1. $x = 3$ 2. $y = -1$ 3. $x = 1$ 4. $t = -7$ 5. $x = \dfrac{1}{3}$ 6. $x = 6$ 7. $a = -3$ 8. $x = 2$

9. $x = \dfrac{1}{2}$ 10. $b = \dfrac{20}{3}$

Self-Study Exercises 8–5

1   **1.** $x + 4 = 12; x = 8$ **2.** $2x - 4 = 6; 5$ **3.** $x + x + (x - 3) = 27;$ **4.** $24 + x = 60;$
Two parts weigh 10 lb.   36 gal
Third part weighs 7 lb.

5. $4.03 - 3.97 = x;$ **6.** $x + 3x = 400$ **7.** $x + 2x + \$70 = \$235$
   0.06 kg   100 ft of solid pipe   Spanish text: \$55
             300 ft of perforated pipe   Calculator: \$110

8. $C = \$14(4)(150) = \$8{,}400$ **9.** $x + 2x = 325$ **10.** $A = \dfrac{1}{2}bh;$ 12 ft

tank 1: $216\dfrac{2}{3}$ gal $60 = \dfrac{1}{2}(b)(10)$

tank 2: $108\dfrac{1}{3}$ gal

11. $B14 = B5 + B6 + B7 + B8 + B9 + B10 + B11 + B12$ $E5 = D5 \div D14 \times 100$
$D14 = D5 + D6 + D7 + D8 + D9 + D10 + D11 + D12$ $E6 = D6 \div D14 \times 100$
$C5 = B5 \div B14 \times 100$ $E7 = D7 \div D14 \times 100$
$C6 = B6 \div B14 \times 100$
$C7 = B7 \div B14 \times 100$ $\vdots$
$\vdots$ $E12 = D12 \div D14 \times 100$
$C12 = B12 \div B14 \times 100$ $F5 = (D5 - B5) \div B5 \times 100$
     $F6 = (D6 - B6) \div B6 \times 100$
     $F7 = (D7 - B7) \div B7 \times 100$
     $\vdots$
     $F12 = (D12 - B12) \div B12 \times 100$

12. See the figure.

	A	B	C	D	E	F
1	The 7th Inning: Budget Operating Expenses and Actual Expenses					
2						
3	Expense	Budget Amount	Percent of Total Budget	Actual Expenses	Percent of Actual Total Expense	Percent Difference from Budget
4						
5	Salaries	$ 45,000.00	17.93%	$ 42,000.00	16.57%	−6.7%
6	Rent	$ 37,000.00	14.74%	$ 36,000.00	14.20%	−2.7%
7	Depreciation	$ 12,000.00	4.78%	$ 14,000.00	5.52%	+16.7%
8	Utilities and phone	$ 13,000.00	5.18%	$ 10,862.56	4.28%	−16.4%
9	Taxes and Insurance	$ 15,000.00	5.98%	$ 13,583.29	5.36%	−9.4%
10	Advertising	$ 2,000.00	0.80%	$ 2,847.83	1.12%	+42.4%
11	Purchases	$ 125,000.00	49.80%	$ 132,894.64	52.41%	+6.3%
12	Other	$ 2,000.00	0.80%	$ 1,356.35	0.53%	−32.2%
13						
14	Total	$ 251,000.00	100.01%	$ 253,544.67	99.99%	

Self-Study Exercises 8–6

1   **1.** 28   **2.** −11   **3.** $\dfrac{18}{7}$ **4.** 0   **5.** $-\dfrac{27}{5}$ **6.** $-\dfrac{5}{27}$ **7.** $\dfrac{8}{5}$ **8.** $\dfrac{11}{15}$ **9.** −350   **10.** 48

11. 27   **12.** $\dfrac{1}{3}$ **13.** 13   **14.** $\dfrac{25}{8}$ **15.** 0   **16.** $\dfrac{1}{9}$ **17.** 576   **18.** 4   **19.** 28   **20.** −68

21. $\dfrac{27}{35}$ **22.** $\dfrac{108}{5}$ **23.** $\dfrac{217}{24}$ **24.** 8   **25.** −5   **26.** $\dfrac{3}{10}$ **27.** 70   **28.** 0   **29.** $\dfrac{32}{63}$ **30.** $\dfrac{32}{5}$

31. $-\dfrac{36}{11}$ **32.** $-\dfrac{1}{108}$ **33.** 21   **34.** $\dfrac{4}{25}$ **35.** $\dfrac{2}{21}$ **36.** $\dfrac{9}{7}$ **37.–42.** Answers to 12, 14, 16, 18, 25, and 32 should be checked.

2 **43.** 80 fixtures   **44.** $1\frac{1}{3}$ hours   **45.** $3\frac{1}{13}$ hr   **46.** $3\frac{3}{7}$ days   **47.** $2\frac{1}{3}$ min

Self-Study Exercises 8–7

1 **1.** 2   **2.** 0.8   **3.** 6.03   **4.** 16   **5.** 4.7   **6.** 0.08   **7.** 0.035   **8.** 0.34   **9.** -3.3
10. 38.8   **11.** 3.3   **12.** -0.2 **13.** 0.6   **14.** 1.128   **15.** 16

2 **16.** 87.5 lb   **17.** 4.71 in.   **18.** 12 hr   **19.** 6.5 hr   **20.** 8.3 Ω **21.** \$5.30   **22.** 156.25 V
23. \$264   **24.** \$136   **25.** \$1,925   **26.** 12.25%   **27.** 1.2%   **28.** 18.5%   **29.** \$2,777.78
30. $\frac{1}{2}$ yr or 6 mo

Assignment Exercises, Chapter 8

1. $\boxed{15x} - \boxed{\dfrac{3x}{7}} + \boxed{\dfrac{x-7}{5}}$ **3.** Five more than a number is two.   **5.** A number divided by 8 is seven.

7. The product of seven more than a number times three is negative three.   **9.** $2x + 7 = 11$

11. $2(x + 8) = 40$ **13.** $x = 7$ **15.** $b = -\dfrac{15}{2}$ **17.** $y = 7$ **19.** $x = -8$ **21.** $x = 15$

23. $x = 64$ **25.** $x = -49$ **27.** $x = 84$ **29.** $x = 5$ **31.** $b = -2$ **33.** $x = 7$ **35.** $x = -4$
37. $x = 3$ **39.** $x = -1$ **41.** $a = 5$ **43.** $x = 5$ **45.** $x = 7$ **47.** $x = 4$ **49.** $x = 2$

51. $x = -3$ **53.** $x = \dfrac{1}{3}$ **55.** $y = -12$ **57.** $y = -3$ **59.** $x = \dfrac{9}{2}$ **61.** $y = 7$ **63.** $x = 9$

65. $x = \dfrac{20}{13}$ **67.** $x = \dfrac{7}{20}$ **69.** $c = -\dfrac{9}{32}$ **71.** $y = 1.5$ **73.** $R = 0.61$ **75.** $y = -1$

77. $x = -9$ **79.** $x = 1$ **81.** $x = 3$ **83.** $x = 2$ **85.** $x = -2$ **87.** $x = 2$ **89.** $x = 0$ **91.** $x = 3$

93. $x = 3$ **95.** $x = -1$ **97.** $x = -\dfrac{1}{8}$ **99.** $x - 6 = 8; x = 14$ **101.** $5(x + 6) = 42 + x; x = 3$

103. $x + (x - 3) = 51; 27$ hr, 24 hr   **105.** $6 = \pi d$; no; $D = 1.9$ in.   **107.** $A = x(2x); 120$ ft $\times$ 60 ft

109.

1	A	B	C	D	E	F	G
2	Employee	Date	Hourly Rate	Hours Worked for Week	Regular Pay	Overtime Pay	Total Gross Pay
3	Gayden, Bertha	8/19	\$ 9.25	40	\$ 370.00	\$ 0	\$ 370.00
4	Harrover, Roy	8/19	\$ 13.60	42	\$ 544.00	\$ 40.80	\$ 584.80
5	Kearney, Claude	8/19	\$ 16.50	48	\$ 660.00	\$ 198.00	\$ 858.00
6	Oswalt, Alisa	8/19	\$ 6.45	40	\$ 258.00	\$ 0	\$ 258.00
7	Stapleton, Iven	8/19	\$ 9.15	45	\$ 366.00	\$ 68.63	\$ 434.63
8	Total Gross Payroll				\$2,198.00	\$ 307.43	\$2,505.43

111. Answers will vary. An example would be, Three times the sum of a number and 5 is increased by 2 and the result is 80. What is the number?
113. $\dfrac{5}{6}$ **115.** $-\dfrac{4}{15}$ **117.** $\dfrac{49}{18}$ **119.** $-\dfrac{60}{13}$ **121.** $\dfrac{8}{3}$ **123.–127.** Answers to 114, 116, 118, and 130 should be checked.   **129.** $-\dfrac{7}{4}$ **131.** $\dfrac{16}{7}$ **133.** $\dfrac{10}{7}$ **135.** 6   **137.** $\dfrac{31}{39}$ **139.** $2\dfrac{1}{10}$ hr
141. 25 fixtures   **143.** 1.9   **145.** 2   **147.** 77.2   **149.** -11.8 **151.** -0.8 **153.** 1.5
155. 0.44   **157.** 0.02   **159.** 3.4   **161.** 17 Ω **163.** 110 V   **165.** 375 lb

Trial Test, Chapter 8

1. $\boxed{5x^2}$ **3.** $x + 5 = 35$ **5.** Eight more than a number is twenty-one.   **7.** 6   **9.** -15 **11.** 3
13. 2   **15.** 2   **17.** 2   **19.** Henderson: 1,500; Brinks: 1,625   **21.** 16   **23.** $-\dfrac{22}{7}$ **25.** $-\dfrac{56}{3}$

27. $\dfrac{7}{25}$ **29.** 6.17   **31.** 254.24   **33.** 1.33 psi   **35.** 11.429 Ω

Self-Study Exercises 9–1

1 **1.** $\dfrac{1}{3}$ and $\dfrac{4}{12}$ **2.** $\dfrac{2}{5}$ and $\dfrac{4}{10}$ **3.** $\dfrac{4}{8}$ and $\dfrac{5}{10}$ **4.** $\dfrac{18}{24}$ and $\dfrac{9}{12}$ **5.** $7 \times 24 = 168$ **6.** $3 \times 20 = 60$
$8 \times 21 = 168$ $5 \times 12 = 60$

7. $3 \times 24 = 72$ **8.** $4 \times 10 = 40$
$8 \times 6 = 48$ $5 \times 8 = 40$
No proportion

2 **9.** 3   **10.** 2   **11.** 29   **12.** $\dfrac{5}{8}$ **13.** $\dfrac{34}{5}$ **14.** $\dfrac{17}{14}$ **15.** 3   **16.** 21   **17.** 4   **18.** $-\dfrac{4}{5}$

Self-Study Exercises 9–2

1 **1.** \$5.90   **2.** \$74.13   **3.** 48 engines   **4.** 15 headpieces   **5.** 1,425 mi   **6.** 3,360 mi
7. 550 lb   **8.** 1.7 gal

2 **9.** 900 rpm   **10.** 200 in.^{3} **11.** 20 in.   **12.** 18 painters   **13.** 10 hr   **14.** 50 rpm   **15.** 5 in.
16. 4 hr   **17.** 5 machines   **18.** 120 rpm   **19.** 4 helpers (5 workers total)   **20.** 9 in.

3 **21.** $\angle A = \angle F, \angle B = \angle E, \angle C = \angle D$ **22.** $PQ = ST, \angle P = \angle S, \angle Q = \angle T$

23. $JL = MP, LK = NP, \angle L = \angle P$ **24.** $a = 20, d = 9$ **25.** $DC = 6, DE = 5\dfrac{1}{3}$ **26.** 65 ft

Assignment Exercises, Chapter 9

1. $20 \neq 24; 120 = 120$ **3.** This is a proportion   **5.** This is a proportion   **7.** 2   **9.** $\dfrac{7}{6}$ **11.** $\dfrac{1}{2}$

13. $\dfrac{49}{12}$ **15.** $-\dfrac{21}{2}$ **17.** $\dfrac{3}{14}$ **19.** $-\dfrac{21}{23}$ **21.** 23 machines   **23.** $2\dfrac{1}{2}$ hr   **25.** 4,500 women

27. 1,500 rpm   **29.** 3 days   **31.** $4\dfrac{1}{5}$ ft   **33.** 129.7 gal   **35.** 5.0 hr   **37.** 1,351 ft   **39.** 60 rpm

41. 3 machines   **43.** $\dfrac{AB}{RT} = \dfrac{BC}{TS} = \dfrac{AC}{RS}$ **45.** 10   **47.** 100 teeth   **49.** $2\dfrac{6}{7}$ gal water, $7\dfrac{1}{7}$ gal antifreeze

Trial Test, Chapter 9

1. no; no; yes   **3.** $Q = -2\dfrac{1}{5}$ or -2.2 **5.** $y = -\dfrac{22}{7}$ or $-3\dfrac{1}{7}$ **7.** $x = 0.30$ or $\dfrac{8}{27}$

9. $x = \dfrac{675}{4}$ or $168\dfrac{3}{4}$ rpm   **11.** $x = 54.7$ L   **13.** $x = 107$ lb   **15.** $x = 5$

Self-Study Exercises 10–1

1 **1.** x^7 **2.** m^4 **3.** a^2 **4.** x^{10} **5.** y^6 **6.** a^2b **7.** a^5b^7 **8.** x^{11}

2 **9.** y^5 **10.** x^4 **11.** $\dfrac{1}{a}$ **12.** b **13.** 1   **14.** $\dfrac{1}{x^2}$ **15.** y^4 **16.** n^7 **17.** $\dfrac{1}{x^{10}}$ **18.** $\dfrac{1}{n}$

19. x^7 **20.** $\dfrac{1}{x^5}$ **21.** $\dfrac{x}{y^6}$ **22.** $\dfrac{a^3}{b^2}$ **23.** $\dfrac{x}{y^2} + y^2$ **24.** $a - \dfrac{1}{b}$

3 **25.** 4,096   **26.** x^6 **27.** 1   **28.** x^{40} **29.** x^{21} **30.** $-\dfrac{1}{8}$ **31.** $\dfrac{4}{49}$ **32.** $\dfrac{a^4}{b^4}$ **33.** $8m^6n^3$

34. $\dfrac{x^6}{y^3}$　**35.** x^6y^{12}　**36.** $4a^2$　**37.** x^6y^3　**38.** $-27a^3b^6$　**39.** $-16{,}807x^{10}$　**40.** $16x^4y^{16}$

Self-Study Exercises 10–2

1　**1.** $7a^2$　**2.** $3x^3$　**3.** $-2a^2 + 3b^2$　**4.** $2a - 2b$　**5.** $-5x^2 - 2x$　**6.** $5a^2$　**7.** $-x^2 - 2y$
8. $2m^2 + n^2$　**9.** $9a + 2b + 10c$　**10.** $-2x + 3y + 2z$

2　**11.** $14x^3$　**12.** $2m^3$　**13.** $-21m^2$　**14.** $-2y^6$　**15.** $2x^2$　**16.** $-\dfrac{a}{2}$　**17.** $\dfrac{1}{2x^2}$　**18.** $-\dfrac{3x^3}{4}$
19. $3x^2 - 18x$　**20.** $12x^3 - 28x^2 + 32x$　**21.** $-8x^2 + 12x$　**22.** $10x^2 + 4x^3$　**23.** $3x - 2$
24. $4x^3 - 2x - 1$　**25.** $7x - \dfrac{1}{x}$　**26.** $4x^2 + 3x$　**27.** $\dfrac{x^2}{3} + 15x^4$　**28.** $5ab^2 - 3b - \dfrac{7}{ab}$
29. $36a^5x^6$　**30.** $96x^{13}y^9$

Self-Study Exercises 10–3

1　**1.** 37　**2.** 1,820　**3.** 0.56　**4.** 1.42　**5.** 780,000　**6.** 62　**7.** 0.00046　**8.** 0.61　**9.** 0.72
10. 42　**11.** 10^{10}　**12.** 10^{-7} or $\dfrac{1}{10^7}$　**13.** 10^{-3} or $\dfrac{1}{10^3}$　**14.** 10　**15.** 10^3　**16.** 10^2　**17.** $\dfrac{1}{10^3}$
18. 1　**19.** $\dfrac{1}{10^5}$　**20.** $\dfrac{1}{10}$

2　**21.** 430　**22.** 0.0065　**23.** 2.2　**24.** 73　**25.** 0.093　**26.** 83,000　**27.** 0.0058　**28.** 80,000
29. 6.732　**30.** 0.00589

3　**31.** 3.92×10^2　**32.** 2×10^{-2}　**33.** 7.03×10^0　**34.** 4.2×10^4　**35.** 8.1×10^{-2}
36. 2.1×10^{-3}　**37.** 2.392×10^1　**38.** 1.01×10^{-1}　**39.** 1.002×10^0　**40.** 7.21×10^2
41. 4.2×10^5　**42.** 3.26×10^4　**43.** 2.13×10^1　**44.** 6.2×10^{-6}　**45.** 5.6×10^1

4　**46.** 2.144×10^7　**47.** 5.6×10^3　**48.** 2.36×10^{-2}　**49.** 4.73×10^{-3}　**50.** 7×10^2
51. 6.5×10^{-5}　**52.** 7×10^2　**53.** 8×10^{-3}　**54.** 3.2285×10^{13}　**55.** 4.2×10^2 or 420

Self-Study Exercises 10–4

1　**1.** binomial　**2.** binomial　**3.** monomial　**4.** monomial　**5.** monomial　**6.** monomial
7. trinomial　**8.** trinomial　**9.** monomial　**10.** binomial　**11.** binomial　**12.** binomial

2　**13.** 1　**14.** 2　**15.** 2, 1, 0　**16.** 3, 1, 0　**17.** 1, 0　**18.** 2, 0　**19.** 0　**20.** 0　**21.** 1, 0
22. 2, 0　**23.** 2　**24.** 3　**25.** 6　**26.** 5　**27.** 2　**28.** 1

3　**29.** $-3x^2 + 5x$; 2; $-3x^2$; -3　**30.** $-x^3 + 7$; 3; $-x^3$; -1　**31.** $9x^2 + 4x - 8$; 2; $9x^2$; 9
32. $5x^2 - 3x + 8$; 2; $5x^2$; 5　**33.** $7x^3 + 8x^2 - x - 12$; 3; $7x^3$; 7　**34.** $-15x^4 + 12x + 7$; 4; $-15x^4$;
-15　**35.** $8x^6 - 7x^3 - 7x$; 6; $8x^6$; 8　**36.** $-14x^8 + x + 15$; 8; $-14x^8$; -14

Self-Study Exercises 10–5

1　**1.** $x^{\frac{3}{5}}$; $\sqrt[5]{x^3}$　**2.** $x^{\frac{3}{5}}$; $(\sqrt[5]{x})^3$　**3.** $x^{\frac{4}{2}}$; $\sqrt{x^4}$　**4.** $x^{\frac{2}{6}}$; $(\sqrt[6]{x})^2$　**5.** $x^{\frac{3}{4}}$; $\sqrt[4]{x^3}$　**6.** $x^{\frac{4}{2}}$; $(\sqrt{x})^4$　**7.** $a^{\frac{1}{4}}$

8. $m^{\frac{1}{3}}$　**9.** $(3x)^{\frac{1}{5}}$　**10.** $(7a)^{\frac{1}{4}}$　**11.** $(xy)^2$　**12.** $3a^2$　**13.** $32^{\frac{1}{3}}x^{\frac{7}{3}}$　**14.** $1{,}296x^6y^8$　**15.** $16^{\frac{2}{3}}a^{\frac{8}{3}}b^{\frac{14}{3}}$

2　**16.** a^3　**17.** $b^{\frac{4}{5}}$　**18.** $64xy^{12}$　**19.** $32mn^{10}$　**20.** $a^{\frac{2}{3}}$　**21.** $\dfrac{1}{b^{\frac{2}{5}}}$　**22.** $m^{\frac{1}{8}}$　**23.** $x^{\frac{7}{6}}$　**24.** $a^{\frac{5}{3}}$

25. $2a^{\frac{17}{5}}$　**26.** $\dfrac{1}{3a^{\frac{13}{3}}}$　**27.** $\dfrac{2}{9a^{\frac{9}{8}}}$　**28.** $a^{\frac{5}{3}} = 3.175$　**29.** $64ab^6$ or 2^{19} or 529,288　**30.** $c^{\frac{7}{20}} = 1.756$

31. $\dfrac{b^{\frac{5}{3}}}{2}$ or $2^{\frac{7}{3}}$　**32.** $a^{7.2}$　**33.** $a^{1.4}$　**34.** $27a^{3.6}b^6$　**35.** $4^{10}c^{13}d^3$

Self-Study Exercises 10–6

1   **1.** $5i$ **2.** $6i$ **3.** $8xi$ **4.** $4y^2i\sqrt{2y}$

2   **5.** i **6.** 1   **7.** 1   **8.** 1

3   **9.** $15 + 0i$ **10.** $0 + 33i$ **11.** $5 + 2i$ **12.** $8 + 4i\sqrt{2}$ **13.** $7 - i\sqrt{3}$ **14.** $0 + 7i$
15. $-4 + 0i$

4   **16.** $11 + 5i$ **17.** $(2\sqrt{3} + 2\sqrt{2}) - 8i\sqrt{3}$ **18.** $4 + i$ **19.** $12 + 20i$ **20.** $1 + 9i$

Self-Study Exercises 10–7

1   **1.** ± 5 **2.** ± 5 **3.** ± 3.464 **4.** ± 0.667 **5.** ± 3 **6.** ± 0.926 **7.** ± 1.155 **8.** ± 4
9. ± 2.449 **10.** $\pm 5i$ **11.** ± 3.207 **12.** ± 3

2   **13.** $x = 81$ **14.** $y = 192$ **15.** $y = 66$ **16.** $x = \pm 5.292$ **17.** $x = -1.979$ **18.** $x = \pm 2$
19. $x = 50$ **20.** $x = \pm 4.899$ **21.** $y = \pm 2.105$ **22.** $y = \pm 2.9$ **23.** $x = \pm 3.606$ **24.** $x = 9$

Assignment Exercises, Chapter 10

1. x^{10} **3.** x^3 **5.** $\dfrac{x}{y^3}$ **7.** x^{12} **9.** x^{15} **11.** $-27x^6$ **13.** $\dfrac{1}{y} - y^2$ **15.** $-2x^4 - 4x^3 + 13x$

17. $11x^2 - 11y^2$ **19.** $21x^6$ **21.** $-\dfrac{2x^3}{3}$ **23.** $10x^3 + 15x^2 - 20x$ **25.** $2x^2 - 4x + 7$ **27.** $\dfrac{x^3}{2} - 6x^6$

29. 10^{12} or $1,000,000,000,000$ **31.** $10^{-3} = \dfrac{1}{1,000} = 0.001$ **33.** $8,730$ **35.** 5.2×10^4

37. 1.7×10^{-4} **39.** 4.368×10^{126} **41.** 3.38×10^{10} **43.** No. The term $-3x^{-2}$ has a negative

exponent.   **45.** $x^{\frac{7}{2}}; \sqrt{x^7}$ **47.** $x^{\frac{2}{3}}; (\sqrt[3]{x})^2$ **49.** $x^{\frac{1}{2}}$ **51.** $x^{\frac{4}{5}}$ **53.** $x^{\frac{4}{3}}y^{\frac{4}{3}}$ **55.** $\sqrt{7}$ **57.** $\sqrt[5]{y^3}$

59. $x^{\frac{1}{3}}$ **61.** $(4y)^{\frac{1}{5}}$ **63.** $2b^4$ **65.** a^2 **67.** y **69.** $27x^{\frac{3}{4}}y^6$ **71.** $64a^3x^{\frac{3}{2}}$ **73.** $x^{\frac{1}{2}}$ **75.** $a^{\frac{7}{6}}$

77. $\dfrac{1}{x^{\frac{1}{8}}}$ **79.** $a^{\frac{8}{3}}$ **81.** $2a^{\frac{7}{2}}; 22.627$ **83.** $\dfrac{3}{2a^{\frac{22}{5}}}; 0.071$ **85.** $a^{6.3}; 78.793$ **87.** $10i$ **89.** $\pm 2y^3i\sqrt{6y}$

91. -1 **93.** i **95.** $0 + 15i$ **97.** $0 - 12i$ **99.** $7 - 4i$ **101.** $11 + i$ **103.** ± 6 **105.** ± 2

107. $\pm 2i$ **109.** ± 3 **111.** $\dfrac{3\sqrt{6}}{2}$ or 3.674 **113.** 46 **115.** ± 10.262 **117.** ± 8.888 **119.** ± 1

121. ± 7.874 **123.** $20 + 15i$

Trial Test, Chapter 10

1. x^5 **3.** $\dfrac{16}{49}$ **5.** $\dfrac{x^4}{y^2}$ **7.** $12a^3 - 8a^2 + 20a$ **9.** 10^6 **11.** $42,000$ **13.** 2.4×10^2

15. 7.83×10^{-3} **17.** 3.5×10^2 **19.** 1.5×10^6 ohms   **21.** $3x^5$ **23.** $x^2y^3z^6$ **25.** $5x^{\frac{1}{6}}y^2$ **27.** $2x$
29. $-i$ **31.** $-3 + 5i$ **33.** ± 9 **35.** ± 1 **37.** 49 **39.** 9 **41.** $\pm 7i$ or no real solutions

Self-Study Exercises 11–1

1   **1.** 25 **2.** 15 **3.** $1,565$ **4.** 9 **5.** 4.5 **6.** \$310   **7.** 14.25% **8.** 3 years   **9.** \$2,600
10. 90 **11.** 8% **12.** \$3,378.38   **13.** 13.6 **14.** 17% **15.** 66 in.   **16.** 9.0 cm   **17.** \$10.50

18. $\dfrac{1}{4}$ mi   **19.** 153.9 in.^{2} **20.** 6.25 km^2 **21.** \$48.75

Self-Study Exercises 11–2

1   **1.** $X = E - I$ **2.** $K = M - A$ **3.** $r = \dfrac{S}{2\pi h}$ **4.** $b = \dfrac{m}{x + y}$ **5.** $y = \dfrac{A - mx}{2m}$ **6.** $T_2 = \dfrac{T_1 V_2}{V_1}$

7. $r = \sqrt{\dfrac{V}{\pi h}}$ 8. $X = \dfrac{m^2}{Y}$ 9. $R = \dfrac{100P}{B}$ 10. $b = \dfrac{P - 2s}{2}$ 11. $r = \dfrac{C}{2\pi}$ 12. $l = \dfrac{A}{w}$

13. $C = \dfrac{R}{A - B}$ 14. $s = \sqrt{A}$ 15. $R = \dfrac{D}{T}$ 16. $D = P - S$ 17. $r = \sqrt{\dfrac{A}{\pi}}$

18. $C = \sqrt{a^2 + b^2}$ 19. $I = A - P$ 20. $T = \dfrac{I}{PR}$

Self-Study Exercises 11–3

1 1. 35   2. 0   3. 45   4. 5   5. 15   6. 10   7. 65   8. 50   9. 80   10. 120

2 11. 158   12. 59   13. 113   14. 122   15. 68   16. 419   17. 590   18. 770   19. 365
20. 32

Self-Study Exercises 11–4

1 1. $P = 209.3$ mm   2. $P = 52.9$ ft   3. 348 in.^{2} 4. 260 ft^2 5. 8,220 ft^2 6. 98 ft
  $A = 5,053.05$ mm^2 $A = 162$ ft^2

2 7. $P = 28.6$ in.   8. $P = 48$ cm   9. 15.75 cm^2 10. 72 ft; 216 ft^2 11. 9.75 ft^2 12. 30 ft^2
  $A = 33$ in.^{2} $A = 96$ cm^2
13. 24.75 in.^{2} 14. \$175.50   15. 8 ft   16. 10 ft 8 in.

3 17. 14.9707 m   18. 3.7123 in.   19. 24.3837 cm   20. 26.7216 cm

4 21. $AC = 24$ cm   22. $BC = 10$ mm   23. $AB = 17$ yd   24. 6.403 cm   25. 6.325 m
26. 12 ft   27. 30 ft   28. 39.783 cm   29. 26 ft 10 in.   30. 12.806 ft   31. 13 cm   32. 45 dm
33. 10.607 mm   34. 61.083 ft   35. $E_a = 170.294$ V

5 36. $LSA = 28$ units^2 37. $LSA = 848.23$ units^2 38. 3,244 ft^2 39. 39.4 in.^{2}
  $TSA = 432$ units^2 $TSA = 1,357.17$ units^2
40. 14,700 mm^2 41. 37.3 in.^{2} 42. 600 in.^{2} 43. 69.1 in.^{2}

6 44. $V = 576$ units^3 45. $V = 3,817.04$ units^3 46. 23.3 in.^{3} 47. 18,850 ft^3 48. 320 cm^3
49. 55.36 in.^{3} 50. 236 ft^3

7 51. 314.2 cm^2 52. 904.8 in.^{3} 53. 6,361.7 ft^2 54. 356,892.8 gal   55. 50.3 ft^2
56. 225.6 gal

8 57. $LSA = 188.5$ ft^2; $TSA = 301.6$ ft^2; $V = 301.6$ ft^3 58. 4,712.4 ft^3 59. 589.0 cm^2
60. 50.3 L   61. 1,649.3 ft^2 62. 10.0 ft   63. \$347

Self-Study Exercises 11–5

1 1. 4.8 Ω 2. 61.6%   3. 35 miles per hour   4. 1.5 amperes   5. 6 ft^3 6. 3 amperes
7. 2 cylinders   8. 1,600 rpm   9. 71.71 in.   10. 4.4 Ω

Assignment Exercises, Chapter 11

1. 12.5   3. 3   5. 1.25   7. 9.6%   9. 10.8   11. \$193.60   13. 2 years   15. 50 in.

17. \$89.50   19. 95.0 in.^{2} 21. $F = D + m + n$ 23. $w = \dfrac{V}{lh}$ 25. $a = \dfrac{c}{h + b}$ 27. $r = \sqrt{\dfrac{B}{cx}}$

29. $B = \dfrac{A}{P}$ 31. $r = \dfrac{I}{Pt}$ 33. $r = s + d$ 35. $t = \dfrac{V_0 - V}{32}$ 37. $t = \dfrac{A - P}{Pr}$ 39. $P = S + D$

41. 203°   43. 104°   45. 185°   47. $P = 12$ in.   49. $P = 339$ mm   51. $A = 8.125$ ft^2
  $A = 6$ in.^{2} $A = 5,899.5$ mm^2

53. $s = 12$; $A = 27.71$ ft^2 **55.** $s = 23$; $A = 99.68$ in.^{2} **57.** $s = 61.5$; $A = 708.6$ in.^{2} **59.** 11.25 ft
61. 19.5 ft^2 **63.** Two sheets   **65.** 48 ft   **67.** 199.5 ft^2 **69.** 15 in.   **71.** 7.141 ft   **73.** 8 yd
75. 20.248 mi   **77.** 30 cm   **79.** 21.633 in.   **81.** 4.243 in.   **83.** 600 cm^3 **85.** 616 cm^2
87. 2,733.186 cm^2 **89.** 102 yd^3 **91.** 60 ft^2 **93.** 348,962 gal   **95.** 1,017.9 m^2 **97.** 7,238.2 ft^3
99. 169.6 cm^2 **101.** 377.0 in.^{3} **103.** 2 gal   **105.** 116.6 ft^2 **107.** 243.3 in.^{2} **109.** 2,094 ft^3
111. 2.75 amperes   **113.** 10 ft^3 **115.** 3.2 amperes   **117.** 1,400 rpm   **119.** 2 cylinders
121. Answers will vary.

Trial Test, Chapter 11

1. $L = \dfrac{RA}{P}$ **3.** $r = \sqrt{\dfrac{d}{\pi s n}}$ **5.** 361   **7.** $s = 16.5$; $A = 36.98$ m^2 **9.** 250 ft^2 **11.** 10 dm
13. 50 in.^{2} **15.** 6,768 gal   **17.** 3,848.5 ft^3 **19.** 2,080.7 cm^2 **21.** 10.9 ft^3 **23.** 1,497.05 lb
25. 12,000 calories

Self-Study Exercises 12–1

1   **1.** $7(a + b)$ **2.** $m(m + 2)$ **3.** $3x(2x + 1)$ **4.** $5(ab + 2a + 4b)$ **5.** $2ax(2x + 3a)$
  6. $a(5 - 7b)$ **7.** $3(4a^2 - 5a + 2)$ **8.** $3x(x^2 - 3x - 2)$ **9.** $2ab(4a + 7b^2)$ **10.** $3m^2(1 - 2m)$

Self-Study Exercises 12–2

1   **1.** $a^2 + 11a + 24$ **2.** $x^2 + x - 20$ **3.** $y^2 - 10y + 21$ **4.** $2a^2 + 5a - 12$
  5. $3a^2 - 8ab + 4b^2$ **6.** $5cx - 25xy - cy + 5y^2$ **7.** $6x^2 - 17x + 12$ **8.** $2a^2 - 7ab + 5b^2$
9. $21 - 52m + 7m^2$ **10.** $40 - 21x + 2x^2$ **11.** $x^2 + 11x + 28$ **12.** $y^2 - 12y + 35$
13. $m^2 - 4m - 21$ **14.** $3bx + 18b - 2x - 12$ **15.** $12r^2 - 7r - 10$ **16.** $35 - 22x + 3x^2$
17. $4 - 14m + 6m^2$ **18.** $6 + 13x + 6x^2$ **19.** $2x^2 + x - 15$ **20.** $20x^2 - 13xy - 21y^2$
21. $14a^2 + 19ab - 3b^2$ **22.** $30a^2 - 13ab - 10b^2$ **23.** $27x^2 + 30xy - 8y^2$ **24.** $20x^2 - 47xy + 24y^2$
25. $21m^2 + 29mn - 10n^2$

2   **26.** $a^2 - 9$ **27.** $4x^2 - 9$ **28.** $a^2 - y^2$ **29.** $16r^2 - 25$ **30.** $25x^2 - 4$ **31.** $49 - m^2$
32. $4 - 12x + 9x^2$ **33.** $9x^2 + 24x + 16$ **34.** $Q^2 + 2QL + L^2$ **35.** $a^4 + 2a^2 + 1$
36. $4d^2 - 20d + 25$ **37.** $9a^2 + 12ax + 4x^2$ **38.** $9x^2 - 49$ **39.** $36 + 12Q + Q^2$ **40.** $y^2 - 25x^2$
41. $16 - 24j + 9j^2$ **42.** $9m^2 - 4p^2$ **43.** $m^4 + 2m^2p^2 + p^4$ **44.** $4a^2 - 49c^2$
45. $81 - 234a + 169a^2$

Self-Study Exercises 12–3

1   **1.** difference   **2.** not difference   **3.** difference   **4.** difference   **5.** not difference
  6. difference   **7.** $(y + 7)(y - 7)$ **8.** $(4x + 1)(4x - 1)$ **9.** $(3a + 10)(3a - 10)$
10. $(2m + 9n)(2m - 9n)$ **11.** $(3x + 8y)(3x - 8y)$ **12.** $(5x + 8)(5x - 8)$ **13.** $(10 + 7x)(10 - 7x)$
14. $(2x + 7y)(2x - 7y)$ **15.** $(11m + 7n)(11m - 7n)$ **16.** $(9x + 13)(9x - 13)$

2   **17.** not perfect square   **18.** perfect square   **19.** not perfect square   **20.** not perfect square
21. perfect square   **22.** perfect square   **23.** $(x + 3)^2$ **24.** $(x + 7)^2$ **25.** $(x - 6)^2$ **26.** $(x - 8)^2$
27. $(2a + 1)^2$ **28.** $(5x - 1)^2$ **29.** $(3m - 8)^2$ **30.** $(2x - 9)^2$ **31.** $(x - 6y)^2$ **32.** $(2a - 5b)^2$

Self-Study Exercises 12–4

1   **1.** $(x + 3)(x + 2)$ **2.** $(x - 7)(x - 4)$ **3.** $(x + 6)(x + 2)$ **4.** $(x - 3)(x - 1)$
5. $(x + 7)(x + 1)$ **6.** $(x + 5)(x + 2)$ **7.** $(x - 4)(x + 3)$ **8.** $(y - 5)(y + 2)$
9. $(y - 3)(y + 2)$ **10.** $(a + 4)(a - 3)$ **11.** $(b + 3)(b - 1)$ **12.** $(7 + b)(2 - b)$
13. $(x - 4)(x - 3)$ **14.** $(x - 6)(x + 5)$ **15.** $(x + 9)(x + 2)$ **16.** $(x - 6)(x - 3)$
17. $(x - 9)(x + 2)$ **18.** $(x + 18)(x - 1)$ **19.** $(x + 5)(x + 4)$ **20.** $(x - 10)(x - 2)$
21. $(x - 8)(x - 2)$ **22.** $(x - 16)(x - 1)$ **23.** $(x - 14)(x + 1)$ **24.** $(x - 7)(x + 2)$

2 **25.** $(x + y)(x + 4)$ **26.** $(3x + 2)(2x - y)$ **27.** $(3x + 5)(m - 2n)$ **28.** $(6x - 7)(5y - 6)$
29. $(x - 2)(x + 8)$ **30.** $(3x - 1)(2x - 7)$ **31.** $(x - 4)(x + 1)$ **32.** $(2x - 1)(4x + 3)$
33. $(x - 5)(x + 4)$ **34.** $(x - 2)(3x + 5)$

3 **35.** $(3x + 1)(x + 2)$ **36.** $(3x + 2)(x + 4)$ **37.** $(3x + 2)(2x + 3)$ **38.** $(x - 2)(x - 16)$
39. $(3x - 4)(2x - 3)$ **40.** $(2x - 5)(x - 2)$ **41.** $(3x - 5)(2x - 1)$ **42.** $(p + 9)(p - 4)$
43. $(6x - 5)(x - 1)$ **44.** $(4x + 3)(2x + 5)$ **45.** $(3x - 5)(5x + 1)$ **46.** $(y - 12)(y - 3)$
47. $(2x - 7)(x + 1)$ **48.** $(6x - 5)(2x + 3)$ **49.** $(5x + 3)(2x - 1)$ **50.** $(Q - 11)(Q + 4)$

4 **51.** $(6x - 5y)(x + 2y)$ **52.** $(3a + 2b)(2a - 7b)$ **53.** $(6x - 5)(3x + 2)$ **54.** $(5x - 4y)(4x + 3y)$
55. $(x + 8)(x - 7)$ **56.** $(a + 19)(a + 2)$

5 **57.** $4(x - 1)$ **58.** $(x + 3)(x - 2)$ **59.** $(2x + 3)(x - 1)$ **60.** $(x + 3)(x - 3)$
61. $4(x + 2)(x - 2)$ **62.** $(m + 5)(m - 3)$ **63.** $2(a + 2)(a + 1)$ **64.** $(b + 3)^2$ **65.** $(4m - 1)^2$
66. $(x + 7)(x + 1)$ **67.** $(2m + 1)(m + 2)$ **68.** $(2m + 1)(m - 3)$ **69.** $(2a - 5)(a + 1)$
70. $(3x - 2)(x + 4)$ **71.** $(3x + 5)(2x - 3)$ **72.** $(4x - 1)(2x + 3)$

Assignment Exercises, Chapter 12

1. $5(x + y)$ **3.** $4(3m^2 - 2n^2)$ **5.** $2a(a^2 - 7a - 1)$ **7.** $5x(3x - 4)(x + 1)$ **9.** $6a^2(3a + 2)$
11. $x^2 + 11x + 28$ **13.** $m^2 - 4m - 21$ **15.** $12r^2 - 7r - 10$ **17.** $4 - 14m + 6m^2$
19. $2x^2 + x - 15$ **21.** $14a^2 + 19ab - 3b^2$ **23.** $27x^2 + 30xy - 8y^2$ **25.** $21m^2 + 29mn - 10n^2$
27. $36x^2 - 25$ **29.** $49y^2 - 121$ **31.** $64a^2 - 25b^2$ **33.** $64x^2 - y^2$ **35.** $9y^2 - 16z^2$
37. $x^2 + 18x + 81$ **39.** $x^2 - 6x + 9$ **41.** $16x^2 - 120x - 225$ **43.** $64 + 112m + 49m^2$
45. $16x^2 - 88x + 121$ **47.** not difference   **49.** difference   **51.** difference   **53.** not perfect-square
trinomial   **55.** perfect-square trinomial   **57.** perfect-square trinomial   **59.** $(5y - 2)(5y + 2)$
61. *NSP*, this is a sum of two squares, not a difference.   **63.** $(a + 1)^2$ **65.** $(4c - 3b)^2$ **67.** $(n - 13)^2$
69. $(6a + 7b)^2$ **71.** $(7 - x)^2$ **73.** *NSP*, this is a sum of two squares, not a difference.   **75.** *NSP*, the
middle term needs a y factor.   **77.** $(7 + 9y)(7 - 9y)$ **79.** $(3x + 10y)(3x - 10y)$ **81.** $(3x - y)^2$

83. $(3xy + z)(3xy - z)$ **85.** $(x + 2)^2$ **87.** $\left(\frac{2}{5}x + \frac{1}{4}y\right)\left(\frac{2}{5}x - \frac{1}{4}y\right)$ **89.** $(x + 8)(x + 3)$

91. $(x + 10)(x + 3)$ **93.** $(x - 8)(x - 1)$ **95.** $(x - 13)(x + 2)$ **97.** $(x + 8)(x - 3)$
99. $(6x + 1)(x + 4)$ **101.** $(5x - 4)(x - 6)$ **103.** $(3x + 7)(2x - 5)$ **105.** $(7x + 8)(x - 3)$
107. $(3a + 10)(3a - 10)$ **109.** $(2x + 1)(x - 2)$ **111.** $(a + 9)(a - 9)$ **113.** $(y - 7)^2$
115. $(b + 5)(b + 3)$ **117.** $(13 + m)(13 - m)$ **119.** $(x - 8)(x + 4)$ **121.** $(x + 20)(x - 1)$
123. $2(x - 4)(x + 2)$ **125.** $2x(x - 6)(x + 1)$ **127.** This enables us to factor more rapidly and easily.
129. $SA = 375 + 40x + x^2$ **131.** Decrease each side by five. $SA = (25 + x)(15 + x)$. Answers will vary.
200 ft^2

Trial Test, Chapter 12

1. $3x + 6y$ **3.** $14x^3 + 21x^2 - 35x$ **5.** $m^2 - 49$ **7.** $a^2 + 6a + 9$ **9.** $2x^2 - 11x + 15$
11. $x(7x + 8)$ **13.** $7ab(a - 2)$ **15.** $(x + 8)(x + 1)$ **17.** $(x - 3)(x + 8)$ **19.** $(3x + 5)(x + 6)$
21. $(a + 8b)^2$ **23.** $(b - 5)(b + 2)$ **25.** $(3x + 4)(x - 1)$ **27.** $(3m - 2)(m - 1)$ **29.** Prime

Self-Study Exercises 13–1

1 **1.** $7x^2 - 4x + 5 = 0$ **2.** $8x^2 - 6x - 3 = 0$ **3.** $7x^2 - 5 = 0$ **4.** $x^2 - 6x + 8 = 0$
5. $x^2 - 9x + 8 = 0$ **6.** $x^2 - 4x - 8 = 0$ **7.** $3x^2 - 6x + 5 = 0$ **8.** $x^2 - 6x - 5 = 0$
9. $x^2 - 16 = 0$ **10.** $8x^2 - 7x - 8 = 0$ **11.** $8x^2 + 8x - 10 = 0$ **12.** $0.3x^2 - 0.4x - 3 = 0$

2 **13.** $x^2 - 5x = 0$ **14.** $3x^2 - 7x + 5 = 0$ **15.** $7x^2 - 4x = 0$
     $a = 1, b = -5, c = 0$ $a = 3, b = -7, c = 5$ $a = 7, b = -4, c = 0$
16. $3x^2 - 5x + 8 = 0$ **17.** $x^2 - 5x + 6 = 0$ **18.** $11x^2 - 8x = 0$
     $a = 3, b = -5, c = 8$ $a = 1, b = -5, c = 6$ $a = 11, b = -8, c = 0$
19. $x^2 - x = 0$ **20.** $9x^2 - 7x - 12 = 0$ **21.** $x^2 - 5 = 0$
     $a = 1, b = -1, c = 0$ $a = 9, b = -7, c = -12$ $a = 1, b = 0, c = -5$

22. $x^2 + 6x - 3 = 0$
$a = 1, b = 6, c = -3$

23. $5x^2 - 0.2x + 1.4 = 0$
$a = 5, b = -0.2, c = 1.4$

24. $\frac{2}{3}x^2 - \frac{5}{6}x - \frac{1}{2} = 0$

$a = \frac{2}{3}, b = -\frac{5}{6}, c = -\frac{1}{2}$

25. $1.3x^2 - 8 = 0$
$a = 1.3, b = 0, c = -8$

26. $\sqrt{3}\,x^2 + \sqrt{5}\,x - 2 = 0$
$a = \sqrt{3}, b = \sqrt{5}, c = -2$

27. $8x^2 - 2x - 3 = 0$ **28.** $x^2 + 3x = 0$ **29.** $5x^2 + 2x - 7 = 0$ **30.** $2.5x^2 - 0.8 = 0$

Self-Study Exercises 13–2

1 **1.** $3, -\dfrac{2}{3}$ **2.** $3, -4$ **3.** $\dfrac{11}{5}, -1$ **4.** $3, 3$ **5.** $3, -2$ **6.** $\dfrac{3}{4}, -\dfrac{1}{2}$ **7.** $1.84, -10.84$

8. $-0.18, -1.82$ **9.** $3.14, -0.64$ **10.** $2.39, 0.28$ **11.** $-0.75 \pm 0.97i$ **12.** $0.5 \pm 1.32i$

2 **13.** width = 5.55 cm, length = 8.55 cm
14. width = 4 in., length = 10 in.
15. 4 m
16. width = 11 ft, length = 16 ft
17. length = 110 ft, width = 70 ft (nearest ft)
18. 111.5 kg

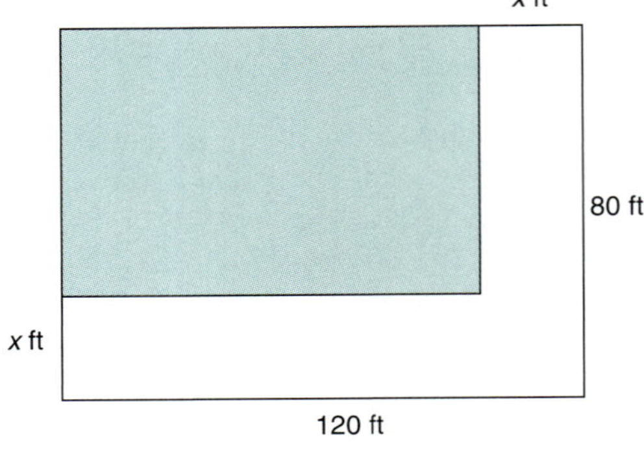

Self-Study Exercises 13–3

1 **1.** $x = \pm 3$ **2.** $x = \pm 7$ **3.** $x = \pm\dfrac{8}{3}$ **4.** $x = \pm\dfrac{7}{4}$ **5.** $y = \pm 3$ **6.** $x = \pm\dfrac{9}{2}$

7. $x = \pm\sqrt{5}$ or ± 2.236 **8.** $x = \pm 1$ **9.** $x = \pm 2$ **10.** $x = \pm 1.732$ **11.** 23 ft **12.** 82 yd
13. Isolate the squared letter, then take the square root of both sides. **14.** opposites

Self-Study Exercises 13–4

1 **1.** $x = 3, 0$ **2.** $x = 0, \dfrac{7}{3}$ **3.** $x = 0, 2$ **4.** $x = 0, -\dfrac{1}{2}$ **5.** $x = 0, \dfrac{1}{2}$ **6.** $x = 0, 3$

7. $x = 0, 6$ **8.** $y = 0, -4$ **9.** $x = 0, -\dfrac{2}{3}$ **10.** $x = 0, \dfrac{4}{3}$ **11.** 8 or 0 **12.** 15 units

13. An incomplete quadratic equation is missing the constant or number term while a pure quadratic equation is missing the linear term. **14.** Yes, the common factor of x will be set equal to zero.

Self-Study Exercises 13–5

1 **1.** $-3, -2$ **2.** $3, 3$ **3.** $7, -2$ **4.** $3, -6$ **5.** $-4, -3$ **6.** $5, 3$ **7.** $14, -1$ **8.** $3, 6$

9. $\dfrac{1}{2}, 3$ **10.** $-\dfrac{1}{3}, -4$ **11.** $\dfrac{3}{5}, -\dfrac{1}{2}$ **12.** $-\dfrac{1}{3}, -\dfrac{3}{2}$ **13.** $-\dfrac{3}{2}, -5$ **14.** $\dfrac{4}{3}, 2$ **15.** $\dfrac{1}{6}, -3$

16. $w = 5$ ft **17.** $w = 18$ in.
$x = 11$ ft $\quad x = 21$ in.

Self-Study Exercises 13–6

1 **1.** $x = \dfrac{2}{3}, -1$ **2.** $x = \dfrac{3 \pm \sqrt{5}}{2}$ or $2.62, 0.38$ **3.** $x = \dfrac{-1 \pm \sqrt{17}}{4}$ or $0.78, -1.28$

4. no real solutions or $x = \dfrac{1 \pm i\sqrt{2}}{3}$ or $0.33 \pm 0.47i$ **5.** $x = \dfrac{3 \pm \sqrt{37}}{2}$ or $4.54,\ -1.54$

6. $x = \dfrac{-5 \pm \sqrt{97}}{6}$ or 0.81 or -2.47 **7.** If a is positive and c is negative, the equation will have real roots.   **8.** If $b^2 - 4ac$ is a perfect square and is positive, the roots will be real, rational, and unequal.

Assignment Exercises, Chapter 13

1. pure   **3.** pure   **5.** incomplete   **7.** pure   **9.** complete   **11.** $a = 1$, $b = -2$, $c = -8$ **13.** $a = 1$, $b = 3$, $c = -4$ **15.** $a = 1$, $b = -3$, $c = 2$

17. $9, -1$ **19.** $2, -4$ **21.** $2, -\dfrac{1}{2}$ **23.** $1.78, -0.28$ **25.** $-0.23, -1.43$ **27.** $w = 11$ ft, $l = 22$ ft

29. $w = 14$ in., $l = 42$ in.   **31.** $x = \pm 10$ **33.** $x = \pm\dfrac{3}{2}$ **35.** $y = \pm 1.740$ **37.** $x = \pm 2.828$

39. $x = \pm 2.236$ **41.** $x = \pm 2$ **43.** $x = \pm 4.123$ **45.** $y = \pm 3.055$ **47.** $x = \pm 4$ **49.** $x = \pm 8$

51. 16.4 cm   **53.** $0, 5$ **55.** $0, 2$ **57.** $0, -\dfrac{1}{2}$ **59.** $0, 5$ **61.** $0, -\dfrac{2}{3}$ **63.** $0, -3$ **65.** $0, 9$

67. $0, -8$ **69.** $0, \dfrac{5}{3}$ **71.** $0, \dfrac{1}{2}$ **73.** $x = 0$ or 4 **75.** $3, 1$ **77.** $-5, 2$ **79.** $-6, -1$ **81.** $4, 2$

83. $-\dfrac{2}{3}, \dfrac{3}{2}$ **85.** $-\dfrac{2}{5}, \dfrac{5}{2}$ **87.** $-\dfrac{3}{4}, -1$ **89.** $\dfrac{3}{4}, -\dfrac{1}{3}$ **91.** $-21, 2$ **93.** $\dfrac{2}{3}, -1$ **95.** $3, 2$

97. $6, -3$ **99.** $-6, -5$ **101.** width $= 12$ ft, length $= 19$ ft   **103.** real, rational, equal   **105.** real, irrational, unequal

107. real, rational, unequal

Trial Test, Chapter 13

1. pure   **3.** incomplete   **5.** ± 9 **7.** $\pm\dfrac{4}{3}$ **9.** ± 2.33 **11.** $0, 2$ **13.** $3, 2$ **15.** $\dfrac{3}{2}, 4$

17. 169.79 mils   **19.** 1.98 amps   **21.** 6

Self-Study Exercises 14–1

1   Values chosen for table of values may vary.

1.

x	$f(x)$
-2	-13
-1	-10
0	-7
1	-4
2	-1

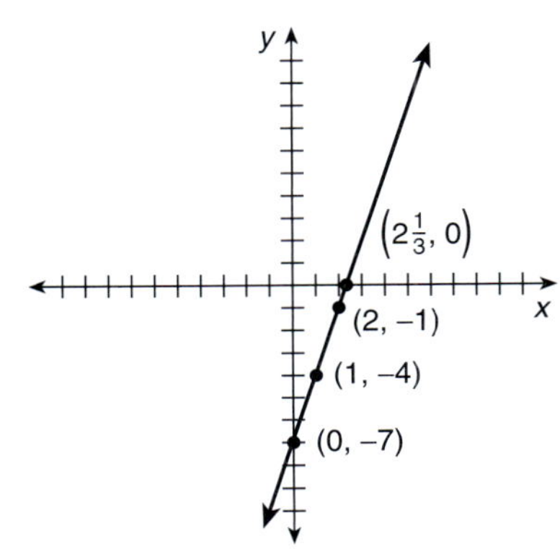

2. $f(x) = 39 + 0.1x$

3.

x	$f(x)$
0	39
50	44
100	49
150	54
200	59

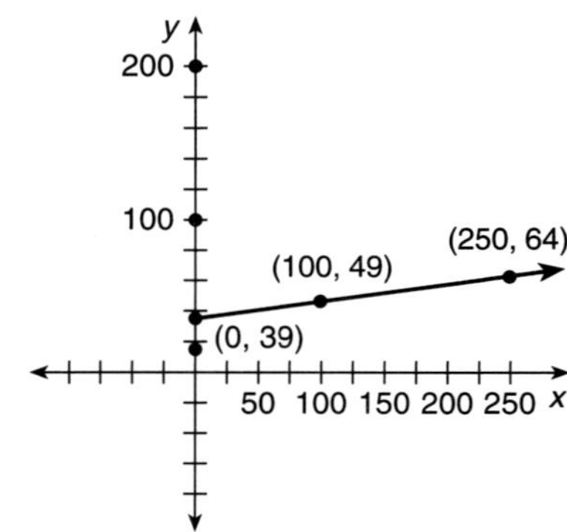

4. $79

5.

x	$f(x)$
-2	17
-1	10
0	5
1	2
2	1
3	2
4	5

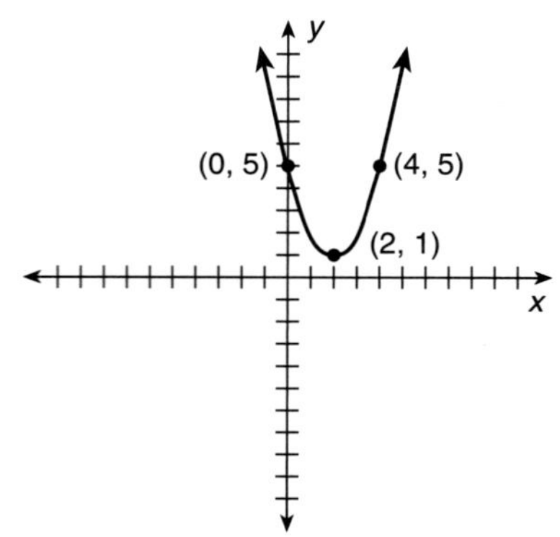

6.

x	$f(x)$
-2	$\frac{1}{9}$
-1	$\frac{1}{3}$
0	1
1	3
2	9

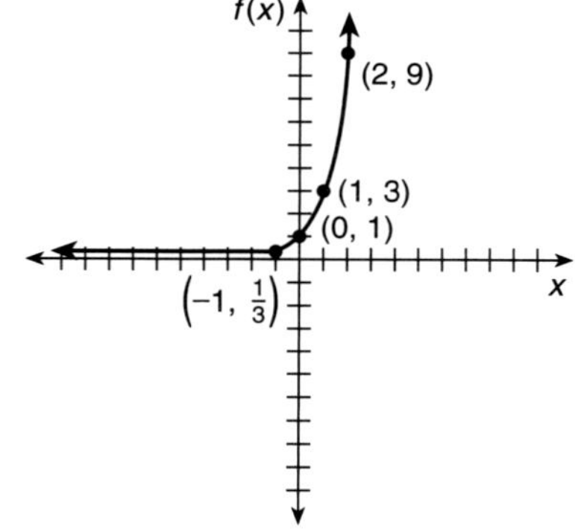

7.

x	$f(x)$
0.1	-1.2
0.3	0
0.5	0.4
1	1.1
2	1.8
3	2.2
4	2.5
5	2.7
8	3.2

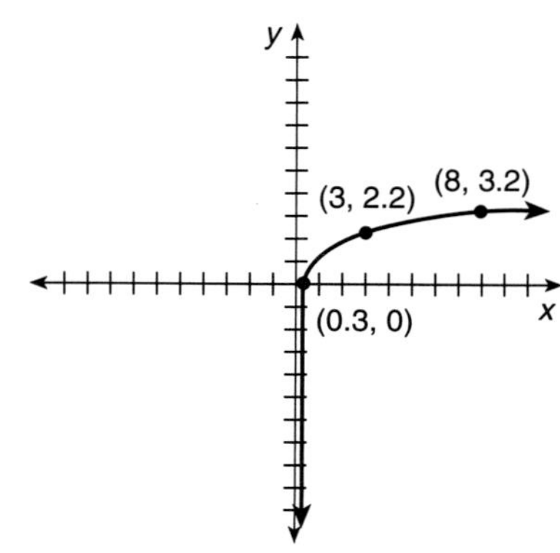

Selected Answers to Student Exercise Material

8. $f(x) = 3x - 5$

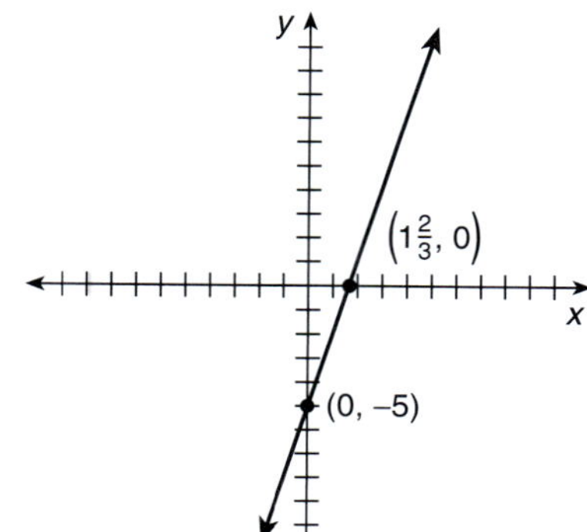

9. $f(x) = 5x - 6$

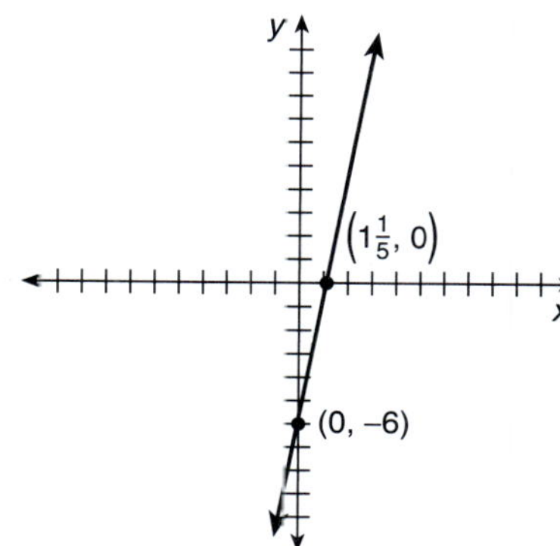

10. $f(x) = 6x - 5$

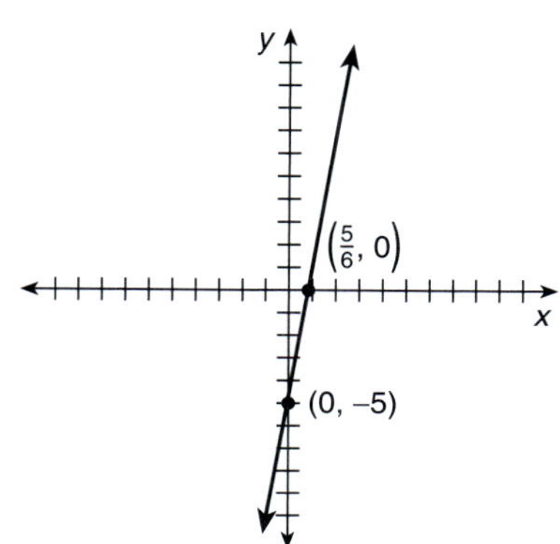

11. $f(x) = x - 1$

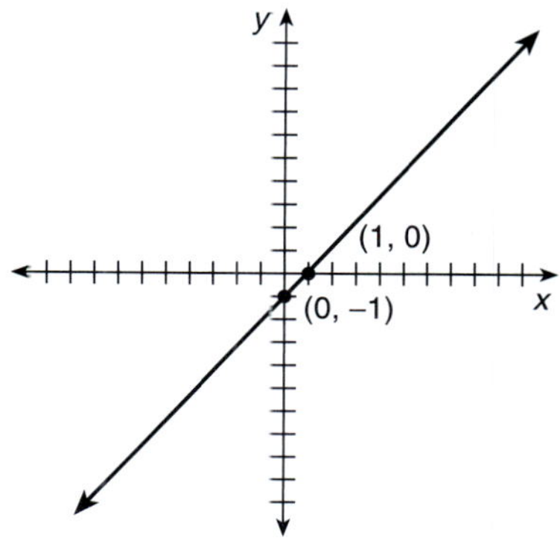

12. $f(x) = x + 9$

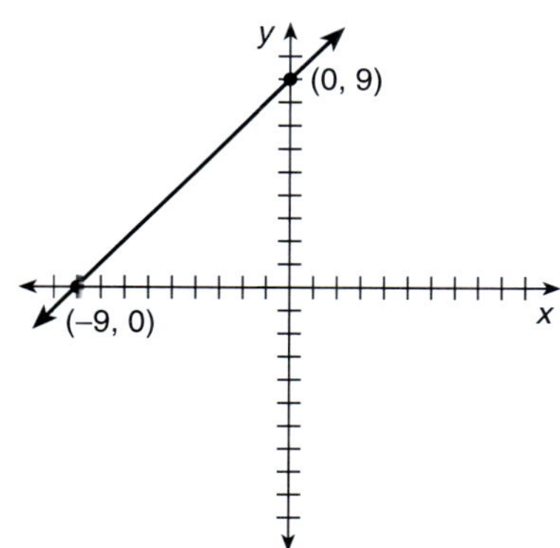

Self-Study Exercises 14–2

1   **1.** no   **2.** yes   **3.** yes   **4.** no   **5.** yes   **6.** yes   **7.** yes   **8.** no   **9.** yes   **10.** no
11. yes   **12.** yes   **13.** yes   **14.** no   **15.** no   **16.** no   **17.** yes   **18.** no

2   **19.** $y = 5$ **20.** $y = 5$ **21.** $x = -6$ **22.** $x = 21$ **23.** $y = 15$ **24.** $y = 0$ **25.** $x = 8$
26. $x = 3$ **27.** $y = 2$ **28.** $y = 1$ **29.** $x = 12$ **30.** $y = -12$ **31.** $x = 18$ **32.** $y = 12$
33. $(1, 3); (1, -2); (1, 0); (1, 1)$ **34.** $(-1, 4); (3, 4); (0, 4); (2, 4)$ **35.** $(2, -7); (-3, -7); (0, -7);$
$(5, -7)$ **36.** $(9, -2); (9, 3); (9, 0); (9, 2)$ **37.** \$12   **38.** \$2   **39.** \$136   **40.** \$13,800

 **1.** *x*-intercept, (5, 0)
  y-intercept, (0, 5)

2. *x*-intercept, (5, 0)
  y-intercept, $\left(0, \dfrac{5}{3}\right)$

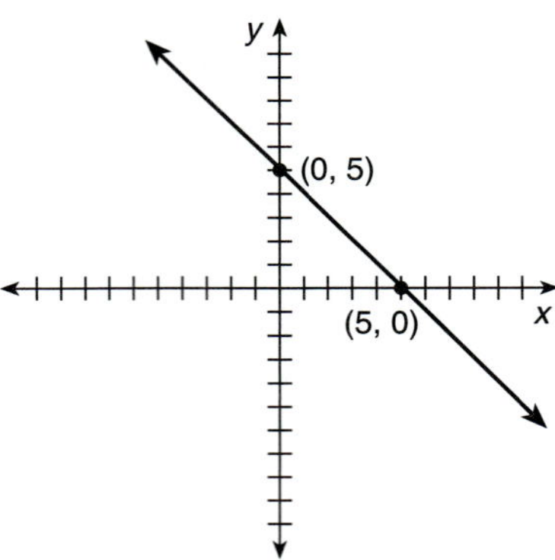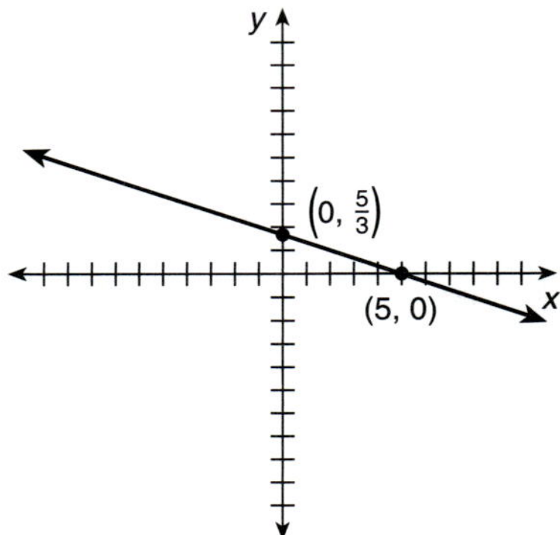

3. *x*-intercept, (−4, 0)
  y-intercept, (0, 8)

4. *x*-intercept, $\left(\dfrac{1}{3}, 0\right)$
  y-intercept, (0, −1)

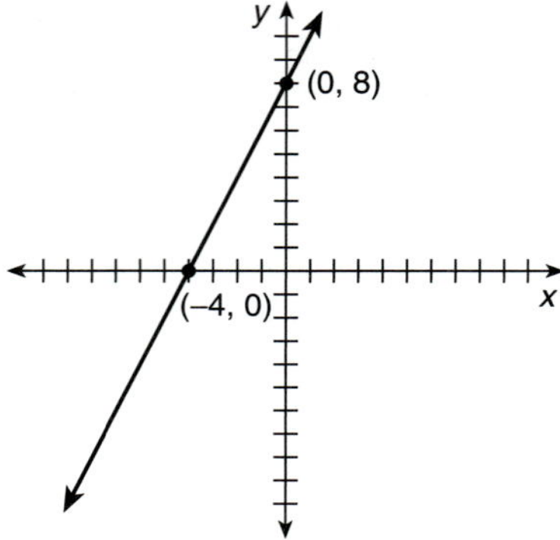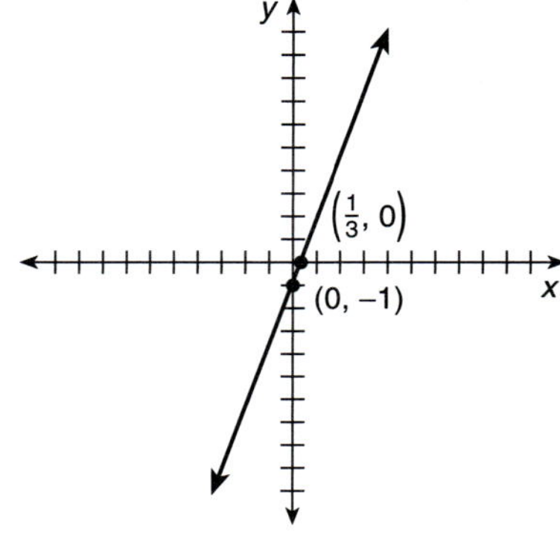

5. *x*-intercept, $\left(\dfrac{2}{5}, 0\right)$
  y-intercept, (0, −2)

6. *x*-intercept, (3, 0)
  y-intercept, $\left(0, \dfrac{3}{2}\right)$

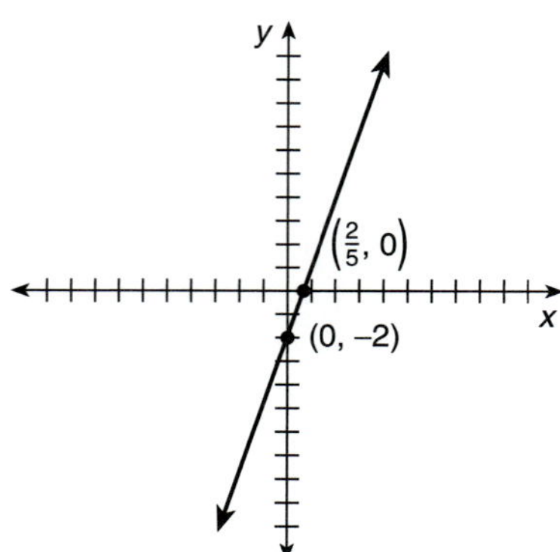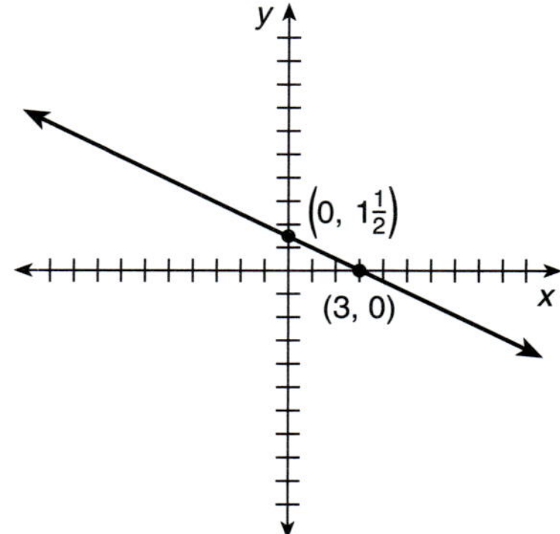

2 **7.** slope $= \dfrac{2}{1}$; *y*-intercept, (0, −3)

8. slope $= -\dfrac{1}{2}$; *y*-intercept, (0, −2)

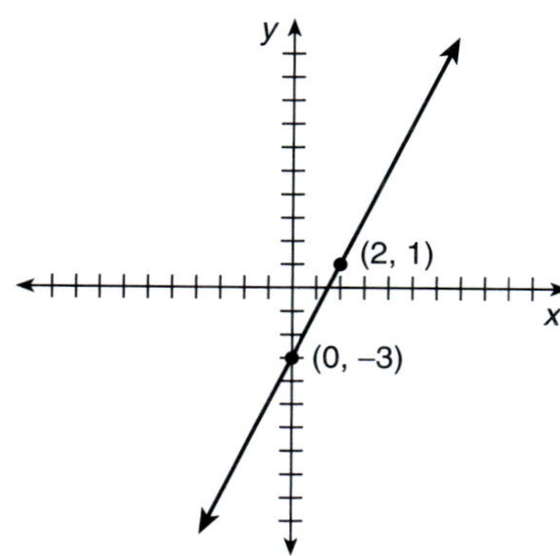

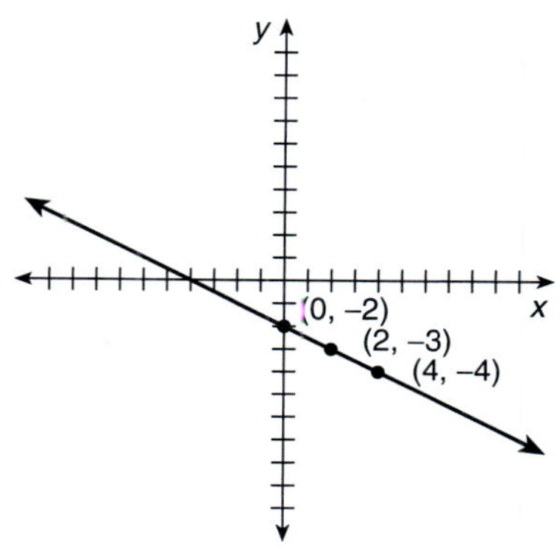

9. slope $= -\dfrac{3}{5}$; y-intercept, $(0, 0)$

10. slope $= \dfrac{1}{2}$; y-intercept, $\left(0, -1\dfrac{1}{2}\right)$

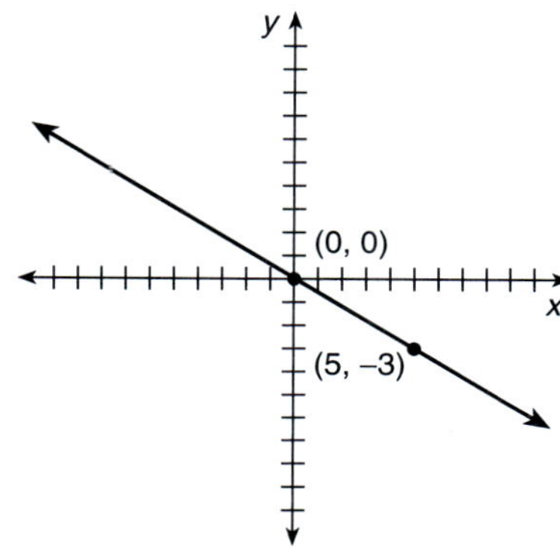

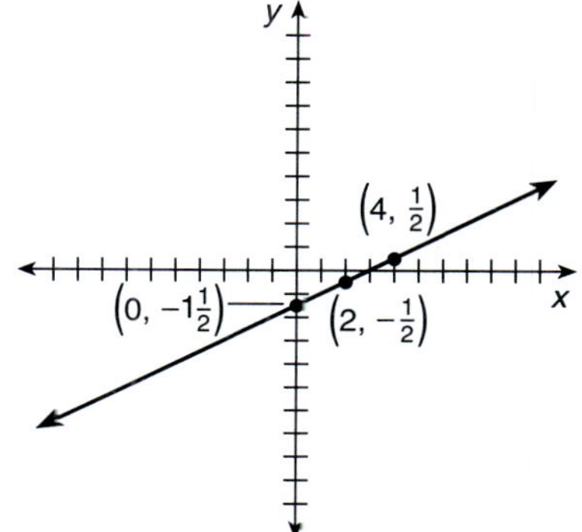

11. slope $= -\dfrac{2}{1}$; y-intercept, $(0, 1)$

Self-Study Exercises 14–4

1   **1.** $\dfrac{1}{2}$ **2.** $-\dfrac{5}{3}$ **3.** 3 **4.** 1 **5.** $\dfrac{3}{4}$ **6.** 0 **7.** $-\dfrac{5}{4}$ **8.** -1 **9.** no slope   **10.** -3

2   **11.–20.** Exercise 6 is horizontal, slope is zero. Exercise 9 is vertical, slope not defined.

3   **21.** $y = \dfrac{2}{3}x + \dfrac{25}{3}$ **22.** $y = -\dfrac{1}{2}x + 3$ **23.** $y = 2x + 1$ **24.** $y = x - 1$

4 **25.** $y = -\dfrac{5}{3}x + \dfrac{38}{3}$ **26.** $x = -1$ **27.** $y = -3$ **28.** $x = -4$ **29.** $y = 2x$ **30.** $y = 0$

5 **31.** slope = 4, y-intercept = 3   **32.** slope = -5, y-intercept = 6   **33.** slope = $-\dfrac{7}{8}$, y-intercept = -3 **34.** slope = 0, y-intercept = 3

35. $y = 2x - 5$, slope = 2, y-intercept = -5 **36.** $y = \dfrac{5}{2}$, slope = 0, y-intercept = $\dfrac{5}{2}$ **37.** $y = -2x + 6$, slope = -2, y-intercept = 6

38. $y = 2x + 5$, slope = 2, y-intercept = 5   **39.** $x = 4$, no slope, no y-intercept   **40.** $x = 9$, no slope, no y-intercept

6 **41.** $y = \dfrac{1}{4}x + 7$ **42.** $y = -8x - 4$ **43.** $y = -3x + 1$ **44.** $y = \dfrac{2}{3}x - 2$ **45.** $y = 4$

46. $y = 2x + 3$ **47.** $x = 10, y = 13$

Self-Study Exercises 14–5

1 **1.** $x + y = -1$ **2.** $2x + y = 9$ **3.** $x - 3y = 4$ **4.** $3x - y = -7$ **5.** $x + 3y = 7$
6. $3x - y = -3$ **7.** $2x + 3y = 5$ **8.** $3x + 2y = 6$ **9.** $2x - 5y = 11$ **10.** $6x - 8y = 3$

2 **11.** $x - y = -1$ **12.** $x - 2y = -13$ **13.** $3x + y = 12$ **14.** $x + 3y = -9$ **15.** $3x - y = 11$
16. $2x - y = -7$ **17.** $3x - 2y = 11$ **18.** $x - 4y = 0$ **19.** $2x - 2y = -3$ **20.** $x + 5y = 4$

Self-Study Exercises 14–6

1 **1.** linear   **2.** quadratic   **3.** quadratic   **4.** other   **5.** linear   **6.** other

2 **7.** $y = x^2$

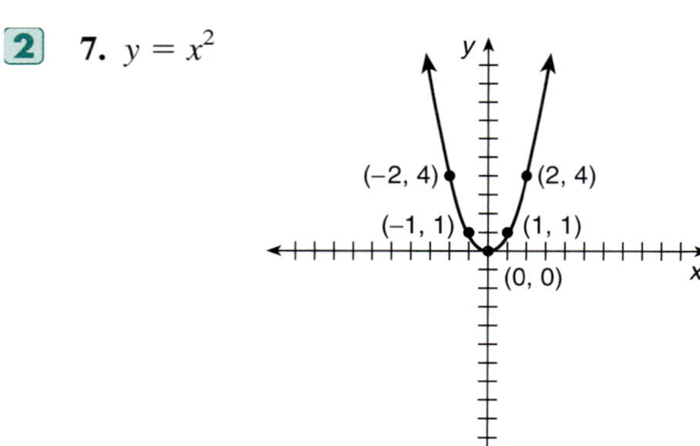

8. $y = 3x^2$

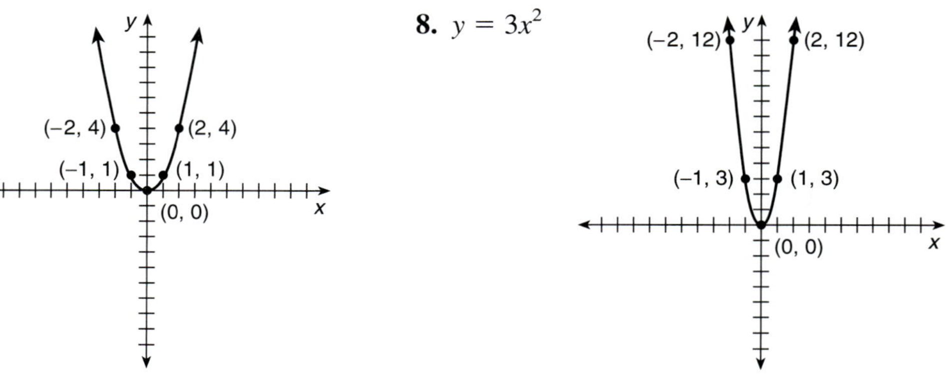

9. $y = \dfrac{1}{3}x^2$

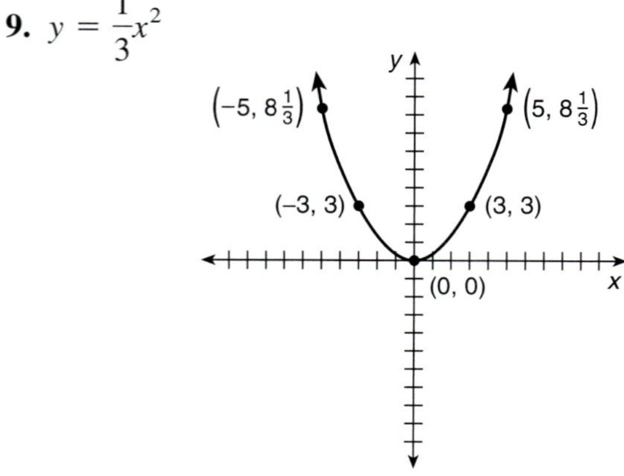

10. $y = -4x^2$

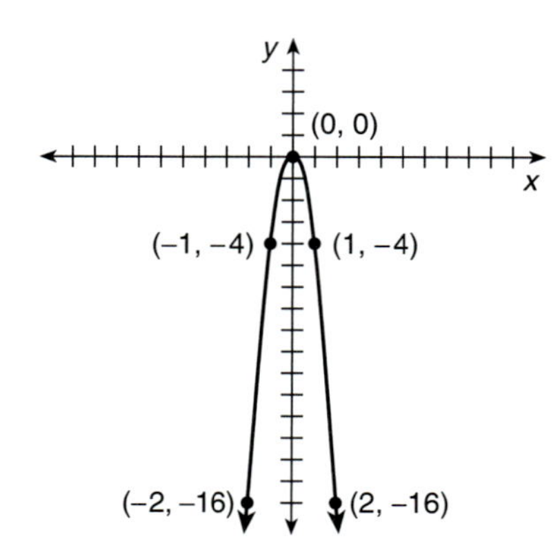

11. $y = -\dfrac{1}{4}x^2$

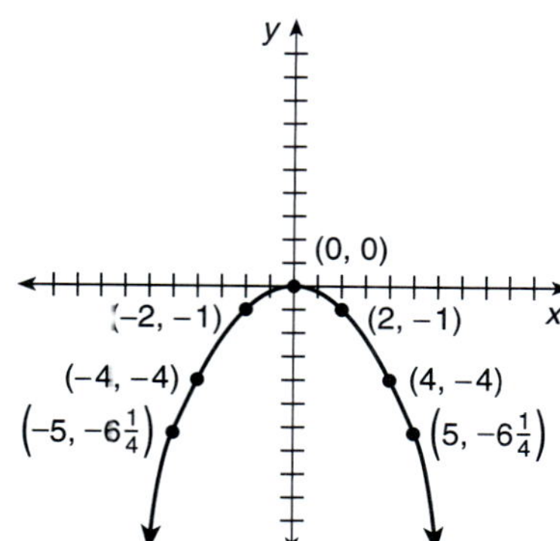

12. $y = x^2 - 4$

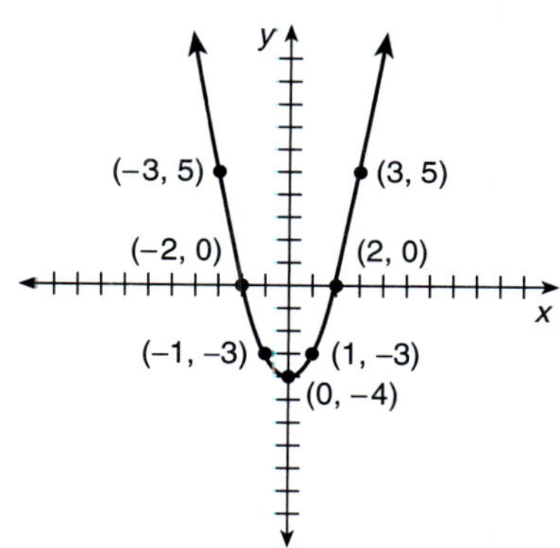

13. $y = x^2 + 4$

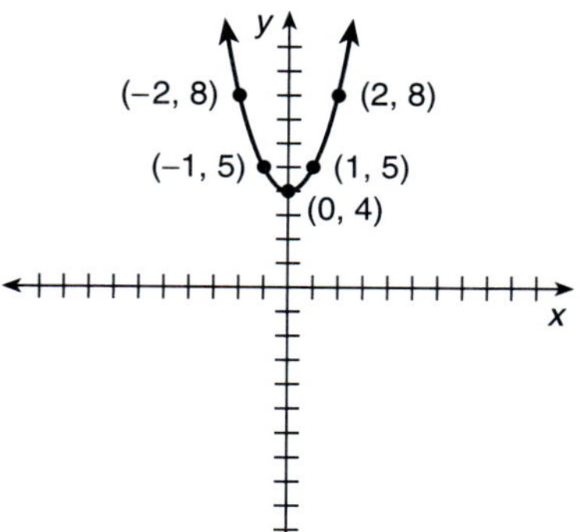

14. $y = x^2 - 6x + 9$

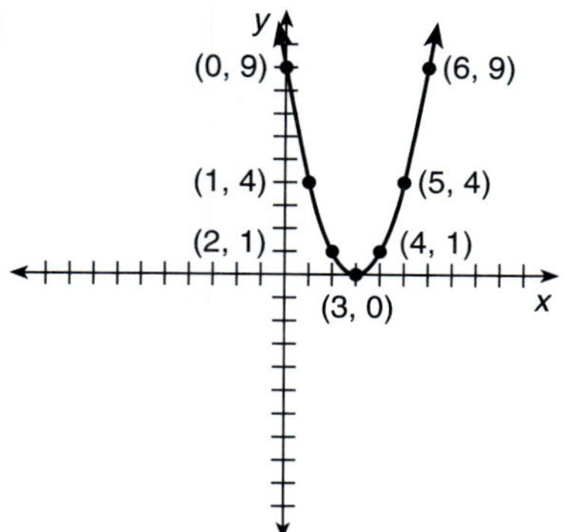

15. $y = -x^2 + 6x - 9$

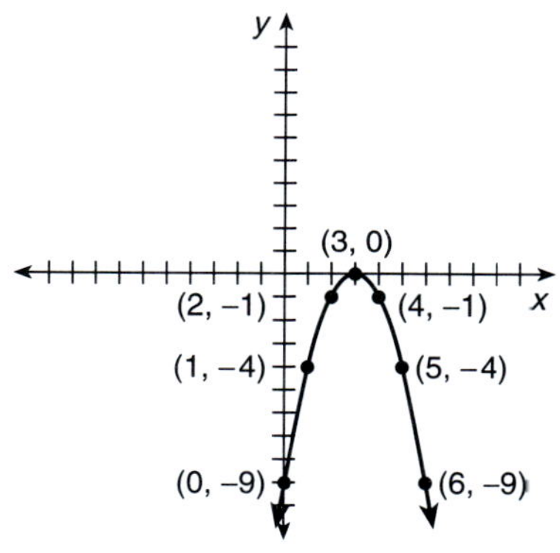

16. $y = 3x^2 + 5x - 2$

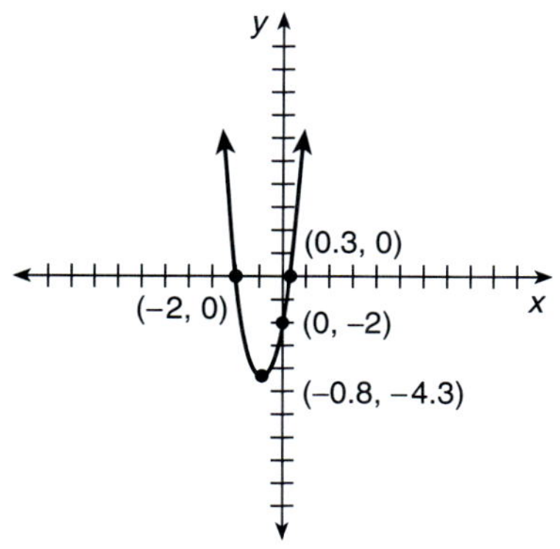

17. $y = (2x - 3)(x - 1)$

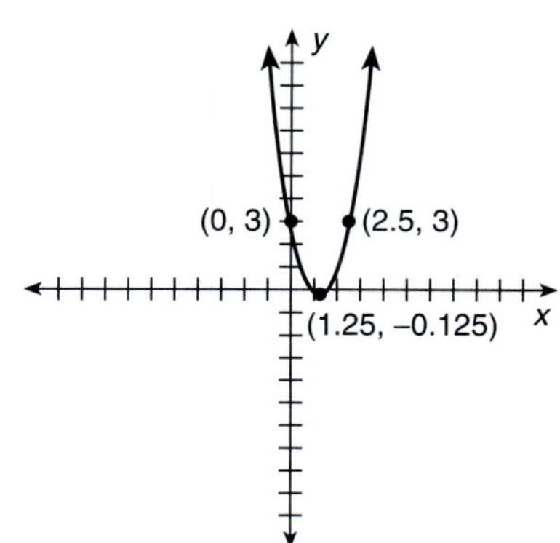

18. $y = 2x^2 - 9x - 5$

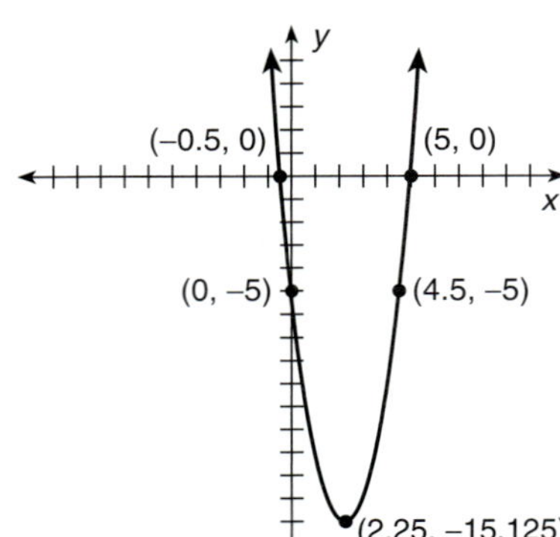

19. $y = -2x^2 + 9x + 5$

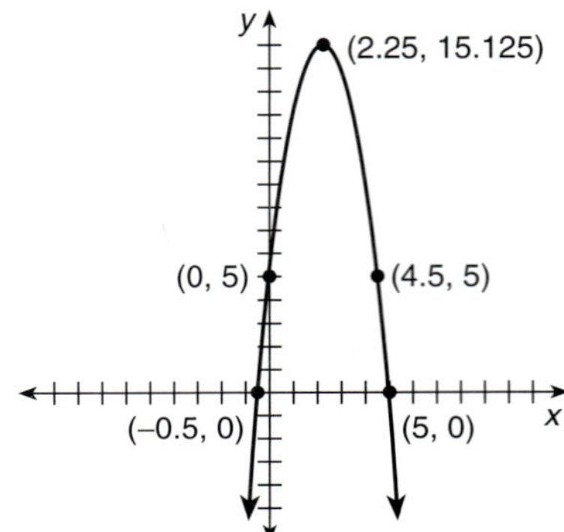

Assignment Exercises, Chapter 14

Since plotted solutions will vary, check graphs by comparing *x*- and *y*-intercepts.

1.

x	$f(x)$
-1	-5
1	-1
3	3

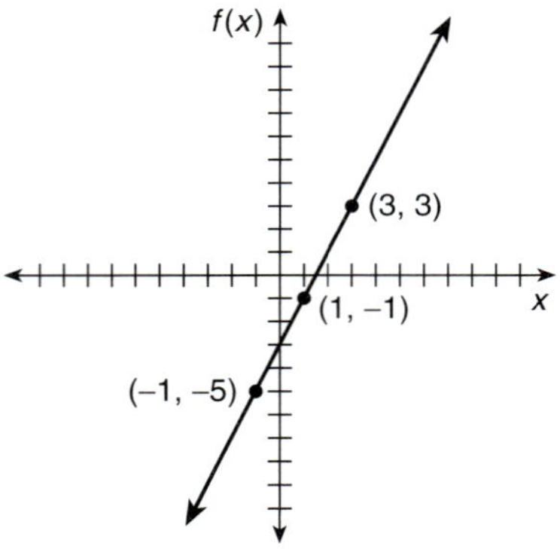

3.

x	$f(x)$
-1	-3
0	0
1	3

5.

x	$f(x)$
-1	3
0	0
1	-3

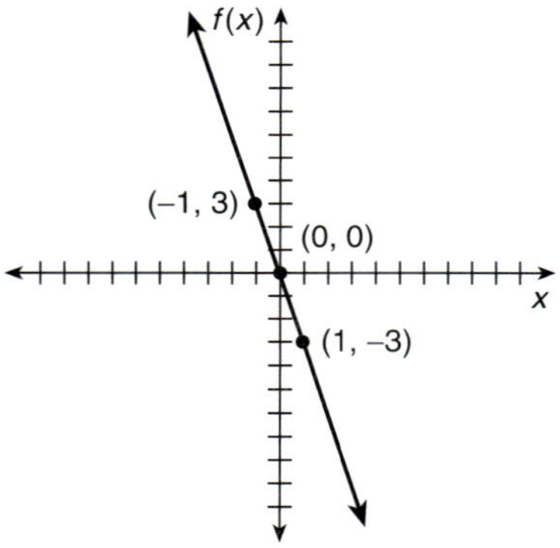

7.

x	$f(x)$
-1	-1
0	1
1	3

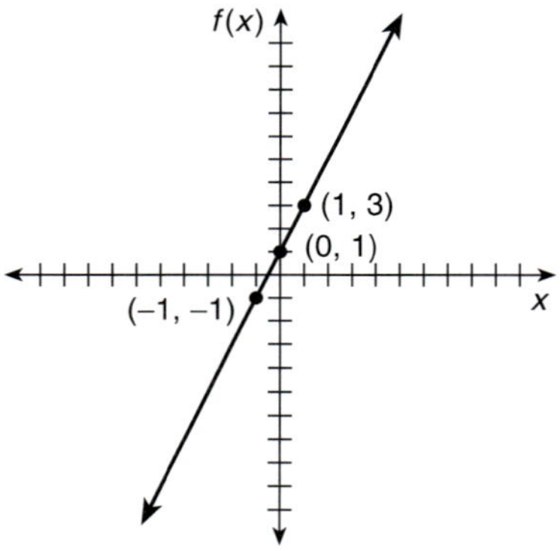

9.

x	$f(x)$
-5	7
-3	-5
-1	-9
1	-5
3	7

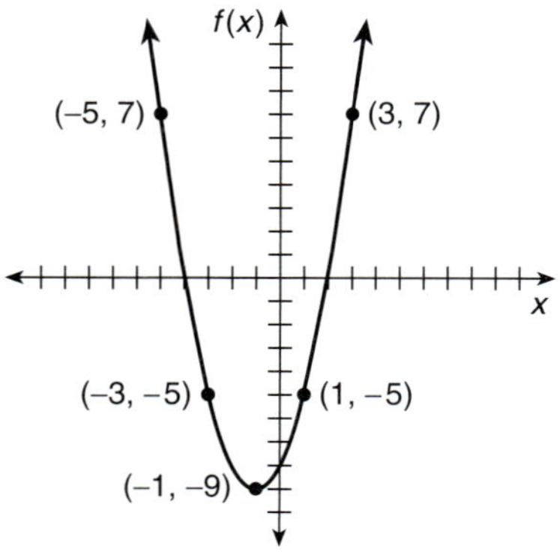

11.

x	$f(x)$
0.1	-2.3
1	0
3	1.1
10	2.3

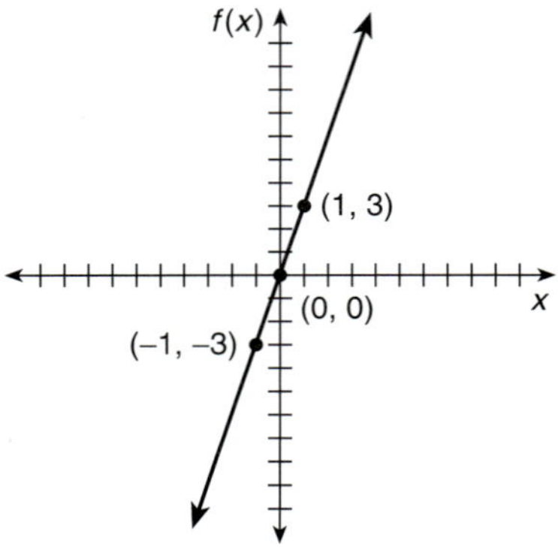

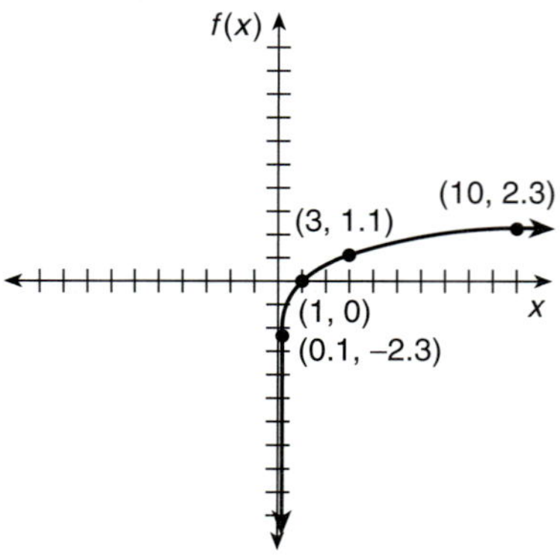

Selected Answers to Student Exercise Material

13. $f(x) = x - 6$

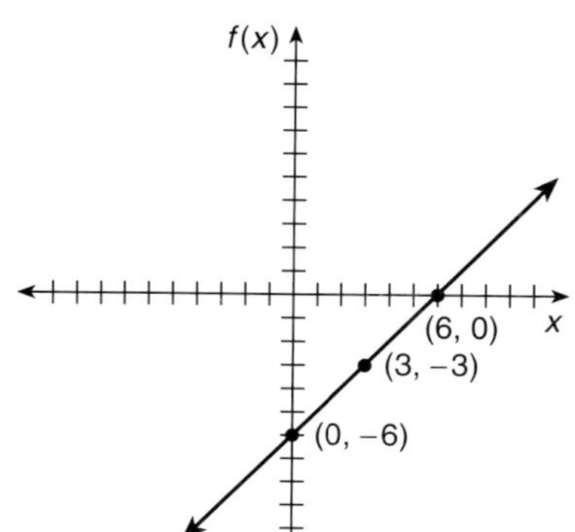

15. $f(x) = 3x + 6$

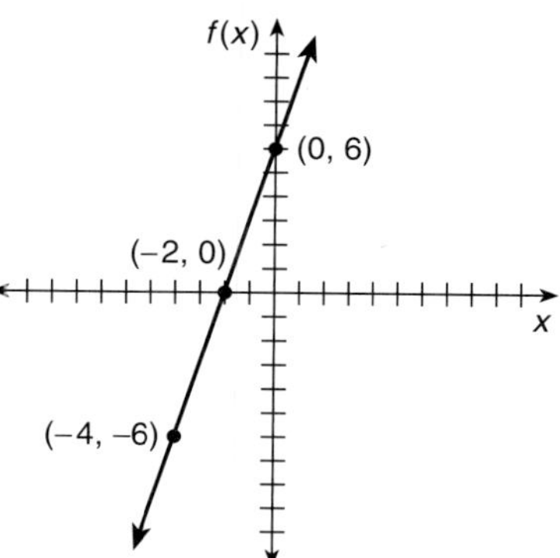

17. $f(x) = 2x - 2$

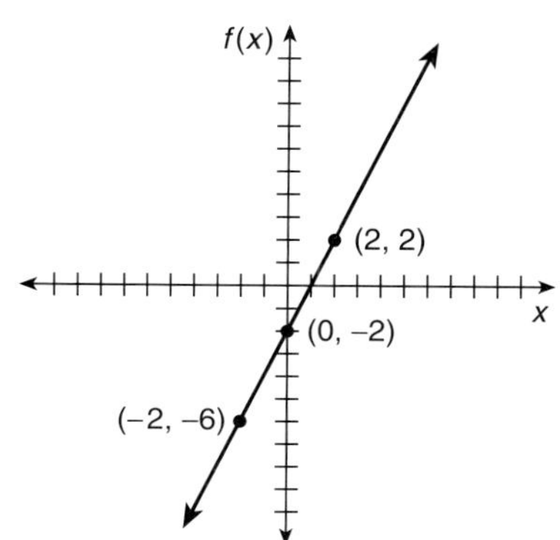

19. no   **21.** yes   **23.** yes   **25.** $y = 1$

27. x-intercept, $(-1, 0)$; y-intercept, $\left(0, -\dfrac{1}{4}\right)$

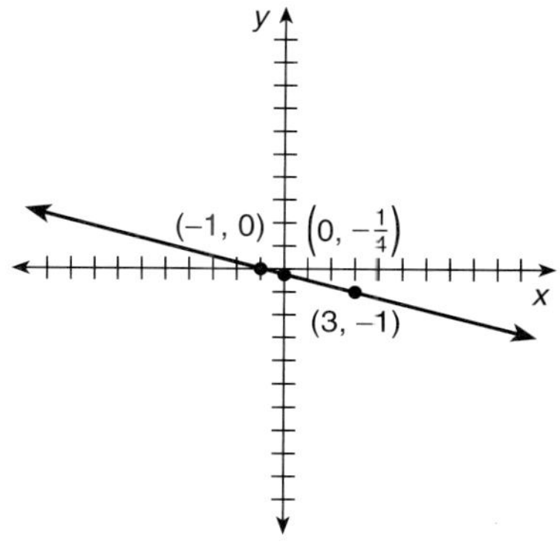

29. $3x - y = 1$; x-intercept, $\left(\dfrac{1}{3}, 0\right)$; y-intercept, $(0, -1)$

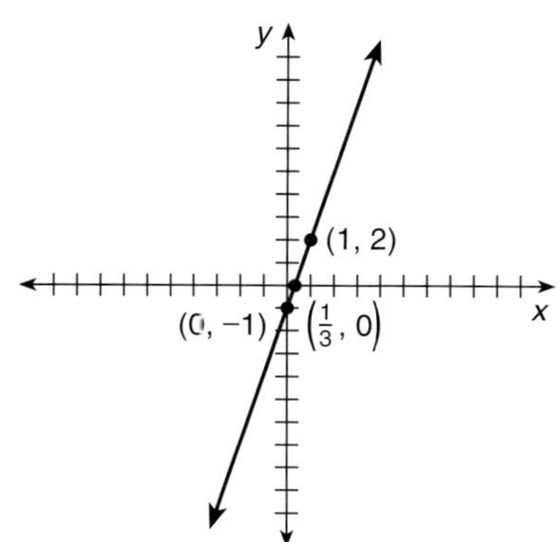

31. $y = 5x - 2$, $m = \dfrac{5}{1}$, $b = -2$

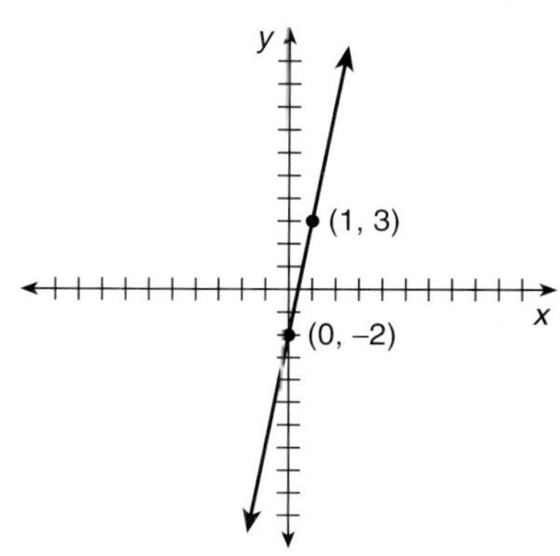

33. $y = -3x - 1, m = \dfrac{-3}{1}, b = -1$

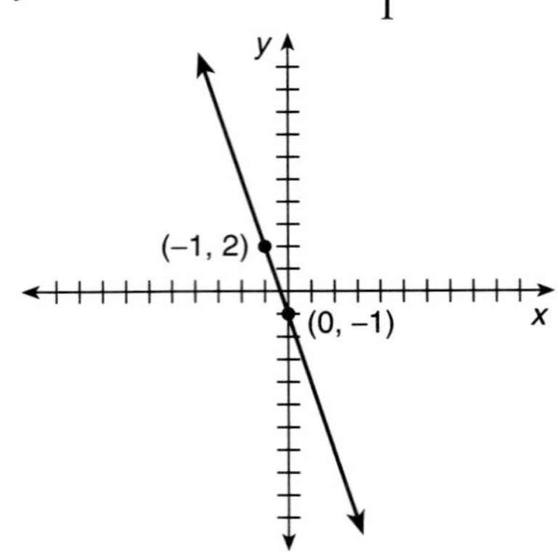

35. $y = x - 4, m = \dfrac{1}{1}, b = -4$

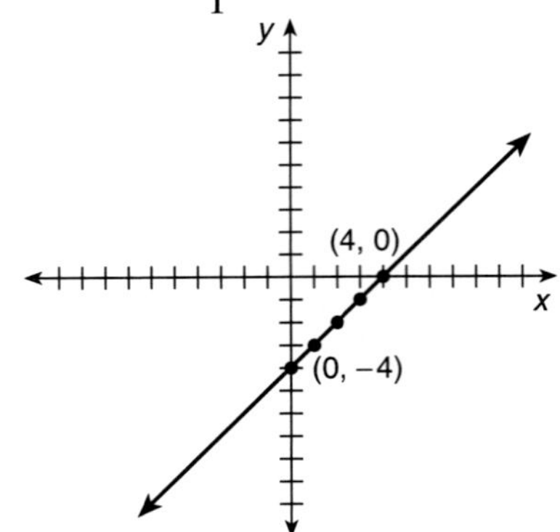

37. $y = \dfrac{1}{2}x + \dfrac{1}{2}, m = \dfrac{1}{2}, b = \dfrac{1}{2}$

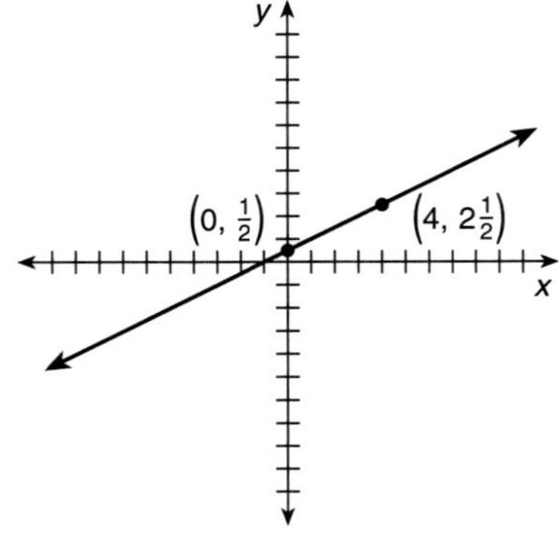

39. $\dfrac{1}{3}$ **41.** 2   **43.** $\dfrac{5}{8}$ **45.** $-\dfrac{4}{3}$ **47.** no slope   **49.** $-\dfrac{4}{7}$ **51.** $\dfrac{7}{2}$ **53.** 0   **55.** no slope   **57.** 1

59. $(-1, 2)(3, 2)$ Answers will vary.   **61.** $y = \dfrac{1}{3}x + 4$ **63.** $y = \dfrac{3}{4}x - 3$ **65.** $y = 4x - 5$

67. $y = -\dfrac{2}{3}x - \dfrac{2}{3}$ **69.** $y = -\dfrac{1}{11}x + \dfrac{17}{11}$ **71.** $y = \dfrac{7}{4}x + \dfrac{5}{4}$ **73.** $y = \dfrac{9}{5}x + \dfrac{3}{5}$ **75.** $y = x - 3$

77. $y = x - 1$ **79.** $y = -2$ **81.** $m = 3$, $b = \dfrac{1}{4}$ **83.** $m = -5$, $b = 4$ **85.** no slope, no y-intercept   **87.** $m = \dfrac{1}{8}$, $b = -5$

89. $y = -2x + 8$, $m = -2, b = 8$ **91.** $y = \dfrac{3}{2}x - 3$, $m = \dfrac{3}{2}, b = -3$ **93.** $y = \dfrac{3}{5}x - 4$, $m = \dfrac{3}{5}, b = -4$ **95.** $y = \dfrac{5}{3}$, $m = 0, b = \dfrac{5}{3}$

97. $y = 3x - 2$ **99.** $y = 2x - 2$ **101.** $x + y = 7$ **103.** $x - 2y = 8$ **105.** $x - 3y = 20$
107. $x + 3y = -10$ **109.** $3x - 4y = -7$
111. $x - y = -4$ **113.** $2x - y = -4$
115. $x - 5y = -11$ **117.** $2x + 10y = 31$
119. $x + 4y = 2$ **121.** Linear   **123.** Quadratic
125. other nonlinear   **127.** $y = -x^2 - 1$

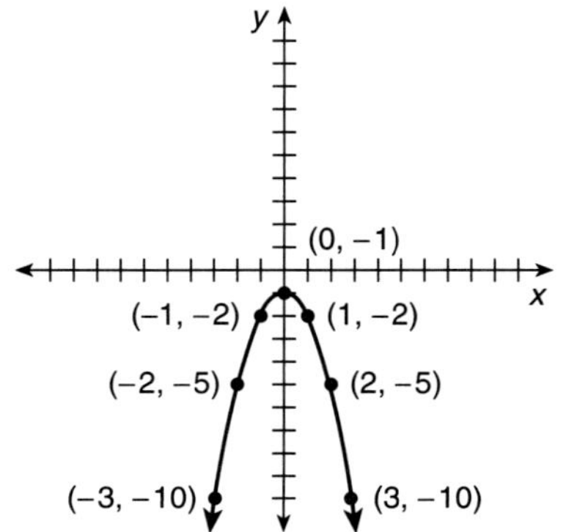

129. $y = x^2 - 6x + 8$

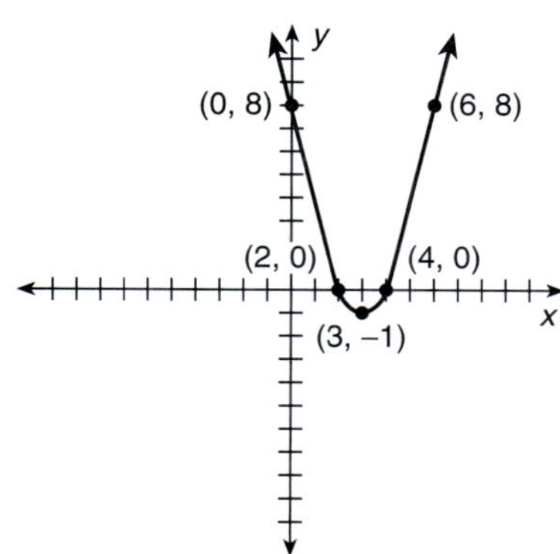

131. $y = -x^2 + 2x - 1$

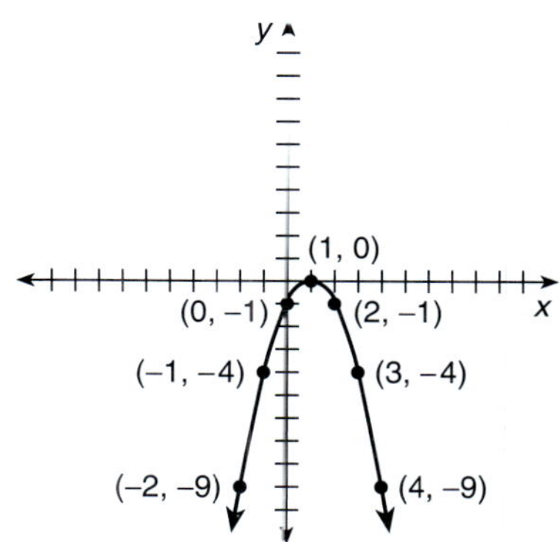

133. $y = -x^2 + 4x - 4$

135. a)

gal	ft²
1	400
2	800
3	1,200
4	1,600
5	2,000
6	2,400
7	2,800
8	3,200
9	3,600
10	4,000

b)

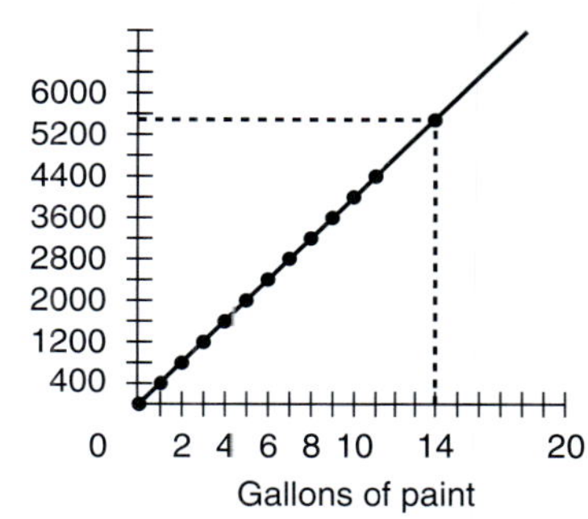

c) 14 gal
d) $279.30
e) $302.34

137.

Lipsticks x	Cost y
0	5,000
100	5,453
1,000	9,530
2,000	14,060

139.

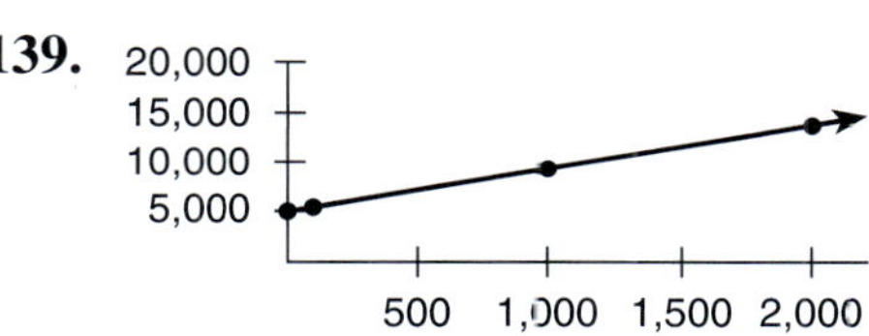

141. 1,122 lipsticks

Trial Test, Chapter 14

Since plotted solutions will vary, check graphs by comparing x- and y-intercepts.

1.

x	$f(x)$
−2	−1
0	0
2	1

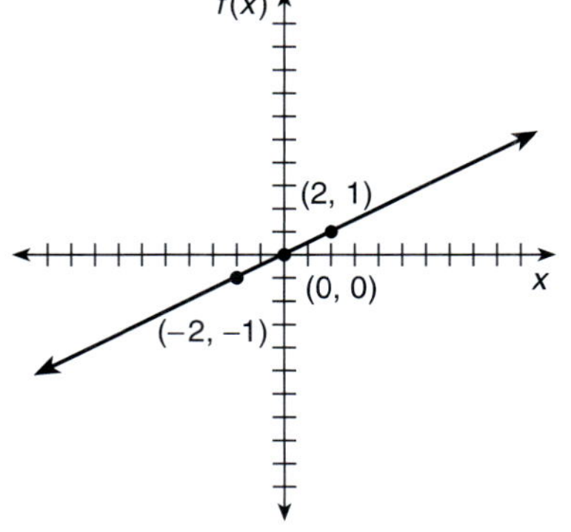

3.

x	$f(x)$
−1	−6
0	−4
1	−2
2	0

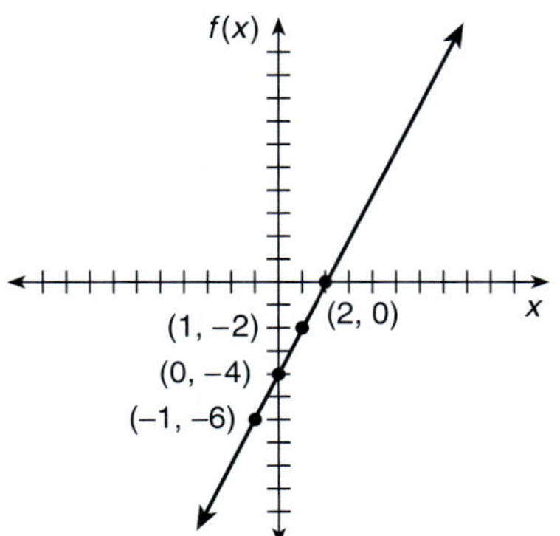

5. $(2, 5)$

7. x-intercept: $(2, 0)$

y-intercept: $\left(0, -\dfrac{1}{2}\right)$

9.

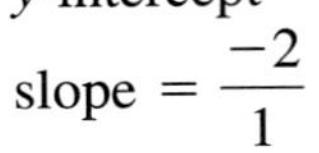
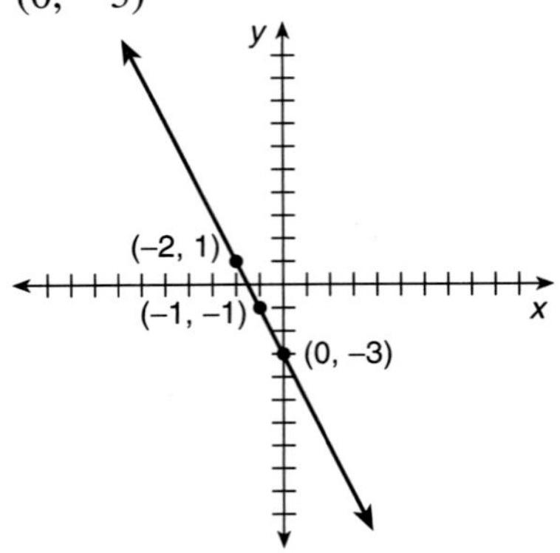

11. x-intercept $(8, 0)$
y-intercept $(0, 4)$

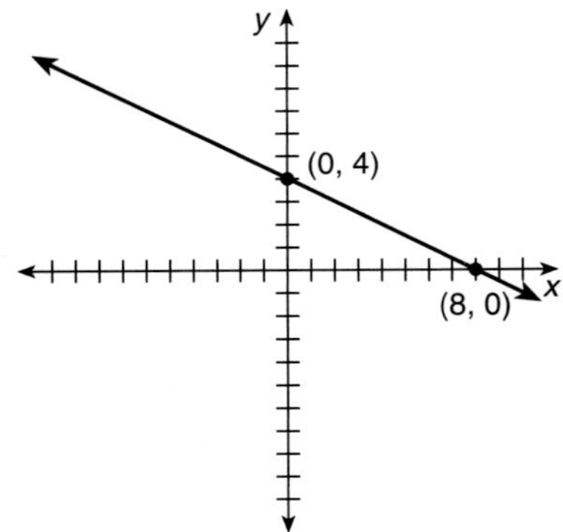

13. y-intercept $= (0, -3)$

slope $= \dfrac{-2}{1}$

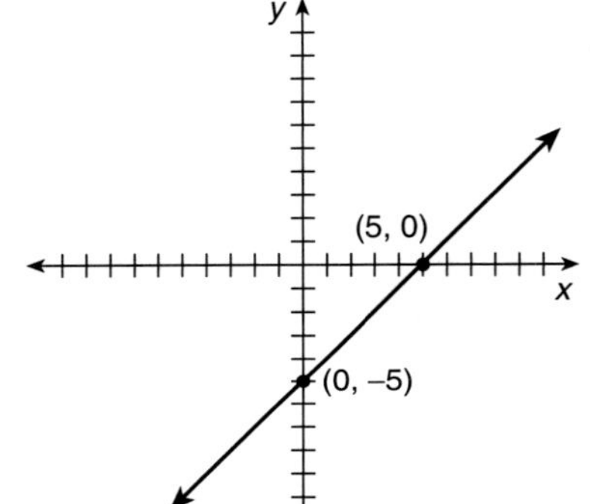

15. y-intercept $= (0, -5)$

slope $= \dfrac{1}{1}$

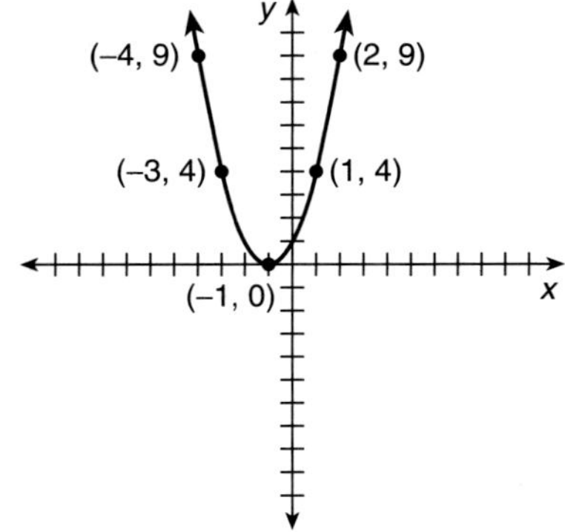

17. $-\dfrac{2}{3}$ **19.** $\dfrac{5}{2}$ **21.** 1 **23.** $y = \dfrac{2}{3}x - 7$ **25.** $y = \dfrac{2}{3}x + \dfrac{7}{3}$ **27.** $y = 2$ **29.** $m = 3$
$b = -22$

31. $y = \dfrac{1}{4}x$
$m = \dfrac{1}{4}, b = 0$

33. $y = -6x + 3$
$m = -6, b = 3$

35. $y = \dfrac{3}{2}x + 3$

37. $2x + y = 5$

39. $x - 2y = 10$

41. $y = x^2 + 2x + 1$

1   **1.** $x = 7, y = 5$

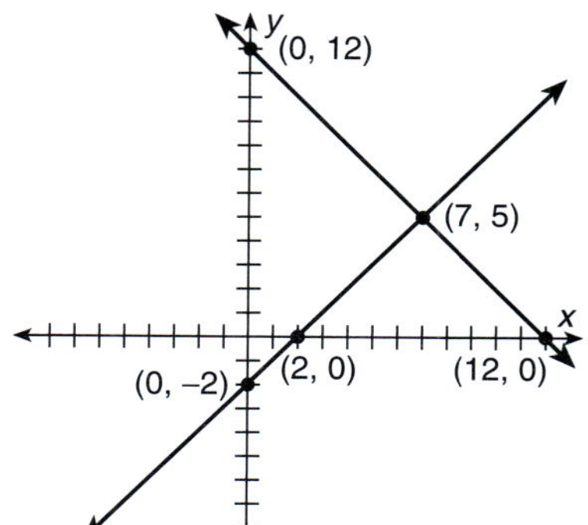

2. $x = 3, y = 3$

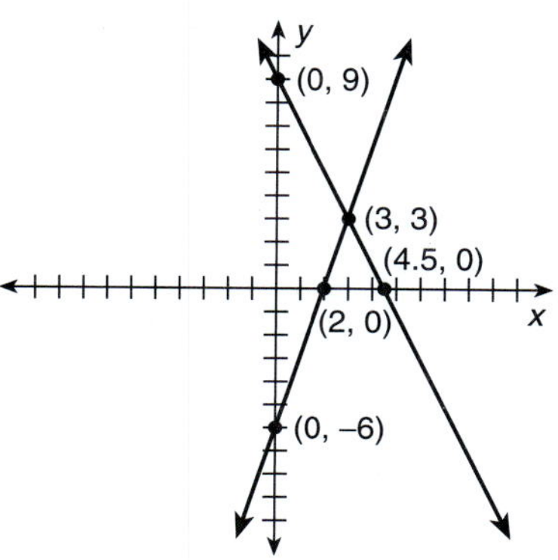

3. no solution, no intersection

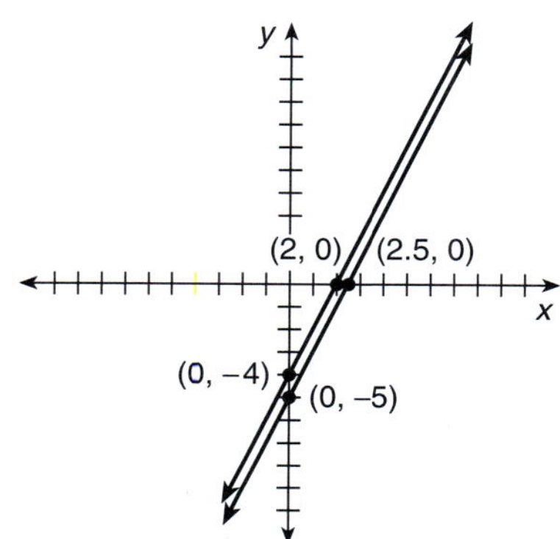

4. many solutions, lines coincide

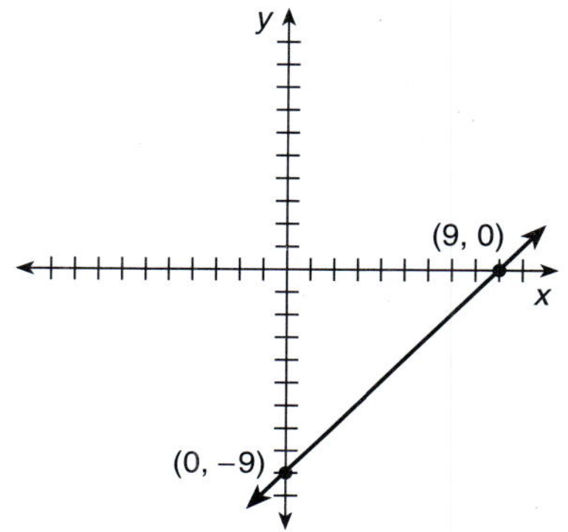

5. $x = 4, y = -2$

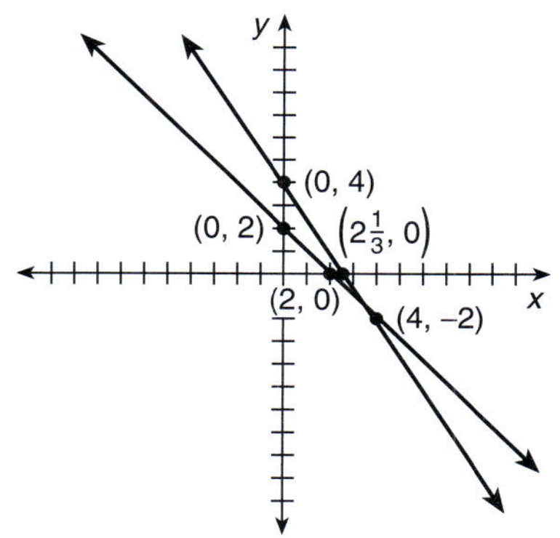

Self-Study Exercises 15–2

1   **1.** $a = 5, b = -1$ **2.** $m = 4, n = -1$ **3.** $x = 1, y = -1$ **4.** $a = 3, b = -3$
5. $x = 2, y = \dfrac{3}{2}$

2   **6.** $x = 3, y = 0$ **7.** $x = 1, y = 5$ **8.** $a = 2, b = \dfrac{8}{3}$ **9.** $x = -1, y = 2$ **10.** $a = 6, y = 1$

3   **11.** inconsistent; no solution   **12.** $x = -3, y = 6$ **13.** dependent; many solutions
14. dependent; many solutions   **15.** $a = 5, b = 1$

Self-Study Exercises 15–3

[1] **1.** $a = 20, b = 10$ **2.** $r = 5, c = 7$ **3.** $x = 13, y = 11$ **4.** $x = 7, y = 5$ **5.** $p = \dfrac{1}{2}, k = \dfrac{1}{3}$

6. $x = 3, y = 1$ **7.** $x = 13, y = 5$ **8.** $x = -2, a = -7$

Self-Study Exercises 15–4

[1] **1.** short 15.5 in.   **2.** \$18,000 at 4%   **3.** \$11 per shirt   **4.** (7) 8-cylinder jobs
long 32.5 in.   \$17,000 at 5%   \$ 6 per hat   (3) 4-cylinder jobs
5. resistor \$0.25   **6.** rate of plane = 130 mph   **7.** \$3,000 at 7%   **8.** rate of current = 1.67 mph
capacitor \$0.30   rate of wind = 10 mph   \$5,000 at 9%   rate of motorboat = 11.67 mph
9. dark roast, \$2.50   **10.** 75% mixture, 123 lb   **11.** scientific \$11.00   **12.** \$4,000 at 10%
with chicory, \$1.90   10% mixture, 77 lb   graphing \$30.00   \$6,000 at 15%
13. reserved \$20.00   **14.** 4 pints at 75%
general \$15.00   4 pints at 25%

Assignment Exercises, Chapter 15

1.

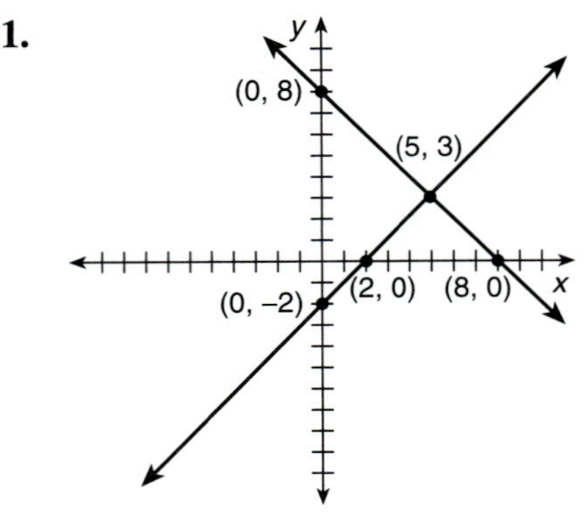

3.

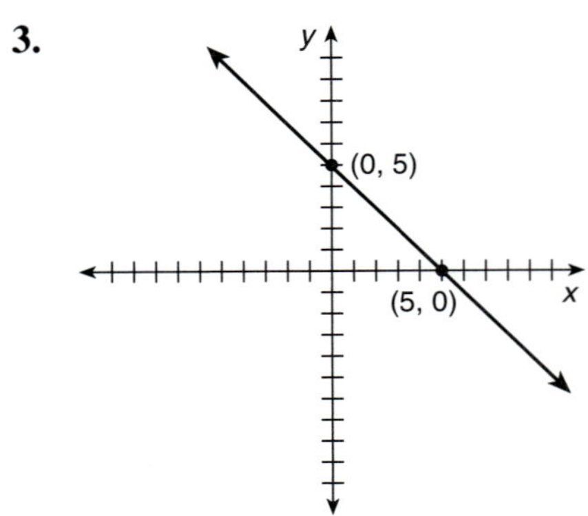

5.

7. $(3, 0)$ **9.** $(-2, 4)$ **11.** $(2, -2)$ **13.** $(6, 3)$ **15.** dependent; many solutions   **17.** $(5, 1)$

19. $\left(2, \dfrac{8}{3}\right)$ **21.** $(2, 0)$ **23.** $(3, 2)$ **25.** $(2, 1.5)$ **27.** $(4, 4)$ **29.** $(7, 5)$ **31.** $(4, -6)$

33. $(3, -4)$ **35.** $\left(\dfrac{4}{5}, \dfrac{2}{5}\right)$ **37.** $(28, 44)$ **39.** $\left(-3, \dfrac{5}{2}\right)$ **41.** $(4, -6)$ **43.** $(12, 2)$ **45.** $(1, -2)$

47. electrician \$75   **49.** shellac \$2.50   **51.** 119°, 56°   **53.** \$2,000 at 10%   **55.** Colombian = \$5
apprentice \$35   thinner \$3.50   \$3,000 at 12%   blended = \$4
57. Ohio = \$1.95   **59.** name brand = \$320   **61.** telephone = \$15,000   **63.** 4 gal 75%
Alaska = \$2.10   generic = \$175   showroom = \$25,000   4 gal 25%
65. In line 4, $-6c$ should be $-9c$.   **67.** Answers will vary.
$c = 3$
$d = 1$

    Selected Answers to Student Exercise Material

1. $(5, 0)$

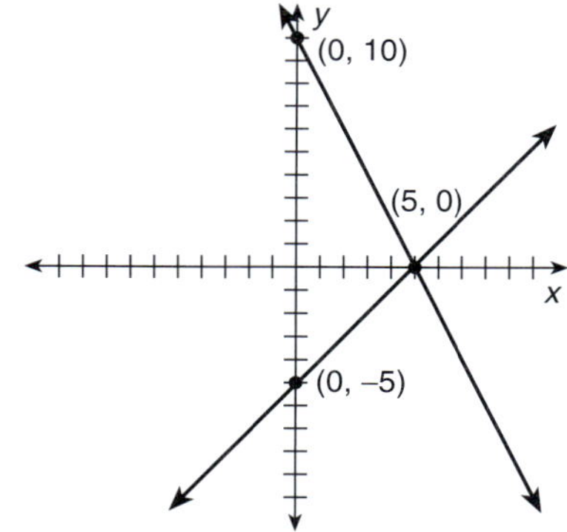

3. $(14, -9)$

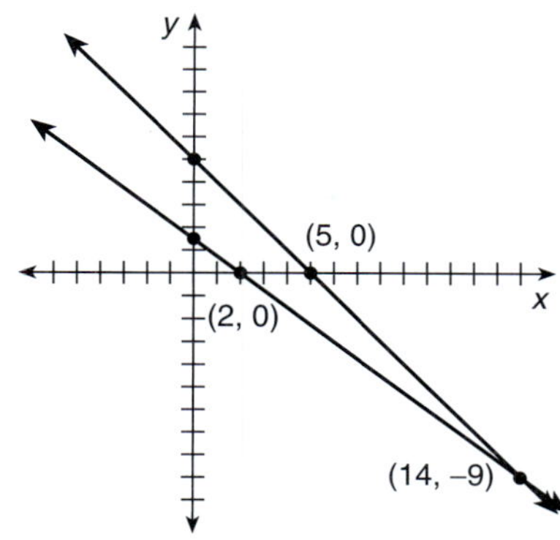

5. $(-6, -6)$

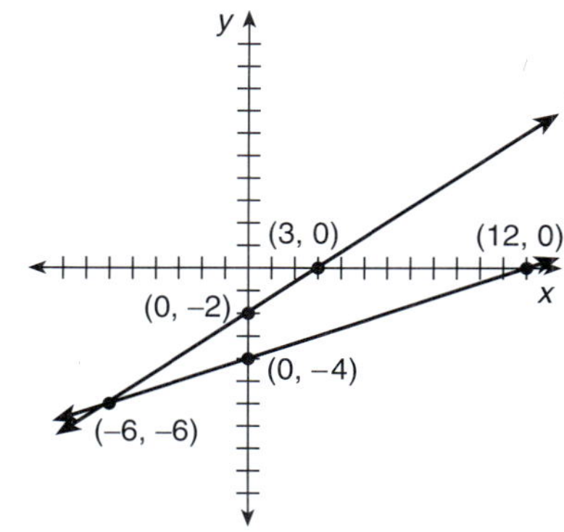

7. $(0, 0)$ **9.** $(1, 2)$ **11.** $(2, 2)$ **13.** $(-3, 15)$ **15.** $(-9, -11)$ **17.** $20A, 15A$
19. $L = 51$ in. **21.** \$20,000 at 12.5% **23.** 0.000215 F
    $W = 34$ in.   \$ 5,000 at 11%   0.000055 F

Self-Study Exercises 16–1

1 **1.** $\overleftrightarrow{PQ}, \overleftrightarrow{QR}, \& \overleftrightarrow{PR}$ **2.** $\overrightarrow{PQ}, \overrightarrow{PR}$ **3.** $\overline{QR}$ **4.** Yes **5.** No. End points are different. **6.** No
7. Yes **8.** Yes **9.** Yes. $\overline{XY}$ and $\overline{YX}$ contain same points. **10.** No

2 **11.** parallel **12.** intersect **13.–14.** coincide or intersect **15.** $\overleftrightarrow{EF}, \overleftrightarrow{GH}$

3 **16.** $\angle DEF, \angle FED$ **17.** $\angle E$ **18.** $\angle 1$ **19.** $\angle a$ **20.** $\angle ABC$ or $\angle CBA$ **21.** $\angle B$

4 **22.** $360°$ **23.** $180°$ **24.** $90°$ **25.** acute **26.** obtuse **27.** right **28.** obtuse **29.** acute
30. straight **31.** acute **32.** obtuse **33.** Yes **34.** No **35.** No **36.** No **37.** Yes **38.** Yes
39. No **40.** $90°$ **41.** congruent

Self-Study Exercises 16–2

1 **1.** $54°$ **2.** $46°8'18''$ **3.** $37°52'56''$ **4.** $74°3'12''$ **5.** $16°59'45''$ **6.** $54°44'12''$
7. $64°24'46''$

2 **8.** $0.7833°$ **9.** $0.01°$ **10.** $0.0872°$ **11.** $0.1708°$ **12.** $29.7°$

3 **13.** $21'$ **14.** $12'$ **15.** $7'12''$ **16.** $12'47''$ **17.** $13'54''$

4 **18.** $7°30'$ **19.** $11°25'43''$ **20.** $38.1908°$ **21.** $12°38'50''$

Self-Study Exercises 16–3

1 **1.** scalene **2.** equilateral **3.** isosceles

2   **4.** longest $\overline{TS}$, shortest $\overline{RS}$ **5.** longest k, shortest l **6.** $\angle C, \angle B, \angle A$ **7.** $\angle X, \angle Z, \angle Y$

3   **8.** $\angle A = \angle F, \angle C = \angle D, \angle B = \angle E$ **9.** $\angle P = \angle S, \angle Q = \angle T, QP = TS$
10. $\angle L = \angle P, PM = LJ, LK = PN$

4   **11.** $BC = 12$ cm, $AB = 16.971$ cm   **12.** $AC = 7.071$ m, $BC = 7.071$ m
13. $AC = 9.899$ m, $AB = 14.0$ m   **14.** $AC = 8$ m, $BC = 8$ m   **15.** $AC = 14.697$ mm, $BC = 14.697$ mm
16. $25'5''$ **17.** 37.830 ft

5   **18.** $AB = 12$ cm, $BC = 10.392$ cm   **19.** $AC = 9.0$ mm, $BC = 15.588$ mm
20. $AC = 4.619$ in., $AB = 9.238$ in.   **21.** $AC = 5.715$ cm, $AB = 11.431$ cm   **22.** $AB = 5'6'', BC = 4'9''$
23. $35'$ **24.** rafter $= 18'0''$, run $= 15'7''$ **25.** 4.5 cm

Self-Study Exercises 16–4

1   **1.** $x = 16', y = 22'$ **2.** $x = 6', y = 25'6''$

2   **3.** $245'6'' = P$ **4.** 96 mm $= P$ **5.** $p = 58'8''$ **6.** 20 yd^2 @ \$16.50 = \$330   **7.** 24 lb
        $1{,}942.5$ ft$^2 = A$ 662.4 mm$^2 = A$
8. $1{,}086$ ft^2, $\angle = 135°$ **9.** 63 in.^{2} **10.** 625 dm^2 **11.** 33 in.^{2}

3   **12.** equilateral triangle, $\angle 60°$ **13.** square, $\angle 90°$ **14.** regular pentagon, $\angle 108°$
15. regular hexagon, $\angle 120°$

Self-Study Exercises 16–5

1   **1.** 120.64 cm^2 **2.** 198.44 mm^2 **3.** 2.43 ft^2 **4.** 14.18 in.^{2} **5.** 2.75 m^2 **6.** 530.14 ft^2
7. 75.40 in.^{2} **8.** 17.0 in.^{2}

2   **9.** 28.27 mm   **10.** 13.09 in.   **11.** 3.20 ft   **12.** 7.06 cm   **13.** 1.64 m   **14.** 15.71 in.

3   **15.** 16.01 cm^2 **16.** 48.98 in.^{2} **17.** 54.83 in.^{2} **18.** 51.99 cm^2 **19.** 91.29 in.^{2}
20. 561.81 mm^2 **21.** 257.93 ft^2

Self-Study Exercises 16–6

1   **1.** $60°$ **2.** $30°$ **3.** $90°$ **4.** 11.5 in.   **5.** 11.5 in.   **6.** 13.33 in.   **7.** 57.5 in.^{2} **8.** 5.75 in.
9. 6.67 in.   **10.** 41.89 in.   **11.** 2.9 cm   **12.** 1 in.

2   **13.** $90°$ **14.** $45°$ **15.** 8.49 in.   **16.** $45°$ **17.** 6 in.   **18.** 0.5 m   **19.** 0.35 m   **20.** $90°$
21. 0.35 m   **22.** 0.29 in.   **23.** 1.06 in.   **24.** 63.62 dkm^2

3   **25.** 34.64 mm   **26.** $120°$ **27.** $60°$ **28.** 40 mm   **29.** $90°$ **30.** 10 mm   **31.** 20 mm
32. $60°$ **33.** $30°$ **34.** 3 in.   **35.** 39.97 mm   **36.** 2.84 in.

Assignment Exercises, Chapter 16

1. $\overleftrightarrow{AB}$ **3.** $\overrightarrow{AB}$ **5.** $\overrightarrow{NO}$ **7.** No   **9.** No   **11.** $\overleftrightarrow{AB}$ & $\overleftrightarrow{CD}$ **13.** $\overleftrightarrow{EF}$ & $\overleftrightarrow{GH}$ **15.** Yes
17. $\angle EDF$ or $\angle FDE$ **19.** $\angle P$ **21.** $\angle M$ **23.** right   **25.** acute   **27.** straight   **29.** obtuse
31. neither   **33.** supplementary   **35.** neither   **37.** equal   **39.** vertical   **41.** $31°08'11''$
43. $34°37'56''$ **45.** $60°40'18''$ **47.** $0.4833°$ **49.** $0.1261°$ **51.** $45'$ **53.** $13'3''$ **55.** $23°20'$
57. $18°39'50.5''$ **59.** isosceles   **61.** longest TS, shortest RS **63.** $\angle C, \angle B, \angle A$ **65.** $\angle C = \angle D,$
$\angle B = \angle E, CB = DE$ **67.** $\angle M = \angle J, JL = PM, JK = MN$ **69.** $RS = 10.607$ cm, $ST = 10.607$ cm
71. $RS = 9.0$ hm, $ST = 9.0$ hm   **73.** $AB = 13.856$ dm, $BC = 6.928$ dm
75. $AC = 17.321$ in., $AB = 20.0$ in.   **77.** $AB = 46'10'', BC = 23'5''$ **79.** $8'4''$ **81.** 19.630 mm
83. 115 in., 908.5 in.^{2} **85.** regular pentagon, $108°$ **87.** 130 ft   **89.** 83 ft^2 **91.** 72 bricks
93. 2.62 cm^2 **95.** $1{,}610.28$ mm^2 **97.** 2.18 ft^2 **99.** 6.66 mm^2 **101.** 6.96 cm^2 **103.** 179.9 mm^2
105. 232.27 ft^2 **107.** 44.50 in.^{2} **109.** 31.42 in.   **111.** 36.74 mm   **113.** $AE = 5$
115. $\angle GJO = 45°$ **117.** $IJ = 20$ **119.** $\angle MOP = 30°$ **121.** 5 mm   **123.** 3.54 cm

125. area pentagon $<$ area circle $\pi(5)^2 = 78.54$ cm^2
area pentagon $>$ area square 50.00 cm^2

Trial Test, Chapter 16

1. $\overline{BC}$ **3.** acute **5.** parallel **7.** perpendicular **9.** 15°45′29″ **11.** 18′45″ **13.** largest $\angle A$, smallest $\angle B$ **15.** 13.657 cm **17.** 155′ **19.** 135° **21.** 43.20 mm **23.** 25.13″ **25.** 1.57 in.

Self-Study Exercises 17–1

1   **1.** 6.45 in. **2.** 14 cm **3.** 4.25 rad **4.** 3.49 rad **5.** 7 cm **6.** 4.5 in.

2   **7.** 17.12 in.^{2} **8.** 4.32 cm^2 **9.** 2.24 cm **10.** 0.85 rad

3   **11.** 0.79 rad **12.** 0.98 rad **13.** 1.36 rad **14.** 2.44 rad **15.** 0.38 rad **16.** 3.10 rad **17.** 0.78 rad **18.** 0.18 rad

4   **19.** 45° **20.** 30° **21.** 143.2394° **22.** 80.2141° **23.** 28°38′52″ **24.** 22°30′ **25.** 42°58′19″ **26.** 63°1′31″ **27.** 1.53 cm **28.** 61° **29.** 47.50 cm^2 **30.** 40.1070°

Self-Study Exercises 17–2

1   **1.** $\dfrac{5}{13}$ **2.** $\dfrac{12}{13}$ **3.** $\dfrac{5}{12}$ **4.** $\dfrac{12}{13}$ **5.** $\dfrac{5}{13}$ **6.** $\dfrac{12}{5}$ **7.** $\dfrac{9}{15} = \dfrac{3}{5}$ **8.** $\dfrac{12}{15} = \dfrac{4}{5}$ **9.** $\dfrac{9}{12} = \dfrac{3}{4}$

10. $\dfrac{12}{15} = \dfrac{4}{5}$ **11.** $\dfrac{9}{15} = \dfrac{3}{5}$ **12.** $\dfrac{12}{9} = \dfrac{4}{3}$ **13.** $\dfrac{16}{34} = \dfrac{8}{17}$ **14.** $\dfrac{30}{34} = \dfrac{15}{17}$ **15.** $\dfrac{16}{30} = \dfrac{8}{15}$

16. $\dfrac{30}{34} = \dfrac{15}{17}$ **17.** $\dfrac{16}{34} = \dfrac{8}{17}$ **18.** $\dfrac{30}{16} = \dfrac{15}{8}$ **19.** $\dfrac{9}{16.64} = 0.5409$ **20.** $\dfrac{14}{16.64} = 0.8413$

21. $\dfrac{9}{14} = 0.6429$ **22.** $\dfrac{14}{16.64} = 0.8413$ **23.** $\dfrac{9}{16.64} = 0.5409$ **24.** $\dfrac{14}{9} = 1.5556$

2   **25.**
$$\sin A = \frac{12}{20} = \frac{3}{5} \qquad \sin B = \frac{16}{20} = \frac{4}{5}$$
$$\cos A = \frac{16}{20} = \frac{4}{5} \qquad \cos B = \frac{12}{20} = \frac{3}{5}$$
$$\tan A = \frac{12}{16} = \frac{3}{4} \qquad \tan B = \frac{16}{12} = \frac{4}{3}$$
$$\csc A = \frac{20}{12} = \frac{5}{3} \qquad \csc B = \frac{20}{16} = \frac{5}{4}$$
$$\sec A = \frac{20}{16} = \frac{5}{4} \qquad \sec B = \frac{20}{12} = \frac{5}{3}$$
$$\cot A = \frac{16}{12} = \frac{4}{3} \qquad \cot B = \frac{12}{16} = \frac{3}{4}$$

26.
$$\sin A = \frac{8.15}{9.73} = 0.8376 \qquad \sin B = \frac{5.32}{9.73} = 0.5468$$
$$\cos A = \frac{5.32}{9.73} = 0.5468 \qquad \cos B = \frac{8.15}{9.73} = 0.8376$$
$$\tan A = \frac{8.15}{5.32} = 1.5320 \qquad \tan B = \frac{5.32}{8.15} = 0.6528$$
$$\csc A = \frac{9.73}{8.15} = 1.1939 \qquad \csc B = \frac{9.73}{5.32} = 1.8289$$
$$\sec A = \frac{9.73}{5.32} = 1.8289 \qquad \sec B = \frac{9.73}{8.15} = 1.1939$$
$$\cot A = \frac{5.32}{8.15} = 0.6528 \qquad \cot B = \frac{8.15}{5.32} = 1.5320$$

Self-Study Exercises 17–3

1   **1.** 0.3584   **2.** 0.9981   **3.** 1.0724   **4.** 0.6088   **5.** 0.5225   **6.** 0.9304   **7.** 0.4068
8. 0.9336   **9.** 0.8039   **10.** 0.9163   **11.** 1.0990   **12.** 0.8843   **13.** 0.9757   **14.** 0.4950
15. 3.3191   **16.** 0.9446   **17.** 0.6845   **18.** 0.9377   **19.** 0.3960   **20.** 0.2199

2   **21.** $20.0°$ **22.** $22.5°$ **23.** $67°$ **24.** $62.5°$ **25.** $57.3°$ **26.** $34.8°$ **27.** $51.7°$ **28.** $82.1°$
29. $39.0°$ **30.** $21.2°$ **31.** 0.5585 rad   **32.** 0.6807 rad   **33.** 0.8028 rad   **34.** 1.1694 rad
35. 0.2537 rad   **36.** 1.4802 rad   **37.** 0.3061 rad   **38.** 1.1804 rad   **39.** 0.2802 rad   **40.** 1.1362 rad

3   **41.** 2.1943   **42.** 1.4945   **43.** 1.0306   **44.** 1.0187   **45.** 2.8824   **46.** 1.8341   **47.** 0.0524
48. 1.3563   **49.** 4.7174   **50.** 1.9661   **51.** $67.0°$ **52.** $56.9°$ **53.** $15.6°$ **54.** $83.0°$
55. $30.0°$

Self-Study Exercises 17–4

1   **1.** $27.8°$ **2.** 6.889 in.   **3.** 28.21 m   **4.** 16.15 ft   **5.** $43.2°$ **6.** $49.6°$
7. $S = 51.1°$ **8.** $S = 45°$
     $U = 38.9°$ $t = 6.647$ m
     $u = 11.31$ yd $s = 4.700$ m
     $18^2 = 14^2 + (11.31)^2$ $(6.647)^2 = (4.7)^2 + (4.7)^2$
     $324 \doteq 323.9161$ $44.18 = 44.18$
9. $U = 55.5°$ **10.** $U = 74°$
     $s = 4.814$ mm $t = 50.79$ m
     $u = 7.005$ mm $u = 48.82$ m
     $(8.5)^2 \doteq (4.814)^2 + (7.005)^2$ $(50.79)^2 \doteq (48.82)^2 + (14)^2$
     $72.25 \doteq 23.17 + 49.07$ $2{,}579.62 \doteq 2{,}383.39 + 196$
     $72.25 \doteq 72.24$ $2{,}579.62 \doteq 2{,}579.39$

2   **11.** $35.7°$ **12.** 10.05 cm   **13.** 31.50 ft   **14.** 35.49 dm   **15.** 15.88 mm
16. 14.12 yd   **17.** $R = 45.9°$ **18.** $Q = 44°$ **19.** $R = 16.5°$ **20.** $Q = 30.5°$
            $Q = 44.1°$ $r = 12.23$ cm $r = 4.147$ dkm $r = 13.58$ m
            $r = 16.52$ ft $q = 11.81$ cm $s = 14.60$ dkm $s = 15.76$ m

3   **21.** $28.6°$ **22.** 4.075 m   **23.** 5.979 ft   **24.** $23.5°$ **25.** 0.04663 cm
26. 0.01212 m   **27.** $D = 34.8°$ **28.** $E = 48°$ **29.** $D = 16.5°$ **30.** $D = 54.0°$
            $E = 55.2°$ $d = 6.303$ ft $d = 5.963$ in. $E = 36.0°$
            $f = 5.604$ m $f = 9.419$ ft $f = 20.99$ in. $f = 13.60$ ft

Self-Study Exercises 17–5

1   **1.** 150.1 m   **2.** 11.91 ft   **3.** 9.950 cm   **4.** 46.71 cm   **5.** 21.09 in.

2   **6.** 22.38 in.   **7.** 53.62 ft   **8.** $59.0°$ **9.** 39.23 ft   **10.** $14.9°$ **11.** 23.25 ft   **12.** 10 ohms
13. 17.32 ohms   **14.** 7.654 cm   **15.** $35.8°$

Assignment Exercises, Chapter 17

1. 1.05 radians   **3.** 5.24 radians   **5.** 1.74 radians   **7.** $150°$ **9.** $97.4028°$ **11.** $67°30'$

13. 1.61 cm   **15.** 1.71 ft   **17.** 3.38 in.   **19.** 2.09 in.   **21.** 15.70 cm^2 **23.** $\dfrac{15}{25} = \dfrac{3}{5}$ **25.** $\dfrac{15}{20} = \dfrac{3}{4}$

27. $\dfrac{25}{20} = \dfrac{5}{4}$ **29.** $\dfrac{20}{25} = \dfrac{4}{5}$ **31.** $\dfrac{20}{15} = \dfrac{4}{3}$ **33.** $\dfrac{25}{15} = \dfrac{5}{3}$ **35.** $\dfrac{2 \text{ ft}}{2 \text{ ft 2 in.}} = \dfrac{24 \text{ in.}}{26 \text{ in.}} = \dfrac{12}{13}$

37. $\dfrac{2 \text{ ft}}{10 \text{ in.}} = \dfrac{24 \text{ in.}}{10 \text{ in.}} = \dfrac{12}{5}$ **39.** $\dfrac{10 \text{ in.}}{2 \text{ ft}} = \dfrac{10 \text{ in.}}{24 \text{ in.}} = \dfrac{5}{12}$ **41.** $\dfrac{7}{12.62} = 0.5547$ **43.** $\dfrac{7}{10.5} = 0.6667$

45. $\dfrac{12.62}{10.5} = 1.2019$ **47.** $\dfrac{10.5}{7} = 1.5000$ **49.** 0.4540   **51.** 0.3057   **53.** 1.0724   **55.** 0.7585
57. 0.8403   **59.** 0.1708   **61.** 52.5°   **63.** 45.3°   **65.** 16.7°   **67.** 1.2188 rad   **69.** 1.2101 rad
71. 0.3313 rad   **73.** 1.0576   **75.** 1.1326   **77.** $A = 38.1°, B = 51.9°, b = 13.15$ in.
79. $G = 48.5°, f = 1.681$ mm, $h = 2.537$ mm   **81.** $X = 52^c, y = 5.469$ m, $z = 8.883$ m   **83.** $D = 28.3°,$
$E = 61.7°, f = 14.76$ ft   **85.** $G = 53°, g = 25.21$ m, $k = 31.57$ m   **87.** 32.2°   **89.** 11.57 in.
91. $B = 60.5°, b = 3.641$ m, $c = 4.183$ m   **93.** 48.6° and 41.4°   **95.** $B = 56°, a = 88.91$ ft, $b = 131.8$ ft
97. yes, 45°   **99.** 12 steps   14   steps
            8 inches high (rise)   6.86 inches high
            14 inches flat (run)   12   inches flat

Trial Test, Chapter 17

1. 0.61 radians   **3.** 5.50 radians   **5.** 15.4167° = 0.27 radian   **7.** 16.2° = 0.28 radian   **9.** 112.5°
11. 68°45′18″   **13.** 1 in.   **15.** 1.71 m   **17.** $\dfrac{10}{26} = \dfrac{5}{13}$ **19.** $\dfrac{26}{10} = \dfrac{13}{5}$ **21.** $\dfrac{11.5}{12.54} = 0.9171$
23. 0.7986   **25.** 0.9135   **27.** 16.0°   **29.** 19.5°   **31.** 0.8726 rad   **33.** 0.3150 rad   **35.** 1.3673
37. 48.8°   **39.** 15.41 ft   **41.** 42.9°   **43.** 6.420 cm   **45.** 62.9°   **47.** $B = 65.5°, b = 30.72$ m,
$c = 33.76$ m   **49.** 38.7°   **51.** 48 in.   **53.** 40.06 A   **55.** 84.3°   **57.** 44.7°   **59.** $X_L = 9.64$ ohms;
$R = 11.49$ ohms

Self-Study Exercises 18–1

1   **1.** 13   **2.** 15   **3.** 7.280   **4.** 8.544   **5.** 2.746

2   **6.** 67.38°   **7.** 53.13°   **8.** 20.56°   **9.** 68.20°   **10.** 75.96°

3   **11.** 19; 42°   **12.** 8; 72°   **13.** 1; 225°   **14.** 6; 75°   **15.** $12 + 5i$; 13; 22.62°
16. $6 + 4i$; 7.211; 33.69°

Self-Study Exercises 18–2

1   **1.** 60°   **2.** 15°   **3.** 70°   **4.** 15°   **5.** 32°   **6.** 70°   **7.** 32°   **8.** 62°   **9.** 0.96 rad
10. 0.44 rad

2   **11.** all positive   **12.** sin, csc positive; others negative   **13.** tan, cot positive; others negative
14. cos, sec positive; others negative   **15.** cos, sec positive; tan, sin zero; cot, csc undefined

3   **16.** −0.5000   **17.** −0.8391   **18.** −0.8011   **19.** −0.6536   **20.** −0.8660   **21.** −0.2910
22. −0.1736   **23.** 0.5000   **24.** 146.3°   **25.** 3.61° rad

Self-Study Exercises 18–3

1   **1.** $C = 63°; b = 40.00; c = 45.22$ **2.** $A = 50°; a = 6.504; c = 8.075$
3. $C = 33°; b = 56.37; c = 35.45$ **4.** $B = 30°; a = 5.543; b = 3.200$
5. $C = 50°; a = 20.84; c = 16.03$

2   **6.** $B = 21.5°; C = 118.5°; c = 57.42$ **7.** $C = 10.8°; A = 99.2°; a = 15.76$ **8.** $A = 43.2°;$
$B = 116.8°; b = 23.47$ or $A = 136.8°; B = 23.2°; b = 10.35$ **9.** $B = 14.5°; C = 135.5°; c = 11.21$

3   **10.**   **11.** 805.9 ft

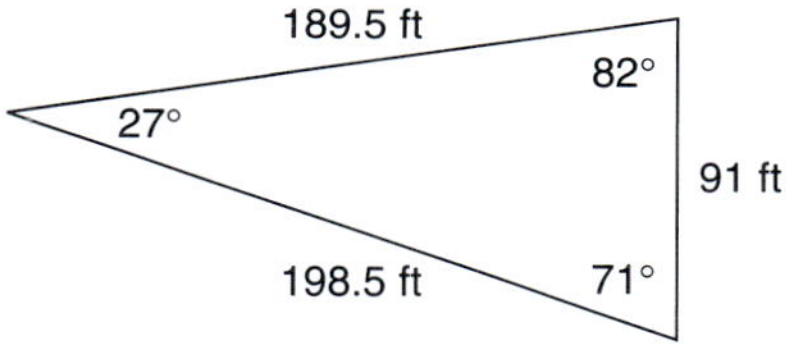

189.5 ft
82°
27°
91 ft
198.5 ft
71°

1 **1.** $A = 30.8°$ **2.** $D = 71.7°$
        $B = 125.1°$ $E = 62.9°$
        $C = 24.1°$ $F = 45.4°$

2 **3.** $r = 32.42$ **4.** $j = 18.68$
        $S = 49.7°$ $K = 37.5°$
        $T = 58.3°$ $L = 34°$

3 **5.** 49.27 ft **6.** 60.8°
                            44.1°
                            75.1°

Assignment Exercises, Chapter 18

1. 7.211; 56.3° **3.** 6.083; 9.5° **5.** 13.89; 59.7° **7.** 0.1016 rad **9.** 41° **11.** 0.8832 rad
13. 32°15′10″ **15.** 0.8632 **17.** −0.3420 **19.** 0.3420 **21.** 1 **23.** 2.83; 2.36 rad
25. $A = 40°, b = 10.8; c = 4.3$ **27.** $A = 30.3°; B = 104.7°; b = 9.6$
29. 1st − $A = 39.4°; C = 112.6°; c = 13.4$ **31.** 42.0 ft
    2nd − $A = 140.6°; C = 11.4°; c = 2.9$

33. 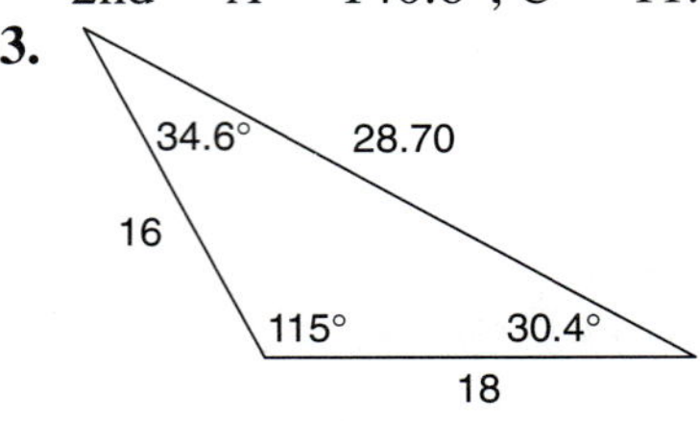

35.

37.

39. 97.98 ft **41.** 343.3 ft; 665.4 ft

Trial Test, Chapter 18

1. 0.8192 **3.** −0.9397 **5.** 135° **7.** 101.3° **9.** 8.991 ft **11.** 131.8° **13.** 6.830 m
15. 150.9 ft **17.** 66.88 ft^2 **19.** 6.4 milliamps **21.** 30 ohms

Glossary and Index

Billion, 3

Binary operation: an operation that involves working with two numbers at a time, 19, 80

Binomial: a polynomial containing two terms. Examples include $x + 3$, $x + y/4$, 419, 488

Binomial square, 489

Calculation, 347

Calculator

addition of whole numbers and decimals, 22–24

angle calculations, 609

area and perimeter, 277

continuous sequence, 60, 178, 306

converting fractions to decimals, 123

decimal part of degree, 811–813

decimal point, 23

decimal to fraction, 123

division of whole numbers and decimals, 49–50

exponents, 56

fraction key, 154

general power key, 57

graphing linear equations in two variables, 544

graphing quadratic equations, 563

graphing systems of equations, 579

hand clearing decimals versus calculator, 365

integers, 95

logarithms, 487

multiplication of whole numbers and decimals, 39–40

negative key, 95

order of operations, 60–61, 93

percents, 180

powers and roots, 56–58

power key, 57

proportions, 178

radians, 659

root(s), 56, 335

scientific notation, 416

shift key, 154

sign change key, 95

square key, 56

square root key, 57

subtraction of whole numbers and decimals, 30–31

toggle key, 95, 154

trigonometric functions of an angle, 668

trigonometric values, 667–670

value of, 287

Calibrate, 238

Caliper, 243

Canceling: dividing a numerator and denominator by the same amount to reduce the terms of the fraction, 141, 409

Capacity, 201, 221

Career applications

aerospace technology, asteroids and comets, 165

analyzing nutrition labels, 392

astronomy, planet alignment, 119

civil engineering, freeway supports, 526

consumer interest

stock market prices, 97

tornado-chasing expeditions, 284, 688

electronics

Kirchhoff's laws, 368

Career applications, *continued*

mesh currents, 594

Ohm's law, 476

percentage error, 193

phase angle, 641

prefixes in technical fields, 435

series and parallel circuits, 684

using slope to find resistance, 570

health sciences, body mass index, 232

horticulture, insect mating, 433

information systems, security coding, 504

interior design, paint, wallpaper, and border, 296

law enforcement, vehicle vaults, 721

nursing, drug dosage and IV concentration, 264

statistics, predicting college class size, 569

teaching, statistics and classroom performance, 319

Carry, 20

Cell, 351

Celsius, 449

Center: the point that is the same distance from every point on the circumference of the circle, 285

Central tendency, 315

Centi-: the prefix used for a unit that is 1/100 of the standard unit, 218

Centimeter: the most common unit used to measure objects less than 1 meter long; it is 1/100 of a meter, 220

Central angle, 654, 657

Chord: a line segment joining two points on the circumference of a circle, 630

Circle: a closed curved line whose points lie in a plane and are the same distance from the center of the figure, 285

arc, 630

arc length, 629

area, 287

center, 303

chord, 630

circumference, 285

diameter, 285

inscribed and circumscribed polygons, 633

radius, 285

sector, 628

segment, 630

semicircle, 285

semicircular-sided figure, 292

see also Geometry

Circle graph: the graph that uses a divided circle to show pictorially how a total amount is divided into parts, 304

Circuit, 369

Circumference: the perimeter or length of the closed curved of the line that forms the circle, 285

Circumscribed polygon: a polygon is circumscribed about a circle when the polygon is outside the circle and all its sides are tangent to (intersecting in exactly one point) the circumference. The circle is said to be inscribed in the polygon, 633, 634

Class frequency: the number of values in each class interval, 308

Class intervals: the groupings of a large amount of data into smaller groups of data, 308

Class midpoint: the middle value of each class interval, 308

Coefficient: each factor of a term containing more than one factor is the coefficient of the remaining factor(s). The

number factor, also called the numerical coefficient is
often implied when the term coefficient is used, 330,
407

Cofunction, 666

Coincide, 603

Common denominator: when comparing two or more fractions, a denominator that each denominator will divide
into evenly, 128

Common denominator in decimals, 11

Common factor: a factor that appears in each of several terms
in an expression, 486

Common fraction: A common fraction is the division of one
integer by another, 105

Commutative property of addition: the property that states
that values being added may be added in any order, 19

Commutative property of multiplication: this property permits numbers to be multiplied in any order, 33

Compare decimal numbers, 10, 129, 130

Complementary angles: two angles whose sum is one right
angle or 90°, 606

Complete quadratic equation: a quadratic equation and that
has a quadratic term, a linear term, and a constant term,
510

Complex fraction: A complex fraction has a fraction or mixed
number in its numerator or denominator or both, 108,
148

Complex notation, 701

Complex numbers: a number that can be written in the form
$a + bi$, where a and b are real numbers and i is $\sqrt{-1}$,
427, 700

Composite figure: A composite figure is a geometric figure
made up of two or more geometric figures, 289, 622

Composite number: a whole number greater than 1 that is not
a prime number, 115

Cone: a right cone is the three-dimensional figure whose base
is a circle and whose side surface tapers to a point,
called the vertex or apex, and whose height is a perpendicular line between the base and the apex, 466

Congruent, 387

Congruent triangles: triangles in which the corresponding
sides and angles are equal. Two triangles are congruent
if they have the same size and shape. Each angle of one
triangle is equal to its corresponding angle in the other
triangle. Each side of one triangle is equal to its corresponding sides in the other triangle, 387, 614

Conjugate pairs, 489

Constant of direct variation: a conversion factor that is found
by solving a function for the coefficient of the independent variable, 381

Constant of inverse (indirect) variation: a conversion factor
for finding inverse variation, 385

Constant term or number term: a term that has degree 0. A
term that contains only a number, 330, 420, 510

Conversion factor: a standard value used to convert one measure to another. Some frequently used conversion factors
convert a metric measurement to a U.S. customary
measurement, or a U.S. customary measurement to a
metric measurement, 206, 230

Corresponding sides, 387

Cosecant: the reciprocal function of the sine; the cosecant of
an angle = hypotenuse/the side opposite the angle. It is
abbreviated csc, 664

Cosine of an angle: the side adjacent to the angle/hypotenuse.
Abbreviated cos, 661, 674

Cost function, 531

Cotangent: the reciprocal function of the tangent; the cotangent of an angle = the side adjacent to the angle/the
side opposite the angle. It is abbreviated cot, 664

Coulomb, 418

Cross products: in a proportion, the cross products are the
product of the numerator of the first fraction times the
denominator of the second, and the product of the denominator of the first fraction times the numerator of
the second. In the proportion $a/b = c/d$, the cross products are $a \times d$ and $b \times c$, 175, 377

Cube: a number raised to the third power, 52

Cubed: another term for "raised to the third power," 52

Cubic polynomial: a polynomial that has degree 3, 420

Cubic term: a term that has degree 3, 420

Cup, 202

Current, 368

Cutting speed, 292

Cylinder: a three-dimensional figure with a curved surface
and two circular bases such that the height is perpendicular to the bases, 461

Data analyzing and interpreting, 303

Deci-: the prefix used for a unit that is 1/10 of the standard
unit, 218

Decimal: a fractional notation based on place values for
fractions with a denominator of 10 or a power of 10,
6, 359

   clearing decimals in equations, 364

   comparing decimal numbers, 10, 129, 130

   divide, 42

   exponents, 426

   fraction, 6, 108

   fraction-decimal conversions, 123

   point, 6

   repeating, 138

   signed decimal numbers, 165

Decimal fraction: a fraction whose denominator is always
ten or some power of 10; a fractional notation that
uses the decimal point and the place values to its right
to represent a fraction whose denominator is 10 or
some power of 10, such as 100, 1000, and so on. A
decimal fraction is also referred to as a decimal, a
decimal number, or number using decimal notation,
6, 108

Decimal number system: the system of numbers that uses ten
individual figures called digits, 0, 1, 2, 3, 4, 5, 6, 7, 8, 9
and place values of powers of 10, 3

Decimal number: an alternate name for a decimal, 6

Decimal point: the symbol (period) placed between the ones
and the tenths place to identify the place value of each
digit, 6, 108

Decrease, 187

Deductive reasoning, 516

Degree: a unit for measuring angles. A degree represents
1/360 of a complete rotation around the vertex of the
angle, 604, 610, 654, 657–658

Degree of a polynomial: the degree of a polynomial that has
only one variable and only positive integral exponents
is the degree of the term with the largest exponent, 420